# THE ARCTIC SEAS

## Climatology, Oceanography, Geology, and Biology

*Edited by*

Yvonne Herman

VAN NOSTRAND REINHOLD COMPANY
*New York*

This work relates to Department of the Navy Grant N00014-85-G-0252 issued by the Office of Naval Research. The United States Government has a royalty-free license throughout the world in all copyrightable material contained herein.


Library of Congress Catalog Card Number 88-33800
ISBN 0-442-23171-7



Printed in the United States of America

Designed by Beehive Production Services

Van Nostrand Reinhold
115 Fifth Avenue
New York, New York 10003

Van Nostrand Reinhold (International) Limited
11 New Fetter Lane
London EC4P 4EE, England

Van Nostrand Reinhold
480 La Trobe Street
Melbourne, Victoria 3000, Australia

Nelson Canada
1120 Birchmount Road
Scarborough, Ontario M1K 5G4, Canada

16 15 14 13 12 11 10 9 8 7 6 5 4 3 2 1



**Library of Congress Cataloging in Publication Data**
The Arctic Seas.
Includes index.
1. Oceanography—Arctic Ocean. 2. Geology—Arctic Ocean. I. Herman, Yvonne.
GC401.A76 1989 551.46′8 88—33800
ISBN 0-442-23171-7

*For Anyu*

# Contents

# Preface

The Arctic region has long held a fascination for explorers and scientists of many countries. Despite the numerous voyages of exploration, the nature of the central Arctic was unknown only 90 years ago; it was believed to be a shallow sea dotted with islands. During Nansen's historic voyage on the polarship *Fram,* which commenced in 1893, the great depth of the central basin was discovered.

In the Soviet Union, investigation of the Arctic Ocean became national policy after 1917. Today research at several scientific institutions there is devoted primarily to the study of the North Polar Ocean and seas.

The systematic exploration of the Arctic by the United States commenced in 1951. Research has been conducted year-round from drifting ice islands, which are tabular fragments of glacier ice that break away from ice shelves. Most frequently, ice islands originate off the northern coast of Ellesmere Island. These research platforms are occupied as weather stations, as well as for oceanographic and geophysical studies. Several international projects, conducted by Canadian, European, and U.S. groups, have been underway during the last three decades. Although much new data have accumulated since the publication of the *Marine Geology and Oceanography of the Arctic Seas* volume in 1974 (Yvonne Herman, ed.), in various fields of polar research—including present-day ice cover, hydrography, fauna, flora, and geology—many questions remain to be answered. The paleoceanography, ice cover history, and the past atmospheric circulation patterns are open to controversy, as are the structure and evolution of the crust beneath the Arctic Ocean. Furthermore, the origin and mode of dispersal of sediments are not fully understood, nor is there a consensus concerning the origin and age of the modern Arctic biota. According to some climatologists, unipolar glaciation existed until the Pleistocene, when the Arctic Ocean became ice covered; the Antarctic ice sheet developed initially approximately 38 Ma. The sedimentary record of the Arctic basin is documented only for the last 4.5–5 My; its history prior to this time will remain open to speculation until sedimentary cores spanning the Cenozoic and possibly the Mesozoic are recovered. The current status of many of these problems is discussed in *The Arctic Seas: Climatology, Oceanography, Geology, and Biology.* It seemed timely to bring together in a single volume important current studies from several disciplines. Because of the wide topical range and the multi-authored nature of this volume, it was not possible to insure precise balance; moreover, certain subjects could not be covered because of the physical limitations of a single book. However, it is hoped that this book will provide an insight into the current status of Arctic research, in particular of studies underway in the Soviet Union inaccessible to date to the West, and that it will also serve as a reference for investigators studying the polar and subpolar seas.

It is with pleasure that I acknowledge the following people for helping in various ways during the preparation of this book: P. E. Rosenberg, R. G. Barry, G. L. Johnson, R. Horner, C. Sancetta, P. Herman, O. Herman, B. R. Heimdal, S. Gartner, M. A. Shnoo-Kee, L. Joachim, A. E. Aksu, C. A. Repenning, A. P. Lisitzin, L. P. Zonenshain, S. Dorobek, W. L. Gates, V. T. Frolov, I. S. Gramberg, D. Bukry, J. England, P. Von Pitzpon, D. E. Strohl, and C. Hooper. I also wish to acknowledge the cooperation I received from all contributing authors. Special appreciation is given to the Office of Naval Research for support in the preparation of this book under Grant N00014-85-G-0252 and to C. S. Hutchinson, Jr., for providing assistance with the redrafting and retyping of the Russian chapters.

YVONNE HERMAN

# Contributors

**O. G. Norden Andersen,** Zoological Museum, Universitetsparken 15, Copenhagen DK-2100, Denmark

**L. Anderson,** Department of Analytical and Marine Chemistry, Chalmers University of Technology and University of Goteborg, Goteborg S-412 96, Sweden

**N. A. Anisimova,** Laboratory of Marine Research, Zoological Institute, USSR Academy of Sciences, Leningrad 199164, USSR

**R. G. Barry,** Cooperative Institute for Research in Environmental Sciences, Campus Box 449, University of Colorado, Boulder, Colorado 80309-0449

**A. N. Belyaeva,** Shirshov Institute of Oceanology, USSR Academy of Sciences, 23 Krasikova, Moscow 117218, USSR

**I. D. Danilov,** Department of Geology, Moscow State University, Moscow 119899, USSR

**A. I. Daniushevskaya,** VNIIOceangeologiya, 120 Moika, Leningrad 190121, USSR

**D. A. Darby,** Department of Geological Sciences, Old Dominion University, Norfolk, Virginia 23529-0496

**N. V. Denisenko,** Laboratory of Benthos, Murmansk Marine Biological Institute, USSR Academy of Sciences, Dalnye Zelentsky, USSR

**M. J. Dunbar,** Climate Research Group, Department of Meteorology, McGill University, 805 Sherbrooke Street West, Montreal, Quebec, Canada H3A 2K6

**D. Dyrssen,** Department of Analytical and Marine Chemistry, Chalmers University of Technology and University of Goteborg, Goteborg S-412 96, Sweden

**M. A. Faustova,** Institute of Geography, USSR Academy of Sciences, Moscow 109017, USSR

**V. V. Fedyakov,** Zoological Institute, USSR Academy of Sciences, Universitetskaya nab. 1, Leningrad 199034, USSR

**G. Gard,** Department of Geology, University of Stockholm, Stockholm S-106 91, Sweden

**A. N. Golikov,** Laboratory of Marine Research, Zoological Institute, USSR Academy of Sciences, Leningrad 199164, USSR

**V. I. Gontar,** Laboratory of Marine Research, Zoological Institute, USSR Academy of Sciences, Leningrad 199164, USSR

**B. Heimdal,** Department of Marine Biology, University of Bergen, N-5065 Biomsterdalen, Norway

**Y. Herman,** Department of Geology, Washington State University, Pullman, Washington 99164-2812

**W. D. Hibler, III,** Thayer School of Engineering, Dartmouth College, Hanover, New Hampshire 03755

**L. L. Isayeva,** Institute of Geography, USSR Academy of Sciences, Moscow 109017, USSR

**R. A. Horner,** 4211 N.E. 88th Street, Seattle, Washington 98115

**G. A. Jones,** Department of Geology and Geophysics, Woods Hole Oceanographic Institute, Woods Hole, Massachusetts 02543

**S. Krugilkova,** Shirshov Institute of Oceanology, USSR Academy of Sciences, 23 Krasikova, Moscow 117218, USSR

**I. A. Mel'nikov,** Shirshov Institute of Oceanology, USSR Academy of Sciences, 23 Krasikova, Moscow 117218, USSR

**T. C. Mowatt,** 9905 Amchitka Circle, Eagle River, Alaska 99577

**A. S. Naidu,** Institute of Marine Sciences, University of Alaska, Fairbanks, Alaska 99701

**L. M. Napatov,** Shirshov Institute of Oceanology, USSR Academy of Sciences, Moscow 117218, USSR

**A. D. Naumov,** Laboratory of Marine Research, Zoological Institute, USSR Academy of Sciences, Leningrad 199164, USSR

**D. B. Oreshkin,** Institute of Geography, USSR Academy of Sciences, Moscow 109017, USSR

**J. K. Osmond,** Department of Geology, Florida State University, Tallahassee, Florida 32306

**V. V. Petryashov,** Laboratory of Marine Research, Zoological Institute, USSR Academy of Sciences, Leningrad 199164, USSR

**E. I. Polyakova,** Department of Geography, Moscow State University, Moscow 117234, USSR

**E. A. Romankevich,** Shirshov Institute of Oceanology, USSR Academy of Sciences, 23 Krasikova, Moscow 117218, USSR

**Yu. G. Samoilovich,** Kola Branch, USSR Academy of Sciences, Apatity, USSR

**O. A. Scarlato,** Marine Research Laboratory, Zoological Institute, USSR Academy of Sciences, Leningrad 199164, USSR

**P. Sharma,** Department of Physics, University of Pennsylvania, Philadelphia, Pennsylvania 19104

**B. L. K. Somayajulu,** Physical Research Laboratory, Navrangpura, Ahmedabad 380 009, India

**S. D. Stepanjants,** Universitetskaya nab. 1, Zoological Institute, USSR Academy of Sciences, Leningrad 199034, USSR

**S. V. Vassilenko,** Laboratory of Marine Research, Zoological Institute, USSR Academy of Sciences, Leningrad 199164, USSR

**A. A. Velitchko,** Institute of Geography, USSR Academy of Sciences, Moscow 109017, USSR

**G. Vilks,** Atlantic Geoscience Center, Geological Survey of Canada, Department of Energy, Mines, and Resources, Bedford Institute of Oceanography, P.O. Box 1006, Dartmouth, Nova Scotia, Canada B2Y 4A2

**J. R. Weber,** Department of Energy, Mines and Resources, Earth Physics Branch, Gravity and Geodynamics Division, 1 Observatory Crescent, Ottawa, Ontario K1A OY3, Canada

**V. S. Zarkhidze,** VNIIOceanologia, 120 Moika, Leningrad 190121, USSR

**L. P. Zonenshain,** Shirshov Institute of Oceanology, USSR Academy of Sciences, 23 Krasikova, Moscow 117218, USSR

# 1: The Present Climate of the Arctic Ocean and Possible Past and Future States

R. G. Barry

## INTRODUCTION

### Atmosphere-Ice-Ocean Processes

The climate of the north polar region is dominated by the year-round presence of the Arctic pack ice, which largely insulates the atmosphere from the waters of the Arctic Ocean. Recently, several lines of climatological research have refocused attention on the long-term stability of this ice cover—a topic considered earlier by Budyko (1962), Donn and Ewing (1966), and Donn and Shaw (1966, 1967). Disagreement over the permanence of the sea ice cover during the late Neogene period remains to be resolved (Herman and Hopkins, 1980; Clark, 1982*a*) and, while this topic has its own intrinsic interest, the problem has acquired new significance in the context of the possible climatic effects of carbon dioxide-induced warming in high latitudes. The sedimentological record of past ice cover, as well as numerical modeling studies, therefore, provide potentially valuable information on the magnitude of global climatic perturbations that can and cannot be sustained by the Arctic sea ice. These topics form the focus of this contribution, but in order to understand the complexity of the problems it is necessary to discuss the nature of the atmosphere–ice–ocean processes and interactions in the north polar region.

### The Physical Setting

The Arctic Ocean and its bordering seas occupy an area of 14 million km$^2$, which is almost all ice covered in late winter. During the summer (September) the minimum ice extent is approximately 8 million km$^2$ (Fig. 1-1). Much of the adjacent land is low lying, which facilitates air mass exchanges with middle latitudes, except for the Greenland ice sheet, the ice-covered mountains of Ellesmere and Axel Heiberg islands, and the ranges in Alaska and northeastern Siberia. The major entrance zone for atmospheric cyclonic systems and warm ocean waters into the central Arctic is the Norwegian–Barents Sea sector.

Major physical features of the Arctic Ocean are the Canadian Basin, separated by the submarine Lomonosov Ridge from the smaller Eurasian Basin, the extensive

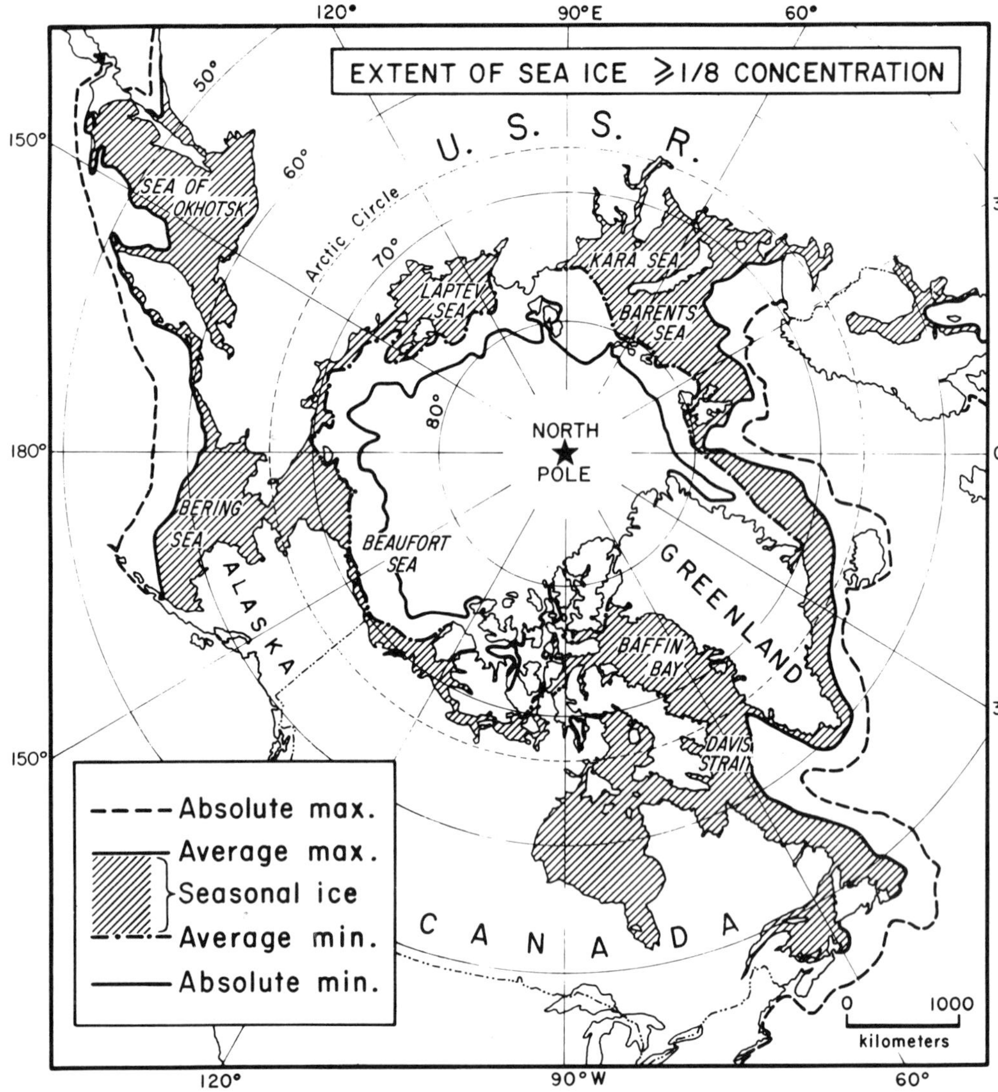


Figure 1-1. The average and extreme seasonal limits of Arctic sea ice extent for ice concentrations ≥ ⅛. (After *Atlas of the Polar Regions,* Central Intelligence Agency, 1978.)

continental shelves of the Eurasian side that occupy about 35 percent of the entire ocean, and the single deep passage of the Fram Strait between Greenland and Svalbard connecting the Arctic Ocean to the Atlantic.

Most river runoff enters the Eurasian shelf seas, particularly the Barents, Kara, Laptev, and East Siberian seas. This freshwater runoff and low-salinity water from the Bering Sea maintain a density stratification in the surface water (to 200 m depth) that prevents deep convective overturning and upward heat flux from the underlying warmer Atlantic water (SCOR, Working Group 58, 1979). The uppermost Polar Mixed Layer, 30–50 m thick, has a strong salinity stratification in summer, resulting from ice melting. During this season, the surface layer of low salinity reduces the downward mixing of heat from solar radiation absorbed in leads in the ice, thereby enhancing ice decay, whereas in autumn its presence aids in ice formation (Carmack, 1986). These upper waters have an estimated residence time of 10 years, compared with about 5 years, on average, for the ice (Ostlund and Hut, 1984). The deep water of the "Arctic Mediterranean" (the Arctic Ocean, Greenland, Iceland, and Norwegian seas) circulates internally, constrained by the Greenland–Scotland ridges (Aagaard, Swift, and Carmack, 1985).

## THE ARCTIC OCEAN ICE REGIME

The pack ice is a mixture of young and old floes that are highly variable in thickness as a result of the complex annual cycle of growth and decay and of the ice motion. Over the full course of a growth season, thermodynamic processes enable first-year ice to attain a thickness of about 2 m (Koerner, 1973), although leads that have frozen more recently will have thinner ice. Winter snow accumulation, which averages about 40 cm on level ice but is drifted up to 1 m or so in depth around hummocks and ridges, modifies the rate of ice growth through its effect on heat conductivity. Motion causes ridging through ice pileup and also fracture through deformation and shear of ice fields. New ice formation takes place, particularly in winter, in the leads that are so formed in the ice. This heterogeneous ice cover is subject to ablation and continued ice dynamical effects during the summer melt season from June through early September. Floes that survive this season thicken further in the following winter unless they are transported out of the Arctic. Sonar studies in the Fram Strait (Wadhams, 1981, 1983) indicate that multiyear ice has drafts ranging between 4 m and 16 m.

Total annual ice growth in the Arctic Basin averages about 1.1 m, which is balanced by 0.7 m of ablation and 0.5 m (equivalent) of removal by export (Koerner, 1973). Three-quarters of the mass exchange from the Arctic and almost all of its heat exchange is handled by the Fram Strait, east of Greenland (Aagaard and Greisman, 1975). The annual flux of ice area is of the order of 0.9 to $1.3 \times 10^6$ km$^2$ according to a recent estimate by Wadhams (1983).

Interannual fluctuations and longer-term trends of Arctic ice extent have received much attention, but until recently only conventional shipboard and coastal observations, supplemented by aerial reconnaissance, were available (Zakharov, 1981). Satellite imagery has greatly improved the data coverage, in particular the use of all-weather, year-round, passive microwave data from Nimbus 5 (the single-channel ESMR) and Nimbus 7 (the dual-polarized multifrequency SMMR).

Many studies demonstrate that interannual variations in ice extent are primarily regional in occurrence and are related to interannual changes in atmospheric circulation characteristics (Campbell et al., 1981; Johnson, 1980; Overland and Pease, 1982). A well-known antiphase relationship exists between summer ice conditions in the Barents–Kara seas and the East Siberian–Chukchi seas on both an annual and a 10-year time scale (Itin, 1936; Zakharov, 1976*a*).

An unexpected result from the recently available SMMR data is the existence of significant interannual variability in the estimated multiyear ice fraction in the central Arctic in the winters of 1979, 1980, and 1981 (Cavalieri and Zwally, 1985). The multiyear ice edge advanced and retreated substantially over these three years in the Alaskan and Eurasian sectors, with opposite patterns in the Laptev–East Siberian seas, on one hand, and in the Beaufort–Chukchi and the Norwegian seas, on the other.

The mean atmospheric wind-forcing sets up the oceanic surface circulation and ice drift. The most significant elements of the ice drift are the Beaufort Sea anticyclonic gyre and the Transpolar Drift

Stream (TPDS), depicted in a recent synthesis of all available observational data by Colony and Thorndike (1984). Mean velocities are of the order of 2 km day$^{-1}$, giving a mean circuit time of ice in the gyre of about 5 years and in the TPDS of about 3 years (Thorndike, 1986). The major exit for ice is through the Fram Strait between Greenland and Svalbard into the East Greenland drift. This topic is discussed in detail by Hibler (Chap. 2).

At the end of summer, the extent of open water in the ice-covered part of the Arctic Ocean amounts to about 26 percent of the area according to calculations from 4 years of passive microwave data (Carsey, 1982). A new analysis of transpolar ice draft data from sonar records, taken by the *USS Nautilus* in early August 1958 and the *USS Queenfish* on the same dates in 1970, indicate an overall Arctic average of 3–4 percent open water or new ice (<30 cm) and about 10 percent of ''skylights'' 0–100 cm thick (McLaren, 1986). In one segment of the *Nautilus* track from 79°–83°N (approximately), these values reached maxima of 5.4 percent open water or thin ice, and 17.9 percent for skylights. Hence, Carsey's estimate seems high, even for the end of summer. In addition, there are extensive melt pools on the bare ice, which at their maximum in mid-July may cover 25 percent of the ice (Nazintsev, 1964; Carsey, 1985) and significantly reduce the albedo. In winter, the fraction of open water and newly frozen leads is less certain. The percentage cover of ice with a draft less than 50 cm has been estimated from submarine sonar data in the Beaufort Sea in April 1976 to be about 1–6 percent (Wadhams and Horne, 1980). Despite this seemingly small area, leads play an important role in the wintertime surface energy balance, as discussed below.

Analysis of ice motion from data provided by the Arctic drifting buoys (1979–1985) has revealed an unexpected pattern of recurrent reversals in the Beaufort Gyre during late summer in response to persistent slow-moving cyclone systems (McLaren, Serreze, and Barry, 1987). These cyclones cause divergent ice motion of about 0.5 percent day$^{-1}$ and associated reductions of ice concentration. Such decreases in ice concentration have previously been inferred from ESMR passive microwave data, as reviewed in Parkinson et al. (1987), but are also now confirmed in visible-band images as well as SMMR data (Maslanik and Barry, 1988). This evidence suggests that changes in atmospheric forcing can have important effects on ice amount within the Arctic Basin.

## THE POLAR ENERGY BUDGET

Global atmospheric circulation is basically driven by the equator to pole gradient between the low-latitude heat sources and the polar heat sinks. The Arctic atmosphere has an annual deficit of net radiative energy, although rather remarkably the net radiation at the surface is close to zero over the year. Figure 1-2 summarizes the major components of the annual energy budget for the earth–atmosphere system over the Arctic. The planetary albedo averages 64 percent. The net incoming solar radiation is 79 W m$^{-2}$ and the outgoing infrared radiation is $-179$ W m$^{-2}$; the net radiative deficit at the top of the atmosphere is $-100$ W m$^{-2}$. The net radiative cooling for the total atmospheric mass of the polar cap, north of 70°N, amounts to 1.3°C day$^{-1}$ in December, January, and February, decreasing to 0.1°C day $^{-1}$ in June, July, and August. The radiative deficit is offset mainly by the horizontal transport of energy poleward in the atmosphere, supplemented by fluxes from the ice–ocean surface to the atmosphere in winter. In summer there is heat flux into the oceans and ice.

Table 1-1 summarizes the energy budget of the earth–atmosphere system of the polar cap, based on data for 1963–1973 analyzed by Nakamura and Oort (1988). The

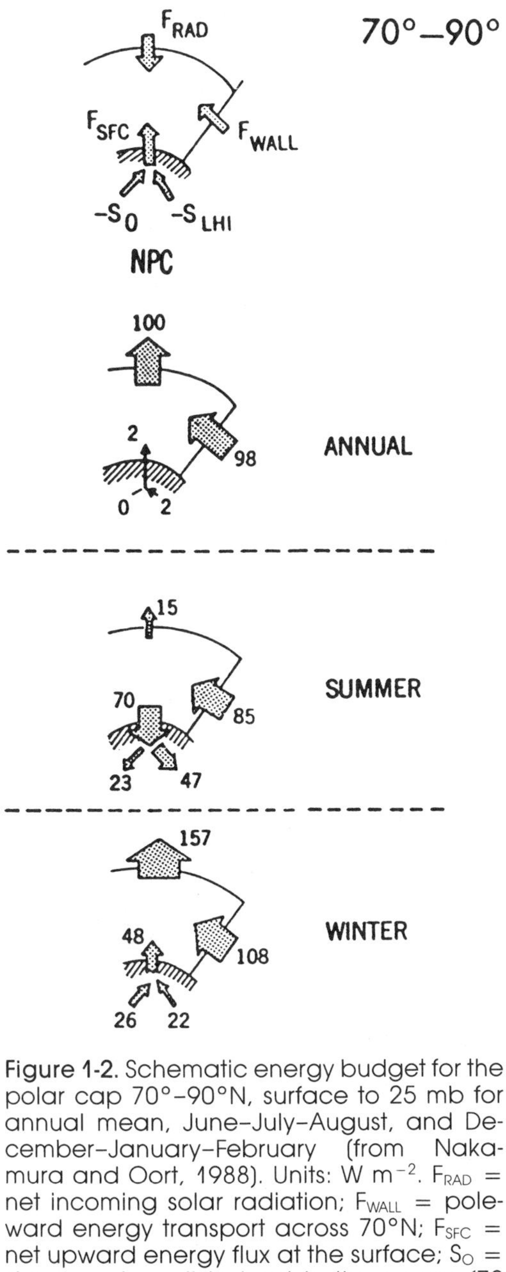


Figure 1-2. Schematic energy budget for the polar cap 70°–90°N, surface to 25 mb for annual mean, June–July–August, and December–January–February (from Nakamura and Oort, 1988). Units: W m$^{-2}$. $F_{RAD}$ = net incoming solar radiation; $F_{WALL}$ = poleward energy transport across 70°N; $F_{SFC}$ = net upward energy flux at the surface; $S_O$ = storage of sensible heat in the ocean (72 percent of the area); $S_{LHI}$ = storage of latent heat in snow and ice.

horizontal transports by the eddy components in the atmosphere, primarily synoptic-scale traveling disturbances, account for almost 55 percent of the available energy annually. The mean meridional plane (south–north) cells depicted in the classic meteorological literature play a minor role except in summer. Sensible heat is the primary mode of energy transport, since the atmospheric vapor content is low. However, in terms of surface to air fluxes, latent heat flux accounts for a larger proportion, especially in summer when these in situ fluxes provide about 20 percent of the total energy to the atmosphere.

The structure of ocean heat transport, which is of secondary importance in high northern latitudes, is not known in detail. The Fram Strait appears to play the predominant role, handling about 90 percent of the heat exchange. Table 1-2 indicates that the total oceanic advective heat gain by the Arctic Ocean is $132 \times 10^{12}$ W, 52 percent via the West Spitsbergen Current and 25 percent in the East Greenland outflow of ice (representing a gain to the Arctic through export of the latent heat of fusion). Distributed uniformly over the Arctic Ocean, the net heat gain is equivalent to an annual heat loss to the atmosphere of 8 W m$^{-2}$, in agreement with earlier estimates of Vowinckel and Orvig (1970). This figure is not directly comparable with the estimate of heat gain from the ocean given in Table 1-1, since the latter includes upward conductive heat flux, latent heat release during ice formation, and heat stored in the snow (Langleben, 1985). Not all of this energy is transferred to the atmosphere.

Assessments of the annual *surface* energy budget of the Arctic Ocean are hampered by the discontinuous nature of the measurements and the question of spatial representativeness. For any given year there are only one or two ice stations and the pack ice varies interannually and spatially in age and thickness. Most station observations are taken over multiyear ice and such data for two Soviet stations (NP 4 and NP 23) are illustrated in Table 1-3. The complete energy balance equation is

$$Rn + H + LE + G + \Delta S + M = 0,$$

Table 1-1. Energy Budget of the Earth–Atmosphere System of the Polar Cap, North of 70°N for 1963–1973

| | *Annual* | *Dec–Jan–Feb* | *June–July–Aug* |
|---|---|---|---|
| Total energy transport across 70°N | | | |
| By transient eddies | 53 | 53 | 50 |
| By stationary eddies | 13 | 30 | 3 |
| By meridional cells | 32 | 21 | 31 |
| Total* | 98 | 108 | 85 |
| Surface to atmosphere flux (residual estimate) | 2 | 48 | −67 |
| Net radiation at top of atmosphere | −101 | −157 | −15 |
| Equivalent cooling (K $day^{-1}$) | −0.8 | −1.3 | −0.1 |

*Source:* After Nakamura and Oort, 1988.
*Smoothed over neighboring months 0.25, 0.50, 0.25.

where $Rn$ is the net all-wavelength radiation; $H$ is the sensible heat flux; $LE$ is the latent heat flux; $G$ is the heat conduction through the ice from the ocean; $\Delta S$ is the absorbed short-wave radiation stored in leads, in the ice and in the water below the ice; and $M$ is the latent heat of snow and ice melt. In Table 1-3, $Rn$ in the winter months is essentially determined by the net long-wave (infrared) radiation, $Ln$, and the heat conduction term, $G$, which includes the latent heat of fusion released by freezing. Hence, the appropriate balance equation is

$$Ln + H + LE + G = 0$$

In winter, thin-ice areas transfer large amounts of turbulent heat to the atmosphere; Maykut (1978, 1982) estimates that half of the annual turbulent heat flux in the Beaufort Sea occurs through ice 20–60

Table 1-2. Mean Annual Heat Budget for the Arctic Ocean

| *Source* | *Heat Transport* | *($10^{12}$ W)* |
|---|---|---|
| Bering Strait: | | |
| Water | 3.8 | |
| Ice | | −1.7 |
| Arctic Archipelago | 13.4 | |
| East Greenland Current | | |
| Polar water | 8.4 | |
| Atlantic water | | −13.4 |
| Ice | 33.5 | |
| West Spitsbergen Current | 68.3 | |
| Spitsbergen–Franz Josef Land | | −1.3 |
| Franz Josef Land–Novaya Zemlya | 2.9 | |
| Runoff | 2.1 | |
| Total advective heat gain | 132.4 | |
| Total advective heat loss | | −16.4 |
| Net Exchange | 116.0 | |

*Source:* After Aagard and Greisman, 1975; SCOR, Working Group 58, 1979.

Table 1-3. Mean Monthly Energy Fluxes (W $m^{-2}$) and Meteorological Parameters over Multiyear Ice

| Station | Month | Ln | H | G | LE | $T_1$(°C) | $U_1$(m/s) |
|---|---|---|---|---|---|---|---|
| NP-23, 76–77°N | D | −27.9 | 4.9 | 7.0 | — | −31 | 2.6 |
| (1976/77) | J | −20.2 | 7.0 | 11.2 | — | −26 | 4.5 |
| | F | −18.1 | 7.0 | 14.7 | — | −32 | 4.7 |
| | M | −20.2 | 7.0 | 16.8* | — | −33 | 3.5 |
| | A | −8.4 | 8.4 | 9.1† | — | −25 | 4.2 |
| NP-4, 85°N | D | −15.4 | 7.7 | 9.1 | −0.7 | | |
| (1956/57) | J | −20.2 | 10.5 | 13.3 | −1.4 | | |
| | F | −26.5 | 10.5 | 14.7 | −0.7 | | |
| | M | −20.2 | 7.7 | 7.0 | −0.7 | | |
| | A | −9.8 | 4.2 | 4.9 | −0.7 | | |

*Source:* After Makshtas, 1984.
$T_1$ = air temperature at 1 m.
$U_1$ = wind speed at 1 m.
*The corresponding value for first-year ice is 47.5 W $m^{-2}$ (measured between 22 cm and 47 cm, but undergoes only small changes with depth); $L_f$ = 34.9 W $m^{-2}$ and the ocean flux is only about 14 W $m^{-2}$. The multiyear ice figure is for 2–82 cm depth.
†The corresponding value for first-year ice is 24.4 W $m^{-2}$; $L_f$ = 8.4 W $m^{-2}$. The ocean heat flux is similar to March (14 W $m^{-2}$, allowing for measurement errors).
*Note:* Positive values are toward the surface.

cm thick, which accounts only for about 6 percent of the ocean area in winter. This is illustrated by the measured values of G on first-year ice in March 1977 and April 1977 at NP-23; these were, respectively, 3 and 2.5 times the figures of multiyear ice (see footnotes to Table 1-3).

Regional averages for the turbulent fluxes nevertheless show a downward flux to the ice. It appears, from special studies over freezing leads, that a substantial fraction of the turbulent heat flux from such leads is transferred only a metre or so above the surface (Andreas et al., 1979) and relayed back to the ice immediately downwind (condensation, ice crystal precipitation, and sensible heat flux to the ice), as well as through long-wave radiation from clouds formed downwind of the lead and via the enhanced surface temperature inversion (Allison, 1982). Major leads can have significant effects, however, transferring condensate in plumes up to several kilometers altitude and extending tens of kilometers or more downwind (Barry et al., 1988). Large-scale advection must also play a part in the overall downward heat flux to the ice, especially in the marginal seas. There is no surface melt during November–April, and Table 1-4 shows an energy balance dominated by net radiation, conductive heat flux, and sensible heat. In summer and autumn the regime is complicated by the melt of snow and ice and by temporary storage of absorbed solar radiation penetrating the ice, $\Delta S$. Maykut's calculations show that about 27 percent of $\Delta S$ is absorbed in leads, 34 percent in thick ice, and 39 percent below the ice. Snow melt of an average 40 cm cover on thick ice and zero on thin ice (as assumed by Maykut, personal communication, 1983) accounts for 27 percent of the M term, and the ice melt represents an average ablation of about 36 cm.

The data in Table 1-4 exclude melt-pond effects, but Maykut (1982) estimates that a 30 percent cover of melt ponds in July would give a surface albedo of 49 percent (instead of 65 percent) and increase the net radiation by 30 W $m^{-2}$. Observations of melt ponds at various North Pole stations

Table 1-4. Regional Energy Budget Estimate for the Beaufort Sea (W m$^{-2}$)

| *Ice Thickness* | *Period* | *Rn* | *H* | *LE* | *G* | *ΔS* | *M* |
|---|---|---|---|---|---|---|---|
| | November–April | | | | | | |
| 0–0.8 m | | −11.1 | −35.0 | −4.4 | 53.2 | | |
| >0.8 m | | −122.2 | 64.4 | −1.3 | 59.2 | | |
| Total | | −133.3 | 29.4 | −5.7 | 112.4 | −2.8 | |
| | May–October | 221.5 | −44.9 | −55.8 | 39.9 | −103.8 | −56.9 |
| | Annual | 88.2 | −15.5 | −61.5 | 152.3 | −106.6 | −56.9 |

*Source:* After Maykut, 1982.
*Note:* Positive values are toward the surface.

(Nazintsev, 1964) indicate a mean coverage of 25 percent of the ice in mid-July and a local maximum of 45 percent; the corresponding minimum surface albedo values, based on aircraft observations for different stages of melting by Buzuev, Shesterikov, and Timerev (1965), are 48 percent and 28 percent, respectively (Table 1-5). Using the latter value, net radiation could be increased by 54 W m$^{-2}$ in July. Maykut shows that the extra ice melt (0.33 m of ice) resulting from his calculated increase in available energy is partially offset by the retention of meltwater in undrained puddles, which later refreeze, and by the effects of increased brine volume in the ice. The net mass loss may be only 0.15 m. However, the revised net radiation estimate for July could increase this figure to about 0.25 m.

## ARCTIC ATMOSPHERIC CIRCULATION

In the middle and upper troposphere, between about 3 km and 10 km, the circulation in high latitudes is dominated by a circumpolar cyclonic vortex (Barry and Hare, 1974). The general westerly air circulation around this cold-cored low-pressure area is a result of the large-scale equator to pole temperature gradient and the Earth's rotation. Even at sea level, low-pressure centers have a greater than 2:1 frequency over high centers, in contrast to the traditional idea of a persistent polar anticyclone. The latter exists as a climatological feature with pressures of ≥1016 mb only in March–June and in September when high pressure is centered west of the Canadian Arctic Archipelago (Fig. 1-3*B*) (Markham, 1981). Knox and Hay (1985) identify frequent blocking in the Arctic from about 90°W westward to 40°E. The annual mean pressure field for 1979–1982, based on buoy data, shows a mean high over the Beaufort Sea (Colony and Thorndike, 1984) reflecting this tendency, and the dominance of a ridge of high pressure in winter months between Alaska and Siberia across the Beaufort and East Siberian seas (Fig. 1-3*A*). During the mid-winter months, anticyclones centered between about 75–85°N, 100°W–170°E, dominate the mean monthly surface pressure field in about 40–50 percent of years (Zakharov, 1976*b*, p. 34). In summer, the mean pressure gradients are slack and generally indeterminate (Figure 1-3*C*).

Analysis of synoptic pressure maps for January, April, July, and October 1958–1977 indicates the occurrence of approximately one cyclone per 5° latitude, −10° longitude box per month over the Arctic (Whittaker and Horn, 1984). However, cyclones were not counted in the same box more than once, and this survey, therefore, does not allow for the persistence tendency of large-scale pressure patterns in the Arctic noted by Wilson (1967); this tendency is especially apparent from December through May.

Table 1-5. Areal Percentage of Melt-Puddle Extent from "North Pole" Station Observations and Estimated Albedos

| | *June* | | | *July* | | | *August* | | |
|---|---|---|---|---|---|---|---|---|---|
| | *1–10* | *11–20* | *21–30* | *1–10* | *11–20* | *21–31* | *1–10* | *11–20* | *21–31* |
| Mean area | 0 | 2 | 7 | 18 | 25 | 23 | 15 | 5 | 1 |
| Maxima (any station) | 0 | 15 | 30 | 36 | 45 | 37 | 30 | 16 | 14 |
| Albedo for mean puddling | 74 | 72 | 65 | 53 | 48 | 50 | 56 | 68 | 73 |
| Albedo for maximum puddling | 74 | 56 | 43 | 36 | 28 | 35 | 43 | 55 | 57 |

*Source:* From Nazintsev, 1964.

a

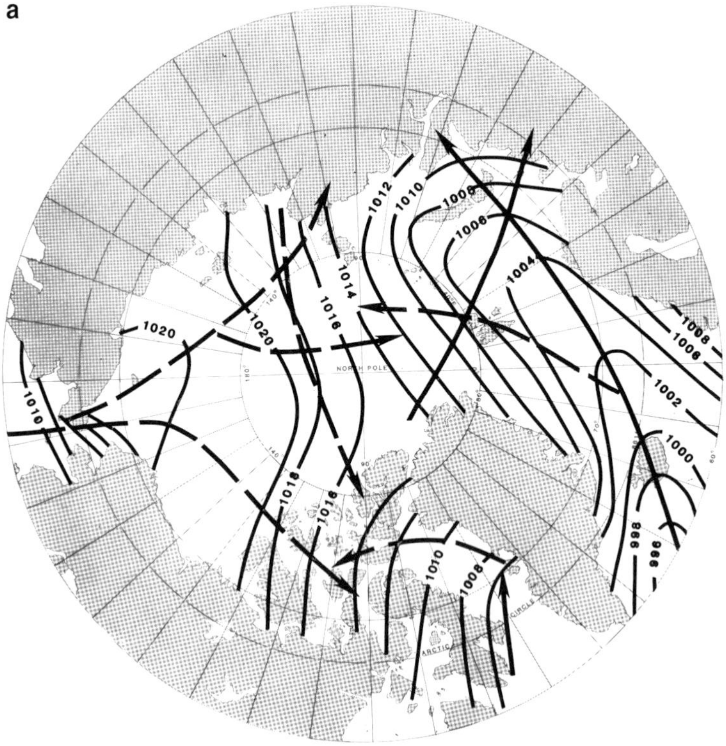


b

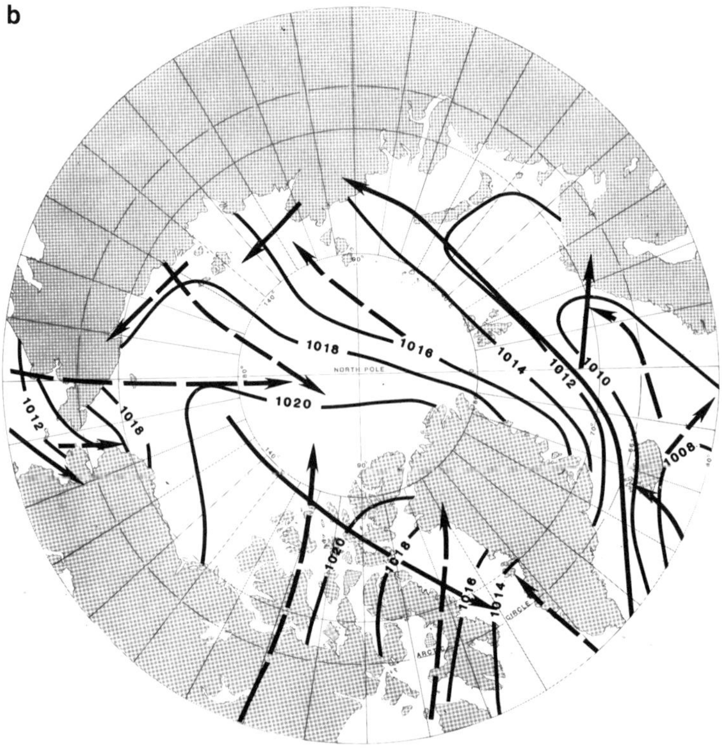


Figure 1-3. The mean MSL pressure distribution (mb) and cyclone tracks in *(A)* January, *(B)* April, *(C)* July, *(D)* October. Primary (solid arrows) and secondary (dashed arrows) cyclone tracks are indicated. (After Gorshkov, 1983.)

c

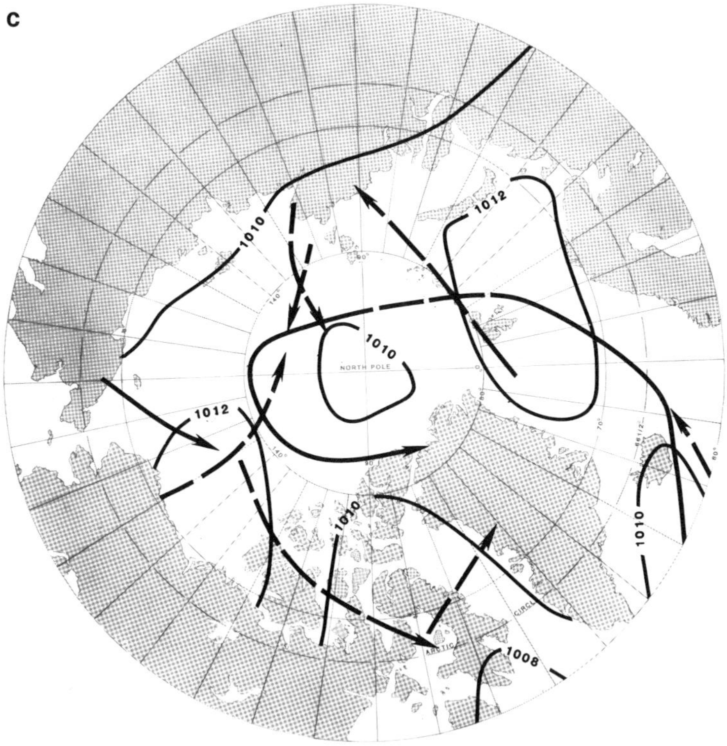


d

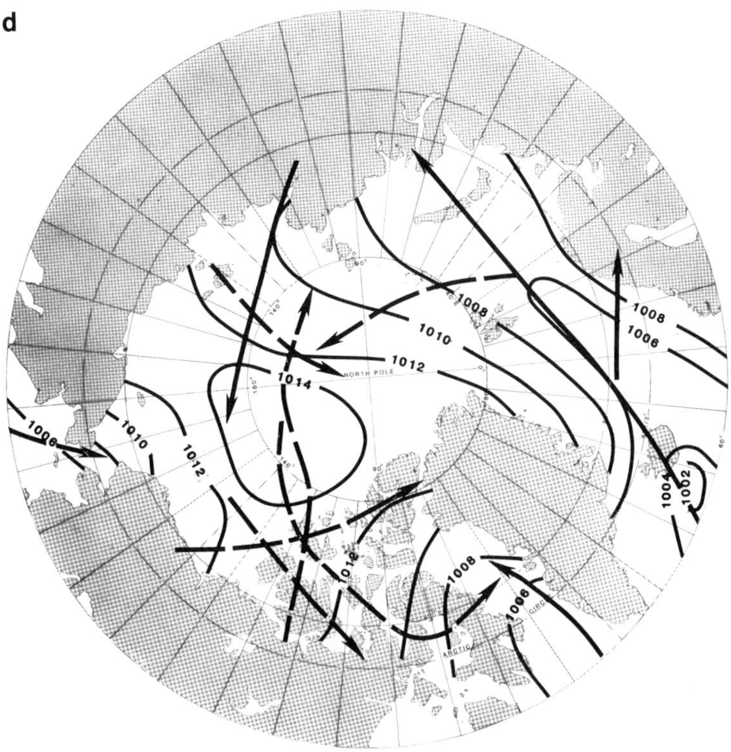


Figure 1-3. Continued

LeDrew (1983, 1985), in an analysis of National Meteorological Center daily height grids at 850 mb for January–March and June–August 1975 and 1976 for the Asiatic and North American sectors of the Arctic, finds cyclones in 80 percent of the summer cases and high-pressure features in 32 percent. Two-thirds of the lows were in the Asiatic sector. For the winters, lows had a 73 percent frequency, with most occurring again in the Asiatic sector. Interestingly, the lows are deeper in winter than summer, but less intense in terms of calculated vertical velocities at 850 mb. This result contradicts the idea that summer cyclones tend to be weak, stagnating systems; Gaigerov (1967) also reports systems with well-defined, upper frontal zones and extensive cloud and precipitation. LeDrew considered the role of surface diabatic heat sources in cyclonic activity. Surprisingly, in both winter and summer, the effects of sensible heat flux from open-water areas were found to be minor. In five case studies, the sensible heat flux was found to be significant to vertical motion and heating rates at 850 mb in only one cyclone over the Laptev Sea in October 1979 (LeDrew, 1984). Instead, large-scale advection of vorticity and thickness* play the major role in both low- and high-pressure systems. Surface friction is an additional factor, especially in the Beaufort Sea sector and, locally, orography is also important.

Various sources show general agreement that the main zone of sea-level cyclonic activity within the higher latitudes in winter (Fig. 1-3*A*) is in the Norwegian–Barents Sea with a separate center of frequent cyclones in Baffin Bay. Serreze and Barry (1988) analyze winters 1979–1985 using pressure data incorporating information from the Arctic Ocean Buoy Program network. Keegan (1958) uses maps for five winters, 1952–1957, and Gaigerov (1967) for January 1957–1961. The latter indicates that in both these areas the systems are also cyclonic at 500 mb, implying that they are mainly tropospheric cold lows (Flohn, 1952). Low-pressure systems from southwest Greenland often rejuvenate and enter the Norwegian Sea, whereas those in the North Pacific less frequently enter the Arctic Basin via the Bering Strait in winter. Typical Arctic cold lows exhibit much less cloud and precipitation in winter than in summer (Reed, 1962), in line with the diagnostic analysis of LeDrew (1985).

In summer, there is more divergence between the Soviet data and other sources. Reed and Kunkel (1960) and Serreze and Barry (1988) show frequent sea-level cyclones in the central Arctic (80–85°N, 140–180°W), over the Kara Sea, and in Baffin Bay, whereas the Gaigerov (1967) map shows maxima only along the Norwegian coast and in the Bering Sea. The Kara Sea is shown as being equally anticyclonic and cyclonic on the sea-level map of Gaigerov. He notes that in summer 1955, nearly all cyclones (84 percent) were upper-level systems and his 5-year average July map for the 500 mb level shows maximum frequencies of lows from Ellsmere Island to the pole, over northern Hudson Bay, and in the Norwegian Sea.

For August–October 1974–1979, LeDrew (1984) traced major cyclones in the Asiatic and North American (Beaufort Sea) section of the Arctic Ocean to their entry points. For 30 systems noted in the Asiatic sector, 50 percent entered from the North Atlantic and 40 percent from the USSR. During the same interval, there were 47 systems in the Beaufort Sea sector of which 75 percent entered via Alaska, 15 percent via the Bering Strait, and 10 percent via eastern Canada (cf. Fig. 3C). The mean minimum pressure of the Asiatic systems was 995 mb, compared with 1,002 mb for the Beaufort Sea systems.

*Thickness refers to the vertical depth of the atmospheric layer between two surfaces of constant pressure; it is proportional to the mean temperature of the layer.

## PRESENT CLIMATIC REGIME

The climatic character of the Arctic Ocean is shaped primarily by the annual cycle of solar radiation, on the one hand, and by the year-round presence of sea ice, on the other.* During the polar night, radiative cooling of the surface forms a persistent surface-based temperature inversion, capped by an isothermal layer, with typically a 10°C contrast between the surface and the inversion base. This is located at a height of 0.5–1.5 km according to Bilello (1966), and Vowinckel and Orvig (1967). Acoustic radar measurements indicate that mechanical mixing in the planetary boundary layer extends to between 100 m and 400 m (Carsey, 1978). The strength of the inversion varies considerably; it weakens in response to winds in excess of 10–12 m $s^{-1}$ and to the radiative effects of ''warm'' medium-level clouds associated with cyclone systems from middle latitudes. However, frontal systems often move above the inversion, and surface temperature changes depend more on cloud cover and associated infrared radiation fluxes than on changes of air mass. Under cloudless skies in January, the net surface IR flux in the central Arctic is about $-60$ W $m^{-2}$, compared with $-32$ W $m^{-2}$ for actual cloud conditions (Herman, 1980).

Heat flux through the ice prevents the occurrence of extreme temperature minima in the Arctic such as those recorded in Siberia and on the Antarctic Plateau. Mean daily air temperatures in midwinter are around $-30$°C to $-35$°C (Prik, 1959) with only about 5 percent of hourly readings in January below $-40$°C (Hastings, 1961).

Daily short-wave radiation totals increase rapidly in spring, especially as cloudiness remains quite low, but the high surface albedo (0.8–0.9) of the snow-covered ice permits little absorption. This factor is much more significant for the delayed temperature rise than the commonly cited effect of the long slant path through the atmosphere (large optical depth) traversed by the solar beam in high latitudes. The low atmospheric vapor content, amounting to only 4 mm of precipitable water in April (Burova, 1981*b*), further limits the atmospheric absorption of outgoing infrared radiation from the surface.

The diurnal temperature amplitude is characteristically small in winter and midsummer, but reaches a maximum in April. This so-called ''Fram'' type of seasonal trend of diurnal temperature amplitude is attributable primarily to the large diurnal range of absorbed solar radiation, relative to the damping introduced by the turbulent heat fluxes between surface and atmosphere (Ohmura, 1984). In April, only sensible heat flux is involved, whereas in summer the increased amplitude of absorbed solar radiation is offset in part by increases in the daily range of turbulent heat fluxes related to melting. A further contribution to smaller amplitudes in summer arises through the $(T)^{-4}$ term in the Stefan-Boltzmann equation because mean temperatures $(T)$ are higher. The Stefan-Boltzmann equation relates the radiance flux density $(F)$ from a black body source to the fourth power of its absolute temperature.

$$F = \sigma T^4$$

where $\sigma$ is the Stefan-Boltzmann constant.

Cloudiness over the Arctic Ocean increases considerably in May (Huschke, 1969), mainly as a result of increased large-scale advection; there is little melt or breakup in the central Arctic until late June (Barry et al., 1987). The cloud cover counteracts the seasonal increase in incoming solar radiation to a high degree. It is a striking fact that the total clear sky solar radiation as a percent of the extraterrestrial value is about the same from May through

*Climatic conditions in the tundra are referred to only incidentally here; they are discussed fully elsewhere (Barry and Hare, 1974; Dingman et al., 1980; Ohmura, 1982).

August at 90°N as at 70°N, whereas the fraction actually absorbed at the surface is substantially smaller at the pole (Doronin, 1970, p. 29). This absence of a latitudinal gradient of potential radiation arises because the day-length effect offsets the solar altitude factor. At 90°N, the calculated values of absorbed solar radiation at the surface remain virtually constant from May through August as a fraction of extraterrestrial radiation (Table 1-6).

Significant melting of the snow cover and the development of puddle formation on the ice takes place only in late June in the central Arctic. The start of snow melt corresponds to a mean air temperature of about −1.2°C (Doronin, 1970, p. 107). Table 1-5, based on observations from Soviet "North Pole" drifting stations, illustrates the development of puddling and the associated albedo changes. The increase in evaporation due to the surface melt and turbulent mixing create a surface mixed layer, or isothermal structure, although a weak elevated advection inversion is still present about 40 percent of the time and surface inversions have a 15–20 percent frequency (Vowinckel and Orvig, 1967).

Conventional observations indicate that cloud cover averages 70–80 percent during July–August (Huschke, 1969), but analysis of satellite imagery for summers 1977 and 1979 by Robinson, Kukla, and Serreze (1986) shows episodes of clear skies over the Arctic in June and less frequent thick stratus than earlier sources. For the central Arctic, north of 80°N, thick, multilayered cloud averaged 36 percent frequency for June through mid-August, thin cloud averaged 34 percent, very thin cloud averaged 10 percent, and clear skies averaged 16 percent. The predominant summer cloud type is stratocumulus and stratus. However, the cloud base is generally below the inversion and commonly the cloud is thin. Soviet aircraft studies over the Kara and Barents seas indicate an average thickness of about 200–400 m (Crane and Barry, 1984). Data from extensive flights by the Soviet Flying Observatory during April–September 1949–1955 show thicknesses of 700 m or less in 76 percent of all observations; greatest thicknesses, up to about 2 km, occur in summer (Zavarina and Romasheva, 1960). Over the Beaufort Sea, single-layer stratiform cloud averages 250–300 m in thickness, double layers between 100 m and 700 m for the lower layer, 400 m and 700 m for the upper layer (Tsay and Jayaweera, 1984).

Precipitation amounts in the Arctic are typically low during the winter half of the year. The precipitable water content in the atmosphere (surface to 5 km) averages less than 2 mm in February and 10–12 mm in August (Burova, 1981*a*), reflecting the low temperatures in winter, which limit the absolute humidity and saturation mixing ratio, and also the relatively infrequent advection of moisture by cyclonic activity. Over the Queen Elizabeth Islands in summer, the low rainfall is determined by the few occasions when the lower troposphere is near saturation (Jackson, 1961), which in turn depends on the occurrence of active frontal cyclones and associated vertical motion.

In the central Arctic, about 20 cm of snow accumulates in the autumn, followed by a slower increase to 30 cm by February and 40 cm by early May (Loshchilov, 1964). On relatively level ice surfaces in the Kara Sea, there is a snowpack of 30–35 cm by April–May, but amounts diminish eastward to 14–22 cm in the Laptev–East Siberian seas (Buzuev, Romanov, and Fedyakov, 1979). According to these authors, depths vary considerably on a local scale in relation to the degree of hummocking and ridging, and also to the age of the ice floes, although Hanson (1980) reported similar mean depths on three-month-old and seven-month-old floes, as a result of wind drifting. Mean wind speeds in January are about 6 m $s^{-1}$ (Burova, 1981*a*). Values are higher in areas where cyclones are more frequent and the surface inversion is less intense and persis-

Table 1-6. Potential and Actual Solar Radiation at the Surface (as a Percentage of Extraterrestrial Radiation)

| *Latitude* | *Parameter* | *Feb.* | *Mar.* | *Apr.* | *May* | *June* | *July* | *Aug.* | *Sept.* | *Oct.* |
|---|---|---|---|---|---|---|---|---|---|---|
| 90° | Clear sky | | | 76 | 80 | 79 | 75 | 71 | 52 | |
| | Actual incoming | | | 57 | 62 | 54 | 42 | 39 | 24 | |
| | Absorbed | | | 14 | 22 | 24 | 24 | 20 | 10 | |
| 70° | Clear sky | 68 | 73 | 80 | 79 | 76 | 74 | 72 | 70 | 66 |
| | Actual incoming | 46 | 62 | 65 | 61 | 50 | 45 | 40 | 38 | 34 |
| | Absorbed | 17 | 24 | 26 | 29 | 34 | 36 | 33 | 31 | 19 |

*Source:* From Vowinckel and Orvig, 1962.

tent. Thus, in the Barents Sea, there is a 40 percent frequency of winds over 11 m $s^{-1}$ in December–February, compared with about 10 percent over the Pole (Hastings, 1961). Newly fallen snow begins to drift with winds of 2–3 m $s^{-1}$, and above 5–8 m $s^{-1}$ snow is lifted off the ground.

A stable snow cover is established by 10–15 September, on average, in the Franz Josef Archipelago and Severnaya Zemlya, 20–30 September in the Kara Sea, Taimyr Peninsula, Laptev–East Siberian seas (Dolgin, 1959). The end of stable snow cover occurs at the end of May in the East Siberian Sea, 5–15 June in the Laptev Sea, and not until the end of June in the Franz Josef Archipelago and Severnaya Zemlya. In the central Arctic, snow melt begins around 10 June at latitude 80°N and about 30 June near the Pole, with melt puddles beginning to form about 10 days later, according to Yanes (1966). The melt onset appears to be determined by synoptic events and its progression is not simply latitudinal (Crane, Barry, and Zwally, 1982). The spatio-temporal pattern of snow melt, mapped from satellite imagery, shows substantial year-to-year variations in early summer. These are large enough to impact the surface energy budget significantly (Scharfen et al., 1987; Barry et al., 1988). The minimum albedo values attained in August in the central Arctic seem not to be greatly affected by these fluctuations. However, there may be reactions in the degree of ice retreat in the marginal Arctic seas. By mid-July when most of the snow has gone, extensive pools up to 70–80 cm deep form on the surface (Chukanin, 1954).

Precipitation in July–August generally falls as rain. At stations along the Arctic coasts rainfall represents up to 40–50 percent of measured annual totals, but over the Arctic Ocean a figure of 20–25 percent seems more likely. Gauge measurements are known to underestimate winter snowfall by at least 60 percent and summer precipitation is also underestimated about 10 percent as a result of the frequent occurrence of trace readings that are recorded as a zero value (Dingman et al., 1980). When winter and summer measurements are adjusted accordingly, snowfall accounts for about 65 percent of annual precipitation in coastal Alaska.

For the central Arctic Ocean, the adjusted annual precipitation is about 20 cm and the evaporation is 7 cm (Bryazgin, 1976; Khrol, 1976). The corresponding figures for the entire Arctic Ocean are 36 cm and 22 cm, respectively (Burova, 1981*a*). Figure 1-4 shows the driest area to be west of the Canadian Arctic Archipelago.

## RECENT VARIABILITY IN ARCTIC CLIMATE

The observational record in high northern latitudes is very limited, although in some locations around the North Atlantic there are long-term proxy records of sea ice and atmospheric parameters from ice cores. The instrumental data from the land areas show a significant warming in high northern latitudes between the 1890s and about 1940, of the order of 2°C for annual temperatures, three times the hemisphere mean values (Kelly et al., 1982) (Fig. 1-5). Temperatures declined after 1940, up to the 1960s, with some subsequent recovery. The 1880s to 1930s–1940s warming was greatest (~2.5°C) in winter and only 1.3°C in summer. Kelly and colleagues note that the major trends affected the whole Arctic, but were strongest in northwest Greenland and the Kara–Barents seas, whereas short-period fluctuations were most pronounced near major sectors of sea-ice growth/decay in the western North American Arctic, the Siberian Arctic, and the Barents Sea.

Correlation analysis of the August ice area (Zakharov, 1981) with summer temperatures for 65°–85°N (Kelly and Jones, 1981), where both series were smoothed by a 5-year binomially weighted moving

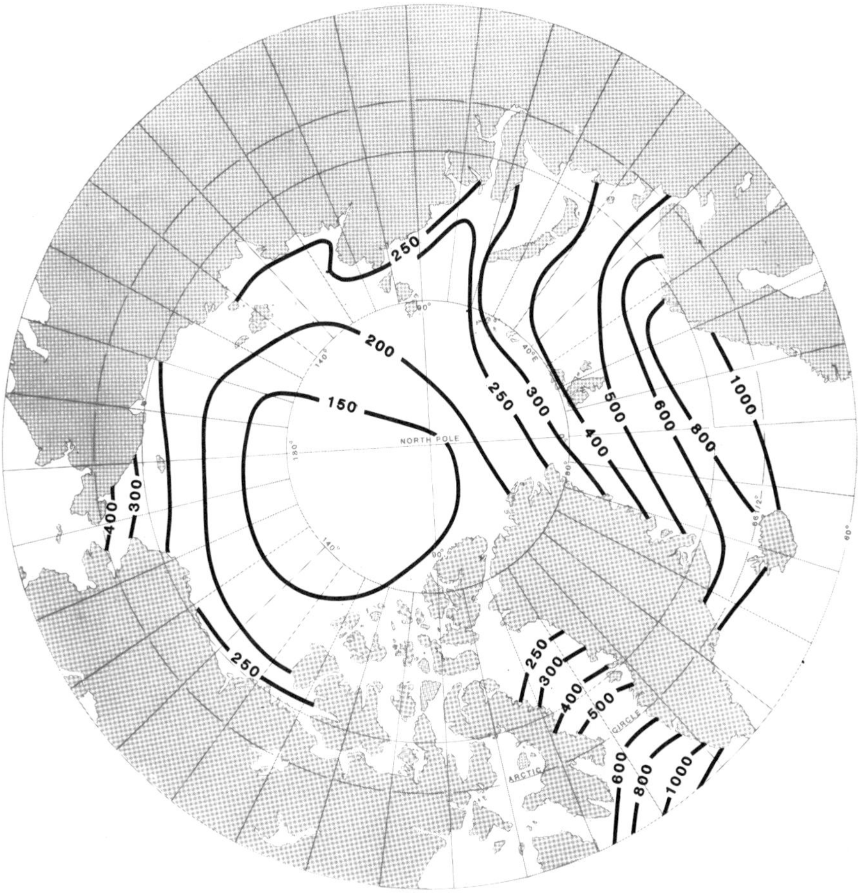


Figure 1-4. Mean annual precipitation (mm), adjusted for bias. (After Gorshkov, 1983.)

average, shows an *r*-value of $-0.81$ for 1951–1980, but the correlation for the full record, 1924–1980, is only $-0.13$ (Barry, 1983). It is not clear whether this breakdown of the close inverse association evident over the last 30 years is solely a result of data inadequacies pre-1935 or is a manifestation of some independent underlying factor such as oceanic changes. The range in minimum summer ice extent is about $10^6$ $km^2$ (15 percent of the mean value), corresponding approximately to a 1°C change in summer temperature, although temperatures and ice extent in individual years show little correlation. The strongest tendency for ice anomalies to respond to atmospheric forcing, with lags of up to several months, is apparent between late February and late July (Walsh and Johnson, 1979). Anomalies of ice extent in spring appear to be primarily regional in extent and driven by anomalies in atmospheric circulation. Soviet workers, for example, have long recognized a tendency for antiphase relationships between ice in the Barents–Kara seas and ice in the Chukchi–East Siberian seas (Barry, 1983).

Long-term records for the East Greenland ice off Iceland, originally analyzed by Koch (1945), have been reexamined by Ogilvie (1984). The broad pattern remains the same: severe ice conditions during the 1690s and 1740s, and less severe conditions during 1640–1670, just prior to a peak

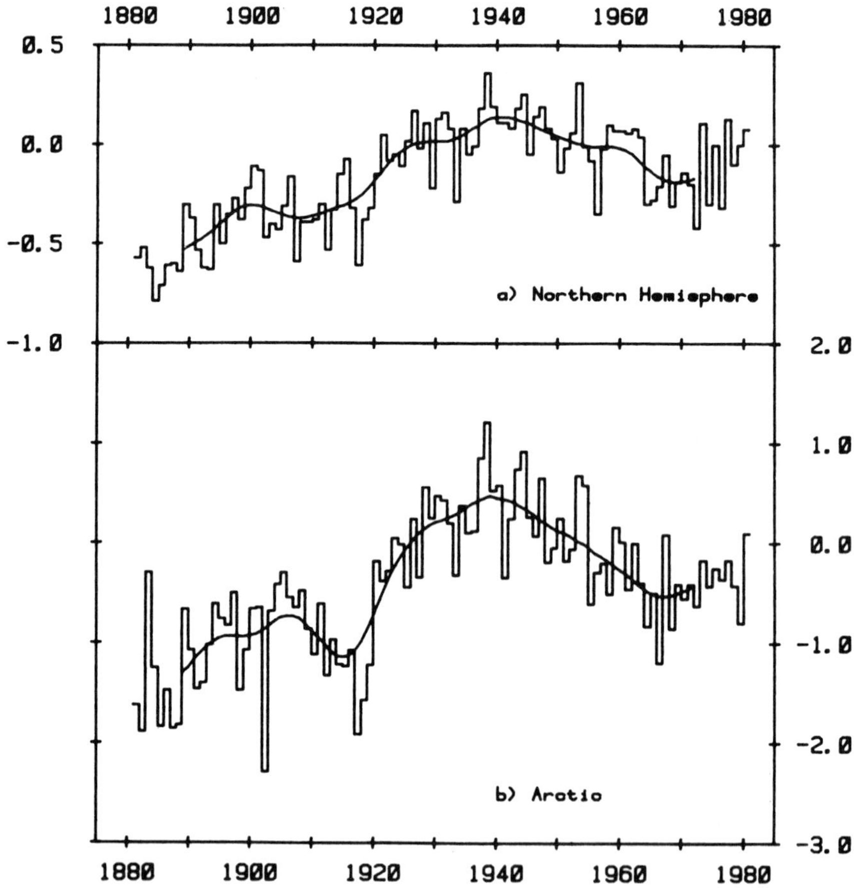


Figure 1-5. Annual temperature (°C) as departures from the 1946–1960 reference: *(a)* Northern Hemisphere (0°–85°N); *(b)* Arctic (65°–85°N). (From Kelly et al., 1982.)

of Little Ice Age severity in Europe. The 1750s are shown to have been exceptionally cold but with less ice than reported by Koch. Ogilvie finds broad agreement between a sea-ice index (the number of seasons—winter, spring, and summer—with ice off the coast of Iceland, weighted by the number of regions, per decade) and a winter/spring thermal index (the number of mild seasons minus the number of severe seasons per decade, weighted by the number of regions, divided by the sum of these two weighted numbers); for the 18 decades of 1601–1780, the correlation is −0.57, similar to nineteenth- and twentieth-century values. However, she notes that severe climatic conditions do not necessarily imply the presence of sea ice (as shown for the 1630s and 1730s) and vice versa (as noted for 1701–1710). Further work by Kelly et al. (1987) demonstrates that the Icelandic sea ice record is a useful indicator of atmospheric and ocean circulation north of Iceland, but is not representative of hemispheric ice or climate conditions.

## PAST CLIMATIC MODES: EVIDENCE AND CONCEPTUAL MODELS

Consideration of past climatic and ice regimes in the north polar region is greatly hampered by the limited observational evidence. This section examines the available evidence and the models proposed for Arctic climatic and ice history.

By the early Cenozoic, the North Pole was located within the Arctic Basin, but global climate underwent major changes during the Tertiary period that are, at least in part, attributable to the effects of continental drift and tectonic evolution with the associated shifts in ocean circulation and maritime/continental climatic influences. Europe began to separate from Greenland in the late Cretaceous, but land linked Greenland and Svalbard until 38 mya. There was probably no significant opening for Atlantic–Arctic water exchange in the Greenland Sea until after 27 mya and even during the Pliocene the Greenland-Svalbard passage remained narrow (Talwani and Edholm, 1977). The Iceland–Faeroes Ridge also separated a shallow Norwegian Sea from the Atlantic. Similarly, despite an opening of Baffin Bay 53 mya (Sclater, Hellinger, and Tapscott, 1977), Baffin Bay remained separated from the deep Labrador Sea 36 mya by volcanic rocks in Davis Strait. This may account for observed difference between marine faunas in the Arctic and Labrador seas (Hopkins and Marincovich, 1984).

The occurrence of broad-leaved, mainly deciduous, nemoral* flora in high northern latitudes during the early Tertiary has long been recognized, although their paleoclimatic significance remains controversial. Two key questions important to climatic reconstructions for the Tertiary concern the paleolatitudes of fossil assemblages (Donn, 1982) and the determination of appropriate analogs in modern vegetation and climatic zones. Wolfe (1980) postulates that high latitudes during the Paleocene and Eocene had mean annual temperatures of 10°–20°C, a low mean annual temperature range of 10°–15°C, and abundant precipitation. However, he recognizes the problem posed by the occurrence of long, dark winters in terms of the proposed equability and forest growth. A possible solution to this may exist if atmospheric $CO_2$ levels were several times their present values during the Tertiary as suggested by Budyko and Efimova (1981). Barron (1985) has proposed that higher $CO_2$ concentrations may account for the warmth of the Cretaceous period with a decrease in concentration contributing to the Tertiary cooling. However, recent methods for estimating past $CO_2$ levels from geological records are open to revision and the sensitivity of global temperatures to several-fold increases of $CO_2$ remains to be modeled in detail (Berner and Barron, 1984). There may be no necessity for special arguments to explain forests and vertebrates in high latitudes during the Cretaceous and Tertiary (Axelrod, 1984) because trees and shrubs now occur poleward of 66°N in areas that were ice covered only 8000–10,000 years ago. Taxa are limited more by summer temperatures than by light and a temperate, equable climatic regime is considered by Axelrod to be sufficient explanation. He suggests that Eocene vertebrates probably hibernated in the season of darkness or low light. Mammals may have been nocturnal or crepuscular in activity.

The implied temperate regime in high northern latitudes during the early Tertiary has served as the basis for hypotheses of an open Arctic Ocean (Flohn, 1982; Thiede, 1983), although at least the timing of Arctic Ocean freezing (2.4 mya according to Flohn) is controversial (Clark, 1982*a*, 1982*b*). Climate deteriorated after the mid-Eocene in the northern North Atlantic (Gradstein and Srivastava, 1980), but hardwood deciduous forest was present on Banks Island (72°N) during the Miocene (Vincent et al., 1983) and coniferous forest was present at 70°N in eastern Siberia and in Iceland according to Wolfe (1980). Wolfe infers a mean annual temperature of 3°–5°C, with a range up to ap-

*A flora adapted to a mild-temperate, moist climate such as the modern midlatitude deciduous forests.

proximately 30°C at 70°N. Hence it seems unlikely that tundra vegetation and ice sheets had yet developed, in contrast to the major east Antarctic ice sheet that formed in the mid-Miocene (14 mya or even 30 mya) (Kerr, 1984). Simulations with Adem's thermodynamic model suggest that Arctic temperatures did not drop below freezing until the land masses had drifted to near their present locations 10–15 mya (Donn and Shaw, 1977). However, it has recently been proposed that summer temperatures in the North Atlantic sector fell during the Eocene as the Norwegian Sea opened, and Greenland separated from North America and moved northward. This separation could have led to the formation of a Greenland ice sheet as early as 30 mya (North and Crowley, 1984; Crowley et al., 1986), although the results are at variance with most conventional paleogeographic reconstructions for the Arctic. Nevertheless, the concept that land–sea distribution exerts a primary control on the annual temperature cycle and particularly summer warmth merits further study.

Löve and Löve (1974) propose that arctic cryophytes replaced the nemoral flora in the Pliocene. Wolfe (1985) suggests that taiga appeared north of the Arctic circle only about 5 mya, and tundra appeared 2–3 mya in the late Pliocene after the glacial mode of water flow from the Arctic to the Atlantic, through the Labrador Strait, became fully established (Gradstein and Srivastava, 1980). The Bering Strait link to the Arctic has existed during interglacial times of high sea level since at least 2.2 mya (Hopkins and Marincovich, 1984). The final closure of the Panama Isthmus about 3.2 mya strengthened the Gulf Stream system and, together with a renewed link between the Mediterranean Sea and the North Atlantic, which increased the salinity of the Atlantic, the closure favored the onset of glaciation on the northern continents. Oxygen isotope evidence (Shackleton and Opdyke, 1977) shows that there was no large-scale continental glaciation in the Northern hemisphere prior to 3.1 mya. Blanc, Fontugne, and Duplessy (1983) identify the first boreal ice caps as forming in Iceland, presumably Greenland, and parts of North America, such as the Sierra Nevada in California. Northern Europe became drier rather than colder at this time. Ice-rafted detritus reached the Hatton–Rockall plateau 2.4 mya (Shackleton et al., 1984), but the west–east climatic differentiation across the North Atlantic disappeared only about 2.3 mya when the oceanic polar front became more zonally oriented and a major cooling phase occurred with ice developing in Scandinavia (Blanc, Fontugne, and Duplessy, 1983). Similar timing is proposed for the Bering Sea and Gulf of Alaska where diatom biostratigraphy indicates significant ice rafting beginning about 2.5 mya near the Matuyama–Gauss magnetic reversal boundary (Rea and Schrader, 1985).

It is not known when the Neogene high-latitude cooling led to the formation of Arctic sea ice. There are currently at least three conflicting views of conditions during the Arctic in the Quaternary (and prior to it):

1. The area was covered during glacial intervals by a floating ice sheet 1000 m thick (Broecker, 1975; Williams, Moore, and Fillon, 1981; Fillon 1984) and/or by ice shelves (Denton and Hughes, 1981).
2. The Arctic Ocean was seasonally open until approximately 0.9 mya to 0.7 mya, after which time ice-free and ice-covered periods alternated (Herman, 1970, 1974, 1983*b*; Herman and Hopkins, 1980; Herman and Osmond, 1984).
3. Sea ice was present before 2 mya and has been since 0.7 mya, with alternating intervals of more open thin-ice or ice-free conditions during the intermediate period (Gilbert and Clark, 1983). Clark (1971) originally proposed that the Arctic Ocean has been continuously frozen since the mid-Pliocene and subsequently suggested that initial freezing

occurred 30 mya in the mid-Cenozoic (Clark, 1980).

These divergent views exist as a result of the incomplete sedimentary record of the Arctic Basin, problems of interpretation, and the difficulties of dating the material. Most workers suggest low sedimentation rates (1–5 mm/$10^3$ yr$^1$) in the deep basins, whereas results of Sejrup et al. (1984) would imply rates of a more rapid order of magnitude. For example, 265 sediment cores studied by Clark et al. (1980) are interpreted from paleomagnetic information to span approximately 5 million years and the dating of at least the upper (Brunhes epoch) segment has been confirmed from continental glacial and interglacial records on Banks Island, NWT (Clark et al., 1984). In contrast, Sejrup et al. (1984) suggest that selected core sections thought to represent 3 million years may span <200 thousand years according to amino-acid geochronology. Darby et al. (Chapter 24) confirm the general time scale established by Lin'kova (1965) and confirmed by Hunkins et al. (1971); they interpret the lowest polarity reversal as Matuyama/Gauss (ca. 1.67–1.87 million years). Their work and new radiometric data by Herman, Osmond, and Somayajulu (Chapter 23) also conflict with the high sedimentation rates postulated by Sejrup and co-workers; the need for further research to resolve this question definitively is apparent.

Reexamination by Witte and Kent (1988) of the magnetostratigraphy of ocean sediment cores collected in the Canada Basin, from ice island T-3, confirms the low sedimentation rates of about 1 mm $10^3$ yr previously reported. Sedimentation rates in the Eurasian Basin are much higher. Thiede (1988) reports values of 1–4 cm $10^3$ yr north of Svalbard, apparently related to the different ice regime and eolian sources over the past 300,000 years.

One rationale for the ice-sheet hypothesis is the need to account for a discrepancy between glacial sea-level lowering calculated from glacio-eustatic records of emergent coastlines and the larger global ice volume indicated by $\delta^{18}O$ records from benthic foraminifera. Thick floating ice shelves grown from atmospheric precipitation could explain this discrepancy. However, uncertainty over the relative roles of ocean temperature and continental ice volume effects on the $\delta^{18}O$ composition of sea water, and overestimates of the isotopic contribution of a floating Arctic ice cap (Dodge et al., 1983) weaken this case. Denton and Hughes (1981) argue independently for a marine ice sheet grounded on the continental shelves, subsequently spreading as floating ice shelves over the deep basins during glacial maxima. This derives from their concept of the necessity for buttressing of ice sheets with marine termini. Lindstrom and MacAyeal (1986) provide support for this argument by modeling the dynamics of ice-shelf buttressing in the Norwegian–Greenland seas for a grounded marine ice sheet in the Barents Sea. Elverhøi and Solheim (1983) find evidence of such an ice sheet at least once during the Pleistocene, but they could not date this feature and discounted the claim for an ice shelf. The Arctic Ocean sedimentary record (if the dating is reliable) shows no evidence of the expected barren intervals that would be associated with thick floating ice (Clark, 1982*a*).

The hypothesis of fluctuating ice regimes during the late Cenozoic is based on analysis of sediment cores. Herman (Chapter 23) identifies three major climatic units. Unit III, deposited prior to 2.5 million years, consists of red clays with fairly well-sorted manganese concretions. Unit II, approximately 2.5–0.9 million years, is mainly a foram-poor zone with at least seasonally ice-free conditions; the foraminifers are dominated by subpolar species. Unit I, spanning the last 0.9 million years, represents alternating ice-free and ice-covered periods. It is proposed that the Arctic Ocean was cold, but ice-free, with strong vertical mixing from about 5 mya to 2.5–3 mya (Herman, 1983*b*); alternatively, there may have been winter coastal ice (Spjeldnaes, 1981). Shelf ice and icebergs also appear to have been present after 4 mya. Approximately 2.5 mya, the waters are thought to have become vertically stratified, possibly with seasonal ice cover. Herman proposes that surface productivity was reduced as a result of limited nutrient availability and lower surface salinity, attributed to somewhat warmer global conditions with meltwater from a deglacial mode affecting the Arctic Ocean about 2.5 mya.

Poorly sorted olive-green and olive-gray "glacial marine" sandy clays of the type deposited in the Arctic between ~3 and ~0.9 mya and dominated by subpolar *Globigerina egelida* cover the present floor of northern Baffin Bay, possibly providing a modern analog. Herman (1983*a*) states that this area is "generally ice-free year-round" at present but, except for the small North Water polynya (~78°N), northern Baffin Bay is in fact ice-covered from November through May (Markham, 1981), and heavy ice lingers off Baffin Island and in the central part of Baffin Bay through July, into early August, in most years. Herman believes that perennial Arctic sea ice formed about 0.9 mya, although fluctuations in abundance of foraminifera are taken to imply intervals of more or less open ice conditions. Foram-poor layers are interpreted as mild, low-salinity deglacial intervals; whereas Clark considers the ice to have been thickest at these times, O'Neill (1981) has disputed the evidence for high productivity before 2.7 million years; arenaceous benthic foraminifera occurring during the early Pliocene (3.5 mya) suggest poorly oxygenated bottom waters, in contrast to the Herman and Hopkins (1980) model of residual sediments deposited in well-aerated basins. O'Neill postulates that conditions ameliorated as a result of increased water flow from the Atlantic and that a sharp increase in planktonic forms about 1.2 mya represents "thinner" (presumably less extensive) ice cover. Cores spanning the last 1.25 million years have generally high abundances of planktonic foraminifera and calcareous benthic assemblages. Low abundance of planktonic species prior to about 1.25 mya could be a result of low productivity (associated with ice cover), dissolution, or both (O'Neill, 1981).

Clark (1980, 1982*a*, 1982*b*) has claimed considerable uniformity of sedimentation processes since at least the late Miocene, implying an essentially "steady-state" Arctic with perennial ice cover. Sedimentation by sea ice and iceberg rafting is dated from at least 3.5 mya on the Alpha Cordillera (Minicucci and Clark, 1983). The sediment deposition process in the Arctic is examined by Clark and Hanson (1983). Fine-grained sediment is transported by sea ice whereas coarser material is rafted principally by glacial ice. At present, icebergs and ice islands (from the Ward Hunt ice shelf, northern Ellesmere Island) are a very minor component of the Arctic ice cover, but they might have been more frequent with ice caps over the Queen Elizabeth Islands, Svalbard, Novaya Zemlya, and the Franz Josef Archipelago. As shown by the present iceberg distribution in the Labrador Sea, they may travel long distances from their source. The existence of a Barents Sea ice sheet (Hughes, Denton, and Grosswald, 1977; Elverhøi and Solheim, 1983) and similarly one on the continental shelf in the Kara Sea is still debated (Gros'vald, 1983), although Boulton (1979) and Velichko et al. (1984) consider a distribution of small ice caps on islands and plateaus of the shelves is more likely. Clark and Hanson note alternations in the deposition of coarser sediments, although these are not dated. Clark (1982*a*) also suggests that somewhat more favorable conditions than at present may have existed between 2.0 and 0.7 mya. This argument is extended by Gilbert and Clark (1983) using other evidence from calcareous dinoflagellates. Their life cycle is unknown, but both dinoflagellates and foraminifera are generally rare during times of ice-rafted deposition of coarse sediment. A tentative interpretation is offered that the Arctic Ocean was probably ice-covered before 2 mya, and since 0.7 mya, with intervals of more open, thin ice or even ice-free conditions alternating with conditions similar to the present between 2 and 0.7 mya.

Indirect evidence that has an important bearing on this problem is available from the Norwegian Sea (Vogt, 1986). Kellogg (1977) postulated that the modern Norwegian Current existed only during the Last Interglacial and not throughout the remainder of the last 450,000 years. Thus, except for these intervals, sea ice must have been present in the Norwegian Sea. For the same area, Belanger (1981) identifies three oceanic modes during the Last Glacial cycle. Full interglacial conditions, with convective overturning, were present only during oxygen isotope stage 5e (~ 125–115 kya), as proposed earlier by Ruddiman and McIntyre (1977), whereas full glacial conditions with year-round sea ice existed in stages 2 and 3 (~ 60–10 kya). Intermediate conditions, with at least seasonal sea ice similar to today, were present during stages 5d, 5a, most probably stage 4, and stage 1 (post-10 kya). In light of the permanent Arctic ice cover at present (stage 1 conditions), and the limited additional hemispheric warmth of the Last Interglacial (stage 5e), it seems unlikely that the Arctic became seasonally ice-free during such stages, which in any case were remarkably brief. Recent analysis of cores in the Fram Strait (Markussen, Zahn, and Thiede, 1985) also indicates no evidence for ice-free conditions in the Arctic Basin during the Last Glacial cycle although it is proposed that iceberg transport of sediment occurred during the isotope stage 3/2 and 2/1 transitions. Ruddiman and McIntyre (1981) postulate that a well-stratified surface layer was present in the North Atlantic during the last deglaciation due to meltwater influx from the Laurentide ice sheet. This low-salinity surface water permitted extensive winter sea ice to form and prevented upward mixing of nutrient-rich deep water, resulting in low biotic productivity (Ruddiman et al., 1980). This argument is similar to Herman's inferences for the Arctic Ocean, although she interprets the foram-poor layers there as mild ice-free intervals.

In the Bering Sea, Gros'vald (1983; Gros'vald and Vozovik, 1984) postulated the existence of a marine-based ice sheet and shelf complex, but cores from the Aleutian Basin in the Bering Sea indicate only a prolonged annual sea ice cover and a short ice-free season during the two glacial maxima of the Wisconsin (Sancetta, 1983; Sancetta and Robinson, 1983). Conditions were apparently similar to the present day in the Sea of Okhotsk. In mid-Wisconsin time there was a greater seasonal contrast than now in the southern Bering Sea, with some winter sea ice. This reconstruction leaves open the nature of the regime in the Arctic Ocean, since the two seas should have been linked in the mid-Wisconsin interstade.

Another source of evidence of conditions in the Arctic Ocean is provided by the late Pleistocene land records. A reconstruction of the glacial ice cover of Greenland and of the Canadian Arctic Islands, based on ice sheet flow modeling, indicates only small changes from the present ice flow patterns (Reeh, 1984), supporting the earlier observational studies of Weidick (1976) and others for limited Late Wisconsin ice extent. This implies a local origin for the ice dome over the Canadian Arctic Islands and it seems unlikely, therefore, that there was a significant moisture source in the Polar Ocean throughout the Last Glacial cycle. This argument is elaborated by Andrews and Barry (1978) and Boulton et al. (1985), who note that high-latitude glacier expansion was out of phase with that of the southern margins of the northern hemisphere continental ice sheets. Ice buildup in high latitudes is primarily dependent on poleward moisture advection, and once ice sheets accumulate in the sub-Arctic and midlatitudes, the moisture flux is reduced as cyclonic activity along the ice margin is shifted southward (Williams, 1975; Carleton, 1985). However, spectral analysis of North Atlantic $\delta^{18}O$ records shows latitudinal dependence in the 23,000- and 41,000-year orbital signals (Ruddiman and McIntyre, 1984). The former appears to reflect midlatitude and the latter high-latitude ice fluctuations. Accordingly, neither accumulation (moisture source effects) nor summer temperature (albedo-feedback effects) dominate the ice sheet responses, since the $\delta^{18}O$ fluctuations change their phase relationships over time.

The paleoclimatic record in Beringia has received particular scrutiny in the context of the "steppe-tundra" or arctic steppe question (Hopkins et al., 1982). Synthesis of the diverse paleoenvironmental records and consideration of plausible meteorological regimes suggest that the mean airflow across Alaska in summer was easterly to southeasterly during the last glacial interval. This pattern accounts for an asymmetry of glaciation on the northern and southern flanks of the east–west ranges in Alaska—a pattern characteristic of middle and late Pleistocene glaciations. However, in the earlier Pleistocene large ice bodies over north Alaska–northeast Siberia suggest at least a seasonally open Arctic Ocean, according to Hopkins and Marincovich (1984). Sea otter remains from the Alaskan Arctic coastal plain, dated 2.4–3 million years (Carter et al., 1986), also suggest that the limit of perennial sea ice may have been 1,600 km or more further north than at present in late Pliocene.

There is no indication of seasonally open water along the Alaskan coast during the Last Glacial Maximum. Winters were cold and dry; summers were mostly dry and somewhat cooler than at present (Barry, 1982; Hopkins, 1982). A specific paleotemperature reconstruction for the Alaskan Arctic coastal plain by amino acid analysis of mollusks of (assumed) last interglacial age suggests that glacial-age mean permafrost temperature averaged 8°C lower than present (Brigham and Miller, 1983). This implies that mean annual air temperatures were between −19°C and −26°C, and summers were probably cooler than at present. A recent model simulation of conditions at 18,000 B.P. compared with present by Manabe and Broccoli (1985) uses the Geophysical Fluid Dynamics Laboratory (GFDL) general circulation model in conjunction with a simple mixed-layer ocean model. Here sea surface temperatures are calculated, although cloudiness is prescribed. They show that the Laurentide ice sheet could have caused a split flow pattern (in winter). At the surface this results in a strong westerly flow north of the ice sheet, although the upper jet is weak. The calculated surface air temperature is lowered 8°–12°C over the Arctic in winter, except over northern Beringia where there is slight warming compared with the control case. For summer, strong cooling is indicated near the northern margins of the ice sheets, but small changes are indicated over the central Arctic Ocean.

Baulin and Danilova (1984) estimate an 11°–12°C cooling during the Zyranska and Sartan* glacial stages in northern Siberia, with permafrost temperatures still 8°C below present values (−7°C) during the intervening Karginsky "interglacial" (interstadial). However, a reconstruction for the Mikulino (Kazantsevo) or last interglacial indicates July mean temperatures of 14°C for most of the Siberian Arctic coast, including Taimyr, where the estimated departure from present values is 8°C (Vel-

*These stages correspond to the early (pre-50,000 yr B.P.) and late (post-25,000 yr B.P.) glacial phases of the Last Glacial cycle.

ichko, 1984); the anomalies are only 2°C in Kolyma and Chukotka. These temperatures suggest that the seasonal ice limit would have been displaced poleward several hundred kilometers from its present location in late summer (Rogers, 1978; Barry, 1983, Fig. 5).

Interglacial records from Banks Island (dated older than 61 thousand years) indicate a climate slightly warmer than today, but sediments from the preceding interglacial interval contained pieces of coniferous wood (Vincent et al., 1983). This evidence would be compatible with inferred seasonally open water in the southern Beaufort Sea during at least some warm interglacials. Recently, Boyd et al. (1984) performed spectral analyses of three central Arctic Ocean sediment cores. They identify a strong ~100,000-year (presumed Earth-orbital) signal, similar to that in midlatitude oceans, implying thinner and more mobile ice in the Arctic during full interglacial conditions. The authors assume that such conditions would permit deposition of coarse deglacial sediments and rather more abundant foraminifera. Evidence for increased ice mobility during the warmer intervals, despite generally weaker atmospheric circulation in high latitudes, is provided by Volkov and Zakharov (1977). They cite average net drift rates in the Arctic Basin of 1.8 km $day^{-1}$ prior to the 1930s and again during 1955–1970, whereas the "Sedov" drifted at 3.7 km $day^{-1}$ in 1938–1939. The long-term mean is about 2.4 km $day^{-1}$. Uplifted beaches in the Canadian Arctic and on Svalbard, dated from 6,500–4,000 years old, suggest greater wave fetch and more open water conditions (Hopkins and Marincovich, 1984; but, see also Barry, 1983) that may provide an analog for earlier interglacial intervals. Nevertheless, the magnitude of such reduced ice extent (and thickness?) is uncertain. Driftwood logs incorporated in the raised beaches have been attributed both to more open sea ice conditions (Stewart and England, 1983) and to heavier ice concentrations (Haggblom, 1982). The latter argument is based on the limited buoyancy of timber in sea water (18 months or less), compared with an estimated 5 years required for wood to drift from Siberia to Svalbard. Consequently, the paleoclimatic interpretation of this evidence remains somewhat ambiguous.

Re-examination by Witte and Kent (1988) of the magnetostratigraphy of ocean sediment cores collected in the Canada Basin, from ice island T-3, confirms the low sedimentation rates of ca. 1 mm $10^3$ yr previously reported.

## POSSIBLE MODES OF ARCTIC OCEAN CLIMATE: MODELS

### Global Considerations

Variables that affect the mean state of the global climate system are both external and internal. External variables include changes in extraterrestrial solar radiation as a result of fluctuations in solar constant or periodic changes in orbital parameters, continental drift, and volcanic activity. Internal factors include atmosphere–surface feedbacks, atmospheric composition, and cloudiness.

New attention is being given to the problem of whether regimes with polar ice sheets represent aberrations from the mean global state (Frakes, 1979; Crowley, 1983; Lloyd, 1984) or whether they are the norm (Hunt, 1984). Until recently, the former view has generally been accepted, with the late Cenozoic, Permo-Carboniferous, Ordovician, and several pre-Cambrian glacial periods seen as the anomalies. However, Barron, Thompson, and Schneider (1981) suggest that it is easier to account for cooling of the Earth than warming.

The presence of continents in polar latitudes has been proposed as a necessary, but not a sufficient, condition for ice sheet buildup (Cox, 1968; Crowell and Frakes, 1970; Sellers and Meadows, 1975), although a clear-cut relationship between the latitudinal distribution of land, total surface albedo, and ice ages between the Cambrian and the Triassic periods is lacking according to Burrett (1982). Nevertheless, an experiment for the Cretaceous period (approximately 135–165 mya), with the NCAR Community Climate Model (CCM) incorporating reconstructed continental positions, topography, and reduced land area (20 percent less than now), resulted in a global average temperature 4.8°C higher than in the present-day control (Barron and Washington, 1984). The warming in the northern hemisphere was attributed primarily to the changes in continental position while that in the southern hemisphere was mainly a result of the deglaciation of Antarctica. Even this calculated warming, however, is insufficient to account fully for the paleoclimatic data. The simulations used a "swamp" ocean and therefore lacked the effects of ocean heat storage and transport. Poleward oceanic heat flux probably contributed to the warmth of the Cretaceous, but a severalfold increase in atmospheric $CO_2$ may be required to explain the discrepancy (Schneider, Thompson, and Barron, 1985).

The Milankovitch theory of glacial–interglacial cycles enters into this wider problem, since it is considered that when the global mean climatic state is close to the threshold temperature of 0°C in higher latitudes, small perturbations in incoming solar radiation can be amplified by the effects of ice-albedo feedback. However, this leaves unanswered the problem of the synchrony of the last interglacial–glacial cycle in the two hemispheres (Mercer, 1984). Deep ocean circulation changes seem to be involved since the production of northern Atlantic deep water will be reduced during northern hemisphere glacial intervals (Corliss, 1982), which subsequently affects the heat budget of the Southern Ocean and the extent of the Antarctic sea ice. It is not clear, however, whether this mechanism would account for the apparent synchrony of the global glacial cycles. Moreover, some overall mechanism is required to shift the global mean state from a higher to a lower temperature level, or vice versa. The possibility of large variations in atmospheric $CO_2$ content over time, as implied by the widespread occurrence of limestone sediments and coal deposits in geologic strata, is one mechanism (Budyko and Efimova, 1981; Berner, Lasaga, and Garrels, 1983) but there are other candidates such as cosmic dust clouds.

Simple climatic models indicate that for various types of forcing (changes in solar constant, atmospheric $CO_2$ content, etc.) the global climate may exhibit two or three states, each existing for a wide range of conditions but undergoing rapid transitions between states for a narrow range of threshold conditions (Barry, 1979). For simple energy budget models (EBMs) with one-parameter equations, the occurrence of solutions with multiple equilibria is now well recognized (Ghil, 1984). Lorenz (1968) first described this possibility for the atmospheric circulation as almost intransivity. A *transitive* system has a unique probability distribution for each set of external forcings, whereas an *intransitive* system permits more than one state for a given set of forcings. The almost-intransitive system is oscillating even for constant forcing. While simple EBMs are useful for investigations of climatic sensitivity to a particular forcing, they cannot account for oscillatory behavior of the climate system. Such behavior can occur, even in the absence of periodic forcing, for a system of two or more differential equations with first-order time derivatives (viz. the time-dependent interactions between two feedbacks, such as ice-albedo and precipitation–temperature) (Ghil, 1984). It is considered that such models can contribute useful insight into the complexities of the global climate system. Results from EBMs and more complex general circulation models (GCMs) are discussed in the following section.

## The Arctic

Two general lines of research have addressed the possible past/future climatic conditions over the Arctic Ocean. One type of study has been concerned with the stability of the present sea ice cover to natural or human-induced climatic perturbations and the other has attempted to model the conditions anticipated with an ice-free ocean. Results obtained in both areas to date must be considered as tentative, primarily because fully coupled, time-dependent numerical models of the atmosphere–ocean–ice system have not yet been developed. We consider these two questions in turn.

### *Stability of the Present Arctic Sea Ice Cover*

The susceptibility of the Arctic ice to a global warming was first pointed out by Budyko (1962). Based on energy budget calculations of ice ablation and temperature estimates of ice accretion, Budyko (1966) concluded that, for the present mean Arctic climate, a 4°C rise in summer temperature would be sufficient to melt ice 4 m thick completely within 4 years. However, the ice would reform seasonally even for a 10°C winter warming.

Simple EBMs have been used to examine the climatic response to a 1 percent increase in solar constant or a $CO_2$ doubling; both forcings give nearly identical responses, not only for geographical and temporal patterns of the responses, but also for their amplitudes in the case of temperature (Robock, 1983). The simplicity of these models, especially with regard to land/sea differences and the climatic effects of ocean processes, necessitates that results from them be interpreted with care. For example, Robock shows that the ice line in the Budyko-type of EBM is too sensitive to external forcings, primarily as a result of the high albedo value adopted poleward of the ice line. Nevertheless, EBMs can reproduce results similar to GCMs if they incorporate a sufficient range of parameterizations. Thus, an EBM incorporating snow/ice feedback shows seasonal melting of polar ice for a $CO_2$ doubling experiment (Robock, 1983), resembling the more detailed calculations of the GFDL GCM shown in Figure 1-6 (Manabe and Stouffer, 1980). However, this does not necessarily mean that either simulation is correct!

Early studies of Arctic sea ice were undertaken by Maykut and Untersteiner (1971), with a detailed 1-D thermodynamic column model that incorporated the effects of snow cover, ice salinity, and internal heating due to penetration of solar radiation. The model predicted an equilibrium ice thickness of 2.88 m and annual surface ablation of 0.40 m, close to ob-

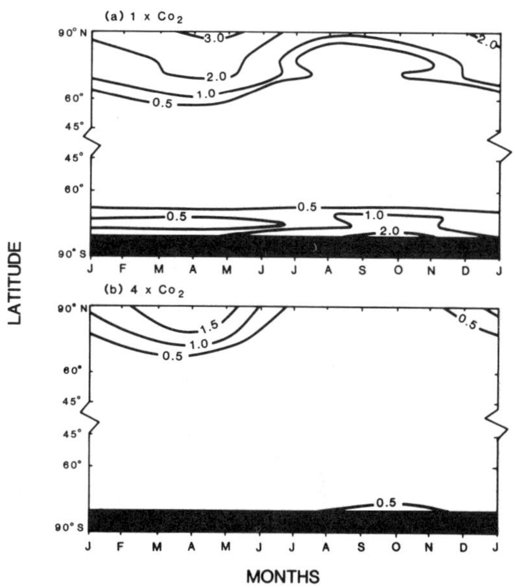


Figure 1-6. Sea ice thickness (m) and extent simulated by the GFDL general circulation model for *(A)* a control case and *(B)* a perturbation experiment with 4 × $CO_2$. (From Manabe and Stouffer, 1980.)

served values. It showed that a twofold to threefold increase in annual ocean heat flux (from about 2 W $m^{-2}$), or a reduction of surface albedo from 64 percent in July to <50 percent, would lead to the ice vanishing in a few years. The model of Maykut and Untersteiner was subsequently simplified by Semtner (1976) and this was incorporated in a single-layer version into a spatially dependent formulation by Parkinson and Washington (1979). By allowing the ice to be forced by mean monthly atmospheric temperature and wind fields, Parkinson and Washington obtained reasonable simulations of the annual ice cycle. When a 5°C warming is incorporated, the Arctic ice disappears in summer (after 5 model years) but reforms in winter (Parkinson and Kellogg, 1979). However, in this treatment the ice limit is unrealistically forced by monthly mean air temperature with no allowance for feedback effects.

More recently, Semtner (1984*a*) has shown that the effects of a warming on the pack ice are highly sensitive to the details of the ice parameterization. This point is also elaborated by Shine and Henderson-Sellers (1985), using Semtner's three-layer model with variable surface albedos and heat fluxes. The Semtner model includes the effects of latent heat storage (in brine pockets in the ice) and of the cold content of the snow and ice in spring. Simulations with this model, for a $CO_2$ quadrupling, do *not* lead to total summer melt of the sea ice at 85°N, in contrast to the corresponding results of Manabe and Stouffer (1980) with the GFDL GCM. The latter experiment incorporated a seasonal cycle, a simple ocean mixed layer, and realistic geography, but has prescribed cloudiness and a simplified surface albedo and sea-ice thermodynamics treatment. In contrast, studies by Semtner (1987), using the thermodynamic/dynamic ice model of Hibler (Chapter 2) coupled to a detailed ocean model, and incorporating monthly atmospheric forcing and river runoff, give different results. For a perturbation experiment with atmospheric temperatures increased by 2°C in all months, the Arctic ice is found to disappear for one month in late summer, although it reforms in winter.

The question of maintaining an ice-free Arctic must also be considered from a large-scale oceanographic point of view. While this cannot be treated in detail here, an outline of the major considerations is essential. The early energy budget studies of an ice-free Arctic (Fletcher, 1965; Donn and Shaw, 1966) focus on the surface albedo effect. However, the GFDL experiments of Hunt (1984) show that a simple change of high-latitude albedo from sea ice (0.55) to water (0.19) gives a warming of only 4.5°C, insufficient to raise the surface temperature above the freezing point of seawater. A second experiment, including ocean heat-capacity effects, gave a smaller warming and regrowth of the ice in less than 100 days. These results are contrary to the earlier energy-budget assessments. In addition, Stigebrandt (1981) notes that the Arctic Ocean receives most of its freshwater input in summer and this limits the depth of the radiational heating.

Whether or not ice will reform in winter depends basically on the depth and density of this layer. It was pointed out by Aagaard and Coachman (1975) that an ice-free Arctic implies destruction of the stably stratified layer above the halocline. At present, the salt released by annual freezing of 0.3 m ice can only drive convection down to about 30 m depth, whereas the present halocline extends to about 200 m. To drive convection to this depth by salt release would require 7.5 m of ice to form annually over the Arctic. Both the salinity structure of the Arctic Ocean and the stability of its ice cover are critically influenced by lateral mixing processes across the continental shelves of the marginal seas (Carmack, 1986) (Fig. 1-7). According to Rudels (1984), low-salinity outflow from the shelves enters the Polar Mixed Layer, while more saline shelf water replenishes

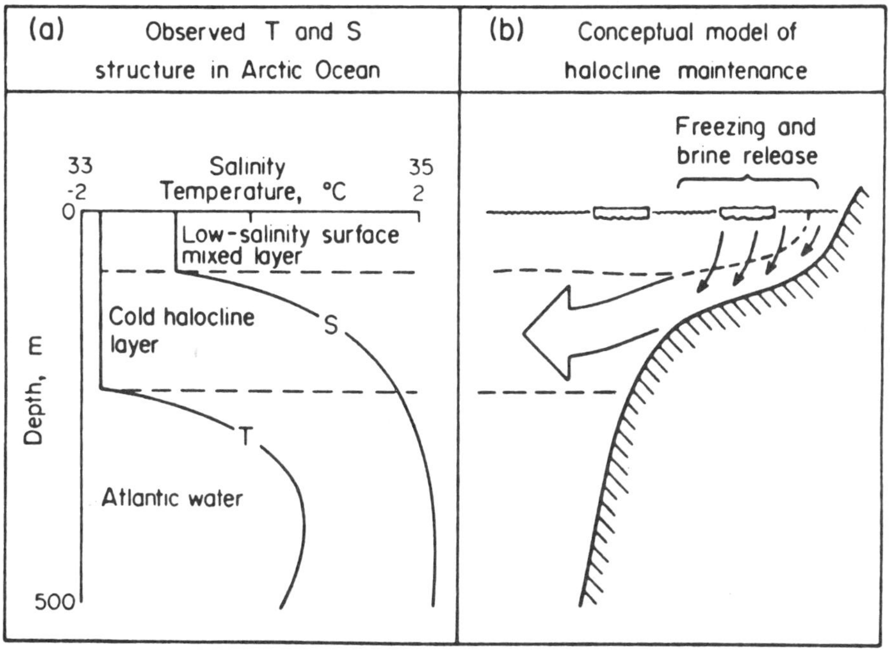


Figure 1-7. *(a)* Temperature and salinity structure in the Arctic Ocean and *(b)* a conceptual model of halocline maintenance. (From Aagaard, Coachman, and Carmack, 1983).

the pycnocline and goes into the deep water. Rudels proposes that mixing is bimodal, comprising the present state that is dominated by shelf-water mixing, and a different extreme where the shelf water becomes sufficiently saline due to reduced runoff and the stratification does not prevent mixing. Heat and salt are mixed upward from the Atlantic waters below the halocline by mechanical turbulence, but Rudels considers the ice cover to be essentially stable with respect to changes in Atlantic inflow.

Fresh water in the Arctic Ocean is provided primarily by river runoff and by the low-salinity Bering Strait inflow (these sources and net precipitation minus evaporation represent 0.25 Sverdrups of fresh water); it has a mean residence of about 10 years. However, in the southern part of the Eurasian Basin, the salinity of the upper layers is about 34.27‰ with a residence time of only 2–3 years (Aagaard and Coachman, 1975). Moreover, at 78°N in the Kara Sea, Micklin (1981) reports a surface salinity as low as 10‰ due to ice melt and river flow. Consequently, if the freshwater input in this sector, principally from the Yenisei and Ob rivers, were drastically reduced on a continuing basis it would be feasible for the sensible heat of the underlying Atlantic water to reach the surface. The St. Anna Canyon, north of Novaya Zemlya, where the Atlantic water is forced upward by the bathymetry, shows late freezing and thinner ice at present according to Aagaard and Hanzlick (1980). Olausson (1985) has argued for many years that the Arctic Ocean was open during the Last Glacial Maximum as a result of reduction of continental runoff due to ice sheet growth, removing the salinity stratification. This mechanism is contrary to the views of Clark and Ruddiman and McIntyre noted earlier, and also different from Herman's argument for mild, low-salinity, ice-free intervals during the Brunhes epoch (Chapter 23).

A detailed ocean circulation model with 13 vertical levels and 110 km horizontal resolution has been used by Semtner (1984*b*, 1987) to study the effects on sea ice extent of reduced river inflow into the entire Arctic Ocean. For the control case a geographically distributed continental runoff totaling 3,225 $km^3\ yr^{-1}$ is assumed; scenarios are then simulated for reductions in runoff of 12 percent and 37 percent (corresponding to the total diversion of the Yenisei, Ob, Pechora, and Dvina rivers). In both cases there is a negligible effect on the halocline in the central Arctic, as postulated by Aagaard and Coachman. For a 37 percent reduction of runoff, Semtner finds that the vertical stability in the Kara and Barents seas is considerably decreased, but not to the point of convective overturning. Ice extent remains almost unchanged. The initial results were considered to be tentative, since the model was run with annual mean forcing. However, further experiments using monthly atmospheric forcing and river runoff confirm negligible overall changes in sea ice with small local increases in extent and thickness in the Barents–Kara seas, for a one-third decrease in Eurasian river runoff (Semtner, 1987).

Empirical data on river flow and ice extent for 1953–1977 are used by Holt, Kelly, and Cherry (1984) to assess the evidence for any changes in ice extent as a result of reduced river flow. Following the conceptual model for ice–ocean–atmosphere interactions (Fig. 1-8) outlined by Micklin (1981), Holt, Kelly, and Cherry test for local effects related to changes in the estuaries and shelves, and basinwide changes. In the first category, they note that reduced runoff may (1) delay quick ice breakup in spring that would give locally increased sea ice; (2) lower the heat content of the surface waters, allowing increased summer ice cover; and (3) cause reduced freshening that would tend to decrease the ice cover in autumn and winter. On the basin scale, reduced runoff from the Siberian rivers could cause a weaker pycnocline, allowing increased upward heat flux and decreased ice in the Eurasian Basin; also, it could weaken shelf gradient currents, causing reduced inflow of warm Atlantic and Pacific water, thereby leading to increased ice in the Barents-Kara and Chukchi seas, respectively. Holt, Kelly, and Cherry confirm significant positive correlations between flow of the Ob River and ice in southeastern Kara Sea in October, negative correlations between flow of the Lena and ice in the Chukchi Sea in August–September, and negative correlations between flow of the Yenisei and ice in the Barents Sea in the next year. These results provide a basis for comparison with modeling studies using seasonal forcing, such as those of Semtner (1987).

Experiments with a simpler two-layer ocean model developed by Stigebrandt (1981, 1985) suggest that for the present Arctic to become ice-free would require elimination of all river runoff and some increase in evaporation. However, the ice cover would not reform under the present climatic regime if it were once removed, in contrast to Semtner's results. He also shows that a large buffering to change is provided at present by the influx of Pacific water via Bering Strait. Closure of the Bering Strait, wholly or partially (corresponding to half of the present buoyancy), together with a reduction of freshwater supply from runoff of 45 percent, or 55 percent, respectively, would be needed to maintain an ice-free Arctic. Stigebrandt notes that "such reductions might have occurred in the past when glaciations led to a lower sea level" and also when there was essentially no river runoff. Additionally, sea-level lowering would have exposed much of the continental shelves, with further impacts on mixing and circulation (Sancetta, 1983). These glacial-age conditions have not yet been explored quantitatively.

A further large component (~30 percent) of the advective heat balance in the Arctic is the ice export of heat deficit (see Table 1-2). In Stigebrandt's model, a 50 percent increase in this export, accom-

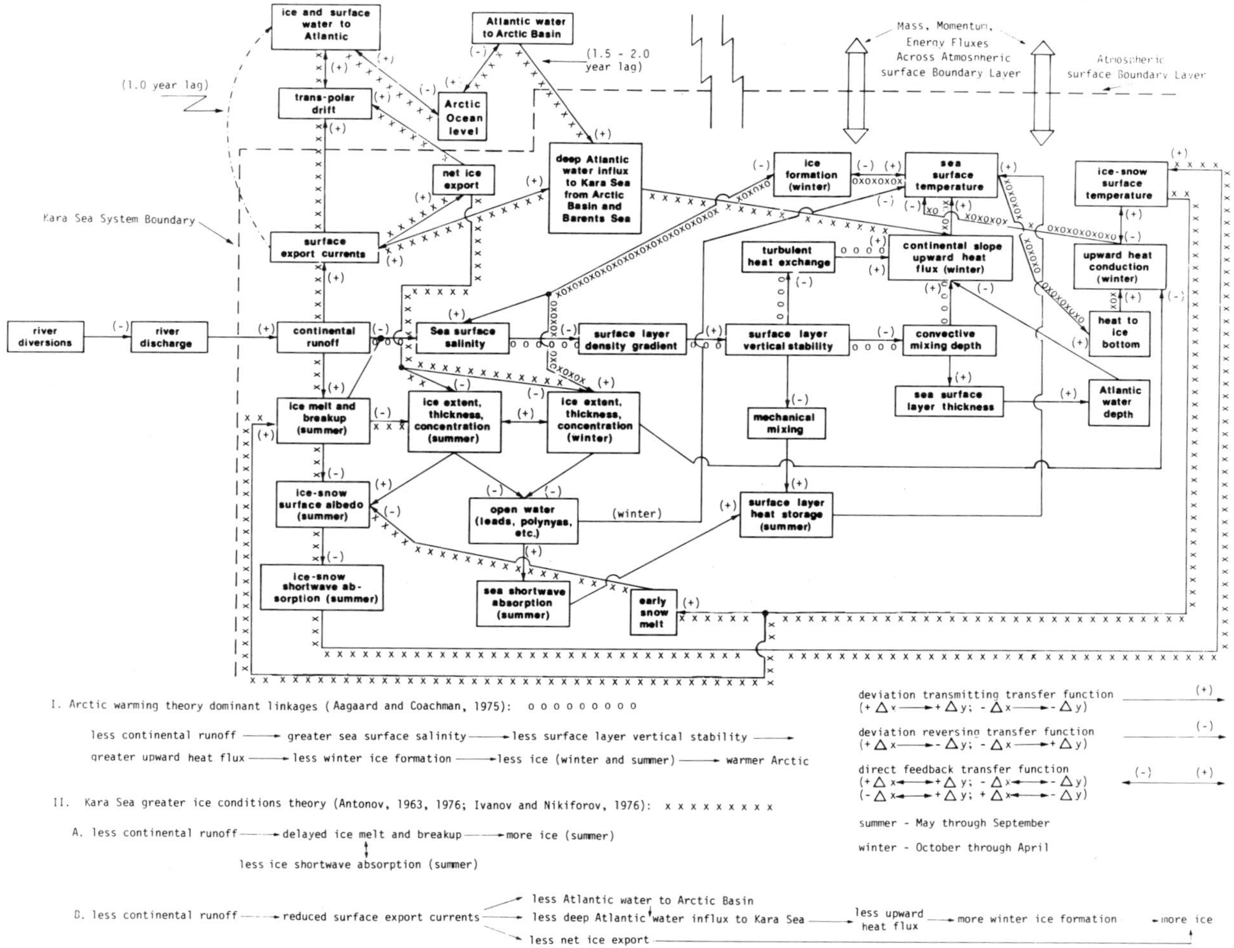


**Figure 1-8.** A conceptual model of the ocean–sea ice linkages in the Kara Sea in the context of changing river inputs and Arctic warming. (From Micklin, 1981.)

panied by reduced flow through the Bering Strait and a 35 percent reduction of freshwater runoff, would also lead to an ice-free state. Ice export is determined primarily by the atmospheric and oceanic circulations and hence is largely independent of the other factors causing changes in the heat and salt balance. Computed interannual variations of ice outflow in the Fram Strait, using a thermodynamic/dynamic ice model forced with three years of actual atmospheric data, are consistent with estimated fluctuations of heat input via the West Spitzbergen Current (Hibler and Walsh, 1982). The interannual variability of this current is about 35 percent. Hence, significant fluctuations in ice export are entirely plausible. The mix of first-year and multiyear ice that is exported will be essentially random, but this affects the ice thickness distribution and hence the open-water fraction in the Arctic Ocean. Stigebrandt's model suggests a range of ice thickness where a small increase in open water increases the turbulent heat losses more than the gain in absorbed solar radiation, resulting in ice growth and a stable condition. However, for another range of ice thickness, such a perturbation may be amplified if the absorbed radiation exceeds the increased heat loss. The critical thickness separating these two states is approximately 1.5 m, similar to the estimates of Maykut and Untersteiner (1971) for a just-persisting annual ice cover.

Ice cover in the subpolar seas (Greenland, Iceland, Norwegian, and Barents seas) depends on the distribution of surface buoyancy (Stigebrandt, 1985). A less dense surface layer could result, for example, from a reduced strength and frequency of northerly airflow off east Greenland, promoting spreading of the east Greenland current surface waters, or from increased frequency of icebergs. A freshwater supply of 0.12 Sv (similar to that currently entering the Arctic from river inflow), with Atlantic water temperatures $\leq 5°C$ would result in perennial ice cover in the subpolar seas, according to Stigebrandt. He postulates such a regime for the deglacial phase of a glacial cycle.

It is appropriate here to consider also the problem of the transition from an open to an ice-covered Arctic. As discussed earlier, the timing of this remains indeterminate. Moreover, the modeling results indicate that both states tend to be stable. Sea-ice cover requires and at the same time maintains the presence of a stratified ocean, whereas an open ocean is characterized by deep vertical mixing. The key to this apparent paradox most probably lies in the characteristics of the shallow shelf seas, particularly their interaction with penetrating Atlantic water and river runoff (Aagaard, Swift, and Carmack, 1985).

Seasonal ice forms readily in the shallow-shelf seas since there is no opportunity for warmer deep waters to penetrate onto the shelves. Stratification is provided by continental river runoff even in the absence of ice melt during the summer. For sea ice cover to develop in the Arctic Ocean, it seems plausible that submergence of the continental margins by downwarping may have occurred in northern Siberia (D. M. Hopkins, personal communication, 1985). Seasonal ice would recur in these initially narrow-shelf seas. During the course of a winter, this ice extended over deeper water, as happens at present in the subarctic Sea of Okhotsk and Baffin Bay. Polynyas occurring along the Siberian coast due to offshore winds and ocean circulation serve as "ice factories," generating large quantities of ice that are advected out over the deeper basins (cf. Barry, 1983). This process permits at least a seasonal Arctic Ocean ice cover to develop. As the shelf seas became more extensive by downwarping, the transition to a permanent ice cover occurred, given some years of thicker winter ice cover and/or less summer melt, and ice export by the wind-driven ocean circulation. It is important in order for thick ice to form not to have deep snow cover, which slows the conductive heat loss and therefore the growth rate (Walsh, Hibler, and Ross, 1985). Shallow snow peaks require a diversion of depression tracks away from the regions of extending sea ice, although Manabe and Broccoli (1985) suggest from their simulations of the Last Glacial Maximum with a coupled model that an apparent cyclone track was established north of the Laurentide ice sheet.

A further possibility involves enhanced flow of fresh water from the North Pacific via Bering Strait. Walin (1985) speculates that the observed asymmetry of the salinity fields in the Atlantic and Pacific oceans may be transient and could play a role in glaciation cycles by affecting the extent of ice in the Arctic Basin.

### *Climatic Reconstructions for an Open Arctic*

Whether or not the Arctic Ocean has been ice-free, at least seasonally, during the last several million years, concern has been expressed that it could become so in the future as a result of $CO_2$-induced global warming. Hence we examine various modeling estimates of the climatic effects of an open Arctic.

An estimate that mean annual air temperature in the central Arctic in the absence of ice would rise to 5°C was made by Budyko (1966). He inferred that ice decreases the north polar temperatures by 30°–35°C in winter and by 5°C in summer, and suggested that mean temperatures at 70°N during the transition to ice-free conditions would rise to 10°C in the warmest month and to −10°C in the coldest. A further estimate for zonal mean temperature, based on calculations of vertical radiation fluxes and vertical and horizontal heat fluxes by Rakipova (1966), indicates sea-level temperatures of 5°C at 70°N and 2°C at 90°N in summer and of −1°C at 70°N and −3°C at 90°N in winter, for ice-free conditions. She concluded that the zonal atmospheric circulation would weaken and summer/winter circulation differences would diminish. Similar work with a zonally symmetric circulation model was performed by MacCracken (1970).

There have subsequently been various sensitivity experiments with GCMs to analyze the possible consequences of an open Arctic Ocean and, accordingly, these results are of relevance in the context of both past and future climatic modes. Table 1-7 summarizes the basic characteristics of the models employed in sensitivity tests of the effects of an ice-free Arctic Ocean. The experiments for winter conditions typically prescribe a water surface at −1°C. All of the experiments suggest that an open Arctic would produce marked warming of the lower troposphere poleward of about 65°N, decreased vertical stability, and a reduced meridional temperature gradient, compared with present ice conditions. Fletcher et al. (1973) and Warshaw and Rapp (1973) find a 10°–15°C polar warming, while Newson (1973; Hills 1983) shows a 30°–40°C warming, with a 2°–6°C cooling over the continents near 50°N. As a result of the reduced meridional temperature gradient, the zonal westerlies are weakened and blocking types of circulation become more frequent. Fletcher et al. note that the Aleutian low is deeper for ice-free conditions with an accompanying pressure rise over Siberia. Warshaw and Rapp point out that geopotential heights are decreased over the Arctic, but increased in the upper troposphere over midlatitudes, implying a dynamic rather than a hydrostatic response. The experiment of Shaw and Donn (1971) with Adem's thermodymic model incorporates ocean heat storage within a 25–50 m homogenous layer. For ice-free conditions a surface albedo of 0.10 is used with 90 percent cloudiness. Heat fluxes and surface temperatures are calculated, starting from

Table 1-7. Model Experiments for an Ice-free Versus an Ice-covered Arctic

| *Model* | *Horizontal Grid* | *Vertical Resolution* | *Boundary Conditions* | *References* |
|---|---|---|---|---|
| Adem thermodynamic model | 512 points N. of ca. 15°N | One 10 km atmospheric layer including clouds and a 50 m ocean layer | Hemispheric, prescribed clouds, snow cover, sensible and latent heat fluxes; ocean heat storage monthly time steps | Donn and Shaw, 1972; Adem, 1964 |
| UCLA/Mintz–Arakawa GCM | 4° × 5° | 2 levels | Continents, orography, prescribed annual mean sea-surface temperature and monthly snow cover. | Fletcher et al., 1973; Warshaw and Rapp, 1973 |
| U.K. Meteorological Office GCM | 3° × 5° to 81° latitude (mean values within polar cap) | 5 levels | Hemispheric, continents, orography prescribed sea-surface temperature, snow and ice albedo = 0.8 | Newson, 1973; Hills, 1983; Corby, Gilchrist, and Newson, 1972 |
| GFDL GCM | Spectral (1,200 points/hemisphere) | 18 levels | Hemispheric, no mountains, global swamp surface, no land–sea contrast; annual mean, nondiurnal state, thermodynamic sea-ice model incorporated | Hunt, 1984 |

an arbitrary 0°C value. After 3 years of model time, surface temperatures range from 2.8°C in March to 9.3°C in August (Fig. 1-9). For a different experiment with an ice-free ocean but glacierized continents, the temperatures are lowered to −1.6°C in March and 4.0°C in August. They consider that cold air advection from the continents could then cause refreezing. They also show that the open Arctic in the model is insensitive to variations of cloudiness between 60 percent and 90 percent.

It should be noted that these experiments were performed with early atmospheric models (except Hunt, 1984) that lacked realistic treatment of many processes. The control experiment results of these models frequently gave somewhat unrealistic simulations of *present* conditions; for example, Gilchrist, Corby, and Newson (1973) note that the U.K. Meteorological Office's five-level model underestimates the Siberian winter anticyclone, while the Icelandic low is too deep and too far south. Such biases may correspondingly distort the results of the ice-free Arctic experiments, but appropriate details are lacking from most of the accounts and hence it is not possible to make complete assessments of the results.

## DISCUSSION AND CONCLUSIONS

Knowledge of the Arctic Ocean's surface climate is still remarkably limited. In particular, there is considerable uncertainty concerning the cloud conditions, and the overall surface energy budget. The former problem is a result of the inadequacy of present remote sensing data for discriminating cloud from snow cover. The latter pertains to the limited knowledge of sea-ice extent and thickness, including the timing and spatial distribution of snow melt. Synoptic and possible subsynoptic systems in the Arctic are inadequately characterized and their genesis is incompletely understood. There are also important interactions between the marginal ice zone (MIZ) and cyclogenetic events, including mesoscale polar lows that need to be examined in terms of their possible influence on the ice margin and on weather in the MIZ.

There are major unresolved problems relating to the paleoclimatic record and consequently our ability to assess potential future modes of Arctic climate. The most serious difficulty is the absence of an accepted framework for the paleooceanographic history of the Arctic Basin and especially its sea-ice cover. The evidence from adjacent land areas and the Norwegian Sea indicates that the ice cover has been less extensive than at present, at least in summer, during some interglacial episodes. Given the present limited role of oceanic compared with atmospheric energy advection, it seems that the latter would be a more likely cause. However, it is not clear how much warmer the northern hemisphere high latitudes would have to be in order to trigger such a shift. It is also not clear when the Arctic pack ice and the Greenland ice sheet first formed following the temperate conditions of the early Tertiary. The most likely date for the growth of a major ice sheet on Greenland is about 2.2 mya. There are three conflicting theories concerning the late Neogene history of Arctic sea ice. A definitive interpretation of the ocean sedimentary record and its dating are still lacking and consequently any paleoclimatic history must remain speculative. Sea-ice cover seems to have been present by mid-Pleistocene times in view of the contrast in north–south patterns of glaciation across the Alaskan mountain systems between the early and late Pleistocene, but the possibility of more open and mobile ice regimes during certain interglacial intervals cannot yet be resolved. Model studies demonstrate that long-term variations of river runoff and Bering Strait inflow would be key factors in changing ocean stratification

a

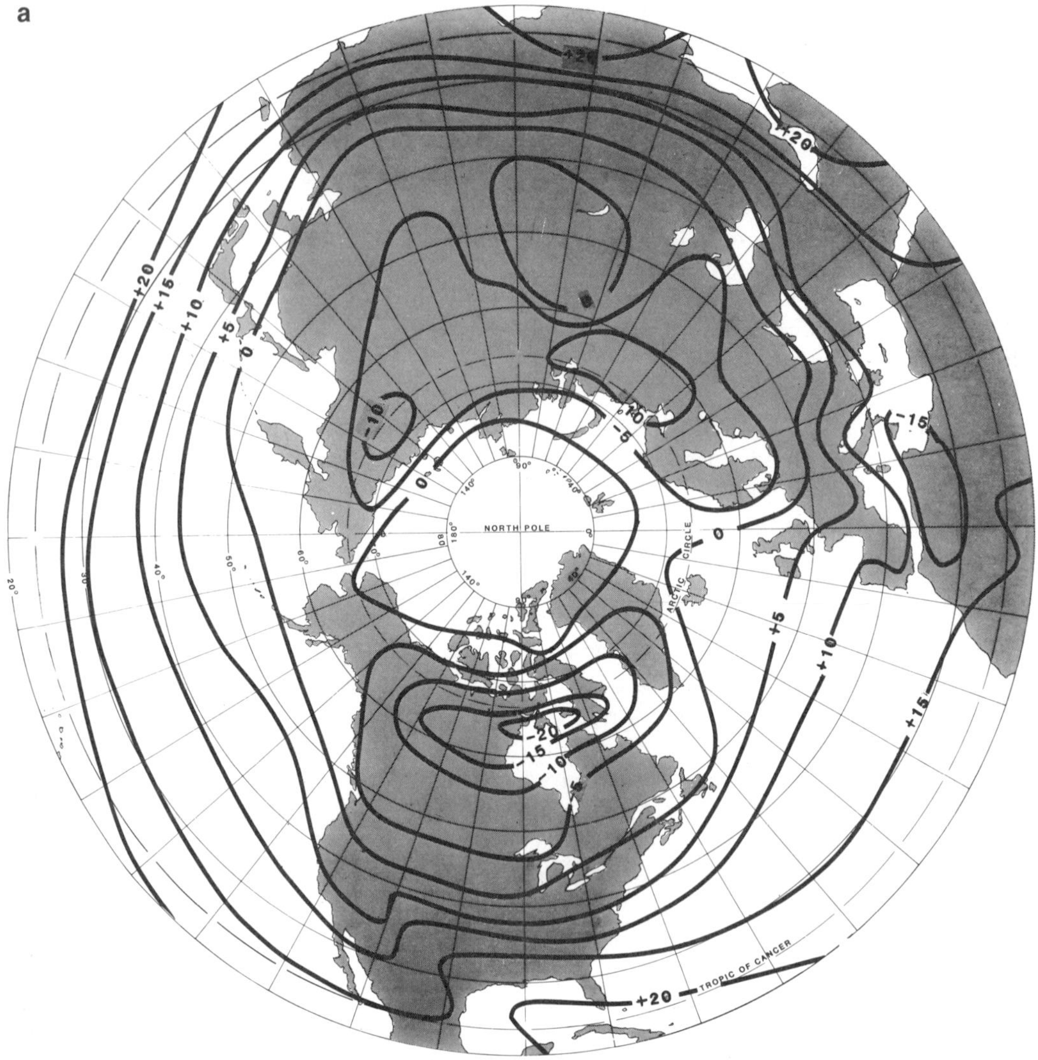


Figure 1-9. The surface temperature distribution (°C) in the Northern Hemisphere calculated with Adem's thermodynamic model for an ice-free Arctic Ocean: *(A)* winter *(B)* summer. (After Donn and Shaw, 1972.)

and consequently the ice regime, but model results cannot provide the time control.

The stability of the present sea-ice cover to a specified climate perturbation is also not well defined. This is a result of the shortcomings in the parameterization of sea ice in existing coupled ocean–atmosphere models. The possible changes in Arctic cloudiness in association with any general warming are also unknown. The historical record suggests an extreme interannual to 10-year variability of about 15 percent in minimum summer ice extent. The controls of such anomalies have not been specifically determined, although they include differences in wind stress (Walsh, Hibler, and Ross, 1985) and cloud amounts (Shine and Crane, 1984). The duration of any potential transition between a permanently and a seasonally ice-covered ocean is uncertain because any in-b

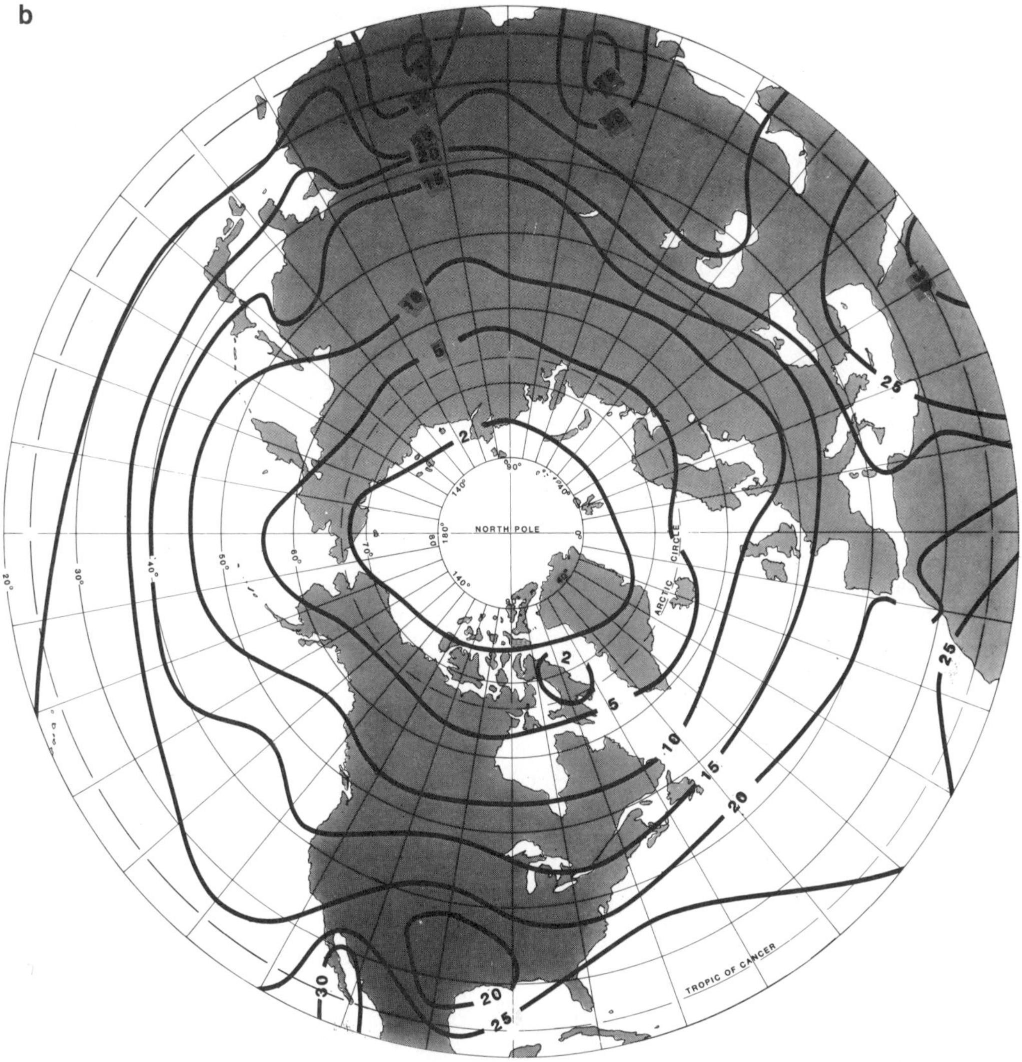


Figure 1-9. Continued

creased summer melt that produces thinner ice tends to be counteracted by a corresponding increase in winter freezing rates through the increased heat losses. The modes of both an open and an ice-covered Arctic will tend to be sustained by negative feedbacks.

The possible effects of an open Arctic Ocean on hemispheric climate have been modeled only by rather primitive sensitivity experiments. Further study of this boundary condition is required using a coupled GCM, although the general indications of considerable warming and some precipitation increase obtained so far are qualitatively realistic. Conditions would resemble those of the present Norwegian Sea.

Fuller understanding of the nature of the past and possible future states of the Arctic Ocean climate offers many challenging research problems. Part of the key will be provided by closer analysis of the present ice-ocean-atmosphere regime through empirical and modeling studies.

## ACKNOWLEDGMENT

I am indebted to D. M. Hopkins, University of Alaska, Fairbanks, for his comments and insight into the paleoclimatic problems, to Margaret Strauch for typing the manuscript, and to T. Wiselogel for drafting. This work was supported in part by National Science Foundation grant DPP 82-17265, NASA Climate Program grant NAG-5-417, and Office of Naval Research contract N00014-86-K-0695 under a University Research Initiative program to Dartmouth College, N.H. and the University of Colorado, Boulder.

## REFERENCES

Aagaard, K., and L. K. Coachman. 1975. Toward an ice-free Arctic Ocean. *EOS* **56**:484–486.

Aagaard, K., and P. Greisman. 1975. Towards new mass and heat budgets for the Arctic Ocean. *J. Geophys. Res.* **80**:3821–3827.

Aagaard, K., and D. Hanzlick. 1980. Freshwater and Atlantic water in the Kara Sea. *J. Geophys. Res.* **85**(C9):4937–4942.

Aagaard, K., L. K. Coachman, and E. Carmack. 1983. On the halocline of the Arctic Ocean. *Deep-Sea Res.* **28A**:529–545.

Aagaard, K., J. H. Swift, and E. C. Carmack. 1985. Thermohaline circulation in the Arctic Mediterranean seas. *J. Geophys. Res.* **90**(C3): 4833–4846.

Adem, J. 1964. On the physical basis for the numerical prediction of monthly and seasonal temperatures in the troposphere-ocean-continent system. *Mon. Weather Rev.* **92**:91–103.

Aksu, A. E., and P. J. Mudie. 1985. Magnetostratigraphy and palynology demonstrate at least 4 million years of Arctic Ocean sedimentation. *Nature* **318**:280–283.

Allison, I. 1982. The role of sea ice in climatic variation. In: *Report of the WMO/CAS-JAS-CCCO Meeting of Experts on the Role of Sea Ice in Climatic Variations,* World Meteorological Organization, Geneva, World Climate Programme, pp. 27–50.

Andreas, E. L., C. A. Paulson, R. M. Williams, R. W. Lindsay, and J. A. Businger. 1979. The turbulent heat flux from Arctic leads. *Boundary-Layer Meteorol.* **17**:57–91.

Andrews, J. T., and R. G. Barry. 1978. Glacial inception and disintegration during the last glaciation. *Ann. Rev. Earth Planet. Sci.* **6**:205–228.

Axelrod, D. I. 1984. An interpretation of Cretaceous and Tertiary biota in polar regions. *Palaeogeogr., Palaeoclimatol., Paleoecol.* **45**:105–148.

Barron, E. J. 1985. Explanation of the Tertiary cooling trend. *Palaeogeogr., Palaeoclimatol., Palaeoecol.* **50**:271–284.

Barron, E. J., S. L. Thompson, and S. H. Schneider. 1981. An ice-free Cretaceous? Results from climate model simulations. *Science* **212**:501–508.

Barron, E. J., and W. M. Washington. 1984. The role of geographic variables in explaining paleoclimates. Results from Cretaceous climate model sensitivity studies. *J. Geophys. Res.* **89**:1267–1279.

Barry, R. G. 1979. Recent advances in climate theory based on simple climate models. *Prog. Phys. Geog.* **3**:119–131.

Barry, R. G. 1982. Approaches to reconstructing the climate of the steppe-tundra biome. In D. M. Hopkins, J. V. Matthews, Jr., C. E. Schweger, and S. B. Young (eds.), *Paleoecology of Beringia.* Academic Press, New York, pp. 195–204.

Barry, R. G. 1983. Arctic Ocean ice and climate: Perspectives on a century of polar research. *Am. Geogr. Assoc. Ann.* **73**:485–501.

Barry, R. G., and F. K. Hare. 1974. Arctic climate. In J. D. Ives and R. G. Barry (eds.), *Arctic and Alpine Environments.* Methuen, London, pp. 17–54.

Barry, R. G., R. G Crane, A. Schweiger, and J. P. Newell. 1987. Arctic cloudiness in spring from satellite imagery. *J. Climatol.* **7**:423–451.

Barry, R. G., M. W. Miles, R. C. Cianflone, G. Scharfen, and R. C. Schnell. 1988. Characteristics of Arctic sea ice from remote sensing data and their relationship to atmospheric processes. *Annals Glaciol.* **12** (in press).

Baulin, V. V., and N. S. Danilova. 1984. Dynamics of late Quaternary permafrost in Siberia. In H. E. Wright, Jr. and C. W. Barnosky (eds.), *Late Quaternary Environments of the Soviet Union.* University of Minnesota Press, Minneapolis, pp. 69–77.

Belanger, P. E. 1981. Paleo-oceanography of the Norwegian Sea during the past 130,000 years: Coccolithophorid and foraminiferal data. *Boreas* **11**:29–36.

Berner, R. A. and E. J. Barron. 1984. Factors affecting atmospheric $CO_2$ and temperature over the past 100 million years. *Am. J. Sci.* **284**:1183–1192.

Berner, R. A., A. C. Lasaga, and R. M. Garrels. 1983. The carbonate-silicate geochemical cycle and its effect on atmospheric carbon dioxide over the last 100 million years. *Am. J. Sci.* **283**:641–683.

Bilello, M. A. 1966. Survey of Arctic and subarctic temperature inversions. *U.S. Army, Cold Regions Research and Engineering Laboratory, Technical Report 161,* Hanover, N.H.

Blanc, P. L., M. R. Fontugne, and J. C. Duplessy. 1983. The time-transgressive initiation of boreal ice-caps: Continental and oceanic evidence reconciled. *Palaeogeogr., Palaeoclimatol., Palaeoecol.* **42**:211–224.

Boulton, G. S. 1979. A model of Weichselian glacier variation in the North Atlantic region. *Boreas* **8**:373–395.

Boulton, G. S., G. D. Smith, A. S. Jones, and J. Newsome. 1985. Glacial geology and glaciology of the last mid-latitude ice sheets. *Geol. Soc. London J.* **142**:447–474.

Boyd, R. F., D. L. Clark, G. Jones, W. F. Ruddiman, A. McIntyre, and N. G. Pisias. 1984. Central Arctic Ocean response to Pleistocene earth-orbital variations. *Quat. Res.* **22**:121–128.

Brigham, J. K., and G. H. Miller, 1983. Paleotemperature estimates of the Alaskan coastal plain during the last 125,000 years. In *Permafrost. Fourth International Conference Proceedings.* National Academy Press, Washington, D.C., pp. 80–85.

Broecker, W. S. Floating glacial ice caps in the Arctic Ocean. *Science* **188**:1116–1118.

Bryazgin, N. N. 1976. Srednegodovoe kolichestvo osadkov v Arktike s uchetom pogreshnostev osadkomerov. *Trudy Arkt. Antarkt. Nauch.-Issled. Inst.* **323**:40–74.

Budyko, M. I. 1962. Polyarnye l'dy i klimat. *Izvestiya Akad. Nauk SSR, Ser. Geogr.,* **6**:5–36.

Budyko, M. I. 1966. Polar ice and climate. In J. O. Fletcher (ed.), *Proceedings of the Symposium on the Arctic Heat Budget and Atmospheric Circulation.* Memorandum RM-5233-NSF, Rand Corporation, Santa Monica, CA, pp. 3–21.

Budyko, M. I., and N. A. Efimova. 1981. Influence of carbon dioxide on climate. *Soviet Meteorol. Hydrol.* **2**:1–10.

Burova, L. P. 1981*a*. Vodnye resursy atmosfery nad Arkticheskim basseinom. *Trudy Arkt. Antarkt. Nauch. Issled. Inst.* **370**:91–110.

Burova, L. P. 1981*b*. Vlagosoderzhanie inversionnyx sloev atmosfery v Arktike. *Trudy Arkt. Antarkt. Nauch.-Issled. Inst.* **370**:111–130.

Burrett, C. F. 1982. Phanerozoic land-sea and albedo variations as climate controls. *Nature* **296**:54–56.

Buzuev, A. Ya., I. P. Romanov, and V. E. Fedyakov. 1979. Variability of snow distribution on Arctic Ocean ice. *Soviet Met. and Hydrol.* **9**:957–964.

Buzuev, A. Ya., N. P. Shesterikov, and A. A. Timerev. 1965. Albedo l'da v Arkticheskikh Moriakh po dannym nabliudenii s samoleta. *Problemy Arktiki i Antarktiki* **20**:49–54.

Campbell, W. J., R. O. Ramseier, H. J. Zwally, and P. Gloersen. 1981. Structure and variability of Bering and Okhotsk sea ice cover by satellite microwave imagery. *Am. Assoc. Petrol. Geol.* **1981**:343–354.

Carleton, A. M. 1985. Synoptic cryosphere-atmosphere interactions in the Northern Hemisphere from DMSP image analysis. *Internat. J. Remote Sensing* **6**:239–261.

Carmack, E. C. 1986. Circulation and mixing in ice-covered waters. In N. Untersteiner (ed.), *The Geophysics of Sea Ice.* Plenum Press, New York, pp. 641–712.

Carsey, F. J. 1978. Character of Arctic PBL structure as determined by acoustic radar. *AIDJEX Bull.* (University of Washington, Seattle), **38**:132–150.

Carsey, F. J. 1982. Arctic sea ice distribution at end of summer 1973–1976 from satellite microwave data. *J. Geophys. Res.* **87**:5809–5835.

Carsey, F. J. 1985. Summer arctic sea ice character from satellite microwave data. *J. Geophys. Res.* **90**(C3):5015–5034.

Carter, L. D., J. Brigham-Grette, L. Marincovitch, Jr., V. L. Pease, and J. W. Hillhouse. 1986. Late Cenozoic Arctic Ocean sea ice and terrestrial paleoclimate. *Geology* **14**:675–678.

Cavalieri, D. J., and H. J. Zwally. 1985. Satellite observations of sea ice. In G. Ohring and H. J. Bolle (eds.), *Space Observations for Climate Studies, Adv. in Space Res.* **5**(6):247–255.

Chernigovskiy, N. T. 1967. Radiation regime of the central Arctic Basin. In *Polar Meteorology,* Tech. Note No. 87. World Meteorological Organization, Geneva, pp. 107–115.

Chukanin, K. I. 1954. Aerometeorology, Volume III, Section 8, 217p. In *Observational Data of the Scientific-Research Drifting Station of 1950–1951,* Vols. I–III, M. M. Somov (ed.), (Izdat. Morskoi Transport, Leningrad), American Meteorological Society, trans. by E. R. Hope (NTIS Document AD117139).

Clark, D. L. 1971. Arctic Ocean ice cover and its late Cenozoic history. *Geol. Soc. Am. Bull.* **82**:3313–3324.

Clark, D. L. 1980. Geology and geophysics of the American Basin. In A. E. M. Nairn, M. Curkin, Jr., and F. G. Stehli (eds.), *The Ocean Basins and Margins, vol. 5, The Arctic Ocean.*

Plenum Publishing Corporation, New York, pp. 599–634.

Clark, D. L. 1982*a*. Origin, nature and world climate effect of the Arctic Ocean ice-cover. *Nature* **300:**321–325.

Clark, D. L. 1982*b*. Commentary (on Flohn 1982). In W. C. Clark (ed.), *Carbon Dioxide Review: 1982.* Clarendon Press, Oxford, pp. 181–184.

Clark, D. L., and A. Hanson. 1983. Central Arctic Ocean sediment texture: A key to ice transport mechanisms. In B. F. Molnia (ed.), *Glacial-Marine Sedimentation.* Plenum, New York, pp. 301–330.

Clark, D. L., J.-S. Vincent, G. A. Jones, and W. A. Morris. 1984. Correlation of marine and continental glacial and interglacial events, Arctic Ocean and Banks Island. *Nature* **311:**147–149.

Clark, D. L., R. R. Whitman, K. A. Morgan, and S. D. Mackey. 1980. Stratigraphy and glacial-marine sediments of the Amerasian Basin, central Arctic Ocean. *Geol. Soc. Am. Spec. Pap. 181,* pp. 1–57.

Colony, R., and A. S. Thorndike. 1984. An estimate of the mean field of Arctic ice motion. *J. Geophys. Res.* **89**(C6):10,623–10,629.

Corby, G. A., A. Gilchrist, and R. L. Newson. 1972. A general circulation model of the atmosphere suitable for long-period integrations. *Quart. J. Roy. Met. Soc.* **98:**809–832.

Corliss, B. H. 1982. Linkage of North Atlantic and Southern Ocean deep water circulation during glacial intervals. *Nature* **298:**458–460.

Cox, A. 1968. Polar wandering, continental drift and the onset of Quaternary glaciation. In J. M. Mitchell, Jr. (ed.), *Causes of Climatic Change, Meteorological Monograph 8(30),* pp. 112–125.

Crane, R. G., and R. G. Barry. 1984. The influence of clouds on climate with a focus on high latitude interactions. *J. Climatol.* **4:**71–93.

Crane, R. G., R. G. Barry, and H. J. Zwally. 1982. Analysis of atmosphere-sea ice interaction in the Arctic basin using ESMR microwave data. *Internat. J. Remote Sensing* **3:**259–276.

Crowell, J. C., and L. A. Frakes. 1970. Phanerozoic ice ages and the causes of ice ages. *Am. J. Sci.* **268:**193–224.

Crowley, T. J. 1983. The geologic record of climatic change. *Rev. Geophys. Space Phys.* **21:**828–877.

Crowley, T. J., D. A. Short, J. G. Mengel, and G. R. North. 1986. Role of seasonality in the evolution of climate during the last 100 million years. *Science* **231:**579–584.

Denton, G., and T. J. Hughes. 1981. The Arctic ice sheet: An outrageous hypothesis. In G. Denton and T. J. Hughes (eds.), *The Last Great Ice Sheets.* Wiley and Sons, New York, pp. 437–467.

Dingman, S. L., R. G. Barry, G. Weller, C. Benson, E. F. LeDrew, and C. W. Goodwin. 1980. Climate, snow cover, microclimate and hydrology. In J. Brown, P. C. Miller, L. L. Tieszen, and F. L. Bunnell (eds.), *An Arctic Ecosystem: The Coastal Tundra at Barrow, Alaska.* Dowden, Hutchinson and Ross, Stroudsburg, Pa., pp. 30–65.

Dodge, R. E., R. G. Fairbanks, L. K. Bennings, and F. Maurrasse. 1983. Pleistocene sea levels from raised coral reefs of Haiti. *Science* **219:**1423–1425.

Dolgin, I. M. (ed.). 1959. Rezultaty nablyudeniy nad snezhnym pokrovom na polyarnykh stantsiyakh 1936/37–1953/54 gg. *Trudy, Arkt. Antarkt. Inst.* **227:**1–167.

Donn, W. L. 1982. The enigma of high-latitude paleoclimate. *Palaeogeogr., Palaeoclimatol., Palaeoecol.* **40:**199–212.

Donn, W. L., and M. Ewing. 1966. A theory of ice ages. III. *Science* **152**:1706–1712.

Donn, W. L., and D. M. Shaw. 1966. The heat budget of an ice-free and an ice-covered Arctic Ocean. *J. Geophys. Res.* **71:**1087–1093.

Donn, W. L., and D. M. Shaw. 1967. The maintenance of an ice-free Arctic. *Prog. Oceanog.* **4:**105–116.

Donn, W. L., and D. M. Shaw. 1972. Discussion. In *Beneficial Modifications of the Marine Environment,* National Academy of Sciences, Washington, D.C., pp. 49–61.

Donn, W. L., and D. M. Shaw. 1977. Model of climate evolution based on continental drift and polar wandering. *Geol. Soc. Am. Bull.* **88:**390–396.

Doronin, Yu. P. 1970. *Thermal Interaction of the Atmosphere and the Hydrosphere in the Arctic.* Israel Program for Scientific Translations, Jerusalem, 244 p.

Elverhøi, A., and A. Solheim. 1983. The Barents Sea ice sheet—a sedimentological discussion. *Polar Res.* **1:**23–32.

Fillon, R. H. 1984. Ice-age Arctic Ocean ice-sheets: A possible direct link with insolation. In A. L. Berger, J. Imbrie, J. Hays, G. Kukla, and B. Saltzman (eds.), *Milankovitch and Climate.* D. Reidel Publishing Company, Dordrecht, pp. 223–240.

Fletcher, J. O. 1965. *The Heat Budget of the Arctic Basin and Its Relation to Climate.* Report R-444-PR, Rand Corporation, Santa Monica, California.

Fletcher, J. O., Y. Mintz, A. Arakawa, and T. Fox. 1973. Numerical simulation of the influence of Arctic sea ice on climate. In *Energy*

*Fluxes over Polar Surfaces,* WMO Tech. Note, No. 129. World Meteorological Organization, Geneva, pp. 181–218.

Flohn, H. 1952. Zur Aerologie der Polargebiete. *Met. Rundsch.* **5:**81–87, 121–128. (Translated in *The Aerology of Polar Regions,* Appendix 1, to Status Report 6, Contract AF19(604)-1141. Arctic Meteorology Res. Group, McGill University, Montreal, 1955.)

Flohn, H. 1982. Climate change and an ice-free Arctic Ocean. In W. C. Clark (ed.), *Carbon Dioxide Review: 1982.* Clarendon Press, Oxford, pp. 145–179.

Frakes, L. A. 1979. *Climates Through Geologic Time.* Elsevier, Amsterdam, 310p.

Gaigerov, S. G. 1967. *Aerology of the Polar Regions.* Israel Program for Scientific Translations, Jerusalem (Gidrometeoizdat, Moscow, 1964), 280p.

Ghil, M. 1984. Climate sensitivity, energy balance models, and oscillatory climate models. *J. Geophys. Res.* **89**(D1):1280–1284.

Gilbert, M. W., and D. L. Clark. 1983. Central Arctic Ocean paleoceanographic interpretations based on Late Cenozoic calcareous dinoflagellates. *Mar. Micropaleontol.* **7:**385–401.

Gilchrist, A., G. A. Corby, and R. L. Newson. 1973. A numerical experiment using a general circulation model of the atmosphere. *Quart. J. Roy. Met. Soc.* **99:**2–34.

Gorshkov, S. G., editor-in-chief. 1983. *World Ocean Atlas, vol. 3, Arctic Ocean.* Pergamon Press, Oxford, 184 plates. (Department of Navigation and Oceanography, Ministry of Defence, USSR, 1980.)

Gradstein, F. M., and S. P. Srivastava. 1980. Aspects of Cenozoic stratigraphy paleoceanography of the Labrador Sea and Baffin Bay. *Palaeogeog., Palaeoclimatol., Palaeoecol.* **30:**261–295.

Gros'vald, M. G. 1980. Late Weichselian ice sheet of northern Eurasia. *Quat. Res.* **13:**1–32.

Gros'vald, M. G. 1983. *Pokrovnye ledniki kontinetal'nykh shel'fov.* Izdatel'stovo Nauka, Moscow, 216p.

Gros'vald, M. G. 1984. Glaciation of continental shelves (Part II). *Polar Geogr. Geol.* **8:**287–351.

Gros'vald, M. G., and Yu. N. Vozovik. 1984. A "marine" ice cap in South Beringia (a working hypothesis). *Polar Geogr. Geol.* **8:**128–146.

Haggblom, A. 1982. Driftwood in Svalbard as an indicator of sea ice conditions. *Geogr. Annal.* **64A:**81–94.

Hanson, A. M. 1980. The snow cover of Arctic sea ice during the Arctic Dynamics Joint Experiment, 1975 to 1976. *Arctic Alp. Res.* **12:**215–226.

Hare, F. K., and S. Orvig. 1958. The Arctic circulation, a preliminary review. Arctic Meteorology Research Group, *Publication in Meteorology No. 12.* McGill University, Montreal, 211p.

Hastings, A. D., Jr. 1961. *Atlas of the Arctic Environment.* U.S. Army Rep. PER-33, HQ Quartermaster Pres. Eng. Command, Environ. Protection Res. Div., Natick, Mass., 22p.

Herman, G. F. 1980. Thermal radiation in Arctic stratus clouds. *Roy. Meteorol. Soc. Quart. J.* **106:**771–780.

Herman, Y. 1970. Arctic paleo-oceanography in late Cenozoic time. *Science* **169:**474–477.

Herman, Y. 1974. Arctic Ocean sediments, microfauna and the climatic record of late Cenozoic time. In Y. Herman (ed.), *Marine Geology and Oceanography of the Arctic Seas.* Springer-Verlag, New York, pp. 283–348.

Herman, Y. 1983*a*. Baffin Bay: Present-day analog of the central Arctic during late Pliocene to mid-Pleistocene time. *Geology* **11:**356–359.

Herman, Y. 1983*b*. Arctic paleoceanography in Late Neogene time and its relationship to global climate. *Oceanology* **23**(1):81–87.

Herman, Y., and D. M. Hopkins, 1980. Arctic oceanic climate in Late Cenozoic time. *Science* **209:**557–562.

Herman, Y. and J. K. Osmond. 1984. Late Neogene Arctic paleoceanography: Micropaleontology and chronology. In A. Berger, J. Imbrie, J. Hays, G. Kukla, and B. Saltzman (eds.), *Milankovitch and Climate.* D. Reidel, Dordrecht, pp. 241–249.

Hibler, W. D., III. 1980. Sea ice growth, drift and decay. In S. Colbeck (ed.), *Dynamics of Snow and Ice Masses.* Academic Press, New York, pp. 141–209.

Hibler, W. D. III, and J. E. Walsh. 1982. On modelling seasonal and interannual fluctuations of Arctic sea ice. *J. Phys. Oceanogr.* **12:**1514–1523.

Hills, T. S. 1983. Sensitivity of general circulation models to changes in sea-ice cover. In A. Ghazi (ed.), *Palaeoclimatic Research and Models.* D. Reidel, Dordrecht, pp. 181–192.

Holt, T., P. M. Kelly, and B. S. G. Cherry. 1984. Cryospheric impact of Soviet river diversion schemes. *Ann. Glaciol.* **5:**61–68.

Hopkins, D. M. 1982. Aspects of the paleogeography of Beringia during the late Pleistocene. In D. M. Hopkins et al., *Paleoecology of Beringia.* Academic Press, New York, pp. 3–28.

Hopkins, D. M. and L. Marincovich, Jr. 1984. Whale biogeography and the history of the Arctic Basin. In *Arctic Whaling: Proceedings of the International Symposium.* University of Groningen, Works of the Arctic Center, no. 8, Groningen, Netherlands, pp. 7–24.

Hopkins, D. M., J. V. Matthews, Jr., C. E. Schweger, and S. B. Young (eds.) 1982. *Paleoecology of Beringia.* Academic Press, New York, 489p.

Hughes, T., G. H. Denton, M. G. Grosswald. 1977. Was there a Late-Wurm Arctic ice sheet? *Nature* **266:**596–602.

Hunkins, K., A. W. H. Bé, N. D. Opdyke, and G. Mathieu, 1971. The late Cenzoic history of the Arctic Ocean. In K. K. Turekian (ed.), *The Late Cenozoic Glacial Ages.* Yale University Press, New Haven, pp. 215–237.

Hunt, B. G. 1984. Polar glaciation and the genesis of ice ages. *Nature* **308**(5954):48–51.

Huschke, R. E. 1969. Arctic cloud statistics from "air-calibrated" surface weather observations. RAND Corporation Memorandum RM-6173, Santa Monica, Calif., 79p.

Itin, V. 1936. Kolebania ledovitosti. *Sovietskaia Arktika* **1:**74–82; **3:**80–92.

Jackson, C. I. 1961. Summer precipitation in the Queen Elizabeth Islands. *Folia Geogra. Danica* **9:**140–153.

Johnson, C. M. 1980. Wintertime arctic sea ice extremes and the simultaneous atmospheric circulation. *Mon. Weather Rev.* **108:**1782–1791.

Keegan, T. J. 1958. Arctic synoptic activity in winter. *J. Meteorol.* **15:**513–521.

Kellogg, T. B. 1977. Paleoclimatology and paleo-oceanography of the Norwegian and Greenland Seas: The last 450,000 years. *Marine Micropaleontol.* **2:**235–249.

Kelly, P. M., and P. D. Jones. 1981. Summer temperatures in the Arctic, 1881–1981. *Climate Monitor* (Univ. of East Anglia, Climatic Research Unit) **10:**66–67.

Kelly, P. M., C. M. Goodess, and B. S. G. Cherry. 1987. The interpretation of the Icelandic sea ice record *J. Geophys. Res.* **92:**10,835–10,844.

Kelly, P. M., P. D. Jones, C. B. Sear, B. S. G. Cherry, and R. K. Tavakol. 1982. Variations in surface air temperatures: Part 2, Arctic regions, 1881–1980. *Mon. Weather Rev.* **110:**71–83.

Kerr, R. A. 1984. Ice cap of 30 million years ago detected. *Science* **244**(4645):141–142.

Khrol, V. P. 1976. Isparenie s poverkhnosti Severnogo Ledovitogo okeana. *Trudy Arkt. Antarkt. Inst.* **323:**148–155.

Knox, J. L., and J. E. Hay. 1985. Blocking signatures in the Northern Hemisphere: Frequency distribution and interpretation. *J. Climatol.* **5:**1–16.

Koch, L. 1945. The east Greenland ice. *Medd. Gronland* **130:**1–374.

Koerner, R. M. 1973. The mass balance of the sea ice of the Arctic Ocean. *J. Glaciol.* **12**(65):173–185.

Langleben, M. P. 1985. Oceanic heat flux as a component of the heat budget of sea ice. *Ann. Glaciol.* **6:**170–173.

LeDrew, E. F. 1983. The dynamic climatology of the Beaufort to Laptev sector of the Polar Basin for the summer of 1975 and 1976. *J. Climatol.* **3:**335–359.

LeDrew, E. F. 1984. The role of local heat sources in synoptic activity within the Polar Basin. *Atmosphere-Ocean* **22:**309–327.

LeDrew, E. F. 1985. The dynamic climatology of the Beaufort to Laptev sector of the Polar Basin for the winters of 1975 and 1976. *J. Climatol.* **5:**253–272.

Lindstrom, D. R., and D. R. MacAyeal. 1986. Paleoclimatic constraints on the maintenance of possible ice-shelf cover in the Norwegian and Greenland Seas. *Paleoceanogr.* **1:**313–337.

Lin'kova, T. I. 1965. Some results of paleomagnetic study of Arctic Ocean floor sediments. In E. R. Hope (trans.), *The Present and Past of the Geomagnetic Field.* Nauka, Moscow, pp. 279–291.

Lloyd, C. R. 1984. Pre-Pleistocene paleoclimates: The geological and paleontological evidence: Modeling strategies, boundary conditions, and some preliminary results. *Adv. Geophys.* **26:**35–140.

Lorenz, E. N. 1968. Climatic determinism. In J. M. Mitchell, Jr. (ed.), *Causes of Climatic Change, Meteorological Monographs* **8**(30):1–3.

Loshchilov, V. S. 1964. Snezhnyi pokrov na l'dakh tsentral'noi Arktiki. *Problemy Arktiki i Antarktiki* **17:**36–45.

Löve, A., and D. Löve. 1974. Origin and evolution of the Arctic and alpine floras. In J. D. Ives and R. G. Barry (eds.), *Arctic and Alpine Environments.* Methuen, London, pp. 571–603.

MacCracken, M. C. 1970. *Tests of Ice Age Theories Using a Zonal Atmospheric Model.* UCRL-72803, University of California, Livermore, Lawrence Radiation Laboratory, 58p.

McLaren, A. S. 1986. Analysis of the under-ice topography in the Arctic Ocean, as recorded by USS Nautilus in 1958 and USS Queenfish in 1970. Ph.D. dissertation, Department of Geography, University of Colorado, Boulder, 152p.

McLaren, A. S., M. C. Serreze, and R. G. Barry. 1987. Seasonal variations of sea ice motion in the Canada Basin and their implications. *Geophys. Res. Lett.* **14:**1123–1126.

Makshtas, A. P. 1984. *Teplovoi balans arkticheskikh l'dov v zimnii period* (Heat Balance of Arctic Ice in the Winter Period). Gidrometeoizdat, Leningrad, 67p.

Manabe, S., and A. J. Broccoli. 1985. The influence of continental ice sheets on the climate of an Ice Age. *J. Geophys. Res.* **90:**2167–2190.

Manabe, S., and R. J. Stouffer. 1980. Sensitivity

of a global climate model to an increase of $CO_2$ concentration in the atmosphere. *J. Geophys. Res.* **85**:5529–5554.

Margolis, S. V., and Y. Herman. 1980. Northern Hemisphere sea-ice and glacial development in the Late Cenozoic. *Nature* **286**:145–149.

Markham, W. E. 1981. *Ice Atlas. Canadian Arctic Waterways.* Atmospheric Environment Service, Environment Canada, 198p.

Markussen, B., R. Zahn, and J. Thiede. 1985. Late Quaternary sedimentation in the eastern Arctic Basin: Stratigraphy and depositional environment. *Palaeogeogr., Palaeoclimatol., Palaeoecol.* **50**:271–284.

Marshunova, M. S., and N. T. Chernigovskiy. 1978. *Radiation Regime of the Foreign Arctic.* NSF-TT-72-51034, 189p. (NTIS, PB 286492) (first published Gidrometeoizdat, Leningrad, 1971).

Maslanik, J. A., and R. G. Barry. 1988. Short-term interactions between atmospheric synoptic conditions and sea ice behaviour in the Arctic. *Annals Glaciol.* **12.**

Maykut, G. A. 1978. Energy exchange over young sea ice in the central Arctic. *J. Geophys. Res.* **83**:3446–3458.

Maykut, G. A. 1982. Large-scale heat exchange and ice production in the central Arctic. *J. Geophys. Res.* **87**:7971–7984.

Maykut, G. A., and N. Untersteiner. 1971. Some results of a time-dependent thermodynamic model of sea ice. *J. Geophys. Res.* **76**:1550–1575.

Mercer, J. H. 1984. Simultaneous climatic change in both hemispheres and similar bipolar warming: Evidence and implications. In J. E. Hansen and T. Takahashi (eds.), *Climate Processes and Climate Sensitivity.* Geophys. Monogr. 29, American Geophysical Union, Washington, D.C., pp. 307–313.

Micklin, P. F. 1981. A preliminary systems analysis of impacts of proposed Soviet river diversions on Arctic sea ice. *EOS* **62**:489–493.

Minicucci, D. A., and D. L. Clark. 1983. A late Cenozoic stratigraphy for glacial-marine sedimentation of the eastern Alpha Cordillera, central Arctic Ocean. In B. F. Molnia (ed.), *Glacial-Marine Ocean Sedimentation.* Plenum Press, New York, pp. 331–365.

Nakamura, N., and A. H. Oort. 1988. Atmospheric heat budgets of the polar regions. *J. Geophys. Res.* **93**(D8):9510:9524.

Nazintsev, Yu. L. 1964. Teplovoi balans poverkhnosti mnogoletnego ledianogo pokrova v tsentral'noi Arktike. *Arkticheskii Antarkticheskii nauchno-issledovatelskii Institut, Trudy* **267**:110–126.

Newson, R. L., 1973. Response of a general circulation model of the atmosphere to removal of the Arctic ice cap. *Nature* **241**:39–40.

North, G. R., and T. J. Crowley. 1985. Modelling glacial climates. *Geol. Soc. London J.* **142**:475–482.

Ogilvie, A. E. J. 1984. The past climate and sea-ice record from Iceland, Part I. Data to A.D. 1970. *Climatic Change* **6**:131–152.

Ohmura, A. 1982. Climate and energy balance on the Arctic tundra. *J. Climatol.* **2**:65–84.

Ohmura, A. 1984. On the cause of "Fram" type seasonal change in diurnal amplitude of air temperature in polar regions. *J. Climatol.* **4**:325–338.

Olausson, E. 1985. The glacial oceans. *Palaeogeogr., Palaeoclimatol., Palaeoecol.* **50**:291–301.

O'Neill, B. J. 1981. Pliocene and Pleistocene benthic foraminifera from the central Arctic Ocean. *J. Paleontol.* **55**:1141–1170.

Oort, A. H. 1975. Year-to-year variations in the energy balance of the Arctic atmosphere. In G. Weller and S. A. Bowling (eds.), *Climate of the Arctic.* Geophysical Institute, University of Alaska, Fairbanks, pp. 68–75.

Oort, A. H., and J. Peixoto. 1983. Global angular momentum and energy balance requirements from observations. *Adv. Geophys.* **25**:355–490.

Osmond, J. K. et al. 1986. Radiometric chronology in Arctic deep sea sediments (this volume).

Ostlund, H. G., and G. Hut. 1984. Arctic Ocean water mass balance from isotope data. *J. Geophys. Res.* **89**(C4):6373–6381.

Overland, J. E., and C. H. Pease. 1982. Cyclone climatology of the Bering Sea and its relation to sea ice extent. *Mon. Weather Rev.* **110**: 5–13.

Parkinson, C. L., and W. M. Kellogg. 1979. Arctic sea ice decay simulated for a $CO_2$-induced temperature rise. *Climatic Change* **2**:149–162.

Parkinson, C. L., and W. M. Washington. 1979. A large-scale numerical model of sea ice. *J. Geophys. Res.* **84**:311–337.

Parkinson, C. P., J. C. Comiso, H. J. Zwally, D. J. Cavalieri, P. Gloersen, and W. J. Campbell. 1987. *Arctic Sea Ice 1973–1976: Satellite Passive-Microwave Observations.* NASA, Washington, D.C., 296p.

Prik, Z. M. 1959. Mean position of surface pressure and temperature distribution in the Arctic. (In Russian.) *Trudy Arkt. Antarkt. Nauch-Issled. Inst.* **217**:5–34.

Rakipova, L. R. 1966. The influence of the Arctic ice cover on the zonal distribution of atmospheric temperature. In J. O. Fletcher (ed.), *Proceedings of the Symposium on the Arctic Heat Budget and Atmospheric Circulation.* Memoran-

dum RM-5233-NSF, Rand Corporation, Santa Monica, Calif., pp. 411–441.
Rea, D. K., and H. Schrader. 1985. Late Pliocene onset of glaciation: Ice-rafting and diatom stratigraphy of North Pacific DSDP cores. *Palaeogeogr., Palaeoclimatol., Palaeoecol.* **49:**313–325.
Reed, R. J. 1962. *Arctic Forecast Guide.* U.S. Navy Weather Research Facility, NWRF-16-0462-058, Norfolk, Va.
Reed, R. J., and B. A. Kunkel. 1960. The Arctic circulation in summer. *J. Meteorol.* **17:**489–506.
Reeh, N. 1984. Reconstruction of the glacial ice covers of Greenland and the Canadian Arctic Islands by three-dimensional perfectly plastic ice-sheet modelling. *Ann. Glaciol.* **5:**115–121.
Robinson, D. A., G. J. Kukla, and M. Serreze. 1986. Arctic summer cloudiness. *Sixth Conf. Atmospheric Radiation Proc.* (Williamsburg, Va.). American Meteorological Society, Boston, Mass., pp. 176–179.
Robock, A. 1983. Ice and snow feedbacks and the latitudinal and seasonal distribution of climate sensitivity. *J. Atmos. Sci.* **40:**986–997.
Rogers, J. C. 1978. Meteorological factors affecting interannual variability of summertime ice extent in the Beaufort Sea. *Mon. Weather Rev.* **106:**890–897.
Ruddiman, W. F., and A. McIntyre. 1977. Late Quaternary surface ocean kinematics and climatic change in the high-latitude North Atlantic. *J. Geophys. Res.* **82:**3877–3888.
Ruddiman, W. F., and A. McIntyre. 1981. The North Atlantic Ocean during the last deglaciation. *Palaeogeogr., Palaeoclimatol., Palaeoecol.* **35:**145–214.
Ruddiman, W. F. and A. McIntyre. 1984. An evaluation of ocean-climate theories on the North Atlantic. In A. L. Berger, J. Imbrie, J. Hays, G. Kukla, and B. Saltzman (eds.), *Milankovitch and Climate.* D. Reidel Publishing Company, Dordrecht, pp. 671–686.
Ruddiman, W. F., B. Molfino, A. Esmay, and E. Pokras. 1980. Evidence bearing on the mechanism of rapid deglaciation. *Climatic Change* **3:**65–87.
Rudels, B. 1984. The importance of the freshwater discharge and the Atlantic outflow for the maintenance of the Arctic ice cover. *Report of the Meeting of Experts on Sea Ice and Climate Modelling.* World Climate Programme WCP-77, World Meteorological Organization, Appendix F, 19p.
Sancetta, C. 1983. Effect of Pleistocene glaciation upon oceanographic characteristics of the North Pacific Ocean and Bering Sea. *Deep-Sea Res.* **30**(8A):851–869.
Sancetta, C., and S. W. Robinson. 1983. Diatom evidence on Wisconsin and Holocene events in the Bering Sea. *Quat. Res.* **20:**232–245.
Scharfen, G., R. G. Barry, D. A. Robinson, G. Kukla, and M. Serreze. 1987. Large-scale patterns of snow melt on Arctic sea ice mapped from meteorological satellite imagery. *Ann. Glaciol.* **9:**200–205.
Schneider, S. H., S. L. Thompson, and E. J. Barron. 1985. Mid-Cretaceous continental surface temperatures: Are high $CO_2$ concentrations needed to simulate above freezing winter conditions? In E. T. Sundquist and W. S. Broecker (eds.), *The Carbon Cycle and Atmospheric $CO_2$: Natural Variations, Archean to Present.* Geophysical Monograph 32, American Geophysical Union, Washington, D.C., pp. 554–559.
Sclater, J. G., S. Hellinger, and C. Tapscott. 1977. The paleobathymetry of the Atlantic Ocean from the Jurassic to the present. *J. Geol.* **85:**509–552.
SCOR, Working Group 58, 1979. *The Arctic Ocean Heat Budget.* University of Bergen, Geophysical Institute, Rep. No. 52, p. 98.
Sejrup, H. P., G. H. Miller, J. Brigham-Grette, R. Lovlie, and D. Hopkins. 1984. Amino acid epimerization implies rapid sedimentation in Arctic Ocean cores. *Nature* **310:**772–775.
Sellers, A., and A. Meadows. 1975. Long-term variations in the albedo and surface temperature of the earth. *Nature* **254:**44.
Semtner, A. J., Jr. 1976. A model for the thermodynamic growth of sea ice in numerical investigations of climate. *J. Phys. Oceanogr.* **6:**379–389.
Semtner, A. J., Jr. 1984*a*. On modelling the seasonal thermodynamic cycle of sea ice in studies of climatic change. *Climatic Change* **6:**27–37.
Semtner, A. J., Jr. 1984*b*. The climatic response of the Arctic Ocean to Soviet river diversions. *Climatic Change* **6:**109–130.
Semtner, A. J., Jr. 1987. A numerical study of sea ice and ocean circulation in the Arctic. *J. Phys. Oceanogr.* **17:**1077–1099.
Serreze, M. C., and R. G. Barry. 1988. Synoptic activity in the Arctic Basin 1979–1985. *J. Climate* **1:**1276–1295.
Shackleton, N. J., and N. D. Opdyke. 1977. Oxygen isotope and paleomagnetic evidence for early Northern Hemisphere glaciation. *Nature* **270:**216–219.
Shackleton, N. J., et al. 1984. Oxygen-isotope calibration of the onset of ice-rafting and history of glaciation in the North Atlantic region. *Nature* **307:**620–623.

Shaw, D. M., and W. L. Donn. 1971. A thermodynamic study of Arctic palaeoclimatology. *Quat. Res.* **1**:175–187.

Shine, K. P., and R. G. Crane. 1984. The sensitivity of a one-dimensional thermodynamic sea ice model to a change in cloudiness. *J. Geophys. Res.* **89**(C6):10,615–10,622.

Shine, K. P., and A. Henderson-Sellers. 1985. The sensitivity of a thermodynamic sea ice model to changes in surface albedo parameterization. *J. Geophys. Res.* **90**:2243–2250.

Spjeldnaes, N. 1981. Ice-rafting, an indication of glaciation? *Nature* **214**:687–688.

Stephens, G. L., G. G. Campbell, and T. H. Vonder Haar. 1981. Earth radiation budgets. *J. Geophys. Res.* **86**(C10):9739–9760.

Stewart, T. G., and J. England. 1983. Holocene sea ice variations and paleoenvironmental change, northernmost Ellesmere Island, N.W.T., Canada. *Arct. Alp. Res.* **15**:1–17.

Stigebrandt, A. 1981. A model for the thickness and salinity of the upper layer in the Arctic Ocean and the relationship between the ice thickness and some external parameters. *J. Phys. Oceanogr.* **11**:1407–1422.

Stigebrandt, A. 1985. On the hydrographic and ice conditions in the northern North Atlantic during different phases of a glaciation cycle. *Palaeogeogr., Palaeoclimatol., Palaeoecol.* **50**:303–321.

Talwani, M., and O. Edholm. 1977. Evolution of the Norwegian-Greenland Sea. *Geol. Soc. Am. Bull.* **88**:969–999.

Thiede, J. 1983. History of the north polar seas during the past 5 million years. In A. Ghazi (ed.), *Palaeoclimatic Research and Models.* D. Reidel, Dordrecht, pp. 178–180.

Thiede, J. (compiler) 1988. Scientific report of Arctic expedition ARK IV/3. *Berichte zur Polarforschung* **43**:171–185.

Thorndike, A. S. 1986. Kinematics of sea ice. In N. Untersteiner (ed.), *The Geophysics of Sea Ice.* Plenum Press, New York, pp. 489–549.

Tsay, S.-C., and K. Jayaweera. 1984. Physical characteristics of Arctic stratus clouds. *J. Climate Appl. Met.* **23**:584–596.

Velichko, A. A. 1984. Late Pleistocene spatial paleoclimatic reconstructions. In A. A. Velichko (ed.) and H. E. Wright, Jr. and C. W. Barnosky (eds., English-language edition), *Late Quaternary Environments of the Soviet Union.* University of Minnesota Press, Minneapolis, pp. 261–285.

Velichko, A. A., L. L. Isayeva, V. M. Makeyev, G. G. Matishov, and M. A. Faustova. 1984. Late Pleistocene glaciation of the Arctic Shelf, and the reconstruction of Eurasian ice sheets. In A. A. Velichko (ed.) and H. E. Wright, Jr. and C. W. Barnosky (eds., English-language edition), *Late Quaternary Environments of the Soviet Union.* University of Minnesota Press, Minneapolis, pp. 35–41.

Vincent, J. S., S. Occhietti, N. Rutter, G. Lortie, J. P. Guilbault, and B. de Boutray. 1983. The late Tertiary-Quaternary stratigraphic record of the Duck Hawks Bluffs, Banks Island, Canadian Arctic Archipelago, *Can. J. Earth Sci.* **20**:1694–1712.

Vogt, P. R. 1986. Sea floor topography, sediments and paleoenvironments. In B. G. Hurdle (ed.), *The Nordic Seas,* Springer-Verlag, New York, pp. 237–410.

Volkov, N. A., and V. F. Zakharov. 1977. Evolution of the ice cover in the Arctic as related to climatic changes. *Soviet Meteorol. Hydrol.* **7**:37–43.

Vowinckel, E., and S. Orvig. 1962. *Relation between Solar Radiation Income and Cloud Type in the Arctic.* Publications in Meteorology No. 53, McGill University, Montreal, 26p.

Vowinckel, E., and S. Orvig. 1967. *The Inversion over the Polar Ocean.* World Meteorological Organization, Geneva, Technical Note 87, pp. 39–59.

Vowinckel, E., and S. Orvig. 1970. The climate of the North Polar basin. In S. Orvig (ed.), *Climates of the Polar Regions, World Survey of Climatology.* Vol. 14 (H. E. Landsberg, editor-in-chief). Elsevier, Amsterdam, pp. 129–252.

Wadhams, P. 1981. The ice cover in the Greenland and Norwegian Seas. *Rev. Geophys. Space Phys.* **19**:345–393.

Wadhams, P. 1983. Sea ice thickness distribution in Fram Strait. *Nature* **305**:108–111.

Wadhams, P., and R. J. Horne. 1980. An analysis of ice profiles obtained by submarine sonar in the Beaufort Sea. *J. Glaciol.* **25**(93):401–424.

Walin, G. 1985. The thermohaline circulation and the control of ice ages. *Palaeogeogr., Palaeoclimatol., Palaeoecol.* **50**:323–332.

Walsh, J. E., W. D. Hibler, III, and B. Ross. 1985. Numerical simulation of Northern Hemisphere sea ice variability, 1951–1980. *J. Geophys. Res.* **88**:4847–4865.

Walsh, J. E., and C. M. Johnson. 1979. Interannual atmospheric variability and associated fluctuations in Arctic sea ice extent. *J. Geophys. Res.* **84**(C11):6915–6928.

Warshaw, M., and R. R. Rapp. 1973. An experiment on the sensitivity of a global circulation model. *J. Appl. Meteorol.* **12**:43–49.

Weidick, A. 1976. Glaciations of northern Greenland—new evidence. *Polarforschung* **46**:26–33.

Whittaker, L. M., and L. H. Horn. 1984. Northern hemisphere extratropical cyclone activity for four mid-season months. *J. Climatol.* **4**:297–310.

Williams, D. F., W. S. Moore, and R. H. Fillon. 1981. Role of glacial Arctic Ocean ice sheets in Pleistocene oxygen and sea level records. *Earth Planet. Sci. Lett.* **56**:157–166.

Williams, J. 1975. The influence of snow-cover on the atmospheric circulation and its role in climatic change: An analysis based on results from the NCAR global circulation model. *J. Appl. Meteorol.* **14**:137–152.

Wilson, C. W. 1967. *Cold Regions Science and Engineering. Part 1. Environment Section A3: Climatology. Introduction. Northern Hemisphere I.* Cold Regions Research and Engineering Laboratory, Hanover, N.H., 141p.

Witte, W. K., and D. V. Kent. 1988. Revised magnetostratigraphies confirm low sedimentation rates in Arctic Ocean cores. *Quat. Res.* **29**:43–53.

Wolfe, J. A. 1980. Tertiary climates and floristic relationships at high latitudes in the Northern Hemisphere. *Palaeogeogr., Palaeoclimatol. and Palaeoecol.*, **30**:313–323.

Wolfe, J. A. 1985. Distribution of major vegetational types during the Quaternary. In E. T. Sundquist and W. S. Broecker (eds.), *The Carbon Cycle and Atmospheric $CO_2$: Natural Variations, Archean to Present,* Geophysical Monograph 32, American Geophysical Union, Washington, D.C., pp. 554–559.

Worsley, T. R. and Y. Herman. 1980. Episodic ice-free Arctic Ocean in Pliocene and Pleistocene time: Calcareous nannofossil evidence. *Science* **210**:323–325.

Yanes, A. V. 1966. Melting of snow and ice in the central Arctic. In *Problems of the Arctic and Antarctic,* No. 11, N. A. Ostenso (ed.), Arctic Institute of North America, pp. 1–13.

Zahn, R., B. Markkussen, and J. Thiede. 1985. Stable isotope data and depositional environments in the late Quaternary Arctic Ocean. *Nature* **314**(6010):433–435.

Zakharov, V. F. 1976*a*. Pokholodanie Arktiki i ledyanoi pokrov Arkticheskikh Morei. *Trudy, Ark. Antarkt. Inst.* **337**:1–96.

Zakharov, V. F. 1976*b*. Ledoviai oppozitsiia. *Trudy, Ark. Antarkt. Inst.* **320**:147–152.

Zakharov, V. F. 1981. *L'ydy Arktiki i Sovremennie Prirodnye Protsessy,* Gidrometeoizdat, Leningrad.

Zavarina, M. V., and M. K. Romasheva. 1960. *Extent of Clouds over the Arctic Seas and Central Arctic USSR.* U.S. Joint Publ. Res. Service Translation, L-2019-D, Office of Technical Services, U.S. Department of Commerce, Washington, D.C. (First published in *Problemy Arktiki* **2**:127–132, 1957.)

# 2: Arctic Ice–Ocean Dynamics

W. D. Hibler III

## INTRODUCTION

A dominant characteristic of sea ice in the polar regions is its almost constant drift and deformation. Because of this deformation, a typical 100 $km^2$ patch of sea ice will contain a variety of ice thicknesses, ranging from open water to very thick ice, to ridges extending possibly 30 m or more below the surface. On top of this variably thick matrix there is often a relatively thin (typically about 30 cm) snow cover. Although thin, this snow cover can cause substantial insulation of the ice and reduce its growth rate.

While the main driving forces that move this ice cover come from wind and currents, the ice does not just move as a passive tracer, but rather has a motion significantly affected by ice/ice interaction. The magnitude of this interaction is especially pronounced near the coast, as manifested by very heavy ridging that often forms there. Far from shore the effects of interaction are more subtle, but they are still considerable in that significant ice buildup is prevented by ice pressure. For comparison purposes, ice pressure is comparable in magnitude to the buildup of surface pressure in the ocean by sea-surface tilt. Ice stresses that are typically used in large-scale models, for example (see, e.g., Hibler, 1980*a;* Hibler and Walsh, 1982), are of the order of $10^4$ N $m^{-2}$, which is approximately equivalent to the pressure force of 1 m of water.

The presence of this dynamic ice cover substantially alters heat, salt, and momentum transfers between the atmosphere and ocean, and hence has the potential to alter atmospheric and oceanic circulation. In these modifications sea-ice dynamics plays a pivotal role inasmuch as leads and ridging substantially change the air–sea heat exchanges as compared to a constant thickness ice cover. In addition, ice transport allows net local imbalances in salt and heat to exist. An important point here is that while volume transports of ice are small, heat and salt transports are magnified by the latent heat of the ice and the difference in salinity between ice and seawater of about $30^0/_{00}$.

The purpose of this paper is to examine the dynamic character of the Arctic ice–ocean system. Toward this end the basic observational characteristics of sea ice are reviewed, and the physics of the dynamics of the ice-ocean system are discussed. This discussion also includes results from numerical modeling experiments and, because of the close coupling of sea-ice dynamics to the ocean, must inevitably include some description also of ocean circulation.

## OBSERVATIONAL CHARACTERISTICS OF SEA-ICE DYNAMICS

A dominant feature of the Arctic ice cover is its relatively landlocked character: Most of the ice is contained in the Arctic Basin (Fig. 2-1). This constrained character contrasts with the Antarctic, where there are relatively few impediments to the free expansion of the ice cover. This restriction of the ice cover leads to greater ridging in the Arctic than in the Antarctic (Weeks, 1968), and thicker ice. The main outlet for this relatively thick ice is the Fram Strait (between Greenland and Spitsbergen), and as a consequence considerable ice is present off the east coast of Greenland, even during times of minimum ice extent.

With regard to areal coverage, Walsh and Johnson (1979) estimate the maximum Arctic ice cover to be about $15 \times 10^6$ km$^2$. This maximum occurs in February, while the minimum, a reduction of about 50 percent, occurs in August. Of this areal extent, over half is in the Arctic Basin and Barents Sea region.

### Sea Ice Drift

The general pattern of the ice cover circulation in the Arctic Basin has been known for some time. The general characteristics (Fig. 2-2) consist of a gyre in the Beaufort Sea and a transpolar drift stream. The transpolar drift stream carries ice from the East Siberian Sea (and even from the Bering Strait), passing it across the North Pole and down the east coast of Greenland. Because of the recirculating character of the Beaufort Gyre, it is possible for ice that survives the summer melt to stay in the gyre for a large number of years (e.g., Koerner, 1973). Consequently, some of the thickest multiyear ice is found in this region.

Ice stations in the Arctic typically have been found to drift at a rate of about 6 km day$^{-1}$ (Reed and Campbell, 1962; Dunbar and Wittman, 1962), but drift rates can vary from day to day and from year to year. One-day drifts of over 20 km have been observed (Thorndike and Colony, 1980). It is also possible for the ice to remain effectively motionless for several weeks. Such behavior was documented by Thorndike and Colony (1980) in the spring of 1976 in the Beaufort Sea. However, except for ice very near shore, this motionless behavior appears to be the exception rather than the rule.

The general drift rates and variability of overall drift patterns from year to year can be deduced from the mean monthly drifting station positions plotted in Figure 2-2. Two notable features illustrated by this figure are the presence of substantial drift (even in winter) parallel to the Alaskan coast and a relatively small component of motion perpendicular to the Canadian Archipelago near the pole. A feature of drift tracks not shown in this figure is the meandering nature of the drift, with week-long reversals in direction being common. Additional perspective on the drift rates in the gyre may be gained by noting that it took the ice island T-3 about 10 years to make a full circuit (Dunbar and Wittmann, 1962). As drifting stations approach the Greenland–Spitsbergen passage, the drift rates begin to increase and meandering decreases. This phenomenon continues after the ice exits from the basin, with the drift rates being as large as 25 km day$^{-1}$ in the East Greenland region (Vinje, 1976). While this general pattern of drift still holds, analysis of more recent Arctic buoy data (see, e.g., Colony, 1984) show a less pro-

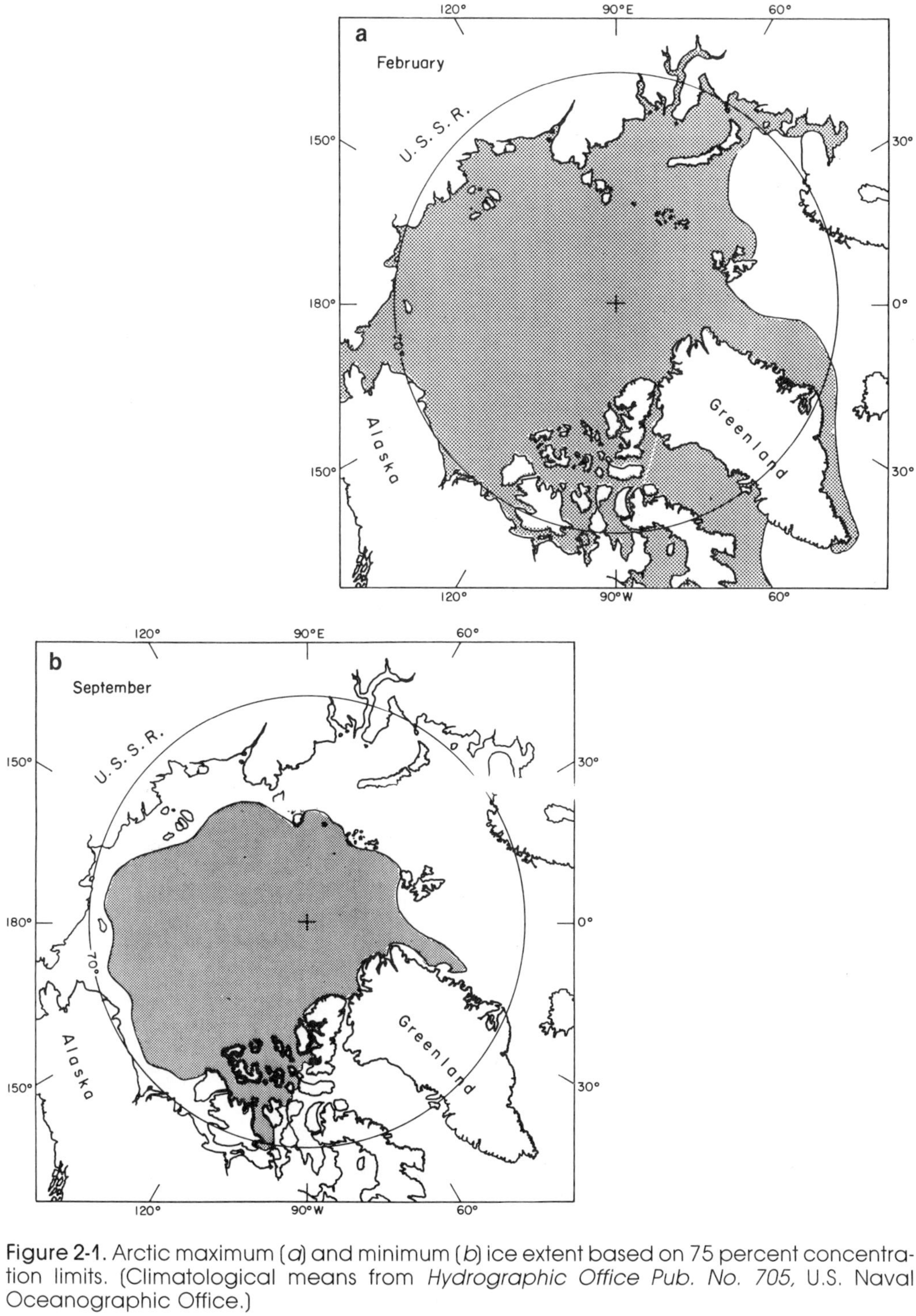


**Figure 2-1.** Arctic maximum (*a*) and minimum (*b*) ice extent based on 75 percent concentration limits. (Climatological means from *Hydrographic Office Pub. No. 705,* U.S. Naval Oceanographic Office.)

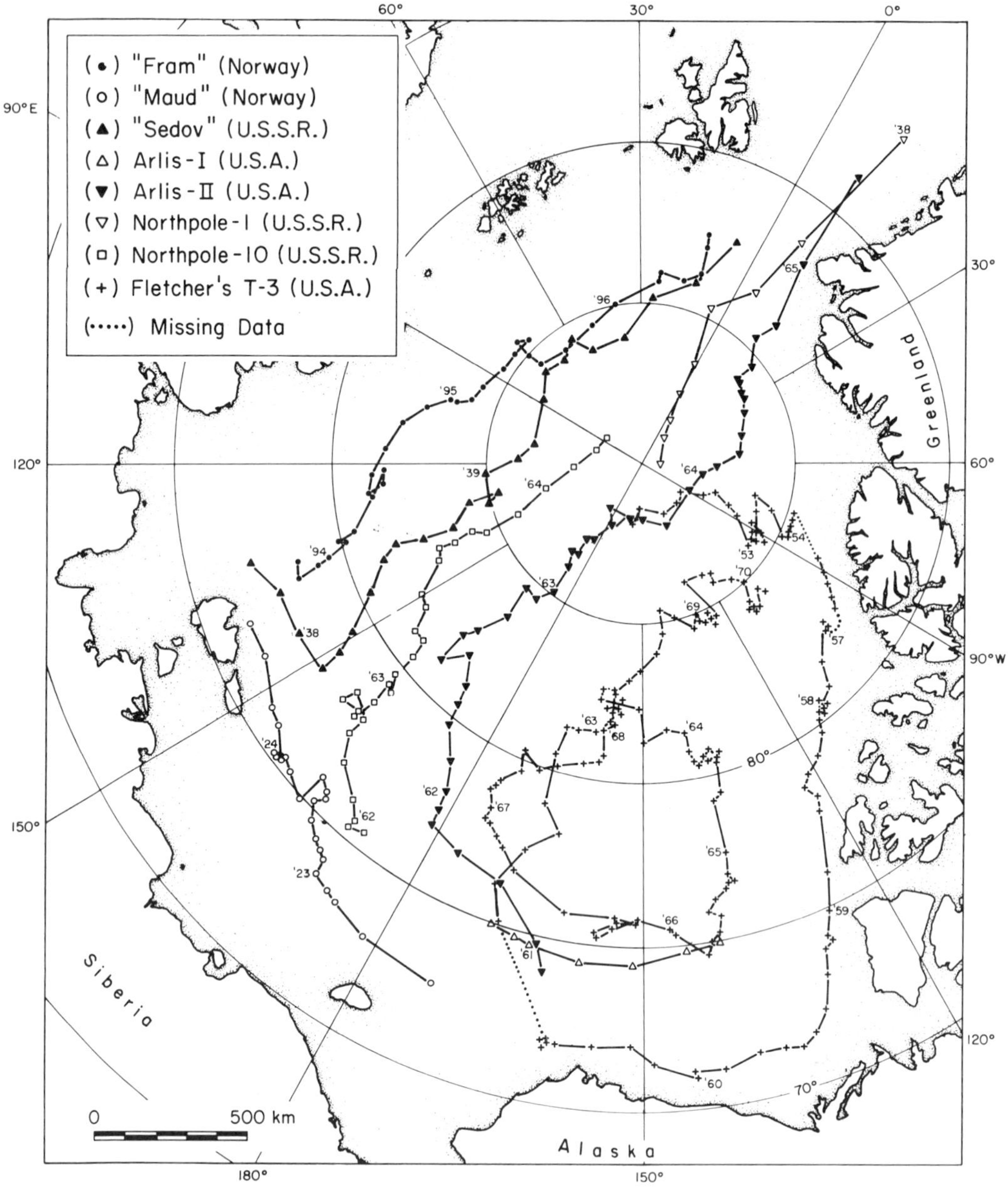


Figure 2-2. Tracks of several drifting stations in the Arctic Ocean. Monthly positions are plotted, with the year of each January 1st position indicated. (From Hastings, 1971.)

nounced transpolar drift stream and a more elongated Beaufort Sea Gyre (along the 60°–120° latitude line). This general character also appears in model simulated drift rates during recent years (Walsh, Hibler, and Ross, 1985; Hibler and Bryan, 1984, 1987), which largely tend to follow the wind.

When compared with wind and currents, the ice station data are not accurate enough to allow detailed examination of the temporal and spatial variations of ice drift. More recent measurements (Thorndike and Colony, 1980) using satellite navigation equipment and laser surveying equipment (Hibler et al., 1974*a*) have supplied a more complete description of the temporal and spatial variation in ice drift. These data show that while the ice velocity has the greatest variation at low frequencies (less than about ⅕ cycle day$^{-1}$), significant velocity perturbations do occur at higher frequencies. These high-frequency motions are dominated by a peak at about 2 cycles day$^{-1}$. This peak is prominent during the summer months, but is substantially suppressed throughout the winter months. Hunkins (1967) and McPhee (1978) have explained these 12-hour oscillations as inertial motions arising from the balance between inertia and the Coriolis force.

When compared with wind and currents, the overall observational characteristic of ice drift is that on the short time scale ice tends to follow the wind, with the drift approximately following the geostrophic wind with about $1/50$ of the magnitude (Zubov, 1943). This general feature of ice drift has been known for many years, beginning with Nansen's expeditions to the Arctic. Since wind variations are larger than currents on a short time scale, this also means that, except for shallow regions, fluctuations in ice drift will be dominated more by wind than by currents. However, on a long time scale, the more steady currents can play a significant role.

These characteristics are illustrated in Figure 2-3, which shows short-term and long-term drift rates using a linear ice-drift model (Hibler and Tucker, 1979). As can be seen, including current effects has only a minor effect on short-term variation. In examining cumulative drift over several years, however, significant differences occur if currents and ocean tilt (see the next section) are neglected, basically while smaller, current effects are steady. On the other hand, wind effects, while large, tend to fluctuate over a long time period, leaving a smaller constant value. This general behavior will be discussed in more detail below in conjunction with a nonlinear ice–ocean model (Hibler and Bryan, 1987).

## Sea Ice Deformation

In addition to drift, the pack ice regularly undergoes deformation. Measurements of deformation on scales of 10–20 km (Hibler et al., 1974*a*) and on scales of 100–500 km in the central Arctic pack (Thorndike and Colony, 1980) show typical magnitudes of the strain rate to be about 1 percent day$^{-1}$. These strain rates generally have little correlation with the ice velocity.

In the Arctic pack on scales smaller than about 20 km the strain rates become very erratic, since motion will often be due to a single lead. A typical view of the matrix of leads and flaws in sea ice in the central pack is shown in Figure 2-4 (top). Basically, in the central pack, sea ice takes on more the appearance of a continuum with leads interspersed. Because of the paucity of flaws in the pack ice (especially in winter), across which motion can occur, it becomes difficult to adequately define the deformation except statistically. Typical observed deviations of the ice velocity field from a linear spatial variation have varied from 0.05 km day$^{-1}$ (Hibler et al., 1974*a*) to about 1.0 km day$^{-1}$ (Thorndike and Colony, 1980).

In the marginal ice zone (region within a hundred kilometers or so of the ice edge), the ice deformation rates tend to be considerably larger, especially as one approaches the edge. In this region the ice tends to resemble a granular medium with relatively small floes, often densely packed (Fig. 2-4, bottom). Measurements of ice deformation in the marginal ice zone on scales of 20 km (Lepparanta and Hibler, 1986) show a much more active deformation, with rates typically an order of magnitude higher, ($\approx$10 percent day$^{-1}$) even

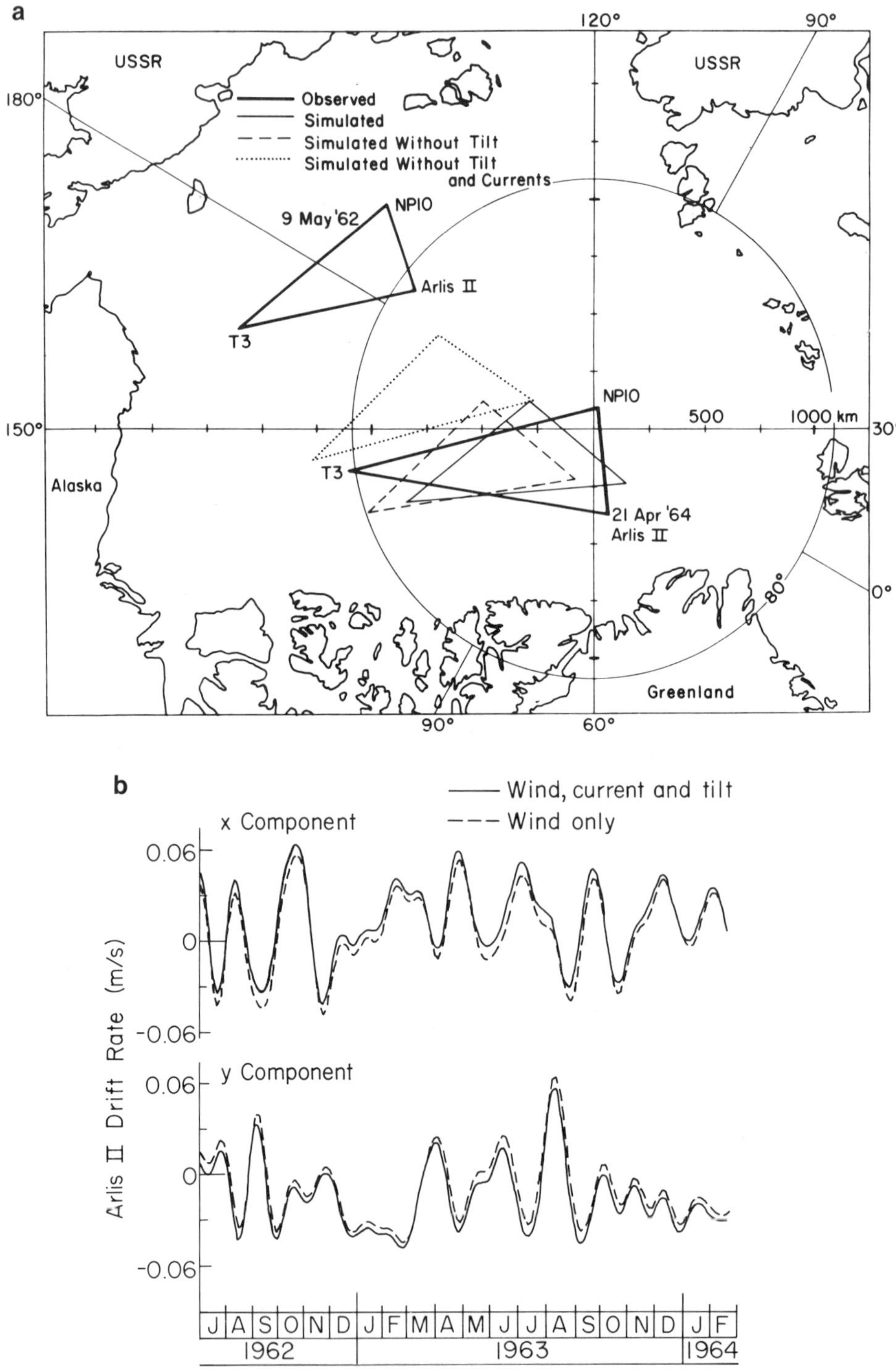


Figure 2-3. (*a*) Predicted and observed net drift of three drift stations, with and without current and ocean tilt effects. (*b*) Effect of currents and ocean tilt on the 25-day smoothed predicted drift rate of the Arlis II ice island. (From Hibler and Tucker, 1979.)

**Figure 2-4.** Typical aerial views of pack ice in the central Arctic in spring (top) and the marginal ice zone in summer (bottom), in which the ship shown is about 50 m in length. In the top view, the lead is about 300 m wide.

under relatively locked "plug flow" conditions, than those obtained in the central pack. Also, under loose conditions the deformation rates can become an additional order of magnitude higher. Because of the accuracy and sampling rate of this particular set of measurements, it was possible to measure the spectral density at very high frequencies. An interesting feature of this spectrum (Fig. 2-5) is the relative red noise character even out to very high frequencies.

A particularly important facet of ice deformation in the Arctic is the formation of a relatively discontinuous shear zone off the North Slope of Alaska and the Canadian Arctic. Winter drift rates very near shore are often substantial in this region. This introduces a discontinuity in the velocity fields very close to the coast. Analysis of this shear zone suggests that shearing is sporadic and may be preceded by a reduction in onshore stress caused by offshore motion. In March 1973, a Landsat study by Hibler et al. (1974*b*), off Barrow, for example, showed the development of strong shearing motion to be initially accompanied by an offshore motion of the pack. The initial and final locations of a series of ice points for a one-day interval from the study are shown in Figure 2-6.

Drift measurements taken further to the east in the same year indicate that a shear zone was also well defined there. In particular, Reimnitz, Toimil, and Barnes (1978) found a relatively persistent shearing motion off Prudhoe Bay from the middle of March to mid-May. This motion tended to occur on the seaward side of a heavily ridged and partially grounded zone approximately 20 km wide. This zone, which was termed a "stamukhi" zone, tends to straddle the 10–30-m depth range. An important observation of both studies (Hibler et al., 1974*b;* Reimnitz, Toimil, and Barnes, 1978) was that the seaward ice next to the shear zone often moves westward (3–10 km day$^{-1}$) at a faster rate than the ice farther from the coast. Thus the ice motion takes on the characteristics of a large, ro-

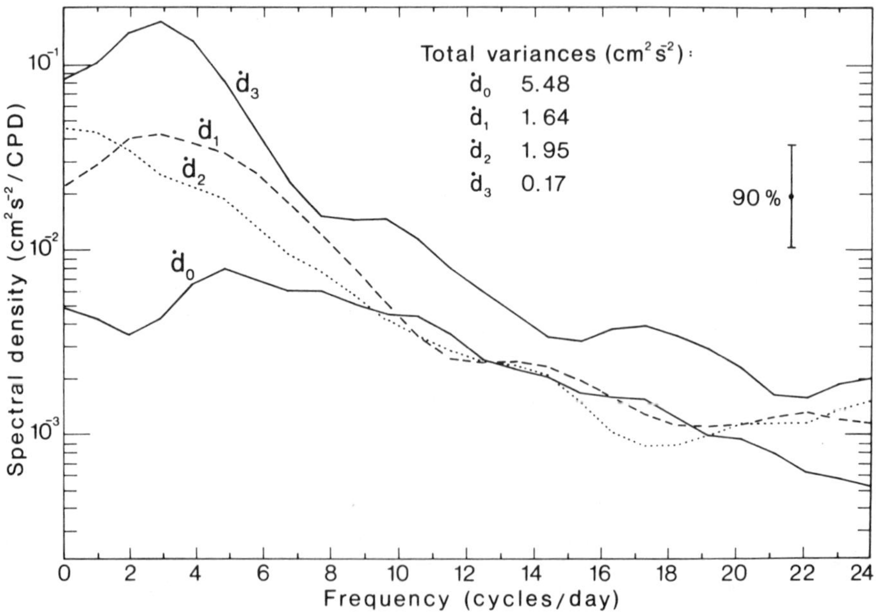


Figure 2-5. Spectra of distance fluctuations between ice floes in the east Greenland marginal ice zone in summer. The distances between floes ranged from 3–6 km. (From Lepparanta and Hibler, 1986.)

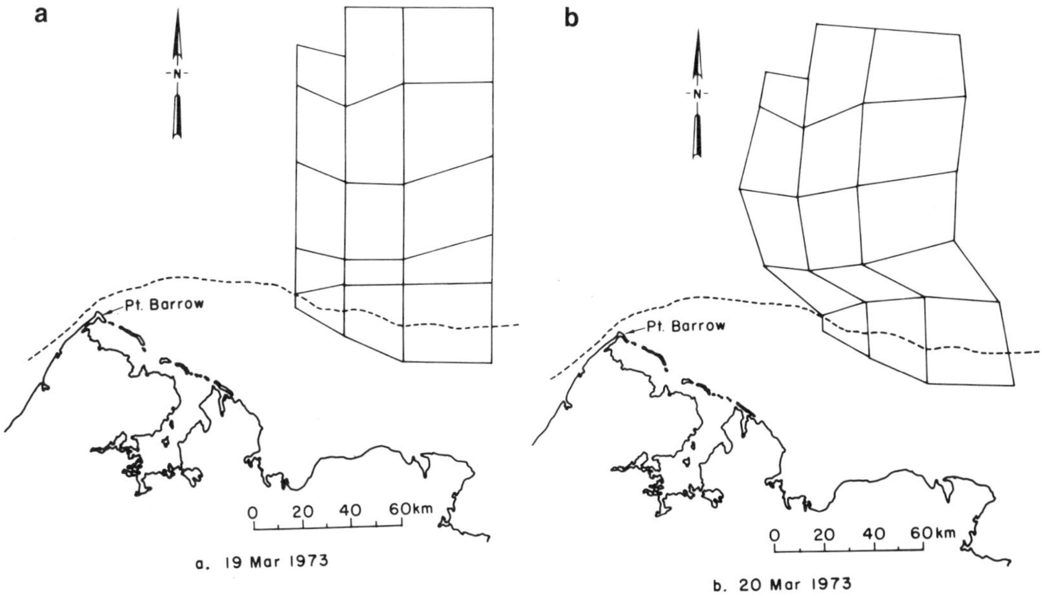


Figure 2-6. Observed ice motion over a one-day interval during shear activity in the near-shore region: (*a*) 19 March 1973; (*b*) 20 March 1973. The observations were made from Landsat imagery and the dashed line marks the boundary between the shore-fast ice and the pack ice. (From Hibler et al., 1974*b*.)

tating, cohesive wheel with slippage at the boundary.

More recent measurements, however, indicate that a well-defined shear zone does not always exist. During spring of 1976 and 1977, detailed measurements of the near-shore ice regions up to about 30 km from shore were made near Prudhoe Bay (Weeks et al., 1977; Tucker et al., 1980). These measurements show a near-shore ice cover that undergoes few large motions (typically one large deformation event per month). During such an event the pack ice may move several kilometers away from the near-shore ice for a day or so and then return. However, in the absence of such large events the ice did exhibit high-frequency fluctuations of about 10 m.

## Ice Thickness and Ridging Characteristics

Due to almost constant motion, formation, and decay, sea ice has a highly variable thickness, which is significantly affected by the amount of ridging. The thin ice plays a particularly important role because of its low strength and high growth rates. Probably the most valuable data for estimating ice thickness characteristics are submarine sonar profiles of the underside of the pack ice (for references see Hibler, 1980b, p. 155). Because about 90 percent of the ice thickness is under water, these profiles can be used to estimate the distribution and spatial variability of thickness. While considerable data have been taken, few have been analyzed in detail. However, in the Arctic Basin a sufficient amount of data have been analyzed (see, e.g., Wadhams, 1981) to infer the general shape of the thickness distribution and spatial variations in thicknesses.

Some typical sonar thickness distributions are shown in Figure 2-7. A salient characteristic of the ice thickness distribution illustrated by Figure 2-7 is the marked bimodal shape. Basically the ice can be thought of as a combination of a thin ice

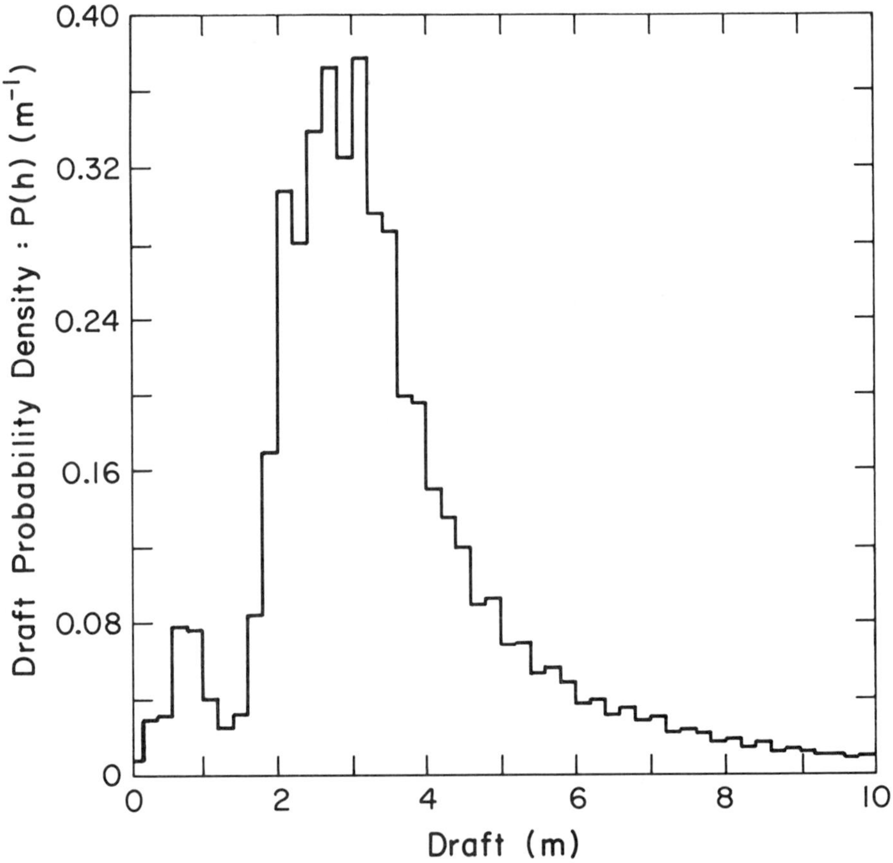


Figure 2-7. Distributions of ice draft taken from two 200-km sections of submarine track intersecting at 72°50′N and 144°17′W in April 1976. (Drawn from data reported in Wadhams and Horne, 1978.)

distribution (up to about 2m) plus a thick ice distribution. The thin ice mode consists of young first-year ice, while the thicker mode is composed of level multiyear ice and ridged ice (both first-year and multiyear). In the early fall, the thin ice consists mostly of ice formed after the cessation of summer melt conditions. However, ice can be added to the thin mode throughout the winter by the creation of leads via diverging deformation. The presence of a substantial amount of ice less than 1 m thick is especially apparent in Figure 2-7 taken by the *USS Gurnard* in April 1976 in the Beaufort Sea (e.g., Wadhams and Horne, 1978).

While the unridged young first-year ice is normally quite level, it should be noted that the undeformed thicker multiyear ice is not. The thicker multiyear ice tends to have a characteristic undulating surface that varies on about a 30 m horizontal scale. Detailed measurements of the thickness and roughness variation of "level" multiyear ice show typical standard deviations of thickness of about 0.6 m. The reason for this variability is the presence of summer melt ponds, which create low spots (which sometimes melt through the ice completely). Concomitant with this thickness variation is a salinity and density variation, with the thicker ice being less salty and less dense.

Very little methodical analysis of the

large-scale spatial variations of thickness has been done. However, L. A. LeShack (private communication, 1979) has analyzed selected portions of submarine tracks over the Arctic Basin. By averaging the mean thickness of the ice-covered portions of different submarine tracks (regardless of season), approximate contours of mean ice thickness for part of the Arctic Basin can be constructed (see Fig. 2-8). The dominant characteristic of this observed variation is greater thickness off the Canadian Archipelago and lesser thickness off the Siberian coast and North Slope.

A particularly important part of the ice thickness distribution are sea-ice pressure ridges. These ridges form when areas of thin ice that have formed in response to wind fields are compressed. This opening and closing action allows considerably greater quantities of ice to be formed than is possible with relatively flat, unridged ice. In the central Arctic Basin, where ridging is relatively light, ridges account for 2–20 percent of the ice mass (Koerner, 1973). In more heavily ridged areas, such as off the Canadian Archipelago, ridges can account for more than 50 percent of the ice mass (Williams et al., 1975).

Field observation (Kovacs, 1972; Weeks, Kovacs, and Hibler, 1971) shows the cross sections of ridge keels and sails to be approximately triangular. For multiyear ridges the keel depth is consistently about 3.2 times the sail height. For first-year ridges the ratio tends to be larger ($\approx 4.5$) and more variable (Kovacs and Mellor, 1974).

In an attempt to classify pressure ridges, statistical distribution models have been developed (Hibler, Weeks, and Mock, 1972; Wadhams, 1981) and used to examine the spatial and temporal variability of ridges. Specifically, Hibler, Weeks,

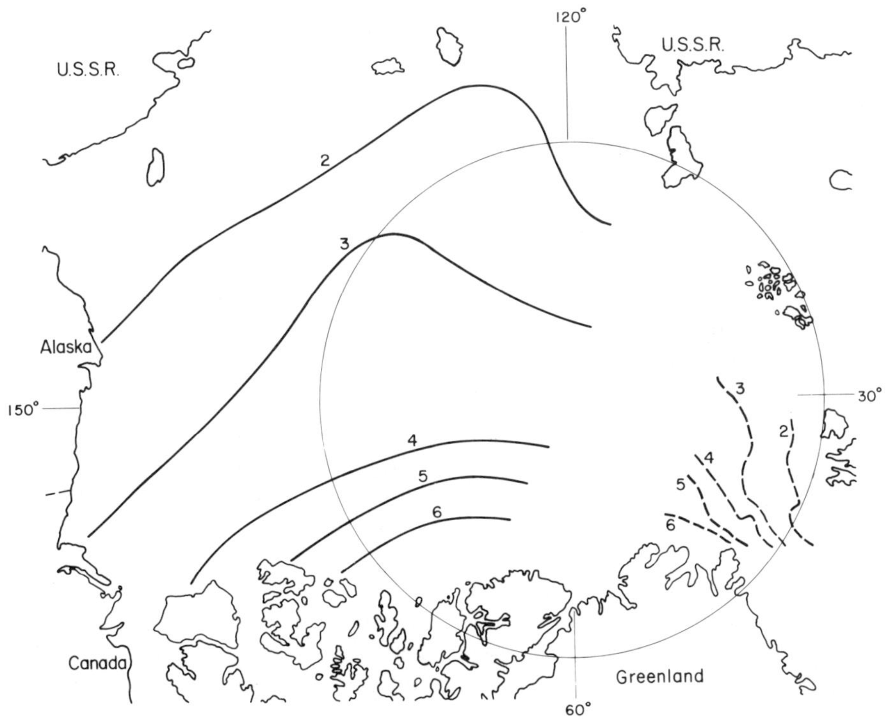


Figure 2-8. Contours (in meters) of observed mean ice thickness values estimated from submarine sonar data. (LeShack, private communication, 1978.)

and Mock (1972) derived height and spacing distributions using general assumptions of randomness and assuming ridge keels and sails to have triangular cross sections. Later modifications to these distributions based on the finite spacing between ridges have been made. These distributions have been found to fit ridge keels well. They also work well for ridge sail heights, except at greater ridge heights where a simpler exponential height distribution (Wadhams, 1981) performs better.

Studies of the spatial and temporal variations of ridging in the western Arctic Basin (Hibler, Mock, and Tucker, 1974) generally show the spatial variability of ridging to correlate well with the thickness contours shown previously. In addition to these basin-wide variations, near-shore ridge studies (Tucker, Weeks, and Frank, 1979) off the North Slope of Alaska and Canada show a buildup of ridging near the coast that decays several hundred kilometers offshore. Field studies show this region to be highly deformed, with substantial amounts of rubble pileup (Kovacs, 1976). While these relative geographical ridge characteristics remain more or less constant, the magnitude of the ridging can vary from year to year, especially near shore. In particular, Hibler, Mock, and Tucker (1974) found the ridging intensity (approximately equivalent to the mean thickness of ridged ice) off the Canadian Archipelago to be twice as large in the winter of 1972 as in 1971. Other regions, however, exhibited less change. Also, Tucker, Weeks, and Frank (1979) found the nearshore ridging intensity off Prudhoe Bay to be substantially less in the winter of 1978 than in 1977.

Spatial variations in thickness can also affect the rate of accumulation and ablation of ice. Using data taken while crossing the ice by dogsled over a period of a year and a half, Koerner (1973) estimated that in the central Arctic Basin there is a total ice growth of about 1.1 m per year, which is dissipated by about 0.6 m of ablation and an equivalent of 0.5 m per year of outflow. The greatest ablation, however, occurred in new (first-year) ridges, and to a lesser degree in melt ponds. These areas had melt rates 1.7–2.6 times larger than level multiyear ice. Old multiyear hummocks or ridges, on the other hand, ablated more slowly (by a factor of 0.6) than level ice.

Koerner also found that about 20 percent of the ice production ends up in ridges, that annual snow accumulation was typically 0.1 m or less and decreased toward the center of the Arctic Basin, and that multiyear ice covered 70–80 percent of the area surveyed. To examine ice thickness characteristics, Koerner made a variety of ice thickness measurement in both first-year and multiyear ice. For first-year ice he found a maximum thickness of 2.04 m at the end of the growth season. For multiyear ice he found a wide range of thicknesses that, while they peaked at 2.5 m, demonstrated no well-defined thickness mode. Also, he was unable to discern any significant old ice thickness differences between the Pacific Gyre and the transpolar drift.

One of the more important features of the mass balance is outflow of ice from the basin (felt to be dominated by outflow through the Greenland–Spitsbergen passage). Estimates of this parameter by other authors (e.g., Vowinckel and Orvig, 1970) are generally lower than Koerner's, but are usually in the neighborhood of 0.1 Sverdrup (1 Sv $= 10^6\ m^3\ s^{-1}$). While this is not large in terms of oceanic transport, it does represent a substantial heat flux because of the latent heat of ice. Estimates of the heat budget of the Arctic Ocean (Aagaard and Greisman, 1975) indicate that this flux accounts for about one-third of the total oceanic advective heat budget. This ice outflow is also undoubtedly an important factor in the ice edge conditions in the Greenland Sea.

## PHYSICS OF SEA-ICE DYNAMICS

The essential physics of sea-ice dynamics may be characterized by the following elements:

1. A momentum balance describing ice drift in response to air and water stresses, Coriolis force, internal ice stress, inertial forces and ocean currents
2. An ice rheology that relates the ice stress to the ice deformation and strength
3. An effective mass balance describing the dependence of ice thickness characteristics on growth, decay, and ice drift
4. An ice strength dependent on the ice thickness and floe size.

In discussing these four elements below, sea ice will be assumed to be a two-dimensional continuum. While such an approximation is reasonable, the fact that pack ice, in some locations, consists of large individual floes causes substantial fluctuations in large-scale motion (Hibler et al., 1974*a*). As a consequence, in some cases continuous motion has to be considered a statistically defined quantity for sea ice. It should be noted, however, that such a continuum description does not rule out representing rapidly changing ice conditions. Nonlinear ice rheologies alone, for example, can introduce discontinuities into the ice velocity field, even though the forcing fields and sea-ice cover may be initially rather smooth.

### Momentum Balance

#### *Governing Equations*

The momentum balance describes the forces that determine drift and deformation of sea ice. Considering a sufficiently large collection of floes so that the average drift may be well defined, to a good approximation the momentum balance is given by

$$m\,D_t\,\mathbf{u} = -\,mf\,\mathbf{k}\times\mathbf{u} + \tau_a + \tau_w + \mathbf{F} - mg\nabla H \quad (2\text{-}1)$$

where $D_t = (\partial/\partial_t + \mathbf{u}\cdot\nabla)$ is the substantial time derivative, $m$ is the ice mass per unit area, k is a unit vector normal to the surface, $\mathbf{u}$ is the ice velocity, $f$ is the Coriolis parameter, $\tau_a$ and $\tau_w$ are the forces due to air and water stresses, $H$ is the elevation of the sea surface with respect to the geoid, and $\mathbf{F}$ is the force due to variations in internal ice stress. In this formulation, $\tau_w$ denotes the frictional drag due to the relative movement between the ice and the underlying ocean. In applications where one is concerned with ice drift over times large compared with inertial periods, the air and water stresses can be estimated from integral boundary layers with constant turning angles (McPhee, 1979; Brown, 1980; Leavitt, 1980):

$$\tau_a = c_a'\,(\mathbf{U}_g\cos\phi + \mathbf{k}\times\mathbf{U}_g\sin\phi) \quad (2\text{-}2)$$

$$\tau_w = c_w'[(\mathbf{U} - \mathbf{u})\cos\theta + \mathbf{k}\times(\mathbf{U}_w - \mathbf{u})\sin\theta] \quad (2\text{-}3)$$

where $\mathbf{U}_g$ is the geostrophic wind (assumed to be much larger than $\mathbf{u}$), $\mathbf{U}_w$ is the geostrophic ocean current, $c_a'$ and $c_w'$ are air and water drag constants, and $\phi$ and $\theta$ are air and water turning angles. In the deep ocean both the currents and the sea surface tilt can be well estimated from geostrophic considerations by setting $H$ equal to the dynamic height and computing currents by $\mathbf{U}_w = gf^{-1}\mathbf{k}\times\nabla H$, where $g$ is the acceleration due to gravity. In general, both $c_a'$ and $c_w'$ are nonlinear functions of the winds and currents. The most commonly used formulation is a quadratic parameterization where

$$c_a' = \rho_a\,c_a\,|\mathbf{U}_g| \quad (2\text{-}4)$$

$$c_w' = \rho_w\,c_w\,|\mathbf{U}_w - \mathbf{u}| \quad (2\text{-}5)$$

with $c_a$ and $c_w$ being dimensionless drag coefficients (with typical values of 0.0012 and 0.0055, respectively; McPhee, 1980). There are also a variety of other nonlinear boundary layer formulations.

Observations and dimensional analysis show that the force balance is largely between the air and water stresses, the Coriolis force, and the ice interaction force. Measurements show water and air stresses to be about 0.1 N m$^{-2}$. For 3-m-thick ice moving at 0.1 m s$^{-1}$, the Coriolis force is about 0.05 N m$^{-2}$, and is commensurate with the wind and water stresses. Inertial forces, in contrast, are smaller. However, the acceleration term can be significant in inertial oscillations where the acceleration and Coriolis force balance. For most applications the momentum advection term ($\mathbf{u} \cdot \nabla \mathbf{u}$) is very small, since velocity changes of 0.1 m s$^{-1}$ per kilometer would be needed for a 0.1 N m$^{-2}$ force.

The relative roles of wind and ocean currents depend on the time scale of interest. Over the span of several days or less, steady current effects are normally quite small (several percent relative to the wind-driven components) (Hibler and Tucker, 1979; Thorndike and Colony, 1982). However, current and tilt effects are significant over several months since the wind effects tend to average out (Hibler and Tucker, 1979). Also, in relatively shallow regions barotropic currents can significantly affect ice drift over short time periods.

The remaining term in the momentum balance is the ice interaction. It is difficult to directly measure the ice stress, so numerical model calculations for both the Arctic (Hibler, 1979; Hibler and Walsh, 1982; Hibler and Bryan, 1984) and Antarctic (Hibler and Ackley, 1983) must be relied on to show the ice stress to be a significant term in the momentum balance, even far from shore. This conclusion is consistent with residual estimates in the Beaufort Sea by Hunkins (1975) where all the terms in the momentum balance were estimated from direct measurements. As discussed below, the highly nonlinear nature of ice interaction does not allow a simple correlation between ice stress and winds, ocean currents, sea surface slope, or acceleration.

### Free Drift

As will be discussed below the internal ice stress term **F** plays a significant role in ice drift, especially when one considers long-term drift and ice buildup. However, it is instructive to first examine a special case of the momentum balance in which there is no ice interaction. This case is usually referred to as free drift or wind-driven drift. To further simplify matters, only steady free drift will be considered so that the inertial terms can also be removed.

Following McPhee (1980) and Thorndike and Colony (1982) it is convenient to use complex notation to analyze this case. Writing all vectors in complex form, that is, $\mathbf{A} \Leftrightarrow Ae^{i\gamma}$, and noting that the force on the ice due to sea surface tilt is given by $-imf\mathbf{U}_w$, the steady-state momentum balance can be written in the form

$$\rho_a c_a U_g \mathbf{U}_g e^{i\phi} = imf\mathbf{U} + \rho_w c_w U \, \mathbf{U} \, e^{i\theta} \qquad (2\text{-}6)$$

where **U** is the ice velocity relative to the steady current, $\mathbf{U} = \mathbf{u} - \mathbf{U}_w$. Letting $\delta$ be the clockwise angle from geostrophic wind to ice drift direction, equation 2-6 can be rewritten

$$U_g^2 \, e^{i\delta} = U^2 \, e^{-i\phi} \, \frac{\rho_w c_{-w}}{\rho_a c_a} \left( e^{i\theta} + \frac{mf}{\rho_w c_w U} \, e^{i\pi/2} \right) \qquad (2\text{-}7)$$

Several facts are clear by inspection of equation 2-7. For example, if the ice mass is small enough to make the Coriolis term negligible, then the ice will drift at a rate equal to a fixed fraction of the geostrophic wind with a drift angle $\theta - \phi$ relative to the geostrophic wind. As the mass increases the ice will tend to drift further to the right of the geostrophic wind. Similar arguments also apply to the ice speed, with larger ice speeds causing the ice to drift nearly parallel to the geostrophic wind. Referring back to equation 2-6, a final point of note is that for linear boundary layers the ratio and angle between the ice drift and the wind will be fixed for a given ice mass, with increasing ice mass causing the ice to drift further to the right of the wind.

A more quantitative illustration of the nonlinear boundary layer case is given in Figure 2-9, which shows solutions of equation 7 in the form of ice speed and turning angle versus geostrophic wind speed for different ice masses. In this figure, drag coefficients and Coriolis values similar to those used by Hibler (1979) were employed ($\rho_w = 10^3$ kg m$^{-3}$, $\rho_a = 1.3$ kg m$^{-3}$, $c_w = 0.0055$, $c_a = 0.0012$, $f = 1.46 \times 10^{-4}$ s$^{-1}$, $\theta = \phi = 25°$). One interesting result of this plot is the very linear relation between the wind speed and ice drift for all but very small speeds, despite the fact that boundary layer drags are nonlinear. This situation arises because both boundary layers are quadratic, and hence largely balance one another except at low speeds.

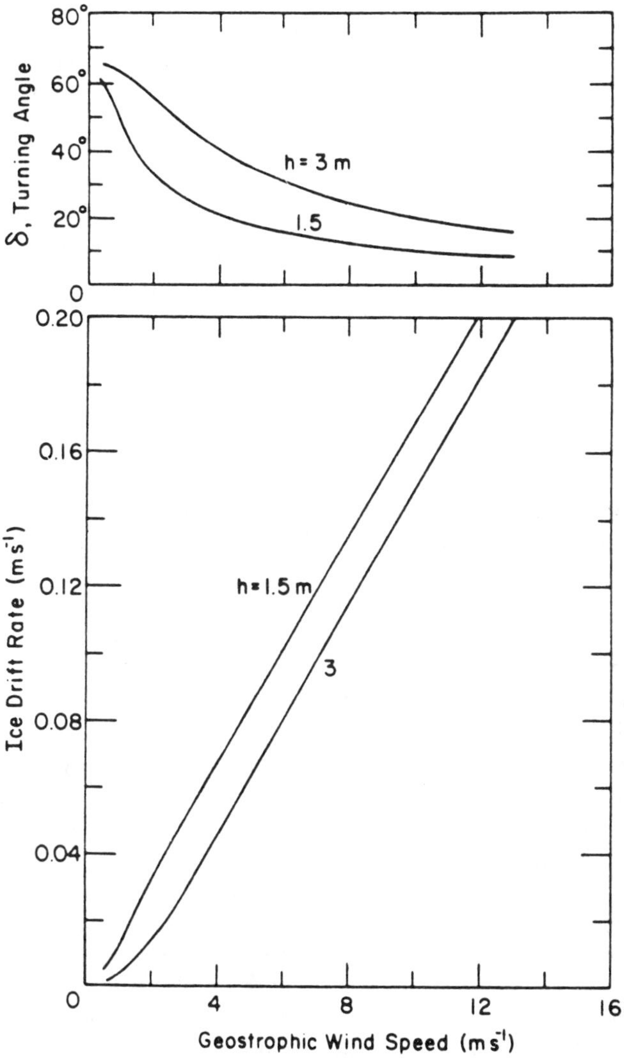


Figure 2-9. The ice drift rate versus the geostrophic wind speed, and the clockwise turning angle from the geostrophic wind to the ice velocity. These curves were calculated assuming free drift and quadratic boundary layers. Curves for different ice thicknesses are shown.

However, the turning angle does show a somewhat greater sensitivity to wind speed. Also note that although $U$ and $U_g$ are linearly related, the best linear fit does not go through the origin.

## Ice–Ice Interaction

An important facet of an ice model is the constitutive law. Considering sea ice to be a two-dimensional isotropic continuum, the most general constitutive law applicable to nonelastic deformation (see, for example, Malvern, 1969) is

$$\sigma_{ij} = 2\eta \dot{\epsilon}_{ij} + [(\zeta - \eta)(\dot{\epsilon}_{kk}) - P]\,\delta_{ij} \qquad (2\text{-}8)$$

where $\sigma_{ij}$ is the two-dimensional stress tensor and $\dot{\epsilon}_{ij}$ is the two-dimensional strain rate tensor, and $\zeta$, $\eta$, and $P$ are functions of the two invariants of the strain rate tensor $\dot{\epsilon}_{ij}$. In constructing a law beginning with this general framework (sometimes called the Reiner Rivlin equation), one would like to take into account the following observational characteristics and/or intuitively reasonable assumptions: discontinuous slippage near shore; rigid ice motion and/or lack of ice motion under considerable wind forcing; small or zero tensile strength, both for uniaxial divergence as well as for areal two-dimensional dilation; and high compressive strengths. These last two points basically say that sea ice should offer very little resistance to being pulled apart, while at the same time offering considerable resistance to being pushed together. In addition to these characteristics, viewed from above (Fig. 2-4), sea ice clearly takes on the appearance of a closely packed, granular medium. Based on granular medium studies in other materials, this suggests some type of constitutive law where the stresses are largely affected by contact stresses between floes rather than stresses transmitted by eddy diffusion.

As a practical modeling problem, it also became clear from theoretical and numerical modeling studies in the early 1970s that some of the basic phenomena of ice dynamics and ice deformation, discussed above, could not be explained by linear rheologies. For example, using a linear viscous model employing both bulk and shear viscosities, it was found that best-fit viscosities of the order of $10^{11}$ to $10^{12}$ kg s$^{-1}$ were needed to model deformation (Hibler, 1974) and drift (Hibler and Tucker, 1979) far from shore. Near shore, however, investigations of satellite imagery (see Fig. 2-6) near coastal boundaries indicated viscosities as small as $10^8$ kg s$^{-1}$ (Hibler et al., 1974*b*), while model simulations of tidal-induced drift indicated that in very small channels near-shore viscosities may be as small as $10^5$ kg s$^{-1}$ (Sodhi and Hibler, 1980). Such variations indicate that some type of nonlinear behavior was necessary.

One way to formulate a nonlinear rheology consistent with the above characteristics is to make use of plasticity theory. This was first noted by Coon (1974), who based much of this proposition on the similarities between sea ice and a granular medium. In the central Arctic it is also possible to base plasticity on the physics of pressure ridging in that mechanistic pressure-ridge models (Parmerter and Coon, 1972) show the work needed to build ridges is relatively independent of how fast the ridge is formed. By dimensional considerations this leads to a relatively rate-independent stress. It is easier to visualize the essential differences between different rheologies in one dimension (Fig. 2-10). As can be seen from this figure, the linear viscous rheology differs essentially from a plastic rheology in that the stress is rate-independent. This figure also sketches the viscous-plastic rheology approach introduced by Hibler (1979) for the purpose of modeling nonlinear flow. Here rigid flow is approximated by a state of very slow creep. Note also that a collision-induced rheology (Shen, Hibler, and Lepparanta, 1986), while nonlinear, has the stress more rapidly increasing with convergence.

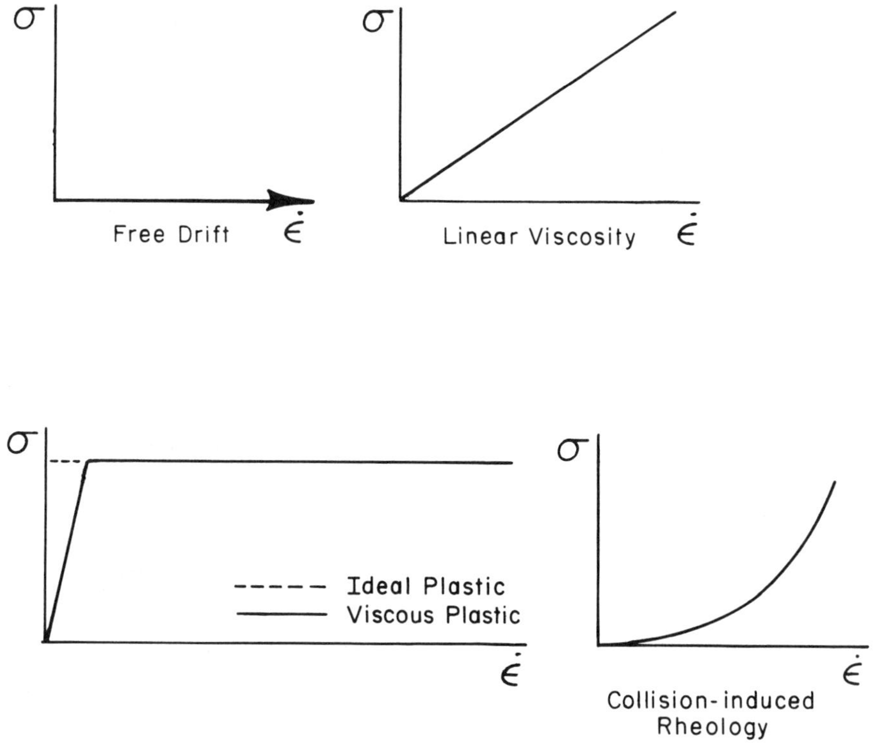


Figure 2-10. Comparison of sea ice rheologies in one dimension (a positive $\dot{\epsilon}$ here represents convergence).

In two dimensions the basic idea in a plastic rheology is to introduce some type of yield curve in stress space, where the stresses lie on the yield curve when flow is occurring (Fig. 2-11). The ratio, but not the magnitude, of the principal components of the strain rate tensor dictates where on the ice yield curve the stress lies. Based on analogies to elastic plastic systems, the associated flow rule has often been used, which specifies that the ratio of principal components of the strain rate tensor is the same as a normal to the yield curve. Since elastic behavior is doubtful for sea ice, such a flow rule is not absolutely necessary. However, as shown below, application of this rule does yield an intuitively acceptable rheology.

Probably the best available smooth yield curve for sea ice is an ellipse as proposed by Hibler (1977) (see Fig. 2-11). The area within this curve where rigid motion oc-

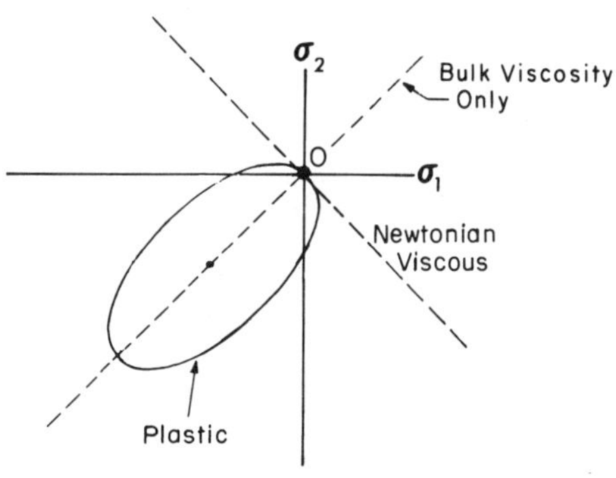


Figure 2-11. Allowable stress states for a Newtonian viscous rheology with only a shear viscosity, a linear viscous rheology with only a bulk viscosity, and an ideal plastic rheology with an elliptical yield curve. The stress states are plotted as a function of the principal components of the two-dimensional stress tensor. In the linear viscous cases, the stress states can lie anywhere on the dotted lines, whereas in the plastic case the stress state is restricted to be inside or on the yield curve.

curs allows only small tensile stress and limited compressive stress. When the stress reaches the yield curve, sea-ice deformation begins to occur. Applying the normal flow rule to this curve yields (Hibler, 1977) values for $\eta$, $\zeta$, and $P$ in equation 2-8 of

$$P = P^*/2 \qquad (2\text{-}9)$$

$$\zeta = P^*/2\Delta \qquad (2\text{-}10)$$

$$\eta = \zeta/e^2 \qquad (2\text{-}11)$$

where

$$\Delta = [(\dot{\epsilon}_{xx}^2 + \dot{\epsilon}_{yy}^2)(1 + \frac{1}{e^2}) + \frac{4}{e^2}\dot{\epsilon}_{xy}^2 + 2\dot{\epsilon}_{xx}\dot{\epsilon}_{yy}(1 - \frac{1}{e^2})]^{1/2} \qquad (2\text{-}12)$$

and where $-P^*$ is the maximum compressive stress allowed and $e$ is the ratio of the semimajor and semiminor axes of the ellipse. Note that a given stress can be produced by various combinations of strain rates, so that strain rate is not a unique function of stress. In simple physical terms, the yield curve of Figure 2-11 encloses the stress regime that a large-scale sea-ice field can maintain without tensile rupture, compressive failure, or other forms of deformation.

One of the strongest justifications for using a plastic law is the behavior of ice slippage near the shore. As discussed earlier (see Fig. 2-6), the basic nature of the deformation is effectively an almost discontinuous slippage near the coast with a more rigid motion further out. This type of behavior is practically impossible to simulate with a linear viscous law. In particular, simulating this behavior requires that the viscosities in equation 2-8 vary as a function of deformation rate. Analysis of seasonal dynamic–thermodynamic simulations (Hibler, 1979) shows this type of discontinuous slip behavior is nicely modeled by a plastic model.

Another strength of a plastic rheology is the capability for simulating relatively stationary behavior under considerable wind forcing. Such behavior was modeled in the southern Beaufort Sea with good agreement to observations by Pritchard, Coon, and McPhee (1977). This type of stationary behavior is also typical of the ice in relatively enclosed regions such as the Bay of Bothnia. Lepparanta (1981), for example, has estimated that ice strengths in the neighborhood of $10^{11}$ N m$^{-1}$ are required to maintain this rigidity.

## Thickness–Strength Coupling

In the cases analyzed above, the behavior of ice dynamics under constant strength conditions has been examined. However, in reality the ice strength is related to the ice thickness characteristics in some way. The thickness characteristics, in turn, affect the air–sea heat exchanges (see, e.g., Maykut, 1978). By coupling an ice thickness distribution and ice strength to a plastic rheology it is possible to carry out an equilibrium seasonal simulation in which ice thickness changes reach an equilibrium seasonal change in three to five years. This can be done using a two-level ice thickness distribution, a procedure used in the Hibler (1979) study. However, it is also possible using a multilevel formulation. This procedure was used by Hibler (1980*a*), in conjunction with a full heat-budget-based thermodynamic model, for a seasonal simulation of the Arctic ice cover. Both methods are briefly described below.

### *Ice Thickness Distribution Formulations*

A simple two-level formulation has been proposed by Hibler (1979) that captures many of the essential features of thickness–strength coupling and modification by dynamics of the air–sea heat exchange. The basic concept is that because of its low strength and high growth rates, many of the dynamic and thermodynamic characteristics of sea ice are dominated by the thin ice. Consequently, the ice thickness distribution may be approximately characterized by breaking it into two parts (thick and thin). This is done by dividing the ice cover into an area $A$ (called the compactness), which is covered by thick ice, and a remaining area (1-$A$), which is covered by thin ice.

For the mean ice thickness per unit area, $h$, and compactness, $A$, the following continuity equations apply:

$$\partial h/\partial t = -\partial(uh)/\partial x - \partial(vh)/\partial y + S_h \tag{2-13}$$

$$\partial A/\partial t = -\partial(uA)/\partial x - \partial(vA)/\partial y + S_A \tag{2-14}$$

where $A < 1$, $u$ is the $x$ component of the ice velocity vector, $v$ is the $y$ component of the ice velocity vector, and $S_h$ and $S_A$ are thermodynamic terms. While equation 2-13 is a simple continuity equation for ice mass (characterized by the mean thickness $h$), with thermodynamic source and sink terms, equation 2-14 is somewhat more complex. By including the restriction that $A \leq 1$, a mechanical sink term for the areal fraction of ice has been added to a simple continuity equation for the ice concentration. Note that this restriction does not affect the conservation of ice mass, which is explicitly guaranteed by equation 2-13.

In these equations, the thermodynamic terms $S_A$ and $S_h$ are given by

$$S_h = f(h/A)A + (1 - A)f(0), \tag{2-15}$$

$$S_A = \begin{cases} (f(0)/h_o)(1 - A) & \text{if } f(0) > 0 \\ 0 & \text{if } f(0) < 0 \end{cases} + \begin{cases} 0 & \text{if } S_h < 0 \\ (A/2h)S_h & \text{if } S_h < 0 \end{cases} \tag{2-16}$$

where $f(h)$ is the growth rate of ice of thickness $h$, and $h_o$ is a fixed demarcation thickness between thin and thick ice.

The $S_h$ term specifies that the net ice growth or melt is given by the sum of the ice grown on open water plus the additional growth over the portion of the cell covered by thick ice, estimated to be ice of constant thickness $h/A$. For melting conditions the same sum is used, which is equivalent to assuming that the heat absorbed by open water will horizontally mix and melt additional ice until the mixed layer returns to freezing. The $S_A$ term characterizes the way in which growth and decay change the relative areal extents of thin and thick ice. The idea is that the areal fraction of thin ice will decrease rapidly under freezing conditions, and increase slowly under melting conditions. For freezing conditions, the fraction of open water $(1 - A)$ is allowed to decay exponentially with a time constant of $h_o/f(0)$, which gives the first term in equation 2-16. In Arctic simulations, $h_o = 0.5$ m is used. The second term in the $S_A$ equation accounts for melting. Its magnitude is derived by assuming that the thick ice is uniformly distributed between 0 and 2 $h/A$ in thickness, and all melts at the same rate.

To couple the ice strength to the ice thickness characteristics the ice strength $P^*$ is taken to be a function of compactness and thickness according to

$$P^* = P_o h \exp[-C(1 - A)], \tag{2-17}$$

where $P_o$ and $C$ are fixed empirical constants and $h$ is in meters. This formulation makes the strength strongly dependent on the amount of thin ice (characterized by the compactness $A$), while also allowing the ice to more slowly strengthen as it becomes thicker. The constant $C$ is taken to be 20, which causes the strength to be very small when the fraction of open water is 15 percent or larger.

In a multilevel thickness distribution, one uses the governing equation for ice thickness distribution (following Thorndike et al., 1975):

$$\frac{\partial g}{\partial t} = -\nabla \cdot (ug) - \frac{\partial (fg)}{\partial h} + \psi \quad (2\text{-}18)$$

where $gdh$ is the fraction of area covered by ice with thickness between $h$ and $h + dh$, $f$ is the vertical growth rate of ice of thickness $h$, and $\psi$ is a redistribution function (depending on $h$ and $g$) that describes the creation of open water and transfer of ice from one thickness to another by rafting and ridging (see Hibler, 1984*a*, for a discussion of different formulations of $\psi$). In this multilevel formulation, ice strength can be deduced by specifying that work done in building ridges is equal to deformation work (first noted in two dimensions by Rothrock, 1975) via the equation

$$C \int_{0}^{\infty} h^2 \psi(h) dh = \Sigma_{ij}\, \sigma_{ij}\, \dot{\epsilon}_{ij} \quad (2\text{-}19)$$

where C is related to the gravitational and buoyancy forces on ice.

### *Role of Rheology in Simulating Ice Buildup*

Thickness–strength coupling causes large-scale spatial variations of thickness to often be determined more by mechanical considerations than by thermodynamics. This feature is illustrated by Figures 2-12 and 2-13, which are taken from a multiple-thickness seasonal equilibrium simulation (Hibler, 1980*a*) of the Arctic Basin ice cover. Very similar results are obtained using a two-level thickness distribution (Hibler, 1979). As can be seen from Figure 2-12, the dynamics introduces a marked change in spatial contours of ice thickness as compared with thermodynamic simulations. Basically, the ice velocity field (whose mean annual values are shown in Figure 2-13) tends to thicken ice by convergence and concomitant ridging off the Canadian Archipelago while thinning the ice off the Alaskan and Siberian coasts. (It should be noted that daily velocities will differ substantially from the mean annual values.) The shape and magnitudes of the simulated thickness contours (Fig. 2-12) agree well with observed estimates (Fig. 2-8). The results considering only thermodynamic effects are very similar to those obtained by Washington et al. (1976) and by Parkinson and Washington (1979), and do not agree well with observations. (An additional obvious shortcoming of thermodynamic-only models is that no outflow from the basin, as we know occurs, is simulated.)

The essential physics of the buildup is that as the ice thickens due to convergence it becomes stronger and reduces the ice velocity field convergence. If no ice interaction were assumed, the ice would build up essentially without bound. The main issue for climatic studies here is that the maximum thicknesses are determined more by mechanical than by thermodynamic considerations. An important footnote is that this reduction in ice velocity convergence is not necessarily concomitant with a reduction in ice drift rate.

With regard to ridge buildup, while the model does not explicitly keep track of ridging, it is possible with the multilevel model to determine the volume of deformed ice over an annual cycle at a fixed location. Figure 2-14 shows fifth-year contours of this quantity. The general shape of the roughness contours agrees well with the surface roughness observations in the western Arctic Basin (Weeks, Kovacs, and Hibler, 1971; Hibler, Mock, and Tucker, 1974). The two major features are a heavy buildup of ridging off the Canadian Archipelago and less ridging in the Beaufort Sea than near the pole. Note, however, that there is a zone of heavy ridging just off the North Slope, in agreement with observations by Tucker, Weeks, and Frank (1979). (This feature is masked in Figure 2-14 by the large contouring interval. Finer contouring intervals show a peak in the volume of deformed ice per unit area between the two 0.2 m contours in Figure 2-14 off the North Slope.) The other major feature

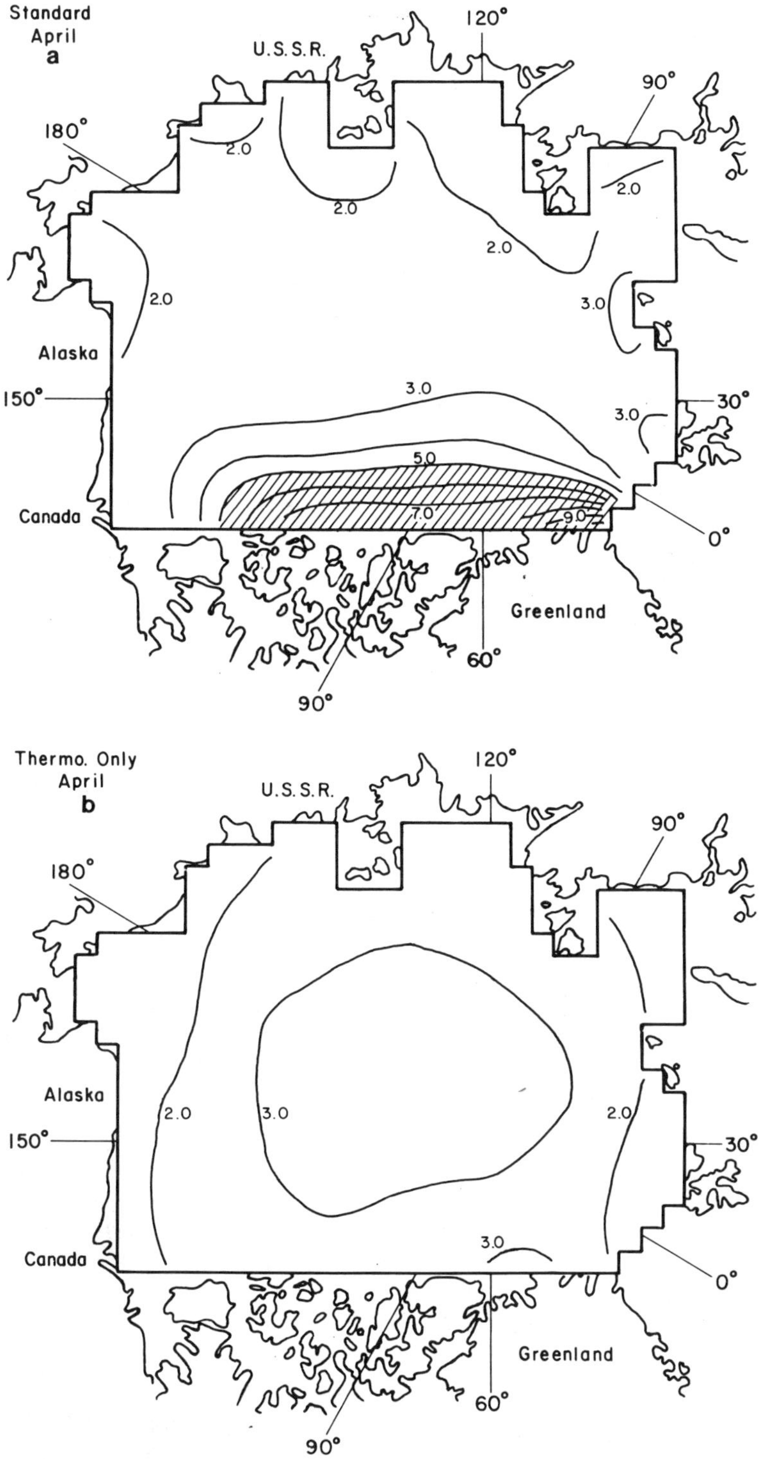


**Figure 2-12.** Average April ice thickness contours (m) for (*a*) a full multilevel dynamic–thermodynamic sea-ice simulation, and (*b*) a "thermodynamics only" simulation. (From Hibler, 1980*a*.)

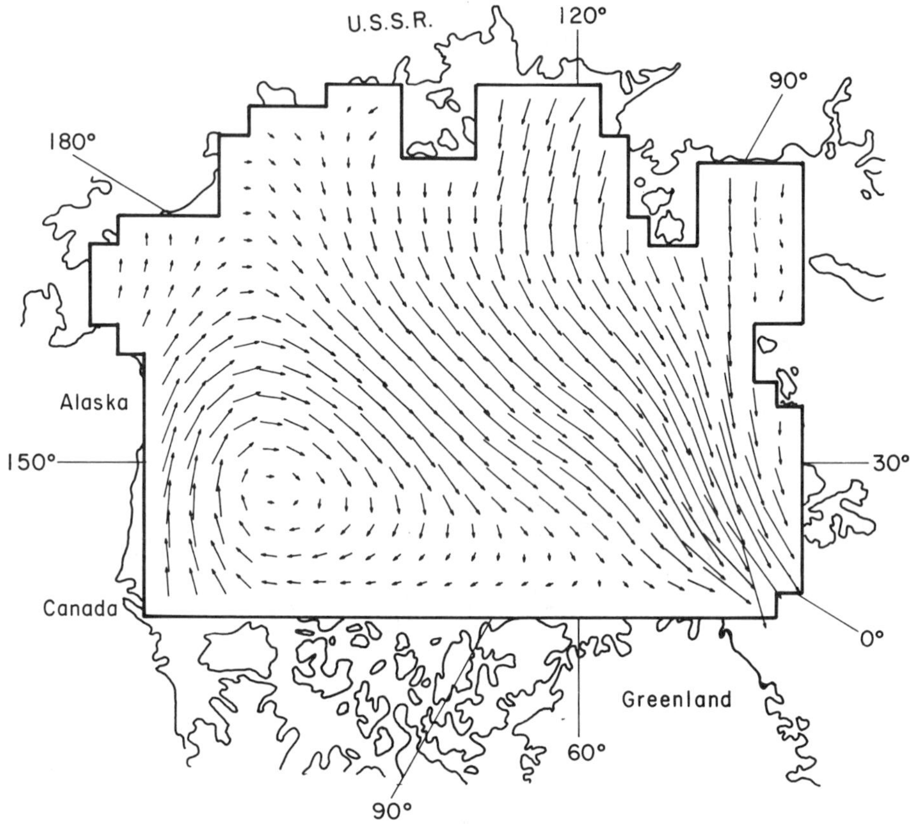


Figure 2-13. Average annual ice velocity field for the fifth year of a multilevel dynamic–thermodynamic sea-ice simulation. (From Hibler, 1980*a*.)

is a tongue of high ridging further offshore near the tip of Greenland. Roughness data obtained from submarine sonar profiles by LeShack (private communication, 1980) also show such a tongue.

In addition to ice buildup, on smaller scales thickness–strength coupling can lead to a number of fluctuating effects, such as kinematic waves (Hibler, Udin, and Ullerstig, 1983). Some of these effects depend on the nonlinear ice interaction and the nonlinear coupling between ice strength and thickness characteristics.

### *Necessity for Ice Interaction in Seasonal Simulations*

An example of the crucialness of thickness–strength coupling is graphically illustrated if one carries out a seasonal simulation with free drift. As discussed above, observationally we know that on short time scales ice drift tends to follow the wind. While it is not obvious, analysis of numerical and analytic ice models shows that this feature is a characteristic of a plastic rheology. However, thickness buildup results are totally different with a plastic rheology as compared to free drift (where the ice is merely taken to follow the wind), with free drift yielding totally unacceptable ice buildup results for climatic studies.

To demonstrate the necessity for using an ice interaction in seasonal simulations, a comparison series of coupled dynamic–thermodynamic simulations was carried out for the Arctic Basin and Greenland–

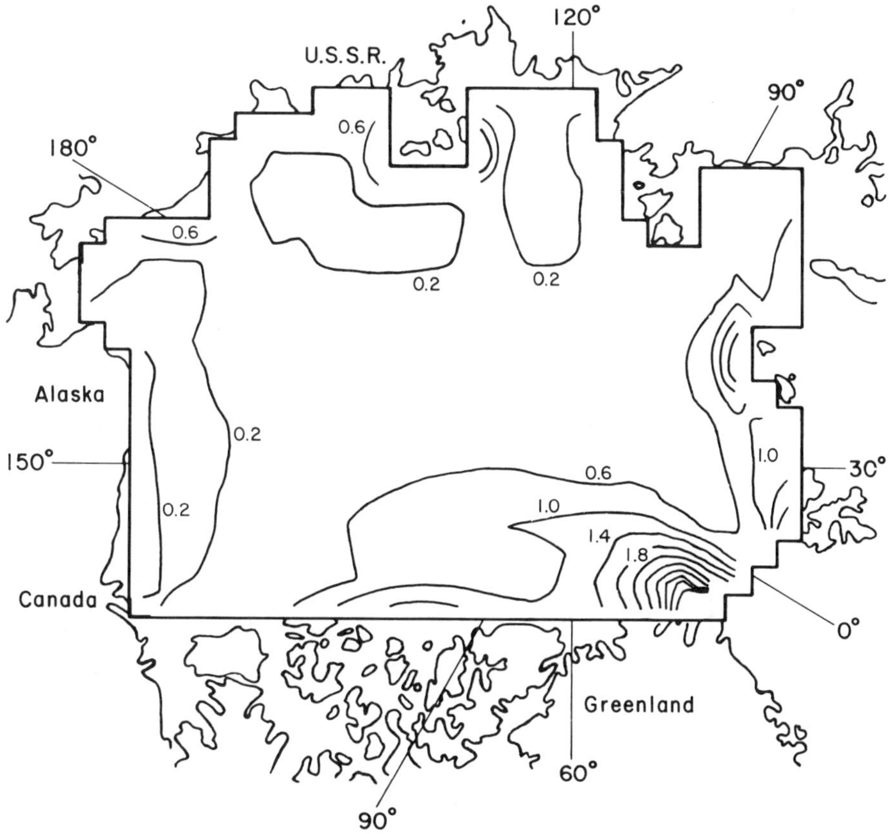


Figure 2-14. Contours (in meters) of simulated volume of deformed ice per unit area for the fifth year of a multilevel dynamic–thermodynamic sea-ice simulation. A description of the model is given by Hibler (1980*a*).

Norwegian seas employing a two-level dynamic–thermodynamic model (Hibler, 1985). The ice strength constant $P^*$ and grid are the same as used by Hibler and Bryan in a coupled ice–ocean simulation to be discussed later. As can be seen from the velocity fields in Figure 2-15, the average annual ice velocity looks rather similar in both the simulations. However, the ice buildup in the free-drift case is totally unacceptable. In particular, even after only one year the free-drift simulation results in unrealistic buildups of over 19 m off the Canadian Archipelago, together with excessive ice convergence in the center of the Beaufort Gyre. The strong resistance to convergence in the plastic model, on the other hand, yields a relatively realistic buildup. Basically, the deformation field of the ice velocity is a much more sensitive test of a model's performance. The above results imply that to optimally aid in model development, satellite remote sensing either should be capable of resolving ice deformation or provide some type of measure of thickness or ridging. These results also illustrate the danger of just using the statistics of predicted ice drift as a test of model performance. If a model yields good drift statistics but unrealistic ice buildup, for example, then its utility and physical basis are clearly limited.

Also shown in Figure 2-15 are the results from a plastic model without shear

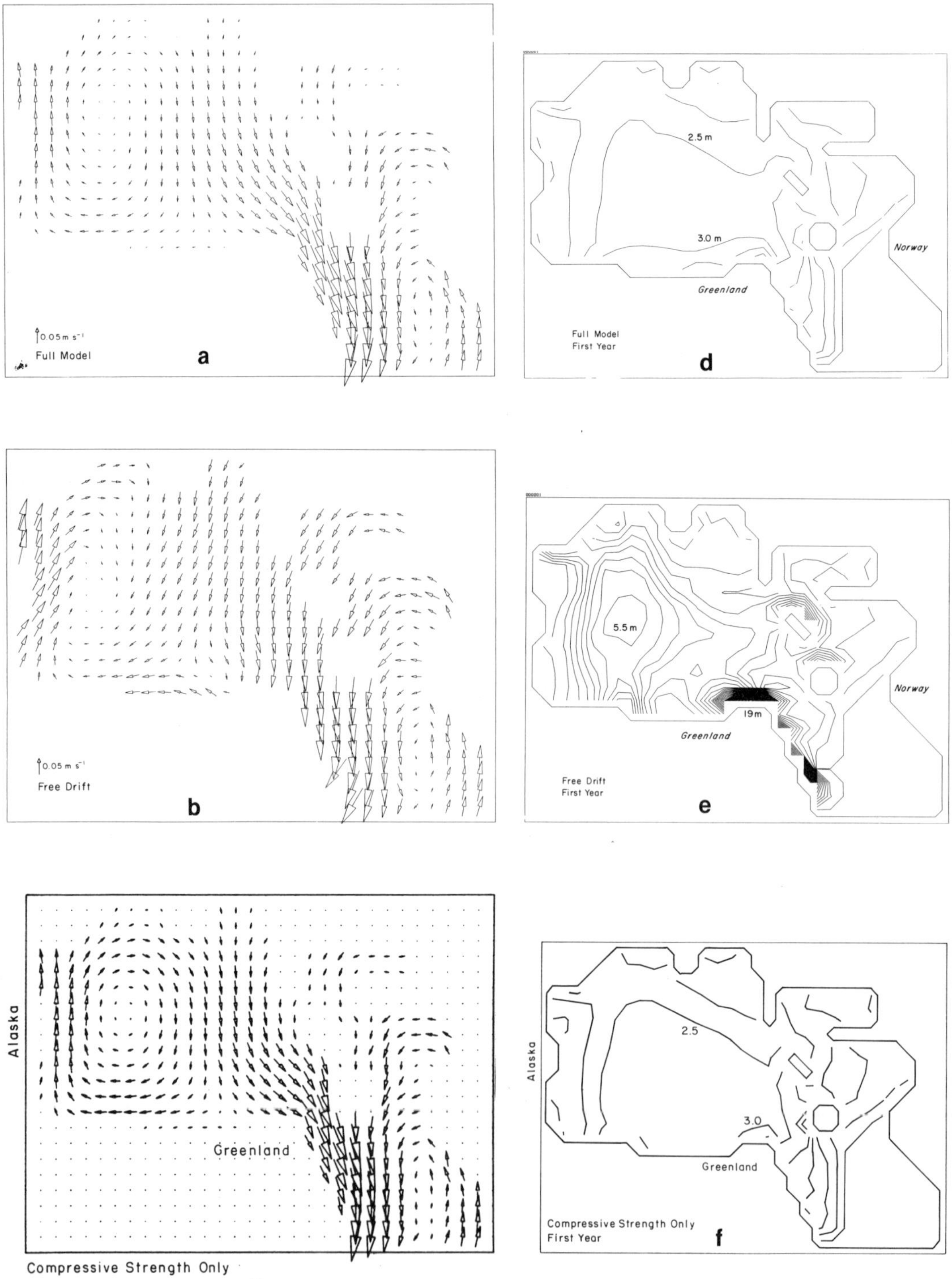


**Figure 2-15.** Simulated annual average ice velocities (*a, b, c*) and ice thicknesses at the end of December (*d, e, f*) for a plastic model versus a free-drift model, and for a plastic model without shear strength.

strength. This model was obtained by taking an elliptical yield curve with the same $P^*$ as in the full model but with an $e$ of 100 instead of 2, so that the ice has effectively no shear strength. This special case of no compressive strength would also be possible to model by utilizing only a nonlinear bulk viscosity. As can be seen in the compressive-strength-only case, the ice velocities are somewhat larger, and although similar in character, the ice buildup off the Canadian Archipelago is less pronounced.

As can be inferred from Figure 2-15, the plastic model offers a consistent way to construct a weakly damped ice velocity field without excessive convergence. The computer time needed for this is not excessive. The year-long integration at one-day time steps for Figure 2-15, for example, required 6 minutes of Cyber 205 computer time, of which about 70 percent was used for solution of the momentum equation. Considering that the equilibrium time scale of the ice model is about 3–5 years, the total computer time needed for equilibrium simulation of both polar covers is substantially less than the equilibrium computer time of an atmospheric or oceanic GCM.

## ICE–OCEAN CIRCULATION

### Barotropic and Baroclinic Currents

Since the ocean currents play an important role in driving the ice drift, it is necessary to understand the basic characteristics of the Arctic Ocean circulation in order to understand ice drift. A particularly critical feature of the Arctic Ocean circulation is the rather complete baroclinic adjustment and the role played by the bottom torque term in the mean circulation. An essential difference between the Arctic and mid-latitude seas is the absence of any beta effect leading to a western intensification of the currents. Instead, in the Arctic the barotropic flow is largely dictated by the bottom depth contours. If the currents were only barotropic, this would lead to extremely small surface currents. However, the baroclinic adjustment is rather complete, so that a strong surface current can occur. This density adjustment also couples with the bottom topography to yield a bottom torque term on the integrated flow that is very significant.

It is important to understand the substantial magnitude of the bottom torque to understand the nature of the Arctic Ocean circulation. Properly simulating this term is also crucial in large-scale ice–ocean modeling.

The essential physics can be outlined as follows. Considering the ocean circulation in the absence of nonlinear momentum advection terms and viscous terms, the governing equations in Cartesian coordinates, together with the Boussinesq approximation, are given by

$$\rho_o \frac{\partial u}{\partial t} = -\frac{\partial P}{\partial x} + \tau_x \delta(0) + \rho_o f v \quad (2\text{-}20)$$

$$\rho_o \frac{\partial v}{\partial t} = -\frac{\partial P}{\partial y} + \tau_y \delta(0) - \rho_o f u \quad (2\text{-}21)$$

with

$$\frac{\partial w}{\partial z} + \nabla \cdot \mathbf{u} = 0 \quad (2\text{-}22)$$

and

$$P = P_o + g \int_{-z}^{0} \rho \, dz = P_o + p' \quad (2\text{-}23)$$

In the momentum equations the vertical and horizontal viscosity terms have been omitted as they are not essential to this argument. Considering a variable bottom depth of $h$, and a coordinate system with $z$ being positive in a direction upward from the ocean bottom, the continuity equation may be integrated, yielding

$$\int_{-h}^{z} \frac{\partial w}{\partial z'} dz' = w(z) - w(-h)$$

$$= -\int_{-h}^{z} \nabla \cdot \mathbf{u}\, dz$$

$$= -\nabla \cdot \int_{-h}^{z} \mathbf{u}\, dz + \nabla h \cdot \mathbf{u} \qquad (2\text{-}24)$$

If we specify that the ocean currents at the bottom run parallel to the bottom slope, then $w(-h) = -\nabla h \cdot u$ and

$$w(z) = -\nabla \cdot \int_{-h}^{z} \mathbf{u}\, dz \qquad (2\text{-}25)$$

Considering that the system has a rigid lid at $z = 0$ so that $w(0) = 0$, then

$$-\nabla \cdot \int_{-h}^{0} \mathbf{u}\, dz = 0 \qquad (2\text{-}26)$$

and it is therefore possible to define a vertically integrated stream function for the system

$$\frac{\partial \psi}{\partial x} = \int_{-h}^{0} \rho_o v\, dz \qquad (2\text{-}27)$$

$$\frac{\partial \psi}{\partial y} = -\int_{-h}^{0} \rho_o u\, dz \qquad (2\text{-}28)$$

Vertically dividing by the depth, integrating the momentum equation, taking the curl and substituting in the definition of stream function for $u$ and $v$, we obtain

$$\frac{1}{h}\frac{\partial}{\partial t}(\nabla^2 \psi) - \left[\frac{\partial}{\partial x}\frac{1}{h}\frac{\partial \psi}{\partial x}\right] + \left[\frac{\partial}{\partial y}\frac{1}{h}\frac{\partial \psi}{\partial x}\right]$$

$$= -\frac{\partial}{\partial y}\left[\frac{1}{h}\int_{-h}^{0}\frac{\partial p'}{\partial x}dz\right] + \frac{\partial}{\partial x}\left[\frac{1}{h}\int_{-h}^{0}\frac{\partial p'}{\partial y}\right]dz$$

$$+ \frac{\partial}{\partial y}\left(\frac{1}{h}\tau_x\right) - \frac{\partial}{\partial x}\left(\frac{1}{h}\tau_y\right) \qquad (2\text{-}29)$$

The terms on the right-hand side involving the pressure are the bottom torque terms, and are clearly nonzero only when there is bottom topography (i.e., $h \neq$ constant).

As a demonstration of how large this term is, Figure 2-16 shows a numerical solution of equation 2-29 for the Arctic Ocean (with a time-invariant surface wind and forcing taken from the 30-year averaged wind fields in Walsh, Hibler, and Ross (1985), both with and without the density term included. In the numerical solution, a rather large constant eddy viscosity of $A_m = 8 \times 10^9\ \text{cm}^2\ \text{s}^{-1}$ was also used. While this quantitatively affects the results, it does not affect the main characteristics of the bottom torque term or the essential argument. As can be seen, without the baroclinic forcing (taken from Levitus, 1982) the flow essentially follows the contour of bottom topography with a greatly reduced flow. In the full baroclinic adjustment case, the flow is much more vigorous, perhaps 30 times as strong as in the barotropic case. This figure is a graphic illustration of the importance of properly simulating the baroclinic adjustment in ocean models, and also gives a good feeling for why the wind-driven component of the current fields is not critical in Arctic ice drift calculations over the deep ocean regions.

With respect to the surface currents from the simulation, Figure 2-16*c* shows the surface pressures from the simulations, which to a good approximation are streamlines for the surface currents. Basically the surface current structure yields a flow pattern resembling observations by Coachman and Aagaard (1974) with a pronounced surface gyre and a transpolar drift pattern.

A question of some interest with regard

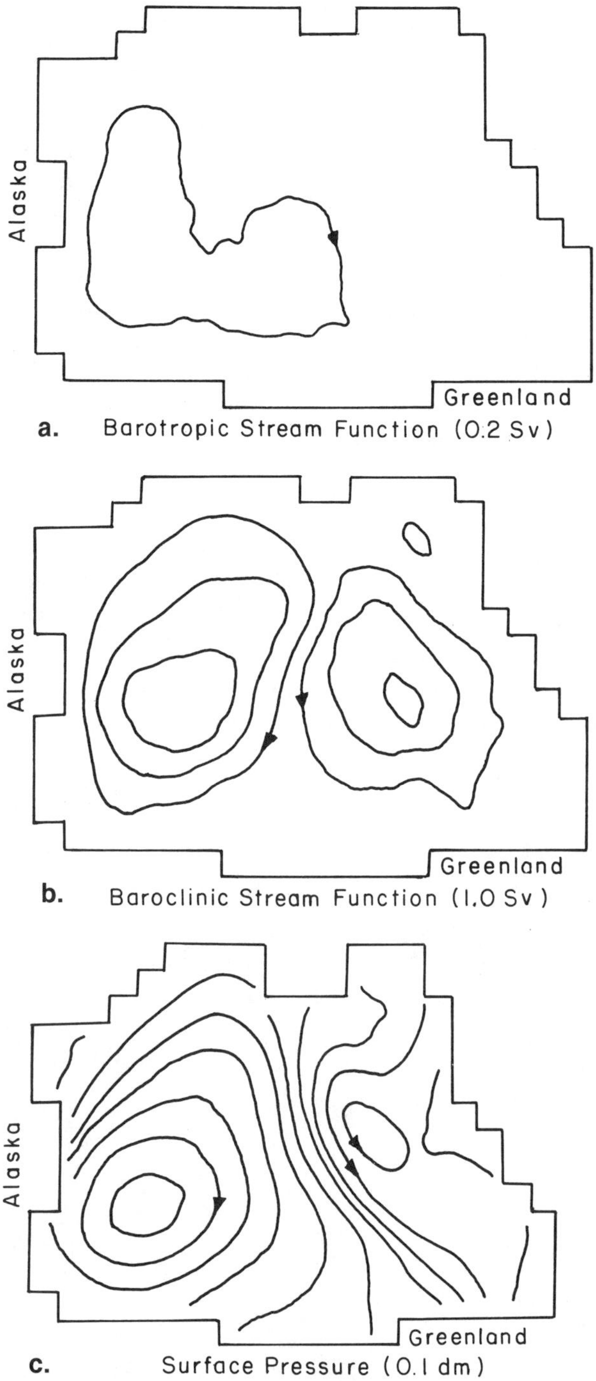


**Figure 2-16.** Contours of vertically integrated flow for the Arctic Ocean for (*a*) a barotropic model, and (*b*) a baroclinic model with fixed density forcing. The contour intervals were 0.2 Sv for the barotropic case and 1.0 Sv for the baroclinic case. Part (*c*) shows the surface pressure for the baroclinic simulation with a contour interval of 0.10 dynamic m. In these model simulations, the bottom topography was resolved using differing numbers of up to 14 vertical levels.

to such large-scale ice–ocean simulations is how rapidly the baroclinic currents change in response to wind forcing and how large the barotropic currents are. These characteristics were investigated by Hibler and Bryan (1987) in a diagnostic simulation of the Greenland–Norwegian Sea, some results from which are shown in Figures 2-17 and 2-18. (A somewhat more complete description of this model is given below.) Shown in Figure 2-17 are the daily RMS fluctuations in surface ocean current speed versus the mean ocean speed averaged for a year on a transect across the Arctic Basin. As can be seen, in the central basin the mean current is substantially larger than the fluctuating current. However, in the very shallow regions the fluctuating barotropic currents can be very large, as in the Barents Sea.

While changing much more slowly than the barotropic currents, the baroclinic currents also undergo an evolution due to the average wind stress change, as shown in Figure 2-18. The final stabilized current field in Figure 2-18 is largely due to diagnostic source terms in the Hibler–Bryan ice–ocean model that relax the model density field to observed mean climatology with a three-year relaxation time. However, the variation over the first few years gives an indication of the magnitude of changes that can occur from one year to the next if the wind forcing is variable. As a consequence, it is clear that significant amounts of interannual variability in the baroclinic currents can occur as the average wind stress forcing changes from year to year. Basically, in Figure 2-18 the seasonal change in the current magnitude is mainly a barotropic effect, while the year-to-year forcing is a baroclinic drift. This seasonal variation is substantially smaller in the central basin.

## Differences Between Ice Drift and Surface Currents

Results from the ice–ocean simulation mentioned above also give some insight into the essential differences between the drift and ocean surface currents. It is notable, for example, that even near the coast where the current fluctuations are larger, the ice motion variations are much larger than the current variations. This character-

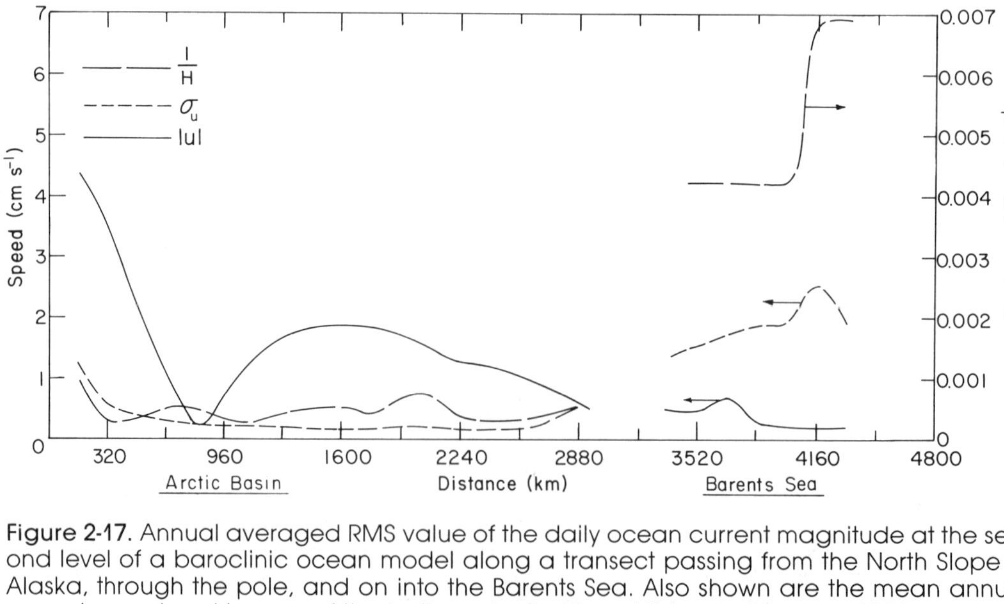


Figure 2-17. Annual averaged RMS value of the daily ocean current magnitude at the second level of a baroclinic ocean model along a transect passing from the North Slope of Alaska, through the pole, and on into the Barents Sea. Also shown are the mean annual current speed and inverse of the bottom depth. (From Hibler and Bryan, 1987.)

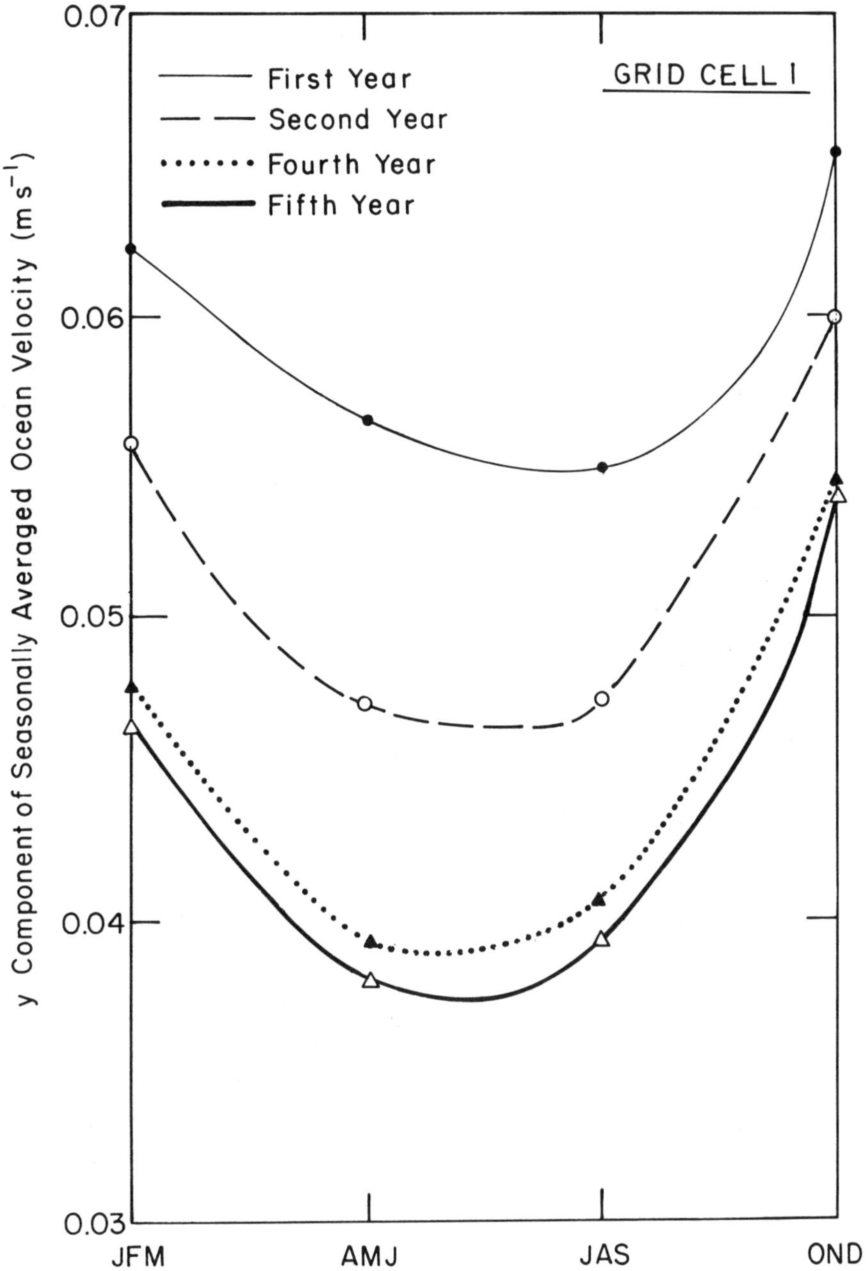


**Figure 2-18.** Seasonally averaged component of ocean surface current of a diagnostic ice–ocean model off the coast of Alaska during different times in the spin-up process. (From Hibler and Bryan, 1987.)

istic is illustrated in Figure 2-19, which compares the mean seasonal $y$ velocity component with the $y$ component of surface ocean velocity (with, in this case, the same scale used for both plots) off the North Slope of Alaska. While there is some variation in the ocean currents, it is smaller than the ice motion.

In order to give some idea of the long-term effects of ocean currents on ice drift,

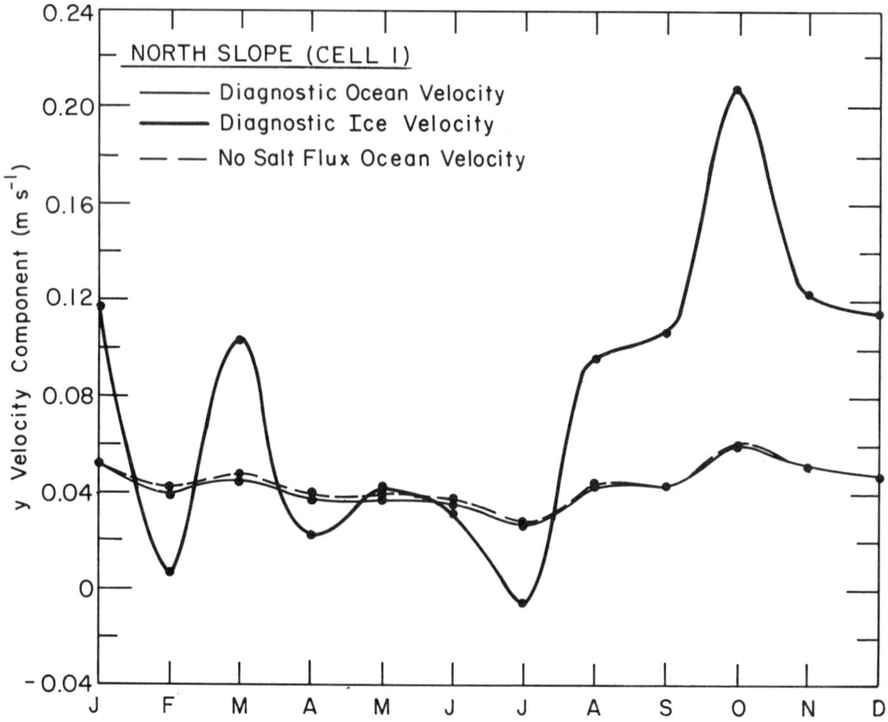


**Figure 2-19.** Time series of *y* velocity component of ice drift and ocean current velocities off the North Slope of Alaska for a coupled ice–ocean model. Also shown (dashed line) is the velocity from a one-year sensitivity study where the salt fluxes due to ice melt and growth were set equal to zero. (From Hibler and Bryan, 1987.)

Figure 2-20 compares observed ice buoy (Thorndike and Colony, 1982) drift to simulated results, both with and without the ocean model included. Although of considerable magnitude, the ocean currents are smaller than the ice drift, and in buoy-drift comparisons they account for less than 50 percent of the net ice drift. Note also that inclusion of ocean currents tends to substantially modify the direction of the ice drift.

An important concept in ice–ocean coupling is the fact that the ice interaction can modify the wind stress that is transferred into the ocean. This effect is particularly pronounced on long time scales since large fluctuations in wind stress tend to average out. As shown in Figure 2–20, the ice interaction at the buoy locations has a magnitude comparable to the mean wind stress. Also, while in most cases the ice stress opposes the wind stress, it can combine with the wind stress or be at right angles to the ice drift due to the configuration of the land boundary and the ice edge.

While these results tend to contradict the intuitive view that ice interaction does not play a significant role far from shore, they are quite reasonable if one considers the theory of a plastic rheology.

In a one-dimensional constant strength plastic system (see, e.g., Hibler, 1984*a*), the simulated ice drift will asymptotically approach the free-drift value for large wind speeds but will be significantly different for small wind speeds. Thus, short-term fluctuations in ice drift will tend to follow the wind, while longer-term variations can be more strongly affected by ice interaction. This is one possible reason why the wind and ice drift have such a high coherence at high frequencies, and, as a conse-

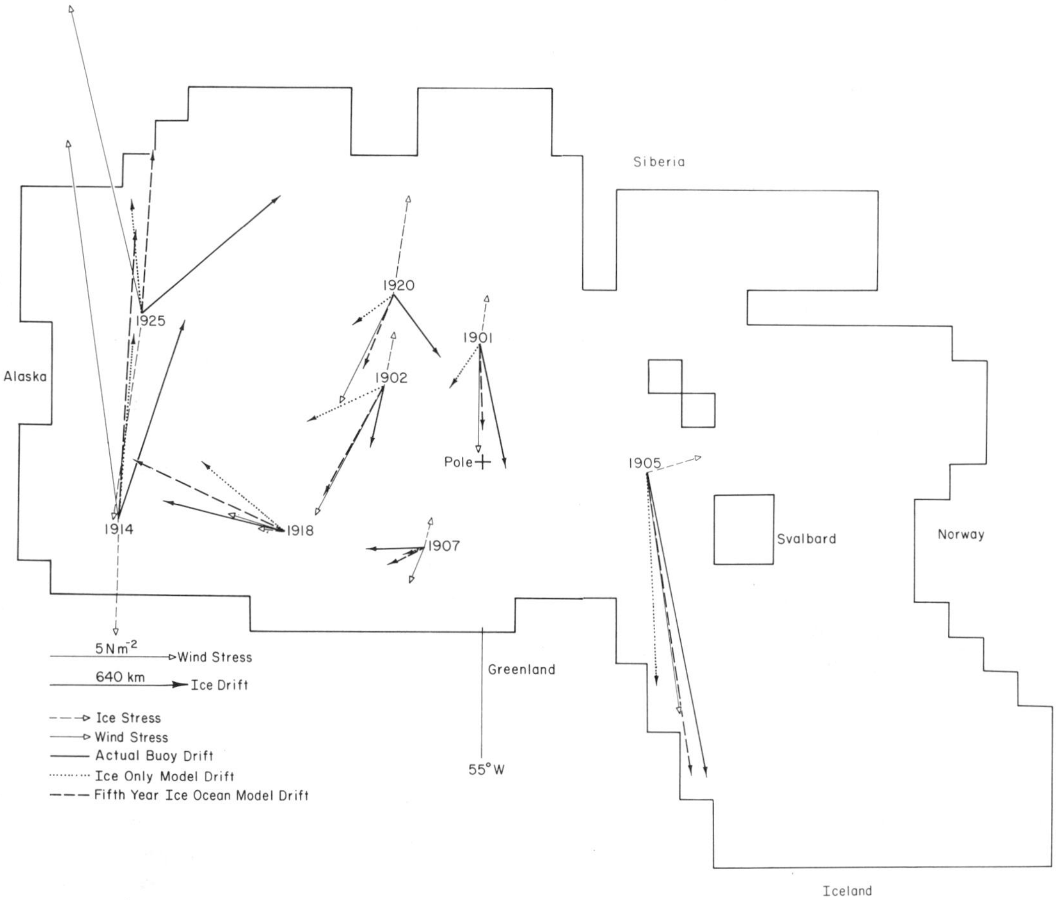


**Figure 2-20.** Long-term averages of simulated and observed drift rates of ice buoys. Drift rates have been converted to distances by multiplication by time. Also shown are the time-averaged wind stress and force due to ice interaction at the locations of the ice buoys. The time intervals used were February–November for buoys 1901 and 1902, March–September for buoy 1905, May–November for buoy 1907, and March–November for the remaining buoys.

quence such a high temporal correlation (Thorndike and Colony, 1982). In general, these remarks highlight the need for more detailed sensitivity studies of this type of ice–ocean model, with verification based on more extensive data sets.

## ICE EDGE AND COMPACTNESS CHARACTERISTICS

As discussed above, many aspects of sea-ice dynamics in the Arctic Basin may be described by considering the ocean currents to be relatively steady and considering the ocean to have a relatively small effect on the ice in terms of heat fluxes. However, when one considers ice edge variations, especially in the east Greenland region, thermal processes and ocean coupling become considerably more important.

### Thermal Processes

To understand the fluctuations of this ice edge it is necessary to consider the overall thermal processes dictating the advance and retreat of the ice cover. To a first approximation we can think of the ice cover as a continuum with spatially varying thickness embedded in an oceanic mixed layer. Analysis of the thermodynamics of individual ice floes shows that once ice reaches 1/2 m or so, the conduction of heat into the atmosphere is greatly reduced (Maykut and Untersteiner, 1971). Studies of these individual ice floes also show that the presence of salinity via brine pockets in the ice substantially changes the thermodynamic characteristics (see, e.g., Hibler, 1980*b*, Section III-B).

However, because of the variable thickness nature of this continuum, this heat conduction can still take place through the leads and thin ice, so that as a practical matter the opening or closing of the ice cover can greatly modify the energy exchanges. Under such growth conditions, most of the heat budget is dominated by the air temperature via sensible heat exchanges.

Under melting conditions, the variable thickness character of the ice cover is also important in that open water regions can absorb radiation, which is then used to melt ice floes either from the side via lateral ablation or by increasing the overall heat content of the mixed layer, which can then melt ice from below. Under these melting conditions, analysis of the heat budget components shows short-wave radiation to be the dominant term, so that the albedo difference between ice and water becomes very important. Because of the importance of the short-wave radiation, the presence of clouds near the ice edge and the amount of snow on the ice become especially relevant during the melting period.

In addition to these ice considerations, the nature of the oceanic heat flux is critical. Cooling conditions at the surface can cause surface overturning, thus bringing up warmer water from below. However, if the surface waters have a much lower salinity than the deeper waters, such convective overturning will be greatly reduced. As a consequence, the amount of melting that occurs at the ice edge can be particularly important in the ice edge advance and retreat.

## Simulation of Seasonal and Interannual Sea-Ice Fluctuations

To get some idea of the relative role of the different processes it is useful to carry out a hierarchy of simulations with differing degrees of ice dynamics and ocean circulation. The simplest first case is to consider a dynamic ice model embedded in a fixed-depth motionless mixed layer with a constant heat flux from below. Such a simulation was carried out by Hibler and Walsh (1982) using the two-level thickness distribution discussed above. Atmospheric forcing for the model consisted of daily atmospheric winds, temperatures, and humidities over the three-year period from 1973–1975. In these model simulations, the treatment of the thermodynamic and oceanic processes consists of estimating vertical growth rates (for use in the $S_A$ and $S_h$ terms in equations 2-13 and 2-14) from (1) heat-budget calculation at the top and bottom surfaces of the ice, and (2) heat storage in a motionless oceanic boundary layer. Growth rates are calculated by using a time-independent thermodynamic sea-ice model (Semtner, 1976) together with the surface heat budget formulation of Parkinson and Washington (1979). The basic surface energy budget (see Hibler, 1980*b*, Appendix A) includes incoming short-wave and long-wave radiation, and sensible and latent heat fluxes formulated in terms of bulk transfer coefficients.

Snow is parameterized by allowing the ice surface albedo to be that of snow (0.75) when the calculated surface temperature is below the melting point and that of snow-free ice (0.66) when it is at the melting point. The treatment of the ocean is based on a motionless mixed layer assumed to be of fixed depth (30 m). When open water is absorbing heat, all of the absorbed heat is assumed to mix underneath any ice in the grid cell, thereby modifying the vertical growth rates. Any heat remaining after all the ice is melted is used to warm the mixed layer. Under growth conditions, the mixed layer is cooled to freezing before ice forms. As a consequence of these constraints, the temperature of the mixed layer is permitted to rise above the melting point if and only if no ice is present.

Some of the main results from this study are shown in Figure 2-21, which shows simulated and observed ice edges in the east Greenland region. Overall, the maximum ice edge position is largely controlled by the thermodynamics, as is apparent from comparing the full model with the thermo-only simulation. However, as discussed earlier, the ice thickness characteristics, especially the spatial variation in ice thickness, are quite different when dynamics is included. In the summer ice edge the dynamics causes a more rapid melt-back, due to the presence of leads, which increase the absorption of shortwave radiation and hence increase the ice melt rate.

Clearly the overall problem is that the simple boundary layer formulation used here yields an excessive ice edge in winter. Considering the physics of the Greenland–Norwegian Sea region, these results are not totally unexpected inasmuch as large-scale oceanic circulation effects would be expected to be particularly pronounced in this region due to warm northward currents encountering rapidly cooling atmospheric conditions and southward-advancing sea ice. This northward heat transport in the ocean ultimately reaches the ice through vertical convection. The amount of heat flux reaching the upper mixed layer through such convective processes can be very substantial. To examine the role of these processes more completely requires a more complete inclusion of ocean effects.

## The Role of Ocean Circulation in Ice Edge Variations

In order to examine the dominant effects of the full three-dimensional ocean circulation on the simulated ice margin and to examine ice–ocean coupling effects, Hibler and Bryan (1984, 1987) constructed a diagnostic ice–ocean model and used it to carry out a series of seasonal simulations of the Arctic, Greenland, and Norwegian seas. The term *diagnostic* refers to the fact that the ocean model contains a linear term that damps the ocean's temperature and salinity toward climatology. However, as discussed below, this damping is done in a relatively weak manner so that seasonal and month-to-month variation in the temperature and salinity due to ice–ocean coupling effects is prognostically simulated by the model.

This ice–ocean model was constructed by coupling the two-thickness-level thermodynamic sea-ice model (Hibler, 1979) with a multilevel baroclinic ocean model (Bryan, 1969). The sea-ice model supplies heat flux, salt flux, and momentum exchange boundary conditions for the top of the ocean. The ocean model, in turn, supplies current and heat exchange information to the ice model. Since the concern here was to examine the effect of ocean circulation on sea ice, observed oceanic temperature and salinity data (Levitus, 1982) were used to weakly force the ocean model so that its equilibrium time scale is similar to that of the ice model (3–5 years). This diagnostic method allows the ocean model to be forced to available climatological ocean data, while at the same time allowing considerable adjustment in the upper ocean due to the effects of ice–ocean interaction. In addition, the barotropic mode of the ocean is fully simulated so that temporally varying currents due to surface stress

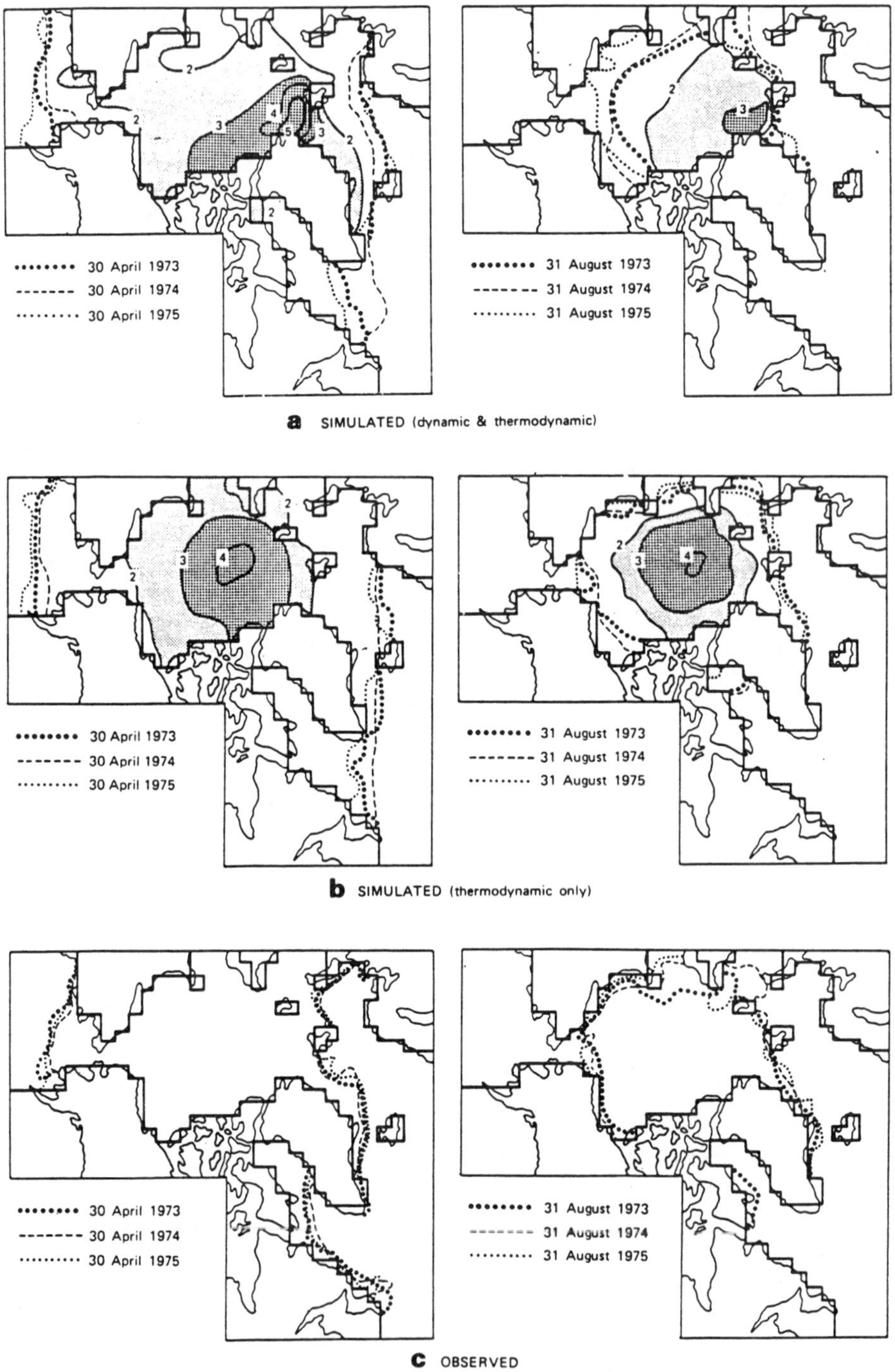


**Figure 2-21.** Positions of simulated (*a, b*) and observed (*c*) ice edges at the end of April and August, 1973–1975. Simulated results are based on model runs containing (*a*) both dynamics and thermodynamics and (*b*) thermodynamics only. Three-year means of simulated ice thicknesses (m) for April and August are also shown in *a* and *b*. (From Hibler and Walsh, 1982.)

fluctuations are part of the model predictions. The basic idea here is to specify the ocean circulation over climatic time scales while modeling the shorter-term fluctuations.

Probably the most noticeable effect of the modeled ocean circulation is to greatly improve the winter ice margin simulation (Fig. 2-22*a*). This improved ice margin is due to the large oceanic heat flux from the deeper ocean into the upper mixed layer in this region (Fig. 2-23). The heat flux occurs primarily in winter; analysis shows that much of the enhancement is due to deep convection, which brings up warm water and prevents ice formation in early winter. This convection explains the absence of ice in the full model simulation far from the ice margin.

Near the ice margin a similar physics applies, but the precise location of the ice edge becomes more sensitive to the surface salt balance and to lateral effects in the ocean circulation. The sensitivity simulations shown in Figure 2-22*b*, for example, indicate that the freshwater flux from melting at the advancing ice edge tends to seal off the ice margin to upward oceanic heat flux and allows a farther advance than would otherwise occur. A sensitivity simulation without lateral motion in the ocean (Fig. 2-22*b*) also tends to produce a more excessive edge at the end of one year, since the lateral transport of heat to both the deep and upper layers is missing. This latter result demonstrates the difficulty of simulating the seasonal cycle of sea ice when only a one-dimensional mixed-layer model is used.

As can be deduced from the above figures, the ice edge position in the east Greenland Sea is largely determined by a

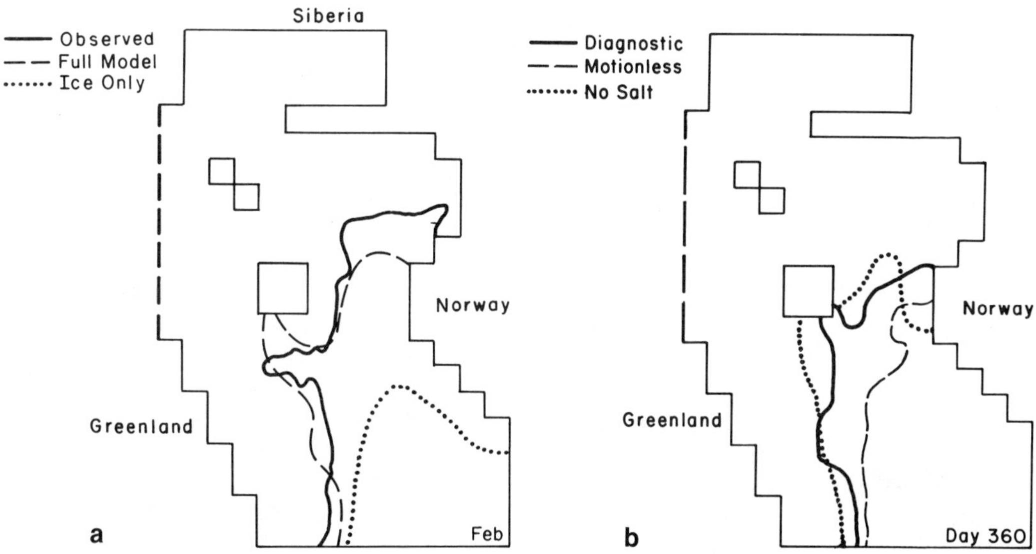


Figure 2-22. (*a*) February 50 percent concentration limits from a full coupled ice–ocean model and for an ice-only model, which includes a fixed-depth mixed layer. The observed limit is taken from Fleet Weather charts (Suitland, Maryland, U.S. Navy) for the end of February. (*b*) Simulated 0.05 m thickness contours for day 360 for the full model, a coupled model with a "motionless" ocean, and a coupled ice–ocean model with no salt fluxes due to ice freezing and melting. The "motionless" model and the no-salt simulations were run for only one year. The starting points for these simulations were the ice thickness and ocean fields at the end of the fourth year of the full ice–ocean model simulation. (From Hibler and Bryan, 1984.)

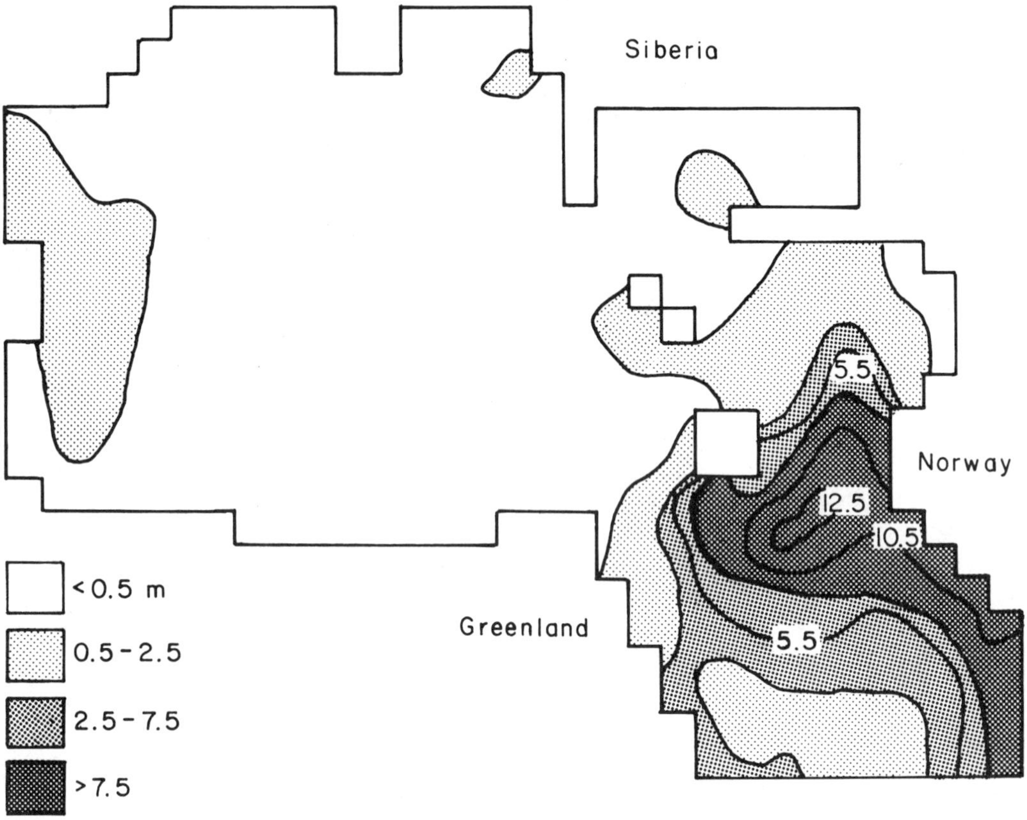


Figure 2-23. Average annual heat gained by the upper layer of the ocean from the deeper ocean and by lateral heat transport during the fifth year of a coupled ice–ocean simulation. The contours are in terms of the melting capacity of the heat (in meters of ice per year). A melt rate of 1 m of ice per year is equivalent to 9.57 W $m^{-2}$. (From Hibler and Bryan, 1984.)

competition between ice advection from the north and heat advection from the south, which ultimately reaches the ice through vertical convection. Melting can cause cold fresh water to cap off the warm salty water, but a lot of melting is required to do this. If the ice production increases we expect the front to shift southward.

While this overall view is largely correct, analysis of the temporal variations of vertical salt and heat fluxes at the ice margin shows the ice advance to be substantially complicated by the local ice dynamics and boundary layer processes occurring at the ice margin. Figure 2-24, for example, shows the time series of monthly averaged ice thickness, freshwater flux, ocean-to-ice heat flux, and atmospheric heat gain at a location just inside the ice edge several hundred kilometers south of Spitsbergen. In this region the ice advance has created a more stably stratified upper ocean than further away from the edge. Hence vertical convection is possible to a lesser degree than further east. What occurs here is a very close oscillatory balance between fluctuations of the oceanic heat flux and freshwater flux, with an increase in heat flux coinciding with an increase in melting. This increased melting then helps stratify the system so that the heat flux decreases. Subsequent freezing or lateral advection

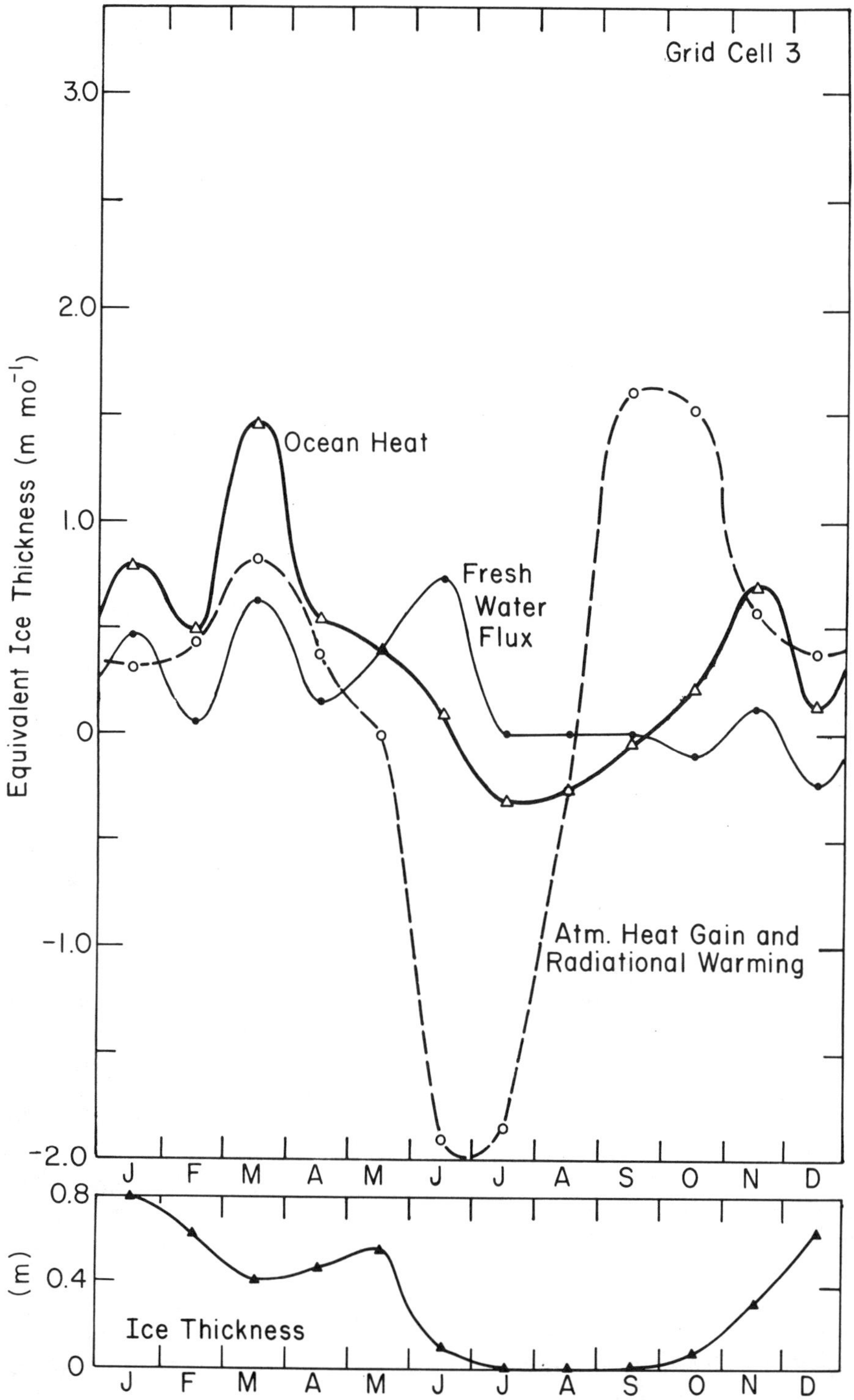


**Figure 2-24.** Time series of monthly averaged heat budget components near the east Greenland ice edge (for a diagnostic ice–ocean model) together with the ice thickness and freshwater flux. The fluxes are in units of meters of ice per month. For comparison, 1 meter of ice per month is equivalent to 115 W $m^{-2}$. (From Hibler and Bryan, 1987.)

can restore the unstable possibilities. It is particularly notable that even though there is a substantial mean ice thickness, the monthly mean ice growth is always negative in winter.

Based on such finely balanced results, even in such a coarse resolution model (160-km grid sizes), it seems quite plausible that details of ice edge processes can play a significant role in the advance of the ice margin on a larger scale. In particular, a closer examination of the fluxes, whose monthly means are shown in Figure 2-24, indicates that open water can be freezing while thicker ice is melting. Consequently, the net salt flux, and hence convection, can hinge on the local ice deformation and advection. Specifically, under large wind-induced divergence the ice growth can dominate, which in turn will lead to more convective overturning and to a reduced stratification. With less divergence and more ice advection, on the other hand, the feedback may favor ice melt, which could cause a rapidly advancing ice edge. Once it formed, the concomitant stratification might be very difficult to remove, and hence the large-scale circulation could be altered.

It should be emphasized that these calculations have not included ocean eddy effects, which may play an important role in both the lateral and the vertical heat exchanges. It is also quite possible for the ice melting and/or freezing to substantially modify the boundary layer drag characteristics and hence accelerate the ice divergence. In addition, the differing wind drag on ice and water can lead to a complex array of ice edge currents and fluctuations, which may in turn affect the ice edge advance and retreat. Clearly, modeling experiments with more complete vertical and horizontal parameterization of ocean processes, together with remote and shipboard oceanographic measurements, are needed to clarify the role of different processes in the ice edge advance and retreat.

## ARCTIC SEA-ICE DYNAMICS AND CLIMATE

The fact that the Arctic ice cover is dynamic rather than static has significant ramifications for climatic change. Specifically, the ice motion and attendant deformation cause heat, salt, and momentum exchanges between the atmosphere and the ocean to be quite different from those that would occur if the ice cover were static. Overall, the main characteristics of a dynamic ice cover that distinguish it, vis-à-vis climate, from a static ice cover are its variable thickness, the presence of ice transport, and ice interaction. How these individual aspects of a dynamic ice cover can affect climate is discussed below. However, it should be cautioned that when all these different components act simultaneously the net result may be more complicated than when the individual components alone are considered.

The variable thickness term refers to the fact that over any region of a dynamic ice cover there will be a variety of ice thicknesses. These different thicknesses are produced by ice deformation, which creates ridged and rafted ice while simultaneously creating leads. Because thin ice and open water grow much faster than thick ice, the growth rate (and hence heat gain by the atmosphere) will often be dominated by the fraction of thin ice rather than the main ice thickness. An example of the magnitude and character of this effect is given in Figure 2-25, taken from a variable thickness dynamic–thermodynamic simulation of the Arctic Basin ice cover by Hibler (1980*a*). In this model simulation an eight-level ice thickness distribution was used together with a dynamic–thermodynamic model employing a plastic rheology. A heat budget and fixed depth mixed-layer calculation similar to that described in the ice edge section is used to calculate the growth rates for each thickness category. Figure 2-25 shows the bas-

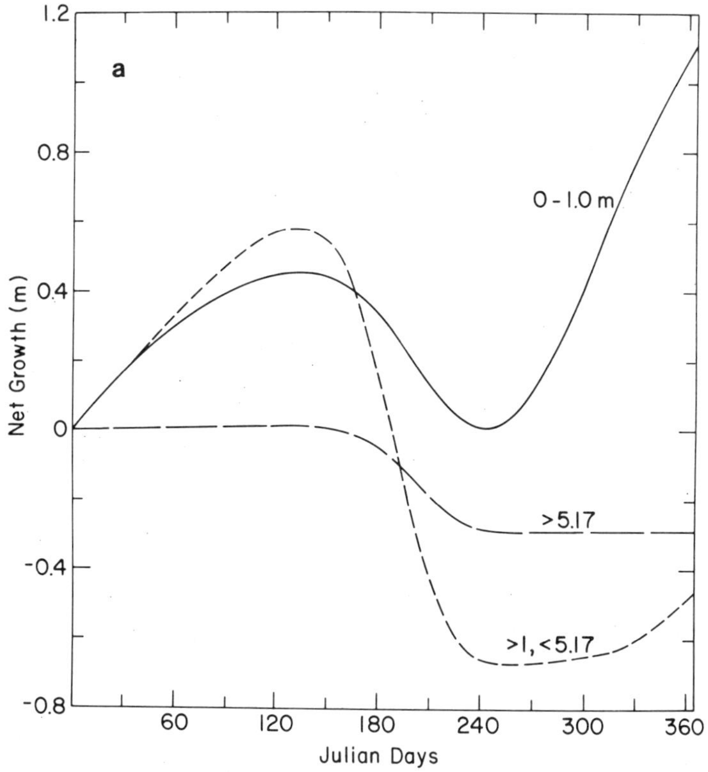


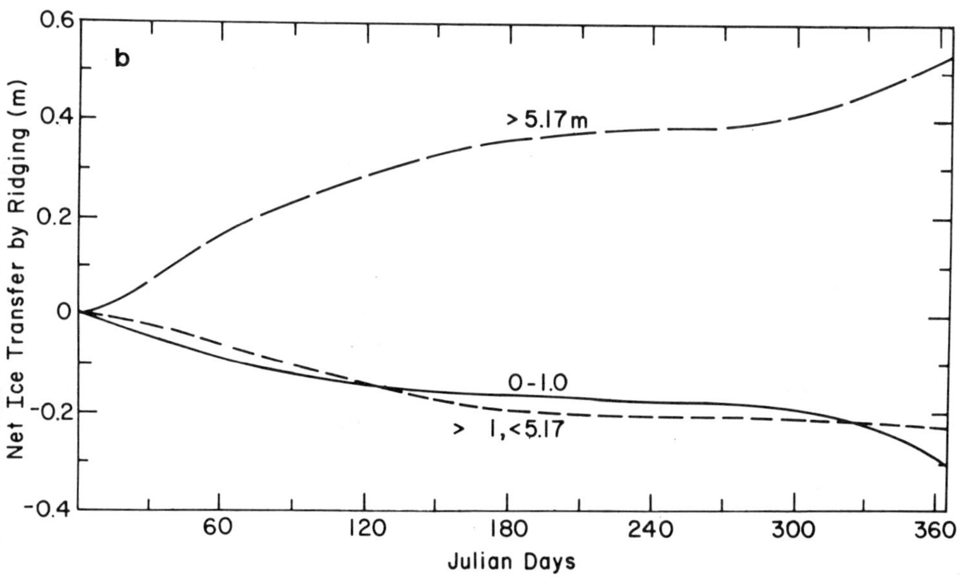


Figure 2-25. Arctic Basin-averaged net ice growth (*a*) and net ice transfer by ridging (*b*) for three categories of ice from a multilevel dynamic-thermodynamic simulation of the Arctic Basin ice cover. (From Hibler, 1980*a*).

in-averaged net growth and net transfer by ridging among three combined thickness categories. On a basin-averaged basis open water creation must be balanced by pressure ridging that formally transfers thin ice to thick ice categories. Consequently, the large amount of fall and winter ridging shown in Figure 2-25 indicates that a large amount of open water is occurring then. Overall, this figure illustrates the important role ridging plays in the ice mass balance. On the average, ridging transfers thin and intermediate ice to thicker categories, thus making room for thin ice and hence greater growth. Note that on an annual basis there is a net ablation of thick and intermediate ice.

This dominance of the growth rates by thin ice can substantially change the way the sea-to-air heat exchange responds to a climatic change, such as a $CO_2$-induced warming. With a constant thickness ice cover (as used by Manabe and Stouffer, 1980, in a numerical $CO_2$ warming experiment), warming will decrease the ice cover thickness, and hence the sea-to-air heat exchange can increase in the polar regions. With a dynamic ice cover, on the other hand, the sensitivity would be reduced or changed in sign, since over leads a warming will decrease sea-to-air heat transfer.

While also arising from the ice motion, the effects of ice transport are somewhat different from the effects of lead formation due to ice deformation. In particular, the creation of leads primarily modifies the local heat budget, while ice transport acts, in effect, to move heat laterally. By means of ice transport, heat can effectively be transferred from a portion of the ocean (where ice is melting) to the atmosphere above a different portion (from where the ice is being transported). Hence, southward ice drift represents a northward heat transport.

The transport also means that the net ice growth, and hence net salt flux, at a given location will differ substantially from zero. The size of this effect is illustrated in Figure 2-26 (also taken from Hibler, 1980*a*), which shows net growth contours from a seasonal equilibrium dynamic–thermodynamic simulation of the Arctic ice cover. This net imbalance in local salt fluxes can be especially important in ice–ocean coupling and modeling, as net melt will tend to stabilize the ocean, whereas net freezing will lead to increased overturning and convection (see also Fig. 2-24).

Ice transport can also play a role in the expansion of an ice edge and in ice decay. In the case of ice decay, irregularities in an ice edge together with lateral transport of ice can enhance ice decay by moving ice into warm water to be melted. This mechanism may be particularly important in the Antarctic due to the strong zonal winds.

An example of how the ice transport and the variable thickness feature can combine so as to negate one another is illustrated by part of Figure 2-27. This figure shows the basin-averaged ice thickness for the full variable thickness model and a single-thickness static–thermodynamic model. At first glance, it appears that the inclusion of dynamics and variable thickness has not changed the thickness variations. However, in fact what has happened is that the increase in growth due to ice dynamics has been offset by an average ice transport from the basin of about 0.06 Sv (which amounts to about half a meter of ice per year). Simulation with only a two-level model produces less thickness increase, and the mean ice thickness will therefore decrease when dynamics is included (see, e.g., Hibler, 1979). This figure also illustrates some of the danger of only comparing ice thicknesses when thermodynamic models are used: There is always the possibility of getting the right answer for the wrong reason.

The third feature of the dynamics is the ice interaction. As discussed above, the ice interaction can play a key role in ice motion by reducing ice convergence while causing little resistance to divergence. As a consequence, the deformation field of pack ice is quite different from that of a

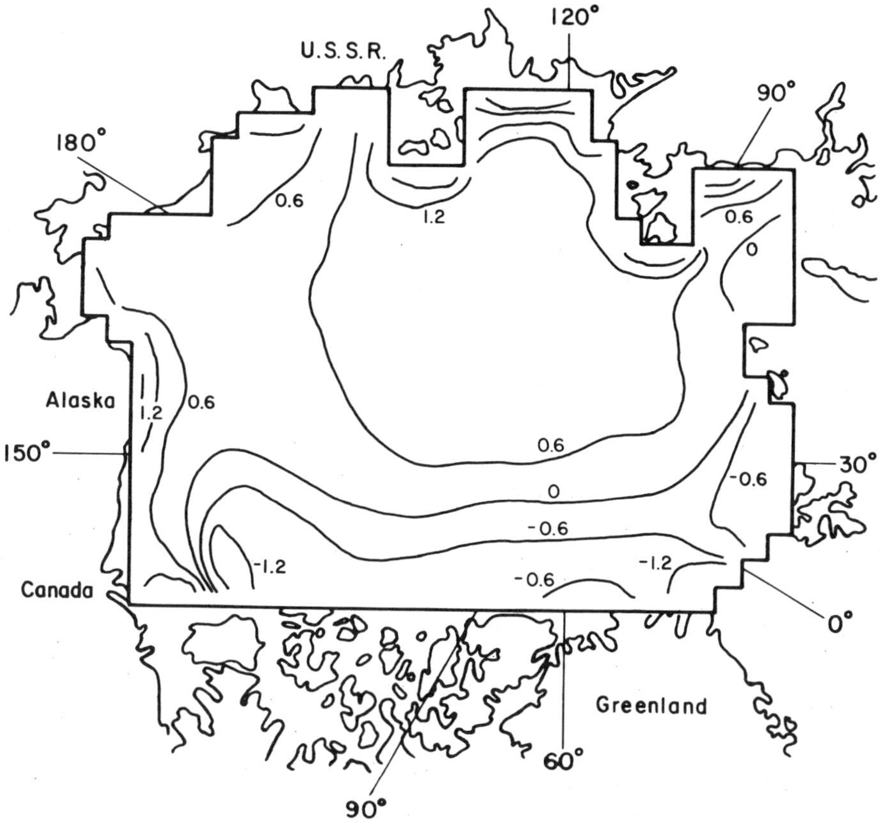


Figure 2-26. Annual net growth contours (at 0.2 m intervals) for a multilevel dynamic–thermodynamic simulation of the Arctic Basin ice cover. (From Hibler, 1980*a*.)

passive tracer moving at the mercy of wind and water currents.

However, probably the most important feature of the ice interaction is that it supplies a mechanical rather than thermodynamic limit on the ice thickness. As noted earlier, this feature is especially important relative to the ice buildup along the Canadian Archipelago, and Figure 2-12 illustrates how critical an effect the strength plays in this buildup. Sensitivity experiments with a multilevel sea ice model (Hibler, 1980*a*) show that increasing the strength causes a substantial decrease in the ice thicknesses off the archipelago. The important point here for climatic change is that since in such heavily ridged areas the ice thickness is dictated by mechanical rather than by thermodynamic considerations, changes in climate (due to $CO_2$ warming, for example) may have considerably less effect on a dynamic ice cover than would be expected from thermodynamic considerations alone. As of yet numerical simulations to test these concepts have not been completed for the Arctic, but a detailed set of studies of the sensitivity to atmospheric warming of the Weddell Sea pack ice in the Antarctic were carried out by Hibler (1984*b*). In terms of the thickness sensitivity, a result of these studies was that the effect of ice interaction tended to decrease the sensitivity to atmosphere warming, both by modifying the velocity field (by ice interaction) to produce fewer leads under warming, and by having the ice thickness in buildup areas be relatively insensitive to atmospheric warming. The

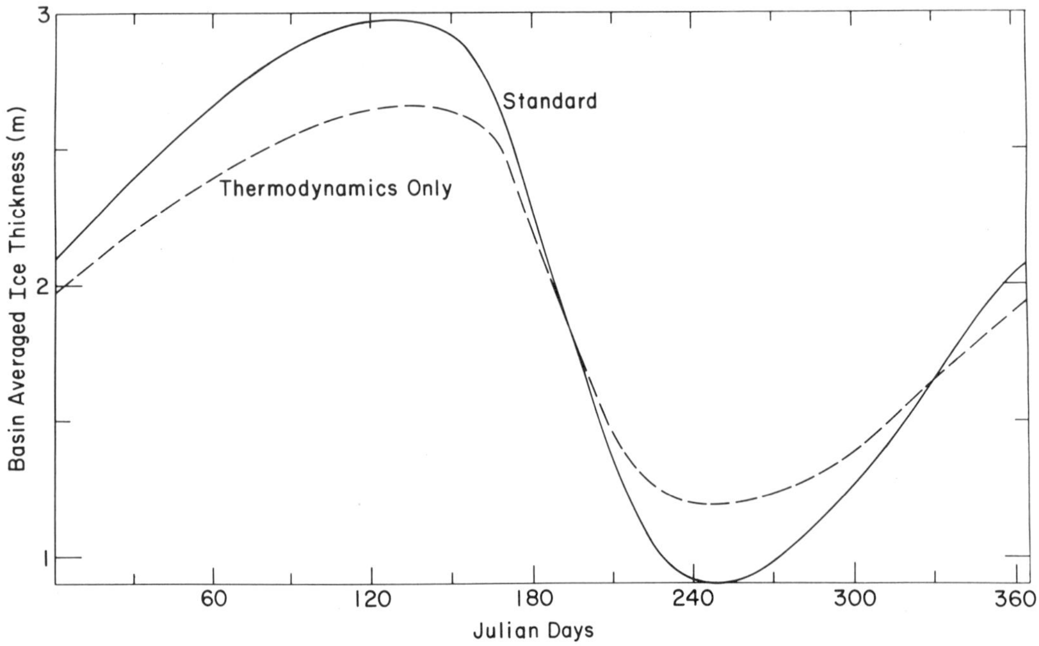


Figure 2-27. Basin-averaged ice thickness time series for a multilevel dynamic–thermodynamic simulation and a thermodynamics-only simulation of the Arctic Basin ice cover. (From Hibler, 1980*a*.)

main point here is that because of the interactive nature of ice dynamics, it is very hard to guess the correct ad hoc dynamical modification to make to climate models. Rather, it seems better to simply include more fully parameterized ice dynamics in climate studies, especially since such models are now well documented and are much faster computationally than oceanic or atmospheric GCMs.

## CONCLUDING REMARKS

A theme of this review has been that the Arctic ice cover is very dynamic rather than passive, and that this dynamic character critically affects air–sea interaction in the polar regions via the variable thickness nature of sea ice. A second more subtle theme is that the motion of the ice cover is not just a passive tracer but that the ice has its own independent circulation due to ice interaction effects. This independent ice circulation can also lead to modifications in the ocean circulation. Such a nonpassive character is particularly important in thickness buildup characteristics and ocean circulation in the Arctic. A third theme is that while moving in response to ocean currents and atmospheric winds, the ice does not simply follow them, nor does it simply follow the Ekman drift. Instead, since the ice sits at the very top of the boundary layer, it drifts at an angle to both the surface currents and Ekman drift, with the angle and drift speed varying with wind speed and ice mass quite apart from ice interaction considerations or boundary layer modifications.

Because of such complexities it is gradually becoming clear that as a geophysical fluid dynamics problem, inclusion of a full dynamic ice cover on an ocean can lead to a wide variety of coupled phenomena, especially in the upper ocean and lower atmosphere. While these phenomena at first glance appear relatively local, because of feedback effects many may well have more global climatic impact. Investigation of such phenomena is only just beginning.

The essential thrust of this review has been to identify many of the key aspects of the physics of sea-ice dynamics and the coupling of the sea ice with the ocean, and to discuss how different aspects may be relevant on a more global level. Since sea-ice dynamics, until recently, has received relatively little attention in the ocean community, appreciation for the complexity and unique nature of this polar geophysical fluid, apart from its thermodynamic insulating character and high albedo, has been slow to develop. It is hoped that this chapter will enhance appreciation of this unique geophysical fluid and substantiate the need for paying increased attention to sea ice at institutions engaged in research related to the atmosphere and oceans.

## ACKNOWLEDGMENTS

This review was made possible by continued encouragement from Yvonne Herman and consistent technical editing advice and aid from Steve Bowen. Figure preparation by Ed Perkins and typing by Donna Harp are also gratefully acknowledged. Finally, it is a pleasure to acknowledge financial support from the Office of Naval Research, Arctic Programs, which has made possible not only this review but much of the research discussed here.

## REFERENCES

Aagaard, K., and P. Greisman. 1975. Toward new mass and heat budgets for the Arctic Ocean. *J. Geophys. Res.* **80**:3821–3827.

Brown, R. A. 1980. Planetary boundary layer modeling for AIDJEX. In R. S. Pritchard (ed.), *Sea Ice Processes and Models.* University of Washington Press, Seattle, pp. 387–401.

Bryan, K. 1969. A numerical method for the study of the circulation of the world oceans. *J. Comput. Phys.* **4**:347–376.

Coachman, H. K., and K. Aagaard. 1974. Physical oceanography of Arctic and Subarctic seas. In Y. Herman (ed.), *Marine Geology and Oceanography of the Arctic Seas.* Springer-Verlag, New York, pp. 1–72.

Colony, R. 1984. Mean ice motion in the Arctic Ocean. *MIZEX Bulletin III.* USA Cold Regions Research and Engineering Laboratory, Special Report 84-7, pp. 91–94.

Coon, M. D. 1974. Mechanical behavior of compacted arctic ice floes. *J. Petrol. Technol.* **26**:466–470.

Dunbar, M., and W. Wittmann. 1962. Some features of ice movement in the Arctic Basin. In *Proceedings of the Arctic Basin Symposium.* Arctic Institute of North America, Washington, D.C., pp. 90–108.

Hastings, A. D. Jr. 1971. *Climatology Atlas of the Arctic Ocean.* U.S. Army Engineering Topography Laboratory, Report ETL-TR-71-5, Fort Belvoir, Va.

Hibler, W. D., III. 1974. Differential sea ice drift. II: Comparison of mesoscale strain measurements to linear drift theory predictions. *J. Glaciol.* **13**:457–471.

Hibler, W. D., III. 1977. A viscous sea ice law as a stochastic average of plasticity. *J. Geophys. Res.* **82**:3932–3938.

Hibler, W. D., III. 1979. A dynamic thermodynamic sea ice model. *J. Phys. Oceanogr.* **9**:815–846.

Hibler, W. D., III. 1980*a*. Modeling a variable thickness sea ice cover. *Mon. Weather Rev.* **108**:1943–1973.

Hibler, W. D., III. 1980*b*. Sea ice growth, drift and decay. In S. Colbeck (ed.), *Dynamics of Snow and Ice Masses.* Academic Press, New York, pp. 141–209.

Hibler, W. D., III. 1984*a*. *Ice Dynamics.* USA Cold Regions Research and Engineering Laboratory, Monograph 84-3, 52p.

Hibler, W. D., III. 1984*b*. The role of sea ice dynamics in modeling $CO_2$ increases. In *Climate Processes and Climate Sensitivity.* American Geophysical Union, Geophysical Monograph 29, Maurice Ewing Volume 5, pp. 238–253.

Hibler, W. D., III. 1985. Sea ice models and remote sensing. in *Proceedings of a Conference on the Use of Satellite Data in Climate Models, Alpbach, Austria.* ESA SP-244, Sept. 1985, pp. 87–94.

Hibler, W. D., III, and S. F. Ackley 1983. Numerical simulation of the Weddell Sea pack ice. *J. Geophys. Res.* **33**:2372–2887.

Hibler, W. D., III, and K. Bryan. 1984. Ocean circulation: Its effect on seasonal sea ice simulations. *Science* **224**:489–492.

Hibler, W. D., III, and K. Bryan. 1987. A diagnostic ice-ocean model. *J. Phys. Oceanogr.* **17**:987–1015.

Hibler, W. D., III, S. J. Mock, and W. B. Tucker. 1974. Classification and variation of sea ice ridging in the western Arctic Basin. *J. Geophys. Res.* **79**:2735–2743.

Hibler, W. D., III, and W. B. Tucker. 1979. Some results from a linear viscous model of the Arctic ice cover. *J. Glaciol.* **22**:293–304.

Hibler, W. D., III, I. Udin, and A. Ullerstig. 1983. On forecasting mesoscale ice dynamics and buildup. *Ann. Glaciol.* **4**:110–115.

Hibler, W. D., III, and J. E. Walsh. 1982. On modeling seasonal and interannual fluctuations of arctic sea ice. *J. Phys. Oceanogr.* **12**:1514–1523.

Hibler, W. D., III, W. F. Weeks, and S. J. Mock. 1972. Statistical aspects of sea ice ridge distributions. *J. Geophys. Res.* **77**:5954–5970.

Hibler, W. D., III, W. F. Weeks, A. Kovacs, and S. F. Ackley. 1974*a*. Differential sea-ice drift. I: Spatial and temporal variations in sea-ice deformation. *J. Glaciol.* **13**(69):437–455.

Hibler, W. D., III, S. F. Ackley, W. K. Crowder, H. W. McKim, and D. M. Anderson. 1974*b*. Analysis of shear zone ice deformation in the Beaufort Sea using satellite imagery. In J. C. Reed and J. E. Sater (eds.), *The Coast and Shelf of the Beaufort Sea.* Arctic Institute of North America, pp. 285–296.

Hunkins, K. 1967. Inertial oscillations of Fletcher's Ice Island (T-3). *J. Geophys. Res.* **72**:1165–1174.

Hunkins, K. 1975. The oceanic boundary layer and stress beneath a drifting ice floe. *J. Geophys. Res.* **80**:3425–3433.

Koerner, R. M. 1973. The mass balance of the sea ice of the Arctic Ocean. *J. Glaciol.* **12**:173–185.

Kovacs, A. 1972. On pressured ice. In T. Karlsson (ed.), *Sea Ice, Proceedings of an International Conference.* National Research Council, Reykjavik, Iceland, pp. 276–298.

Kovacs, A. 1976. *Grounded Ice in the Fast Ice Zone along the Beaufort Sea Coast of Alaska.* USA Cold Regions Research and Engineering Laboratory, CRREL Report 76-32, 31p.

Kovacs, A., and M. Mellor. 1974. Sea ice morphology and ice as a geologic agent in the southern Beaufort Sea. In J. L. Reed and J. E. Sater (eds.), *The Coast and Shelf of the Beaufort Sea.* Arctic Institute of North America, pp. 113–161.

Leavitt, E. 1980. Surface-based air stress measurements made during AIDJEX. In R. S. Pritchard (ed.), *Sea Ice Processes and Models.* University of Washington Press, Seattle, pp. 419–429.

Lepparanta, M. 1981. *On the Structure and Mechanics of Pack Ice in the Bothnian Bay.* Finnish Marine Research Report No. 248, Institute of Marine Research Helsinki, 86p.

Lepparanta, M., and W. D. Hibler, III. 1986. Mesoscale sea ice deformation in the east Greenland marginal ice zone. *J. Geophys. Res.* **92**:7060–7070.

Levitus, S. 1982. *Climatological Atlas of the World Ocean.* NOAA Publ. 13, U.S. Department of Commerce, 173p.

McPhee, M. 1978. A simulation of inertial oscillation in drifting pack ice. *Dyn. Atm. and Oceans* **2**:107–122.

McPhee, M. G. 1979. The effect of the oceanic boundary layer on the mean drift of pack ice: Application of a simple model. *J. Phys. Oceanogr.* **9**:388–400.

McPhee, M. G. 1980. Analysis of pack ice drift in summer. In R. S. Pritchard (ed.), *Sea Ice Processes and Models.* University of Washington Press, Seattle, pp. 62–75.

McPhee, M. G. 1982. *Sea Ice Drag Laws and Simple Boundary Layer Concepts, Including Application to Rapid Melting.* CRREL Report 82-4, USA Cold Regions Research and Engineering Laboratory.

Malvern, L. E. 1969. *Introduction to the Mechanics of a Continuous Media.* Prentice Hall, Englewood Cliffs, New Jersey.

Manabe, S., and R. Stouffer. 1980. Sensitivity of a global climate model to an increase of $CO_2$ concentration in the atmosphere. *J. Geophys. Res.* **85**:5529–5554.

Maykut, G. A. 1978. Energy exchange over young sea ice in the central Arctic. *J. Geophys. Res.* **83**:3646–3658.

Maykut, G. A., and N. Untersteiner. 1971. Some results from a time dependent, thermodynamic model of sea ice. *J. Geophys. Res.* **76**:1550–1575.

Parkinson, C. L., and W. M. Washington, 1979. A large-scale numerical model of sea ice. *J. Geophys. Res.* **84**:311–337.

Parmerter, R. R., and M. D. Coon. 1972. Model of pressure ridge formation in sea ice. *J. Geophys. Res.* **77**:6565–6575.

Pritchard, R. S., M. D. Coon, and M. G. McPhee. 1977. Simulation of sea ice dynamics during AIDJEX. *J. Press. Vess. Tech.* **99J**:491–497.

Reed, R. J., and W. J. Campbell. 1962. *J. Geophys. Res.* **67**:281–297.

Reimnitz, E., L. Toimil, and P. Barnes. 1978. Arctic continental shelf morphology related to sea ice zonation, Beaufort Sea, Alaska. *Marine Geol.* **28**:179–210.

Rothrock, D. A. 19075. The energetics of the plastic deformation of pack ice by ridging. *J. Geophys. Res.* **80**:4514–4519.

Semtner, A. J., Jr. 1976. A model for the thermodynamic growth of sea ice in numerical investigations of climate. *J. Phys. Oceanogr.* **6**:379–389.

Shen, H. H., W. D. Hibler, III, and M. Lepparanta. 1986. On applying granular flow theory to a deforming broken ice field. *Acta Mecanica* **63**:143–160.

Sodhi, D. S., and W. D. Hibler, III. 1980. Nonsteady ice drift in the Strait of Belle Isle. In R. S. Pritchard (ed.), *Sea Ice Processes and Models.* University of Washington Press, Seattle, pp. 177–186.

Thorndike, A. S., and R. Colony. 1980. Large-scale ice motion in the Beaufort Sea during AIDJEX, April 1975–April 1976. In R. S. Pritchard (ed.), *Sea Ice Processes and Models.* University of Washington Press, Seattle, pp. 249–260.

Thorndike, A. S., and R. Colony. 1982. Sea ice motion in response to geostrophic winds. J. Geophys. Res. **87**:5845–5852.

Thorndike, A. S., D. A. Rothrock, G. A. Maykut, and R. Colony. 1975. The thickness distribution of sea ice. *J. Geophys. Res.* **80**:4501–4513.

Tucker, W. B., W. F. Weeks, and M. Frank. 1979. Sea ice ridging over the Alaskan continental shelf. *J. Geophys. Res.* **84**:4885–4897.

Tucker, W. B., W. F. Weeks, A. Kovacs, and A. J. Gow. 1980. Nearshore ice motion at Prudhoe Bay, Alaska. In R. S. Pritchard (ed.), *Sea Ice Processes and Models.* University of Washington Press, Seattle, pp. 261–272.

Vinje, T. E. 1976. Sea ice conditions in the European sector of the marginal seas of the Arctic, 1966–1975. In *Norsk Polar-Institutt Arbok 1975.* Oslo, pp. 163–174.

Vowinckel, E., and S. Orvig. 1970. The climate of the North Polar Basin. In S. Orvig (ed.), *Climates of the Polar Regions.* World Survey of Climatology, Vol. 14. Elsevier, Amsterdam, 370p.

Wadhams, P. 1981. Sea ice topography of the Arctic Ocean in the region 70°W to 25°E. *Roy. Soc. Philos. Trans.* **A302**:1464–1504.

Wadhams, P., and R. J. Horne. 1978. *An Analysis of Ice Profiles Obtained by Submarine Sonar in the AIDJEX Area of the Beaufort Sea.* Technical Report 78-1, Scott Polar Research Institute, Cambridge.

Walsh, J. E., and C. M. Johnson. 1979. An analysis of Arctic sea ice fluctuations, 1953–77. *J. Phys. Oceanogr.* **9**:580–591.

Walsh, J. E., W. D. Hibler, III, and B. Ross. 1984. A model simulation of 20 years of Northern Hemisphere sea ice fluctuations. *Ann. Glaciol.* **5**:170–176.

Walsh, J. E., W. D. Hibler, III, and B. Ross. 1985. Numerical simulation of Northern Hemisphere sea ice variability, 1951–1980. *J. Geophys. Res.* **90**:4847–4865.

Washington, W. M., A. J. Semtner, C. Parkinson, and L. Morrison. 1976. On the development of a seasonal change sea-ice model. *J. Phys. Oceanogr.* **6**:679–685.

Weeks, W. F. 1968. Characteristics of Arctic and Antarctic sea ice. In *SCAR/SCOR/IAPO/IUPS Symposium on Antarctic Oceanography, Santiago, Chile, 1966.* Scott Polar Research Institute, Cambridge, pp. 173–190.

Weeks, W. F., A. Kovacs, and W. D. Hibler, III. 1971. Pressure ridge characteristics in the Arctic coastal environment. In *Proceedings of the First International Conference on Port and Ocean Engineering Under Arctic Conditions.* Technical University of Norway, Trondheim, Vol. 1, pp. 152–183.

Weeks, W. F., A. Kovacs, S. J. Mock, W. B. Tucker, W. D. Hibler, III, and A. J. Gow. 1977. Studies of the movement of coastal sea ice near Prudhoe Bay, Alaska, USA. *J. Glaciol.* **19**(81):533–546.

Williams, E., C. W. M. Swithinbank, and G. DeQ. Robin. 1975. A submarine sonar study of the Arctic pack ice. *J. Glaciol.* **15**:349–362.

Zubov, N. N. 1943. *Arctic Ice.* National Technical Information Service (AD426972), 491p.

# 3: Chemical Oceanography of the Arctic Ocean

Leif Anderson
and David Dyrssen

## INTRODUCTION

The American GEOSECS program followed by the Transient Tracers in Oceanography (TTO) program has supplied oceanography with a vast amount of high-quality chemical data and stressed the importance of the use of tracers. Östlund's measurements of H bomb-produced tritium (HTO) have given us two "snapshots" (1972 and 1981) of how the deep Atlantic water is formed in the Greenland and Norwegian seas and penetrates toward the south Atlantic (Östlund, 1985). During the Ymer-80 expedition an attempt was made to obtain similar high-quality data for the eastern Arctic Ocean (Anderson and Dyrssen, 1981; Carmack and Rudels, 1983), but in spite of the ice camp work by the Canadians and Russians more work needs to be carried out before we have a good picture of the Arctic Ocean. The main objectives of future research were discussed at a meeting on the ventilation of the Arctic Ocean (Melling, 1985). Before this meeting a symposium in Reykjavik on the use of chemical tracers in chemical oceanography (Stefansson, 1985) stressed the advantage of these tools in the understanding of the transport of chemical constituents in the oceans, the main goal being the role of the oceans (not least the Arctic Ocean) in the circulation of carbon dioxide.

In this entry we have made an attempt to summarize the most important contributions to our present knowledge of the chemical oceanography of the Arctic together with our view on how the determination of chemical constituents (both natural and anthropogenic contaminants) can be used to understand the transport and mixing of Arctic water masses.

## CHEMICAL CONSTITUENTS

### Conservative Constituents

The concentrations of the conservative constituents in moles or grams per kilogram seawater are proportional to the salinity (S). Such constituents are lithium, sodium, potassium, rubidium, cesium, magnesium, calcium, fluoride, chloride, bromide, sulfate, alkalinity, and presence of some trace metals (Bruland, 1983). Precise titrations of the calcium and alkalinity ratios, Ca/S and $A_t$/S, may however reveal small shifts due to the formation and dissolution of calcium carbonate and the input of high-alkalinity river water.

## Nutrient-type Constituents

The concentration of nutrient-type constituents is changed during photosynthesis and microbial decomposition of organic matter. Total carbonate, nitrate, phosphate, and silica are decreased upon formation of organic matter while the oxygen concentration is increased. Simple stoichiometric relations may be explained by the following reaction:

$$(x \cdot CO_2) + (x \cdot H_2O) + (y \cdot HNO_3) + (y \cdot H_2O) + (z \cdot MgHPO_4) \rightarrow (CH_2O)_x(NH_3)_y(MgHPO_4)_z + (x + 2y) \cdot O_2$$

Since $HNO_3$ + $HPO_4^{2-}$ are removed, a small shift in alkalinity will occur, $\Delta A_t = y - z$. Most of the Pacific and Atlantic seawater has a ratio $x$:$y$:$z$ close to 106:16:1 and the ratio of oxygen to nitrate is $(x + 2y)/y = 8.625$.

Some trace metals like nickel, zinc, and cadmium, as well as trace elements like germanium, arsenic, selenium, and iodine (Bruland, 1983), behave as nutrient-type constituents. Thus they are depleted in the surface waters and enriched below the euphotic zone.

## Other Natural Constituents

Trace metals that are hydrolyzed at the pH of seawater generally have short residence times in the sea, and elements like scandium, the rare earths, and thorium show surface depletion. The depletion is, however, not only dependent on the removal process but also on the rate of supply (source function). Lead, for example, does not show surface depletion in the north Pacific and Atlantic because of the wide contamination of the northern hemisphere. Strontium shows a slight surface depletion that most likely is due to the extensive production of zooplanktonic shells (calcium carbonate) while barium belongs to the nutrient-type elements (Bruland, 1983). Uranium has a long residence time in seawater due to the formation of a carbonate complex, $UO_2(CO_3)_3^{4-}$, which has almost no affinity to descending particulate matter. It therefore behaves almost as a conservative element. Furthermore, uranium generates a whole series of natural radioactive elements (Fig. 3-1). The depth profiles of lead and lead-210 are quite different and this difference constitutes a good example of the influence of the source function on the depth profile of an element.

## Contaminants

The testing of heavy nuclear devices in the beginning of the sixties gave large inputs of carbon-14 and tritium in the atmosphere. This testing led to the contamination of the sea surface by $^{14}CO_2$ and HTO. The half-life of $^{14}CO_2$ in the troposphere is about seven years and that of HTO is a few weeks. The testing of nuclear weapons also produced chlorine-36 by the capture of neutrons by chloride in the seawater. Thus the $^{36}Cl$ bomb pulse has been measured in Greenland ice core (Elmore et al., 1982). Other radioactive contaminants have been supplied by the nuclear fuel reprocessing plants in La Hague and Sellafield (windscale). Due to malfunctioning the outlets from the Sellafield works into the Irish Sea vary, which may be important when tracing Irish Sea water into the Arctic Ocean (Livingston et al., 1984).

Another type of contaminant is the halocarbons (Fogelqvist, 1985), which can be detected in extremely low amounts by the electron capture detector. Halocarbons are separated by gas chromatography. Large amounts of freon-11 ($CCl_3F$, bp 23.82), freon-12 ($CCl_2F_2$, bp −29.79), carbon tetrachloride ($CCl_4$, bp 76.54) and tetrachloro-

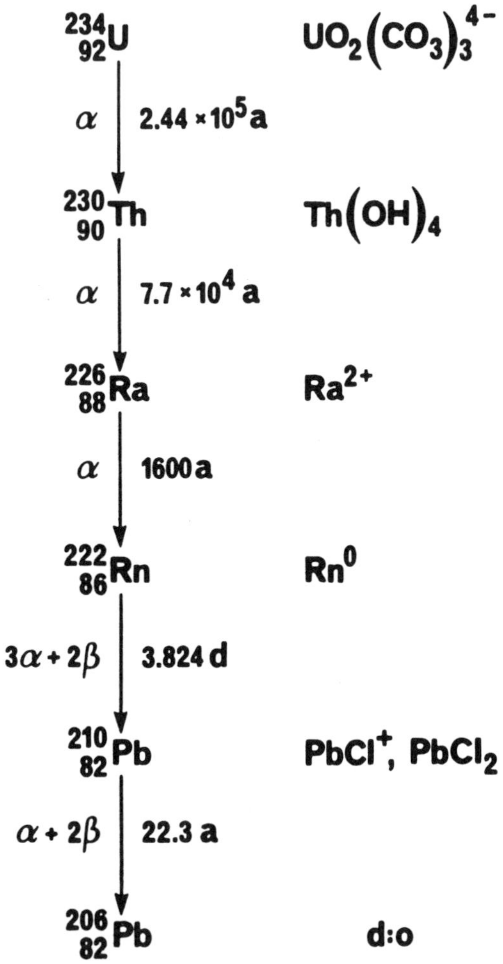


Figure 3-1. Part of the decay series for natural uranium. Some main species are given.

ethylene ($CCl_2CCl_2$, bp 121) have been released into the atmosphere and the sea has subsequently become contaminated. Bromoform ($CHBr_3$, bp 149.5) is naturally produced by belts of red algae (Fig. 3-2), but can also be a contaminant when produced during chlorination of seawater containing organic matter.

## INFLUENCE OF BIOLOGICAL PROCESSES

Biological production, consumption, and microbial decay affect the following constituents of seawater:

Total carbonate
Alkalinity
Oxygen
Nitrate
Phosphate
Calcium
Silica
Some trace metals

The total carbonate is influenced by the production and decomposition of carbohydrates:

$$CO_2 + H_2O \rightarrow CH_2O + O_2$$

The alkalinity is mostly influenced by the uptake of nitrate and release and microbial oxidation of organic nitrogen (proteins):

$$HNO_3 + H_2O \rightarrow NH_3(org) + 2O_2$$
$$NH_3(org) \rightarrow NH_3$$
$$NH_3 + 2O_2 \rightarrow HNO_3 + H_2O$$

The phosphate is essential for the photosynthesis and is usually taken up and released in a ratio to nitrate of between 1:15 and 1:17 (Takahashi, Broecker, and Langer, 1985). However, in Arctic waters the ratio is lower, 1:14 (Anderson and Dyrssen 1981) or 1:9.6 (Jones et al., 1984). The calcium and silica are used to produce the biogenic calcium carbonate and opal. The calcium carbonate production and dissolution is also associated with shifts in the total carbonate and alkalinity:

$$Ca^{2+} + CO_3^{2-} \rightarrow CaCO_3(s)$$

Among the trace metals, cadmium shows a marked correlation with phosphate (Danielsson and Westerlund, 1983; Mart, Nürnberg, and Dyrssen, 1984).

The photosynthetic production is associated with the formation of both particulate and dissolved organic matter. The rate of plankton production is measured by the uptake of carbonate labeled with $NaH^{14}CO_3$. The respiration rate can be measured by the determination of the elec-

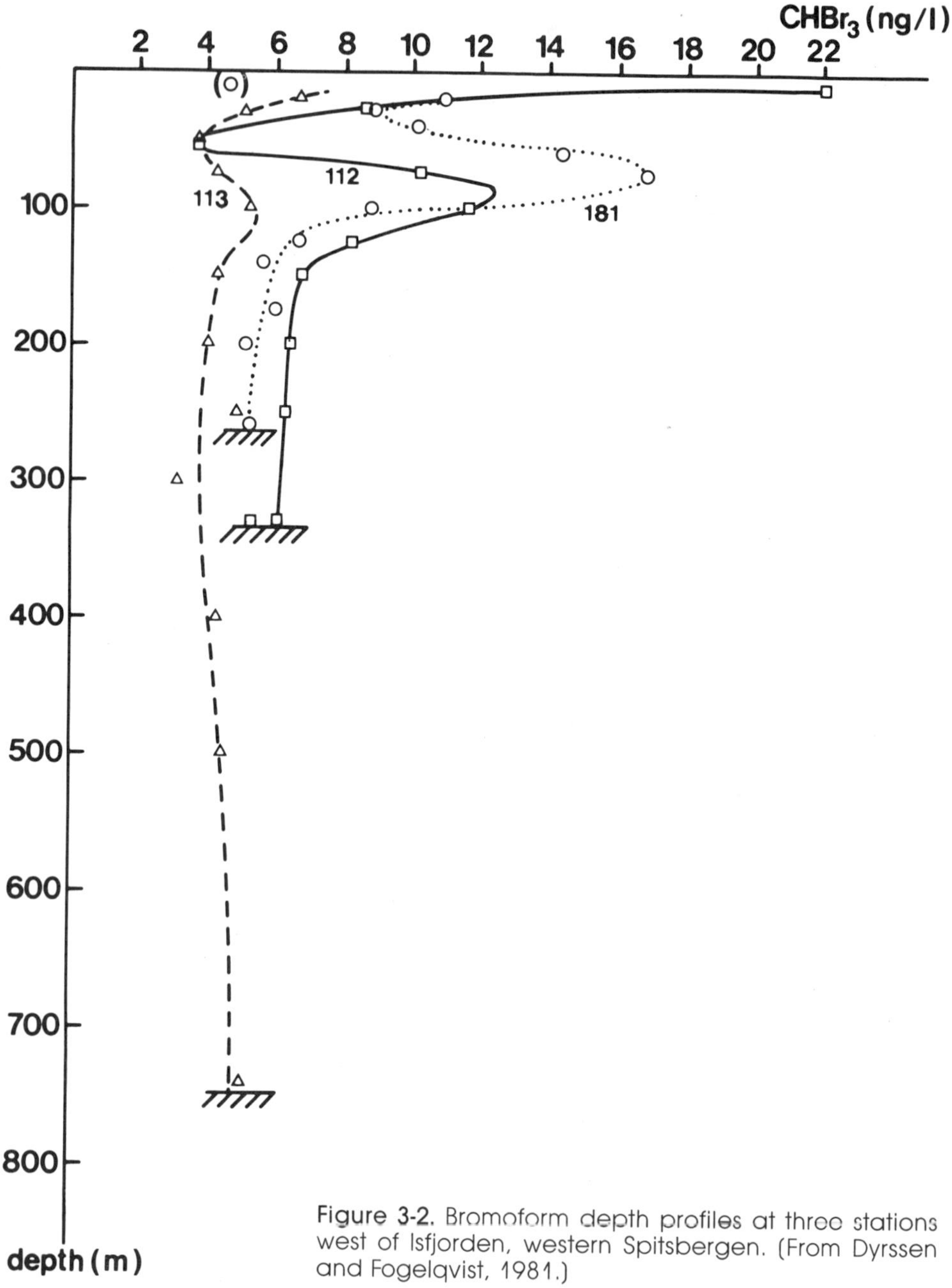


Figure 3-2. Bromoform depth profiles at three stations west of Isfjorden, western Spitsbergen. (From Dyrssen and Fogelqvist, 1981.)

tron transport system (ETS) activity (Garfield, Codispoti, and Packard, 1982). Belts of red benthic algae can produce bromoform (Dyrssen and Fogelqvist, 1981), which can be used as a tracer for water passing through such belts.

## WATER MASSES

The Arctic Ocean has a deep central part (depth maximum 4,880 m) that is divided into four basins by underwater ridges. Surrounding the deep basins is a huge area of continental shelves, mainly on the eastern side. The deep basin is connected to the Atlantic Ocean through Fram Strait (threshold depth 2,600 m) between Greenland and Svalbard and to the Pacific Ocean through Bering Strait (threshold depth 40–50 m). There are also several communications through the Canadian Archipelago to Baffin Bay, the deepest being Nares Strait of approximately 250 m and the widest being Lancaster Sound. Besides these connections, water is added by several rivers entering the Arctic Ocean.

One result of the broad variations in the composition of the waters entering the Arctic Ocean is well-stratified water masses in the upper layers. Due to circulation and a residence time of about ten years, the water masses have a remarkably homogeneous horizontal distribution.

Figure 3-3 summarizes the water exchange, schematically describes the vertical distribution of the water masses, and points to some of the processes influencing their chemical composition. In the top 50 m there is a well-mixed surface layer with temperatures close to the freezing point and a salinity of around 31. Below

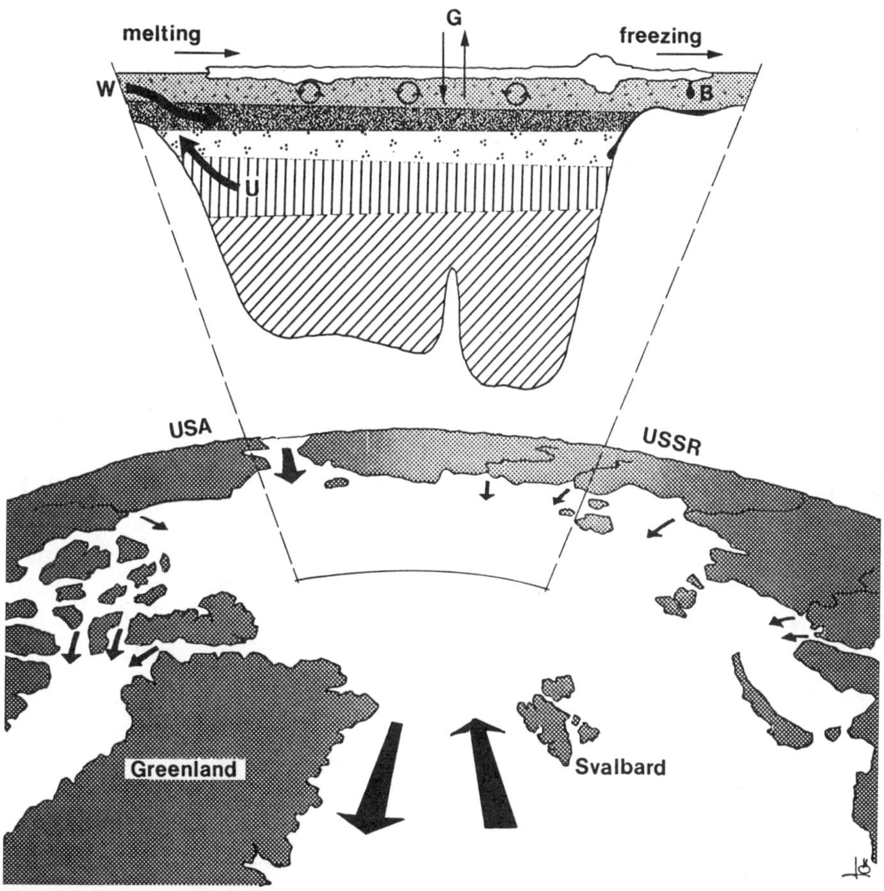


Figure 3-3. The Arctic Ocean with inflows and outflows. The schematic depth section illustrates the different main water masses from the surface; the surface mixed layer; Bering Sea summer water; water of shelf origin, including sea-ice-produced brine; Atlantic water; deep waters. G denotes gas transfer; W, inflow of Bering Strait winter water; B, brine drainage; U, upwelling of Atlantic water.

this layer the salinity increases dramatically and the temperature shows a local maximum at 75 m depth. Deeper, concentrated around 130 m, there is a water mass close to freezing with a salinity of just over 33. Below this water mass the temperature and salinity diagram (Fig. 3-4) indicates a water mass of salinity 34.3 followed by the warm Atlantic water. The deep water was earlier thought of as one homogeneous mass, but has recently been divided into two components (Aagaard, Swift, and Carmack, 1985).

The low salinity of the surface mixed layer together with the ice cap was thought to prevent any major exchange with the deep water within the central ocean. However, recent investigation has stressed the importance of the large shelf areas in the production of high-salinity waters as a result of the annual sea-ice production. These waters are transported over the shelf sediments into the deep basins where they penetrate downward, entraining water and mixing into the basins at different depths. Thereby we have a process to transport different chemical constituents from the surface waters to the sediments and into the deep basin. The exact nature of these processes needs to be elucidated by future research (Jones, 1985).

## INPUTS AND OUTPUTS

Anderson et al. (1983) have tried to estimate the inputs and outputs of salt, fresh water, alkalinity, and silica in the Arctic Ocean. Such estimates are not very precise and are based on uncertain volume flows. In Table 3-1 the inputs and outputs are

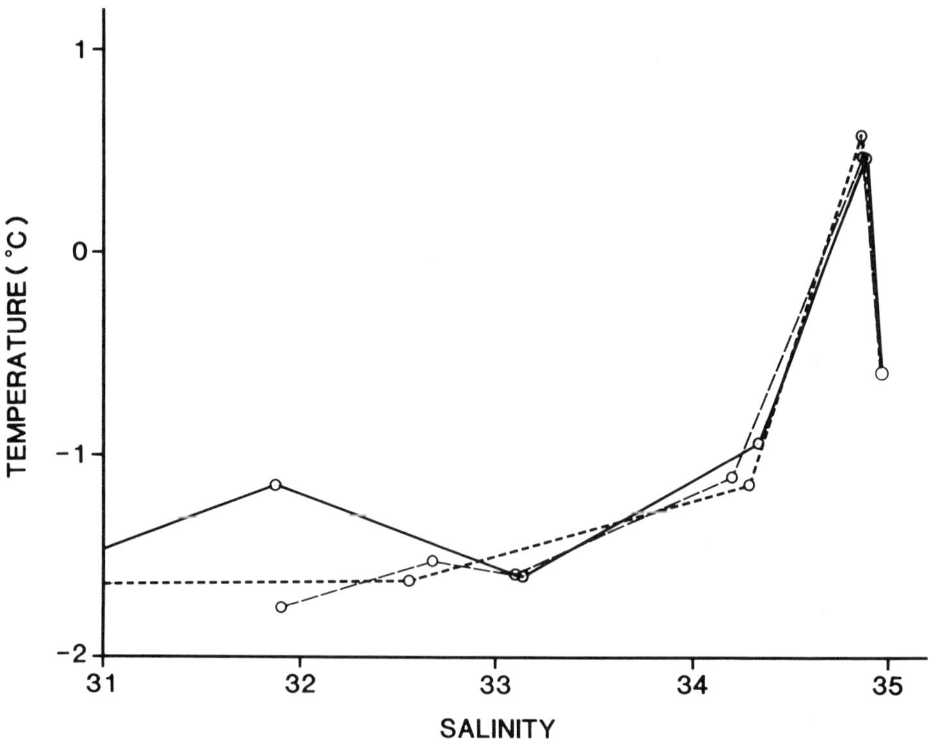


Figure 3-4. Temperature–salinity diagram from the experiments T3 (Kinney, Arhelgeg, and Burrell, 1970), LOREX (Moore, Lowings, and Tan, 1983), and CESAR (Jones and Anderson, 1986). T3—— line, LOREX--- line, CESAR——— line.

Table 3-1*a*. Salt Budget for the Arctic Ocean (in $10^6$ kg $s^{-1}$)

| | | |
|---|---|---|
| Inputs: | Bering Strait | +43.1 ± 0.4 |
| | Atlantic inflow above 200 m | +17.0 ± 0.14 |
| | Atlantic inflow below 200 m | +70.0 – |
| | River water | Negl. |
| | Total | +130.1 ± 0.54 |
| Outputs: | Canadian Archipelago | −67.8 ± 0.6 |
| | East Greenland and outflow | −61.9 ± 0.6 |
| | Sea ice | −1.0 ± 0.1 |
| | Total | −130.7 ± 1.3 |

Table 3-1*b*. Freshwater Budget for the Arctic Ocean (in $10^6$ kg $s^{-1}$)

| | | |
|---|---|---|
| Inputs: | Bering Strait | +146 ± 12 |
| | Atlantic inflow above 200 m | +7 ± 4 |
| | Atlantic inflow below 200 m | — |
| | River water | +110 ± 11 |
| | Precipitation | +30 ± 1 |
| | Total | +293 ± 22 |
| Outputs: | Canadian Archipelago | −109 ± 17 |
| | East Greenland outflow | −74 ± 15 |
| | Sea ice | −71 ± 1 |
| | Evaporation | −20 ± 1 |
| | Total | −274 ± 34 |

Table 3-1*c*. Alkalinity Budget for the Arctic Ocean (in $10^6$ mol $s^{-1}$)

| | | |
|---|---|---|
| Inputs: | Bering Strait | +2.98 ± 0.03 |
| | Atlantic inflow above 200 m | +1.14 ± 0.01 |
| | Atlantic inflow below 200 m | +4.60 ± 0.04 |
| | River water | +0.12 ± 0.01 |
| | Total | +8.84 ± 0.09 |
| Outputs: | Canadian Archipelago | −4.57 ± 0.04 |
| | East Greenland and outflow | −4.22 ± 0.04 |
| | Sea ice | −0.05 ± 0.002 |
| | Total | −8.85 ± 0.08 |

Table 3-1*d*. Silica Budget for the Arctic Ocean (in $10^3$ mol $s^{-1}$)

| | | |
|---|---|---|
| Inputs: | Bering Strait | +31.1 ± 12.2 |
| | Atlantic inflow above 200 m | +2.4 ± 0.5 |
| | Atlantic inflow below 200 m | +9.6 ± 2.0 |
| | River water | +13.2 ± 2.0 |
| | Total | +56.3 ± 16.7 |
| Outputs: | Canadian Archipelago | −41.5 ± 2.0 |
| | East Greenland outflow | −18.4 ± 2.6 |
| | Sea ice | −0.1 ± 0.1 |
| | Total | −60.0 ± 4.7 |

based on an inflow of seawater through Bering Strait of $1.35 \cdot 10^6\ m^3\ s^{-1}$ and an inflow through Fram Strait and Barents Sea of $2.43 \cdot 10^6\ m^3\ s^{-1}$. The estimated outflows were $2.0 \cdot 10^6\ m^3 s^{-1}$ through the Lancaster, Jones, and Smiths sounds into the Baffin Bay and $1.8 \cdot 10^6\ m^3\ s^{-1}$ through the Fram Strait by the east Greenland current. The inflow of river water was 105 600 $m^3\ s^{-1}$ and the outflow of sea ice was 100 000 $m^3\ s^{-1}$. The freshwater input in saline waters was calculated by multiplying the volume flow by (35 − S)/35. Table 3-1 shows that the inputs and outputs agree within the estimated uncertainties. In the case of the budgets for salt (Table 3-1*a*) and alkalinity (Table 3-1*c*), the river contributions are small or negligible. The agreement is then mainly due to the balance in the volume inflows and outflows ($3.8 \cdot 10^6\ m^3\ s^{-1}$). The freshwater budget was used for a sensitivity test. The silica budget is rather rough and does not describe the influence of the production and dissolution of biogenic opal or the circulation of silica. In spite of the uncertainties, it may be concluded from Table 3-1*d* that the Bering Strait inflow supplies the Arctic Ocean with considerable amounts of silica, which also has been stressed by Codispoti (1979). It also supplies the Arctic Ocean with the other nutrients, nitrate and phosphate, both dissolved and particulate. According to Sambrotto, Goering, and McRoy (1984), the main source of these nutrients is the Bering Shelf–Anadyr water.

## PRECIPITATION AND EVAPORATION

The mean annual precipitation over the Arctic Basin is low compared to the runoff and is estimated to be about $0.03 \cdot 10^6\ m^3/s$. The values given for the evaporation varies and range from $0.012 \cdot 10^6$ to $0.09 \cdot 10^6\ m^3/s$. Hence, in a freshwater budget for the Arctic Ocean the precipitation and evaporation may be neglected (Aagaard and Greisman, 1975; Anderson et al., 1983). However, in the treatment of tritium and oxygen-18 data one has to be aware of these fluxes.

Östlund (1982) has discussed the distribution pattern of bomb-produced tritium (HTO) in rain and atmospheric water vapor in order to estimate the residence time of the freshwater component in the Arctic Ocean. When calculating the freshwater components using the oxygen-18 distribution, the source values in runoff and sea-ice meltwater are important (see the section on sea-ice formation and melting). Special care has to be taken when the concentrations in the sea ice are determined since layers of snow are often included in sea ice. The problems related to these questions have been thoroughly discussed by Östlund and Hut (1984).

## AIR–SEA GAS TRANSFERS

Even though large areas of the Arctic Ocean are covered with ice, the transfer of gases between the atmosphere and the sea plays an important role in modifying the chemistry of the surface waters. The transfer takes place not only in open-water areas as it has been shown that sea ice is permeable to gases (Gosink, Pearson, and Kelly, 1976). The migration rate was shown to be two to six times faster along the grain boundaries than perpendicular to them, indicating that the gases are diffusing through the brine channels of sea ice.

The quantitatively most important gases in the Arctic ocean air–sea transfer are oxygen and carbon dioxide. The transfer is affected by biological activity, changing the concentration in the surface water, and by seawater temperature and salinity, changing the solubility of the gases in the seawater. The temperature effect for carbon dioxide in the Barents and Kara seas has been overrated by Kelly (1970). The cooling of inflowing Atlantic water increases the solubility of carbon dioxide, but cannot cause an undersaturation of over 100 ppm.

The oxygen and carbon dioxide content of the waters in Fram Strait, corrected for biological activity, has been presented by Anderson and Dyrssen (1981). For carbon dioxide, the data fall close to the saturation value within the range of temperatures and an atmospheric carbon dioxide concentration of 330 ppm. However, the oxygen concentrations are close to the saturation value at 3°C regardless of the sample temperature and depth. This might be explained by the fact that oxygen has been lost to the atmosphere while the water was in the photic zone. These results agree with the slow exchange rate of carbon dioxide and comparably fast rate of oxygen. The resulting deviation from the saturation value has to be considered in calculations of performed values where a 100 percent saturation value for oxygen is assumed as the initial concentration.

In two papers Lyakhin and Rusanov (1980, 1983) summarized surface water oxygen and carbon dioxide data, both summer and winter, and estimated the flux of these compounds through the air–sea interface. Although they stress the uncertainties, the fluxes match the estimated primary production. The resulting annual input of carbon dioxide to the Arctic seas is calculated to be around $25 \cdot 10^{12}$ g C.

Except for the major gases there are a number of gases with low concentrations that are of interest in the study of the Arctic Ocean. Among these are those of anthropogenic origin, for example, freons and other halocarbons, and those of natural origin, for example, the noble gases. The freons (F-11 and F-12) have been investigated at the ice station CESAR (Canadian Expedition to Study the Alpha Ridge) and the data were used to make some simple modeling of the water transports including time estimates (Wallace and Moore, 1985).

The halocarbons, carbon tetrachloride, tetrachloroethylene, and trichloroethane have been investigated in the Fram Strait region by Fogelqvist (1985) and calculations were carried out to see if the Arctic Ocean was acting as a source or sink. Fogelqvist found an undersaturation in the surface waters of the Arctic Ocean together with a net outflow through Fram Strait.

Neon and helium have been determined in the Arctic Ocean by Top, Clarke, and Moore (1983), and the most striking feature of their results is the high Ne/He ratios in shallow waters. They explain this by the different solubilities of the gases in sea ice. This effect, that many gases have low or practically no solubility in ice, is important when evaluating their behavior in Arctic Ocean surface waters. In general, gases are expelled when sea ice is formed.

## SEA-ICE FORMATION AND MELTING

Even if practically no variation in the annual mean ice thickness of the Arctic Ocean occurs, the seasonal freezing and melting play an important role in the physical and chemical properties of Arctic Ocean water masses. The summer melting will add large amounts (of the same order as the runoff) of relatively fresh water, thereby increasing stratification and inhibiting mixing of the upper layers. During the freezing period brine is drained out of the ice, which drives convection and transports salts downward.

### Deuterium and Oxygen-18

Sea-ice meltwater has almost the same deuterium and oxygen-18 content as the seawater from which the ice was formed. This is in contrast with the content in meteoric water where fractionation during evaporation gives a higher concentration of the lighter isotopes (i.e., $^{1}H_2{}^{16}O$). The ratios of the isotopes in seawater, $^2H/^1H = 1.5 \cdot 10^{-4}$ and $^{18}O/^{16}O = 20 \cdot 10^{-4}$, are each very constant in the bulk ocean. Hence, it is convenient to express the concentrations (isotopic shifts) by a notation $\delta$ (delta), which designates the per mille by which the ratio differs from that of a standard mean ocean water (SMOW):

$$\delta^2H(\text{sample}) = (((^2H/^1H)_{\text{sample}})/((^2H/^1H)_{\text{SMOW}}) - 1) \cdot 1000\ ‰$$

$$\delta^{18}O(\text{sample}) = (((^{18}O/^{16}O)_{\text{sample}})/((^{18}O/^{16}O)_{\text{SMOW}}) - 1) \cdot 1000\ ‰$$

In a $\delta S$ diagram each sample will have a value as a result of the relative mixing of the three source components SMOW (here Atlantic seawater), meteoric water, and sea-ice meltwater (Fig. 3-5). Included in meteoric water is both precipitation and runoff.

From the mass balance equations

$$F_{\text{SMOW}} + F_M + F_{SI} = 1$$

$$F_{\text{SMOW}} \cdot S_{\text{SMOW}} + F_M \cdot S_M + F_{SI} \cdot S_{SI} = S_{\text{Sample}}$$

$$F_{\text{SMOW}} \cdot \delta_{\text{SMOW}} + F_M \cdot \delta_M + F_{SI} \cdot \delta_{SI} = \delta_{\text{sample}}$$

the fractions, $F_{\text{SMOW}}$, $F_M$ (meteoric), and $F_{SI}$ (sea ice), of the different sources can be calculated. A negative $F_{SI}$ will mean that salt rejected during ice formation is added to the sample; see point B in Figure 3-5. Redfield and Friedman (1969) presented statistically treated data of deuterium in samples, both ice and seawater, mainly from the Canadian Arctic. Using the data, they calculated excesses of salinity in different deep waters that they attributed to brine added from sea-ice production.

Vetshteyn, Malyuk, and Rusanov (1974) presented the oxygen-18 distribution in

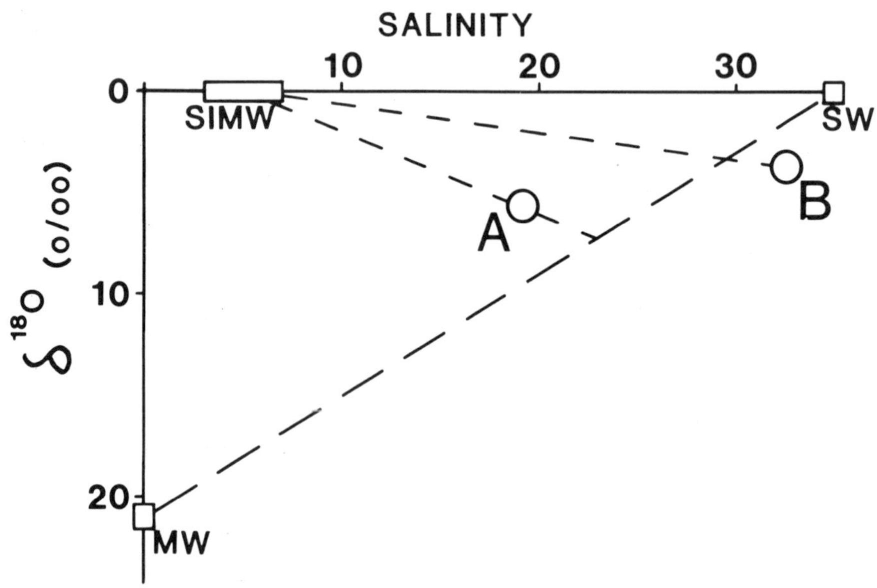


Figure 3-5. The oxygen-18 distribution, expressed as $\delta^{18}$, in the Arctic Ocean seawater system. MW = meteoric water, SIMW = sea-ice meltwater, and SW = seawater. A seawater sample with the $\delta^{18}$ property of *A* consists of all three source waters, while sample *B* is a mixture of MW and SW to which salt has been added as brine or water has been taken away as sea ice.

the central Arctic Ocean, and using the source values by Östlund and Hut (1984) we have calculated the relative fractions of the source components, Atlantic water, runoff, and sea-ice meltwater. A depth plot of the mean values is presented in Figure 3-6. Note that the negative sea-ice values mean that ice has been removed or that salt has been added to the seawater. A calculation like this is mainly sensitive to the accuracy of the primary data, and then especially to the $\delta^{18}O$ value of the meteoric water. A decrease of $\delta^{18}O_M$ to −25‰ will roughly cut the brine addition in half. However, from Figure 3-6 we can conclude that during the winter season the only freshwater source in the area is runoff; within the accuracy reported (±0.3‰ in

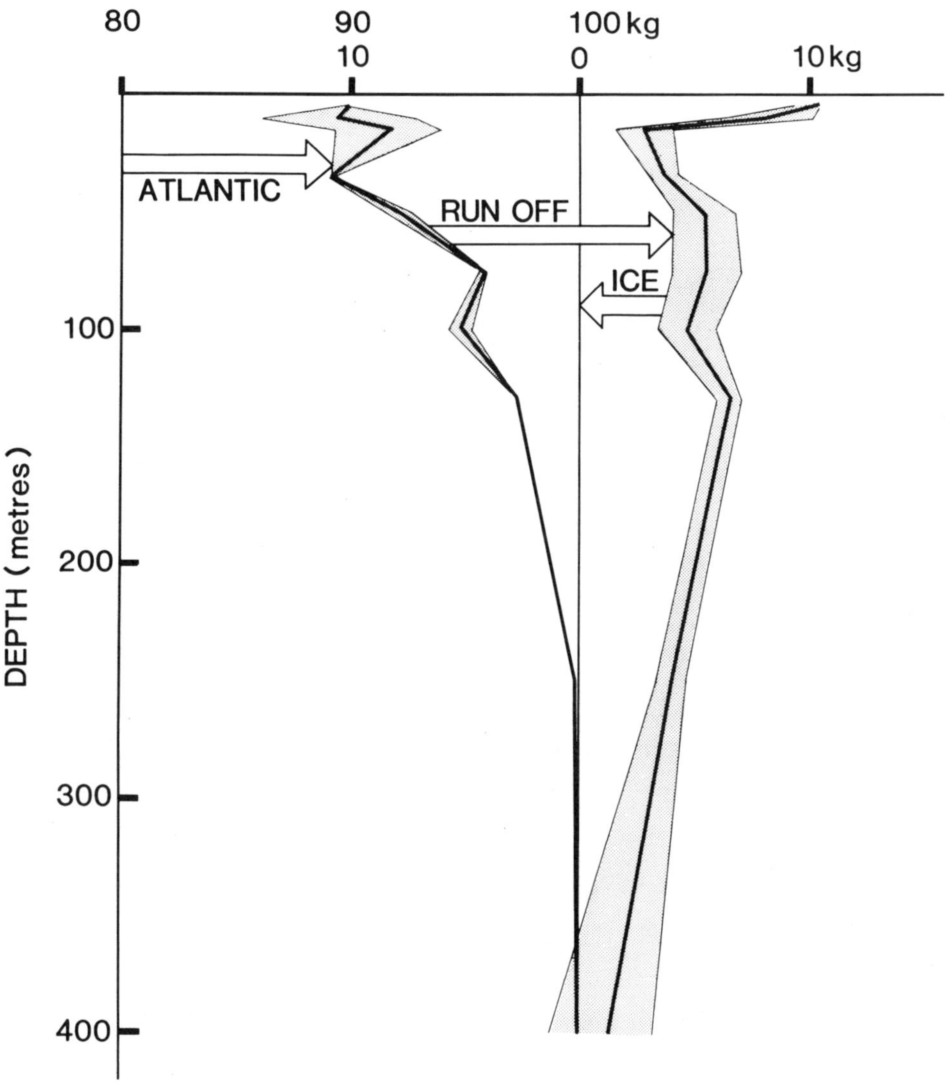


**Figure 3-6.** The mean fractions of the components of Atlantic water, runoff and sea ice, in 100 kg of central Arctic Ocean seawater, calculated using the oxygen-18 data by Vetshteyn, Malyuk, and Rosanov (1974) and the source water values by Östlund and Hut (1984). The shaded area reflects the inaccuracy of the oxygen-18 data, ± 0.3‰.

$\delta^{18}O$) there is a significant addition of salt in the form of brine from sea-ice production in the water above the Atlantic core, which has its temperature maximum at a depth of around 400 m. The data reported by Moore, Lowings, and Tan (1983) for the central Arctic Ocean also suggest an addition of brine at some depths above 150 m.

Östlund and Hut (1984) made the same type of calculations using high-accuracy data from the Fram Strait. They present beautiful sections across the strait that clearly show the dominant freshwater component in the eastern part is sea-ice meltwater, while runoff is the only source in the western part. In the western part there is even a negative sea-ice meltwater component, which means that brine has been added to these waters. Having a good estimate of the fraction of runoff in the sample, Östlund and Hut, using tritium data, calculated the time since the runoff entered the sea, giving a mean residence time in the Arctic Ocean of 10 ± 1 years for the surface and halocline waters.

The distribution of the freshwater inputs in the Canadian Arctic and Baffin Bay has been described by Tan and Strain (1980) and Bédard et al. (1981) using the oxygen-18 technique. From the sea-ice meltwater result Tan and Strain (1980) calculated a mean sea-ice thickness of 1.5 m for Baffin Bay.

With the help of oxygen-18 data, Tan, Dyrssen, and Strain (1983) investigated to what extent the excess alkalinity in the east Greenland current is due to chemical fractionation within the sea ice. They concluded that the alkalinity source mainly is an input from the rivers and not from the sea ice.

## Precipitation of Salts in Sea Ice during Formation and Aging

When seawater starts to freeze it loses more than half of its salt concentration. The rest is trapped as brine in pockets between the ice crystals. The lower the temperature the ice is exposed to, the more concentrated the brine that is obtained. At given temperatures different salts start to precipitate. The salts that precipitate at temperatures above −40°C are as follows:

| *Salt Composition* | *Temperature of Initial Salt Formation in Sea-Ice Brine (°C)* |
|---|---|
| $CaCO_3 \cdot 6H_2O$ | −2.2 |
| $NaSO_4 \cdot 10H_2O$ | −8.2 |
| $CaSO_4 \cdot 2H_2O$ | −10.0 |
| $NaCl \cdot 2H_2O$ | −22.9 |
| KCl | −36.8 |

A phase diagram drawn from the data by Richardson (1976) shows the relative composition of sea ice at different temperatures (Fig. 3-7). Several authors have determined and discussed these solid salts in sea ice produced in the laboratory (e.g., Ringer, 1906; Nelson and Thompson, 1954; Thompson and Nelson, 1956; Assur, 1958; Richardson, 1976). In order for the brine and sea-ice meltwater to have a chemical composition different from that of seawater, the solid salts have to stay in the ice when the brine is leaching out during aging. If this is the case, we would expect to find high concentration, relative to the salinity, especially of total carbonate (or total alkalinity), sulfate, and calcium in natural sea ice. Although several investigators have found some samples with enrichment (e.g., Tsurikov, 1974; Reeburgh and Springer-Young, 1983; Overgaard, Wadhams, and Leppäranta, 1983; Anderson and Jones, 1985) and Sinha (1977) has shown the existence of salt crystals in brine pockets, it does not seem likely that this fractionation process will have a significant effect on the chemical composition of the mean sea-ice meltwater of the Arctic Basin. However, in local areas the effect might be important; for example, in the Sea of Okhotsk, Lyakhin (1970) has reported low specific total alkalinity in brine, and in Baffin Bay, Jones, Coote, and Levy (1983) explain the high normalized alkalinity of the upper 50 m by input of sea-ice

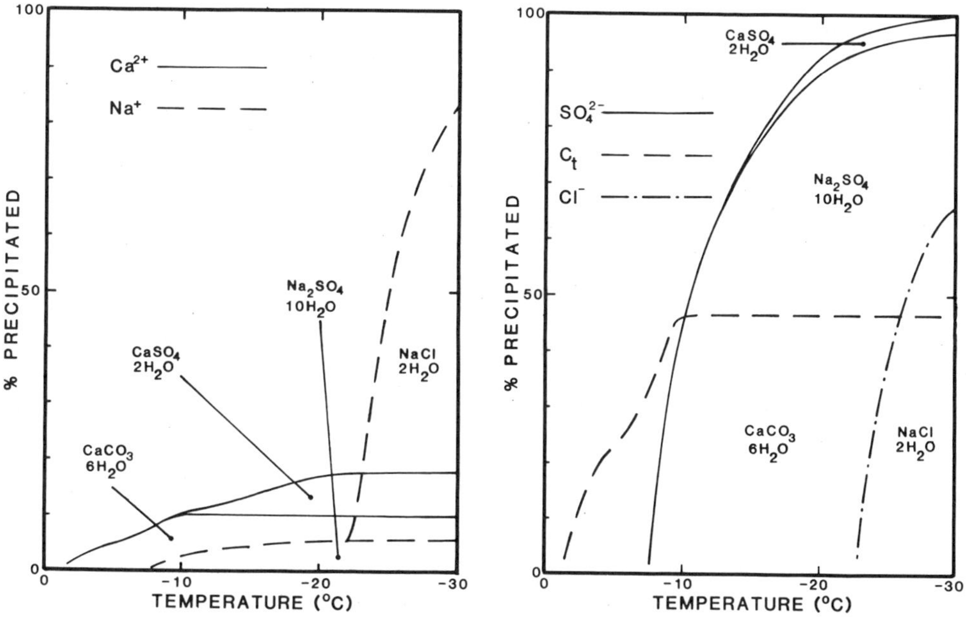


Figure 3-7. The relative precipitation for anions and cations in standard seawater at subfreezing temperatures. (Data from Richardson, 1976.)

meltwater having excess calcium carbonate.

## Release of Brine from Sea Ice

As mentioned above, the seawater loses about two-thirds of its salinity during the freezing process. The brine trapped in the sea ice is slowly draining out, changing the salinity of the sea ice from an initial value of about 12‰ to typically 5‰ after one year. Even if the chemical composition of the brine is not different from that of seawater, the rejected brine is very important in the transport of chemical constituents in the Arctic Ocean, as will be discussed in the next section.

## SHELF PROCESSES

### Organic Matter

The Arctic Ocean shelves cover nearly the same area as the deep Arctic Basin. Although the biological production in the central Arctic Ocean is low (e.g., Subba Rao and Platt, 1984), areas with very high productivity can be found primarily over the shelves (e.g., Codispoti and Richards, 1971; Belyayeva and Romankevich, 1982). As the shelves are mainly shallow and the bacterial activity is low in low-temperature waters, it is probable that a large part of the biogenic organic matter ends up in the shelf sediments. It should be noted that even if Bering Sea is not included in the Arctic Ocean, a lot of the organic matter produced in this very high productive area (e.g., Koike et al., 1982) is transported by currents through Bering Strait into the Arctic Ocean, where to a large extent it settles on the shelf (e.g., Walsh, 1983; Sambrotto, Goering, and McRoy, 1984). The sediment–seawater interaction will, hence, play an important role in the cycling of chemical constituents in the Arctic Ocean.

### Brine Produces High-Salinity Bottom Water

Different from more southerly regions, in the Arctic Ocean there is a process of brine drainage from sea ice, which forms high-salinity, low-temperature bottom water that slowly flows over the shelf sediments into the deeper basin. The process of brine drainage from sea ice during its growth has been discussed previously. The existence of the resulting high-salinity, low-temperature water has been shown by Aagaard, Coachman, and Carmack (1981) in the Chukchi Sea, by Campbell (1964) in Foxe Basin, and by Midttun (1985) in the eastern Barents Sea. However, Nansen (1906) first pointed to the importance of brine rejection from sea ice in the production of high-density bottom waters in the eastern Barents Sea.

A simplified model of the process involved in the transfer of chemical constituents from the shelf water into the deep basin is shown in Figure 3-8. The depths to which the high-salinity water penetrates depends on its density. Since the temperature is rather uniform the density is mainly a function of the initial salinity and residence time on the shelves during which it is to some extent mixed with overlaying water. Nansen (1906) believed that salt originating from sea-ice brine contributes to the Arctic Ocean deep water in the Nansen basin, and this has since been confirmed (e.g., Swift, Takahashi, and Livingston, 1983; Anderson and Jones, 1985).

## Shelf Process Contributing to the Nutrient Maximum of the Deep Basins

Source waters of salinity in the range 33.0 to 34.3 and with low temperatures are clearly seen in a temperature–salinity plot

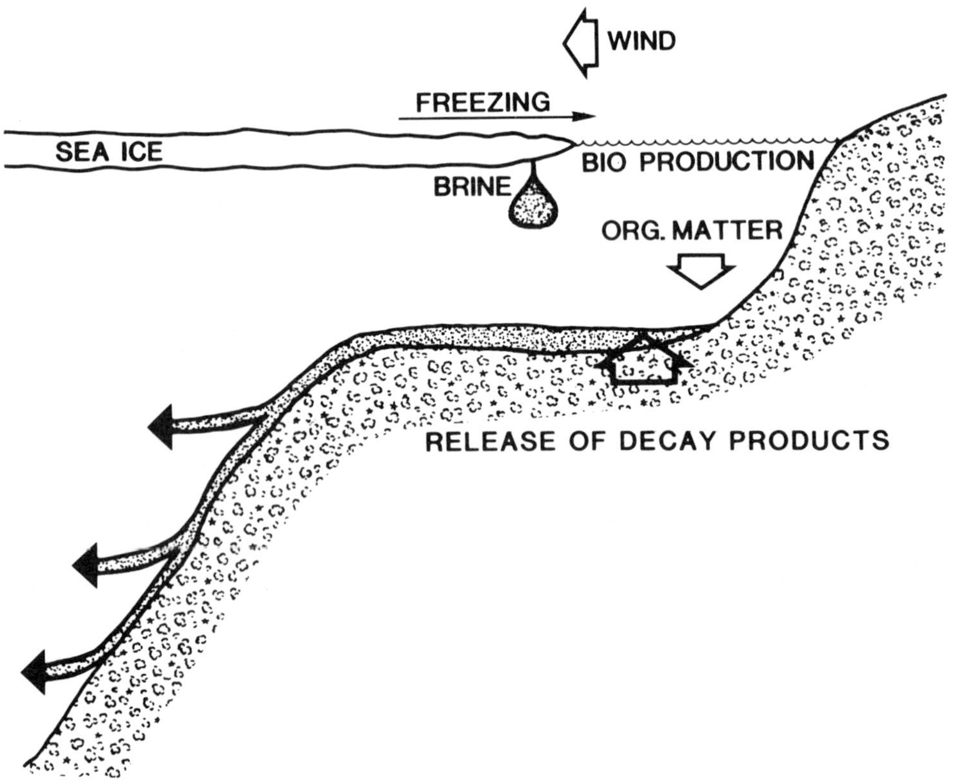


Figure 3-8. Processes involved in the transfer of chemical constituents from the surface water into the deeper waters. Note that the brine production is enhanced by winds that keep waters ice-free.

of a depth profile in the Arctic Basin not too close to Fram Strait (Fig. 3-4). Different origins of these source waters are suggested. One is sea-ice brine added, another is upwelling and cooling of Atlantic water (Aagaard, Coachman, and Carmack, 1981; Melling and Lewis, 1982), and a third is input of Bering Sea winter water (Kinney, Arhelger, and Burrell, 1970).

High nutrient and low oxygen concentrations are observed at a salinity of around 33 in the Canadian Basin (Fig. 3-9) (Nikiforov, Belysheva, and Blinov, 1966; Kinney, Arhelger, and Burrell, 1970; Melnikov and Pavlov, 1978; Moore, Lowings, and Tan, 1983). The nutrient maximum was assumed by Kinney, Arhelger, and Burrell (1970) to have a Bering winter water origin. This view was also considered by Rusanov (1974) and Rusanov and Shapaykher (1979) using the dissolved silica signal. Moore, Lowings, and Tan (1983) mention the possibility that even though the Bering Strait region is the original source, some addition of brine with nutrient enrichment can contribute to the nutrient maximum.

Data from the 1983 CESAR ice camp, including nutrients, total alkalinity, and total carbonate, show clearly that mineralization processes have significantly affected the concentrations of the nutrient maximum (Jones and Anderson, 1986). This is especially distinct in the alkalinity to total carbonate ratio (Fig. 3-10), which differs significantly from what has been reported just north of Bering Strait (Anderson and Lowings, 1982) or in the Bering Sea at the most northern GEOSECS Station, 219 (Broecker, Spencer, and Craig, 1982). There are no reasons to believe otherwise than that the mineralization processes take place on the shelves at the sediment seawater interface. This explanation does not exclude the Bering Sea as the major source for the water mass. However, the chemical signature has been modified after the entrance into the Arctic Ocean.

No release rates of these constituents from Arctic sediments are reported in the literature, but Anderson et al. (1986) measured the release rates from organic rich sediments off the Swedish west coast at low temperatures (−1°C). A rough calculation using these rates, the thickness and excess of the nutrient maximum, and the area ratio between that of the deep basin and the shelves give residence times that are of the same order (10–30 years) as reported elsewhere (Coachman and Barnes, 1961; Östlund, 1982).

The Atlantic warm water that is upwelled on the shelves will be cooled down and mixed with some surface water, thereby decreasing its salinity. As this process is relatively fast, and the cooled water has not necessarily been in contact with the sediment, we do not expect this water to have elevated nutrient concentrations.

The processes producing the low-temperature water, in the salinity range 33.0 to 34.4 over the deep basin, do not produce the same density of waters for all production sites. Most data from the Arctic Ocean are collected from ice stations, thus sampling a fixed moment in time and space. The resulting temperature–salinity diagram will show very straight mixing lines between the water masses. However, in Figure 3-4, data from several stations are included that clearly show different temperature–salinity properties of the source waters. The differences in densities are a result of the quantity of brine added to the water, how much mixing is taking place on the shelf, topographic conditions in the area where the water is formed, and so on. Variations in the nutrient maximum from May 1975 to October 1975 are described by Melnikov and Pavlov (1978).

This discussion points to the importance of the shelves in the transport of chemical constituents, for example, carbon dioxide (Walsh et al., 1981; Anderson, Dyrssen, and Jones, 1989), from the surface waters into the deeper layers. Until recently this transport was considered un-

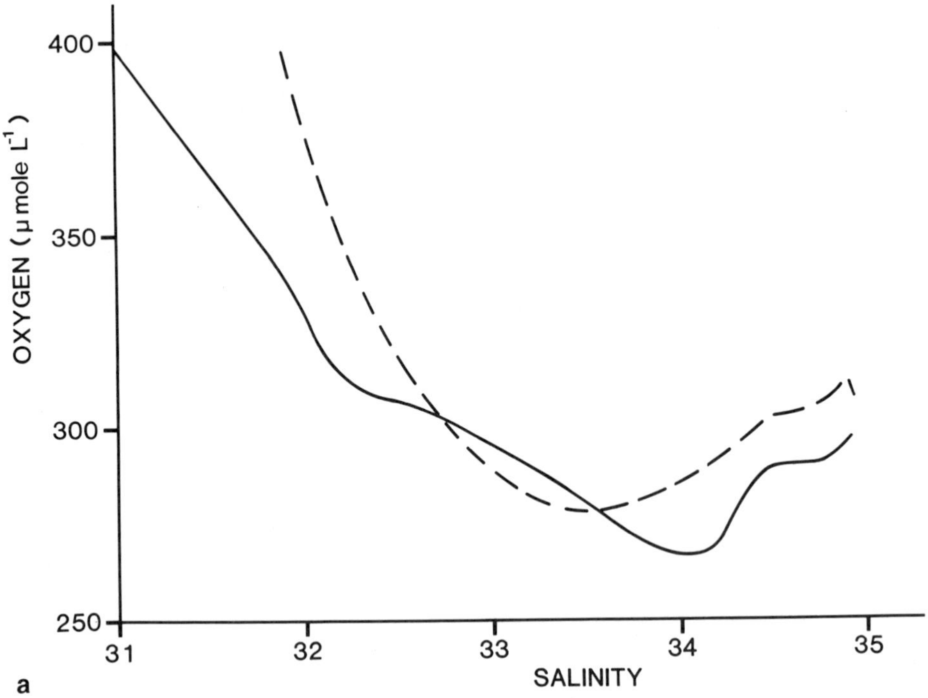


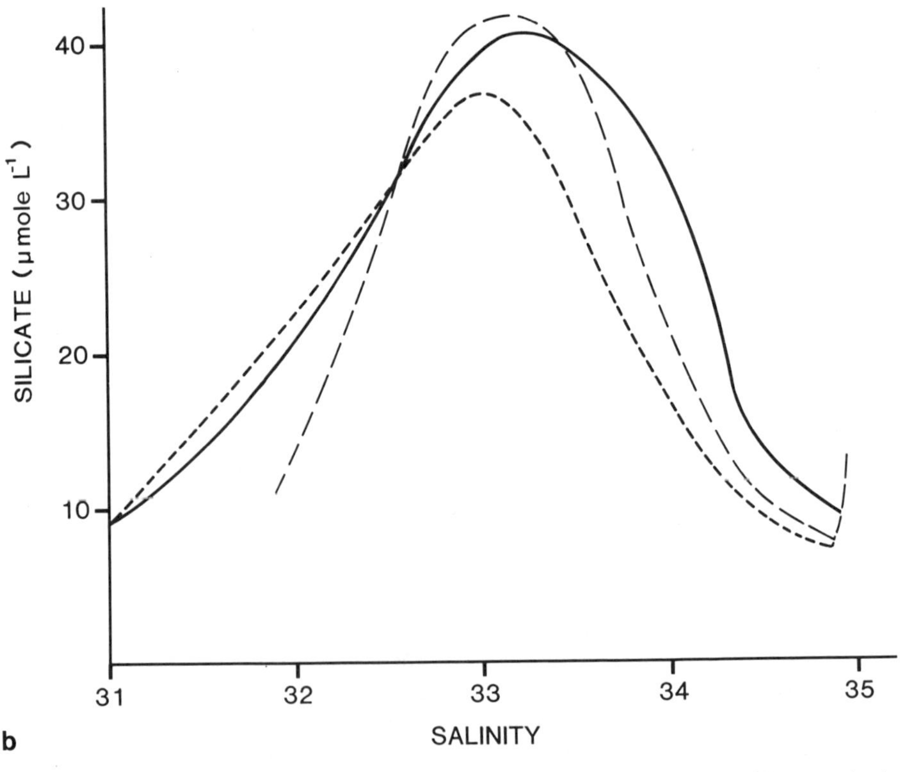


**Figure 3-9.** Oxygen versus salinity (*a*) and silicate versus salinity (*b*) for the central Arctic Ocean. T3 —— line, LOREX --- line, CESAR ——— line. (Data from the experiments T3, LOREX (only silicate), and CESAR.)

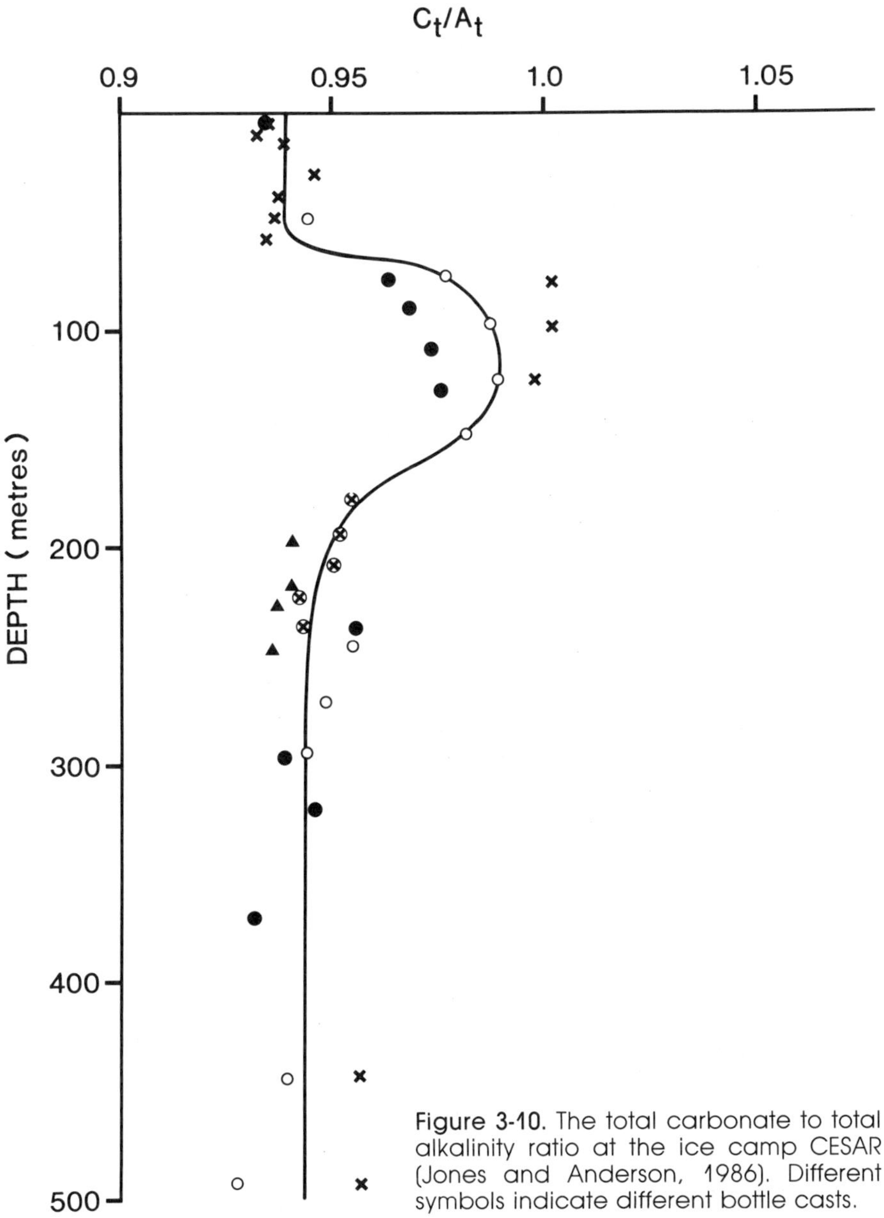


Figure 3-10. The total carbonate to total alkalinity ratio at the ice camp CESAR (Jones and Anderson, 1986). Different symbols indicate different bottle casts.

likely because of the stratified top 200 m of the Arctic Ocean.

## THE CENTRAL BASINS

Our present knowledge of the chemical oceanography of the Arctic Ocean deep basin is founded on few investigations, mostly based on ice camp experiments. Except for investigations close to Fram Strait, all show a pronounced maximum in nutrient concentrations and minimum in oxygen concentration at a depth of 100–200 m (Fig. 3-11) (Nikiforov, Belysheva, and Blinov, 1966; Melnikov and Pavlov, 1978; Kinney, Arhelger, and Burrell, 1970; Moore, Lowings, and Tan, 1983; Jones and Anderson, 1986). The maxima coincide with a bend in the temperature–salinity diagram,

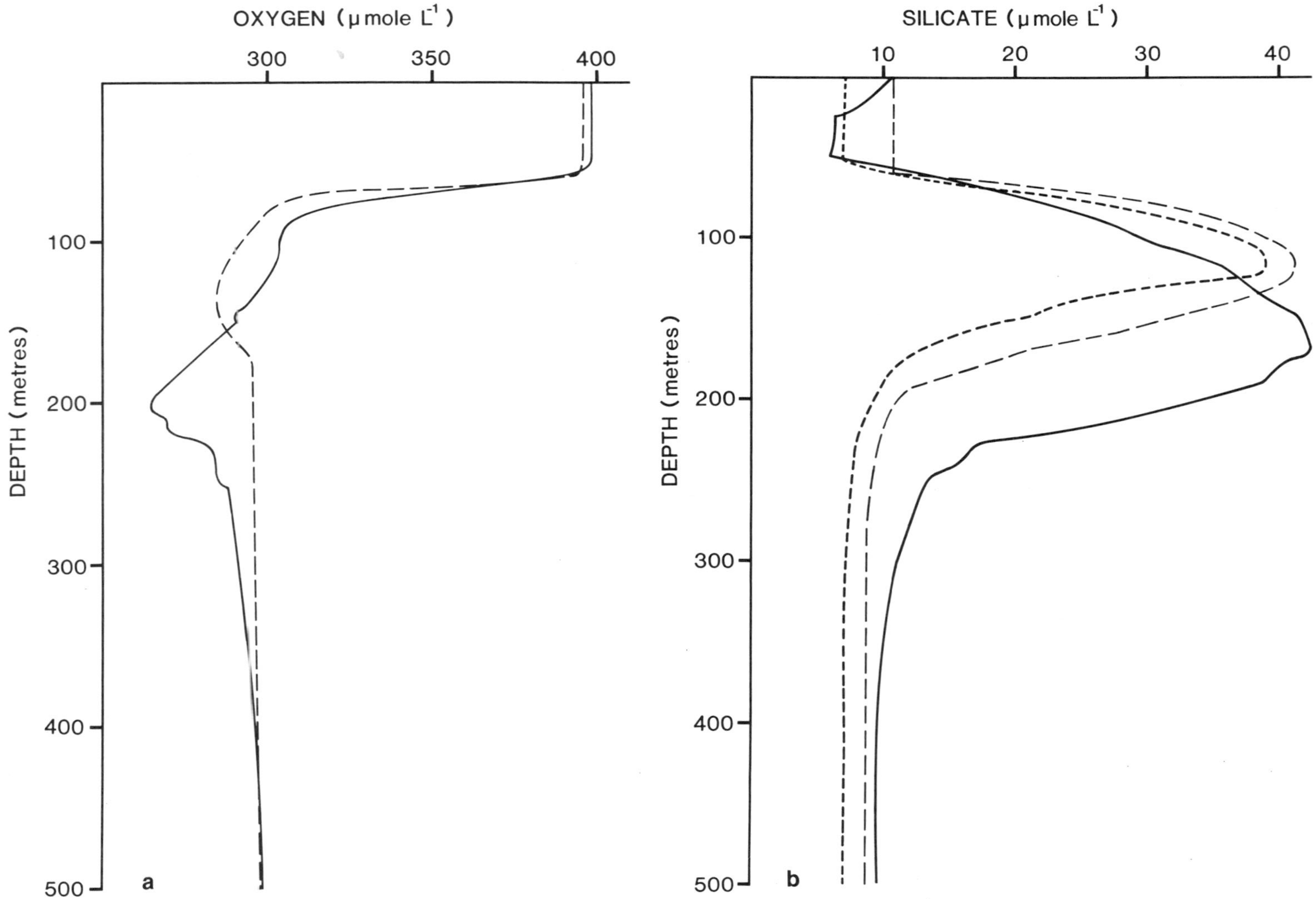


**Figure 3-11.** Depth profiles for oxygen (*a*) and silicate (*b*) in the top 500 m of the central Arctic Ocean. T3 —— line, LOREX --- line, CESAR - - - - line. (Data from the experiments T3, LOREX (only silicate), and CESAR.)

which indicates an input of a water mass having the corresponding nutrient and oxygen concentrations. The origin of this water mass is described previously and points to the importance of sediment–water interactions for the elevated concentrations.

The biochemical in situ modification taking place in the water column over the deep basin is probably of little importance. There are two reasons for this conclusion. First, the biological production within the central basin is low, reflected in the very low particular organic carbon (POC) values throughout the whole depth column (Kinney, Arhelger, and Burrell, 1970; Gordon and Cranford, 1985). Second, the decomposition rate is slow, reflected both in the relatively high dissolved organic carbon (DOC) values (Kinney, Arhelger, and Burrell, 1970; Gordon and Cranford 1985) and the low respiration rates determined from ETS activity measurements at ice station Fram III in the deep waters north of Svalbard (Garfield, Codispoti, and Packard, 1982).

It is not possible to state whether decomposition within the deep Arctic Ocean sediment is of any importance. The only published depth profiles penetrating deep into any of the basins are those from the T3 ice island (Kinney, Arhelger, and Burrell, 1970). A thorough look at these nutrient and oxygen profiles reveals a slight increase of the nutrient concentration together with a decrease in the oxygen concentration with depth. These changes with depth can be a result of in situ decomposition and/or decomposition at the sediment surface with subsequent advection into the basin. However, taking the estimated age of these waters into consideration (more than a few hundred years [H. G. Östlund, personal communication, 1986]), there is another possible explanation. Some of the high-salinity bottom waters produced on the shelves might have densities that allow them to penetrate below the halocline into the deep waters. That this takes place in the Eurasian Basin has been shown by Swift, Takahashi, and Livingston (1983) and by Anderson and Jones (1985); however, to what extent this also occurs in the Canadian Basin, which is surrounded by very large shallow shelves, is unknown.

## SUMMARY

The Arctic Ocean is unique regarding its ice cover and limited water exchange with surrounding oceans. However, some of the deep waters formed in the Greenland and the Norwegian seas are transported through Fram Strait into the Eurasian Basin, where they undergo chemical modification before being transported back into the Norwegian–Greenland sea system. The source areas for most of the chemical modifications taking place in the Arctic Ocean are the huge shelves where sea-ice and brine formations take place and where sediment–seawater interactions play an important role. The large biological production occurring over the shelves supplies the sediments with fresh material and is thus of great significance in the transport of chemical constituents between the different water masses.

## REFERENCES

Aagaard, K. 1981. On the deep circulation in the Arctic Ocean. *Deep-Sea Res.* **28:**251–268.

Aagaard, K., L. K. Coachman, and E. Carmack. 1981. On the halocline of the Arctic Ocean. *Deep-Sea Res.* **28:**529–545.

Aagaard, K., and P. Greisman. 1975. Toward new mass and heat budgets for the Arctic Ocean. *J. Geophys. Res.* **80:**3821–3827.

Aagaard, K., J. H. Swift, and E. C. Carmack. 1985. Thermohaline circulation in the Arctic Mediterranean seas. *J. Geophys. Res.* **90:**4833–4846.

Anderson, L., and D. Dyrssen. 1981. Chemical constituents of the Arctic Ocean in the Svalbard Area. *Oceanologica Acta* **4:**305–311.

Anderson, L. G., D. Dyrssen, and E. P. Jones. 1989. An assessment of the transport of atmospheric $CO_2$ into the Arctic Ocean. *J. Geophys. Res.*

Anderson, L. G., and E. P. Jones. 1985. Measurements of total alkalinity, calcium and sulphate in natural sea ice. *J. Geophys. Res.* **90:**9194–9198.

Anderson, L. G., and E. P. Jones. 1986. Water masses and their chemical constituents in the western Nansen Basin of the Arctic Ocean. *Oceanologica Acta* **9:**277–283.

Anderson, L., and M. Lowings. 1982. *Constituent Data for the Chukchi and Northern Bering Sea.* Report on the Chemistry of Seawater, XXX, Department of Analytical and Marine Chemistry, Chalmers University of Technology and University of Göteborg, S-412, 96, Göteborg, Sweden.

Anderson, L. G., D. W. Dyrssen, E. P. Jones, and M. G. Lowings. 1983. Inputs and outputs of salt, fresh water, alkalinity, and silica in the Arctic Ocean. *Deep-Sea Res.* **30:**87–94.

Anderson, L. G., P. O. J. Hall, Å. Iverfeldt, M. M. Rutgers van der Loeff, B. Sundby, and S. F. G. Westerlund. 1986. Benthic respiration measured by total carbonate production. *Limnol. Oceanogr.* **31:**319–329.

Assur, A. 1958. Composition of sea ice and its tensile strength. *Arctic Sea Ice, N.A.S./N.R.C. Publ. 598,* pp. 106–138.

Bédard, P., C. Hillaire-Marcel, and P. Pagé. 1981. $^{18}O$ modeling of freshwater inputs in Baffin Bay and Canadian Arctic coastal waters. *Nature* **293:**287–289.

Belyayeva, A. N., and Ye. A. Romankevich. 1982. Organic matter in a water column of the Barents Sea. *Oceanology* **23:**188–194.

Broecker, W. S., D. W. Spencer, and H. Craig. 1982. *GEOSECS Pacific Edition, vol. 3, Hydrographic Data 1973–1974.* U.S. Government Printing Office, Washington, D.C.

Bruland, K. W. 1983. Trace metals in sea-water. In J. P. Riley and R. Chester (eds.), *Chemical Oceanography,* vol. 8. Academic Press, London, pp. 157–220.

Campbell, N. J. 1964. The origin of cold high-salinity water in Foxe Basin. *J. Fish. Res. Bd. Canada* **21:**45–55.

Carmack, E., and B. Rudels. 1983. Report on the meeting, The Physical and Chemical Oceanography of the Arctic Ocean. *Report on the Chemistry of Seawater, XXXI.* Department of Analytical and Marine Chemistry, Chalmers University of Technology and University of Göteborg, S-412 96 Göteborg, Sweden.

Coachman, L. K., and C. A. Barnes. 1961. The contribution of Bering Sea water to the Arctic Ocean. *Arctic* **14:**147–161.

Codispoti, L. A. 1979. Arctic Ocean processes in relation to the dissolved silicon content of the Altantic. *Marine Sci. Comm.* **5**(6):361–381.

Codispoti, L. A., and F. A. Richards. 1971. Oxygen supersaturations in the Chukchi and East Siberian seas. *Deep-Sea Res.* **18:**341–351.

Danielsson, L-G., and S. Westerlund. 1983. Trace metals in the Arctic Ocean. In C. S. Wong, E. Boyle, K. W. Bruland, J. D. Burton, and E. D. Goldberg (eds.), *Trace Metals in Sea Water.* Plenum Press, New York and London.

Dyrssen, D. 1982. Tracing mixing processes in the Arctic Ocean. *Thalassia Jugoslavica* **18:**45–60.

Dyrssen, D., and E. Fogelqvist. 1981. Bromoform concentrations of the Arctic Ocean in the Svalbard area. *Oceanologica Acta* **4:**313–317.

Elmore, D., L. E. Tubbs, D. Newman, X. Z. Ma, R. Finkel, K. Nishiizumi, J. Beer, H. Owschger, and M. Andree. 1982. $^{36}Cl$ bomb pulse measured in a shallow ice core from Dye 3, Greenland. *Nature* **300:**735–737.

Fogelqvist, E. 1985. Carbon tetrachloride, tetrachloroethylene, 1,1,1-trichloroethane, and bromoform in Arctic seawater. *J. Geophys. Res.* **90**(C5):9181–9193.

Fonselius, S., and G. Östlund. 1959. Natural radiocarbon measurements on surface water from the North Atlantic and Artic Sea. *Tellus* **1:**77–82.

Garfield, N., L. A. Codispoti, and T. T. Packard. 1982. Chemical and biochemical data obtained by Bigelow Laboratory Investigators: FRAM III, 1981. *Technical Report No. 26.* Bigelow Laboratory for Ocean Sciences, McKown Point, West Boothbay Harbor, Maine 04575.

Gordon, D. C., and P. J. Cranford. 1985. Vertical distribution of dissolved and particulate organic matter in the Arctic Ocean at 86°N and comparison with other regions. *Deep-Sea Res.* **32**(10):1221–1232.

Gosink, T. A., J. G. Pearson, and J. J. Kelly. 1976. Gas movement through sea ice. *Nature* **263:**41–42.

Jones, E. P. 1985. A report on the chemical oceanography of the central Arctic Ocean. Ventilation of the Arctic Ocean. National Science Foundation. 29–31 October 1984, at the Rosentell School of Marine and Atmospheric Sciences, University of Miami, Florida.

Jones, E. P., and L. G. Anderson. 1986. On the origin of the chemical properties of the Arctic Ocean halocline. *J. Geophys. Res.* **91:**10759–10767.

Jones E. P., A. R. Coote, and E. M. Levy. 1983. Effect of sea ice meltwater on the alkalinity of seawater. *J. Mar. Res.* **41:**43–52.

Jones, E. P., D. Dyrssen, and A. R. Coote. 1984. Nutrient regeneration in deep Baffin Bay with consequences for measurements of conservative tracer NO and fossil fuel $CO_2$ in the oceans. *Can. J. Fish. Aquat. Sci.* **41:**30–35.

Kelly, J. J. 1970. Carbon dioxide in the surface waters of the north Atlantic Ocean and the Barents and Kara seas. *Limnol. Oceanogr.* **15:**80–87.

Kinney, P., M. E. Arhelger, and D. C. Burrell. 1970. Chemical characteristics of water masses in the Amerasian Basin of the Arctic Ocean. *J. Geophys. Res.* **75:**4097–4104.

Kinney, P. J., T. C. Loder, and J. Groves. 1971. Particulate and dissolved organic matter in the Amerasian Basin of the Arctic Ocean. *Limnol. Oceanogr.* **16:**132–137.

Koike, I., K. Furuya, H. Otobe, T. Nakai, T. Memoto, and A. Hattori. 1982. Horizontal distribution of surface chlorophyll a and nitrogenous nutrients near Bering Strait and Unimak Pass. *Deep-Sea Res.* **29:**149–155.

Livingston, H. D., S. L. Kupferman, V. T. Bowen, and R. M. Moore. 1984. Vertical profile of artifical radionuclide concentrations in the central Arctic Ocean. *Geochim. Cosmochim. Acta* **48:**2195–2203.

Lyakhin, Yu. I. 1970. Saturation of water of the sea of Okhotsk with calcium carbonate. *Oceanology* **10:**789–795.

Lyakhin, Yu. I., and V. P. Rusanov. 1980. Oxygen and carbon dioxide exchange between the Arctic Basin and the atmosphere. *Oceanology* **20:**411–418.

Lyakhin, Yu. I., and V. P. Rusanov. 1983. Exchange of oxygen and $CO_2$ between water and atmosphere in the Arctic Seas. *Oceanology* **23:**722–726.

Mart, L., H. W. Nürnberg, and D. Dyrssen. 1984. Trace metals in the eastern Arctic Ocean. *Sci. Total Environ.* **39:**1–14.

Melling, H. 1985. *Ventilation of the Arctic Ocean.* A report on a workshop funded by the division of Polar Programs, National Science Foundation. 29–31 October 1984, at the Rosentell School of Marine and Atmospheric Sciences, University of Miami, Florida.

Melling, H., and E. L. Lewis. 1982. Shelf drainage flows in the Beaufort Sea and their effect on the Arctic Ocean pychnocline. *Deep-Sea Res.* **29:**967–985.

Melnikov, I. A., and G. L. Pavlov. 1978. Characteristics of organic carbon distribution in the waters and ice of the Arctic Ocean. *Oceanology* **18:**163–167.

Midttun, L. 1985. Formation of heavy bottom water in the Barent Sea. *Deep-Sea Res.* **32**(10):1233–1241.

Moore, R. M., M. G. Lowings, and F. C. Tan. 1983. Geochemical profiles in the central Arctic Ocean: Their relation to freezing and shallow circulation. *J. Geophys. Res.* **88:** 2667–2674.

Nansen, F. 1906. Northern waters: Captain Roald Amundsen's oceanographic observations in the Arctic Seas in 1901. *Vitensk.-Selsk. Skr., I, Math.-Naturv. Kl. No. 3,* 145p.

Nelson, K. H., and T. G. Thompson. 1954. Deposition of salts from seawater by frigid concentration. *J. Mar. Res.* **13:**166–182.

Nikiforov, Ye. G., Ye. Y. Belysheva, and N. I. Blinov. 1966. The structure of water masses in the eastern part of the Arctic Basin. *Oceanology* **6:**59–64.

Östlund, H. G. 1982. The residence time of the freshwater component in the Arctic Ocean. *J. Geophys. Res.* **87:**2035–2043.

Östlund, H. G. 1985. *Rit Fiskideildar* **9:**13–19.

Östlund, H. G., and G. Hut. 1984. Arctic Ocean water mass balance from isotope data. *J. Geophys. Res.* **89:**6373–6381.

Overgaard, S., P. Wadhams, and M. Leppäranta. 1983. Ice properties in the Greenland and Barents seas during summer. *J. Glaciol.* **27**(101):142–164.

Redfield, A. C., and I. Friedman. 1969. The effect of meteoric water, meltwater and brine on the composition of polar seawater and of deep waters of the ocean. *Deep-Sea Res.* **16**(suppl.):197–213.

Reeburgh, W. S., and M. Springer-Young. 1983. New measurements of sulphate and chlorinity in natural sea ice. *J. Geophys. Res.* **88**(C5):2959–2966.

Richardson, C. 1976. Phase relationships in sea ice as a function of temperature. *J. Glaciol.* **17:**507–519.

Ringer, W. E. 1906. De veranderinger im samenstelling van zeewater bij het bevriezen. *Chem. Weekbl.* **3:**223–249.

Rusanov, V. P. 1974. Silicon distribution in the surface waters of the Arctic Basin in winter. *Oceanology* **14:**669–674.

Rusanov, V. P., and A. D. Shpaykher. 1979. Advection of dissolved silica in the Chukchi Sea. *Oceanology* **19:**418–421.

Sambrotto, R. N., J. J. Goering, and C. P. McRoy. 1984. Large yearly production of phytoplankton in the western Bering Strait. *Science* **225:**1147–1150.

Sinha, N. K. 1977. Technique for studying structure of sea ice. *J. Glaciol.* **18:**315–323.

Stefansson, U. 1985. Chemical tracers for studying water masses and physical processes in the sea. *Rit Fiskideildar* **9:**1–186.

Subba Rao, D. V., and T. Platt. 1984. Primary

production of Arctic waters. *Polar Biol.* **3:**191–201.

Swift, J. H., T. Takahashi, and H. D. Livingstone. 1983. The contribution of the Greenland and Barents seas to the deep water of the Arctic Ocean. *J. Geophys. Res.* **88:**5981–5986.

Takahaski, T., W. S. Broecker, and S. Langer. 1985. Redfield ratio based on chemical data from isopycnal surfaces. *J. Geophys. Res.* **90**(C4):6907–6924.

Tan, F. C., and P. M. Strain. 1980. The distribution of sea ice meltwater in the eastern Canadian Arctic. *J. Geophys. Res.* **85:**1925–1932.

Tan, F. C., D. Dyrssen, and P. M. Strain. 1983. Sea ice meltwater and excess alkalinity in the east Greenland current. *Oceanologica Acta* **6:**283–288.

Thompson, T. G., and K. H. Nelson. 1956. Concentration of brines and deposition of salts from water under frigid conditions. *Am. J. Sci.* **254:**227–238.

Top, Z., W. B. Clarke, and R. M. Moore. 1983. Anomalous neon–helium ratios in the Arctic Ocean. *Geophys. Res. Lett.* **10:**1168–1171.

Tsurikov, V. L. 1974. Statistics of salt composition in sea ice. *Oceanology* **14**(3):360–367.

Vetshteyn, V. Ye., G. A. Malyuk, and V. P. Rusanov. 1974. Oxygen-18 distribution in the central Arctic Basin. *Oceanology* **14:**514–519.

Wallace, D. W. R., and R. M. Moore. 1985. Vertical profiles of $CCl_3F$ (F-11) and $CCl_2F_2$ (F-12) in the central Arctic Ocean. *J. Geophys. Res.* **90:**1155–1166.

Walsh, J. J. 1983. Death in the sea: Enigmatic phytoplankton losses. *Prog. Oceanogr.* **12:**1–86.

Walsh, J. J., G. T. Rowe, R. L. Iverson, and C. P. McRoy. 1981. Biological export of shelf carbon is a neglected sink of the global $CO_2$ cycle. *Nature* **291:**196–201.

# 4: Polar Marine Ecosystem Evolution

M. J. Dunbar

From the mid-1960s to the late 1980s, several papers have analyzed polar marine ecosystems (Holdgate, 1967; Hedgpeth, 1977; Nemoto and Harrison, 1981; Dunbar, 1982, 1985), and there have been studies and speculations concerning the evolution of these systems (Dunbar, 1968, 1972, 1977). In the 1980s, new ideas have appeared about the basic mechanisms of evolution beyond the classical neo-Darwinian system of selection at the individual or specific level (Johnson, 1981; Ott, 1981; Wiley and Brooks, 1982). It is not intended here to go over this ground once more but rather to explore other ideas as they affect our understanding of the evolution of polar marine ecosystems, including (1) a recent paper by Jablonski (1986) and (2) ideas on the basic problem of how biological evolution began, and on what terms and conditions, and therefore of how it continued. High-latitude ecosystem evolution is only one aspect of the global pattern, but it happens to be within the focus of this book.

The abiding importance of the work of Darwin and Wallace is not in question; nevertheless the hold that they have had on biologists is quite extraordinary. Just before his famous final paragraph (''tangled bank''), which has an ecosystem flavor to it, Darwin (1859) wrote: ''Natural selection works solely by and for the good of each being,'' but he did not claim that his natural selection was the only process involved in evolution. The insistence by so many students of evolution today in answering those suspected of the heresy of an ecosystem approach, that no matter or phenomenon has yet been raised that cannot be explained by strict individual selection, is beside the point; that narrow view should have been abandoned long ago—is it Occamism that causes this dogmatism? If so, is individual selection really the simplest possible theory?

The total emphasis on the individual, or on the individual species, since the beginning of evolutionary thought has to be explained. The answer must be that the sense of ecosystem did not appear until late in the history of biology. The view put forward here is that evolution of the whole system, as a whole, is to be expected, and that the emphasis of theory should be on the system, not on the species. If the Greek biologists had been aware of the ecosystem, or had been "ecosystem-minded," evolutionary theory would have developed very differently. The concept of evolution at the ecosystem level removes difficulties in understanding the evolution of highly complex social (system) organizations (ant slaves; insect societies, in general; parasitism; commensalism; symbiosis). Woodbury (1954) made an advance in grouping all these phenomena under the heading of "consortism," a grouping that was more than just word juggling, for it was a step in getting rid of school-text insistence on classifying these patterns by fine distinctions. The names of Bates, Emerson, Ott, Wilson, and Darnell are all associated with this more general approach. Marx (1985), reviewing the Sixth International Symposium on Nitrogen Fixation, emphasizes the intricate interplay between the plant and the nodulating rhizobial bacterium required to produce the nitrogen-fixing nodules.

*Ecosystem* has been variously defined, but the sense of all of the definitions can be summed up in the statement by Dickie (1971): "A system is a set of units combined by nature or art to form an integral, organic and organized whole; an orderly working totality." The behavior of ecosystems has impressed several authors within this century by their ability to evolve in a cohesive or unified manner; it is useful at least once per generation to remind ourselves of the classical statement by Lotka (1925) on the subject of this ecological organization:

> The picture we must keep before us, then, is that of a great world engine or energy transformer composed of a multitude of subsidiary units, each separately, and all together as a whole, working in a cycle. It seems, in a way, a singularly futile engine, which, with a seriousness strangely out of keeping with the absurdity of the performance, carefully and thoroughly churns up all the energy gathered from the source. It spends all its work feeding itself and keeping itself in repair, so that no balance is left over for any imaginable residual purpose. Still, it accomplishes one very remarkable thing; it *improves* itself as it goes along, if we may employ this term to describe those progressive changes in its composition and construction which constitute the evolution of the system.

McGowan (1974) introduced a new way of looking at oceanic ecosystems that has general application in nature. His Pacific systems are described as semiclosed and ancient, as "highly tuned to climate but not to weather," and as having developed over millennia in association with the circulation pattern of the Pacific, which has maintained its present form since at least the Miocene. The implication is that large environmental changes set the scene for ecosystem changes.

Woodger's (1929) *Biological Principles* contain thoughts that are perhaps seminal to the most recent ecological approaches to evolution. The first sentence of the first chapter is "Biology is a science of antitheses"; the three fundamental antitheses being (1) structure and function, (2) organism and environment, and (3) preformation and epigenesis. On the second antithesis he wrote: "An organism without an environment is never an object of experience and to a biologist is unthinkable in the sense of unmeaning." It follows that when we consider organisms as evolving entities, we are in fact considering "organism plus environment," and that environment itself consists of other "organisms plus environments." Therefore, the smallest unit of natural selection must be the organism plus its environment, which evolves within a system of similar entities. From that point it is not far to reach the concept of the ecosystem as an evolving unit in its own right.

Ecosystems are remarkably tough and resilient, and adjust to natural disasters and to all but the most devastating perturbations put upon them by humankind.

This quality of resilience was well described by Hutchinson (1948) in the earlier literature on ecosystems:

> It is well known from mathematical theory, . . . that circular paths often exist which tend to be self-correcting within certain limits, but which break down, producing violent oscillations, when some variable in the system transgresses limiting values. When a breakdown of the self-correcting system takes place in nature, it may be expected to end in disaster for some element in the system which consequently disappears. The original system is thus destroyed, to be replaced by another in which the lost element plays no part. It is, therefore, usual to find in natural circular systems various mechanisms acting to damp oscillations, and self-correcting mechanisms may be introduced at several points in the circular path . . . The resulting stability, which appears to characterize ecological systems, while it permits their persistence, makes investigation difficult.

Paleoecology is likely to prove as fruitful to ecosystem evolution study as paleontology proved to classical evolution. Consider the ''primal soup'' and the origin of life on this planet. If we stick to Darwinian thinking alone, it is not easy to conceive why the first organisms diversified at all. What is the advantage to a humble chemosynthetic or photosynthetic organism to split off other forms that will compete with it? And yet if other ecosystem elements did not develop, the original species must have been doomed to ultimate extinction by the concentration of its own population and the failure of the nutrient supply. Bacteria at least would have been essential and a herbivorous population of some sort would have been an asset.

Darwin had intended to include a discussion of the origin of life in his *Origin of Species*, but abandoned the idea as being too difficult (Gruber and Barrett, 1974). Perhaps it was too difficult because of his obsession with individual selection, which prevented the concept of the primal appearance not of one kind of organism but of several kinds at the same time in ecological equilibrium. Stanley (1973) suggested that the first appearance of heterotrophs made further evolution possible. Quoting from the *New Scientist* (1973) describing Stanley's work:

> All the organisms (in the Precambrian seas) were autotrophic; . . . the number of removable nutrients was small; and hence only a few types predominated. That being so, there was very little possibility of evolutionary development. Eventually, however, forms developed that were heterotrophic. The first ones obviously would have been herbivorous, but carnivorous types, feeding on the first herbivores, would soon have followed.

This is an interesting scenario, but I wonder why *eventually* was inserted in this quotation; why should the heterotrophs have appeared later? Why not at the same time as the autotrophs? In other words, why should not a whole simple ecosystem have appeared from the beginning? It would have been no more remarkable than the appearance of life as such.

Once the evolutionary process began, it appears to have progressed very slowly indeed during very great periods of time, to be accelerated only during periods of significant environmental change. The literature on this point is now voluminous, involving such phrases as ''punctuated equilibrium.'' Lewin (1984), reporting on workshop proceedings on paleoclimates and evolution, concludes that the workshop discovered that ''examination of major events in the climatic and faunal records begins to identify environmental change as an engine of evolution.'' This is indeed the re-invention of the wheel.

Environmental change was recognized by both Wallace and Darwin as an ''engine of evolution,'' and before that, Lamarck based his whole evolutionary theory on environmental change. An example is the case of the mummified animals found in Thebes and in Memphis, which had been examined by Cuvier and by Geoffroy St.-Hilaire, both of whom had pointed out that the animals in question had not changed at all when compared with contemporary specimens; therefore, no evolution. In a lecture published in 1803, Lamarck wrote:

> I have seen them, these animals, and I believe in the conformity of their resemblance with the individuals of the same species which live today. . . . But it would assuredly be very singular if this should be otherwise; for the position of Egypt and its climate are still the same or very nearly the same as at former times. Therefore the animals which live there have not been compelled to change their habits. (quoted from Packard, 1901)

Lamarck went on to point out that the observation proved only that the animals had remained the same for 3,000 years, a very short period of evolutionary time.

As effective as environmental change must be on the changes in direction of evolutionary development at the level of the species, it very probably has a much more direct and rapid effect on the ecosystem, in the form of extinction of elements, simplication of pattern (or the reverse), and change in the pattern of energy flow through the system. This would be especially marked if the change involved the pattern of seasonality. Seasonality (the normal variation in different regions between summer and winter conditions) has been identified as the single most important environmental variable determining the nature of the ecosystem in different climatic regions (McLaren, 1964; Dunbar, 1968).

Present polar climatic conditions and other conditions are the result of a general history of cooling beginning in the Miocene. There is no evidence that such polar wandering as may have taken place during the Cenozoic was sufficient to seasonally change the distribution of light in the present high latitudes; but nevertheless light conditions at the surface of the earth, and especially below the sea surface, must have changed very considerably, owing to increase in cloud cover as the result of cooling climate, and with the development of sea ice. Environmental conditions in the seas in polar and subpolar regions, therefore, changed in the direction of lower water temperatures and decrease in total light energy available for marine life. It is also probable that changes occurred in the pattern of upwelling, and therefore in the nutrient supply to the euphotic zone; but it is not useful or even reasonable to speculate on this point. We must take the present distribution of water column stability or instability as a given, although we may suspect that overall the total primary productivity of the mid- and high latitudes in the sea has decreased somewhat, and certainly in the Arctic ocean itself.

From a biological standpoint, these environmental changes (temperature, light, productivity) are interpreted in this model to have the following general effects:

1. Shorter seasonal periods of phytoplankton and other algal development.
2. Lower growth and metabolic rates, at least in ectotherms. (Since this paper is a discussion of ecosystem evolution, not individual evolution, I shall not address questions of individual physiological problems.)
3. Longer life spans, especially in zooplankton, to adjust both to the lower temperature and to the more seasonal primary production (see Dunbar [1968] and later papers), leading to the development of life cycles of one, two, or more years in the zooplankton and the appearance of overlapping broods.
4. Necessity for fewer species to ensure survival of specific populations to the next phytoplankton bloom; therefore extinction of other species and development of lower diversity.
5. Probable changes in the plankton/benthos biomass ratio to the advantage of the benthos.

Item 5 above would be expected to follow from the Thorson effect (reduction or disappearance of planktonic larvae from the life cycles of benthonic species), itself the result of the shorter phytoplankton bloom. A change in the same direction would follow also from the slower turnover rates in the plankton as a result of longer life cycles and slowing down of growth rates. This is the effect described by Petersen and Curtis (1980) as follows:

> Primary planktonic production increases from the subarctic to the tropics, but the transfer of this energy to higher trophic levels appears most efficient in the high latitude ecosystem. Although the zooplankton comprises a dominant link within the food web, energy flow in the subarctic may be more equally partitioned between the benthos and the zooplankton. Consequently, pelagic organisms form the basis for a fishery in the tropical study area while demersal forms are of greater importance in the subarctic.

This effect can also be referred to as seasonality, since it is the high and constant turnover of energy at the primary level in the tropics that produces immediate food for pelagic fishes year-round. In higher latitudes, the seasonal accumulation of zooplankton biomass spills over, and in the sedimentation and sinking process it benefits the benthos and the demersal fish.

There are other results of the high seasonality. The necessity to produce a large enough specific population in the zooplankton to ensure survival to the next period of high primary production has encouraged the development of very large numbers of eggs per female, which in turn has selected in favor of large body size, since there is a nonlinear relation between body size and egg number; hence, the well-known large size of high-latitude zooplankton. Finally, high seasonality overrides other effects on metabolism and growth rates, controlling them at the ultimate level (evolutionary exigencies). This is implied in item 3 discussed previously.

Longhurst (1985) has reviewed the structure and evolution of plankton communities on a global scale, and has produced results relevant to the present subject, but that will probably take some time to digest and interpret. He finds that copepods form only 50 percent of the biomass of all plankton, but they represent 70 percent of biomass in polar seas, compared with 35 percent in tropical seas; also there are fewer predators among the polar plankton. The dominance of copepods in cold water may be correlated with the abundance of diatoms rather than other phytoplankters. Diatoms are larger in body size, especially in polar regions, than in low latitudes, and there may be some relation to the development of diatom populations within the sea ice. The smaller numbers of predators may well be associated with the smaller number of species of all ecological levels on the principle brought forward by Paine (1966, 1980) concerning the relation between food web complexity and species diversity.

The vertebrate fauna, riding on top of the ecological substructure, had to adapt to the changing situation as best as they could during the Miocene to Holocene process. The endotherms (mammals and birds) had an obvious advantage over the ectotherms (fishes) in adapting to cooling conditions, and this may perhaps (but only perhaps) be enough to explain the dominance of the endotherms among the larger fauna. Fish species are remarkably few in number in the Arctic water (water formed in the upper 200 m of the Arctic ocean), and are not abundant in the mixed, or sub-Arctic, water in the peripheral seas (Dunbar, 1968).

In a stimulating discussion on the ecological role of sea mammals, Ray (1981) makes the point that "such large animals as marine mammals exert important effects on their ecosystems." This is clearly so; the great whales of the Southern Ocean have in the past maintained a level of food supply in such a way as to control the population levels of their competitors—seals, birds, and fishes—and their serious depletion during the present century has allowed the accumulation of a surplus of food (especially the euphausiids) and a consequent, or concomitant, increase in the abundance of the competitors. Nevertheless, in the present context, that of ecosystem evolution, the top predators must be recognized as "cream skimmers," and therefore theoretically expendable. Ecosystems adjust very easily to the loss of a few top predators. Many of them (whales and birds) are seasonal migrants, using the northern food supplies in the summer only.

Jablonski (1986) has brought something new into the study of evolutionary history, ably reviewed by Lewin (1986), which is directly relevant to the present matter. During the periods of mass extinctions, Jablonski finds the rules of selection are not the same as those that hold for the long background periods when the environment remains stable and the pace of evolution slackens. During background periods normal Darwinian selection is the rule, so that selection is in fact the conservative process that Darwin imagined it to be for most of the time. But during the more turbulent periods of large environmental change "perfectly good adaptations might be eliminated . . . for reasons quite unrelated to their utility" (Lewin, 1986). The variables involved (for marine bivalves and gastropods, the material of Jablonski's study) are (1) presence or absence of planktotrophic larval development, (2) species diversity within a higher taxon, and (3) geographic range. To quote from Jablonski's summary:

> During background times, traits such as planktotrophic larval development, broad geographic range of constituent species, and high species richness enhanced survivorship of species and genera. In contrast, during the end-Cretaceous and other mass extinctions these factors were ineffectual, but broad geographic deployment of an entire lineage, regardless of the ranges of its constituent species, enhanced survivorship.

The same applies to Jablonski's summary as to the Longhurst (1985) paper quoted above, namely that it will take time to estimate the summary's importance with respect to the present polar marine systems, but it can at least be claimed that the Tertiary and Quaternary cooling was a large climatic change and also a period of considerable widespread extinction. Jablonski points out that during the Pliocene molluscan extinction (Stanley, 1982) "species survivorship was unrelated to traits that enhanced background survivorship," and that as in the end-Cretaceous extinction, endemic taxa (with small geographic range) became disproportionately extinct relative to widespread taxa regardless of species richness. High-latitude faunas are remarkable for the abundance of circumpolar genera, which has in the past been put down simply to similarity of environmental conditions and to the lack of barriers to dispersal. The polar fauna is also marked by the apparent absence of rare species, which may also very well be a result of the Jablonski effect. For other taxonomic peculiarities shown by polar faunas, see Dunbar (1977).

Presumably the most recent development in polar marine ecosystems is that of the sea-ice ecosystem, based on algal growth within the ice itself, which can reach remarkable concentrations. The literature on this subject is now voluminous, recently summarized by Horner (1985). This special subsystem is an example of the invasion and exploitation of a new habitat that came into existence with the beginning of sea-ice formation in Pleistocene or pre-Pleistocene time. The growth of ice volume during the process of climatic cooling and its relation to seasonality have recently been considered by Crowley et al. (1986).

This has been, and is, an outspokenly

speculative essay, as an essay on evolution at the ecosystem level has to be, for the moment. There is no testable hypothesis, no mathematical model, nor are there statistical data (see Parsons, 1985; Dunbar, 1980). Ecosystem evolution, as defined here, may be considered either as an ''unconvincing possibility'' or as a ''likely impossibility.'' Aristotle, in his *Poetics*, considered the latter to be always preferable to the former.

## REFERENCES


Crowley, T. J., D. A. Short, J. G. Mengel, and G. R. North. 1986. Role of seasonality in the evolution of climate during the last 100 million years. *Science* **231**(4738):579–584.

Darwin, C. R. 1859. *The Origin of Species by Means of Natural Selection, or the Preservation of Favoured Races in the Struggle for Life.* Murray, London, 490p.

Dickie, L. M. 1971. *The Gulf as a Biological Production System.* Second Gulf of St. Lawrence Workshop, Bedford Institute, Dec. 3, 1970, pp. 88–110. E. M. Hassan, Coordinator. Dartmouth, Nova Scotia.

Dunbar, M. J. 1968. *Ecological Development in Polar Regions.* Prentice-Hall, Englewood Cliffs, N.J., 119p.

Dunbar, M. J. 1972. The ecosystem as a unit of natural selection, *Conn. Acad. Arts and Sci. Trans.* **44:**113–130.

Dunbar, M. J. 1977. The evolution of polar ecosystems. In G. A. Llano (ed.), *Adaptations within Antarctic Ecosystems.* Smithsonian Institution, Washington, D.C., pp. 1063–1076.

Dunbar, M. J. 1980. The blunting of Occam's razor, or to hell with parsimony. *Canadian J. Zool.* **58**(2):123–128.

Dunbar, M. J. 1982. Arctic marine ecosystems. In L. Rey (ed.), *The Arctic Ocean.* Macmillan Press, London, pp. 233–262.

Dunbar, M. J. 1985. The Arctic marine ecosystem. In F. R. Engelhardt (ed.), *Petroleum Effects in the Arctic Environment.* Elsevier Applied Science Publishers, London and New York, pp. 1–35.

Gruber, H. E., and P. H. Barrett. 1974. *Darwin on Man.* Woldwood House, London, 495 p.

Hedgpeth, J. W. 1977. The Antarctic marine ecosystem. In G. A. Llano (ed.), *Adaptations within Antarctic Ecosystems.* Third SCAR Symposium on Antarctic Biology, Washington, 1974. Smithsonian Institution, pp. 2–10.

Holdgate, M. W. 1967. The Antarctic ecosystem. *Roy. Soc. London Philos. Trans,* ser. B., **252:**363–383.

Horner, R. A. 1985. *Sea Ice Biota.* CRC Press Inc., Boca Raton, Fla., 240p.

Hutchinson, G. E. 1948. Circular causal systems in ecology. *N.Y. Acad. Sci. Ann.* **50**(4): 221–246.

Jablonski, D. 1986. Background and mass extinctions: The alternation of macroevolutionary regimes. *Science* **231**(4734):129–133.

Johnson, L. 1981. The thermodynamic origin of ecosystems. *Can. J. Fish. and Aquat. Sci.* **38**(5):571–590.

Lewin, R. 1984. The paleoclimatic numbers game. *Science* **226**(4671):154–156.

Lewin, R. 1986. Mass extinctions select different victims. *Science* **231:**219–220.

Longhurst, A. R. 1985. The structure and evolution of plankton communities. *Progr. Oceanogr.* **15**(1):1–35.

Lotka, A. J. 1925. *Elements of Physical Biology.* Williams and Wilkins, Baltimore, Md. (Reprinted by Dover, New York, 1956.)

McGowan, J. A. 1974. The nature of oceanic ecosystems. In C. B. Miller (ed.), *Biology of the Oceanic Pacific.* Oregon State University Press, Corvallis, Ore., pp. 9–28.

McLaren, I. A. 1964. Marine life in Arctic waters. In I. N. Smith (ed.), *The Unbelievable Land.* Queen's Printer, Ottawa, pp. 93–97.

Marx, J. L. 1985. How Rhizobia and Legumes get it together. *Science* **230:**157–158.

Nemoto, T., and G. Harrison. 1981. High latitude ecosystems. In A. R. Longhurst (ed.), *Analysis of Marine Ecosystems.* Academic Press, London and New York, pp. 95–127.

New Scientist, 1973. Did the first predators make ancient life burgeon? *New Scientist,* 14 June 1973, p. 664.

Ott, J. A., 1981. Adaptive strategies at the ecosystem level: Examples from two benthic marine systems. *Marine Ecol.* **2**(2):113–158.

Packard, A. S. 1901. *Lamarck, the Founder of Evolution: His Life and Work.* Longmans, Green & Co., London and New York.

Paine, R. T. 1966. Food web complexity and species diversity. *Am. Nat.* **100:**65–75.

Paine, R. T. 1980. Food webs: Linkage, interaction, strength and community infrastructure. *J. Anim. Ecol.* **49:**667–685.

Parsons, T. R. 1985. Hypothesis testing and rigorous statistics as criteria for marine research proposals. *La Mer* **23:**109–110.

Petersen, G. H., and M. A. Curtis. 1980. Differences in energy flow through major components of subarctic, temperate and tropical marine shelf ecosystems. *Dana* **1:**53–64.

Ray, G. Carleton. 1981. The role of large organisms. In A. R. Longhurst (ed.), *Analysis of Marine Ecosystems.* Academic Press, London and New York, pp. 397–416.

Stanley, S. M. 1973. An ecological theory for the sudden origin of multicellular life in the late Precambrian. *Nat. Acad. Sci. (USA) Proc.* **72:**646–650.

Stanley, S. M. 1982. Species selection involving alternative character states: An approach to macroevolutionary analysis. *Third N. Am. Paleontol. Conv. Proc.* **2:**505–510.

Valentine, J. W. 1972. Conceptual models of ecosystem evolution. In T. J. M. Schopf (ed.), *Models in Paleobiology.* Freeman, Cooper and Co., San Francisco, pp. 192–215.

Wiley, E. O., and D. R. Brooks. 1982. Victims of history—a nonequilibrium approach to evolution. *Syst. Zool.* **31**(1):1–24.

Woodbury, A. M., 1954. *Principles of General Ecology.* The Blakiston Company Inc., New York and Toronto, 503p.

Woodger, J. H., 1929. *Biological Principles.* Kegan Paul, Trench, Trubner and Co., London, 498p.

# 5: Arctic Sea-Ice Biota

Rita A. Horner

Ice is an important physical feature of the environment in polar regions and strongly affects the plants and animals living in these areas. Ice reduces the amount of light reaching the water column, and also reduces heat and gas exchange. A salinity minimum directly beneath the ice and reduced mixing of the water combine to enhance the vertical stability of the water column. At the same time, the upper surface of the ice provides a place for marine mammals and birds to rest, feed, and bear their young, while the lower surface provides a substrate for microalgae, meiofauna, epifauna, and some fish. The ecological importance of these complex communities associated with sea ice is gradually becoming better known. In this chapter, I will review the information currently available on those organisms associated with the bottom few centimeters of ice at the ice–seawater interface, in particular, the microalgae, meiofauna, and epifauna.

## TERMINOLOGY

The microalgae that live in close association with the sea ice, either attached to ice crystals or in the interstitial water between ice crystals, are called ice algae. In the Arctic, algal cells are found scattered throughout the ice thickness in winter and concentrated in the bottom few centimeters of ice in spring, although cells may also be found elsewhere in the ice. This bottom ice community is often called the epontic community (Bunt and Wood, 1963), but epontic (*epi* = on, to, upon; *ontos* = being) is perhaps not a very good term because it has also been used to describe cells attached to plants, animals, rocks, and debris (Crosby and Wood, 1959; Wood, 1964).

A number of other terms have been used to describe the ice algal community, including plankton ice (Matsuda, 1961), psychrophilic (Bunt, Owens, and Hoch, 1966), cryophilic (Usachev, 1949), cryophyton (Round, 1981), cryobiont (Kol, 1942; Bursa, 1963), and sympagic (Whitaker, 1977). Some of these names imply association with ice, while others imply only an affinity for low temperature. Others suggest that the organisms are planktonic in origin, which may not always be true. Recently, Ackley, Buck, and Taguchi (1979) introduced the terms surface, interior, and bottom for ice algal communities depending on their location in the ice. These terms indicate not only the origin of the cells in the ice but also where in the ice the cells are found. That terminology is used here, but see Horner et al. (1988).

In the Arctic, only the bottom ice community has been studied and, although there are reports of interior (J. Burns, personal communication, 1970) and surface (Booth, 1984) ice communities, they are not thought to be widespread. In the Antarctic, however, interior and surface communities are found in a number of localities (Ackley, Buck, and Taguchi, 1979; Hoshiai, 1977; Meguro, 1962; McConville and Wetherbee, 1983); thus the necessity for terminology to separate these communities from the bottom ice communities.

## HISTORY

The history of ice algal investigations has been extensively reviewed (Horner, 1976, 1977, 1985). In the Arctic, microalgae found in sea ice were first mentioned by Ehrenberg (1841, 1853). The early studies on ice algae were primarily during voyages of exploration and are mainly lists of species.

Sutherland (1852) described a ''greenish slimy-looking substance'' located in small cavities on the bottom of the ice and reported ''minute vegetable forms.'' Because the same species were found from a number of places, including the Kara Sea (Cleve and Grunow, 1880); Cape Wankarema, Siberia; Baffin Bay and Davis Strait; Franz Josef Land; between Greenland and Spitsbergen; and north of Jan Mayen Island (Cleve, 1883, 1896, 1898, 1899, 1900), Cleve (1896) suggested that they originated in the Bering Strait region and then drifted across the Arctic to northern Greenland. Here some cells continued to drift along the east coast of Greenland, while others drifted in the Labrador Current. Nansen (1897) used this information to help formulate his theory of Arctic Ocean currents that led to the First Norwegian North Polar Expedition, which, in turn, provided the most extensive early description of the ice algal community (Gran, 1904; Nansen, 1897, 1906). Additional lists of species were provided by Meister (1930) and Usachev (1938). Later, Usachev (1949) reviewed the Soviet work on ice algae.

Experimental studies on ice algae began during the International Geophysical Year (IGY) 1957–1958 (Apollonio, 1961, 1965), but it was not until the early 1970s that adequate field techniques were devised to allow for the in situ measurement of primary productivity (Clasby, Horner, and Alexander, 1973; Alexander, Horner, and Clasby, 1974). Even in the mid-1980s, there are still no techniques available to adequately measure the in situ nutrient and salinity concentrations in the brine channels where the microalgae live and the technique for the in situ measurement of primary production also has its limitations (Schrader, Horner, and Smith, 1982).

Apollonio (1961, 1965) first collected ice core samples and measured chlorophyll *a* concentrations at Devon Island in the Canadian Arctic. An important discovery during this study was finding a relationship between snow cover, light, and ice algal biomass as measured by chlorophyll *a*. Light is rapidly attenuated by snow (Thomas, 1963), so that under even a relatively shallow snow cover, less than 1 percent of the surface irradiance may reach the bottom of the ice where the algae live (Apollonio, 1965; Horner and Schrader, 1982). With no snow cover, more than 75 percent of the surface irradiance may reach the bottom of the ice (Apollonio, 1965). This means that ice algae are shade adapted to live at these low light intensities.

Apollonio (1965) also found very small amphipods on the bottom of one ice core where they were probably feeding on the ice algae. This concentrated source of food offered optimum grazing conditions for the herbivores and the ice algae were clearly a more significant source of food than phytoplankton early in the spring (Apollonio, 1965).

Other IGY studies included those of Mohr and Tibbs (1963), who found no evidence of invertebrates obtaining food from the pack ice and no organisms in the ice, and English (1961), who reported films of organic material on the underside of the ice and observed amphipods feeding on the films. These studies were carried out in the permanent pack ice over deep water and away from shore, and suggested to English (1961) that the films were probably not very important in the overall Arctic ecosystem.

Later, and closer to shore, Meguro, Ito, and Fukushima (1966, 1967) found diatoms in Chukchi Sea ice. They thought the diatoms were frozen into the ice in the fall and formed cysts in order to survive the winter. Nutrients in the ice were provided by brine drainage within the ice, by microbial regeneration in the ice, and by penetration of nutrients from seawater under the ice, with brine drainage considered to be the most important.

In the nearshore Chukchi Sea, Horner, Alexander, and their co-workers provided the most extensive study of the ice algal community and also followed the annual cycle of ice algae from freeze-up in the fall to ice breakup in the spring. They found dinoflagellates, green flagellates, chrysophytes, and euglenoids, as well as diatoms (Horner and Alexander, 1972). Using a combination sampler–incubation chamber (Clasby, Horner, and Alexander, 1973), they were able to measure in situ $^{14}C$ uptake and also to collect samples for pigment, salinity, nutrient concentrations, species determinations, and cell concentrations. A similar chamber (Schrader, Horner, and Smith, 1982) was used in the nearshore Beaufort Sea (Horner and Schrader, 1982).

At both sites, Barrow and Narwhal Island, Alaska, cells were present throughout the ice in winter with the cells concentrating in the bottom few centimeters of ice in late March–early April. Primary productivity began when light levels were 2–9 $\mu E\ m^{-2}\ s^{-1}$ and maximum production was in late May when light levels were 10–65 $\mu E\ m^{-2}\ s^{-1}$. Species composition was similar at the two sites (Horner and Schrader, 1982).

Ice algae have also been studied at a number of sites in the Canadian Arctic. Dunbar and Acreman (1980) compared chlorophyll *a*, standing crops of cells, nutrient concentrations, and salinity from several high Arctic sites with the Gulf of St. Lawrence. Chlorophyll *a* concentrations were 1–2 orders of magnitude higher at the high Arctic sites and 96 percent of the diatoms were benthic, pennate species, while in the Gulf of St. Lawrence, only 57 percent of the diatoms were benthic species. Also in the Gulf of St. Lawrence, Demers, Therriault, and Descolos-Gros (1984) found two chlorophyll peaks, one in the interior of the ice and one at the bottom with higher chlorophyll concentrations in the bottom frazil ice. Taxonomic composition of the ice flora was different during the two years of this study, with benthic diatoms comprising about 50 percent of the community one year and pelagic microflagellates dominating the second year (Demers, Therriault, and Descolos-Gros, 1984).

Cells were present in the ice in the fall in the Eskimo Lakes region (Hsiao, 1980) and nutrients accumulated in the ice during winter in Frobisher Bay (Grainger, 1977). Growth of the ice algae was slow during winter, increased rapidly in spring, reached a peak in late spring and early summer, and declined rapidly as the ice melted. More than 200 species, mostly diatoms, were identified, with other organisms sometimes being abundant, including green flagellates that were relatively numerous in the bottom 5 cm from December to June. At Pond Inlet on Baffin Island, ice algae were patchy during the spring (Cross, 1982). Of the 77 algal species identified, 10 species of pennate diatoms accounted for more than 95 percent of the total number of cells present. Large numbers of meiofauna and macrofauna were associated with the ice algae as masses of the algae became detached from the ice (Cross, 1982).

Elsewhere in the Arctic, Andersen (1977) estimated primary production of the ice community to be about 0.15 g C $m^{-2}$ $yr^{-1}$ at Disko Island, West Greenland, and in the ice edge zone of Davis Strait, <0.10 g C $m^{-2}$ $yr^{-1}$ can be calculated from Booth's (1984) ice data.

Recent Soviet work in the Arctic has included determination of the chemical composition of ice. Oradovskiy (1972) has sampled ice, snow from ice floes, newly fallen snow, and seawater near ice and suggested that sea ice may serve as an additional source of nutrients to the euphotic zone in the water column. Mel'nikov and Pavlov (1978) found organic carbon to be higher in ice than in the water column and considered the drifting ice to be an allochthonous source of organic carbon. Maximum suspended organic carbon (25–45 $\mu g$ C $l^{-1}$) was present in surface water during ice melt from June through August. At other times, the suspended organic carbon (~ 5 $\mu g$ C $l^{-1}$) was evenly distributed in the water column. In the central Arctic Basin, Mel'nikov (1979) found the upper ice surface melted from June to August. Meltwater that drained through the ice formed a lens of fresh water at the ice–water interface that was enriched with organic carbon. Three species of green algae, originally found in meltwater at the upper ice surface, were found in this freshwater layer. It was thought that they formed spores and survived in new ice formed on the lower ice surface. Based on measurements of ice melting and refreezing, these species probably reached the upper surface of the ice in 2–3 yr. Surface melt ended in late August and then the bottom surface of the ice melted, but there was less melting of the bottom surface (Mel'nikov, 1979).

## SPECIES COMPOSITION

Over 300 species belonging to about 75 genera of microalgae from 10 algal classes have been reported from Arctic sea ice. These include members of the Bacillariophyceae, Chrysophyceae, Chlorophyceae, Cryptophyceae, Cyanophyceae, Dictyochophyceae, Dinophyceae, Euglenophyceae, Prasinophyceae, and Prymnesiophyceae (Hsiao, 1983; Horner, 1985). A list of some of the most common microalgae found in Arctic sea ice is given in Table 5-1.

Diatoms (Bacillariophyceae), with nearly 300 species or groups of species, are usually the most abundant organisms in sea ice with pennate species being more abundant than centric species. There are exceptions, however, particularly in lower latitudes, such as the Gulf of St. Lawrence and also in some marginal ice zones, where centric diatom species may predominate. Reasons for this are unclear, but may include the shorter period of time that ice is present at lower latitudes and the presence of thinner ice and therefore higher light intensities in and under the ice (Dunbar and Acreman, 1980). The structure of the ice may also be a factor, as in the frazil ice layer in the Antarctic where centric diatoms are abundant.

Cyanophyceae are not common in Arctic marine waters, but may be abundant in fresh water. They have been reported from freshwater ice that probably formed in the Yenisey River (Usachev, 1938) and from sea ice at Frobisher Bay in February and May (Grainger and Hsiao, 1982). Freshwater members of the Chlorophyceae have also been reported from the Yenisey River (Usachev, 1938) and from melt ponds on the upper surface of sea ice near Barrow, Alaska (Bursa, 1963).

Microflagellates from several algal

Table 5-1. Most Common Microalgae Found in Arctic Sea Ice

| *Species* | *References* |
|---|---|
| *Amphiprora paludosa* var. *hyperborea* (Grunow) Cleve (*Entomoneis paludosa* var. *hyperborea* (Grunow) Poulin et Cardinal | Usachev, 1949; Alexander, Horner, and Clasby, 1974; Horner and Schrader, 1982; Hsiao, 1983; Poulin and Cardinal, 1983 |
| *Cylindrotheca closterium* (Ehrenberg) Reimann et Lewin | Alexander, Horner, and Clasby, 1974; Horner and Schrader, 1982; Hsiao, 1983; Pett et al., 1983; Booth, 1984 |
| *Gomphonema exiguum* var. *arcticum* (Grunow) Cleve | Usachev, 1949; Meguro, Ito, and Fukushima, 1967; Horner and Schrader, 1982 |
| *Melosira arctica* (Ehrenberg) Dickie* | Gran, 1904; Usachev, 1949; Hsiao, 1983; Pett et al., 1983; Booth, 1984 |
| *Navicula directa* (W. Smith) Ralfs | Usachev, 1949; Meguro, Ito, and Fukushima, 1967; Horner and Schrader, 1982; Poulin and Cardinal, 1982*b*; Hsiao, 1983; Pett et al., 1983; Booth, 1984 |
| *Navicula marina* Ralfs | Alexander, Horner, and Clasby, 1974; Horner and Schrader, 1982; Hsiao, 1983 |
| *Navicula spicula* (Hickie) Cleve | Horner and Schrader, 1982; Hsiao, 1983; Pett et al., 1983 |
| *Navicula transitans* Cleve | Usachev, 1949; Meguro, Ito, and Fukushima, 1967; Alexander, Horner, and Clasby, 1974; Horner and Schrader, 1982; Hsiao, 1983; Pett et al., 1983 |
| *Navicula transitans* var. *derasa* (Grunow) Cleve | Usachev, 1949; Meguro, Ito, and Fukushima, 1967; Alexander, Horner, and Clasby, 1974; Poulin and Cardinal, 1982*b*; Hsiao, 1983; Pett et al., 1983; Booth, 1984 |
| *Nitzschia cylindrus* (Grunow) Hasle [*Fragilaria cylindrus* Grunow; *Fragilariopsis cylindrus* (Grunow) Helmcke et Krieger] | Usachev, 1949; Horner and Schrader, 1982; Hsiao, 1983; Pett et al., 1983; Poulin and Cardinal, 1983 |
| *Nitzschia frigida* Grunow | Usachev, 1949; Alexander, Horner, and Clasby, 1974; Horner and Schrader, 1982; Hsiao, 1983; Pett et al., 1983; Poulin and Cardinal, 1983; Booth, 1984 |
| *Nitzschia grunowii* Hasle [*Fragilaria arctica* Grunow; *F. oceanica* Cleve; *Fragilariopsis oceanica* (Cleve) Hasle] | Usachev, 1949; Alexander, Horner, and Clasby, 1974; Hsiao, 1983; Pett et al., 1983 |
| *Pinnularia quadratarea* (A. Schmidt) Cleve (including numerous varieties; see Hsiao, 1983) | Usachev, 1949; Meguro, Ito, and Fukushima, 1967; Horner and Schrader, 1982; Poulin and Cardinal, 1982*a*; Hsiao, 1983; Pett et al., 1983 |
| *Pleurosigma stuxbergii* Cleve et Grunow | Usachev, 1949; Meguro, Ito, and Fukushima, 1967; Alexander, Horner, and Clasby, 1974; Poulin and Cardinal, 1982*a*; Hsiao, 1983 |
| Unidentified flagellates from various algal classes | Usachev, 1949; Alexander, Horner, and Clasby, 1974; Horner and Schrader, 1982; Hsiao, 1983; Pett et al., 1983; Booth, 1984 |

*Note:* All are pennate diatoms except *Melosira arctica* and unidentified flagellates. (For more comprehensive lists see Hsiao, 1983; Horner, 1985.)

*This species also forms huge mucilaginous masses in the water column around drifting ice floes in spring and summer

classes often dominate the sea-ice community in winter and may be numerous at other times as well (Grainger and Hsiao, 1982; Demers, Therriault, and Descolos-Gros, 1984). Marine chlorophycean flagellates have been reported from bottom ice (Alexander, Horner, and Clasby, 1974; Grainger and Hsiao, 1982; Pett et al., 1983). Euglenoids are present, but never abundant, in and near the ice, and are rare in the water column when ice is not present (Horner and Alexander, 1972; Horner and Schrader, 1982; Horner, unpublished observations, 1974). Members of the Prasinophyceae, such as *Phaeocystis pouchetii* (Har.) Lagerh., are often more abundant in the water column in spring, but this species has also been reported from ice (Hsiao, 1983; Booth, 1984). *Chrysochromulina birgeri* Hällfors and Niemi has been reported from slushy snow and ice in the Baltic (Hällfors and Niemi, 1974). Unidentified members of the Prasinophyceae have been reported from sea ice in Davis Strait (Booth, 1984). Dinoflagellates (Dinophyceae) may be relatively common in the ice and at least 15 species or species groups have been reported from Arctic sea ice (Horner, 1985). Both photosynthetic and nonphotosynthetic species may be present in the ice.

Unidentified members of the Cryptophyceae are often present in the ice (Alexander, Horner, and Clasby, 1974; Horner and Schrader, 1982). The skeletons of *Distephanus speculum* (Ehrb.) Haeckel, a silicoflagellate (Dictyochophyceae), common in the water column in the Arctic (Tibbs, 1967), have also been found in sea ice (Usachev, 1949; Booth, 1984; Horner, unpublished observations, 1972). Members of the Craspedophyceae (choanoflagellates), reported from sea ice in the Antarctic (Buck, 1980; Takahashi, 1981), have not yet been reported from Arctic sea ice, although they are relatively common in the water column (Horner, unpublished observations, 1972, 1973). Archaeomonads have also been reported from sea ice in the Antarctic (Mitchell and Silver, 1982; Mitchell, 1982), and they are thought to be siliceous, fossil, marine chrysophyte cysts that cannot be matched to living species (Adam and Mahood, 1981). Archaeomonads have also been found in surface sediments from the Bering Sea (Mitchell and Silver, 1982), and perhaps they will be found in sea ice there as well. They are thought to be indicators of ice environments (Mitchell and Silver, 1982).

Of the species present in the ice, usually only a few make up the bulk of the biomass. Horner and Schrader (1982) reported only 6 of 55 species accounted for more than 10 percent of the cells counted at Narwhal Island and Cross (1982) found that 10 of 77 species accounted for more than 95 percent of the total cells counted at Pond Inlet. At Narwhal Island, species changes occurred during the course of the ice algal bloom (Horner and Schrader, 1982). *Nitzschia cylindrus* (Grun.) Hasle was dominant in April and early May, but by mid-May, other diatoms and flagellate species that had been rare or not identified earlier became more common and diversity and chlorophyll *a* increased. Environmental factors, including increasing light or nitrate levels and decreasing salinity, may have contributed to these species changes.

Some species are found in sea ice over a wide geographic area (Cleve, 1896), but others have been reported from only one or a few small areas. Species differences may occur in the same area in successive years (Alexander, Horner, and Clasby, 1974; Demers, Therriault, and Descolos-Gros, 1984) or over relatively narrow geographic areas in the same year (Alexander, Horner, and Clasby, 1974). Reasons for these differences are not known, but may be related to environmental conditions.

The same species are generally not found in both the ice and the water column in most Arctic areas, including Barrow (Alexander, Horner, and Clasby, 1974; Matheke and Horner, 1974; Horner, 1984) and Narwhal Island (Horner and Schrader, 1982), Alaska, and Eskimo Lakes (Grainger and Evans, 1982) and Frobisher Bay (Grainger, 1977), Northwest Territories. Ice diatoms found in the water column are usually unhealthy in appearance with shrunken and faded chloroplasts, suggesting that they do not survive very long in the low-salinity meltwater usually found immediately under the ice. Also, pennate diatoms are more abundant in the ice, while centric diatoms are more common in the water column. An important exception is the *Nitzschia cylindrus-N. grunowii* Hasle complex of pennate diatoms that may be common in both the ice and water column.

## ECOLOGY

### Distribution

#### *Present-day Distribution of Ice Algae*

The distribution of ice algae is not well known. Ice algae are apparently a regular feature on the bottom of annual ice in both polar regions, and also in some more temperate areas including the Baltic Sea and the Gulf of St. Lawrence. Their distribution in ice more than one year old is not known, although Gran (1904) reported no ice algae when the *Fram* was well into the permanent pack and away from shore during the First Norwegian North Polar Expedition. Schell (personal communication, 1980) has found ice algae 160 km offshore from Prudhoe Bay, Alaska, but always in relatively thin new ice in refrozen leads. I have found *Thalassiosira* sp. spores and *Distephanus speculum* skeletons in sea ice approximately 7 m thick off northern Greenland in April, but no live diatom cells (Horner, unpublished observations, 1972).

Ice algae are also commonly found in the Bering Sea (Alexander and Chapman 1981; Schandelmeier and Alexander, 1981), where their distribution based on chlorophyll *a* measurements is extremely patchy on both local and widespread geographic scales (Alexander and Chapman, 1981). Neither the seasonal distribution nor the length of the productive period are known for ice algae in the Bering Sea. McRoy and Goering (1974) suggested that the ice algal community has a marked seasonal cycle, with the ice algae not being present throughout the winter but developing in late winter and increasing until the ice disappears in spring. That would be in contrast to the Chukchi and Beaufort seas where cells are present in the ice from the time the ice forms in the fall.

Sancetta (1981*a*) reported a "Sea Ice Assemblage," primarily composed of *Thalassiosira nordenskioeldii* Cleve, *Nitzschia grunowii,* and *N. cylindrus* in surface sediments from the northern Bering Sea Shelf and the Chukchi Sea where ice is present for about six months of the year. This assemblage is a combination of the spring ice algal community (*Nitzschia* spp.) and the spring phytoplankton bloom in the water column (*Thalassiosira* sp.). It indicates a shelf water depth of 50–100 m, a long period of ice cover, and a highly productive spring bloom during ice melt (Sancetta, 1981*a*).

In the ice itself, the ice algae may be found concentrated on the upper surface of the ice at the snow–ice interface or on ice in melt ponds on the ice surface. They may be found in the interior of the ice or in the bottom few centimeters of ice at the ice–water interface. Algal cells may be found throughout the ice thickness as well but usually not in large numbers. As already noted, the surface and interior ice communities occur in the Antarctic, but are generally not present in the Arctic, although cells may be found in melt ponds on the ice surface (Bursa, 1963).

### Ice Algae as Possible Past Environmental Indicators

There is some evidence to suggest that ice algae may be possible indicators of past environmental conditions (Zhuse, Kozlova, and Mukhina, 1967; Sancetta, 1981*a,b*; 1982; Mitchell, 1982; Mitchell and Silver, 1982; Dale, 1983). The Antarctic complex described by Zhuse, Kozlova, and Mukhina (1967) consisted primarily of endemic ice algae. Sancetta (1981*a,b*; 1982) found diatom assemblages in Bering and Okhotsk sea sediments that were consistent with present hydrographic and productivity regimes, suggesting that they can be useful in determining past conditions, such as the duration and extent of sea ice during glacial periods.

Mitchell (1982) and Mitchell and Silver (1982) found archaeomonads in sea ice from the Ross, Weddell, and Bering seas. Their modern distribution in sea ice suggested to these authors that the cysts are produced during periods of environmental stress, thus indicating that the fossil forms might be useful indicators of sea ice. Dinoflagellates that produce cysts under adverse environmental conditions are also found in sea ice and there is some evidence that dinoflagellate cysts are abundant in Quaternary and Recent sediments near sea ice (Dale, 1983).

## Modes of Adaptation to the Ice Environment

Knowledge of both the physical features of the environment and the biological characteristics of the organisms themselves is necessary in order to understand how the ice algae have adapted to their environment.

### Environmental Factors

Organisms living in sea ice are affected by a number of environmental factors and have adapted to living under rather extreme conditions. These environmental factors include light, temperature, nutrients, and salinity, with light usually considered to be the most important factor influencing the development and distribution of the ice algal community.

*Light.* Relatively few light measurements have been made at the same time that productivity has been measured in the field (Clasby, Horner, and Alexander, 1973; Alexander, Horner, and Clasby, 1974; Horner and Schrader, 1982; Booth, 1984), and some light measurements have been made along with chlorophyll *a* determinations (Apollonio, 1965; Schell, 1980*a*). These, and similar measurements made in the Antarctic (Bunt, 1963, 1964; Sullivan et al., 1982), all indicate that the ice algae are extremely shade adapted, often growing at light intensities near 0.1 percent of surface irradiances. Generally, the 1 percent light depth is considered to be the bottom of the euphotic zone in the pelagic environment.

Snow on the ice surface severely attenuates the amount of light reaching the ice algae. When no snow was present, about 2 percent of the surface irradiance reached the algal layer through 180 cm of ice, but when 3–4 cm of snow was present, the light level was reduced to 1 percent and 20 cm reduced the light to $<0.1$ percent (Horner and Schrader, 1982). A sediment layer or layers in the ice can also reduce the amount of light reaching the ice algae (Schell, 1980*a,b*; Horner and Schrader, 1982), but they may be a serious problem only in shallow lagoon systems such as those found along the North Slope of Alaska. Sediment layers have not been reported in other areas where ice algae have been studied.

The cells in the bottom of the ice are

probably living at the highest light levels available to them and are able to use this light more efficiently than cells in the water column (Meguro, Ito, and Fukushima, 1967). Furthermore, the layer of ice algae effectively reduces the amount of light reaching the phytoplankton even more. The extremely low light levels reaching the water column and the sea floor thus prevent phytoplankton and benthic microalgal growth until the snow and ice algae disappear, melt ponds appear on the ice surface, and the ice begins to break up.

*Nutrients and Salinity.* Nutrient concentrations and salinity are usually measured on melted ice cores (Grainger, 1977; Dunbar and Acreman, 1980; Booth, 1984). This dilutes the hypersaline interstitial water where the organisms are living and does not give an accurate indication of their environment. Grainger (1977) found nitrate and phosphate accumulated in the ice in winter with low levels in the upper part of the ice and high levels at the bottom. Dunbar and Acreman (1980) reported higher salinity and nutrient values at the top and bottom of the ice, and Booth (1984) thought that nutrient concentrations could be 10–15 times higher than her reported values because of dilution from melted ice. Alexander, Horner, and Clasby (1974) allowed cores of bottom ice to drain, but not melt, before salinity and nutrient determinations were made. They found salinity to be near that of the underlying seawater, ranging from 20.4–34.1 ‰, while nutrient concentrations were variable, being either similar in both habitats or considerably higher in the ice. When bottom ice cores were allowed to melt completely, salinity in the ice was 1–2 ‰ lower than in the underlying seawater with nutrient concentrations generally similar in both environments (Horner and Schrader, 1982).

Salinity of the underlying seawater was the major environmental factor controlling the distribution and abundance of species in southeastern Hudson Bay (Poulin, Cardinal, and Legendre, 1983) where a surface water salinity gradient controlled the rate of ice formation, its thickness, and its structure, and therefore the biomass of the microalgae. The species composition of the bottom ice community was controlled by the ice salinity.

Under laboratory conditions, diatoms isolated from Chukchi and Beaufort sea ice grew over a salinity range of 5–60 ‰, averaging 0.4–0.8 divisions $d^{-1}$ in the range of 10–50 ‰. Greater mortality and longer lag periods occurred at lower salinities (Grant and Horner, 1976), suggesting greater cell stress. This is supported by the presence of large numbers of dead or unhealthy cells in low salinity meltwater in spring (Horner and Schrader, 1982). Salinity might also help to explain why cells concentrate in the bottom layer of ice in the spring.

*Temperature.* Few measurements have been made of the effect of temperature on the growth of ice algae and none have been made in the Arctic. In the Antarctic, net carbon fixation at 14 $\mu E\ m^{-2}\ s^{-1}$ was markedly temperature dependent in the range $-1.5$ to 5.0°C, with maximum carbon uptake from 10–15°C (Bunt, 1964), while maximum growth of *Fragilaria sublinearis* Van Heurck (now *Nitzschia sublineata* Hasle), an obligate psychrophile, occurred at a temperature higher than that of sea ice (Bunt, 1968), both suggesting that temperature may control the maximum growth rate of the ice algae (McConville, 1985).

Motility and shape of flagellates living in sea ice may also be affected by temperature because microtubules in the cytoplasm and flagellar apparatus often depolymerize at 4°C (Burch and Marchant, 1983). However, some organisms are able to accumulate polyols and sugars, including glycerol and mannitol, that tend to stabilize the microtubules and may also aid in osmoregulation. These compounds have not yet been identified in organisms from sea ice, but they have been found in closely related organisms (Burch and Marchant, 1983).

### Biochemistry and Physiology

Knowledge of the physiological state of microalgae found in sea ice is important in order to determine the environmental factors that limit their growth and enable them to survive during the winter dark period. Ratios of certain chemical components of the cells, including pigments and storage products, may provide indirect estimates of growth rates and thus, of the physiological conditions of the cells (McConville, 1985). These indirect estimates of growth are promising because use of $^{14}C$-uptake rates to measure primary production of ice algae is difficult and may lead to underestimates due to possible sources of error in the methods (Clasby, Horner, and Alexander, 1973; Schrader, Horner, and Smith, 1982; Horner and Schrader, 1982).

*Pigments.* The dominant light harvesting pigments of microalgal classes found in sea ice are given in Table 5-2. The diatoms with chlorophyll *c* and fucoxanthin that absorb in the blue-green region of the spectrum are well adapted to live in the ice where attenuation of short-wave radiation (blue-green region) is less than that of long-wave radiation (McConville, 1985; Perovich and Grenfell, 1981). Later in the spring, when the ice begins to melt and incident radiation increases, green algae (Chlorophyceae, Prasinophyceae) with chlorophyll *a* and *b* that absorb in the red region may become more numerous (McConville, 1985).

Microalgae in the ice are said to be shade adapted because of their ability to live at the low light levels available to them. Changes must occur in the photosynthetic apparatus to enable the cells to be photosynthetically efficient at these low light levels. The nature of these changes is not known, but may involve changes in the activities of the enzymes having to do with carbon uptake, changes in photosynthetic–irradiance profiles, or changes in the number and size of the photosynthetic units (Falkowski, 1980). It is known that shade adaptation is more pronounced and cells are not able to respond to short-term ($<24$ h) variations in irradiance at low temperatures (Hitchcock, 1980).

*Dark Adaptation and Survival.* Microalgae are found in sea ice from the time the ice forms in the fall and the cells are usually scattered throughout the ice thickness in winter. The cells must be able to survive the prolonged period of winter darkness at temperatures near $-1.8°C$ in order to provide the seed population for the spring bloom in the ice.

Palmisano and Sullivan (1982) have proposed three possible mechanisms for dark survival: (1) facultative heterotrophy; (2) storage and utilization of carbon reserves at reduced metabolic rates; and (3) formation of resting spores. Heterotrophy has been suggested as the means of winter survival for phytoplankton under the ice in subarctic lakes (Rodhe, 1955), for Arctic seaweeds (Wilce, 1967), and for ice algae (Allen, 1971). Horner and Alexander (1972) found that heterotrophic metabolism by ice algae was negligible, but their study was done in March when light was already available to the cells and may have triggered a switch from heterotrophy to photosynthesis. Bunt and Lee (1972) also failed to show that heterotrophy was important for dark survival of several Antarctic ice algae. Palmisano and Sullivan (1982), however, found that some ice algae could take up in the dark less than 1 percent of the carbon needed for growth and suggested that this might be sufficient for maintenance metabolism during the winter.

Table 5-2. Algal Classes found in Arctic Sea Ice, Their Major Light-Harvesting Pigments,* and Reserve Food Material†

| | | *Chlorophylls* | | | | | *Major Accessory Light-Harvesting Pigments* | | |
|---|---|---|---|---|---|---|---|---|---|
| *Class* | *Common Name* | *a* | *b* | $c_1$ | $c_2$ | *β-carotene* | *Xanthophylls* | *Biliproteins* | *Food Reserves* |
| Bacillariophyceae | Diatoms | + | | + | + | + | Fucoxanthin | | Chrysolaminarin, lipid |
| Chrysophyceae | Golden-brown flagellates | + | | + | + | + | Fucoxanthin | | Chrysolaminarin |
| Prymnesiophyceae (Haptophyceae) | Golden-brown flagellates with haptonema, including coccolithophorids | + | | + | + | + | Fucoxanthin | | Chrysolaminarin |
| Dictyochophyceae | Silicoflagellates | + | | + | + | + | Fucoxanthin | | Chrysolaminarin |
| Dinophyceae | Dinoflagellates | + | | | + | + | Peridinin | | Starch, lipid |
| Chlorophyceae | Green algae | + | + | | | + | Lutein | | Starch |
| Prasinophyceae | Scaly green flagellates | + | + | | | + | Siphonoxanthin (some species) | | Starch, mannitol |
| Cryptophyceae | Cryptomonads | + | | | + | + | Alloxanthin | + | Starch |
| Cyanophyceae | Blue-green algae (Cyanobacteria) | + | | | | + | Myxoxanthin | + | Myxophycean starch |
| Euglenophyceae | Euglenoids | + | + | | | + | | | Paramylon |

*From Jeffrey and Vesk, 1981; Richardson, Beardall, and Raven, 1983.
†From Price, 1981.

Ice diatoms often contain large bodies of reserve food material and the utilization of these stored lipids and carbohydrates at reduced rates may provide enough carbon for maintenance metabolism in winter. Dark respiration by *Fragilaria sublinearis* at 3°C was less than 50 percent that at 10°C and would probably be even lower at −1.8°C (Bunt, Owens, and Hoch, 1966). In a simulated summer–winter transition experiment, ice diatoms were found to store and subsequently utilize cellular lipids and carbohydrates (Palmisano and Sullivan, 1982). These observations suggest that slow metabolism may provide enough energy for the cells during winter.

Resting spores are heavily silicified cells with a morphology and metabolism different from growing vegetative cells. Morphologically distinct resting cells have been reported for only a small number of pennate diatoms and there are few, if any, pennate diatoms in polar waters that produce resting spores (Hargraves and French, 1983). Species of the centric diatoms *Chaetoceros* and *Thalassiosira* commonly form resting spores and these may be present in the ice, but centric diatoms are usually not dominant members of the ice community in the Arctic, although they may be abundant in lower latitudes such as the Gulf of St. Lawrence (Dunbar and Acreman, 1980). There is little evidence to suggest that the formation of resting spores is important for winter survival, but there is the possibility that resting cells may not be morphologically different from vegetative cells and are therefore not recognized (Fryxell, Villareal, and Doucette, 1981). It is likely that all of these mechanisms may be used, either singly or in combination, to enable cells in the ice to survive the winter dark period.

***Storage Products.*** The major storage products of microalgae are lipids and carbohydrates. Lipid storage frequently occurs in microalgae during nutrient limitation, particularly silica (Werner, 1977) and nitrate (Shifrin and Chisholm, 1981) deficiencies. Ice algae often accumulate oil droplets in late spring, which may indicate the cells are nutrient limited. Lipid accumulation may also occur at low temperatures (Palmisano and Sullivan, 1982; Smith and Morris, 1980), but this is not always the case (Li and Platt, 1982), and may be due to variation in species composition rather than to low temperature (Li and Platt, 1982).

Carbohydrates are accumulated when carbon is in excess of metabolic needs. At low light levels and in rapidly growing, nutrient-sufficient cultures, the photosynthetic rate is low relative to growth rate and cellular levels of carbohydrates are low (McConville, 1985). Carbohydrates may accumulate during the day and when the growth rate is less than the photosynthetic capacity of the cells. Accumulation is dependent on light intensity and may occur at low temperatures, high salinities, or nutrient limitation (McConville, 1985). In ice diatoms, carbohydrate accumulation occurred during a simulated summer–winter transition with temperature, salinity, and light all influencing carbohydrate metabolism (Palmisano and Sullivan, 1982).

## Origin of the Arctic Bottom Ice Algal Community

The origin of the ice algae is still not known. Early investigators (Østrup, 1895) thought the cells were carried to the ice by wind or rivers. The marine forms were thought to be washed onto the ice during storms or frozen into the ice as the ice formed. Ross (1954) and Meguro, Ito, and Fukushima (1967) thought diatoms formed spores or cysts in the fall and these were incorporated into the ice as it formed. The resting cells then germinated in the spring when the environmental conditions were favorable. However, as previously discussed, morphologically distinct resting spores or cysts have been reported for only a few pennate diatoms and few, if any, endemic polar pennate diatoms produce resting cells (Hargraves and French, 1983). It is possible, however, that the so-called resting cells are not morphologically distinct from vegetative cells and thus are not recognized as special spores or cysts.

Cells are frozen into the ice as it forms in the fall. Cells from the benthos may be brought into the water column during fall storms, or cells from the water column may be frozen into the ice, but in some areas at least, there is no significant species correlation between either the benthic or planktonic diatom floras and the ice flora. Microflagellates from several algal classes are often the most abundant organisms in the ice during fall and winter, although fall blooms of diatoms have been reported in the ice from Narwhal Island, Alaska (Schell, personal communication, 1980), and the Eskimo Lakes region, Northwest Territories (Hsiao, 1980). These fall blooms may be remnant populations left from the previous spring bloom because the Narwhal Island bloom contained the same species complement as was seen the previous spring (Horner and Schrader, 1982).

Mel'nikov (1976) reported that freezing ice in the central Arctic concentrates biogenous and organic forms present in the water column and redistributes them within the ice as the ice grows. The ice then contains several times more organic matter than the water from which it formed.

Other possible sources of ice algae in shallow water may be a subsurface chlorophyll maximum in the water column or the formation of a thick, soft ice layer on the underside of the ice. In the Bering Sea, a subsurface chlorophyll maximum has been reported (Alexander and Chapman, 1981) that may be a remnant of the spring ice diatom bloom (Saito and Taniguchi, 1978). It is not known if these cells remain viable and are brought back into the euphotic zone in the fall either by fall storms or by coming into contact with and being collected by frazil ice crystals that are then incorporated into the newly forming ice.

The soft ice layer on the underside of the ice is apparently a widespread phenomenon in the Bering, Chukchi, and Beaufort seas and is associated with fall freezing events (Larsen, 1980). It is not known how this layer forms, but it may be a reservoir for concentrating organic material and organisms into the ice canopy, although no explanation was given for how the cells and organic matter actually become trapped in the ice.

Ice scavenging and nucleation are other possible mechanisms for concentrating cells in the ice (Ackley, 1982; Garrison, Ackley, and Buck, 1983). This has been suggested for the Antarctic where frazil ice crystals moving up in the water column collide with and collect algal cells from the water column (scavenging) or where frazil ice crystals nucleate on algal cells (nucleation). This harvesting mechanism may explain the origin of cells in areas where frazil ice is continuously formed and where the same species are found in the ice and water column as in the Antarctic, but in nearshore areas of the Arctic where few cells are found in the water column during ice formation other mechanisms for concentrating cells may be necessary.

The origin of freshwater algae in the ice is also not known. Cleve (1900) thought the cells were transported from land by winds. Other investigators who found freshwater algae in the ice, including river ice (Usachev, 1938), and on the surface of sea ice (Bursa, 1963) did not speculate on how the cells actually got into the ice.

## Annual Cycle of Arctic Ice Algae

Most of the studies on Arctic ice algae have been done in the spring when the ice community is growing and at its peak production. Few studies have followed the annual cycle of ice algae from the time the ice forms in the fall, through the winter dark period, and through the spring increase and decline as the ice melts in late spring–early summer. The annual cycle is relatively well known only from Barrow, Alaska (Alexander, Horner, and Clasby, 1974), so much of the following discussion will be based on that location with other studies added where relevant.

Cells are present in the ice from the time the ice forms. Microflagellates are the dominant organisms in the fall, but pennate diatoms are also present. The fall bloom in November has not been found at Barrow, but at Narwhal Island the bloom of pennate diatoms contained $1 \times 10^8$ cells $m^{-2}$, which was comparable to the level found the previous spring in the same area (Horner and Schrader, 1982). In the Eskimo Lakes region, the fall population of pennate diatoms contained $3.9 \times 10^3$ cells $m^{-2}$ (Hsiao, 1980).

Cells remain scattered throughout the ice in winter (Horner, 1976; Hsiao, 1980). Chlorophyll *a* concentrations throughout the ice are low and variable from January to mid-March (Alexander, Horner, and Clasby, 1974; Horner and Schrader, 1982). The algae gradually become concentrated in the relatively soft bottom few centimeters of ice during late March and by early April a golden-brown layer becomes visible on the bottom of the ice. The concentration mechanism for the cells is not known, but may be a combination of brine drainage and active migration of cells through brine channels in the ice (Horner, 1985). The cells may be moved passively by the brine as it drains toward the ice–seawater interface or many of the pennate diatoms are capable of a slow gliding movement and may move down the brine channels toward more favorable environmental conditions.

Increasing light levels in the spring trigger the growth of the ice algae. At Barrow, productivity was $<0.1$ mg C $m^{-2}$ $h^{-1}$ until a threshold light level was reached (Alexander, Horner, and Clasby, 1974). This threshold level at Barrow was 2.3 and 9.3 $\mu E$ $m^{-2}$ $s^{-1}$ in successive years (Alexander, Horner, and Clasby, 1974) and at Narwhal Island was 7.7 $\mu E$ $m^{-2}$ $s^{-1}$ (Horner and Schrader, 1982). Growth continues during spring, reaching maximum levels in mid- to late May when the ice is near maximum thickness, the snow cover is gone, and light levels are 10–65 $\mu E$ $m^{-2}$ $s^{-1}$ (Alexander, Horner, and Clasby, 1974; Horner and Schrader, 1982). Growth may be slowed or interrupted by adverse weather conditions such as periods of fog or storms that decrease the light intensity. By the end of May, the bottom ice containing the algae becomes softer and gradually deteriorates. Weak water currents under the ice wash the algal layer away, although the decaying ice containing the cells may remain associated with the harder ice above for a few days. At Pond Inlet, Baffin Island, divers observed thin sheets of algae partially detached from the underside of the ice that persisted for some time as strands and amorphous masses (Cross, 1982). At Barrow and Narwhal Island, the brown layer on the underside of the ice is no longer visible by early June.

Annual production of ice algae in the Arctic has been calculated from $^{14}C$ uptake experiments and from chlorophyll *a* concentrations (Table 5-3). Errors are present with both methods, with the $^{14}C$ method probably underestimating production because it is difficult to ensure adequate mixing of the isotope throughout the sample and because some isotope may diffuse upward and be lost during the in situ incubation period, both caused by the porous nature of the ice. Production calculations from chlorophyll *a* measurements may be misleading unless carbon:chlorophyll ratios are carefully determined for the population being studied (Bunt, 1968; Banse, 1974). In most areas, ice algal production amounts to about 2 percent of the total annual primary production although the 5 g C $m^{-2}$ $y^{-1}$ at Barrow was estimated to be as much as 25–30 percent of the total annual primary production in that area (Alexander, 1974).

Table 5-3. Annual Ice Algal Production (g C m$^{-2}$) in the Arctic Calculated from In Situ $^{14}C$ Uptake Experiments and Chlorophyll *a* Concentrations

| *Location* | *Production* | *Reference* |
|---|---|---|
| Barrow, Alaska | 5.0 | Alexander, 1974 |
| Stefansson Sound, Alaska | 0.9 | Horner and Schrader, 1982 |
| | 1.4* | Schell, 1980*a* |
| Simpson Lagoon, Alaska | 0.18* | Schell, 1980*a* |
| Offshore Beaufort Sea | 1.7* | Schell, 1980*a* |
| Narwhal Island, Alaska | 0.7 | Horner and Schrader, 1982 |
| Davis Strait (Disko Island) | 0.15 | Andersen, 1977 |
| Davis Strait | <0.10 | Booth, 1984 |
| Bering Sea | 0.22–0.5 | McRoy and Goering, 1974 |

*Calculated from chlorophyll *a* concentrations.

## Fate of the Arctic Ice Algal Community

A number of conflicting views concerning the fate of the ice algal community have been suggested and the literature has recently been reviewed (Horner, 1984). The most commonly held hypothesis is that the ice algae produce the spring phytoplankton bloom in the water column. This is not the case in the nearshore Chukchi Sea at Barrow, Alaska, where the ice algal bloom and the spring phytoplankton bloom are clearly separated by time and by the species present (Alexander, Horner, and Clasby, 1974; Horner and Alexander, 1972; Clasby, Horner, and Alexander, 1973; Matheke and Horner, 1974). The ice algal bloom consists primarily of pennate diatoms that are usually not found in the water column during the phytoplankton bloom, while the phytoplankton bloom consists primarily of centric diatoms and occurs some time after the end of the ice algal bloom when adequate light reaches the water column (Horner, 1984). Even though no data are available for the spring phytoplankton bloom at Narwhal Island, Alaska, few live phytoplankton cells were present in the water column and phytoplankton production was low when the ice algal bloom ended (Horner and Schrader, 1982). Primary productivity peaks in the water column at the end of the ice algal bloom at both Barrow (Clasby, Horner, and Alexander, 1973) and Narwhal Island (Horner and Schrader, 1982) were short-lived and were probably due to healthy cells just released from the ice before they were adversely affected by the low-salinity meltwater immediately beneath the ice.

Offshore in the Chukchi Sea, Meguro, Ito, and Fukushima (1966, 1967) thought the high chlorophyll *a* concentration in the water column as the ice melted was caused by ice diatoms being liberated into the water column. They apparently did not sample diatoms in the water column around the melting ice, but they did comment that the ice species were commonly found in the water as well. Hameedi (1978) found higher chlorophyll *a* concentrations in the water column at stations close to the ice than at open water stations and concluded that cells released from the ice were responsible for the increases. He did not sample the ice, nor did he determine what species were present in the water column, so it is not possible to say with certainty that the ice algae produced the higher chlorophyll *a* concentrations.

In the Bering Sea, ice algae were released into the water column as the ice melted and, at the same time, the spring bloom occurred in the surface layer that was vertically stabilized by low-salinity meltwater (Saito and Taniguchi, 1978). As a result, both ice diatoms and spring bloom diatoms were present in the surface water at the same time. During the spring bloom, the ice diatoms gradually sank to some subsurface depth and the planktonic diatoms dominated in the surface layer.

A layer of photosynthetically active cells was found near the bottom at 30–40 m depth near the ice edge and over a wide area of the Bering Sea (Alexander and Chapman, 1981) that might have been the remnant of the ice or ice edge bloom. No list of species was given for this layer so it is not possible to determine if it was, in fact, from the ice. A similar high chlorophyll *a* layer in the eastern Chukchi Sea was composed primarily of centric diatom species that are common in the water column in spring (Horner, unpublished observations, 1974).

In the marginal ice zone of the Bering Sea, Alexander and Chapman (1981) found the same species in slush ice, ice cores, and seawater; high primary productivity for cells from brash ice; and high photosynthetic rates for ice algae incubated at surface light intensities. These authors concluded that ice algae are important in the ice-edge phytoplankton bloom in the water column. Few planktonic species may be present in the water column early in spring and perhaps the ice algae are more important in the water column at that time. There is still the question of whether the marginal ice phytoplankton community is the same as one found under the ice well away from the ice edge. Also, are the same species present in the ice in the marginal ice zone and in the more stable ice away from the ice edge?

In the Canadian Arctic, ice algae were released from the ice more or less at the beginning of the phytoplankton bloom in the Eskimo Lakes region, but Grainger and Evans (1982) said this did not imply that the phytoplankton bloom was initiated by the ice algae. At Frobisher Bay, there may be a time lag between the release of the ice algae and the beginning of the phytoplankton bloom (Grainger, 1977), while in Hudson Bay, there is a simultaneous release of the ice algae and increased phytoplankton production as the ice melts, all initiated by increased solar energy (Legendre, Ingram, and Poulin, 1981).

It is not known how long algal cells, especially diatoms, from the ice can survive in the water column. A number of factors probably affect survival, but from the number of unhealthy or dead ice algal cells in the low-salinity meltwater immediately beneath the ice, salinity may be the most important one (Grant and Horner, 1976; Horner and Schrader, 1982). These unhealthy cells are easily distinguished from the healthy, golden-brown cells by their shrunken, blue-green chloroplasts.

Another possible fate for the ice algae is incorporation into the sediments. Sancetta (1981*a,b*) reported an assemblage of diatoms associated with melting ice in the surface sediments of the Bering Sea, but in shallow, nearshore areas of the Chukchi and Beaufort seas, few ice diatoms were found in the sediments (Matheke and Horner, 1974; Horner and Schrader, 1982).

Utilization of ice algae by invertebrates has also been suggested as a possible fate for the ice algae and will be discussed later in this chapter.

## ICE FAUNA

Sea ice provides a solid substrate for a diverse group of animals ranging from protozoans to birds and marine mammals, and an extensive food web is supported by the ice algae on the undersurface of the ice (Carey, 1985).

## Terminology

As with the ice algae, terminology for the animals that live associated with sea ice is ambiguous. Andriashev (1968) defined two faunal assemblages associated with the fast ice in the Antarctic. The ice fauna lives at least temporarily in the bottom layer of sea ice, while the sub-ice fauna does not live directly in the ice, but is associated with it in varying degrees as part of the food web. Carey (1985) has chosen the word *sympagic,* meaning "with ice," as a broad descriptive term to include the meiofauna and macrofauna of the true ice community that originates in the benthos and the plankton.

## History

Sutherland (1852) was apparently the first person to report finding "infusory animalcules" in sea ice and Nansen (1906) described and figured "infusoria" from sea ice collected in the East Greenland Sea in 1882, Denmark Strait in 1888, and the Arctic Ocean in 1894. Usachev (1949) also reported on a number of protozoans found in the ice. MacGinitie (1955) found large numbers of the amphipod *Apherusa glacialis* living under ice floes during the summer at Barrow, while Horner and Alexander (1972) reported an extensive meiofauna community consisting of flagellates, heliozoans, ciliates, turbellarians, nematodes, polychaete larvae, and copepods from the same area. Amphipods were also seen grazing on the underside of the ice and their fecal pellets contained fragments of ice diatoms (Horner, 1977). Other early observations came from investigators working on drifting ice stations (English, 1961; Mohr, 1959; Mohr and Tibbs, 1963; Apollonio, 1965; Barnard, 1959; George and Paul, 1970). None of these early observations was quantitative.

Quantitative studies on ice fauna have been done only since the 1970s. Golikov and Averincev (1977) used scuba diving to study the ice fauna at a number of sites in the Siberian Arctic islands where they found specific faunal assemblages associated with the bottom of the ice or in cavities in the ice. Animals in these assemblages included *Gammarus setosus, G. wilkitzkii, Apherusa glacialis,* and *Boreogadus saida.*

In the Canadian Arctic, ice fauna have been studied at Pond Inlet (Cross, 1982; Bradstreet and Cross, 1982), Resolute Bay (Green and Steele, 1975), Frobisher Bay (Grainger and Hsiao, 1982), and McKinley Bay (Pett et al., 1983). Ice fauna have also been studied off northern Alaska (Carey and Montagna, 1982; Kern and Carey, 1983; Carey, 1982; and Carey and Boudrias, 1987) and at the edge of the pack ice in the Barents Sea (Gulliksen, 1984).

## True Ice Fauna

### Macrofauna

The true ice macrofauna (>0.5 mm) is almost entirely composed of gammarid amphipods (Carey, 1985). Amphipods apparently congregate on the undersurface of the ice to feed and also to release their young. There are four main macrofaunal ice communities separated by water depth and ice type: (1) the protected lagoon/bay fast-ice community dominated by *Gammaracanthus loricatus;* (2) the shallow-water, coastal, fast-ice community dominated by *Pseudalibrotus litoralis;* (3) a deeper-water, seasonal, fast-ice zone community dominated by *Apherusa glacialis;* and (4) a multiyear ice community dominated by *A. glacialis* and *Gammarus wilkitzkii* (Carey, 1985).

### Meiofauna

The meiofauna community (62–500 $\mu$m) is composed of a number of permanent and temporary organisms that includes nematodes, harpacticoid, cyclopoid, and calanoid copepods, rotifers, and larvae of benthic polychaetes, pelecypods, gastropods, tunicates, turbellarians, and cirripedes. Nematodes are the dominant meiofaunal group comprising about 50 percent of the population by number. Copepods, comprising 30 percent to 40 percent of the population, are the next most abundant group (Carey, 1985).

### Microfauna

Animals <62 μm comprise the microfauna assemblage. Little is known about this group, but unpigmented flagellates, heliozoans, and ciliates are known to be present (Nansen, 1906; Usachev, 1949; Horner and Alexander, 1972; Grainger and Hsiao, 1982).

### Sub-ice Fauna

This group of animals consists primarily of fish species, including *Boreogadus saida*, *Arctogadus glacialis*, and *Ammodytes hexapterus*, that are trophically linked to the true ice fauna and feed extensively on *Apherusa glacialis* (Andriashev, 1968; McAllister, 1975; Alexander, 1980; Lowry and Frost, 1981). The ice may also be a source of protection from predators for these animals.

### Interactions with the Benthic and Pelagic Regimes

The ice fauna interact with the benthic and pelagic environments, with water depth and length of the ice-covered season affecting both the source of the ice fauna and the fate of detrital material produced by it (Carey, 1985). In shallow water, the source of the ice meiofauna derives from the water column and the bottom sediments (Carey, 1985) and is composed of benthic organisms, such as nematodes and meroplanktonic larvae of benthic organisms, including polychaetes and cirripedes. The macrofauna in this environment is derived primarily from the benthos. In the shallow oceanic region, the ice macrofauna is dominated by benthic amphipods, while offshore, the deep-water, fast-ice community is pelagic in origin (Carey, 1985).

## THE ICE-ALGAL-BASED FOOD WEB

Carey (1985) has summarized the little available data on the ice-algal-based food web, and pointed out that there is little quantitative data and temporal and spatial coverage is scant. The ice algae would be an especially important food source early in spring when little or no primary production occurs in the water column and the ice algae are concentrated at the ice–water interface. We know that animals from a number of invertebrate groups live in or on the sea ice (Horner and Alexander, 1972; Grainger and Hsiao, 1982; Cross, 1982; Carey, 1985). While some authors (Meguro, Ito, and Fukushima, 1967) believed that little grazing occurred in the ice, the abundant meiofauna (Horner and Alexander, 1972; Nansen, 1906; Carey and Montagna, 1982; Kern and Carey, 1983) and the large number of amphipods and fish on the undersurface of the ice or in cracks and brine channels (English, 1961; Mohr, 1959; McAllister, 1975; Green and Steele, 1975; Apollonio, 1965; Horner and Schrader, 1982; Horner, 1976; Cross, 1982; Carey and Montagna, 1982; Kern and Carey, 1983; Holmquist, 1958) all suggest that grazing does occur. For some of these groups, at least, we know they are grazing on the ice algae based on observations of gut contents and fecal pellets (Horner and Alexander, 1972; Horner, unpublished observations, 1972).

Grazing of the ice algae probably also occurs in the water column. In the Chukchi and Beaufort seas, where ice algae are released from the ice in a single pulse over a short time period, utilization is probably rapid, leaving few cells in the water column. For example, Cross (1982) found large numbers of meio- and macrofauna associated with masses of algae that were partially detached from the ice. Cells that are not grazed in the water column may sink to the bottom and either be utilized by benthic invertebrates or be incorporated into the sediments. The amount of material reaching the benthos, however, may depend on the structure of the grazing community. In the southeastern Bering Sea, Niebauer, Alexander, and Cooney (1981) found that when the ice edge was over oceanic waters during cold years, a diverse assemblage of large copepods was present that could ingest a large portion of the ice algae as they were released from the ice, but when the ice edge was over the shelf in warm years, the numerically dominant small copepods could not utilize the ice algae as efficiently and more organic material settled to the bottom.

## SUMMARY AND CONCLUSIONS

The sea-ice community that consists of microalgae, macrofauna, and meiofauna, plus sub-ice fauna that apparently utilize the ice community for food, is very complex. Even providing adequate terminology to describe the ice community is difficult. The ice algal community is common in annual sea ice in both polar regions and in some more temperate areas such as the Gulf of St. Lawrence and the Baltic Sea, but its distribution in sea ice more than one year old is not known.

The annual cycle of ice algae is well documented only off the north coast of Alaska (Barrow and Narwhal Island), and factors affecting its production have been studied primarily in coastal areas throughout the Arctic. Light is the most important factor affecting the growth and production of the ice algal community, but other factors such as temperature, salinity, nutrient concentrations, and grazing pressure may also be significant. The origin of the ice community is not known, although it is possible that cells are collected and concentrated by ice crystals forming in the water column. The fauna associated with the ice has been shown to have connections with both the benthic and pelagic communities depending to some extent on local water depth.

It is not known how the ice algae survive the winter dark period, but heterotrophy, slow metabolism of accumulated food reserves, and formation of resting spores have been suggested. Rates of production are difficult to determine because of sampling problems and providing natural conditions for incubations. The fate of the ice algal community is not well known, but it may initiate the spring bloom in the water column, be grazed either in situ or after the cells are released from the ice, or the cells may settle to the bottom where they are either utilized by benthic invertebrates or become incorporated into the sediments.

About 300 species have been found in the ice, but few comparisons have been made with the water column and especially with the benthos. These comparisons would help determine the fate of the ice algal community. Changes in community structure may occur during the course of the spring ice bloom.

The geographic extent of the ice communities is not known. This is true for both the algae and the invertebrates and should be determined on both local and broad geographic scales. Additional food web studies are needed, particularly of the smaller-size groups, to define the roles of the macrofauna and meiofauna.

There is some evidence that ice algae may be possible indicators of past environmental conditions. The effects of pollution on the ice community are not known.

## REFERENCES

Ackley, S. F. 1982. Ice scavenging and nucleation: Two mechanisms for incorporation of algae into newly formed sea ice. *EOS*, **63**:54.

Ackley, S. F., K. R. Buck, and S. Taguchi. 1979.

Standing crop of algae in the sea ice of the Weddell Sea region. *Deep-Sea Res.* **26A:**269–281.

Adam, D. P., and A. D. Mahood. 1981. Chrysophyte cysts as potential environmental indicators. *Geol. Soc. Am. Bull.* **92:**839–844.

Alexander, V. 1974. Primary productivity regimes of the nearshore Beaufort Sea, with reference to potential roles of ice biota. In J. C. Reed and J. E. Sater (eds.), *The Coast and Shelf of the Beaufort Sea.* The Arctic Institute of North America, Arlington, pp. 609–635.

Alexander, V. 1980. Interrelationships between the seasonal sea ice and biological regimes. *Cold Regions Sci. Tech.* **2:**157–178.

Alexander, V., and T. Chapman. 1981. The role of epontic algal communities in Bering Sea ice. In D. W. Hood and J. A. Calder (eds.), *The Eastern Bering Sea Shelf: Oceanography and Resources.* University of Washington Press, Seattle, pp. 773–780.

Alexander, V., R. Horner, and R. C. Clasby. 1974. *Metabolism of Arctic Sea Ice Organisms.* Rep. R74-4, Institute of Marine Science, University of Alaska, Fairbanks. 120p.

Allen, M. B. 1971. High latitude phytoplankton. *Ann. Rev. Ecol. Syst.* **2:**261–276.

Andersen, O. G. N. 1977. Primary production associated with sea ice at Godhaven, Disko, West Greenland. *Ophelia* **16:**205–220.

Andriashev, A. P. 1968. The problem of the life community associated with the Antarctic fast ice. In R. I. Currie (ed.), *Symposium on Antarctic Oceanography.* W. Heffer & Sons Ltd., Cambridge, pp. 147–155.

Apollonio, S. 1961. The chlorophyll content of Arctic sea-ice. *Arctic* **14:**197–200.

Apollonio, S. 1965. Chlorophyll in Arctic sea ice. *Arctic* **18:**118–122.

Banse, K. 1974. Determining the carbon-to-chlorophyll ratio of natural phytoplankton. *Mar. Biol.* **41:**199–212.

Barnard, J. L. 1959. Epipelagic and under-ice Amphipoda of the central Arctic Basin. In V. Bushnell (ed.), *Scientific Studies at Fletcher's Ice Island T-3 (1952–1955).* Geophysical Research Paper 63, pp. 115–152.

Booth, J. A. 1984. The epontic algal community of the ice edge zone and its significance to the Davis Strait ecosystem. *Arctic* **37:**234–243.

Bradstreet, M. S., and W. E. Cross. 1982. Trophic relationships at high Arctic ice edges. *Arctic* **35:**1–12.

Buck, K. R. 1980. *Morphology and Distribution of the Acanthoecidae (Choanoflagellata) from the Weddell Sea during the Austral Summer, 1977.* Report 80-16, Cold Regions Research and Engineering Laboratory, Hanover, N.H., 26p.

Bunt, J. S. 1963. Diatoms of Antarctic sea-ice as agents of primary production. *Nature* **199:**1255–1257.

Bunt, J. S. 1964. Primary productivity under sea ice in Antarctic waters. 2: Influence of light and other factors on photosynthetic activities of Antarctic marine microalgae. *Antarct. Res. Ser.* **1:**27–31.

Bunt, J. S. 1968. Some characteristics of microalgae isolated from Antarctic sea ice. *Antarct. Res. Ser.* **11:**1–14.

Bunt, J. S., and C. C. Lee. 1972. Data on the composition and dark survival of four sea-ice microalgae. *Limnol. Oceanogr.* **17:**458–461.

Bunt, J. S., and E. J. F. Wood. 1963. Microalgae and Antarctic sea-ice. *Nature* **199:**1254–1255.

Bunt, J. S., O. van H. Owens, and C. Hoch. 1966. Exploratory studies on the physiology and ecology of a psychrophilic marine diatom. *J. Phycol.* **2:**96–100.

Burch, M. D., and H. J. Marchant. 1983. Motility and microtubule stability of Antarctic algae at sub-zero temperatures, *Protoplasma* **115:**240–242.

Bursa, A. S. 1963. Phytoplankton in coastal waters of the Arctic Ocean at Point Barrow, Alaska. *Arctic* **16:**239–262.

Bursa, A. S. 1965. *Discoasteromonas calciferus* n. sp., an Arctic relict secreting *Discoaster* Tak [sic] Sin Hok 1927. *Grana Palynol.* **6:**147–165.

Bursa, A. S. 1971. Morphogenesis and taxonomy of fossil and contemporary Dinophyta secreting discoasters. In A. Farinacci (ed.), *Proceedings II Planktonic Conference Roma, 1970,* vol. 1. Edizioni Technoscienza, Rome, pp. 129–143.

Carey, A. G., Jr. 1982. Shallow southwest Beaufort Sea macrofauna: Interactions with the epibenthic community. *EOS* **3:**949.

Carey, A. G., Jr. 1985. Marine ice fauna: Arctic. In R. Horner (ed.), *Sea Ice Biota,* CRC Press, Boca Raton, pp. 173–190.

Carey, A. G., Jr., and M. A. Boudrias. 1987. Feeding ecology of *Pseudalibrotus (Onisimus) litoralis* Kröyer (Crustacea: Amphipoda) on the Beaufort Sea inner continental shelf. *Polar Biol.* **8:**29–33.

Carey, A. G., Jr., and P. A. Montagna. 1982. Arctic sea ice faunal assemblages: First approach to description and source of underice meiofauna. *Mar. Ecol. Prog. Ser.* **8:**1–8.

Clasby, R. C., V. Alexander, and R. Horner 1976. Primary productivity of sea-ice algae. In D. W. Hood and D. C. Burrell (eds.), *Assessment of the Arctic Marine Environment: Selected Topics.* Institute of Marine Science, University of Alaska, Fairbanks, pp. 289–327.

Clasby, R. C., R. Horner, and V. Alexander. 1973. An in situ method for measuring primary productivity of Arctic sea ice algae. *J. Fish. Res. Board Can.* **30**:835–838.

Cleve, P. T. 1883. Diatoms collected during the Expedition of the *Vega*. *Vega-Expeditionens Vetenskapliga Iakttagelser* **3**:455–517.

Cleve, P. T. 1896. Diatoms from Baffin Bay and Davis Strait. *Bih. K. Sven. Vetenskapsakad. Handl.* **22,** Afd. III(4):1–22.

Cleve, P. T. 1898. Diatoms from Franz-Josef Land collected by the Harmsworth-Jackson Expedition. *Bih. K. Sven. Vetenskapsakad. Handl.* **24**, Afd. III (2). 3–26.

Cleve, P. T. 1899. Mikroskopiskundersökning at stoft, funnet på drifis i Ishafvet. *Oefv. K. Vetenskapsakad. Foreh.* **1899**(3):123–130.

Cleve, P. T. 1900. Microscopical examination of dust from drift-ice north of Jan Mayen. *Oefv. K. Vetenskapsakad. Foreh.* **1900**(4):393–397.

Cleve, P. T., and A. Grunow. 1880. Beiträge zur Kenntniss der arctischen Diatomeen. *K. Sven. Vetenskapsakad. Handl.* **17**(2):1–122.

Crosby, L. H., and E. J. F. Wood. 1959. Studies on Australian and New Zealand diatoms. II: Normally epontic and benthic genera. *R. Soc. N.Z. Trans.* **86**:1–58.

Cross, W. E. 1982. Under-ice biota at the Pond Inlet ice edge and in adjacent fast ice areas during spring. *Arctic* **35**:13–27.

Dale, B. 1983. Dinoflagellate resting cysts: "Benthic plankton." In G. A. Fryxell (ed.), *Survival Strategies of the Algae.* Cambridge University Press, Cambridge, pp. 69–136.

Demers, S. J., C. Therriault, and C. Descolos-Gros. 1984. Biomasse et composition spécifique de la microflore des glaces saisonnières: Influences de la lumière et de la vitesse de congélation. *Marine Biol.* **78**:185–191.

Dunbar, M. J., and J. C. Acreman. 1980. Standing crops and species composition of diatoms in sea ice from Robeson Channel to the Gulf of St.Lawrence. *Ophelia* **19**:61–72.

Ehrenberg, C. G. 1841. Einen Nachtrag zu dem Vortrage über Verbreitung und Einfluss des mikroskopischen Lebens in Süd- und Nord-Amerika. *Akad. Wiss., Berlin, Monatsber.* **1841**:202–209.

Ehrenberg, C. G. 1853. Über neue Anschauungen des kleinsten nördlichen Polarlebens, *Akad. Wiss., Berlin, Monatsber.* **1853**:522–529.

English, T. S. 1961. *Some Biological Oceanographic Observations in the Central North Polar Sea, Drift Station Alpha, 1957–1958.* Sci. Rep. No. 15, The Arctic Institute of North America, Washington, D.C. 79p.

Falkowski, P. G. 1980. Light-shade adaptation in marine phytoplankton. In P. G. Falkowski (ed.), *Primary Productivity in the Sea.* Plenum Press, New York, pp. 99–119.

Fryxell, G. A., T. A. Villareal, and G. J. Doucette. 1981. Diatom resting spores and Agulhas collections. *Antarct. J. U.S.* **16**:128–130.

Garrison, D. L., S. F. Ackley, and K. R. Buck. 1983. A physical mechanism for establishing algal populations in frazil ice. *Nature* **306**:363–365.

George, R. Y., and A. Z. Paul 1970. *USC-FSU Biological Investigations from the Fletcher's Ice Island T-3 on Deep-sea and Under-ice Benthos of the Arctic Ocean.* Tech. Rep. No. 1, University of Southern California Arctic Project, Los Angeles, 73p.

Golikov, A. N., and A. G. Averincev. 1977. Distribution patterns of benthic and ice biocenoses in the high latitudes of the Polar Basin and their part in the biological structure of the World Ocean. In M. J. Dunbar (ed.), *Polar Oceans.* The Arctic Institute of North America, Calgary, pp. 331–364.

Grainger, E. H. 1977. The annual nutrient cycle in sea-ice. In M. J. Dunbar (ed.), *Polar Oceans.* The Arctic Institute of North America, Calgary, pp. 285–299.

Grainger, E. H., and M. S. Evans. 1982. Seasonal variations in chlorophyll and nutrients in a Canadian Arctic estuary. *Estuaries* **5**:294–301.

Grainger, E. H., and S. I. C. Hsiao. 1982. A study of the ice biota of Frobisher Bay, Baffin Island 1979–1981. *Can. Manuscr. Rep. Fish. Aquat. Sci. No. 1647,* 128p.

Gran, H. H. 1904. Diatomaceae from the ice-floes and plankton of the Arctic Ocean. *Sci. Res. Norw. N. Polar Exped.* **4**(11):3–74.

Grant, W. S., and R. Horner. 1976. Growth responses to salinity variation in four Arctic ice diatoms. *J. Phycol.* **12**:180–185.

Green, J. M., and D. H. Steele. 1975. Observations on marine life beneath sea ice, Resolute Bay, N.W.T. *Proceedings Circumpolar Conference on Northern Ecology,* vol. 2. National Research Council of Canada, Ottawa, pp. 77–86.

Gulliksen, B. 1984. Cryopelagic fauna from Svalbard waters. *Sarsia* **69**:17–23.

Hällfors, G., and Å. Niemi. 1974. A *Chrysochromulina* (Haptophyceae) bloom under the ice in the Tvärminne Archipelago, southern coast of Finland. *Mem. Soc. Fauna Flora Fenn.* **50**:89–104.

Hameedi, M. J. 1978. Aspects of water column primary productivity in the Chukchi Sea during summer. *Marine Biol.* **48**:37–46.

Hargraves, P. E., and F. W. French. 1983. Diatom resting spores: Significance and strategies. In G. A. Fryxell (ed.), *Survival Strategies of the Algae.* Cambridge University Press, Cambridge, pp. 49–68.

Hitchcock, G. L. 1980. Diel variation in chlorophyll *a*, carbohydrate and protein content of the marine diatom *Skeletonema costatum. Marine Biol.* **57**:271–278.

Holmquist, C. 1958. An observation on young specimens of *Ammodytes dubius. Medd. Grønl.* **159**(2):9–14.

Horner, R. 1976. Sea ice organisms. *Oceanogr. Mar. Biol. Ann. Rev.* **14**:167–182.

Horner, R. 1977. History and recent advances in the study of ice biota. In M. J. Dunbar (ed.), *Polar Oceans.* The Arctic Institute of North America, Calgary, pp. 269–283.

Horner, R. 1984. Do ice algae produce the spring phytoplankton bloom in seasonally ice-covered waters? A review of recent literature. In D. Mann (ed.), *Proceedings of the Seventh Symposium on Living and Fossil Diatoms.* J. Cramer, Braunschweig, pp. 401–409.

Horner, R. 1985. History of ice algal investigations. In R. Horner (ed.), *Sea Ice Biota,* CRC Press, Boca Raton, Fla., pp. 1–19.

Horner, R. and V. Alexander. 1972. Algal populations in Arctic sea ice: An investigation of heterotrophy. *Limnol. Oceanogr.* **17**:454–458.

Horner, R., and G. C. Schrader. 1982. Relative contributions of ice algae, phytoplankton, and benthic microalgae to primary production in nearshore regions of the Beaufort Sea. *Arctic* **35**:484–503.

Horner, R., E. E. Syvertsen, D. P. Thomas, and C. Lange. 1988. Proposed terminology and reporting units for sea ice algal assemblages. *Polar Biol.* **8**:249–253.

Hoshiai, T. 1977. Seasonal change of ice communities in the sea ice near Syowa Station Antarctica. In M. J. Dunbar (ed.), *Polar Oceans.* The Arctic Institute of North America, Calgary, pp. 307–317.

Hsiao, S. I. C. 1980. Quantitative composition, distribution, community structure and standing stock of sea ice microalgae in the Canadian Arctic. *Arctic* **33**:768–793.

Hsiao, S. I. C. 1983. A checklist of marine phytoplankton and sea ice microalgae recorded from Arctic Canada. *Nova Hedwigia* **37**:225–313.

Jeffrey, S. W., and M. Vesk. 1981. The phytoplankton—systematics, morphology and ultrastructure. In M. N. Clayton and R. J. King (eds.), *Marine Botany: An Australasian Perspective.* Longman Cheshire, Melbourne, pp. 138–179.

Kern, J. C., and A. G. Carey, Jr. 1983. The faunal assemblage inhabiting seasonal sea ice in the nearshore Arctic Ocean with emphasis on copepods. *Mar. Ecol. Prog. Ser.* **10**:159–167.

Kol, E. 1942. The snow and ice algae of Alaska. *Smithson. Misc. Collect.* **101**(16):1–36.

Larsen, L. 1980. Sediment-laden sea ice: Concepts, problems, and approaches. Outer Continental Shelf Environmental Assessment Program, BLM/NOAA. *Arctic Project Bull.* **29**:59–73.

Legendre, L., R. G. Ingram, and M. Poulin. 1981. Physical control of phytoplankton production under sea ice (Manitounuk Sound, Hudson Bay). *Can. J. Fish. Aquat. Sci.* **38**:1385–1392.

Li, W. K. W., and T. Platt. 1982. Distribution of carbon among photosynthetic end-products in phytoplankton of the eastern Canadian Arctic. *J. Phycol.* **18**:466–471.

Lowry, L. F., and K. J. Frost. 1981. Distribution, growth and food of Arctic cod *(Boreogadus saida)* in the Bering, Chukchi, and Beaufort seas. *Can. Field Nat.* **95**:186–191.

McAllister, D. E. 1975. Ecology of the marine fishes of arctic Canada. *Proceedings Circumpolar Conference on Northern Ecology,* vol. 2. National Research Council of Canada, Ottawa, pp. 49–65.

McConville, M. J. 1985. Chemical composition and biochemistry of sea ice microalgae. In R. Horner (ed.), *Sea Ice Biota.* CRC Press, Boca Raton, Fla., pp. 105–129.

McConville, M. J., and R. Wetherbee. 1983. The bottom-ice microalgal community from annual ice in the inshore waters of East Antarctica. *J. Phycol.* **19**:431–439.

MacGinitie, G. E. 1955. Distribution and ecology of the marine invertebrates of Point Barrow, Alaska. *Smithson. Misc. Collect.* **128**:1–201.

McRoy, C. P., and J. J. Goering. 1974. The influence of ice on the primary productivity of the Bering Sea. In D. W. Hood and E. J. Kelley (eds.), *Oceanography of the Bering Sea.* Institute of Marine Science, University of Alaska, Fairbanks, pp. 403–421.

Matheke, G. E. M., and R. Horner. 1974. Primary productivity of the benthic microalgae in the Chukchi Sea near Barrow, Alaska. *J. Fish. Res. Board Can.* **31**:1779–1786.

Matsuda, T. 1961. Biological investigation during Antarctic Expedition voyage and at East and West Ongul Islands in Antarctica. *Nankyoku Shiryo* **11**:159.

Meguro, H. 1962. Plankton ice in the Antarctic Ocean. *Antarct. Rec.* **14**:1192–1199.

Meguro, H., K. Ito, and H. Fukushima. 1966. Diatoms and the ecological conditions of their growth in sea ice in the Arctic Ocean. *Science* **152**:1089–1090.

Meguro, H., K. Ito, and H. Fukushima. 1967. Ice flora (bottom type): A mechanism for primary production in polar seas and the growth of diatoms in sea ice. *Arctic* **20:**114–133.

Meister, F. 1930. Mission du "Porquoi-Pas?" en 1929 sous le Commandement du Dr. J.-B. Charcot. Diatomes récoltées par R.-Ph. Dollfus sur un glace flottante. *Bull. Mus. Nat. Hist. Nat.* (ser. 2), **2:**329–330.

Mel'nikov, I. A. 1976. Hydrobiological investigations in the central Arctic Ocean. *Oceanology* **16:**314–315.

Mel'nikov, I. A. 1979. Cryobiological observations in the Central Arctic Basin (method and some results of studies). *Oceanology* **19:**93–96.

Mel'nikov, I. A., and G. L. Pavlov. 1978. Characteristics of organic carbon distribution in the waters and ice of the Arctic Basin. *Oceanology* **18:**163–167.

Mitchell, J. G. 1982. Sea ice environments are indicated by archaeomonads. *EOS* **63:**47.

Mitchell, J. G., and M. W. Silver. 1982. Modern archaeomonads indicate sea-ice environments. *Nature* **296:**437–439.

Mohr, J. L. 1959. Marine biological work. In V. Bushnell (ed.), *Scientific Studies at Fletcher's Ice Island T-3 (1952–1955).* Geophysical Research Paper 63, pp. 82–103.

Mohr, J. L., and J. Tibbs. 1963. Ecology of ice substrates. In *Proceedings of the Arctic Basin Symposium.* Tidewater Publishing Corp., Centreville, Maryland, pp. 245–249.

Nansen, F. 1897. *Farthest North,* 2 vols. Harper & Brothers, Publishers, New York, 729p.

Nansen, F. 1906. Protozoa from the ice-floes of the North Polar Sea. *Sci. Res. Norw. N. Polar Exped.* **5**(16)1–22.

Niebauer, H. J., V. Alexander, and R. T. Cooney. 1981. Primary production at the eastern Bering Sea ice edge: The physical and biological regimes. In D. W. Hood and J. A. Calder (eds.), *The Eastern Bering Sea Shelf: Oceanography and Resources,* vol. 2. University of Washington Press, Seattle, pp. 763–772.

Oradovskiy, S. G. 1972. Studies on the composition of nutrients in the ice of the Barents Sea. *Trudy vses. nauchno-issled. Inst. morsk. ryb. Khoz. Okeanogr.* **75:**65–73. (In Russian; English summary.)

Østrup, E. 1895. Marine diatomeer fra østgrønland. *Medd. Grønl.* **18:**395–476.

Palmisano, A. C., and C. W. Sullivan. 1982. Physiology of sea ice diatoms. I. Response of three polar diatoms to a simulated summer-winter transition. *J. Phycol.* **18:**489–498.

Perovich, D. K., and T. Grenfell. 1981. Laboratory studies of the optical properties of young sea ice. *J. Glaciol.* **27:**331–346.

Pett, R. J., G. P. Vickers, J. C. Acreman, and A. G. Ethier. 1983. *A Seasonal Study of the Epontic and Planktonic Communities Near the Issungnak Artificial Island and McKinley Bay in the Canadian Beaufort Sea.* Contract ISZ81-00026 for Department of Supply and Services Canada by Arctic Laboratories Ltd., Sydney. 89p.

Poulin, M., and A. Cardinal. 1982*a*. Sea ice diatoms from Manitounuk Sound, southeastern Hudson Bay (Quebec, Canada). I. Family Naviculaceae. *Can. J. Bot.* **60:**1263–1278.

Poulin, M., and A. Cardinal. 1982*b*. Sea ice diatoms from Manitounuk Sound, southeastern Hudson Bay (Quebec, Canada). II. Naviculaceae, genus *Navicula. Can. J. Bot.* **60:**2825–2845.

Poulin, M., and A. Cardinal. 1983. Sea ice diatoms from Manitounuk Sound, southeastern Hudson Bay (Quebec, Canada). III. Cymbellaceae, Entomoneidaceae, Gomphonemataceae, and Nitzschiaceae. *Can. J. Bot.* **61:**107–118.

Poulin, M., A. Cardinal, and L. Legendre. 1983. Réponse d'une communauté de diatomées de glace à un gradient de salinité (baie d'Hudson). *Marine Biol.* **76:**191–202.

Price, I. R. 1981. Plants of the marine environment. In M. N. Clayton and R. J. King (eds.), *Marine Botany: An Australasian Perspective.* Longman Cheshire, Melbourne, pp. 15–24.

Richardson, K., J. Beardall, and J. A. Raven. 1983. Adaptation of unicellular algae to irradiance: An analysis of strategies. *New Phytol.* **93:**157–191.

Rodhe, W. 1955. Can plankton production proceed during winter darkness in subarctic lakes? *Verh. Internat. Verein. Limnol.* **12:**117–1–2.

Ross, R. 1954. Algae: Planktonic. *Bot. Rev.* **20:**400–416.

Round, F. E. 1981. *The Ecology of Algae.* Cambridge University Press, Cambridge, 653p.

Saito, K., and A. Taniguchi. 1978. Phytoplankton communities in the Bering Sea and adjacent seas. II. Spring and summer communities in seasonally ice-covered areas. *Astarte* **11:**27–35.

Sancetta, C. 1981*a*. Oceanographic and ecologic significance of diatoms in surface sediments of the Bering and Okhotsk seas. *Deep-Sea Res.* **28:**789–817.

Sancetta, C. 1981*b*. Diatoms as hydrographic tracers: Example from Bering Sea sediments. *Science* **211:**279–281.

Sancetta, C. 1982. Distribution of diatom species in surface sediments of the Bering and Okhotsk seas. *Micropaleontology* **28:**221–257.

Schandelmeier, L., and V. Alexander. 1981. An analysis of the influence of ice on spring phytoplankton population structure in the southeast Bering Sea. *Limnol. Oceanogr.* **26**:935–943.

Schell, D. M. 1980*a*. Foodweb and nutrient dynamics studies in nearshore Alaskan Beaufort Sea waters. Outer Continental Shelf Environmental Assessment Program, BLM/NOAA. *Environmental Assessment of the Alaskan Continental Shelf, Annual Reports of Principal Investigators* **2**:467–515.

Schell, D.M. 1980*b*. Chemical overview of biological-physical process interactions. Outer Continental Shelf Environmental Assessment Program, BLM/NOAA. *Arctic Project Bull.* **29**:25–31.

Schrader, G. C., R. Horner, and G. F. Smith. 1982. An improved chamber for in situ measurement of primary productivity by sea ice algae. *Can. J. Fish. Aquat. Sci.* **39**:522–524.

Shifrin, N. S., and S. W. Chisholm. 1981. Phytoplankton lipids: Interspecific differences and effects of nitrate, silicate and light-dark cycles. *J. Phycol.* **17**:374–384.

Smith, A. E., and I. Morris. 1980. Pathways of carbon assimilation in phytoplankton from the Antarctic Ocean. *Limnol. Oceanogr.* **25**:865–872.

Sullivan, C. W., A. C. Palmisano, S. Kottmeier, and R. Moe. 1982. Development of the sea ice microbial community in McMurdo Sound. *Antarct. J. U.S.* **17**:155–157.

Sutherland, P. C. 1852. *Journal of a Voyage in Baffin's Bay and Barrow Straits, in the Years 1850–1851, Performed by H. M. Ships "Lady Franklin" and "Sophia," under the Command of Mr. William Penny, in Search of the Missing Crews of H. M. Ships "Erebus" and "Terror,"* 2 vols. Longman, Brown, Green, and Longmans, London. 869p.

Takahashi, E. 1981. Loricate and scale-bearing protists from Lützow-Holm Bay, Antarctica. I. Species of the Acanthoecidae and the Centrohelida found at a site selected on the fast ice. *Antarct. Rec.* **73**:1–22.

Thomas, C. W. 1963. On the transfer of visible radiation through sea ice and snow. *J. Glaciol.* **4**:481–484.

Tibbs, J. F. 1967. On some planktonic Protozoa taken from the track of Drift Station ARLIS I, 1960–61. *Arctic* **20**:247–254.

Usachev, P. I. 1938. Biological analysis of ice-floes. *C. R. (Dokl.) Acad. Sci. URSS* **19**:645–648.

Usachev, P. I. 1949. The microflora of polar ice. *Tr. Inst. Okeanol., Akad. Nauk SSSR* **3**:216–259. (In Russian.)

Werner, D. 1977. Silicate metabolism. In D. Werner (ed.), *The Biology of Diatoms.* Blackwell Scientific Publications, Oxford, pp. 110–149.

Whitaker, T. M. 1977. Sea ice habitats of Signy Island (South Orkneys) and their primary productivity. In G. A. Llano (ed.), *Adaptations within Antarctic Ecosystems.* Gulf Publishing Co., Houston, pp. 75–82.

Wilce, R. T. 1967. Heterotrophy in arctic sublittoral sea weeds. *Bot. Mar.* **10**:185–197.

Wood, E. J. F. 1964. Studies in microbial ecology of the Australasian region. V. Microbiology of some Australian estuaries. *Nova Hedwigia* **8**:461–527.

Zhuse, A. P., O. G. Kozlova, and V. V. Mukhina. 1967. Vidovoi sostav i zonal'noe rasprelelenie diatomei v poverkhnostnom sloe osodkov Tikhogo Okeana. *Dokl. Akad. Nauk SSSR* **172**:1183–1186. (In Russian.)

# 6: Primary Production, Chlorophyll, Light, and Nutrients Beneath the Arctic Sea Ice

Ole G. N. Andersen

## INTRODUCTION

Determinations of primary production and related factors beneath the Arctic sea ice have been few. A thorough treatment, however, of all available data shall not be attempted here. The main topic concerns conditions of primary production in the Polar Basin and adjacent ice-laden areas, with additional information from more southerly waters, where winter ice is more or less a regular occurrence. Production by phytoplankton, ice algae, and benthic microalgae is discussed.

## PHYTOPLANKTON

The 1972 FAO map showing the global distribution of marine phytoplankton production has the ice-laden Arctic Polar Basin marked as a low production area with $<100$ mg C m$^2$ d in the productive season as opposed to rich ice-free coastal upwelling areas with $>500$ mg C m$^2$ d. The Arctic, however, encompasses more than the Polar Basin, and it can be divided into a number of areas that are very different with respect to hydrography and primary production in the pelagic realm.

The least productive area is the Central Polar Basin, which along with some fjords and straits in northern Canada and Greenland, may be termed high-Arctic due to the perennial ice cover of 85–100 percent. More productive are some marginal seas and inshore waters. Still higher in production is the Arctic currents with their marginal ice zones (MIZ) leaving the Polar Basin. Highest in production are the polar fronts, as shown in Nielsen's 1958 map (Fig. 6-1). This map also shows that production can be very high inshore in the Arctic. A special case is glacier fjords and waters carrying great concentrations of icebergs away from them, such as the Vaigat north of Disko and parts of Disko Bay. Far les productive are turbid nonglacier fjords and estuarine waters, in spite of their being ice-free during part of the summer.

## ICE ALGAE

Production is largely restricted to first-year ice, and is greatest in neritic shallow areas

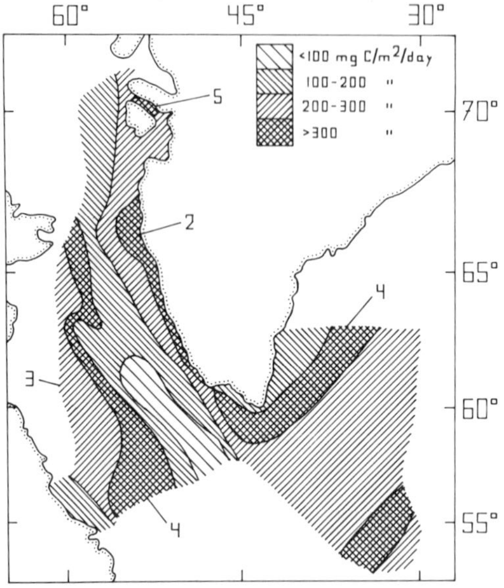


Figure 6-1. Gross primary production in the waters around Greenland in July–August 1954. (After Nielsen, 1958.)

with nutrient replenishment from the sediment and from land.

## BENTHIC MICROALGAE

Production is restricted to shallow waters. For primary production to occur, several conditions must be met. First, a seed of algal cells or spores (cysts) must be present. These need light and nutrients to subsist, grow, and multiply. The algal cells also need time and sufficient light, that is, light for daily periods long enough to cover respiratory energy expenditures. For the stationary ice and benthic algae the question of daily light is simple, whereas for the planktonic forms vertical stability is of vital importance.

### Seeding

If light, nutrients, and time are favorable, algal seeding is not a problem. Algal cells or spores are always present in the water column. Even in darkness, beneath the snow-covered pack ice of the high-Arctic Polar Sea, algae are present and ready to multiply when the opportunity arises. They rest on the sea bottom, are found scattered in the water, or occur often in great numbers inside or on the lower surface of the ice. In April 1979, a few hours after a newly formed lead had frozen over in the Polar Basin, chlorophyll could be found in the ice (Andersen, unpublished). The same was the case in Disko Bay in March 1976, with chlorophyll concentrations 3–66 times greater in the lower 4.5 cm of ice and 3–7 times greater in the 4.5 cm of unfrozen water next to the ice than 1 m below it. This was in spite of conditions being highly unstable in the upper 40 m. (Andersen, 1977*b*; see section on phytoplankton and ice algal production).

### Light

While light in tropical and temperate latitudes is at no time totally limiting to primary production—that is, the daily surface irradiance is always sufficient to cover respiration loss—in the Arctic the opposite is true. The solar irradiance is on the average 1.5 g cal cm² min for the Earth, distributed as shown in Fig. 6-2. If no clouds were present, the mean solar irradiance and the annual insolation at the North Pole would be one-fourth the values at the equator, and at 60°N latitude they would be one-half the values, due to differences in solar elevation and day length.

The transmission of irradiance through the sea surface is dependent on solar

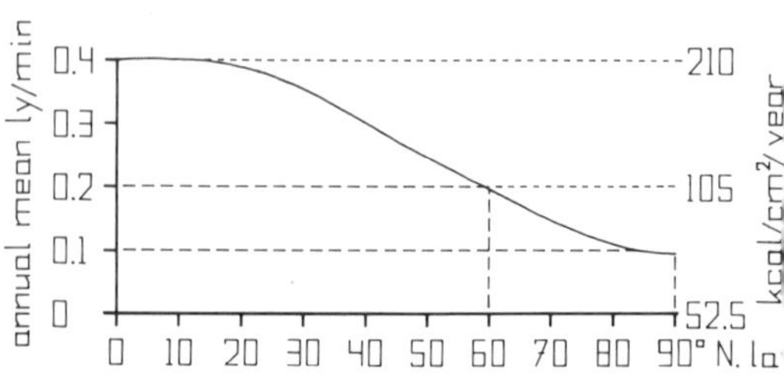


Figure 6-2. Global annual insolation and mean irradiance, given no clouds.

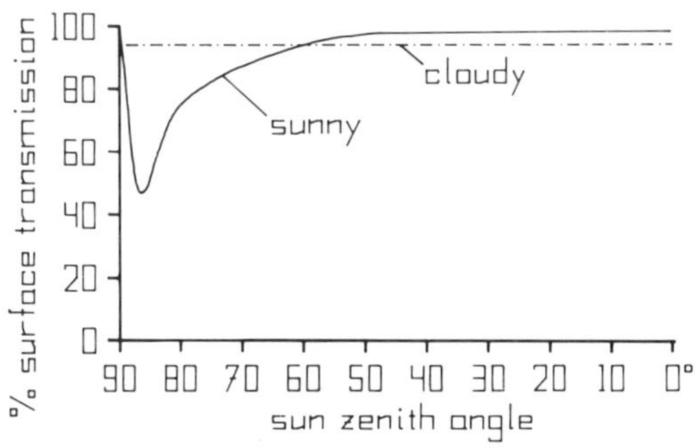


Figure 6-3. Sea-surface light transmission related to zenith angle of the sun, with and without clouds.

elevation, as shown in Figure 6-3. The elevation of the sun has little influence on surface transmission under overcast conditions, whereas surface transmission is reduced to a minimum of 50 percent at a 5° elevation of the sun. However, it is only when the sun is below about 30° that a reduction in surface transmission is noteworthy. This occurs in temperate and Arctic latitudes. In Jørgen Brønlund Fj. (JBF), North Greenland (82° N lat.), the maximum elevation of the sun is precisely 30°. Here surface transmission under clear skies is 6.5 times greater at a sun elevation of 30° than at 10° (Andersen, 1977*a*).

Figure 6-4 shows the daily amplitude of the movements of the sun above the horizon in JBF. Nearly 7.5 months elapse from the first appearance of the sun in late February till it is seen for the last time in early October. If the plankton needs 8 percent of what can be produced at light saturation and it is assumed that saturation is attained at least at midday under clear skies, then production cannot take place before early April and will last no longer than 5.3 months, ending in early September. Under overcast conditions production can, at best, last for 4.7 months. While the theo-

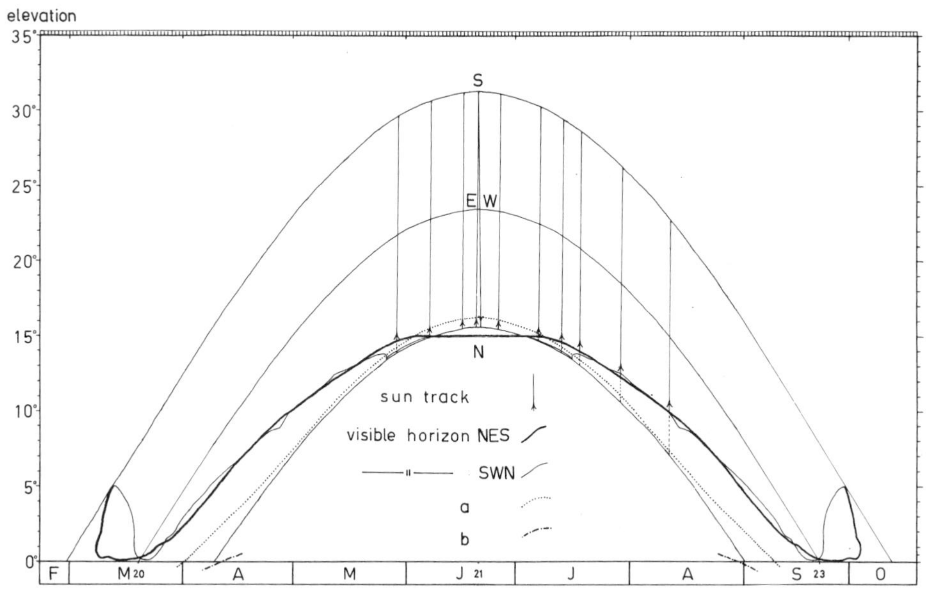


Figure 6-4. The amplitude described by the sun above the true horizon at Jørgen Brønlund Fjord in the period with sun is shown by lines representing the sun angle when the sun is to the north, to the east or west, and to the south, respectively. Sun tracks mark the primary production experiments performed in 1968. The curve *a*, representing maximum production on a sunny day, has been derived by subtracting 8 percent of midday illumination (to compensate for respiration) from illumination at any time throughout the 24-hour cycle. The curves *b* are the end points of a similar curve representing cloudy conditions. (From Andersen, 1977*a*.)

retical period of production in the high-Arctic JBF, calculated on the basis of insolation alone, is about 5 months, actual production is hampered by the thick cover of ice during the first 3 of these months. This leaves, at best, a production period of 2 months or even less due to freezing over already in late August. However, JBF is far better off than the Polar Basin with its perennial cover of pack ice. The situation under near-shore fast ice is even worse.

Ice and especially snow-covered ice is the major factor restricting the penetration of light into Arctic waters. While the midsummer albedo from open water in the Arctic Ocean is quite constantly about 4 percent (Hanson, 1961), the reflection due to ice and snow is extremely varied. Ice with a cover of snow reflects up to 93 percent of direct incident solar radiation (Langleben, 1969). In the spring, when most of the ice is covered with snow, the albedo is about 80–90 percent. Melting snow reflects as little as 50 percent, and melting sea ice has an albedo of about 45–70 percent, depending on wetness and degree of contamination (Zubov, 1943; Hanson, 1961; Langleben, 1969, 1971). Where puddling occurs, the albedo is reduced linearly to a low of 0–20 percent where the surface is entirely flooded. Where puddles are frozen over, the albedo increases to 50 percent (Rothrock, 1978).

Mid-summer albedo over the Polar Basin is 46 percent if the ice cover is 85 percent and puddles cover 30 percent of the ice (Hanson, 1961). Puddle cover in the Polar Basin probably does not exceed 50 percent (Hanson, 1961), at least not for very long. However, almost total flooding does occur where sufficient amounts of fresh water from land and from the surface of the melting sea ice come into contact with the cold lower surface of the ice. Here the fresh water freezes and seals off the bottom of the ice, as well as tidal cracks and holes bored through the ice, so meltwater from the major part of the ice cannot seep through. Andersen (1977*a*) describes this melting process in JBF (Fig. 6-5) where the entire lower surface of the approximately 2-m-thick ice cover became penetrable to meltwater on June 20, and the totally flooded surface was transformed overnight into a dazzling white, highly reflecting, snowlike surface consisting of closely set vertical ice needles, several centimeters long, of the congealed ice (Fig. 6-6). Puddling again occurred, diminishing the albedo, which reached a minimum when at a thickness of about 1 m the ice became totally drenched and dark. Thus, the disappearance of all snow for the summer does not necessarily imply that the albedo is permanently reduced.

Snow is a decisive factor, nevertheless. Ten percent of surface irradiance passes 5 cm of fresh snow, and, while an additional 10 cm lower transmission to 1 percent, it takes the same 5 cm of snow plus 3 m of ice to lower transmission similarly to 0.5–1 percent. However, 5 cm of partially melted snow and 15 cm of icy snow transmit 23 percent and 7 percent respectively (Zubov, 1943). In the Antarctic, Bunt and Lee (1970) measured a transmission of 27 percent of visible irradiance (380–750 nm) through 1 m of ice, whereas in West Greenland only 8–15 percent of green light passed 15–34 cm of ice with 2–3 cm of crusty snow (Andersen, 1977*b*). English (1961) measured less than 1 percent of visible direct surface irradiance (actually 300–3,000 nm) beneath 3 m of level polar ice with a few centimeters of snow. In the Antarctic, Bunt and Lee (1970) similarly measured less than 1 percent (380–750 nm) beneath all thicknesses of ice with snow. Approximately 0.1 percent passed 3 m of thick ice, and 0.01 percent passed 4.5 m of thick ice with snow. Quantum measurements of photosynthetically available energy (400–700 nm) in the Polar Basin (84°N) give an even gloomier picture, since only 0.04–0.13 percent passed 2.0–2.5 m of thick ice with 3–5 cm of snow in

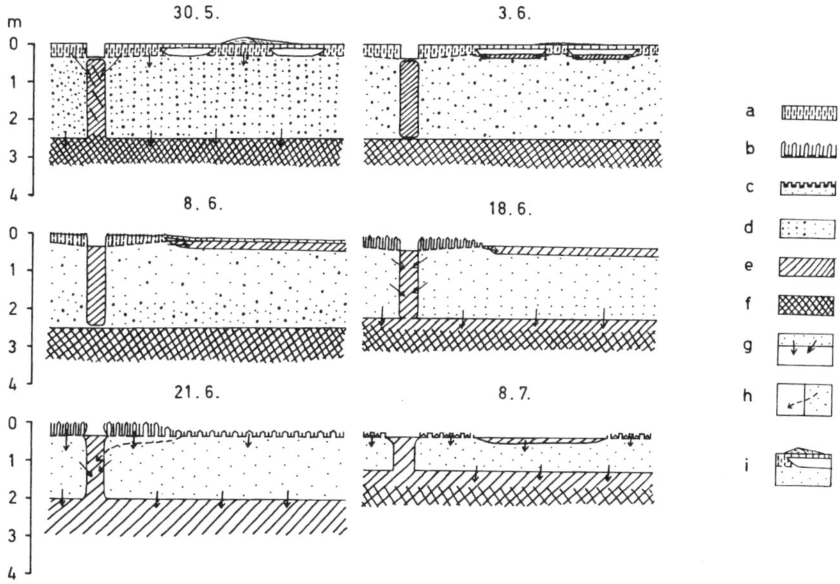


Figure 6-5. Vertical sections of the fjord ice in Jørgen Brønlund Fjord in 1968. From June 21 onward the lower surface is not level as shown. *a:* smooth ice with air bubbles; *b:* snowlike ice; *c:* snowlike ice soaked in water; *d:* porous ice below the water line with a stronger (left) and a weaker (right) brine; *e:* fresh water; *f:* salt water; *g:* seeping and melting out of brine; *h:* meltwater runoff; *i:* snow drift on smooth ice with bubbles of air and a larger air-filled cavity.

Figure 6-6. Ice needles from snowlike ice at Jørgen Brønlund Fjord 1968.

late April 1979 (Andersen, unpublished). On T-3 extinction coefficients for snow and ice were found to be 0.29 cm and 0.012 cm respectively (Weller and Parker, 1972). According to these coefficients, only 20 cm of snow reduces illumination to 0.1 percent of incident, while ice for every half meter reduces illumination approximately to the following: 55 percent (0.5 m), 30 percent (1 m), 16 percent (1.5 m), 9 percent (2 m), 5 percent (2.5 m), 2.5 percent (3 m), 1.5 percent (3.5 m), 0.8 percent (4 m), 0.44 percent (4.5 m), and 0.24 percent (5 m).

### Pack Ice

Ice growing without disturbance in the Polar Basin has been estimated from direct observations to reach an equilibrium thickness of approximately 3.5 m in May at the end of the accretion period, and approximately 3 m in September at the end of the summer ablation period (Untersteiner, 1963). A model of Maykut and Untersteiner (1971) predicts a maximum equilibrium thickness of slightly over 3 m and a thickness in late October of approximately 2.7 m. Wadhams' (1981*a*) measurements in the central Polar Basin in October 1976 gave a greater mean thickness of level ice of 3.3–3.5 m, probably due to a greater snowfall than foreseen in the model. The central polar pack, however, consists of only 28 percent first-year ice (Koerner, 1973), and only about 50 percent consisted of level ice in October 1976, making the mean thickness 4.75 m, while pressured ice near Greenland was thicker, and ice nearer the marginal ice zone was thinner.

The polar pack ice is not consolidated. It is broken up into pieces of many sizes moving about and forming pressure ridges and leads that only stay open during the summer except in polynial areas. Up to 20 percent, but usually only 1–5 percent, of the Polar Basin consists of leads, which, in October when freezing over occurs quickly, are defined as ice less than 1 m thick (Wadhams 1981*a*).

### Fast Ice

In coastal areas, in fjords and straits, where currents are not severe enough to keep the surface free from ice, consolidated fast ice is formed. In arid, high-Arctic fjords with almost no snow, such as JBF, first-year ice attains a thickness of 2.5 m before summer ablation sets in, whereas in Dumbell Bay at the same latitude off Ellesmere Island, where 15 cm of snow covers the ice in June, the ice attains a thickness of only 1.8 m. In some straits and fjords the ice survives for many years. Ice at least 39 years old has been found, measuring 10 m in thickness (Serson, 1972). Maykut and Untersteiner's (1971) model, modified by Walker and Wadhams (1979), which incorporated a yearly snow fall of 1 m, produced an ice thickness of 12 m after 65 years. Floes of a similar thickness are encountered in the drifting pack. The equilibrium thickness of this version of the model is 20 m attained after 200–300 years.

What light intensities exist under the ice, and what are the requirements of the algae living there? In order to compare the various light data, these are all converted to $\mu$E m$^2$ s PAR ($6.02 \cdot 10^{17}$ quanta m$^2$ s photosynthetically active radiation [400–700 nm] according to Li-Cor standards).

Vollenweider (1974) converts 1 lux = 0.0929 incident radiation to $956 \cdot 10^{-10}$ gcal foot-candles cm$^2$ s = $956 \cdot 10^{-10}$ ly s PAR. Jitts et al. (1976) found for a wide range of solar elevations (17–80°) and total irradiances (80–1,100 W m$^2$) that PAR was equal to $1.176 \cdot 10^{18}$ quanta s = 1.95 $\mu$E s per W of total irradiance. Morel and Smith (1974) found for a wide range of sun altitudes and meteorological conditions that PAR was equal to $2.77 \cdot 10^{18}$ quanta s = 4.60 $\mu$E s per W of PAR. 1.176/2.77 = 0.425 is the ratio of PAR to total irradiance. For white fluorescent light, 1 lux = $4 \cdot 10^{-6}$ ly min PAR (Vollenweider, 1974). Ninety-three percent of daylight fluorescent light and 98 percent of cool white fluorescent light is PAR, according to Jitts et al. (1964), who used filtered tungsten with about 98 percent PAR measured as ly s. For unfiltered tungsten, 1 lux = 8 • $10^{-6}$ ly min = 9 percent PAR (Vollenweider, 1974). One foot-candle cool white fluorescent ≃ 11,800 • $10^{-10}$ ly s PAR, according to Bunt (1967), while Bunt and Lee (1970) used white fluorescent light (380–750 nm), where 1 foot-candle was equivalent to 80,500 • $10^{-6}$ μE $m^2$ s PAR (Table 6-1).

Maximum incident PAR in the Polar Basin north of the Arctic Circle calculated using Brock's (1981) equation is shown in Figure 6-7. In Dumbell Bay at 82°30'N, Apollonio (1980), however, recorded a maximum mid-summer daily mean of about 950 μE $m^2$ s PAR, which corresponds to a midday maximum of about 1,320 μE $m^2$ s PAR. The summer, however, is a season with frequent fog and overcast days. Especially from July onward, open water promotes the formation of fog and clouds, which reduce mean daily incident radiation to 200 μE $m^2$ s PAR. In the central Polar Basin, the mid-summer mean is about 615–750 μE $m^2$ s PAR (English, 1961), corresponding to a midday maximum of about 770–940 μE $m^2$ s PAR (at 85°N), whereas closer to open water the mid-summer mean can be about 478 μE $m^2$ s PAR (Apollonio, 1980), corresponding perhaps to a midday maximum of 600 μE $m^2$ s PAR.

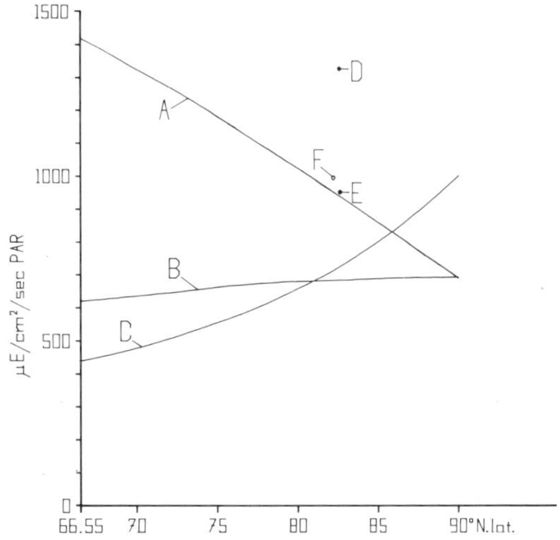


**Figure 6-7.** Theoretical ground illumination curves based on Brock (1981): *A:* mid-summer midday maximum; *B:* mid-summer daily mean maximum; *C: B* as percentage of *A.* Actual measurements at Dumbell Bay after Apollonio (1980); *D:* mid-summer midday maximum; *E:* mid-summer daily mean maximum. Estimate at Jørgen Brønlund Fjord; *F:* mid-summer midday maximum.

The mean thickness of Polar Basin ice (4.75 m) with 10 cm of snow on top permits 0.01 percent of incident radiation to pass (refer to coefficients of Weller and Parker, 1972). With mid-summer daily means of 200–950 μE $m^2$ s incident PAR,

**Table 6-1. Light Parameters Converted to μE m2 PAR**

| *Reference Unit* | *lux* | *ly s PAR* | *W m² PAR* | *μE m² s PAR* |
|---|---|---|---|---|
| 1 f.c. tot. | 10.764 | 10290 • $10^{-10}$ | 43065 • $10^{-6}$ | 198157 • $10^{-6}$ |
| 1 lux tot. | | 956 • $10^{-10}$ | 4001 • $10^{-6}$ | 18409 • $10^{-6}$ |
| 1 ly s tot. | | 0.425 | 17786 | 81840 |
| 1 ly s PAR | | | 41850 | 192566 |
| 1 W m² tot. | | | 0.425 | 1.956 |
| 1 W m² PAR | | | | 4.601 |
| 1 μE m² s tot. | | | | 0.425 |
| 1 f.c. white fluoresc. | 10.764 | 7176 • $10^{-10}$ | 30033 • $10^{-6}$ | 138183 • $10^{-6}$ |
| 1 f.c. white fl. (380–750) | 10.764 | 4181 • $10^{-10}$ | 17500 • $10^{-6}$ | 80519 • $10^{-6}$ |
| 1 f.c. cool white fl. | 10.764 | 11800 • $10^{-10}$ | 49385 • $10^{-6}$ | 227224 • $10^{-6}$ |
| 1 ly s cool white fl. | | 0.98 | 41014 | 188711 |
| 1 ly s daylight fl. | | 0.93 | 38922 | 179083 |
| 1 f.c. tungsten | 10.764 | 14348 • $10^{-10}$ | 60050 • $10^{-6}$ | 276297 • $10^{-6}$ |
| 1 ly s filtered tungst. | | 0.98 | 41014 | 188711 |

this gives sub-ice mean values of 0.02–0.10 $\mu$E m$^2$ s PAR with a mean of about 0.05 $\mu$E m$^2$ s PAR. The maximum midday PAR at Dumbell Bay was 0.13 $\mu$E m$^2$ s. The level ice that covers about 50 percent of the Polar Basin and has a mean thickness of 3.3–3.5 m will, with a snow cover of 8 cm, similarly allow 0.1 percent of incident radiation to pass. This corresponds to 0.2–1 $\mu$E m$^2$ s with a mean of 0.5 $\mu$E m$^2$ s PAR and a maximum midday value of about 1.3 $\mu$E m$^2$ s PAR.

### Phytoplankton

For planktonic algae to be able to reproduce at all in the constantly changing light field beneath unablated Polar Basin ice, they probably must have compensation points below the approximate 0.05 $\mu$E m$^2$ s PAR. No planktonic algae do (Table 6-2). The lowest recorded compensation point in Arctic waters is 0.22 $\mu$E m$^2$ s PAR for nannoplankton in Bothnia Bay. This value could well apply to the nannoplankton

**Table 6-2. Light Compensation, Saturation, and Inhibition Values for Phytoplankton Under Ice (+), in Open Water (−), and in Partly Open Water (±)**

| *$\mu$E m$^2$s PAR* | | | | | |
|---|---|---|---|---|---|
| *Compensation* | *Saturation* | *Inhibition* | *Ice* | *Location* | *Source* |
| ca. 6 | ca. 325 | | + | Polar Basin | English, 1961 |
| | 123–136 | | + | Polar Basin | Hameedi, 1978 |
| <20 | ≥60 (pre-bloom) | | + | Arctic Ocean | Apollonio, 1980 |
| | ≥80 (bloom) | | ± | (Dumbell Bay) | |
| ≥0.45 (daily mean) | | | + | Arctic Ocean | Andersen, unpubl. |
| ≥0.60 (daily max.) | | | + | (Fram Strait) | |
| ≤0.22 (nannoplankton) | | | + | Bothnia Bay | Müller-Haeckel, 1982 |
| | 102 | | ± | Chukchi Sea | Hameedi, 1978 |
| ca. 1 | ca. 70 | ca. 250 | + | Antarctica | Bunt, 1964 |
| 0.7–1.4 | | | + | mountain lake | Pechlaner, 1971 |
| | 642 | | − | Chukchi Sea | Dawson, 1965 |
| | 411 | | − | Chukchi Sea | Hameedi, 1978 |
| | 60 | ≥77 (division) | − | Frobisher Bay | Hsiao, 1985 |
| 30 | | | − | Baffin Bay | Harrison, Platt, and Irwin, 1982 |
| 110 | 580 (50% light depth) | | − | Baffin Bay | Platt et al., 1982 |
| 115 | 354 (1% light depth) | | − | Baffin Bay | |
| | 322 | | − | Can. E. Arctic | Platt, Gallegos, and Harrison, 1980 |
| 18.5–350 | 87–1450 | | − | Can. E. Arctic | Subba Rao and Platt, 1984 |
| 133 (mean) | 561 (mean) | | − | Can. E. Arctic | |
| ca. 3–7 (compensation depth) | | | − | Norwegian Sea | Sverdrup, 1953 |
| ca. 3 (compensation depth) | | | − | Barents Sea | Marshall, 1958 |
| 114–142 (starts spring bloom) | | | − | Barents Sea | Rey and Loeng, 1985 |
| 13.8–46 | 83–220 | | − | Antarctica | Jacques, 1983 |
| 70 | | | − | temperate seas | Jenkins, 1937 |

fraction of Arctic plankton in general, a fraction that has been found to be appreciable, especially outside the major blooms (Reynolds, 1974; Andersen, 1981; Horner and Schrader, 1982; Smith et al., 1985). Inshore, where perennial ice is found, or a great amount of snow falls every year, light conditions are even more severe. But, where level ice dominates and snow is scarce, and where the ice is thinner, as in the marginal areas of the Arctic Ocean, conditions are better suited for phytoplankton production.

In Jørgen Brønlund Fjord (82°N), level ice that forms every year attains a thickness of about 2.5 m. This ice is largely free of snow, and until surface ablation sets in during June, making the surface snowlike, approximately 5 percent of surface radiation should be able to penetrate the ice. With a theoretical maximum midday irradiance of about 1,000 $\mu$E $m^2$ s PAR (Andersen, 1977*a*), the midday maximum is 50 $\mu$E $m^2$ s PAR and the maximum daily mean is 35 $\mu$E $m^2$ s PAR beneath the ice. These values are well above compensation points found for sub-ice phytoplankton and even for most open-water plankton (Table 6-2). The very low compensation point of 0.22 $\mu$E $m^2$ s PAR should be reached even as the sun appears above the horizon by February in JBF, whereas the much higher compensation point of 6 $\mu$E $m^2$ s PAR found by English (1961) in the Polar Basin should be attained at noon by mid-March, and should equal the daily mean in early April (not measured). In spite of this projection, primary production measured with $C^{14}$ was barely detectable when first attempted in late May in JBF, possibly due to scattered snowdrifts and/or terrestrial material in snow and ice. Even small amounts of sediment in the ice can prevent under-ice spring bloom (Schell, 1980*a,b;* Horner and Schrader, 1982).

Under first-year ice in the Polar Basin, where contamination can be ruled out, a daily mean of 0.22 $\mu$E $m^2$ s PAR was measured in early April, and by the end of April a mean sub-ice irradiance of only 0.45 $\mu$E $m^2$ s PAR was accompanied by a slight increase in chlorophyll *a* in the upper 5 m next to the ice (Andersen, unpublished).

The disappearance of snow and the formation of pools of meltwater on the ice raises the mean sub-ice irradiance in the uneven pack ice. But only when ablation creates a thin layer of stabilizing freshened surface water beneath the thinnest level ice is phytoplankton subjected to much greater mean light intensities. Where melt pools lower the albedo and have thinned the ice, optimal or even inhibitory intensities can occur, like in the open-water areas.

### Ice Algae and Benthic Algae

For these algae, light conditions are different, since the only short-term light fluctuations are those caused by variations in insolation. Epontic algae living on the lower surface and in the lower few centimeters of ice have compensation points that compare with those for Arctic sub-ice phytoplankton. The lowest values, based on in situ measurements of chlorophyll in the Polar Basin (Andersen, unpublished), are close to the very low values found experimentally in the Antarctic (Table 6-3). Should this daily mean of 0.4 $\mu$E $m^2$ s PAR represent the compensation point for ice algae of the entire Polar Basin, then epontic production is possible at mid-summer under at least 25 percent of the Polar pack (about 50 percent of the level ice that constitutes about 50 percent of the pack), where mean surface transmission at midsummer is about 0.5 $\mu$E $m^2$ s PAR. If 28 percent of the pack is first-year ice (Koerner, 1973), production can already commence by late April under 28 percent of the pack. Most compensation values based on chlorophyll measurements, however, are higher (1.2–9.3 $\mu$E $m^2$ s PAR in Table 6-3). These values are theoretically obtainable in mid-summer in the Polar Basin under first-year ice (2.5 m) with 5–10 cm of snow.

Table 6-3. Light Compensation, Saturation, and Inhibition Values for Ice Algae

| μE m² s PAR | | | | | |
|---|---|---|---|---|---|
| *Compensation* | *Saturation* | *Inhibition* | *Community* | *Location* | *Source* |
| 0.4 (daily mean) | | | epontic | Arctic Ocean | Andersen, unpubl. |
| 0.6 (daily max.) | | | epontic | (Fram Strait) | |
| 1.2 | | | epontic | Chukchi Sea | Clasby, Horner, and Alexander, 1973 |
| 7.7 | 10–25 | | epontic | Beaufort Sea | Horner and Schrader, 1982 |
| 2.3–9.3 | 16–65 | | epontic | Beaufort Sea | |
| 7.6 | | | epontic | Hudson Bay | Gosselin et al., 1985 |
| | 1.8 | >2 | epontic | Davis Strait | Booth, 1984 |
| ≤12 | | | epontic | Disko Bay | Andersen, 1977*b* |
| ca. 0.5–1 | ca. 23 | >50–70 | epontic | antarctica | Bunt, 1964 |
| | (15°C optimum) | | epontic | Antarctica | Bunt and Lee, 1970 |
| | ca. 16 | | epontic | Antarctica | Bunt, 1967 |
| 0.45–1.36 | 23 (<0°C, suboptimal), sp. 1 | | epontic | Antarctica | |
| ca. 1.6 | >45 (−2°C, after 10 days), sp. 1 | | epontic | Antarctica | |
| | 15 (0°C, suboptimal), sp. 2 | | epontic | Antarctica | |
| | 63 (5–7°C, optimal), sp. 2 | | epontic | Antarctica | |
| | 32 (4–5°C, optimal), sp. 3 | | epontic | Antarctica | |
| | >45 (after 7 days), sp. 3 | | epontic | Antarctica | |
| ≤0.3 | | | epontic | Antarctica | Palmisano and Sullivan, 1982 |
| | 200–400 (ca. 30% at 20) | | snow | Antarctica | Burkholder and Mandelli, 1965 |

The wide range of saturation intensities, from 1.8 μE m² s PAR for Davis Strait ice algae (Booth, 1984) to 1,450 μE m² s PAR for Canadian East Arctic phytoplankton (Subba Rao and Platt, 1984), may indicate that what we are dealing with are more or less parallel upward shifts in inhibition, saturation, and perhaps compensation values that occur as daily mean irradiances increase. These shifts may proceed gradually due to adaptation and/or more or less stepwise due to species succession. Harrison and Platt (in press) (Table 6-4) state that PBI (production per biomass per irradiance) is predominantly controlled by light history, and accordingly increases with latitude on a yearly basis. What seems to be a compensation intensity in the laboratory may not, however, mark the initiation of a bloom in the field, since adaptation to an increase in irradiance can take from 4–6 hr (Gallegos et al., 1983) to up to 3 days (Nielsen and Park, 1964).

Bunt (1967), working with individual species of Antarctic ice algae, found an upward shift in inhibitory intensities after prolonged (7–10 days) subjection to optimal light intensities. He also found, as Hsiao (1985) did, for mixed populations of Arctic plankton, that cell growth was inhibited at much lower light intensities than was photosynthesis. Seasonal differences in susceptibility to photoinhibition may also be species dependent (Glibert, Dennett, and Goldman, 1985). For example, at Vineyard Sound a difference in susceptibility could be due to a shift in abundance from a diatom-dominated winter population to a cryptomonad-dominated summer population (Goldman and Dennet, 1984). Thus the fact that Apollonio (1980) found algae thriving at <20–50 μE m² s PAR did not bloom or respond to elevated light intensities until they had been subjected to intensities above 75 μE m² s PAR for approximately 2 days may well be explained by adaptation and/or changes in species composition. The same may apply to the findings of Gosselin et al. (1985) in Hudson Bay, where ice algae gradually seemed to change from shade-adapted to light-adapted as irradiance was increased above 7.6 μE m² s PAR.

## Dark Survival

Arctic plankton must be able to survive long periods of darkness or low light intensity. Antia (1976) has shown that cold-water strains of plankton algae survive longer in darkness at low temperatures than warm-water strains, and vice versa. The bulk of the more resistant strains qualified as benthic types, 70 percent of which tolerated 11–12 months of darkness while the rest tolerated 5–6 months. A few tolerated 3 years of darkness. Bunt and Lee (1972) found that two species of Antarctic ice algae survived 3 months of darkness without any major influence on pigment ratios, while a third species did not remain viable. There was no indication of heterotrophic activity as found in previous experiments, although organic supplements appeared to affect pigment levels and ratios. Horner and Alexander (1972) found heterotrophic activity in Arctic ice algae to be negligible. Palmisano et al. (1985) found heterotrophy to be responsible for less than 0.3 percent of production under 1.5 m ice with 7 cm snow. According to Palmisano and Sullivan (1982), a simulated transition to winter by reducing the daily light sum brought about an increase up to 60-fold in heterotrophic potential in ice diatoms. Another consequence was a reduction in metabolism, growth rate, and photosynthetic rate, and an increase in indogeneous carbon reserves.

Survival of ice algae is enhanced by the formation of spores. These increase their content of carbon and chlorophyll *a*, while kept in a light–dark cycle, and ensure a better survival in darkness by retaining more of their carbon and chlorophyll *a* (Doucette and Fryxell, 1983). Horrigan (1981) reports 0.11 mg C m³ d chemoautotrophic activity and 0.025 μg C m³ d hetero-

Table 6-4. Photosynthetic Parameters for Phytoplankton Under Ice (+), in Marginal Ice Zones (MIZ), in Open Water (−), and in Partly Open Water (±)

| P/B = mgC mg chlorophyll a | | P/B μE m² h | | | |
|---|---|---|---|---|---|
| *per hour* | *per day* | *per hour* | *Ice* | *Location* | *Source* |
| 0.6 | | | 3 m | Polar Basin | English, 1961 |
| 3.6–7.2 | | | + | Polar Basin | Smith and English, 1973 |
| 3.8 | | | + | Polar Basin | Pautzke, 1974 |
| 1 (max.) | | | 0.5–1.8 m | Arctic Ocean | Apollonio, 1980 |
| 0.15–4.2 (net) | | | 0–1.8 m | (Dumbell Bay) | |
| (0.08) | 0.2 (mean) | | + | Arctic Ocean | Harrison et al., 1982 |
| (0.17) | 4.0 (mean) | | ± | Baffin Bay | |
| (0.28) | 6.7 (mean) | | ± | up to 80°N | |
| (2.25) | 54 (max) | | ± | Chukchi Sea | Hameedi, 1978 |
| 0.4 | | | + | Bering Sea | McRoy and Goering, 1974 |
| 0.04 | | | MIZ | Bering Sea | |
| (0.3 mean) | 8 (max.) | | 0.5 m | Disko Bay | Andersen, 1981 |
| 0.2–0.6 (0–4.5 cm under ice) | | | 0.35 m | Disko Bay | |
| 0.09–0.3 (0.2–50 m under ice) | | | 0.35 m | Disko Bay | |
| <1.7 (Sept.–Oct.) | | | MIZ | N. Svalbard | Heimdal, 1983 |
| 0.3–2 | | | − | Chukchi Sea | Hameedi and Shaw, 1975 |
| 1.37–2.9 | | 0.01–0.02 | − | N. Foxe Basin | Smith et al., 1985 |
| 0.31–3.8 | | | − | W. Baffin Bay | Hsiao and Trucco, 1980 |
| 0.79–1.22 | | | − | Baffin Bay | Platt et al., 1982 |
| 0.8–1.4 (pre-spring) | | | − | Davis Strait | MacLaren Marex, 1979*c* |
| 0.4–0.5 (spring) | | | − | Davis Strait | |
| 0.1 (Aug.) | | | − | Davis Strait | |
| 0.11–2.41 | | <0.002–0.03 | − | Can. E. Arctic | Gallegos et al., 1983 |
| 0.4 (−1.5°C) | | | − | Can. E. Arctic | Li, Smith, and Platt, 1984 |
| 2.0 (0–8°C, general max.) | | | − | Can. E. Arctic | |
| 3.2 (0.2°C, max.) | | | − | Can. E. Arctic | |
| 4 (max.) | | | − | Disko Bay | Andersen, 1977*b* |
| 0.09–8.9 | 2–212 | | − | Disko Bay | Andersen, 1981 |
| ca. 6.5 (mean) | 160 (max) | | − | Disko Fjord | |
| 0.09–2.66 | | | − | Barents Sea | Vedernikov and Solov'yeva, 1972 |
| 1–1.5 | | | − | Arctic | Nielsen and Hansen, 1959 |
| 0.11–10.33 (max.) | | 0.00087–0.05 | − | Arctic | Subba Rao and Platt, 1984 |
| 1.96 (max. mean) | | 0.015 (mean) | − | Arctic | |
| 1–1.5 (mean) | | | − | Arctic | |

trophic activity beneath Antarctic shelf ice 412 m thick.

## NUTRIENTS

As in any other ocean, the photic zone of the ice-covered Arctic Ocean receives nutrients directly from the atmosphere, from precipitation, from riverine input, and from upwelling or admixture from below. Additional sources are sea ice and glacial ice, and associated with ice are phenomena causing nutrient replenishment.

Looking at the nutrient profiles of ice-covered areas as well as open-water areas north of Svalbard, Dyrssen (1982) found that the effect of photosynthesis reached to depths of 150–200 m. Mean deficits down to 150 m of total $CO_3$, $NO_3$, $PO_4$, and $SiO_4$ were 67, 11, 0.8, and 5.6 μM/l, respectively, that is, 84:14:1:7. This deficit of course does not correspond to what is biogenically fixed in the Polar Basin. It is an overestimate, since the upper 150 m contain an admixture of already nutrient-poor Atlantic water in the area mentioned. The overestimate may, however, be lessened or even changed to an underestimate, due to the more near surface inflow of nutrient-rich water from the Pacific Ocean through Bering Strait.

T-3 and other data from the Amerasian Basin (Kinney, Arhelger, and Burrell, 1970) show that the nutrient-poor Atlantic water has its core at about 500 m, above which nutrient-rich water from Bering Strait is evident between 50 m and 250 m, where a core at 75 m is postulated Bering summer water. Rusanov (1980) has divided the Polar Basin into six regions on the basis of hydrographic characteristics, including P and Si (Fig. 6-8; Table 6-5). Regions I, II, and V, covering a sector between the North Pole and northern Europe and Siberia and influenced by the inflow of nutrient-poor Atlantic water, have no subsurface nutrient maxima except for a $SiO_4$ maximum in region V due to riverine input from Siberia. In regions III, IV, and VI, subsurface maxima of both $SiO_4$ and $PO_4$ tell of Bering water with highest values in region IV, and the maxima nearest to the surface in region VI, due to upwelling off the Canadian Archipelago as shown by Apollonio (1980).

Many authors have looked at the $SiO_4$ budget of the Arctic Ocean (Codispoti and Lowman, 1973; Rusanov, 1974; Codispoti and Owens, 1975; Codispoti, 1979a; Jones and Coote, 1980; Dyrssen, 1982) (Table 6-6). Dyrssen's deficit of 5.4 kmol s in the table is less than those of Codispoti and Owens (1975) (17 kmol s) and Jones and Coote (1980) (7 kmol s) due to differences in volume transport. Despite the fact that most of the $SiO_4$ from the Bering inflow (84 percent) is considered to exit via the Canadian Arctic Archipelago, the Pacific water is the main contributor to the $SiO_4$ pool of the Polar Basin. The input from the Siberian rivers entering the photic zone probably supplies most of the $SiO_4$, which is biogenically fixed in the Polar Basin.

Table 6-5. Distribution of Silicates and Phosphates in the Polar Basin

| *Region* | *I* | *II* | *V* | *III* | *VI* | *IV* |
|---|---|---|---|---|---|---|
| $SiO_4$ μM at 5 m | 2 | 13 | 11 | 11 | 15 | 18 |
| $SiO_4$ μM maximum | 5 | — | 23 | 35 | 36 | 48 |
| $SiO_4$ depth of max. in m | 250 | — | 120 | 140 | 90 | 120 |
| $PO_4$ μM at 5 m | 0.60 | 0.60 | 0.85 | 1.30 | 1.50 | 1.05 |
| $PO_4$ μM maximum | 0.55 | 0.70 | — | 1.70 | 1.55 | 1.75 |
| $PO_4$ depth of max. in m | 350 | 150 | — | 100 | 65 | 170 |

*Source:* After Rusanov, 1980.
*Note:* Roman numerals refer to the zones in Figure 6–8.

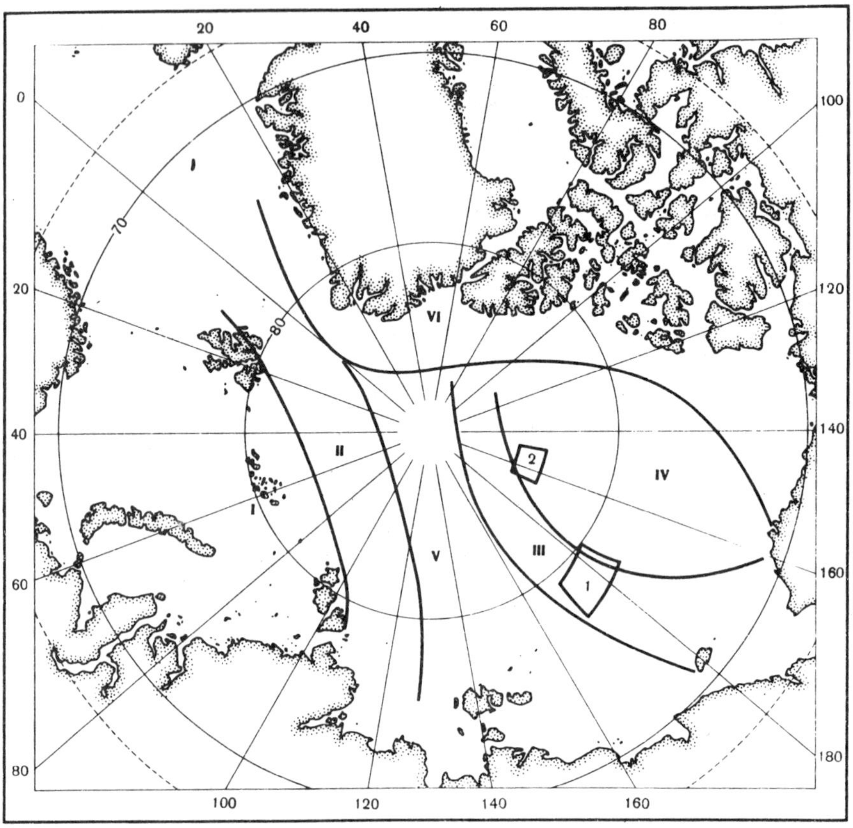


Figure 6-8. Surface chemical-oceanographic zones of the Arctic Basin. I: Spitzbergen zone; II: North Land zone; III: Wrangel zone; IV: Canadian zone; V: Central zone; VI: Canadian–Greenland zone. See Table 6-5. (After Rusanov, 1980.)

Table 6-6. Polar Basin Silicate Budget

| | *µM* $SiO_3$ | *Water Flow %* | *Si Flow %* | *kmol s Flow* |
|---|---|---|---|---|
| *Inflows* | | | | |
| Atlantic | 5 ( 5) | 83 (84.5) | 37.5 (38.7) | + 39.0 |
| Pacific | 35 ( 35) | 16 (14.1) | 50.5 (45.2) | + 52.5 |
| Riverine | 125 (125) | 1.1 (1.4 ) | 12.0 (16.0) | + 12.5 |
| *Outflows* | | | | +104.0 |
| Atl. water | 5 7.5 (8) | 57.5 | 27.4 | − 27.0 |
| E. Greenl. Polar water | 15 (10) | 19.1 (84.6) | 27.3 (67.7) | − 27.0 |
| ice | 5 | 1.1 | 0.5 | − 0.5 |
| Canadian Archipel. | 21 (21) | 22.3 (15.4) | 44.7 (32.3) | − 44.1 |
| Total | | | | − 98.6 |
| Deficit | | | | + 5.4 |

Based on concentration and volume flow data from Anderson and Dyrssen (1981) and Codispoti (1979). Values are in parentheses.

Nutrient replenishment from precipitation, primarily from snow, is negligible in the relatively unpolluted high-Arctic (Grainger, 1979).

Due to the perennial ice cover, direct wind-induced mixing is virtually nonexistent in the Polar Basin. In spite of this, the characteristic hydrographic profile—as first described by Nansen (1902)—consists of a well-mixed surface layer of polar water with low salinity and temperatures at freezing or almost freezing overlying a steep pycnocline with rising salinity and a more or less corresponding rise in temperature due to intruding water, primarily of Atlantic origin and to some degree of Pacific origin. The low salinity of the homogeneous surface layer is due equally to the melting of sea ice and to the riverine input that together amount to about 2 percent of the mixed layer each year, corresponding to a layer about 1 m thick. The mean residence time of this fresh water in the Polar Basin is approximately 13 years (Dyrssen, 1982).

The surface layer under the polar pack is rapidly mixed due to the wind-induced movements of the ice, and its thickness coincides closely with the frictional boundary layer observed during maximum wind speeds (McPhee and Smith, 1976). A thickness of 40–50 m in the central Polar pack has been reported by several authors (Gudkovich, 1955; Farlow, 1958; Collin, 1959; English, 1961; Kinney, Arhelger, and Burrell, 1970). Variations, however, are many. In the Eurasian Basin northeast of Greenland, more than 500 km from the ice edge north of Svalbard, the mixed layer can vary from 5–80 m in thickness (McPhee, 1981). Here a temperature minimum found in April in the steep pycnocline beneath the shallow surface layer suggest intrusion of relatively nutrient-poor (Andersen, unpublished) Atlantic water, which must have been cooled near the surface to the south.

The mixed layer generally has a lower nutrient content than the pycnocline in which a nutrient maximum usually constitutes a substantial but not immediately accessible nutrient reserve. The rate of salinity increase with time in the mixed layer shows that salt is supplied both from the water freezing to ice and from admixture of water from below (Solomon, 1973). A mean thickness of about 1.1 m of ice is formed each year. Approximately half of this (0.6 m) melts again, and the remainder (0.5 m) drifts out of the Polar Basin (Dyrssen, 1982).

The nutrients found in the sea ice originate solely in the water from which they froze. Therefore, melting ice does not serve as a nutrient replenisher as such. During freezing the sea ice initially loses about 70 percent of its salt content, and subsequent loss during the period of ice growth lowers the residual content from 30 percent to 10 percent of the content in seawater due to the seeping out of brine at a rate of 3–4 percent of the original water content per month (Grainger, 1977). Since the ion ratios in the brine trapped in the ice are close to that of seawater (Addison, 1977), the nutrient content is correspondingly lowered. However, the nutrient content of the water, which is frozen during the latter part of the period with ice growth, is the highest of the year. At the same time, the brine pockets in the ice become more numerous near the lower surface of the ice due to seeping (Meguro, Ito, and Fukushima, 1967). Therefore, very high nutrient contents can be found in the lower regions (few cm) of ice in the spring (see Table 6-7).

Although these concentrations can be several times higher than in the water below at the time of melting, the main body of ice is poorer in nutrients than the water. Therefore, a total melting of the ice would only cause a replenishment of nutrients amounting to a few percent in the photic zone. Grainger (1977) calculated a replenishment of 2–5 percent of the $PO_4$ and 1–2 percent of the $NO_3$ available in the mixed layer at the time of melting in Frobisher

Table 6-7. Maximum Values Found for Nutrient Contents of Sea Ice

| Maximum µM/l | | | Locality | Source |
|---|---|---|---|---|
| $PO_4$ | $NO_3$ | $SiO_2$ | | |
| | 19 | | Polar Basin | Apollonio, 1958 |
| 1.5 | 10 | 22 | Beaufort Sea | Horner and Schrader, 1982 |
| 2.7 | | 14–75 | Chukchi Sea | Meguro, Ito, and Fukushima, 1967 |
| 3.0 | | | Jones Sound | Apollonio, 1961 |
| 8.5 | 15 | | Frobisher Bay | Grainger, 1977 |
| 3.1 | >30 | 20.9 | Davis Strait | Booth, 1984 |
| 3.5 | 15.5 | | Disko Bay | Andersen, 1977*b* |
| | 13 | | Barents Sea | Oradovskii, 1972 |

Bay. This calculation is, however, offset to the positive by the fact that the melting of the lower portions of ice where the main nutrient deposit is located occurs over a short period of time, namely up to and during the sub-ice phytoplankton bloom. Here any nutrient replenishment as near to the surface as possible would be advantageous. Brine drainage, however, can be so intense that local downward flows of brine transversing the mixed layer can rob it of any nutrient benefit (Green and Steele, 1977).

Meguro, Ito, and Fukushima (1967) found that algae living in the ice are capable of using up any nutrients found there, and actually must rely on nutrients from the water below to be able to continue growing to any extent. Gosselin et al. (1985) found that production by epontic algae in Hudson Bay largely depended on an input of nutrients from the mixed surface layer.

An upwelling of 6–10 m each year (Solomon, 1973) is estimated to balance the runoff from the Polar Basin via the East Greenland Current and through the Canadian Arctic Archipelago. The nutrient increase with depth per 10 m at the top of the pycnocline is up to 500 percent. If 10 m upwelling and a corresponding runoff after mixing is assumed, this will give a maximum annual replenishment of nutrients in the mixed layer of 500:11 = 45 percent. According to nutrient data from four drift stations, this should give maximum annual replenishments as shown in Table 6-8.

The phytoplankton in the Polar Basin, however, does not seem to lack nutrients. The very low summer production does not lower the mixed-layer nutrient content to any great extent (English, 1961). Kawamura (1967) states, on the basis of ARLIS 2 data, that a nutrient cycle exists with respect to $PO_4$ but not $SiO_4$. The profiles

Table 6-8. Nutrient Gradients in the Upper Part of the Pycnocline Immediately Beneath the Mixed Polar Surface Water

| µM/l Change in Upper 10 m of Nutrient Gradient | | | Basin | Drift Station |
|---|---|---|---|---|
| $PO_4$ | $NO_3$ | $SiO_2$ | | |
| 0.31–0.5 | 2.5–5 | 10–12 | Amerasian | Alpha |
| 0.45 | 5 | 7.5 | Amerasian | T-3 |
| 0.55 | | 13 | Eurasian | ARLIS 2 |
| 0.23 | 2.7–3 | 4.36 | Eurasian | Fram 1 |

from ARLIS 2 (Kusunoki et al., 1966) show that the peak concentrations of $PO_4$ were found in June–July at the start of the ablation period, indicating that the extra $PO_4$ could at least in part have come from the ice. This is, however, questionable, as argued above. Moreover, the rhythm in $PO_4$ concentrations mentioned by Kawamura (1967) can be questioned due to apparent analytical difficulties evidenced by irregular profiles and shifts in the profiles more or less paralleling the changes in the mixed layer. The lack of nutrient depletion in the Polar Basin is also indicated by the fact that English (1961) found that adding nutrients ($PO_4$, $PO_4$ + $NO_3$, $NO_3$) did not enhance production.

Where the polar surface layer is left undisturbed—that is, away from contacts with other water masses, as at polar fronts, or where bottom topography or hydrographic phenomena induce upwelling—it will eventually be depleted of nutrients if it receives sufficient light. In this respect it is no different from any other marine area. But the pycnocline forms an exceptionally rigid barrier to nutrient transport to the photic zone from below. It acts as a "false" bottom, making the polar surface water appear as a shallow well-mixed neritic regime. In such an area the concentration and thickness of ice and snow, governing the penetration of light, is the major factor determining the onset and intensity of primary production (Slagstad, 1982). When the ice gets thin and/or scattered enough, or when it is swept away by winds (MacLaren Marex Inc., 1979c), the polar water is soon depleted. After that, summer production will largely be confined to the top of the pycnocline, where shade-adapted algae subsist on a compromise between light and nutrients. This situation is characteristic of the Arctic, for example, in Ogac Lake (62°52′ N) (McLaren, 1969), in Disko Bay (69°30′ N) (Andersen, 1981b), in the northern Barents Sea (75° N) (Rey and Loeng, 1985), and where Polar water from the East Greenland Current reaches Iceland (67° N) (Thordardottir, 1984).

The thinner the mixed layer is, the quicker is it depleted. Excessive meltwater from the ice can cause a narrow surface layer to become depleted in the upper part of the mixed polar water. Then the polar water above the pycnocline acts as a backup for the surface meltwater, for a time supplying nutrients and ensuring against excessive mixing and sinking. In the northern Barents Sea, for instance (Rey and Loeng, 1985), meltwater from the ice eventually causes the entire surface layer to become extremely isolated above the pycnocline, and depletion of $NO_3$ but not of $SiO_4$ or $PO_4$ is complete by early May. Depletion of $SiO_4$ follows gradually, indicating that diatoms thrive almost exclusively on remineralized N, and after thorough depletion, the meltwater still contains small amounts of phytoplankton mainly consisting of small flagellates, which in this oligotrophic situation subsist on remineralized nutrients.

In Baffin Bay, Harrison, Platt, and Irwin (1982) found no obvious signs of severe N limitations despite very low concentrations. They concluded that since $NO_3$ uptake/$NO_3$ + $NH_4$ uptake was 0.5, about 50 percent of N requirements were met by regeneration, and that nutrients may play a less important role in controlling primary production in ice-free Arctic waters than previously assumed. Hsiao (1985) states that light is the sole limiting factor to primary production in Frobisher Bay. This would apply even more to the ice-covered Polar Basin. Goldman et al. (1985), however, found that a phagotrophic microflagellate, feeding equally well on phytoplankton and bacteria, regenerated 50 percent of N excreted as $NH_4$, regenerated only 15–30 percent under exponential growth, and regenerated only 8 percent when prey was grown under N limitation. Regeneration was negligible without the flagellates. Therefore, to rule out the limiting effect of nutrients is perhaps inappropriate.

Along the polar fronts, conditions are markedly different. The polar fronts not only extend around the world but they also penetrate deep, to a depth of 250 m between Svalbard and Greenland (Dyrssen, 1982). The fronts are often referred to as upwelling areas, bringing nutrients to the surface. When Atlantic water meets polar water, the heavier, more saline Atlantic water is forced down, forming a sloping front beneath the polar water. Upwelling and spreading of deeper-lying polar water over the Atlantic water can be greatly enhanced by certain ice-edge wind conditions, as has been demonstrated by satellite imagery in the East Greenland Current, where the ice edge follows the continental slope (Electromagnetics Institute, Technical University, Denmark, unpublished). Wind-induced upwelling along ice edges, bringing water from at least 150 m to the surface and spreading it to over 12 km from the ice edge, has been described from Svalbard (Buckley et al., 1979).

Upwelling of nutrient-rich water is, however, of little avail if some degree of stabilization does not occur. Slagstad (1984, 1985), in fact, postulates that the high and prolonged production, lasting for 2–3 months in the frontal areas where Atlantic water meets polar water in the northern Barents Sea, is not due to sinking of Atlantic water and upwelling of nutrient-rich polar water, as postulated by Zenkevitch (1963). What happens according to a beautifully generated model by Slagstad is that turbulent Atlantic water, relatively rich in nutrients, stabilizes upon meeting the polar water, which has been freshened by the melting of the ice. The admixture of Atlantic water from beneath should afford a continuous input of nutrients to the mixed layer and a simultaneous deepening of this layer, ultimately effecting a prolongation of the productive period. In real life, however, as indicated in Slagstad's figure 8, which does not quite match the model simulations, some wind-generated ice-edge upwelling is probably also involved. An exceptional area in this respect is the western Bering Strait, where a continuous cross-shelf flow of nutrient-rich water from the Bering Sea continental slope moves north beneath the south-going polar current (Sambrotto, Goering, and McRoy, 1984).

## GLACIERS AND NUTRIENTS

Vibe (1939) expressed the view that glaciers far surpass precipitation as an erosive factor procuring the inorganic material necessary for organic life. Cooper (1952) cites Sverdrup for suggesting that Antarctic waters should receive much silica from the bedrock scraped by the immense glaciers of the Antarctic continent. Hartley and Dunbar (1938) discussed upwelling of nutrient-rich water—brown zones—adjacent to glaciers terminating in the sea. Petersen (1964, 1977) comments on the disruptive effect of calving glaciers and icebergs (called ''tagsaq'' by the Greenlanders) and on the upwelling occurring around icebergs in West Greenland. Wherever icebergs encounter water above freezing, the surface will at some time begin to melt, and the fresh meltwater will move upward as a brackish current, bringing with it nutrients from the surrounding saltier water to the photic zone. Despite the negative effect on surface temperatures, this constant upwelling of nutrients has a markedly positive effect on primary production, as shown by Andersen (1981).

Neshyba (1977) has calculated that the melting of a 1 km$^2$ Weddel Sea iceberg extending 250 m into the water in the course of one year is equivalent to 3.3 km$^2$ of Oregon coastal upwelling, the rate of which is

$10^{-2}$cm/s. On a yearly basis, this value is 13.2 km$^2$, and on an annual nutrient basis it is equivalent to 26.3 km$^2$, since the Antarctic upwelling due to melting is irreversible and continual as opposed to the Oregon coastal upwelling, which occurs for only 3 months each year, and since only 50 percent of the nutrients there are utilized due to remixing. According to Parker et al. (1978), Antarctic icebergs contain about 1.4 $\mu$M $NO_3$ and 1.0 $\mu$M $NH_4$ per liter. This $NO_3$ content on the average is 7.2 times as high as that of the Antarctic circumpolar water. The average annual input of $NO_3$ from melting ice to the Antarctic circumpolar surface water amounts to about 7 percent of the total mixed-layer reservoir. The beneficial effect from melting as well as from upwelling is probably less in the Arctic.

Limestones contain up to about 2.5 ppm mineral N in the form of soluble salts retained by physical entrapment (Chalk and Keeney, 1971). These are normally released by weathering. Glacial erosion, abrasion, and scouring of such rocks have a similar effect, causing release of the limiting nitrates as well as the less limiting silicates into nearby seawater (Apollonio, 1973). No significant dissolution of phosphate from limestones results from glacial activity.

## SALINITY EFFECTS

Ice algae, which are known to tolerate a wide range of salinities (5–60 ‰ S) (Grant and Horner, 1976), have been found to thrive optimally at about 13–18 ‰ S (Bunt, 1964).

## TEMPERATURE EFFECTS

Arctic and Antarctic algae are not specifically adapted to low temperatures. Fogg (1977) writes that there is no adaptation to low temperatures, thus enabling higher growth rates. Lowering the temperature lowers the respiration rate, thus enabling longer survival in the darkness and slow growth at low light intensities due to a lowered compensation point. Low temperatures also lower saturation and inhibition intensities, which are lowered even more if nutrients are at a minimum (Jitts et al., 1964).

There is no evidence of increased enzyme content, as shown by Nielsen and Jørgensen (1968) and Jørgensen (1968), since there was a difference in the physiological state of their algae dependent on temperature (Morris and Glover, 1974). Low temperature reduces cell division rates relative to carbon fixation (Fogg, 1977). Optimal temperatures are often higher than the environmental temperatures (Jitts et al., 1964). Li, Smith, and Platt (1984) found that production rates increased fivefold in Arctic plankton with a temperature increase from $-1.5$°C to 0°C, and maximum was attained at 0.2°C (Table 6-4). In Baffin Bay and in the Labrador Current, the increase was twofold to threefold from 0.2°C to 2.1°C (in situ). Bunt (1964, 1967) found that one ice diatom normally living at $-2$°C grew optimally at 5–15°C, while another thrived best at 3–7°C and died at 9°C.

Another effect of low temperatures is increased incorporation of carbon into lipids. Smith and Morris (1980) report from the Antarctic that up to 80 percent carbon is incorporated into lipids at $-1.8$°C to $-2$°C, while the amount is less than 20 percent at temperatures above 0°C. In the Canadian East Arctic (Li and Platt, 1982), the amount is only 18 percent and only up to 30 percent at $-1$°C to 6°C. Smith and Nelson (1985) additionally attributed a high carbon/chlorophyll *a* ratio of 118 and a high silicon/carbon ratio of 0.62 in the Antarctic to low temperatures compared to oceanic values of 25–40 and 0.3, respectively.

## PRIMARY PRODUCTION

In the high Arctic and in other places where blooming in the plankton is greatly delayed due to winter darkness and to the light-limiting effect of ice and snow, and where wind-driven vertical mixing is impaired due to a cover of ice, up to 10 distinct "communities" of unicellular primary producers, each inhabiting a characteristic regime, have been recognized. Their species composition, however, have not all been disclosed. They include 3 ice floras, 6 phytoplankton floras and 1 benthic flora. In ice-covered areas spring and summer epontic ice floras are found at the bottom and a surface of snow-ice flora is found at the surface of the ice, and a winter flora and a spring–summer surface halocline flora is found beneath the sea ice. In sub-Arctic areas, which become more or less free from ice, the phytoplankton follows a characteristic sequence including a spring bloom flora, a summer upper oligotrophic flora, a summer deep pycnocline flora, and a fall bloom flora. Finally, a benthic microflora inhabits shallow bottoms.

### Ice Floras

The spring ice flora with very low compensation points is the first to bloom. The composition of this flora can vary greatly. In the Canadian Arctic, it is primarily composed of pennate diatoms of which 168 species have been identified compared to 21 centrics, 3 flagellates, 2 dinoflagellates, and 2 chrysophytes (Hsiao, 1980). In the east Canadian Arctic, Dunbar and Acreman (1980) found 96–99 percent pennates, mainly consisting of benthic forms (50 percent of species). From Davis Strait, Booth (1984) reports 30.9–100 percent pennates (mostly >90 percent). In the Gulf of St. Lawrence, Dunbar and Acreman (1980) found only 57 percent pennates, including 5 percent benthics, and the rest were centric planktonic forms, while Demers, Theriault, and Descolas-Gros (1984) found 50 percent benthic diatoms one year and nearly 100 percent planktonic forms, mostly flagellates, another year. In Isefjord, Denmark, a 5 $\mu$m chrysophyte that came from the plankton dominated the ice flora—ultimately as cysts (Thomsen, 1985)—while in the neighboring Kattinge Vig (Bay), Grøntved (1950) found more than 1 million diatoms as well as some flagellates per milliliter of ice.

Fifty percent of the total photosynthate of pennate diatoms is excreted, forming a mucilage of mono- plus polysaccharides that has a cryoprotective as well as an adhesive function and probably forms the basis of a food chain involving bacteria, which are positively correlated with chlorophyll *a* in the ice (Sullivan and Palmisano, 1984). In Antarctic sea ice, Iizuka, Tanabe, and Meguro (1966) found 74 million brevibacteria in 1 kg of ice, and Sullivan and Palmisano (1984) found up to $10^{12}$ bacteria per $m^2$ of ice. A mean of $14 \cdot 10^{10}$ corresponds to 9.8 mg C $m^2$. Ninety-three percent of the biomass in the lower 20 cm of 1.3–2.5 m thick ice consisted of bacteria. Seventy percent lived freely in the ice, while 30 percent were associated with detritus and algae.

Pennate diatoms and flagellates are present in the ice as soon as it forms in the fall, and a limited fall bloom may even set in (Schell, 1980*a*). In low-Arctic areas without winter darkness, algae from the subice winter bloom (see below) can penetrate and cause an intense coloration of the lower 8–10 cm of ice, as in the Isefjord, Denmark (Thomsen, 1985). The later the ice forms, the quicker it will freeze due to the low temperature. This limits the chances for the algae to survive. Therefore, the percentage of phaeopigments is high in late-formed ice (Hoshiai and Fukuchi, 1981).

Seeding of the ice can occur throughout the winter (Bunt and Lee, 1972; Andersen, 1977*b*, unpublished). In the Antarctic, many algae are assumed to be transported to the surface by ice platelets and incorporated into the so-called frazil ice, since high chlorophyll values often correspond to high salinity values, indicating frazil ice (El-Sayed and Taguchi, 1977). While Antarctic ice can consist of 50 percent frazil ice, only 5–10 percent of Arctic ice (top 10–20 cm) usually consists of frazil ice in which brine drainage is impaired (Ackley et al., 1981; Martin, 1979). In the remaining 90–95 percent congelation ice, brine drainage, and possibly active movement by the algae gradually concentrate them in the lower few centimeters of the ice and on the lower surface. Upwelling and platelet formation adjacent to icebergs and glaciers (Petersen, 1977) may also contribute to seeding in the Arctic. Ice algae start growing on nutrients entrained in the ice (Meguro, Ito, and Fukushima, 1967), but they depend on nutrients from the water below to actually bloom (Gosselin et al., 1985).

Biomass and production in different localities are shown in Tables 6-9 through 6-12. Although spring epontic chlorophyll concentrations (up to 1,460 mg chlorophyll *a* $m^3$) can be up to 560 times that in the water beneath (Fukushima and Meguro, 1966) and several times greater per $m^2$ (up to 30.5 mg chlorophyll *a* $m^2$) at a given time, and daily production in the ice (up to 193 mg C $m^3$ d and at least 350 mg C $m^2$ day) can outweigh the simultaneous planktonic production, the annual epontic production is usually negligible (0.14 to at least 10 g C $m^2$) compared to the planktonic production. Less than 1–15 percent have been measured in areas with seasonal ice cover, while in the arid high-Arctic, Welch and Kalff (1977) have considered 33 percent as realistic (excluding benthic production). If their 15 g C $m^2$ yr is correct, and even if 5–10 g C $m^2$ yr should be representative for the central Arctic Ocean, then spring epontic production outweighs plankton production. This is certainly the case if plankton production is 0.6 g C $m^2$ yr as found by English (1961), and even if total production (ice plus plankton) is 9 g C $m^2$ yr (Subba Rao and Platt, 1984). The high annual rates computed by these authors by multiplying hourly rates by 24 hours and 120 days are, however, often out of proportion. For example, the 164 g C $m^2$ yr obtained by them from Matheke and Horner (1974) is based on a mid-summer peak of 57 mg C $m^2$ h and in fact applies to benthic microflora.

A summer epontic flora of light-adapted algae, the production of which is unknown, has been recognized as a distinct flora by Apollonio (1985). *Melosira spp.* form massive mucilaginous accumulations noted already in 1952 by Dickie in the Canadian Arctic, and described by Usachev (1949) and Mel'nikov (1980) from the Polar Sea. These centric diatoms form up to 1.5 m long tufts hanging down from the undersurface of the ice. The mucilaginous nature of these "summer ice algae" suggests that they thrive in a nutrient-deficient milieu (according to Miller, 1962 and Fisher and Schwarzenbach, 1978). Different from this is a summer ice flora community of pennate diatoms described by Meguro, Ito, and Fukushima (1967) from the Chukchi Sea, and also seen littering upturned highly productive summer pack ice in the north Barents Sea (author, unpublished). Perhaps it is more like the snow or slush community described from the Antarctic (see below).

A snow or slush algal community has been described by Meguro (1962) and Burkholder and Mandelli (1965). This community is in fact a plankton community that seeped into the snow when the weight of snow forced the ice below the surface. It is a light-adapted community living under very stable conditions. It is found in snow and in slush layers up to 1 m thick. With a production of about 190 mg C $m^2$ d it is about 30 times more productive than the surrounding plankton community.

Table 6-9. Chlorophyll *a* Concentrations of Sea Ice and of the Water Beneath Ice (+) or Scattered Ice (+ −) in the Arctic

| *Epontic Ice Flora* | | *Sub-Ice Plankton Flora* | | | | | | | |
|---|---|---|---|---|---|---|---|---|---|
| *mg chlorophyll a m³* | *mg chlorophyll a m²* | *mg chlorophyll a m³* | *mg chlorophyll a m²* | *Ice* | *Part of Ice* | *Part of Water* | *Time* | *Locality* | *Source* |
| | | 0.37 (max.) | 21 (max.) | 4.5 m | | | | Polar Basin | Apollonio, 1959 |
| | | | 66 (max.) | 3 m | | 4–64 m m | | Polar Basin | English, 1961 |
| | | 2.3 (max.) | | 3 m | | 6 m | | Polar Basin | |
| | | 5.7 (max.) | | lead | | 2 m | | Polar Basin | |
| | | 8.2 (max.) | 42 (max.) | pack | | 5 m | Aug. | Dumbell Bay | Apollonio, 1980 |
| | 1–>26 | 0.1–0.4 | | + | | 7m | | Beaufort Sea | Horner and Schrader, 1982 |
| | 30.5 (max.) | | | + | | | | Chukchi Sea | Clasby, Horner, and Alexander, 1973 |
| | 22–30 (max.) | | | + | | | | Chukchi Sea | Horner, 1976 |
| 10–427 | | | | + | colored | | | Chukchi Sea | Meguro, Ito, and Fukushima, 1967 |
| 120 (mean) | | | | + | layer | | | Chukchi Sea | |
| | | 8.8 (max.) | | MIZ | | | | Chukchi Sea | Hameedi, 1978 |
| | 0.34–3.0 | | | + | | | | Bering Sea | McRoy and Goering, 1974 |
| | | 35 | | MIZ | | | | Bering Sea | Alexander and Niebauer, 1981 |
| 25.9 | | 0.09 | | + | lower 3 cm | | | Okhotsk Sea | Hoshiai and Fukuchi, 1981 |
| 33–60 | | | | + | colored ice | | | Okhotsk Sea | |
| 12.8 | | 0.2 | | + | lower 3 cm | | | Okhotsk Sea | |
| 14 (max.) | | | | + | | | | Sea of Japan | Kuznetsov, 1981 |
| 19 (max.) | | 23.8 | | + | | | | (Amur Bay) | Kuznetsov, 1983 |
| 30–1,460 | 3–23 | | | 1.65 m | colored | | | Jones Sound | Apollonio, 1965 |
| | 12 (mean) | | | 1.65 m | layer | | | Jones Sound | |
| | 10.2–22 | | 0.2–0.5 | + | | 0–50 m | | Resolute Bay | Welch and Kalff, 1977 |
| | 15.2 (mean) | | | + | | | | Resolute Bay | |
| 300 (max.) | 7.7 (max.) | | | + | | | | Frobisher Bay | Hsiao, 1980 |
| $160 \cdot 10^9$ (max.)* | $34 \cdot 10^9$ (max.)* | | | + | | | | Frobisher Bay | |

| | | | | | | | | | |
|---|---|---|---|---|---|---|---|---|---|
| | | 1–2 | 2–4 | + – | | 2–3 m halocline | | Ogac Lake | McLaren, 1969 |
| | | 1.5 (max.) | | + | | | May | Hudson Bay | Legendre, Ingram, and Poulin, 1981 |
| 60–202 (max.) | 10–25 (max.) | | | + | | | | Can. E. Arctic | Dunbar and Acreman, 1980 |
| 160 (max.) | | 0.36 | | + | lower 6 cm | | | Davis Strait | Booth, 1984 |
| 10.24 (mean) | | 10.4–27.5 (max.) | | + | lower 6 cm | | | Davis Strait | |
| | 0.06–0.76 (mean) | | | + | total floe | next to ice | | Davis Strait | |
| 160 (max.) | 16.25 (max.) | | 2–10 | + | lower 6 cm | | | Davis Strait | MacLaren Marex Inc., 1979*c* |
| 205 (max.) | 3.69 (mean) | | | + | lower 1 cm | | | Davis Strait | |
| $27 \cdot 10^9$ (max.)* | | | | + | | | | Davis Strait | Booth, 1984 |
| 7.6 (max.) | 0.5 (max.) | 0.25 (max.) | | 0.5m | lower 4.5 cm | | | Disko Bay | Andersen, 1977*b* |
| | | 0.25 (max.) | | 1 m | | | | Disko Fjord | Andersen, 1981 |
| 25 (max.) | | 0.5 (max.) | | + | | | | St. Lawrence Gulf | Demers, Therriault, and Descolas-Gros, 1984 |
| 13 (max.) | 3 (max.) | | | + | | | | St. Lawrence Gulf | Dunbar and Acreman, 1980 |
| | | 2.2 (max.) | | MIZ | | | Sept.–Oct. | N of Svalbard | Heimdal, 1983 |
| 8.6 (tot. max.) | ca. 0.7 (tot. max.) | 66.4 (tot. max.) | | 0.35 m | lower 8 cm | next to ice | 12 Feb. | Isefjord, Denmark | Andersen, unpubl. |
| | | 54.6 (<20$\mu$m max.) | | 0.35 m | lower 8 cm | next to ice | 12 Feb. | Isefjord, Denmark | |
| 62.3 (tot. max.) | ca. 6 (tot. max.) | 23.6 (tot. max.) | | 0.38 m | lower 9.5 cm | 2 m | 19 Mar. | Isefjord, Denmark | |
| 42.4 (<20$\mu$m max.) | ca. 4 (<20$\mu$m max.) | 4.0 (<20$\mu$m max.) | | 0.38 m | lower 9.5 cm | 2 m | 19 Mar. | Isefjord, Denmark | |
| | | $>75 \cdot 10^9$* | | 0.35 m | | next to ice | 12 Feb. | Isefjord, Denmark | Thomsen, 1985 |
| $>1100 \cdot 10^9$* | | | | 0.7 m | | | | Kattinge Vig, Denm. | Grøntved, 1950 |

*Note:* * = cell counts; MIZ = marginal ice zone.

Table 6-10. Chlorophyll *a* Concentrations of Sea Ice and of the Water Beneath Ice (+) in the Antarctic

| *Epontic Ice Flora* | | *Sub-Ice Plankton Flora* | | | | | | |
|---|---|---|---|---|---|---|---|---|
| *mg chlorophyll a m³* | *mg chlorophyll a m²* | *mg chlorophyll a m³* | *mg chlorophyll a m²* | *Ice* | *Part of Ice* | *Part of Water* | *Locality* | *Source* |
| ca. 153 | | | | + | interstitial | | Antarctica | Bunt, 1963 |
| ca. 617 | | | | + | on crystals | | Antarctica | |
| ca. 248 | | | | + | total | | Antarctica | |
| 7830 ml (max.) | | 14 ml (max.) | | + | | | Antarctica | Fukushima and Meguro, 1966 |
| ca 100 | | | | + | | | Antarctica | Bunt and Lee, 1970 |
| >1000 | | | | 1.3m | | | Antarctica | Hoshiai, 1977 |
| 2–4.5 | 0.4–0.8 | 0.8 (max.) | | 1m | | 10m | Antarctica | Fukuchi and Sasaki, 1981 |
| | | | 3 | 1m | | 0–150m | Antarctica | |
| 86–921 | | | | + | | | Antarctica | Hoshiai and Fukuchi, 1981 |
| 250 | | 55 (max.) | | + | | 30m | McMurdo Sound | Bunt, 1968 |
| 0.31–4.54 (max.) | | | | + | | | Weddel Sea | Ackley, Buck, and Taguchi, 1979 |

| *Snow Ice Flora* | | *Sub-Ice Plankton Flora* | | | | |
|---|---|---|---|---|---|---|
| *mg chlorophyll a m³* | *mg chlorophyll a m²* | *mg chlorophyll a m³* | *Ice* | *Part of Ice* | *Locality* | *Source* |
| 670 | 97 | | + | upper 15–20 cm | Antarctica | Meguro, 1962 |
| 300–407 | | | + | upper 30 cm | Antarctica | Burkholder and Mandelli, 1965 |
| 80 ml (max.) | | 14 ml (max.) | + | | Antarctica | Fukushima and Meguro, 1966 |

Table 6-11. Primary Production per m³ of Sea Ice and of the Water Beneath Ice (+) or Scattered Ice (+ −) in the Arctic and Antarctic

| *Epontic Ice Flora* | | | *Sub-Ice Plankton Flora* | | | | | | | |
|---|---|---|---|---|---|---|---|---|---|---|
| *mg C m³ h* | *mg C m³ d* | *g C m³ yr* | *mg C m³ h* | *mg C m³ d* | *g C m³ yr* | *Ice* | *Part of Ice* | *Part of Water* | *Locality* | *Source* |
| Arctic | | | | | | | | | | |
| | | | 0.12–0.78 | | | 3 m | | 4–64 m | Polar Basin | English, 1961 |
| | | | 4.9 (max.) | | | 3 m | | 0.5 m lead | Polar Basin | |
| | | | | 110 (max. net) | | 0.5 m pack | | 5 m | Dumbell Bay | Apollonio, 1980 |
| | | | | 158 (max. gross) | | 0.5 m pack | | 5 m | Dumbell Bay | |
| | | | 6.58 (max. mean) | | | 0.5 m pack | | 5 m | Dumbell Bay | Subba Rao and Platt, 1984 |
| | | | 0.5 (max. mean) | 12.1 (max.) | | 1.7 m | | next to ice | J. Brøn-lund Fj. | Andersen, 1977*a* |
| | | | 20 (max.) | 476 | | MIZ | | 16 m | Chukchi Sea | Hameedi, 1978 |
| | 44–95 | | | | | + | | | Bering Sea | McRoy and Goering, 1974 |
| | 193 (max.) | | | 1,600 (max.) | | 0.6 m | | | Amur Bay | Kuznetsov, 1981 |
| | | | | 20–40 | 0.8–2.2 | + − | | 2–3 m halocl. | Ogac Lake | McLaren, 1969 |
| measured: | 45 (max.) | ca. 1 | | 20 (max.) | | 0.5 m | lower 4.5 cm | 0–4.5 cm | Disko Bay | Andersen, 1977*b* |
| estimated: | 122 (max.) | ca. 2.9 | | 4.5 (max.) | | 0.5 m | lower 4.5 cm | 1 m | Disko Bay | |
| | | | 0.1 (max.) | 1 (max. mean) | | 0.5 m | | 0–50 m | Disko Bay | |
| | | | 0.05 (max.) | <0.5 (max.) | | 1 m | | 0–50 m | Disko Fjord | Andersen, 1981 |

Table 6-11. *(continued)*

| *Epontic Ice Flora* | | | *Sub-Ice Plankton Flora* | | | | | | | |
|---|---|---|---|---|---|---|---|---|---|---|
| *mg C m³ h* | *mg C m³ d* | *g C m³ yr* | *mg C m³ h* | *mg C m³ d* | *g C m³ yr* | *Ice* | *Part of Ice* | *Part of Water* | *Locality* | *Source* |
| Arctic | | | | | | | | | | |
| | | | 2.5 (max.) | | | MIZ | | | N. Svalbard | Heimdal, 1983 |
| | | | 0.02–14.01 | | | + − | | | Arctic seas | Subba Rao and Platt, 1984 |
| | | | 2.73 (mean) | | | + − | | | Arctic seas | |
| Antarctic | | | | | | | | | | |
| 17 | | | | | | + | | | Antarctica | Bunt, 1968 |
| | | | 82 (max.) | | | pack | | 0–9 m | Antarctica | El-Sayed, 1971*a* |
| | | | | 0.11 (chemoautotroph) | | 412 m | | | Antarctica | Horrigan, 1981 |
| | | | | $25 \cdot 10^{-6}$ (heterotroph) | | 412 m | | | Antarctica | |
| Snow Ice Flora | | | | | | | | | | |
| Antarctic | | | | | | | | | | |
| 749–1,080 (max.) | | | | | | | | | Antarctica | Burkholder and Mandelli, 1965 |

*Note:* MIZ = marginal ice zone.

## Plankton Floras

A winter bloom of microflagellates growing at very low light intensities (0.22 $\mu$E $m^2$ s) has been described by Müller-Haeckel (1982) from the Gulf of Bothnia (60°N). More southerly in Peter the Great Bay, Sea of Japan, at 44°N (Vyshkvartsev and Konolova, 1979; Kuznetsov, 1981, 1983) and in Isefjord, Denmark, at 56°N (Thomsen, 1985; Andersen unpublished), an intense winter bloom has been recorded under newly formed ice under less severe light conditions (max. 60 $\mu$E $m^2$ s in Isefjord). These winter sub-ice blooms occur as a direct consequence of the formation of fast ice, which immediately eliminates wind mixing. Hereby the phytoplankters are given a chance to keep close to the surface where light conditions are best at the time. Also, any nutrient-rich outflow of fresh water from land will stay close to the surface. Therefore, in the eutrophied Amur Bay, Sea of Japan (Kuznetsov, 1981), winter sub-ice production is much greater than production in the ice (up to 1.6 g C $m^3$ d), and in Isefjord, Denmark, chlorophyll concentrations immediately beneath the ice reached 66.4 mg chlorophyll *a* $m^3$, while at 2 m concentrations were only 3 mg chlorophyll *a* $m^3$, at 4 m they were 1.9, at 6 m they were 0.44, and at 8 m they were 0.39 mg chlorophyll *a* $m^3$. The main constituent was a 5 $\mu$m chrysophyte numbering 75 million per litre.

From under high-Arctic (63°–82°30′ N) fast ice and nearshore pack ice, which does not move about like the pack ice of the Polar Basin and the polar currents, Apollonio (1985) recognizes a distinct "halocline" flora (McLaren, 1969). It is perhaps identical to the winter floras mentioned above, since as suggested by Apollonio, it may consist mainly of small chlorophytes and chrysophytes, as found by Bursa (1963) under the Chukchi Sea ice, and of euglenoids. This halocline bloom occurs 1–3 m beneath the ice in the steep halocline present at that level. Chlorophyll concentration is remarkably constant at 1–2 mg chlorophyll *a* $m^3$, as if in dynamic equilibrium, perhaps controlled by nutrient diffusion from below. It can persist for as long as 6–8 weeks, producing 40–80 mg C $m^2$ d or 1.7–4.5 g C $m^3$ yr. These last two phytoplankton blooms of restricted vertical extent have probably often been missed in phytoplankton production studies because of their restricted occurrence in space and/or time, or they have been included but not recognized as distinct from the "normal" phytoplankton bloom.

Normal vernal phytoplankton blooming—defined as a threefold increase above initial concentrations (Slagstad, 1982)—in shallow and well-mixed water is most dependent on the concentration and thickness of ice and snow, governing the penetration of light (Slagstad, 1984). The mixed polar surface water, with its pycnocline acting as a false bottom, can be viewed as one such area. Normally shallow mixed-surface layers (<10 m according to Slagstad, 1982) delay the bloom due to sinking. Also (in ice-free waters) vertical stability is the most important factor determining the onset and extent of blooming. Deep mixed layers give delayed blooms but greater total production if not initiated too late in the year. In sheltered waters, however, as under ice, an early bloom can start in the top of a seemingly deep mixed layer like a winter sub-ice bloom. Beneath and adjacent to the polar pack, the mixed polar surface layer (20–50 m thick) acts as a backup for any minor surface mixed layer of meltwater, supplying nutrients and ensuring against excessive mixing and sinking. The shallower this mixed layer of polar water above the steep pycnocline is, however, the more short-lived the bloom will be and the less total production will be (Andersen, 1981).

In the permanently ice-covered offshore high-Arctic, the conditions never surpass those of vernal blooming. Therefore, the annual production cycle has only one maximum (Apollonio, 1959). The bloom inten-

Table 6-12. Primary Production per $m^2$ of Sea Ice and of the Water Beneath Ice (+) and Scattered Ice (+ −) in the Arctic and Antarctic

| *Epontic Ice Flora* | | | *Sub-Ice Plankton Flora* | | |
|---|---|---|---|---|---|
| *mg C m² h* | *mg C m² d* | *g C m² yr* | *mg C m² h* | *mg C m² d* | *g C m² yr* |
| Arctic | | | | | |
| | | | | 24 (max.) | 0.6 |
| | | | | 5–6 (max.) | |
| | | | 5–19.6 | (• 24 = 120–470) | |
| | | | | 100 net (max.) | |
| | | | | 550 net (max.) | |
| | | | | 830 gross (max.) | |
| | | | | | 6–9 net |
| | | | | | 11–12 gross |
| | | | | 33.3 exp. (max.) | |
| | | | | 88.7 est. (max.) | |
| | | | | | ca. 1.5 |
| | 350 | 15 est. (= 33%) | | | |
| <0.2–2.6 | | 0.7 | | | 0.38 |
| | | | | | 0.01 |
| | | 0.18–1.7 | | | |
| | | 5–10 | | | |
| 4.56–14.9 (max.) | (• 24 = 110–358) | | | | |
| 0.3–7.67 (mean) | (• 24 = 7.2–184) | (• 120 = 0.9–22) | | | |
| 4.5 (max.) | ca 100 (max.) | ca. 3 (max.) | 9.5 (max.) | ca. 200 (max.) | ca. 2 |
| | | | ca. 10 ? (max.) | ca. 250 (max.) | ca. 3 |
| | | | | 6000 (max.) | |
| | | | 0.6 | (• 24 = 14.4) | |
| | 2.2–4.8 (max.) | | | 15–20 | |
| | | 10–15% | | | |
| 0.22 (max.) | (• 24 = 5.28) | <1% | | | |
| | 0.003–2.4 (max.) | | | | |
| | 2 exp. (max.) | 0.044 exp. (max.) | | 0.9 (max.) | |
| | 5.5 est. (max.) | 0.14 est. (max.) | | | |
| | | | | 15 (max.) | |
| | | | | 8 (max.) | |
| Antarctic | | | | | |
| | | | | 0.6–8.9 | |
| | | | | 4.2 | 1.5 |
| | | | | 0.001 | 0.00038 |

*Note:* exp = experimental; est = estimated.

sity is limited by low illumination, and it can only occur as a result of the steep pycnocline beneath the mixed surface layer. Chlorophyll *a* concentrations up to 2.3 mg $m^3$ at 6 m, 5.7 mg $m^3$ at 2 m in a lead, and 66 mg $m^2$ at 4–64 m have been recorded (English, 1961). Annual production is low. The 0.6 g C $m^2$ yr found by Apollonio (1959) could, however, be an underestimate, since he may have missed a possible bloom immediately beneath the ice. Subba Rao and Platt (1984) consider 9 g C $m^2$ as a mean estimate of offshore production including the epontic contingent. How they arrive at this figure is, however, not clear. Attention should perhaps also be paid to chemoautotrophic and heterotrophic production, which under Antarctic shelf ice

Table 6-12.
*(continued)*

| *Ice* | *Depth* | *Time* | *Locality* | *Source* |
|---|---|---|---|---|
| 4.5 m | | | Polar Basin | Apollonio, 1959 |
| 4.5 m pack | 0–30 m | | Polar Basin | English, 1961 |
| + – | | | Polar Basin | Subba Rao and Platt, 1984 |
| 1.5 m fast | | | Polar Basin | Apollonio, 1980 |
| 0.5 m pack | | | Dumbell Bay | |
| 0.5 m pack | | | Dumbell Bay | |
| 1.8–0.5 m | | | Dumbell Bay | |
| 1.8–0.5 m | | | Dumbell Bay | |
| 2.3 m fast | 0.27.6 m | June | J. Brønlund Fj. | Andersen, 1977*a* |
| 2.3 m fast | 0–34.3 m | June | J. Brønlund Fj. | |
| 2.5–1 m fast | | | J. Brønlund Fj. | |
| 2 m fast | | | Resolute Bay | Welch and Kalff, 1977 |
| + | | spring | Beaufort Sea | Horner and Schrader, 1982 |
| + | benthos | spring | Beaufort Sea | |
| + | | | Beaufort Sea | Schell 1980*b* |
| + | | | Beaufort Sea | Alexander, 1974 |
| + | | | Chukchi Sea | Clasby, Horner, and Alexander, 1973 |
| + | | | Chukchi Sea | |
| + – | | | Chukchi Sea | Matheke and Horner, 1974 |
| + – | benthos | | Chukchi Sea | |
| pack edge | 0–32 m | | Chukchi Sea | Hameedi, 1978 |
| + | | | Chukchi Sea | Booth, 1984 |
| + | | | Bering Sea | McRoy and Goering, 1974 |
| + | | | Bering Sea | McRoy and Goering, 1976 |
| 0.3–2.7 m | | | Davis Strait | Booth, 1984 |
| 1–1.6 m (max.) | | | Davis Strait | |
| 0.35 m | 0–0.045 m | | Disko Bay | Andersen, 1977*b* |
| 0.35 m | 0–0.045 m | | Disko Bay | |
| 0.5 m | 0–50 m | | Disko Bay | Andersen, 1981 |
| 1 m | 0–50 m | | Disko Fjord | |
| pack | | | Antarctica | Burkholder and Mandelli, 1965 |
| 412 m shelf (chemoautotroph) | | | Antarctica | Horrigan, 1981 |
| 412 m shelf (heterotroph) | | | Antarctica | |

has been estimated to be 1.5 g and 0.38 mg C $m^2$ yr, respectively (Horrigan, 1981). Consistent with a very low level of production in the Polar Basin, Packard (1982) was unable to detect any catabolic metabolism in particulate matter below 125 m in the Polar Basin. This furthermore supports the case for sedimenting particulate matter formed in the Polar Basin in playing the dominant role in fueling deep metabolism as opposed to organic matter entrained with deep-ocean water masses during their formation.

In high-Arctic nearshore waters influenced by terrestrial runoff and winds blowing from land, especially in fjords, the annual production cycle exhibits two maxima (Andersen, 1977*a;* 1981; Apollonio,

1980). The spring bloom starting in late May to late June does not attain any great maximum, since it is checked by an increase in surface turbidity accompanying the freshwater outflow from land. Although these fjords become ice-free or have their ice broken up in July–August, allowing more light to reach the water, production does not increase markedly until turbidity decreases as runoff from land is reduced in mid- to late August. Then a greater bloom sets in, with up to 830 mg C d gross production, giving up to 8.2 mg chlorophyll *a* $m^3$, but this is only in a shallow stratum where the chlorophyll *a* per $m^2$ only amounts to 42 mg (Apollonio, 1980) (Table 6-9). Production is, however, soon brought to a stop by new ice and snow and reduced insolation. The result is, nonetheless, a greater annual production of about 6–14 g C $m^2$ in turbid high-Arctic fjords (Andersen, 1977*a;* Apollonio, 1980) than in the offshore Polar Basin. Newbury (1983) estimates that about one-third of annual primary production in the nearshore waters of the Beaufort Sea occurs under the landfast ice in the spring (including 5 percent epontic). Adding to the primary production, Schell (1980*b*) has calculated that an equal amount of carbon entering the nearshore marine food web apparently is of terrestrial origin coming from peat banks and rivers. About 90 percent of this material is probably first assimilated by bacteria (Griffiths and Morita, 1981).

In and off glacier fjords turbidity is also a major factor; however, it is not restricted to the spring and summer period. Nor does the turbidity originate in surface outflow alone. Upwelling in front of glaciers, as was described by Hartley and Dunbar (1938), and around icebergs scraping the bottom (Petersen, 1964, 1977) continually brings turbid material and nutrient-rich water to the surface, possibly aided by a density-dependent outgoing surface current of lighter (fresher) water. In Disko Bay, off Jakobshavn Isfjord, at least the upper 50 m are constantly cooled and diluted by icebergs. Immediately below a steep surface density gradient with high turbidity, depleted of nutrients, a dense productivity layer is found concentrated at 5–10 m where it thrives on a plentiful supply of upwelling nutrients. Primary production probably amounts to more than 1 g C $m^2$ d throughout the summer (1.1 g measured in mid-August [Andersen, 1981]).

The usual primary production cycle in Arctic waters, which become free from ice early enough for summer depletion of nutrients and are not subjected to excessive turbidity, exhibit two major blooms, a spring bloom and a fall bloom reaching close to the surface. In between these blooms more or less oligotrophic conditions exist in the surface near-water layers, sometimes interrupted by smaller intermediate blooms caused by tide-generated or other upwelling. In Disko Bay, for example, one such mid-summer bloom occurs each year (Andersen, 1981). Otherwise a deeper shade-adapted flora upholds a lower level of mid-summer production based on nutrient diffusion or upwelling from beneath.

Spring bloom chlorophyll *a* concentrations in Arctic waters of up to 18 mg $m^3$ and 340 mg $m^2$ have been measured under normal conditions, while ice-edge values reach 40 mg $m^3$ and 492 mg $m^2$ (Table 6-13). Subba Rao and Platt (1984) consider a production of 14 mg C $m^3$ h to be maximum for Arctic seas, but report 21 mg C $m^3$ h as a maximum in Disko Bay, corresponding to a maximum daily mean of 7 mg C $m^3$ h. An ice-edge production of up to 44 mg C $m^3$ h has been measured in the Chukchi Sea (Hameedi, 1978) (Table 6-14), while 82 mg C $m^3$ h has been recorded from the Antarctic (El-Sayed, 1971*a*) (Table 6-11). In Disko Bay, a spring maximum of 168 mg C $m^3$ d has been recorded at 5 m, while the highest 0–50 m mean of about 38 mg C $m^3$ d was measured on the same day. This corresponds to 1,920 mg C $m^2$ d, which is close to the maximum of 2,256 mg

## Table 6-13. Chlorophyll *a* in Arctic and Antarctic Waters Free from Ice

| *mg chlorophyll a* / $m^3$ | *mg chlorophyll a* / $m^2$ | *Part of Water* | *Time* | *Locality* | *Source* |
|---|---|---|---|---|---|
| Arctic | | | | | |
| 7.4 (max.) | | 3.5 m | | Prudhoe Bay | Horner, Coyle, and Redburn, 1974 |
| 0.1–2.0 | | | | Beaufort Sea | Grainger and Evans, 1980 |
| 10.2 (max.) | | | | Beaufort Sea | Hsiao and Trucco, 1980 |
| 0.4–17.8 (mean) | 128 (mean) | | | Chukchi Sea | Hameedi, 1978 |
| 40 (max.) | | | | Chukchi Sea | |
| 4 (max.) | | surface | | N. Bering Strait | Koike et al., 1982 |
| 3.5 (max.) | | surface | | E. Bering Sea | |
| 4.9 (max.) | | | | Murman coast | Sokolova and Solov'-yeva, 1971 |
| 15 (max.) | 25 (mean) | | | Resolute Bay | Welch and Kalff, 1977 |
| | 100 (max.) | | | Resolute Bay | |
| 9.5 (max.) | | 12 m | July–Aug. | Frobisher Bay | Hsiao, 1985 |
| 7.5 (max) | | 5 m | Aug.–Sept. | Frobisher Bay | |
| 5 (max.) | | | | Frobisher Bay | Grainger, 1979 |
| 13.6 (max.) | | 25 m | | Lancaster Sound | LGL, 1976 |
| 2.47 (mean) | | 25 m | | Lancaster Sound | |
| $2 \cdot 10^7$ cells (max.) | | 10 m | | Lancaster Sound | |
| 5 (max.) | | | | N. Baffin Bay | Herman, 1983 |
| 1.26 (mean) | 26–108 | photic zone | | Baffin Bay | Harrison, Platt, and Irwin, 1982 |
| 18 (max.) | | pycnocline | | E. Can. Arctic | Boorstad and Gover, 1984 |
| 8 (max.) | | eddy | | E. Can. Arctic | |
| | 69 (mean) | 0–35 m | | E. Can. Arctic | |
| 15 (max.) | | | | E. Can. Arctic | Subba Rao and Platt, 1984 |
| | 492 (max.) | | | Davis Strait | MacLaren Marex Inc., 1979*c* |
| 10 (max.) | 340 (mean) | | spring | Disko Bay | Andersen, 1981 |
| 1.6 (max.) | 31 (mean) | | | Barents Sea | Vedernikov and Solov'yeva, 1972 |
| $>10 \cdot 10^9$ cells | | | | Barents Sea | Reynolds, 1974 |
| Antarctic | | | | | |
| | 17.4 (mean) | | | Antarctica | El-Sayed and Turner, 1977 |
| 11–27 (max.) | | | | Antarctica | Krebs, 1983 |
| 118.35 (max.) | | surface | | Antarctica Atl. | El-Sayed, 1971*b* |
| 5.8 (max.) | | surface | | Antarctica Pac. | |

Table 6-14. Primary Production per $m^3$ in Arctic Waters Free from Ice

| *mg C m³ h* | *mg C m³ d* | *Part of Water* | *Locality* | *Source* |
|---|---|---|---|---|
| 1.8 (max. mean) | 44.2 (max.) | 1.35 m | J. Brønlund Fj. | Andersen, 1977*a* |
| 8.8 (max.) | | | Beaufort Sea | Hsiao, Foy, and Kittle, 1977 |
| 8.0 (max.) | | | Beaufort Sea | Hsiao and Trucco, 1980 |
| 6.4 (max.) | | | Beaufort Sea | Grainger and Evans, 1982 |
| 18.4 (max.) | | | Beaufort Sea | Duval, 1977 |
| 44.2 (max.) | | surface | Chukchi Sea | Hameedi, 1978 |
| 1 (max.) | | mixed layer | NE Baffin Bay | Herman, 1983 |
| 4 (max.) | | 12 m | NE Baffin Bay | |
| 1.7 (max.) | | 17 m | NE Baffin Bay | |
| | 7.4 (max. mean) | photic zone | Baffin Bay | Harrison, Platt, and Irwin, 1982 |
| 4.2 (max.) | 50–100 (max.) | | Frobisher Bay | Grainger, 1979 |
| | 40 (max. mean) | 0–50 m | Disko Bay | Andersen, 1981 |
| 7 (max. mean) | 168 (max.) | 5 m | Disko Bay | |
| 21 (max.) | | | Disko Bay | Subba Rao and Platt, 1984 |
| 6 (max. mean) | 150 (max.) | | Disko Fjord | Andersen, 1981 |
| 7 (max.) | | | SW Greenland | Petersen, 1979 |
| 6 (max.) | | | W Iceland | Thordardottir, 1973 |
| 13.5 (max.) | | | Norway | Throndsen and Heimdal, 1976 |
| 8.6 (max.) | | | W Barents Sea | Corlett, 1958 |
| 2.4 (max.) | | | Barents Sea | Vedernikov and Solov'yeva, 1972 |
| 1.4 (max.) | | | Murman coast | Sokolova and Solov'yeva, 1971 |

C $m^2$ d measured in Davis Strait, the 2,300 mg C $m^2$ d from Svalbard, and the 2,400 mg C $m^2$ d from Godthaab, West Greenland (Table 6-15). Most other Arctic areas, as well as Antarctic areas (Table 6-16), have lower maximum values. Theoretically much higher values of production are recorded from Lancaster Sound and west Baffin Bay (4,750 and 4,890 mg C $m^2$ d in Table 6-15), and even higher values have actually been measured in the upwelling areas of the Chukchi Sea and Bering Strait (up to 6,000 and more than 4,000 mg C $m^2$ d). These areas are the most productive in the world, with annual rates of 500 g C $m^2$ or more. Ice-edge production as such cannot compete in these areas in spite of high daily rates, since ice-edge production zones are always on the move away from depletion. Only where upwelling occurs along the ice edge, or in the Atlantic water, which constantly submerges beneath polar water (as in the north Barents Sea and the Greenland Sea), or in the Bering Strait, can the proliferation of phytoplankton be prolonged within a restricted geographical area.

## Benthic Floras

Few measurements of benthic microalgal production have been made in the Arctic. At Barrow, the production rate was no more than 1 mg C $m^2$ h at 5 m before breakup in late June–early July. A high of 57 mg C $m^2$ h was measured in late July after breakup, and an annual mean of less than 5 mg C $m^2$ was estimated (Matheke and Horner, 1974). This is twice the epontic estimate and eight times that for the plankton, the production of which, however, had not peaked when experiments stopped by 21 August. The estimated annual production of 43.8 g C $m^2$ at 5 m at Barrow is about three times an estimate of 15 g C $m^2$ at 7–11 m in Resolute Bay (Welsh and Kalff, 1977).

## CONCLUSIONS

A review of primary production of Arctic phytoplankton, ice algae, and benthic microalgae, including their magnitude and relationship to light, darkness, nutrients, sea ice, glaciers, salinity, and temperature, is presented. Up to six distinct phytoplankton floras have been recognized, restricted in time and space beneath and adjacent to the ice. The stabilizing effect of an ice cover on the water column permits production to occur very near to the surface at low light intensities. The total annual production per $m^2$ by phytoplankton ranges from one to a few g C under the polar pack to more than 500 g C along the polar fronts.

Three ice floras have been recognized. An epontic spring flora is estimated to produce up to 15 g C $m^2$ yr or possibly one-third of total annual production in the central Polar Basin, excluding production by surface-type or snow algae and by summer epontic algae, the production of which is largely unknown. The biomass of the summer epontic algae, however, can be many times greater than that of the spring community. Benthic microalgal production in the near-shore shallow-ice-covered high-Arctic equals epontic spring production. The primary limiting factor is light, which under average central Polar Basin ice 4.75 m thick is only 0.01 percent of surface irradiance. This corresponds to a mid-summer daily mean of about 0.05 $\mu$E $m^2$ s PAR, which is less than the lowest recorded compensation point of 0.22 $\mu$E $m^2$ s PAR. Under only about 25 percent of the central polar pack does mean mid-summer irradiance reach an assumed compensation intensity of about 0.4 $\mu$E $m^2$ s PAR. Primary production by phytoplankton as well as by spring ice algae in the Polar Basin sets in at about this irradiance level in late April. In nearshore arid areas it may already occur in early March. Saturation intensities range from 1.8 $\mu$E $m^3$ s PAR for ice algae

Table 6-15. Primary Production per m² in Arctic Waters with (+) or without (−) Ice Cover

| *mg C m² h* | *mg C m² d* | *g C m² yr* | *Ice* | *Depth* | *Time* | *Locality* | *Source* |
|---|---|---|---|---|---|---|---|
| | | 15.7 (mean) | + − | incl. ice + benthos | | Arctic Ocean | Subba Rao and Platt, 1984 |
| | | 27 | + − | shelf | | Arctic Ocean | |
| | | 9 | + − | offshore | | Arctic Ocean | |
| | 75 (mean) | | + − | incl. ice + benthos | | Arctic waters | |
| | 225 | | + − | shelf | | Arctic waters | |
| | 20–260 | | + − | offshore | | Arctic waters | |
| | 140 exp. (max.) | | − | 0–13.5 m | Aug. | J. Brønlund Fj. | Andersen, 1977*a* |
| | 187 est. (max.) | | − | 0–13.5 m | Aug. | J. Brønlund Fj. | |
| | | 7.4–13.7 | 2.5–0 m | | | J. Brønlund Fj. | |
| | 300 (mean) | 30 | 2–0 m | | | Resolute Bay | Welch and Kalff, 1977 |
| | | +15 | 2–0 m | ice, offshore | | Resolute Bay | |
| | | or 15 | 2–0 m | benthos, inshore | | Resolute Bay | |
| | | 45 (total) | 2–0 m | | | Resolute Bay | |
| | 750–1170 (max.) | 41–70 | + − | | | Frobisher Bay | Grainger, 1979 |
| | 450 | (• 120 = 55.9) | + − | | | Frobisher Bay | Harrison, Platt, and Irwin, 1982 |
| | 220 | | + − | | | High Arctic | Nemoto and Harrison, 1981 |
| | 40–80 | ca 1.7 | + − | 2–3 m halocline | | Ogac Lake | McLaren, 1969 |
| | | 12 | + − | | | Ogac Lake | |
| 45–198 | (• 24 = 1090–4750) | | − | | | Lancaster Sound | Gallegos et al., 1983 |
| | 200 | | − | | | Foxe Basin | Smith et al., 1985 |
| 2–8 | (• 24 = 58–190) | | − | | | Can. E. Arctic | Platt and Herman, 1983 |

| | | | | | | | |
|---|---|---|---|---|---|---|---|
| | 80–860 | | – | | | Can. E. Arctic | Harrison and Platt, in press |
| | 227 (mean) | 27 | – | | | Baffin Bay | Harrison, Platt, and Irwin, 1982 |
| | 140–260 | | – | | | Baffin Bay | Herman, 1983 |
| | 2210–4890 | | – | | | W. Baffin Bay | Hsiao and Trucco, 1980 |
| | 55–180 | | – | | | Beaufort Sea | Alexander, 1974 |
| | 280 | | – | | | Beaufort Sea | Alexander, 1976 |
| | 10–15 | | – | | | Beaufort Sea | Alexander et al., 1975 |
| | 15–40 | | – | | | Beaufort Sea | Subba Rao and Platt, 1984 |
| | 120–550 | | – | | | Beaufort Sea | |
| 0.4–120 | (• 24 = 10–2540) | | – | | | Beaufort Sea | Grainger and Evans, 1982 |
| | 100–1190 | | – | | | S. Beaufort Sea | Hsiao, Foy, and Kittle, 1977 |
| | 4 | | – | | | N. Beaufort Sea | Horner and Schrader, 1982 |
| | | 18–28 | – | | | NE. Chukchi Sea | Carey, 1978 |
| <5 (ann. mean) | (• 24 = 120) | (• 360 = 43.8) | + – | benthic 5 m | | Chukchi Sea | Matheke and Horner, 1974 |
| 0.5–57 | (• 24 = 12–1368) | (• 120 = 1.4–164) | – | benthic 5 m | Aug. | Chukchi Sea | |
| 32 (max.) | ca 800 (max.) | | – | | | Chukchi Sea | |
| 23.8 | (• 24 = 572) | | – | | summer | Chukchi Sea | Booth, 1984 |
| | 2000 (mean) | | MIZ | | | Chukchi Sea | Hameedi, 1978 |
| | 150–6000 | | MIZ | | | Chukchi Sea | |
| | >4000 | | – | | | Bering Strait | McRoy, Goering, and Shields, 1972 |
| | 2700 | (• 120 = 324) | – | | | Bering Strait | Sambrotto, Goering, and McRoy, 1984 |
| | 89 (max.) | | MIZ | | | Bering Sea | McRoy and Goering, 1974 |

Table 6-15. *(continued)*

| *mg C m² h* | *mg C m² d* | *g C m² yr* | *Ice* | *Depth* | *Time* | *Locality* | *Source* |
|---|---|---|---|---|---|---|---|
| | 15–21 | | – | | | Bering Sea | Subba Rao and Platt, 1984 |
| | | 45% | – | | spring | Bering Sea | McRoy and Goering, 1976 |
| | 243 (mean) | 18% | – | | summer | Bering Sea | |
| | 18–867 | | – | | summer | Bering Sea | |
| | | 25% | – | | fall | Bering Sea | |
| | | 2% | + | | winter | Bering Sea | |
| | | 9% | + | ice | | Bering Sea | |
| | | 2% | + | limnic | | Bering Sea | |
| | 160–630 | | – | | summer | Bering Sea | Taniguchi, 1969 |
| | 100–150 | | – | | summer | Bering Sea | Koblentz-Minshke, 1965 |
| | 11 | | – | | summer | Bering Sea | Holmes, 1953 |
| | 458 | 89 | – | shelf | summer | Bering Sea | Taguchi, 1972 |
| | 330 | 71 | – | central | summer | Bering Sea | |
| | 399 (total) | 85 (total) | – | | summer | Bering Sea | |
| | | 605 | + – | shelf | | Bering Sea | Ivanenkov, 1961 |
| | | 230 | + – | oceanic | | Bering Sea | |
| | | 387 (total) | + – | | | Bering Sea | |
| | 210 (max.) | | + – | | prespring | Davis Strait | Booth, 1984 |
| | 2256 (max.) | | – | | spring | Davis Strait | |
| | 54 (max.) | | – | | Aug. | Davis Strait | |
| | 20 (mean) | | + – | | | Davis Strait | Subba Rao and Platt, 1984 |
| | 1018 (max.) | 52 | – | | | Disko Bay | Petersen, 1964 |
| | 1920 (max.) | 75–104 | – | 0–50 m | | Disko Bay | Andersen, 1981 |
| | 510 (max.) | 32 | – | 0–50 m | | Disko Fjord | |
| | 900 | | – | | | Godthaab Fjord | Petersen, 1977 |

| | | | | | | | |
|---|---|---|---|---|---|---|---|
| | 350 (max.) | 50 | – | | | Godthaab inner | Smidt, 1979 |
| | 2400 (max.) | 160 | – | | | Godthaab outer | |
| | >500 | | – | coastal | | E. Greenland | Nielsen, 1958 |
| | ca 150–300 | | – | offshore | | (55–70°N) | |
| | 170–500 | | – | | | W. Iceland | Thordardottir, 1973 |
| | 380–2300 | | – | coastal | | Svalbard | Throndsen and Heimdal, 1976 |
| 18–37 | (• 24 = 430–890) | | MIZ | | Oct. | N. Svalbard | Heimdal, 1983 |
| | 980–1300 | | – | | | Bear Island | Marshall, 1957 |
| | 45–594 | | – | coastal | | Barents Sea | Vedernikov and Solov'yeva, 1972 |
| | 304 (mean) | | – | | | Barents Sea | |
| | 280–1330 | | – | | | W. Barents Sea | Corlett, 1959 |
| | 7–262 | | + – | | | Murman coast | Sokolova and Solov'yeva, 1971 |
| | 101 (mean) | 18.5 | + – | | | White Sea | Fedorov and Bobrov, 1977 |
| | 290 (max.) | | + – | | | White Sea | |
| | 30–40 | | + – | | | Gulf of Finland | Dunbar, 1982 |
| | 1100 (max.) | 76–80 | + – | | | Bothnia Bay | Hagström, Andersson, and Öquist, 1983 |

*Note:* exp = experimental; est = estimated; MIZ = marginal ice zone.

Table 6-16. Primary Production per m² in Antarctic Waters with (Pack) or without (–) Ice Cover

| *mg C m² d* | *mg C m² yr* | *Ice* | *Depth* | *Locality* | *Source* |
|---|---|---|---|---|---|
| 1560 | | pack | | Antarctica | El-Sayed, 1971*a* |
| ca. 800 | | – | shelf | Antarctica | Subba Rao and Platt, 1984 |
| 200–500 | | – | offshore | Antarctica | |
| 20–170 | | – | offshore | Antarctica | Saijo and Kawashima, 1964 |
| 40–294 | | – | offshore | Antarctica | El-Sayed and Turner, 1977 |
| 134 (mean) | | – | offshore | Antarctica | |
| 220–424 | | – | offshore | Antarctica | Jennings, Gordon, and Nelson, 1984 |
| | 130 | – | | Signy Island | Horne, Fogg, and Eagle, 1969 |
| <500 | 86–289 | – | offshore | Signy Island | Whitaker, 1982 |

to 1,450 μE m² s PAR for some open-water Arctic phytoplankton.

Some Arctic algae are able to survive periods of darkness far exceeding the extent of the polar night. There are always algae present, ready to start up production and to colonize the ice. Nutrients become limiting only where open water occurs in summer. Glaciers supply extra nutrients by melting and by causing upwelling, and they aid production by stabilizing the surface waters. Arctic algae in the plankton as well as in the ice are not living at their optimum temperature, but they (especially the ice algae) are able to tolerate very low temperatures and a wide range of salinities.

## REFERENCES

Ackley, S. F., K. R. Buck, and S. Taguchi. 1979. Standing crop of algae in the sea ice of the Weddell Sea region. *Deep-Sea Res.* **26A**:269–281.

Ackley, S. F., A. J. Gow, K. R. Buck, and K. M. Golden. 1981. Sea ice studies in the Weddell Sea aboard USCGC Polar Sea. *Antarc. J. U. S.* **15**(5):84–86.

Addison, J. R. 1977. Impurity concentrations in sea ice. *J. Glaciol.* **18**(78):117–127.

Alexander, V. 1974. Primary productivity regimes of the nearshore Beaufort Sea, with reference to potential roles of ice biota. In J. C. Reed and J. E. Sater (eds.), *The Coastal Shelf of the Beaufort Sea.* Arctic Institute of North America, Calgary, Alberta, pp. 609–632.

Alexander, V., and H. J. Niebauer. 1981. Oceanography of the eastern Bering Sea ice-edge zone in spring. *Limnol. Oceanogr.* **26**(6):1111–1125.

Alexander, V., D. C. Burrell, J. Chang, R. T. Cooney, C. Coulon, J. J. Crane, J. A. Dygas, G. E. Hall, P. J. Kinney, D. Kogl, T. C. Mowat, A. S. Naidu, T. E. Osterkamp, D. M. Schell, R. D. Siefert, and R. W. Tucker. 1975. *Environmental Studies of an Arctic Estuarine System. Final Report.* U.S. Environmental Protection Agency ORD, Ecological Research Series, 542p.

Andersen, O. G. N. 1977*a*. Primary production, illumination and hydrography in Jørgen Brønlund Fjord, North Greenland. *Medd. Grønland* **205**(4):27p.

Andersen, O. G. N. 1977*b*. Primary production associated with sea ice at Godhavn, Disko, West Greenland. *Ophelia* **16**(2):205–220.

Andersen, O. G. N. 1981. The annual cycle of phytoplankton primary production and hydrography in the Disko Bugt area, West Greenland. *Medd. Grønland, Biosci.* **6**:65p.

Anderson, G., and D. Dyrssen. 1981. Chemical constituents of the Arctic Ocean in the Svalbard area. *Oceanologica Acta* **4**(3):305–311.

Antia, N. J. 1976. Effects of temperature on the darkness survival of marine microplanktonic algae. *Microb. Ecol.* **3**:41–54.

Apollonio, S. 1958. *Hydrobiological Measurements on T-3, 1957–58.* Woods Hole Oceanographic Institute manuscript, 7p.

Apollonio, S. 1959. Hydrobiological measurements on IGY drifting station Bravo. *Am. Geophys. Union Trans.* **40**(3):316–319 & IGY Bull. 27:316–319.

Apollonio, S. 1961. The chlorophyll content of Arctic sea-ice. *Arctic* **14**(3):197–200.

Apollonio, S. 1965. Chlorophyll in Arctic sea ice. *Arctic* **18**:118–122.

Apollonio, S. 1973. Glaciers and nutrients in Arctic seas. *Science* **180**:491–493.

Apollonio, S. 1980. Primary production in Dumbell Bay in the Arctic Ocean. *Marine Biol.* **61**:41–51.

Apollonio, S. 1985. Arctic marine phototrophic systems: Functions of sea ice stabilization. *Arctic* **38**(3):167–173.

Boorstad, G. A., and J. F. R. Gower. 1984. Phytoplankton chlorophyll distribution in the eastern Canadian Arctic. *Arctic* **37**(3):224–233.

Booth, J. A. 1984. The epontic algal community of the ice edge zone and its significance to the Davis Strait ecosystem. *Arctic* **37**(3):234–243.

Brock, T. D. 1981. Calculating solar radiation for ecological studies. *Ecol. Model.* **4**:1–19.

Buckley, J. R., T. Gammelsrød, J. A. Johannesen, O. M. Johannesen, and L. P. Røed. 1979. Upwelling: Oceanic structure at the edge of the Arctic pack in winter. *Science* **203**:165–167.

Bunt, J. S. 1963. Diatoms of Antarctic sea ice as agents of primary production. *Nature* **199**:1254–1257.

Bunt, J. S. 1964. *Primary Productivity Under Sea Ice in Antarctic Waters. 2. Influence of Light and Other Factors on Photosynthetic Activities of Antarctic Marine Microalgae.* Antarctic Res. Ser. I. pp. 27–31.

Bunt, J. S. 1967. *Some Characteristics of Microalgae Isolated from Antarctic Sea Ice.* Antarctic Res. Ser. 11, Biol. Antarct. Seas III, pp. 1–14.

Bunt, J. S. 1968. Microalgae of the Antarctic pack ice zone, in *Symposium on Antarctic Oceanography,* R. I. Currie (ed.), pp. 198-218.

Bunt, J. S., and C. C. Lee. 1970. Seasonal primary production in Antarctic sea ice at McMurdo Sound in 1967. *J. Marine Res.* **28**:304–320.

Bunt, J. C., and C. C. Lee. 1972. Data on the composition and dark survival of four sea-ice microalgae. *Limnol. Oceanogr.* **17**:458–461.

Burkholder, P. R., and E. F. Mandelli. 1965. Productivity of microalgae in Antarctic sea ice. *Science* **149**:872–874.

Bursa, A. 1963. Phytoplankton in coastal waters of the Arctic Ocean at Point Barrow, Alaska. *Arctic* **16**(4):239–262.

Carey, A. G. (ed.) 1978. Marine biota. In *Environmental Assessment of the Alaskan Continental Shelf, Interim Synthesis: Beaufort/Chukchi.* National Oceanic and Atmospheric Administration, Boulder, Colorado pp. 174–237.

Chalk, P. M., and D. R. Keeney. 1971. Nitrate and ammonium contents of Wisconsin limestones. *Nature* **229**:42.

Clasby, R. C., R. Horner, and V. Alexander. 1973. An in situ method for measuring primary productivity of Arctic sea ice algae. *J. Fish. Res. Bd. Can.* **30**:835–838.

Codispoti, L. A. 1979. Arctic Ocean processes in relation to the dissolved silicon content of the Atlantic. *Marine Sci. Com.* **5**(6):361–381.

Codispoti, L. A., and D. Lowman. 1973. A reactive silicate budget for the Arctic Ocean. *Limnol. Oceanogr.* **18**(3):448–456.

Codispoti, L. A., and T. G. Owens. 1975. Nutrient transport through Lancaster Sound in relation to the Arctic Ocean's reactive silicate budget and the outflow of Bering Strait waters. *Limnol. Oceanogr.* **20**:115–119.

Collin, A. E. 1959. Canadian oceanographic activities on IGY drift station "Bravo." *Fish. Res. Bd. Can.* Manus. Rep. Ser. (Oceanogr. Limnol.) 40.

Cooper, L. H. N. 1952. Factors affecting the distribution of silicate in the North Atlantic Ocean and the formation of North Atlantic deep water. *J. Mar. Biol. Assoc. U.K.* **30**:511–526.

Corlett, C. 1958. Measurements of primary production in the western Barents Sea. Rapp. P.-v. Réun. Cons. perm. int. Explor. Mer. **144**:76–78.

Dawson, W. A. 1965. *Phytoplankton from the Chukchi Sea.* Technical Report. Department of Oceanography, University of Washington, 117: pp. 1-123.

Demers, S., J.-C. Therriault, and C. Descolas-Gros. 1984. Biomass and specific composition of the seasonal sea-ice microflora: Influences of light and the rate of freezing. *Marine Biol.* **78**:185–191. (In French; English summary.)

Dickie, G. 1852. Notes on the algae. In P. C. Sutherland (ed.), *Journal of a Voyage in Baffin's Bay in search of H. M. Ships Erebus and Terror,* vol. 2. Longman, Brown, Green and Longmans, London, pp. CXCI–CC.

Doucette, G. J., and G. A. Fryxell. 1983. *Thalassiosira antarctica:* Vegetative and resting stage chemical composition of an ice-related marine diatom. *Marine Biol.* **78:**1–6.

Dunbar, M. J. 1982. Arctic marine ecosystems. In L. Rey (ed.), *The Arctic Ocean.* Comité Arctique International, Monaco, pp. 233–261.

Dunbar, M. J., and J. C. Acreman. 1980. Standing crops and species composition of diatoms in sea ice from Robeson Channel to the Gulf of St. Lawrence. *Ophelia* **19**(1):61–72.

Duval, W. S. 1977. Environmental assessment of construction and construction support activities on plankton communities of the Mackenzie estuary related to the proposed ten year Beaufort Sea offshore exploration. Part 4. In Environmental Assessment of Construction and Construction Support Activities Related to the Proposed Ten Year Beaufort Sea Offshore Exploration Program 2. Unpublished report prepared by F. F. Slaney and Co. Ltd. for Imperial Oil Ltd., Calgary.

El-Sayed, S. Z. 1971*a*. Biological aspects of the pack ice ecosystem. In *SCAR Symposium on Antarctic Ice and Water Masses, 1970,* pp. 35–54.

El-Sayed, S. Z. 1971*b*. On the productivity of the southern ocean. In W. M. Holdgate (ed.), *Antarctic Ecology.* Academic Press, New York, pp. 115–135.

El-Sayed, S. Z., and S. Taguchi. 1977. Phytoplankton studies in the water column and in the pack ice of the Weddell Sea. *Antarctic J. U. S.* **12:**35–36.

El-Sayed, S. Z., and J. T. Turner. 1977. Productivity of the Antarctic and subtropical regions: A comparative study. In M. J. Dunbar (ed.), *Polar Oceans.* 1974 Proc. SCOR/SCAR Polar Oceans Conf., McGill University, Montreal, pp. 463–504.

English, T. S. 1961. Some biological oceanographic observations in the central North Polar Sea. Drift Station ALPHA, 1957–1958. *Arctic Inst. N. Amer. Res. Pap. 13,* 79p.

FAO. 1972. *Atlas of the Living Resources of the Seas,* 3rd ed. VIIIp and 20p and 62 maps.

Farlow, J. S. 1958. *Project Ice Skate Oceanographic Data.* Woods Hole Oceanographic Institute Ref. No. 58–28, 18p.

Fedorov, V. D., and Yu. A. Bobrov. 1977. Seasonal changes of the indices of productivity of the phytoplankton of the White Sea. *Becth. Mock. yh-ta, Cep. duor. No. 1.*

Fisher, N. S., and R. P. Schwarzenbach. 1978. Fatty acid dynamics in *Thalassiosira pseudonna* (Bacillariophyceae): Implications for physiological ecology. *J. Phycol.* **14**(2):143–150.

Fogg, F. R. S. 1977. Aquatic primary production in the Antarctic. *R. Soc. London Philos. Trans.,* series B, **279:**27–38.

Fukuchi, M., and H. Sasaki. 1981. Phytoplankton and zooplankton standing stocks and downward flux of particulate material around fast ice edge of Lützov-Holm Bay, Antarctica. *Natl. Inst. Polar Res. Mem.,* ser. E., *Biol. and Med. Sci.,* **4:**13–43.

Fukushima, H., and H. Meguro. 1966. The plankton ice as basic factor of the primary production in the Antarctic Ocean. *Antarct. Rec.* **27:**99–101.

Gallegos, C. L., T. Platt, W. G. Harrison, and B. Irwin. 1983. Photosynthetic parameters of Arctic marine phytoplankton: Vertical variations and time scales of adaptations. *Limnol. Oceanogr.* **28**(4):698–708.

Glibert, P. M., M. R. Dennett, and J. C. Goldman. 1985. Inorganic carbon uptake by phytoplankton in Vineyard Sound, Massachusetts. II. Comparative primary productivity and nutritional status of winter and summer assemblages. *J. Exp. Marine Biol. Ecol.* **86:**101–118.

Goldman, J. C., and M. R. Dennett. 1984. Effect of photoinhibition during bottle incubations on the measurement of seasonal primary production in a shallow coastal water. *Marine Ecol. Program Ser.* **15:**169–180.

Goldman, J. C., D. A. Caron, O. K. Andersen, and M. R. Dennett. 1985. Nutrient cycling in a microflagellate food chain: I. Nitrogen dynamics. *Marine Ecol. Program Ser.* **24:**231–242.

Gosselin, M., L. Legendre, S. Demers, and R. G. Ingram. 1985. Responses of sea-ice microalgae to climatic and fortnightly tidal inputs (Manitounuk Sound, Hudson Bay). *Can. J. Fish. Aquat. Sci.* **42:**999–1,006.

Grainger, E. H. 1977. The annual nutrient cycle in sea-ice. In M. J. Dunbar (ed.), *Polar Oceans.* 1974 Proc. Polar Oceans Conf., McGill University, Montreal, pp. 285–299.

Grainger, E. H. 1979. Primary production in Frobisher Bay, Arctic Canada. In M. J. Dunbar (ed.), Marine production mechanisms. IBP **20**(2):9–30.

Grainger, E. H., and M. S. Evans. 1982. Seasonal variations in chlorophyll and nutrients in a Canadian Arctic estuary. *Estuaries* **5:**294–301.

Grant, W. S., and R. A. Horner. 1976. Growth responses to salinity variations in four Arctic ice diatoms. *J. Phycol.* **12:**180–185.

Green, J. M., and D. H. Steele. 1977. Observations on marine life beneath sea ice, Resolute Bay, N. W. T. *Proc. Circumpolar Conf. Northern Ecol. Sept. 15–18, 1975, Ottawa II,* pp. 77–86.

Griffiths, R. P., and R. Y. Morita. 1981. Study of microbial activity and crude oil–microbial interactions in the waters and sediments of Cook Inlet and the Beaufort Sea. In *Environmental Assessment of the Alaskan Continental Shelf.* Final report of principal investigator, Res. Unit 190. NOAA-OCSEAP, Rockville, Md., 361p.

Grøntved, J. 1950. Phytoplankton studies 1. *Nitzschia frigida* Grun., an Arctic–inner–Baltic diatom found in Danish waters. *K. Danske Vidensk. Selsk. Biol. Medd.* **XVIII**(12):1–19.

Gudkovich, Z. M. 1955. Results of a preliminary analysis of the deep-water hydrological observations. In M. M. Somov (ed.), *Observational Data of the Scientific Research Drifting Station of 1950–1951,* vols. 1-3, 170p. (English translation, ASTIA Document AD 117133, American Meteorological Society, Boston, Mass.)

Hagström, Å. A. Andersson, and G. Öquist. 1983. Algal and bacterial production in a coastal area (Norrbyn) of the northern Bothnian Sea. *Aquilo Ser. Zool.* **22**:131–137.

Hameedi, M. J. 1978. Aspects of water column primary productivity in the Chukchi Sea during summer. *Marine Biol.* **48**:37–46.

Hameedi, M. J., and S. R. Shaw. 1975. Bioacoustic studies in the Chukchi Sea. *Arctic Bull.* **2**:8–10.

Hanson, K. J. 1961. The albedo of sea-ice and ice islands in the Arctic Ocean Basin. *Arctic* **14**(3):188–196.

Harrison, W. G., and T. Platt. in press. Photosynthetic characteristics of phytoplankton in the Eastern Canadian Arctic. -manus. 41 pp.

Harrison, W. G., T. Platt, and B. Irwin. 1982. Primary production and nutrient assimilation by natural phytoplankton populations of the Eastern Canadian Arctic. *Can. J. Fish. Aquat. Sci.* **39**(2):335–345.

Hartley, C. H., and M. J. Dunbar. 1938. On the hydrographic mechanism of the so-called brown zones associated with tidal glaciers. *J. Marine Res.* **1**(4):305–311.

Heimdal, B. R. 1983. Phytoplankton and nutrients in the waters north-west of Spitzbergen in the autumn of 1979. *J. Plank. Res.* **5**:901–918.

Horne, A. J., G. E. Fogg, and D. J. Eagle. 1969. Studies in situ of the primary production of an area of inshore Antarctic Sea. *J. Marine Biol. Assoc. U.K.* **49**:393–405.

Horner, R. A. 1976. Sea ice organisms. *Oceanogr. Marine Biol. Ann. Rev.* **14**:167–182.

Horner, R. A. 1977. History and recent advances in the study of ice biota. In M. J. Dunbar (ed.), *Polar Oceans.* 1974 Proc. Polar Oceans Conf., McGill University, Montreal, pp. 269–283.

Horner, R., and V. Alexander. 1972. Algal populations in Arctic sea ice: An investigation of heterotrophy. *Limnol. Oceanogr.* **17**:454–458.

Horner, R. A., K. O. Coyle, and D. R. Redburn. 1974. *Ecology of the Plankton of Prudhoe Bay, Alaska.* Univ. Alaska Inst. Mar. Sci. Rep. R74-2. Sea Grant Rep. 73-15, 78p.

Horner, R., and G. C. Schrader. 1982. Relative contributions of ice algae, phytoplankton, and benthic microalgae to primary production in nearshore regions of the Beaufort Sea. *Arctic* **35**(4):485–503.

Horrigan, S. G. 1981. Primary production under the Ross Ice Shelf, Antarctica. *Limnol. Oceanogr.* **26**(2):378–382.

Hoshiai, T. 1977. Seasonal changes in ice communities in the sea ice near Syowa Station, Antarctica. In M. J. Dunbar (ed.), *Polar Oceans.* 1974 Proc. Polar Oceans Conf., McGill University, Montreal, pp. 307–317.

Hoshiai, T., and M. Fukushi. 1981. Sea ice colored by ice algae in a lagoon, Lake Saroma, Hokkaido, Japan. *Antarct. Rec.* **71**:113–120.

Hsiao, S. I. C. 1980. Quantitative composition, distribution, community structure and standing stock of sea ice microalgae in the Canadian Arctic. *Arctic* **33**(4):768–793.

Hsiao, S. I. C. 1985. The growth of Arctic marine phytoplankton in Frobisher Bay. *Arctic* **38**(1):31–38.

Hsiao, S. I. C., M. G. Foy, and D. W. Kittle. 1977. Standing stock, community structure, species composition, distribution and primary production of natural populations of phytoplankton in the southern Beaufort Sea. *Can. J. Bot.* **55**:685–694.

Hsiao, S. I. C., and R. Trucco. 1980. A marine biological study of Brevoort Harbour and nearby waters of Eastern Baffin Island phytoplankton. *Can. Mar. Rep. Fish. Aquat. Sci. 1557.*

Iizuka, H., I. Tanabe, and H. Meguro. 1966. Microorganisms in plankton-ice of the Antarctic Ocean. *J. Gen. Appl. Microbiol.* **12**(1):101–102.

Ivanenkov, V. N. 1961. Primary production of the Bering Sea. *Tr. Inst. Okeanol. AN USSR* **55.** (In Russian.)

Jacques, G. 1983. Some ecophysiological aspects of the Antarctic phytoplankton. *Polar Biol.* **2**:27–33.

Jenkins, P. M. 1937. Oxygen production by the diatom *Coscinodiscus excentricus* Ehr. in relation to submarine illumination in the English Channel. *J. Marine Biol. Assoc. U.K.* **22**:301–342.

Jennings, J. C., Jr., L. I. Gordon, and D. M.

Nelson. 1984. Nutrient depletion indicates high primary productivity in the Weddell Sea. *Nature* **309:**51–54.

Jitts, H. R., C. D. McAllister, K. Stephens, and J. D. H. Strickland. 1964. The cell division rates of some marine phytoplankters as a function of light and temperature. *J. Fish. Res. Bd. Can.* **21:**139–157.

Jitts, H. R., A. Morel, and Y. Saijo. 1976. The relation of oceanic primary production to available photosynthetic irradiance. *Aust. J. Mar. Freshwater Res.* **27:**441–454.

Jones, E. P., and A. R. Coote. 1980. Nutrient distribution in the Canadian Archipelago: Indicators of summer water mass flow characteristics. *Can. J. Fish. Aquat. Sci.* **37:**589–599.

Jørgensen, E. G. 1968. The adaptation of plankton algae. II. Aspects of the temperature adaptation of *Skeletonema costatum. Physiol. Plant.* **21:**423–427.

Kawamura, A. 1967. Observations of phytoplankton in the Arctic Ocean in 1964. *Inf. Bull. Planktol. Japan Commemoration number of Dr. Y. Matsue:* **71–88.**

Kinney, P., M. E. Arhelger, and D. C. Burrell. 1970. Chemical characteristics of water masses in the Amerasian Basin of the Arctic Ocean. *J. Geogr. Res.* **75**(21):4097–4104.

Koblentz-Minshke. 1965. Primary production in the Pacific. *Okeanol.* **5:**325–337. (In Russian.) (English translation, *Oceanology* **5:**104–116.)

Koerner, R. M. J. 1973. The mass balance of the sea ice of the Arctic Ocean. *J. Glaciol.* **12:**173–185.

Koike, I., K. Furuya, H. Otobe, T. Nakai, T. Memoto, and A. Hattori. 1982. Horizontal distributions of surface chlorophyll *a* and nitrogeneous nutrients near Bering Strait and Unimak Pass. *Deep-Sea Res.* **29**(2A):149–155.

Krebs, W. N. 1983. Ecology of neritic marine diatoms, Arthur Harbor, Antarctica. *Micropaleontology* **29**(3):267–297.

Kusunoki, K., T. Minoda, K. Fujino, and A. Kawamura. 1966. Description of oceanographic observations at Drift Station ARLIS II in 1964–1965. *Arctic Inst. N. Amer. Tech. Rep. ONR-335:1,* 107p.

Kuznetsov, L. L. 1981. Chlorophylls and primary production associated with ice of Amur Bay, Sea of Japan. *Sovyet J. Marine Biol.* **6**(5):297–299.

Kuznetsov, L. L. 1983. Chemical parameters and chlorophyll *a* in the ice and subglacial waters of the Amur Bay, Sea of Japan. *Biol. Morya* **1983**(2):59–61. (In Russian with English abstract.)

Langleben, M. P. 1969. Albedo and degree of puddling of a melting cover of sea ice. *J. Glaciol.* **8**(54):407–412.

Langleben, M. P. 1971. Albedo of melting sea ice in the southern Beaufort Sea. *J. Glaciol.* **10**(58):101–104.

Legendre, L., R. G. Ingram, and M. Poulin. 1981. Physical control of phytoplankton production under sea ice (Manitounuk Sound, Hudson Bay). *Can. J. Fish. Aquat. Sci.* **38:**1385–1392.

LGL Ltd Environmental Research Associates. 1976. Biological oceanographic studies in Lancaster Sound. For Norland Petroleum Ltd, Calgary, Alberta.

Li, W. K. W. and T. Platt. 1982. Distribution of carbon among photosynthetic end-products in phytoplankton of the Eastern Canadian Arctic. *J. Phycol.* **18:**466–471.

Li, W. K. W., J. C. Smith, and T. Platt. 1984. Temperature responses of photosynthetic capacity and carboxylase activity in Arctic marine phytoplankton. *Marine Ecol. Progr. Ser.* **17:**237–243.

McLaren, I. A. 1969. Primary production and nutrients in Ogac Lake, a land-locked fiord on Baffin Island. *J. Fish. Res. Bd. Canada* **26:**1561–1576.

MacLaren Marex Inc. 1979*a*. *Report on Cruise 78-1. Environmental Studies in the Pack Ice Zone of Davis Strait for ESSO Resources Canada Ltd., Aquitaine Co. of Canada Ltd. and Canada Cities Services Ltd.* Arctic Petroleum Operators Association Project 146, Jan. 1979, 300p.

MacLaren Marex Inc. 1979*b*. *Report on Cruise 78-2, August 1978. Environmental Studies in the Davis Strait for ESSO Resources Canada Ltd., Aquitaine Co. of Canada Ltd. and Canada Cities Services Ltd.* Arctic Petroleum Operators Association Project 146, Feb. 1979, 92p.

MacLaren Marex Inc. 1979*c*. *Primary Productivity Studies in the Water Column and Pack Ice of the Davis Strait. April, May, and August 1978.* Department of Indian Northern Affairs and Supply and Services, Ottawa, Canada, 102p.

McPhee, M. G. 1981. Upper ocean temperature, salinity and density in the vicinity of Arctic drift station Fram I, March to May, 1979. Special Report 81-5, Office of Naval Research, Arlington, Va., 20p.

McPhee, M. G., and J. D. Smith. 1976. Measurements of the turbulent boundary layer under pack-ice. *J. Phys. Oceanogr.* **6:**696–711.

McRoy, C. P., and J. J. Goering. 1974. The influence of ice on primary productivity of the Bering Sea. In D. W. Hood and E. J. Kelley (eds.), *Oceanography of the Bering Sea.* University of Alaska Occasional Pub. 2, pp. 403–421.

McRoy, C. P., and J. J. Goering. 1976. Annual

budget of primary production in the Bering Sea. *Marine Sci. Comm.* **2**(5):255–267.

McRoy, C. P., J. J. Goering, and W. E. Shields. 1972. Studies of primary productivity of the Bering Sea. In A. Y. Takenouti (ed.), *Biological Oceanography of the Northern North Pacific Ocean.* Idemitsu Shoten, Tokyo, pp. 199-216.

Marshall, P. T. 1958. Primary production in the Arctic. *J. Cons. Int. Explor. Mer.* **23:**173–177.

Martin, S. 1979. A field study of brine drainage and oil entrapment in the first-year sea ice. *J. Glaciol.* **22**(88):473–502.

Matheke, G. E. M., and R. Horner. 1974. Primary productivity of the benthic microalgae in the Chukchi Sea near Barrow, Alaska. *J. Fish. Res. Bd. Can.* **31:**1779–1786.

Maykut, G. A., and N. Untersteiner. 1971. Some results from a time dependent thermodynamic model of sea ice. *J. Geophys. Res.* **76:**1550–1575.

Meguro, H. 1962. Plankton ice in the Antarctic Ocean. *Antarct. Rec.* **14:**72–79 (1192–1199).

Meguro, H., K. Ito, and H. Fukushima. 1967. Ice flora (bottom type): A mechanism of primary production in Polar seas and the growth of diatoms in sea ice. *Arctic* **20:**114–133.

Mel'nikov, I. A. 1980. The ecosystem of Arctic pack ice. In M. E. Vinogradov and I. A. Mel'nikov (eds.), *The Biology of the Central Arctic Basin.* Nauka Press, Moscow, pp. 61–111. (In Russian. English translation, Dept. Secretary of State, Canada. Multilingual Services Division. Translation Bureau No. 1035889, 66p., 1982. Available from Nat. Res. Council Canada. Can. Inst. Scient. Tech. Inf., Ottawa.)

Miller, J. P. A. 1962. Fats and steroids. In R. A. Lewin (ed.), *Physiology and Biochemistry of Algae.* Academic Press, New York, 927p.

Morel, A., and R. C. Smith. 1974. Relation between total quanta and total energy for aquatic photosynthesis. *Limnol. Oceanogr.* **19:**591–600.

Morris, I., and H. E. Glover. 1974. Questions on the mechanism of temperature adaptation in marine phytoplankton. *Marine Biol.* **24:** 147–154.

Müller-Haeckel, A. 1982. Nanoplankton under ice and snow in a coastal area of the Bothnia Sea. In K. Müller (ed.), *Coastal Research in the Gulf of Bothnia.* Junk Publishers, The Hague, pp. 225–231.

Nansen, F. 1902. Oceanography of the North Polar Basin. The Norwegian North Polar Expedition 1893–96. *Sci. Results* **3**(9):427p.

Nemoto, T., and G. Harrison. 1981. High latitude ecosystems. In A. R. Longhurst (ed.), *Analysis of Marine Ecosystems.* Academic Press, London, pp. 95–126.

Neshyba, S. 1977. Upwelling by icebergs. *Nature* **267:**507–508.

Newbury, T. K. 1983. Under landfast ice. *Arctic* **36**(4):328–340.

Nielsen, E. S. 1958. A survey of recent Danish measurements of the organic productivity of the seas. Rapp. p.-v. Cons. int. *Explor. Mer.* **144:**92–95.

Nielsen, E. S., and V. K. Hansen. 1959. Light adaptation in marine phytoplankton populations and its interrelation with temperature. *Physiol. Plant.* **12:**353–370.

Nielsen, E. S., and E. G. Jørgensen. 1968. The adaptation of algae. I. General part. *Physiol. Plant.* **21:**401–413.

Nielsen, E. S., and T. S. Park. 1964. On the time course in adaptation to low light intensities in marine phytoplankton. *J. Cons. Perm. Int. Explor. Mer.* **29:**19–24.

Oradovskii, S. G. 1972. Studies on the composition of nutrients in the ice of the Barents Sea. *VNIRO Trudy* **75:**65–73. (In Russian.)

Packard, T. T. 1982. Respiratory enzyme activity and associated biochemical properties of the water column below the Arctic pack ice. *EOS* **63**(3):47 (21 B-8).

Palmisano, A. C., and C. W. Sullivan. 1982. Physiology of sea ice diatoms. I. Response of three Polar diatoms to a simulated summer-winter transition. *J. Phycol.* **18:**489–498.

Palmisano, A. C., S. T. Kottmeier, R. L. Moe, and C. W. Sullivan. 1985. Sea ice microbial communities. IV. The effect of light perturbation on microalgae of the sea-ice–water interfaces in McMurdo Sound, Antarctica. *Marine Ecol. Program Ser.* **21:**37–45.

Parker, B. C., L. E. Heiskell, W. J. Thompson, and E. J. Zeller. 1978. Non-biogenic fixed nitrogen in Antarctica and some ecological implications. *Nature* **271:**651–652.

Pautzke, C. G. 1974. Stimulating light and primary production below Arctic Ocean pack ice. Research report, Master's degree, Department of Oceanography, University of Washington, Seattle, 90p (unpublished).

Pechlaner, R. 1971. Factors that control the production rate and biomass of phytoplankton in high-mountain lakes. *Mitt. Int. Verein. Theor. Angew. Limnol.* **19:**125–145.

Petersen, G. H. 1964. The hydrography, primary production, bathymetry, and "Tagsâq" of Disko Bugt, West Greenland. *Medd. Grønland* **159**(10):45p.

Petersen, G. H. 1977. Biological effects of sea-ice and icebergs in Greenland. In M. J. Dunbar (ed.), *Polar Oceans.* 1974 Proc. Polar

Oceans Conf., McGill University, Montreal, pp. 319–329.

Petersen, G. H. 1979. On the analysis of dark fixation in primary production computations. *J. Cons. Int. Explor. Mer.* **38:**326–330.

Platt, T., C. L. Gallegos, and W. G. Harrison. 1980. Photoinhibition of photosynthesis in natural assemblages of marine phytoplankton. *J. Marine Res.* **38:**687–701.

Platt, T., and A. W. Herman. 1983. Remote sensing of phytoplankton in the sea: Surface-layer chlorophyll as an estimate of water-column chlorophyll and primary production. *Int. J. Remote Sensing* **4:**343–351.

Platt, T., W. G. Harrison, B. Irwin, E. P. Horne, and C. L. Gallegos. 1982. Photosynthesis and photoadaptation in marine phytoplankton in the Arctic. *Deep-Sea Res.* **29**(10A):1159–1170.

Rey, F., and H. Loeng. 1985. The influence of ice and hydrographic conditions on the development of phytoplankton in the Barents Sea. In J. S. Gray and M. E. Christiansen (ed.), *Marine Biology of Polar Regions and Effects of Stress on Marine Organisms.* Wiley, New York, pp. 49–63.

Reynolds, N. 1974. What matters in Arctic phytoplankton. *Brit. Phycol. Soc. J.* **9**(2):222–223.

Rothrock, D. A. 1978. Modeling sea ice features and processes. Symposium on dynamics of large ice masses, 21–25 Aug. 1978, Ottawa. 38p (unpublished manuscript).

Rusanov, V. P. 1974. Silicon distribution in the surface waters of the Arctic Basin in winter. *Oceanology* **14:**669–674.

Rusanov, V. P. 1980. Hydrochemical characteristics of the surface waters of the Arctic Basin. In M. E. Vinogradov and I. A. Melnikov (eds.), *Biology of the Central Arctic Basin.* Nauka, Moscow, pp. 15–35. (In Russian.)

Saijo, Y., and T. Kawashima. 1964. Primary production in the Antarctic Ocean. *Oceanogr. Soc. Japan J.* **19:**190–196.

Sambrotto, R. N., J. J. Goering, and C. P. McRoy. 1984. Large yearly production in western Bering Strait. *Science* **225:**1147–1150.

Schell, D. M. 1980*a*. Chemical overview of biological-physical process interactions. Outer continental shelf environmental assessment program. *Arctic Project Bull.* **29:**25–31.

Schell, D. M. 1980*b*. Foodweb and nutrient dynamics studies in nearshore Alaskan Beaufort Sea waters. Environmental assessment of the Alaskan continental shelf. Ann. Rep. 2:467–513.

Sekerak, A. D., R. A. Buchanan, W. B. Griffiths, and M. G. Foy. 1976. *Biological Oceanographic Studies in Lancaster Sound, 1976.* LGL Ltd., for Norlands Petroleums Ltd., Calgary, Ottawa, 246p.

Serson, H. V. 1972. *Investigation of a Plug of Multi-Year Ice in the Mouth of Nansen Sound.* Tech. Note 72-6, Def. Res. Estab. Ottawa, Ontario.

Slagstad, D. 1982. A model of phytoplankton growth—effects of vertical mixing and adaptation to light. *Modeling Ident. Control* **3**(2):111–130.

Slagstad, D. 1984. A model of phytoplankton production in the marginal sea-ice zone of the Barents Sea. *Modeling Ident. Control* **5**(3):129–140.

Slagstad, D. 1985. A model of phytoplankton in the marginal sea-ice zone of the Barents Sea. In J. S. Gray and M. E. Christiansen (eds.), *Marine Biology of Polar Regions and Effects of Stress on Marine Organisms.* Wiley, New York, pp. 35–48.

Smidt, E. L. B. 1979. Annual cycles of primary production and of zooplankton of Southwest Greenland. *Greenland Biosci.* **1:**52p. (Formerly *Medd. Grønland,* now *Medd. Grønland, Biosci.* with Smidt, 1979 starting as no. 1.)

Smith, A. E., and I. Morris. 1980. Pathways of carbon assimilation in phytoplankton from the Antarctic Ocean. *Limnol. Oceanogr.* **25**(5):865–872.

Smith, J. C., T. Platt, W. K. W. Li, E. P. W. Horne, W. G. Harrison, D. V. Subba Rao, and B. D. Irwin. 1985. Arctic marine photoautotrophic picoplankton. *Marine Ecol. Progr. Ser.* **20:**207–220.

Smith, T. D., and T. S. English. 1973. Studies of the primary productivity under the Arctic pack ice. Department of Oceanography, University of Washington, Seattle, 24p (unpublished).

Smith, W. O., and D. M. Nelson. 1985. Phytoplankton bloom produced by a receding ice edge in the Ross Sea: Spatial coherence with the density field. *Science* **227:**163–166.

Sokolova, S. A., and A. A. Solov'yeva. 1971. Primary production in Dal'nezelenetskaya Bay (Murman coast) in 1867. *Oceanology* **11:**386–395.

Solomon, H. 1973. Wintertime surface convection in the Arctic Ocean. *Deep-Sea Res.* **20:**269–283.

Subba Rao, D. V. and T. Platt. 1984. Primary production of Arctic waters. *Polar Biol.* **3:**191–201.

Sullivan, C. W., and A. C. Palmisano. 1984. Sea ice microbial communities: Distribution, abundance, and diversity of ice bacteria in McMurdo Sound, Antarctica in 1980. *Appl. Environ. Microbiol.* **47**(4):788–795.

Sverdrup, H. U. 1953. On the conditions for the vernal blooming of phytoplankton. *J. Cons. Int. Explor. Mer.* **18**(3):287–295.

Taguchi, S. 1972. Mathematical analysis of primary production in the Bering Sea in summer. In A. Y. Takenouti (ed.), *Biological Oceanography of the Northern North Pacific Ocean.* Idemitsu Shoten, Tokyo, pp. 253–262.

Taniguchi, A. 1969. Regional variations of surface primary productivity in the Bering Sea in summer and the vertical stability of water affecting the production. *Bull. Fac. Fish., Hokkaido Univ.* **20**:169–179.

Thomsen, H. A. 1985. Ice-related occurrence of *Pedinella tricostata* (Chrysophyceae): An ultrastructural investigation of the flagellated stage and the cyst. *Second Internat. Phycol. Congr., Copenhagen 4–10 Aug. 1985. Book of Abstracts,* p. 160.

Thordardottir, T. 1973. Successive measurements of primary productivity and composition of phytoplankton at two stations west of Iceland. *Nor. J. Bot.* **20**:257–270.

Thordardottir, T. 1984. Primary production north of Iceland in relation to water masses in May–June 1970–1980. *Cons. Int. Explor. Mer C.M. 1984/620.*

Throndsen, J., and B. R. Heimdal. 1976. Primary production, phytoplankton and light in Straumsbukta near Tromsø. *Astarte* **9**:51–60.

Untersteiner, N. 1963. Ice budget of the Arctic Ocean. In *Proc. Arctic Basin Symp., Oct. 1962.* Arctic Institute of North America, pp. 219–226.

Usachev, P. I. 1949. The microflora of Polar ice. *Trudy Inst. Okeanol., Moscow* **3**:216–259. (In Russian. *Fish. Res. Bd. Canada Transl.,* ser. 1305, 93p, 1969).

Vedernikov, V. I., and A. A. Solov'yeva. 1972. Primary production and chlorophyll in the coastal waters of the Barents Sea. *Oceanology* **12**:559–565.

Vibe, C. 1939. Preliminary investigations on shallow water animal communities in the Upernavik—and Thule—districts (northwest Greenland). *Medd. Grønland* **124**(2):43p.

Vollenweider, R. A. 1974. *A Manual on Methods for Measuring Primary Production in Aquatic Environments.* IBP Handbook No. 12, 2nd ed., 225p.

Vyshkvartsev, D. I., and G. V. Konovalova. 1979. Mechanism of winter "blooming" of microalgae in Peter the Great Bay (Sea of Japan). *Sovyet J. Marine Biol.* **4**(1):53–54.

Wadhams, P. 1981. Sea-ice topography of the Arctic Ocean in the region 70°W to 25°E. *R. Soc. London Philos. Trans.,* series A, **302**(1464):45–85.

Walker, E. R., and P. Wadhams. 1979. On the thickness of sea-ice floes. *Arctic* **32**(2):140–147.

Welch, H. E., and J. Kalff. 1977. Marine metabolism at Resolute Bay, Northwest Territories. *Proc. Circumpolar Conf. Northern Ecol., Sept. 15–18 1975,* vol. 2. Ottawa, pp. 67–75.

Weller, G., and S. Parker. 1972. Radiation regime of the Arctic Basin—studies at T-3, AIDJEX sites and Barrow. *Tech. Rep. Geophys. Inst. Univ. Alaska* **2**:4–11.

Whitaker, T. M. 1982. Primary production of phytoplankton off Signy Island, South Orkneys, the Antarctica. *R. Soc. London Proc.,* series B, **214**:169–189.

Zenkevitch, L. 1963. *The Biology of the Seas of the USSR.* George, Allen & Unwin Ltd., London, 955p.

Zubov, N. N. 1943. *Arctic Ice.* U.S. Navy Oceanographic Office and American Meteorological Society, 491p.

# 7: Arctic Ocean Phytoplankton

B. R. Heimdal

## INTRODUCTION

All organisms (bacteria, plants, and animals) that live unattached to the bottom of the sea and that have limited powers of locomotion for maintaining their position against the water currents are referred to as *plankton* (Hensen, 1887). The members of the phytoplankton are predominantly microscopic, unicellular, or colonial algae; are able to synthesize organic matter by using solar energy (*autotrophic*); and are the major producers in the sea. The diameter of the individual algal cells varies from about 1 $\mu$m to > 1,000 $\mu$m and the cell volume varies a millionfold. The algal cells are often classified into size groups that are in part related to the mesh sizes of the usual plankton nets. Thus, organisms less than 2 mm, but large enough to be caught by a net with a mesh size of 20 $\mu$m, form the *netplankton*. Cells between 2 $\mu$m and 20 $\mu$m form the *nanoplankton* and unicellular organisms smaller than 2 $\mu$m are collectively known as *picoplankton* (Sieburth, Smetacek, and Lenz, 1978). It should be noted, however, that these distinctions are arbitrary ones and that not all workers use 20 $\mu$m as the dividing point between the netplankton and nanoplankton fractions.

In the sea, the most abundant representatives of the netplankton are diatoms and dinoflagellates. The nanoplankton consist of small, often motile unicells belonging to several other algal classes, but small diatoms and dinoflagellates are of course also included. According to recent discoveries the marine picoplankton include small (0.2–2 $\mu$m) coccoid cyanobacteria and minute eukaryotic algae (mostly members of the algal classes Prymnesiophyceae and Loxophyceae, see Christensen, 1966). At present, however, our knowledge of these subjects is very limited and more work is needed to understand the composition of the nanoplankton and picoplankton in the ocean.

The geographical distribution of plankton associated with the oceans and with the coastal waters is characterized by the terms *oceanic* and *neritic,* whereas plankton organisms that produce resting spores or have a sedentary stage are known as *meroplankton* as opposed to *holoplankton,* passing the complete life cycle in the pelagic state (see, e.g., Smayda, 1958*a*). There are also distinct cold-water and warm-water phytoplankton species (Hasle, 1976), although very many of the common species are cosmopolitans.

Much of the earlier phytoplankton work from the Arctic Ocean included microscopic examination and the floral composition of the netplankton is fairly well known, at least in some areas and at certain times of the year. With the exceptions of Throndsen (1970) and Reynolds (1973) studying the composition and abundance of flagellate plankton in the Barents Sea, the smaller planktonic algae have, unfortunately, been greatly overlooked until now, because their size and morphology in most cases make them very difficult to capture and preserve with conventional methods. This gap is particularly serious since evidence continues to accumulate that nanoplanktonic algae are an important component of the phytoplankton populations and responsible for a major portion of the phytoplankton biomass and production in the sea (Throndsen, 1976; Eppley and Weiler, 1979; Malone, 1980; Hallegraeff, 1981). The ecological importance of the picoplankton as oceanic primary producers has yet to be fully assessed, but indications are that picoplanktonic algae contribute significantly to total primary production in open-ocean systems (Platt, Subba Rao, and Irwin, 1983). Their contribution to the rate of turnover of phytoplankton biomass in Arctic waters is not known.

Arctic planktonic algae have been studied for more than a century. The early studies were primarily of a taxonomic and biogeographic nature (for references, see Heimdal, 1983). But by the 1930s, when factors affecting the growth of the individual algae and the density of the phytoplankton in the euphotic layer (light, nutrient concentrations, temperature, grazing, sinking, and vertical transport) were generally adopted in the study of the phytoplankton, these factors were also applied to the conditions in the Arctic Ocean (see, e.g., Kreps and Verjbinskaya, 1930). The majority of the Arctic phytoplankton and primary production studies so far are, however, from the subarctic regions of the Pacific, while data from the eastern Canadian Arctic and the Eurasian sector of the Atlantic are more scarce (Nemoto and Harrison, 1981). For all these areas, the studies have principally been confined to the development of the phytoplankton during late spring and summer.

The purpose of this chapter is to summarize important concepts of Arctic phytoplankton as revealed by representative data from the literature. It is not claimed to be exhaustive either in the sense of covering all aspects of Arctic phytoplankton or in treating all of the available literature on those aspects that are covered. In particular, the author regrets that she has not been able to give the Soviet literature the full attention it deserves. Investigations such as those of Horner (1976, 1977) have drawn the attention to the importance of algae associated with ice and snow in polar seas. Since this topic is given a detailed discussion in Chapter 5, it is excluded from this review even though the ice flora in arctic waters consists mainly of diatoms with a number of species common to the phytoplankton community (Raymont, 1980).

## ALGAE OF THE PLANKTON

For readers not familiar with the various groupings of algae mentioned above, their main distinguishing features are summarized in the following section. The textbooks by Round (1973), Bold and Wynne (1978), and Christensen (1980) are recommended for those wishing to explore the classification of algae beyond the present treatment.

## Diatoms (Class Bacillariophyceae)

All diatoms are solitary or colonial. Cell size ranges from 2 μm to 2 mm. The silica wall consists of two parts (valves) with one (epitheka) overlapping the other (hypotheka) in the girdle region. The siliceous box or frustule encloses a layer of cytoplasm that lines it internally and extends into all cavities within the wall, the vacuole, and the nucleus. The chloroplasts occur within the cytoplasm and may be either laminate or lobed structures (1–2 per cell) or many disc-shaped bodies. The pigments of the diatoms (including chlorophylls and certain carotenoid pigments, e.g., fucoxanthin) give the cells a yellow-brown color. The diatoms store chrysolaminaran, lipids, and polyphosphates. The casing is perforated, allowing entry of nutrients from outside, and is sculptured in each species into a different perforated pattern. In some diatoms there is a long slit-like opening in the valve surface (the raphe) that allows contact between the cytoplasm and the medium and enables the cell to make gliding movements along a solid surface. The cell wall of the elongated *pennate* diatoms are typically ornamented with bilateral symmetry, while the representatives of the *centric* group usually have a circular outline with an obvious radial symmetry arranged around a centric or eccentric point.

Valve shape, frustule structure, and raphe morphology are important in the classification. Reproduction is generally by simple binary fissions. When asexual multiplication has continued for a long time, the resulting cells will be of very different sizes. Full size is restored by sexual reproduction, which may be oogameous (centric diatoms, male gametes with one flagellum) or isogameous (pennate diatoms, nonflagellated gametes). Some diatoms can form cysts (resting spores) with heavy, silicified walls. Remains of centric diatoms are found from the Jurassic, while the pennates, the present predominant group, date back to the Tertiary.

The diatoms frequently form the dominant phytoplankton of both temperate and high latitudes (Plate 7-I).*A fuller account of their distribution and relative abundance in the Arctic will be given in the following text.

## Dinoflagellates (Class Dinophyceae)

The dinoflagellates constitute a diverse group of algae, containing both monadoid (*monad* = single cell with flagella), amoeboid, and filamentous forms from approximately 5 μm to > 2 mm in size. The majority of them, however, are unicellular and biflagellate. A typical dinoflagellate cell (Plate 7-II, fig. 1) has a longitudinal furrow (the sulcus) on the ventral side and a transverse furrow (the cingulum or girdle), dividing the cell into an anterior part (the epicone) and a posterior part (the hypocone). The two flagella causing cell rotation and forward movements lie within these furrows and are attached where the furrows meet. The monadoid forms either have no cell covering (theka) or they have a carbohydrate layer inside the plasma membrane consisting mainly of cellulose with a precise pattern of plates for each species. The number and the arrangement of the thecal plates are accorded taxonomic significance at various hierarchical levels. Prominent horns and winglike extensions of the theka in some species have been described as floating devices. The most common method of reproduction for dinoflagellates is by binary fission. In addition,

*Plates and explanations appear as an appendix to the chapter.

sexual reproduction is known for some monadoid species.

Dinoflagellates are important members of the plankton throughout the seas of the world, but may predominate in the tropics in the open seas. They often follow the abundance of diatoms in the seasonal cycle in temperate seas. Most dinoflagellates are free-living, but parasitic forms and some that live as symbionts in animals are known. Many species are phagotrophic. In autotrophic species the brownish chloroplasts are variable in number and form. Heterotrophic species are frequently pink. Photosynthetic pigments include chlorophylls and carotenoids. Of the carotenoids, which appear to be peculiar to the Dinophyceae, a special xanthophyll called peridinin greatly increases the photosynthetic efficiency of this group. The few dinoflagellates lacking peridinin have fucoxanthin as their major xanthophyll pigment. Starch is the common storage product in autotrophic species. Phagotrophic forms store fatty oils, often as reddish or yellowish droplets. Most species have trichocysts, which are organelles capable of throwing out mucilaginous projectiles. Several dinoflagellate species have an eyespot (stigma) located near the flagellar bases on the ventral surface. The dinoflagellate nucleus is very large in proportion to cell volume with permanently condensed chromosomes that are readily distinguished under a phase-contrast microscope. Many dinoflagellates are known to produce cysts. The cysts, well represented in fossil records as far back as the Triassic, have the greatest interest for the paleontologists, as it is the cysts only that are fossilized. The sole exception is the genus *Actiniscus,* which is represented in the fossil records by its internal siliceous skeleton in vegetative cells.

A number of dinoflagellates are able to migrate vertically in the water, apparently guided by the light together with an internal rhythm. Their characteristic red-brown pigments often result in the discoloration of the seas (*red tides*) when the cell numbers temporarily become very great, usually near the shore. The main problem with such blooms (a bloom is a very dense population of planktonic algae, and is sometimes used to describe the seasonal maxima of production) is that they generally consist of species that produce toxins. The toxins either kill the animals directly (mass fish mortality) or they accumulate in the living filter-feeding bivalves and cause paralytic shellfish poisoning of those consuming the bivalves. To date, toxic dinoflagellate blooms that cause shellfish poisoning and/or fish deaths have not been reported from the Arctic. Some dinoflagellates are luminescent, and bioluminescence in the sea due to members of this class is often associated with red tides.

## Coccolithophorids (Class Prymnesiophyceae)

Coccolithophorids are monadoid or coccoid (*coccum* = berry, without flagella), mostly $< 30\ \mu m$ and many $< 10\ \mu m$. A typical coccolithophorid cell is spherical or oval and bears two flagella equal in length. Between the flagella is usually a third threadlike appendage called the haptonema. The cell surface is covered by organic scales encrusted with calcium carbonate (the coccoliths). The coccoliths have a characteristic architecture in each species and coccolithophorids are classified largely on structural features of the coccoliths. The calcium carbonate present in the coccoliths occurs almost always in the crystal form of calcite. Holococcoliths are constructed of rhombohedral units in which the crystal form of calcite is unmodified. In heterococcoliths, the crystals are also basically rhombohedral, but the shape may be modified so that their characteristic crystal faces and angles become partly or wholly supressed. The holococcoliths are more fragile in their particular form of construction than the heterococcoliths. It is therefore to be expected that holococcoliths are less commonly found in sediments or fossil deposits (e.g., Haq, 1978; Wind and Wise, 1978; Tappan, 1980), and unlike the heterococcolithophorids, which have an extensive literature on fossil forms, most reports on holococcolithophorids have dealt with only extant species.

The intracellular fine structure of the coccolithophorids resembles that of other examined representatives of the class. The two or more (rarely one or four) chloroplasts are golden-brown, differing only in minor details of pigmentation from the prymnesiophycean type. Likewise, chrysolaminaran appears to be a common storage product. Coccolithophorids multiply vegetatively by binary fission. Two planktonic species show a dimorphic life cycle involving a motile flagellated stage and a nonmotile stage. In some neritic species, a coccolith-covered flagellated stage alternates with a filamentous sessile stage lacking coccoliths. Under unfavorable conditions some species are observed to produce calcareous cysts.

Although predominantly oceanic, some of the coccolithophorid species are neritic. Coccolithophorids and dinoflagellates are characteristic of marine tropical regions, but coccolithophorids occur also in cooler waters where diatoms are more important numerically (Halldal and Halldal, 1973; Manton, Sutherland, and Oates, 1976, 1977; Thomsen, 1981). The most abundant living species, *Emiliania huxleyi*, is a cosmopolitan that has been reported in numbers of 115 million cells per litre on the southern part of the Norwegian west coast, giving the water a milky green color (Berge, 1962). In the Southern Ocean, McIntyre and Bé (1967) observed a monospecific flora of this species south of the Antarctic Convergence to about 65°S and delineated by the 2°C isotherm. An extension of the temperature range of *E. huxleyi* to include negative temperatures was supported by observations in the waters northwest of Spitsbergen (Heimdal, 1983).

Coccoliths have been found in deposits as far back as the Triassic, and are abundant since the Jurassic.

### Other Phytoflagellates

A variety of nanoplankton flagellates representing several algal classes generally contributes significantly to the phytoplankton of the ocean world (e.g., Malone, 1980). Because earlier studies of the phytoplankton from the Arctic Ocean used tow nets for collection, or harsh preservatives for fixation, such flagellates have only rarely been identified by species. The most common one is the prymnesiophyte *Phaeocystis pouchetii* (Plate 7-II, fig. 6), which is somewhat exceptional in that the colonial, gelatinous stage can be large enough to be visible to the naked eye. It is widely distributed and often occurs in vast numbers in the Arctic.

## MEASURING PHYTOPLANKTON BIOMASS AND PRIMARY PRODUCTION IN AQUATIC ENVIRONMENTS

In the sea, phytoplankton distribution is usually nonrandom or patchy. In most seas, the patches exist in a wide range of scales and make it difficult to collect a representative sample from a given area and also to tell to what extent the results of an investigation are characteristic of the area being sampled. In addition, the method of collection and the frequency and regularity of the sampling program may influence the biomass assessment greatly.

Phytoplankton samples are collected using nets of synthetic gauze, water bottles, and pumps. Net samples cannot be used for quantitative studies because the nets do not retain all the organisms quantitatively. They may, however, be very useful in providing material for morphological and taxonomic work. Pumps and water bottles sample the entire spectrum of planktonic algae. Furthermore, the sample volume can be accurately determined and several depths may be sampled simultaneously with water bottles. They do not, however, concentrate the organisms as the nets do. This must be done after collection, most often after preservation with buffered formalin or an iodine fixative. The purpose of the fixative is to stabilize the cells well enough to permit identification at some later date.

Neutral formalin preserves diatoms, thecate dinoflagellates, and coccolithophorids in identifiable conditions, but distorts the cell shape of "naked" flagellates (i.e., the flagellates in which the external cell cover is submicroscopic or absent) and causes many of the flagellated forms to shed their flagella, whereas the iodine method in many cases preserves these organisms better. The naked flagellates are studied by concentration of the original sample (centrifugation or membrane filtration) or by means of a serial dilution technique (Throndsen, 1969). While centrifugation and filtration seldom give adequate cell numbers for studying such things as cell shape, color, and motility, the culture method provides many cells of each species and thus better opportunity for reliable identification of species. It should not be overlooked, however, that this method is highly selective, recording only species that are able to grow under the culture conditions offered, and that centrifugation and filtration have demonstrated the presence of species that seldom or never occur in dilution cultures.

Phytoplankton populations are measured either by the number of cells or the biomass (e.g., chlorophyll *a*, total carbon) per unit volume of water or under a given unit area, usually 1 $m^2$. Prior to enumeration and identification, the preserved material is concentrated by centrifugation or sedimentation of a known volume of sample. The organisms are examined by using an inverted microscope when settled to the bottom of the counting chamber. Cell numbers are useful for documenting community structure and for comparative studies, but may be very misleading as a measure of biomass due to the extreme variation in size among planktonic algae.

Although the chlorophyll content of an algal cell is dependent on species and growth conditions (see, e.g., Nusch, 1980) and thus is by no means a perfect measure of biomass, it has proved its merits in comparison with other biomass parameters as the most selective one. It is also easy and rapid to measure compared with the more time-consuming cell counting. For chlorophyll measurements, the particulate material is concentrated, usually by filtration, and extracted in an organic solvent (acetone, methanol). The absorbance of the extract is measured in a spectrophotometer or a fluorometer and converted to $\mu$g chlorophyll per litre (or mg per $m^3$).

The productivity of planktonic algae is measured using the radiocarbon method (Steemann Nielsen, 1952). The isotope, $^{14}C$, is added as bicarbonate to water samples in light and dark bottles. The bottles are incubated in situ, or in a constant-temperature bath on the deck of the ship, using light filters to simulate illumination at various depths, or in the laboratory with an artificial light source. At the end of the incubation period, the sample is filtered and the radioactivity contained on the filter is counted with a suitable instrument, most often a liquid scintillation counter. To calculate the amount of carbon assimilated, one must know the total number of counts added and the content of total inorganic carbon present in the sample. The total inorganic carbon may be determined from pH and alkalinity measurements (Parsons, Maita, and Lalli, 1984) or by using an infrared analyzer.

## FACTORS CONTROLLING PHYTOPLANKTON GROWTH IN THE SEA

Phytoplankton growth depends on several factors such as light, temperature, supply of nutrients, and grazing by pelagic animals. Because the energy source for photosynthesis is light between 350 nm and 700 nm wavelength, corresponding to the visible part of the spectrum, it is obvious that both quality and quantity of the ambient light will affect the photosynthetic production. In lakes and seas, the boundary between the *euphotic zone* (the illuminated surface waters where there is sufficient light for photosynthesis) and the *aphotic zone* (where there is not) depends on several factors such as location (latitude), time of day/year, weather conditions, and transparency of the water.

Planktonic algae show the usual relationship between biological activity and temperature by increased growth rate with increasing temperature up to some optimum after which growth rate declines, often sudden, to zero (e.g., Marshall Darley, 1982). Temperature also has an important indirect effect on the mixing processes and the stability of the water column.

Algal growth is dependent on the supply of biologically important elements from the environment. Most of these elements are present in abundance relative to what is needed, whereas lack of others such as nitrogen, phosphorus, and silicon (used by diatoms to form the siliceous cell walls) often limit production. The limiting nutrient in the sea is usually assumed to be nitrogen. The potentially limiting role of phosphorus as essential for phytoplankton growth in the marine environment is, however, still less well understood. *New* nitrogen in the form of nitrate mixed in from the waters below the euphotic zone and *regenerated* nitrogen mainly as ammonia produced in the euphotic zone by zooplankton excretion and bacterial breakdown of organic matter are the most important nitrogen sources for algae.

## THE ARCTIC MARINE ENVIRONMENT

The Arctic Ocean covering an area of about 14 million km$^2$ (Sakshaug and Holm-Hansen, 1984) is the smallest ocean of the globe. Located within the Arctic Circle (66°33′N latitude) and occupying the region around the North Pole, it remains frozen throughout the year except in the fringe areas. Almost completely encircled by the land masses of North America, Eurasia, and Greenland, the Arctic Ocean is connected with the Pacific Ocean and the Atlantic Ocean through the Bering Strait and the Greenland Sea, respectively. The connection with the Pacific is very shallow and narrow and the influx from the Pacific is of minor importance. Its principal exchange of water is therefore with the Atlantic Ocean. There are, however, factors in the Arctic Ocean that make its processes significantly different from those in the Atlantic. Most notable are the sea-ice cover that varies seasonally (Fig. 7-1) and from one year to another, the extreme cycles of incident light on both the seasonal and daily time scales, and the low seawater temperature. Sea ice reflects a large portion of the incoming radiation, reduces the penetration of sunlight needed for the photosynthetic processes of the marine plants, supresses wind stress, imposes an upper limit on the surface temperature, and impedes evaporation. Its thickness varies between thin (0–2 m) first-year ice and thicker (up to 30 m) multilayer ice (Weeks, 1976).

All waters of the Arctic Ocean are cold. In areas with drifting pack ice the temperature of the surface layer is about −1.8°C year round; however, the water warms up to several degrees above 0°C in ice-free areas in summer. Although very little is known about the temperature response of

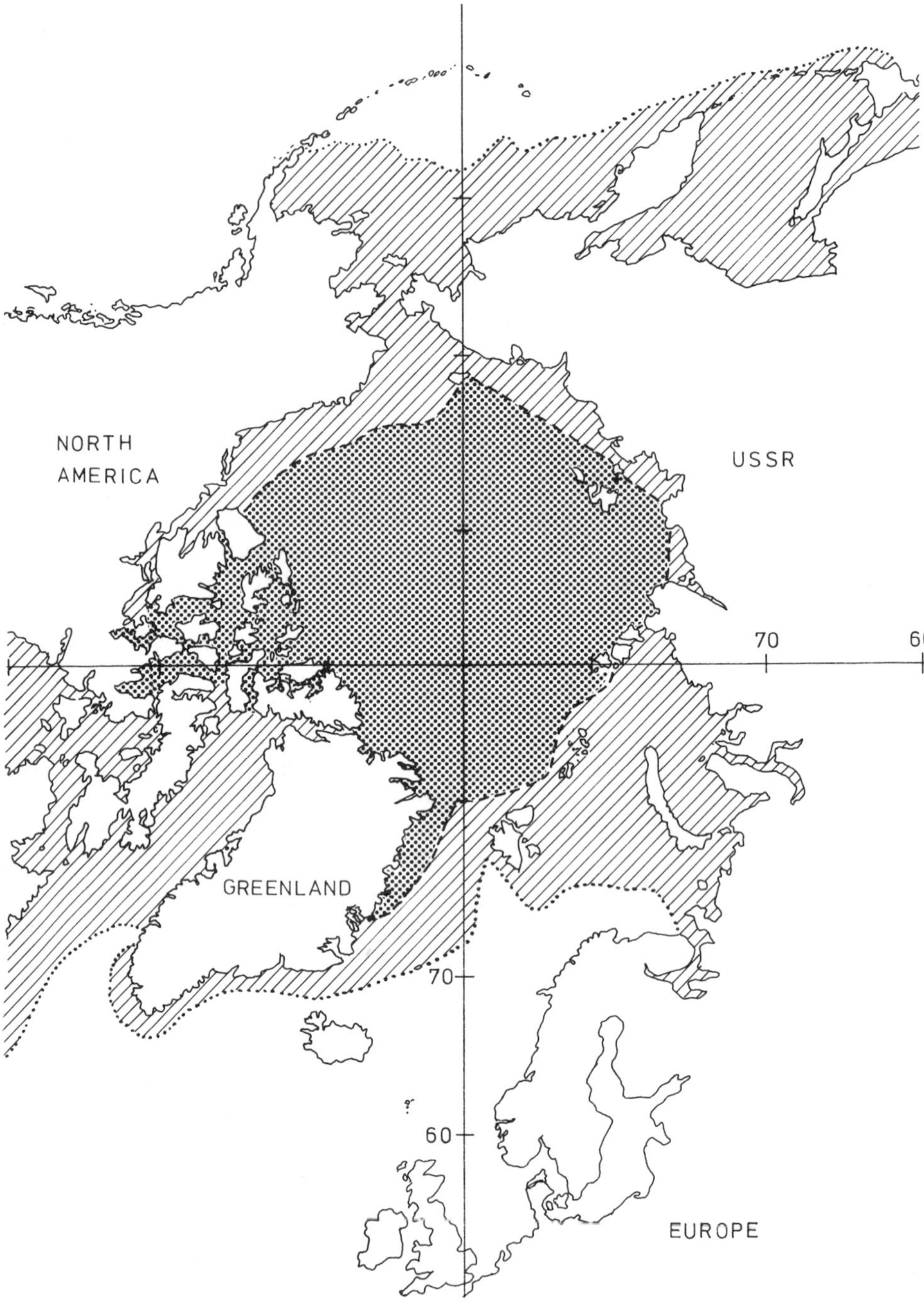


**Figure 7-1.** Map of Arctic waters. Hatched and shaded areas denote the maximum extent of sea ice in winter and summer. (Redrawn after Sakshaug and Holm-Hansen, 1984.)

polar planktonic algae (Sakshaug, 1985), most species have apparently no particular mechanisms that compensate for the low temperature they experience and they do not grow faster than temperate species at low temperatures but instead exhibit a unique tolerance of strong light at temperatures below 0°C. Even if temperature itself does not seem to be a barrier to growth (Raymont, 1980), the indirect effect of the impact on water column stability can be extremely important.

Yearly total radiation reaching the sea surface at different latitudes in the northern hemisphere varies (Fig. 7-2), with remarkable seasonal changes at higher latitudes. For plants in the sea, however, it is the light penetrating the water that is of vital importance and this again is influenced by weather conditions, sun height, and distribution of sea ice. The fraction of underwater light converted to chemical energy in the form of algal biomass varies because the algae have to compete with other light-absorbing components of the system (the water itself and nonphotosynthesizing particles). Except for periods with dense phytoplankton blooms or local influence of particles from melting ice polar waters are, however, rather transparent (Sakshaug and Holm-Hansen, 1984).

The growth of phytoplankton depends also to a large extent on how deep turbulence is active. As long as the depth to which the water column is mixed is so great that the loss due to respiration of the algae exceeds or equals that gained by assimilation, net primary production is impossible. For effective production the depth of the mixed layer must be less than the *critical depth* (Sverdrup, 1953), which is the depth at which the total photosynthesis of the algal cells in water column equals their total respiration.

In polar latitudes with seasonal ice cover, the formation of the ice will result in a more stable water column than found in temperate seas during winter. Snow covering the ice will prevent what little there might be of light (if any at all) to reach the water. When the snow melts, the increased light penetration supports some growth in the water under the ice (e.g., Sakshaug and Holm-Hansen, 1984). Dense populations of phytoplankton do not occur until the ice breaks up and melting water stabilizes the upper layer of the water column. Thus nutrients and plank-

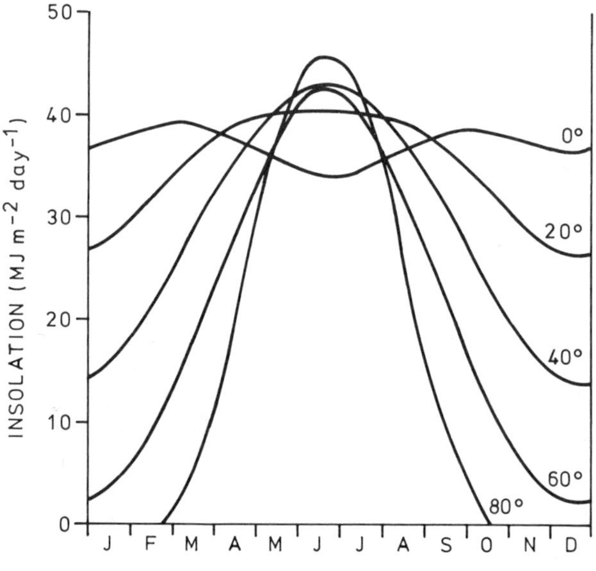


**Figure 7-2.** Seasonal variation in the daily insolation at different latitudes in the northern hemisphere. (Redrawn after Kirk, 1983.)

tonic algae are isolated from those in the layer below, and given enough light, a rapid phytoplankton growth can commence, closely followed by a peak in herbivores.

After some weeks, production is limited by depletion of nutrients above the seasonal pycnocline and/or grazing. The potentially limiting role of nutrients as essential for phytoplankton growth in these waters is, however, still discussed (Subba Rao and Platt, 1984; Sakshaug, 1985). Harrison, Platt, and Irwin (1982) suggest that temperature and light may be more important. According to their calculations, only half the nutrient requirements for phytoplankton growth in Arctic waters appear to be supported by vertical diffusion processes. Regeneration supplies the balance.

In the fringe area new nutrients are at maximum abundance in the surface layers before the bloom of phytoplankton. Their maximum concentrations in the Eurasian sector of the Arctic Ocean and the Bering Sea show distinctive differences. Sakshaug (1985) summarized maximum values for various North-East Atlantic and Barents Sea waters to be generally in the range of 9–14 $\mu$mol N litre$^{-1}$, 0.5–0.6 $\mu$mol P litre$^{-1}$ and 4–5 $\mu$mol Si litre$^{-1}$, while the Bering Sea waters contained about twice as much nitrogen and phosphorus and seven times as much dissolved silica (Codispoti, 1983). This difference probably reflects the nature of the water below the euphotic zone, which is younger and less fertilized in the North-East Atlantic and the Barents Sea than in the Bering Sea.

In Arctic waters, the amount of light is only sufficient for a single plankton bloom during summer (e.g., Parsons, Takahashi, and Hargrave, 1977) as opposed to the North Atlantic where there are two peaks during the year (Fig. 7-3). The Arctic bloom will in principle follow the retreating ice edge as it moves northward in summer (confirmed by the observations by Rey and Loeng, 1985) until it has reached its northern limit (Fig. 7-1). Conditions in these areas are, however, very variable with the ice cover strongly affected by the wind. Furthermore, other factors that frequently complicate the hydrography at the ice edge (e.g., upwelling, downwelling, and formation of eddies; Buckley et al., 1979; Johannessen et al., 1983; McPhee, 1984) introduce nutrients to the euphotic zone but at the same time mix the water below the critical depth and thus inhibit the phytoplankton growth. It should also be noted that the position of the ice is extremely dynamic and that its movements in response to local forcing will introduce short-time (e.g., in the order of days; Smith, 1985) variability into the biological processes.

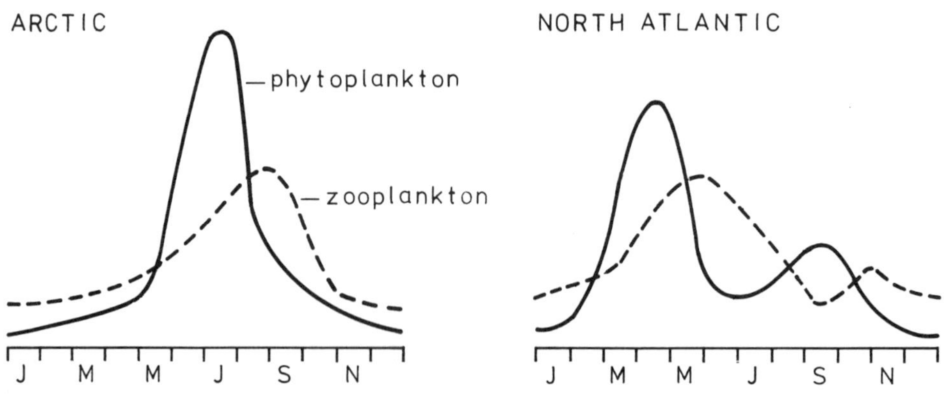


Figure 7-3. Geographical variation in the seasonal cycle of phytoplankton and herbivorous zooplankton.

The Arctic plankton blooms are usually dominated by fairly large chain-forming diatoms (Plates 7-I and 7-III) and the prymnesiophycean flagellate *Phaeocystis pouchetii* (Plate 7-II, fig. 6), which may be identified in preserved material. With the exception of Throndsen (1970), studies of other phytoflagellates have, as already mentioned, lagged seriously behind those of diatoms and thekate dinoflagellates and little is known yet about their occurrence and abundance in Arctic waters. By using serial dilution cultures from five oceanic stations in the Barents Sea, Throndsen (1970) identified several smaller flagellates not previously reported from this area. Estimates for autotrophic flagellates were a maximum $10^5 - 10^6$ cells litre$^{-1}$ and the predominant algal classes were Prasinophyceae and Prymnesiophyceae. The very small prasinophyte *Micromonas pusilla* (cell length $1.5 - 2.5$ $\mu$m) seemed to be ubiquitous and occurred in numbers from $10^3$ to $10^6$ cells litre$^{-1}$. In Norwegian coastal waters the highest concentrations reported for this species are 13 million cells litre$^{-1}$ (Throndsen, 1969).

While Throndsen (1970) drew attention to the species composition of the flagellate plankton in the Barents Sea, Reynolds (1973) focused his work on methods of assessing the amounts of the smaller members of the phytoplankton. He found that over a wide area of the Barents Sea and also in some other northern waters, smaller planktonic algae, mostly under 10 $\mu$m, at times contributed more than 90 percent of the biomass expressed as total chlorophyll *a* in the water.

## HISTORY OF INVESTIGATIONS

### Early Studies

Many classical phytoplankton studies of a taxonomic or biogeographical nature have been made in Arctic seas. The first report on diatoms was given by Ehrenberg in 1841 (see Cleve, 1883). The limited number he observed was considerably increased by O'Meara (1860), who listed 85 diatom species from Arctic waters. Only a few of these were, however, marine phytoplankters. Later Cleve (1867, 1873, 1883) and Cleve and Grunow (1880) examined diatom material collected during several expeditions to Greenland, Spitsbergen, and the Siberian Sea and found a considerable number of planktonic species in addition to those from the ice.

In 1892–1893, Vanhöffen (1893) carried out some investigations in the Karajak Fjord of West Greenland. It was covered with ice from early December until the beginning of June (Gran, 1904). There were very few diatoms in the plankton from November to February. In March, Vanhöffen (1893) found a rich vegetation of epontic (under ice) diatoms, and the same species were also found living planktonically at that time as well as later on in May. Østrup (1895) studied tow nets and ice samples collected by N. Hartz from east Greenland. According to Hartz (in Østrup, 1895) the diatoms were missing in the tow nets during winter and early spring. The planktonic samples from the summer (Østrup, 1895) consisted almost entirely of *Chaetoceros* species and *Thalassiosira nordenskioeldii* (centrics) accompanied by another centric, *Melosira nummuloides* var. *hyperborea* (valid name *M. arctica*, the names at that time being in many cases others than accepted at present; see Table 7-1), and the pennate *Fragilaria oceanica* (valid name *Nitzschia grunowii*).

In 1897, Cleve proposed classifying the plankton of the Atlantic Ocean and its adjacent seas into various plankton types characteristic of particular water masses of definite salinity and temperature. The plankton off the coasts of Greenland and in Baffin Bay and the ice edge region of the Arctic Ocean proper were considered as a kind of neritic plankton. Typically, it was almost free from animals and contained as its most characteristic species *Chaetoceros*

Table 7-1. List of Synonyms Cited in the Text

| Synonym | Valid Name | Reference |
|---|---|---|
| Diatoms | | |
| *Corethron hystrix* | *C. criophilum* | Hendey, 1937 |
| *Fragilaria cylindrus* | *Nitzschia cylindrus* | Hasle, 1972 |
| *F. nana* | *N. pseudonana* | Hasle, 1974 |
| *F. oceanica* | *N. grunowii* | Hasle, 1972 |
| *Lauderia fragilis* | *Bacterosira fragilis* | Gran, 1905 |
| *Melosira nummuloides* var. *hyperborea* | *M. arctica* | see Hustedt, 1930 |
| Dinoflagellates | | |
| *Exuviaella baltica* | *Prorocentrum balticum* | Loeblich, 1969 |
| *Peridinium* spp. with three girdle plates, in general marine species | *Protoperidinium* spp. | Balech, 1974 |
| Coccolithophorids | | |
| *Coccolithus huxleyi* | *Emiliania huxleyi* | Hay et al., 1967 |

*furcellatus, C. socialis, Lauderia fragilis* (valid name *Bacterosira fragilis,* centric), and *Fragilaria oceanica.* As it touched other plankton types, it frequently became mixed with a number of neritic species from other regions (Cleve, 1899).

Later, Cleve (1900) described the summer plankton (July) in the region of the drift ice, close to the east coast of Greenland, as a part of his report on the plankton collected during the Swedish expedition to Greenland in 1890. There was little plankton in the surface waters. According to Cleve (1900), the organisms (mainly diatoms) were mainly Arctic species mixed with a few other forms. Among the Arctic forms, a considerable number consisted of neritic or brackish water species. He classified several *Chaetoceros* species, *Eucampia groenlandica, Fragilaria oceanica* (Plate 7-I), *Lauderia fragilis, Melosira arctica,* and other diatoms as Arctic neritic and *Phaeocystis pouchetii* (Plate 7-II, fig. 6) as Arctic oceanic. Samples collected in the fjords of East Greenland in August or near the coast of Greenland were very poor in plankton, partly due to the intrusion of oceanic water.

Also Gran (1897*a,b*) analyzed the plankton from the coast of Greenland. He (Gran, 1897*b*) discussed in detail registrations of the same type of plankton vegetation, mainly meroplanktonic diatom species such as *Fragilaria oceanica* and *Thalassiosira hyalina* (Plate 7-I, fig. 9; Plate 7-III, fig. 2), on the east coast in early July 1896 as Vanhöffen (1893) and Gran (1897*a*) found in the Karajan Fjord and Cleve (1896) found in Baffin Bay in May some years before. Resting spores were frequent in Gran's (1897*b*) material. He assumed that one way for arctic planktonic diatoms to survive unfavorable conditions is to form resting spores that may settle in the ice and later germinate and repopulate overlying waters when conditions again become favorable.

Gran (1900) examined a number of plankton samples collected during the Norwegian Polar Expedition 1893–1896. The majority of these samples were collected in the summer of 1894 at about 81°N and 125°E. He classified the diatom species from these waters into four groups: true plankton forms, species living both in plankton and on ice, species that are known from ice floes only or also as littoral forms, and freshwater species. While the last-mentioned group played a very unimportant role in his material, the number of true planktonic diatoms was fairly large. Many of the neritic species were recorded on the ice in the form of spores (e.g., *Chaetoceros* and *Thalassiosira* species; Plate 7-I, figs. 1, 3, 6, 7; Plate 7-III, fig. 3). Gran (1904) reported a changing diatom flora from the same area with varying proportions of Arctic neritic, Arctic boreal neritic, boreal neritic, temperate Atlantic neritic, and oceanic species, thus reflecting the incursion of various water masses. All of his observations clearly show the mixed biogeographic character of the plankton community in polar seas.

Ostenfeld (1910), Paulsen (1910), and Ostenfeld and Paulsen (1911) studied the plankton vegetation from the east Greenland Sea north to 73°N latitude more extensively. The last-mentioned authors characterized the plankton of Denmark Harbor (76°46′N, 18°43′W) as an Arctic coast plankton with a short flowering period in later summer. It was fairly poor in species and it consisted mainly of neritic diatoms having a wide distribution in the northern seas, also outside the Arctic. Only four centric (*Bacterosira fragilis, Biddulphia arctica, Chaetoceros gracilis, Eucampia groenlandica*) and two pennate (*Navicula septentrionalis, N. vanhoeffenii*) diatoms were classified as true Arctic species. The coastal waters and the pack ice region were divided into three regions according to their vegetation: the innermost plankton region of the coastal waters being characterized by diatoms (*Chaetoceros* species, *Coscinodiscus subbuliens* [centric], *Fragilaria oceanica*) and by a copepod, *Calanus finmarchicus;* the plankton region of the pack ice characterized by a ciliate, namely *Cyttarocylis denticulata,* some diatoms (*Chaetoceros* and *Rhizosolenia* species), and dinoflagellates (*Peridinium,* valid name see Table 7-1), as well as a golden-brown flagellate belonging to the genus *Dinobryon* (Chrysophyceae); the plankton region of the open ocean with small quantities of two dinoflagellates, *Ceratium arcticum* (Plate 7-II, fig. 3) and *Peridinium ovatum.*

## Research Since 1930

The early phytoplankton studies from Arctic seas reviewed so far were naturally of a systematic/taxonomic and biogeographic nature. No research on the dynamics of phytoplankton growth as a function of the Arctic environment was made until the 1930s. So far, most phytoplankton production studies from the Arctic Ocean have been concentrated in the Canadian Arctic, while data from the Eurasian sector are still scarce (for references see Subba Rao and Platt, 1984 and references cited therein).

## Eurasian Waters

Measurements of nutrients (phosphate, nitrate, nitrite, silicate) and chlorophyll and their seasonal variations were first presented by Kreps and Verjbinskaya (1930) from the central Barents Sea. Their data showed that the waters of different origin also differed with respect to their relative richness in nutrients. Thus Atlantic water was somewhat poorer in nutrients than Arctic water, but richer than inshore waters. In May, the Atlantic water still showed winter conditions characterized by uniform chemical conditions throughout the depths, while the inshore waters and the Arctic water showed depletion of nutrients and a rise in the H-ion concentration in the upper 25 m, both resulting from the spring bloom of phytoplankton. The decrease in nitrate and phosphate during spring and summer caused by the planktonic algae was proportional to the relative content of chlorophyll in the plankton samples from the upper 25 m. Intense vertical mixing of the water during winter reestablished a uniform vertical distribution of nutrients with depth.

Later, Braarud (1935) studied the phytoplankton and the conditions of plant production in the Atlantic, Arctic and Polar water masses of Denmark Straits between Iceland and Greenland from June to August. He also included some observations on net samples from the east coast of Greenland and from waters in the Spitsbergen–Franz Josef Land area, listed all species he observed, and commented on their distribution and the taxonomy of some of the species. At the border of the east Greenland Polar Current in Denmark Straits, he observed a rich diatom vegetation that consisted of components from both Polar and Atlantic waters (*Achnanthes* [pennate], *Nitzschia* and *Thalassiosira* species). Since the main part of the population was of the same character as that of the Polar water, Braarud assumed that the large production was a result of a mixture of Polar waters fairly rich in phytoplankton and poor in nutrients, with nutrient-rich Atlantic water poor in phytoplankton. While the Polar waters were covered with ice, they had, according to Braarud, a moderate supply of nutrients, were slightly subsaturated in oxygen, and had a scanty plankton numerically dominated by flagellates, mainly brown and colorless dinoflagellates, but also green flagellates and some coccolithophorids. Most of them were very small forms ($< 20\ \mu m$) and the diatoms were only represented by six species in the quantitative samples from this period. When the ice became more open, fresh water from the melting ice caused a pycnocline and the diatom population increased rapidly. Since the phytoplankton production in Polar waters took place as soon as the ice opened and without any mixture of Atlantic waters, and regardless of the low temperature of Polar waters ($-1.8°C$ to $0°C$), Braarud assumed that it was the lack of light under the ice and not the low temperature that prevented a large plant production from taking place in the areas with very dense pack ice. The systematic observations of the nonmotile stage of the cold water coccolithophorid *Coccolithus pelagicus* (Plate 7-IV, fig. 3) within the east Greenland Polar Current from dense hummocked ice to the more open ice on the outskirts of the current (Braarud, 1935, 1979) indicate an extension of the temperature range of this species to include negative temperatures.

Marshall (1957) and Corlett (1958) studied the submarine light penetration and the production of phytoplankton during spring and summer in the western Barents Sea. The measurements of primary production were made at three stations in the Bear Island area in May. The production was highest ($1.3\ g\ C\ m^{-2}d^{-1}$) in the cold Arctic water on the Spitsbergen Shelf (75°15′N, 25°00′E; Corlett, 1958). The number of diatom cells under 1 $m^2$ was $6.23 \times 10^9$, of which the main constituents were the two centric species, *Thalassiosira gravida* (40 percent) and *T. hyalina* (24 percent) (Plate 7-I, figs. 8 and 9; Plate 7-III, figs. 1 and 2), and the pennate *Fragilaria islandica* (20 percent). Marshall (1957) discussed the conditions for phytoplankton production in this area in terms of the mixed layer and the critical depth (Sverdrup, 1953). The results of the primary production measurements supported his theory that phytoplankton production might be expected to start in April, or as soon as the water becomes ice-free, on the Bear Island Bank and in June in the deeper Atlantic water. The wave of production apparently following the retreating ice edge in spring and summer was accounted for by a shallower and mixed surface layer at the ice edge than the critical depth, so that the phytoplankton could increase quantitatively there.

Smayda's (1958*b*) phytoplankton studies from the waters around Jan Mayen Islands between Iceland and Spitsbergen from mid-March through April represented the first quantitative study of Arctic phytoplankton so early in the growth season. At that time, Jan Mayen was completely surrounded by waters with surface temperatures between −1°C and −2°C. The phytoplankton vegetation was in its initial growth phase when the investigations started. During April a considerable increase in the populations of diatoms (*Fragilaria nana,* valid name is *Nitzschia pseudonana,* accompanied by a variety of more subordinate species) and green flagellates were observed, but no real bloom was recorded at any of the stations. According to Braarud (1935), the presence of green flagellates indicated melting ice conditions.

At the end of April, *Thalassiosira* and *Chaetoceros* species had increased in composition and abundance. Of the factors controlling the phytoplankton growth during the investigated period, Smayda found that light was of prime importance. The community consisted of a mixture of meroplanktonic and holoplanktonic species. According to Smayda, the presence of meroplanktonic species was readily explained by their being mainly Arctic species whose resting spores are commonly found in the ice. When such ice melts, the spores become free and germinate, thereby giving rise to local populations, also capable of seeding other areas. The occurrence of holoplanktonic forms so early in the growth season was somewhat more problematic. But Smayda (1958*b*) suggested that these species represented local populations capable of active growth at the low temperatures observed; hydrographic observations supported this view.

As part of his investigations on the Norwegian Sea, Ramsfjell (1960) studied the vegetation and the conditions of phytoplankton growth in the waters north to northeast of Jan Mayen, off Bear Island and along the west coast of Spitsbergen in June 1952 and 1953. As observed by Smayda (1958*b*), the spring increase seemed to start fairly early in the waters north of Jan Mayen and was over in June. The vegetation in June 1952, characterized by the occurrence of oceanic species of the centric diatom genus *Chaetoceros,* consisted generally of fewer species occurring in smaller numbers than in 1953. Also, the diatom cells were poorly silicified in 1952 and the 25 m population was considerably larger than at lower depths. These conditions indicate a later stage of the annual cycle than in 1953. The causes of the later spring development in this area in 1953, as compared with 1952, was explained by the different ice and wind conditions in the two years. The investigations on the water masses off Bear Island and west of Spitsbergen indicated that a more advanced stage in the phytoplankton cycle also had been reached in these areas in 1952 than in 1953.

Fewer Arctic species were recorded in 1952 and the most important ones, *Chaetoceros furcellatus* and *Fragilaria oceanica,* occurred in smaller numbers than in 1953 when the fairly abundant vegetation was dominated by Arctic boreal neritic diatoms (*C. furcellatus, C. socialis, F. oceanica,* and *Thalassiosira nordenskioeldii*). The predominant species in the cold water masses off Spitsbergen were different from those recorded from the east Greenland Polar Current in Denmark Straits during the corresponding period of the annual phytoplankton cycle, which reflected the distinct water masses within these areas.

The importance of diatoms and *Phaeocystis pouchetii* in these waters was also focused on by Zenkevitch (1963). He found that in the central Arctic Basin diatoms formed some 75 percent of the species with dinoflagellates accounting for nearly 20 percent. In biomass the difference was even more pronounced; diatoms comprised 98 percent of the phytoplankton biomass near the pole. In the Barents Sea and partly in the Kara Sea flagellates were more important, *Phaeocystis pouchetii* and the chrysophyte *Dinobryon* developing rapidly and sometimes giving the appearance of a bloom.

Since the 1960s, Soviet and Norwegian scientists have studied more extensively the variation in the assimilation number (which means the maximum photosynthetic rate per unit of chlorophyll *a* and has been used as an index of the physiological state of the natural phytoplankton) in different parts of the Barents Sea (Vedernikov, 1975), and the phytoplankton biomass and primary production in the Arctic Ocean (see Table 7-2 and references cited therein). According to Sokolova and Solov'yeva (1971), the coastal algal communities were composed of a diversity of species although diatoms were most abundant. Along the Murman coast the centric diatom *Skeletonema costatum* was the most common species during the summer months, while Halldal and Halldal (1973) found that two golden-brown chrysophytes, *Meringosphaera tenerrima* and *Apedinella spinifera,* dominated the plankton flora in Kings Bay in July. The motile form of the coccolithophorid *Coccolithus pelagicus, Crystallolithus hyalinus* (Plate 7-IV, fig. 4), was also very common in their samples, indicating the influence of Atlantic waters.

Mel'nikov (1976*a,b,* 1978) and Mel'nikov and Tsinovskiy (1978) studied the spatial distribution of Arctic phytoplankton and the primary production in open water and in puddles from the start of melting to the start of freezing by using a drifting research station. Diatoms characterized the flora of open water (Mel'nikov, 1976*a*) where the primary production varied from 0.4–0.7 mg C $l^{-1}d^{-1}$ (cf. Table 7-2 and comments therein). They indicate that the vegetation period in the central Arctic Basin lasts at least three months from mid-June (Mel'nikov and Tsinovskiy, 1978).

Rey and Loeng (1985) concluded from observations of the spring phytoplankton bloom in the central Barents Sea that the bloom will start at different times depending on the vertical stability and the light conditions. They assume that the phytoplankton at the ice edge and near the Norwegian coast will bloom more or less simultaneously, sometime during April, with a slightly earlier start at the ice edge if the maximum extent of the ice reaches to the south of the Polar Front (Rey and Loeng, 1985, fig. 1) and comes in contact with warmer Atlantic water so that melting can commence earlier. In the region dominated by Atlantic water, between the Polar Front (or the ice edge) and the Norwegian Coastal Current, the spring bloom usually takes place in late May when the warming of the upper layers has given the necessary stability (Rey and Loeng, 1985). Also, Rey and Loeng stress the importance of *Phaeocystis pouchetii* and several diatoms (*Chaetoceros, Nitzschia* and *Thalassiosira* species) in these waters. Other important constituents of the plankton included *Dinobryon balticum* (Plate 7-II, fig. 5) and the craspedophyte *Monosiga marina* var. *minima* (choanoflagellate, heterotrophic).

In the meltwater chlorophyll *a* was evenly distributed with concentrations between 5 and 6 mg $m^{-3}$ in early May 1981 and the nitrate was almost exhausted, indicating that the phytoplankton bloom had already reached its maximum. Other results from the same area (Sakshaug and Hegseth, 1985) show that the spring bloom had ended at the ice edge in late May 1984. In July and August 1984, the surface layer, characterized by meltwater, contained only 0.1–1 mg chlorophyll *a* $m^{-3}$ in contrast to a marked chlorophyll maximum (Table 7-2) of a few metres thickness just below the pycnocline at 25–50 m. This was also the case in the summer of 1985 (Loeng, 1986). When analyzing these data it must be remembered, however, that marine planktonic algae respond to variations in light intensities by changing the content and the ratios of the photosynthetic pigments (for a summary of light–shade adaptation, see Falkowski, 1980). The deep communities reported by Sakshaug and Hegseth (1985) and Loeng (1986) were probably light-limited and chlorophyll *a* may therefore be even more questionable as a measure of phytoplankton biomass in these cases. In the entire area studied in 1985 (Loeng, 1986), about 78 percent of the

Table 7-2. Maximum Values of Biomass and Primary Production from the Eurasian Sector of the Arctic Ocean and Some Northern Norwegian Fjords Situated Well North of the Arctic Circle

| *Region* | *Reference* | *Season of Investigation* | *Chlorophyll* a *mg m⁻³* | *Primary Production mg C m⁻³ h⁻¹* |
|---|---|---|---|---|
| W. Barents Sea | Corlett, 1958 | May | | 8.60 |
| Murman Coast | Sokolova and Solov'yeva, 1971 | April–Oct. | 4.85 | 1.30 |
| Murman Coast | Vedernikov and Solov'yeva, 1972 | June–Aug. | 1.59 | 2.40 |
| Kings Bay, Spitsbergen | Halldal and Halldal, 1973 | July | 5.45 | |
| Barents Sea | Reynolds, 1973 | Aug.–Sept. | 1.00 | |
| Central Arctic Ocean | Mel'nikov, 1976*a* | June–Aug. | | 0.70* |
| Van Mijenfjord, Spitsbergen | Schei et al., 1979 | July | | 23.90 |
| Barents Sea | Ellertsen et al., 1981 | June–July | 12 | |
| Northwest of Spitsbergen | Heimdal, 1983 | Sept.–Oct. | 2.20 | 2.50 |
| Barents Sea | Rey and Loeng, 1985 | May–July | 5–6 | |
| | | Aug. | >12 | |
| Barents Sea | Sakshaug and Hegseth, 1985 | July–Aug. | 17 | |
| Barents Sea | Hassel, Loeng, and Skjoldal, 1986 | Jan. | 0.01 | |
| Barents Sea | Loeng, 1986 | Aug. | >8 | |
| *Northern Norwegian fjords* | | | | |
| Straumsbukta | Throndsen and Heimdal, 1976 | All seasons | | 13.50 |
| Ramfjorden | Bech, 1982 | All year | 8.02 | 18.24 |
| Tromsøysundet | Bech, 1982 | All year | 8.42 | 8.80 |
| Balsfjorden | Eilertsen and Taasen, 1984 | All year | | >20 |

*mg C $l^{-1}d^{-1}$, measured by the oxygen method.

chlorophyll *a* was found in the size fraction of algae smaller than 10 μm. According to Hewes (1986), the average biomass for autotrophic nanoplankton was 42 μg C litre$^{-1}$ as compared to 35 μg litre$^{-1}$ heterotropic biomass (grazers).

As summarized above, parts of the seasonal phytoplankton cycle in the Barents Sea and its dynamics appear to be fairly well described, whereas information on the phytoplankton from these waters during autumn and winter is more scarce. Heimdal (1983) studied the biological and the chemical properties of the water in the ice edge region and the open sea in the Arctic Ocean northwest of Spitsbergen in September–October 1979. Supplementary observations on water samples from a few stations located further south were also reported to compare the Polar pattern with those in the Greenland Sea and the Barents Sea. Concentrations of phosphate were quite low (maximum 0.55 μmol litre$^{-1}$). Nitrate and silicate ranged from no measurable quantities to 5.7 μmol litre$^{-1}$ and 3.8 μmol litre$^{-1}$, respectively, and the nutrient concentrations were usually highest at the maximum depth of sampling (32 m). The standing stock of phytoplankton measured as chlorophyll *a* was noticeably higher (maximum 2.2 mg m$^{-3}$) in the ice edge region northwest of Spitsbergen than in the more homogenous water column further south (0.1 mg chlorophyll *a* m$^{-3}$) where the salinity and temperature differences were 0.05 ‰ and 0.05°C, respectively, between the surface and a depth of 40 m. Nearly all measurable primary production took place within the uppermost 12–16 m. The highest rate, equivalent to 37 mg m$^{-2}$ h$^{-1}$, occurred on a sunny day in an open lead in the ice on September 19.

More than 70 taxa of diatoms and dinoflagellates were identified in Heimdal's (1983) material, as well as about 10 forms belonging to other algal classes and 2 species of planktonic choanoflagellates. Diatoms were predominant and certain neritic species (*Chaetoceros socialis, C. subsecundus, Eucampia groenlandica, Leptocylindrus danicus, Nitzschia 'cylindrus'* [comprising *N. cylindrus* and *N. grunowii*], *Thalassiosira 'antartica'* [probably containing two closely related species], *T. bioculata,* and *T. nordenskioeldii*) and *N. seriata,* which may be indigenous to both coastal and oceanic waters (Hasle, 1965), were particularly important. At some stations *L. danicus* (centric) formed >90 percent of the total diatom population, but wide differences, especially in the relative abundance of the most numerous species, were observed. It is remarkable that this species was neither observed in the summer plankton from the east coast of Greenland and the Spitsbergen–Franz Josef Land area (Braarud, 1935; Halldal and Halldal, 1973; Schei et al., 1979) nor from the ice-edge region of the Barents Sea (Ellertsen et al., 1982), whereas Grøntved and Seidenfaden (1938) found that *L. danicus* was one of the most common diatom species in the West Greenland Current in August. They recorded *L. danicus* up to the edge of the ice as far north as 78°N latitude, and it was most numerous in the northernmost studied areas. In northern Norwegian coastal waters it is abundant in the summer plankton (e.g., Heimdal, 1974).

It is also remarkable that the cosmopolitan diatom *Skeletonema costatum* was absent in the waters northwest of Spitsbergen in autumn (Heimdal, 1983), in accordance with previous studies in both the Arctic and Antarctic by Braarud (1935) and Hasle (1969), respectively. Resting spores of *Chaetoceros socialis* were numerous in the ice zone, but otherwise absent. Spores of other diatoms were less frequent. The dinoflagellates were rich in species at all stations studied, but the species identified were quantitatively insignificant compared with the diatoms. Fairly large populations of unidentified athecate dinoflagellates (of <30 μm) were, however, observed throughout the area. Coccolithophorids were present in small numbers only. Numerically *Emiliania huxleyi* (Plate 7-IV, fig.

2) was most important. Depth distributions of the diatom and dinoflagellate populations derived from cell counts were reasonably consistent with those obtained by the chlorophyll *a* measurements. The discrepancy was most pronounced in the vicinity of the ice edge where resting spores of *C. socialis* dominated the diatom populations at the maximum depths of sampling.

A winter situation is described by Hassel, Loeng, and Skjoldal (1986), who carried out physical and biological investigations in the northern part of the ice-free areas of the Barents Sea in January 1985. The concentrations of chlorophyll *a* were extremely low (about 0.01 mg $m^{-3}$). The microscopical examinations showed, however, a fairly high number of unidentified organisms smaller than 10 $\mu m$ (maximum $1.6 \times 10^6$ cells $litre^{-1}$). This would suggest a large heterotrophic component. Diatoms were scarce, maximum 4,000 cells $litre^{-1}$ in the completely mixed water on the shallow Central Bank. This is of considerable interest because until recently overwintering strategies were very poorly understood.

## Canadian Arctic and Associated Waters

While the whole Norwegian coast and the Barents Sea south of 73°N and west of 35°E remain permanently ice-free during winter due to the influx of warm Atlantic water (e.g., Sakshaug and Holm-Hansen, 1984), the maximal extent of sea ice reaches south to about 55°N latitude in the Canadian region (Fig. 7-1). Furthermore, Arctic and Atlantic waters meet and mix in the Davis Strait, thus forming a complex system of circulation that affects the development of plankton in these waters.

The plankton flora of the waters west of Greenland has been studied for nearly a century (see Grøntved and Seidenfaden, 1938 and the summaries of previous investigations given therein). Though primarily concerned with the taxonomic problems, Grøntved and Seidenfaden (1938) also discussed the composition of the various plankton communities in the different water masses of the Labrador Sea and in part at different seasons. Along the coast the biological spring started when the ice broke up with a *Fragilaria* bloom, succeeded by *Thalassiosira* and *Chaetoceros* species in the order stated. The centric diatom *Rhizosolenia hebetata* f. *semispina* was the only important diatom in late summer together with some *Peridinium* species (dinoflagellates). The central eddy of the Labrador Sea was characterized by *C. decipiens* (oceanic) and a more oceanic temperate *Thalassiothrix* species (pennate) accompanied by a number of neritic species originating from the coast of Iceland. The cold-water masses from the northern part of the Canadian coast were very poor in plankton but rich in nutrients, so when passing northward they attained a rich plankton production somewhat similar in species (*Chaetoceros, Leptocylindrus, Rhizosolenia,* and *Peridinium*) to that of the surrounding waters, but dominated by the Arctic dinoflagellate *Ceratium arcticum* (Plate 7-II, fig. 3).

Holmes also (1956) focused his studies on the mixed biogeographic character of the phytoplankton communities and the successional pattern in the Labrador Sea, while Pavshtiks (1968) concentrated his work on the pattern of the development of the biological spring in the same waters (Fig. 7-4). Holmes's (1956) material was collected at a sub-Arctic, oceanic locality (56°30′N, 51°00′W) with an annual temperature minimum of about 3°C. The phytoplankton population was composed of varying proportions of Arctic, boreal, temperate, oceanic, and neritic species, reflecting the incursions of various water masses into the area (*Chaetoceros atlanticus, C. borealis,* and *Thalassiothrix longissima* characteristic of the Irminger–Atlantic portion of the West Greenland Current; *Fragilaria oceanica, C. debilis, C. decipiens, Rhizosolenia styliformis,* and *Thalassiosira gravida* of the mixed water of the West Greenland Cur-

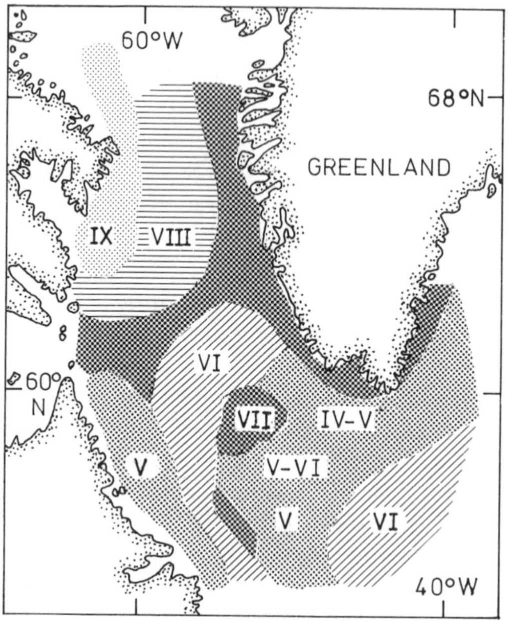


Figure 7-4. The biological spring in the Davis Strait. Roman numerals indicate the months as the development extends northward and out over deeper water. (Redrawn after Pavshtiks, 1968.)

rent; *Coscinodiscus centralis, Ceratium arcticum,* and *Peridinium depressum* of the Labrador Current; solitary *Chaetoceros* spp., *Corethron hystrix, Fragilaria nana,* and *Peridinium* spp. of the Central Eddy). The annual cycle was dominated by diatoms (numerically most abundant *F. nana*) and the dinoflagellate *Exuviaella baltica*. Coccolithophorids were scarce.

Bursa (1961) studied the annual phytoplankton cycle and its hydrographical background on the coastal shelf of the northern Foxe Basin (Igloolik, 69°20.5′N, 81°43.5′W). The main limiting factor for marine life in these waters is the sea ice, which strongly affects salinity, temperature, and light penetration during most of the year. The biological winter at Igloolik lasted from November until April. When the ice broke up in late April, spring started with a rich phytoplankton, mainly of *Fragilaria cylindrus*. Bursa (1961) classified the pennate diatoms found beneath the ice at Igloolik in April as a shade flora in contrast to a sun flora of centrics (*Chaetoceros* and *Thalassiosira* species), which was most abundant during the open-water period and with a maximum in early August.

In recent years, phytoplankton and primary production in Canadian Arctic and associated waters have been most extensively studied by Canadian scientists (for references see the reviews of Nemoto and Harrison, 1981; and Subba Rao and Platt, 1984). Data on biomass (0.05–15.0 mg chlorophyll *a* $m^{-3}$), primary production (0.02–14.01 mg C $m^{-3}$ $h^{-1}$), and assimilation number (0.11–10.33 mg C • mg chlorophyll *a* $h^{-1}$) in the eastern Canadian Arctic and adjacent areas (Subba Rao and Platt, 1984) are comparable to values from a wider geographical area including Eurasian waters (cf. Table 7-2). Daily integrated production in the Arctic waters, based on an extended photoperiod (24 h), ranged according to Subba Rao and Platt (1984) between 0.004 and 4.89 g C $m^{-2}$. Usually the higher values were observed in the coastal waters. The maximum rate from eastern Baffin Island is of similar magnitude as the rates recorded from some of the most productive seas (see Platt and Subba Rao, 1975).

Converted to annual production the values from the Arctic Ocean, based on a 120-day growth season, range from 12–98 g C $m^{-2}$. These rates are considerably lower than the 86–289 g C $m^{-2}y^{-1}$ reported from the Antarctic by Subba Rao and Platt (1984). More than 50 percent of the annual production in Arctic waters may depend on new nutrients (observations from the Bering Sea; McRoy and Goering, 1974). Microscopic analysis of preserved samples from the Baffin Bay showed a wide variety of planktonic algae belonging to the Bacillariophyceae, Dinophyceae, Chrysophyceae, Cryptophyceae, and Cyanophyceae (Platt et al., 1982). Diatoms (*Chaetoceros, Rhizosolenia,* and *Thalassiosira* as well as

*Nitzschia* species) were dominant and often accounted for over 90 percent of the numerical abundance.

## SUMMARY AND CONCLUSIONS

The Arctic planktonic flora has been studied for more than a century. This chapter reviewed important concepts of this flora as revealed by data from the literature. A surprisingly large number of planktonic algae can thrive in Arctic waters (e.g., Heimdal, 1983), but there seem to be relatively few truly Arctic species. The blooms are usually dominated by chain-forming diatoms, particularly species of the genera *Chaetoceros, Eucampia, Leptocylindrus, Nitzschia, Rhizosolenia,* and *Thalassiosira,* and the prymnesiophycean flagellate *Phaeocystis pouchetii* (Platt et al., 1982; Heimdal, 1983), while dinoflagellates and other nondiatom planktonic algae are less important.

According to Gran (1905), Hustedt (1930), Smayda (1958*a*), Hasle (1976), and Syvertsen and Hasle (1984) there are ten diatoms (*Achnanthes taeniata, Bacterosira fragilis, Chaetoceros furcellatus, C. septentrionalis, Fragilaria oceanica, Navicula pelagica, N. vanhoeffenii, Nitzschia frigida, Thalassiosira bulbosa, T. hyalina*) and one dinoflagellate (*Ceratium arcticum*) classified as Arctic species. In addition, *Eucampia groenlandica,* listed as a cold-water diatom by Syvertsen and Hasle (1983), may have a similar distribution, while *N. cylindrus, Porosira glacialis,* and *T. antarctica* appear to be bipolar (Hasle, 1976). The most important cosmopolitans from Arctic seas are probably *Leptocylindrus danicus* (diatom) and *Prorocentrum balticum* (dinoflagellate). Apparently the very common cosmopolitan diatom *Skeletonema costatum* is absent in these waters.

In Arctic waters, sufficient light for phytoplankton growth is only available during 5 or 6 months of the year. For the early and late part of this period, light intensities are so low and the critical depth is so near the surface that mixing effectively prevents any increase in biomass. Even with the increase in light during spring, the vernal bloom of phytoplankton is dependent to a large extent on how far turbulence is active. At the ice edge, reduced salinity of the upper layers may produce stability relatively early in the year. This stability, coupled with an adequate supply of light, provides the necessary conditions for an increase in the standing stock, which follows the melting ice northward in summer until it has reached its northern limit (e.g., Rey and Loeng, 1985). In the Atlantic water, between the Polar Front (or the ice edge) and the Norwegian Coastal Current, the mixed layer becomes shallower than the critical depth only in late May, and the spring bloom is correspondingly delayed.

A wide range of daily integrated production values (0.004–4.89 g C $m^{-2}$ $d^{-1}$) has been reported from the Arctic waters (Subba Rao and Platt, 1984). Generally speaking, the values are higher in the coastal bays and fjords. But considerable variations in the magnitude of phytoplankton production apparently also exist within the neritic areas and more data are needed from all regions. In the Arctic Ocean, assuming a 120-day growth season, the annual production range from about 12–98 g C $m^{-2}$.

# APPENDIX

## Plate 7-I

Centric (1–10) and pennate (11–13) diatoms common in Arctic seas. Scale bar = 10 $\mu$m. For synonyms see Table 7-1.

1. Part of chain of *Chaetoceros furcellatus* with resting spores. Northern neritic species.
2. *C. septentrionalis.* Arctic species very common and abundant, especially in the ice zone.
3. *C. socialis.* Northern temperate neritic species. *a:* Typical colony kept together by some long setae and slime. *b:* Part of chain with resting spores.
4. *Detonula confervacea.* Neritic species that belongs to a complex of species (*Bacterosira fragilis, Nitzschia grunowii, N. seriata, Thalassiosira constricta, T. hyalina,* and *T. nordenskioeldii*) indigenous to the cold water of the northern hemisphere (Hasle, 1976). *a:* Part of a chain of vegetative cells. *b:* Part of a chain with resting spores.
5. *Eucampia groenlandica.* Curved chain. Cold-water species in coastal and oceanic waters. No resting spores known (Syvertsen and Hasle, 1983).
6. *T. antarctica. a:* Typical chain. Bipolar species that forms resting spores (*b*) and is assumed to be dependent on the coast or ice for survival.
7. *T. constricta.* Neritic species. *a:* Chain showing dividing cells. *b:* Pair of resting spores split up.
8. *T. gravida.* Bipolar species (biogeographic classification questionable, see Hasle, 1976).
9. *T. hyalina.* Arctic neritic species, from about 80°N in the ice north of Spitsbergen to about 63°N off the Norwegian west coast and to 50°N in the colder western part of the North Atlantic Ocean (Hasle, 1976). Very often accompanied by *N. grunowii* and *T. nordenskioeldii,* which have, however, a more southern range.
10. *Porosira glacialis.* Bipolar species that forms resting spores.
11. *N. grunowii. a:* Part of band-shaped colony in girdle view. *b:* Valve view.
12. *N. cylindrus.* Bipolar species. Resting spores unknown. *N. cylindrus, P. glacialis,* and *T. antarctica* seem all to belong to extreme inshore waters and/or associated with ice where another Arctic diatom, *T. bulbosa,* frequently predominates the plankton flora (Syvertsen and Hasle, 1984). *a:* Girdle view. *b:* Valve view.
13. *Navicula vanhoeffenii.* An Arctic neritic species found occasionally together with another Arctic diatom, *N. pelagica. a:* Part of colony in girdle view. *b:* Valve view showing the raphe.

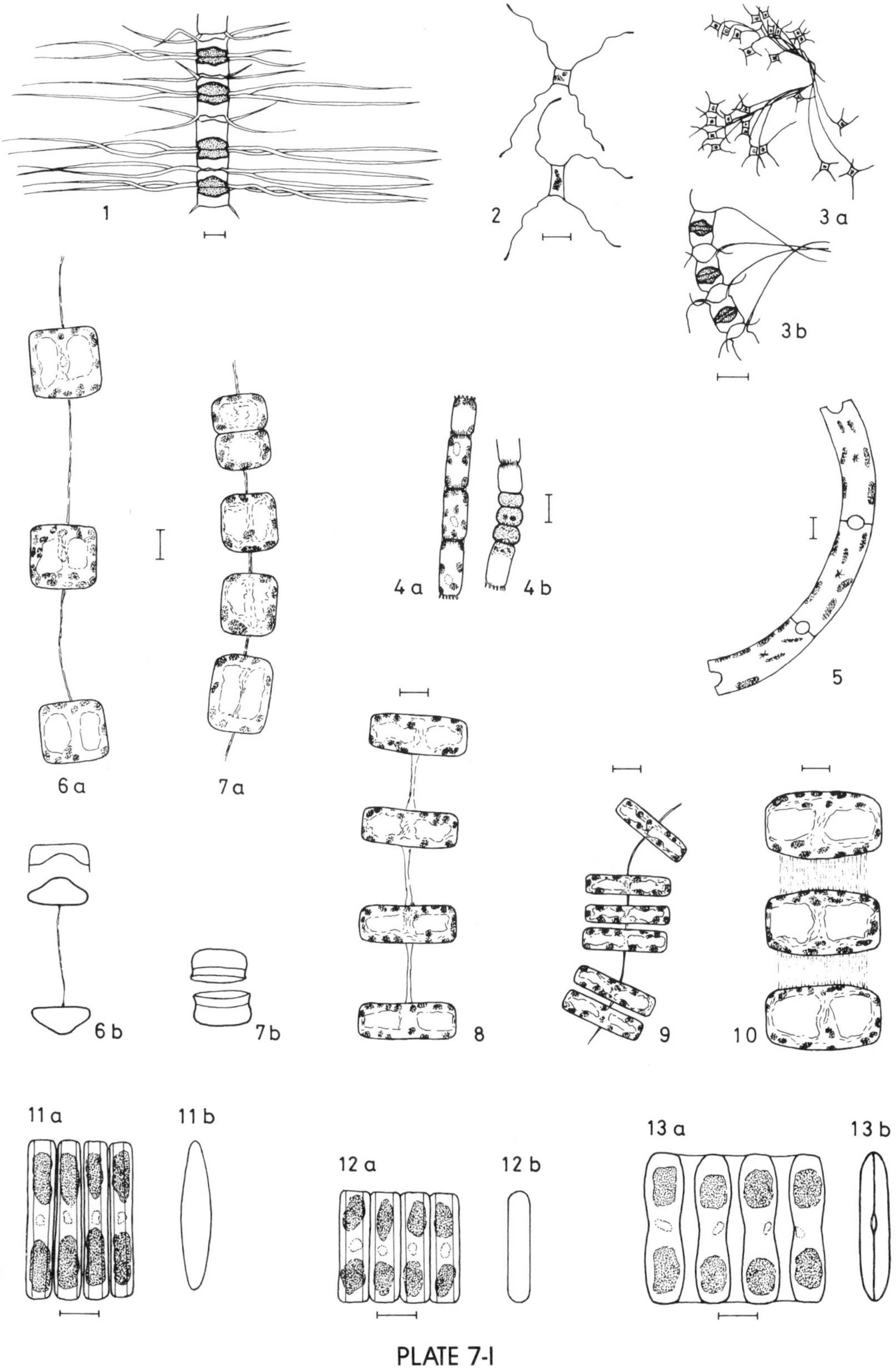


PLATE 7-I

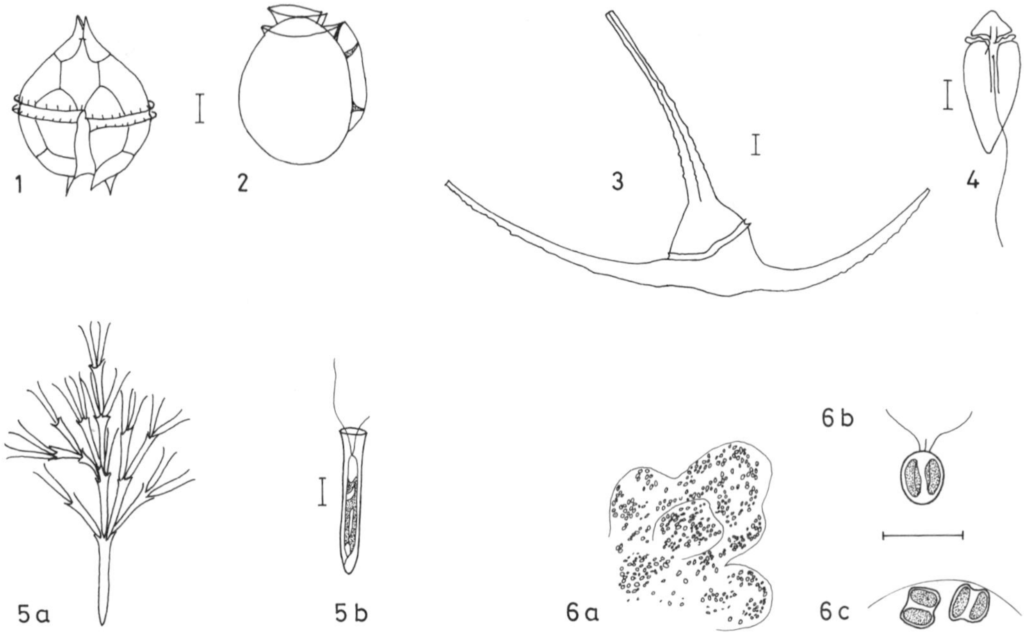


## Plate 7-II

Some common phytoflagellates in Arctic seas. Scale bar = 10 $\mu$m. For synonyms see Table 7-1.

1. *Protoperidinium pellucidum.* Neritic and oceanic dinoflagellate.
2. *Dinophysis acuminata.* Neritic dinoflagellate, widely recorded.
3. *Ceratium arcticum.* Arctic dinoflagellate.
4. *Oxytoxum* sp. Dinoflagellate.
5. *Dinobryon balticum.* Chrysophyte recorded from the Baltic Sea in addition to the northern waters. *a:* Colony. *b:* Lorica with attached cell.
6. *Phaeocystis pouchetii.* Prymnesiophyte, predominant species in northern seas. *a:* Part of lobed colony. *b:* Flagellated cell. *c:* Larger magnification of nonmotile cells.

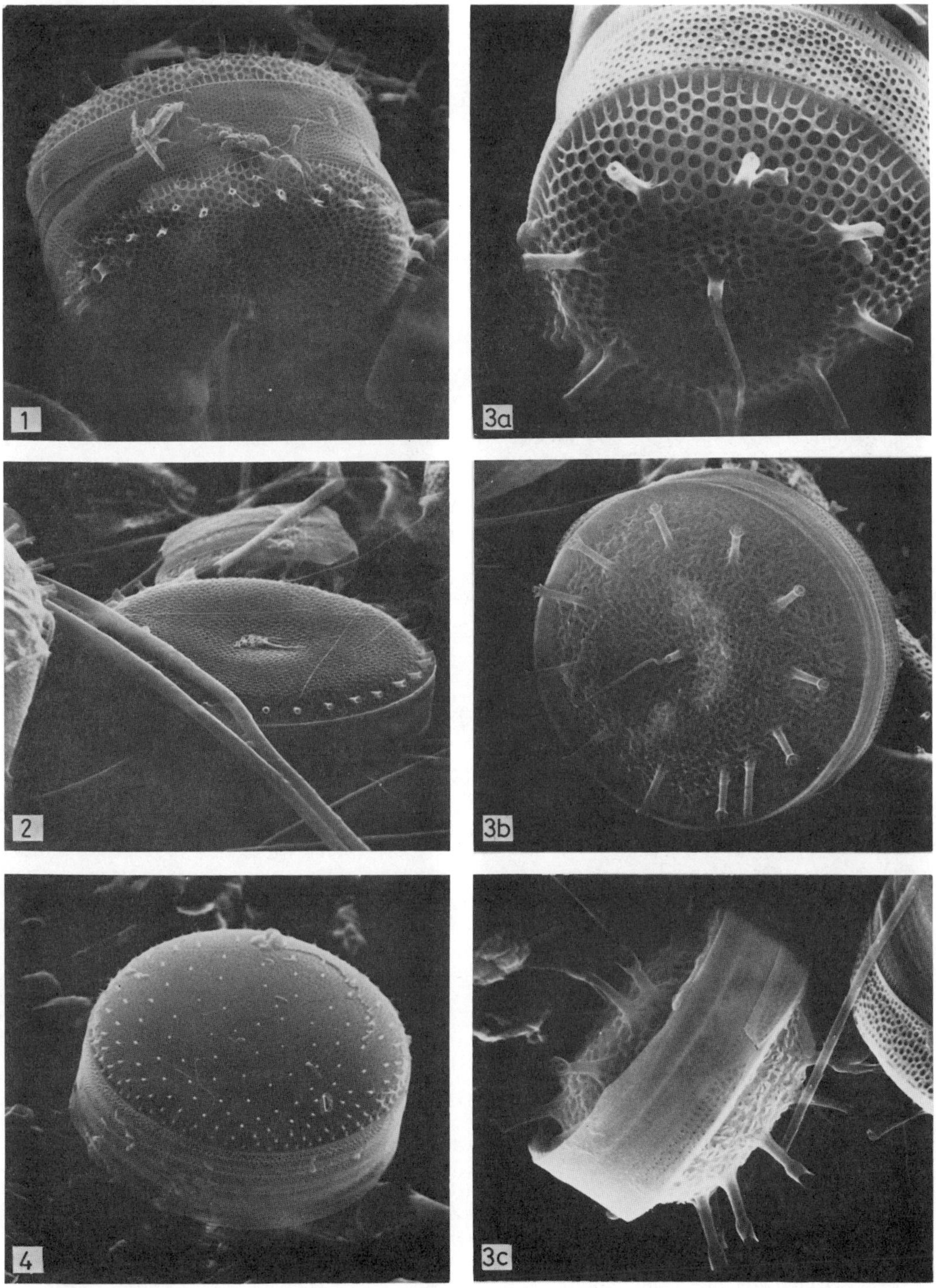


## Plate 7-III

Common centric diatoms in Arctic plankton.

1. *Thalassiosira gravida* (× 2,100).
2. *T. hyalina* (× 2,800).

3*a*. *T. nordenskioeldii* (× 6,300).

3*b*. *T. nordenskioeldii*. Resting spore, valve view (× 3,150).

3*c*. *T. nordenskioeldii*. Resting spore, lateral view (× 4,900).

4. *Porosira glacialis* (× 1,400).

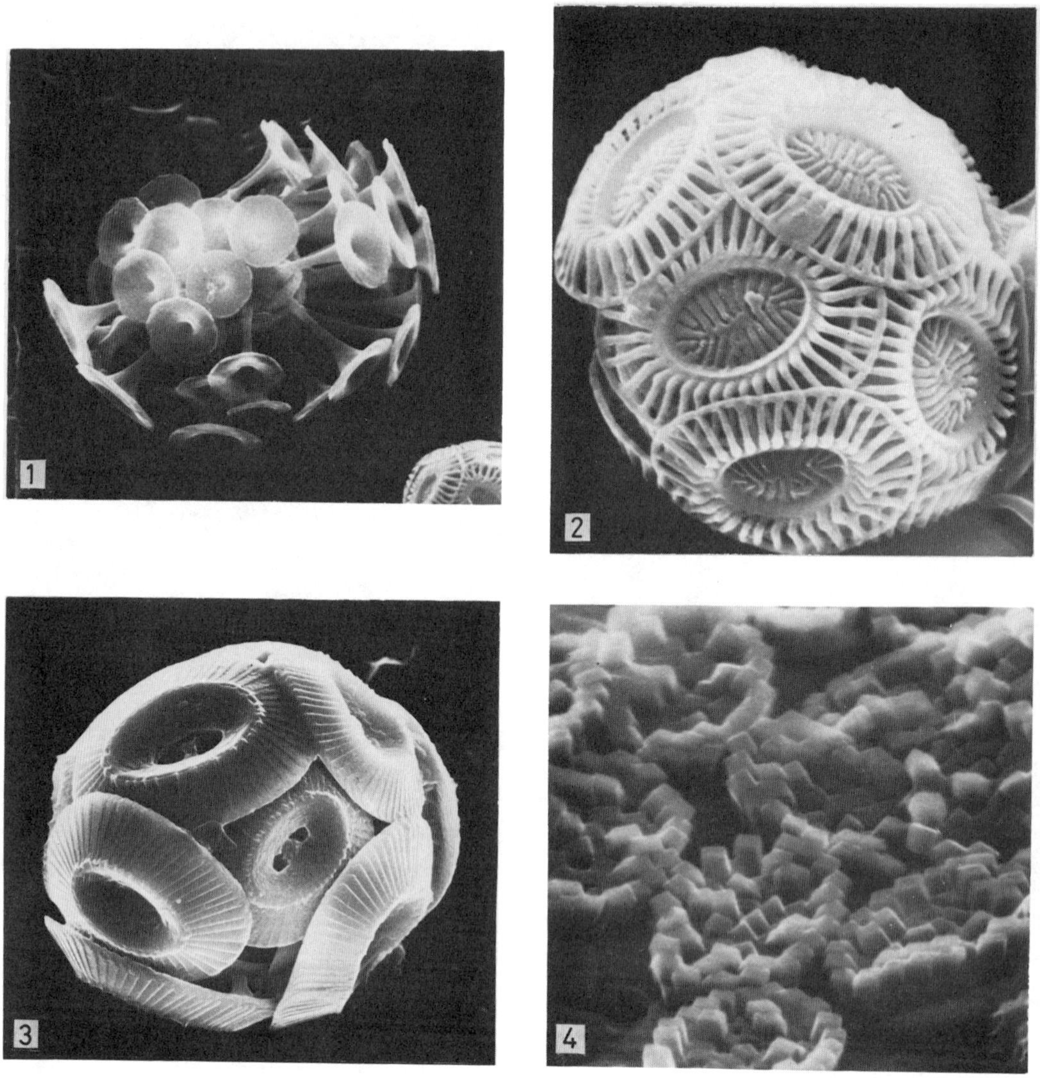


## Plate 7-IV

Coccolithophorids encountered in samples from Arctic waters (Halldal and Halldal, 1973; Braarud, 1979; Heimdal, 1983).

1. *Discosphaera tubifer* (× 3,780). Widely distributed in warmer seas.
2. *Emiliania huxleyi* (× 14,700). Cosmopolitan.
3. *Coccolithus pelagicus* (× 2,100). Nonmotile stage. Cold-water species showing a dimorphic life cycle involving a motile flagellated stage with holococcoliths and a nonmotile stage with heterococcoliths (Parke and Adams, 1960).
4. Holococcoliths of the flagellated stage, *Crystallolithus hyalinus* (× 16,800).

## REFERENCES


Balech, E. 1974. El genero *Protoperidinium* Bergh, 1881 (''*Peridinium*'' Ehrenberg, 1831, partim). *Rev. Mus. Arg. C. Nat. Bernardino Rivadavia Inst. Hidrobiol.* **4:**1–79.

Bech, P. -A. 1982. Planteplankton od primaerproduksjon i Ramfjorden og Tromsøysundet, 1980. *Cand. real. thesis,* University of Tromsø, 163p.

Berge, G. 1962. Discoloration of the sea due to *Coccolithus huxleyi* ''bloom.'' *Sarsia* **6:**27–40.

Bold, H. C., and M. J. Wynne. 1978. *Introduction to the Algae. Structure and Reproduction.* Prentice-Hall, Englewood Cliffs, N.J., 706p.

Braarud, T. 1935. The ''Øst'' expedition to the Denmark Strait 1929. II. The phytoplankton and its conditions of growth (including some qualitative data from the Arctic in 1930). *Hvalråd. Skr.* **10:**1–173.

Braarud, T. 1979. The temperature range of the non-motile stage of *Coccolithus pelagicus* in the North Atlantic Region. *Br. Phycol. J.* **14:**349–352.

Buckley, J. R., T. Gammelsrød, O. M. Johannessen, and L. P. Røed. 1979. Upwelling: Oceanic structure at the ice edge of the Arctic pack ice in winter. *Science* **203:**165–167.

Bursa, A. S. 1961. The annual oceanographic cycle at Igloolik in the Canadian Arctic. II. The phytoplankton. *J. Fish. Res. Bd. Canada* **18:**563–615.

Christensen, T. 1966. Alger. In *Botanik,* vol. 2. Systematisk Botanik, No. 2, 2nd ed. Munksgaard, København, 180p.

Christensen, T. 1980. *Algae. A Taxonomic Survey.* Fasc. 1. AiO Tryk as, Odense. 228p.

Cleve, P. T. 1867. Diatomaceer från Spetsbergen. *Öfvers. K. VetenskAkad. Förh.* **24:**661–670.

Cleve, P. T. 1873. On diatoms from the Arctic Sea. *Bih. K. Svenska VetenskAkad. Handl.* **1:**1–28.

Cleve, P. T. 1883. Diatoms collected during the expedition of the Vega. *Vega-Exped. Iakttagelser* **3:**455–517.

Cleve, P. T. 1896. Diatoms from Baffins Bay and Davis Strait. *Bih. K. Svenska VetenskAkad. Handl.* **22:**1–22.

Cleve, P. T. 1899. Plankton collected by the Swedish Expedition to Spitzbergen in 1898. *K. Svenska VetenskAkad. Handl.* **32:**1–51.

Cleve, P. T. 1900. Report on the plankton collected by the Swedish expedition to Greenland in 1899. *K. Svenska VetenskAkad. Handl.* **34:**1–21.

Cleve, P. T., and A. Grunow. 1880. Beiträge zur Kenntniss der Arctischen Diatomeen. *K. Svenska VetenskAkad. Handl.* **17:**1–122.

Codispoti, L. A. 1983. On nutrient variability and sediments in upwelling regions. In E. Suess and J. Thiede (eds.), *Coastal Upwelling.* Plenum Publishing Corporation, pp. 125–145.

Corlett, J. 1958. Measurements of primary production in the western Barents Sea. *Rapp. P.-v. Réun. Cons. Perm. Int. Explor. Mer* **144:**76–78.

Eilertsen, H. C., and J. P. Taasen. 1984. Investigations on the plankton community of Balsfjorden, northern Norway. The phytoplankton 1976–1978. Environmental factors, dynamics of growth, and primary production. *Sarsia* **69:**1–15.

Ellertsen, B., A. Hassel, H. Loeng, F. Rey, S. Tjelmeland, and D. Slagstad. 1982. Økologiske undersøkelser naer iskanten i Barentshavet somrene 1979 og 1980 (Ecological investigations in the marginal ice zone in the Barents Sea the summers 1979 and 1980). *Fisken Hav.* **3:**31–83. (In Norwegian; English summary.)

Ellertsen, B., H. Loeng, F. Rey, and S. Tjelmeland. 1981. ''Lodda på sommerbeite.'' Feltobservasjoner i 1979 og 1980 (The feeding conditions of capelin during summer. Field observations in 1979 and 1980). *Fisken Hav.* **3:**1–68. (In Norwegian; English summary.)

Eppley, R. W., and C. S. Weiler. 1979. The dominance of nanoplankton as an indicator of marine pollution: A critique. *Oceanologica Acta* **2:**241–245.

Falkowski, P. G. 1980. Light-shade adaptation in marine phytoplankton. In P. G. Falkowski (ed.), *Primary Productivity in the Sea.* Plenum Press, New York and London, pp. 99–119.

Gran, H. H. 1897*a*. Bacillariaceen vom kleinen Karajakfjord. *Bibl. Bot.* **42:**13–24.

Gran, H. H. 1897*b*. Bemerkungen über das Plankton des Arktischen Meeres. *Ber. Dt. Bot. Ges.* 15:132–136.

Gran, H. H. 1900. Diatomaceae from the icefloes and plankton of the Arctic Ocean. *Norw. North Polar Exped. 1893–1896, 1904* **4:**1–74.

Gran, H. H. 1904. Die diatomeen der Arktischen Meeres. 1. Teil: Die diatomeen des planktons. *Fauna Arct.* **3:**509–554.

Gran, H. H. 1905. Diatomeen. *Nord. Plankt., Bot. Teil* **19:**1–146.

Grøntved, J., and G. Seidenfaden. 1938. The phytoplankton of the waters west of Greenland. *Medd. Grönland* **82:**1–380.

Halldal, P., and K. Halldal. 1973. Phytoplankton, chlorophyll, and submarine light conditions in Kings Bay, Spitsbergen, July 1971. *Norw. J. Bot.* **20:**99–108.

Hallegraeff, G. M. 1981. Seasonal study of phytoplankton pigments and species at a coastal station off Sydney: Importance of diatoms and the nanoplankton. *Marine Biol.* **61**:107–118.

Haq, B. U. 1978. Calcareous nannoplankton. In B. U. Haq and A. Boersma (eds.), *Introduction to Marine Micropaleontology*. Elsevier, New York, pp. 79–107.

Harrison, W. G., T. Platt, and B. Irwin. 1982. Primary production and nutrient assimilation by natural phytoplankton populations of the eastern Canadian Arctic. *Can. J. Fish. Aquat. Sci.* **39**:335–345.

Hasle, G. R. 1965. *Nitzschia* and *Fragilariopsis* species studied in the light and electron microscopes. II. The group Pseudonitzschia. *Skr. norske Vidensk-Akad. I . Mat.-naturv. Kl. N. S.* **18**:1–45.

Hasle, G. R. 1969. An analysis of the phytoplankton of the Pacific Southern Ocean: Abundance, composition, and distribution during the Brategg Expedition, 1947–1948. *Hvalråd. Skr.* **52**:1–168.

Hasle, G. R. 1972. *Fragilariopsis* Hustedt as a section of the genus *Nitzschia* Hassall. *Nova Hedw., Beih.* **39**:11–119.

Hasle, G. R. 1974. Validation of the names of some marine planktonic species of *Nitzschia* (Bacillariophyceae). *Taxon* **23**:425–428.

Hasle, G. R. 1976. The biogeography of some marine planktonic diatoms. *Deep-Sea Res.* **23**:319–338.

Hassel, A., H. Loeng, and H. R. Skjoldal. 1986. Marinøkologiske undersøkelser i Barentshavet i januar 1985 (Ecological investigations in the Barents Sea in January 1985). *Havforskningsinstituttet. Rep. FO 8604,* 33p.

Hay, W. W., H. P. Mohler, P. H. Roth, R. R. Schmidt, and J. E. Boudreaux. 1967. Calcareous nannoplankton zonation of the Cenozoic of the Gulf coast and Caribbean-Antillean area, and Transoceanic correlation. *Gulf Coast. Assoc. Geol. Soc. Trans.* **17**:428–480.

Heimdal, B. R. 1974. Composition and abundance of phytoplankton in the Ullsfjord area, North Norway. *Astarte* **7**:17–42.

Heimdal, B. R. 1983. Phytoplankton and nutrients in the waters north-west of Spitsbergen in the autumn of 1979. *J. Plankton Res.* **5**:901–918.

Hendey, N. I. 1937. The plankton diatoms of the Southern Seas. *Discovery Rep.* **16**:1–317.

Hensen, V. 1887. Über die bestimmung des planktons. *Ber. Kommis. Wiss. Untersuch. deutschen meere in Kiel, Jahre 1882–6,* **12–16**:1–108.

Hewes, C. D. 1986. Nanoplankton biomass in the Barents Sea. In H. Loeng (ed.), *Ecological Investigations in the Barents Sea, August 1985.* Rep. PRO MARE-Cruise No. 5, pp. 63–68.

Holmes, R. W. 1956. The annual cycle of phytoplankton in the Labrador Sea, 1950–51. *Bingham Oceanogr. Coll. Bull.* **16**:1–74.

Horner, R. 1976. Sea ice organisms. *Oceanogr. Marine Biol. Ann. Rev.* **14**:167–182.

Horner, R. 1977. History and recent advances in the study of ice biota. In M. J. Dunbar (ed.), *Polar Oceans.* Arctic Institute of North America, pp. 269–282.

Hustedt, F. 1930. Die kieselalgen deutschlands, Østerreichs und der schweiz. *Kryptogamenflora* **7**(1):1–920.

Johannessen, O. M., J. A. Johannessen, J. Morison, B. A. Farrelly, and E. A. S. Svendsen. 1983. Oceanographic conditions in the marginal ice zone north of Svalbard in early fall 1979 with an emphasis on mesoscale processes. *J. Geophys. Res.* **88**:2755–2769.

Kirk, J. T. O. 1983. *Light and Photosynthesis in Aquatic Ecosystems.* Cambridge University Press, 401p.

Kreps, E., and N. Verjbinskaya. 1930. Seasonal changes in the phosphate and nitrate content and in hydrogen ion concentration in the Barents Sea. *J. Cons. Perm. Int. Explor. Mer* **5**:329–346.

Loeblich, A. R., III. 1969. The amphiesma or Dinoflagellate Cell Covering. *North American Paleontological Convention Proceedings, part G.* pp. 867–929.

Loeng, H. (ed.). 1986. *Ecological Investigations in the Barents Sea, August 1985.* Rep. PRO MARE-Cruise No. 5, 125p.

McIntyre, A., and A. W. H. Bé. 1967. Modern coccolithophoridae of the Atlantic Ocean. I. Placoliths and cyrtoliths. *Deep-Sea Res.* **14**:561–597.

McPhee, M. G. 1984. MIZEX 84 upper ocean turbulence experiment. *EOS* **65**:935.

McRoy, C. P., and J. T. Goering. 1974. The influence of ice on the primary productivity of the Bering Sea. In D. Hood and E. Kelley (eds.), *The Oceanography of the Bering Sea.* Institute of Marine Science, University of Alaska, pp. 403–421.

Malone, T. C. 1980. Size-fractionated primary productivity of marine phytoplankton. In P. G. Falkowski (ed.), *Primary Productivity in the Sea.* Plenum Press, New York and London, pp. 301–319.

Manton, I., J. Sutherland, and K. Oates. 1976. Arctic coccolithophorids: Two species of *Turrisphaera* gen. nov. from west Greenland, Alaska, and the Northwest Passage. *R. Soc. London Proc.,* ser. B, **194**:179–194.

Manton, I., J. Sutherland, and K. Oates. 1977. Arctic coccolithophorids: *Wigwamma arctica* gen. et sp. nov. from Greenland and Arctic Canada, *W. annulifera* sp. nov. from South

Africa and S. Alaska and *Calciarcus alaskensis* gen. et sp. nov. from S. Alaska. *R. Soc. London Proc.*, ser. B, **197:**145–168.

Marshall, P. T. 1957. Primary production in the Arctic. *J. Cons. Perm. Int. Explor. Mer* **23:**173–177.

Marshall Darley, W. 1982. Algal biology: A physiological approach. In J. F. Wilkinson (ed.), *Basic Microbiology*, vol. 9. Blackwell Scientific Publications, Oxford, London, pp. 1–168.

Mel'nikov, I. A. 1976*a*. Hydrobiological investigations in the central Arctic Ocean. *Oceanology* **16:**314–315.

Mel'nikov, I. A. 1976*b*. Hydrobiological investigations in the central part of the Arctic Ocean (spring 1976). *Oceanology* **16:**647.

Mel'nikov, I. A. 1978. Hydrobiological research in the central Arctic basin (spring 1978). *Oceanology* **18:**748.

Mel'nikov, I. A., and V. D. Tsinovskiy. 1978. Hydrobiological studies in the northern Arctic Ocean on NP-23 (May–October 1977). *Oceanology* **18:**246–247.

Nemoto, T., and G. Harrison. 1981. High latitude ecosystems. In A. R. Longhurst (ed.), *Analysis of Marine Ecosystems.* Academic Press, London, pp. 95–126.

Nusch, E. A. 1980. Comparison of different methods for chlorophyll and phaeopigment determination. *Arch. Hydrobiol. Ergebn. Limnol.* **14:**14–36.

O'Meara, E. 1860. On some diatomaceous forms from the Arctic regions, collected during the late voyage of the Fox. *J. Royal Dublin Soc.* **3:**59–60.

Ostenfeld, C. H. 1910. Marine plankton from the east-Greenland Sea (W. of 6° W. long. and N. of 73°30'N. lat.) collected during the "Danmark Expedition" 1906–1908. I. List of diatoms and flagellates. *Medd. Grönland* **43:**259–285.

Ostenfeld, C. H., and O. Paulsen. 1911. Marine plankton from the east-Greenland Sea (W. of 6°W. long. and N. of 73°30'N. lat.) collected during the "Danmark Expedition" 1906–1908. IV. General remarks on the microplankton. *Medd. Grönland* **43:**321–336.

Østrup, E. 1895. Marine diatoméer fra Østgrønland. *Medd. Grönland* **18:**395–476.

Parke, M., and I. Adams. 1960. The motile (*Crystallolithus hyalinus* Gaarder & Markali) and non-motile phases in the life history of *Coccolithus pelagicus* (Wallich) Schiller. *J. Marine Biol. Ass. U. K.* **39:**263–274.

Parsons, T. R., Y. Maita, and C. M. Lalli. 1984. *A Manual of Chemical and Biological Methods of Seawater Analysis.* Pergamon Press, Oxford, 173p.

Parsons, T. R., M. Takahashi, and B. Hargrave. 1977. *Biological Oceanographic Processes,* 2nd ed. Pergamon Press, Oxford, 332p.

Paulsen, O. 1910. Marine plankton from the east-Greenland Sea (W. of 6°W. long. and N. of 73°30'N. lat.) collected during the "Danmark Expedition" (1906–1908). III. Peridiniales. *Medd. Grönland* **43:**303–318.

Pavshtiks, E. A. 1968. The influence of currents upon seasonal fluctuations in the plankton of the Davis Strait. *Sarsia* **34:**383–392.

Platt, T., and D. V. Subba Rao. 1975. Primary production of marine microphytes. In J. P. Cooper (ed.), *Photosynthesis and Productivity in Different Environments.* Cambridge University Press, Cambridge, pp. 249–280.

Platt, T., D. V. Subba Rao, and B. Irwin. 1983. Photosynthesis of picoplankton in the oligotrophic ocean. *Nature* **301:**702–704.

Platt, T., W. G. Harrison, B. Irwin, E. P. Horne, and C. L. Gallegos. 1982. Photosynthesis and photoadaptation of marine phytoplankton in the Arctic. *Deep-Sea Res.* **29:**1159–1170.

Ramsfjell, E. 1960. Phytoplankton distribution in the Norwegian Sea in June, 1952 and 1953. *FiskDir. Skr. Ser. Havunders.* **12:**1–112.

Raymont, J. E. G. 1980. *Plankton and Productivity in the Oceans. 1. Phytoplankton,* 2nd ed. Pergamon Press, Oxford, 489p.

Rey, F., and H. Loeng. 1985. The influence of ice and hydrographic conditions on the development of phytoplankton in the Barents Sea. In J. S. Gray and M. E. Christiansen (eds.), *Marine Biology of Polar Regions and Effects of Stress on Marine Organisms.* John Wiley & Sons Ltd., pp. 49–63.

Reynolds, N. 1973. The estimation of the abundance of ultraplankton. *Br. Phycol. J.* **8:**135–146.

Round, F. E. 1973. *The Biology of the Algae.* Edward Arnold, London, 278p.

Sakshaug, E. 1987. *The Physiological Ecology of Polar Phytoplankton.* Contr. No. 239. The Biological Station, Trondheim, 48 p.

Sakshaug, E., and E. N. Hegseth. 1985. Algae in Ice and Polynyas. *PRO MARE Annual Rep. 1984,* pp. 75–77.

Sakshaug, E., and O. Holm-Hansen. 1984. Factors governing pelagic production in polar oceans. In O. Holm-Hansen, L. Bolis, and R. Gilles (eds.), *Marine Phytoplankton and Productivity. Lecture Notes on Coastal and Estuarine Studies,* vol. 8. Springer-Verlag, Berlin, pp. 1–18.

Schei, B., H. C. Eilertsen, S. Falk-Petersen, B. Gulliksen, and J. P. Taasen. 1979. Marinbiologiske undersøkelser i Van Mijenfjorden (Vest-Spitsbergen) etter oljelekkasje ved Sveagruva 1978. *Tromura* **2:**1–50.

Sieburth, J. M., V. Smetacek, and J. Lenz. 1978. Pelagic ecosystem structure: Heterotrophic compartments of the plankton and their relationship to plankton size fractions. *Limnol. Oceanogr.* **23**:1256–1263.

Smayda, T. J. 1958*a*. Biogeographical studies of marine phytoplankton. *Oikos* **9**:158–191.

Smayda, T. J. 1958*b*. Phytoplankton studies around Jan Mayen Island March–April, 1955. *Nytt Mag. Bot.* **6**:75–96.

Smith, W. 1985. MIZEX biological oceanography meetings. *MIZEX Newsl.* **3**:5–6.

Sokolova, S. A., and A. A. Solov'yeva. 1971. Primary production in Dal'nezelenetskaya Bay (Murmansk coast) in 1967. *Oceanology* **11**:386–395.

Steemann Nielsen, E. 1952. The use of radioactive carbon ($C^{14}$) for measuring organic production in the sea. *J. Cons. Perm. Int. Explor. Mer.* **17**:117–140.

Subba Rao, D. V., and T. Platt. 1984. Primary production of Arctic waters. *Polar Biol.* **3**:191–201.

Sverdrup, H. U. 1953. On conditions for the vernal blooming of phytoplankton. *J. Cons. Perm. Int. Explor. Mer.* **18**:287–295.

Syvertsen, E. E., and G. R. Hasle. 1983. The diatom genus *Eucampia:* Morphology and taxonomy. *Bacillaria* **6**:169–210.

Syvertsen, E. E., and G. R. Hasle. 1984. *Thalassiosira bulbosa* Syvertsen, sp. nov., an Arctic marine diatom. *Polar Biol.* **3**:167–172.

Tappan, H. 1980. *The Paleobiology of Plant Protists.* W. H. Freeman and Company, San Francisco, 1,028p.

Thomsen, H. A. 1981. Identification by electron microscopy of nannoplanktonic coccolithophorids (Prymnesiophyceae) from west Greenland, including the description of *Papposphaera sarion* sp. nov. *Br. Phycol. J.* **16**:77–94.

Throndsen, J. 1969. Flagellates of Norwegian coastal waters. *Nytt Mag. Bot.* **16**:161–216.

Throndsen, J. 1970. Flagellates from Arctic waters. *Nytt Mag. Bot.* **17**:49–57.

Throndsen, J. 1976. Occurrence and productivity of small marine flagellates. *Norw. J. Bot.* **23**:269–293.

Throndsen, J., and B. R. Heimdal. 1976. Primary production, phytoplankton and light in Straumsbukta near Tromsø. *Astarte* **9**:51–60.

Vanhöffen, E. 1893. Frühlingsleben in Nord-Grönland. *Verh. Ges. Erdk. Berl.* **20**:454–469.

Vedernikov, V. I. 1975. Dependence of the assimilation number and concentration of chlorophyll *a* on water productivity in different temperature regions of the world ocean. *Oceanology* **15**:484–485.

Vedernikov, V. I., and A. A. Solov'yeva. 1972. Primary production and chlorophyll in the coastal waters of the Barents Sea. *Oceanology* **12**:559–565.

Weeks, W. F. 1976. Sea ice conditions in the Arctic. *AIDJEX Bull.* **34**:173–206.

Wind, F. H., and S. W. Wise. 1978. Mesozoic holococcoliths. *Geology* **6**:140–142.

Zenkevitch, L. 1963. *Biology of the Seas of the USSR.* Allen and Unwin Ltd., London, 995p.

# 8: Foraminifera and Pteropoda Beneath the Arctic Sea Ice: New Distributions

Yvonne Herman and
Ole G. N. Andersen

## INTRODUCTION

The usefulness of planktonic foraminifera and pteropods as biological indicators has long been established. Due to their specific environmental requirements, most single-species populations, as well as species assemblages, characterize various water masses and circulation patterns. Ecological biogeographers have mapped the limits of ranges of numerous taxa and have drawn broad ecological inferences. Being planktonic, both foraminifera and pteropods float passively, or nearly so, with currents. Consequently, low-latitude species are transported by currents such as the Gulf Stream and its ramifications to distant areas beyond their habitat. Knowledge of pteropod and planktonic foraminiferal distribution in polar–subpolar regions based on studies from limited geographic areas is scanty.

*Thecosomatous Pteropods:* Data on thecosome distribution was summarized by van der Spoel (1967). A literature review published more recently (Bé and Gilmer, 1977) reveals that comparatively little has been published on high-latitude species in the last decade. Even approximate assessments of geographic distribution, temperature–salinity or depth ranges are difficult for many species based on data from the literature. Some of the difficulties encountered stem from the different sampling devices and sampling intervals used by various investigators, enormous areal gaps in data, as well as the different methods of presenting results.

One thecosomate pteropod, *Limacina helicina* (Plate 8-I) is known to inhabit the Arctic and Antarctic oceans (Meisenheimer, 1905*a;* Chen, 1966). This species has also been reported from subpolar regions: In the North Pacific it is common north of ~45°N latitude in the Subarctic Water mass, the Gulf of Alaska water, and the cold California Current (Tesch, 1948; McGowan, 1960, 1963). McGowan recognizes intraspecific variations in *L. helicina,* with varieties A and B, each having distinct geographic and depth distributional ranges in the North Pacific. In the North Atlantic, *L. helicina* was reported by several authors, north of the Grand Banks in the Norwegian, Greenland, and Labrador seas, as well as in the Hudson and Baffin bays and Davis Strait (Bonnevie, 1913; Kerswill, 1940; Kramp, 1961; Chen, 1962). In the Southern Hemisphere it is most abundant south of the Antarctic Polar Front (Meisenheimer, 1906), becoming gradually less common; its northern limit coincides roughly with the Subtropical Convergence (Chen, 1966; van der Spoel, 1967; Boltovskoy, 1971*a,b*, 1973). The presence of *L. helicina* in subtropical–tropical zones may be due to its ability to inhabit deep cold water (submergence) or to current transport (Meisenheimer, 1905*b*; van der Spoel, 1967).

*Limacina retroversa retroversa* and *L. retroversa balea* are subpolar–temperate epiplanktonic forms (Plate 8-I) (Chen, 1966; van der Spoel, 1967 and references therein), the latter adapted to lower temperatures. The geographic and depth ranges of both formae are given by several authors and are summarized by van der Spoel (1967). *Clio pyramidata antarctica* (Plate 8-I) is found north of the Antarctic Polar Front in the Subantarctic Water mass; its depth distribution and temperature range have not been firmly established (van der Spoel, 1967). *Clio pyramidata sulcata* (Plate 8-I) is able to tolerate lower temperatures than *C. pyramidata antarctica,* inhabiting Antarctic and sub-Antarctic waters (Chen, 1966; van der Spoel, 1967 and references therein); its depth distribution and temperature range have not been conclusively determined (van der Spoel, 1967). Pteropod distribution in Quaternary sediments of the world ocean has been studied by Herman (1971).

*Planktonic Foraminifera:* Knowledge of the distribution of Arctic foraminifera is based on studies from a limited geographic area, collected along the track of drifting ice stations in the Amerasian Basin (Bé, 1960; Tibbs, 1967; Vilks, 1970). Only one species, sinistral (left-coiling), *Globigerina pachyderma* (Plate 8-II), was reported from water samples. The same species has been recorded from the Antarctic Ocean (Chen, 1966; Bé, 1969). Subpolar species are dominated by *Globigerina bulloides.* Other subpolar taxa are *Globigerina quinqueloba,* dextral (right-coiling) *G. pachyderma, Globigerinita uvula,* and *Globigerinita glutinata* (Plate 8-II) (Bé, 1977 and bibliography therein).

*G. pachyderma* is the best indicator species of polar water. The percentage ratios of left- versus right-coiling individuals are apparently temperature dependent. Ericson (1959) discovered that sinistral tests constitute at least 90 percent of the *G. pachyderma* fauna in the modern sediments of the Arctic, Labrador, and Norwegian–Greenland seas, while dextral populations predominate in the Subarctic and Transition zones (Bé, 1977). Left- and right-dominant coiling provinces in North Atlantic fossil assemblages are separated by the present-day 7.2°C April surface water isotherm (Ericson, 1959). Similar distributions of living populations in the Northern and Southern hemispheres were found by Bé and Tolderlund (1971) and Bé (1977).

## MATERIAL AND METHODS

The plankton samples were collected in 1979 during the FRAM I Expedition from drifting sea ice off northeast Greenland at approximately 84°24′N and 06°00′W (Table 8-1). After an initial northward movement, the ice station drifted mainly southward, covering nearly 300 km between March 11 and May 13. The zooplankton was sampled between early April and May 1979 (Tables 8-1 and 8-2) by drilling holes with a 15 cm "snail" drill through the 2–2.5-m-thick platform while the ice floe drifted roughly southward. The plankton was sampled quantitatively with a Jabsco rotary pump, with pre-pump filtration (80 $\mu$m) of 1–11 $m^3$. The specimens were preserved in 2 percent formaldehyde buffered with borax (pH ~8.4). Sixty-one aliquots were prepared for foraminiferal and pteropodal studies. Samples were collected from water depths ranging between 2.5

Table 8-1. Station Data and Number of Pteropods per Sample

| Coordinates: | 84°39'N | 84°35'N | 84°33'N | 84°29'N | 84°31'N | 84°31'N | 84°19'N | 84°18'N | 84°17'N | 84°17'N | 84°10'N | 83°50'N |
|---|---|---|---|---|---|---|---|---|---|---|---|---|
| | 9°30'W | 9°4'W | 8°46'W | 8°55'W | 9°8'W | 9°37'W | 8°20'W | 8°12'W | 8°1'W | 7°58'W | 7°30'W | 6°55'W |
| Date: | 4/4-6/79 | 4/11/79 | 4/12/79 | 4/13/79 | 4/14-15/79 | 4/18/79 | 4/23/79 | 4/24/79 | 4/26/79 | 4/26-28/79 | 4/30/79-5/4/79 | 5/6-7/79 |
| | A / B / C | A / B / C | A / B / C | A / B / C | A / B / C | A / B / C | A / B / C | A / B / C | A / B / C | A / B / C | A / B / C | A / B / C |
| | | | | | | | | | | | | |
| Depth (m) | | | | | | | | | | | | |
| 2, 5 | 20 / 2 / 2 | | | | 10 / 5 / 5 | | | 36 / 11 / 7 | | | 5 / 3 / 3 | |
| 5 | | | | | 10 / 1 / 1 | | | | | 6 / 1 / 0 | 3 / 2 / 0 | 40 / 10 / 5 |
| 7, 5 | 5 / 1 / 0 | | | | | | | | | | | |
| 10 | | | | | 20 / 2 / 2 | | | | | 35 / 7 / 7 | | |
| 12, 5 | 10 / 1 / 0 | | | | | | | | | | | |
| 15 | | | | | | 6 / 7 / 5 | | | | | | |
| 17, 5 | 20 / 2 / 2 | | | | | | | | | | | |
| 20 | | | | | 20 / 2 / 2 | | 10 / 1 / 0 | | | | 20 / 2 / 2 | 33 / 5 / 5 |
| 22, 5 | 23 / 7 / 1 | | | | | | | | | | | |
| 25 | | 0 | 7 / 1 / 1 | 2 / 2 / 2 | | | | | | | | |
| 30 | | | | | | | 10 / 1 / 1 | | | | | |
| 32, 5 | 10 / 2 / 2 | | | | | | | | | | | |
| 40 | | | ? / 1 / 1 | | | | | | | | | |
| 42, 5 | | | | | | | | | | | | |
| 45 | | | | | | | | | | | | |
| 50 | | 0 | | | 8 / 1 / 1 | | | 2 / 1 / 1 | | | | |
| 52, 5 | | | | | | | | | | | | |
| 60 | | | | | | | | | | | | |
| 70 | | | | | | | | | | | | |
| 80 | | | | | | | | | | | | |
| 90 | | | | | | | | | | | | |
| 100 | | | | | | | | | | | | |
| | | | | | | | | | | | | |
| Legend: | A. number of specimens per 10 m3 | | | | | | | | | | | |
| | B. number of specimens seen | | | | | | | | | | | |
| | C. number of specimens in aliquot | | | | | | | | | | | |

Table 8-2. Station Data and Number of Foraminifera per Sample

| Date: | 4/4-6/79 | 4/11/79 | 4/12/79 | 4/13/79 | 4/14/79 | 4/18/79 | 4/23/79 | 4/24/79 | 4/26/79 | 4/26-28/79 | 4/30/79-5/3/79 | 5/4-7/79 |
|---|---|---|---|---|---|---|---|---|---|---|---|---|
| | A / B / C | A / B / C | A / B / C | A / B / C | A / B / C | A / B / C | A / B / C | A / B / C | A / B / C | A / B / C | A / B / C | A / B / C |
| Water Depth (m) | | | | | | | | | | | | |
| 2, 5 | 0 | | | | | | | 0 | | 0 | 5 / 3 / 2 | |
| 5 | | | | | 20 / 2 / 1 | | 0 | | | 0 | 22 / 13 / 5 | 20 / 5 / 2 |
| 7, 5 | 4.8 / 1 / 1 | | | | | | | | | | | |
| 10 | | | | | 20 / 2 / 1 | | 0 | | | 60 / 3 / 3 | 12 / 4 / 4 | 80 / 10 / 1 |
| 12, 5 | 0 | | | | | | | | | | | |
| 15 | | | | | | 0 | | | | | | |
| 17, 5 | 0 | | | | | | | | | | | |
| 20 | | | | | 10 / 1 / 1 | | 0 | | | 10 / 1 / 1 | 10 / 1 / 1 | 6.7 / 1 / 1 |
| 22, 5 | 21 / 7 / 1 | | | | | | | | | | | |
| 25 | | D | 0 | ? / ? / 1 | | | | | | | | |
| 30 | | | | | 33 / 5 / 4 | | 0 | | | 5 / 1 / 1 | 0 | 0 |
| 32, 5 | 10 / 2 / 1 | | | | | | | | | | | |
| 40 | | | | | ? / ? / 2 | | 0 | | | 0 | 0 | 0 |
| 42, 5 | 0 | | | | | | | | | | | |
| 45 | | | | | 13 / 3 / 3 | | | | | | | |
| 50 | | D | | | 0 | 1.4 / 1 / 1 | 0 | ? / ? / 1 | 5.7 / 2 / 2 | 0 | 0 | 50 / 5 / 4 |
| 52, 5 | 0 | | | | | | | | | | | |
| 60 | | | | | | | | | | 10 / 1 / 1 | 0 | 0 |
| 70 | | | | | | | | | | 0 | 20 / 2 / 2 | 0 |
| 80 | | | | | | | | | | ? / ? / 2 | 20 / 3 / 3 | 5 / 1 / 1 |
| 90 | | | | | | | | | | 10 / 1 / 1 | 20 / 2 / 2 | |
| 100 | | | | | | | | | | 16 / 2 / 2 | 0 | 25 / 5 / 5 |

Legend:
A. number of specimens per 10m3
B. number of specimens seen
C. number of specimens in aliquot
D. samples not studied
Coordinates are the same as in Table 1

and 100 m, and from ice cores drilled through sea ice. Their distribution is presented in Tables 8-1 through 8-4 and Figure 8-1*a,b*. The abundances of planktonic foraminifera range from 2 to 80 per 10 $m^3$ and that of pteropods from 2 to 40 per 10 $m^3$.

## HYDROGRAPHY AND PHYTOPLANKTON IN THE STUDY AREA

A typical high-Arctic sub-ice profile with a surface layer of 40–50 m above steep thermoclines and haloclines was present. This layer was periodically disturbed by eddies of colder, clearer, less saline water of lower nutrient content, first detected at 40–70 m in mid-April. This finally replaced the surface layer down to ~50 m by late April–early May, when chlorophyll concentrations increased in the upper few meters. Ice salinities were 4.0–6.7 ‰ from top to bottom.

Primary production proved too low to be measurable with the technique used (in situ, $^{14}C$); chlorophyll increased slightly during late April and early May, especially in the upper 10–20 m (Fig. 8-1*c*). In the lower 4.5 cm of the sea ice, chlorophyll showed an appreciable increase, especially from late April on, rising from 0.5 mg chlorophyll *a* $m^2$ on April 11 to 2.7 mg $m^2$ on May 2.

## ZOOPLANKTON

As mentioned previously, only one pteropod, *L. helicina,* and one planktonic foraminifer, sinistral *G. pachyderma,* have been previously reported in Arctic Ocean water.

In the samples collected for the present study we found several species of planktonic foraminifera, one pteropod, and one benthic foraminifera in surface water beneath the permanent ice cover and in the sea ice (Tables 8-1 through 8-4); these are sinistral *Globigerina cryophila;* sinistral *Globigerina egelida;* dextral *Globigerina pachyderma;* sinistral *G. pachyderma* and *G. quinqueloba;* one pteropod, *L. helicina;* and one benthic foraminifer, *Cibicides lobatulus* (Tables 8-3 and 8-4; Plates 8-I and 8-II). The samples were collected from water depths ranging between 2.5 and 100 m, and from ice cores drilled through sea ice. The abundance of planktonic foraminifera range from ~2 to 80 per 10 $m^3$ and that of pteropods from 2 to 40 per 10 $m^3$. The average number of foraminifers above 50 m increases in samples collected after April 25. The same is true for pteropods whose numbers also increase after April 24, coinciding with the increase in chlorophyll *a* concentration.

Sinistral *G. pachyderma* occurs in many samples taken between the first and last days of collection (April 5–May 7) in water depth ranging between 5 and 100 m; >55 percent of the examined specimens were juvenile (Table 8-3). The second most common species is sinistral *G. cryophila,* which occurs in samples taken between April 27 and May 7, in water depth ranging from 2.5 to 100 m (Table 8-3). All the specimens are adult, suggesting different reproductive and growth periods for these two closely related taxa (Herman, 1974). Sinistral *G. egelida* ranged third in abundance. Only juvenile individuals were observed; they occur in samples collected between April 26 and May 5 in water depth ranging between 5 and 10 m. A juvenile sinistral *G. quinqueloba* was present in a sample that was taken on April 5 from a depth of 32.5 m (Table 8-3). *G. egelida* was absent in the sample in which *G. quinqueloba* was found, also suggesting that these two closely related taxa may have different reproductive and growth seasons. Dextral adult and juvenile *G. pachyderma* were observed in several samples (Table 8-3). A juvenile sinistral *G. pachyderma* was found in the bottom of an ice core sample, while two specimens of the benthic foraminifer *Cibicides lobatulus*

Table 8-3. Foraminifera; Eurasian Basin, Arctic Ocean

| Date | Water Depth (m) | *Globigerina cryophila(s)* | *Globigerina pachyderma(s)* | *Globigerina pachyderma(d)* | *Globigerina egelida(s)* | *Globigerina quinqueloba(s)* | Globigerina sp. (juvenile) | *Cibicides lobatulus* |
|---|---|---|---|---|---|---|---|---|
| 4/5/79 | 7.5 | | 1(J) | | | | | |
| 4/5/79 | 22.5 | | | | | | 1(J) | |
| 4/5/79 | 32.5 | | | | | 1(J) | | |
| 4/23/79 | 25 | | | | | | 1(J) | |
| 4/14/79 | 5 | | | | | | 1(J) | |
| 4/14/79 | 10 | | | | | | 1(J)s | |
| 4/14/79 | 30 | | 1(J) | | | | 2(J)s | |
| 4/14/79 | 40 | | 1(J) | | | | 1(J)s | |
| 4/14/79 | 45 | | 2(J) | | | | | |
| 4/18/79 | 50 | | 1(J) | | | | | |
| 4/26/79 | 50 | | | 1(J) | | | | |
| 4/27/79 | 80 | 1(A) | 1(J) | | | | | |
| 4/27/79 | 90 | | | 1(J) | | | | |
| 4/27/79 | 100 | 1(A) | 1(A) | | | | | |
| 4/28/79 | 20 | | 1(J) | | | | | |
| 4/28/79 | 30 | | | | | | | |
| 4/28/79 | 60 | 1(A) | | | | | | |
| 4/30/79 | 5 | 1(A) | | | | | | |
| 4/30/79 | 90 | | | | | | | |
| 5/1/79 | 5 | 1(A) | 3(A) | | | | | |
| 5/2/79 | 70 | | 1(J) | | | | | |
| 5/2/79 | 80 | 1(A) | | | | | | |
| 5/3/79 | 5 | | 2(A) | 2(A&J) | 1(J) | | | |
| 5/3/79 | 10 | | | | 2(J) | | 1(J)s | |
| 5/3/79 | 20 | | | | | | | |
| 5/4/79 | 2.5 | 1(A) | | | | | | |
| 5/5/79 | 80 | | | | | | | |
| 5/5/79 | 100 | | | | | | 3(J)s | |
| 5/7/79 | 5 | | | | 1(J) | | | |
| 5/7/79 | 10 | | | | | | | |
| 5/7/79 | 20 | | | | | | | |
| 5/7/79 | 50 | 2(A) | 2(A) | | | | | |
| Bottom few cm of ice | | | 1(J) | | | | | |
| Upper 2m of ice 4/18 | | | | | | | | 2(J) |
| | | | | | | | | |
| Legend: | A = Adult | | | | | | | |
| | B = Juvenile | | | | | | | |
| | s = Sinistral | | | | | | | |
| | d = Dextral | | | | | | | |

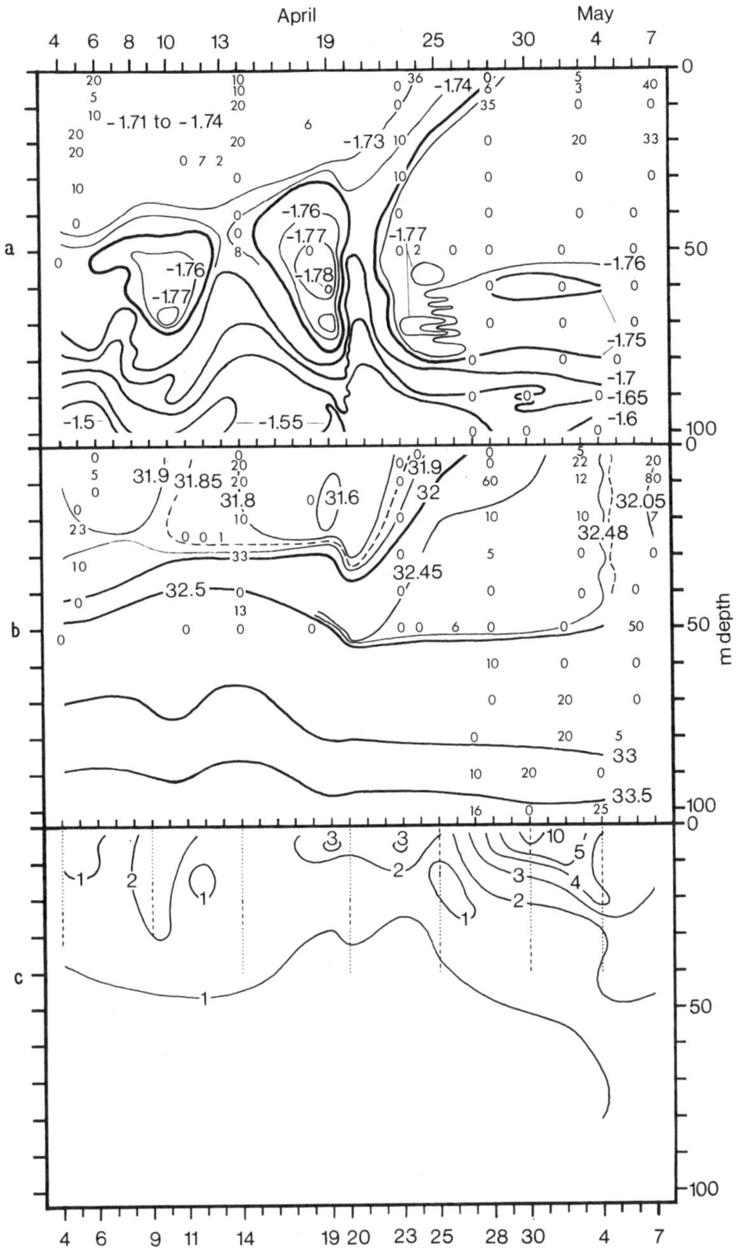


Figure 8-1. *a:* Isopleth diagram of temperatures (°C) from CTD casts made on the dates indicated on the abscissa at the top, with numbers of pteropod *Limacina helicina* per 10 $m^3$ quantitative pump samples; *b:* Same for salinities (‰) with numbers of foraminifera per 10 $m^3$; *c:* Isopleth diagram of chlorophyll *a* (10 $\mu g \cdot m^{\div 3}$), with extinction coefficients for light quanta (400–700 nm) indicated (. . . = >0.1; --- <0.1). Dates for chlorophyll measurements are indicated on the abscissa at the bottom of the diagram.

Table 8-4. *Limacina helicina*, Eurasian Basin, Arctic Ocean

| *Date* | *Water Depth* | *Number of Specimens Recorded* |
|---|---|---|
| 4/5/79 | 22.5 | 4 (J) |
| 4/5/79 | 32.5 | 1 (J) |
| 4/6/79 | 2.5 | 1 (J) |
| 4/6/79 | 7.5 | 1 (J) |
| 4/6/79 | 17.5 | 1 (J) |
| 4/12/79 | 25.0 | 1 (J) |
| 4/12/79 | 40.0* | 1 (A) |
| 4/13/79 | 25.0 | 2 (J) |
| 4/14/79 | 5.0 | 1 (J) |
| 4/14/79 | 10.0 | 1 (J) |
| 4/14/79 | 20.0 | 2 (J) |
| 4/14/79 | 50.0 | 1 (J) |
| 4/15/79 | 2.5 | 4 (J) |
| 4/18/79 | 15.0 | 5 (J) |
| 4/23/79 | 30.0 | 1 (J) |
| 4/24/79 | 2.5 | 7 (J) |
| 4/24/79 | 50.0 | 1 (J) |
| 4/26/79 | 50.0 | 1 (J) |
| 4/28/79 | 10.0 | 7 (J) |
| 5/3/79 | 20.0 | 2 (J) |
| 5/4/79 | 2.5 | 3 (J) |
| 5/7/79 | 5.0 | 5 (J) |
| 5/7/79 | 20.0 | 5 (J) |
| | bottom few cm of ice | 1 (J) |

*Drift net used; actual depth 0–40 m.
(J) = juvenile; (A) = adult.

were present in the uppermost 2 m of the ice core. Eleven unidentified juvenile globigerinids were counted (Table 8-3).

Thecosomate pteropods are exemplified by *L. helicina,* which was present in all samples, in water depths ranging from 2.5 to 50 m (Table 8-1). Over 97 percent of the specimens are juvenile; one immature pteropod was recovered from the bottom few centimeters of an ice core (Table 8-4).

## DISCUSSION AND CONCLUSIONS

In previous studies only one thecosomate pteropod, *L. helicina,* and one planktonic foraminifera, sinistral *G. pachyderma,* were found in Arctic Ocean water. Their reported absence in polar plankton tows may be due to the different (coarser) plankton nets used by other workers. Our discovery of several species of planktonic foraminifera considered subpolar, namely *G. egelida, G. quinqueloba,* and dextral *G. pachyderma,* in surface water beneath the ice can be explained either by their wider geographic range than hitherto suggested or to their transport from lower latitudes via the West Spitsbergen Current into the Arctic Ocean, where they are refluxed westward to the East Greenland Current, the region under study. We believe that the second alternative accounts for the observed occurrences. Until adequate data on polar species composition and distributional patterns in various water masses and during different seasons become available, the presence of these "lower-latitude" fauna in the Arctic will remain elusive.

Pteropods constitute >33 percent by number of the calcareous zooplankton in our samples, but only up to 5 percent in the shallowest Alpha Rise Recent sediments (1,521 m; Herman, 1974), suggesting that dissolution occurs during descent through the water column and/or on the seafloor. The distribution of solution-susceptible *G. egelida, G. quinqueloba,* and juvenile globigerinids in surface sediments cored from various physiographic provinces of the central Arctic does not change consistently with water depth (Herman, unpublished) and therefore cannot be compared with the plankton distribution of the present study.

## ACKNOWLEDGMENTS

We thank S. van der Spoel for discussions of the pteropod fauna, R. Cifelli (deceased) for reading an earlier draft of this manuscript and for offering valuable suggestions, and L. K. Coachman for data concerning present-day surface-water circulation patterns. The FRAM I Expedition was organized as a U.S. station with the participation of Danish and Norwegian scientists. It was supported by the Office of Naval Research, the Commission for Scientific Research in Greenland, the Danish Natural Science Research Council, and the Norwegian Polar Research Institute. We are indebted to G. Leonard Johnson for making this project possible. We also thank R. E. Garrison, UCSC, for making the scanning electron microscope available and Lamont Doherty Geological Observatory (LDGO) for providing the core samples. Support for the curating facilities of the LDGO Deep-Sea Sample Repository was provided by the National Science Foundation through Grant OCE 85-00232 and the Office of Naval Research through Grant N00014-84-C-0132.

## APPENDIX

### Plate 8-I

1*A*. *Limacina helicina antarctica* forma *antarctica*.
1*B*. *Limacina helicina* forma *helicina*.
1*C*. *Limacina helicina helicina* forma *pacifica*.
1*D*. *Limacina helicina antarctica* forma *rangii*.
1*E*. *Limacina helicina helicina* forma *acuta*. (Magnification not given. Shell measurements are: height up to 6 mm; diameter up to 8 mm.) (From van der Spoel, 1967.)
2*A*-*C*. *Limacina retroversa* forma *balea*; × 9.
2*D*,*E*. Transitional forms between *Limacina retroversa* forma *balea* and *Limacina retroversa* forma *retroversa*; × 9.
2*F*,*G*. *Limacina retroversa* forma *retroversa*; × 9. (From Tesch, 1946.)
3*A*. *Clio pyramidata* forma *antarctica*; shell length 14 mm, dorsal view.
3*B*. Same shell, embryonic portion; higher magnification.
3*C*. Same shell, lateral view. (After Tesch, 1948.)
4*A*. *Clio pyramidata* forma *sulcata*; shell length 18 mm, dorsal view.
4*B*. Embryonic portion of *C. pyramidata* forma *sulcata*, higher magnification. (From van der Spoel, 1967.)

### Plate 8-II

1*A*. *Globigerina cryophila*, dextral; spiral view; 1*B*, side view; 1*C*, umbilical view; × 60.
2*A*. *Globigerina pachyderma*, dextral, spiral view; 1*B*, side view; 1*C*, umbilical view; × 60. (From Cifelli and Smith, 1970.)
3*A*. *Globigerina pachyderma*, dextral, umbilical view; × 271, SEM, from Core T3-67-9-220 cm.
4*A*. *Globigerina bulloides*, spiral view; 4*B*, umbilical view; × 45.
5*A*. *Globigerina egelida*, spiral view; 5*B*, side view; 5*C*, umbilical view; × 101. (Figs. 4 and 5 from Cifelli and Smith, 1970.)
6*A*. *Globigerina egelida*, umbilical view; × 338, SEM, from Core T3-67-12-192 cm.
6*B*. *Globigerina egelida* (same as fig. 6*A*); × 938.
7*A*. *Globigerina egelida*, umbilical view; × 405, SEM, from Core T3-66-S-8-8-10 cm.
8*A*. *Globigerina quinqueloba*, umbilical view; × 293, SEM, from Core T3-66-S-8-8-10 cm.
8*B*. *Globigerina quinqueloba* (same as fig. 8*A*); × 405, SEM.

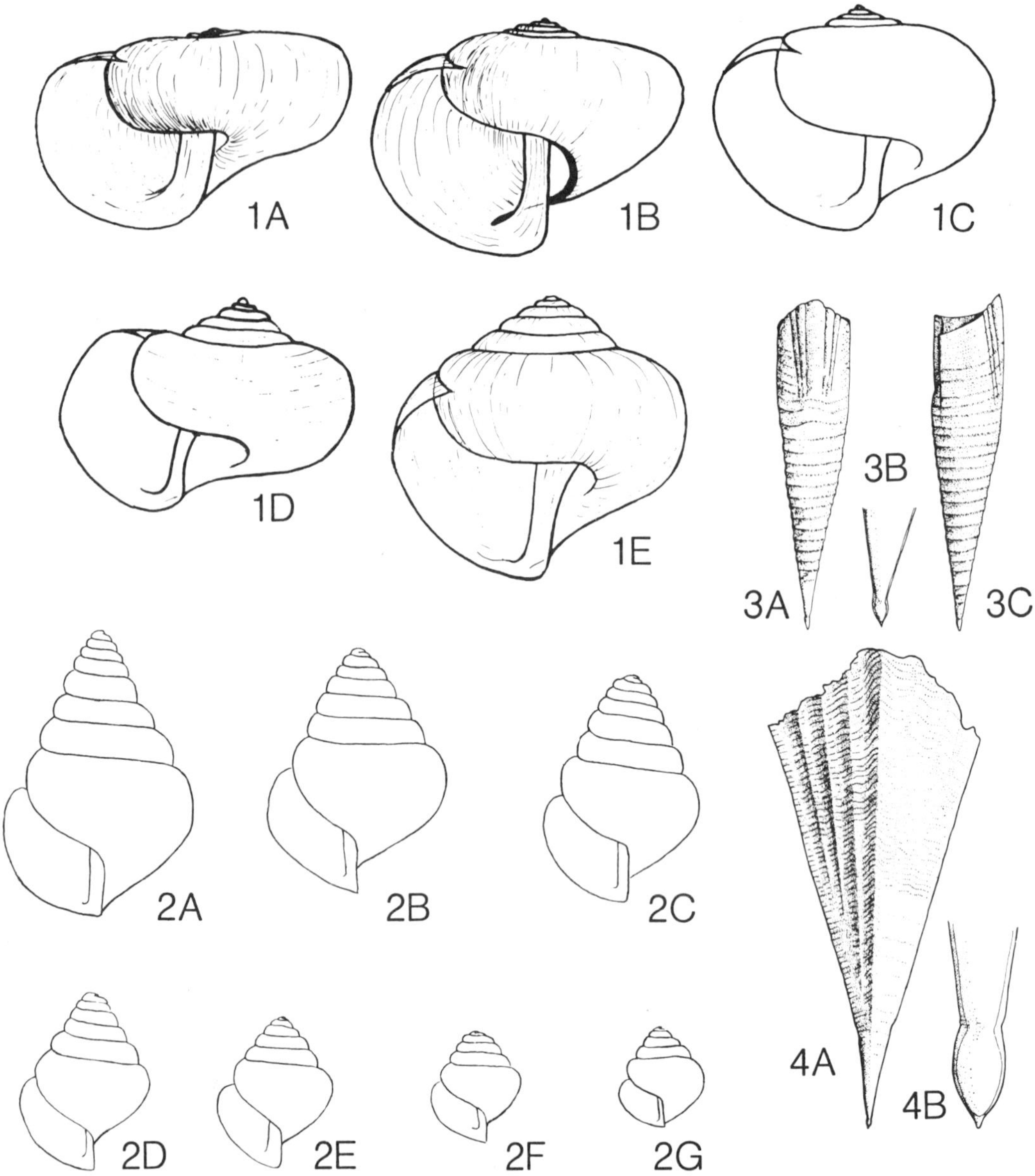


PLATE 8-I

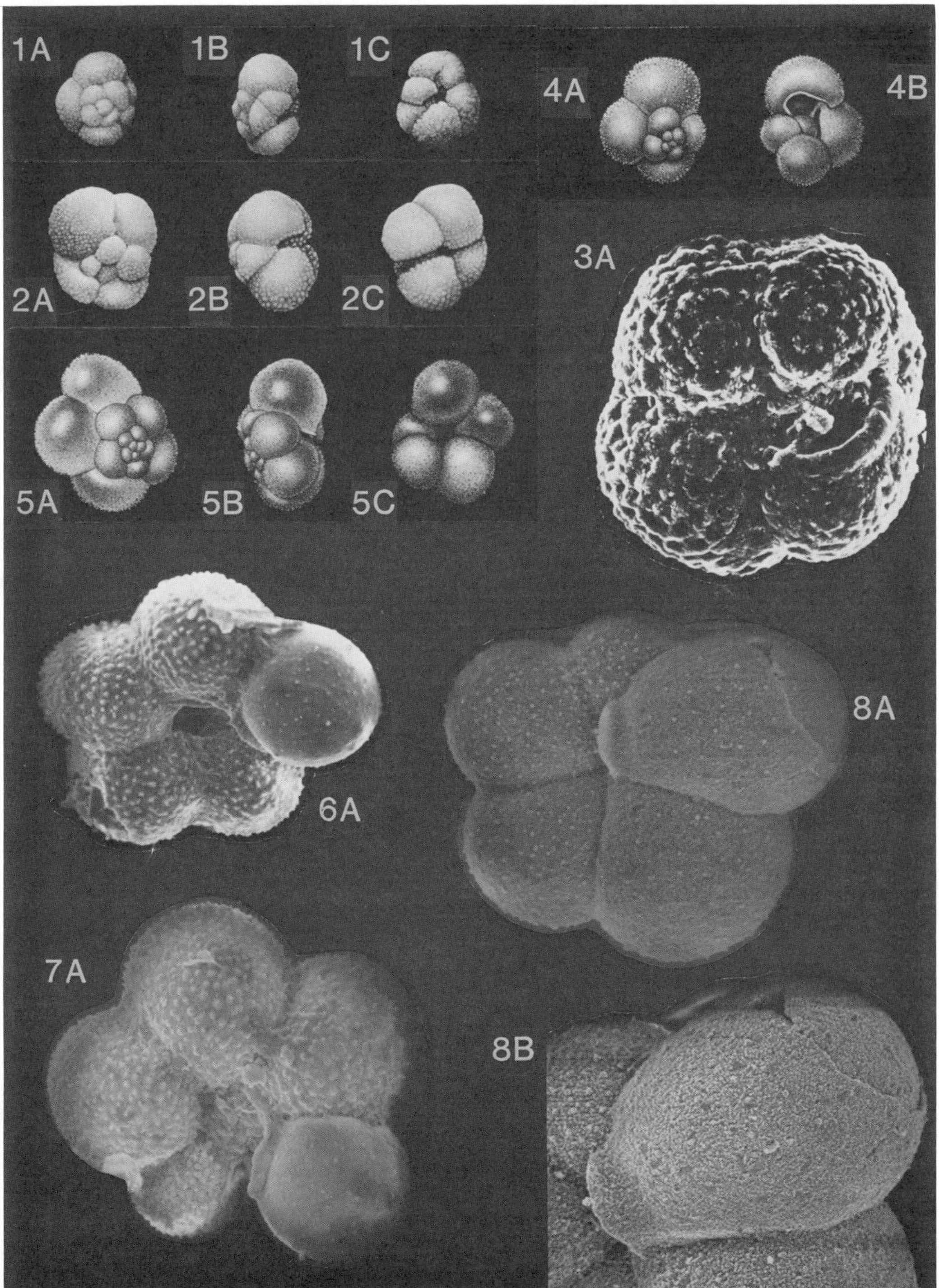


PLATE 8-II

## REFERENCES


Bé, A. W. H. 1960. Some observations on Arctic planktonic foraminifera. *Contrib. Cushman Found. Foram. Res.* **11**:64–68.

Bé, A. W. H. 1969. Planktonic foraminifera. In *Folio 11: Distribution of Selected Groups of Marine Invertebrates in Waters South of 35°S Latitude.* Antarctic Map Folio Series, *American Geographical Society,* New York, pp. 9–12.

Bé, A. W. H. 1977. An ecological zoogeographic and taxonomic review of recent planktonic foraminifera. In A. T. S. Ramsay (ed.), *Oceanic Micropaleontology.* Academic Press, London, pp. 1–100.

Bé, A. W. H., and R. W. Gilmer, 1977. A zoogeographic and taxonomic review of euthecosomatous pteropods. In A. T. S. Ramsay (ed.), *Oceanic Micropaleontology.* Academic Press, London, pp. 733–808.

Bé, A. W. H., and D. S. Tolderlund. 1971. Distribution and ecology of living planktonic foraminifera in surface waters of the Atlantic and Indian Oceans. In B. M. Funnell and W. R. Riedel (eds.), *The Micropaleontology of Oceans.* Cambridge University Press, Cambridge, pp. 105–149.

Boltovskoy, D. 1971*a*. Contribución al conocimiento de los pteropodos Thecosomados sobre la plataforma continental bonaerense. *Revista del Museo de La Plata (Neuva Serie), Seccion Zoologia* **11**:121–136.

Boltovskoy, D. 1971*b*. Pteropodos thecosomados del Atlantico suboccidental. *Malacologia* **11**:121–140.

Boltovskoy, D. 1973. Los euthecosomados del Mar Argentino y adayacencias. *Universidad de Buenos Aires. Physis Section A* **32**:511–526.

Bonnevie, K. 1913. *Pteropoda. Rept. Sci. Res. "Michael Sars," North Atlantic Deep-Sea Exped., 1910* **3**(1):1–69.

Chen, C. 1962. Ecological studies of shell-bearing pteropods from the western part of the North Atlantic Ocean. Ph.D. Dissertation, Boston University, Boston, 177p.

Chen, C. 1966. Calcareous zooplankton in the Scotia Sea and Drake Passage. *Nature* **212**(5063):678–681.

Cifelli, R., and R. K. Smith. 1970. Distribution of planktonic foraminifera in the vicinity of the North Atlantic Current. *Smithsonian Contrib. Paleont. No. 4,* pp. 1–52.

Ericson, D. B. 1959. Coiling direction of *Globigerina pachyderma* as a climatic index. *Science* **130**:219–220.

Herman, Y. 1971. Vertical and horizontal distribution of pteropods in Quaternary sequences. In B. M. Funnell and W. R. Riedel (eds.), *The Micropaleontology of Oceans.* Cambridge University Press, Cambridge, pp. 463–486.

Herman, Y. 1974. Arctic Ocean sediments, microfauna and the climatic record in late Cenozoic time. In Yvonne Herman (ed.), *Marine Geology and Oceanography of the Arctic Seas.* Springer-Verlag, New York, pp. 283–348.

Kerswill, C. J. 1940. The distribution of pteropods in the waters of eastern Canada and Newfoundland. *J. Fish. Res. Bd. Can.* **5**:23–31.

Kramp, P. L. 1961. Pteropoda. *Medd. Grønland* **81**:1–12.

McGowan, J. A. 1960. The systematics, distribution and abundance of the Euthecosomata of the North Pacific. Ph.D. Dissertation, University of California, San Diego, 197p.

McGowan, J. A. 1963. Geographic variation in *Limacina helicina* in the North Pacific. *Syst. Ass. Publ. No. 5. Speciation in the Sea.* pp. 109–128.

Meisenheimer, J. 1905*a*. Die arktischen Pteropoden. In F. Romer and F. Schaudinn (eds.), *Fauna Arctica,* vol. 4. Jena, Fisher, pp. 407–430.

Meisenheimer, J. 1905*b*. Pteropoda. *Wiss. Ergebn. Tiefsee Exped. "Valdivia"* **9**:1–314.

Meisenheimer, J. 1906. Die Pteropoden der deutschen Süd-polar Expedition 1901–1903. *Deutsch. Südpol. Exp. IX, 1901–1903, (Zool.),* **1**:92–152.

Tesch, J. J. 1946. The Thecosomatous Pteropods. I, The Atlantic. *Dana Report No. 24.* Oxford University Press, London, pp. 1–82.

Tesch, J. J. 1948. The Thecosomatous Pteropods II, the Indo-Pacific. *Dana Report No. (5) 30.* Oxford University Press, London, pp. 1–45.

Tibbs, J. F. 1967. On some planktonic protozoa taken from the track of Drift Station ARLIS I, 1960–61. *Arctic* **20**:247–254.

van der Spoel, S. 1967. Euthecosomata, a group with remarkable development stages (Gastropoda, Pteropoda). *J. Noorduyn en Zoon.* N.V., Gorinchem, 375p.

Vilks, G. 1970. Circulation of surface waters in parts of the Canadian Arctic Archipelago based on foraminiferal evidence. *Arctic* **23**:100–111.

# 9: Ecology of Arctic Ocean Cryopelagic Fauna

I. A. Mel'nikov

## INTRODUCTION AND PREVIOUS STUDIES

The Arctic Ocean and surrounding epicontinental seas have a perennial ice cover in the central sector and throughout most of the year in their peripheral parts. Paleoceanographic reconstructions have shown that the perennial Arctic sea ice developed about 0.9 million years ago (Herman, 1970; Herman and Hopkins, 1980). During the past 0.9 my, the sea-ice cover has repeatedly melted and formed again (Maksimov, 1972). The last phase of the ice-free ocean was in early Würm. According to Olausson and Jonasson (1969) the Arctic Ocean has had a sea-ice cover during the last 11,000 years. The ice thickness did not exceed 3–4 m (Nesis, 1983) and was much like the present sea-ice cover. It is possible that the general features of Arctic sea-ice ecology have developed since the Last Glaciation.

The early descriptions of the high-latitude ice flora (Ehrenberg, 1853; Dickie, 1880; Cleve and Grunov, 1880; Gran, 1904; Palibin, 1925; Usachev, 1949) suggested to investigators that the fauna developing on the underside of sea ice is trophically related to the ice flora. Such fauna was first observed in the Arctic (Mohr, 1959; Barnard, 1959; Apollonio, 1961) and later in the Antarctic (Peckham, 1961; Ray, 1966; Gruzov, Propp, and Pushkin, 1967). Along with crustaceans, various fish taxa were found. Both in the Arctic and Antarctic regions the ecological niches were always occupied by the same "performers" (term coined by L. Zenkevich), mainly amphipods. Cryopelagic organisms are presently believed to form a link in the transport of energy from primary production (epontic ice algae, bottom-type ice flora, phytoplankton) to sea birds and mammals living in Arctic areas and to deep-sea plankton in the open ocean (Dunbar and Wittman, 1963).

The investigation of species composition of the sea-ice communities undoubtedly is of great ecological significance; hence the various observations concerning the Arctic cryofauna (Shoemaker, 1955; Mohr, 1959; Mohr and Tibbs, 1963; Bar-

*Note:* This chapter has been reviewed, edited, and rewritten by the Editor, Yvonne Herman.

nard, 1959; George and Paul, 1970; Horner, 1976; Golikov and Averinzev, 1977; Mel'nikov and Kulikov, 1980; Cross, 1982; Gulliksen, 1984). Based on these investigations it was established that two species of the cryophilic taxa, the amphipods *Gammarus wilkitzkii* and *Apherusa glacialis,* are found both in the open ocean and on the shelf (Barnard, 1959; George and Paul, 1970; Golikov and Averinzev, 1977; Pavshtiks, 1980). In addition to the main components of the biocenosis, we find in the central Arctic Ocean great numbers of *Pseudalibrotus nanseni* and *Gammaracanthus loricatus* (Barnard, 1959; George and Paul, 1970), while on the shelf (Barrow, Alaska), *Pseudalibrotus litoralis, P. birulai,* and *Gammarus setosus,* are present (Mohr and Tibbs, 1963). The latter were found by Golikov and Averinzev (1977) as dominant forms in the ice biocenosis on the shelf of Rudolf Island (Franz Josef Land). In samples obtained off Point Barrow, Horner (1976) found amphipods, polychaete larvae, and turbellarians associated with the underside of the ice.

The continuous drift of the ice cover in the Arctic Ocean unites in a sense the open ocean with the shelf. Because of this, the planktonic fauna inhabiting these zones are included permanently or temporarily in the composition of the drifting sea-ice community. Hence, it is of importance to distinguish both the true sea-ice species, that is, the permanent inhabitants (aborigines) of the ecosystem, and temporary taxa (immigrants), which occur near the sea ice as a result of certain ecological factors.

The primary production of the sea-ice flora is now believed to be a first trophic level in the Arctic ecosystem (Meguro, Ito, and Fukushima, 1967; Clasby, Horner, and Alexander, 1972; Horner, 1976; Mel'nikov, 1980, and others) and in the Antarctic sea ice (Bunt and Wood, 1963; Andrijashev, 1970). The consumers of the sea-ice flora, the cryopelagic fauna, represent the second trophic level of the food chain of the sea-ice ecosystem. Our knowledge of this important trophic level is very limited. The present chapter synthesizes data on the ecology of the cryophilic fauna of the central Arctic Ocean. We wanted to ascertain whether the species composition of the fauna had populations whose character was similar to those of related species of zooplankton. The present study had two principal goals: (1) to determine both the qualitative and quantitative composition as well as the structure of the cryopelagic faunal communities and (2) to consider the peculiarities in the distribution and the behavior of the dominant species of the cryopelagic fauna living under the drifting Arctic sea ice. We collected the fauna under undeformed sea-ice cover together with vertical net tows of plankton to assess the planktonic contribution to the composition of the sea-ice community. The samples were taken during the following drifting North Pole (NP) expeditions:

1. NP-23 in the summer of 1977 (Mel'nikov and Tzinovsky, 1978) and in the spring of 1978 (Mel'nikov, 1978)
2. NP-22 and NP-24 in the winter of 1980 (Mel'nikov, 1981)
3. NP-22 in the spring of 1981 (Mel'nikov, 1982)

## MATERIAL AND METHODS

This research is based largely on material obtained in the region off the east Siberian shelf up to the Lomonosov Ridge, not far from the geographic north pole. All samples were collected from a 100 × 100 m pack-ice floe located 350–400 m from the station base in order to avoid sample contamination.

Cryopelagic fauna were caught by diving under the drifting ice cover. Dives were carried out through a separate hydrohole made in the pack ice. Samples were collected by scraping a plankton net (mesh size 180 $\mu$m) with a 40 × 20-cm-wide flat top in the opening frame for 40 m along the under-ice surface. During the catch of cryofauna the plankton net was kept in front of the diver to protect the under-ice surface from bubbles from the diver's air supply. After the tow catch the plankton net was closed in the water in order to protect the sample from planktonic fauna. In the laboratory the samples were washed on nylon gauze (mesh size 180 $\mu$m). The organisms were preserved in 4 percent formalin.

Catches of cryofauna were simultaneously coupled with (1) control horizontal tows of fauna at a depth of 5 m using the same plankton net, and (2) vertical zooplankton tows made with a Juday net (mesh size 180 $\mu$m).

Counts of organisms were made using Bogorov's device. The sample was concentrated to 5–10 ml volume and the entire sample was counted. In samples obtained by Juday net, organisms found on the underside of the ice were noted. Counts are given as specimens per $m^3$. The data concerning the cryofauna and the zooplankton catches are given in Tables 9-1 and 9-2.

In view of the small number of cryopelagic organisms under a $m^2$ of drifting sea ice, all animals were counted. In all cases the drifting sea ice was normalized to an area of 16 $m^2$, that is, 40 m the distance of the catch and 0, 4 m the diameter of the net aperture. This manipulation allowed us to express the quantity of individuals in specimens per 16 $m^2$.

## TERMINOLOGY

The sea-ice terminology used in this paper follows the International Nomenclature of Sea Ice (Volkov and Treshnikov, 1969).

The under surface of different types of sea ice is known to be inhabited by a fauna that Mohr and Tibbs (1963) identified as *anakatobenthos,* meaning "upside down benthic community or bottom-type fauna." Andrijashev (1970) used the term *cryopelagic* for fishes that swim actively near the ice surface and associate in some way with drifting or fast ice during their life cycle. Golikov and Scarlato (1973) included invertebrates in the term *cryopelagic* and described *cryopelagic biocenoses* with amphipods such as *Apherusa glacialis* and *Gammarus wilkitzkii.* Bradstreet and Cross (1982) used the term *epontic* for the fauna that lives on, or just within, the under surface of the ice. Gulliksen (1984) named the organisms associated with sea ice as *under-ice fauna.* In the present study, the term *cryopelagic* suggested by Andrijaschev in 1970 is adopted by the author.

## RESULTS

### Species Composition

Systematic analysis of samples obtained from the underside of the drifting sea ice identified about 50 species belonging to the following taxonomic groups: Foraminifera, Rotatoria, Nematoda, Polychaeta, Chaetognatha, Copepoda, Mysidacea, Isopoda, Amphipoda, Decapoda, Pteropoda, and Appendicularia. Species data for cryopelagic fauna and zooplankton species are given in Tables 9-3 and 9-4 (pages 240–241 and 242–243), respectively.

Continuous finding of several species in samples obtained from the underside of the sea ice and the absence of these species in samples collected at a depth of 5 m and in the deep sea allow me to distinguish two general ecological groups:

1. Autochthonous organisms permanently living on or near the underside of the sea ice
2. Allochthonous organisms temporarily related to the underside of the sea ice

Criteria for distinguishing these two ecological groups were as follows:

For the first group:

1. Preferential occurrence of animals of both sexes of all developmental stages

Table 9-1. Station Data for the Cryopelagic Fauna from NP-23 (1977–1978)

| Sample No. | Depth of Catch (m) | Length of Catch (m) | Date | Position | | Depth (m) |
|---|---|---|---|---|---|---|
| | | | | Latitude (N) | Longitude (E) | |
| | | | 1977 r. | | | |
| 1 | 0 | 40 | 3.VI | 78°32′8 | 161°34′9 | 180 |
| | 5 | 40 | | | | |
| 2 | 0 | 40 | 16.VI | 79 01 4 | 161 22 1 | 260 |
| 3 | 0 | 2 | 19.VI | 78 59 0 | 162 03 3 | 285 |
| 4 | 0 | 4 | 29.VI | 79 49 7 | 160 05 6 | 1,800 |
| | 5 | 40 | | | | |
| 5 | 0 | 20 | 3.VII | 80 06 1 | 159 01 9 | 2,000 |
| 6 | 0 | 20 | 4.VII | 80 09 8 | 157 46 1 | 2,016 |
| | 5 | 20 | | | | |
| 7 | 0 | 20 | 7.VII | 80 18 9 | 157 45 1 | 2,333 |
| 8 | 0 | 15 | 10.VII | 80 28 9 | 158 16 8 | 2,565 |
| 9 | 0 | 15 | 17.VII | 81 10 0 | 157 57 8 | 2,800 |
| | 0 | 15 | | | | |
| 10 | 0 | 15 | 20.VII | 81 17 7 | 158 29 7 | 2,800 |
| 11 | 0 | 15 | 23.VII | 81 27 3 | 158 45 8 | 2,800 |
| 12 | 0 | 10 | 26.VII | 81 34 3 | 150 11 6 | 2,800 |
| | 5 | 15 | | | | |
| 13 | 0 | 10 | 26.VII | 81 34 3 | 159 11 6 | 2,800 |
| | 5 | 20 | | | | |
| 14 | 0 | 20 | 27.VII | 81 36 3 | 158 57 9 | 2,800 |
| | 5 | 40 | | | | |
| 15 | 0 | 50 | 4.VIII | 81 50 8 | 159 19 6 | 2,810 |
| | 5 | 50 | | | | |
| 16 | 0 | 30 | 12.VIII | 82 05 3 | 156 09 7 | 2,800 |
| | 5 | 30 | | | | |
| 17 | 0 | 40 | 28.VIII | 82 38 2 | 152 22 1 | 2,770 |
| | 5 | 40 | | | | |
| 18 | 0 | 50 | 6.IX | 82 52 5 | 154 19 1 | 2,600 |
| | 5 | 50 | | | | |
| 19 | 0 | 40 | 24.IX | 83 28 0 | 151 37 6 | 2,480 |
| | 5 | 40 | | | | |
| 20 | 0 | 50 | 9.X | 83 30 2 | 150 42 1 | 2,467 |
| | 5 | 50 | | | | |
| 21 | 0 | 40 | 20.X | 83 35 0 | 150 21 0 | 2,430 |
| | 5 | 40 | | | | |
| | | | 1978 r. | | | |
| 22 | 0 | 10 | 20.IV | 88 38 9 | 146 36 8 | 2,520 |

on the underside of the sea ice during the year
2. Absence of these species in the plankton
3. Constant relationships of these species with the ice during the ice movement through different parts of the Arctic Ocean

For the second group:

1. Temporary inhabiting of several species on the underside of the ice during reproduction and growth of animals during different seasons
2. Presence of these species at depth

### *Autochthonous Cryophilic Fauna*

The following species belong to this ecological group: *Gammarus wilkitzkii, Mysis*

Table 9-2. Station Data for Plankton Samples from NP-23 (1977–1978)

| *Station No.* | *Date* | *Position* N | E | *Depth of Catches (m)* | *Layer of Catch (m)* |
|---|---|---|---|---|---|
| | | | 1977 | | |
| 1 | 4.VI | 78°35′3 | 161°17′5 | 177 | 25–0,50–27, 100–49, 150–99, 172–151 |
| 2 | 1.VII | 79 56 5 | 158 56 7 | 1833 | 25–0, 50–24, 100–50, 150–97, 142–200, 145–300, 293–500, 1,000–487, 1,800–997 |
| 4 | 1.VIII | 81 45 9 | 158 43 5 | 2821 | 25–0, 50–24, 100–49, 150–94, 200–144, 300–196, 500–285, 1,000–499 |
| 6 | 1.IX | 82 40 7 | 153 04 1 | 2775 | 25–0, 50–25, 100–49, 200–100, 300–197, 500–297, 1,000–499, 2,700–1,003 |
| 8 | 13.X | 83 38 2 | 149 38 5 | 2403 | 25–0, 50–25, 100–50, 200–99, 500–203, 1,000–501, 2,350–1,009 |
| | | | 1978 r. | | |
| 9 | 18.IV | 88 19 3 | 144 02 9 | 2550 | 25–0, 50–25, 100–50, 200–97, 500–218, 1,000–503, 2,500–989 |

*polaris, Jaschnovia johnsoni, Tisbe furcata,* and *Harpacticus superflexus. Gammarus wilkitzkii* inhabits the underside of the sea ice year-round. Here we found all its generations: juveniles, subadults, adults, females, and males. Gravid females occurred in winter and juveniles were in water in July and August. Twenty-three of 25 individuals from all collections were caught at 9 stations on the underside of the ice. Although *Gammarus wilkitzkii* was absent in other collections, it was grouped with the autochthonous taxa since it was continuously observed within the under-ice surface, and its absence in some samples was due to the high mobility of the animals attempting to avoid the plankton net during sampling. This species was not observed in water deeper than 1–2 m below the ice surface.

The same observation was made concerning *Mysis polaris;* 86 out of 94 collections were observed near the undersurface of the ice. Among them we found specimens belonging to both sexes and all generations. This species was not found in plankton tows. *Jaschnovia johnsoni* was found in 13 out of 21 stations during the year. Between 0 and 5 m depths, 232 out of 242 individuals were caught near the ice surface. Probably *J. johnsoni* lives within the thin "water-ice" boundary and has been considered as an epibenthic form. The maximum number of specimens were observed between September and October when both copepodites and adults of this species were present.

*Harpacticus superflexus* was found at 9 out of the 21 stations and 39 out of 45 individuals were collected on the undersurface of the ice. In vertical catches it was found only once at 0–25 m. *Tisbe furcata* was caught at 16 out of 21 stations; 86 out of the 95 individuals of the total collection were

Table 9-3. Cryopelagic Fauna from NP-23 (1977–1978)

| | | Sample No. and Depth of Catches (m) | | | | | | | | | | | | |
|---|---|---|---|---|---|---|---|---|---|---|---|---|---|---|
| | | 1 | | 2 | 3 | 4 | | 5 | 6 | | 7 | 8 | 9 | |
| Species | | 0 | 5 | 0 | 0 | 0 | 5 | 0 | 0 | 5 | 0 | 0 | 1 | 2 |
| *Globigerina* sp | | – | – | – | – | – | – | – | – | – | 2 | – | – | – |
| Radiolaria gen. sp. | | – | – | – | – | – | – | – | – | – | – | – | – | – |
| Nematoda gen. sp. | | – | – | – | – | – | – | – | – | – | – | – | – | – |
| *Antinoella sarsi* | | – | – | – | – | – | – | – | 2 | – | – | – | – | – |
| *Synchaeta* sp. | | – | – | – | – | – | – | – | – | – | – | – | – | – |
| *Parasagitta arctica* | | – | – | – | + | – | – | – | – | – | – | 5 | – | – |
| *Euarohnia hamata* | | – | – | – | – | – | 3 | – | 2 | – | 2 | – | – | – |
| *Conchoecia* sp. | | – | – | – | – | – | – | – | – | – | – | – | – | – |
| *Calanus glacialis* | cop. II–IV | – | 1 | – | + | – | 1 | – | 4 | – | 2 | – | – | – |
| | cop. V | – | – | – | – | – | – | – | – | – | – | – | – | – |
| | ♀ | – | – | – | – | – | – | – | – | – | – | – | – | – |
| *C. hyperboreus* | cop. IV | – | – | – | – | – | – | – | – | – | – | – | – | – |
| | ♀ | – | – | – | – | – | – | – | – | – | – | – | – | – |
| *Derjuginia af. tolli* | cop. III–V | – | – | 39 | + | – | – | – | 8 | 2 | 12 | – | – | – |
| | ♀ | – | – | 1 | – | – | – | – | – | – | – | – | – | – |
| | ♂ | – | – | – | – | – | – | – | – | – | – | – | – | – |
| *Pseudocalanus minutus* | ♀ | – | – | – | – | – | 14 | – | 38 | 2 | 24 | – | 3 | – |
| | ♂ | – | – | – | – | – | 1 | – | – | – | – | – | – | – |
| *P. elongatus* | ♀ | 1 | 2 | 6 | + | + | 3 | 2 | 26 | – | 22 | – | 3 | – |
| | ♂ | – | – | – | + | – | – | – | – | – | 4 | – | – | – |
| *Pseudocalanus* sp. | cop. III–V | 2 | 7 | 14 | + | – | 9 | 2 | 34 | 4 | 6 | – | 8 | – |
| *Microcalanus pygmaeus* | ♀ | – | – | – | – | – | – | 4 | – | – | – | – | – | – |
| *Drepanopus bungei* | ♀ | – | – | – | – | – | – | – | – | – | – | – | – | – |
| *Spinocalanus magnus* | ♀ | – | – | – | – | – | – | – | – | – | – | – | – | – |
| *Lucicutia polaris* | ♀ | – | – | – | – | – | – | – | – | – | – | – | – | – |
| *Metridia longa* | cop. IV–V | – | – | – | – | – | – | 4 | – | – | – | – | – | – |
| | ♀ | – | – | – | – | – | – | – | – | – | – | – | – | – |
| *Mormonilla polaris* | ♀ | – | – | – | – | – | – | – | – | – | – | – | – | – |
| *Acartia* sp. | cop. V | – | – | – | – | – | – | – | 2 | – | – | – | – | – |
| *Oithona similis* | cop. III–V | – | 2 | – | + | + | – | 12 | – | – | 24 | 3 | 3 | 3 |
| | ♀ | – | 2 | – | – | – | 2 | – | 2 | 2 | – | – | – | – |
| | ♂ | 1 | – | | – | – | – | – | – | – | – | – | – | – |
| *Cyclopina schneideri* | cop. III–V | – | – | – | – | – | – | – | – | – | – | – | – | 3 |
| | ♀ | – | – | – | – | – | – | – | – | – | – | – | – | – |
| *Oncaea borealis* | cop. III–V | – | – | – | – | – | – | – | – | – | – | – | – | – |
| | ♀ | – | – | – | – | – | – | – | – | – | – | – | – | – |
| *Tisbe furcata* | cop. III–V | – | – | – | + | – | 1 | – | 2 | – | 2 | – | 8 | – |
| | ♀ | – | – | 3 | + | – | – | – | 6 | – | 2 | – | 13 | – |
| | ♂ | – | – | – | – | – | – | – | – | – | 2 | – | 5 | – |
| *Harpacticus superflexus* | cop. III–V | – | – | – | – | – | – | 2 | 12 | – | 2 | 5 | – | – |
| | ♀ | – | – | – | – | – | – | – | 2 | – | – | – | – | – |
| | ♂ | – | – | – | – | – | – | – | – | – | – | 3 | – | – |
| *Microsetella norvegica* | ♀ | – | – | – | – | – | – | – | – | – | – | – | – | – |
| Copepoda (naupl.) | | – | – | – | + | – | 1 | 2 | 2 | – | 2 | – | 8 | – |
| *Mysis polaris* | juv. | – | – | 10 | – | – | – | – | 8 | 2 | 14 | – | 11 | – |
| | ♀ | 1 | – | 3 | – | – | – | – | – | 2 | – | – | 13 | – |
| | ♂ | – | – | 2 | – | – | – | – | – | – | – | – | 3 | – |
| *Dayus mysidis* | | – | – | 3 | – | – | – | – | – | – | – | – | 243 | [illegible] |
| *Parathemisto abyssorum* | | – | – | – | – | – | – | – | 2 | – | – | 3 | 8 | – |
| *P. libellula* | | – | – | – | – | – | – | – | – | – | – | – | – | – |
| *Pseudalibrotis glacialis* | juv. | – | – | – | – | – | – | – | – | 2 | – | – | – | – |
| | ♀ | – | – | – | – | – | – | – | – | – | – | – | – | – |

| 10 | 11 | 12 | | 13 | | 14 | | 15 | | 16 | | 17 | | 18 | | 19 | | 20 | | 21 | | 22 |
|---|---|---|---|---|---|---|---|---|---|---|---|---|---|---|---|---|---|---|---|---|---|---|
| 0 | 0 | 0 | 5 | 0 | 5 | 0 | 5 | 0 | 6 | 0 | 5 | 0 | 5 | 0 | 5 | 0 | 5 | 0 | 5 | 0 | 5 | 0 |
| – | – | – | – | – | – | – | – | – | – | – | – | – | – | – | – | – | – | – | – | – | – | – |
| – | – | – | – | – | – | – | – | – | – | – | – | 1 | – | – | 1 | – | – | – | – | – | – | – |
| – | – | – | – | – | – | – | – | – | – | – | – | – | – | 11 | – | 7 | – | – | – | – | – | – |
| – | – | – | – | – | – | – | – | – | – | – | – | – | – | – | – | – | – | – | – | – | – | – |
| – | – | – | – | – | – | – | – | – | – | – | – | – | – | – | – | – | – | – | – | – | – | – |
| – | – | 4 | 5 | – | 2 | – | – | – | – | – | 3 | – | – | – | – | – | – | – | 5 | 1 | 4 | – |
| – | – | 4 | – | – | – | – | – | – | 1 | – | – | – | – | – | – | – | – | – | – | – | – | 3 |
| – | – | – | – | – | – | – | – | – | – | – | – | – | – | – | – | – | – | – | 1 | – | 1 | – |
| – | – | 36 | 61 | 80 | 138 | 8 | 4 | – | 2 | 4 | 16 | – | 26 | – | – | 1 | – | 6 | 142 | – | 1 | – |
| – | – | 48 | 150 | 76 | 500 | 4 | – | – | – | 2 | 20 | 1 | 40 | – | – | – | – | – | – | – | – | – |
| – | – | – | 16 | 15 | 96 | – | – | – | – | – | 3 | – | 1 | – | – | – | – | – | – | – | – | – |
| – | – | – | – | – | – | – | – | – | 1 | – | 2 | – | – | – | – | – | – | – | – | – | – | – |
| – | – | – | – | – | – | – | – | – | – | – | – | 1 | – | – | – | – | – | – | – | – | – | – |
| 5 | 4 | – | – | – | – | 4 | – | – | – | 8 | 2 | 9 | 7 | 11 | – | 32 | – | 28 | – | 1 | – | – |
| – | – | – | – | – | – | – | – | – | – | – | – | – | – | 3 | – | 4 | – | 15 | – | 4 | – | 2 |
| – | – | – | – | – | – | – | – | – | – | – | – | – | – | – | – | 1 | – | 2 | – | 1 | – | – |
| 3 | – | – | – | 8 | – | – | – | – | 3 | 3 | 10 | 1 | 6 | – | – | 1 | 1 | 3 | 4 | – | 1 | – |
| – | – | – | – | – | – | – | – | – | – | – | – | – | – | – | – | – | – | – | – | – | – | – |
| 3 | 3 | – | 10 | 20 | 16 | – | 1 | – | 2 | – | 3 | – | 5 | – | – | – | – | – | 1 | – | – | – |
| – | – | – | – | – | – | – | – | – | – | – | – | – | – | – | – | – | – | – | – | – | – | – |
| 6 | – | 4 | 9 | 29 | – | – | – | – | 1 | – | 5 | 18 | 98 | – | 3 | 14 | 21 | 888 | 542 | 30 | 184 | 9 |
| – | – | – | 2 | – | – | – | – | – | – | – | – | 1 | – | – | – | – | – | – | – | – | – | – |
| – | – | – | – | – | – | – | – | – | – | – | 2 | – | – | – | – | – | – | – | – | – | – | – |
| – | – | – | – | – | – | – | – | – | – | – | – | 1 | – | – | – | – | – | – | – | – | – | – |
| – | – | – | – | – | – | – | – | 1 | – | – | – | – | – | – | – | – | – | – | – | – | – | – |
| – | – | – | – | – | – | – | – | – | – | – | – | – | – | – | – | – | – | – | – | – | 22 | – |
| – | – | – | – | – | – | – | – | – | – | – | – | – | – | – | – | – | – | – | – | – | 11 | – |
| – | – | – | – | – | – | – | – | 2 | – | – | – | – | – | – | – | – | – | – | – | – | – | 10 |
| – | – | – | – | – | – | – | – | – | – | – | – | – | – | – | – | – | – | – | – | – | – | – |
| 3 | – | – | 5 | – | – | – | – | 2 | – | 3 | – | – | – | – | – | – | – | 2 | 4 | – | – | 14 |
| – | – | – | – | – | – | – | – | – | 6 | – | 3 | 1 | – | 2 | 2 | 2 | 1 | – | 1 | – | 1 | 10 |
| – | – | – | – | – | – | – | – | – | – | – | – | – | – | – | – | – | – | – | – | – | – | – |
| 3 | – | – | – | – | – | – | – | – | – | – | – | – | – | – | – | 4 | – | – | – | – | – | – |
| – | 1 | – | – | – | – | – | – | – | 2 | 3 | – | – | – | 1 | – | – | – | – | – | – | – | – |
| – | – | – | – | – | – | – | – | 1 | – | – | – | – | – | 1 | – | – | – | – | – | – | – | 10 |
| – | – | – | 2 | – | – | – | – | – | – | – | – | – | – | – | – | – | – | – | – | – | 2 | 2 |
| – | – | – | 2 | – | – | – | – | – | – | – | – | – | – | 8 | – | 4 | – | – | – | – | 2 | 2 |
| – | 12 | – | 2 | 4 | – | 4 | 2 | – | – | – | 5 | 2 | – | 9 | – | 6 | – | 8 | 1 | – | – | 17 |
| – | 4 | – | – | – | – | – | – | – | – | – | – | – | – | – | – | – | – | – | – | – | – | – |
| – | – | – | – | 8 | – | – | 1 | – | – | – | – | – | – | – | – | – | – | – | – | – | – | – |
| – | 4 | – | 2 | 64 | 4 | 8 | 1 | – | 1 | – | – | – | – | – | – | – | – | – | – | – | – | – |
| – | – | – | – | 4 | – | – | – | 1 | – | – | – | – | – | – | – | – | – | – | – | – | – | – |
| – | – | – | – | – | – | – | – | 1 | – | – | 2 | – | 5 | 39 | – | 12 | – | – | – | – | – | – |
| – | – | – | – | – | – | 4 | – | – | 1 | 15 | 2 | 15 | – | 9 | 2 | 4 | – | 3 | – | – | – | – |
| 3 | – | – | – | – | – | – | – | – | – | 2 | – | 2 | 1 | – | – | 1 | – | 2 | – | – | 1 | – |
| – | – | – | – | – | – | – | – | – | – | 2 | – | – | – | – | – | – | – | 1 | – | – | – | – |
| – | – | – | – | – | – | – | – | – | – | – | – | 1 | – | – | – | – | – | 1 | – | – | – | – |
| – | – | – | – | – | – | – | – | – | 1 | – | – | – | – | – | – | – | – | – | 2 | – | 1 | – |
| – | 1 | – | – | – | – | – | 3 | – | – | – | – | – | 2 | – | – | – | – | – | – | – | – | – |
| – | – | – | 14 | – | – | – | 1 | – | – | 2 | – | – | – | – | – | – | – | 2 | – | – | – | – |
| – | – | – | – | – | – | – | – | – | – | – | – | – | – | – | – | – | – | 1 | – | – | – | – |
| – | – | – | – | 4 | – | – | – | – | – | – | – | – | – | – | – | – | – | 1 | – | – | – | – |

Table 9-3. *(continued)*

| | | *Sample No. and Depth of Catches (m)* | | | | | | | | | | | | |
|---|---|---|---|---|---|---|---|---|---|---|---|---|---|---|
| | | *1* | | *2* | *3* | *4* | | *5* | *6* | | *7* | *8* | *9* | |
| *Species* | | *0* | *5* | *0* | *0* | *0* | *5* | *0* | *0* | *5* | *0* | *0* | *1* | *2* |
| *P. nanseni* | juv. | 1 | – | 1 | + | – | – | – | – | – | 2 | – | – | – |
| | ♀ | – | – | – | + | – | – | – | – | – | – | – | – | – |
| | ♂ | – | – | – | – | – | – | – | – | – | – | – | – | – |
| *Gammarus wilkitzkii* | juv. | – | – | – | + | – | – | – | – | – | – | – | – | – |
| | ♀ | – | – | – | – | – | – | – | – | – | – | – | 3 | – |
| | ♂ | – | – | – | + | – | – | 2 | – | – | – | – | 3 | – |
| *Gammaracanthus loricatus* | juv. | 1 | – | – | + | – | – | – | – | – | – | – | – | – |
| | ♀ | – | – | – | – | – | – | – | – | – | – | – | – | – |
| | ♂ | – | – | – | – | – | – | – | – | – | – | – | – | – |
| *Weyprechtia pinguis* | juv. | 1 | – | – | – | – | – | – | – | – | – | – | – | – |
| *Eusirus holmi* | juv. | – | – | – | – | – | – | – | – | – | – | – | – | – |
| | ♂ | – | – | – | – | – | – | – | – | – | 2 | – | – | – |
| *Neopleustes* sp. | juv. | 1 | 1 | 10 | + | – | – | – | 2 | – | – | – | – | – |
| | ♀ | – | – | – | + | – | – | – | – | – | – | – | – | – |
| *Metopa longirama* | juv. | 2 | 1 | 3 | + | – | – | – | – | – | – | – | – | – |
| | ♀ | – | – | 2 | + | – | – | – | – | – | – | – | – | – |
| | ♂ | – | – | – | + | – | – | – | – | – | – | – | – | – |
| *Apherusa glacialis* | juv. | 1 | – | 2 | + | – | – | – | 12 | – | 8 | 5 | 8 | – |
| | ♀ | 2 | – | 16 | + | – | – | 2 | 2 | – | – | – | 11 | – |
| | ♂ | – | – | 1 | + | – | – | – | – | – | – | – | – | – |
| *Eualus gaimardi belcheri* | ♀ | – | – | 1 | – | – | – | – | – | – | – | – | – | – |
| *Hymenodora glacialis* | | – | – | – | – | – | – | – | – | – | – | – | – | – |
| *Clione limacina* | | – | – | – | – | – | – | – | 4 | – | 6 | – | 3 | – |
| *Limacina helinica* | | – | – | – | – | – | 5 | – | 8 | – | – | – | – | – |
| *Oikopleura* sp. | | – | – | – | – | – | – | – | – | – | – | – | – | – |
| *Fritillaria borealis* | | 1 | – | – | – | – | – | – | – | – | – | – | – | – |

caught on the underside of the ice. Both copepodites and adults were found. This species was sometimes found at 0–25 m.

The animals described above belonging to this ecological group have, strictly speaking, a nekto-benthic way of life, but only near the underside of the sea ice. For this reason, occasionally they were absent in samples taken from the underside of the ice. However, during the year in this biotop all developmental stages and both sexes of these species were found. It may be suggested that their life cycle is closely associated with the underside of the ice and their populations are a constant component of the cryopelagic community of the drifting permanent ice cover in the Arctic Ocean.

### *Allochthonous Cryophilic Fauna*

Animals of this group are found near the underside of the ice as a result of hydrodynamical, geographical, or biological factors. *Apherusa glacialis* is the dominant species among the amphipods found on the underside of the ice in summer. Near the ice 172 out of 182 individuals in 18 out of 21 collections were caught. Only 10 specimens were found in horizontal catches at 5 m and none were found in vertical plankton tows. In spite of its constant presence within the ice surface, this species was included with the second ecological group because all of the 172 individuals found on the underside of the ice were juveniles.

It is likely that *A. glacialis* grows during

| 10 | 11 | 12 | | 13 | | 14 | | 15 | | 16 | | 17 | | 18 | | 19 | | 20 | | 21 | | 22 |
|---|---|---|---|---|---|---|---|---|---|---|---|---|---|---|---|---|---|---|---|---|---|---|
| 0 | 0 | 0 | 5 | 0 | 5 | 0 | 5 | 0 | 6 | 0 | 5 | 0 | 5 | 0 | 5 | 0 | 5 | 0 | 5 | 0 | 5 | 0 |
| – | – | – | – | – | – | – | – | – | – | – | – | – | 1 | – | – | – | – | 2 | – | 1 | – | – |
| – | – | – | – | – | – | – | – | – | – | – | – | – | – | – | – | – | – | – | – | 1 | – | – |
| – | – | – | – | – | – | – | – | 1 | – | – | – | – | – | – | – | – | – | – | – | 3 | – | 126 |
| – | – | – | 2 | – | – | – | – | – | – | – | – | 1 | – | – | – | – | – | – | – | 2 | – | 2 |
| – | 4 | – | – | – | – | – | – | 3 | – | – | – | 1 | – | 1 | – | – | – | – | – | 2 | – | – |
| – | – | – | 2 | – | – | – | – | – | – | – | – | – | – | – | – | 1 | – | – | – | – | – | – |
| – | – | – | – | – | – | – | – | – | – | – | – | – | – | – | – | – | – | – | – | – | – | – |
| – | 1 | – | – | – | – | – | – | – | – | – | – | – | – | – | – | – | – | 1 | – | – | – | – |
| – | – | – | – | – | – | – | – | – | – | – | – | – | – | – | – | – | – | – | – | – | – | – |
| – | – | – | – | – | – | – | – | – | – | – | – | – | – | – | – | – | – | 1 | – | 3 | – | – |
| – | – | – | – | – | – | – | – | – | – | – | – | – | 1 | – | – | – | – | – | – | – | – | – |
| – | – | – | – | – | – | – | – | – | – | – | – | – | – | – | – | 4 | – | 2 | – | 1 | – | 14 |
| – | – | – | – | – | – | – | – | – | – | – | – | – | – | – | – | – | – | – | – | – | – | – |
| – | – | – | – | – | – | – | – | – | – | – | – | – | – | 1 | – | 4 | – | 7 | – | 1 | – | 12 |
| – | – | – | – | – | – | – | – | – | – | – | – | – | – | – | – | 1 | – | – | – | – | 2 | – |
| – | – | – | – | – | – | – | – | – | – | – | – | – | – | – | – | – | – | – | – | – | – | – |
| – | – | – | – | – | – | – | 1 | 2 | – | 19 | – | – | – | 2 | – | – | – | – | – | 1 | – | – |
| 8 | 1 | 16 | – | 32 | 10 | 26 | 2 | 2 | – | 13 | – | – | – | 3 | – | – | – | 2 | – | 2 | – | – |
| 3 | – | 28 | – | 68 | 4 | 2 | – | – | – | – | – | – | – | – | – | – | – | 1 | – | – | – | – |
| – | – | – | – | – | – | – | – | – | – | – | – | – | – | – | – | – | – | – | – | – | – | – |
| – | – | – | – | – | – | – | – | – | – | – | – | – | 1 | – | – | – | – | – | – | – | – | – |
| – | – | – | – | – | – | – | – | – | 1 | – | – | – | – | – | – | – | – | – | 1 | – | – | – |
| 3 | – | – | – | – | – | – | – | – | – | – | – | – | – | – | – | – | – | – | – | – | – | – |
| – | – | – | – | – | – | – | – | – | – | – | – | – | – | – | – | – | – | – | – | – | 1 | – |
| – | – | – | – | – | – | – | – | – | – | – | – | – | – | – | – | – | – | – | – | – | – | – |
| – | – | – | – | – | – | – | – | – | – | – | – | – | – | – | – | – | – | – | – | – | – | – |

the summer and in the dark season migrates to greater depths to reproduce. In winter this species stayed within the lowest layer of the Arctic surface water masses at a depth of 175–300 m (Harding, 1966; Pavshtiks, 1980). In my winter observations at NP-22 during 1979–1980, I observed a conspicuous rise in *A. glacialis* to the ice surface in January, although until that time the species was absent. In contrast to *A. glacialis,* two small amphipods, *Metopa longirama* and Neopleustes, were always found near the ice surface in June, disappearing later and reappearing in the fall and apparently staying near the ice during the entire winter.

The occurrence of *Pseudalibrotus nanseni, P. glacialis,* and *Gammaracanthus loricatus* on the underside of the ice also had a cyclic character: They were often found in the spring and fall but very seldom during the summer. Adults of both sexes as well as juveniles were present. It has been established (Barnard, 1959; George and Paul, 1970) that *P. nanseni* spawns in winter. Adults are found near the ice in autumn when the polar night begins. I observed a large number of specimens of this species in mid-November 1975 at NP-22 when the station was located over the Canadian Abyssal Plain, and in January–March 1980 at NP-22 and simultaneously at NP-24 when the first station was drifting along the east Siberian shelf and the second was not far from the geographic north pole (unpublished data). *P. nanseni* seems to be a very

**Table 9-4. Planktonic Fauna from NP-23 (1977–1978)**

| Species | | Layer of Catch (m) and Station No. | | | | | | | | | | | | | | | | | |
|---|---|---|---|---|---|---|---|---|---|---|---|---|---|---|---|---|---|---|---|
| | | 25–0 | | | | | | 50–25 | | | | | | 100–50 | | | | | |
| | | 1 | 2 | 4 | 6 | 8 | 9 | 1 | 2 | 4 | 6 | 8 | 9 | 1 | 2 | 4 | 6 | 8 | 9 |
| *Globigerina* sp. | | – | – | – | + | – | – | – | – | – | – | – | – | – | – | 1 | – | – | – |
| Radiolaria gen. sp. | | – | – | – | + | – | 1 | – | – | – | + | – | – | – | – | + | – | – | + |
| *Parasagitta arctica* | | 1 | 7 | 2 | 1 | 1 | – | 2 | 4 | 2 | – | 1 | + | 1 | + | + | – | 1 | – |
| *Eucrohnia hamata* | | – | – | – | – | – | 4 | – | – | – | – | 5 | 10 | – | 10 | 9 | – | 8 | 8 |
| *Calanus glacialis* | cop. III | 1 | 13 | 1 | – | 4 | – | 7 | 2 | – | – | 1 | – | 4 | – | – | – | 1 | – |
| | cop. IV | + | 20 | 1 | – | – | – | 1 | 18 | – | 1 | 2 | + | – | + | + | 3 | 3 | 1 |
| | cop. V | – | 3 | – | 2 | 1 | – | – | 7 | + | 8 | 28 | + | – | 1 | + | 4 | 13 | 7 |
| | ♀ | – | + | + | 1 | – | – | – | 9 | 2 | 17 | 23 | 2 | – | 2 | 2 | 9 | 7 | 13 |
| | ♂ | – | – | – | – | – | – | – | – | – | – | + | – | – | – | – | – | – | – |
| *C. hyperboreus* | cop. III | – | – | 2 | – | – | – | – | – | 2 | – | – | – | – | – | – | – | – | – |
| | cop. IV | – | – | 6 | 4 | – | – | – | 4 | 2 | 4 | – | – | – | 1 | – | 1 | + | – |
| | cop. V | – | – | + | 3 | – | – | – | 10 | 1 | 4 | – | – | – | 2 | + | 1 | – | – |
| | ♀ | – | – | – | 2 | – | – | – | 1 | – | 1 | – | – | – | + | + | 1 | 1 | – |
| *Pseudocalanus minutus* | ♀ | – | – | 2 | 2 | 2 | – | – | – | – | – | – | – | – | – | + | + | – | – |
| *P. elongatus* | ♀ | 15 | 49 | 1 | – | 1 | – | 10 | 18 | + | + | – | – | 4 | + | – | + | – | – |
| | ♂ | 5 | 6 | – | – | – | – | 5 | + | – | – | – | – | 3 | – | – | – | – | – |
| *Pseudocalanus* sp. | cop. II–III | 33 | 6 | – | – | 298 | 1 | 20 | – | – | – | 19 | – | 8 | – | – | – | 2 | – |
| | cop. IV–V | 73 | 18 | 2 | 1 | 269 | 2 | 56 | – | – | – | 59 | – | 21 | – | + | – | 7 | – |
| *Microcalanus pygmaeus* | cop. II–III | – | – | 17 | – | 29 | + | – | – | 4 | 20 | 37 | 2 | – | – | 11 | 20 | 12 | [illegible] |
| | cop. IV–V | 6 | – | 11 | 3 | 38 | 6 | 3 | 5 | 12 | 53 | 106 | 16 | 1 | 8 | 45 | 54 | 56 | [illegible] |
| | ♀ | 2 | 6 | – | – | 1 | 2 | 1 | 10 | 6 | 2 | 2 | 20 | 1 | 26 | 7 | 5 | 4 | [illegible] |
| | | – | – | – | – | – | – | – | – | 1 | 1 | – | – | + | 2 | 1 | + | – | + |
| *Spinocalanus magnus* | cop. V | – | – | – | – | – | – | – | – | – | – | – | – | – | – | – | – | – | – |
| | | – | – | – | – | – | – | – | – | – | – | – | – | – | – | – | – | – | – |
| *Metridia longa* | cop. II | – | – | – | – | – | – | – | – | – | + | 43 | + | – | – | 1 | – | 29 | – |
| | cop. III | – | – | – | – | – | – | – | + | – | + | 3 | 3 | – | – | 1 | + | 4 | [illegible] |
| | cop. IV | – | + | 1 | – | – | + | – | 1 | + | + | 2 | + | – | + | 2 | 2 | 1 | |
| | cop. V | – | – | + | 2 | 2 | – | – | + | 1 | 15 | 13 | 4 | – | + | 2 | 3 | 2 | [illegible] |
| | ♀ | – | – | – | – | – | + | – | 3 | – | 5 | 17 | 1 | – | 2 | 2 | 2 | 3 | [illegible] |
| | | – | – | – | – | – | – | – | – | + | – | 1 | – | – | + | + | – | + | – |
| *Lucicutia polaris* | ♀ | – | – | – | – | – | – | – | – | – | – | – | – | – | – | – | – | – | – |
| | ♂ | – | – | – | – | – | – | – | – | – | – | – | – | – | – | – | – | – | – |
| *Mormonilla polaris* | ♀ | – | – | – | – | – | 1 | – | – | – | – | + | + | – | – | + | – | – | + |
| | ♂ | – | – | – | – | – | – | – | – | – | – | – | – | – | – | – | – | – | – |
| | cop. III–V | – | + | – | – | – | – | – | – | – | – | – | + | – | – | – | – | – | – |
| *Oithona similis* | cop. III–V | 16 | 27 | 372 | 152 | 1248 | 46 | 18 | 32 | 179 | 52 | 350 | 69 | 12 | 8 | 96 | 94 | 38 | |
| | ♀ | 47 | 78 | 84 | 214 | 102 | 16 | 43 | 121 | 105 | 136 | 72 | 45 | 34 | 27 | 97 | 37 | 41 | [illegible] |
| | ♂ | – | + | 8 | 8 | – | – | + | 1 | – | 6 | 2 | – | + | 1 | 1 | 1 | 5 | – |
| *Cyclopina schneideri* | cop. III–V | – | – | – | – | 3 | – | – | – | – | – | 1 | – | – | – | – | – | – | – |
| *Oncaea borealis* | ♀ | + | 29 | 20 | 2 | 1 | 1 | – | 29 | 5 | 3 | 6 | 4 | – | 16 | 51 | 10 | 11 | |
| *Tisbe furcata* | cop. III–V | 1 | 2 | + | – | – | – | – | – | – | – | – | – | – | – | – | – | – | – |
| | ♀ | + | – | – | + | – | + | – | – | – | – | – | – | – | – | – | – | – | – |
| | ♂ | – | – | 1 | – | – | – | – | – | – | – | – | – | – | – | – | – | – | – |
| *Harpacticus superflexus* | ♀ | – | – | – | – | – | – | – | – | – | 1 | – | – | – | – | – | + | – | – |
| *Microsetella norvegica* | ♀ | – | – | – | + | + | – | – | – | – | – | – | – | – | – | – | – | – | – |
| *Parathemisto abyssorum* | | – | + | 2 | 1 | – | – | – | 1 | + | + | – | – | – | 1 | + | – | + | – |
| *P. libellula* | | – | – | – | – | – | – | – | – | – | – | – | – | – | – | – | – | – | – |
| *Cyclocaris guilelmi* | juv. | – | – | – | – | – | – | – | – | – | – | – | – | – | – | – | – | – | – |
| *Eusirus holmi* | ♀ | – | – | – | – | – | – | – | – | – | – | – | – | – | – | – | – | – | – |
| *Hymenodora glacialis* | | – | – | – | – | – | – | – | – | – | – | – | – | – | – | – | – | – | |
| *Limacina helicina* | | – | 11 | – | – | – | – | + | 11 | – | – | – | – | 1 | + | – | – | – | |
| *Clione limacina* larv. | | – | 2 | – | 1 | – | + | – | 1 | 2 | 1 | – | 1 | – | + | 1 | 1 | 2 | |

wide-ranging species in the Arctic Ocean, and probably, *P. nanseni*, *P. glacialis*, and *G. loricatus* are occupying the logical niche of *A. glacialis*, which develops at depth during the winter. The following animals were found near the ice due to geographic factors.

The amphipods *Anonyx nugax* and *Weyprechtia pinguis* and the decapod *Eualus belcheri* were collected by means of conventional baited wire-mesh traps using various kinds of meat when NP-23 was drifting over the east Siberian shelf. They are typical Arctic–boreal species (Bushueva,

| *150–100* | | | *200–100* | | | *300–200* | | | *500–200* | | *500–300* | | | *1000–500* | | | | | AHO–1000 | | | |
|---|---|---|---|---|---|---|---|---|---|---|---|---|---|---|---|---|---|---|---|---|---|---|
| *1* | 2 | *4* | 6 | *8* | 9 | 2 | 4 | 6 | 8 | 9 | 2 | 4 | 6 | 2 | 4 | 6 | *8* | 9 | 2 | 6 | 8 | 9 |
| — | — | — | — | — | — | — | — | — | — | — | — | — | — | — | — | — | — | — | — | — | — | — |
| — | — | + | 1 | 1 | 4 | — | — | + | + | + | — | 2 | + | — | — | — | + | — | — | + | 1 | — |
| + | — | — | + | — | — | — | — | — | — | — | — | — | — | — | — | — | — | — | — | — | — | — |
| — | 5 | 4 | 2 | 3 | 14 | 1 | 3 | 1 | 3 | 10 | 2 | + | + | + | — | — | 1 | + | — | — | — | + |
| 2 | — | — | — | 2 | — | — | — | — | + | — | — | — | — | — | — | — | — | — | — | — | — | — |
| 2 | + | 1 | + | 1 | + | + | — | + | + | — | — | + | — | — | + | — | — | — | — | — | — | — |
| — | 1 | 1 | + | 1 | 2 | — | + | + | + | — | — | — | + | — | — | + | — | + | — | — | — | — |
| — | 1 | + | + | 1 | 2 | + | + | — | + | — | — | — | + | — | — | — | — | — | — | — | — | — |
| — | — | — | — | — | — | — | — | — | — | — | — | — | — | — | — | — | — | — | — | — | — | — |
| — | — | 1 | — | + | + | — | — | — | — | + | — | + | — | + | — | + | + | + | — | — | — | — |
| 1 | — | — | + | 1 | + | + | + | + | + | + | + | — | + | + | + | + | + | 1 | + | + | + | + |
| — | — | — | 1 | + | + | + | — | + | + | + | — | — | + | + | + | + | + | + | + | + | + | + |
| — | + | + | 1 | + | + | + | + | + | + | + | + | + | + | + | — | + | + | + | + | + | + | + |
| — | — | + | + | — | + | — | + | — | — | — | — | — | — | — | — | — | — | — | — | — | — | — |
| 1 | + | + | + | — | — | — | — | — | — | — | — | — | — | — | — | — | — | — | — | — | — | — |
| 1 | — | — | — | — | — | — | — | — | — | — | — | — | — | — | — | — | — | — | — | — | — | — |
| 1 | — | — | — | — | — | — | — | — | — | — | — | — | — | — | — | — | — | — | — | — | — | — |
| 3 | — | — | — | + | — | — | — | — | + | — | — | — | — | — | — | — | + | — | — | — | — | — |
| — | — | 7 | 11 | 6 | 5 | — | 5 | 6 | 2 | 1 | — | 1 | 1 | — | + | — | 1 | 1 | — | — | — | + |
| + | — | 29 | 24 | 25 | 18 | — | 16 | 9 | 7 | 6 | — | 4 | 3 | + | + | + | 4 | 5 | + | + | — | 2 |
| + | 9 | 6 | 6 | 4 | 3 | — | 3 | 3 | 3 | 1 | 1 | 3 | 1 | 1 | 1 | 2 | 5 | 2 | 1 | + | — | 2 |
| + | 1 | + | 1 | 1 | 1 | — | 1 | 1 | 1 | + | — | + | + | + | + | + | 1 | + | + | — | — | + |
| — | — | — | — | — | — | + | 1 | — | — | — | + | + | — | + | + | + | + | + | — | + | + | + |
| — | — | — | — | — | — | + | + | — | — | — | + | + | + | + | + | + | + | — | — | + | — | + |
| — | — | + | — | 8 | + | — | + | — | 1 | — | — | — | — | — | — | — | + | — | — | — | — | — |
| — | — | 1 | + | 3 | 2 | 3 | 3 | + | + | — | 1 | + | — | + | — | — | + | + | — | — | — | — |
| — | — | 1 | 1 | — | 2 | 3 | 2 | + | + | + | 2 | + | + | + | — | — | — | + | + | — | — | + |
| — | — | 1 | 1 | + | + | 1 | + | + | + | + | + | + | + | + | — | — | + | + | + | — | — | + |
| — | 1 | + | 1 | + | 2 | + | + | — | — | — | + | — | + | — | + | — | — | — | + | — | + | + |
| — | + | — | — | + | — | 1 | + | — | + | + | 1 | + | + | + | — | — | — | + | + | — | — | — |
| — | — | — | — | — | — | — | — | — | — | — | — | — | — | — | — | — | — | — | — | + | + | + |
| — | — | — | — | — | — | — | — | — | — | — | — | — | — | — | — | — | — | — | — | + | — | — |
| — | — | + | + | + | + | + | 3 | 2 | 1 | 1 | + | 1 | 1 | + | + | + | 1 | + | + | + | + | + |
| — | — | — | + | — | + | — | 1 | 2 | 1 | 1 | 1 | 1 | 1 | + | 1 | 1 | 1 | 1 | + | + | + | + |
| — | — | + | + | — | — | — | — | — | + | + | — | — | — | — | + | — | — | — | — | — | — | — |
| 1 | 17 | 24 | 11 | 22 | 10 | — | 1 | 1 | 1 | + | — | + | + | — | + | + | 1 | + | — | + | — | + |
| 27 | 8 | 6 | 3 | 6 | 20 | — | 1 | + | 1 | + | — | — | + | — | + | — | + | + | — | + | — | + |
| + | — | 1 | 2 | + | — | — | + | 1 | + | — | — | — | — | — | — | — | — | — | — | — | — | — |
| — | — | — | — | — | — | — | — | — | — | — | — | — | — | — | — | — | — | — | — | — | — | — |
| 1 | 10 | 3 | 11 | 5 | 6 | 7 | 1 | 2 | 2 | 1 | 1 | — | + | + | + | — | + | + | + | + | — | — |
| — | — | — | — | — | — | — | — | — | — | + | — | — | — | — | — | — | — | — | — | — | — | — |
| — | — | — | — | — | — | — | — | — | — | — | — | — | — | — | — | — | — | — | — | — | — | — |
| — | — | — | — | — | — | — | — | — | — | — | — | — | — | — | — | — | — | — | — | — | — | — |
| — | — | — | — | — | — | — | — | — | — | — | — | — | — | — | — | — | — | — | — | — | — | — |
| — | — | — | — | — | — | — | — | — | — | — | — | — | — | — | — | — | — | — | — | — | — | — |
| — | + | — | — | — | — | + | + | — | — | — | + | — | — | + | — | — | — | — | — | — | — | — |
| — | — | + | — | — | — | — | — | — | — | — | — | — | — | — | — | — | — | — | — | — | — | — |
| — | — | — | — | — | — | — | — | + | — | — | — | — | — | + | — | — | — | — | — | — | — | — |
| — | — | — | — | — | — | — | — | — | + | — | — | — | — | — | — | — | — | + | — | — | — | — |
| — | — | — | — | — | — | — | — | — | — | — | — | — | — | + | — | — | — | — | — | — | — | — |
| 2 | + | — | — | — | — | — | — | — | — | — | — | — | — | — | — | — | — | + | — | — | — | — |
| — | + | 3 | 1 | 1 | + | 2 | — | — | + | — | + | — | — | — | — | — | — | — | — | — | — | — |

1977) rising to the ice from the bottom. When the station was off the shelf and near the central sector of the Arctic Ocean, these species were not found. Single occasional occurrences of foraminifera, radiolarians, chaetognaths, and ostracods near the ice were recorded. These taxa were observed when the direction of the ice drift changed and when intensive mixing of surface Arctic water took place. The occurrence of common Arctic copepods, *Calanus glacialis, C. hyperboreus,* Pseudocalanus spp., *Oithona similis, Spinocalanus magnus, Lucicutia polaris, Metridia longa,* and *Mor-*

*monilla polaris,* near the surface of the ice is caused by the ontogenetic migration of these species. For example, populations of *C. glacialis* and *C. hyperboreus,* which inhabit the Intermediate Arctic and Atlantic water masses in summer, migrate upward to the undersurface of the ice where they spawn and grow (Pavshtiks, 1980). In June–July, *C. glacialis* was found in horizontal tows and in sample 26.06 (Table 9-3); the number of specimens was the highest for the entire year. Such abundance of *C. glacialis* was observed near the ice only once. The next day (28.06) their number decreased and between August and October the abundance of both species did not exceed several specimens per $m^3$. The migration of *C. hyperboreus* was the same as that of *C. glacialis.* More specimens of *Pseudocalanus elongatus* and *P. minutus* near the undersurface of the ice were observed simultaneously with increased abundances of these species in vertical plankton tows.

## Distribution and Behavior of the Cryofauna

The distribution of the cryopelagic fauna under the sea ice is related to certain environmental factors, and most important to

1. Type of ice
2. Morphological features of the underside of the ice
3. Hydrological conditions within the ice–water boundary layer

In the Arctic Basin, the open ocean is covered with 71–81 percent multiyear ice, 9–10 percent hummocked ice, 8–17 percent undeformed first-year ice and 0.2–1 percent leads in the ice (Koerner, 1973).

Based on a survey at NP-22, 23, and 24 between 1975 and 1981, the multiyear sea ice was considered as (1) a concentrator and transporter of substances in space and time (Mel'nikov and Pavlov, 1978), and (2) a biological substrate (Mel'nikov and Kulikov, 1980). In the ice three different biotops were distinguished subject to differences in the interacting factors at the water–ice and ice–atmosphere boundaries and within the ice itself. Each biotype is characterized by the composition of characteristic cryophilic fauna and flora. In spite of the mobility and variability of the substrate, the multiyear sea ice is inhabited by an unchanged cryophylic biota in space and time. Based on these criteria it is considered a steady-state ecological system (Mel'nikov, 1980).

The multiyear sea ice accumulates large quantities of organic matter produced by the ice flora during the summer. This matter represents a reserve of energy for the organisms of the cryopelagic community living during the polar nights. The multiyear sea ice is a substrate for the cryophylic fauna. Neither the underside of hummocked ice nor the undersurface of the first-year ice are inhabited by the cryopelagic fauna as is the multiyear ice (Table 9-5). The bottom of hummocked ice is 4–15 m deeper than the bottom of the multiyear ice (Grichenko, 1980), occurring in a turbulent zone where the velocity of water currents is high (Belakov, 1974).

First-year ice forms in winter when there is no photosynthesis and therefore there is no organic matter as compared with the multiyear sea ice that has concentrations 10–15 times greater than the first-year ice (Mel'nikov, 1980). Thus, the macrodistribution of the cryopelagic fauna within the Arctic Ocean is found mainly near the multiyear sea ice. The microdistribution of the cryopelagic fauna is related to the morphology of the ice underside as well as to the direction and velocity of the under-ice currents. The average thickness of the floes of the permanent ice pack has been estimated to be 3–5 m (Timofeev, 1960). The macrorelief of the underside of the multiyear sea ice consists of elements of tens to hundreds of meters long formed year-round during hummocking, melting, and growing. Elements of microrelief sev-

Table 9-5. Individuals of the amphipods *Gammarus wilkitzkii* and *Pseudalibrotus nanseni* and Polychaete *Antinoella sarsia* under m$^2$ of Ice

| | NP-22 Winter 1980 | | | NP-23 Summer 1977 | | | NP-24 Winter 1980 | | |
|---|---|---|---|---|---|---|---|---|---|
| *Species* | *m* | *f* | *h* | *m* | *f* | *h* | *m* | *f* | *h* |
| *G. wilkitzkii* | 10 | — | 1 | 15 | — | 2 | 11 | — | — |
| *A. sarsia* | 3 | — | — | 5 | — | 1 | 2 | — | — |
| *P. nanseni* | 18 | 1 | 2 | 14 | — | 1 | 21 | — | 3 |

h = hummocked ice; m = multiyear ice; f = first-year ice.

eral meters to several tens of centimeters cover the surface of the macrorelief and are formed by thermodynamic actions.

The type of macrorelief (in this case the underside of the hummocked part of the pack ice—named ''bummock'') and the under-ice currents form calm and turbulent zones on the underside of the drifting ice (Fig. 9-1). Their locations vary due to current direction. Abundant groups of animals are found in hydrologically calm zones. This distribution was characteristic of the species of the cryopelagic community that form swarms under the ice (mainly amphipods). Observations in the summer of 1977 at NP-23 and in the winter of 1980 at NP-22 were designed to show the swarming behavior of the amphipod *Parathemisto libellula*.

It was shown that (1) out of the swarm *P. libellula* was practically absent, (2) the swarms are obviously distributed in hydrologically calm locations (Fig. 9-1*a*), and (3) in times of high-velocity currents under the ice the swarms were destroyed and reappeared in calm zones (Fig. 9-1*b*). The shape and size of the swarm and the behavior of the animals varied sharply between seasons. In summer a swarm represents a group of actively moving individuals, but in winter *P. libellula* behaves more quietly: The motion is slow and individuals are immobile. The animals' heads are oriented toward the current.

The distance between individuals in summer swarms is about 2–3 cm and the number of animals is 15,000–30,000 individuals m$^3$ while in winter the distance between animals is approximately 10–15 cm and the number of animals is 500–1,000 individuals m$^3$. Winter swarms (6–8 m in diameter) are wider in space than summer ones, and the shapes vary from round to various ellipsoid forms. Swarm formation is a well-known phenomenon for many planktonic animals in temperate and tropical latitudes. Swarms are formed by some calanoid copepods (Hamner and Carleton, 1979), mysids (Clutter, 1969), euphausids (Brown, Barker, and Gaskin, 1979) decapods (Berenboim and Popkov, 1980), medusae, ctenophores, and pteropods (Harbison and Madin, 1979). Information about the swarming of the under-ice cryopelagic fauna was not available until recently. It is interesting to note that masses of the diatom swarms of *Melosira arctica* (Mel'nikov, 1980) were observed to be surrounded by calm zones under the sea ice. When the velocity of the ice drift increased, the masses were generally destroyed and pulled apart by the strong current under the ice.

The distribution of cryofauna is affected by cracks, fractures, and polynyas formed during shearing and fracturing of the ice cover as well as by the macrorelief caused by the bummocks. Their influence on the distribution of *Calanus glacialis*, a dominant planktonic species of the Arctic water, was particularly noted. This species has a seasonal vertical ontogenetic migration

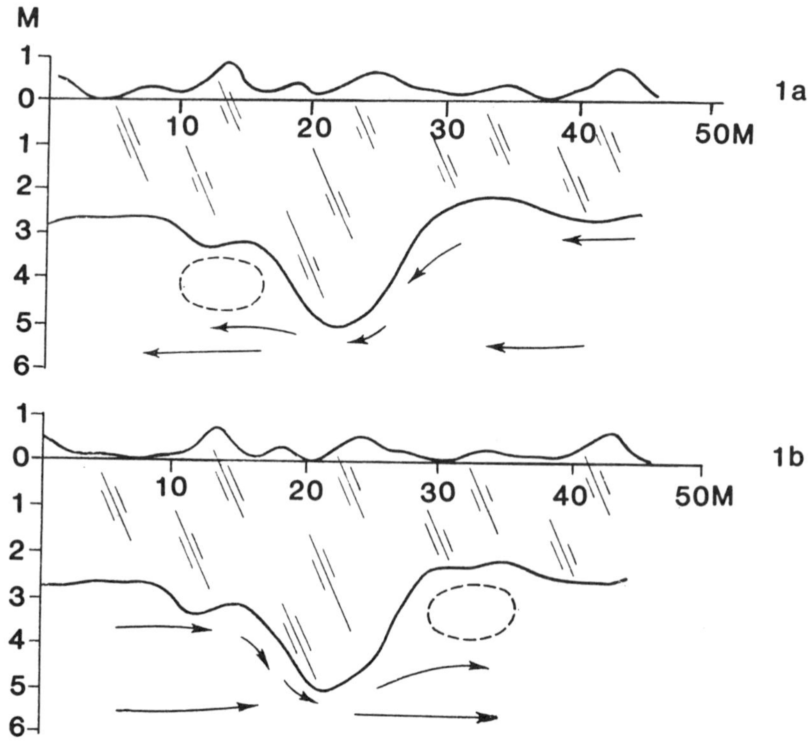


Figure 9-1. Influence of macrorelief on the formation of the hydrodynamic calm zones near the underside of the drifting sea ice; arrows indicate the under-ice currents; ellipsoids show the position of crowd *Parathemisto libellula* in relation to a vector drift.

from depth to the phytoplankton-rich surface waters and sometimes up to the underside of the ice (Pavshtiks, 1980). The highest concentrations of *C. glacialis* were observed in August. Observations have been carried out under multiyear ice (NP-23) and within a lagoon after the development of fractures; observations were made three times at 12-hour intervals. Results are shown in Table 9-6.

Such abundances of *C. glacialis* in fractures along with its scarcity under ice (catches 23.06 and 4.08; Table 9-6) may be interpreted as follows. On the one hand,

Table 9-6. *Calanus glacialis* in Samples Obtained from the Underside of the Ice and in a Lagoon (NP-23, Summer 1977)

| | *Date of Sampling and Depth of Catches (m)* | | | | | | | | | |
|---|---|---|---|---|---|---|---|---|---|---|
| | *23.07* | | *26.07* | | | | *27.07* | | *4.08* | |
| *Individuals* | *0* | *5* | *0* | *5* | *0* | *5* | *0* | *5* | *0* | *5* |
| Copepodits II–IV | — | — | 36 | 61 | 80 | 138 | 8 | 4 | — | 2 |
| Copepodits V | — | — | 48 | 150 | 76 | 500 | 4 | — | — | — |
| Females | — | — | — | 16 | 15 | 96 | — | — | — | — |

Catches on 26 July and 27 July were made in a lagoon. On 26 July catches were made twice at 12-hour intervals. Catches on 23 July and 4 August were carried out under the multiyear ice.

there is a light signal for animals moving toward the surface even when the ice cover is permanent, and so the incident solar radiation is decreased 3–5 orders of magnitude by the ice (Smith, 1973). The main population of the *C. glacialis* at this time is distributed at depths of 25–150 m. In the water adjacent to the ice, as well as at 5 m, not more than 2 individuals were found (Table 9-6). After the fracturing of the ice cover, the light penetrates deeper into the water column. In the dark field of vision for the crustaceans, a light patch or an irritant appear to be a signal for migration toward the surface and at that time part of the population rises toward the surface. On the other hand, there are calm hydrodynamic conditions in a lagoon: Currents are very sluggish and drift of the ice cover occurs. After a heavy storm (26–27 July), the swarm was destroyed and the number of *C. glacialis* abruptly decreased both in tows near the ice and at 5 m (Table 9-6). Apparently the animals were dispersed by a strong under-ice current, but some specimens were able to stay within the lagoon.

In the water–ice zone interface, cryoorganisms are settled among small cracks, crystals, and other morphological features that form the microrelief of the underside of the multiyear ice. It is well known (Belakov, 1974; Shirshov, 1938) that the minimum velocity of the under-ice current occurs under the surface of the ice and the maximum velocity occurs at a 20–30 m depth. Probably these differences in current velocity may produce better hydrodynamic conditions at the underside of the ice than at depth. However, in summer when the ice is melting and in winter when the ice is growing, animals of the cryopelagic community, the autochthonous fauna *Gammarus wilkitzkii, Mysis polaris, Jaschovia johnsoni, Tisbe furcata,* and *Harpacticus superflexus,* continue to live on the underside of the multiyear ice.

During under-ice dives certain behavioral peculiarities of *G. wilkitzkii* and *M. polaris* near the ice surface were observed. An animal must leave the ice surface for periods of time to look for food or for other reasons. The change in location occurs by locomotion on the ice surface, active swimming, or passive transportation by the under-ice currents. The latter happens as follows: An animal takes off from the substrate at a distance of 15–20 cm, and at this moment it is caught by the under-ice current, which carries it from one place to another, where the animal again fastens to the ice. If the new place is already occupied by other animals or is not suitable, the animal repeats its jump once more. Such image-jumping change-of-place is characteristic of *Mysis polaris. Gammarus wilkitzkii* is a more active swimmer, but it also moves in this manner.

In addition to living between rough features that form the microrelief of the underside of the ice, cryopelagic fauna also settle on the vertical walls of the hummocked ice. Summer observations at NP-23 (1977) indicated the occurrence of the typical plankter *Calanus hyperboreus* near the ice. On the vertical walls of hummocked ice there were many cavities that contained numerous individuals of *C. hyperboreus.* The animals did not move, but they were alive because they began to swim when they were turned out of the cavities. It is interesting that the form of cavities resembled the crustacean shape. Probably after finding a free place not occupied by other animals, *C. hyperboreus* lies on the ice and apparently the metabolic heat of the individual and the sun radiation heat absorbed by the crustacean's carapaces were sufficient to melt the surrounding ice. This phenomenon is of significance as well for other groups of the Arctic fauna including *Gammarus wilkitzkii* and *Pseudalibrotus nanseni.* During dives at NP-23 (summer 1977) and NP-22 (winter 1980) we found repeatedly thawed patches containing both crustaceans on the vertical walls of the hummocked ice. The form of the thawed patches was the same as the half-moon shape of the crustaceans. Apparently this behavior of crustaceans is one form of adaptation to conditions of the dynamic ice substrate in the Arctic Ocean.

## DISCUSSION

Andrijashev (1967) considered the ice fauna in the Antarctic and suggested two ecological groups of animals: (1) the ice fauna proper—animals living at least temporarily in the lower part of the ice, and (2) the sub-ice fauna—organisms that do not penetrate directly into the friable ice, but are connected trophically with the underside of the ice. The first group includes Polychaeta (2 species), Calanoida (the young stage), Cyclopoida (1 species), Harpacticoida (3 species), Amphipoda (2 species), and one fish species (*Trematomus bochgrevinki*). The second group includes planktonic crustaceans Euphausiidae, Hyperiidae, Gammaridea, Calanoida, and some fish species. Andrijashev (1967) mentioned that the criteria for identifying the sub-ice species are less well-founded than those for the ice fauna proper. The accurate boundary between these ecological groups is not available and it is possible that some animals, mainly crustaceans, which belong to the second group, may belong to the first association.

The absence of an accurate boundary is explained by the ice structure. The friable ice but not the solid lower part of the ice are permeable to all organisms of the herbivorous plankton. Meguro (1962) introduced the term *plankton ice* to characterize this type of ice. When it becomes colder, the freezing from friable to solid structures takes place. At this time, the fauna of the first ecological group either leaves the brines of the plankton ice or stays in part in the ice. Observations on the components of the cryopelagic fauna, which are frozen into fast ice (Andrijashev and Gruzov, 1974), showed that they include not only planktonic crustaceans but some benthic invertebrates such as actinias, sea urchins, and starfishes. After the melting of the ice, these animals fell out into the water or to the bottom and continued their development.

Quite different conditions occur in Arctic, where the cryophilic fauna develop under conditions of a moving and always growing substrate (Mel'nikov, 1980). The drifting multiyear ice grows constantly upward with rates depending on the ice thickness. For example, in winter if the ice thickness is 3–4 m it grows downward at a rate of 1, 1–0, 15 cm/day and results in a thickening of 30–70 cm per winter (Grichenko, 1980; Mel'nikov, 1978). The rate of growth and consequently the thickness of young ice are greater than for old ice. The melting of the underside of the ice is very slow and occurs only in the young layer of the ice that forms on the undersurface of the old ice as a result of the melting water, which is about 5–25 cm (Maykut and Grenfell, 1975; Grichenko, 1980; Mel'nikov, 1978).

In conditions of a continuously growing substrate, the probability of animals returning to the water from the ice is very small. Microscopic analysis of a seston filtered from ice samples showed that adults and juveniles are not found in the ice. Only the interestitial fauna such as nematodes, turbellarians, and protozoans were found in ice. Because of their size, form, and manner of locomotion, these animals are most suitable for living in the narrow space of brine pockets between ice crystals (Mel'nikov, unpublished data).

One of the most outstanding features of the cryopelagic fauna of the drifting sea ice is that all organisms are actively moving in contrast to "pagon's animals" (term coined by Zernov, 1928). Cryophilic orgaisms do not have an anabiotic stage, but pagon's animals survive in ice, and in spring when ice melting begins, they return into the water. This ecological feature of the development of some water invertebrates is characteristic of both freshwater

(Zernov, 1928) and seawater taxa (Andrijashev and Gruzov, 1974). Zernov (1928) listed 117 species of planktonic animals that can be frozen into the ice in winter and come to life in spring after the ice melts.

The absence of a resting stage for organisms living during the polar night, the stage when photosynthesis does not occur, is not profitable from the point of view of the energy flow throughout the biotope. This seems to suggest that at this time the cryophilic organisms have decreased locomotory activities, especially organisms belonging to the first ecological group. Experiments carried out with *Gammarus wilkitzkii* (George and Paul, 1970) show that this species seems to have long resting periods followed by periods of constant swimming. Individuals can be stationary without exhibiting locomotory or swimming movements for periods up to six hours.

The populations of *Gammarus wilkitzkii* in the Central Arctic Ocean attain dimensions larger than populations inhabiting the surrounding marginal seas: 40–46 mm (Kulikov, 1980) and 30–36 mm, respectively (Tzvetkova, 1977), following Bergman's rule (Geptner, 1937). Increasing size and decreasing locomotion are considered a consequence of lowering metabolic rates in conditions of scarcity of organic matter during polar nights. In the present author's view, this is a form of adaptation to the extreme Arctic conditions.

Finally, it is necessary to consider some biogeographical features of the cryopelagic faunal distribution in the Arctic Ocean. It has been shown (Gordienko, 1958; Dunbar and Wittman, 1963) that in the Arctic Ocean there are two macroscale circulation patterns of ice: (1) along the 180° meridian, toward the North Pole, and along the great circle route from the New Siberian Islands to Spitzbergen, and (2) an anticyclonic gyre in the Canadian subbasin between the East Siberian, Laptev, Lincoln, Beaufort, and Chukchi seas. From this general scheme it may be seen that the animals of the cryopelagic community, related to the drifting ice in the gyre, are staying in theory infinitely within the gyre's area.

By contrast, the cryofauna of the Arctic Ocean and the continental shelves of the surrounding seas are advected into the North Atlantic Ocean by the trans-Arctic ice drift. Biological evidence for the penetration of cryofauna through the Fram Strait into the North Atlantic Ocean by the trans-Arctic ice drift has been reported for several groups and expatriate species have been used to indicate the degradation of the ice ecosystem (Mel'nikov, 1980). As all pack ice is eventually carried in the trans-Arctic drift, we may consider the continuous loss of some part of the population. There is the loss of the autochthonous cryofauna, because in this region there is no reverse flow from the Fram Strait into the Arctic Ocean. From a biogeographical point of view (Beklemishev, 1969) the entire functional structure of the cryopelagic biotope may be as follows:

1. The zone of the anticyclonic gyre—this is where the cryofauna remains for a long period.
2. The zone of the trans-Arctic drift—the unsterile region of expatriation where populations of all species are maintained by the import of cryofauna from other regions (e.g., from the anticyclonic gyre) for 2–3 years. The population cannot remain here permanently because of the ice drift in one direction (from the Arctic Ocean to the Fram Strait).
3. The zone of the North Atlantic—the sterile region of species expatriation where animals of the cryopelagic community fall into the water after the degradation of the ice ecosystem and are incorproated here in the food chain of boreal communities.

It is well known that micro- and mesoscale eddies occur within the macroscale ice circulation (Gorbunov, Losev, and Kulakov, 1982). Gorbunov and Losev (1979) observed 54 ice eddies simultaneously at the surface over an area of 12,000 km$^2$. Each eddy was 3–5 km in diameter. Probably ice floes are exchanged at the periphery when the eddies move. It is suggested that this form of ice exchange promotes the separation of animals in the basin and, apparently, is the cause of the uniform distribution of the cryopelagic fauna in the Arctic Ocean. Observations carried out for three seasons in different regions (Canada Abyssal Plain, the shelf of the East Siberian Sea, and the eastern slope of the Lomonosov Ridge not far from the North Pole) show that dominant species of the cryopelagic community such as *Gammarus wilkitzkii* and *Mysis polaris* were nearly always found in the Central Arctic Ocean. On these grounds we may conclude that these and probably other components of the cryopelagic population in the basin are stable populations.

One other factor is of importance in maintaining the stability of populations of the allochthonous cryofauna. It has been shown that animals that belong to this ecological group stayed under the ice only during a part of their life cycle and occurred at depth during the rest of their life cycle, which is characteristic for heterotopic hemipopulations of the cryofauna. Animals such as *Apherusa glacialis* and *Pseudalibrotus nanseni* perform annually inverse ontogenetic migrations from the Atlantic water to the underside of the ice. Until now it has not been possible to establish precisely the time of occurrence of animals in each biotope. However, there are indirect data suggesting that independent populations of the cryofauna correspond to the allochthonous ecological group within the transarctic drift of the ice.

Figure 9-2 shows a schematic cross-section along 180°L, a section along the trans-Arctic drift of the ice. The relative scale between the surface Arctic water and the Atlantic water as well as the direction and relative current velocity are shown. The Atlantic water flows into the Arctic Ocean through the Fram Strait (Timofeev, 1960; Coachman and Barnes, 1963), but the surface Arctic water movement corresponds to the trans-Arctic drift of the ice (Fig. 9-2). The velocity of these currents is different: Both the surface water and the drifting ice have speeds $V_1$ = 10–50 cm/c (Stepanov, 1969), and the Atlantic water is $V_2$ = 3 cm/c and $V_2$ = 8 cm/c (Timofeev,

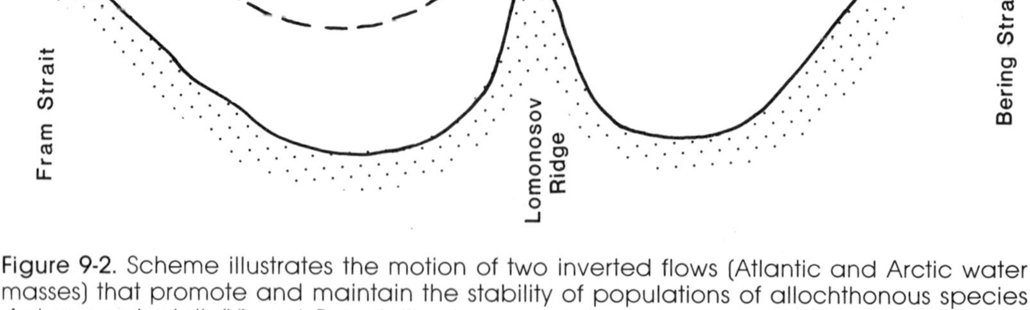


Figure 9-2. Scheme illustrates the motion of two inverted flows (Atlantic and Arctic water masses) that promote and maintain the stability of populations of allochthonous species *Apherusa glacialis* (X) and *Pseudalibrotus nanseni* (O) in the Arctic Ocean.

1960), respectively. The time of finding animals near the ice is $T_1$ and in the Atlantic water it is $T_2$. Index *X* is related to *Apherusa glacialis* and *O* is related to *Pseudalibrotus nanseni*.

When in t. *A. glacialis* begins to rise to the surface from the Atlantic water, *P. nanseni* sinks. Because of the inverted currents, the animals are transported by the first current out of the basin, but the second returns them. For the maintenance of the independent populations of these species, it is necessary to observe certain conditions, for example, $V_1T_1 = V_2T_2$. As $V_1 > V_2$, so $T_2$ must be longer than $T_1$, that is, the residence of animals at depth must be longer than near the ice surface. Data from NP-23 (Table 9-3) and in winter 1980 at NP-22 (unpublished data) show that *A. glacialis* as well as *P. nanseni* spend the winter (approximately 5 months) near the ice and the rest of the time in the Atlantic water ($T_2 > T_1$). Based on the above data it may be concluded that the heterotopic unity of the allochthonous cryofauna and the hydrodynamic conditions within the trans-Arctic zone promote the retention of animals and consequently maintain the population's stability in this region of the Arctic Ocean.

## CONCLUSION

On the basis of material obtained from NP-22, NP-23, and NP-24 (1977–1980), it has been shown that the underside of the sea ice acts as a substrate for the cryopelagic fauna of approximately 50 species. The cryopelagic community consists of two ecological groups: autochthonous and allochthonous faunas. Animals of the first group live during the year near the underside of the ice. On this surface we found animals of both sexes and of all developmental stages. Populations are stable throughout different regions of the basin, but only within the Arctic Ocean.

The abundance of allochthonous fauna varies with season. The variations have been attributed to environmental factors and to the life histories of the animals, caused by vertical ontogenetic migration of planktonic species. Since the ice cover is known to move geographically and relative to the water layers in the Arctic Ocean, spatial planktonic species changes are large enough to change the species composition of the cryopelagic biocenosis during the year. The apparent uniformity of the enviroment beneath the permanent ice pack suggests that the distribution and behavior of the cryopelagic fauna obtained in this study can be applied with some confidence to other areas of the Arctic Ocean.

## REFERENCES


Andrijashev, A. P. 1967. Microflora and fauna associated with the Antarctic fast ice, *Zool. Zhurnal* **44**(10):1585–1594.

Andrijashev, A. P. 1970. Cryopelagic fishes of the Arctic and Antarctic and their significance in polar ecosystems, *Antarct. Ecol.* **1**:294–303.

Andrijashev, A. P., and E. N. Gruzov. 1974. Biological communities near shore Antarctica. In *Nauka i Chelovechestvo*. Znanie, Moscow, pp. 101–121.

Apollonio, S. 1961. The chlorophyll content of Arctic sea ice, *Arctic* **14**(3):197–200.

Barnard, J. L. 1959. Epipelagic and under-ice Amphipoda of the Central Arctic basin, *Geophysical Res. Pap. 63,* **1**:115–129.

Beklemishev, C. W. 1969. *Ecology and Biogeography of the Open Ocean*. Nauka, Moscow, 291p.

Belakov, L. N. 1974. Drift currents under ice in the Arctic basin. *Oceanology* **14**(2):256–263.

Berenboim, B. I., and G. V. Popkov. 1980. Some results on the underwater observations for the behavior of shrimp Pandalus borealis (Kr) in Barents Sea. *Trudy Polar Res. Inst. Marine Fish. Oceanogr. Murmansk* **44**:56–63.

Bradstreet, N. W., and W. E. Cross. 1982. Trophic relationships at high Arctic ice edge. *Arctic* **35**:1–12.

Brown, R. G. B., S. P. Barker, and D. E. Gaskin. 1979. Raytime surface by Meganyctiphanes norvegica (M. Sars) (Crustacea, Euphausiacea) off Brier island, Bay of Fundy. *Canadian J. Zool.* **57**(12):2,285–2,291.

Bunt, J. S., and E. J. F. Wood. 1963. Microalgae

and Antarctic sea ice, *Nature* **199**(4,900): 1,254–1,255.

Bushueva, I. V. 1977. About ecology and distribution of Amphipoda-Gammaridea in shoal waters of the Novaya Zemlya and Franz-Josef Land. *Explor. Fauna Seas* **14**(22):277–290.

Clasby, R., R. Horner, V. Alexander. 1972. Ecology and metabolism of sea-ice organisms. *Inst. Marine Scin. (Univ. Alaska, Fairbanks), Rep. R72,* 56p.

Cleve, P. T., and A. Grunov. 1880. Beitrage zur Kenntniss der arctischen Diatomeen. *K. Sv. Vetesk. Akad. Handl.* **17**(2):1–122.

Clutter, R. T. 1969. The microdistribution and social behavior of some pelagic mysid shrimps. *J. Exp. Marine Biol Ecol.* **3**:125–155.

Coachman, L. K., and C. A. Barnes. 1963. The movement of Atlantic water in the Arctic Ocean. *Arctic* **16:**9–16.

Cross, W. E. 1982. Under-ice biota at the Pond Inlet ice edge and in adjacent fast ice areas during spring. *Arctic* **35:**13–27.

Dickie, G. 1880. On the algae found during the Arctic expedition. *J. Linn. Soc. Bot.* **17:**6–12.

Dunbar, M., and W. Wittman. 1963. Some features of ice movement in the Arctic basin. In *Proceedings of the Arctic Basin Symposium, 1962.* Arctic Institute of North America, Arlington, Va., pp. 90–108.

Ehrenberg, C. G. 1853. Uber neue Anschauungen des kleinsten nordlichen Polarlebens, *Monatsber. d. Berl. Akad.* **1,853:**522–529.

George, R. Y., and A. Z. Paul. 1970. USC-FSU biological investigations from the Fletcher's ice island T-3 on deep-sea and under-ice benthos of the Arctic Ocean. University of Southern California, Department of Biological Sciences, Tech. Report 1.

Geptner, V. G. 1937. *General Zoogeography.* Biomedgiz, Moscow, 358p.

Golikov, A. N., and V. G. Averinzev. 1977. Biocoenoses of the upper regions of the shelf of Franz-Josef Land archipelago and some regularities of their distribution. In O. A. Scarlato (ed.), *Biocoenoses of the Shelf of Franz-Josef Land and the Fauna of Adjacent Waters.* pp. 5–54.

Golikov, A. V., and O. A. Scarlato. 1973. Comparative characteristics of some ecosystems of the upper regions of the shelf in tropical, temperate and Arctic waters. *Helgolander Wiss. Meersunters.* **24**:219–234.

Gorbunov, Yu. A., and S. M. Losev. 1979. Some features of local distribution of the sea ice in relationship with its drift. *Trudy AANII* **364:**64–69.

Gorbunov, Yu. A., S. M. Losev, and I. Yu. Kulakov. 1982. Structure features of the sea ice drift in Arctic ocean. *Material of Glaciol. Invest. Moscow* **45:**51–56.

Gordienko, P. A. 1958. Arctic ice drift. In *Arctic Sea Ice.* National Research Council Pub. 598, National Academy of Science, Washington, D.C.

Gran, E. N. 1904. Diatomaceae from the ice-floes and plankton of the Arctic Ocean. In F. Nansen (ed.), *Norwegian N. Polar Expedition 1893–98.* Scientific Research of the Norwegian N. Polar Expedition, vol. 4, series 11, pp. 3–74.

Grichenko, V. D. 1980. About morphology of the sea ice surfaces of the Arctic Ocean. In *Biology of the Central Arctic Basin.* Nauka, Moscow, pp. 35–56.

Gruzov, E. N., M. V. Propp, and A. F. Pushkin. 1967. Biological communities of the near-shore regions of the Davis sea (on the results of underwater diving observations). *Inform. Bull. SAE* **65:**124–141.

Gulliksen, B. 1984. Under-ice fauna from Svalbard waters. *Sarsia* **69**(1):17–23.

Hamner, W. M., and J. H. Carleton. 1979. Copepod swarms: Attributes and role in coral reef ecosystems. *Limnol. Oceanogr.* **24**(1):1–14.

Harbison, G. R., and L. P. Madin. 1979. Diving—a new view on plankton biology. *Oceanus* **22**(2):18–27.

Harding, G. C. H. 1966. Zooplankton distribution in the Arctic Ocean with notes on life cycles, Master's thesis. McGill University.

Herman, Y. 1970. Arctic paleo-oceanography in late Cenozoic time. *Science* **169:**474–477.

Herman, Y., and D. M. Hopkins. 1980. Arctic Ocean climate in late Cenozoic time. *Science* **209**(4,456):557–562.

Horner, R. 1976. Sea ice organisms. *Oceanogr. Marine Biol. Ann. Rev.* **14:**167–182.

Koerner, R. M. 1973. The mass balance of the sea ice of the Arctic Ocean. *J. Glaciol.* **12**(65):123–140.

Kulikov, A. S. 1980. About ecology of two species of gammarids (Amphipoda, Gammaridae) and mysid (Mysidacea) in cryopelagic biocoenosis of the Central Arctic Ocean. In *Biology of the Central Arctic Ocean.* Nauka, Moscow, pp. 111–118.

Maksimov, E. V. 1972. Problems of glaciation of the Earth and rhythms in nature. Nauka, 296p.

Maykut, G. A., and T. C. Grenfell. 1975. The spectral distribution of light beneath first-year sea ice in the Arctic Ocean. *Limnol. Oceanogr.* **20:**554–563.

Megu, O. H. 1962. Plankton ice in the Antarctic Ocean. *Antarct. Rec.* **14:**1192–1199.

Meguro, H., K. Ito, and H. Fukushima. 1967. Ice flora (bottom type): A mechanism of primary production in Polar seas and the growth of diatoms in sea ice. *Arctic* **20**(2):114–133.

Mel'nikov, I. A. 1978. Hydrobiological studies in the Central Arctic Ocean (spring, 1978). *Oceanology* **18**(6):1,132.

Mel'nikov, I. A. 1979. Cryobiological observations in two Central Arctic Ocean (methods and some results of the studies). *Oceanology* **19**(1):150–156.

Mel'nikov, I. A. 1980. Ecosystem of the arctic drifting sea ice, In *Biology of the Central Arctic Ocean.* Nauka, Moscow, pp. 61–97.

Mel'nikov, I. A. 1981. Hydrobiological studies in the Arctic Ocean (NP-22, December 1979–April 1980). **21**(2):397–398.

Mel'nikov, I. A. 1982. Hydrobiological studies at SP-22 in the Arctic Ocean (spring, 1981). *Oceanology* **22**(1):158.

Mel'nikov, I. A., and A. S. Kulikov. 1980. Cryopelagic fauna of the Central Arctic ocean. In *Biology of the Central Arctic Ocean.* Nauka, Moscow, pp. 97–111.

Mel'nikov, I. A., and G. L. Pavlov. 1978. Peculiarities of organic carbon distribution in the waters and ice of the Arctic basin. *Oceanology* **18**(2):248–254.

Mel'nikov, I. A., and V. D. Tsynovsky. 1978. Hydrobiological studies in the Arctic Ocean (May–October, 1977). *Oceanology* **18**(2):378–379.

Mohr, J. L. 1959. Marine biological work. In *Scientific Studies at Fletcher's Ice Island, T-3, 1952–1955.* Geophysical Research Papers 3, vol. 1, pp. 82–103.

Mohr, J. L., and J. Tibbs. 1963. Ecology of ice substrates, In *Proceedings of the Arctic Basin Symposium, October, 1962.* Arctic Institute of North America, Arlington, Va., pp. 245–249.

Nesis, K. N. 1983. Panarctic ice sheet during the Pleistocene? *Marine Biol. Vladivostok* **6:**11–20.

Olausson, E., and U. C. Jonasson. 1969. The Arctic Ocean during the Wurm and early Flandrian. *Geologika Foreningens i Stockholm Forbandlingar* **91:**185–200.

Palibin, I. V. 1925. Microorganisms as a destroyer of the polar sea ice. *Izvestija Zentr. Gidrometeorol. Buro* **25.**

Pavshtiks, E. A. 1980. About some regularities in plankton life of the Central Arctic Ocean. In *Biology of the Central Arctic Ocean.* Nauka, Moscow, pp. 142–154.

Peckham, V. 1964. Year-round SCUBA diving in the Antarctic. *Polar Rec.* **12**(77:)143–146.

Ray, C. 1966. Stalking seals under Antarctic ice. *Geogr. Mag.* **129:**54–65.

Shirshov, P. P. 1938. Oceanological observations. *Acad. Nauk USSR (Doklady)* **19**(8):569–580.

Shoemaker, C. 1955. Amphipoda collected at the Arctic Laboratory, office of Naval Research, Point Barrow, Alaska, by G. E. MacGinitie. *Smithsonian Misc. Colln.* **128**(1):1–78.

Smith, R. C. 1973. Optical properties of the Arctic upper waters. *Arctic* **26**(4):63–67.

Stepanov, V. N. 1969. About specific features of water vertical circulation in the Arctic Ocean. *Acad. Nauk USSR (Doklady)* **187**(1):187–189.

Timofeev, V. T. 1960. *Water Masses of the Arctic Ocean.* Hydrometeoizdat, Leningrad, 191p.

Tzvetkova, N. L. 1977. Some peculiarities of ecology, growth and production of two species of gammarids (Amphipoda, Gammaridae) in high latitudes of the Arctic. In *Explor. Fauna Seas* **14**(22):291–298.

Usachev, P. I. 1949. The microflora of polar ice. *Akad. Nauk USSR, Inst. Oceanol. (Trudy)* **3:**216–259.

Volkov, N. A., and A. F. Treshnikov. 1969. About modern international nomenclature of the sea ice. *Problemi Arctiki i Antarctiki* **32:**132–143.

Zernov, S. A. 1928. About hibernation of the water organisms in the ice thickness and in frozen ground of the materials of N. V. Boldireva, P. P. Smirnova, and V. D. Shmeleva. *Russian J. Hydrobiol.* **7**(1–2):1–8.

# 10: Evolution of Arctic Ecosystems During the Neogene Period

A. N. Golikov and
O. A. Scarlato

## INTRODUCTION AND PREVIOUS STUDIES

In spite of continued studies, there is no general agreement concerning the evolution of the Arctic Ocean and its ecosystems. Some authors believe that the Arctic Ocean was originally a bay of the Atlantic Ocean and this determined the composition and recent distribution of its flora and fauna (Zenkevich, 1939; Nesis, 1961; Kussakin, 1973; Kafanov, 1979; Beklemishev, 1980). Other authors regard the Arctic as a separate biogeographic region, Pacific invaders having contributed greatly to its faunal and floral composition (Soot-Ryen, 1932; Gurjanova, 1934, 1939, 1949, 1957, 1970, 1974; Ekman, 1935; Diakonov, 1945; Zinova, 1974; Levenstein, 1981; Golikov, Scarlato, and Pavshtiks, 1983).

No less conflicting are the views on the role of the mixed sub-Arctic (low Arctic) waters and their recent biogeographic position in the formation of the populations of the Arctic Ocean. Some authors attach great importance to these waters in the formation of the ecosystems of the Arctic Basin and distinguish them as a separate zone or biogeographic region (Derjugin, 1915; Zernov, 1934; Dunbar, 1951, 1953, 1972; Zenkevich, 1963), whereas other researchers, because of the absence of characteristic species in these waters (Filatova, 1938), treat the sub-Arctic waters as an ecotone without a biogeographic rank.

It is not an easy task to gain insight into the mechanism of tropho-dynamic processes occurring in the Arctic Ocean. Primary production of bottom-dwelling macrophytes and microphytes as well as of the planktonic algae is very low because of the lack of light during much of the year and the severe ice conditions. However, the biomass of bottom heterotrophic organisms is high in some regions (Golikov and Averincev, 1977; Golikov, Scarlato, and Pavshtiks, 1983).

The aim of the present study is to examine the above problems analyzing the evolution of the ecosystems in the Arctic Ocean. For this purpose we have used geological, paleogeographical, paleontological, and paleoclimatic data (Krishtafov-

*Note:* This chapter has been reviewed, edited, and rewritten by the Editor, Yvonne Herman.

ich, 1932; Saks, 1953; Sinitsin, 1962, 1967; Golikov, 1963, 1980; Emiliani and Flint, 1963; Belousov, 1968; Velichko, 1973; Lindberg, 1972; Kaplin, 1973; Suzdalsky, 1976; Krasilov, 1977; Monin, 1977; Gladenkov, 1978; Scarlato, 1981; Grosswald, 1983), ecological principle of evolutionary reconstructions (Golikov and Tzvetkova, 1972), and data from under-ice diving biological research, off the Franz Josef Archipelago, off Novaya Zemlya, on the New Siberian Shoals, and off Wrangel Island (Golikov, and Averincev, 1977*b;* Golikov, Averincev, and Sheremetyevsky, 1974; Golikov et al., 1978).

During the Paleogene (67–24 mya) (Golikov, 1985), the Arctic Ocean was inhabited by relatively thermophilic species, differing radically from today's cold-water morphotypes. The Canadian Arctic waters were inhabited by crocodiles and turtles as well as by coral reefs extending as far as Greenland (Krishtafovich, 1932; Swarzbach, 1955). The warm Siberian Sea covered most of Middle Asia (Sinitsin, 1962, 1967). This basin and the Norwegian and Greenland seas were apparently isolated in the Eocene and the Oligocene from the Atlantic Ocean. The radiolarians that inhabited the pseudoabyssal region* of the Norwegian and Greenland seas at that time differed from the species inhabiting the adjacent Atlantic waters (Petrushevskaya, 1976). In the Early Miocene (22–23 mya), the Norwegian Sea and the adjacent Arctic Basin, by this time connected with the Atlantic Ocean, were inhabited by the same tropical thermophilic radiolarians (Petrushevskaya, 1976, 1979). Approximately 16–18 mya, thermophilic subtropical mollusks predominated in the southern Chukchi Sea (Petrov, 1976).

Approximately 14 mya, volcanic activity increased; this increase may have been related to the Arctic ''tectono-magnetic'' Alpine orogeny (Monin, 1977).* As a result of tectonic activity in the Drake Passage, the circum-Antarctic circulation was established and the glaciation of Antarctic increased (Barker and Burrell, 1977; Verbitskiy and Kvasov, 1980). In our view this glaciation was one of the most significant causes of cooling of the Northern Hemisphere, especially of the North Pacific region. Increased volcanic activity occurred at that time off northern Iceland (McDougall, 1984). Sea level rose (Vail and Hardenbol, 1979; Herman, 1979) and temperature fell (Shackleton and Opdyke, 1976).

Approximately 14 mya, following the above-mentioned dramatic tectonic and climatic events, numerous genera and species evolved in the North Pacific Ocean (Golikov, 1963, 1980; Golikov and Tzvetkova, 1972; Scarlato, 1981); thus, the first boreal biogeographic region was established. Today these taxa exclusively inhabit the boreal (temperate) regions. They are widely distributed Pacific boreal species. To our knowledge, Northern Hemisphere species did not originate in the Atlantic 14 mya. Between ≈ 14 my BP and the Pliocene, subtropical temperatures prevailed in the North Atlantic as evidenced by its subtropical fauna† (Sorgenfrei, 1958; Glibert, 1959, 1963). The geological, climatic, and biological events of the period between 14 mya and 5 mya justify our suggestion of defining it as a separate geological epoch that should be referred to as the Boreocene (Golikov, 1985).

Several authors assume that the first migrations of faunas between the Pacific and Atlantic oceans were via the northern waters about 12 mya (Hopkins, 1959; MacNeil, 1965; Durham and MacNeil, 1967; Einarsson, Hopkins, and Doel, 1967). However, data on gastropod evolution

*Pseudoabyssal (Derjugin, 1915; Andriyashev, 1964) is a depression in the bathyal zone below 2,000 m, not connected with the World Ocean floor, at depths of 2,000–5,000 m, for example, in the North Polar Ocean.

*This statement is not supported by geological data; see, for example, Chapter 30 [Editor's note].

†This statement is not supported by DSDP results [Editor's note].

(Golikov, 1963, 1980) and the presence of subtropical mollusks only of Mediterranean–Lusitanian origin in northern Europe sediments (Sorgenfrei, 1958; Glibert, 1959, 1963) do not support this point of view. The transgression of that period may have resulted in the opening of the Bering Strait, but was not sufficient to cover the Thule Ridge (Strauch, 1970, 1972; Gladenkov, 1978). The Pacific boreal fauna could have dispersed in the parts of the Arctic Ocean that were covered with water, but did not penetrate the North Atlantic. Water temperatures of the latter were 2°–3°C higher than those of the Pacific Ocean. Deep-water temperatures in high latitudes were 6°–7°C, while surface temperatures off Alaska were about 14°C (Emiliani, 1956, 1961). These slight differences in the temperatures of the deep and surface waters may have determined the penetration of Pacific taxa not only in shallow regions but in the deep sectors of the Arctic Basin as well (Andriyashev, 1951). This cooling brought about the regional extinction of subtropical species in the Arctic Ocean.

The Arctic taxa may have been derived from Pacific species with wide distributional ranges. Such gastropod species include *Littorina squalida, Buccinum rossicum, B. ochotense, B. percrassum, Neptunea pribillofensis,* and *N. lyrata*. Bivalve species such as *Panomya ampla, Crassicardia crassidens, Keenocardium californiense, Peronidia lutea, Spisula voyi, Cyclocardia crebricostata,* and *Protothaca staminea* may have been predominant in some ecosystems of the Arctic Basin at that time. The warmer waters of the North Atlantic supported a subtropical fauna and flora. Approximately 10 mya, sea level dropped significantly (Vail and Hardenbol, 1979). A major cooling and coincident evolution commenced about 5 mya; it marked the boundary between the Boreocene and the Pliocene.

In early Pliocene, North Atlantic surface water temperatures were 8°–12°C during the winter and 20°–22°C in summer (Funell, 1970). The first Atlantic boreal species of Mediterranean–Lusitanian origin developed at that time (Harmer, 1914; Gladenkov, 1978). During the Gilbert Reversed magnetic epoch regression, the Beringian Land and Thule Ridge isolated almost completely the Arctic from the Pacific and Atlantic oceans (Lindberg, 1962; Einarsson, Hopkins, and Doel, 1967; Durham and MacNeil, 1967), and land existed in the area now covered by the Siberian seas. The Arctic Ocean was ice-free (Herman, 1970, 1983; Margolis and Herman, 1980; Herman and Hopkins, 1980). This created favorable conditions for the most eurybiontic widely distributed Pacific boreal species, which had invaded the Arctic Ocean earlier (Golikov, 1963, 1980; Golikov and Tzvetkova, 1972).

A major transgression 3.5 mya led to an unrestricted connection between Pacific, Atlantic, and Arctic oceans. Climates were somewhat milder than today, the Arctic Ocean was ice-free, and boreal conditions prevailed (Einarsson, Hopkins, and Doel, 1967; Herman, 1970; Gladenkov, 1978; Herman and Hopkins, 1980). A significant part of the Siberian seas was covered by sea, Novaya Zemlya separated from the continent, and a large bay was formed at the site of the White Sea (Suzdalsky, 1976). However, the regions of the New Siberian Island and Wrangel Island remained relatively highly elevated land areas (Panov, 1946). Because of this transgression, numerous Pacific boreal species invaded the Arctic Ocean and its seas and penetrated into the North Atlantic. As a result of the Coriolis force, the migrations of organisms from the Pacific were toward the northeast along the sea-covered coast of Canada (Diakonov, 1945; Golikov, 1963, 1980). Reverse migrations, if any, were slight during this period. This was due (1) primarily to the fact that numerous eurybiontic boreal genotypes capable of inhabiting the high latitudes did not inhabit the North Atlantic, and (2) to the existence of a partial land barrier in the region of the New Siberian Island. However, Atlantic boreal species of Mediterranean–Lusitanian origin that had already evolved by that time must have penetrated the Arctic Ocean from the Norwegian Sea (Gurjanova, 1939, 1949, 1957; Golikov and Tzvetkova, 1972).

The Thule Land was apparently submerged at that time and was not a barrier for the migrations of marine organisms (Strauch, 1970, 1972; Gladenkov, 1978; Golikov, 1963, 1980). The lower layers of the Red Crags of England that continued the Coralline Crags at the onset of the Gauss Epoch already contained remnants of Atlantic boreal species that were derived mainly from Pacific invaders, and only to a small degree from the local Mediterranean–Lusitanian species (Golikov and Tzvetkova, 1972). The species diversity in the Arctic Ocean increased sharply. The biocenoses of the Arctic Basin must have still been dominated (though to a smaller degree) by widespread boreal species of Pacific origin that were distributed in the temperate Pacific waters as well. These were the first groups to occupy the Arctic Ocean. High boreal species of Pacific origin and sometimes of Atlantic origin either occupied vacant ecological niches, or coadapting with the widespread boreal species, entered the existing ecosystems (Golikov and Averincev, 1977*a,b*).

About 3 mya, cooling intensified. By the end of the Gauss Epoch the glaciation of Greenland was underway and ice sheets appeared in Iceland (Einarsson, Hopkins, and Doel, 1967; Zakharov, 1978). Simultaneously the Antarctic ice sheet was expanding (Mercer, 1978). Icebergs calved into the Arctic Ocean, though the ocean was generally ice-free (Herman, 1970, 1983). About 2.5 mya, numerous boreal Arctic species evolved from eurybiontic boreal Pacific species in the Northern Bering Sea and fewer boreal Arctic taxa originated in the Norwegian Sea from Atlantic Ocean ancestors. They became widely distributed in the Arctic Ocean. This view is supported by the adaptation of these taxa to their environment (Golikov and Tzvetkova, 1972), and by the vast fossil evidence of that period from Kamchatka, Alaska, Chukchi Peninsula, and to a lesser degree, from Iceland (Golikov, 1963, 1980; Gladenkov, 1978; Scarlato, 1981). The boreal species of the Arctic Ocean, which could not adapt to the harsh environment, declined and eventually became extinct (Gladenkov, 1978; Petrov, 1982). The vacant ecological niches were occupied by boreal-Arctic species. These gastropods, such as *Neptunea communis, N. ventricosa, Buccinum ciliatum, B. elatior, B. glaciale,* and *B. polare,* became an important part in the heterotrophic food chain. The bivalves *Leionucula tenius, Nuculana pernula, Musculus niger, Chlamys islandicus,* and *Tridonta borealis* probably dominated the biocenoses. At about the same time, the tundra landscape evolved on land. The dramatic events that occurred 2.5–3 mya strongly suggest that this threshold was the boundary between the lower and upper Pliocene.

Local ice sheets and a tundra landscape zone emerged on the Pribilof Islands (Hopkins, 1972). Volcanic activity and ice development increased in Iceland. It is noteworthy that in the region of the New Siberian and Wrangel islands a land barrier persisted that may have prevented circum-Arctic species migrations, whereas north of Canada the sea was ice-free most of the year (Herman, 1970). By the onset of the Olduvai and Gilsa events that marked the beginning of the Pleistocene, the temperatures of the surface water, even in the temperate latitudes, dropped to 3°–6°C (Savin, Douglas, and Stehli, 1975), and in the deep water dropped to 2°C (Emiliani and Flint, 1963). The Calabrian age of high sea level was replaced by a marked regression by 560–750 m* (Krishtafovich, 1932; Saks, 1953; Einarsson, Hopkins, and Doel, 1967; Kotenev, 1970). The Bering Land and the Thule Land*

*This statement is not supported by geological data [Editor's note].

were possibly above sea level and isolated the Arctic Ocean from the Atlantic and Pacific waters (Scott, 1926; Panov, 1946; Rukhin, 1962; Strauch, 1970, 1972). An Arctic water mass may have formed at that time; it subsequently filled the pseudoabyssal zone.

Endemic bathyal and pseudoabyssal genera and species, mostly of Pacific origin, evolved in the Canada Basin, while those of Atlantic and secondary Atlantic origin evolved in the Norwegian–Greenland Basins. The boreal elements probably became extinct by and the inhabitants of the ecosystems that survived in the Arctic Basin were represented by boreal Arctic and newly evolved Arctic species. During this regression, water salinity was about 35‰ and temperatures were between −0.4°C and −0.8°C. Most deep parts of the Arctic Ocean basins and the abyssal depths in some other regions of the World Ocean must have developed during this regression and are believed by some authors to account for these conditions* (Belousov, 1968; Lindberg, 1970). However, the Lomonosov and the Wyville–Thompson ridges, characterized by continental crust and depths less than 1,000 m (Treshnikov, 1967; Dementskaya and Karasik, 1971; Vogt, 1972), could already have separated the bathyal Eurasian and Amerasian subbasins, which in turn were isolated from the deep North Atlantic and Pacific oceans. The development of water with well-defined physical and chemical characteristics in the deep "pseudoabyssal" basins of the Arctic Ocean led to the development of genera and species in these waters; since they were unable to extend their range, these taxa became endemic to the Arctic Ocean. Species endemism in the pseudoabyssal zone of the Arctic Ocean is about 70–90 percent (Gorbunov, 1946; Gurjanova, 1957, 1970; Filatova, 1957; Clarke, 1963; Koltun, 1964; Bernard, 1979; Bouchet and Waren, 1979; Bilyard and Carey, 1980). Many of these endemic species comprise abyssal ecosystems only in the Eurasian and Amerasian subbasins, while those in the Canada deep basin are of Pacific origin (Levenstein, 1981).

In our view, most boreal species that inhabited the Arctic continental shelves were wiped out during the first glaciation. Only the most eurybathic boreal Arctic taxa survived. The major regression was not protracted and was followed by a transgression; sea level stood ≈200 m below its present stand, possibly for a relatively long time (Suzdalsky, 1976), and shaped the modern continental shelves.

The surface Arctic water mass was formed at the onset of this transgression under relatively cold conditions. It had reduced salinities of ≈30–32 ‰ and annual temperatures ranged from −0.1°C to −1.5°C. About 1.2–1.4 mya, the genus Portlandia, which is endemic to the Arctic Basin, originated; its type species, *P. arctica,* is an indicator of Arctic water. Arctic species became widely distributed in the ecosystems of the Arctic Basin, but predominant in only a few of them, since earlier invaders of eurybiontic boreal Arctic species already dominated most regions.

Before the Brunhes–Matuyama boundary, a marked transgression is suggested by Lindberg (1972). This resulted in the influx of relatively warm Atlantic water into the Arctic Ocean, and a rise of temperature by about 2°–3°C above the present temperature (Troitsky, 1961). The intermediate Atlantic–Arctic water mass of the Arctic Ocean may have been formed at that time. This water mass was characterized by 35 ‰ salinity and temperatures above zero (about 2°–3°C). Atlantic waters sank that had higher density and salinity and were relatively warm near surface Arctic water, extending along the lower edge of the shelf in the upper parts of the bathyal zone. Today a similar pattern can be observed; however, the present-day surface Arctic water is colder. A number of charac-

*This view is not supported by geological data [Editor's note].

teristic species (e.g., *Cryptonatica bathybii*) emerged in the intermediate Atlantic–Arctic water mass in the Arctic Basin.

After the Arctic water was formed, a sub-Arctic or low-Arctic hydrological regime developed in the zone of mixing with the boreal water (Derjugin, 1915; Zernov, 1934; Dunbar, 1951, 1953, 1972; Beklemishev, 1980; Filatova, 1957*a,b;* Zenkevich, 1963). This water of mixed origin and its fauna is distinguished by the above-mentioned authors as a specific zone, a biogeographic region, or a subregion. However, analyses of the bivalve faunal distribution (Filatova, 1938) and numerous subsequent data on most other groups of plants and animals have shown that the mixed waters do not contain unique species. This fact and the variable hydrological regime suggest strongly that the sub-Arctic (i.e., low-Arctic) water and its fauna were an ecotone between the Arctic and the boreal regions, having no zonal or biogeographic significance.

The temperature rise in the early Pleistocene resulted in a temporary extension of the ranges of numerous boreal species. Thus, *Macoma balthica* and *Mytilus edulis* were distributed in the shallow water of the Eurasian subbasin eastward up to Taymyr. The beginning of the Brunhes was accompanied by the onset of a major regression and rapid cooling. Some of the deep basins in the Arctic Basin and in the Pacific Ocean were formed at that time,* and sea level fell 200–300 m below present-day stand (Lindberg, 1970, 1972). The Beringian Land connected Asia and America (Lindberg, 1962). The Thule Land, extending as during previous regressions nearly up to 30°N, was partially reestablished.* Siberian Sea shelves were exposed in some parts as far as 78°N. Alluvial, lake, and terrigenous deposits on the Siberian Sea shelves suggest that they were land area during most of the Quaternary (Suzdalsky, 1976; Tolmachov and Yurzev, 1970). A drastic temperature decrease 0.7–0.9 mya resulted in the inception of a 3–4-m-thick perennial sea ice over the Arctic Ocean (Herman, 1970, 1983; Herman and Hopkins, 1980). This may have caused the formation of cryopelagic species (some species of diatoms, the Amphipoda *Apherusa glacialis* and *Gammarus wilkitskii,* and probably the Pisces *Boreogadus saida*) that formed a peculiar biocenoses on the lower surface of the ice. In our view, the onset of the last normal polarity epoch and the sharp and frequent climatic fluctuations 0.6–0.4 mya mark the boundary between early Pleistocene and Pleistocene proper.

The suggestion that the Lomonosov Ridge may have been emergent in the Pleistocene is questioned (Gurjanova, 1949, 1957; Demenitskaya and Karasik, 1971; Golikov, 1976*a;* Nesis, 1983*a*). However, in our view, the existence of land area far northward in the region of the New Siberian Islands is beyond doubt. This must have affected the water circulation and caused the differences of their regimes in the Eurasian and Amerasian subbasins of the Arctic Ocean. As a result, western and eastern Arctic species originated from boreal-Arctic and circum-Arctic taxa. The reasons for their limited distribution were the past existence of the New Siberian land with cold climates throughout most of the Pleistocene, and the present extremely severe hydrological regime and low productivity of the water off the New Siberian Islands induced by turbid water. Species that originated in the Eurasia and Amerasian subbasins may not have attained circumpolar distribution in 10,000 years, since the onset of the recent distribution of fauna on the Arctic Ocean shelf. Arctic species with limited range appeared when the ecological niches were occupied by boreal and circumpolar species and are generally of no essential importance for the bioenergetics of the Arctic Ocean ecosystems.

Regression and temperature decrease

*This statement is not supported by geological data [Editor's note].

led to the Günz (Nebraskan, Varangian) Glaciation. In the European plains, the ice sheet extended to 50°–53°N and air temperature fell in summer to 5°–6°C. It is postulated by some workers (Grosswald, 1983) that during periods of glaciation a single pan-Arctic ice sheet was formed as a result of the confluence of the Laurentian, Innitian, Greenland, Iceland, British, Scandinavian, Barents Sea, and Kara ice sheets, which descended below 800–850 m, with ice thickness above open water exceeding tens of meters. Such an assumption is groundless, since Atlantic waters penetrated into the Arctic Ocean uninterruptedly, though to a lesser extent than today (Nesis, 1983*b*). Furthermore, cryopelagic and surface water Arctic species, which evolved earlier, survived throughout this time interval. Most probably the ice sheets covered the continental shelf up to Taymyr, but did not reach the bathyal zone and were not as thick as suggested by the above-mentioned authors. The ice thickness apparently was not greater than 3–4 m (Herman, 1970) and allowed light penetration and development of primary production.

The Günz Glaciation and regression were followed, about 0.6–0.5 mya, by the Cromerian (Aftonian, Byelovezhskian) transgressive phase with a sea level about 50 m above the present sea-level stand (Kaplin, 1973). This transgression was accompanied by a considerable freshening of the sea water, so that very large interconnected brackish water areas were formed in the estuarine parts of rivers (Sinitsin, 1962; Lindberg, 1970, 1972). Salinities in these areas ranged from 2–16‰. Mixing of melting ice with cold surface Arctic water led to the formation of a peculiar estuarine Arctic water mass and the development of its fauna, represented chiefly by mobile crustaceans (Golikov and Averincev, 1977). Dominating the estuarine–Arctic ecosystems are usually *Saduria entomon, Gammaracanthus loricatus,* and *Mysis oculata relicta.* Species of the estuarine Arctic complex penetrated the Caspian and Baltic seas and acquired interzonal distribution. The warming and transgression were followed ≈0.47 mya by cooling, the Mindel, Kansas, Elster, Oka, Demyanovskoe Glaciation, and regression to 50–100 m below present sea level.

In Europe, ice sheets advanced to 50°N (Grosswald, 1983). The temperature of the Arctic Ocean surface water must have been close to zero or below it. The shelves of its seas became emergent as during previous glaciations. At that time, as during the previous Pleistocene glaciations, Arctic species may have been formed either in the Eurasian or in the Amerasian parts of the Arctic Ocean. These include *Neptunea denselirata, Buccinum nivale, B. hydrophanum,* and *Montacuta spitzbergensis;* their ancestors may have originated in the Atlantic sector of the Arctic region. *Buccinum fringillum* and *Periploma alaskensis* originated in the Pacific sector of the Arctic region and were distributed only in the west Siberian and Chukchi seas.

About 0.4–0.35 mya, temperatures rose (Shackleton and Opdyke, 1976), ice sheets melted, and the Holsteinian (Lichvinian, Yarmouthian) transgression began; this resulted in the dispersal of marine fauna on the shelves and the wide distribution of the estuarine Arctic fauna. Air temperature rose to 20°C in summer in Europe (Woldstedt, 1954) and the temperature of the Arctic Ocean surface water rose to 3°–4°C (Troitsky, 1961). The White and Baltic seas were connected and the Gulf of Anadyr of the Bering Sea were connected with the Penjin Bay of the Okhotsk Sea.

About 0.24–0.25 mya, a severe cooling commenced (the Riss, Saalian, Illinoian, Dnieper–Moscow Glaciation), accompanied by a very marked regression, with sea level falling 150 m below its present stand (Lindberg, 1972). This glaciation, separated by an interstadial into two phases, may have been the coldest during the Pleistocene. Permafrost was developing in eastern Siberia. The Gulf Stream formed a frontal zone with the near-ice cold waters off the European coasts south of the Bay of Biscay. In spite of these adverse conditions, some taxa survived in the ice-free water and even in water covered by thin ice where primary production can take place. These same species inhabit the Arctic Ocean today. This statement is supported by the fossil remains of mollusks in the adjacent Atlantic Ocean of Riss age; they are almost exclusively species present in the Arctic Ocean (Harmer, 1914; Brogger, 1900–1901; Knipovitsch, 1901). The same is true for the inhabitants of the Canada Basin (Bernard, 1979). The Pleistocene environment was established during early glaciations and interglacials; this resulted in the process of species reorganization. In marine organisms this process was essentially completed during the early phases of the Pleistocene. The same process occurred with the highly mobile groups of terrestrial organisms. Thus, no speciation occurred in birds during the last 0.65 my (Mayr, 1968, 1981).

Approximately 0.19 mya, temperatures increased, gradually melting the ice. Sea level rose once again. The last interglacial, the Eemian (Sangamon, Kazantsevo), was underway. During winter the ice extended north of the Bering Strait according to Hopkins (1972); the Barents Sea was ice-free. The Barents, White, and Baltic seas were connected by wide straits. Surface water temperatures were 2°–3°C higher than today (Velichko, 1973; Vella, 1973). The shelves were once again covered by seawater. Atlantic boreal fauna penetrated as far as the Taymyr lowland, which was flooded by seawater (Lavrova, 1960; Troitsky, 1961). However, the cold water, and to some degree the terrestrial barrier, in the region of the New Siberian Islands were obstacles for organisms to disperse farther. Pacific waters spread eastward only as far as Wrangel Island (Ushakov, 1952).

Another cold cycle commenced about 75 thousand years ago; global temperatures decreased, ice sheets expanded, and sea level fell. This cold interval, the Würmian Glaciation, may be subdivided into several stages. During the earliest, or the first stadial, the Barents, White, and Kara shelves became emergent and are believed to have been covered by ice sheets (Grosswald, 1983). Their biota perished, except for those species able to survive in the bathyal or abyssal zones. The first interstadial, a time of slight warming and sea level rise, resulted in a negligible repopulation of these shelves.

The second interstadial was warmer and sea level rose to approximately 40 m below present sea level stand. Water exchange between the Norwegian Sea and the Atlantic, somewhat restricted during periods of maximum glaciation, became freer (Moyes et al., 1974; Petrushevskaya, 1976). However, even in the Mediterranean Sea surface water temperatures did not exceed −14°–16°C (Barash and Oskina, 1979). Furthermore, coeval sediments from the Sicily region and from areas west of this island contain fossils tolerant of lower temperatures than those prevailing in the western Mediterranean today. Among these species are *Modiolus modiolus, Arctica islandica, Astarte sulcata* (Reyss, 1964), as well as *Littorina littorea, Neptunea antiqua,* and *Buccinum undatum.* At the same time, *Homo sapiens* may have appeared on the eastern coasts of the Mediterranean Sea.

The last and coldest phase of the Würmian Glaciation followed when sea level fell by more than 150 m (Lindberg, 1970, 1972; Kaplin, 1973). Ice sheets extended to the shelf up to Taymyr and may have merged (Grosswald, 1983), but this did not preclude life beneath the floating ice. Approximately 18 thousand years ago, the Würm Glaciation reached its peak. Surface water temperatures even in the tropics dropped by more than 6°C (Emiliani and Flint, 1963) and in the Mediterranean Sea temperatures dived to 7° – 12°C in winter and 13° – 18°C in summer (Shackleton,

1978; Herman, 1980). The Gulf Stream once again was deflected south of the Bay of Biscay (Molina-Cruz and Thiede, 1978).

The melting of the Würm ice sheets, which began about 15 thousand years ago, greatly intensified during Allerod time ≈12 thousand years ago. The melting brought about strong freshening along the margins of the dwindling ice (Berger, Johnson, and Kileingley, 1977). Therefore, the first shelf faunas to reappear were the estuarine-Arctic assemblages that had survived the glaciations in the eastern sector of the Arctic and in brackish water in the vicinity of ice sheets. Species belonging to these assemblages migrated into the White and Caspian seas, as well as into the St. Lawrence Bay (Schafer, 1977) and into some areas of the northwestern American waters (Segerstrale, 1971). Increased salinities and warming of the intermediate water led to an interzonal distribution of these assemblages. About 10 thousand years ago, sea level rose to 20–30 m below its present stand. Arctic species, such as *Portlandia arctica,* penetrated as far south as the Baltic Sea. The most eurybiontic boreal species appeared in the shallow areas of the Atlantic part of the Arctic Ocean. The lack of any significant events 10 thousand years ago, as compared to those of the earlier Pleistocene during deglaciations and subsequent onset of interglacial phases, does not allow us to regard the Holocene as a separate geological epoch.

Approximately 6 thousand years ago, the ocean attained its present level, and 5–4.5 thousand years ago the level was 2.5 m higher than today. Temperatures rose by 2°–3°C as in other interglacial phases. The Littorinian (Tapes) transgression of the Atlantic period began. Species of subtropical origin, such as *Ostrea edulis,* migrated northward as far as the coast of Scotland (Smith, 1972) and boreal species (*Macoma balthica* and *Mytilus edulis*) again penetrated the eastern sector of the Arctic as far as Taymyr (Panov, 1946; Lavrova, 1960; Troitsky, 1961). The Atlantic boreal water mass was formed again in the Arctic Ocean and its inhabitants also acquired their modern high boreal character. The Arctic species in the Barents and Norwegian seas survived only in parts of fjords with cold near-bottom and bottom water. The subsequent slight cooling and climatic fluctuations have led to the present-day distribution of the ecosystems.

## DISCUSSION

The study of evolution of Arctic Ocean ecosystems shows that the evolution of the biological systems is determined by changes in the major environmental physico-chemical factors and is restricted to them in space and time. Geological and biological processes have a distinct sinusoidal pattern. Relatively long periods of slow development are replaced at the boundaries of geological epochs by relatively rapid and sharp geomorphological, climatic, and biological changes. During the Neogene, the phases of rapid development occurred under conditions of major temperature decrease and changes in sea level.

It is surmised that major global coolings enhance rates of speciation (Golikov, 1973). In the Neogene, in the Northern Hemisphere oceans and during the brief phases of global temperature changes, 4 families and 20 genera of gastropods made their evolutionary appearance whereas only 2 genera evolved during the extended periods of stable climatic regimes (Golikov, 1985). In the history of development of the Arctic Ocean fauna and flora, the eurybiontic species evolved more frequently during transgressions whereas stenobiontic taxa originated during regressions (Golikov, 1976*a*, 1980).

It is useful to study the sequence of events during the Neogene that has a regular exponential character and can be approximated by the formula

$$T = a \cdot e^{bn}$$

where $T$ is time of events, $a$ and $b$ are coefficients, and $n$ is sequence of events. In this particular case

$$T = 14 \pm 2 \cdot e^{-(1 \pm 0.01)n}$$

However, as seen in Figure 10-1, a time scale of not less than 1 my is only appropriate for defining epochs in accepted practice. Using a time scale of 0.1 my, the Eupleistocene and Pleistocene proper, and also the inception phases of major glaciations, can be distinguished. With a time scale of thousands of years, the beginning of stadials and the most drastic climatic changes are determined.

The history of ecosystem development has shaped the recent biogeographic structure of the Arctic Ocean (Fig. 10-2). The earliest inhabitants of the Arctic Ocean are Pacific widely distributed boreal species whose northern limit of distribution is defined by the Pacific high boreal water mass (Golikov, 1976*a*). The extent of penetration to the northern regions by different taxa varies and is a function of their morpho–

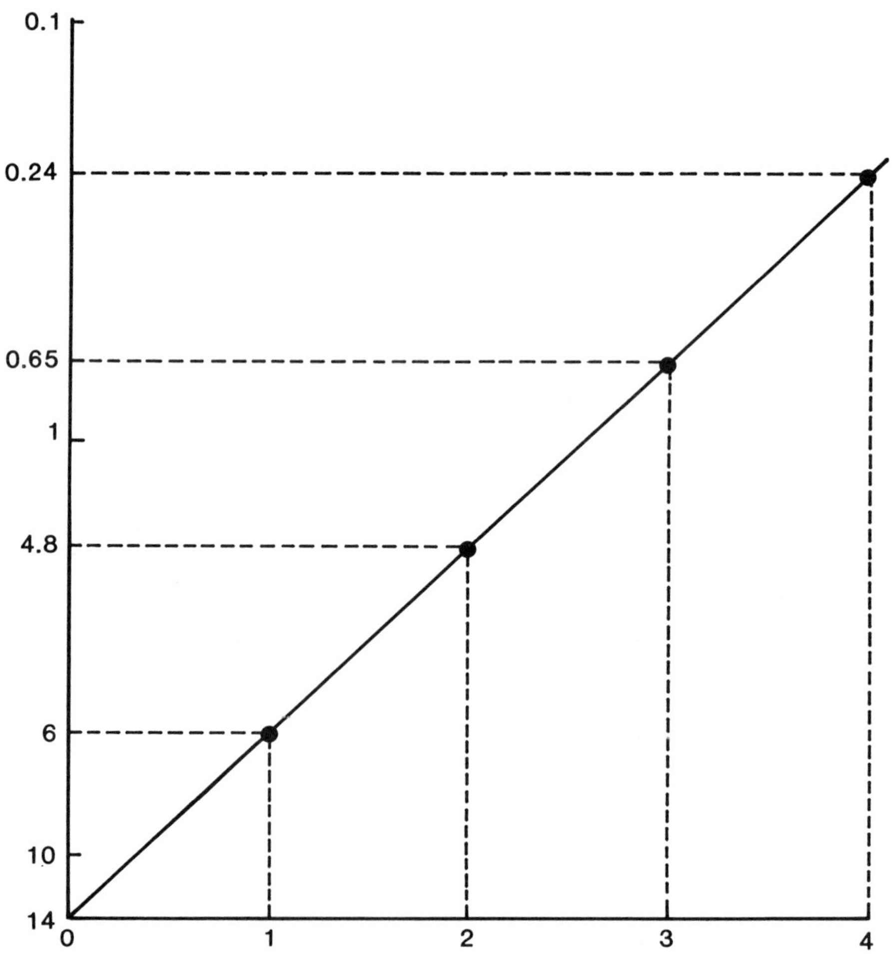


Figure 10-1. Relationship between the time of geological events and the sequence of ecosystems in the Arctic Ocean, Abscissa: sequence of events; ordinate: time, on logarithmic scale.

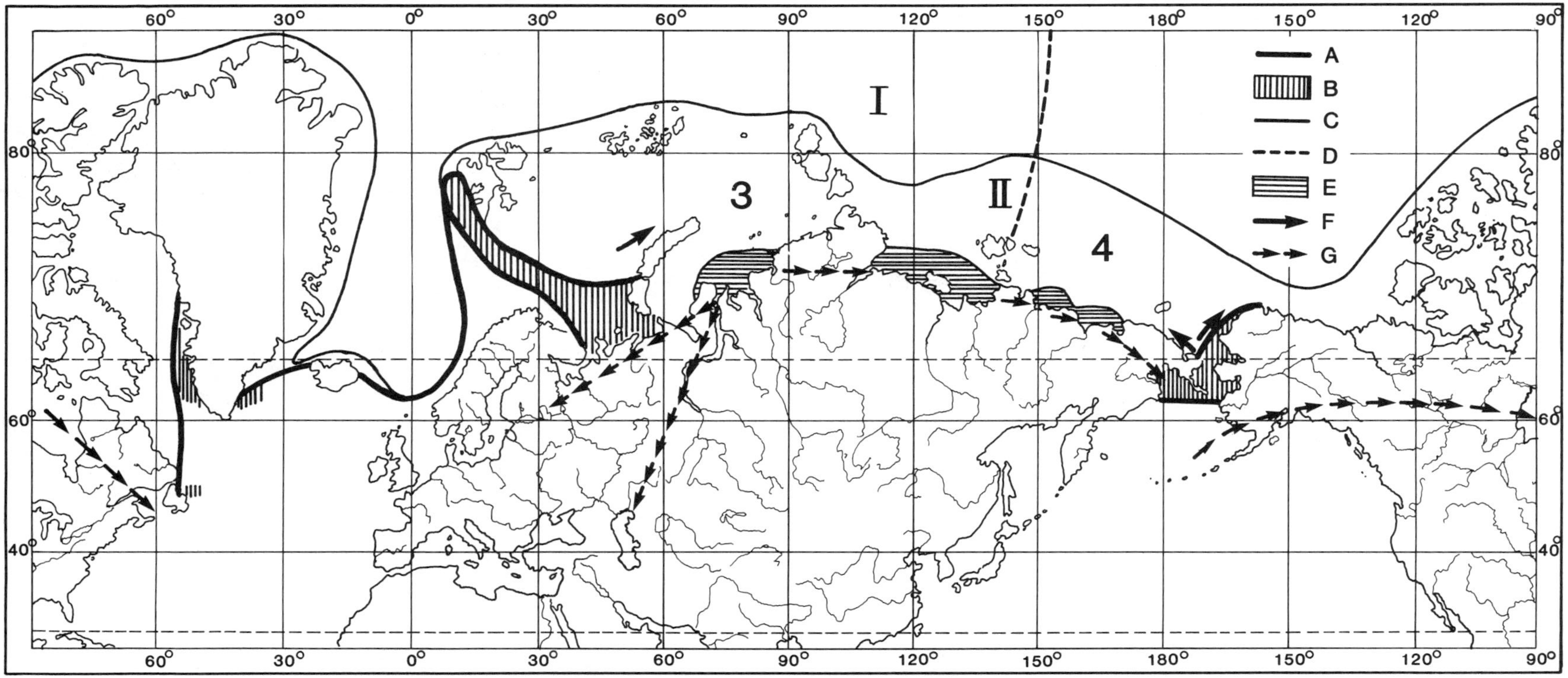


**Figure 10-2.** Division of the Arctic Ocean and adjacent waters into biogeographic regions. *A:* biogeographic regions' boundaries; *B:* ecotones between the Arctic and boreal regions; *C:* boundaries between provinces; *D:* boundaries between sectors; *E:* estuary Arctic province; *F:* recent migration routes; *G:* ancient migration routes; *I:* deep Arctic province; II: shelf Arctic province with Eurasian (3) and Amerasian (4) sectors.

functional and ecological attributes. Highly mobile animals (e.g., crustaceans or fishes) or ones having pelagic larval ontogenetic stages (e.g., bivalves or echinoderms) can penetrate further north and boreal species dominate the Arctic taxa north of the Bering Strait, where the southern boundary of the Arctic biogeographic region is drawn according to the distribution of these groups (Gurjanova, 1939, 1949; Diakonov, 1945; Tzvetkova, 1975; Scarlato, 1981). The southern boundary of taxa with low mobility and (or) with direct ontogenetic development crossed the northern part of the Bering Sea, south of the Gulf of Anadyr, St. Lawrence Island, and Norton Sound. Therefore, the boundary between the Arctic region and the Pacific boreal region is a low-Arctic (sub-Arctic) ecotone with variable hydrological regime, covering the Bering Strait region and the adjacent waters.

The Arctic deep waters have temperatures of −0.4°C to −0.8°C and salinities of 35‰, and their faunas appear to be the most ancient subdivision of the Arctic region proper. The deeper bathyal and pseudoabyssal zones of the Arctic Ocean have several species in common; furthermore, we find species (and perhaps genera) endemic to the Canada Basin. The deep basins of the Arctic are geologically the oldest; therefore, their biota inhabit the "special" deep provinces of the Arctic region that may be subdivided into the Amerasia Basin and the Eurasia–Norwegian–Greenland basins. Preceding them are the shelf faunal elements inhabiting the surface water with temperatures ranging between −0.1°C and −1.8°C and salinities of 30–32‰. In the course of its evolution the province was divided into the Eurasian sector (including the shelves of the Baffin, Greenland, Kara, and Laptev seas as well as the eastern part of the Barents Sea) and the Amerasian sector (including the shelves of the East Siberian, Chukchi, and Beaufort seas), the boundary crossing along the region of the New Siberian Islands (Gurjanova, 1949; Golikov, 1963, 1980). There seems to be reason to regard the Atlantic-Arctic Intermediate waters, extending between 200–400 m and 650–1,000 m, with temperatures of 1.3°–2.3°C and salinities of 35‰, and their inhabitants as a separate province. In spite of the relative paucity of their population, these waters have their characteristic fauna, such as *Cryptonatica bathybii*.

An ecotone serves as a boundary of the Arctic region both in the North Atlantic and in the Pacific Ocean. Off the eastern American coasts, this boundary, fluctuating with the strength of the Labrador Current and northern branches of the Gulf Stream, passes in the region of the Newfoundland Bank, rises off the western Greenland coasts under the influence of the warm Irminger Current north of 75°N, and further descends abruptly toward southeastern Greenland and northern Iceland. In the Norwegian Sea it flows along the southern margin of the Norwegian Basin, forced northward by the warm Nansen Current off western Spitsbergen. In the Barents Sea the boundary of the Arctic region crosses 75°N, descending southward in the region of Cape Sv. Nos on the Kola Peninsula. However, the shoals of the White and Barents seas to depths of 25–30 m retain their boreal characteristics up to the southern end of Novaya Zemlya.

The estuarine Arctic assemblages have a specific interzonal position and should be treated as a separate province. This province is represented in the cold estuarine sectors of the Arctic zone and interzonally in the cold waters of the Baltic and Caspian seas. In the Arctic, the estuarine Arctic assemblage does not descend to depths greater than 10 m in water with summer temperatures of 4°–6°C and salinities of 2–3‰ to 16‰. The Atlantic boreal region in the Arctic Ocean is represented only by the northern parts of the German high boreal province, occupying the shelves and the upper parts of the bathyal zone of the

Norwegian Sea, the western parts of the Barents Sea, and the shallow eastern part of the Barents and White seas. Water temperatures in these areas may rise above 6°–12°C in summer and salinities fluctuate from 25‰ (White Sea) to 34.1‰.

Preliminary data (Golikov, Averincev, and Scarlato, 1983) suggest that as a result of the evolution and migrations the Arctic Ocean flora and fauna comprise over 3,000 species, belonging to 1,200 genera and 930 families, all known phyla, about 90 percent of classes, 60 percent of orders, 30 percent of families, 20 percent of genera, and 0.14–4 percent of the total number of the World Ocean's species. Genetically, the inhabitants of the Arctic Ocean are related only to the boreal water taxa of the North Atlantic and Pacific oceans to which they bear the greatest resemblance. The Arctic environment being the youngest, endemism of its population is low: It does not exceed 36 genera and 540 species, that is, about 3 percent and 18 percent of its flora and fauna, respectively.

The highest number of species and biomass in the Arctic Ocean occurs in those high-rank taxonomic groups that originated in earlier geological periods during times of major global temperature decrease (Golikov, 1975). Among crustaceans these are the Peracarida, which originated during the Carboniferous temperature minima. Amphipoda of this group have the highest number of genera and species in the Arctic. The same pattern and predominance of advanced groups in the Arctic is observed in mollusks (Golikov, 1963, 1980; Golikov and Scarlato, 1977). The evolutionarily advanced taxonomic groups that have originated during cold geological epochs in the Arctic Basin also have the highest rate of diversification, which has a regular parabolic character in closed and semiclosed systems (Golikov, 1976*b*).

The number of species of all taxonomic groups in the Arctic Ocean is the highest in the Norwegian and Barents seas where gene flow from the North Atlantic is particularly extensive and hydrological conditions are most favorable. In temperate water of the Barents Sea there are over 2,000 benthic and 900 planktonic species. Their number declines rapidly northward and eastward and particularly abruptly in the regions influenced by the great Siberian rivers. The Chukchi Sea flora and fauna comprise only about 850 benthic and 400 planktonic species. The small number of Pacific boreal species penetrating into the Chukchi Sea cannot make up for the decline in the total gene pool of the population. This is because the North Atlantic water as the main source of the present hydrological regime of the Arctic Ocean penetrates up to the northern coasts of Canada, gradually loses its strength, and therefore also its biota, which is primarily, or secondarily, of Atlantic origin (Svetovidov, 1940; Tzvetkova, 1975; Bilyard and Carey, 1980).

The species of Pacific origin that were originally distributed in the Arctic Basin because of the partial and/or complete isolation from the Pacific became gradually influenced by the Atlantic waters and became partly transformed into "secondary" Atlantic species. For example, primary Atlantic species originated from Atlantic species and were dispersed in recent times in the Arctic Ocean. They include all representatives of the amphipod Gammarus and the fish Gadus and Clupea. The "secondary" Atlantic species originated from species of Pacific origin. They include all the Atlantic species belonging to Neptunea and Buccinum (Golikov, 1963, 1980). During the time of dispersal of boreal-Arctic species in the Arctic Ocean, about 2 mya, over 80 percent of these species were of Pacific origin.

When the Arctic species began to differentiate into Eurasian and Amerasian taxa about 0.5–0.6 mya, about 75 percent of species had been formed in the Atlantic part of the Arctic Basin and only 25 percent of the Pacific sector (Golikov, Averincev, and Scarlato, 1983). However, even today the surface Arctic water mass contains 40–50 percent Arctic species that are of Pacific origin and the intermediate Atlantic–Arctic water mass supports about 40 percent of these species (Golikov, unpublished).

The quantitative distribution of life in the Arctic Ocean has been studied to a much lesser degree than its qualitative aspects. In the Eurasian sector of the Arctic Basin, quantitative data have been obtained chiefly from the sub-Arctic (low-Arctic) waters (Brotskaya, 1930; Idelson, 1931; Brotskaya and Zenkevich, 1932; Thorson, 1933, 1936; Sparck, 1933; Pergament, 1944; Filatova and Zenkevich, 1957). In the last decades detailed information has been obtained using quantitative diving methods to study the biota in the upper parts of the shelf, off the Franz Josef Land Archipelago, in the Laptev and East Siberian seas, in the Long Strait, and off Wrangel Island (Golikov, Averincev, and Sheremetyevsky, 1974; Golikov et al., 1978; Golikov and Averincev, 1977*a,b*). The distribution of the mean biomass is shown in Figure 10-3.

One can see a decline in the biomass of plankton and benthos occurring rapidly from the Norwegian Sea to the estuarine parts of the Ob and Yenisei rivers in the Kara Sea and then, more slowly, toward the Chukchi Sea. In its eastern part, influ-

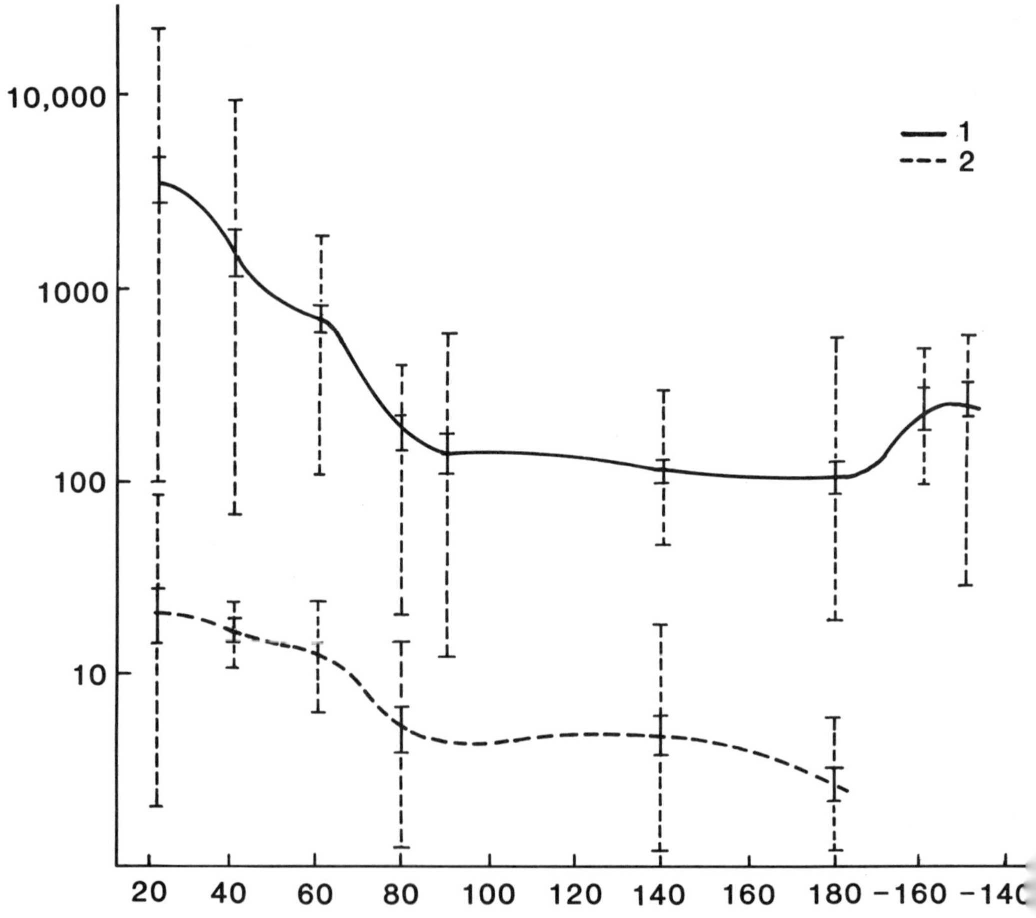


Figure 10-3. Biomass of (1) benthos and (2) plankton in different areas of the Arctic Basin. Abscissa; longitude, degrees; ordinate; biomass in g m$^2$. (After Golikov, Scarlato, and Pavshtiks, 1983.)

enced by the Pacific water, the benthos biomass increases slightly. The benthos biomass declines from the boreal waters of the Barents Sea toward the lower salinity Arctic water of the Kara Sea by nearly one order of magnitude. This decline is not only due to the decrease in velocity of the North Atlantic Drift carrying dissolved and suspended organic matter from the eutrophic temperate latitudes but also to a sharp increase in water turbidity and decline in primary production.

Boreal-Arctic species are dominant in the Arctic Ocean, often comprising up to 90 percent of the total biomass; they are chiefly of Pacific origin and the earliest emigrants into the Arctic waters and the most eurybiontic taxa. The biomass of the Arctic ecosystems is generally dominated by macrobenthic organisms, the proportion of meiobenthos of the total biomass not higher than 6–15 percent. However, where the biomass of macrobenthos is low, that is, near river estuaries, at depths, beneath permanent pack sea ice, and so forth, the role of meiobenthic organisms increases abruptly in biocenoses, sometimes attaining a high biomass. In these regions they dominate the ecosystems. For example, in the northern part of the Kara Sea at depths of ≈400 m, foraminiferal biomass was 200 g $m^2$, the macrobenthos being poorly developed (Filatova and Zenkevich, 1957). This may be related to the ability of meiobenthic organisms to feed on fine-grained and dissolved particles, which are not available to macroorganisms even in the absence of their usual prey. Such ecological replacement is observed in every region with well-defined physico-chemical conditions (Golikov and Averincev, 1977).

The planktonic and benthic biomass in the Siberian Seas' shelves is formed by two assemblages different by their origin and ecology:

1. Hydrobio-assemblage of the surface Arctic water mass, extending near large rivers, generally from 10–20 m to 200–300 m
2. Hydrobio-assemblages of the estuarine-Arctic water mass in shallower water

In the first hydrobio-assemblage, the highest mean biomass values (300 g $m^2$) are attained at salinities greater than 24‰, which is already favorable for euryhaline species, as well as in the periodically ice-free regions. Beneath the pack ice the benthic biomass, devoid or nearly devoid of macrophytes, does not exceed 150–200 g $m^2$. In the estuarine Arctic hydrobio-assemblage, maximal biomass values (100–200 g $m^2$) are attained at salinities of 6–7‰. At salinities of 14–16‰, unfavorable both for Arctic and for estuarine Arctic species, the biomass does not exceed 10–20 g $m^2$ (Golikov and Averincev, 1977).

In the surface Arctic water layer, in regions that are ice-free during summer, macrophyte biomass is 2,000 g $m^2$ (on the average 300 ± 100 g $m^2$) and macrophytes produce up to 1,500 kcal $m^2$ yr (on the average 200 ± 75 kcal $m^2$ yr). The biomass of phytoplankton fluctuates markedly in different parts of the Arctic Basin depending on the ice regime and season and may attain 5–14 g $m^2$ in summer, though on the average it is 0.21–1 g $m^2$. Because of the short time of phytoplankton development, production in the Arctic seas generally does not exceed 60–150 kcal $m^2$ yr. A significant role in the primary production of the photic zone of the Arctic shelf belongs to benthic microalgae capable of producing 18–1,000 kcal $m^2$ yr (Matheke and Horner, 1974).

Heterotrophic organisms, represented chiefly by filter feeders and detritophages, can produce on the average up to 500 kcal $m^2$ yr and consume about 625 kcal $m^2$ yr. These preliminary data suggest that in the ice-free photic zone during 1–2 summer months a trophodynamic balance is possible if the local primary production is sufficient for the consumption by heterotrophic organisms. However, major parts of the Arctic Ocean are covered by nearly continuous 2–4-m-thick pack ice year-round. Here, even on the shoals, macrophytes are either absent or in seasonally ice-free regions they attain a biomass not higher than 90 ± 40 g $m^2$ and production of 60–30 kcal $m^2$ yr. Phytoplankton biomass here exceeds 0.2–0.4 g $m^2$, and production exceeds 40–50 kcal $m^2$ yr (Andersen, 1977). An additional source of primary production is formed on the subsurface of pack ice by diatoms and few blue-green algae of the cryopelagic biocenoses. Under favorable conditions they can produce up to 300 kcal $m^2$ yr. This production, however, is fully consumed by heterotrophic organisms of the cryopelagic biocenoses functioning as a semiclosed tropho-dynamic system, for examples, algae–amphipods (*Apherusa glacialis* and *Gammarus wilkitzkii*) and fishes (*Boreogadus saida*).

Under ice heterotrophic benthic organisms produce 170–250 kcal $m^2$ yr and consume 250–350 kcal $m^2$ yr. In the estuarine Arctic waters with high turbidity and almost total lack of transparency, primary production is reduced to a minimum, and the biomass of the heterotrophic organisms attains on the average 60–80 g $m^2$ (production about 100 kcal $m^2$ yr).

As seen from these calculations, a tropho-dynamic balance is possible in the Arctic seas only under the most favorable conditions and is directly related to local primary production. In most cases, a flow of allochthonous organic matter is indispensable for the existence of abundant heterotrophic biota. The source of this flow are the boreal waters of the Arctic Ocean and the adjacent parts of the North Atlantic. Even in the relatively poor Barents Sea, macrophytes' biomass in the photic layer reaches 10–20 kg $m^2$, production is 6,000–12,000 kcal $m^2$ yr, and production of phytoplankton is usually 1,000–2,500 kcal $m^2$ per yr. This production is not completely consumed by heterotrophic organisms in the North Atlantic and is dispersed in suspended and dissolved form, partly carried by currents together with metabolites of animals into the Arctic Basin.

Analysis of changes in the quantitative biotic distribution with increasing depth shows the relationship between the biomasses of the biocenoses and depth-related factors and the vertical structure of water masses (Fig. 10-4). The biomass in the photic layer of the Atlantic high boreal water mass in the boreal waters of the Arctic Ocean is 2–3 orders of magnitude higher than in the Arctic waters. This is related to the abundant development of macrophytes and such species of animals as *Mytilus edulis* capable of producing biomasses up to 10–15 kg $m^2$.

In the Arctic littoral zone, a certain biomass is produced in the coastal ice by organisms belonging to a specific cryolittoral biocenosis (e.g., Acrosiphonia algae: amphipods; *Gammarus setosus;* some nematodes and other elements of the meiofauna), and because of the mechanical action of floating ice, the quantity of life is low in deeper waters (to 5 m). Below 5 m, we find a rapid decline of biomass in the boreal waters and an increase in the Arctic water. At depths of 20–30 m, free from the continuous mechanical action of ice, the biomass in the Arctic waters reaches maximum values and is comparable to that at equivalent depths in boreal waters. However, the population of the Arctic water does not produce biomasses of more than 500 g $m^2$ at these depths and greater depths, whereas in the Atlantic high boreal waters ecosystems with such dominant species as *Chlamys islandicus* and modiolus (*Modiolus modiolus*) may produce a biomass at depths of 100 m of 1 kg $m^2$.

In the deeper regions the biomass is generally decreasing and remains relatively constant within the Atlantic Intermediate Arctic water mass at depths of 200–800 m. In the deep Arctic water mass the biomass declines rapidly toward the greatest depths. The trophic structure of the benthos in the upper layer of the surface Arctic water mass in the regions that are ice-free in summer and in areas covered with pack ice year-round is shown in Figure 10-5. Shoals to a depth of 10 m are

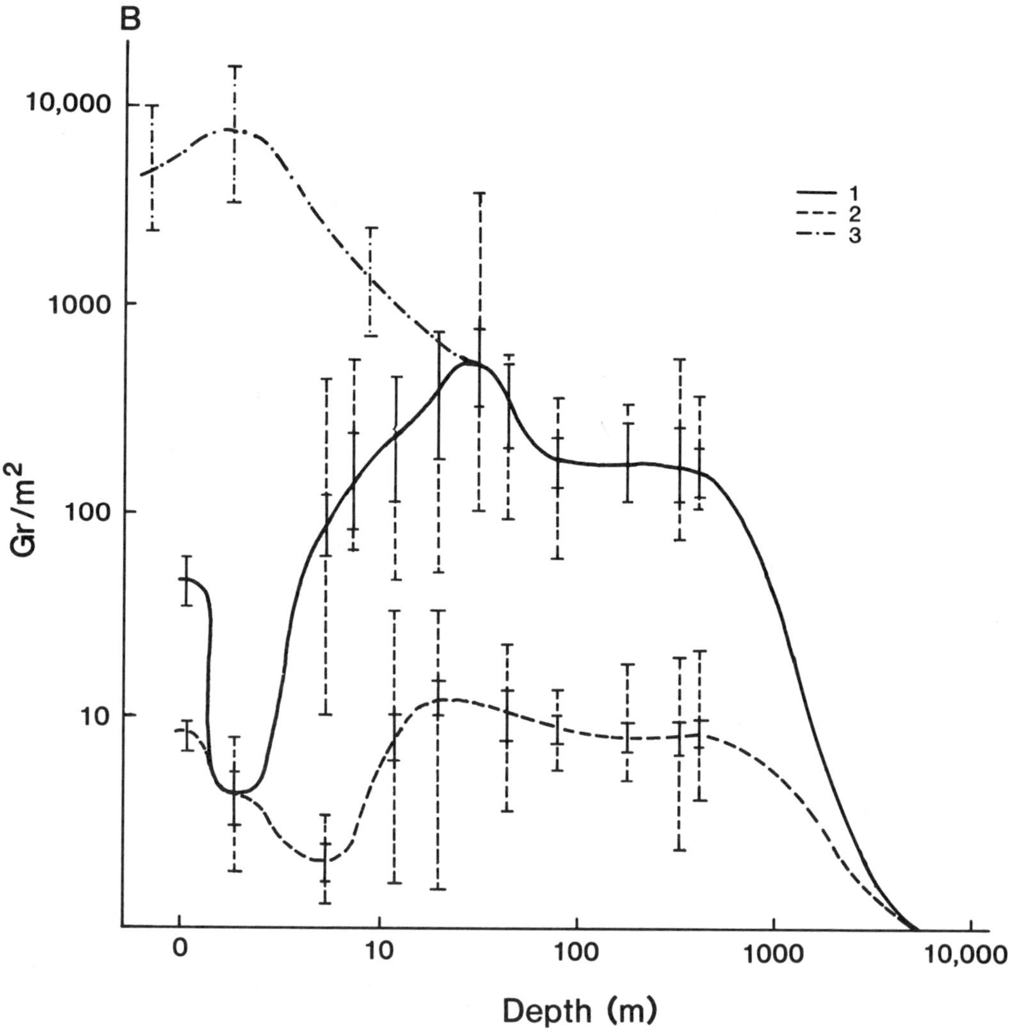


Figure 10-4. Relationship between (1) benthos and (2) plankton biomass and depth as a factor in the Arctic and boreal (3) waters of the Arctic Basin. Abscissa: depth on logarithmic scale in m; ordinate: biomass in g $m^2$.

dominated by predators, highly mobile and capable to avoid mechanical ice action. Macrophytes are relatively abundant in the parts that are ice-free in summer at depths to 30 m and are nearly absent in the ecosystems beneath perennial ice. Detritophages most closely bound with predators by the predator–prey relationship are relatively numerous at depths of 5–15 m, but predominate only in a few ecosystems. Depths of more than 20–30 m are dominated by filtrato that comprise over 50 percent of the average biomass of all ecosystems and are capable of consuming the inflowing organic matter.

Our preliminary studies, based on limited data on the Arctic biota, show that although the majority of Arctic organisms are formed primarily by Pacific immigrants and their descendants, the present-day

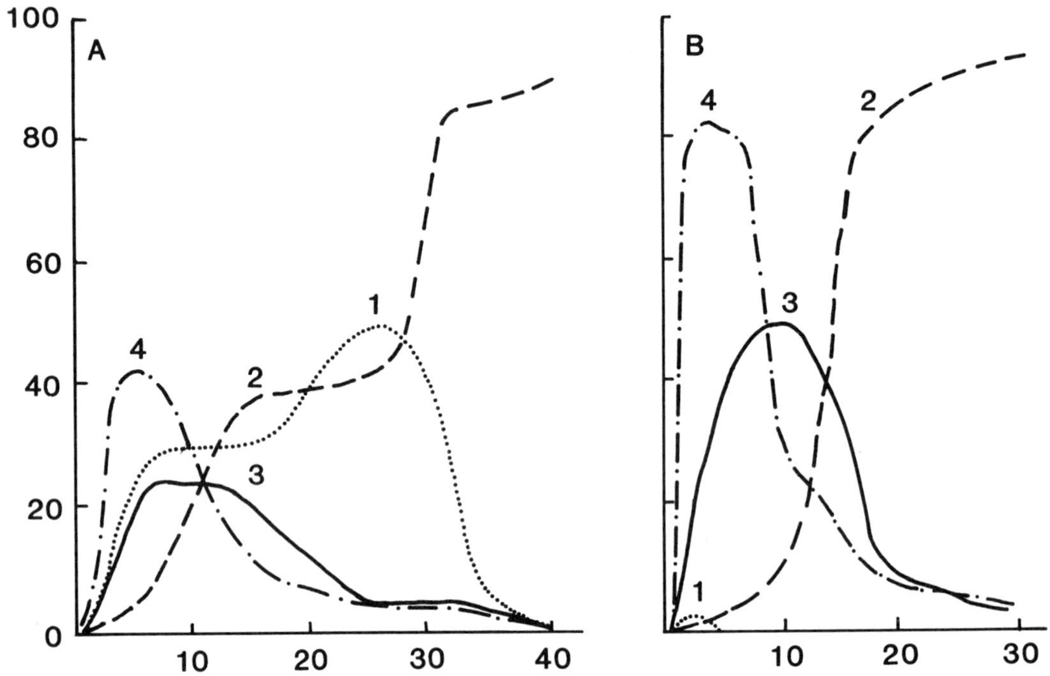


Figure 10-5. Percentage of biomass of benthos trophic groups at different depths in the Arctic Ocean in ice-free areas in summer (*A*) and in perennial sea ice-covered areas (*B*). 1: macrophytes; 2: filter feeders; 3: detritophages; 4: predators. Abscissa: depth in m; ordinate: percentage of trophic group in biocenosis biomass at a given depth.

biological regime of this ocean is almost entirely determined by the North Atlantic waters.

## REFERENCES

Andersen, O. G. 1977. Primary production, illumination and hydrography in Jorgen Bronlund fjord, Northern Greenland. *Medd. Grønland* **205:**2–27.

Andriyashev, A. P. 1964. *Fishes of the Northern Seas of the USSR.* Israel Program for Scientific Translations, Jerusalem, No. 836, 617p. (Originally published in 1954 in Russian.)

Barash, M. S., and N. S. Oskina. 1979. Paleotemperatury Atlanticheskogo okeana 1800 i 4000 let nazad (po planktonnym Foraminifera). *Okeanologia* **19:**98–101. (In Russian.)

Barker, P., and J. Burrell. 1977. The opening of Drake Passage. *Marine Geol.* **25:**15–34.

Beklemishev, K. V. 1980. Biogeograficheskaya struktura subarkticheskikh vod. *Kompleksn. issled. prirody okeana. Moskva* **7**(5):248–254. (In Russian.)

Belousov, V. V. 1968. *Zemnaya Kora i Verkhnyaya Mantiya Okeanov.* Nauka, Moskva, 255p. (In Russian.)

Berger, W. H., R. F. Johnson, and J. S. Kileingley. 1977. Global freshening of the upper layer during deglaciation. *Naturwissenschaften* **64**(12):634–365.

Bernard, F. R. 1979. Bivalve mollusks of the Western Beaufort Sea. *Contrib. Sci. Los Angeles County Mus.* **313:**80p.

Bilyard, G. R., and A. G. Carey. 1980. Zoogeography of western Beaufort Sea, Polychaeta (Annelida). *Sarsia* **65**(1):19–25.

Bouchet, Ph., and A. Waren. 1979. The molluscan fauna of the Norwegian Sea and its relation to other faunas. *Sarsia* **64:**211–243.

Brogger, W. C. 1900–1901. Om de senglaciale og posglaciale nivaforandringer i Kristianiafeltet. *Norges Geol. Undersogelse 31.* Kristiana, 731p.

Brotskaya, V. A. 1930. Materialy po kolichestvennomu uchotu donnoi fauny Storfiorda (Vost. Spitzbergen). *Trudy Morskogo Nauch. Instituta, Moskva* **4**(3):46–57. (In Russian.)

Brotskaya, V. A., and L. A. Zenkevich. 1932. Materialy po kolichestvennomu uchotu donnoi fauny Barentseva, Belogo i Karskogo morei. 6. Kolichesvennyi uchot donnoi fauny Cheshskoi guby. *Trudy Gos. Okeanogr. Instituta, Moskva.* pp. 41–57. (In Russian.)

Chizhov, O. P. 1976. *Oledeneniye Severnoi Polarnoi Oblasti.* Nauka, Moskva, 240p. (In Russian.)

Clarke, A. H. 1963. Arctic archibenthal and abyssal molluscs. II. Molluscs dredged from Drifting Station Charli (Alpha II). *Bull. Nat. Mus. Canada* **185**(7):90–109.

Demenitskaya, R. M., and A. M. Karasik. 1971. Problemy genezisa Severnogo Ledovitogo okeana. In *Istoriya Mirovogo Okeana.* Nauka, Moskva, pp. 58–76. (In Russian.)

Derjugin, K. M. 1915. Fauna Kol'skogo zaliva i usloviya eyo susuchestvovaniya. *Zap. Akad. Nauk, Fiz.-Mat. Otd.*, series VIII, **34**(1):929p. (In Russian.)

Diakonov, A. M. 1945. Vzaimootnosheniye arkticheskikh i tikhookeanskikh faun na osnovye izucheniya zoogeografii iglokozhikh. *Zhurnal Obshchyei Biologii* **6**(2):125–155. (In Russian.)

Donk, J., and G. Mattieu. 1969. Oxygen isotope composition of Foraminifera and water samples from the Arctic Ocean. *J. Geophys. Res.* **74**(13):1,396–1,407.

Dunbar, M. J. 1951. Eastern Arctic waters. A summary of our present knowledge on the physical oceanography of the eastern Arctic area from Hudson Bay to Cape Farewell and from Belle Isle to Smith Sound. *Fish. Res. Bd. Canada Bull.* **88:**1–131.

Dunbar, M. J. 1953. Arctic and subarctic marine ecology: Immediate problems. *Arctic* **6:**75–90.

Dunbar, M. J. 1972. The nature and definition of the marine subarctic with a note on the sea life areas of the Atlantic salmon. *Roy. Soc. Canada Trans.* **1–3**(16):249–257.

Durham, J. W., and P. S. MacNeil. 1967. Cenozoic migrations of marine invertebrates through the Bering Strait region. In *The Bering Land Bridge,* Stanford University Press, pp. 326–333.

Einarsson, T., D. M. Hopkins, and S. E. Doel. 1967. The stratigraphy of Tjornes, northern Iceland and the history of the Bering Land Bridge. In *The Bering Land Bridge.* Stanford University Press, 112p.

Ekman, S. 1935. *Tiergeographic des Meeres.* Leipzig, 542p.

Emiliani, C. P. 1956. Oligocene and Miocene temperature of the equatorial and subtropical Atlantic Ocean. *J. Geol.* **64**(3):281–294.

Emiliani, C. P. 1961. Cenozoic climatic changes as indicated by the stratigraphy and chronology of deep-sea cores of Globigerina—Ooze facies. In *Solar Variations, Climatic Change and Related Geophysical Problems.* New York, pp. 521–536.

Emiliani, C. P., and R. F. Flint. 1963. The Pleistocene record. The sea, 3. *The Earth Beneath the Sea History.* International Publication 7, Wiley, 150p.

Filatova, Z. A. 1938. Kolichestvennyi uchot donnoi fauny yugozapadnoi chasti Barentseva morya, Trudy Polyarnogo nauch-issled. Instituta morsk. rybn. khoz. i okeanografii. Pishchepromizdat, Moskva, Leningrad, 2, 53–59. (In Russian.)

Filatova, Z. A. 1957*a*. Obshchii obzor fauny dvustvorchatykh mollyuskov severnykh morei SSSR. *Trudy Inst. Okeanol., Akad. Nauk SSSR* **20:**53–59. (In Russian.)

Filatova, Z. A. 1957*b*. Zoogeograficheskoye raiyonirovaniye severnykh morei po raspredeleniyu dvustvorchatykh mollyuskov. *Trudy Inst. Okeanol., Akad. Nauk SSSR* **23:**195–215. (In Russian.)

Filatova, Z. A., and L. A. Zenkevich. 1957. Kolichestvennoye raspredeleniye donnoi fauny Karskogo morya. *Trudy Vsesoyuzn. Gidrobiol. Obshchestva* **8:**3–67. (In Russian.)

Funell, B. M. 1970. The Paleogene and early Pleistocene of Norfolk. In *The Geology of Norfolk.* London, 25p.

Gladenkov, Yu. B. 1978. *Morskoi Verkhnii Kainozoi Severnykh Raiyonov.* Nauka, Moskva, 194p. (In Russian.)

Glibert, M. 1959. Gastropodes du Diestein, du Scaldisien et du Mertemien de la Belgique. 3rd note. *Inst. Roy. Sci. Natur. Belg. Bull.* **35**(10):27p.

Glibert, M. 1963. Les Muricaecea et Buccinacea fossiles du Cenozoique etranger. *Des Collections de L'Institut Royal des Sciences Naturelles de Belgique.* Mem. Inst. Sci. Nat. Belg., ser. 2, Fasc. 74, Bruxelles, 179p.

Golikov, A. N. 1963. Bryukhonogiye mollyuski roda Neptunea Bolten. *Fauna SSSR Mollyuski* **5**(1):217p. (In Russian.)

Golikov, A. N. 1973. Species and speciation in Poikilothermal animals. *Marine Biol.* **21**(4): 257–268.

Golikov, A. N. 1975. O paralellizmye i konvergentsii v razvitii morskikh. ekosistem. In *Biologiya Shel'fa. Vsesoyuznoye Soveshchaniye po Biologii Shel'fa.* Vladivostok, pp. 30–31. (In Russian.)

Golikov, A. N. 1976*a*. Znacheniye biogeograficheskogo metoda i ucheniya o krupnykh geogidrokraticheskikh kolebaniyakh urovnya Mirovogo okeana dlya poznaniya zakonomernostei evolyutsii i rasseleniya morskoi fauny. In *Zoogeografiya i Sistematika Ryb.* Leningrad, pp. 24–36. (In Russian.)

Golikov, A. N. 1976*b*. O kolichestvennykh zakonomrrnostyakh protsessa divergentsii. In *Gidrobiologicheskiye Issledovaniya Samoochishcheniya Vodoyomov.* Leningrad, pp. 90–96. (In Russian.)

Golikov, A. N. 1980. Mollyuski Buccininae Mirovogo okeana. *Fauna SSSR* **5**(2):465p. (In Russian.)

Golikov, A. N. 1985. Etapy i rubezhi kainozoya. *Trudy Zool. Inst. AN SSSR, Leningrad* **5**(130):37–52. (In Russian.)

Golikov, A. N., and V. G. Averincev. 1977*a*. Distribution patterns of benthic and ice biocenoses in the high latitudes of the Polar Basin and their part in the biological structure of the World Ocean. *Polar Ocean,* pp. 331–364.

Golikov, A. N., and V. G. Averincev. 1977*b*. Biotsenozy verkhnikh otdelov shel'fa arkhipelaga Zemlya Frantsa Iosifa i nekotorye zakonomernosti ikh raspredeleniya. *Issled. Fauny Morei* **14**(22): 5–55. (In Russian.)

Golikov, A. N., V. G. Averincev, and O. A. Scarlato, 1983. Proiskhozhdeniye i nekotorye zakonomernosti raspredeleniya ekosistem verkhnikh otdelov shel'fa Severnogo Ledovitogo okeana. *Probl. Ekol. Polar. Obl. Trudy Shkoly-Seminara, Murmansk, Apr. 1980.* Moskva, pp. 91–96. (In Russian.)

Golikov, A. N., V. G. Averincev, and A. M. Sheremetyevsky. 1974. Nekotorye zakonomernosti raspredeleniya zhizni na Novosibirskom melkovodye. In *Otchotnaya Sessiya ZIN AN SSSR za 1973.* Nauka, Leningrad, pp. 12–13. (In Russian.)

Golikov, A. N., and O. A. Scarlato. 1965. Gidrobiologicheskiye issledovaniya v zalivye Possjet s primeneniyem vodolaznoi tekhniki. *Issled Fauny Morei* **3**(II):5–21. (In Russian.)

Golikov, A. N., and O. A. Scarlato. 1967. Ecology of bottom biocenoses in Possjet Bay (Sea of Japan) and the peculiarities of their distribution in connection with physical and chemical conditions of their habitat. *Helgolander Wiss. Meeresunters* **15**:193–201.

Golikov, A. N., and O. A. Scarlato. 1977. Sostav, raspredeleniye i ekologiya bryukhonogikh i dvustvorchatykh mollyuskov u arkhipelaga Zemlya Frantsa Iosifa. *Issled Fauny Morei* **14**(22):313–390.

Golikov, A. N., O. A. Scarlato, and E. A. Pavshtiks. 1983. Biologicheskiye resursy Severnogo Ledovitogo okeana. In *Geografiya i Proirodnye Resursy, N 3.* Nauka, Novosibirsk, pp. 62–71. (In Russian.)

Golikov, A. N., and N. L. Tzvetkova. 1972. The ecological principle of evolutionary reconstruction as illustrated by marine animals. *Marine Biol.* **14**(1):1–9.

Golikov, A. N., V. I. Lyulyeyev, O. K. Novikov, V. V. Potin, B. I. Sirenko, and A. M. Sheremetyevsky. 1978. Nekotorye zakonomernosti raspredeleniya zhizni na verkhnikh otdelakh shel'fa ostrova Vrangelya i mysa Schmidta. In *Morfologiya Sistematika i Evolyutsiya Zhivotnykh.* Leningrad, pp. 11–12. (In Russian.)

Gorbunov, G. P. 1946. Donnoye naseleniye Novosibirskogo melkovodiya i tsentral'noi chasti Severnogo Ledovitogo okeana. Trudy dreif. eksp. Glavsevmorputi na L/p "G. Sedov," 1937–1940, 3, pp. 30–138. (In Russian.)

Grosswald, M. G. 1983. *Pokrovnye Ledniki Kontinental'nykh Shel'fov.* Nauka, Moscow, 215p. (In Russian.)

Gurjanova, E. F. 1934. Zoogeograficheskii ocherk fauny Isopoda Arktiki. *Arctica* **2**:127–152. (In Russian.)

Gurjanova, E. F. 1936. K zoogeografii Karskogo morya. *Izv. AN SSSR,* ser. biol., **213**:565–598. (In Russian.)

Gurjanova, E. F. 1939. K voprosu o proiskhozhdenii i istorii razvitiya fauny Polarnogo basseina na osnovye analiza faun Amphipoda i Isopoda. *Izv. AN SSSR,* ser. biol., **5**:671–704. (In Russian.)

Gurjanova, E. F. 1948. Donnaya fauna Vostochno-Sibirskogo morya. *Nauchn. Bull. Len. gos. Univ.* **24**:15–18. (In Russian.)

Gurjanova, E. F. 1949. Fauna Polarnogo basseina i puti eyo obmena s faunami sosednikh raiyonov Mirovogo okeana. *Trudy 2 Vsesoyuzn. Geogr., Syezda, Geografgiz, Moskva,* **3**:202–203. (In Russian.)

Gurjanova, E. F. 1957. K. zoogeografii Arkticheskogo basseina. *Materialy Nablyudenii Nauchno-Dreif.,* stns SP-3 i SP-4, 1954/1955, **1**:343–362. (In Russian.)

Gurjanova, E. F. 1970. Osobennosti fauny Severnogo Ledovitogo okeana i ikh znacheniye dlya ponimaniya istorii ego formirovaniya. In *Severnyi Ledovityi Okean i Ego Poberezhye v Kainozoye.* Nauka, Leningrad, pp. 126–161. (In Russian.)

Gurjanova, E. F. 1974. K zoogeograficheskomu raiyonirovaniyu kholodnykh i umerennykh vod Severnogo polushariya. In *Gidrobiologiya i Biogeografiya Kholodnykh i Umerennykh vod Mirovogo Okeana.* Leningrad, pp. 11–12. (In Russian.)

Harmer, F. W. 1914. *The Pliocene Mollusca of Great Britain. Part I.* London, 200p.

Herman, Y. 1970. Arctic paleoceanography in late Cenozoic time. *Science* **160**:474–477.

Herman, Y. 1979. Plancton distribution in the past. In *Zoogeography and Diversity in Plancton.* Utrecht, pp. 29–49.

Herman, Y. 1980. Paleoclimatic and paleohydrologic record of Mediterranean deep-sea cores based on pteropods, planktonic and benthonic foraminifera. *Revista Española de Micropaleontologia* **13**(2):171-200.

Herman, Y. 1983. Baffin Bay: Present-day analog of the central Arctic during late Pliocene and mid-Pleistocene time. *Geology* **2**:356-359.

Herman, Y., and D. M. Hopkins. 1980. Arctic oceanic climate in late Cenozoic time. *Science* **209**(1):557-562.

Hopkins, D. M. 1959. Cenozoic history of the Bering Land Bridge. *Science* **29**(3362):1519-1920.

Hopkins, D. M. 1972. The paleogeography and climatic history of Beringia during Late Cenozoic time. *Internat. J. Arctic and Nordic Studies,* Paris, **2**:121-130.

Idelson, M. S. 1931. Materialy po kolichestvennomu uchotu donnoi fauny Spitzbergenskoi banki (Barentsevo morye). *Trudy Morskogo Nauchnogo Instituta,* Moscow, pp. 27-46. (In Russian.)

Kafanov, A. I. 1979. O statusye arkticheskoi morskoi biogeograficheskoi oblasti (po paleontologicheskim dannym). *Sb. Rabot Inst. Biol. Morya, Dal'nevost. Nauch Tsentra AN SSSR* **16**:100-129. (In Russian.)

Kaplin, P. A. 1973. *Noveishaya Istoriya Poberizhii Mirovogo Okeana.* Izd. Mosk. Univ., Moscow, 265p. (In Russian.)

Kaplin, P. A. 1976. *Izmenenie Urovnya Mirovogo Okeana v Pleistotsene po Dannym Opredelenii Absolyutnogo Vozrasta Drevnikh Beregovykh Linii. V. sb.: Problemy Paleogidrologii.* Nauka, Moscow, pp. 95-101. (In Russian.)

Knipovitsch, N. M. 1901. Mollusca und Brachiopoda. Zoologische Ergebnisse der russischen Expeditionen nach Spitzbergen. Ezhegodn. *Zool. Muz. Akad. Nauk* **6**:435-558.

Koltun, V. M. 1964. Gidrobiologicheskiye raboty vysokoshirotnykh articheskikh ekspeditsee na 1/r "F. Litke" i d/e "Ob" i "Lena." *Trude Arkt. i Antarkt. Nauch.-Issled. Inst Izd. Transport,* Moscow and Leningrad, **259**:5-13. (In Russian.)

Kotenev, B. N. 1970. Rel'yef, donnykh otlozheniyi i nekotorye osobennosti geologicheskogo stroyeniya materikovogo sklona vostochnykh raiyonov Beringova morya. *Trudy Vsesoyuzn. Nauch. Issled. Inst. Rybn. Khoz i Okeanogr.* **70**:43-69. (In Russian.)

Krasilov, V. A. 1977. *Evolyutsiya i Biostratigrafiya.* Nauka, Moscow, 256p. (In Russian.)

Krishtafovich, A. N. 1932. *Geologicheskii Obzor Stran Dal'nego Vostoka.* WNIGRI, Moscow and Leningrad, 332p. (In Russian.)

Kussakin, O. G. 1973. Biogeograficheskoye raiyonirovaniye kholodnykh i umerennykh vod shel'fovykh zon Severnogo polushariya na materialye po ravnonogim rakoobraznym. Tezisy dokladov 5 soveschaniya geografov Sibiri i Dal'nego Vostoka. *Trkutsk* **1**(1):107-108. (In Russian.)

Lavrova, M. A. 1960. *Chetvertichnaya geologiya Kol'skogo Poluostrova.* Moscow and Leningrad, 234p. (In Russian.)

Levenstein, R. Ya. 1981. Osobennosti rasprostraneniya polikhet sem. Polynoidae Kanadsekoi kotloviny Polarnogo basseina. *Trudy Inst. Okeanologii AN SSSR* **115**:26-36. (In Russian.)

Lindberg, G. U. 1962. O svyazi kontinentov Evropy i Ameriki. In *Sovetskiye Rybokhozyaistvennye Issledovaniya v Severo-Zapadnoi Chasti Atlanticheskogo Okeana.* Moscow, pp. 135-142. (In Russian.)

Lindberg, G. U. 1970. Krupnye kolebaniya urovnya okeana v chetvertichnoye vremya i ikh vliyaniye na bassein Severnogo Ledovitogo okeana i ego organicheskii mir. In *Severnyi Ledovity Okean i Ego Poberezh'ye v Kainozoye.* Leningrad, pp. 101-112. (In Russian.)

Lindberg, G. U. 1972. *Krupnye Kolebaniya Urovnya Okeana v Chetvertichnyi Period.* Nauka, Leningrad, 547p. (In Russian.)

McDougall, V. I. 1984. Magnetostratigraphy and geochronology of northern Iceland. *J. Geophys. Res.* **89**(138):7,029-7,060.

MacNeil, F. S. 1965. Evolution and distribution of the genus Mya and tertiary migration of molluscs. *Geol. Surv. Prof. Pap. 483-6.* U.S. Government Printing Office, Washington, D.C., pp. 1-51.

Margolis, S. V., and Y. Herman. 1980. Northern hemisphere ice and glacial development in the Late Cenozoic. *Nature* **286**:145-149.

Matheke, G. E., and R. Horner. 1974. Primary productivity of the benthic microalgae in the Chukchi Sea near Barrow Alaska. *J. Fish. Res. Bd. Canada* **31**(11):1,779-1,786.

Mayr, E. 1968. *Zoologicheskii vid i Evolyutsiya.* Mir, Moscow, 597p. (In Russian.)

Mayr, E. 1981. Evolyutsiya. In *Evolyutsiya.* Moscow, pp. 11-31. (In Russian.)

Mercer, J. H. 1978. Glacial development and temperature trends in the Antarctic and in South America. In Banker (ed.), *Antarctic Glacial History and World Paleoenvironments.* Rotterdam, pp. 73-81.

Molina-Cruz, A., and J. Thiede. 1978. The glacial eastern boundary current along the Atlantic Eurafrican continental margin. *Deep-Sea Res.* **25**(4):337-356.

Monin, A. S. 1977. *Istoriya Zemli.* Nauka, Leningrad, 228p. (In Russian.)

Moyes, J., J. C. Duplessy, E. Gonthier, C. Latouche, N. Maillet, M. Parra, and C. Pujol. 1974. Les sediments profonds actuels et Pleistocene recent de l'Atlantique nord-oriental et du sud de la mer de Norvege. *Le Colloq. Int. Exploit. Oceans.* Bordeaux, 4, Paris, Box 201, pp. 1–15.

Nesis, K. N. 1961. Puti i vremya formirovaniya razorvannogo areala u amfiboreal'nykh vidov morskikh donnykh zhivotnykh. *Okeanologiya* **1**((5):893–903. (In Russian.)

Nesis, K. N. 1983*a*. Gipoteza o prichinye vozniknoveniya zapadnoi vostochnoarkticheskikh arealov morskikh donnykh zhivotnykh. *Biologiya Morya* **5**:3–13. (In Russian.)

Nesis, K. N. 1983*b*. Sushchestvoval li v pleistotsene panarkticheskii lednikovyi pokrov? *Biologiya Morya* **6**:11–19. (In Russian.)

Panov, D. G. 1946. Paleogeografiya Sovetskoi Arktiki v chetvertichnoye vremya. *Trudy Inst. Geogr. AN SSSR*, pp. 41–54. (In Russian.)

Pergament, T. S. 1944. Benthos of the Kara Sea. *Problems of the Arctic* **1**:102–132.

Petrov, O. M. 1976. Geologicheskaya istoriya Beringova proliva v pozdnem kainozoye. In *Beringiya v Kainozoye*. Vladivostok, pp. 28–32. (In Russian.)

Petrov, O. M. 1982. *Morskiye Mollyuski Antropogena Severnoi Chasti Tikhogo Okeana*. Nauka, Moscow, 142p. (In Russian.)

Petrushevskaya, M. G. 1976. K voprosu o formirovanii fauny Norwezhsko-Grenlandskogo basseina. In *Problemy Zoologii*. Nauka, Leningrad, pp. 69–72. (In Russian.)

Petrushevskaya, M. G. 1979. Paleobiogeografiya Norwezhsko-Grenlandskogo basseina. *Issled. Fauny Morei* **23**:158–168. (In Russian.)

Reyss, D. 1964. Observations faites en soucoupe plongeante dans deux valees sousmaines de la mer Catalane: Le rech du Cap et le rech Lacaze—Dathiers. *Bull. Inst. Oceanogr.* **63**(1,308):1–8.

Rukhin, L. B. 1962. *Osnovy Obshchei Paleogeografii*. Gostoptekhizdat, Leningrad, 628p. (In Russian.)

Saks, V. N. 1953. Chetvertichnyi period v Sovetskoi Arktikye. *Trudy Nauch-Issled. Inst. Geol. Arktiki* **77**:627p. (In Russian.)

Savin, S. M., R. C. Douglas, and F. G. Stehli. 1975. Tertiary marine paleotemperatures. *Geol. Soc. Am. Bull.* **86**:1,499–1,510.

Scarlato, O. A. 1981. Dvustvorchatye mollyuski umerennykh shirot zapadnoi chasti Tikhogo okeana. Opredelitel, po faune SSSR Nauka, Leningrad, 126, 477p. (In Russian.)

Schafer, C. T. 1977. Distribution and depositional history of sediments in Baie des Chaleurs, Gulf of St. Lawrence, Canada. *J. Earth Sci.* **14**:593–605.

Scott, G. 1926. Geographic des Atlantischen Ordeus. In *Ardet, Handbuch der Palaeogeographie*. Berlin, 250p.

Segerstrale, S. G. 1971. The zoological problem involved in the presence of the glacial relict Pontoporeia affinis (Crustacea, Amphipoda) in lake Washington, USA. *J. Res. Bd. Canada* **28**:1331–1334.

Shackleton, N. J. 1978. Ice-age paleo-oceanography of the Mediterranean. *Nature* **276** (5,689):667.

Shackleton, N. J., and N. D. Opdyke. 1976. Oxygen-isotope and paleomagnetic stratigraphy of Pacific core 28-239 late Pliocene to latest Pleistocene. *Geol. Soc. Am. Mem.* **145**:449–464.

Sinitsin, V. M. 1962. Paleogeografiya Azii. *Izd. AN SSSR*. Moscow and Leningrad, 267p. (In Russian.)

Sinitsin, V. M. 1967. *Vvedeniye v Paleoklimatologiyu*. Nedra, Leningrad, 230p. (In Russian.)

Smith, S. M. 1972. Palaeoecology of postglacial beaches in last Lothian. *Scot. J. Geol.* **8**(1):31–49.

Soot-Ryen, R. 1932. Pelecypoda with a discussion of possible migrations of Arctic pelecypoda in Tertiary times. Norwegian Polar expedition "Maund" 1918–1925. *Sci. Results* **5**(12):3–26.

Sorgenfrei, T. 1958. Molluscan assemblages from the marine middle Miocene of South Jutland and their environments. *Danmarks Geol. Undersog. Kovenhavn* **2**:356–564.

Sparck, R. 1933. Contributions to the animal ecology of the Franz Josef Fjord and adjacent east Greenland waters. *Medd. Grønland* **100**(1):1–38.

Strauch, F. 1970. Die Thule-Landbrücke als Wanderweg und Faunenscheide zwischen Atlantic und Skandik in Tertiär. *Sonderdruck aus der Geologischen Rundschau* **60**:381–417. (In German.)

Strauch, F. 1972. Phylogenese adaptation und Migration einiger nordischer mariner Molluskengenera (Neptunea, Panomya, Cyrtodaria and Mya). *Abhandlangen der Senchenberg Naturforsch. Geselsch. Frank. am Main* **531**:50p. (In German.)

Suzdalsky, O. V. 1976. *Paleogeografiya Arkticheskikh Morei SSSR v Neogene i Pleistotsene*. Nauka, Leningrad, 250p. (In Russian.)

Svetovidov, A. N. 1940. O geograficheskom rasprostranenii treskovykh i drugikh semeistv otryada Gadiforms. *Bull. Mosk. Obshch. Ispytatelei Prirody* **49**(1):50–59. (In Russian.)

Swarzbach, M. 1955. *Klimaty Proshlogo*. Mir., Moscow, 283p. (In Russian.)

Thorson, G. 1933. Investigations on shallow water animal communities in the Franz Josef Fjord (east Greenland) and adjacent waters. *Medd. Grønland* **100**:1–68.

Thorson, G. 1936. The larval development, growth and metabolism of Arctic marine bottom invertebrates *Medd. Grønland* **100**(6):1–155.

Tolmachov, A. I., and B. A. Yurzev. 1970. *Istoriya Arkticheskoi Flory i Eyo Svyaz s Istoriyei Polarnogo Basseina. V sb.: Severnyi Ledovityi Okean i Ego Poberezh'ya v Kainozoye.* Nauka, Leningrad. pp. 87–100. (In Russian.)

Treshnikov, A. F. 1967. Geograficheskie naimenovania osnovnych chastei relyefa dna Arkticheskogo basseina. In *Problemy Arktiki i Antarkiki* **27**:15p.

Troitsky, S. A. 1961. Sovremennaya i iskopayemaya Macoma baltica (L.) na poberezh'ye morya Laptevykh. *DAN SSSR* **136**(2):449–452. (In Russian.)

Tzvetkova, N. L. 1975. *Pribrezhnye Gammaridy Severnykh i Dal'nyevostochnykh Morei SSSR i Sopredel'nykh Vod.* Nauka, Leningrad, 257p. (In Russian.)

Udintsev, G. B. 1955. Proiskhozhdeniye rel'yefa dna Okhotskogo morya. *Trudy Inst. Okeanol. AN SSSR* **13:**5–15. (In Russian.)

Ushakov, P. V. 1952. Chukotskoye morye i ego donnaya fauna. In *Krainii Severo-Vostok Soyuza SSSR II.* Nauka, Moscow and Leningrad, pp. 3–78. (In Russian.)

Vail, P. R., and J. Hardenbol. 1979. Sea level changes during the Tertiary. *Oceanus* **22:**71–80.

Velichko, A. P. 1973. *Prirodnyi Protsess v Pleistotsene.* Nauka, Moscow, 256p. (In Russian.)

Vella, P. 1973. *Ocean Paleotemperatures and Oscillations of the Subtropical Convergence Zone on the Eastern New Zealand.* Oceanogr. S. Pacif., 1972, Wellington, pp. 315–318.

Verbitskiy, M. Ya., and D. D. Kvasov. 1980. Prichiny oledeneniya Antarktidy. *Antarktika. Dokl. Komissii* **19:**23–88. (In Russian.)

Vogt, P. R. 1972. The Faeroe-Iceland-Greenland aseismic ridge and the western boundary undercurrent. *Nature* **239**(5,367):79–81.

Woldstedt, D. 1954. *Das Eizeitalter.* Bd. I., Stuttgart, 250p.

Zakharov, B. F. 1978. *Mirovoi Okean v Lednikovye Epokhi Pleistotsena.* Gidrometeoizd. Leningrad, 64p. (In Russian.)

Zenkevich, L. A. 1939. Nekotorye momenty zoogeografii Polarnogo basseina v svyazi s voprosom o ego paleogeograficheskom proshlom. *Zool. Zhurn.* **12**(4):17–34. (In Russian.)

Zenkevich, L. A. 1963. Biologiya morei SSSR. *Izd. AN SSSR* 739p. (In Russian.)

Zernov, S. A. 1934, *Obshchaya Gidrobiologiya.* Biomedgiz, Moscow, 215p. (In Russian.)

Zinova, A. D. 1974. Sostav i fitogeograficheskoye deleniye arkticheskoi vodoroslevoi flory. In *Gidrobiologiya i Biogeografiya Shel'fov Kholodnykh i Umerennykh vod Mirovogo Okeana.* Nauka, Leningrad, pp. 12–13. (In Russian.)

# 11: Distributional Patterns of Echinoderms in the Eurasian Sector of the Arctic Ocean

N. A. Anisimova

Our knowledge of echinoderms, including echinoids, asteroids, and ophiuroids from the Eurasian Arctic seas (Barents, Kara, Laptev, East Siberian, and Chukchi seas), were established with the studies of Diakonov (1926, 1933, 1950, 1952, 1954), Shorygin (1982), Gorbunov (1932, 1933*a,b*), Baranova (1964, 1977), Khodkina (1964), and Moskalyov (1980).

To date the following have been described: 42 species of asteroids (starfishes) belonging to 23 genera, 10 families, and 4 orders; 22 species of ophiuroids (brittle stars) belonging to 14 genera, 7 families, and 2 orders; and 9 species of echinoids (sea urchins) belonging to 7 genera, 7 families, and 3 orders. They include taxa inhabiting various depth zones. The number of taxa (orders, families, genera, and species) increases with decreasing rank, that is, the lowest ranking unit (the species) contains the largest number of taxa. According to Golikov (1976*a*), the relationship between the abundance of a taxa and hierarchical rank is approximated by the parabolic function $N = aT^b$, where $N$ is the number of taxa (number of orders, families, etc.); $T$ is the rank of taxon; $a$, $b$ are the coefficients. If the rank of the taxa in the diagram is expressed in units from 1 (class) to 5 (species), the dependence of the number of taxa on their hierarchy on a logarithmic scale is a straight line. The diagram for starfishes and brittle stars (Fig. 11-1) shows that these two well-known groups are found in the Northern seas of the USSR (lines with circles and straight line nearly coincide).

Figure 11-1*B* expresses the decline in the rate of increase of faunal diversity. Figure 11-1 shows that the fauna of starfishes is richer than that of brittle stars. Figure 11-1*A*, which stands for faunal diversity of orders, expresses to some extent the history of the fauna, its relative age, and rates of evolution. The similarity of the values of these parameters both in starfishes and in brittle stars implies their simultaneous distribution and similar ways of dispersal.

In order to compare the Arctic echinoderms with the faunas of adjacent areas, similar diagrams were made for the British waters (Mortensen, 1927) and the USSR Far Eastern seas (Diakonov, 1949). If we compare these data we see that the latter possess the highest brittle stars and star-

*Note:* This chapter has been reviewed, edited, and rewritten by the Editor, Yvonne Herman.

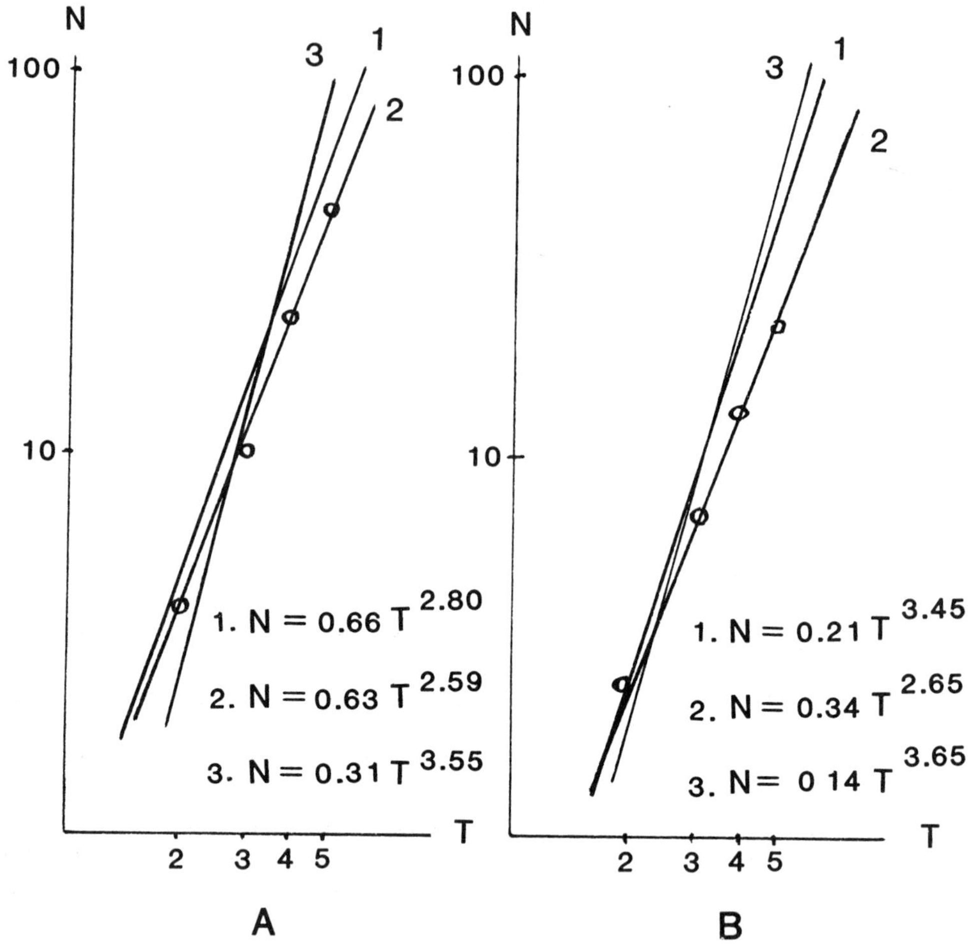


Figure 11-1. Relationship between number of taxa and their systematic rank. Ordinate: logarithm of number of taxa; abscissa: logarithm of rank of taxa designated in figures from 1 (class) to 5 (species). *A:* starfishes; *B:* brittle stars. 1: British waters (Mortensen, 1927); 2: Northern seas of the USSR (references in text); 3: Far Eastern seas of the USSR (Diakonov, in references).

fish diversity as well as the highest rate of increase in the number of taxa. These values are slightly lower in the Atlantic boreal waters and reach a minimum in the Arctic seas. The author believes that this is due to the "old" age of the Pacific Ocean fauna and young age of the Arctic fauna.

Unlike the starfishes and brittle stars, the fauna of sea urchins is very poor. Only three species of sea urchins inhabit the high-Arctic waters (boreal species penetrating the southern region of the Barents and Chukchi seas are not taken into account). In the estuary Arctic water sea urchins do not occur.

## AREAL DISTRIBUTION

Twenty-two biogeographic groups have been recognized to date. The nomenclature and classification proposed by Golikov (1980, 1982) is followed here. Their distributional patterns, based on references mentioned in the previous section as well as on the vast literature including Ekman (1953) and the *Treatise on Invertebrate Paleontology* (1966, 1978), are as follows:

1. Tropical-boreal, wide-ranging species—*Amphipholis squamata* (Ophiuroidea) wide-ranging in tropical and temperate sectors of the Atlantic and Pacific oceans; found off the Australian coasts; also in the Atlantic Ocean its northern limit is the southwestern Barents Sea.
2. Subtropical-Arctic species—*Ctenodiscus crispatus* (Asteroid) in the Arctic Ocean has circumpolar distribution; in the Atlantic Ocean it was observed near Cape Cod and off the North American and Norwegian coasts; in the Pacific Ocean it occurs along the Japanese coast and the American coast as far as the Panama Canal.
3. Atlantic subtropical-boreal wide-ranging species—*Psilaster andromeda* (Asteroid); *Ceramaster granularis* (Ast.), *Ophiacantha abyssicola* (Oph.); *Spatangus purpureus* (Echinoid); and *Brisaster fragilis* (Ech.). These species were recorded between the southern Barents Sea and Cape Hatteras, except the last taxon, which has been recorded on the coasts of Florida and the Antilles and along the Moroccan coast as well as in the Azores.
4. Atlantic European subtropical-boreal species ranging from the southern Barents to the Mediterranean Sea along the European coast: *Ophiura albida* (Oph.) and *Echinus acutus* (Ech.); *Echinocyamus pusillus* (Ech.) as far as the Bay of Biscay; *Ophiactis abyssicola* (Oph.) and *Echinus esculentis* to the coasts of Senegal; and *Asterias rubens* (Ast.), absent from the Mediterranean Sea and found in the White Sea.
5. Amphiboreal wide-ranging species—*Pseudarchaster parelii* (Ast.), *Leptychaster arcticus,* (Ast.), and *Solaster endeca* (Ast.) are known in the Atlantic Ocean from Labrador to Cape Cod along the American coast and from the coast of Great Britain to the southern sector of the Barents Sea (the last species occurs in the White Sea, around Spitsbergen and eastward as far as the southern sector of the Kara Sea); in the Pacific Ocean it occurs from the Bering Strait to the Sea of Japan. *Echinarachnus parma* (Ech.) has a similar distribution; however, it penetrates into the southern Chukchi Sea, and is apparently absent from the Atlantic Ocean off the European coasts.
6. Atlantic boreal wide-ranging species—*Hippasteria phrygiana* (Ast.) present along the American coast and the southern coast of England, and *Poraniomorpha hispida* (Ast.) observed as far north as Cape Hatteras off the American coast and the Skagerrak Strait off the Norwegian coast; the northern limit of both species is the southern Barents Sea.
7. Atlantic European boreal species—*Leptasterias mülleri* (Ast.) and *Henricia sanguinolenta* occur between the European coast and the southern Barents Sea.
8. Pacific boreal wide-ranging species—*Solaster dawsoni* occurs between Cape Barrow in the Chukchi Sea and Monterey Bay along the American coast and the Yellow Sea, the Sea of Japan, the Okhotsk Sea, as well as in the Bering Sea.
9. Pacific Asian boreal species—*Gorgonocephalus caryi* (Oph.), *Amphiodia craterodmeta* (Oph.), and *Ophiura maculata* are known from the Sea of Japan, Okhotsk, Bering, and the southern Chukchi seas.
10. Amphiboreal species—*Leptasterias arcticus* recorded from the Chukchi, Bering, and Okhotsk seas, the western Barents Sea, off the Faroe Islands and off the coast of Iceland.
11. Atlantic European high boreal species—*Amphiura borealis* (Oph.), known from the Norwegian coast, the eastern Barents Sea, and off the Faroe Islands.
12. Pacific high boreal species—*Asterias*

*rathbuni* and *Henricia tumida,* known from the southern Chukchi Sea around Shumanskiya Islands on the American coast and in the Bering and Okhotsk seas as well as off the Aleutian and Kurile islands, and *Leptasterias sibirica* endemic species known from the southern Chukchi Sea.

13. Wide-ranging boreal–Arctic species—*Ophiura sarsi, Crossaster papposus* (Ast.) with circumpolar distribution in the Arctic, *Ophiacantha bidentata, Ophiopholis aculeata* (Oph.), *Pteraster pulvillus* (Ast.), *P. militaris, Strongylocentrotus droebachiensis* (Ech.), *S. pallidus* absent from the Siberian Sea. In the Atlantic and Pacific oceans they are confined to the boreal zone.
14. Atlantic wide-ranging boreal–Arctic species—*Ophiocten sericeum* (Oph.) with circumpolar distribution in the Arctic, *Pontaster tenuispinus* (Ast.), *Ophioscolex glacialis* (Oph.) along the continental slope eastward to the Chukchi Sea, *Ophiura robusta* occurring eastward to the Kara Sea; none of these species in the Atlantic Ocean are found south of Cape Hatteras along the American coast, and the Bay of Biscay along the European coast.
15. Pacific boreal–Arctic wide-ranging species—*Stegophiura nodosa* (Oph.). In the Arctic, nearly circumpolar distribution; it occurs off the western coasts of Greenland, Spitsbergen, Franz Josef Land, southeastern Barents Sea, Kara, Laptev, East Siberian, Chukchi, Bering, Okhotsk seas, and as far south as the coasts of Japan and Korea.
16. High boreal–Arctic species—*Amphiura sundvalli:* Barents, Kara, Laptev seas, as well as the Chukchi Sea, and in the Atlantic Ocean along the American coast; its southern limit along the European coast is Scotland; in the Pacific Ocean, the Bering Sea up to Olyutorskiy Cape. *Stephanasterias albula* (Ast.): Similar to the previous species, but absent from the Laptev, East Siberian, and Chukchi seas, and in the Pacific Ocean present in the Okhotsk Sea. *Pteraster obscurus:* Similar to the previous species, but inhabiting the southern Chukchi Sea also.
17. Atlantic high boreal–Arctic species—*Gorgonocephalus eucnemis:* In the Arctic on the continental slope as far as the Chukchi Sea; *Leptasterias hyperborea* and *Henricia solida* in the western Barents Sea; *Pedicellaster typicus* (Ast.), *Solaster syrtensis,* and *Henricia eschrichtii* to the Kara Sea; *H. knipowitschi, H. scabrior,* and *Lophaster furcifer* (Ast.) as far as the New Siberian Islands. None of the Atlantic species occur south of Cape Cod off the American coast or south of the Faroe Islands off the European coast.
18. Pacific high boreal–Arctic species—*Leptasterias polaris:* In the eastern Arctic as far as the Bennet Island, in the Chukchi Sea; in the Bering Sea as far as Lopatka Cape, the Olyutorskiy Bay, and off the Pribylov Islands.
19. Atlantic Arctic species—*Urasterias lincki* (Ast.) with circumpolar distribution in the Arctic; *Hymenaster pellucidus* (Ast.) occurring along the continental slope as far as the Chukchi Sea; *Icasterias panopla* (Ast.) inhabiting the eastern Arctic as far as the New Siberian Shoals; and *U. lincki* in the White Sea.
20. Pacific Arctic species—*Leptasterias groenlandica* with circumpolar distribution in the Arctic, occurs in the Pacific Ocean along the American coast as far as 54°NL.
21. West Arctic species—*Korethraster hispidus* (Ast.), *Crossaster squamatus, Solaster glacialis, Tylaster willei* (Ast.), *Poraniomorpha bidens, P. tumida, Bathybiaster vexillifer* (Ast.), and *Gorgonocephalus arcticus; Ophiopleura borealis,* (Oph.), *Ophiopus arcticus* (Oph.), and *Amphipholis torelli* (Oph.) in the Atlantic Ocean, adjacent to the Arctic Ocean.
22. Arctic abyssal species—*Pourtalesia jeffreysi* (Ech.) and *Ophiopyren striatum*

(Oph.) inhabiting abyssal regions of the central Arctic Basin.

The biogeographic distribution of faunas in different Arctic regions is shown in Figure 11-2.

The degree of similarity between high taxonomic units in different Arctic regions is illustrated by using the Jaccard coefficient

$$I_J = \frac{C}{D_1 + D_2 - C} \times 100\%$$

(Jaccard, 1901); for lower taxonomic units the Shimkevish–Simpson coefficient is used (Simpson, 1943),

$$I_S = \frac{C}{D_{min}} \times 100\%$$

where $D_1$ and $D_2$ are the numbers of species in the areas compared, $C$ is the number of species in common, and $D_{min}$ is the smaller number of species.

The Shimkevich–Simpson index along with the Jaccard index allows us to evaluate the degree of impoverishment of one fauna relative to another (Pesenko, 1982). The mean value of these two coefficients showing the compared faunas was used to construct similarity dendrograms in various regions of the Arctic (Fig. 11-3).

The Chukchi Sea was divided into northern and southern sectors along the latitude of the Herald Island and Cape Barrow. The boundary between the northern and southern parts of the Barents Sea was drawn along 75°NL corresponding to the

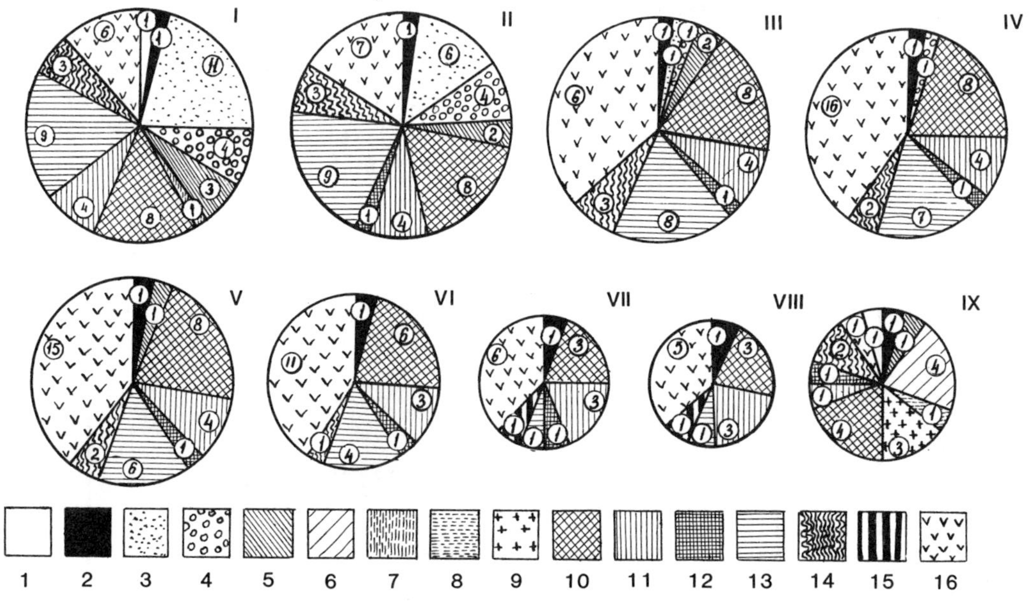


Figure 11-2. Biogeographic distribution of the fauna of sea urchins, starfishes, and brittle stars in different Arctic regions. 1: Tropical boreal wide-ranging species; 2: Subtropical–Arctic circumpolar species; 3: Atlantic subtropical–boreal wide-ranging species; 4: Atlantic boreal wide-ranging species; 5: Amphiboreal wide-ranging species; 6: Pacific boreal wide-ranging species; 7: Atlantic European high boreal species; 8: Amphi-high boreal species; 9: Pacific high boreal species; 10: Atlantic Boreal–Arctic wide-ranging species; 11: Boreal–Arctic wide-ranging species; 12: Pacific boreal–Arctic wide-ranging species; 13: Atlantic high boreal–Arctic species; 14: High boreal–Arctic species; 15: Pacific high boreal–Arctic species; 16: Arctic species. I: Southwestern part of Barents Sea; II: southeastern part of Barents Sea; III: northwestern part of Barents Sea; IV: northeastern part of Barents Sea; V: Kara Sea; VI: Laptev Sea; VII: East Siberian Sea; VIII: northern part of Chukchi Sea; IX: southern part of Chukchi Sea. Arabic numerals within each sector indicate number of species.

**Table 11-1. Comparison of Species Composition of Sea Urchins, Starfishes, and Brittle Stars in Different Arctic Regions**

| | *I* | *II* | *III* | *IV* | *V* | *VI* | *VII* | *VIII* | *IX* |
|---|---|---|---|---|---|---|---|---|---|
| I | 51 | 96 | 73 | 70 | 71 | 74 | 69 | 71 | 40 |
| II | 81 | 45 | 78 | 75 | 76 | 81 | 81 | 78 | 50 |
| III | 52 | 64 | 45 | 98 | 100 | 100 | 94 | 93 | 50 |
| IV | 44 | 55 | 85 | 40 | 95 | 100 | 94 | 93 | 45 |
| V | 43 | 54 | 84 | 86 | 38 | 100 | 94 | 93 | 45 |
| VI | 34 | 44 | 60 | 68 | 71 | 27 | 94 | 93 | 35 |
| VII | 20 | 27 | 33 | 37 | 38 | 54 | 16 | 100 | 44 |
| VIII | 18 | 23 | 28 | 32 | 33 | 46 | 87 | 14 | 43 |
| IX | 13 | 19 | 18 | 18 | 18 | 17 | 24 | 21 | 20 |

Figures below underscores: degree of similarity according to Jaccard; figures above underscores: measure of inclusion of smaller into larger samples (per unit). Roman numerals indicate regions as in Figure 11-2. Underlined numbers indicate number of species in each region.

northern boundary of the North Cape Current and between the western and eastern parts along the Kola Longitude (33°30′E).

As shown in Figure 11-3, in the southern part of the Chukchi Sea, species diversity is high. The fauna of this region is characterized by the prevalence of Pacific boreal species that have penetrated in the southern Chukchi Sea through the Bering Strait (e.g., *G. caryi, S. dawsoni, H. tumida, A. rathbuni, A. craterodmeta, O. maculata,* and *E. parma*). Some taxa extend westward into Kolyuchinskaya Inlet (e.g., *H. tumida* and *A. craterodmeta*), and occasionally occur in Long Strait (e.g., *O. maculata*). The northern limits of boreal species are Herald Island and Cape Barrow. *L. sibirica,* described from the Kolyuchinskaya Inlet and

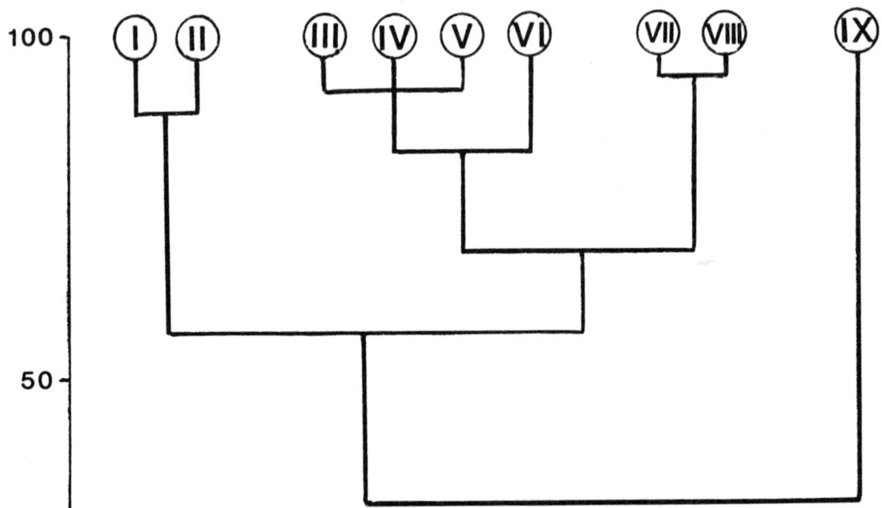


**Figure 11-3.** Similarity of fauna of starfishes, sea urchins, and brittle stars in different Arctic regions. Ordinate: similarity percent (mean value of Jaccard coefficient and measure of inclusion); abscissa: numerals of regions as in Figure 11-2.

the Bering Strait, is endemic to the southern Chukchi Sea. Thus, in the southern Chukchi Sea, endemic and boreal Pacific species constitute 40 percent of the fauna, circumpolar Arctic taxa (e.g., *L. groenlandica*) comprise 25 percent, and boreal–Arctic species (e.g., *C. crispatus*, *C. papposus*, *O. cericeum*, and *O. sarsi*) make up 35 percent. The last group penetrates the Chukchi Sea from the North Pacific, but is found in other parts of the Arctic Ocean as well.

The second region with well-defined species composition is the southern Barents Sea. Numerous boreal Atlantic species penetrate this region via the warm waters of the North Atlantic Drift. Some species are confined to the southwestern sector of the sea (e.g., *P. parelii*, *C. squamatus*, *E. acutus*, *S. purpureus*, *Ophiacantha abyssicola*, *A. squamata*, and *A. borealis*) while others penetrate along the Murmansk coast eastward as far as Cape Svyatoi Nos (e.g., *H. sanguinolenta*, *E. esculentus*, *A. rubens*, *P. hispida*, *B. fragilis*, *L. mulleri*, *H. phrygiana*, *P. andromeda*, and *C. granularis*). *S. endeca* ranges further east as far as the south Kara Sea. Several species occur along the edge of the Barents Sea shelf as far as western Spitsbergen (78°NL) (e.g., *L. arcticus*, *S. endeca*, and *Ophiactis abyssicola*). Thus 37 percent of the species in this region occur in the Norwegian high boreal zone as well; 10 percent are circumpolar boreal–Arctic and Arctic species; and 52 percent are species present both in the southern Barents Sea and in other sectors of the Arctic Ocean, but without circumpolar distribution. Among them only *S. nodosa* penetrates in the Barents Sea from the northern sectors of the Pacific Ocean (along the western route). Other species are either North Atlantic emigrants or have penetrated from the Atlantic adjacent to the Arctic Ocean.

In the southern Barents Sea, boreal species constitute 37 percent of the fauna; boreal–Arctic taxa, 50 percent; and Arctic species, 13 percent of the total number of species. The echinoderm fauna of the southern part of the Barents Sea may be referred to as the Norwegian high boreal province of the Atlantic boreal region. The Arctic and boreal regions are separated by a transitional zone where both Arctic and boreal species occur. Figure 11-4 shows the transitional zone in which both Arctic and boreal species occur and the boundary between boreal and Arctic species. This boundary is based on data collected at 470 stations between 1969–1970 by the Polar Institute of Fisheries and Oceanography. The position of this boundary fluctuates and is related to the fluctuations of the North Atlantic Drift.

The region of the Kaninsko-Kolguevskoyi shoals and the Pechora Sea is extremely poor in echinoderms. Shallow water depths, extensive fishing, strong tidal currents, and the sandy substrate constitute unfavorable conditions for echinoderms. Of the rare species present in this region the most common are *S. nodosa*, *O. cericeum*, *L. groenlandica*, *H. eschrichtii*, *H. knipowitschi*, and *S. endeca*. This province belongs to the transitional zone between the boreal and Arctic regions (Fig. 11-4). Based on sea urchins, starfishes, and brittle stars, the northwestern and northeastern sectors of the Barents, Kara, Laptev, East Siberian, and northern Chukchi seas belong to the Arctic region (Golikov, Scarlato, and Pavshtiks, 1983). The dendogram (Fig. 11-3) shows the western and eastern parts with the boundary along the New Siberian Islands. The following species are endemic to the Arctic region: *O. striatum*, *B. vexillifer*, *K. hispidus*, *T. willei*, *P. bidens*, *S. golikovi*, *S. glacialis*, *O. arcticus*, *O. borealis*, and *P. jeffreysi* (24 percent of the total number of species). The first 8 species are present only in the western sector while *O. borealis* and *P. jeffreysi* have been found east of the above-mentioned boundary on the continental slope as well. Eighty percent of the endemic species inhabit the Atlantic sector of the Arctic region. In the vicinity of the Pacific Ocean, the Arctic Basin has no endemic Arctic echinoderms.

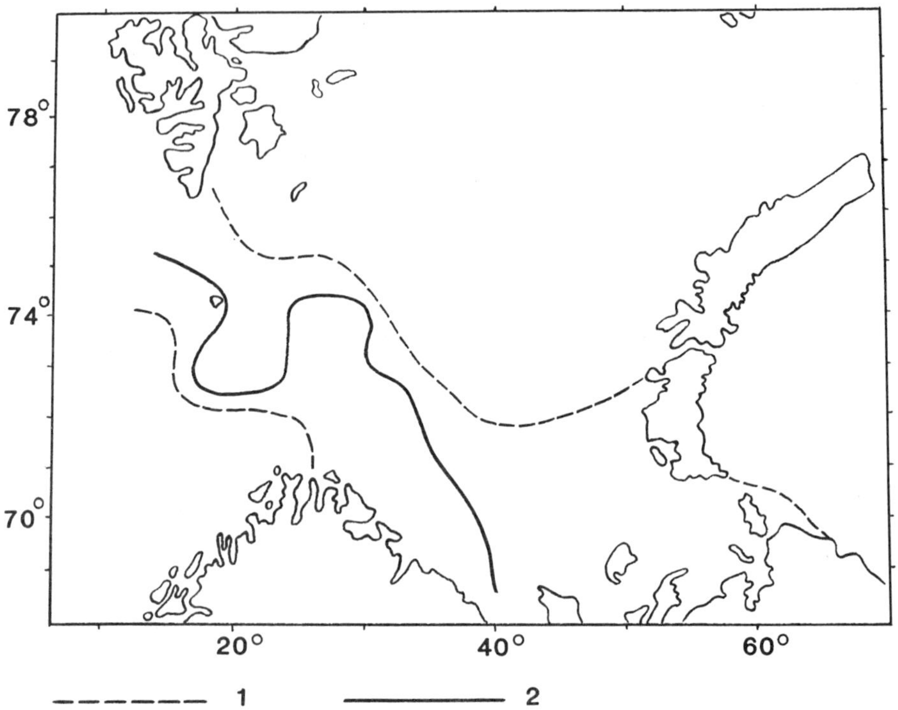


Figure 11-4. Boundaries of transitional zones (1) between the Arctic– and Atlantic-boreal regions and (2) of predominance of boreal over Arctic species.

Five predominantly Arctic species (12 percent of the total number of species) are widespread in the Arctic region but are found in both the Atlantic (*U. lincki, I. panopla, H. pellucidus,* and *G. arcticus*) and in the Pacific (*L. groenlandica*) oceans. In the Arctic region, 27 boreal–Arctic species occur; they comprise 64 percent of the total species number. Of this group, 4 species have circumpolar distribution in the Arctic region, 5 occur in the eastern and western sectors, and 2 species are in the Pacific.

Eastward the species composition is impoverished. The decline in the number of species is most abrupt in the area of the New Siberian Islands.

## VERTICAL DISTRIBUTION

In the Barents and Kara seas, the largest number of species occur between 150–200 m depth; in the Laptev and East Siberian seas, between 50–80 m depth; and in the Chukchi Sea, between 200–600 m depth (Fig. 11-5), except for a maximum at 50 m depth where there is a slight increase in species number. Figure 11-6 illustrates the number of samples required to obtain a representative assemblage, assuming homogenous faunal distribution for a given region. Our results are based on the study of 4,209 samples from the Barents Sea, 1,081 samples from the Kara Sea, 190 samples from the Laptev Sea, 152 samples from the East Siberian Sea, and 217 samples from the Chukchi Sea.

The author believes that the higher diversity of the Barents–Kara seas' faunas between 150 m and 200 m, as compared to their shallower-water counterparts, could be due to the ''older age'' of these taxa inhabiting the Atlantic water. The latter group may have evolved or repopulated these regions following the termination of the Last Glaciation only 10,000–13,000 years ago.

The surface Arctic Water Mass formed

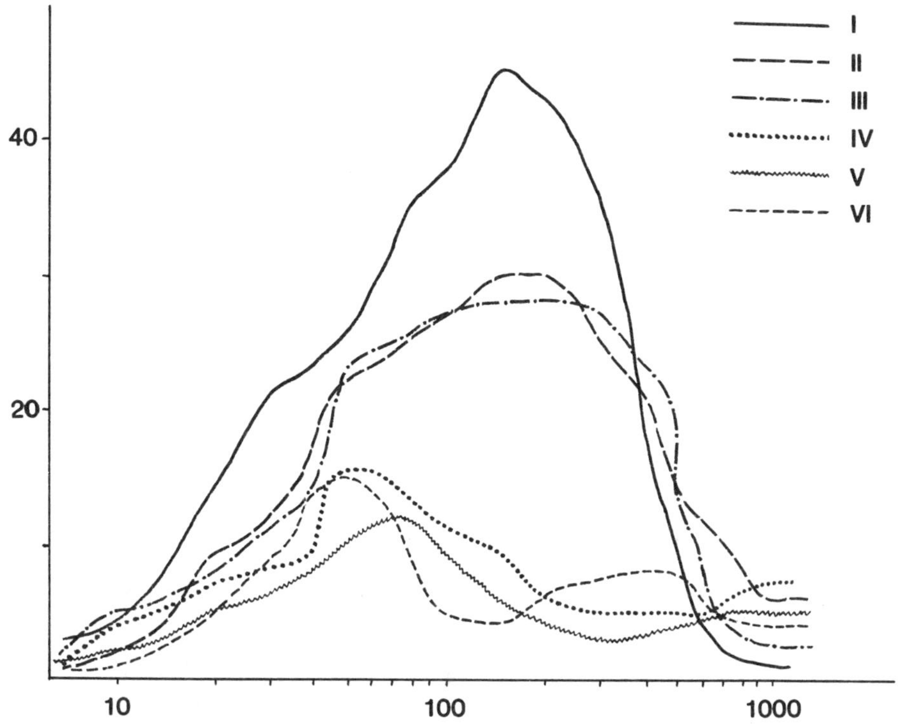


Figure 11-5. Distribution of sea urchins, starfishes, and brittle stars as a function of depth in different regions of the European sector of the Arctic. Ordinate: number of species; abscissa: depth in meters (logarithmic scale). I: Southern part of Barents Sea; II: northern part of Barents Sea; III: Kara Sea; IV: Laptev Sea; V: East Siberian Sea, VI: Chukchi Sea.

in the Arctic is characterized by low salinities (S) and temperatures (T) as well as by pronounced TS seasonal fluctuations (*Oceanographic Atlas of the Arctic,* 1980). This water covers most Arctic shelves except for the southern sectors of the Barents and Chukchi seas. Hence, the dominance of the most eurybiontic boreal–Arctic species on the shelves (Fig. 11-7). All of the Arctic–boreal taxa with circumpolar distribution are confined to this water mass (e.g., *U. lincki, L. groenlandica,* and *I. panopla*). Among them *L. groenlandica* inhabits the shallowest water (10 m). The low species diversity at depths shallower than 20 m is probably due to the adverse environment, particularly the low salinities resulting from summer river discharge and the melting of pack ice and icebergs (Golikov and Averintsev, 1977).

The Chukchi Sea fauna at 50 m depth is dominated by North Pacific elements (81 percent of the fauna) while wide-ranging circumpolar Arctic and boreal Arctic taxa comprise only 19 percent of the echinoderm population. The second species diversity maximum in the Chukchi Sea occurs between 200 m and 500 m. The fauna is composed of Atlantic elements that migrated along the shelf and the upper continental slope (e.g., *P. tenuispinus, O. glacialis, G. eucnemis, H. pellucidus, O. bidentata,* and *O. borealis*). Gurjanova (1957) observed that species living between 600 m and 800 m depths in the regions adjacent to the Atlantic descend to 1,500–2,000 m in the Pacific. Among echinoderms this distributional pattern was observed for *H. pellucidus, O. bidentata,* and *P. tenuispinus.* This phenomenon appears to be related to water mass characteristics.

In the view of the author, the endemic abyssal Arctic echinoderm fauna (including *P. jeffreysi* and *O. striatum*) evolved in

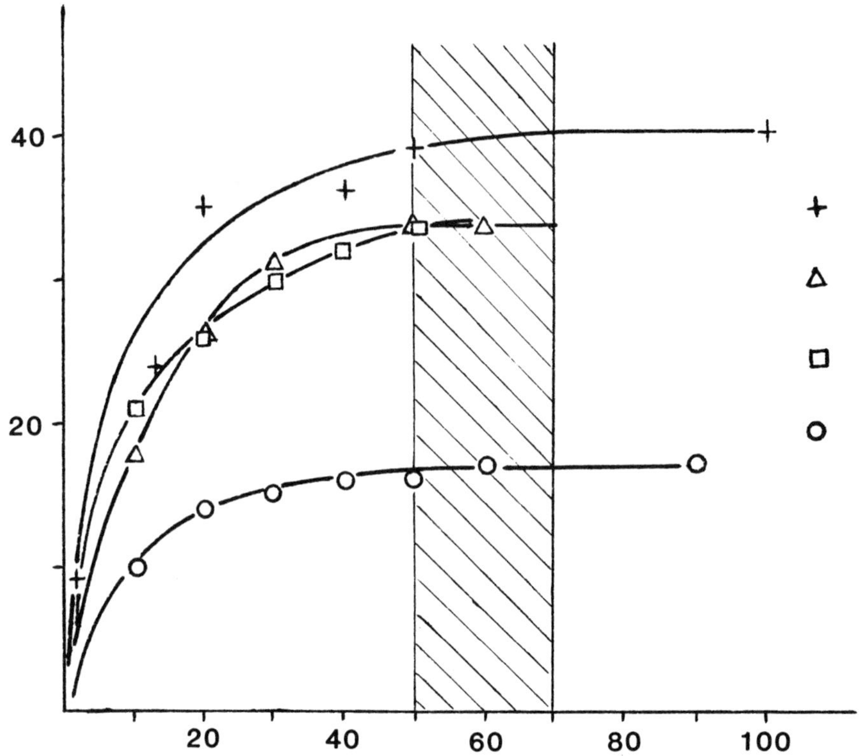


Figure 11-6. Relationship between number of species observed and number of samples taken in different parts of the Barents Sea. Ordinate: number of species; abscissa: number of samples. 1: Northwestern part of Barents Sea; 2: southeastern part of Barents Sea (beyond Cape Svyatoi Nos), 3: southwestern part of Barents Sea (between 23°E up to Cape Svatoi Nos); 4: coastal zone of eastern Murman area (from Kildin Island up to Cape Svyatoi Nos). Dashed zone indicates the number of samples required for obtaining representative faunas.

early Pleistocene time with the initial glaciation, which resulted in the cooling of the world ocean's deep water (Golikov, 1976*b*, 1980) and the simultaneous deepening of the Arctic Basin (Saks, 1953; Einarsson, Hopkins, and Doel, 1967).

Not many echinoderms seem to have preference for any particular type of substrate (Fig. 11-8). Among species tolerant of narrow ranges of substrate type, only *S. nodosa* appears to be confined to silty sands, while *S. endeca, S. droebachiensis, A. rubens, E. acutus,* and *O. aculeata* seem to prefer rocky substrates, and *C. crispatus, P. tenuispinus, H. pellucidus, L. arcticus,* and *O. borealis* occur mainly on silty substrates.

## CONCLUSIONS

The study of recent Arctic echinoderms suggests that species distributional pat-

Figure 11-7. Biogeographic composition of sea urchins, starfishes, and brittle stars at different depths and in different regions of the Eurasian sector of the Arctic. Ordinate: species number in a given biogeographic group as percentage of the total species number; abscissa: depth in meters (logarithmic scale). *A:* Southwestern part of Barents Sea; *B:* northern part of Barents Sea; *C:* Kara Sea and Laptev Sea; *D:* East Siberian Sea; *E:* Chukchi Sea. 1: Arctic species; 2: boreal–Arctic species; 3: boreal species.

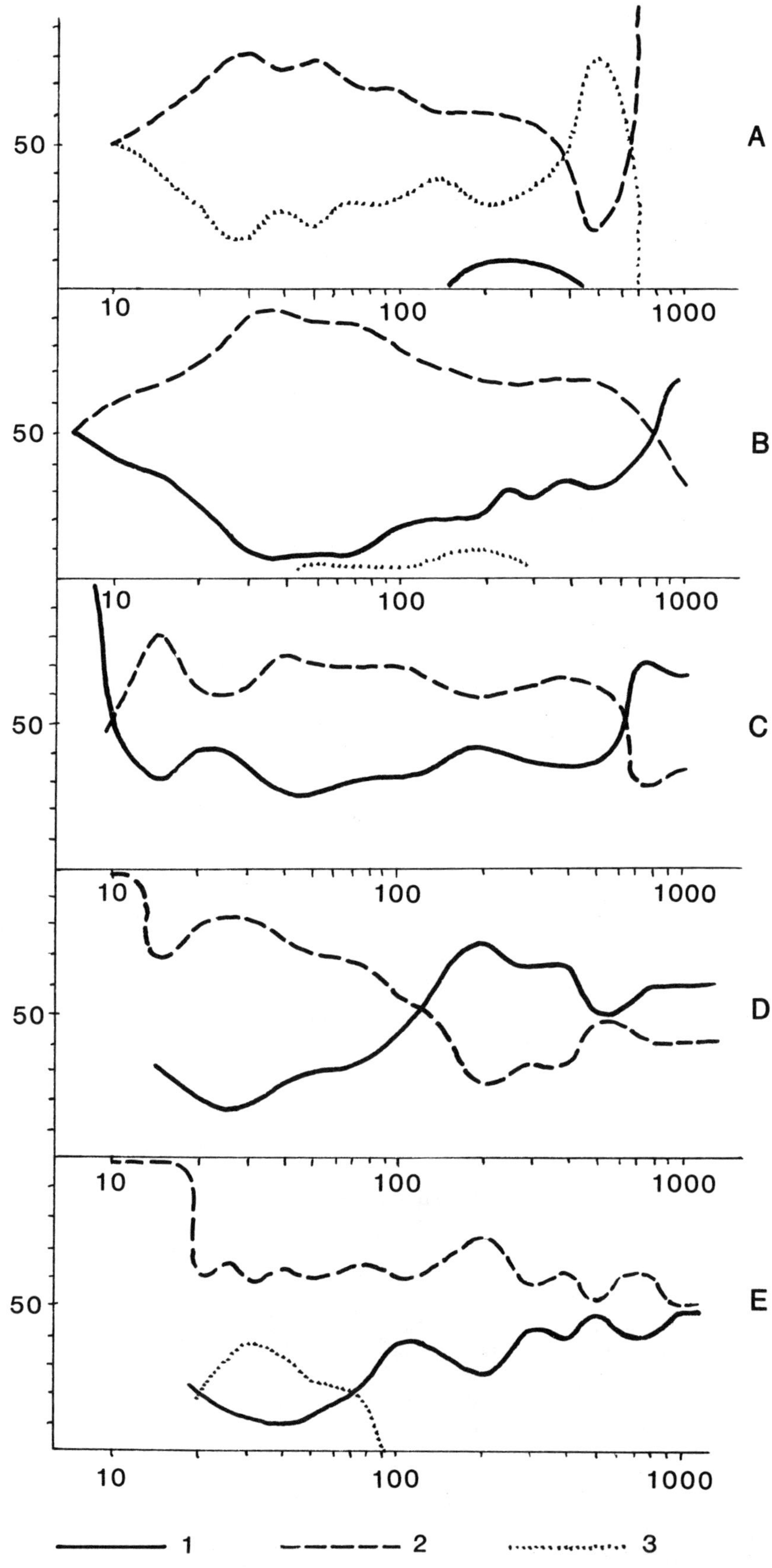

50
A
10
100
1000
50
B
10
100
1000
50
C
10
100
1000
50
D
10
100
1000
50
E
10
100
1000
1
2
3

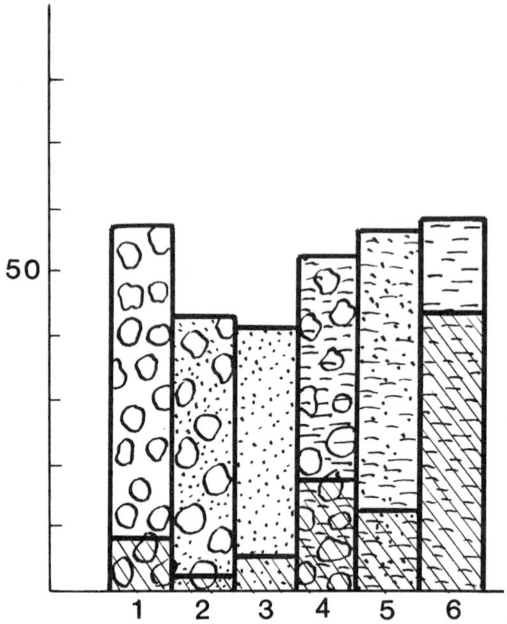


Figure 11-8. Distribution of species of brittle stars, starfishes, and sea urchins on various substrates. Ordinate: number of species occurring on various types of substrate. Dashed lines indicate number of species preferring a certain type of substrate; abscissa: type of substrate 1: Rocky substrate; 2: gravel, sand; 3: sand; 4: gravel and clays; 5: clays; 6: clay–sand, sandy clays.

terns are determined primarily by water-mass characteristics and only secondarily by the type of substrate and by their mode, time, and area of evolution.

The present-day oceanographic regime of the Arctic is principally influenced by the Atlantic water mass; consequently, the Arctic fauna has closest affinities with and probably originated from Atlantic species.

Based on the echinodermal faunas, several biogeographical regions are distinguished. In the Atlantic region, from Cape Svyatei Nos up to northwestern Spitsbergen, and in the Pacific region, from Kolyuchinskaya Bay up to Herald Island and Point Barrow, boreal faunas are recognized. North and south of these boundaries, transitional zones occur. Within the northern boundaries, Arctic and boreal taxa occur. In the Pacific, south of these boundaries, the Aleutian high boreal subregion is recognized, while in the Atlantic, the Norwegian high province occurs. Within the Arctic region proper, faunal depth zonation is much more conspicuous than latitudinal zonation. Three provinces (the last comprising two subprovinces) are recognized in this region: (1) the abyssal Arctic province corresponding to the Arctic Abyssal Water Mass at depths greater than 800–1,000 m, (2) the intermediate Atlantic province within the Atlantic Intermediate Water at 200–800 m depth, and (3) the western Arctic surface subprovince and the eastern Arctic surface subprovince separated by the New Siberian Islands of the surface Arctic water mass.

## APPENDIX

### Plate 11-I

1. *Crossaster papposus* Rec., × 0.3–0.4
2. *Solaster syrtensis* Rec., × 0.2–0.4
3. *Pontaster tenuispinus* Rec., × 0.75
4. *Poraniomorpha tumida* Rec., × 0.75
5. *Ctenodiscus crispatus* Rec., × 1, (*a*) aboral surface, (*b*) oral surface
6. *Strongylocentrotus pallidus* Rec., × 0.75
7. *Gorgonocephalus eucnemis* Rec., × 0.2–0.3
8. *Stegophiura nodosa* Rec., × 1.125
9. *Ophiura sarsi* Rec., × 0.75
10. *Ophiopholis aculeata* Rec., × 1.125
11. *Amphiura sundevalli* Rec., × 0.75

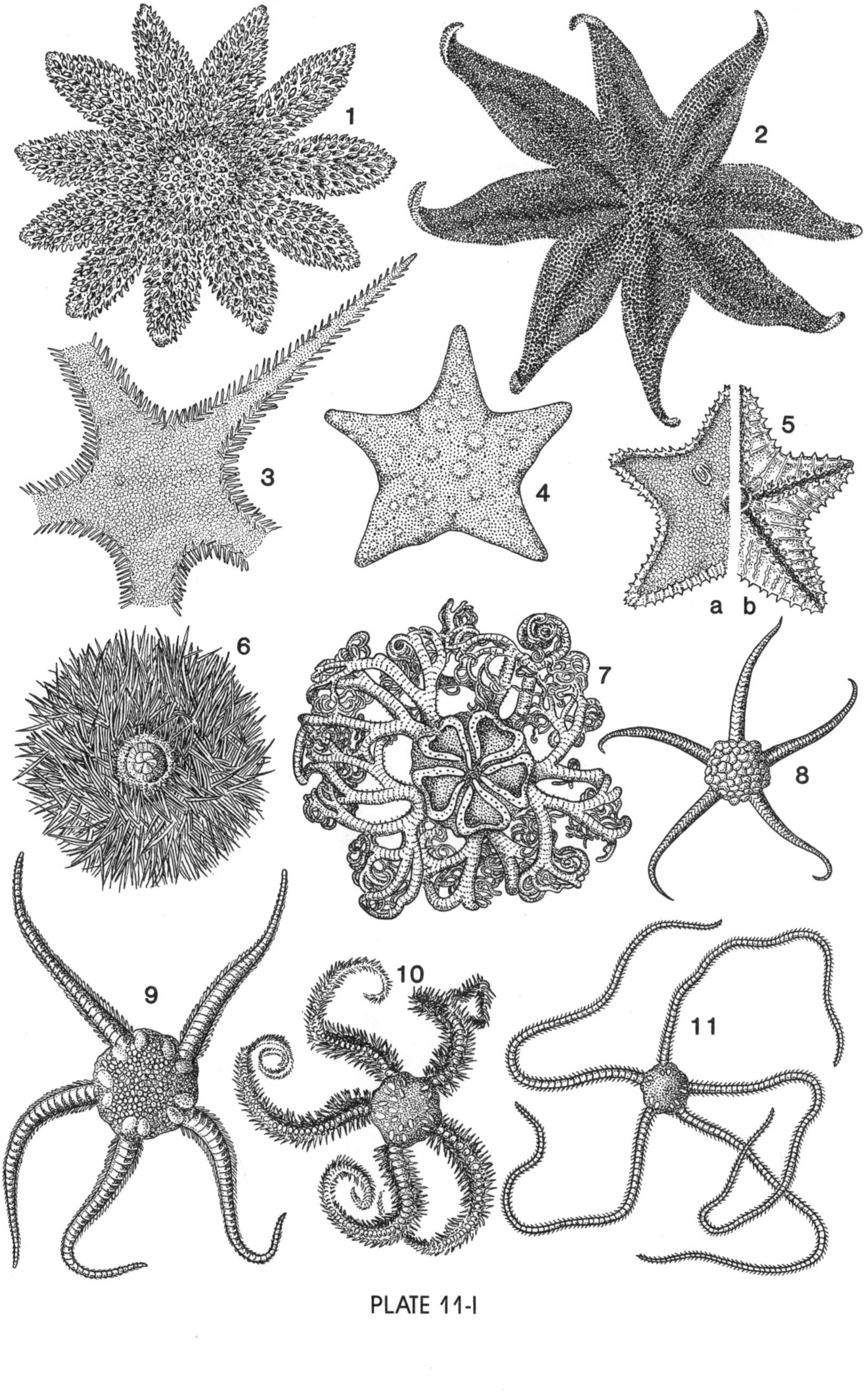


PLATE 11-I

## Distribution of Representative Taxa

Ten maps have been prepared based on data from the author's studies and from the vast literature listed in the References section.

Map 11-1. Distribution of sea urchins *Strongylocentrotus* in the Eurasian sector of the Arctic Ocean.

Map 11-2. Distribution of the starfish *Urasterias lincki* in the Eurasian sector of the Arctic Ocean.

Map 11-3. Distribution of the starfish *Ctenodiscus crispatus* in the Eurasian sector of the Arctic Ocean.

Map 11-4. Distribution of the starfish *Pontaster tenuispinus* in the Eurasian sector of the Arctic Ocean.

Map 11-5. Distribution of the starfish *Crossaster papposus* in the Eurasian sector of the Arctic Ocean.

Map 11-6. Distribution of the brittle star *Stegophiura nodosa* in the Eurasian sector of the Arctic Ocean.

Map 11-7. Distribution of the brittle star *Ophiopleura borealis* in the Eurasian sector of the Arctic Ocean.

Map 11-8. Distribution of the brittle stars (1) *Gorgonocephalus eucnemis* and (2) *G. arcticus* in the Eurasian sector of the Arctic Ocean.

Map 11-9. Distribution of the brittle star *Amphiura sundvalli* in the Eurasian sector of the Arctic Ocean.

Map 11-10. Distribution of the brittle star *Ophiura sarsi* in the Eurasian sector of the Arctic Ocean.

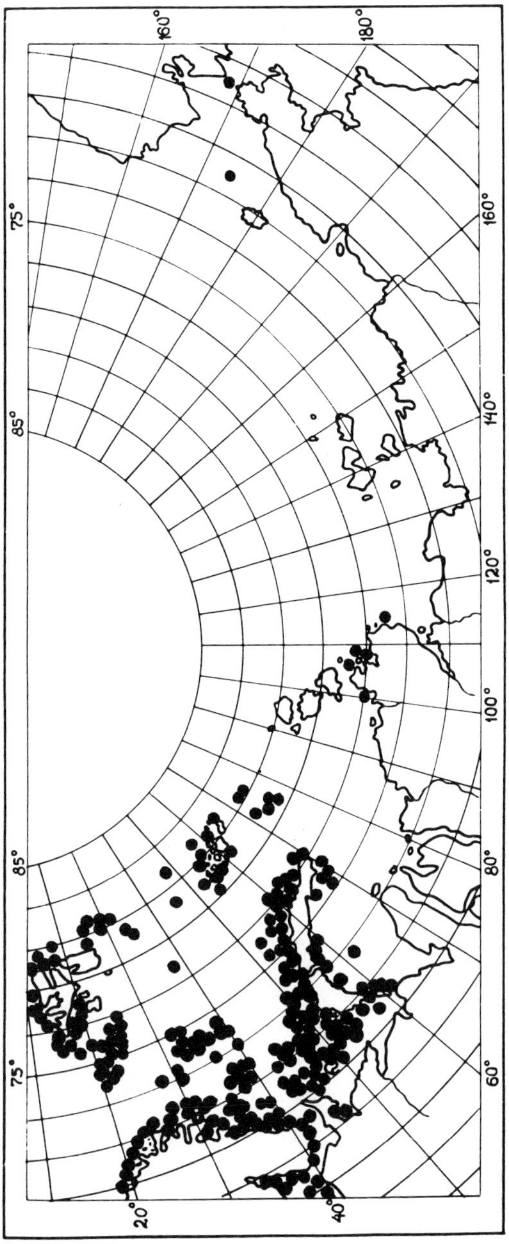


MAP 11-1

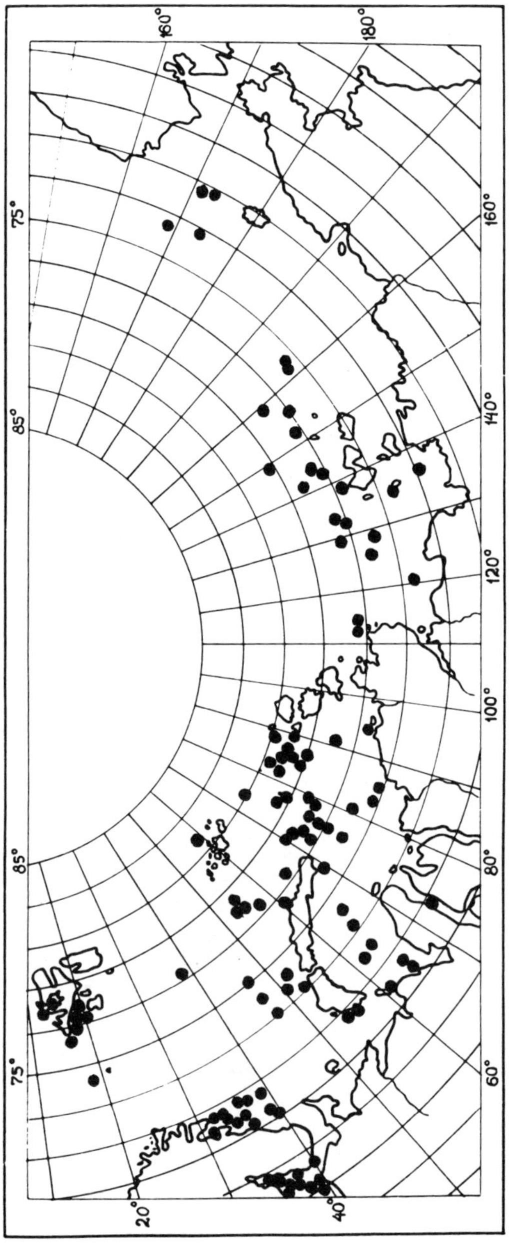


MAP 11-2

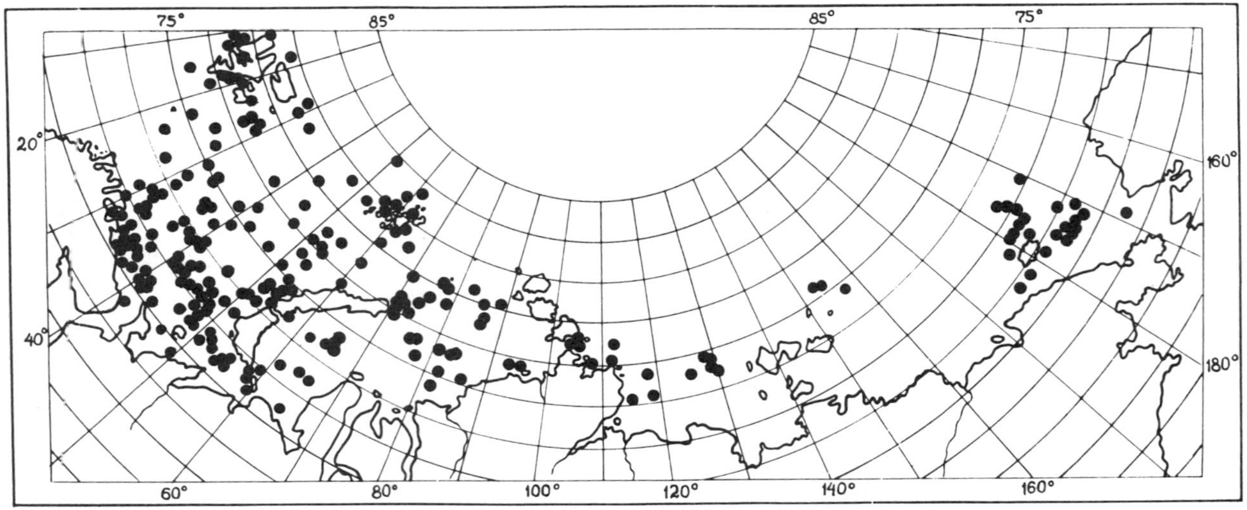


MAP 11-3

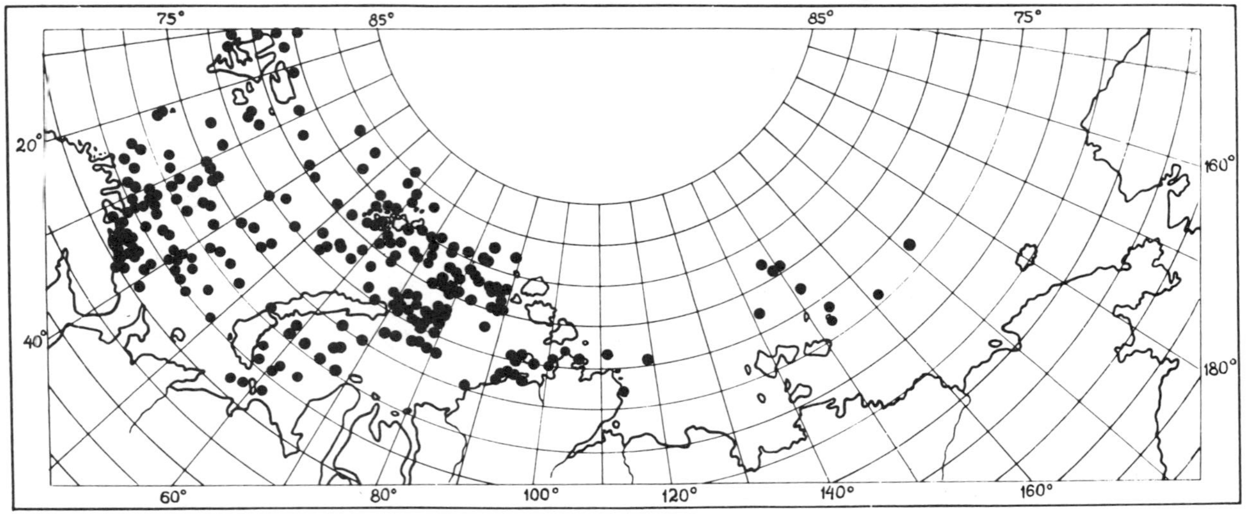


MAP 11-4

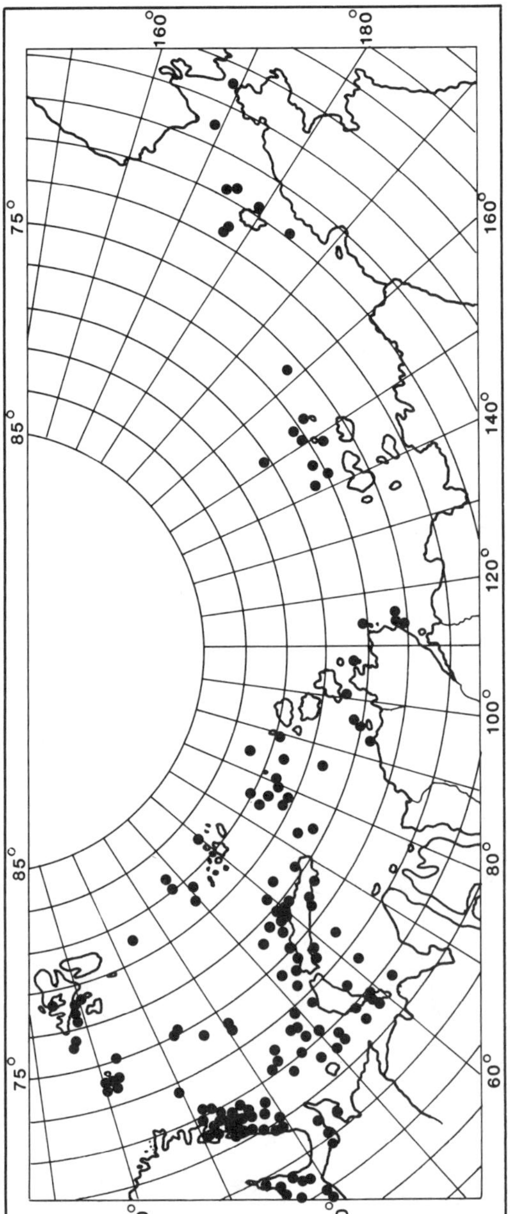


MAP 11-5

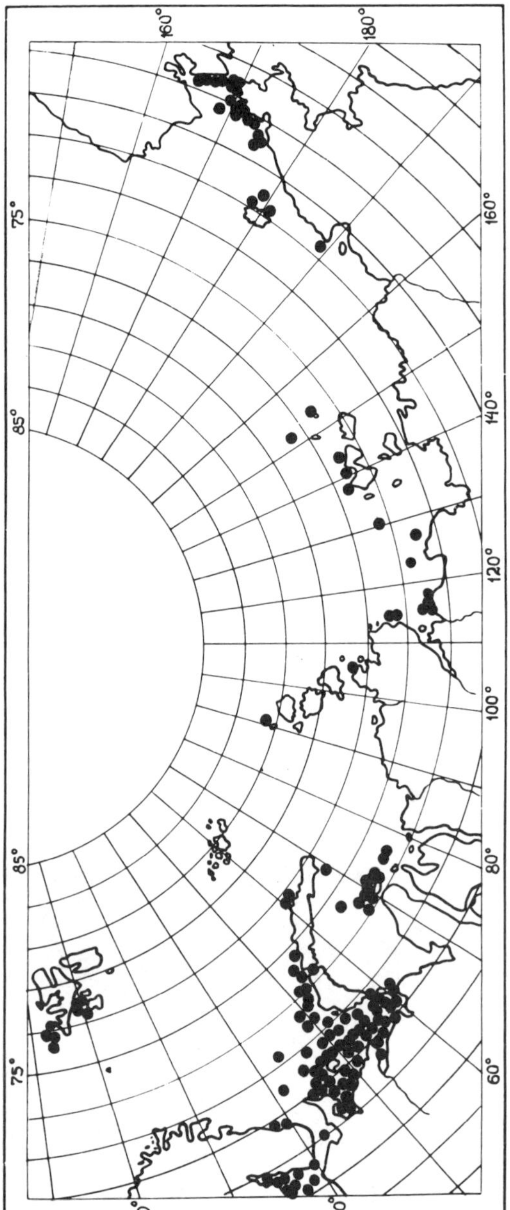


MAP 11-6

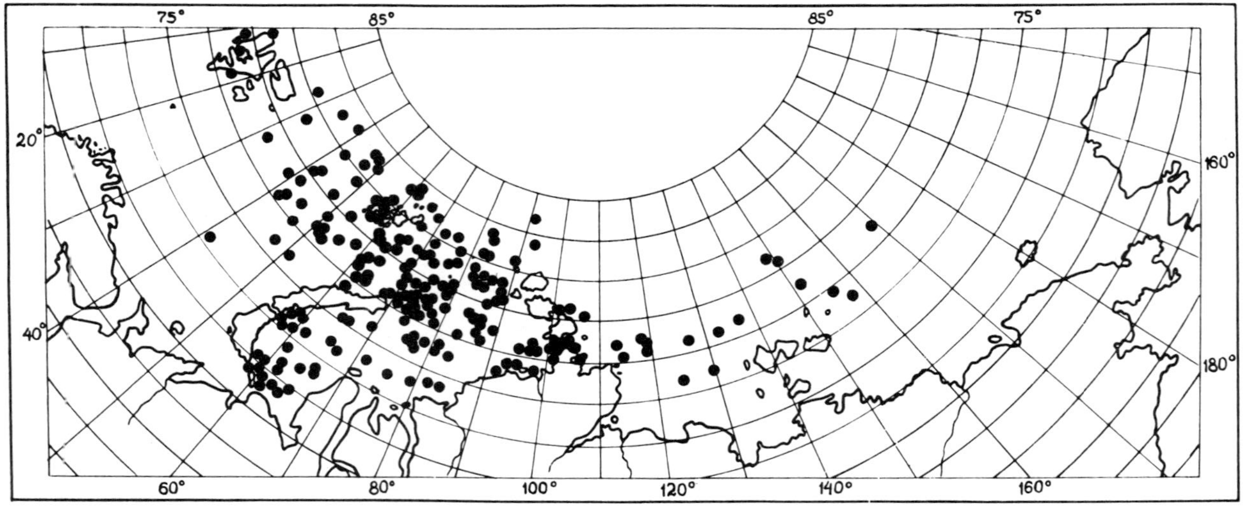


MAP 11-7

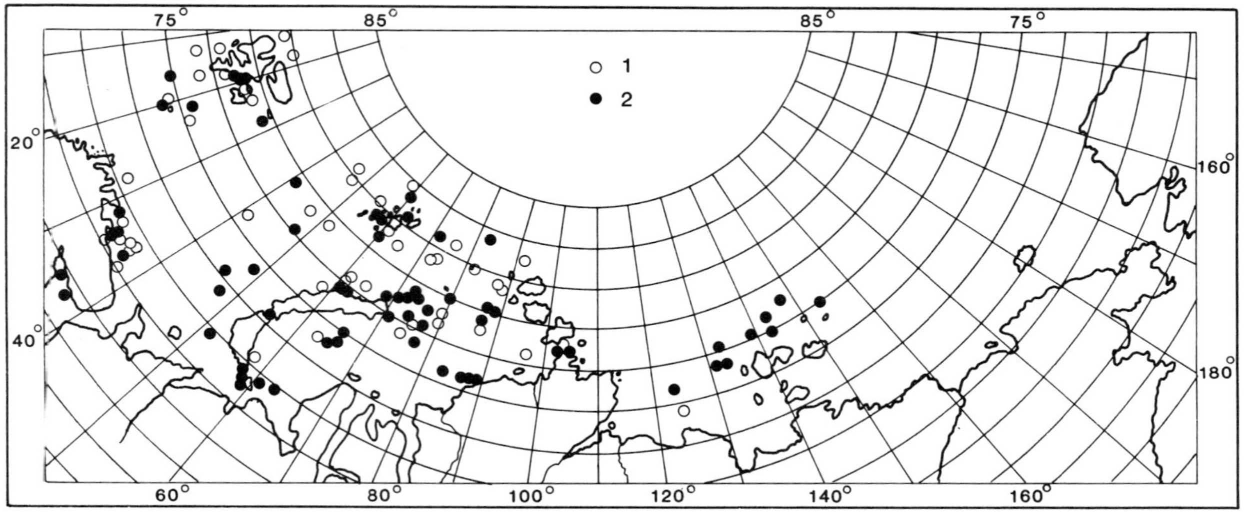


MAP 11-8

MAP 11-9

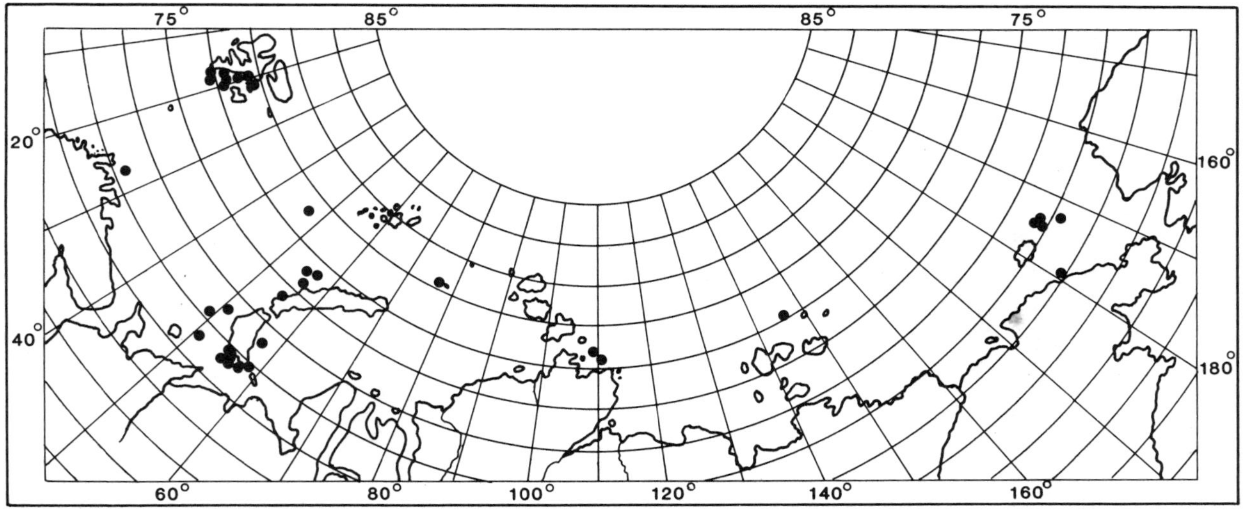


MAP 11-10

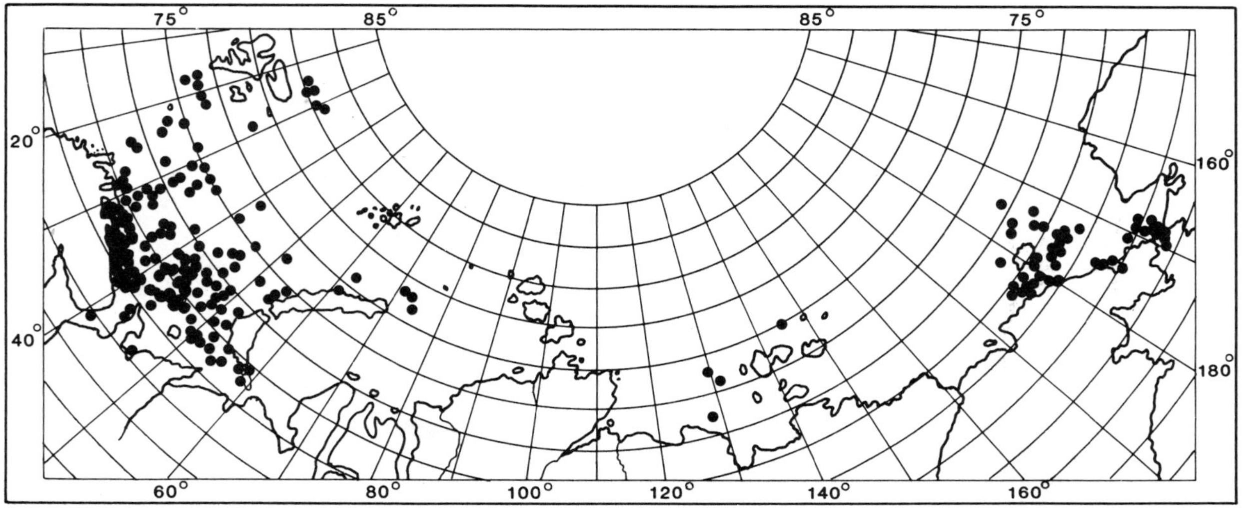

## REFERENCES


Baranova, Z. I. 1964. Iglokozhiye (Echinodermata), sobrannye ekspeditsiyei na L/K "F. Litke" v 1965. *Tr. Arkt. i Antarkt. In-ta* **259**:335–372. (In Russian.)

Baranova, Z. I. 1977. Iglokozhiye Zemli Frantsa Iosifa i sopredel'nykh vod. *Issledovaniya Fauny Morei* **14**(22):435–465. (In Russian.)

Diakonov, A. M. 1926. Iglokozhiye Barentseva, Byelogo i Karskogo morei. *Tr. Len. Obshch.* **56**(2):98–131. (In Russian.)

Diakonov, A. M. 1933. Iglokozhiye severnykh morei. Opred. po faune SSSR, izd. *Zool. In-ta AN SSSR* **8**(1):1–166. (In Russian.)

Diakonov, A. M. 1945. Vzaimootnosheniya arkticheskoi i tikhookeanskoi morskikh fauna primere zoogeographicheskogo analiza iglokozhikh. *Zhurn Obshch. Biol.* **6**(2):125–155. (In Russian.)

Diakonov, A. M. 1949. Opredelitel' iglokozhikh dal'nevostochnykh morei. *Izv. TINRO* **3**:1–127. (In Russian.)

Diakonov, A. M. 1950. Morskie zvoyzdy morei SSSR. Opred. po faune SSSR, izd. *Zool. In-ta AN SSSR* **34**:203. (In Russian.)

Diakonov, A. M. 1952. Iglokozhiye Echinodermata Chukotskogo morya i Beringova proliva. *Krainii Severo-Vostok Soyuza SSR* **2**:286–310. (In Russian.)

Diakonov, A. M. 1954. Ofiury (zmeekhvostki) morei SSSR. Opred. po faune SSSR, izd. *Zool. In-ta AN SSSR* **55**:1–136. (In Russian.)

Einarsson, T., D. M. Hopkins, and S. E. Doel. 1967. The stratigraphy of Tjornes, Northern Iceland and the history of the Bering land bridge. In D. M. Hopkins (ed.), *The Bering Land Bridge*. Stanford, 1–112.

Ekman, S. 1953. *Zoogeography of the Sea*. Sidgwick and Jackson, London, 417p.

Golikov, A. N. 1976*a*. *O Kolichestvennykh Zakonomernostyakh Protsessa Divergentsii*. Gidrobiologicheskiye issledovaniya samoochicheniya vodoyomov, Leningrad, pp. 90–96. (In Russian.)

Golikov, A. N. 1976*b*. Znacheniye biogeograficheskogo metoda i ucheniya o krupnykh geogidrograficheskikh kolebaniyakh urovnya mirovogo okeana dlya poznaniya zakonomernostei evolutsii i rasseleniya morskoi faun. *Zoogeografiya i Sistematika* **1**:24–36. (In Russian.)

Golikov, A. N. 1980. Mollhyuski Buccinine Mirovogo okeana. *Fauna SSSR* **5**(2):465p. (In Russian.)

Golikov, A. N. 1982. O printsipakh raionirovaniya i unifikatsii terminov v morskoi biogeografii. *Morskaya Biogeografiya: Predmet, Metody, Printsipy Raionirovaniya*. Nauka, Moscow, pp. 94–99. (In Russian.)

Golikov, A. N., and V. G. Averintsev. 1977. Biotsenozy verkhnykh otdelov shel'fa Zemli Frantsa Iosifa i nekotorye zakonomernosti ikh raspredeleniya. *Issledovaniya Fauny Morei* **14**(22):5–54. (In Russian.)

Golikov, A. N., O. A. Scarlato, and E. A. Pavshtiks. 1983. Biologicheskiye resursy Severnogo Ledovitogo Okeana. *Geografiya i Prirodnye Resursy* **3**:62–71. (In Russian.)

Gorbunov, G. P. 1932. K poznaniyu fauny iglokozhikh Zemli Frantsa Iosifa i morya Korolevy Viktorii. *Tr. Arkt. i Antarkt. In-ta* **2**:93–193. (In Russian.)

Gorbunov, G. P. 1933*a*. K poznaniyu fauny iglokozhikh poberezh'ya Severnogo ostrova Novoi Zemli. *Tr. Arkt. i Antarkt. In-ta* **7**:41–70. (In Russian.)

Gorbunov, G. P. 1933*b*. Iglokozhiye severnoi poloviny Karskogo morya. *Tr. Arkt. i Antarkt. In-ta* **8**:5–78. (In Russian.)

Grainger, E. H. 1966. Sea stars (Echinodermata: Asteroidea) of Arctic North America. *Fish. Res. Bd. Canada* **152**:70p.

Gurjanova, E. F. 1957. K zoogeografii Arkticheskogo basseina. Materialy nablyudenii nauchno-issledovatelskikh deistvuyushchikh stantsii *Severnyi polus 3* i *Severnyi polyus 4*. 1954/1955 gg., 1, 1., 343–362. (In Russian.)

Gurjanova, E. F. 1970. Osobennosti fauny Severnogo Ledovitogo okeana i ikh znacheniye dlya ponimaniya istorii ego formirovaniya. *Severnyi Ledovityi Okean i Ego Poberezh'ye v Kainozoye*. Leningrad, pp. 126–161. (In Russian.)

Gurjanova, E. F. 1972. Zoogeograficheskoye raionirovaniye morya. - v kn.: *Issledovaniye Fauny Morei*. Nauka, Leningrad. **10**(18):8–21. (In Russian.)

Jaccard, P. 1901. Distribution de la flore alpine dans la bassin de Drunses et dans quelque regions voisines. *Bull. Soc. Vandoise. Sc. Nat. Dr.* **37**:241–272.

Khodkina, N. V. 1964. Iglokozhiye yuzhnoi chasti Barentseva morya. *Novye Issledovaniya Planktona i Bentosa Barentseva Morya*. Nauka, Moscow and Leningrad, pp. 41–75. (In Russian.)

Koehler, R. 1924. *Les Echinodermes des Mers d'Europe*. Librarie Octave Doin, Paris, 360p.

Kuksa, V. I. 1982. *Promezhutochnye Vody Mirovogo Okeana*. Gidrometizdat, Leningrad, pp. 1–272. (In Russian.)

Mortensen, T. H. 1927. *Echinoderms of the British Isles*. Oxford, 461p.

Moskalyov, L. I., 1980. Donnaya fauna vneshnei chasti shel'fa Chukotskogo morya. *Ekologicheskiye Issledovaniya Shel'fa,* Moscow, pp. 72–79. (In Russian.)

*Oceanographic Atlas of the Arctic.* 1980. Moskwa, SSSR, 184P. (In Russian.)

Pesenko, J. A. 1982. *Printsipi i Metody Kolichestvennogo Analiza v Faunisticheskikh Issledovaniyakh.* Nauka, Moscow, pp. 1–288. (In Russian.)

Saks, V. N. 1953. Chetvertichnyi period v Sovetskoi Arktike. *Tr. Nauch. Issled. In-ta Geol. Arktiki* 77(4):627p. (In Russian.)

Shorygin, A. A. 1928. Iglokozhiye Barentseva morya, *Tr. Plavuch. Morsk. Inst.* **33:**107p. (In Russian.)

Simpson, G. G. 1943. Mammals and the nature of continents. *Am. J. Sci.* **241:**1–31.

Tolmachov, A. I., and B. A. Yurtsev. 1970. Istoriya arkticheskoi flory i eyo svyaz' s istoriei Polyarnogo basseina. *Severnyi Ledovityi Okean i Ego Poberezhye v Kainozoye.* Leningrad, pp. 87–100. (In Russian.)

*Treatise on Invertebrate Paleontology.* Part S, 650p (1966); Part T, 1027p (1978); Part U, 695p (1978). R. C. Moore (ed.). University of Kansas Press.

# 12: Marine Bivalvia of the Arctic Ocean

V. V. Fedyakov and
A. D. Naumov

Bivalve mollusks are important constituents of the marine fauna. They are widely distributed in the Arctic Ocean, particularly on continental shelves.

## PREVIOUS STUDIES

The investigation of Arctic bivalves commenced in the eighteenth century. However, the systematic sampling and study of the Arctic seas began only in the second half of the nineteenth century when numerous sea-going expeditions to high-latitude regions were organized. Among the early investigators, the following should be mentioned: Moebius (1874), Dall (1874, 1885), Sars (1878), Leche (1878, 1883), Cattie (1886), Herzenstein (1885, 1893), Melville and Standen (1900), Knipowitsch (1901*a*, 1901*b*, 1902, 1903), Friele and Grieg (1901), Hägg (1904, 1905), etc. These studies were primarily of taxonomic nature.

In the early part of the twentieth century, ecological and biogeographical studies of mollusks were emphasized in an attempt to resolve the problem of evolution of the Arctic fauna. Important contributions in this field were made by Odhner (1951), Derjugin (1915, 1928, 1932), Zenkevich (1927, 1931), Mesyatsev (1931), Soot-Ryen (1932), Madsen (1936, 1940), Thorson (1934, 1935, 1936), and Gorbunov (1941). Advances in ecology were matched by systematic studies (Jensen, 1912; Dall, 1919, 1924; Grieg, 1924; Derjugin and Guryanova, 1926; Thiele, 1928; Mossewitsch, 1928; Richards, 1936). During this period the investigation of the central Arctic Basin was underway. Data were collected on board the Research Vessels *Krasin* (1928), *Sadko* (1935), and *Sedov* (1937–1940), as well as on the Drifting Station *Severnyi Polyus-1* (1937–1938) (Gorbunov, 1946*a*, 1946*b*).

More recent taxonomic studies were published by Ryabinina (1952), Ockelmann (1954), Soot-Ryen (1964, 1966), Kafanov (1974, 1978), Miloslavskaya (1977), Odhner (1970), Clarke (1960, 1962), Bernard (1979), and Bouchet and Waren (1979). Among ecological investigations the following should be mentioned: Thorson (1951), Savilov (1953), Filatova and

*Note:* This chapter has been reviewed, edited, and rewritten by the Editor, Yvonne Herman.

Zenkevich (1957), Matveeva (1976, 1977), Golikov and Scarlato (1977), and Petersen (1978). New concepts concerning the origin and mode of dispersal in the Arctic based on current concepts of the geological evolution of the Arctic were put forth by Filatova (1948, 1951, 1957*a*, 1957*b*), Nicol (1955), Govberg (1975), Lande (1975), Wagner (1977), and Bernard (1979).

The present study is an attempt to synthesize the data on composition and distribution of Arctic Ocean Bivalvia. We have drawn upon the faunal collections deposited at the Leningrad Zoological Institute of the USSR Academy of Sciences as well as on the vast literature mentioned in the previous paragraph and publications by Becher (1886), Jensen (1905), Grieg (1909), Brotskaya and Zenkevich (1939), Guryanova (1948), Madsen (1949), Gorbunov (1952), MacGinitie (1955), Ockelmann (1958), Hülsemann (1962), Koltun (1964), Sparks and Pereyra (1966), and Clarke (1974).

## SPECIES COMPOSITION

Over 180 species of bivalves have been described from the Arctic Ocean and marginal seas (Table 12-1). Faunal diversity in the polar seas is meager as compared to other oceans. From the northwest Pacific 270 species have been described (Scarlato, 1981), as well as 450 species from the northeast Pacific (Dall, 1921) and 730 species from the northwest Atlantic Ocean (Johnson, 1934). The impoverished nature of the Arctic fauna may be due to the fact that only one biogeographic province exists in the Central Polar Sea. Further, we believe that speciation of the Arctic shelf biota might have ceased during the Late Cenozoic glacial temperature minima (glaciation maxima). Only one genus and 25 endemic species are known to have evolved in the Arctic (Table 12-1, species designated "arc"). In our view, at lower latitudes speciation went on uninterruptedly as tectonic events (mainly sea-level fluctuations) and climatic changes were less marked and consequently did not affect faunal evolution to the extent that they did in the Arctic.

The majority of Arctic taxa (Scarlato, 1981) have affinities with Atlantic boreal and subtropical boreal species (Table 12-1). Their widespread occurrence in the Arctic is due to the fact that the German (Norwegian) province of the Atlantic boreal province with its rich fauna is located within the geographical boundaries of the Arctic (see Fig. 12-2). The environment is favorable not only to autochthonous mollusks but also to some subtropical mollusks. The next large group includes wide-ranging boreal Arctic species, which are the most eurybiontic descendants of the earliest Arctic immigrants (Chapter 13).

Atlantic and Pacific water masses can be traced into the Arctic; in these Arctic regions Atlantic boreal Arctic and Pacific boreal species are abundant. Within the Arctic proper, Pacific, and Atlantic temperate water, wide-ranging taxa are poorly represented (e.g., the Mytilidae, Pectinidae, Cardiidae, Mactridae, Pholadidae) and, with the exception of the first family, evolved between the Carboniferous and Cretaceous. About one-third of the families, including endemic Arctic species, are primarily confined to the Arctic Ocean and found in deposits of Ordovician–Devonian age (*Treatise on Invertebrate Paleontology,* 1969). They include the Malletidae, Nuculidae, and Nuculanidae. It is noteworthy that in the Arctic Ocean the most diverse are the ancient groups of Bivalvia, whereas among Gastropoda, evolutionarily young forms predominate (Chapter 13).

Almost all Arctic genera with a large number of species have been found in post late Cretaceous deposits (Fig. 12-1). These taxa may have evolved during the late Cretaceous cooling. This interpretation agrees with Golikov's hypotheses. An alternative explanation for the present-day

Table 12-1. Distribution of Bivalve Mollusks in the Arctic Ocean

| *Species* | *Biogeography* | *Presence* | | | | | | | | | |
|---|---|---|---|---|---|---|---|---|---|---|---|
| | | *a* | *b* | *c* | *d* | *e* | *f* | *g* | *h* | *i* | *j* |
| *Malletia cuneata** | cosm. | + | + | | | | | | | | |
| *M. obtusa* | atl.b. | ⊕ | | + | | | | | | | |
| *Tindaria derjugini* | arc. | + | ⊕ | | | | | | | | |
| *Nucula tumidula* | atl.b. | | | + | | | | | | | |
| *N. zophos* | arc. | | ⊕ | | | | | | | | |
| *Leionucula tenuis** | b.-a. | | | + | + | + | + | ⊕ | + | ⊕ | + |
| *L. delphinodonota* | atl.b. | + | | ⊕ | + | | | | | + | + |
| *Nuculana pernula* | b.-a. | + | + | + | ⊕ | ⊕ | + | | + | + | + |
| *N. minuta* | b.-a. | | | + | + | + | + | + | + | + | + |
| *N. radiata* | b.-a. | | | | | | + | ⊕ | ⊕ | | |
| *N. lamellosa* | arc. | | | | | | ⊕ | ⊕ | | | |
| *Ledella tamara* | arc. | | ⊕ | | | | | | | | |
| *Portlandia arctica* | arc. | | | | ⊕ | ⊕ | | | + | ⊕ | + |
| *P. siliqua* | arc. | | | | | + | ⊕ | + | + | + | |
| *P. aestuariorum* | arc. | | | | + | + | ⊕ | | + | | |
| *P. sulcifera* | arc. | | | | | | | | | + | |
| *Yoldiella lucida* | atl.b. | + | | ⊕ | + | + | | | | | |
| *Y. persei* | arc. | | + | | + | ⊕ | + | | | | |
| *Y. lenticula* | atl.b.-a. | | + | + | + | + | + | + | + | ⊕ | + |
| *Y. fraterna* | atl.b.-a. | | + | + | + | ⊕ | + | | | + | |
| *Y. frigida* | atl.b.-a. | + + | + | + | + | + | + | | + | + | |
| *Y. intermedia* | atl.b.-a. | + | + | + | + | ⊕ | + | | + | ⊕ | + |
| *Y. subaequilatera* | atl.b. | ⊕ | | + | + | | | | | + | |
| *Y. annenkovae* | arc. | | + | | | | | | | | |
| *Y. tenuis* | atl.b. | | + | | | | | | | | |
| *Y. expansa* | atl.b. | + | + | + | | | | | | + | |

*May belong to a species complex.

†Identification errors may exist as a result of uncertain systematics.

††Species of this genus are closely related.

*Note:* arc: endemic Arctic; atl.b.-a.: Atlantic boreal–Arctic; b.-a: widely distributed boreal–Arctic; atl.b.: Atlantic boreal and subtropical boreal; amph.b.: amphiboreal; pac.b.: Pacific boreal; cosm.: cosmopolitan species. Presence (+); (⊕: characteristic species) in *a:* Norwegian–Greenland Basin; *b*: Central Abyssal region; *c*: Norwegian–Iceland region; *d*: Spitsbergen–White Sea region; *e:* Kara Sea region; *f:* Siberian Sea region; *g*: northern part of the Chukchi Sea; *h:* Beaufort Sea and southern part of the Chukchi Sea; *i:* Canadian–Greenland region; *j:* Hudson Bay.

Table 12-1. *(continued)*

| Species | Biogeography | Presence | | | | | | | | | |
|---|---|---|---|---|---|---|---|---|---|---|---|
| | | *a* | *b* | *c* | *d* | *e* | *f* | *g* | *h* | *i* | *j* |
| *Y. pustulosa* | atl.b. | + | | + | | | | | | | |
| *Yoldia amygdalea* | b.-a. | | | + | + | + | + | + | + | + | + |
| *Y. myalis* | amph.b. | | | | | | | | ⊕ | | + |
| *Cnesterium seminuda* | pac.b. | | | | | | | ⊕ | ⊕ | | |
| *Megayoldia thraciaeformes* | amph.b. | | | | | | | | | + | |
| *Limopsis minuta* | atl.b. | | | ⊕ | + | | | | | + | |
| *Crenella decussata* | b.-a. | | | + | ⊕ | + | | | + | + | + |
| *Arvella faba* | atl.b.-a. | | | | | | | | | + | + |
| *Mytilus edulis* | amph.b. | | ⊕ | ⊕ | | | | | | + | + |
| *Musculus discors** | b.-a. | | | + | + | + | + | + | + | + | + |
| *M. niger* | b.-a. | | | + | + | + | + | + | + | + | + |
| *M. corrugatus* | b.-a. | | | + | + | + | + | + | + | + | + |
| *M. marmoratus* | atl.b. | | | + | | | | | | | |
| *Modiolus modiolus* | amph.b. | | | ⊕ | + | | | | | | |
| *Modiolula phaseolina* | atl.b. | | | + | | | | | | | |
| *Dacrydium vitreum* | b.-a. | + | + | ⊕ | + | + | + | + | + | + | + |
| *Bathyarca glacialis* | atl.b.-a. | | | + | + | ⊕ | + | | + | ⊕ | + |
| *B. pectunculoides* | atl.b.-a. | + | + | + | + | + | + | | + | ⊕ | |
| *B. frielei* | arc. | + | + | | | + | | | | + | |
| *Cyclopecten imbrifer* | Atl.b.-a. | + | + | + | + | + | + | | | + | |
| *C. vitreum* | atl.b. | + | | + | + | | | | | | |
| *Delectopecten simile* | atl.b. | | | + | | | | | | | |
| *D. groenlandicum* | atl.b.-a. | + | + | + | + | ⊕ | + | | + | ⊕ | + |
| *Hyalopecten frigidus* | atl.b.-a. | ⊕ | + | | | | | | | | |
| *Pseudamussium septemradiatus* | atl. b. | | | + | | | | | | | |
| *Chlamys islandicus* | b.-a. | | | ⊕ | ⊕ | + | | + | + | + | + |
| *Ch. beringianus* | pac.b. | | | | | | | | + | | |
| *Ch. aratus* | atl.b. | | | + | | | | | | | |
| *Ch. tigrinus* | atl.b. | | | + | | | | | | | |
| *Ch. striatus* | atl.b. | | | + | | | | | | | |
| *Ch. pusio* | atl.b. | | | + | | | | | | | |
| *Aequipecten opercularis* | atl.b. | | | + | | | | | | | |
| *Acesta excavata* | atl.b. | + | | + | | | | | | | |
| *Limatula subauriculata* | atl.b. | | | + | + | + | | | | + | |

| | | a | b | c | d | e | f | g | h | i | j |
|---|---|---|---|---|---|---|---|---|---|---|---|
| *L. subovata* | atl.b. | | + | + | + | | | | | | |
| *L. sarsi* | atl.b. | + | | + | | | | | | | |
| *L. hyperborea* | arc. | + | + | | + | + | + | | + | ⊕ | |
| *Pododesmus macroshisma* | pac.b. | | | | | | | | + | | |
| *Heteranomia squamula* | atl.b.-a. | | | ⊕ | ⊕ | | | | | | |
| *H. aculeata* | atl.b. | | | + | + | | | | | | |
| *Monia patelliformes* | atl.b. | | | + | | | | | | | |
| *Pandora glacialis* | b.-a. | | | + | + | + | + | + | + | + | + |
| *P. grandis* | pac.b. | | | | | | | | + | | |
| *Lyonsia arenosa* | b.-a. | + | + | + | + | + | + | + | + | + | + |
| *Periploma abyssorum* | b.-a. | | | | | | + | + | + | + | + |
| *P. alaskana* | arc. | | | | | | | + | ⊕ | | |
| *Thracia myopsis* | b.-a. | | | + | + | + | + | + | + | + | + |
| *Th. devexa* | atl.b-a. | | | | + | | | | + | + | + |
| *Th. septemtrionalis* | atl.b.-a. | | | + | + | * | | | | + | + |
| *Astarte crenata** | atl.b.-a. | + | + | + | ⊕ | + | + | | | ⊕ | + |
| *A. sulcata* | atl.b. | | | ⊕ | + | + | | | | | |
| *Elliptica elliptica* | atl.b.-a. | | | + | ⊕ | + | | | | + | + |
| *E. alaskensis* | pac.b. | | | | | | | | ⊕ | | |
| *Tridonta borealis** | b.-a. | + | | + | + | + | + | + | + | + | + |
| *Nicania montagui* | b.-a. | | | + | + | + | + | + | + | + | + |
| *Rictocyma asquimalta* | pac.b. | | | | | | | | + | | |
| *Hiatella arctica* | b.-a. | | | ⊕ | ⊕ | + | + | + | + | + | + |
| *Cyrtodaria kurriana* | b.-a. | | | | + | + | ⊕ | + | + | + | + |
| *Panomya arctica* | b.-a. | | | + | + | + | | | | | |
| *P. ampla* | pac.b. | | | | | | | | + | | |
| *Saxicavella plicatilis* | atl.b. | | | + | | | | | | | |
| *Lucina borealis* | atl.b. | | | + | | | | | | | |
| *Thyasira equalis* | arc. | + | + | + | + | ⊕ | + | | | + | |
| *Th. gouldi** | b.-a. | + | | + | + | + | + | + | + | + | + |
| *Th. sarsi* | atl.b.-a. | | | + | + | + | | | | | |

*May belong to a species complex.

†Identification errors may exist as a result of uncertain systematics.

††Species of this genus are closely related.

*Note:* arc: endemic Arctic; atl.b.-a.: Atlantic boreal–Arctic; b.-a: widely distributed boreal–Arctic; atl.b.: Atlantic boreal and subtropical boreal; amph.b.: amphiboreal; pac.b.: Pacific boreal; cosm.: cosmopolitan species. Presence (+); (⊕: characteristic species) in *a:* Norwegian–Greenland Basin; *b*: Central Abyssal region; *c*: Norwegian–Iceland region; *d*: Spitsbergen–White Sea region; *e:* Kara Sea region; *f:* Siberian Sea region; *g*: northern part of the Chukchi Sea; *h:* Beaufort Sea and southern part of the Chukchi Sea; *i:* Canadian–Greenland region; *j:* Hudson Bay.

Table 12-1. *(continued)*

| Species | Biogeography | Presence *a* | *b* | *c* | *d* | *e* | *f* | *g* | *h* | *i* | *j* |
|---|---|---|---|---|---|---|---|---|---|---|---|
| *Th. phrygiana* | arc. | | | | | | + | | | | |
| *Axinulus eumyarius* | atl.b. | | | + | | | | | | | |
| *A. ferruginosa* | atl.b.-a. | + | | + | + | + | + | | | | |
| *A. croulinensis* | atl.b.-a. | | | | + | + | | | | | |
| *A. ottoschmidti* | arc. | | + | | | | | | | | |
| *A. pygmaeus* | atl.b. | ⊕ | | | | | | | | + | |
| *A. careyi* | arc. | | + | | | | | | | | |
| *Axinopsida orbiculata* | b.-a. | | | + | + | + | + | + | + | + | + |
| *Turtonia minuta* | amph.b. | | | ⊕ | ⊕ | | | | | + | |
| *Kellia suborbicularis* | amph.b. | + | | + | | | | | | | |
| *K. pumila* | atl.b. | + | | | + | | | | | | |
| *Pseudopythina compressa* | pac.b. | | | | | | | | + | | |
| *Montacuta*† *substriata* | atl.b.-a. | | | + | | | | | | + | |
| *M. spitzbergensis* | arc. | | | | + | + | | | | | |
| *M. dawsoni* | atl.b.-a. | | | | + | | + | | + | | |
| *M. maltzani* | atl.b.-a. | | | + | + | + | | | | | |
| *Mysella bidentata* | atl.b. | | | + | | | | | | | |
| *M. mölleri* | arc. | | | | | | ⊕ | | | | |
| *M. derjugini* | b.-a. | | | | | | | + | + | | |
| *M. planata* | b.-a. | | | | | | | | + | | |
| *Tellimya ferruginosa* | atl.b. | | | + | | | | | | | |
| *Boreacola vadosa* | arc. | | | | | | | | + | | |
| *Axinodon symmetros* | atl.b. | + | + | | | | | | | | |
| *Cyclocardia borealis* | atl.b. | | | | | | | | | | + |
| *C. crebricostata* | pac.b. | | | | | | | | + | | |
| *C. ventricosa* | b.-a. | | | | | | + | + | ⊕ | | |
| *Crassicardia crassidens* | pac.b. | | | | | | | + | + | | |
| *Parvicardium exiguus* | atl.b. | | | + | | | | | | | |
| *Cardium echinatum* | atl.b. | | | + | | | | | | | |
| *C. faeciatum* | atl.b. | | | + | + | | | | | | |
| *C. minutum* | atl.b. | + | | + | | | | | | | |
| *C. nodosum* | atl.b. | | | + | | | | | | | |
| *Cilialocardium ciliatum* | b.-a. | | | + | + | + | + | + | + | + | + |
| *Keenocardium californiensis* | pac.b. | | | | | | | | ⊕ | | |

| | | | | | | | | | | | |
|---|---|---|---|---|---|---|---|---|---|---|---|
| *Cerastoderma elegantulum* | atl.b. | | | + | | | | | | + | |
| *C. edule* | atl.b. | | | ⊕ | | | | | | | |
| *Serripes groenlandicus* | b.-a. | + | | + | + | + | + | + | + | + | + |
| *S. laperousi* | pac.b. | | | | | | | | + | | |
| *Diplodonta torelli* | atl.b.-a. | | | | + | + | | | | | |
| *D. aleutica* | pac.b. | | | | | | | | + | | |
| *Tellina fabula* | atl.b. | | | + | | | | | | | |
| *T. pusilla* | atl.b. | | | + | | | | | | | |
| *T. tenuis* | atl.b. | | | + | | | | | | | |
| *Peronidia lutea* | pac.b. | | | | | | | | + | | |
| *Macoma calcarea* | b.-a. | | | + | ⊕ | + | + | + | + | ⊕ | + |
| *M. balthica* | amph.b. | | | ⊕ | + | | | | + | + | + |
| *M. torelli* | b.-a. | | | | + | + | ⊕ | ⊕ | + | + | + |
| *M. loveni* | b.-a. | | | | + | + | + | + | + | + | |
| *M. moesta* | b.-a. | | | | + | + | ⊕ | + | + | + | + |
| *M. lama* | pac.b. | | | | | | | | | + | |
| *M. incongrua* | pac.b. | | | | | | | | + | | |
| *M. middendorffi* | pac.b. | | | | | | | | + | | |
| *Psammobia feroensis* | atl.b. | | | + | | | | | | | |
| *Ps. tellinella* | atl.b. | | | + | | | | | | | |
| *Abra prismatica* | atl.b. | | | + | | | | | | | |
| *A. nitida* | atl.b. | | | + | | | | | | | |
| *Arctica islandica* | atl.b. | | | ⊕ | + | | | | | | |
| *Kelliella milliaris* | atl.b. | + | | + | | | | | | | |
| *Saxidomus giganteus* | pac.b. | | | | | | | | + | | |
| *Dosina lincta* | atl.b. | | | + | | | | | | | |
| *D. exoleta* | atl.b. | | | + | | | | | | | |
| *Liocyma fluctuosa* | b.-a. | | | + | ⊕ | ⊕ | ⊕ | + | ⊕ | + | |
| *Venerupis pullastra* | atl.b. | | | + | | | | | | | |
| *Chamelea gallina* | atl.b. | | | + | | | | | | | |
| *Protothaca staminea* | pac.b. | | | | | | | | + | | |

*May belong to a species complex.
†Identification errors may exist as a result of uncertain systematics.
††Species of this genus are closely related.
*Note:* arc: endemic Arctic; atl.b.-a.: Atlantic boreal–Arctic; b.-a: widely distributed boreal–Arctic; atl.b.: Atlantic boreal and subtropical boreal; amph.b.: amphiboreal; pac.b.: Pacific boreal; cosm.: cosmopolitan species. Presence (+); (⊕: characteristic species) in *a:* Norwegian–Greenland Basin; *b*: Central Abyssal region; *c*: Norwegian–Iceland region; *d*: Spitsbergen–White Sea region; *e:* Kara Sea region; *f:* Siberian Sea region; *g*: northern part of the Chukchi Sea; *h:* Beaufort Sea and southern part of the Chukchi Sea; *i:* Canadian–Greenland region; *j:* Hudson Bay.

Table 12-1. *(continued)*

| Species | Biogeography | Presence *a* | *b* | *c* | *d* | *e* | *f* | *g* | *h* | *i* | *j* |
|---|---|---|---|---|---|---|---|---|---|---|---|
| *Timpclea ovata* | atl.b. | | | + | | | | | | | |
| *Varicorbula gibba* | atl.b. | | | + | | | | | | | |
| *Phaxus pellucidus* | atl.b. | | | + | | | | | | | |
| *Ensis ensis* | atl.b. | | | + | | | | | | | |
| *E. siliqua* | atl.b. | | | + | | | | | | | |
| *Siliqua alta* | pac.b. | | | | | | | | + | | |
| *Spisula elliptica* | atl.b. | | | + | + | | | | | | |
| *S. voyi* | pac.b. | | | | | | | | + | | |
| *Mya arenaria* | atl.b. | | | + | + | | | | | | |
| *M. truncata* | b.-a. | | | + | + | + | + | + | + | + | + |
| *M. japonica* | pac.b. | | | | | | | | + | | |
| *Zirfaea crispata* | amph.b. | | | + | | | | | | | |
| *Xylophaga dorsalis* | atl.b. | | | + | | | | | | | |
| *Teredo norvegica* | atl.b. | | | + | | | | | | | |
| *Psiloteredo megotara* | atl.b. | | | + | | | | | | | |
| *Lyonsiella*† *abyssicola* | atl.b.-a. | + | + | + | + | ⊕ | + | | | + | |
| *L. ushakovi* | arc. | | + | | | | | | | | |
| *Policordia cordata* | atl.b.-a. | + | + | | | + | | | | | |
| *P. jeffreysi* | arc. | + | + | | | | | | | | |
| *Cuspidaria arctica*†† | atl.b.-a. | + | + | + | + | ⊕ | + | | | + | |
| *C. obesa** | atl.b.-a. | + | + | + | + | ⊕ | + | | + | + | + |
| *C. centobi* | atl.b. | + | | + | + | | | | | + | |
| *C. exiqua* | atl.b. | + | | | | | | | | + | |
| *C. rostrata* | atl.b. | + | | + | | | | | | | |
| *C. turgida* | arc. | | + | | + | + | + | | | | |
| *C. lamellosa* | atl.b. | + | | + | | | | | | | |
| *Poromya granulata* | atl.b. | + | | + | + | | | | | | |

*May belong to a species complex.

†Identification errors may exist as a result of uncertain systematics.

††Species of this genus are closely related.

*Note:* arc: endemic Arctic; atl.b.-a.: Atlantic boreal–Arctic; b.-a: widely distributed boreal–Arctic; atl.b.: Atlantic boreal and subtropical boreal; amph.b.: amphiboreal; pac.b.: Pacific boreal; cosm.: cosmopolitan species. Presence (+); (⊕: characteristic species) in *a:* Norwegian–Greenland Basin; *b*: Central Abyssal region; *c*: Norwegian–Iceland region; *d*: Spitsbergen–White Sea region; *e:* Kara Sea region; *f:* Siberian Sea region; *g*: northern part of the Chukchi Sea; *h:* Beaufort Sea and southern part of the Chukchi Sea; *i:* Canadian–Greenland region; *j:* Hudson Bay.

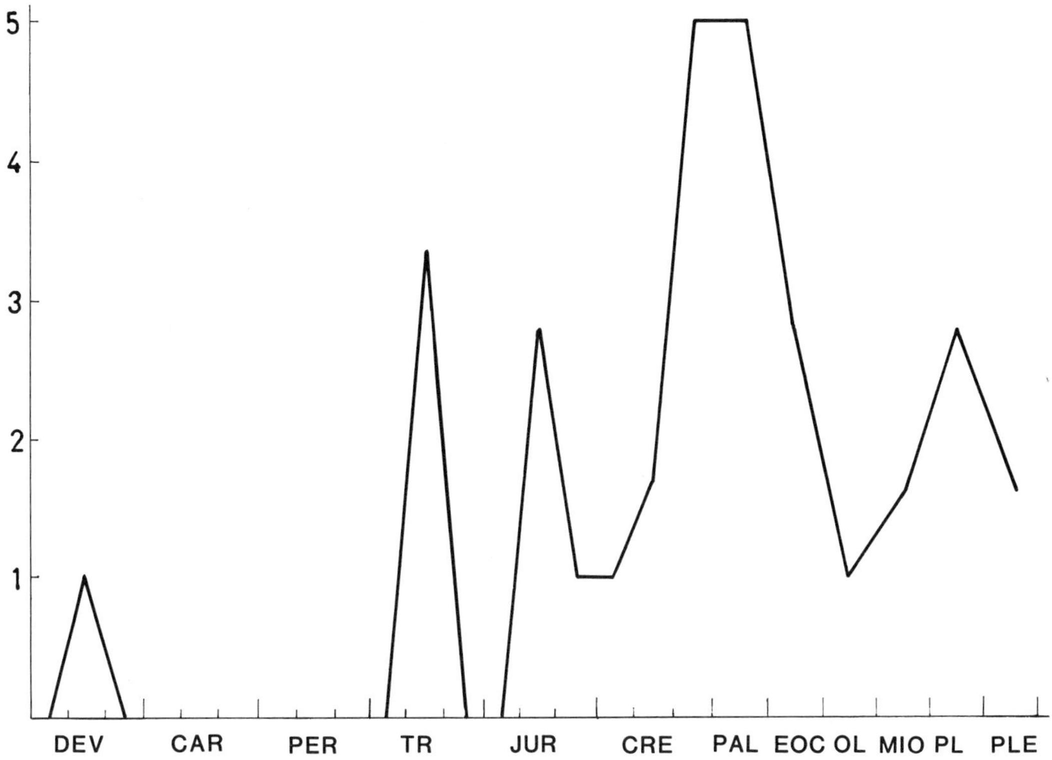


Figure 12-1. Relationship between average number of bivalve species and their evolution. Abscissa: geologic time scale; ordinate: average number of species. Dev, Devonian; Car, Carboniferous; Per, Permian; Tri, Trias; Jur, Jurassic; Cre, Cretaceous; Pal, Paleocene; Eoc, Eocene; Oli, Oligocene; Mio, Miocene; Pli, Pliocene; Ple, Pleistocene.

faunal composition is their adaptive feeding mechanism. Detritophagy and commensalism, a common mode of feeding among Arctic bivalves, might have evolved under the rigorous Arctic environment of low primary production during the late Cenozoic glacial periods.

## PRESENT-DAY DISTRIBUTIONAL PATTERNS

Twenty-three faunal provinces are recognized* (Fig. 12-2). The similarity between provinces was determined using the following formula:

$$K = \frac{N_{++}N_{--} - N_{+-}N_{-+}}{(N_{++} + N_{+-})(N_{++} + N_{-+})(N_{--} + N_{+-})(N_{--} + N_{-+})}$$

where $K$ is a similarity index, $N_{++}$ is the number of regions where two species occur, $N_{--}$ is the number of regions where neither the first nor the second species occur, $N_{+-}(N_{-+})$ is the number of regions inhabited either by the first (second) or the second (first) species. Cluster analyses of similarity indices matrix distinguished 19 Arctic regions. Our data suggest that most recent shelf species are Atlantic Ocean emigrants (Fig. 12-3). At first they colonized the Norwegian coast and later the East Siberian Sea.

*The term *faunal province* is used informally to denote geographical regions.

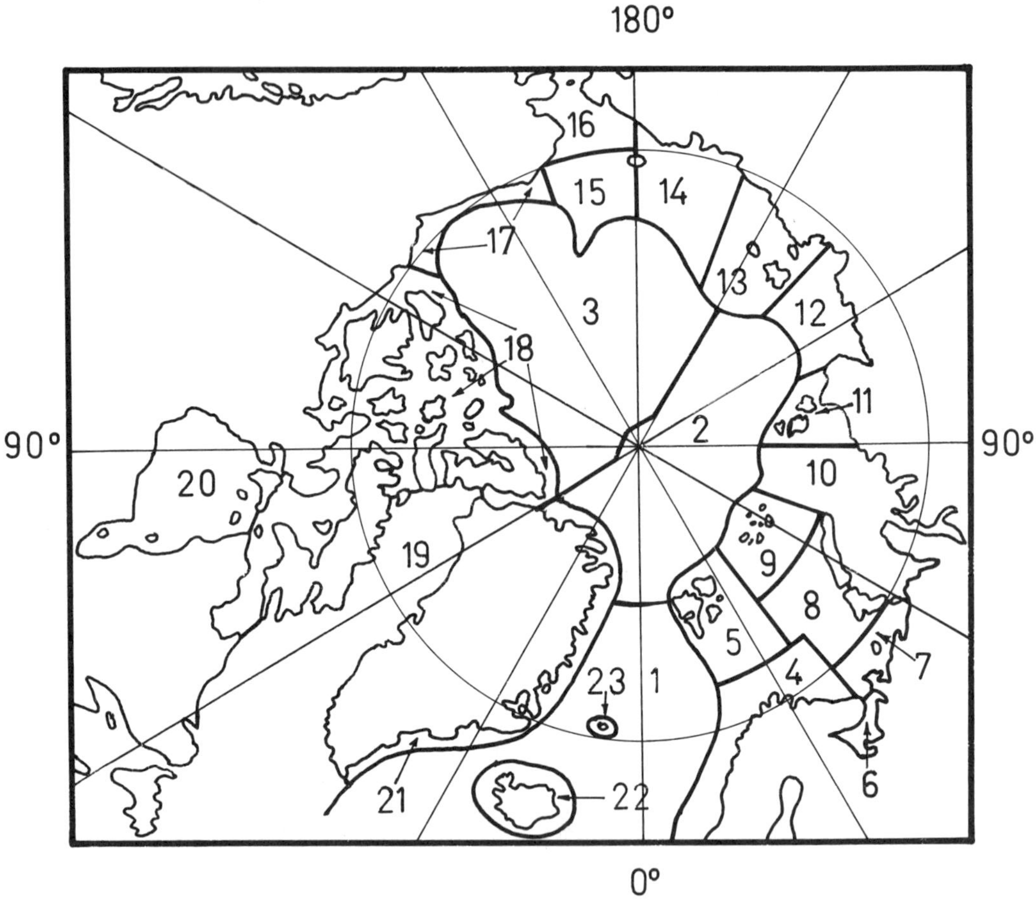


Figure 12-2. Preliminary division of the Arctic region.

The earliest immigrants penetrated the Arctic during the Tertiary. They were isolated from the main Arctic region at the time of the initial late Neogene glaciation (Lindberg, 1972). About two-thirds of these taxa gave rise to Arctic species that inhabit the Laptev Sea and the New Siberian Islands' shelves. During subsequent migrations from adjacent European waters, bivalves reached only the Severnaya Zemlya region, becoming isolated from the main Atlantic province during the Last Glaciation. These species colonized the deeper parts of the Barents and Kara seas.

Thus the most eurybiontic Atlantic boreal species entering the Arctic today have penetrated only as far as the Novaya Zemlya Islands. Their further migration eastward is precluded by the present-day hydrological regime. In our view, a second, less common, dispersal route in the Arctic was via the Baffin Bay, Canadian Arctic Archipelago, East Greenland Sea, and during glacial regression maxima along the Nansen Ridge onto the Spitsbergen shelf. More recently, bivalves extended their distribution to the Beaufort Sea, the region of Franz Josef Land, and the Kara and White seas. Of course, only deep-water species migrated along the Nansen Ridge; they survived the glacial periods in the deep sectors of the Kara Sea. Early Pacific boreal species occupied only the southern part of the Chukchi Sea and restricted regions of the Beaufort Sea. Later, they penetrated the northern sectors of the Chukchi Sea

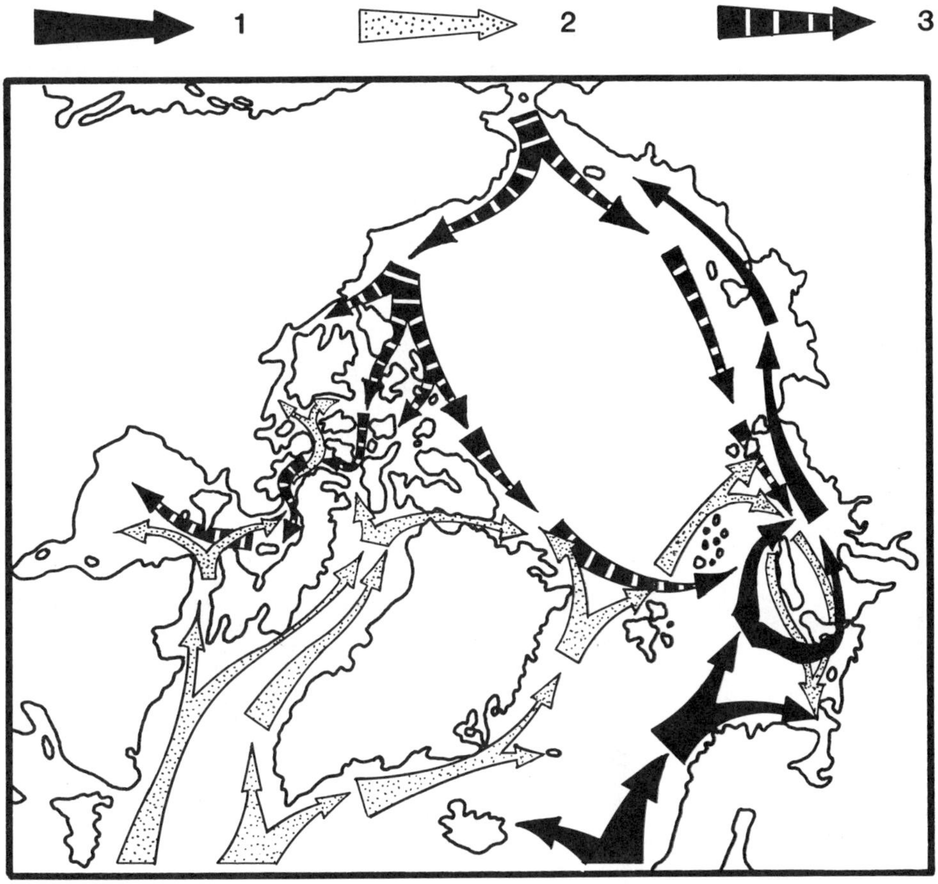


Figure 12-3. Suggested routes of dispersal of bivalve mollusks in the Arctic. 1: migration route from the European Atlantic shelf; 2: migration route from the American Atlantic shelf; 3: migration route from the Pacific shelf.

and the East Siberian Sea, eventually reaching the Canadian Arctic Archipelago and the Laptev Sea (Fig. 12-3).

There is limited information concerning the colonization of Arctic abyssal regions by bivalves during past geologic times. In our view, almost all Arctic bathyal species are Atlantic emigrants settled in a zone bathed by the Atlantic Intermediate Water Mass (Table 12-1).

## FAUNAL PROVINCES OF THE ARCTIC REGION

Species from different sectors of the Arctic Ocean and seas were compared using a cluster analysis on the matrix of similarity coefficients (Jaccard, 1912). Ten regions were distinguished with faunal similarity lower than 50 percent (Fig. 12-4). The largest number of species occur in the Norwegian–Iceland region. The fauna of this region consists principally of Atlantic boreal and subtropic boreal elements (Table 12-2). The total lack of Arctic mollusks in this region can be accounted for by the year-round positive temperatures on the shelf and the upper bathyal zone. This fauna belongs to the Atlantic boreal province. The relative abundance of mollusks at various depth zones derived from different "centers of evolution" vary only slightly (Fig. 12-5*a*). The relative proportion of widely

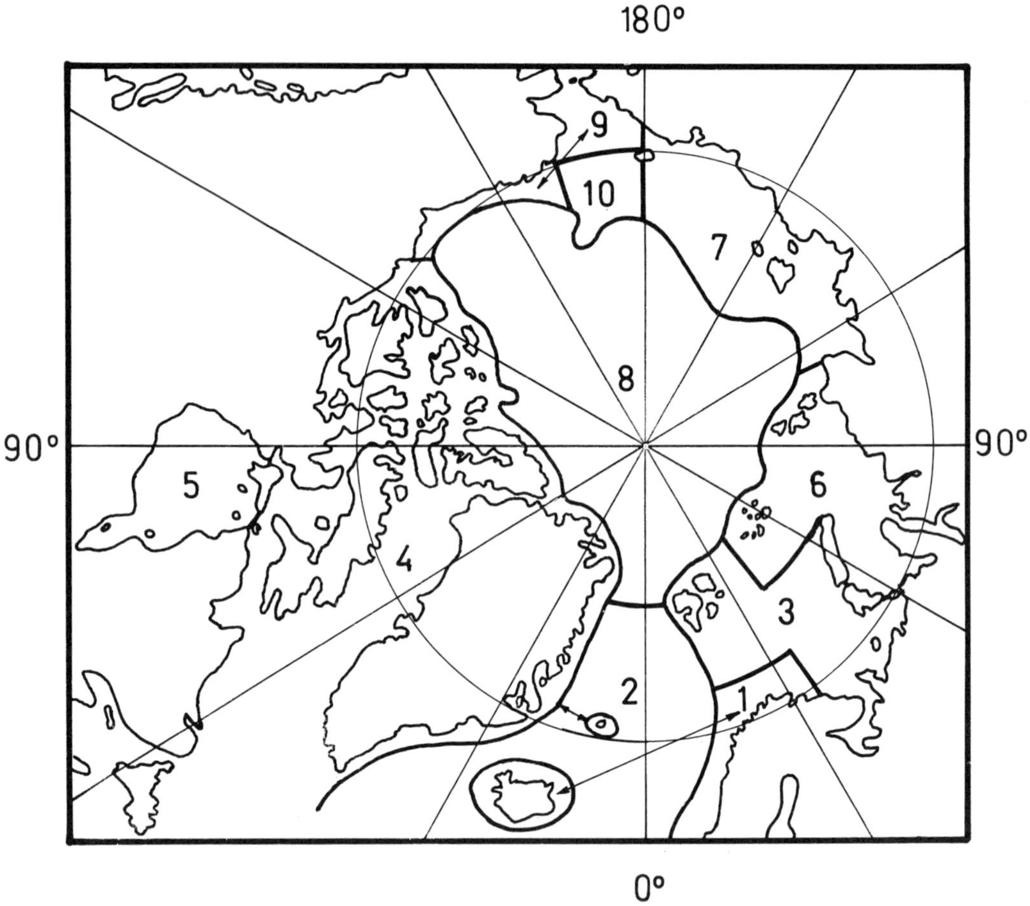


Figure 12-4. Faunal regions of the Arctic according to the distribution of bivalve mollusks. 1: Norwegian–Iceland region; 2: Norwegian–Greenland Basin region; 3: Spitsbergen–White Sea region; 4: Canadian–Greenland region; 5: Hudson Bay; 6: Kara Sea region; 7: Siberian Seas region; 8: Central Arctic Basin region; 9: southern Chukchi Sea and Beaufort Sea region; 10: northern Chukchi Sea region.

distributed eurybiontic boreal–Arctic and amphiboreal species (such as *Mytilus edulis* and *Dacrydium vitreum*) declines with increasing water depth, whereas that of more stenobiontic Atlantic boreal–Arctic and boreal elements (e.g., *Leionucula delphinodonta, Cardium edule,* and *Yoldiella lucida*) increases. This change of biogeographic composition is evidently brought about by a gradual decrease in fluctuation in environmental factors with increasing water depth. Fjord elements are absent from this region (Knipowitsch, 1893; Naumov, 1979 and references therein). This may be due to the different hydrological regimes of the two regions.

The Norwegian–Greenland region appears to be the most similar to the previous area (Table 12-2). Here Atlantic boreal species dominate over the Arctic elements. This does not mean that the entire region should be referred to as the Atlantic boreal province. A depth-related change in the proportion of taxa of different origin is observed (Fig. 12-5*b*): Atlantic boreal species dominate the Arctic elements only down to a depth of about 2,500 m. At greater depths the faunal composition is strongly

Table 12-2. Species Composition in the Arctic Ocean and Adjacent Seas (%)

| *Biogeographic Groups* | *Regions* | | | | | | | | | | |
|---|---|---|---|---|---|---|---|---|---|---|---|
| | *Arctic Ocean* | *Norwegian–Iceland* | *Norwegian–Greenland Depression* | *Spitsbergen–White Sea* | *Canadian–Greenland* | *Hudson Bay* | *Kara Sea* | *Siberian Seas* | *Central Abyssal* | *Southern Chukchi Sea* | *Northern Chukchi Sea* |
| Arctic | 13 | 1 | 12 | 9 | 9 | 2 | 16 | 18 | 38 | 9 | 9 |
| Atlantic boreal–Arctic | 14 | 17 | 29 | 29 | 26 | 26 | 33 | 27 | 38 | 13 | 3 |
| Boreal Arctic | 18 | 21 | 12 | 37 | 43 | 60 | 46 | 55 | 9 | 45 | 82 |
| Atlantic boreal | 37 | 55 | 43 | 19 | 12 | 5 | 3 | 0 | 12 | 0 | 0 |
| Amphi-boreal | 5 | 6 | 2 | 6 | 8 | 7 | 2 | 0 | 0 | 3 | 0 |
| Pacific boreal | 12 | 0 | 0 | 0 | 2 | 0 | 0 | 0 | 0 | 30 | 6 |
| Cosmo-politan | 1 | 0 | 2 | 0 | 0 | 0 | 0 | 0 | 0 | 0 | 0 |

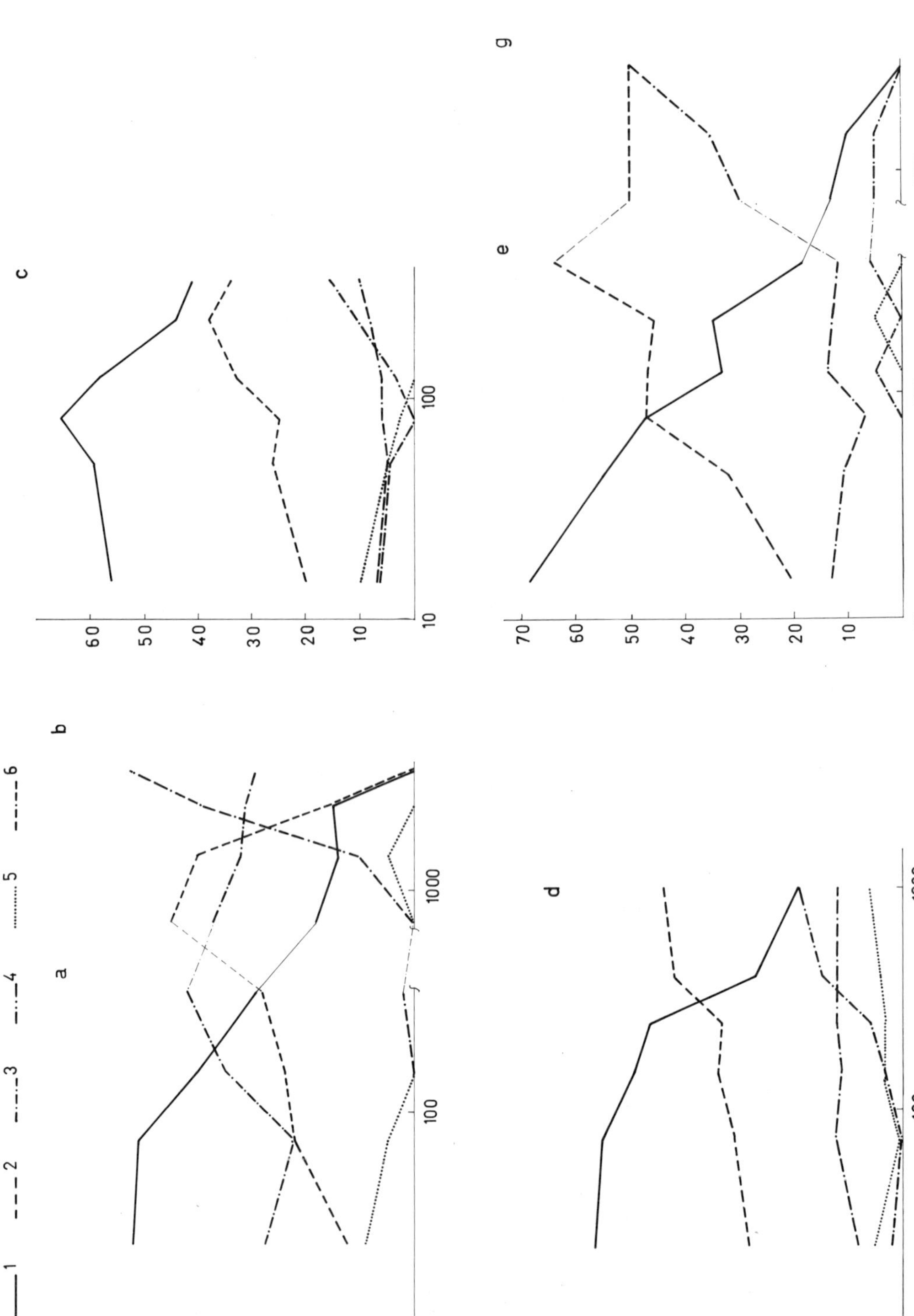

1
2
3
4
5
6
a
b
c
d
e
g
60
50
40
30
20
10
70
10
100
1000

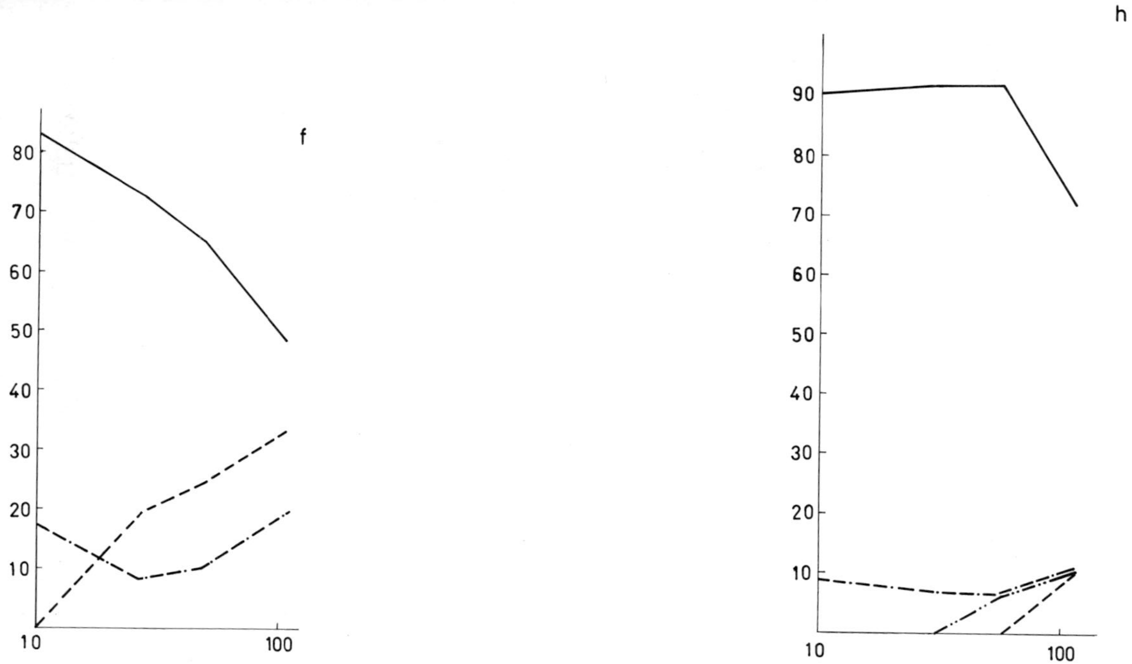


**Figure 12-5.** Relationship between the biogeographic composition of bivalves and depth in the *a,* Norwegian–Iceland Region; *b,* Norwegian–Greenland Basin region; *c,* Spitsbergen–White Sea region; *d,* Canadian–Greenland region; *e,* Kara Sea region; *f,* Siberian Seas region; *g,* Central Arctic Basin region; *h,* northern Chukchi Sea region. Abscissa: depth, m, logarithmic scale; ordinate: percentage of species of different biogeographic groups. 1: Widely distributed boreal–Arctic species; 2: Atlantic boreal–Arctic species; 3: Atlantic boreal and subtropical boreal species; 4: Pacific boreal species; 5: amphiboreal species; 6: Arctic species.

influenced by the Arctic Deep-Water Mass. Consequently, the boundary between the Atlantic boreal and Arctic regions north of Iceland and northwest of Norway is located in the Norwegian–Greenland Basin at ≈2,500 m depth.

The Spitsbergen–White Sea region is characterized by the predominance of wide-ranging boreal-Arctic mollusks (Table 12-2). Generally, the number of boreal taxa is slightly higher than that of Arctic species. However, the proportion of these two biogeographic groups depends greatly on water depth as well as on hydrological characteristics of the various regions. Warm shallow areas are dominated by Atlantic boreal and amphiboreal elements throughout the region. In shallow inlets with local depressions ≈50–70 m deep in the White Sea and 70–100 m in Spitsbergen fjords, species of boreal origin were not found. At the same time, even in the deepest sectors of the central part of the Barents Sea (except for local basins), Arctic mollusks do not occur. The nearly equal proportion of Atlantic boreal and Arctic forms make us classify this region as a transitional one between the two biogeographic provinces.

Some biologists consider the White Sea Basin a distinct region of the Arctic because of its impoverished fauna (Golikov, 1980; Fedyakov, 1983). The depth-related pattern of biogeographic composition is illustrated in Figure 12-5*c*. Figure 12-5*c* also shows the change in relative abundance of eurybiontic species such as *Hiatella arctica* and *Macoma calcarea* and eurybiontic taxa (e.g., *Portlandia arctica* and *Tortonia minuta*). Biogeographically the Canadian–Greenland region (Fig. 12-4) is very similar to the Spitsbergen–White Sea region (Table 12-2). Depth-related faunal changes in these two areas are also comparable (Fig. 12-5*d*).

At great depth the Baffin Bay is filled with the relatively warm Atlantic water, which accounts for the increased proportion of temperate water fauna. The Canadian–Greenland region should be regarded as a transitional zone, with the northern and northwestern parts of the Canadian Arctic Archipelago belonging to the Arctic region. In the Atlantic sector of the Arctic, the Hudson Bay region is regarded as a separate province, with wide-ranging boreal–Arctic species (Table 12-2). Its fauna is transitional between that of the Atlantic boreal and the Arctic regions. The Kara Sea region supports numerous Atlantic elements, its fauna being dominated by wide-ranging boreal–Arctic species. At depths greater than 100 m they inhabit the Intermediate Atlantic Water Mass. Between 50 m and 100 m the relative abundance of Arctic elements is low (Fig. 12-5*e*).

The fauna of the shallow Siberian Sea is dominated by wide-ranging boreal–Arctic species. The relative abundance of Arctic elements increases in this sea as compared to the previous region except for the 20–50 m depth zone (Fig. 12-5*f*). A distinct Estuarine–Arctic province is recognized by Filatova (1957*b*). A gradual increase in Arctic species with increased water depth is noted in this area. The Central Arctic Basin region is characterized by the almost total absence of wide-ranging species (Table 12-2; Fig. 12-5*g*). The southern sector of the Chukchi Sea and the coastal zone of the Beaufort Sea are characterized by the dominance of Pacific boreal species. Arctic elements are subordinate in these regions (Table 12-2; Fig. 12-4). The shallow northern Chukchi Sea region supports a predominantly eurybiontic wide-ranging boreal–Arctic bivalve fauna (Table 12-2). In this region, which we consider a transition zone, the proportion of Pacific boreal versus Arctic elements is even except for the 40–70 m depth zone where Pacific bivalves dominate (Fig. 12-5*h*).

## SUMMARY AND CONCLUSIONS

The following biogeographic regions are recognized in the Arctic: the German (Norwegian) province of the Atlantic boreal region north of Iceland and northwest of Scandinavia at ≈2,500 m depth. The boundary between the Atlantic boreal and the Arctic regions can be drawn roughly between the Spitsbergen–White Sea region, the Norwegian–Greenland Sea and the southern part of the Canadian–Greenland region (Fig. 12-2), and the Pacific boreal region including the southern sector of the Chukchi Sea and the coastal area of the Beaufort Sea at least to Point Barrow.

Within the Arctic region four provinces characterized by endemic bivalves are recognized. The east Arctic province includes most of the Laptev Sea and the entire East Siberian Sea. To the north it is limited by the ≈200 m isobath. A distinct faunal region is the estuarine–Arctic province, which may have developed during an early interglacial period. Today the fauna of this province is restricted to large river mouths; however, the province may have formed a continuous brackish water system during earlier interglacials. Fossil bivalves belonging to this province have been discovered in the Leningrad region, the Baltic region, and in Germany (Fedyakov, unpublished). Another relatively shallow Arctic province may exist near the northern coast of Canada and Greenland where several endemic species such as *Portlandia sulcifera* and *Boreocola vadosa* occur. This assumption is supported by Lubinsky's data (1980). Based on our preliminary data, the Arctic deep-water province includes the Arctic Basin proper at depths greater than 200–500 m and the Norwegian–Greenland Sea at depths greater than 2,500 m. The Lomonosov Ridge does not seem to present a barrier to the distribution of most Arctic endemic abyssal bivalves. The seafloor covered by the Atlantic Intermediate Water supports an Atlantic boreal–Arctic molluskan fauna and is devoid of endemic elements. This area may be regarded as transitional, surrounding the deep-water province of the Arctic province.

## ACKNOWLEDGMENTS

The authors express their sincere gratitude to Professor Dr. A. N. Golikov for valuable discussion and Mr. O. K. Novikov for assistance.

# APPENDIX

## Plate 12-I

The plate shows representative Arctic bivalves, the distributional patterns for which are given in Table 12-1.

1. *Tindaria derjugini.* Specimen from the vicinity of Franz Joseph Land, water depth 2,000 m; Recent.
2. *Nuculana pernula.* Specimen from the East Siberian Sea, water depth 40 m; Recent.
3. *Portlandia arctica.* Specimen from the Kara Sea, water depth 150 m; Recent.
4. *Yoldiella persei.* Specimen from the Kara Sea, water depth 300 m; Recent.
5. *Yoldiella annenkovae.* Specimen from the vicinity of Severnaya Zemlya, water depth 800 m; Recent.
6. *Bathyarca frielei.* Specimen from the vicinity of Novaya Zemlya, water depth 500 m; Recent.
7. *Lyonsiella abyssicola.* Specimen from the Kara Sea, water depth 200 m; Recent.
8. *Cuspidaria turgida.* Specimen from the East Barents Sea, water depth 50 m; Recent.

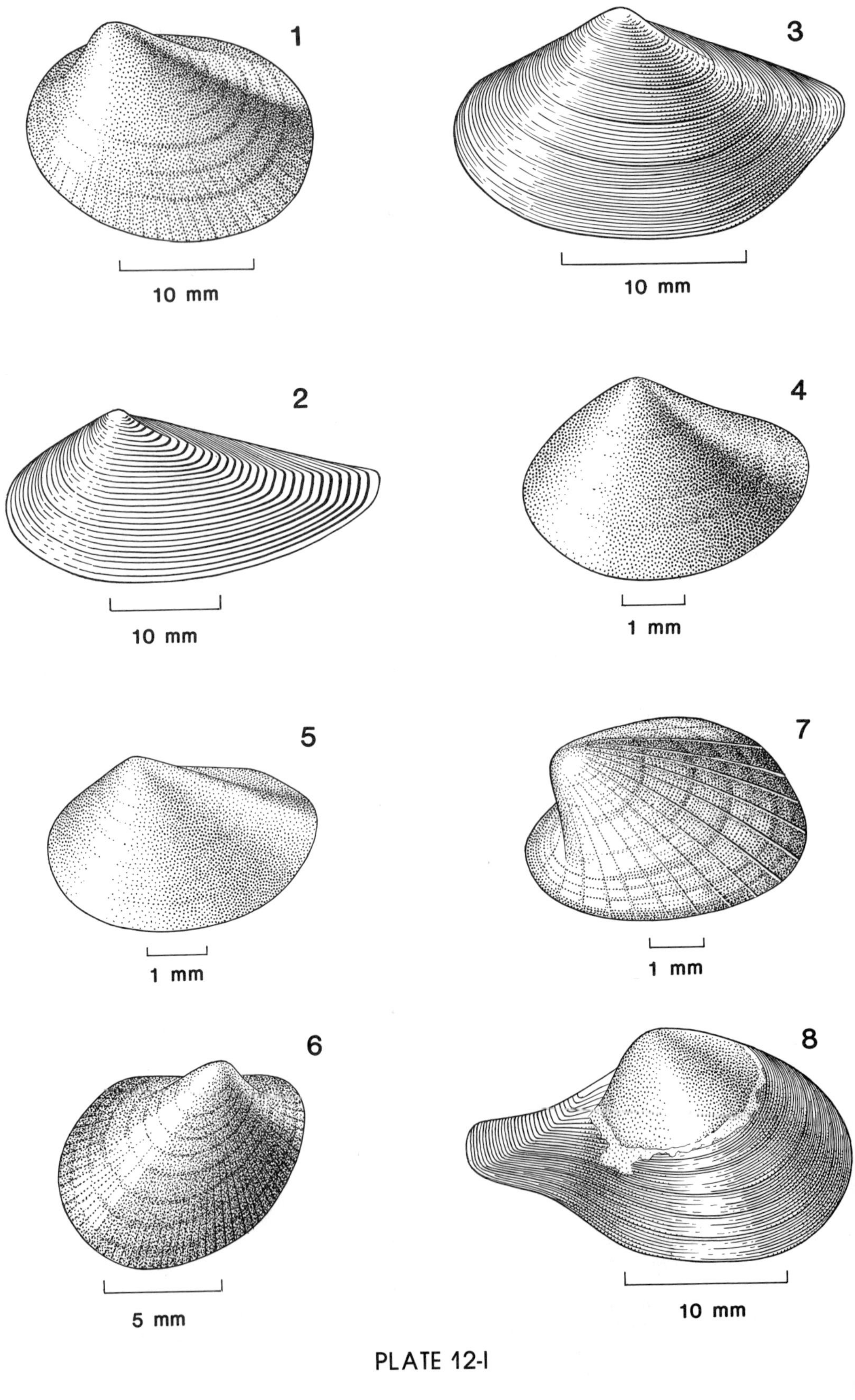


PLATE 12-I

## REFERENCES


Becher, F. 1886. Mollusken von Jan Mayen. *Die Österreich Polarstation Jan Mayen.* Beob.-Ergeb. herausg v. d. Akad. d. Wissensch, B. 3.

Bernard, F. R. 1979. *Bivalve Mollusks of the Western Beaufort Sea.* Contrib. Sci. Los Angeles County Museum, N 313, 80p.

Bouchet, P., and A. Waren. 1979. The abyssal molluscan fauna of the Norwegian Sea and its relation to the other faunas. *Sarsia* **64**(3):211–243.

Brotskaya, V. A., and L. A. Zenkevich. 1939. Kolichestvennyi uchot donnoi fauny Barentseva morya. *Trudy VNIRO* **4**:5–126. (In Russian.)

Cattie, J. Th. 1886. Les Lamellibranches, recueillies dans les course du "Willem Barents" durant les mois de Mai a'Septembre 1880 et 1881. *Bijdr. tot de Dierkunde* 13-e Affr., Amsterdam.

Clarke, A. H. 1960. Arctic archibathal and abyssal mollusks from drifting station Alpha. *Breviora* **119**:17p.

Clarke, A. H. 1962. Artic archibathal and abyssal molluscs—II. Molluscs dredged from drifting station Charlie (Alpha II), *Nat. Mus. Can Bull.* **185:**90–105.

Clarke, A. H. 1974. Molluscs from Baffin Bay and the northern North Atlantic Ocean, *Nat. Mus. Can. Bull.* **7**:23p.

Dall, W. H. 1874. Catalogue of shells from Bering Strait and the adjacent portions of the Arctic Ocean, with descriptions of three new species. *Calif. Acad. Sci. Proc.* **5**:246–253.

Dall, W. H. 1885. Report of the mollusks. *Rep. Intern. Polar Exp. to Point Barrow, Alaska* **4**:177–184.

Dall, W. H. 1919. The Mollusca of the Arctic coast of America, collected by the Canadien Arctic Expedition West from Bathurst Inlet with an appended report on a collection of Pleistocene Fossil Mollusca. *Rep. Can. Arct. Exp. 1913–1918* **8**:3a–29a.

Dall, W. H. 1921. Summary of the marine shell bearing mollusks of the northwest coast of America from San Diego, California, to the Polar Sea, mostly contained in the collection of the United States National Museum, with illustrations of hitherto unfigured species. *Smithsonian Inst. U.S. Nat. Mus. Bull.* **112**:217p.

Dall, W. H. 1924. Supplement to the report of the Canadian Arctic expedition. Mollucan recent and Pleistocene. *Rep. Can. Arct. Exp. 1913–1918.* **8A**:31a–35a.

Derjugin, K. M. 1915. Fauna Kolskogo zaliva i usloviya ego sushchestvovaniya. *Zap. Akad. Nauk* **34**(8):929p. (In Russian.)

Derjugin, K. M. 1928. Fauna Belogo morya i usloviya ego sushchestvovania. *Issled. Morei SSSR* **7–8**:511p. (In Russian.)

Derjugin, K. M. 1932*a*. Bentos estuariya r. Leny. *Issled. Morei SSSR* **15:**63–66. (In Russian.)

Derjugin, K. M. 1932*b*. Iglokozhiye i molluski iz morya Laptevykh. *Issled. Morei SSSR* **15**:147–156. (In Russian.)

Derjugin, K. M., and E. F. Guryanova. 1926. Novyye vidy molluskov iz russkikh severnykh morei. *Tr. Len. Obshch. Estesvt.* **1**:17–23. (In Russian.)

Fedyakov, V. V. 1983. *Dvustvorchatyye i Rakovinnyye Bryuchonogiye Molluski Bentosa Belogo Morya i Zakonomernosti ikh Raspredeleniya.* Avtoref. kand. diss., Leningrad, 18p. (In Russian.)

Filatova, Z. A. 1948. Geograficheskoye rasprostraneniye i nekotorye ekologicheskie osobennosti dvustvorchatykh molluskov iz roda Leda severnykh morei SSSR. *Probl. Arctiki.* **1**:82–99. (In Russian.)

Filatova, Z. A. 1951. Nekotoryye zoogeograficheskiye osobennosti dvustvorchatykh molluskov iz roda Portlandia. *Tr. In-ta Okeanologii AN SSSR* **6**:117–131. (In Russian.)

Filatova, Z. A. 1957*a*. Obshichii obzor fauny dvustvorchatykh molluskov severnykh morei CCCP. *Tr. In-ta Okeanologii AN SSSR* **20**:3–59. (In Russian.)

Filatova, Z. A. 1957*b*. Zoogeograficheskoe raionirovanie severnykh morei po rasprostraneniyu dvustvorchatykh molluskov *Tr. In-ta Okeanologii AN SSSR* **23:**195–215. (In Russian.)

Filatova, Z. A., and L. A. Zenkevich. 1957. Kolichestvennoe raspredelenie donnoi fauny Karskogo moria. *Tr. Vsesoiuzn. Gidrobiol.* **8**:3–67. (In Russian.)

Friele, H., and J. Grieg. 1901. Mollusca III. Den Norske Nordhaus Exp. 1876–1878. *Zool. Christiania* **7**:128p.

Golikov, A. N. 1980. Molluski buccinidae Mirovogo okeana. *Fauna SSSR* **2**:465p. (In Russian.)

Golikov, A. N., and O. A. Scarlato. 1977. Sostav, raspredelenie i ekologiya bruchonogikh i dvustvorchatykh molluskov u arkhipelaga Zemlya Frantsa Iosifa. *Issled. Fauny morei* **14**(22):313–390. (In Russian.)

Gorbunov, G. P. 1941. Dvustvorchatyi mollusk Portlandia arctica (Gray) kak pokazatel raspredelenia materikovykh vod v Sibirskikh moryakh. *Problemy Arktiki* **11**:46–55. (In Russian.)

Gorbunov, G. P. 1946*a*. Novye i interesnye vidy Mollusca i Brachiopoda iz Severnogo Ledovitogo okeana. *Tr. Dreif. Exp. Glavsevmorputi na* *l/p "G. Sedov" 1937–1940.* 5.3, pp. 308–322. (In Russian.)

Gorbunov, G. P. 1946*b*. Donnoe naselenie novosibirskogo melkovodia i tsentralnoi chasti Severnogo Ledovitogo okeana. *Tr. Dreif. Exp. Glavsevmorputi na l/p "G. Sedov" 1937–1940.* 5.3, pp. 30–138. (In Russian.)

Gorbunov, G. P. 1952. Dvustvorchatye molluski (Bivalvia) Chukotskogo morya i Beringova proliva. In *Krainii Severo-Vostok SSSR.* Moscow, t.2, pp. 216–279. (In Russian.)

Govberg, L. I. 1975. Stanovlenie fauny molluscov Belogo morya. In *Biol. Shelfa. Tez. Dokl. Vses. Konf. Vladivostok,* pp. 28–29. (In Russian.)

Grieg, J. A. 1909. Brachiopods and molluscs with a supplement to echinoderms. *Rep. Second Norwegian Arctic Exp. in the "Fram" 1898–1902* **3**(30):3–45.

Grieg, J. A. 1924. Molluscs, brachiopods and echinoderms from Novaya Zemlya. Rep. Sci. Res. Norw. Exp. to Novaya Zemlya 1921. *Kristiania* **26:**1–16.

Guryanova, E. F. 1948. Donnaya fauna Vostochno-Sibirskogo morya. *Nauchn. Bull. Leningradsk. Univers.* **21:**15–18. (In Russian.)

Hägg, R. 1904. Mollusca und Brachiopoda gesammelt von der schwedischen zoologischen Polarexpedition nach Spitzbergen, dem nord-östichen Grønland und Jan Mayen im J. 1900. *Arkiv Zoolog.* **2**(2):1–66.

Hägg, R. 1905. Mollusca und Brachiopoda von der schwedischen Polarexpedition nach Spitzbergen, dem nordostlichen Grønland und Jan Mayen in J. 1900. *Arkiv Zoolog.* **2**(13):1–136.

Herzenstein, S. M. 1885. Beiträge zur Kenntnis der Fauna der Murman-Küste und des Weissen Meeres. 1. Mollusca. *Tr. St-Peterburgsk. Obshch. Estestv.* **16**(2):635–814.

Herzenstein, S. M. 1893. Apercu sur la faune malacologoque de l'Ocean glaciale Russe congress. *Int. Zool. a Moscou* **11:**1–20.

Hülsemann, K. 1962. Marine Pelecypoda from the North Alaskan coast. *Veliger* **5**(2):67–73.

Jaccard, P. 1912. The distribution of the flora in the alpine zone. *New Phytol.* **11:**37–50.

Jensen, A. S. 1905. On the Mollusca of East Greenland. I. Lamellibranchiata. *Medd. Grønland* **29:**289–362.

Jensen, A. S. 1912. Lamellibranchiata. Part I. *The Danish Ingolf Exped.* II, 5, 119p.

Johnson, W. 1934. List of marine Mollusca of the Atlantic coast from Labrador to Texas. *Boston Soc. Nat. Hist. Proc.* **40**(1):204p.

Kafanov, A. I. 1974. Sostav, sistematika i istoria razvitia gruppy Clinocardium (Mollusca, Cardiidae). *Zool. Zhurn.* **53**(10):1466–1476. (In Russian.)

Kafanov, A. I. 1978. Nekotorye aspekty ekologichestkikh vzaimootnoshenii tryokh vidov podsemeistva Clinocardiinae (Bivalvia, Cardiidae). In *Zakonomernosti Raspredelenia i Ekol. Pribrezhn. biotsenozov.* Leningrad, pp. 86–88. (In Russian.)

Knipowitsch, N. M. 1893. Neskolko slov otnositelno fauny Dolgoy guby Solovetskogo ostrova i fiziko-geograficheskikh eyo usloviy. *Vestn. Estestv,* pp. 44–57. (In Russian.)

Knipowitsch, N. M. 1901*a*. Zoologicheskie issledivaniya na ledokole "Ermak" letom 1901 goda. *Ezhegodnik Zool. Mus. Imper. Akad. Nauk* **6:**1–xx. (In Russian.)

Knipowitsch, N. M. 1901*b*. Zoologische Ergebnisse der russischen Expeditionen nach Spitzbergen. Mollusca und Brachiopoda I–IV. *Ezhegodnik Zool. Mus. Imper. Akad. Nauk.* **6:**435–554.

Knipowitsch, N. M. 1902. Zoologische Ergebnisse der russischen Expeditionen nach Spitzbergen. Mollusca und Brachiopoda I–IV. *Ezhegodnik Zool. Mus. Imper. Akad. Nauk.* **7:**355–459.

Knipowitsch, N. M. 1903. Zoologishe Ergebnisse der russischen Expeditionen nach Spitzbergen. Mollusca und Brachiopoda I–IV. *Ezhegodnik Zool. Mus. Imper. Akad. Nauk.* **8:**133–143.

Koltun, V. M. 1964. K izucheniu donnoi fauny Grenlandskogo morya i tsentralnoi chasti Arkticheskogo basseina. *Tr. Arkt. i Antarkt. Nauch.-Issl. In-ta* **259:**13–78. (In Russian.)

Lande, E. 1975. A collection of Pelecypoda from Øre and romsdal, western Norway. *Sarsia* **59:**7–13.

Leche, W. 1878. Øfversig öfver de of Svenska Expeditionerna till Novaja Semlya och Jenisse 1875 och 1876 insamlade Hafs Mollusker. *Sv. Ak. Handl.* 16, 2, 86p.

Leche, W. 1883. Øfversigt öfver de of Vega-expeditionen insamlade arctiska Hafs-Mollusker. I. Lamellibranchiata. *Vega-Exped. Vetensk Iakttagelser* **3:**437–452.

Lindberg, G. U. 1972. *Krupnye Kolebania Urovnya Okeana v chetvertichnyi Period.* Nauka, Leningrad, 548p. (In Russian.)

Lubinsky, I. 1980. Marine bivalve molluscs of the Canadian central and eastern Arctic: Faunal composition and zoogeography. *Rep. Fisheries and Oceans,* 111p.

MacGinitie, G. E. 1955. Distribution and ecology of the marine invertebrates of Point Barrow, Alaska. *Smithsonian Misc. Colln.* **128**(9):201p.

Madsen, H. 1936. Investigation on the shore fauna of east Greenland with a survey of the shores of other Arctic regions. *Medd. Grønland* **100**(8):1–79.

Madsen, H. 1940. A study of the littoral fauna of northwest Greenland. *Medd. Grønland* **124**(3):1–24.

Madsen, H. 1949. Marine bivalvia. *Zool. Iceland* **4**(63):116p.

Matveeva, T. A. 1976. Biologia dvustvorchatogo molluska Turtonia minuta v razlichnykh chastyakh areala. *Biologia Morya* **6**:33–39. (In Russian.)

Matveeva, T. A. 1977. Osobennosti razmnozhenia dvustvorchatykh molluskov semeistva Astartidae. Biotsenozy shelfa Zemli Fratsa Iosifa i fauna sopredelnykh akwatorii. *Issled. Fauny Morei* **14**(22):418–427. (In Russian.)

Melville, J. C., and R. Standen. 1900. Report on the Mollusca of the "Jackson Harmsworth" Expedition to Franz Josef Land (1896–97) and of the "Andrew Coast" Cruise (1898) to Kolguev etc. *Mem. Proc. Manchester Liter Phil. Soc.* **44**:1–14.

Mesyatsev, I. I. 1931. Molluski Barentseva moria. *Tr. GOIN* **1**(1):1–168. (In Russian.)

Middendorf, A. T. 1851. Mollusken. *Reise in der Aussersten Norden und Ostsibirien, Während der Jahre 1843 und 1844.* **2**(1):163–464.

Miloslavskaya, N. M. 1977. Molluski semeistva Thyasiridae (Bivalvia, Lucinidae) arkticheskikh morei SSSR. Biotsenozy shelfa Zemli Fratsa Iosifa i fauna sopredelnykh akwatorii. *Issled. Fauny Morei* **14**(22):391–416. (In Russian.)

Moebius, K. 1874. *Mollusken Die Zweite Deutsche Nordpolarfahrt 1869 u. 70 unter Koldewey. II. Zoologie.* Part 7. Leipzig, pp. 249–250.

Morch, O. A. L. 1869. Catalogue des Mollusques du Spitzberg recueillis par le Dr. H. Kroyer. *Ann. Soc. Malac. Belg.* **4**:7–32.

Mossewitsch, N. A. 1928. Materialy k sistematike, ekologii i rasprostraneniu sowremennoy i iskopaemoy Yoldia arctica Gray. *Mat. Kom. po Izuch. YASSR* **19**:44p. (In Russian.)

Muratov, M. V. 1975. *Proisckozhdenie Materikov i Okeanicheskikh Vpadin.* Nauka, Moscow, 176p. (In Russian.)

Naumov, A. D. 1979. Donnaya fauna guby Lov (Beloe more, Kandalakshskii zaliv) i ee osobennosti. In *Ekologia Donnogo Naselenia Shelfovoi Zony.* Moscow, pp. 128–136. (In Russian.)

Nicol, D. 1955. An analysis of the Arctic marine pelecypod fauna. *Nautilus* **68**(4):115–122.

Ockelmann, W. K. 1954. On the interrelationship and the zoogeography of Northern species of Yoldia Moller s. str. (Mollusca, fam. Lididae). *Medd. Grønland* **107**(7):1–32.

Ockelmann, W. K. 1958. Marine Lamellibranchiata. *Medd. Grønland* **122**(4):256p.

Odhner, N. 1951. Die Molluskenfauna des Eisfjordes. *Kungl. Sv. Vet. Akad. Handl. Bd.* **54**(1):274p.

Odhner, N. 1960. Mollusca. *Rep. Swed. Deep-Sea Exp.* **2**:365–400. (Zool. 22.)

Oskarson, I. 1952. Lamellibranchia. Sleldyrafauna Islands. Reykjavik. 119p.

Petersen, G. 1978. Life cycles and population dynamics of marine benthic bivalves from the Disco Bugt area of west Greenland. *Ophelia* **17**(1):95–120.

Popov, A. M. 1932. Gidrobiologicheskii ocherk morya Laptevykh. *Issled. Morei SSSR* **15**:189–229. (In Russian.)

Richards, H. G. 1936. Recent and Pleistocene marine shells of James Bay. *Am. Midl. Nat.* **17.**

Ryabinina, N. V. 1952. Carditacea Chukotskogo morya i Beringova proliva. *Krainii Severo-Vostok SSSR* **2**:279–286. (In Russian.)

Sars, G. O. 1878. Bidrag til Kundskaben om Norges arktiske Fauna. I. Mollusca regionis arcticae Norvegiae. *Christiania* 466p.

Savilov, A. I. 1953. Rost i ego izmenchivost u bespozvonochnykh Belogo morya Mytilus edulis, Mya arenaria i Balanus balanoides. Ch. I. Mytilus edulis v Belom more. *Tr. In-ta Okeanologii AN SSSR* **7**:198–258. (In Russian.)

Scarlato, O. A. 1981. Dvustvorchatye molluski umerennykh shirot zapadnoi chasti Tikhogo okeana. Opred. po faune SSSR, izd. *Zool. Intom AN SSSR* **126**:480p.

*Severnyi Ledovityi Okean i Ego Poberezhie v Kainozoe.* 1970. Gidrometeoizdat, Leningrad, 562p. (In Russian.)

Soot-Ryen, T. 1932. Pelecypoda, with a discussion of possible migrations of Arctic pelecypods in Tertiary times. *Medd. fra det Zool. Mus. Bergen-Oslo* **5**(12):1–35.

Soot-Ryen, T. 1964. Yoldia Moller, 1842 and Portlandia Morch, 1857: proposed designation of a type-species under the plenary powers with rejection of Yoldia arctica Moller, 1842 A.N.(S.) 1522. *Bull. Zool. Nomenc.* **21**(2):127–129.

Soot-Ryen, T. 1966. Revision of the pelecypods from the Michael Sars North Atlantic Deep-Sea Expedition 1910. *Sarsia* **24**:1–32.

Sparks, A. K., and W. T. Pereyra. 1966. Benthic invertebrates of the south-eastern Chukchi Sea. In *Environment of the Cape Thompson Region, Alaska.* U.S. Atomic Energy Commission, Division of Technical Information, pp. 817–837.

Thiele, J. 1928. Arctische Loricaten, Gastropoden, Scaphopoden und Bivalven. *Fauna Arctica* **5**(2):563–632.

Thorson, G. 1934. Contributions to the animal ecology of the Scoresby Sound fjord complex (east Greenland). *Medd. Grønland* **100**(3):1–68.

Thorson, G. 1935. Biologische Studien über die

Lamellibranchier Modiolaria discors L. und Modiolaria nigra Gray in *Ostgrönland. Zool. Anz. Leipzig* **111**(11–12):297–304.

Thorson, G. 1936. The larval development, growth and metabolism of Arctic marine bottom invertebrates, compared with those of other seas. *Medd. Grønland* **100**(6):1–155.

Thorson, G. 1951. Scaphopoda, Placophora, Solenogastres, Gastropoda, Prosobranchiata, Lammelibranchiata. The Godthaab Expedition 1928. *Medd. Grønland* **81**(2):1–117.

*Treatise on Invertebrate Paleontology*. 1969. R. C. Moore (ed.). Part N, Vols. 1–3, Mollusca 6, 951p.

Ushakov, P. V. 1952. Chukotskoe more i ego donnaya fauna. *Krainii Severo-Vostok SSSR* **2**:5–83. (In Russian.)

Wagner, F. J. E. 1977. Recent mollusc distribution patterns and palaeobathymetry, southeastern Beaufort Sea. *Can. J. Earth Sci.* **14**:9.

Zenkevich, L. A. 1927. Kolichestvennyi uchyot donnoi fauny Pechyorskogo raiona Barentseva morya i Belogo morya. *Tr. Plav. Mor. In-ta* **1**(4):1–64. (In Russian.)

Zenkevich, L. A. 1931. Kolichestvennyi uchyot donnoi fauny prikaninskogo raiona. *Tr. Mor. Nauch. In-ta* **4**:3.

# 13: Arctic Ocean Gastropod Prosobranchs

A. N. Golikov

## INTRODUCTION

The study of 225 species of Arctic prosobranch gastropods has revealed a regular parabolic sequence in the number of members in taxa of different ranks. As related to their rank, the regression coefficient is 2.8.

The geographic distribution of mollusks of different biogeographic groups shows the Norwegian and western Barents seas to be dominated by Atlantic boreal species; the eastern Barents, Kara, and Laptev seas by Atlantic boreal-Arctic and Arctic species; and the East Siberian and Chukchi seas by Pacific boreal-Arctic and Arctic species, replaced off Bering Strait by Pacific boreal taxa. A considerable proportion of species are confined either to the Eurasian or to the Amerasian subbasins with the boundary in the region of the New Siberian Islands. The fauna is dominated by eurybathic species. The lowest similarity in species composition at adjacent depths is observed in the Barents Sea at a depth of 25 m, marking the boundary between the Atlantic high boreal water mass and the surface Arctic water; in the Siberian Seas at 10 m depth, separating the estuarine–Arctic from the Arctic waters; and in all other seas at 200–700 m depth at the boundaries between the Intermediate Atlantic and Arctic water masses. Species belonging to different biogeographic groups are excellent indicators of water masses. However, thermophilic Pacific boreal species inhabit the eastern parts of the Chukchi Sea, the Intermediate Atlantic–Arctic Water Mass, sometimes dominating the Atlantic boreal-Arctic species characteristic of this water mass. Species inhabiting the Arctic region show no considerable decline in growth rate as compared to their temperate water counterparts.

*Note:* This chapter has been reviewed, edited, and rewritten by the Editor, Yvonne Herman.

## PREVIOUS STUDIES—MATERIALS AND RESULTS

Prosobranch gastropods are an excellent group for the study of general patterns of biotic distributions in the Arctic. Their shells are well preserved and therefore they are used widely for biostratigraphic zonation, having well-defined zonal–geographic distributions. Furthermore, they do not contain cosmopolitans, not even species distributed in more than two geographic zones. Their shells bear morphological growth characteristics and features believed to represent environmental characteristics. However, in spite of the vast number of review books (Middendorff, 1849*a*, 1849*b*, 1851; Mörch, 1875; Sars, 1878; Friele, 1882; Herzenstein, 1885; Aurivillius, 1887; Brögger, 1899–1901; Knipowitsch, 1900; Thorson, 1941, 1944, 1951; Filatova and Zatsepin, 1948; McGinitie, 1955; Bousfield, 1960; Macpherson, 1971) and reviews dealing with various taxonomic groups (Derjugin, 1950; Galkin, 1955; Golikov, 1963, 1980), knowledge of the ecology of many species and of gastropod faunal composition in the Arctic Ocean is still incomplete; new genera and species are continuously being discovered. Nevertheless, the collection of the Zoological Institute, assembled over 200 years, as well as the literature data, are sufficient for a preliminary synthesis of the Arctic Ocean prosobranch fauna. In the compilation of faunal distributions, the author has drawn upon his own data as well as on the vast open literature.

The accepted systematic position of the prosobranch mollusks follows the latest classification (Golikov and Starobogatov, 1975), placing the orders Epitoniiformes, Pyramidelliformes (Heterostropha), and Eulimiiformes (Homoeostropha) in a separate subclass. In compliance with the system the prosobranch fauna in the Arctic Ocean is represented by 4 subclasses, 12 orders, 42 families, 95 genera, and about 220 species.

The number of species found in the Arctic Ocean comprise 1.6 percent of the marine prosobranches' world fauna and one-third of the species distributed in the cold and temperate waters of the Northern Hemisphere. The hierarchy of taxa in the regional faunas and floras is approximated by a parabolic sequence of the number of members in taxa related to their rank (Golikov, 1976*a*)

$$N = a\,T^{b}$$

where $N$ is the number of taxa, $T$ is the rank of taxa, and $a$ and $b$ are coefficients.

This sequence for the gastropod prosobranchs in the Arctic Ocean (Fig. 13-1) can be expressed as

$$N = 1.7\,T^{2.8}$$

The changes in species composition in various seas of the Arctic Ocean were analyzed, estimating the degree of similarity between the diverse and less diverse faunas (Jaccard, 1901, 1912):

$$j = \frac{C \cdot 100}{D_1 + D_2 - C}$$

and with the method of analysis of inclusion of less diverse fauna into more diverse fauna

$$i = \frac{C \cdot 100}{D_{\min}}$$

where $C$ is the number of common species, $D_1$ and $D_2$ are the numbers of species in the faunas compared, and $D_{\min}$ is relatively poor fauna. The results are presented in Table 13-1.

There is a rapid decline in the number of species in the Arctic Ocean from the boreal zone, in the vicinity of the Atlantic waters, eastward. A slight increase is observed only in the East Siberian Sea near the boundary between the Eurasian and Amerasian parts of the Arctic region. Generally the boreal gastropod fauna of the Norwegian and Barents seas shows a statistically significant difference from the faunas of other Arctic Ocean seas.

The less diverse faunas of the Arctic seas include a significant number of spe-

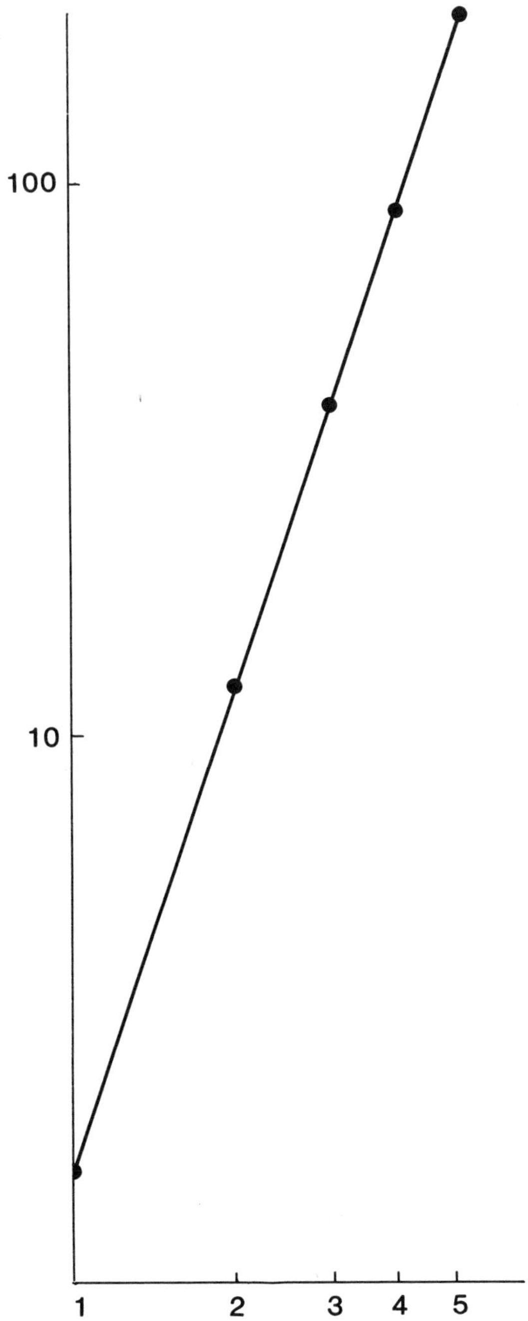


Figure 13-1. Relationship between the number of gastropod prosobranchs and their rank in the Arctic Ocean. Ordinate: number of taxa (on a logarithmic scale); abscissa: rank of taxa (sequence: 2, orders; 3, families; 4, genera; 5, species; on a logarithmic scale).

cies of the shelf and bathyal zones of the Barents Sea. The differences between the species composition of neighboring seas of the Arctic Ocean increase from west to east, with decreasing influence of the North Atlantic fauna. Three-fourths of the Kara Sea fauna are also present in the Barents Sea but are more closely related to the Laptev Sea fauna. The Laptev and Kara seas faunas are more closely related to the fauna of the Barents Sea than to the faunas of the East Siberian and Chukchi seas.

The East Siberian Sea prosobranchs, including both Atlantic and Pacific invaders, display a statistically significant difference in composition from the Barents, Kara, and Laptev seas' prosobranchs, as well as from the Chukchi and Beaufort seas' taxa. The Chukchi Sea fauna is similar to that of the Beaufort Sea, including 20 percent boreal Pacific invaders; it is most peculiar among the Arctic Ocean Seas' faunas.

In order to undertake a biogeographic study of the Arctic Ocean and present the composition of its prosobranch fauna, a list of species is given arranged in biogeographic groups.

**I.** Atlantic subtropical-boreal species (20) distributed in the temperate waters of the North Atlantic and within the range of the subtropical Mediterranean–Lusitanian biogeographic region: *Helcion pellucidus, Tectura virginea, Emarginula fisura, Gibbula cineraria, G. tumida, Ganesa trochoidea, Littorina saxatilis, Mohrensternia interrupta, Alvania cimicoides, Onoba jeffreysi, Skeneopsis planorbis, Pseudopolynices nanus, Capulus hungaricus, Triphora perversa, Nucella lapillus, Taranis morchi, Buccinum humphreyseanum, Balcis stenostoma, Aclis walleri,* and *Omalogyra atomus.*

**II.** Atlantic boreal species (39) are divided into two groups.

1. Atlantic boreal wide-ranging species (15) distributed in the temperate waters of the North Atlantic from the English

Table 13-1. Similarity (below underscored numbers) and Degree of Inclusion (above underscored numbers) (%) from Less Diverse to More Diverse Faunas of Prosobranchs

| | D1 | D2 | D3 | D4 | D5 | D6 |
|---|---|---|---|---|---|---|
| D1 | 120 | 81 | 78 | 70 | 51 | 30 |
| D2 | 50 | 84 | 71 | 75 | 54 | 38 |
| D3 | 33 | 40 | 56 | 75 | 59 | 38 |
| D4 | 27 | 40 | 58 | 53 | 62 | 45 |
| D5 | 11 | 31 | 38 | 39 | 65 | 44 |
| D6 | 10 | 17 | 20 | 26 | 18 | 45 |

*Note:* Western Barents Sea (D1), eastern Barents Sea (D2), Kara Sea (D3), Laptev Sea (D4), Eastern Siberian Sea (D5), Chukchi Sea (D6).

Channel and Cape Cod to the south, up to Labrador, south Greenland, Spitsbergen, and Novaya Zemlya Island in the north: *Pilidium fulvum, Emarginula crassa, Calliostoma occidentalis, Littorina obtusata, Littorina littorea, Hydrobia ulvae, Onoba tenuicostata, Jeffreysina globularis, Lamellaria latens, Beringius turtoni, Colus turgidulus, Buccinum undatum, Boreofusus berniciensis, Chrysallida eximia,* and *Spirotropis carinata.*

2. Atlantic high boreal species (24) distributed similarly to the species of the biogeographic group II, 1, but off the European coasts; they do not penetrate farther south than northern Scotland and Ireland: *Ganesa areolata, G. laevigata, Lacuna neritoidea, Onoba aculeus, Setia tumidula, Pseudosetia turgida, Torellia vestita, Ariadnaria conica, Marsenina groenlandica, Laeocochlis granosa, Eumetula costulata, Liomesus dalli, Siphonorbis dalli, S. ebur, Colus stimpsoni, C. latericeus, C. fusilformes, C. undulatus, Bucci num finmarchianum, Neptunea despecta, Thesbia nana, Oenopota pingeli, O. angulosa,* and *Raphitoma amoena.*

**III.** Amphiboreal species (8) distributed in the temperate waters of the Pacific and Atlantic oceans and absent in the Arctic region: *Puncturella noachina, Margarites helicinus, Epheria vincta, Trichotropis bicarinata, Iphinoe kroyeri, Onchidiopsis groenlandicus, Colus spitzbergensis,* and *Obestoma tenuilirata.*

**IV.** Boreal–Arctic species (76) are divided into three groups.

1. Boreal–Arctic widespread species (28) distributed in the Arctic Ocean and in the adjacent temperate waters of the Pacific and Atlantic oceans: *Erginus rubella, Lepeta coeca, Margarites groenlandicus s.l., M. costalis s.l., M. vahlii, Solariella obscura, S. varicosa, Tachyrhynchus erosus, T. reticulatus, Ariadnaria borealis, Velutina undata, V. velutina, V. plicatilis, Onchidiopsis glacialis, Capulacmaea radiata, Bulbus smithii, Lunatia tenuistriata, L. pallida, Cryptonatica clausa, Buccinum ciliatum s.l., Admete couthouyi, Oenopota pyrameidalis, O. bicarnata, O. trevelliana, O. nobilis, O. exarata, O. schantarica,* and *Boreoscala groenlandica.*
2. Boreal–Arctic Atlantic species (33) distributed in the boreal waters of the North Atlantic and in the Arctic region and absent from the Pacific Ocean. They are subdivided into species with circumpolar (pan-Arctic) distribution (13): *Frigidoalvania janmaeni, Marsenina glabra, Amauropsis islandica, Astyris rosacea, Colus togatus, C. tortuosus, C. islandicus, Boreptrophon truncatus, Oenopota decussata, O. turricula, O. incisula, O. harpularia,* and *O. cancellata;* and species

distributed in the Arctic region only, in the Eurasian part, not farther in the east than the New Siberian Islands and the Lomonosov Ridge (20): *Testudinalia tessellata, Scisurella crispata, Moelleria costulata, Margarites olivaceus, Turritellopsis acicula, Epheria crassior, Onoba castanea, Cerithiella whiteavesii, Beringius ossiani, Volutopsius norvegicus, Colus pubescens, Buccinum nivale, B. fragile, B. cyaneum, B. micropoma, B. hydrophanum, Oenopota cinerea, O. mitrula, O. obliqua,* and *Liostomia eburnea.*

3. Boreal–Arctic Pacific species (15) distributed in the boreal waters of the Pacific Ocean and in the Arctic region, but absent from the Atlantic Ocean. They are also subdivided into species with circumpolar distribution (8): *Margarites gigantea, Pyrulofusus deformis, Plicifuscus kroyeri, Neptunea ventricosa, N. communis, s.l., Buccinum polare, B. elatior,* and *B. glaciale;* and species distributed in the Arctic region, mainly in the Amerasian part, not farther than the Laptev Sea (7): *Margarites vorticifera, M. pribiloffensis, M. frigidus, Beringion beringi, Neptunea beringiana, Buccinum physematum,* and *B. polium obsoletum.*

**V.** Arctic species (42) distributed only in the Arctic region are subdivided into species with circumpolar distribution (14): *Tachirinchus spitzbergensis, Punctulum wyville-thomsoni, Lunatia bathibii, Buccinum belcheri, Turrisipho lachesis, Colus curtus, Buccinum angulosum, B. maltzani, Oenopta arctica, O. gigantea, O. novajasemlensis, O. schmidti, O. impressa,* and *Menestho truncatula;* Arctic, mainly Eurasian species that penetrate to the east, not farther than the New Siberian Islands and the Lomonosov Ridge (23): *Frigidoalvania scorbiculata, F. cruenta, Punctulum moerchi, P. minuta, Onoba verrilli, Lissospira profunda, Ankes undulisculpta, Velutina schneideri, V. lanigera, Cryptonatica septentrionalis, Colus propinquus, Anomalosipho altus, Tacita danielseni, Neptunea denselirata, Mohnia mohni, Boreotrophon fabricii, Taranis amoena, Oenopta harpularioides, O. declivis, O. ovalis, Pleurotomella packardi, Entocolax ludwigi,* and *Arcrisa costulata;* and Arctic, mainly Amerasian species that penetrate to the west, not farther than the New Siberian Islands (5): *Aquilonaria turneri, Anomalosipho martensi, A. verkruzeni, Colus roseus,* and *Buccinum fringillum.*

**VI.** Pacific boreal species that penetrate into the Chukchi and Beaufort seas (13): *Crepidula grandis, Lunatia monterona, Beringius stimpsoni malleatus, Colus capponius, C. hypolispus, Voluptosius stephanssoni, Neptunea lyrata, Buccinum plectrum, Boreotrophon beringi, B. pacificus, Admete regina, Oenopota nazanensis,* and *O. harpa.*

## DISCUSSION

There is an increase in the relative diversity of the evolutionarily advanced species and phylogenetically young taxonomic groups of gastropods from the temperate waters of the Arctic Ocean toward the Arctic water. Thus, the Atlantic boreal species comprise 15–17 percent of the primitive gastropods belonging to the orders Patelliformes and Fissurelliformes, 30–40 percent of the more advanced taxa belonging to the orders Bucciniformes and Coniformes, whereas in the boreal–Arctic and Arctic biogeographic groups the former constitute about 10 percent, and the latter about 60–75 percent. This phenomenon may be explained by the genetic advantage (cold adaptation) of the representatives of taxa that originated in previous geological epochs during global cooling (Golikov, 1975). Apparently, the interdependent protein and nucleotide synthesis bear traces of the original genetic adaptation of the ancestral forms to the comparatively low temperatures. The orders Bucciniformes and Coniformes have originated chiefly in the Cenomanian (Cretaceous, 97.5–91 mya) during a global cooling and their numerous genera (Neptunea, Colus, Buccinum, Oenopota) have achieved their highest development mostly in cold water (Golikov, unpublished).

A comparison between the number of gastropod species of the various biogeographic groups in the different Arctic Ocean seas (Fig. 13-2) has shown that Atlantic boreal species dominate other biogeographic groups in the Norwegian and Barents seas as far as about 33°E longitude. Atlantic boreal–Arctic and widespread boreal–Arctic species become more numerous in the eastern Barents Sea. The number of Arctic species is exceeding that of the boreal taxa off Novaya Zemlya and in the western Kara Sea, suggesting that this region represents the eastern boundary of the Arctic region. Therefore, the region between 33°N and 60°N can be regarded as an ecotone, an area of mixing of the boreal and Arctic hydrobiocomplexes. These waters contain no endemic species or even intraspecific forms of gastropods, whereas the Atlantic boreal and Arctic waters comprise numerous endemic species.

The relative number of Pacific boreal–Arctic species increases in the Arctic Ocean from west to east. In the Laptev Sea, these Pacific boreal–Arctic species are becoming more numerous than the Arctic species, and in the East Siberian and Chukchi seas they dominate the species of other biogeographic groups. There are major differences between the prosobranch faunal composition in the Eurasian and the Amerasian parts of the Arctic Basin, with the boundary in the region of the New Siberian shoals. Beyond the region of the New Siberian Islands, the species of various biogeographic groups may occur due to upwelling (Gorbunov, 1946) and to mixing of Eurasian and Amerasian species. There is a decline in the relative number of Atlantic boreal–Arctic species eastward and of Pacific boreal–Arctic species westward, whereas widespread boreal–Arctic species, more ancient phylogenetically (evolving ≈2.5 mya) show an even distributional pattern.

The more eurybiontic widespread boreal–Arctic species were the first to invade the Arctic and dominate a number of shelf ecosystems (Golikov and Scarlato, 1977). These species became established on the shelves after the Würm regression faster and more efficiently than other groups.

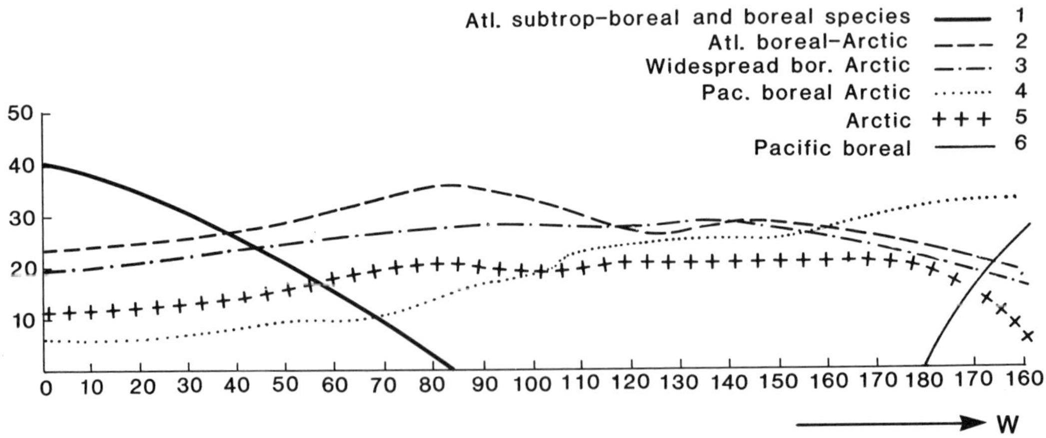


Figure 13-2. Number of species of different biogeographic groups in the Arctic Seas. Ordinate: percentage of species of a given biogeographic group out of the total species number in all areas of the Arctic Ocean; abscissa: longitude in the Arctic Ocean. 1, Atlantic subtropic boreal, widespread boreal, and high boreal species; 2, Atlantic boreal–Arctic species; 3, widespread boreal–Arctic species; 4, Pacific boreal–Arctic species; 5, Arctic species; 6, Pacific boreal species, present in some areas of the Arctic.

For a number of relatively stenobiontic Pacific boreal–Arctic and Amerasian–Arctic species, as well as for Atlantic boreal–Arctic and Eurasian–Arctic species, the rigorous conditions of the New Siberian shoals were an impassable barrier. In the Bering Strait and western Beaufort Sea, Pacific boreal species outnumber the Arctic and even the Atlantic as well as the widely distributed boreal–Arctic taxa, being only less numerous than the Pacific boreal–Arctic species. This means that the waters of the southeastern Chukchi Sea and southwestern Beaufort and Bering seas north of Navarinsks Cape and Norton Sound may be regarded as an ecotone between the Arctic and Pacific boreal biogeographic regions. In contrast with the Eurasian mixed waters they do not contain endemic species.

The biomass of prosobranchs in the Arctic Ocean is low and, much like that of the entire benthos, is decreasing eastward. However, in the temperate coastal waters of the Barents and Norwegian seas, the most abundant periwinkles can reach a biomass of several kg $m^2$, and the Buccinids, most abundant in the Arctic waters, can reach a biomass of several hundred g $m^2$. The vertical distribution of prosobranchs in the Arctic Ocean is shown in Figure 13-3.

The maximum number of species at all depths is observed in the Norwegian and Barents seas, where, primarily in boreal water, the species reach their greatest diversity at depths of 55–100 m. In the Arctic seas the maximum number of species is always confined to a depth of 55 m. A decline in species number is observed in the East Siberian, Laptev, Kara, and Chukchi seas.

A comparison of the degree of similarity between the malacofaunas of adjacent depths (Fig. 13-4) has shown that depths up to 5 m have distinct prosobranch faunas in the temperate and intermediate wa-

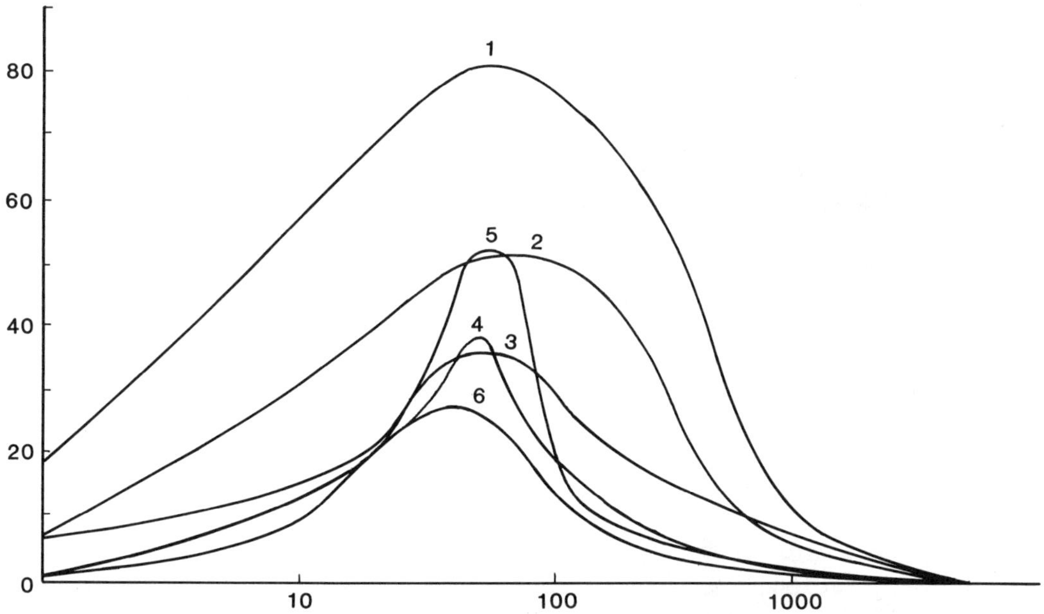


**Figure 13-3.** Vertical distribution of species in (1) the Norwegian Sea and western Barents Sea, (2) the eastern and northern parts of the Barents Sea, (3) the Kara Sea, (4) the Laptev Sea, (5) the East Siberian Sea, and (6) the Chukchi Sea. Ordinate: number of species; abscissa: depth on a logarithmic scale.

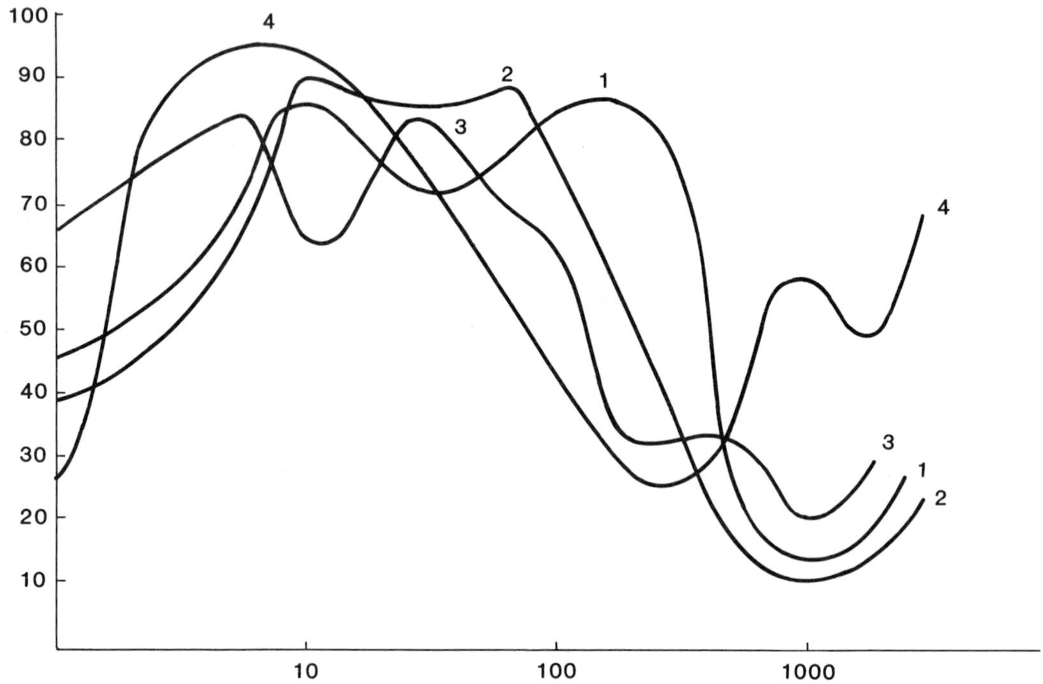


Figure 13-4. Similarity (%) of species composition at (1) adjacent depths in the Norwegian Sea and western Barents Sea, (2) in eastern and northern parts of the Barents Sea, (3) in the Kara Sea and in the Laptev Sea, (4) in the East Siberian and Chukchi Seas. Ordinate: percentage of similarity (after Jaccard); abscissa: depth, m (on a logarithmic scale).

ters both in the Pacific and Atlantic sectors, but are devoid of specificity in the Arctic region.

A marked though not significant increase of differences in malacofaunal composition is observed at depths of about 10 m in the Arctic and at 25 m in the Atlantic water. These represent the lower boundaries of the upper layers of the surface Arctic and Atlantic high boreal water masses. There are statistically significant differences between the malacofauna of the inner shelf and that of the slope. They are as follows: in the Norwegian and western Barents seas at a depth of about 400 m, in the eastern Barents seas at 200 m, in the Kara and Laptev seas at 130–150 m, in the East Siberian and Chukchi seas at 80–90 m. The lowest degree of similarity between the faunas of adjacent depths is observed off the boundaries of the Atlantic Intermediate water mass, at depths of 500–1,200 m in the Norwegian, Barents, Kara, and Laptev seas and 180–500 m in the East Siberian and Chukchi seas. At depths greater than 2,000 m, in the Deep Arctic Water mass, the similarity of malacofaunas increases again. The most complete information on the vertical distribution of water masses is provided by the study of relationships of biogeographic groups of species at various depths (Fig. 13-5).

In the Norwegian and Western Barents seas, down to depths of 500 m, Atlantic boreal species dominate all other taxa (Fig. 13-5*A*). These species evolved 3–5 mya Harmer, 1914; Golikov, 1963, 1980; Gladenkov, 1978). The inner shelf Atlantic boreal–Arctic species are less numerous than the more eurybiontic widespread boreal–Arctic species, but on the shelf edge and on the slope they are increasing in numbers and at depths ranging from 500–2,000 m they dominate the other biogeographic groups and are indicators of the Atlantic-Arctic water mass. The number of Pacific boreal–Arctic species present only on shelves is not high in the Norwegian Sea or in the western part of the Barents Sea. The Arctic species here occur at depths greater than 10 m and increase in numbers rapidly. At 500 m depth, they dominate over the widespread boreal–Arctic species, at between 1,200 m and 2,000 m they dominate the Atlantic boreal-Arctic species, and in the pseudoabyssal zone, at depths in excess of 2,000 m, they are the predominant malacofaunal group.

In the eastern Barents Sea (Fig. 13-5*B*), the Atlantic boreal species dominate only at depths of 10 m, whereas Atlantic and widespread boreal–Arctic species dominate the deeper parts of the shelf and slope. Pacific boreal–Arctic species are scarce and are nearly evenly represented at 10–1,000 m depths. Arctic species exceed the boreal taxa in numbers at depths greater than 25 m, thus defining the lower boundary of the Atlantic high boreal water mass, which is thinner in this region than in the Norwegian and western Barents seas. These species in the eastern part of the Barents Sea dominate the malacofauna at depths greater than 800 m; they are indicators of the Deep Arctic Water Mass.

Atlantic boreal–Arctic species are indices of the Intermediate Atlantic–Arctic Water Mass. They dominate other biogeographic groups in the eastern part of the Barents Sea at depths of 120–800 m. In the Kara and Laptev seas (Fig. 13-5*C*), eurybiontic widespread boreal–Arctic species are dominant over the other biogeographic groups on the shelf, to a depth of 120 m. The boundary between the estuarine–Arctic Water Mass and the surface Arctic Water Mass is marked at a depth of 10 m by the predominance of Atlantic boreal–Arctic species over Pacific boreal–Arctic and Arctic taxa. The Intermediate Atlantic–Arctic Water Mass is defined by Atlantic boreal–Arctic species dominating between 120 m and 800 m depths. The Deep Arctic Water Mass is dominated by the Arctic species. In the East Siberian and Chukchi seas (Fig. 13-5*D*), Pacific boreal–Arctic species are most diverse on the shelf down to depths of 120 m. At greater depths (including the Intermediate Atlantic–Arctic Water Mass) the Chukchi Sea faunas are dominated by Pacific boreal immigrants that are replaced at depths of 800 m in the Deep Arctic Water Mass by Arctic species.

## CONCLUSIONS

An attempt has been made to relate species of different origin inhabiting various depths to water masses. Our data suggest that the formation of water masses with well-defined temperature–salinity regimes is followed by the evolutionary appearance of taxa adapted to the newly formed environment (Golikov, 1973). Eventually the water mass and the species inhabiting it form a hydrobiocomplex characterized by certain spatial–temporal and ecological features. The limits of the Atlantic High Boreal Water Mass distribution are distinctly shown by the predominance of the Atlantic boreal species over most of the shelf, on the upper part of the slope in the Norwegian Sea and in the western Barents Sea, as well as by their more restricted distribution (up to 5 m) off Novaya Zemlya (Fig. 13-6). The proportion of Atlantic boreal–Arctic species in the Arctic Ocean is increasing with depth and they dominate other biogeographic groups in most parts of the ocean between 120 m and 800 m depths, forming a hydrobiocomplex of the Intermediate Atlantic–Arctic Water Mass. In the Chukchi Sea off the Bering Strait, this biogeographic group is in some parts dominated by Pacific boreal species that favor the relatively warm Intermediate Atlantic–Arctic Water Mass.

The Surface Arctic Water Mass in the Eurasian Arctic sector is characterized by boreal–Arctic widely distributed species, while the Amerasian sector supports Pa-

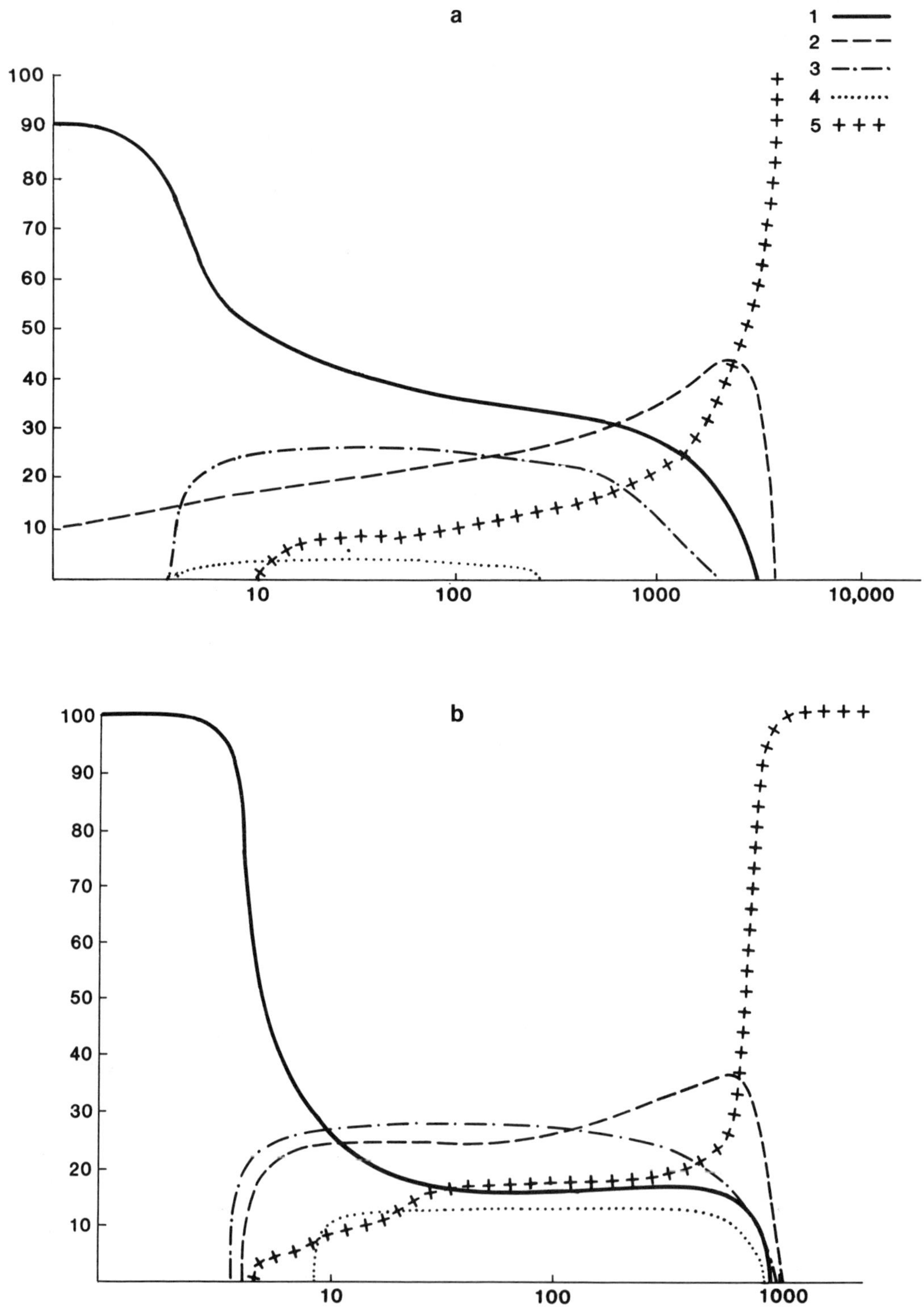


Figure 13-5. Percentages of species in biogeographic groups at different depths. *A:* Norwegian Sea and western Barents Sea; *B:* eastern Barents Sea; *C:* Kara Sea and Laptev Sea; *D:* East Siberian and Chukchi seas. Ordinate: number of species in biogeographic groups at various depths; abscissa: depth, m (on a logarithmic scale). Symbols legend as in Figure 13-2.

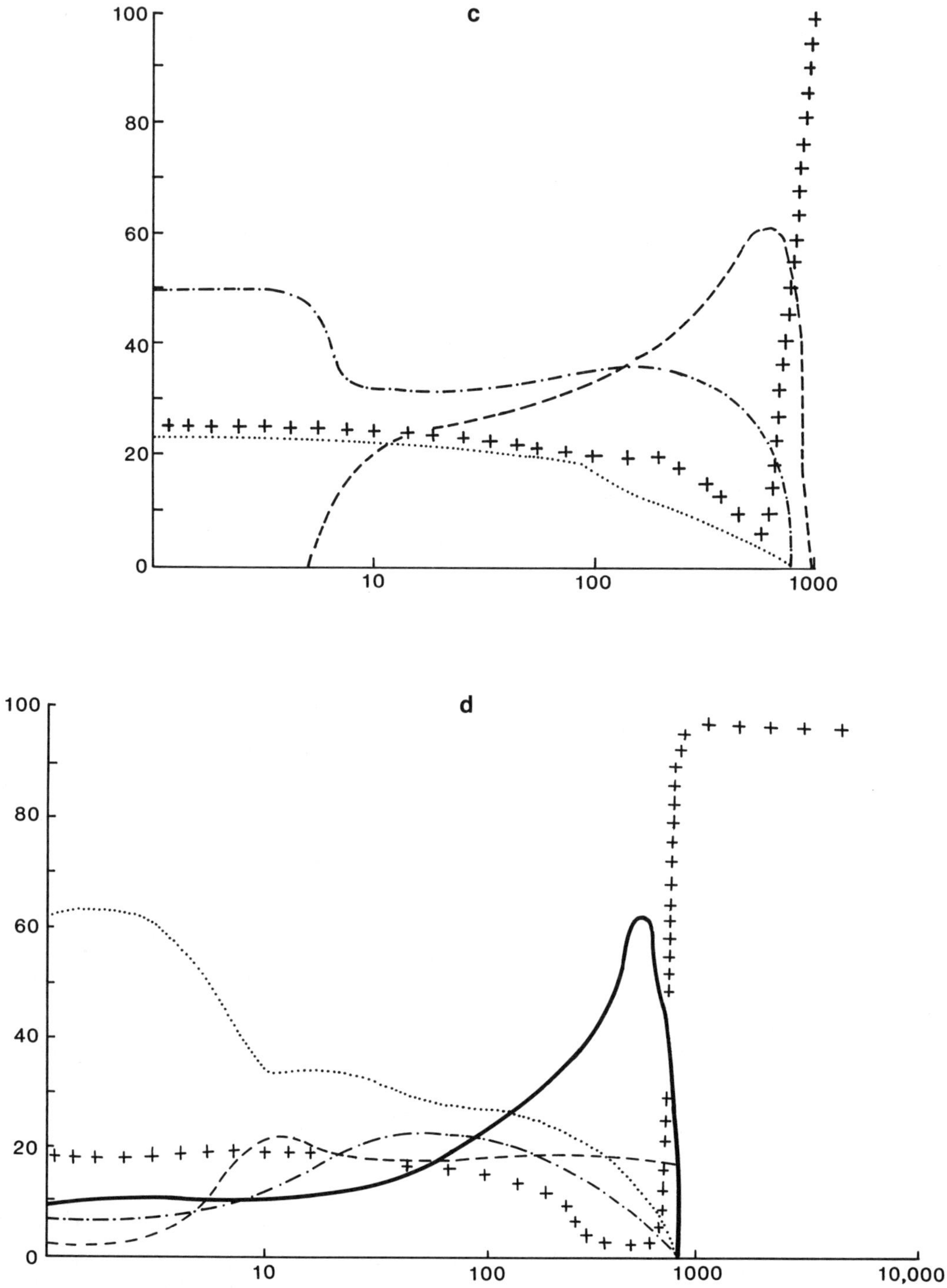


Figure 13-5. (continued)

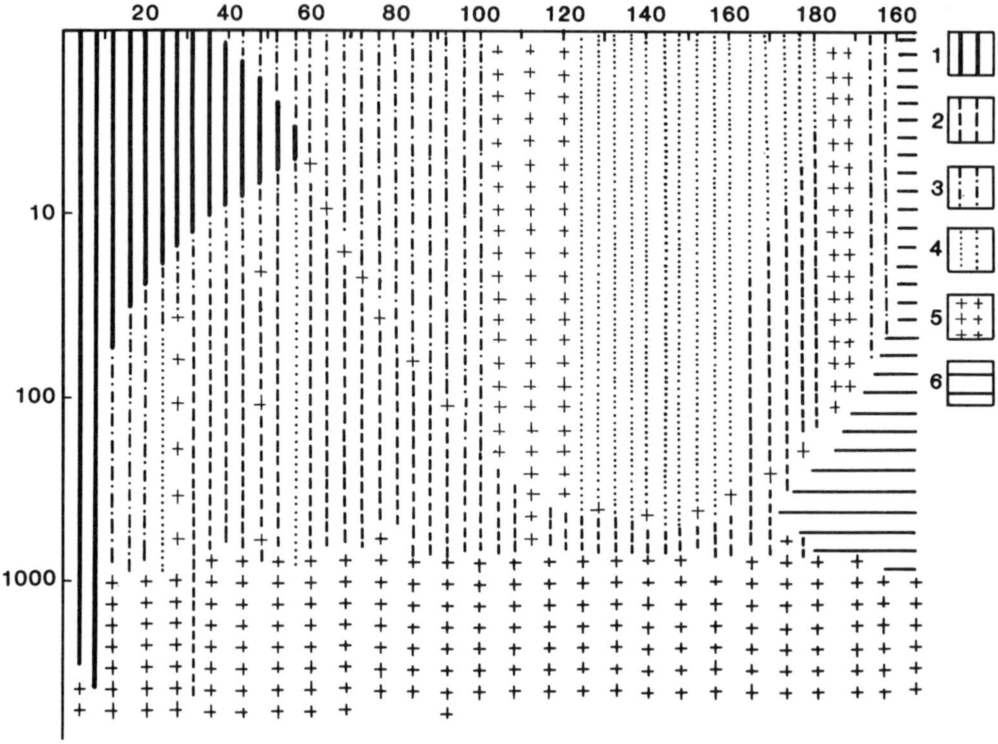


Figure 13-6. Distribution of hydrobiocomplexes in the seas of the Arctic Ocean at different depths. Ordinate: depth (on a logarithmic scale); abscissa: correlation (%) or predominant biogeographic groups: Barents Sea; Kara Sea and Laptev Sea; East Siberian Sea and Chukchi Sea. 1, Species of Atlantic high boreal hydrobiocomplex; 2, species of Atlantic-Arctic hydrobiocomplex; 3, wide-spread boreal-Arctic species; 4, Pacific boreal–Artic species; 5, species of Arctic hydrobiocomplex; 6, species of Pacific high boreal hydrobiocomplex.

cific boreal–Arctic taxa as well. Species indicator of Surface Arctic water in both sectors are numerous Arctic taxa distributed chiefly on the shelf; they include circumpolar, Eurasian, and Amerasian species.

The biocenoses of the Deep Arctic Water Mass is formed primarily by bathyal and abyssal Arctic species, such as *Cryonella wyville-thomsoni, Lunatia bathibii, Turrisipho lachesis, Lissospira profunda, Ankes unduliscuIpta, Tacita danielseni,* and *Mohnia mohni.* A number of species, characteristic of the deep Eurasian subbasin, have not been found to date in the Canada Basin proper.

Gastropods do not play a major role in most ecosystems, being generally detritophages, necrophages, or predators, dominated in the upper parts of the shelf by autotrophic organisms and at depth by sestonophages in their production and biomass. The history of the gastropod's evolution in the Arctic is briefly described in Chapter 10.

Thorson (1935, 1936, 1941, 1944) has shown that most high-latitude prosobranchs do not possess a pelagic stage in their ontogenesis development and their eggs are rich in nutrients. One might assume that ontogenesis does not compensate for the restrictive effect of low temperatures and rigorous trophic conditions on their growth. However, as indicated by the study of growth of several prosobranch

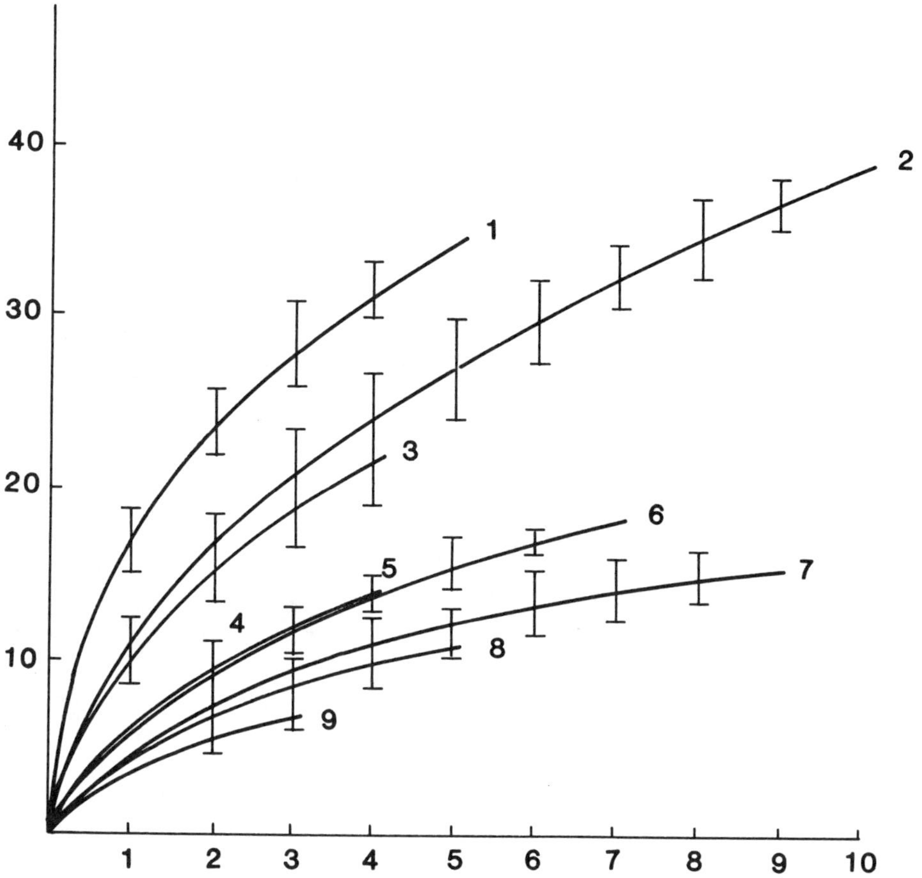


Figure 13-7. Linear growth of individuals of several species in the Arctic Ocean. Ordinate: height of shell in mm; abscissa: age in years. 1, *Buccinum cyaneum*; 2, *Tachyrchynchus erosus*; 3, *T. reticulatus*; 4, *Lacuna neritoidea*; 5, *Epheria vincta*; 6, *Littorina saxatilis*; 7, *Margarites groenlandica umbilivalis*; 8, *Lepeta coeca*, *Problacmaea rubella*.

species in the Arctic (Fig. 13-7), this is not the case, their growth rate being similar to that in species characteristic of temperate latitudes (Golikov, 1976, 1980). The growth rate of prosobranch species is proportional to the maximum size attained by their individuals varying with environmental conditions within the genotype range. The significant decline in growth rate of organisms under Arctic conditions may be the result of a highly efficient utilization of the assimilated food for growth at low temperatures, partly due to the high lipid content (Hochachka and Somero, 1971; Golikov, 1976*b*).

Further study of the Arctic prosobranchs will provide more detailed information on their ecology and further clues to the problems presented in this chapter.

# APPENDIX

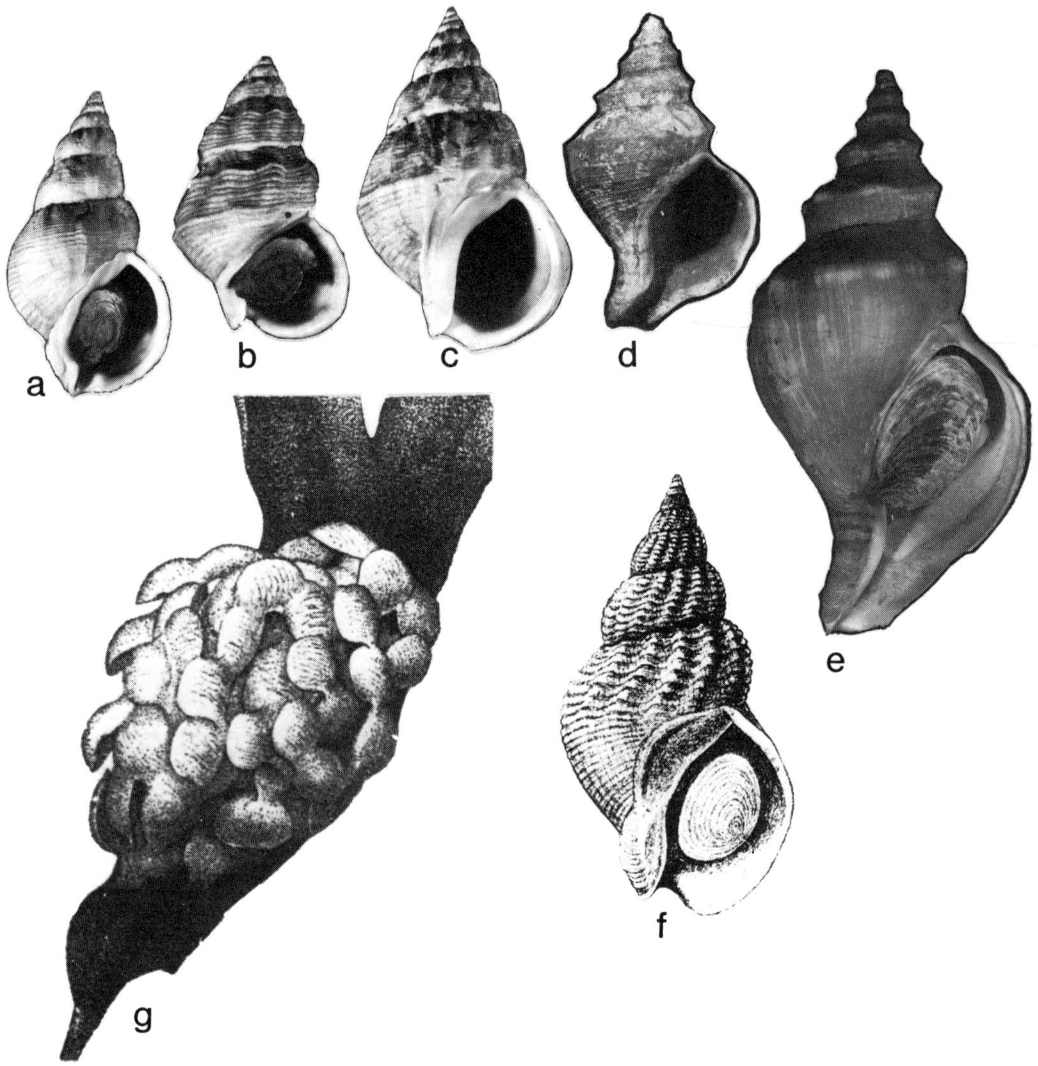


Plate 13-1

*a, b, c.* *Buccinum glaciale L.* Kara Sea, 40 m water depth; substrate: sandy clay and gravel.

*d.* *Neptunea communis* var *borealis.* Kara Sea, 80 m water depth; substrate: clay and rock fragments.

*e.* *Neptunea ventricosa.* Laptev Sea, 50 m water depth; substrate: clay, sand, and rock fragments.

*f.* *Buccinum undatum L.,* shell. Barents Sea, 100 m water depth; substrate: clay and sand.

*g.* Egg clusters. Barents Sea, 100 m water depth; substrate: clay and sand.

## REFERENCES


Aurivillius, C. 1887. Ofversigt ofver de at Vega—Expeditionen insamlade Arktiska Hafsmollusker. II Placophora och Gastropoda. *Vega—Expeditionen Vetenskapliga Jakttagelser* **4**:113–183.

Bousfield, E. L. 1960. Canadian Atlantic sea shells. *Nat. Mus. Canada, Ottawa,* pp. 1–72.

Brögger, W. C. 1900–1901. Om de senglaciale og postglaciale niváforandringer; Kristianiafeltet. Norges Geol. undersögelse. *Kristiania* **31**:731p.

Derjugin, K. M. 1950. Novye dannye po sistematike, morfologii i biogeografii roda Velutina Flem (Mollusca, Gastropoda, Lamellariidae). *Issled. Dal'nevostochnykh Morei SSSR* **2**:2–27. (In Russian.)

Filatova, Z. A., and V. I. Zatsepin. 1948. Klass Gastropoda—Bryukhonogiye mollyuski. In N. S. Gayevskaya (ed.), *Opredelitel' fauny i Flory Severnykh Morei SSSR.* Izd. Sovetskaya Nauka, Moscow, pp. 358–401. (In Russian.)

Friele, H. 1882–1901. The Norwegian North-Atlantic expedition 1876–1978. (Zoology Mollusca, I–III Christiania), 200p.

Galkin, Yu. I. 1955. Bryukhonogiye mollyuski trokhidy dal'nevostochnykh i severnykh morei SSSR (sem. Trochidae). Opredeliteli po faune SSSR. *Izd Zool. Inst. AN SSSR.* **57**:132p. (In Russian.)

Gladenkov, Yu. B. 1978. *Morskoi verkhnii Kainozoi Severnykh Raiyonov.* Nauka, Moscow, 194p. (In Russian.)

Golikov, A. N. 1963. Bryukhonogiye mollyuski roda Neptunea Bolten. *Fauna SSSR Mollyuski* **5**(1): 217p. (In Russian.)

Golikov, A. N. 1973. Species and speciation in poikilothermal animals. *Marine Biol.* **21**:257–268.

Golikov, A. N. 1975. O parallelizme i konvergentsii v razvitii morskikh ekosistem. In *Biologiya shel'fa.* Vladivostok, pp. 30–31. (In Russian.)

Golikov, A. N. 1976*a*. O kolichestvennykh zakonomernostyakh protsessa divergentsii. In *Gidrobiol. Issled. Samoochishcheniya Vodoyomov.* Leningrad, 2, pp. 90–96. (In Russian.)

Golikov, A. N. 1976*b*. Nekotorye zakonomernosti rosta i izmenchivosti na primere mollyuskov. In *Gidrobiol. Issled. Samoochischeniya Vodoyomov.* Leningrad, pp. 97–118. (In Russian.)

Golikov, A. N. 1980. Mollyuski Buccininae Mirovogo okeana. *Fauna SSSR* **5**(2):466p. (In Russian.)

Golikov, A. N., and O. A. Scarlato. 1977. Sostav, raspredeleniye i ekologiya bryukhonogikh i dvustvorchatykh mollyuskov u arkhipelaga Zemlya Frantsa Iosifa. Issled. *Fauny Morei* **14**(22):313–390. (In Russian.)

Golikov, A. N., and Ya. I Starobogatov. 1975. Systematics of Prosobranch Gastropods. *Malacologia* **15**(1):185–232.

Gorbunov, G. P. 1946. Donnoye naseleniye Novosibirskogo melkovod'ya i tsentral'noi chasti Severnogo Ledovitogo okeana. *Trudy Dreif. Exp. Glavsevmorputi na 1/p "G. Sedov" 1937–1940* **3**:30–138. (In Russian.)

Harmer, F. W. 1994. *The Pliocene Mollusca of Great Britain,* Pt. 1. London 200p.

Herzenstein, S. M. 1885. Materialy k faune Murmanskogo berega i Byelogo morya. I. *Mollyuski, Tr. sPb. Obshchestva Estestvoispyt* **2**:635–814. (In Russian.)

Hochachka, P. W., and G. N. Somero. 1971. Biochemical adaptation to the environment. In *Fish Physiology,* vol. 6. New York and London, pp. 99–156.

Jaccard, P. 1901. Distribution de la flore alpine dans le bassin de Drunses et dans quelques regions voisines. *Soc. Vandoise, Sc. Natur Bull.* **37**:241–272.

Jaccard, P. 1912. The distribution of the flora in the alpine zone. *New Phytol.* **11**:37–50.

Knipowitsch, N. 1900. Zur Kentniss der geologischen Geschichte der Fauna des Weissen und des Murman Meeres (postliocaene Mollusken und Brachiopoden). *Verh. Russ. Mineral. Ges.,* ser. 2, **38**(1):169p.

Knipowitsch, N. 1901. Mollusca und Brachiopoda. Zoologische Ergebnisse der russischen Expeditionen nach Spitzbergen. Ezhegodn. *Zool. Muz. Akad. Nauk* **6**(2):435–558.

MacGinitie, G. E. 1955. Distribution and ecology of the marine invertebrates of Point Barrow, Alaska, *Smithsonian Misc. Colln.* **128**(9):201p.

Macpherson, E. 1971. The marine molluscs of Arctic Canada. *Biol. Oceanography, Ottawa* **3**: 149p.

Middendorff, A. G. 1849*a*. *Beiträge zu Einer Malacozoologia Rossica Abt* **2**(3):1–187.

Middendorff, A. G. 1849*b*. Aufzahlung und Beschreibung der zur Meeresfauna Russlands gehörigen zweischaler. *Beiträge zu Einer Malacozoologia Rossica Abt* **6**:1–80.

Middendorff, A. G. 1851. *Mollusken, Reise in den äusserten Norden und Osten Sibiriens Während der Jahre 1843 und 1844.* St.-Petersb., 2, pp. 164–464.

Mörch, O. A. L. 1875. *Prodromus Faunae Molluscorum Gröenlandieae.* Rink's Danish Gröenland, London, 9, pp. 124–135.

Sars, G. O. 1878. Mollusca regions arctica norvegia. Bidrag til Kundskaben om Norges Arktistke Fauna. *Christiania* **1**:466p.

Thorson, G. 1935. Studies on the egg-capsules and development of Arctic marine Prosobranchs. *Medd. Grønland* **100**(5):1–71.

Thorson, G. 1936. The larval development, growth and metabolism of Arctic marine bottom Invertebrata, compared with those of other seas. *Medd. Grønland* **100**(6):1–155.

Thorson, G. 1941, Marine Gastropoda, Prosobranchiata. *Zool. Iceland* **4**(60):1–50.

Thorson, G. 1944. The zoology of East Greenland Marine Gastropoda Prosobranchiata. *Medd. Grønland* **121**(13):1–181.

Thorson, G. 1951. Scaphopoda, Placophora, Solenogastres, Gastropoda, Prosobranchiata, Lamellibranchiata. The Godthaab Expedition 1928. *Medd. Grønland* **8192:** 1–117.

# 14: Arctic Ocean Bryozoa

Valentina I. Gontar and
Nina V. Denisenko

## INTRODUCTION

Bryozoans (Bryozoa, Polyzoa, Ectoprocta) are aquatic colonial animals. A few kinds live in fresh water (class Phylactolaemata) while the majority are marine (class Gymnolaemata and Stenolaemata). The phylum contains over 4,000 living species and about 15,000 extinct species. The colonies are formed by asexual reproduction from one individual, the ancestrula, that arises through metamorphosis of the larva developed by sexual reproduction. The bryozoan colony, the zoarium, is almost always firmly attached to the substratum and can be ramified, lobed, fan-shaped, and very frequently have the appearance of crusts covering the substratum. The size of colonies vary from several millimeters to several decimeters. The colonies consist of individuals, the zooids. Bryozoans are bilaterally symmetrical, coelomic, protostomian animals. As a result of their sessile and colonial way of life, their organization has undergone profound changes in different subclasses.

*Note:* This chapter has been reviewed, edited, and rewritten by the Editor, Yvonne Herman.

The structure of bryozoans is as follows: Every zooid bears a circlet of tentacles. The mouth is surrounded by tentacles. The anus is located near the mouth, but outside the crown of tentacles. In marine bryozoans, the lophophore, a fleshy structure that surrounds the mouth, is ring-shaped. When the anterior part of the body is withdrawn, the tentacles are also retracted into the body known as the tentacle sheath.

Marine bryozoans have neither heart nor vascular system. The inner zooidal cavity is a coelom that is separated from the lophophore cavity by an incomplete partition. The nervous system comprises a nerve ganglion that lies between the mouth and the anus and a plexus in the body wall. From the nerve ganglion, nerves radiate to the tentacles and other organs. The muscular system consists of circular and longitudinal muscles, namely, circular sphincter muscles, transverse parietal muscles, and strong retractor muscles. Reproduction can be both asexual and sexual; colonies are formed mainly by asexual reproduction. Most bryozoans are hermaphrodites.

The zooidal body cavity contains gonads of both sexes that develop simultaneously, although some species show a tendency for protandry (functional male gonad maturation precedes female gonad maturation). Spermatozoa are released from the zooidal cavity either through the terminal pores of tentacles, or through the supraneural pore coelomopore situated at the vertex of the funnel between tentacles. In some species, the coelomopore is transformed into an intertentacular organ in which fertilization occurs. Fertilized eggs can be discharged into the sea, although in many marine taxa their development up to larval stage occurs either in the coelom or the zooid in a special brooding chamber (ovicell). The larvae are close to the trochophore and according to some authors may be significant as a dispersal mechanism. After being set free, the larvae swim for a few hours and eventually settle on a suitable substratum, producing through metamorphosis and budding a primary zooid, the ancestrula, which subsequently initiates a colony.

The above-described complex of organs is termed the polypide; it is encased in the exoskeleton of various shapes, called the zooecium or cystid. It is secreted by the body wall, which consists of cellular layers. The zooidal skeleton can be separated, but its walls are usually in touch with other zooecia or zooecia may have common proximal and distal walls. Connection between separate zooids is through single pores or more complex structures (porous lamellae or septa, porous chamber or dietalia) in the cystid walls. The zooecium bears an orifice through which the anterior end of the polypide is protruded. It can be provided with an operculum. The space between the tips of tentacles withdrawn into the tentacle sheath and the orifice is the atrium or vestibulum. The zooecium can be chitinous, jellylike (other polymer of chitin of Insecta), or calcareous.

In bryozoan colonies the zooids are usually polymorphic, that is, there are in addition to the zooids' typical structures: the autozooids, morphologically modified and physiologically specialized zooids such as the heterozooids, including gonozoids (where maturation of sperm and ooecia occurs), the kenozooids (performing supporting function), the avicularia (defense function), the vibracula (apparently providing water current beneath the colony), and other structures (*Treatise on Invertebrate Paleontology,* 1983). The phylum Bryozoa is divided into three classes.

Class Stenolaemata with 6 orders:

Order Cyclostomata (Paleozoic–Recent)
Order Trepostomata (Paleozoic–Triassic)
Order Cystoporata (Paleozoic)
Order Rhabdomesonata (Paleozoic)
Order Cryptostomata (Paleozoic)
Order Fenestrata (Paleozoic)

Class Gymnolaemata with 2 orders:

Order Ctenostomata (Paleozoic–Recent)
Order Cheilostomata (Jurassic–Recent)

Class Phylactolaemata (Carboniferous–Recent)

Bryozoa are known to have been widespread in the Paleozoic seas (7 orders) and are valuable in stratigraphy; their ancestry and early evolutionary relationships are difficult to infer. Skeleton-bearing bryozoans are believed to have evolved during the lower Ordovician (Baltic province), although the high level of organization of the early Ordovician bryozoans suggests that they might have evolved during the early Cambrian. The evolutionary development of skeleton-bearing Bryozoa is divided into large stages, their boundaries corresponding to boundaries of geologic epochs defined by major evolutionary changes that led to a change in the rank of orders (Morosova and Viskova, 1977). The first stage encompasses the period from early Ordovician up to late Permian; the second stage occurred during the Mesozoic–Cenozoic. During the first stage, 7 orders evolved simultaneously with an abrupt change of genera and families at three boundaries: The first boundary corresponds to the end of the Silurian; the second, to the end of the Devonian; and the third, to the end of the Permian. The second stage covers the Mesozoic–Cenozoic eras. It is noteworthy that the currently accepted systematics of the Bryozoa is far from being adequate, as the paleontological classification is based only on the morphology of skeletal elements.

The Permian is the final period of evolution of Paleozoic bryozoans. Permian taxa were widespread in the World Ocean. However, the Arctic region was rich in bryozoans between the Ordovician and Permian (Astrova, 1965; Morosova, 1981; Morosova and Kruchinina, 1986; Nekchorosheva, 1965, 1966, 1970, 1976). About 300 species described by these authors are used for the stratigraphic division of Paleozoic deposits of this region as well as for dividing the Paleozoic Arctic seas into biogeographic regions.

During the Dzchulphinian and Dorosham times, the bryozoan faunas were declining. They survived only in the Tethys Sea. Nearly all bryozoans had disappeared by the end of the Permian. Only 4 genera (Discritella, Reptonodicava, Paralioclema, and Pseudobatostomella) with few species characterized by simple morphological organization continued to evolve in the Triassic seas (known from the Primorsky Territory and the northeastern USSR) together with a small number of Cyclostomata. The Mesozoic–Cenozoic stage commences in Triassic and is characterized by a different type of bryozoans, which must have given rise to all the younger taxa. Favorable environmental conditions in the Middle Jurassic seas, where the Cyclostomata were probably the only bryozoans, may have triggered their rapid speciation. During the Late Jurassic and throughout the entire Cretaceous period, the Cyclostomata, which were the dominant bryozoans, became ubiquitous in the World Ocean. The penetration of the Cyclostomata to relatively great depths in Campanian and Maestrichtian seas must have been related not only to extensive marine transgressions but also to the rapid development of the Cheilostomata that had evolved in early Jurrasic time (*Pyriporopsis portlandensis*). The Cyclostomata are represented today by relicts of Mesozoic seas.

Another order of recent bryozoans is the Ctenostomata. The fossil record of this group is very scanty because they lack calcareous skeleton; to date 55 genera have been described. They belong to 23 families, 3 of which are Paleozoic while 15 are recent. Most recent species belong to the order Cheilostomata (80 percent of the total number of recent species). In the Early Cretaceous, only the Anasca suborder was represented. Two other suborders, Ascophora and Cribrimorpha, evolved at the beginning of the Late Cretaceous. The end of the Late Cretaceous and the beginning of the Cenozoic are characterized by the extinction of most cribrimorphs and by the diversification of the Anasca and Ascophora. The evolution of bryozoans suggests a parallel evolution of colonies of different phylogenetic branches of the phylum owing to polymorphism. This provided bryozoans with a high degree of adaptation to environmental factors with evolutionary plasticity and led to their flourishing in modern oceans and seas.

Bryozoans, represented by three orders in the World Ocean—the Cyclostomata, Ctenostomata, and Cheilostomata—are a ubiquitous group of aquatic invertebrates. They occur in every climatic zone and inhabit depths ranging from the littoral to the abyssal zones. Bryozoans constitute a part of many marine bottom biocenoses and play an important role in the total economy of the sea. Colonies live on rocks, algae, hydroids, worm tubes, mollusk shells, crab tests, ascidians, corals, colonies of other bryozoans, fragments of sea urchins' skeletons, and also on various sunken objects such as buoys and nets.

## ARCTIC BRYOZOA

The distributional pattern of various biogeographic groups shows that all Arctic seas are dominated by boreal–Arctic* species. The western part of the Barents Sea and areas of the Atlantic Ocean adjoining the Arctic Basin are dominated by Atlantic boreal† species. Boreal species† are replaced by the Arctic taxa‡ in the eastern Barents, Kara, Laptev, and East Siberian seas. Arctic species‡ are replaced by boreal Pacific taxa in the eastern Chukchi Sea and adjoining part of the Bering Sea. North American Arctic waters are dominated by Arctic species. Most species are confined to the Eurasian and Amerasian subbasins with the boundary in the area of the New Siberian Islands. Eurybathyal species are predominant in the western Barents, and the Chukchi Sea fauna exhibits the lowest degree of similarity with the fauna of Siberian seas and the Arctic–American waters.

The majority of bryozoans inhabit rocky substrates. They comprise many marine benthic biocenoses and play an important part in the overall trophic relations of the sea. They are well preserved in sediments of different geological age and are very important in biostratigraphic zonation. There are numerous reviews dealing with the bryozoan fauna of the Arctic Basin (Packard, 1867; Smitt, 1878, 1879; Vigelius, 1882, 1884; Waters, 1900, 1904; Nordgaard, 1906, 1912, 1918, 1923, 1929; Norman, 1906; Osburn, 1913, 1932, 1936; Kluge, 1962; Gostilovskayja, 1978). While the faunal composition in this region has been adequately studied, knowledge of their ecology remains incomplete. Nevertheless, the literature and material collected during the last 250 years and deposited at the Zoological Institute of the USSR Academy of Sciences allows us to summarize their distribution in the Arctic Ocean.

Here we follow the classification of Kluge (1962) with several additions based on recent investigations of Soviet and foreign scientists. In the Arctic Ocean, the phylum is represented by 3 orders, 11 suborders, 41 families, 89 genera, 340 species including 40 varieties and 1 form.

Variations in species composition in different seas of the Arctic Ocean have been analyzed using cluster analysis and estimating the degree of similarity according to Jaccard (1901),

$$j = \frac{c \times 100\%}{D_1 + D_2 - c}$$

and with the inclusion method

$$j = \frac{c \times 100\%}{D_{\min}}$$

Results are presented in Tables 14-1 and 14-2 (see pages 349–363) and Figure 14-1. As shown in Table 14-1, the decline in number of species occurs from boreal Atlantic waters to the Arctic proper. There is a significant difference between the bryozoan fauna of the western Barents Sea of boreal character, and faunas of eastern Barents Sea and adjoining seas. The groups showing the closest affinity are from the eastern Barents, Kara, and Lap-

*Boreal–Arctic species (for their boundaries see map) occur in the boreal and Arctic biogeographic regions.
†Boreal species including amphiboreal, widespread boreal, low boreal, and high boreal taxa are distributed within the boreal biogeographic regions.
‡Arctic species include Eurasian species distributed in the Arctic Ocean off the European and Asian coasts; Euroamerican or Amereuropean species distributed in the Arctic Ocean off the Asian and North American coasts. Circumpolar species are distributed in the Arctic Ocean off the coasts of Europe, Asia, North America, Greenland, Iceland, and Arctic Ocean islands. Amerasian species are distributed in the Arctic Ocean off the Asian and North American coasts. American species are distributed in the Arctic Ocean off the North American coasts.

tev seas. The fauna of these seas contain a larger proportion of Atlantic taxa than of Pacific elements. There is a significant difference in composition between the fauna of the East Siberian Sea, comprising species of Atlantic and Pacific origin, and the fauna of the Eurasian sector of the Arctic and Chukchi seas.

| $D_1$ | $D_2$ | $D_3$ | $D_4$ | $D_5$ | $D_6$ | |
|---|---|---|---|---|---|---|
| 193 | 68 | 63 | 60 | 64 | 62 | $D_1$ |
| 47 | 211 | 93 | 80 | 76 | 84 | $D_2$ |
| 45 | 78 | 186 | 90 | 85 | 94 | $D_3$ |
| 30 | 40 | 55 | 119 | 72 | 68 | $D_4$ |
| 31 | 35 | 46 | 53 | 110 | 56 | $D_5$ |
| 25 | 35 | 45 | 53 | 34 | 93 | $D_6$ |

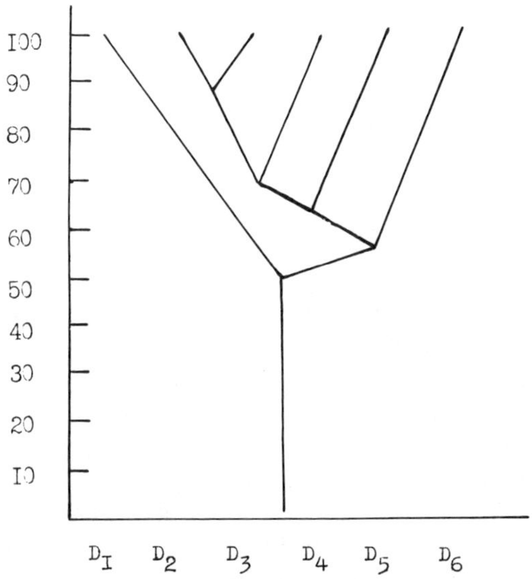


Figure 14-1. Dendrogram expressing similarities; similarities (bottom) and degree of inclusion (top) of bryozoan fauna (%) in Arctic seas. $D_1$, western part of the Barents Sea; $D_2$, eastern part of the Barents Sea; $D_3$, Kara Sea; $D_4$, Laptev Sea; $D_5$, Eastern Siberian Sea; $D_6$, Chukchi Sea.

Correlation of different biogeographic groups in the Arctic seas (Fig. 14-2) shows that the boreal–Arctic species are the most abundant group in all seas. Boreal species predominate in the North Atlantic and western Barents Sea as well as in the eastern Chukchi Sea. The number of Arctic species exceeds that of boreal species in the eastern Barents Sea and western Chukchi Sea, delineating the boundaries of the Arctic region. Major faunal differences exist between the Eurasian and the American sectors of the Arctic Basin; the boundary between the two regions is the New Siberian shoals. Boreal–Arctic species are predominant here; however, there is a decline in species numbers in this region. The western part of the Barents and eastern part of the Chukchi seas may be considered a transitional zone between the Atlantic and the Pacific boreal biogeographic regions and the Arctic region (Fig. 14-3).

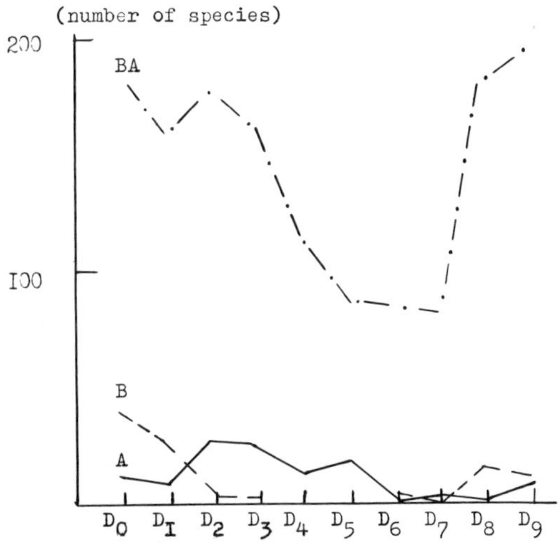


Figure 14-2. Depth distribution of different biogeographic groups of bryozoan species in the Arctic Seas. Legend as in Figure 14-1. $D_0$, northern part of the Atlantic Ocean; $D_6$, eastern part of the Chukchi Sea; $D_7$, western part of the Chukchi Sea; $D_8$, northern part of the Pacific Ocean; $D_9$, Arctic coastal waters of North America; BA, boreal–Arctic species; B, boreal species; A, Arctic species.

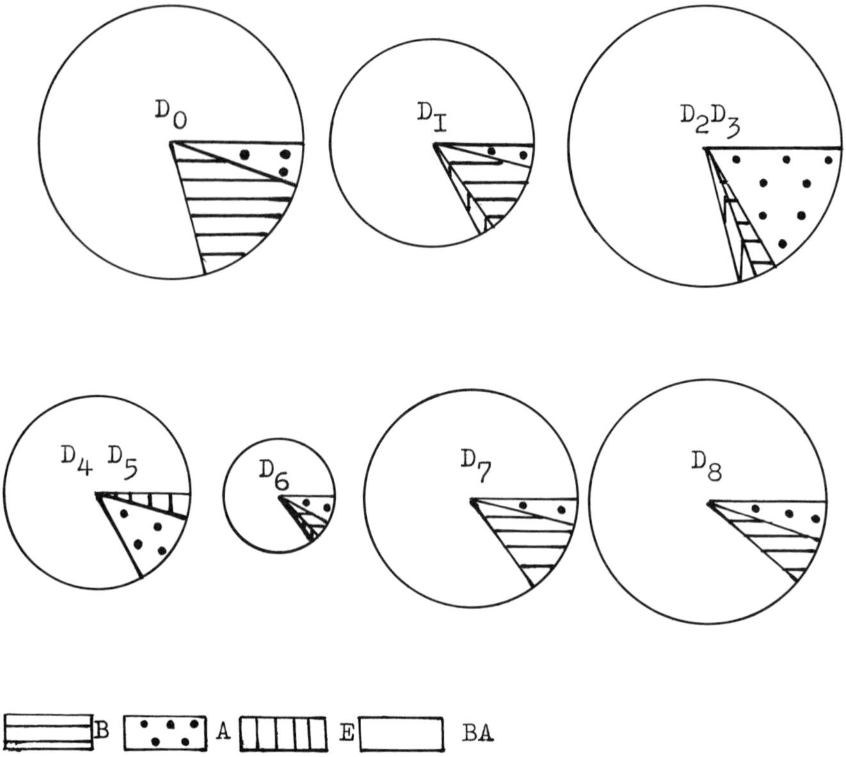


Figure 14-3. Biogeographic distribution of the bryozoan fauna in the Arctic Seas. Legend as in Figures 14-1 and 14-2 except $D_7$, northern part of the Pacific Ocean; $D_8$, Arctic coastal waters of North America; E, endemic species.

The distribution of bryozoans in the Arctic region (Fig. 14-4) indicates that species number is highest between 50 m and 100 m depths. Valuable information concerning the vertical distribution of water masses is provided by analyses of various biogeographic groups (Fig. 14-5). Boreal–Arctic species are predominant to a depth of 500 m with a maximum at 50–100 m depths. The number of boreal species gradually declines with increasing depth, increasing slightly at 50–100 m depths. Species endemic to the Arctic region occur between 0 m and 250 m depths, reaching maximum diversity between 50 m and 100 m depths. The number of Arctic species exceeds that of boreal taxa between 40 m and 450 m depths.

Bryozoa colonies live on a variety of substrates (Fig. 14-6). It should be emphasized that although the number of Bryozoa in different facies depends on various factors, including temperature and currents, generally the most important factor is believed to be the substrate (Hayward and Ryland, 1979). In our opinion, the low number of species on rocky substrates and their relatively high number on sandy and argillaceous facies is principally due to the total area occupied by these types of facies in the Arctic seas.

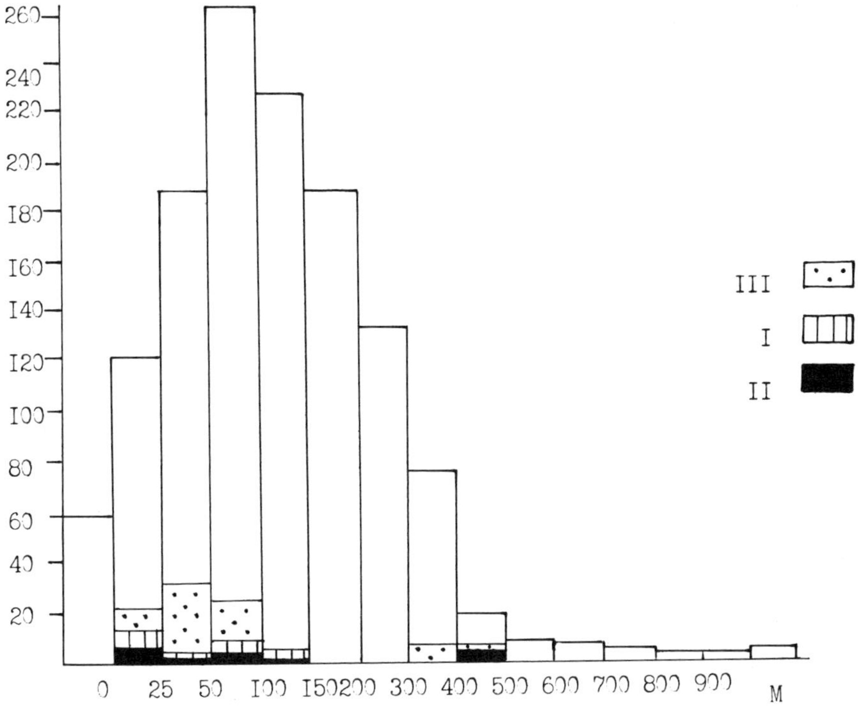


Figure 14-4. Depth distribution of species. Patterns indicate species confined to certain depth ranges. I, Barents Sea; II, western part of the Barents Sea; III, other Arctic seas. Ordinate: number of species; abscissa: water depth in meters.

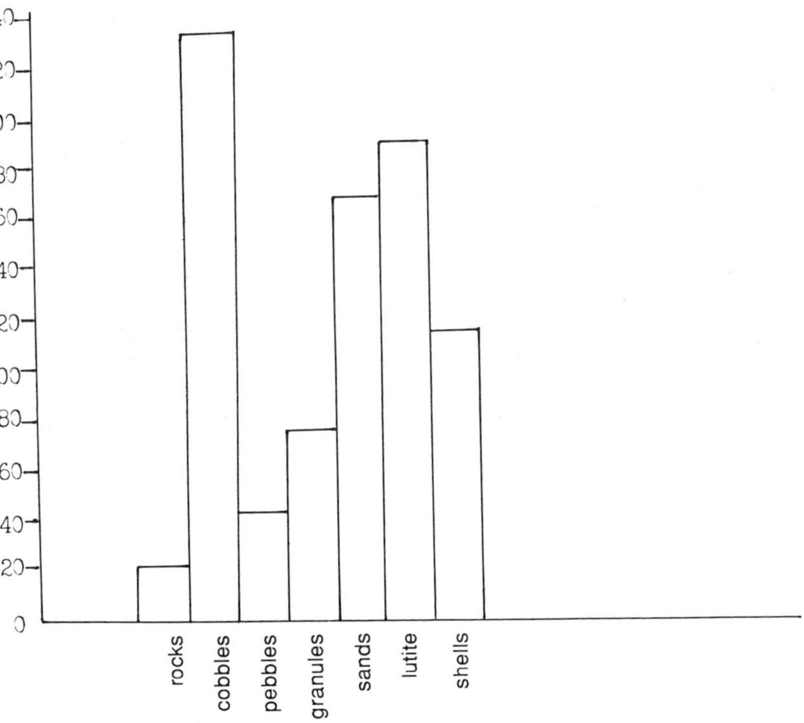


Figure 14-5. Distribution of bryozoans on various substrates. Ordinate: number of species.

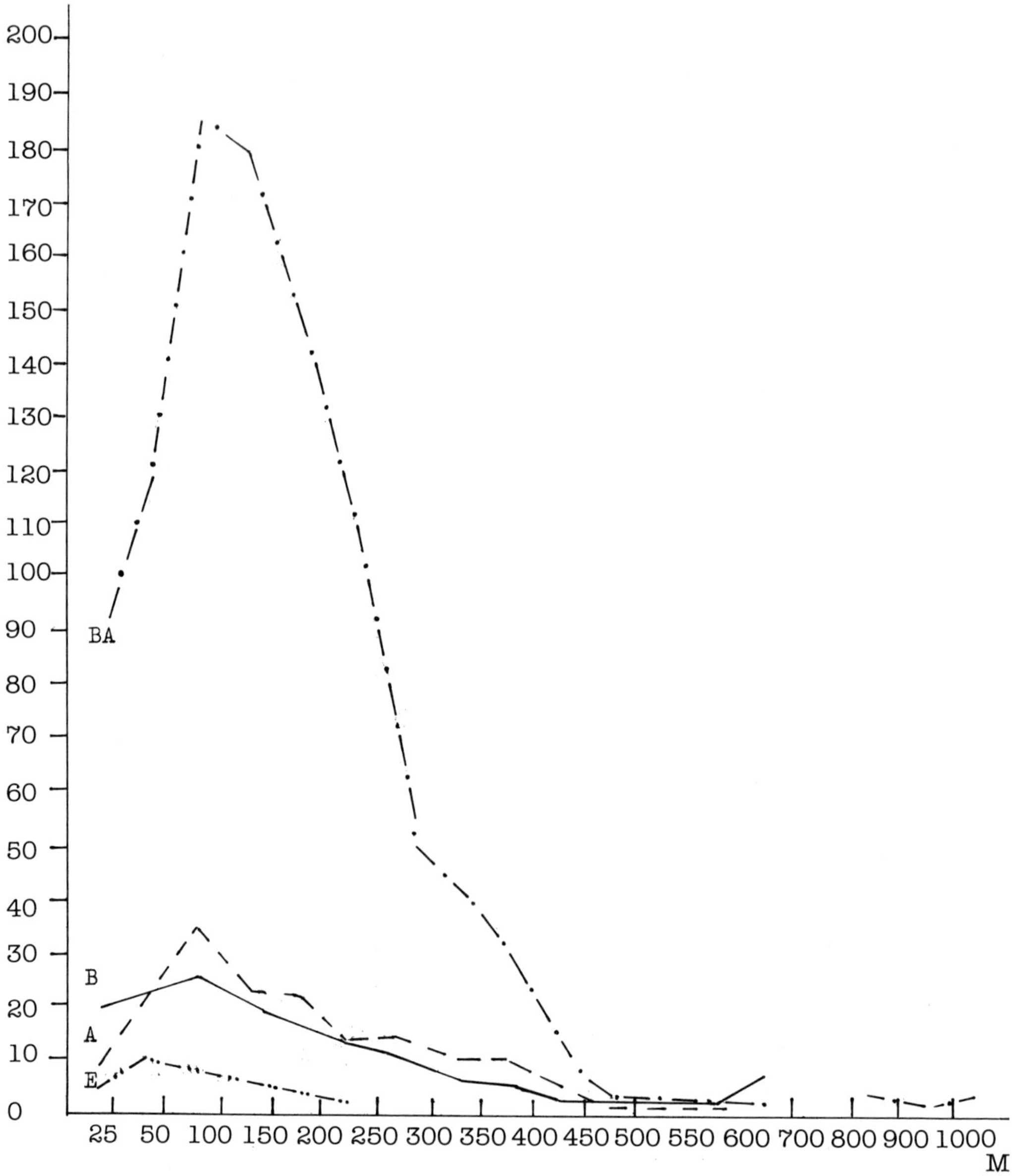


Figure 14-6. Depth distribution of species belonging to different biogeographic groups. Ordinate: number of species. Legend as in Figure 14 2.

Table 14-1. Distribution of Bryozoa in the Arctic Ocean

| Species | Biogeography | 1 | 2 | 3 | 4 | 5 | 6 | 7 | 8 | 9 | 10 | 11 | 12 | 13 | 14 | 15 | 16 | 17 | 18 | 19 | 20 | 21 | 22 | 23 | 24 |
|---|---|---|---|---|---|---|---|---|---|---|---|---|---|---|---|---|---|---|---|---|---|---|---|---|---|
| | | Species Presence | | | | | | | | | | | | | | | | | | | | | | | |
| *Parasmittina trispinosa* (Johnston) | sub.-b. | | | | | | | + | | + | | | | | | | | | | + | | + | | + | |
| *Schizomavella auriculata* (Hassall) | sub.-b. | | | | + | | + | + | + | + | + | + | | + | | | | + | | | | | | | |
| *Proboscina major* (Johnston) | atl.sub.-b. | | | | | + | | | | | | | | | | | | | | + | | | | | |
| *Tubulipora dilatans* (Johnston) | atl.sub.-b. | | | | | | | | + | | | | | | + | | | | + | + | + | | | | |
| *Tubulipora liliacea* (Pallas) | atl.sub.-b. | | | | | | | + | | + | + | | | | | | | + | | + | | | | | |
| *Crisidia cornuta* (Linnaeus) | atl.sub.-b. | | | | | | | | | + | | | | + | | | | | | + | + | | | | |
| *Diplosolen obelia* (Johnston) | atl.sub.-b. | | | | + | | + | + | | + | + | + | | | | | | | | + | | | | | |
| *Stegohornera violacea* (M. Sars) | atl.sub.-b. | | | | | | | | | + | | | | | | | + | | | + | | | | | |
| *Bowerbankia pustulosa* (Ellis et Solander) | atl.sub.-b. | | | | | | | | | | | | | | | | + | | | + | | | | | |
| *Bowerbankia imbricata* (Adams) | atl.sub.-b. | | | | | | | | | | | | | | | | | | + | + | | | | | |
| *Callopora aurita* (Hincks) | atl.sub.-b. | | | | | | + | | | | | | | | | | + | | + | + | | | | | |
| *Tessarodoma boreale* (Busk) | atl.sub.-b. | | | | | | | | | | + | | | | | | | | | + | + | + | | | |
| *Tubulipora penicillata* (Fabricius) | atl.b. | | + | + | | | | | | | + | | | | | | | | | + | | | | | |
| *Bowerbankia gracilis* Leidy | atl.b. | | | | | | | | | | | | | | | | | | + | + | | | | | |
| *Valkeria uva* (Linnaeus) | atl.b. | | | | | | | | | | | | | | | | | | + | + | | | | | |
| *Amphiblestrum flemingii* (Busk)) | atl.b. | | | | | | + | | | | | | | | | | | | | + | | | | | |
| *Amphiblestrum solidum* (Packard) | atl.b. | | | | | | | | | | | | | | | | | | | + | | | | | |
| *Chartella barleei* (Busk) | atl.b. | | | | | | | | | | | | | | | | | | | + | | | | | |

*Note:* sub.-b., subtropical boreal; atl.b., Atlantic boreal widespread; atl.sub.-b., Atlantic subtropical boreal; atl.h.-b., Atlantic high boreal; amph.b., amphiboreal; b.-a., boreal–Arctic widespread; b.-a.atl.circ.-p., boreal–Arctic Atlantic circumpolar; b.-a.atl.eurasian, boreal–Arctic Atlantic Eurasian; b.-a.atl.am.-eur., boreal–Arctic Atlantic American–European; b.-a.pac.circ.-p., boreal–arctic Pacific circumpolar; b.-a-pac.amer.-as, boreal–Arctic Pacific Amerasian; b.-a.pac.eur.-as., boreal–Arctic Pacific Eurasian; arc.circ.-p., Arctic circumpolar; arc.eur.-as., Arctic Eurasian; arc.amer.-as., Arctic Amerasian; arc.eur.-am., Arctic Euroamerican; arc.amer., Arctic American; pac.b., Pacific boreal.

+ = species in: 1, northwestern coast of Alaska; 2, northern coast of North America; 3, the Archipelago of Canadian Islands; 4, Baffin Bay; 5, Davis Strait; 6, Hudson Bay; 7, Gulf of Labrador; 8, Newfoundland; 9, St. Lawrence Gulf; 10, western Greenland; 11, eastern Greenland; 12, Greenland Sea; 13, Iceland; 14, Jan Mayen; 15, Norwegian Sea; 16, North Sea; 17, northern Norway; 18, White Sea; 19, Barents Sea; 20, Kara Sea; 21, Laptev Sea; 22, East Siberian Sea; 23, Chukchi Sea; 24, Bering Strait.

This table is based on the authors' present data and the literature listed.

Table 14-1. *(continued)*

| Species | Biogeography | Species Presence | | | | | | | | | | | | | | | | | | | | | | | |
|---|---|---|---|---|---|---|---|---|---|---|---|---|---|---|---|---|---|---|---|---|---|---|---|---|---|
| | | 1 | 2 | 3 | 4 | 5 | 6 | 7 | 8 | 9 | 10 | 11 | 12 | 13 | 14 | 15 | 16 | 17 | 18 | 19 | 20 | 21 | 22 | 23 | 24 |
| *Membraniporella nitida* (Johnston) | atl.b. | | | | | | | | | | | | | | | | + | | | + | | | | | |
| *Cribrilina punctata* (Hassall) | atl.b. | | + | | | | | | | + | + | | | | | | + | | + | + | | | | | |
| *Cribrilina cryptooecium* Norman | atl.b. | | | | | | | | | | | | | | | | | | | + | | | | | |
| *Escharella laqueata* (Norman) | atl.b. | | | | | | | | | | | | | | | | | + | | + | | | | | |
| *Fenestrulina malusii* (Audouin) | atl.b. | | | | | | | | | | | | | | | | + | | | | | | | | |
| *Cellepora pumicosa* (Linnaeus) | atl.b. | | | | | | | | | | | | | | | | | | | + | | | | | |
| *Turbicellepora avicularis* (Hincks) | atl.b. | | | | | | | | | | | | | | | | | | | + | | | | | |
| *Proboscina gracilis* (Kluge) | atl.h.-b | | | | | | | | | | | | | | | | | | | + | | | | | |
| *Tubulipora marisalbi* Gostilovskaja | atl.h.-b. | | | | | | | | | | | | | | | | | | + | | | | | | |
| *Tubulipora fructuosa* Gostilovskaja | atl.h.-b. | | | | | | | | | | | | | | | | | | + | | | | | | |
| *Tubulipora smitti* Gostilovskaja | atl.h.-b. | | | | | | | | | | | | | | | | | | | + | | | | | |
| *Tubulipora uniformis* Gostilovskaja | atl.h.-b. | | | | | | | | | | | | | | | | | | | + | + | | | | |
| *Tubulipora aperta* Harmer | atl.h.-b. | | | | | | | | | | | | | | | | | | + | | | | | | |
| *Tubulipora murmanica* Kluge | atl.h.-b. | | | | | | | | | | | | | | | | | | | + | | | | | |
| *Defrancia lucernaria* var. prolifera Kluge | atl.h.-b. | | | | | | | | | | | | | | | | | | + | + | | | | | |
| *Alcyonidium excavatum* Hincks | atl.h.-b. | | | | | | | | | | | | | | | | | | | + | | | | | |
| *Alcyonidium irregulare* Kluge | atl.h.-b. | | | | | | | | | | | | | | | | | | | + | | | | | |
| *Alcyonidium proboscideum* Kluge | atl.h.-b. | | | | | | | | | | | | | | | | | | | + | | | | | |
| *Arachnidium simplex* Hincks | atl.h.-b. | | | | | | | | | | | | | | | | | | | + | | | | | |
| *Arachnoidea barentsia* Kluge | atl.h.-b. | | | | | | | | | | | | | | | | | | | + | | | | | |
| *Electra crustulenta* var. *catenulariasimilis* Kluge | atl.h.-b. | | | | | | | | | | | | | | | | | + | | | | | | | |
| *Tegella kildinensis* Kluge | atl.h.-b. | | | | | | | | | | | | | | | | | | | + | | | | | |
| *Callopora derjugeni* (Kluge) | atl.h.-b. | | | | | | | | | | | | | | | | | | | + | | | | | |
| *Ramphonotus minax* (Busk) | atl.h.-b. | | | | | | | | | + | | | | + | | | | | | + | | | | | |
| *Larnacicus corniger* (Busk) | atl.h.-b. | | | | | | | | | | | + | | | | | | | | + | | | | | |
| *Megapora ringens* (Busk) | atl.h.-b. | | | | | | | | | | | | | | | | | | | + | | | | | |

| | | | | | | | | | | | | | | | | | | | | | | | | | |
|---|---|---|---|---|---|---|---|---|---|---|---|---|---|---|---|---|---|---|---|---|---|---|---|---|---|
| *Reussina impressa* (Reuss) | atl.h.-b. | | | | | | | | | | | | | | | | | | | + | | | | | |
| *Dendrobeania murmanica* Kluge | atl.h.-b. | | | | | | | | | | | | | | | | | | | + | | | | | |
| *Bicellarina alderi* (Busk) | atl.h.-b. | | | | | | | | | | | + | | | | | | | | + | | | | | |
| *Notoplites jeffreysii* Norman | atl.h.-b. | | | | | | | | | | | | | | | | | | | + | | | | | |
| *Scrupocellaria intermedia* Norman | atl.h.-b. | | | | | | | | | | | | | | | | | | | + | | | | | |
| *Gephyrotes nitidopunctata* (Smitt) | atl.h.-b. | | | | | | | | | | + | | | | | | | | | + | | | | | |
| *Escharella octodentata* Hincks | atl.h.-b. | | | | | | | | | | | | | | | | | | | + | | | | | |
| *Escharella latodonta* Kluge | atl.h.-b. | | | | | | | | | | | | | | | | | | | + | | | | | |
| *Schizoporella hexagona* Nordgaard | atl.h.-b. | | | | | | | | | | | | | | | | | | | + | | | | | |
| *Schizoporella smitti* Kluge | atl.h.-b. | | | | | | | | | | | | | | | | | + | | + | | | | | |
| *Schizoporella thompsoni* Kluge | atl.h.-b. | | | | | | | | | | | + | | | | | | | | | | | | | |
| *Turbicellepora smitti* Kluge | atl.h.-b. | | | | | | | | | | + | | | | | | | + | | + | | | | | |
| *Berenicea meandrina* (S. Wood) | amph.-b. | | | | | | | | + | | + | | | | | | | | | | | | | | |
| *Filicrisia geniculata* (Milne-Edwards) | amph.b. | | | | | | | | | | | | | | | | | | + | + | | | | | |
| *Lichenopora radiata* (Audouin) | amph.b. | | | | | | | | | | | | | | | | | | | | | | | + | |
| *Flustrellidra hispida* (Fabricius) | amph.b. | | | | | | | | | + | + | | | | | | + | | + | + | | | | | |
| *Bowerbankia composita* (Kluge) | amph.b. | | | | | | | | | | | | | | | | | + | | + | | | | | |
| *Triticella pedicellata* (Alder) | amph.b. | | | | | | | | | | | | | | | | | + | | + | | | | | |
| *Membranipora membranacea* (Linnaeus) | amph.b. | | | | | | | | | | | | | | | | | + | | + | | | | | |
| *Tegella unicornis* (Fleming) | amph.b. | | + | | | | + | | | + | + | | | | | | | | | + | | | | | |
| *Cauloramphus spiniferum* (Johnston) | amph.b. | | | | | | | | | | | | | | | | | | + | + | | | | | |
| *Flustra foliacea* (Linnaeus) | amph.b. | | | | | | | | | | | + | | | | | + | | + | + | + | | | | |
| *Corynoporella tenuis* Hincks | amph.b. | | | | | | | | | + | | | | | | | | | | + | | | | | |
| *Tricellaria peachii* var. *beringia* Kluge | amph.b. | | | | | | | | | | | | | | | | | | + | | | | | | |

*Note:* sub.-b., subtropical boreal; atl.b., Atlantic boreal widespread; atl.sub.-b., Atlantic subtropical boreal; atl.h.-b., Atlantic high boreal; amph.b., amphiboreal; b.-a., boreal–Arctic widespread; b.-a.atl.circ.-p., boreal–Arctic Atlantic circumpolar; b.-a.atl.eurasian, boreal–Arctic Atlantic Eurasian; b.-a.atl.am.-eur., boreal–Arctic Atlantic American–European; b.-a.pac.circ.-p., boreal–arctic Pacific circumpolar; b.-a-pac.amer.-as, boreal–Arctic Pacific Amerasian; b.-a.pac.eur.-as., boreal–Arctic Pacific Eurasian; arc.circ.-p., Arctic circumpolar; arc.eur.-as., Arctic Eurasian; arc.amer.-as., Arctic Amerasian; arc.eur.-am., Arctic Euroamerican; arc.amer., Arctic American; pac.b., Pacific boreal.

+ = species in: 1, northwestern coast of Alaska; 2, northern coast of North America; 3, the Archipelago of Canadian Islands; 4, Baffin Bay; 5, Davis Strait; 6, Hudson Bay; 7, Gulf of Labrador; 8, Newfoundland; 9, St. Lawrence Gulf; 10, western Greenland; 11, eastern Greenland; 12, Greenland Sea; 13, Iceland; 14, Jan Mayen; 15, Norwegian Sea; 16, North Sea; 17, northern Norway; 18, White Sea; 19, Barents Sea; 20, Kara Sea; 21, Laptev Sea; 22, East Siberian Sea; 23, Chukchi Sea; 24, Bering Strait.

This table is based on the authors' present data and the literature listed.

Table 14-1. *(continued)*

| Species | Biogeography | Species Presence | | | | | | | | | | | | | | | | | | | | | | | |
|---|---|---|---|---|---|---|---|---|---|---|---|---|---|---|---|---|---|---|---|---|---|---|---|---|---|
| | | 1 | 2 | 3 | 4 | 5 | 6 | 7 | 8 | 9 | 10 | 11 | 12 | 13 | 14 | 15 | 16 | 17 | 18 | 19 | 20 | 21 | 22 | 23 | 24 |
| *Smittoidea reticulata* (Mac Gillivray) | amph.b. | | | | | | | | | | | | | | | | | | | + | | | | | |
| *Stomachetosella magniporata* (Nordgaard) | amph.b. | | + | | | | | | | | | | | | | | | | | + | | | | | |
| *Microporella ciliata* (Pallas) | amph.b. | + | + | | | | + | + | | + | + | | | | | | | + | + | + | | | | + | |
| *Retepora beaniana* King | amph.b. | | | | | | | | | + | + | | + | | | | | | | + | | | | | |
| *Tubulipora lobulifera* Hastings | amph.b. | | | | | | | | | + | | | | | | | | | | | | | | | |
| *Tubulipora flabellaris* (Flabricius) | b.-a. | + | + | + | | | | | | | | | | | | | | | + | + | + | + | + | + | |
| *Diplosolen obelia* var. *arctica* (Waters) | b.-a. | | + | + | | | | + | | + | + | | | + | | | | | + | + | + | + | + | + | |
| *Crisia eburnea* (Linnaeus) | b.-a. | | | + | | | | + | | + | + | | | + | | | | + | + | + | + | + | + | + | |
| *Crisia klugei* (Ryland) | b.-a. | | + | | | | | | | | + | | | | | | | | | | + | | | + | |
| *Lichenopora verrucaria* (Fabricius) | b.-a. | | + | + | + | | | + | | + | + | + | | | + | | + | + | + | + | + | + | + | + | |
| *Lichenopora hispida* (Fleming) | b.-a. | | + | + | | | + | + | | + | + | + | | | + | | + | | + | + | + | | | | |
| *Alcyonidium gelatinosum* (Linnaeus) | b.-a. | | + | | | + | + | | | + | + | + | | | + | | + | + | + | + | + | + | | | |
| *Alcyonidium mytili* Dalyell | b.-a. | | + | | | | | | | | + | + | | | + | | + | | + | + | + | + | + | + | |
| *Buskia nitens* Alder | b.-a. | | | | | | | | | | | | | | | | | | + | + | + | | + | | |
| *Eucratea loricata* (Linnaeus) | b.-a. | | + | + | + | | + | + | | + | + | + | | + | | | | + | + | + | + | + | + | + | |
| *Electra arctica* (Borg) | b.-a. | | + | + | + | + | + | | | + | + | | | | | | | | + | + | + | + | | + | |
| *Tegella armifera* (Hincks) | b.-a. | | + | + | + | | + | + | | + | + | + | | | + | | | | + | + | + | + | + | + | |
| *Tegella spitzbergensis* (Bidenkap) | b.-a. | + | + | | | | + | | | | + | + | | | | | | | + | + | + | | + | + | |
| *Tegella arctica* (d'Orbigny) | b.-a. | | + | + | + | | + | + | | + | + | + | | | | | | | + | + | + | | | + | |
| *Tegella nigrans* (Hincks) | b.-a. | | + | + | + | | + | | | | + | + | | | | | | | + | + | + | + | + | + | |
| *Callopora lineata* (Linnaeus) | b.-a. | + | + | + | | | + | + | | + | + | + | | | + | | | + | + | + | + | + | + | + | |
| *Callopora craticula* (Alder) | b.-a. | | + | + | + | | + | + | + | + | + | + | | + | + | | | + | + | + | + | + | + | + | |
| *Cauloramphus cymbaeformis* (Hincks) | b.-a. | | + | + | + | | + | + | | + | + | + | | | + | | | | + | + | + | + | + | + | |
| *Doryporella spatulifera* (Smitt) | b.-a. | | + | + | + | + | | | + | + | + | + | | + | + | | | + | + | + | + | | | + | |
| *Carbasea carbasea* (Ellis et Solander) | b.-a. | | + | + | | | + | + | | + | + | | | | + | | | + | | + | + | + | + | + | |
| *Terminoflustra membranaceotruncata* (Smitt) | b.-a. | | | + | + | | | | | + | + | + | | | | | | | + | + | + | + | + | + | |
| *Dendrobeania murrayana* (Bean in Johnston) | b.-a. | + | + | + | + | | + | + | | + | + | | | | | | | | + | + | + | | | + | |

| | | | | | | | | | | | | | | | | | | | | | | | | | |
|---|---|---|---|---|---|---|---|---|---|---|---|---|---|---|---|---|---|---|---|---|---|---|---|---|---|
| *Dendrobeania pseudomurrayana* Kluge | b.-a. | | | + | | | | | | | + | + | | | | | | + | | + | | | | | + |
| *Dendrobeania pseudomurrayana* var. *fessa* Kluge | b.-a. | | | | | + | | + | | + | + | | | + | | | | + | + | + | + | | | | + |
| *Dendrobeania fruticosa* (Packard) | b.-a. | | + | + | | | | + | | + | + | + | | + | + | | | | + | + | + | + | | | + |
| *Dendrobeania fruticosa* var. *quadridentata* (Lowen) | b.-a. | | | + | | + | | + | | | + | + | | | | | | | | + | + | + | | | |
| *Tricellaria peachii* (Busk) | b.-a. | | | + | + | | + | + | | + | + | + | | + | | | | | + | + | + | + | + | + | + |
| *Tricellaria ternata* (Ellis et Solander) | b.-a. | | + | + | | | + | + | + | + | + | + | | + | + | | | | + | + | + | | | | |
| *Tricellaria gracilis* (Van Beneden) | b.-a. | | | + | + | + | + | + | | + | + | + | | + | + | | | + | + | + | + | + | + | + | + |
| *Scrupocellaria scabra* (Van Beneden) | b.-a. | | + | + | + | | + | + | + | + | + | + | | + | + | | + | | + | + | + | | | | + |
| *Scrupocellaria scabra* var. *paenulata* (Van Beneden) | b.-a. | | + | + | | | | + | | | + | + | | | | | | + | + | + | + | + | | | + |
| *Scrupocellaria elongata* (Smitt) | b.-a. | | | + | | | + | | | | + | | | | | | | | + | + | + | | | | |
| *Caberea ellisii* (Fleming) | b.-a. | | + | + | | | + | + | | + | + | | | + | | | + | | + | + | | | | | |
| *Cribrilina annulata* (Fabricius) | b.-a. | + | + | + | + | | + | + | | + | + | + | | | + | | + | | + | + | + | | | | + |
| *Escharella immersa* (Fleming) | b.-a. | | | | | | + | + | | + | + | + | | | + | | + | | + | + | + | | + | | + |
| *Escharella ventricosa* (Hassall) | b.-a. | | + | + | + | | + | + | | + | + | + | | + | + | | | | + | + | + | + | + | | + |
| *Smittina majuscula* (Smitt) | b.-a. | | + | | + | | + | + | | + | + | + | | | | | | + | + | + | + | + | + | | + |
| *Smittina bella* (Busk) | b.-a. | | + | + | + | | | | | + | + | + | | + | | | | | | + | | | | | |
| *Smittina minuscula* (Smitt) | b.-a. | | | + | | | + | + | | | + | + | | + | | | | | + | + | + | + | | | + |
| *Smittina rigida* Lorenz | b.-a. | | | | | + | + | + | | | + | + | | | + | | | | + | + | + | + | | | + |
| *Porella concinna* (Busk) | b.-a. | + | + | + | + | + | + | + | + | + | + | + | | | + | | + | | + | + | + | + | | | + |
| *Porella concinna* var. *belli* (Dawson) | b.-a. | | | | | | | + | | + | + | + | | | | | | + | + | + | | | | | + |
| *Parasmittina jeffreysii* (Norman) | | + | + | + | | | | + | | + | + | + | | | | | + | | + | + | + | + | + | | |
| *Porella compressa* (Sowerby) | b.-a. | | + | + | + | | + | + | | | + | + | | + | + | | | | + | + | + | + | + | + | + |
| *Porella acutirostris* Smitt | b.-a. | | + | + | + | | | | | + | + | + | | | + | | | | | + | + | + | + | + | + |
| *Porella smitti* Kluge | | | | | | | | | | | | | | | | | | | | | | | | | |

*Note:* sub.-b., subtropical boreal; atl.b., Atlantic boreal widespread; atl.sub.-b., Atlantic subtropical boreal; atl.h.-b., Atlantic high boreal; amph.b., amphiboreal; b.-a., boreal–Arctic widespread; b.-a.atl.circ.-p., boreal–Arctic Atlantic circumpolar; b.-a.atl.eurasian, boreal–Arctic Atlantic Eurasian; b.-a.atl.am.-eur., boreal–Arctic Atlantic American–European; b.-a.pac.circ.-p., boreal–arctic Pacific circumpolar; b.-a-pac.amer.-as, boreal–Arctic Pacific Amerasian; b.-a.pac.eur.-as., boreal–Arctic Pacific Eurasian; arc.circ.-p., Arctic circumpolar; arc.eur.-as., Arctic Eurasian; arc.amer.-as., Arctic Amerasian; arc.eur.-am., Arctic Euroamerican; arc.amer., Arctic American; pac.b., Pacific boreal.

+ = species in: 1, northwestern coast of Alaska; 2, northern coast of North America; 3, the Archipelago of Canadian Islands; 4, Baffin Bay; 5, Davis Strait; 6, Hudson Bay; 7, Gulf of Labrador; 8, Newfoundland; 9, St. Lawrence Gulf; 10, western Greenland; 11, eastern Greenland; 12, Greenland Sea; 13, Iceland; 14, Jan Mayen; 15, Norwegian Sea; 16, North Sea; 17, northern Norway; 18, White Sea; 19, Barents Sea; 20, Kara Sea; 21, Laptev Sea; 22, East Siberian Sea; 23, Chukchi Sea; 24, Bering Strait.

This table is based on the authors' present data and the literature listed.

Table 14-1. *(continued)*

| Species | Biogeography | Species Presence 1 | 2 | 3 | 4 | 5 | 6 | 7 | 8 | 9 | 10 | 11 | 12 | 13 | 14 | 15 | 16 | 17 | 18 | 19 | 20 | 21 | 22 | 23 | 24 |
|---|---|---|---|---|---|---|---|---|---|---|---|---|---|---|---|---|---|---|---|---|---|---|---|---|---|
| *Cystisella saccata* (Busk) | b.-a. | | + | + | + | | | + | + | + | + | + | | + | + | | | + | + | + | + | + | + | + | |
| *Umbonula patens* (Smitt) | b.-a. | | | | | | | | | + | + | | | | | | | | | + | | | | + | |
| *Buffonellaria biaperta* Michelin | b.-a. | | + | + | + | | | | | + | + | | | + | + | | | | + | + | + | + | | + | |
| *Myriozoella crustacea* (Smitt) | b.-a. | | + | + | + | | + | + | | + | + | + | | + | + | | | + | + | + | + | + | + | + | |
| *Schizomavella auriculata* var. *lineata* (Nordgaard) | b.-a. | | | | + | | | | | + | + | | | + | | | | + | + | + | + | | | | |
| *Schizomavella porifera* (Smitt) | b.-a. | | + | + | + | | + | + | | + | + | + | | + | + | | | + | + | + | + | + | | | |
| *Stomachetosella incerta* (Kluge) | b.-a. | | | | | | | + | | | + | | | | | | | | + | + | + | + | | | |
| *Stomachetosella sinuosa* (Busk) | b.-a. | + | + | | | + | + | | | + | + | + | | + | + | | | + | + | + | | | | | |
| *Stomachetosella cruenta* (Busk) | b.-a. | | + | | + | | | | | + | + | | | | + | | | + | + | + | + | + | + | + | |
| *Hippoporina propinqua* (Smitt) | b.-a. | | | | | | | + | | | + | + | | + | | | | + | + | + | + | | | | |
| *Myriapora coarctata* (M. Sars) | b.-a. | | | | | + | | + | | + | + | + | | + | + | | | + | + | + | + | | | | |
| *Hippothoa divaricata* var. *arctica* Kluge | b.-a. | | | | + | | | + | | + | + | + | | | | | | + | + | + | + | + | | + | |
| *Hippothoa hyalina* (Linnaeus) | b.-a. | | + | + | + | | | + | | + | + | + | | + | + | | | + | + | + | + | + | + | + | |
| *Hippoponella hippopus* (Smitt) | b.-a. | | + | + | + | | | + | | + | + | + | | | | | | + | + | + | + | | | + | |
| *Rhamphostomella scabra* (Fabricius) | b.-a. | | | + | + | + | | + | | + | + | + | | | | | | + | + | + | + | + | + | + | |
| *Rhamphostomella costata* Lorenz | b.-a. | | | + | + | | + | + | | + | + | + | | | + | | | + | + | + | + | + | + | + | |
| *Rhamphostomella ovata* (Smitt) | b.-a. | + | + | + | + | + | + | + | | + | + | + | | | | | | + | + | + | + | + | + | + | |
| *Rhamphostomella radiatula* (Hincks) | b.-a. | | + | + | | + | | + | | + | + | + | | + | + | | | + | + | + | + | | | | |
| *Ragionula rosacea* (Busk) | b.-a. | | | | | | | | | + | + | | | + | + | | | + | + | + | + | | + | | |
| *Stomatopora granulata* (Milne-Edwards) | b.-a.atl.circ.-p. | | | | | | | | | | | | | | | | | | + | + | + | | | | |
| *Proboscina incrassata* (Smitt) | b.-a.atl.circ.-p. | | | | | | | | | + | | | | | | | | | | + | + | + | | | |
| *Oncousoecia diastoporoides* (Norman) | b.-a.atl.circ.-p. | | | | | | | + | | + | | | | + | | | | | + | + | + | | | + | |
| *Tubulipora ventricosa* Busk | b.-a.atl.circ.-p. | | | + | | | | | | | + | | | | | | | | + | + | + | + | + | | |
| *Idmonea atlantica* Forbes MSS | b.-a.atl.circ.-p. | | + | + | + | | | + | + | + | + | + | | + | + | | | | | + | + | + | | | |
| *Entalophora clavata* (Busk) | b.-a.atl.circ.-p. | | + | | | | | | | | | + | | | + | | | | | + | + | + | + | | |
| *Entalophora harmeri* (Osburn) | b.-a.atl.circ.-p. | | + | + | | | | | | + | + | | | | | | | | | + | + | | | | |
| *Defrancia lucernaria* M. Sars | b.-a.atl.circ.-p. | | | | | | | | | + | + | + | | | + | | | | | + | + | + | + | | |
| *Fasciculiporoides americana* (d'Orbigny) | b.-a.atl.circ.-p. | | | | | | | | + | + | + | | | | | | | | | | | + | + | + | |

| | | | | | | | | | | | | | | | | | | | | | | | | | |
|---|---|---|---|---|---|---|---|---|---|---|---|---|---|---|---|---|---|---|---|---|---|---|---|---|---|
| *Stegohornera lichenoides* (Linnaeus) | b.-a.atl.circ.-p. | | | | | | | | + | + | + | | + | + | | | | | + | + | + | + | | | |
| *Alcyonidium mammilatum* Alder | b.-a.atl.circ.-p. | | + | | | | | | | + | + | | | + | | | | + | + | + | | + | + | | |
| *Nolella dilatata* (Hincks) | b.-a.atl.circ.-p. | + | | | | | | | | + | | | | + | | | | | + | | | | | | |
| *Electra pilosa* (Linnaeus) | b.-a.atl.circ.-p. | + | | | + | | | | + | + | | | + | | | | + | | + | + | + | | | | |
| *Amphiblestrum septentrionalis* (Kluge) | b.-a.atl.circ.-p. | | + | | | | | | + | + | + | | | + | | | | + | + | | | + | | | |
| *Amphiblestrum trifolium* var. *quadrata* (Hincks) | b.-a.atl.circ.-p. | | + | | | | | | + | + | + | | | + | | | + | | + | | | | + | | |
| *Securiflustra securifrons* (Pallas) | b.-a.atl.circ.-p. | | + | | | + | + | + | | + | + | | | + | | + | + | + | + | + | + | | + | | |
| *Sarsiflustra abyssicola* (G.O. Sars) | b.-a.atl.circ.-p. | | | | | | | | | + | + | | + | + | | | | + | + | + | + | | | | |
| *Bugula elongata* (Nordgaard) | b.-a.atl.circ.-p. | | | | | | | | + | | | | | | | | | | + | + | + | | | | |
| *Kinetoskias arborescens* Daniellsen | b.-a.atl.circ.-p. | | + | | | + | | | + | + | | | | | | | | | + | + | + | + | | | |
| *Kinetoskias smitti* Daniellsen | b.-a.atl.circ.-p. | | | | | | | | | | | | | | | | | | + | + | | + | | | |
| *Lepralioides nordlandica* (Nordgaard) | b.-a.atl.circ.-p. | | + | | | | | | + | + | | | | + | | | | | + | + | | + | + | | |
| *Pachyegis groenlandica* Norman | b.-a.atl.circ.-p. | | + | + | | | | | + | + | + | | | | | | + | + | + | + | | | + | | |
| *Retepora cellulosa* (Linnaeus) | b.-a.atl.circ.-p. | | | | | | | | | + | + | | | + | | | | | + | + | + | | | | |
| *Idmonea atlantica* var. *gracillima* Busk | b.-a.atl.circ.-p. | + | + | | | | + | | + | + | + | | + | + | | | | | + | + | + | + | | | |
| *Alcyonidium hirsutum* (Fleming) | b.-a.atl.eur.-as. | | | | | | | | | | | | | | | | | + | + | | | | | | |
| *Alcyonidium gelatinosum* var. *diaphanum* (Farre) | b.-a.atl.eur.-as. | | | | | | | | | | | | | | | | | | + | | | | | | |
| *Alcyonidium albidum* Alder | b.-a.atl.eur.-as. | | | | | | | | | | | | | | | | | | + | + | | | | | |
| *Arachnidium clavatum* Hincks | b.-a.atl.eur.-as. | | | | | | | | | | | | | | | | | | + | | | | | | |
| *Arachnidium hippothoides* Hincks | b.-a.atl.eur.-as. | | | | | | | | | | | | | | | | | | + | | | | | | |

*Note:* sub.-b., subtropical boreal; atl.b., Atlantic boreal widespread; atl.sub.-b., Atlantic subtropical boreal; atl.h.-b., Atlantic high boreal; amph.b., amphiboreal; b.-a., boreal–Arctic widespread; b.-a.atl.circ.-p., boreal–Arctic Atlantic circumpolar; b.-a.atl.eurasian, boreal–Arctic Atlantic Eurasian; b.-a.atl.am.-eur., boreal–Arctic Atlantic American–European; b.-a.pac.circ.-p., boreal–arctic Pacific circumpolar; b.-a-pac.amer.-as, boreal–Arctic Pacific Amerasian; b.-a.pac.eur.-as., boreal–Arctic Pacific Eurasian; arc.circ.-p., Arctic circumpolar; arc.eur.-as., Arctic Eurasian; arc.amer.-as., Arctic Amerasian; arc.eur.-am., Arctic Euroamerican; arc.amer., Arctic American; pac.b., Pacific boreal.

+ = species in: 1, northwestern coast of Alaska; 2, northern coast of North America; 3, the Archipelago of Canadian Islands; 4, Baffin Bay; 5, Davis Strait; 6, Hudson Bay; 7, Gulf of Labrador; 8, Newfoundland; 9, St. Lawrence Gulf; 10, western Greenland; 11, eastern Greenland; 12, Greenland Sea; 13, Iceland; 14, Jan Mayen; 15, Norwegian Sea; 16, North Sea; 17, northern Norway; 18, White Sea; 19, Barents Sea; 20, Kara Sea; 21, Laptev Sea; 22, East Siberian Sea; 23, Chukchi Sea; 24, Bering Strait.

This table is based on the authors' present data and the literature listed.

Table 14-1. *(continued)*

| Species | Biogeography | Species Presence | | | | | | | | | | | | | | | | | | | | | | | |
|---|---|---|---|---|---|---|---|---|---|---|---|---|---|---|---|---|---|---|---|---|---|---|---|---|---|
| | | 1 | 2 | 3 | 4 | 5 | 6 | 7 | 8 | 9 | 10 | 11 | 12 | 13 | 14 | 15 | 16 | 17 | 18 | 19 | 20 | 21 | 22 | 23 | 24 |
| *Electra pilosa* var. *dentata* Ellis et Solander | b.-a.atl.eur.-as. | | | | | | | | | | | | | | | | | | + | + | + | | | | |
| *Electra pilosa* var. *baltica* Borg | b.-a.atl.eur.-as. | | | | | | | | | | | | | | | | + | | | | | | | | |
| *Bugula purpurotincta* Norman | b.-a.atl.eur.-as. | | | | | | | | | | | | | + | | | | | | | | | | | |
| *Smittina peristomata* (Nordgaard) | b.-a.atl.eur.-as. | | | | | | | | | | | | | + | | | | + | + | + | + | | | | |
| *Smittina pseudoacutirostris* Gostilovskaja | b.-a.atl.eur.-as. | | | | | | | | | | | | | | | | | | + | + | | | | | |
| *Haplopoma impressum* (Audouin) | b.-a.atl.eur.-as. | | | | | | | | | | | | | | | | | | | + | | | | | |
| *Bicellariella ciliata* (Linnaeus) | b.-a.atl.am.-eur. | | + | | | | | | | + | | | | | | | | + | | | | | | | |
| *Cribrilina watersi* Andersson | b.-a.atl.am.-eur. | | | | | | | | | | + | + | | + | + | | | + | | + | + | + | + | | |
| *Escharella microstoma* (Norman) | b.-a.atl.am.-eur. | | | + | | | | | | | + | + | | | | | | | + | + | + | | + | | |
| *Phylactella labiata* (Boeck in MS. Smitt) | b.-a.atl.am.-eur. | | | | | | | | | | + | + | | + | | | | + | | + | + | + | | | |
| *Porelloides struma* (Norman) | b.-a.atl.am.-eur. | | | | + | | + | + | | + | + | | | + | | | | | + | + | + | | | | |
| *Porelloides laevis* (Fleming) | b.-a.atl.am.-eur. | | + | | | + | | + | | | | | | | | | | | | + | | | | | |
| *Porella minuta* (Norman) | b.-a.atl.am.-eur. | | + | + | + | | | + | | | | | | | | | | + | + | + | | | | | |
| *Palmicellaria skenei* (Ellis et Solander) | b.-a.atl.am.-eur. | | | | | + | | | | | | | | | | | | | | + | + | + | | | |
| *Palmicellaria skenei* var. *bicornis* (Busk) | b.-a.atl.am.-as. | | | | | | | | | | | | | | | | | | | + | | | | | |
| *Escharina alderi* (Busk) | b.-a.atl.am.-eur. | | | | | | | | | + | | | | | | | | | | + | | | | | |
| *Schizoporella pachystega* Kluge | b.-a.atl.am.-eur. | | | | | | | | | + | + | | | | | | | | | + | + | | + | | |
| *Microporina articulata* (Fabricius) | b.-a.pac.circ.-p. | + | + | + | + | | + | | | | + | | | | | | | | | + | | | | | |
| *Smittina mucronata* (Smitt) | b.-a.pac.circ.-p. | | | | | | + | | | | | | | | | | | + | + | + | + | | | + | |
| *Umbonula arctica* (M. Sars) | b.-a.pac.circ.-p. | + | + | | + | | | | | + | + | + | | + | + | | | + | + | + | + | + | + | + | |
| *Porella fragilis* Levinsen | b.-a.pac.circ.-p. | | | + | | | | + | | + | + | + | | | | | | | + | + | + | + | + | + | |
| *Stomachetosella limbata* (Lorenz) | b.-a.pac.circ.-p. | | | + | + | | | | | + | + | | | | + | | | + | + | + | + | + | | | |
| *Schizobrachiella stylifera* (Levinsen) | b.-a.pac.circ.-p. | | + | + | + | | | | | + | + | + | | | | | | | | + | + | + | | + | |
| *Hippoporina reticulatopunctata* (Hincks) | b.-a.pac.circ.-p. | | + | + | + | + | | + | | + | + | + | | | | | | | + | + | + | + | + | + | |

| | | | | | | | | | | | | | | | | | | | | | | | | |
|---|---|---|---|---|---|---|---|---|---|---|---|---|---|---|---|---|---|---|---|---|---|---|---|---|
| *Porella obesa* Waters | b.-a.pac.circ.-p. | | | | | | + | | | + | | | | | | | | | + | + | + | + | | + |
| *Hippothoa expansa* Dawson | b.-a.pac.circ.-p. | + | + | + | + | + | + | + | | + | + | + | | | + | | | | + | + | + | | | + |
| *Cylindroporella tubulosa* (Norman) | b.-a.pac.circ.-p. | | + | + | + | | | | | + | + | + | | | + | | | + | + | + | + | | | + |
| *Phydolopora elongata* (Smitt) | b.-a.pac.circ.-p. | | + | + | + | + | | + | | + | + | + | | | | | | + | | + | + | + | + | |
| *Lepraliella contigua* (Smitt) | b.-a.pac.circ.-p. | | + | + | + | | + | | | + | + | + | | + | + | | | + | + | + | | | | + |
| *Rhamphostomella hinksi* Nordgaard | b.-a.pac.circ.-p. | | + | + | + | + | + | | | + | + | + | | + | + | | | | | + | + | + | | + |
| *Rhamphostomella plicata* (Smitt) | b.-a.pac.circ.-p. | | | + | + | + | | | | + | + | + | | + | | | | + | + | + | + | + | | + |
| *Escharopsis lobata* (Lamouroux) | b.-a.pac.circ.-p. | | + | + | + | | | + | + | + | + | + | + | + | + | | | + | + | + | + | + | + | + |
| *Celleporina incrassata* Lamarck | b.-a.pac.circ.-p. | | + | + | + | | + | + | + | + | + | | | + | | | | + | + | + | + | | + | + |
| *Celleporina ventricosa* Lorenz | b.-a.pac.circ.-p. | + | + | + | + | + | | + | | + | + | + | | + | + | | | + | | + | + | + | + | + |
| *Filicrisia smitti* (Kluge) | b.-a.pac.am.-as. | | | | | | | | | + | + | | | | | | | | | + | + | | + | |
| *Crisiella producta* (Smitt) | b.-a.pac.am.-as. | | | | | | | | | | + | | | | | | | | + | + | + | | | |
| *Smittina thompsoni* Kluge | b.-a.pac.am.-as. | | | | + | + | | | | + | + | | | | | | | | | | | | | |
| *Myriapora orientalis* (Kluge) | b.-a.pac.am.-as. | | | | + | | | | | | | | | | | | | | | | | + | + | + |
| *Myriapora subgracilis* (d'Orbigny) | b.-a.pac.am.-as. | + | + | + | + | + | + | + | + | + | + | + | | | | | | | | + | + | + | + | + |
| *Eucratea loricata* var. *macrostoma* (Ortmann) | b.-a.pac.eur.-as. | | | | | | | | | | | | | | | | | | | | + | + | | |
| *Tegella arctica* var. *retroversa* Kluge | b.-a.pac.eur.-as. | | | | | | | | | | | | | | | | | | | + | + | | + | + |
| *Cheiloporina sincera* Smitt | b.-a.pac.circ.-p. | + | + | + | + | + | | + | | + | + | + | | + | + | | | + | | + | + | + | + | + |
| *Cauloramphus intermedius* Kluge | b.-a.pac.eur.-as. | + | | | | | | | | | | | | | | | | | + | + | + | | | + |
| *Flustra nordenskjoldi* Kluge | b.-a.pac.eur.-as. | | | | | | | | | | | | | | | | | | | | | + | + | + |
| *Dendrobeania pseudomurrayana* var. *tenuis* Kluge | b.-a.pac.eur.-as. | | | | | | | | | | | | | | | | | | | + | | | | |
| *Dendrobeania flustroides* Levinsen | b.-a.pac.eur.-as. | | | | | | | | | | | | | | | | | | | | + | | | |

*Note:* sub.-b., subtropical boreal; atl.b., Atlantic boreal widespread; atl.sub.-b., Atlantic subtropical boreal; atl.h.-b., Atlantic high boreal; amph.b., amphiboreal; b.-a., boreal–Arctic widespread; b.-a.atl.circ.-p., boreal–Arctic Atlantic circumpolar; b.-a.atl.eurasian, boreal–Arctic Atlantic Eurasian; b.-a.atl.am.-eur., boreal–Arctic Atlantic American–European; b.-a.pac.circ.-p., boreal–arctic Pacific circumpolar; b.-a-pac.amer.-as, boreal–Arctic Pacific Amerasian; b.-a.pac.eur.-as., boreal–Arctic Pacific Eurasian; arc.circ.-p., Arctic circumpolar; arc.eur.-as., Arctic Eurasian; arc.amer.-as., Arctic Amerasian; arc.eur.-am., Arctic Euroamerican; arc.amer., Arctic American; pac.b., Pacific boreal.

+ = species in: 1, northwestern coast of Alaska; 2, northern coast of North America; 3, the Archipelago of Canadian Islands; 4, Baffin Bay; 5, Davis Strait; 6, Hudson Bay; 7, Gulf of Labrador; 8, Newfoundland; 9, St. Lawrence Gulf; 10, western Greenland; 11, eastern Greenland; 12, Greenland Sea; 13, Iceland; 14, Jan Mayen; 15, Norwegian Sea; 16, North Sea; 17, northern Norway; 18, White Sea; 19, Barents Sea; 20, Kara Sea; 21, Laptev Sea; 22, East Siberian Sea; 23, Chukchi Sea; 24, Bering Strait.

This table is based on the authors' present data and the literature listed.

Table 14-1. *(continued)*

| Species | Biogeography | Species Presence 1 | 2 | 3 | 4 | 5 | 6 | 7 | 8 | 9 | 10 | 11 | 12 | 13 | 14 | 15 | 16 | 17 | 18 | 19 | 20 | 21 | 22 | 23 | 24 |
|---|---|---|---|---|---|---|---|---|---|---|---|---|---|---|---|---|---|---|---|---|---|---|---|---|---|
| *Semibugula birulai* Kluge | b.-a.pac.eur.-as. | | | | | | | | | | | | | | | | | | | + | + | + | | | |
| *Scrupocellaria scabra* var. *paenulata forma orientalis* Kluge | b.-a.pac.eur.-as. | | | | | | | | | | | | | | | | | | | | | | + | | |
| *Schizoporella costata* Kluge | b.-a.pac.eur.-as. | | | | | | | | | | | | | | | | | | | + | | | | | |
| *Hippoporina ussowi* (Kluge) | b.-a.pac.eur.-as. | | | | | | | | | | + | | | | | | | | + | + | + | | | | |
| *Stomachetosella hincksi* Powell | b.-a.pac.circ.-p. | | | + | + | | | | | + | + | | | | + | | | + | + | + | + | + | | | + |
| *Rhamphostomella bilaminata* var. *sibirica* Kluge | b.-a.pac.eur.-as.-am. | | | | | | | | | + | + | + | | | | | | | + | + | + | + | + | + | |
| *Celleporina nordenskjoldi* (Kluge) | b.-a.pac.eur.-as. | | | | | | | | | | | | | | | | | | | | + | + | + | | |
| *Cheloporina inermis* (Busk) | b.-a.pac.circ.-p. | | | | + | | | | | + | + | | | | | | | | | + | + | | | + | |
| *Oncousoecia canadensis* (Osburn) | arc.circ.-p. | | | + | + | | | | | + | + | + | | | | | | | + | + | | + | | + | |
| *Oncousoecia polygonalis* (Kluge) | arc.circ.-p. | | | | | | | | | + | | | | | | | | | | + | | | | + | |
| *Diplosolen intricarius* (Smitt) | arc.circ.-p. | | | | + | | | | | | + | | | + | | | | | | + | + | + | | | |
| *Crisia eburneodenticulata* Smitt | arc.circ.-p. | | | | | | | | | + | + | | | | + | | | | + | + | + | + | + | | |
| *Crisia denticulata* var. *borgi* Kluge | arc.circ.-p. | | | | | | | | | | + | | | | + | | | | + | + | + | + | + | | |
| *Lichenopora crassiuscula* (Smitt) | arc.circ.-p. | | | + | | | | | | | + | + | | | | | | | | + | + | + | | + | |
| *Crisia denticulata* var. *arctica* M. Sars | arc.circ.-p. | | | | | | | | | | + | | + | | | | | | | + | + | + | | + | |
| *Alcyonidium mammilatum* var. *erectum* Andersson | arc.circ.-p. | | + | | | | | | | | + | + | | | | | | | + | + | + | + | + | | |
| *Alcyonidium disciforme* Smitt | arc.circ.-p. | | | + | | + | | | | | + | + | | | | | | + | | + | + | + | + | + | |
| *Flustrellidra corniculata* (Smitt) | arc.circ.-p. | | | + | | | | | | | + | | | | + | | | | | + | | | | | |
| *Bowerbankia arctica* Busk | arc.circ.-p. | | | + | | | | | | | + | + | | | | | | | + | + | + | + | + | | |
| *Eucratea loricata* var. *cornuta* (Osburn) | arc.circ.-p. | | | | | | + | | | | | | | | | | | | | | | + | + | + | |
| *Eucratea loricata* var. *arctica* (Kluge) | arc.circ.-p. | | | | | | | | | + | | | | | | | | | | + | + | + | + | + | |
| *Callopora smitti* Kluge | arc.circ.-p. | | + | | | | + | | | | + | | | | | | | | | + | + | + | + | | |
| *Callopora whiteavesi* Norman | arc.circ.-p. | | | | | | | | | + | + | + | | | | | | | + | + | + | + | + | + | |
| *Callopora lata* (Kluge) | arc.circ.-p. | | | | | | | | | | + | | | | + | | | | + | + | + | + | + | | |
| *Flustra serrulata* Busk | arc.circ.-p. | | + | + | + | | | + | + | + | + | | | | | | | | | + | + | + | + | + | |

| | | | | | | | | | | | | | | | | | | | | | | | | | |
|---|---|---|---|---|---|---|---|---|---|---|---|---|---|---|---|---|---|---|---|---|---|---|---|---|---|
| *Notoplites smitti* (Norman) | arc.circ.-p. | | + | + | + | | | | | | + | + | | | | | | + | | + | + | + | | | |
| *Tricellaria gracilis* var. *inermis* Kluge | arc.circ.-p. | | | | | | | | | | + | | | | + | | | | | + | + | | | + | |
| *Escharella abyssicola* (Norman) | arc.circ.-p. | | + | | + | | | | | + | + | + | | | + | | | + | | + | | | | | |
| *Smittina smitti* (Kirchenpauer) | arc.circ.-p. | | + | + | | | | | | + | + | + | | | + | | | + | | + | + | + | + | | |
| *Pseudoflustra solida* (Stimpson) | arc.circ.-p. | | + | | | + | + | | | + | + | + | + | + | + | + | | | | + | + | + | + | | |
| *Pseudoflustra hincksi* Kluge | arc.circ.-p. | | | | | | | | | | | + | | | | | | | | + | + | + | + | | |
| *Pseudoflustra anderssoni* Kluge | arc.circ.-p. | | | | | | | | | | | | + | | | | | | | + | + | + | + | | |
| *Pseudoflustra sinuosa* (Andersson) | arc.circ.-p. | | | | | | | | + | | | + | | | | | | | | + | + | + | + | | |
| *Porella aperta* (Boeck) | arc.circ.-p. | | + | | | + | | | | + | + | + | | | | | | | | + | + | + | | | |
| *Pachyegis producta* (Packard) | arc.circ.-p. | | | + | | | | | | + | + | + | | | | | | + | + | + | + | | | + | |
| *Porella proboscidea* Hincks | arc.circ.-p. | | | + | | | | | | + | + | + | | | + | | | | + | + | + | + | + | | |
| *Porella plana* Hincks | arc.circ.-p. | | | + | | | + | | | | + | + | | + | | | | | | + | + | + | + | | |
| *Palmicellaria skenei* var. *tridens* (Busk) | arc.circ.-p. | | | | | | + | | | + | + | + | | + | | | | + | | + | + | | | | |
| *Hippoporina harmsworthi* (Waters) | arc.circ.-p. | | | | | | | | | + | + | + | | | | | | | + | + | + | + | + | + | |
| *Harmeria scutulata* (Busk) | arc.circ.-p. | | + | + | + | | + | | | | + | | | | | | | | + | + | + | | | | |
| *Microporella ciliata* var. *arctica* Norman | arc.circ.-p. | | + | | | | | + | | | + | + | | | | | | + | + | + | + | | | | |
| *Rhamphostomella costata* var. *cristata* Hincks | arc.circ.-p. | | | | | | + | + | | + | | + | | | | | | | | + | | | | | |
| *Rhamphostomella spinigera* Lorenz | arc.circ.-p. | | + | + | | | + | | | | + | + | | | + | | | | + | + | + | | | | |
| *Rhamphostomella bilaminata* (Hincks) | arc.circ.-p. | + | + | | + | + | + | + | | + | + | + | | | + | | | | + | + | + | | | + | |
| *Celleporina nodulosa* (Lorenz) | arc.circ.-p. | | + | + | | + | | | | | + | + | | | + | | | | + | + | + | + | + | + | |
| *Escharoides jacksoni* (Waters) | arc.circ.-p. | | + | + | | | | | | | + | + | | | | | | | | | + | + | + | + | |
| *Proboscina fecunda* Kluge | arc.eur.-as. | | | | | | | | | | | | | | | | | | | | | | + | | |
| *Tubulipora minuta* Kluge | arc.eur.-as. | | | | | | | | | | | | | | | | | | + | + | | | + | | |

*Note:* sub.-b., subtropical boreal; atl.b., Atlantic boreal widespread; atl.sub.-b., Atlantic subtropical boreal; atl.h.-b., Atlantic high boreal; amph.b., amphiboreal; b.-a., boreal–Arctic widespread; b.-a.atl.circ.-p., boreal–Arctic Atlantic circumpolar; b.-a.atl.eurasian, boreal–Arctic Atlantic Eurasian; b.-a.atl.am.-eur., boreal–Arctic Atlantic American–European; b.-a.pac.circ.-p., boreal–arctic Pacific circumpolar; b.-a-pac.amer.-as, boreal–Arctic Pacific Amerasian; b.-a.pac.eur.-as., boreal–Arctic Pacific Eurasian; arc.circ.-p., Arctic circumpolar; arc.eur.-as., Arctic Eurasian; arc.amer.-as., Arctic Amerasian; arc.eur.-am., Arctic Euroamerican; arc.amer., Arctic American; pac.b., Pacific boreal.

+ = species in: 1, northwestern coast of Alaska; 2, northern coast of North America; 3, the Archipelago of Canadian Islands; 4, Baffin Bay; 5, Davis Strait; 6, Hudson Bay; 7, Gulf of Labrador; 8, Newfoundland; 9, St. Lawrence Gulf; 10, western Greenland; 11, eastern Greenland; 12, Greenland Sea; 13, Iceland; 14, Jan Mayen; 15, Norwegian Sea; 16, North Sea; 17, northern Norway; 18, White Sea; 19, Barents Sea; 20, Kara Sea; 21, Laptev Sea; 22, East Siberian Sea; 23, Chukchi Sea; 24, Bering Strait.

This table is based on the authors' present data and the literature listed.

**Table 14-1.** *(continued)*

| *Species* | *Biogeography* | *Species Presence* 1 | 2 | 3 | 4 | 5 | 6 | 7 | 8 | 9 | 10 | 11 | 12 | 13 | 14 | 15 | 16 | 17 | 18 | 19 | 20 | 21 | 22 | 23 | 24 |
|---|---|---|---|---|---|---|---|---|---|---|---|---|---|---|---|---|---|---|---|---|---|---|---|---|---|
| *Tubulipora eminens* Kluge | arc.eur.-as. | | | | | | | | | | | | | | | | | | | | | | + | | |
| *Tubulipora borgi* Kluge | arc.eur.-as. | | | | | | | | | | | | | | | | | | + | | | | | | |
| *Tubulipora soluta* Kluge | arc.eur.-as. | | | | | | | | | | | | | | | | | | | + | + | + | + | | |
| *Tubulipora nordgaardi* Kluge | arc.eur.-as. | | | | | | | | | | | | | | | | | | | | | | + | | |
| *Tubulipora fruticosa* Kluge | arc.eur.-as. | | | | | | | | | | | | | | | | | | + | + | + | | + | | |
| *Idmonea tumida* (Smitt) | arc.eur.-as. | | | | | | | | | | | | | | | | | | | + | + | | + | | |
| *Idmonea bidenkapi* Kluge | arc.eur.-as. | | | | | | | | | | | | | | | | | | | + | | | | | |
| *Idmoneoides arctoflabellaris* (Kluge) | arc.eur.-as. | | | | | | | | | | | | | | | | | | | + | + | + | + | | |
| *Idmoneoides simplex* Kluge | arc.eur.-as. | | | | | | | | | | | | | | | | | | | + | | | | | |
| *Berenicea arctica* Kluge | arc.eur.-as. | | | | | | | | | | | | | | | | | | | + | + | | | | |
| *Berenicea oblonga* Kluge | arc.eur.-as. | | | | | | | | | | | | | | | | | | | + | + | | | | |
| *Crisiella diversa* (Kluge) | arc.eur.-as. | | | | | | | | | | | | | | | | | | | + | + | | | | |
| *Crisiella complecta* (Kluge) | arc.eur.-as. | | | | | | | | | | | | | | | | | | | + | + | | | | |
| *Crisis aculeata* var. *bathyalis* Kluge | arc.eur.-as. | | | | | | | | | | | | | | | | | | | | + | | | | |
| *Crisia constans* Kluge | arc.eur.-as. | | | | | | | | | | | | | | | | | | | | + | | | | |
| *Stegohornera arctica* (Kluge) | arc.eur.-as. | | | | | | | | | | | | | | | | | | | + | + | | | | |
| *Lichenopora multicentra* Kluge | arc.eur.-as. | | | | | | | | | | | | | | | | | | | | | + | | | |
| *Lichenopora sibirica* Kluge | arc.eur.-as. | | | | | | | | | | | | | | | | | | | | | + | | | |
| *Fungella dalli* Kluge | arc.eur.-as. | | | | | | | | | | | | | | | | | | | | | | | + | |
| *Alcyonidium gelatinosum* var. *anderssoni* Abrikosov | arc.eur.-as. | | | | | | | | | | | | | | | | | | | + | + | + | + | + | |
| *Alcyonidium gelatinosum* var. *pachydermatum* Kluge | arc.eur.-as. | | | | | | | | | | | | | | | | | | | + | + | | | | |
| *Alcyonidium radicellatum* Kluge | arc.eur.-as. | | | | | | | | | | | | | | | | | | | + | + | + | + | | |
| *Tegella armiferoides* Kluge | arc.eur.-as. | | | | | | | | | | | | | | | | | | | | | | + | | |
| *Callopora craticula* var. *sedovi* Kluge | arc.eur.-as. | | | | | | | | | | | | | | | | | | | + | | | | | |
| *Rhamphonotus gorbunovi* Kluge | arc.eur.-as. | | | | | | | | | | | | | | | | | | | | | | + | | |
| *Dendrobeania fruticosa* var. *frigida* (Waters) | arc.eur.-as. | | | | | | | | | | | | | | | | | | | + | | | | | |
| *Bugula tricuspis* Kluge | arc.eur.-as. | | | | | | | | | | | | | | | | | | | + | | | | | |
| *Uschakovia gorbunovi* Kluge | arc.eur.-as. | | | | | | | | | | | | | | | | | | | | + | | + | | |
| *Notoplites sibirica* (Kluge) | arc.eur.-as. | | | | | | | | | | | | | | | | | | | + | + | + | | + | |

| | | | | | | | | | | | | | | | | | | | | | | | | | |
|---|---|---|---|---|---|---|---|---|---|---|---|---|---|---|---|---|---|---|---|---|---|---|---|---|---|
| *Notoplites normani* (Nordgaard) | arc.eur.-as. | | | | | | | | | | | | | | + | | | | + | + | | | | | |
| *Cribrilina spitzbergensis* Norman | arc.eur.-as. | | | | | | | | | + | + | | | | | | | + | + | + | + | + | + | | |
| *Cribrilina spiculifera* Kluge | arc.eur.-as. | | | | | | | | | | | | | | | | | | | | | + | | | |
| *Escharella ventricosa* var. *peristomata* Kluge | arc.eur.-as. | | | | | | | | | | | | | | | | | + | + | | | | | | |
| *Escharelloides cancellatum* (Smitt) | arc.circ.-p. | + | | | | + | | | | | | | | | | | | | + | | | | | | |
| *Escharelloides simplex* Kluge | arc.eur.-as. | | | | | | | | | | | | | | | | | | | + | | | | | |
| *Smittina glaciata* (Waters) | arc.eur.-as. | | | | | | | | | | | | | | | | | | + | + | | | | | |
| *Smittina muliebris* Kluge | arc.eur.-as. | | | | | | | | | | | | | | | | | | | + | | | | | |
| *Pseudoflustra birulai* Kluge | arc.eur.-as. | | | | | | | | | | | | | | | | | | + | + | | + | | | |
| *Umbonula inarmata* Kluge | arc.eur.-as. | | | | | | | | | | | | | | | | | | | | + | | | | |
| *Schizoporella elmwoodia* var. *mammilata* Kluge | arc.eur.-as. | | | | | | | | | | | | | | | | | | + | | | | | | |
| *Hippoporina borealis* (Waters) | arc.eur.-as. | | | | | | | | | | | | | | | | | | + | + | | | | | |
| *Stomachetosella collaris* (Kluge) | arc.eur.-as. | | | | | | | | | | | | | | | | | | | | + | | | | |
| *Retepora beaniana* var. *watersi* Nordgaard | arc.eur.-as. | | | | | | | | | | | | + | | | | | | + | | | | | | |
| *Echaroides bidenkapi* (Kluge) | arc.eur.-as. | | | | | | | | | | | | | | | | + | | + | + | + | + | | | |
| *Echaroides jacksoni* var. *rostrata* (Kluge) | arc.eur.-as. | | | | | | | | | | | | | | | | | | | | | + | | | |
| *Escharoides monstruosa* (Kluge) | arc.amer.-as. | | | | | | | | | | | | | | | | | | | | | | + | | |
| *Idmonea fenestrata* Busk | arc.amer.-as. | | | | | | | | | | + | | | | | | | | + | + | + | | | | |
| *Bugula harmsworthi* Waters | arc.amer.-as. | | + | | + | | | | | + | + | | | + | | | | | | | | | | | |
| *Escharella diimphnae* (Kluge) | arc.amer.-as. | | | | | | | | | | + | | | | | | | | | | | | | | |
| *Escharella indivisa* Levinsen | arc.circ.-p. | | + | + | + | + | | | | + | + | | | | | | | | + | + | + | + | | | |
| *Hippoporina tschukotkensis* Kluge | arc.amer.-as. | | | | | | | | | | | | | | | | | | | | | | + | | |
| *Escharelloides spinulifera* (Hincks) | arc.circ.-p. | + | | + | + | | + | | + | + | | | | | | | | + | + | + | | | | | |
| *Esharelloides stenostoma* (Smitt) | arc.eur.-am. | | | | | | | | | | + | | | | | | | | + | | | | | | |

*Note:* sub.-b., subtropical boreal; atl.b., Atlantic boreal widespread; atl.sub.-b., Atlantic subtropical boreal; atl.h.-b., Atlantic high boreal; amph.b., amphiboreal; b.-a., boreal–Arctic widespread; b.-a.atl.circ.-p., boreal–Arctic Atlantic circumpolar; b.-a.atl.eurasian, boreal–Arctic Atlantic Eurasian; b.-a.atl.am.-eur., boreal–Arctic Atlantic American–European; b.-a.pac.circ.-p., boreal–arctic Pacific circumpolar; b.-a-pac.amer.-as, boreal–Arctic Pacific Amerasian; b.-a.pac.eur.-as., boreal–Arctic Pacific Eurasian; arc.circ.-p., Arctic circumpolar; arc.eur.-as., Arctic Eurasian; arc.amer.-as., Arctic Amerasian; arc.eur.-am., Arctic Euroamerican; arc.amer., Arctic American; pac.b., Pacific boreal.

+ = species in: 1, northwestern coast of Alaska; 2, northern coast of North America; 3, the Archipelago of Canadian Islands; 4, Baffin Bay; 5, Davis Strait; 6, Hudson Bay; 7, Gulf of Labrador; 8, Newfoundland; 9, St. Lawrence Gulf; 10, western Greenland; 11, eastern Greenland; 12, Greenland Sea; 13, Iceland; 14, Jan Mayen; 15, Norwegian Sea; 16, North Sea; 17, northern Norway; 18, White Sea; 19, Barents Sea; 20, Kara Sea; 21, Laptev Sea; 22, East Siberian Sea; 23, Chukchi Sea; 24, Bering Strait.

This table is based on the authors' present data and the literature listed.

Table 14-1. *(continued)*

| Species | Biogeography | *Species Presence* | | | | | | | | | | | | | | | | | | | | | | | |
|---|---|---|---|---|---|---|---|---|---|---|---|---|---|---|---|---|---|---|---|---|---|---|---|---|---|
| | | *1* | *2* | *3* | *4* | *5* | *6* | *7* | *8* | *9* | *10* | *11* | *12* | *13* | *14* | *15* | *16* | *17* | *18* | *19* | *20* | *21* | *22* | *23* | *24* |
| *Hemicyclopora emucronata* (Smitt) | arc.eur.-am. | | | | + | + | | | | | | | | | | | | | | + | + | | | | |
| *Schizoporella bispinosa* Nordgaard | arc.eur.-am. | | | + | | | | | | | | | | | | | | | | + | + | | | | |
| *Schizoporella elmwoodiae* Waters | arc.eur.-am. | | | + | | | | | | | | | | | | | | | + | + | + | | | | |
| *Cellepora canaliculata* Busk | arc.circ.-p. | | | | | + | | | | + | + | + | | + | | | | + | + | + | + | + | + | | |
| *Escharella macrodonta* Levinsen | arc.amer. | | | | | | | | | | | + | | | | | | | | | | | | | |
| *Smittina tuberosa* Kluge | arc.amer. | + | | | | | | | | | | | | | | | | | | | | | | | |
| *Cystisella elegantula* (d'Orbigny) | arc.amer. | | | | + | + | + | + | + | + | + | | | | | | | | | | | | | | |
| *Membraniporella crassicosta* Hincks | arc.amer. | | | | | + | + | | | + | | | | | | | | | | | | | | | |
| *Hippoporina murdochi* Kluge | arc.amer. | + | | | | | | | | | | | | | | | | | | | | | | | |
| *Schizoporella ortmanni* Kluge | arc.amer. | | | | + | | | + | | | | | | | | | | | | | | | | | |
| *Cheiloporina sincera* var. *praelucida* (Hincks) | arc.amer. | | + | + | | | | + | | + | + | | | | | | | | | | | | | | |
| *Vesicularia fasciculata* Soule | arc.amer. | | + | | | | | | | | | | | | | | | | | | | | | | |
| *Alcyonidium vermiculare* Okada | pac.b | | | | | | | | | | | | | | | | | | | | | | | + | |
| *Tegella anguloavicularis* Kluge | pac.b | | | | | | | | | | | | | | | | | | | | | | | + | |
| *Tegella inermis* Kluge | pac.b | | | | | | | | | | | | | | | | | | | | | | | | + |
| *Tegella amissavicularis* (Kluge) | pac.b | | | | | | | | | | | | | | | | | | | | | | | | + |
| *Callopora obesa* Kluge | pac.b | | | | | | | | | | | | | | | | | | | | | | | | + |
| *Dendrobeania pseudolevinseni* Kluge | pac.b | | | | | | | | | | | | | | | | | | | | | | | + | |
| *Bugula tschukotkensis* (Kluge) | pac.b | | | | | | | | | | | | | | | | | | | | | | | + | |
| *Cystisella saccata* var. *beringia* Kluge | pac.b | | | | | | | | | | | | | | | | | | | | | | | + | |
| *Porella tumida* (Kluge) | pac.b | | | | | | | | | | | | | | | | | | | | | | | | + |
| *Schizobrachiella stylifera* var. *perforata* (Kluge) | pac.b | | | | | | | | | | | | | | | | | | | | | | | | + |

| Species | Group | 1 | 2 | 3 | 4 | 5 | 6 | 7 | 8 | 9 | 10 | 11 | 12 | 13 | 14 | 15 | 16 | 17 | 18 | 19 | 20 | 21 | 22 | 23 | 24 |
|---|---|---|---|---|---|---|---|---|---|---|---|---|---|---|---|---|---|---|---|---|---|---|---|---|---|
| *Hippoponella fastigatoavicularis* (Kluge) | pac.b | | | | | | | | | | | | | | | | | | | | | | | | + |
| *Adeonellopsis tuberculata* (Busk) | pac.b | | | | | | | | | | | | | | | | | | | | | | | + | |
| *Smittina beringia* (Kluge) | arc.am. | + | | | | | | | | | | | | | | | | | | | | | | | |
| *Scrupocellaria minor* Kluge | amer. | | | | | | | | | | + | | | | | | | | + | + | | | + | + | |

*Note:* sub.-b., subtropical boreal; atl.b., Atlantic boreal widespread; atl.sub.-b., Atlantic subtropical boreal; atl.h.-b., Atlantic high boreal; amph.b., amphiboreal; b.-a., boreal-Arctic widespread; b.-a.atl.circ.-p., boreal-Arctic Atlantic circumpolar; b.-a.atl.eurasian, boreal-Arctic Atlantic Eurasian; b.-a.atl.am.-eur., boreal-Arctic Atlantic American-European; b.-a.pac.circ.-p., boreal-arctic Pacific circumpolar; b.-a-pac.amer.-as, boreal-Arctic Pacific Amerasian; b.-a.pac.eur.-as., boreal-Arctic Pacific Eurasian; arc.circ.-p., Arctic circumpolar; arc.eur.-as., Arctic Eurasian; arc.amer.-as., Arctic Amerasian; arc.eur.-am., Arctic Euroamerican; arc.amer., Arctic American; pac.b., Pacific boreal.

+ = species in: 1, northwestern coast of Alaska; 2, northern coast of North America; 3, the Archipelago of Canadian Islands; 4, Baffin Bay; 5, Davis Strait; 6, Hudson Bay; 7, Gulf of Labrador; 8, Newfoundland; 9, St. Lawrence Gulf; 10, western Greenland; 11, eastern Greenland; 12, Greenland Sea; 13, Iceland; 14, Jan Mayen; 15, Norwegian Sea; 16, North Sea; 17, northern Norway; 18, White Sea; 19, Barents Sea; 20, Kara Sea; 21, Laptev Sea; 22, East Siberian Sea; 23, Chukchi Sea; 24, Bering Strait.

This table is based on the authors' present data and the literature listed.

## Table 14-2. Biogeographic Distribution of the Bryozoan Fauna of the Arctic Ocean (%)

| *Biogeographic Groups* | *Entire Arctic Ocean* | *Regions* 1 | 2 | 3 | 4 | 5 | 6 | 7 | 8 | 9 | 10 | 11 | 12 | 13 | 14 | 15 | 16 | 17 | 18 | 19 | 20 | 21 | 22 | 23 | 24 |
|---|---|---|---|---|---|---|---|---|---|---|---|---|---|---|---|---|---|---|---|---|---|---|---|---|---|
| sub.-b. | 0.6 | 0 | 0 | 0 | 1.4 | 0 | 1.6 | 2.6 | 5.9 | 1.65 | 0 | 0.8 | 0 | 1.8 | 0 | 0 | 0 | 1.4 | 0 | 0.3 | 0 | 0.85 | 0 | 0.9 | 0 |
| atl.sub.-b. | 2.9 | 0 | 0 | 0 | 1.4 | 2.6 | 3.2 | 2.6 | 5.9 | 2.5 | 2.5 | 0.8 | 0 | 1.8 | 1.4 | 0 | 14.3 | 1.4 | 2.3 | 3.8 | 1.6 | 0.85 | 0 | 0 | 0 |
| atl.b. widespread | 3.9 | 0 | 2.3 | 1 | 0 | 0 | 1.6 | 0 | 0 | 0.8 | 1.2 | 0 | 0 | 0 | 0 | 0 | 14.3 | 1.4 | 2.3 | 4.4 | 0 | 0 | 0 | 0 | 0 |
| atl.h.-b. | 9.2 | 0 | 0 | 0 | 0 | 0 | 0 | 0 | 0 | 0.8 | 1.2 | 2.6 | 0 | 1.8 | 0 | 0 | 0 | 4.1 | 3.1 | 9.5 | 0.5 | 0 | 0 | 0 | 0 |
| amph.-b. | 5.0 | 0 | 2.3 | 2.0 | 0 | 0 | 3.2 | 1.3 | 5.9 | 4.1 | 3.1 | 0.8 | 0 | 1.8 | 0 | 0 | 9.5 | 5.5 | 4.6 | 4.8 | 0.5 | 0 | 0 | 3.6 | 0 |
| b.-a. widespread | 19.2 | 47 | 47.1 | 48.5 | 47.9 | 31.6 | 53.3 | 59.3 | 35 | 40.6 | 38.9 | 41 | 0 | 46.4 | 46.5 | 0 | 47.7 | 43.7 | 44.2 | 23.5 | 32.1 | 33.3 | 31.4 | 43.6 | 0 |
| b.-a.atl.circ.-p. | 7.1 | 0 | 5.8 | 10.1 | 2.8 | 2.6 | 3.2 | 3.9 | 11.8 | 11.6 | 10.5 | 9.4 | 0 | 8.9 | 16.4 | 0 | 9.5 | 4.1 | 6.2 | 8.1 | 10.3 | 10.2 | 10.8 | 6.4 | 0 |
| b.-a.atl.eur.-as. | 3.5 | 0 | 1.2 | 1.0 | 0 | 0 | 0 | 1.3 | 0.8 | 0.5 | 0.5 | 0.8 | 0 | 5.3 | 1.4 | 0 | 4.7 | 1.4 | 3.1 | 3.7 | 2.2 | 0.85 | 0.9 | 0 | 0 |
| b.-a.atl.am.-eur. | 29 | 0 | 3.4 | 2.0 | 2.8 | 5.3 | 1.6 | 3.9 | 0 | 3.3 | 3.1 | 2.5 | 0 | 5.3 | 1.4 | 0 | 0 | 5.5 | 2.3 | 3.7 | 3.3 | 2.6 | 2.0 | 0.9 | 0 |
| b.-a.pac.circ.-p. | 5.9 | 29.4 | 17.2 | 16.2 | 23.9 | 18.4 | 12.9 | 11.6 | 11.8 | 14 | 11.1 | 11.9 | 0 | 16.2 | 15.1 | 0 | 0 | 17.8 | 11 | 7.3 | 9.8 | 9.4 | 6.9 | 14.6 | 0 |
| b.-a.pac.am.-as. | 1.5 | 5.9 | 1.2 | 1.0 | 4.2 | 5.3 | 1.6 | 1.3 | 5.9 | 11.65 | 2.5 | 1.6 | 0 | 0 | 0 | 0 | 0 | 0 | 0.8 | 1.1 | 1.6 | 1.7 | 2.9 | 1.8 | 0 |
| b.-a.pac.eur.-as. | 3.5 | 5.9 | 0 | 0 | 0 | 0 | 0 | 0 | 0 | 0.8 | 1.2 | 0.8 | 0 | 0 | 0 | 0 | 0 | 0 | 2.3 | 2.6 | 3.8 | 4.3 | 4.9 | 4.6 | 0 |
| arc.circ.-p. | 12.4 | 5.9 | 17.2 | 14.2 | 11.4 | 23.7 | 14.6 | 9.3 | 11.8 | 14.9 | 20.9 | 22.2 | 100 | 8.9 | 16.4 | 50 | 0 | 12.3 | 13.1 | 15.1 | 19.6 | 25.6 | 22.6 | 12.7 | 0 |
| arc.eur.-as. | 13.6 | 0 | 0 | 0 | 0 | 0 | 0 | 0 | 0 | 0 | 0.5 | 0.8 | 0 | 1.8 | 0 | 50 | 0 | 1.4 | 3.9 | 10.3 | 12.6 | 9.4 | 17.6 | 3.6 | 0 |
| arc.am.as. | 0.9 | 0 | 0 | 0 | 0 | 0 | 0 | 0 | 0 | 0 | 0.5 | 0.8 | 0 | 0 | 0 | 0 | 0 | 0 | 0 | 0.3 | 0.5 | 0.85 | 0 | 1.8 | 0 |
| arc.eur.-am. | 1.2 | 0 | 0 | 2.0 | 1.4 | 2.6 | 0 | 0 | 0 | 0 | 0.5 | 1.6 | 0 | 0 | 0 | 0 | 0 | 0 | 0.8 | 1.5 | 1.6 | 0 | 0 | 0 | 0 |
| arc.amer. | 3.2 | 11.8 | 2.3 | 2.0 | 2.8 | 7.9 | 3.2 | 3.9 | 0 | 2.5 | 1.8 | 1.6 | 0 | 0 | 1.4 | 0 | 0 | 0 | 0 | 0 | 0 | 0 | 0 | 0 | 0 |
| pac.b. | 3.5 | 0 | 0 | 0 | 0 | 0 | 0 | 0 | 5.9 | 0 | 0 | 0 | 0 | 0 | 0 | 0 | 0 | 0 | 0 | 0 | 0 | 0 | 0 | 5.5 | 100 |

*Note:* Legend of biogeographic groups appears in footnote to Table 14-1.

# APPENDIX

## Plate 14-I

1. Proboscina sp., × 21, Barents Sea, incrusting kelp.
2. Tubulipora sp., × 35, Barents Sea, incrusting kelp.
3. Tubulipora sp., × 21, Barents Sea, incrusting kelp.
4. *Tubulipora soluta* Kluge, × 21, Barents Sea, incrusting kelp.
5. Idmoneoides sp., × 5, Barents Sea, on polychaetes tubes.

5*a*. Idmoneoides sp., fragment of zoarium with oecistome.

6. *Entalophora clavata* (Busk), × 3.5, Barents Sea, on rock.
7. Tubulipora sp., × 14, Barents Sea, incrusting kelp.
8. *Tubulipora eminens* Kluge, × 3.5, East Siberian Sea, on rock and argillaceous sand.

## Plate 14-II

1. Berenicea sp., × 10.5, Chukchi Sea, on rocks.
2. Fasciculiporoides sp., × 10.5, East Siberian Sea, on a hydrozoa.
3. Berenicea sp., × 7, Barents Sea, on polychaetes tubes.
4. *Lichenopora multicentra* Kluge, solitary zoarium, × 3.5, substratum: argillaceous sand with rocks, Laptev Sea.
5. *Lichenopora multicentra* Kluge, compound zoarium, × 6, Laptev Sea, on the legs of sea spiders.
6. *Lichenopora sibirica* Kluge, × 6, Laptev Sea, on hydrozoa, on argillaceous substratum.

## Plate 14-III

1. *Eucratea loricata* var. *arctica* (Kluge), × 17.5, Kara Sea, on rocks; argillaceous rocky substratum.
2. *Notoplires smitti* (Norman), × 14, Barents Sea, on shells; on argillaceous rocky substratum.
3. *Semibugula birulai* Kluge, × 28, Barents Sea, on rocks; on argillaceous rocky substratum.
4. *Dendrobeania fruticosa* (Packard), × 17.5, Barents Sea, on rocks; substratum: rocks, shells, clays.
   *a*. marginal avicularium side view, × 42.
   *b, c*. middle avicularium view from above, × 52.5.
5. *Dendrobeania levinseni* (Kluge), × 17.5, Kara Sea, on rocks.
   *a*. marginal avicularium side view, × 28.
   *b*. middle avicularium side view, × 28.
   *c*. middle avicularium frontal view, × 28.
6. *Flustra serrulata* Busk, Chukchi Sea, × 17.5, on hydrozoa; on argillaceous rocky substratum.

**PLATE 14-I**

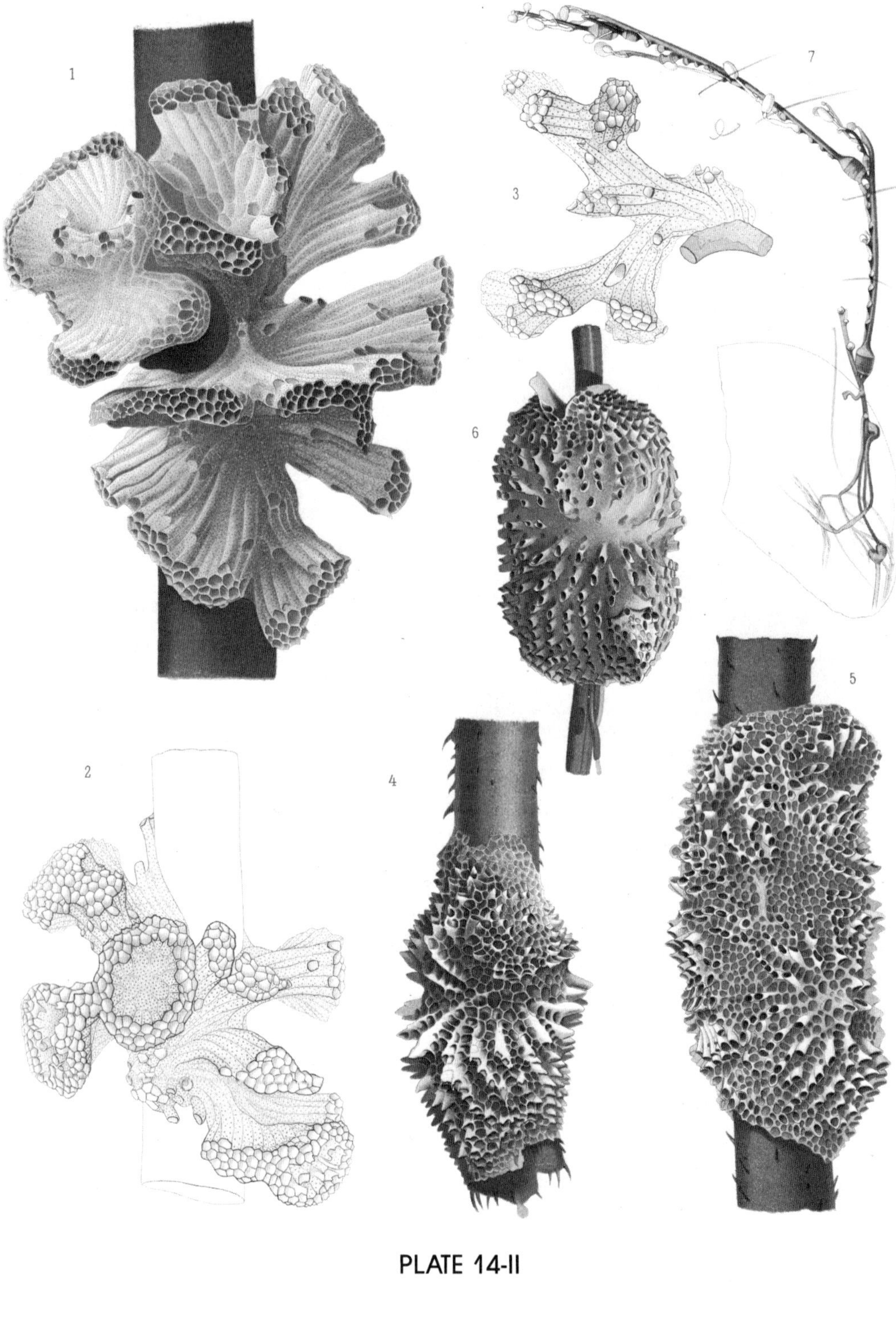


PLATE 14-II

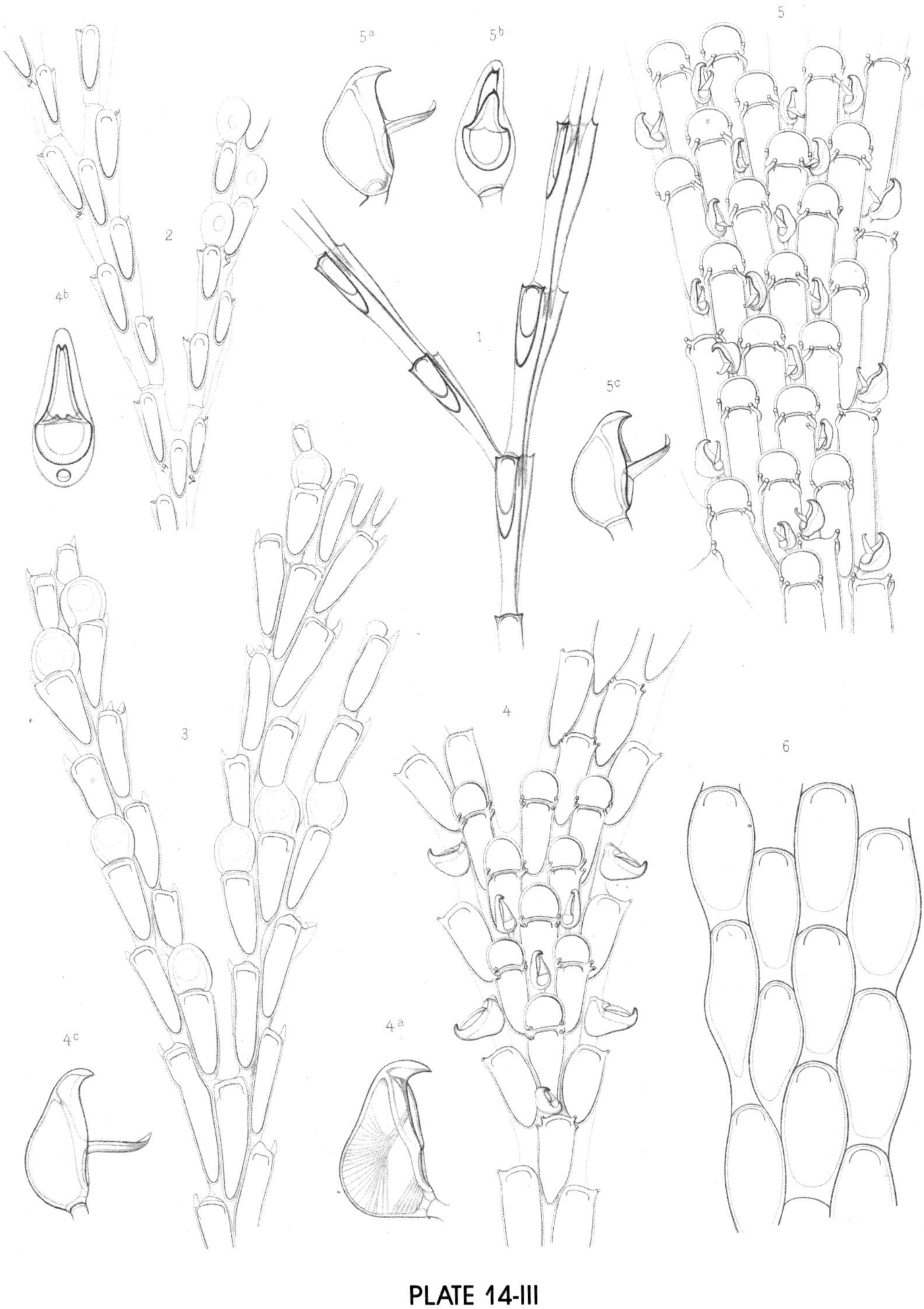


PLATE 14-III

## Plate 14-IV

1. *Callopora smitti* (Kluge), × 21, Kara Sea, on shells; on argillaceous-rocky-sandy substratum.
2. *Stomachetosella incerta* (Kluge), × 21, Barents Sea, on rocks.
3. *Schizoporella pachystega* (Kluge), × 21, Barents Sea, on shells, on argillaceous-rocky-sandy substratum.
4. *Pseudoflustra solida* (Stimpson), × 21, Laptev Sea, on rocks.
5. *Pseudoflustra birulai* (Kluge), × 21, Barents Sea, on rocks, on argillaceous rocky substratum.
6. *Rhamphostomella bilaminata* var. *sibirica* (Kluge), × 21, on shells, Chukchi Sea, on rocky argillaceous substratum.
7. *Rhamphostomella ovata* (Smitt), × 21, Barents Sea, on algae, on rocky argillaceous substratum.
8. *Cellepora canaliculata* Busk, × 28, on shells, Kara Sea.
9. *Celleporina nordenskjoldi* (Kluge), × 17.5, East Siberian Sea, on algae.

All illustrations were drawn by Kluge in 1916 but were never published.

## Map 14-1

Map of the seas of the Northern Hemisphere divided into biogeographic regions, subregions, and provinces.

I. Arctic subregion, Arc–Atlantic region.
II–IV. Atlantic boreal subregion, Arc–Atlantic region.
II. Celtic high boreal province.
III. Delaware low boreal province.
IV. Scandinavian low boreal province.
V. Aleutian high boreal subregion, Pacific boreal region.
VI. Ainian low boreal subregion, Pacific boreal region.
VII. Oregonian low boreal subregion, Pacific boreal region.

## Map 14-2

Distribution of the three Arctic species:

● *Dendrobeania fruticosa* (Packard)
■ *Semibugula birulai* Kluge
▲ *Rhamphostomella ovata* (Smitt)

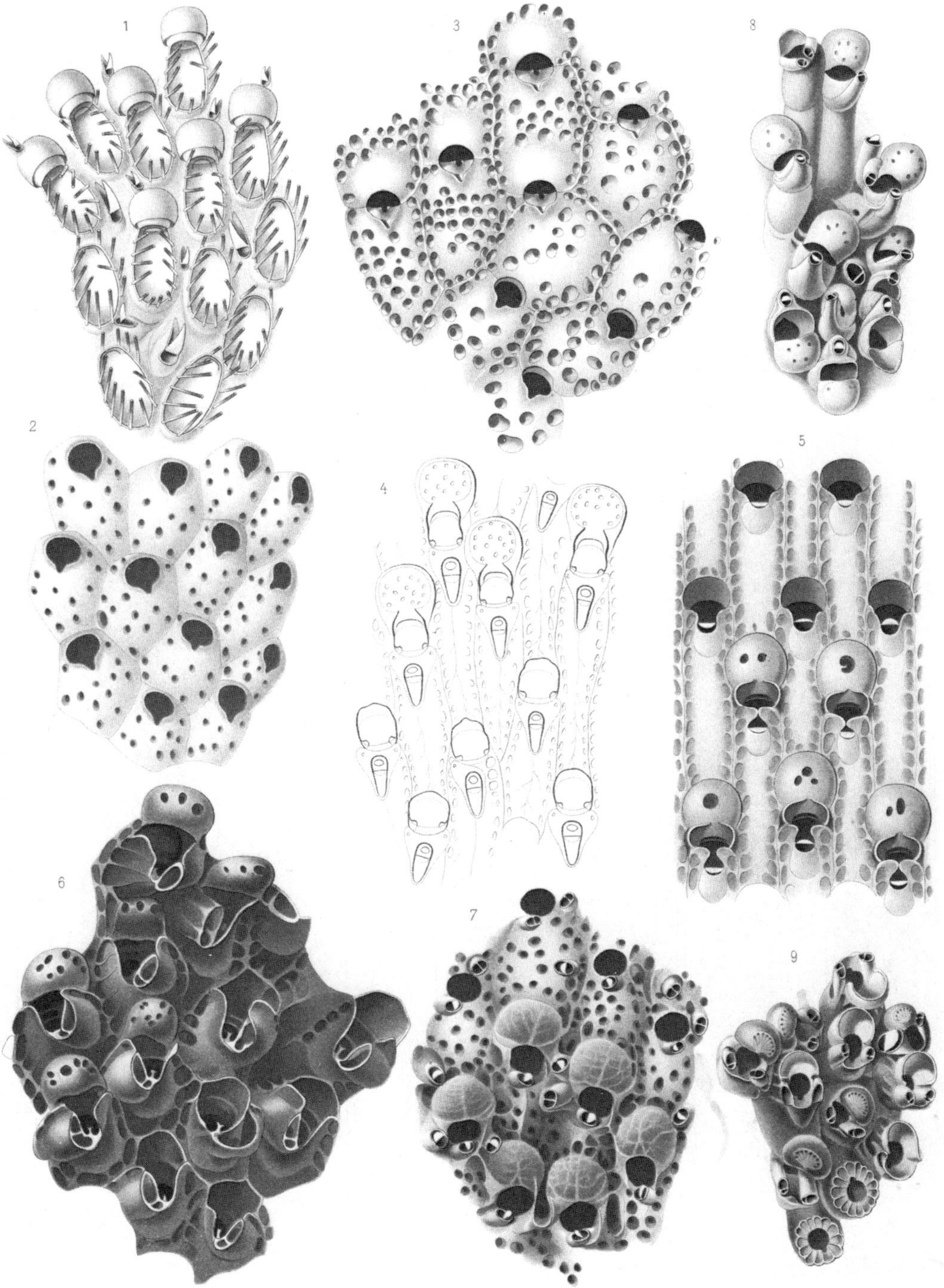


PLATE 14-IV

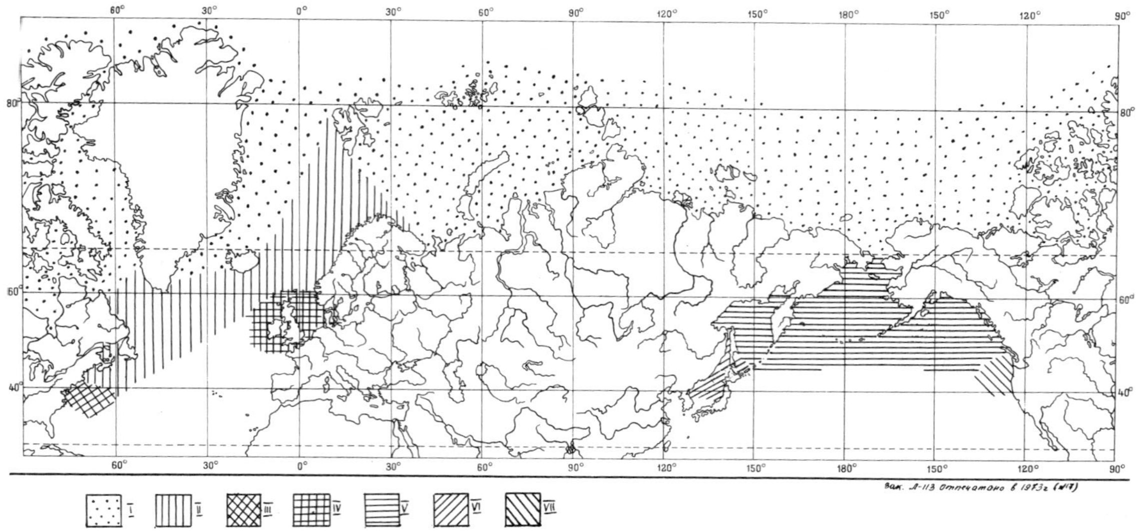


MAP 14-1

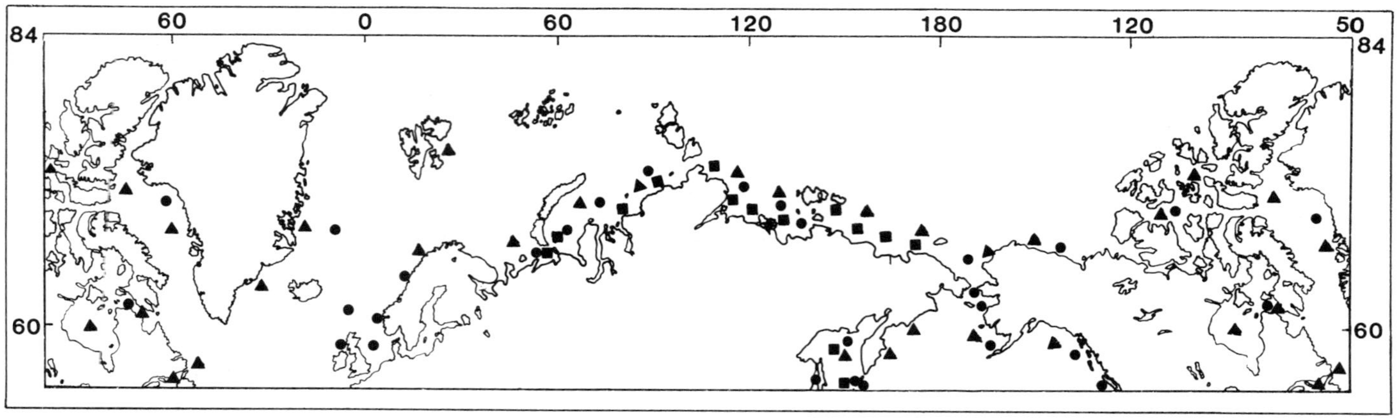


MAP 14-2

## REFERENCES


Gostilovskayja, M. G. 1978. *Opredelitel Mschanok Belogo Morya*. Leningrad, 246p. (In Russian.)

Hayward, P. J., and J. S. Ryland. 1979. British Ascophoran Bryozoa, in D. M. Kermack and R. S. K. Barnes (eds.), *Synopses of the British Fauna* (new series), **14**:312p.

Jaccard, P. 1901. Distribution de la flore alpine dous le Bassin de Drouces et dans quelques regions voisines. *Bull. Soc. Vandoise Sci. Natur.* **37**(140):241–272.

Kluge, G. A. 1961. *Mschanki Severnych Morei SSSR*. Moscow and Leningrad, 582p. (In Russian.)

Nordgaard, O. 1906. Bryozoa from 2nd "Fram" expedition 1898–1902. Rep. *2nd Norweg. Arctic. Exp. of the Fram,* pp. 1–44.

Nordgaard, O. 1912. Bryozoaires. *Duc d'Orleans Campagne Arctique de 1907.* Bruxelles, pp. 1–43.

Nordgaard, O. 1918. Bryozoa from the Arctic regions. *Bromso. Mus. Aarshefter XL (1917)* **1**:1–99.

Nordgaard, O. 1923. Bryozoa. *Rep. Sci. Res. Norweg. Exp. Novaja Zwmlya, 1921* **17**:1–19.

Nordgaard, O. 1929. Bryozoa. The Norweg. North. Polar Exp. with the "Maud" 1918–1925. *Sci. Res. V* **10**:1–10.

Norman, A. 1906. Greenlandic Polyzoa. *Ann. Mag. Nat. Hist.*, ser. 7, **27**:90–93.

Osburn, R. C. 1913. Bryozoa from Labrador, Newfoundland, and Nova Scotia, collected by Dr. Owen Bryant. *U.S. Nat. Mus. Proc.* **43**:275–289.

Osburn, R. C. 1932. Biological and oceanographic conditions in Hudson Bay and Strait. *Contrib. Canada Biol. Fish.* **7**(20):361–376 (1–10).

Osburn, R. C. 1936. Bryozoa collected in the American Arctic by Captain R. A. Barlett. *J. Wash. Acad. Sci.* **26**:538–543.

Packard, A. S. 1867. Observations on the glacial phenomena of Labrador and Maine, with a view of recent invertebrate fauna of Labrador. *Boston Soc. Nat. Hist. Mem.* **Part II**:270–303, pl. 7–8.

Smitt, F. 1879. Recensio systematica animalium Bryozoorum, quae in itineribus, annis 1875 et 1876, and insulas Novaja Zemlja et adostium fluminis jenisei duce, Professore A. E. Nordensrioli. *Ofvers. Kongl. Vetensk. Akad. Forh.* XXXV, **3**:11–26.

*Treatise on Invertebrate Paleontology.* 1983. Bryozoa 1, Raymond C. Moore and R. A. Robinson (eds.). Geological Society of America and University of Kansas Press, 621p.

Vigelius, W. 1882. Catalogue of the Polyzoe collected during the Dutch North Polar Cruises of the "Willem Barents" in 1878 and 1879. *Niederl. Arch. Zool. Suppl. I* **3**:1–20, pl. 1.

Vigelius, W. 1884. Die Bryozoen, gesammelt wahrend der dritten und vierten Polarfahrt des "Willem Barents" in den Jahren 1880–1881. *Bijdr. Dierkunde* **11**:1–104.

Waters, A. W. 1900. Bryozoe from Franz-Josef Land, collected by the Jackson–Harmsworth Expedition, 1896–1897. Part I: Cheilostomata. *Linn. Soc. Zool. London J.* **27**:43–105.

Waters, A. W. 1904. Bryozoe from Franz-Josef Land, collected by the Jackson–Harmsworth Expedition, 1896–1897. Part II: Cyclostomata, Ctenostomata and Endroprocts. *Linn. Soc. Zool. London J.* **29**:161–184.

# 15: Arctic Ocean Mysids *(Crustacea, Mysidacea)*: Evolution, Composition, and Distribution

V. V. Petryashov

## INTRODUCTION

Forty species of mysids belonging to 20 genera, 2 families, and 1 suborder have been described from the Arctic Ocean. These species belong to 5 biogeographic groups. There is a decline in the number of species northward and eastward of the adjacent Atlantic boreal waters (from 27 to 4 or 3 species) and a slight increase in the area of mixing of Arctic and Pacific waters (up to 5 species). Atlantic boreal and amphiboreal species are predominant in the Norwegian Sea, the southwestern part of the Barents Sea, and the Greenland Sea; Pacific boreal species dominate in the Chukchi Sea off the Alaskan coast, and boreal–Arctic and Arctic species dominate in other areas of the Arctic Ocean. Species that belong to different biogeographic groups occur in specific water masses. The distributional patterns of the recent mysid fauna were established at the onset of the Postglacial transgression with the development of present-day water circulation in the Northern Hemisphere.

*Note:* This chapter has been reviewed, edited, and rewritten by the Editor, Yvonne Herman.

## PREVIOUS STUDIES

Studies of the Arctic Ocean Mysidacea (restricted to the area north of the Wyville–Thomson Ridge, the southern coast of the Baffin Bay, the eastern part of the Hudson Strait, and the Bering Strait) commenced in the first half of the nineteenth century. Ross (1835) mentioned *Praunus (= Mysis) flexuosus* in his work on marine invertebrates collected during his second voyage to the Arctic regions. Further studies of mysids from this area were carried out in the second half of the nineteenth century by Walker (1862), Jarzynsky (1870), Sars (1870), and Miers (1877). These investigations dealing with mysids collected mainly in the Norwegian, Barents, and White seas were followed by a study of Mysidacea off the northwestern and northern coast of Alaska (Murdoch, 1885*a,b*). Simultaneously intensive studies of mysids from the Norwegian, Greenland, and Barents seas as well as from the Baffin Bay commenced (Sars, 1885, 1886; Hansen, 1888, 1908; Aurivillius, 1896; Vanhöffen, 1897; Birula, 1897; Stebbing, 1900; Ohlin, 1901; Gerstaecker and Ortmann, 1901; Ortmann, 1901).

The results allowed Zimmer (1904, 1909) to draw the first conclusions concerning the Arctic Mysidacea based on his study of 32 species from the Arctic Ocean within the geographical limits mentioned above. Studies by Linko (1907, 1908) off the northern Russian coast and by Stephensen (1912, 1913, 1918, 1933, 1938) in the seas surrounding Greenland were subsequently published. Investigations by Paulson (1909), Kramp (1913), Björck (1916), Schmitt (1919), Madsen (1936), Bertelsen (1937), Saemundsen (1937), and Dunbar (1940) dealing with mysids or coastal plankton containing Mysidacea from Iceland, Greenland, and the North American coast should also be mentioned. During the last decades the Arctic Ocean mysids were studied by Holmquist (1949, 1958, 1959, 1963, 1982), Rusanova (1962, 1963, 1971), and Kulikov (1980). In addition to the study of Arctic species, the investigation of this group in other regions was underway (Czerniavsky, 1887; Holt and Tattersall, 1905; Illig, 1930; Tattersall, 1951; Mauchline, 1980; Mauchline and Murano, 1977). However, to date the study of the Central Arctic Basin mysids has been inadequate.

## MATERIAL AND METHODS

The results presented in this chapter are based on our studies of Mysidacea from the collection of the Zoological Institute of the USSR Academy of Sciences (710 samples containing 5,500 specimens were analyzed). Most of these samples were collected with a Sigsbee dredge and a few were collected with a Djedy net.

## DISTRIBUTIONAL PATTERNS

Our results and literature data indicate that 40 species of mysids inhabit the Arctic Ocean; they belong to 1 suborder, 2 families, and 20 genera. The number of species inhabiting the Arctic Ocean constitute 5 percent of the world fauna.

Changes in species composition in different sectors of the Arctic Ocean were analyzed. The degree of similarity of faunas in these areas was determined using the Chekanovsky–Sørensen index (Sørensen, 1948):

$$I_{CS} = \frac{2 \cdot C \cdot 100\%}{D_1 + D_2}$$

and the degree of similarity of faunas was determined using the Jaccard index (Jaccard, 1901):

$$I_J = \frac{C \cdot 100\%}{D_1 + D_2 - C}$$

For the degree of inclusion of fauna of one area into the fauna of another region the index of Shimkevich (Simpson, 1943) was used:

$$I_{SZS} = \frac{C \cdot 100\%}{D_{min}}$$

For these three equations, $C$ is the number of species in common, $D_1$ and $D_2$ are the numbers of species in the areas compared, and $D_{min}$ is the minimum number of species in the areas compared ($D_{min} \leq D_{max}$).

The results of our calculations are given in Table 15-1. The data obtained indicate a rapid decline in the number of species northward and eastward from the adjacent Atlantic boreal waters (from 27 to 4 to 3 species) and a slight increase in the area of mixing of Arctic and Pacific waters (up to 5 species). The mysids of the Norwegian and southern regions of the Barents Sea are of boreal character and differ from their counterparts in other regions. The fauna of this area has close affinities with faunas of the coast of eastern Greenland and the southern Baffin Bay (as far as 75°NL); there is a great similarity between the faunas of these areas and the mysids from the continental slope off the Central Arctic Basin from Spitzbergen and up to the northern Laptev Sea. The fauna of most of the remaining area of the coastal Arctic region is rather monotonous. The only exception is the Chukchi Sea: Its northwestern regions support a fauna similar to that of the adjacent East Siberian Sea. Off of the Alaskan coast the mysids closely resemble the faunal composition of taxa from analogous coastal regions of the Siberian Sea and the Canadian Arctic Archipelago.

## BIOGEOGRAPHIC DISTRIBUTION

The following biogeographical groups are recognized:

**I.** Boreal-subtropical species

1. Atlantic upper sublittoral boreal-subtropical species: *Hemimysis lamornae:* in the Atlantic boreal and subtropical as well as in the Mediterranean–Lusitanian areas (Zimmer, 1904).
2. Atlantic bathypelagic boreal-subtropical species found in the adjacent Arctic sector as well: *Boreomysis tridens* (Plate 15-I, *a*), found in the Atlantic boreal and subtropical Mediterranean–Lusitanian areas and off the Greenland coast (Stephensen, 1933; Tattersall, 1951).
3. Amphiboreal-subtropical species, found in the Atlantic sector of the Arctic: *Boreomysis arctica* and *B. microps;* in boreal and subtropical waters of the Northern Hemisphere, in the Atlantic sector of the Arctic, reported off Greenland and off Franz Josef Land (Zimmer, 1904, 1909; Stephensen, 1913; Tattersall, 1951).

**II.** Atlantic boreal species

1. Upper sublittoral ubiquitous boreal species: *Neomysis integer, Praunus flexuosus, P. inermis* (Plate 15-I, *b*), and *Schistomysis ornata,* present in the temperate waters of the North Atlantic from the English Channel in the south to Murman and the southern Novaya Zemlya Island to the north (Zimmer, 1904, 1909; Linko, 1908).
2. Elittoral*-upper bathyal ubiquitous boreal species: *Erythrops microps, E. serrata, Pseudomma affine, Mysideis insignis, Mysidopsis didelphys,* and *Hemimysis abyssicola:* in the boreal waters of the North Atlantic and Arctic Ocean from the English Channel and Cape Cod to the south, up to the southern regions of Baffin Bay (up to 71–75°NL), along the eastern coast of Greenland (up to 76–78°NL), Nordkapp Cape, and West Murman to the north (Zimmer, 1904, 1909; Linko, 1908; Tattersall, 1951).
3. Bathyal ubiquitous boreal species: *Erythrops glacialis, Parerythrops spectabilis,* and *Michthyops parva.* Their distribution is analogous to species of the previous biogeographic group (Vanhöffen, 1897; Zimmer, 1904; Stephensen, 1913; Tattersall, 1951).
4. Sublittoral high boreal.† Species: *Dactylamblyops sarsi,* and *Mysis mixta* with distributions similar to species of the abovementioned biogeographic groups; however, they penetrate southward only as far as the coast of southern Norway, Faroes Islands, Massachusetts Bay, off the eastern coast of Greenland in the

*Elittoral zone of Kjelman (1877) ranging from 55–70 m to 120–180 m, bathyal zone of Appellof (1912) ranging from 200 m to 2000 m (in Golikov, 1985*a*).

†High boreal species are boreal species occurring in the northern parts of the boreal region; in the Atlantic and Arctic they do not penetrate south of Massachusetts Bay, Faroes Islands, Norway (off Bergen). Their northern limits of distribution are the same as those of the above-mentioned biogeographic group.

Table 15-1. Results of the Calculations of Biogeographic Indexes.

| | 1 | 2 | 3 | 4 | 5 | 6 | 7 | 8 | 9 | 10 | 11 | 12 | 13 | 14 | 15 |
|---|---|---|---|---|---|---|---|---|---|---|---|---|---|---|---|
| 1 | 14 | 37.5<br>64.29 | 24.24<br>57.14 | 22.58<br>50 | 30<br>42.86 | 50<br>61.54 | 37.5<br>75 | 44.44<br>66.66 | 28.57<br>100 | 50<br>80 | 31.85<br>71.43 | 20<br>75 | 11.76<br>40 | 5.0<br>14.29 | 13.33<br>100 |
| 2 | 54.55<br>9 | 19 | 43.75<br>73.68 | 34.38<br>57.89 | 22.22<br>42.85 | 45.45<br>76.92 | 17.39<br>50 | 34.78<br>66.66 | 9.52<br>50 | 31.82<br>70 | 18.18<br>57.14 | 9.52<br>50 | 9.09<br>40 | 4.0<br>14.29 | 10<br>66.66 |
| 3 | 39.02<br>8 | 60.87<br>14 | 27 | 88.88<br>100 | 37.67<br>78.57 | 25<br>61.54 | 12.9<br>50 | 21.88<br>58.33 | 0<br>0 | 15.63<br>50 | 9.68<br>42.86 | 0<br>0 | 0<br>0 | 0<br>0 | 0<br>0 |
| 4 | 36.84<br>7 | 57.89<br>11 | 94.12<br>24 | 24 | 40.75<br>78.57 | 27.59<br>61.54 | 14.29<br>50 | 24.14<br>58.33 | 0<br>0 | 17.24<br>50 | 10.71<br>42.86 | 0<br>0 | 0<br>0 | 0<br>0 | 0<br>0 |
| 5 | 42.86<br>6 | 36.36<br>6 | 53.66<br>11 | 57.89<br>11 | 14 | 42.11<br>61.54 | 46.67<br>87.5 | 44.44<br>66.66 | 12.5<br>50 | 33.33<br>60 | 10.53<br>28.57 | 12.5<br>50 | 11.76<br>40 | 5.0<br>14.29 | 13.33<br>66.66 |
| 6 | 59.26<br>8 | 62.5<br>10 | 40<br>8 | 43.24<br>8 | 59.26<br>8 | 13 | 40<br>75 | 66.66<br>83.33 | 13.33<br>50 | 64.29<br>90 | 33.33<br>71.43 | 13.33<br>50 | 12.5<br>40 | 5.26<br>14.29 | 14.29<br>66.66 |
| 7 | 54.55<br>6 | 29.63<br>4 | 22.88<br>4 | 25<br>4 | 63.64<br>7 | 57.14<br>6 | 8 | 33.33<br>62.5 | 33.33<br>75 | 38.46<br>50 | 25<br>42.86 | 33.33<br>75 | 18.18<br>40 | 7.14<br>14.29 | 22.22<br>66.66 |
| 8 | 61.54<br>8 | 51.61<br>8 | 35.9<br>7 | 38.89<br>7 | 61.54<br>8 | 80<br>10 | 50<br>5 | 12 | 23.07<br>75 | 83.33<br>100 | 35.31<br>71.43 | 14.29<br>50 | 13.33<br>40 | 5.56<br>14.29 | 15.38<br>66.66 |

| | | | | | | | | | | | | | | | |
|---|---|---|---|---|---|---|---|---|---|---|---|---|---|---|---|
| *9* | 44.44<br>4 | 17.39<br>2 | 0<br>0 | 0<br>0 | 22.22<br>2 | 23.53<br>2 | 50<br>3 | 37.5<br>3 | 4 | 27.27<br>75 | 57.14<br>100 | 60<br>75 | 28.57<br>50 | 10<br>25 | 75<br>100 |
| *10* | 66.66<br>8 | 48.28<br>7 | 27.03<br>5 | 29.41<br>5 | 50<br>6 | 78.26<br>9 | 55.55<br>5 | 90.91<br>10 | 42.86<br>3 | 10 | 41.66<br>71.43 | 16.67<br>50 | 15.38<br>40 | 6.25<br>14.29 | 30<br>100 |
| *11* | 47.52<br>5 | 30.77<br>4 | 17.65<br>3 | 19.35<br>3 | 19.05<br>2 | 50<br>5 | 40<br>3 | 52.63<br>5 | 72.72<br>4 | 58.82<br>5 | 7 | 37.5<br>75 | 20<br>40 | 7.69<br>14.29 | 42.86<br>100 |
| *12* | 33.33<br>3 | 17.39<br>2 | 0<br>0 | 0<br>0 | 22.22<br>2 | 23.53<br>2 | 50<br>3 | 25<br>2 | 75<br>3 | 28.57<br>2 | 54.55<br>3 | 4 | 50<br>75 | 22.22<br>50 | 40<br>66.66 |
| *13* | 21.05<br>2 | 16.67<br>2 | 0<br>0 | 0<br>0 | 21.05<br>2 | 22.22<br>2 | 30.75<br>2 | 23.53<br>2 | 44.44<br>2 | 26.66<br>2 | 28.57<br>2 | 66.66<br>3 | 5 | 33.33<br>60 | 33.33<br>66.66 |
| *14* | 9.52<br>1 | 7.69<br>1 | 0<br>0 | 0<br>0 | 9.52<br>1 | 10<br>1 | 13.33<br>1 | 10.53<br>1 | 18.18<br>1 | 11.76<br>1 | 14.29<br>1 | 36.36<br>2 | 50<br>3 | 7 | 11.11<br>33.33 |
| *15* | 35.29<br>3 | 18.18<br>2 | 0<br>0 | 0<br>0 | 23.53<br>2 | 25<br>2 | 36.36<br>2 | 26.67<br>2 | 85.71<br>3 | 46.15<br>3 | 60<br>3 | 57.14<br>2 | 50<br>2 | 20<br>1 | 3 |

Arranged Diagonally: Number of Mysid Species. Lower Left Corner Above Chekanovsky–Sørensen Coefficient: Number of Species the Regions Have in Common (bottom). Upper Right Corner: Jaccard Coefficient (top). Shimkevich–Simpson Coefficient (bottom).

Regions: 1, Baffin Bay; 2, eastern Greenland; 3, Norwegian Sea, region of Tronheim; 4, Norwegian Sea, region of Lofoten Islands; 5, southwestern Barents Sea; 6, northern Barents Sea; 7, Barents Sea southwest of the Novaya Zemlya; 8, northern Kara Sea and Novaya Zemlya Kara Sea trough; 9, southeastern Kara Sea and Laptev Sea; 10, northern Laptev Sea; 11, East Siberian Sea; 12, Chukchi Sea, east of Wrangel Island; 13, Chukchi Sea area off Point Barrow; 14, Bering Sea, Anadyr Bay; 15, Straits of the Canadian Arctic Archipelago.

southern parts of the boreal region; in the Atlantic and Arctic from the English Channel, the southern coast of Ireland, and Cape Cod (eastern coast of the United States) to the south; up to Novaya Zemlya, the western coast of Spitzbergen, Iceland, southern parts of the Baffin Bay, and in the Intermediate Water Mass of the Greenland Sea to the north; (up to 72°N) (Zimmer, 1904; Linko, 1908; Tattersall, 1951).

5. Bathypelagic high boreal species distributed in the sector adjacent to the Arctic: *Boreomysis nobilis* is present from the Faroes Islands, Iceland, eastern coast of Canada up to Greenland (up to 77°NL off the eastern coast and up to 75°NL in Baffin Bay) off Spitzbergen (Hansen, 1888; Vanhöffen, 1897; Ohlin, 1901; Zimmer, 1904, 1909) and Franz Josef Land.
6. Meso-bathypelagic amphiboreal species: *Amblyops abbreviata* (Plate 15-I, *c*), present in the temperate waters of the Atlantic, Pacific, and Arctic Ocean but absent from the Arctic water masses (Zimmer, 1904, 1909; Linko, 1908; Tattersall, 1951).

**III.** Sublittoral Pacific boreal species penetrating into the Chukchi and Beaufort Seas: *Neomysis rayii, Exacanthomysis arctopacifica,* and *Xenacanthomysis pseudomacropsis* (Murdoch, 1885*a,b;* Holmquist, 1982).

**IV.** Bathyal-pseudoabyssal* bipolar species: *Birsteiniamysis inermis* (Plate 15-I, *d*), observed in the Antarctic region and Subantarctic, in the North Pacific, North Atlantic, and Arctic (Zimmer, 1904; Stephensen, 1913; Tattersall, 1951).

*Pseudoabyssal of Derjugin (1915) basins in the bathyal zone 2,000–5,000 m depth not connected with the seafloor (in Golikov, 1985*a*).

**V.** Boreal–Arctic species

1. Sublittoral upper bathyal boreal–Arctic species: *Meterythrops robusta, Pseudomma truncatum* (Plate 15-I, *e*) and *Stilomysis grandis* (Plate 15-I, *f*), present in the temperate waters of the Atlantic, Pacific, and Arctic oceans (Zimmer, 1904, 1909; Linko, 1908; Stephensen, 1913; Tattersall, 1951); in the Arctic mainly in the Intermediate Water Mass from Spitzbergen up to the northern sector of the Laptev Sea, seldom penetrating the Arctic surface water.
2. Sublittoral upper bathyal Atlantic boreal–Arctic species: *Erythrops abyssorum, E. erythrophthalma* (Plate 15-I, *g*), *Parerythrops obesa, Pseudomma roseum,* and *Ps. frigidum* have the same distribution as the previous biogeographic groups, but not recorded from the Pacific Ocean (Zimmer, 1904, 1909; Linko, 1908; Stephensen, 1913; Tattersall, 1951).
3. Bathypelagic Atlantic boreal–Arctic species: *Amblyopsoides ohlinii,* found in the northwestern sector of the Atlantic Ocean and in the Norwegian Sea (Zimmer, 1904; Stephensen, 1913; Tattersall, 1951).
4. Sublittoral circumpolar Arctic–boreal species: *Mysis oculata* (Plate 15-I, *h*) and *M. litoralis,* found in the Arctic Ocean and adjacent temperate waters of the Atlantic and Pacific oceans (Zimmer, 1904, 1909; Linko, 1908; Tattersall, 1951; Holmquist, 1982).
5. Estuarine-freshwater circumpolar Arctic–boreal species: *Mysis relicta,* found in estuaries of rivers flowing into the Arctic Ocean and also in a number of rivers and lakes of northern Europe, northern Asia, and North America (Linko, 1908).

**VI.** Arctic species

1. Cryopelagic circumpolar species: *Mysis polaris* (Holmquist, 1959; Kulikov, 1980).

2. Sublittoral upper bathyal species: *Michthyops theeli,* distributed in the adjacent Atlantic sector of the Arctic Ocean from the eastern coast of Greenland (Ohlin, 1901) up to the northern Laptev Sea.
3. Bathypelagic species: *Pseudomysis abyssi,* observed in the Greenland and Norwegian seas (Zimmer, 1904).

## DISCUSSION AND CONCLUSIONS

Distributional patterns of species belonging to different biogeographic groups in the Arctic seas (Figs. 15-1 and 15-2) show that Atlantic boreal and boreal-subtropical species dominate the mysid fauna of the Norwegian Sea, the southwestern areas of the Barents and part of the Greenland seas (up to 76–78°NL); Pacific boreal species dominate the fauna of the Chukchi Sea, northwestern Alaskan coast; boreal–Arctic species dominate the fauna of other parts of the Arctic Ocean. Arctic species do not dominate the mysids of other biogeographic groups in any sector of the Arctic Ocean, if the mysid fauna is regarded as a whole and is not divided by separate water masses.

Areas from the Labrador Peninsula up to 75°NL off the west coast and 78°NL off the east coast of Greenland and from Nordkapp Cape up to the southwest coast of Novaya Zemlya and Franz Josef Land should be considered as an ecotone, a transitional zone where mixing of boreal and Arctic hydrobiocomplexes occur.* The fauna of the latter, in the coastal regions, is dominated by Arctic and sublittoral Arctic–boreal taxa belonging to the genus *Mysis.* Sublittoral upper bathyal and elittoral upper bathyal species dominate the fauna of the upper parts of the continental slope from Spitzbergen up to the northern sectors of the Laptev Sea and the Novaya Zemlya troughs of the Kara Sea. The region from Wrangel Island and Beechey Point (North Alaska) to the north and to Prince of Wales Cape and Olyutorsky Cape (North Kamchatka) to the south can be designated as an ecotone at the boundary with the Pacific boreal fauna.

The vertical distribution of mysids in the Arctic Ocean is presented in Figure 15-3. Maximum number of species is observed in the Norwegian Sea, where species attain greatest diversity between 200 m and 400 m depth. Northward and eastward the number of upper sublittoral, bathyal, and bathypelagic species declines. Thus, boreal and boreal-subtropical bathyal and bathypelagic species can penetrate easily along the coast of east Greenland with the Intermediate Water Mass, whereas boreal upper sublittoral species are not found here and the fauna of the coastal regions is represented by two sublittoral Arctic–boreal species. Bathyal and bathypelagic species are absent from the southwestern parts of the Barents Sea and therefore the maximum number of mysids occurs at 90–200 m depth. Likewise, maximum species diversity is observed off the northwestern Alaskan coast (Chukchi Sea) at 5–30 m depth. In the northern Laptev Sea, due to the predominance of elittoral-bathyal species, the number of taxa is the highest at 200–400 m depth.

The distribution of mysids at various depths is related to the distribution of water masses in the Arctic Ocean. Thus the brackish-freshwater species *Mysis relicta* occurs in water with temperatures not higher than 8°C or 9°C and salinities of 0–7‰. This species occurs in the Arctic Ocean in estuaries at depths of 0–20 m; in fresh water it was observed even at a depth of 117 m (Lake Onega). The cryopelagic species *Mysis polaris* is stenothermal (−1.7°C to +1.1°C) and stenohaline (29.47–34.42‰), living in the surface Arctic Water Mass and on the underside of ice.

*hydrobiocomplex = hydroecosystem

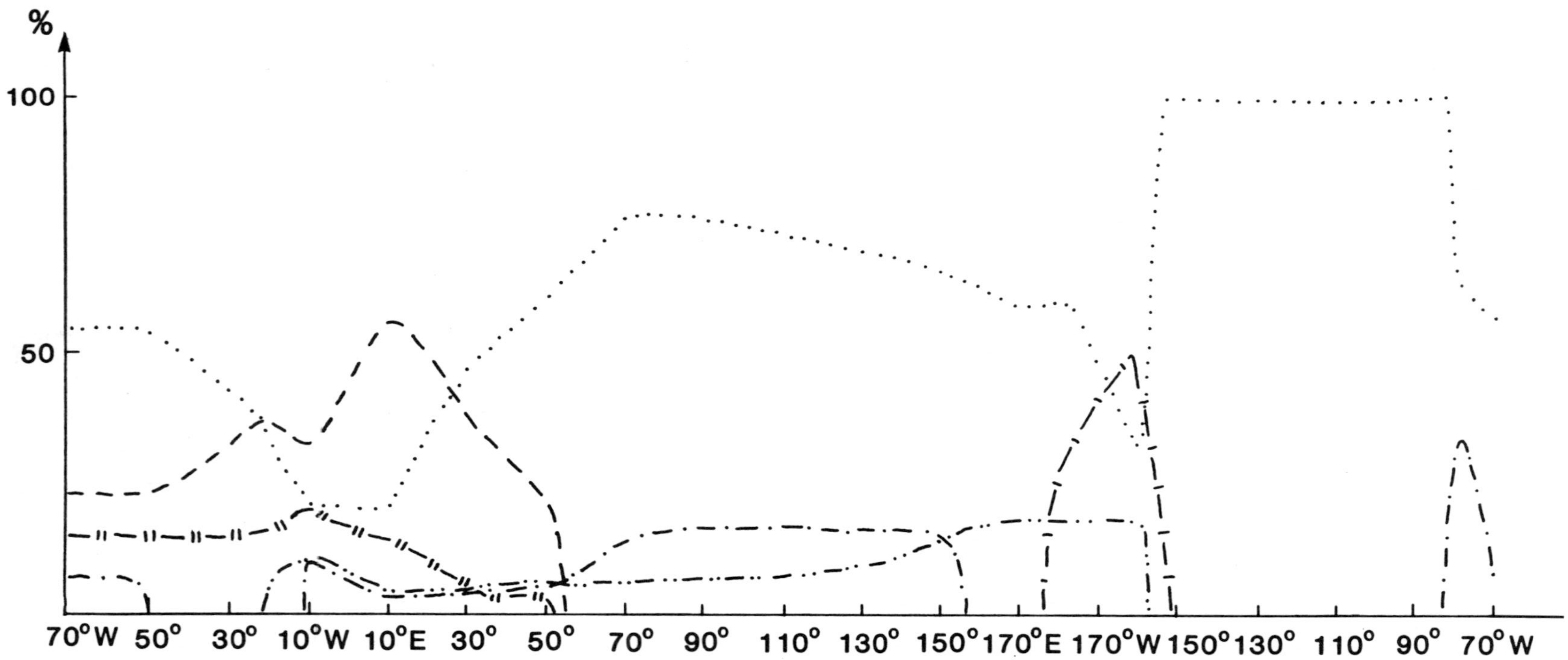


Figure 15-1. Number of mysid species belonging to different biogeographic groups in the Arctic Ocean. Abscissa: longitude; ordinate: percentage of species belcnging to a certain biogeographic group out of the total number of species at a given site (%):

boreal Atlantic species and amphiboreal species
boreal–Arctic species
Arctic species
bipolar species
boreal subtropical species
boreal Pacific species

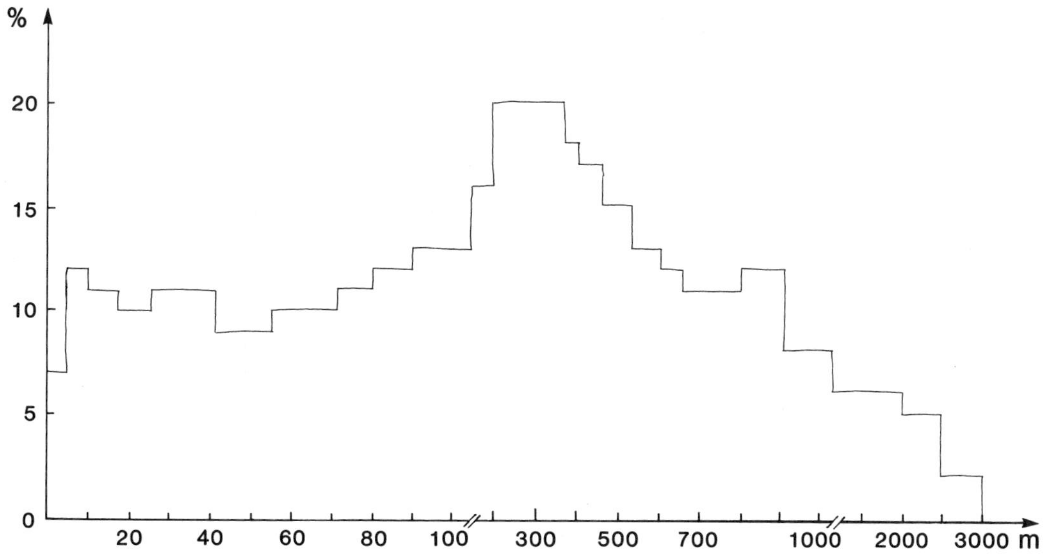


Figure 15-2. Distribution of Arctic Ocean mysids at various depth zones. Abscissa: depth in meters (m); ordinate: number of species (%).

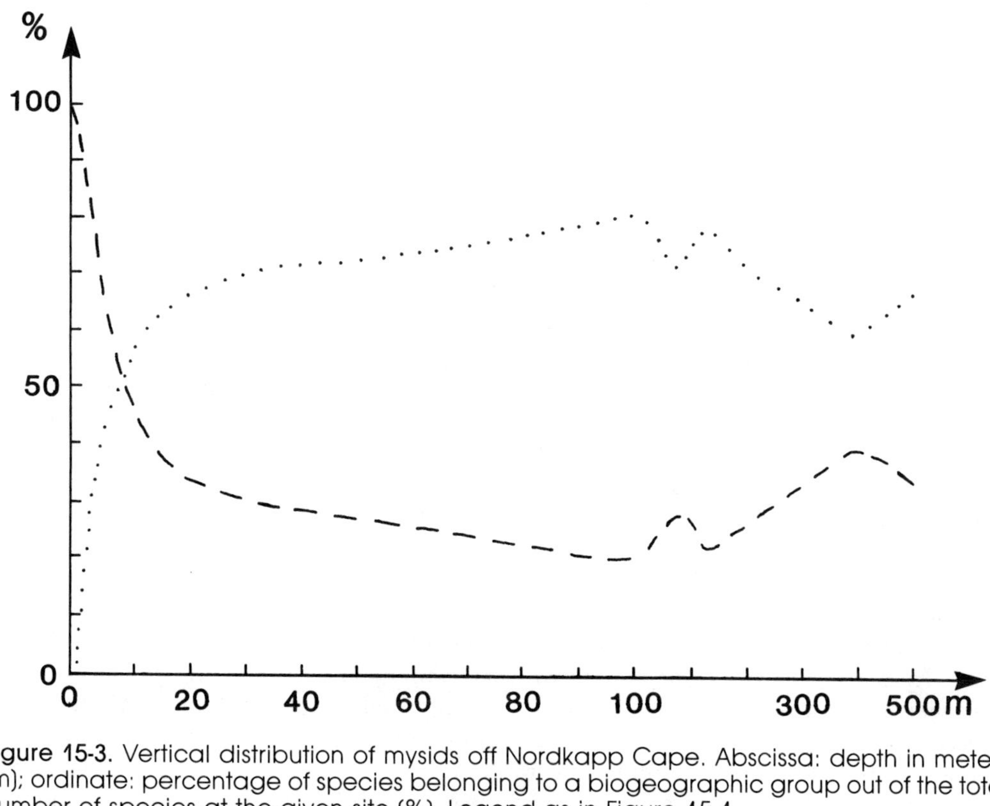


Figure 15-3. Vertical distribution of mysids off Nordkapp Cape. Abscissa: depth in meters (m); ordinate: percentage of species belonging to a biogeographic group out of the total number of species at the given site (%). Legend as in Figure 15-1.

The upper sublittoral species *Exacanthomysis arctopacifica, Hemimysis lamornae, Neomysis integer, N. rayii, Praunus flexuosus, P. inermis, Schistomysis ornata,* and *Xenacanthomysis pseudomacropsis* are eurythermal (−0.4° to +10.08°C) and euryhaline (23–32.66‰); some species occur even at river mouths where salinities are much lower than 23‰ and in the Arctic region where they have been observed in the surface North Atlantic and surface Pacific water masses at depths of 0–51 m. The sublittoral *Mysis litoralis, M. oculata,* and *M. mixta* are eurythermal (*M. mixta* 0.02°–13.68°C and *M. oculata* and *M. litoralis* −1.77°–5.4°C) and euryhaline (6.2–34.93‰); they have been observed when the water was saturated with oxygen (up to 45–96.3 percent). *M. mixta* occurs in the surface Atlantic Water Mass at 1–200 m depth while *M. oculata* and *M. litoralis* have been observed in surface Arctic water, estuarine–Arctic water and even in cool surface Atlantic and Pacific water masses at 1–249 m depth.

The sublittoral upper bathyal Stilomysis grandis, *Erythrops erythrophthalma, Pseudomma truncatum, Misidopis didelphys,* and probably *Dactylamblyops sarsi* can be both stenothermal (e.g., *E. erythrophthalma* and *Ps. truncatum,* −0.2° to −1.82°C, but probably occur at higher temperatures) and eurythermal (e.g., *St. grandis* −1.59° to +6.4°C); however, all are stenohaline (31.15–35‰) and have been observed in water saturated with oxygen (e.g., *E. erythrophthalma* at 53.3–83%, *Ps. truncatum* at 32.3–93.6‰ saturation). These taxa are found in the Arctic Ocean within the Atlantic boreal and intermediate Atlantic–Arctic water masses at 18–580 m depth. Very closely related to this group are the elittoral upper bathyal species *Erythrops abyssorum, E. glacialis, E. microps, E. serrata, Meterythrops robusta, Parerythrops obesa, P. spectabilis, Michthyops theeli, Pseudomma affine, Ps. roseum, Ps. frigidum, Mysideis insignis,* and *Hemimysis abyssicola.* In spite of the differences among these species in their thermopathy *Michthyops theeli* (1.19°C to −1.80°C), *Erythrops abyssorum* (−1.77°–7.22°C), *E. glacialis* (0.9°–1.1°C), *E. microps* (10.7°C), *Meterythrops robusta* (−1.77°–2°C), *Pseudomma roseum* (−1.77°–4.4°C), *Ps. frigidum* (0.6°–1.3°C), *Parerythrops spectabilis* (0.3°–1.0°C) all are stenohaline (31.47–35‰) and were observed in water saturated with oxygen (72.8–109.8%). These taxa occur at 54–900 m depth in the Atlantic boreal and a considerable number have been found in the Intermediate Atlantic–Arctic water masses.

The meso-bathypelagic species *Boreomysis arctica, B. nobilis, Amblyops abbreviata,* and *Michthyops parva* are also eurythermal (−0.82–5.5°C) and stenohaline (34.9–35.55‰), but inhabit depths of 140–1,250 m in the Atlantic boreal and Intermediate Atlantic–Arctic water masses. The distribution of bathypelagic species *Boreomysis microps, B. tridens, Birsteiniamysis inermis, Amblyopsoides ohlinii,* and *Pseudomysis abyssi* at various depths is more complex, although all are stenothermal (+1.0° to −1.0°C) and stenohaline (34.9–34.99‰). *B. microps* and *B. tridens* occur in the Atlantic boreal and the Intermediate Atlantic–Arctic water masses, while *A. ohlinii* is present in the Arctic deep-water mass at depths of 400–2,000 m. *B. inermis* inhabits depths of 900–2,970 m within the Arctic Ocean, and *P. abyssi* lives in the deep-water mass exclusively at 1,406–2,700 m depth.

It should be emphasized that in the Intermediate Atlantic–Arctic waters, which upwell in the Novaya Zemlya trough in the Kara Sea, a number of species occur at uncharacteristic depths. Thus *Erythrops abyssorum* (with a depth habitat of 180–540 m) rises here to 64 m and *Michthyops theeli* (119–575 m) rises here to 27 m. Analogous phenomenon has been observed in this region for other bathyal invertebrates (Zenkevich, 1963).

Information about the distribution of different water masses is provided by the composition of different biogeographic groups at different depths. Atlantic boreal species (Fig. 15-3) dominate at depths shallower than 10 m in the southwestern Barents Sea while boreal–Arctic species dominate at greater depths and the number of Atlantic boreal species slightly increases at depths of 400–600 m. In the southern areas of this sea, east of Kola Bay, boreal–Arctic species dominated within the entire water mass. This region is characterized by a sharp decline in the numbers of boreal taxa in the area between the Varanger fjord and Kola Bay.

Atlantic boreal species penetrate into the southern Barents Sea since the Atlantic Boreal Water Mass extends in this area from the surface to 600 m depth. Sublittoral Arctic-boreal species occur in the cooler regions of this water mass (Nordkapp branch of the North Atlantic Drift) in the southwestern areas of the Barents Sea and further east. This mysid group has been observed here mainly within a limited depth range (10–30 m though sometimes to depths of 200–240 m); the mysids do not dominate species of other biogeographic groups. Atlantic boreal waters do not penetrate the Kara Sea beyond its southern sector; therefore, Atlantic boreal species have not been observed in this sea.

In the Kara Sea and further east along the coastal regions of the Siberian Arctic seas, the surface Arctic and estuarine–Arctic water masses (except for the Chukchi Sea) at depths of 100–200 m are entirely dominated by sublittoral Arctic–boreal species and estuarine–freshwater Arctic–boreal taxa. These groups are replaced by cryopelagic circumpolar Arctic species only in cryobiocenoses. The distributional patterns of mysids off east Greenland are somewhat different (Fig. 15-4). Boreal–Arctic species occur in this area only to a

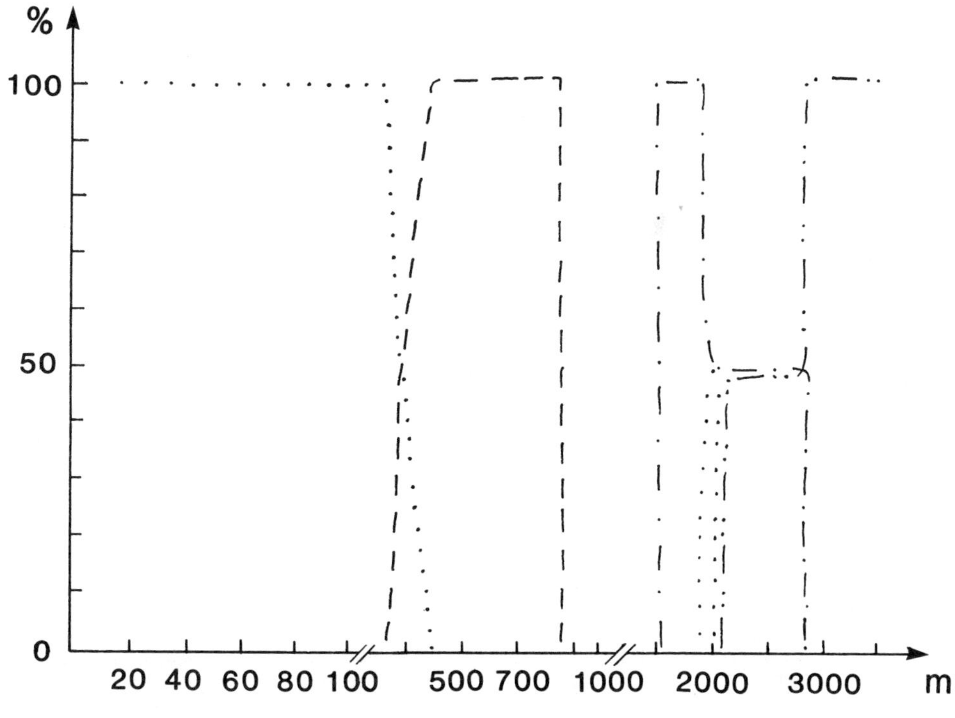


Figure 15-4. Vertical distribution of mysids off the eastern Greenland coast (75°). Abscissa: depth in meters (m); ordinate: percentage of species of a biogeographic group of the total number of species at the given site (%). Legend as in Figure 15-1.

depth of 200 m, in the surface water of the Greenland Sea. In the Intermediate Water Mass these taxa occur only in the uppermost layers, to a depth of 400 m. Atlantic boreal and amphiboreal-subtropical species found in the Arctic Ocean predominate in this water mass. Only two Atlantic boreal species penetrate to depths greater than 600 m, into the cold Arctic Water Mass and even then only in its upper layers (to a depth of 800 m). In the Greenland and Norwegian seas' basins at depths in excess of 1,400 m only the Arctic *Pseudomysis abyssi,* the boreal–Arctic *Amblyopsoides ohlinii,* and the bipolar *Birsteiniamysis inermis* have been observed to date.

The distributional patterns of mysids in different depth zones in the northern areas of Kara and Laptev seas is quite unusual. For example, the Arctic cryopelagic species *Mysis polaris* occurs in the northwestern Laptev Sea (Fig. 15-5) in the uppermost layers (0–5 m), and at depths of 5–40 m in the summer period the fauna is dominated by sublittoral Arctic–boreal species. The mysids of these two groups are typical representatives of the surface Arctic Water Mass. Lower layers of this water mass and the Intermediate Atlantic–Arctic Water Mass are dominated by sublittoral upper bathyal boreal–Arctic species. In the deep-water mass of the Central Arctic Basin, at depths in excess of 900 m, the bipolar *Birsteiniamysis inermis* has been found exclusively. The mysid fauna within the Arctic Ocean is most peculiar in the southeastern parts of the Chukchi Sea where some Pacific boreal species penetrate (Fig. 15-6). The latter dominate mysids off the northwestern Alaskan coast from the Bering Strait up to Cape Barrow at depths of 1–45 m. The great depths and regions not influ-

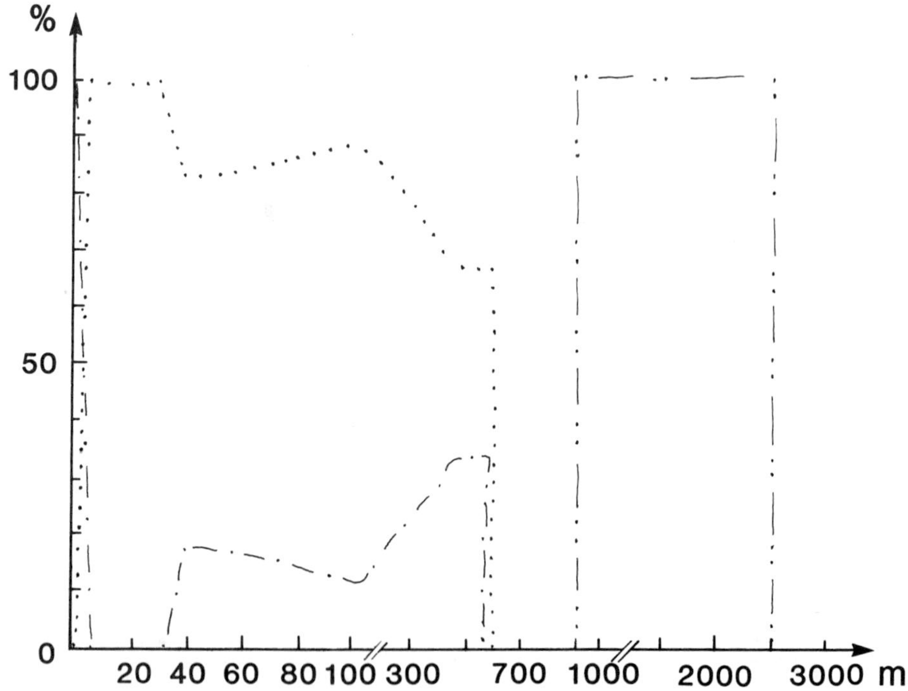


Figure 15-5. Vertical distribution of mysids in the northwestern Laptev Sea. Abscissa: depth in meters (m); ordinate: percentage of species of a biogeographic group out of the total number of species at the given site (%). Legend as in Figure 15-1.

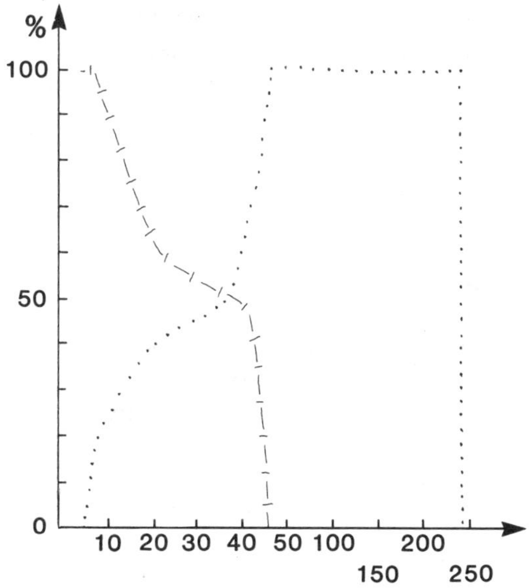


Figure 15-6. Vertical distribution of mysids off the Northwestern Alaskan Coast. Abscissa: depth in meters (m); ordinate: percentage of species of the biogeographic group out of the total number of species at a given site (%). Legend as in Figure 15-1.

enced by the Pacific surface waters are dominated by sublittoral Arctic–boreal species.

## EVOLUTION OF THE MYSID FAUNA

In the view of this author, the formation of the recent mysid fauna of the Arctic Ocean is related to the termination of the Last Glaciation (Würm–Valdai–Wisconsin Glaciation, 10–12 thousand years ago), the onset of the Postglacial transgression, the establishment of recent water circulation in the Northern Hemisphere, and the development of Arctic Ocean water masses with their specific temperature–salinity structure.

Each biogeographic group has its own history of distribution within the study area. The genus *Mysis* holds a particular place in the mysid fauna of the Arctic Ocean. In the Arctic and adjacent waters the species belonging to this genus may well have evolved in three directions: (1) by adaptation to conditions of the coastal Atlantic boreal waters (*M. gaspensis, M. mixta,* and *M. stenolepis*), (2) by adaptation to coastal Arctic regions (*M. oculata, M. litoralis,* and *M. relicta*), and (3) by adaptation to the epipelagic zone and later to the cryopelagic zone (*M. polaris*). The evolution of the cryopelagic species *M. polaris* is related to the onset of ice cover over the Arctic Ocean. Therefore, this species must have existed in the Arctic during the entire late Neogene ice age or at least during the Last Glaciation. Following the melting of ice sheets of the Last Glaciation along the shallow Arctic seas and the subsequent intensive freshening of coastal waters, the estuarine freshwater species *M. relicta* penetrated the coastal zones. At the same time, *M. oculata* and *M. litoralis* penetrated to great depths where salinities were exceeding 6–8‰. With decreased freshwater discharge from the continent, these species became distributed in coastal waters also, and *M. relicta* must have formed isolated populations in the river estuaries of the Arctic seas.

Mysids belonging to the genus Mysis (e.g., *M. relicta*), which originated in the period of intensive melting of glaciers and therefore strong freshening, were euryhaline and were able to penetrate from the White Sea through the area of the recent Onega and Ladoga lakes into the arm of the sea that existed at the locus of the present Baltic Sea (Lavrova, 1960). When this connection ceased, isolated populations of *M. relicta* developed in the deepest lakes, the areas occupied by a former marine basin. It is believed that at that time or during a previous boreal transgression mysids belonging to this genus penetrated from the region south of the recent Ladoga and Onega lakes through river systems or through Rybnoye Lake into the Volga Basin and further into the Caspian Sea (Högbom, 1917; Berg, 1928) where the Mysis fauna (*M. amblyops, M. caspia, M. micro-*

*lepis,* and *M. microphthalma*) endemic to the Caspian Sea originated.

The Atlantic boreal and the genetically related boreal–Arctic taxa (except for Arctic–boreal) is the largest group, including 36 species. Subsequent to the Last Glaciation, species belonging to these groups migrated from the southwestern European coast and the eastern coast of North America northward. This distribution was enhanced by the North Atlantic Drift and its branches that reach far north. Thus, for example, with the Nordkapp branch of the North Atlantic Drift, 12 boreal and boreal–Arctic species reach into the middle Murman region (Kola and Motoka fjord) and some taxa (*Pseudomma truncatum, Stilomysis grandis, Praunus inermis,* and *Mysis mixta*) extend their range to Novaya Zemlya.

The West Spitzbergen branch of the North Atlantic Drift has had a significant influence on the regional faunal composition. East of Spitzbergen this branch descends gradually under the surface Arctic water and in the Arctic Ocean descends under the Intermediate Atlantic–Arctic Water Mass. Near Spitzbergen it flows at a depth of 0–900 m, north of Franz Josef Land at a depth of 150–900 m, and north of Severnaya Zemlya at a depth of 200–900 m (Zenkevich, 1963). With this narrow water mass band, 9 elittoral upper bathyal and meso-bathypelagic species penetrate into the northern areas of the Barents, Kara, and Laptev seas including the Novaya Zemlya trough of the Kara Sea. The boundary of distribution of these species coincides almost entirely with the 0.5°C isotherm at 300 m depth (Zenkevich, 1963). Some boreal and boreal–Arctic species belonging to the main elittoral upper bathyal and meso-bathypelagic zones occur off the eastern Greenland coast bathed by the cool East Greenland Current. This occurrence is related to the formation of Greenland Sea's Intermediate Water Mass at 200–500 m depth. The penetration of Atlantic boreal and boreal–Arctic (except Arctic–boreal) species into Baffin Bay is via the Irminger Current and the relatively warm West Greenland Current.

The distribution of three Pacific boreal species in the Chukchi Sea is related to the development of the Postglacial transgression and the opening of the Bering Strait (Monin, 1980; *Morskaya Geomorphologiya,* 1980). These events led to the penetration of a warm current through the Bering Strait that bifurcates in the Chukchi Sea. The northeastern branch flows toward Point Barrow while the northwestern arm flows toward the region northeast of Wrangel Island. This current influences a restricted area, as compared to the influence of the North Atlantic Drift on the Arctic Ocean. This is due to the shallow depths of the Bering Strait, which limits the volume of warm water carried through the strait to 20,000 $km^3$ as compared to the 150,000 $km^3$ of Atlantic water carried into the Norwegian Sea from the south (Zenkevich, 1963).

Of the two species *Michthyops theeli* and *Pseudomysis abyssi* endemic to the Arctic Ocean, the former most probably is of boreal origin as suggested by its distribution, which is similar to that of a number of Atlantic boreal–Arctic species (Golikov, 1985*b*). The second species *Pseudomysis abyssi* may be an Ice Age relict in the Greenland and Norwegian seas. The genus Pseudomysis is amphiboreal. The ancestors of this taxon must have penetrated into the North Atlantic Ocean about 3 mya from the Pacific Ocean along the Canadian coast during a period when the Arctic was ice-free. The dispersal of other amphiboreal species and genera of mysids from the North Pacific into the North Atlantic may have been the same.

# APPENDIX

## Plate 15-1

Several characteristic mysids of the Arctic Ocean (after Zimmer, 1904): *a, Boreomysis tridens; b, Praunus inermis; c, Amblyops abbreviata; d, Birsteiniamysis inermis; e, Pseudomma truncatum; f, Stilomysis grandis; g, Erythrops erythrophthalma; h, Mysis oculata.*

## Map 15-1

Northern limit of distribution of mysids in the Arctic Ocean.

| | | |
|---|---|---|
| 1. | ——————— | Atlantic upper sublittoral boreal-subtropical species. |
| 2. | — — — — — | Bathypelagic boreal and boreal-subtropical species: |
| | △ | *Boreomysis arctica.* |
| 3. | ═══════ | Atlantic boreal and amphiboreal species: |
| | ○ | *Boreomysis nobilis.* |
| 4. | — — — — | Sublittoral Pacific boreal species. |
| 5. | —··——··— | Bathyal, pseudoabyssal bipolar species. |

## Map 15-2

Northern limits of distribution of mysids in the Arctic Ocean.

| | | |
|---|---|---|
| 1. | ············ | Northern limit of sublittoral, upper bathyal boreal–Arctic species. |
| | - - - - - - - | Northeast limit of *Erythrops abyssorum.* |
| 2. | △ | Bathypelagic Atlantic boreal–Arctic species. |
| 3. | —·—·—·— | Southern limits of sublittoral circumpolar Arctic–boreal species. |
| 4. | ■ | Cryopelagic circumpolar Arctic species. |
| 5. | —~——~— | Sublittoral, upper bathyal Arctic species. |
| 6. | □ | Bathypelagic Arctic species. |

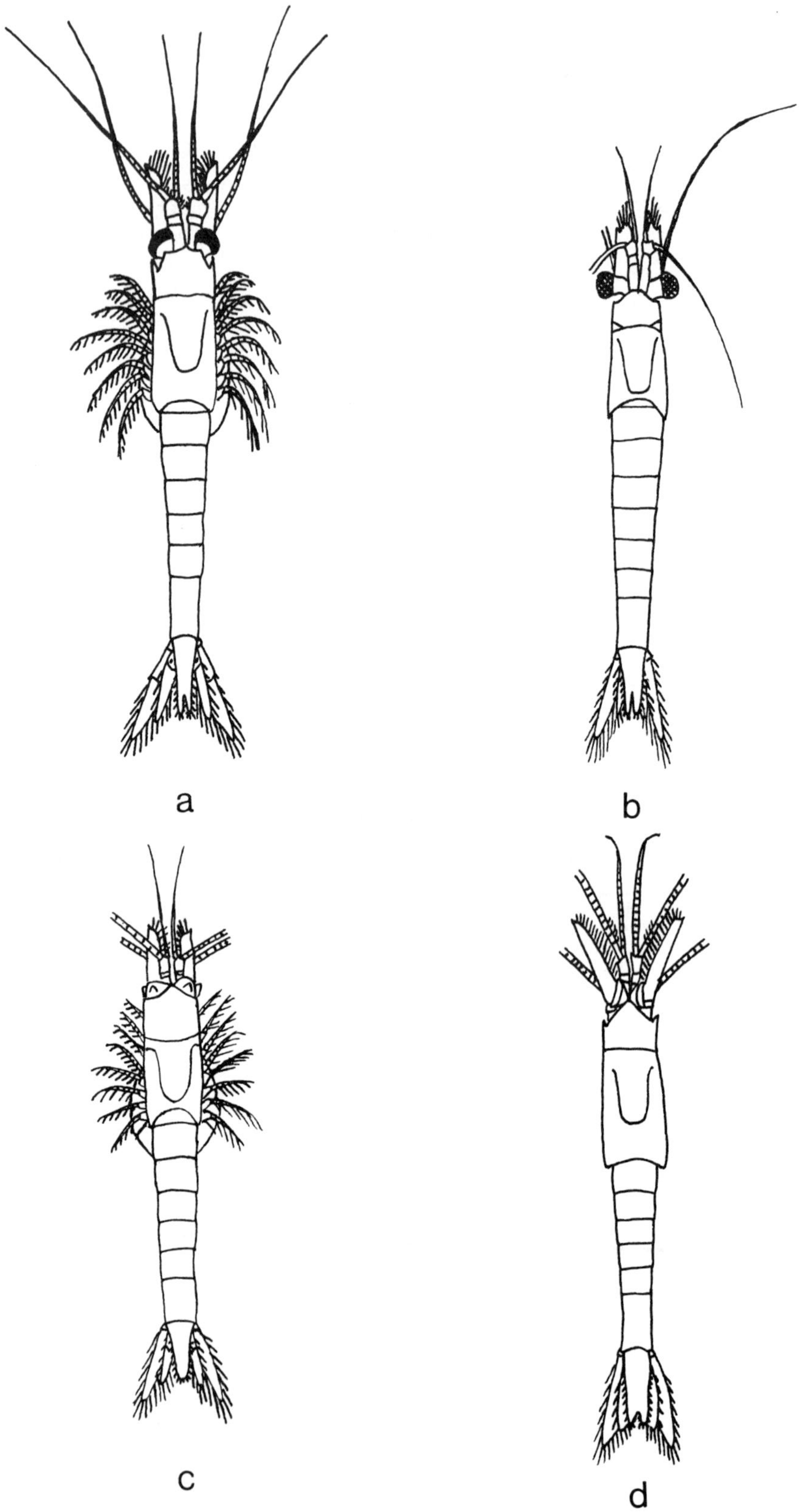


PLATE 15-I

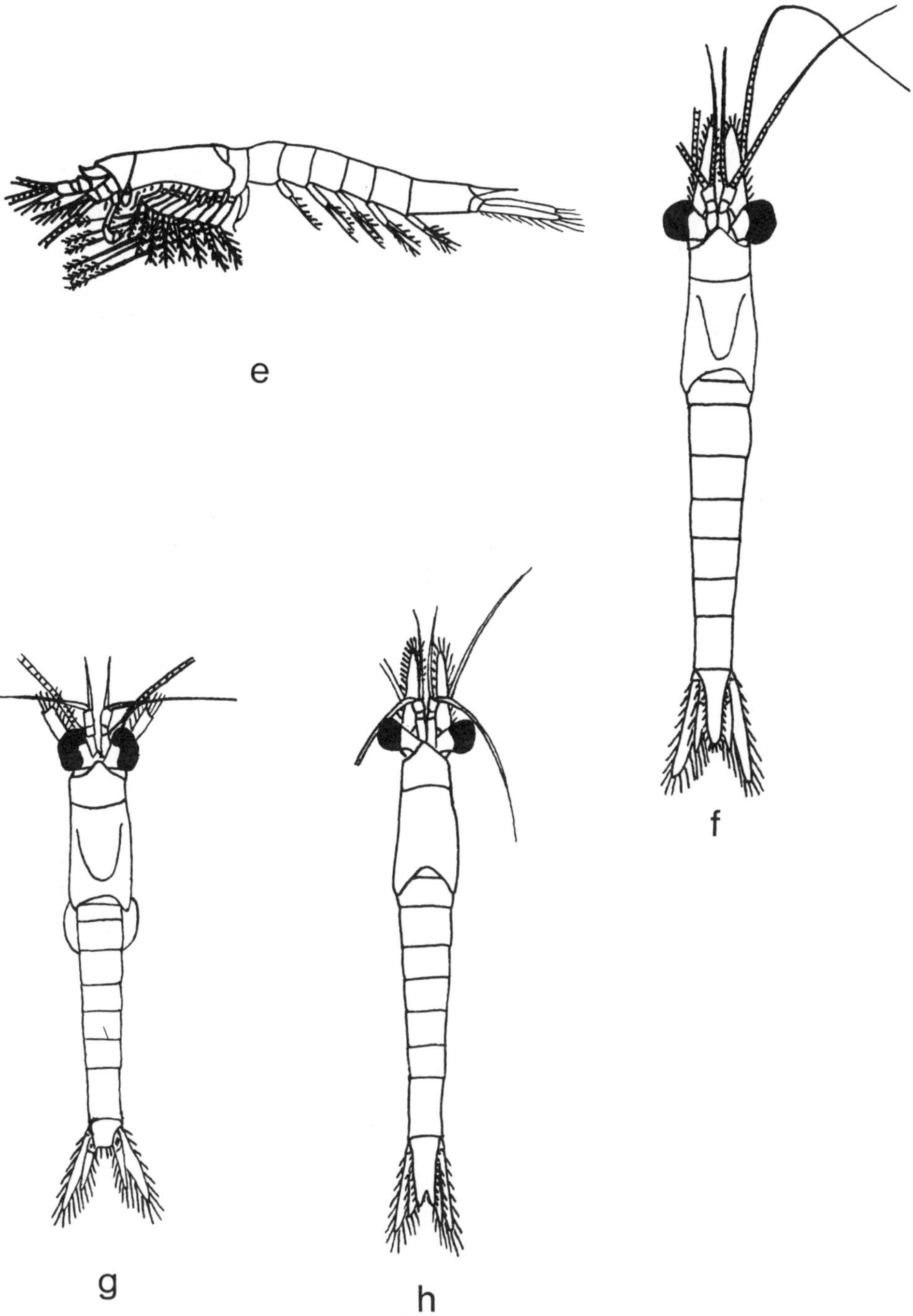


PLATE 15-I

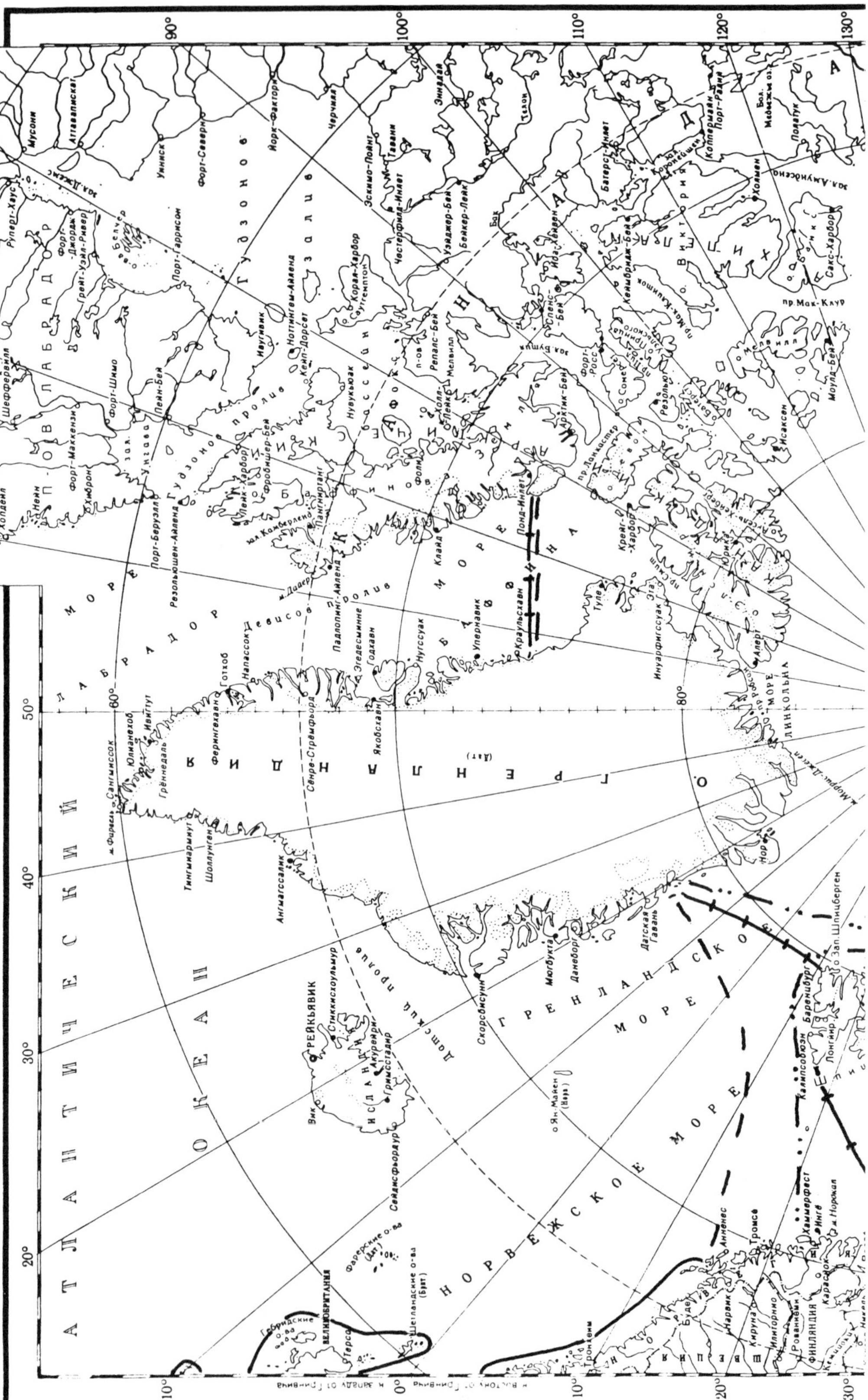
АТЛАНТИЧЕСКИЙ ОКЕАН
МОРЕ ЛАБРАДОР
Гудзонов залив
Гудзонов пролив
Девисов пролив
МОРЕ БАФФИНА
ГРЕНЛАНДИЯ
ГРЕНЛАНДСКОЕ МОРЕ
НОРВЕЖСКОЕ МОРЕ
ИСЛАНДИЯ
РЕЙКЬЯВИК
Датский пролив
ВЕЛИКОБРИТАНИЯ
ФИНЛЯНДИЯ

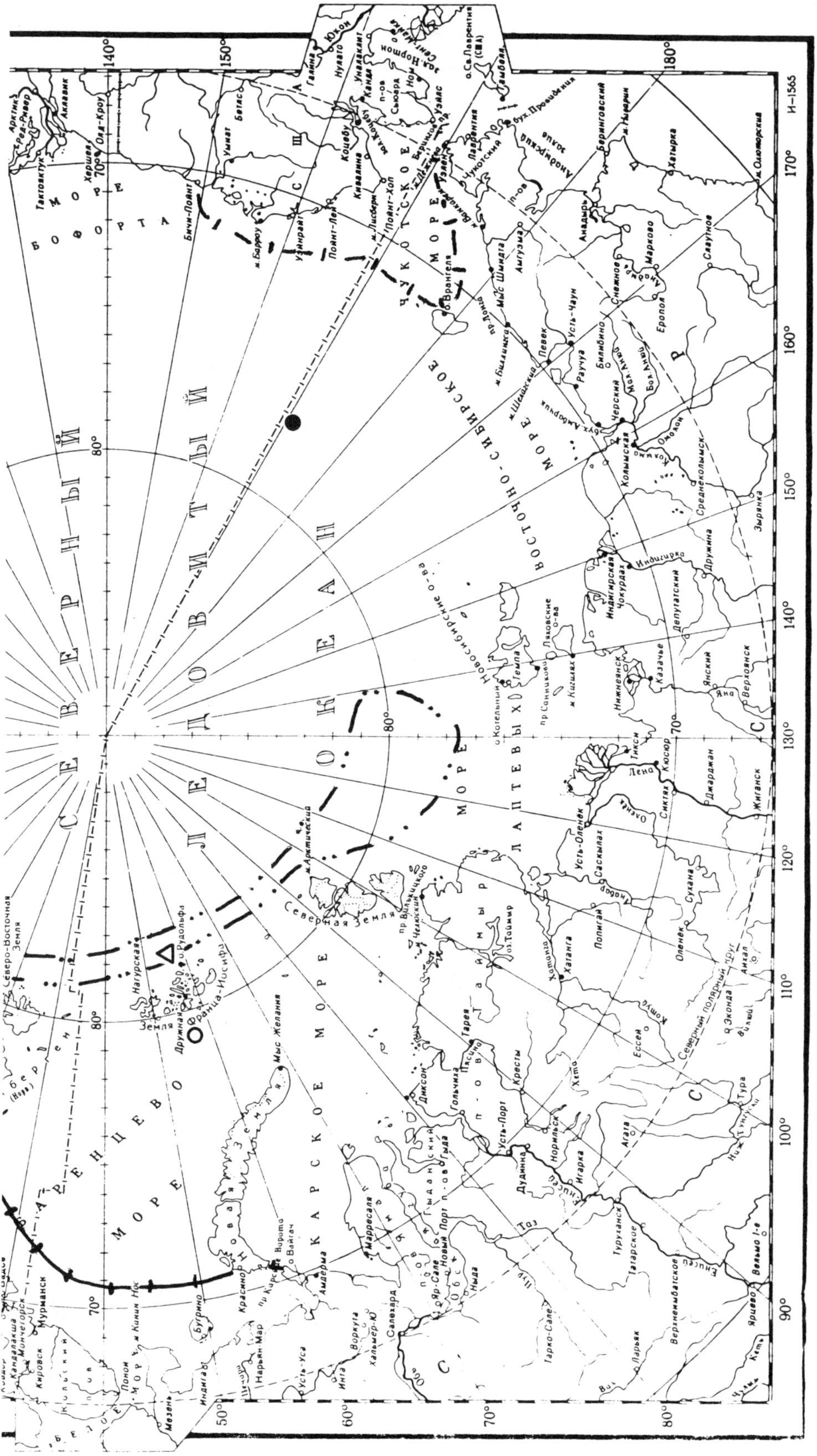
СЕВЕРНЫЙ ЛЕДОВИТЫЙ ОКЕАН
МОРЕ БОФОРТА
ЧУКОТСКОЕ МОРЕ
ВОСТОЧНО-СИБИРСКОЕ МОРЕ
МОРЕ ЛАПТЕВЫХ
КАРСКОЕ МОРЕ
БАРЕНЦЕВО МОРЕ

MAP 15-1

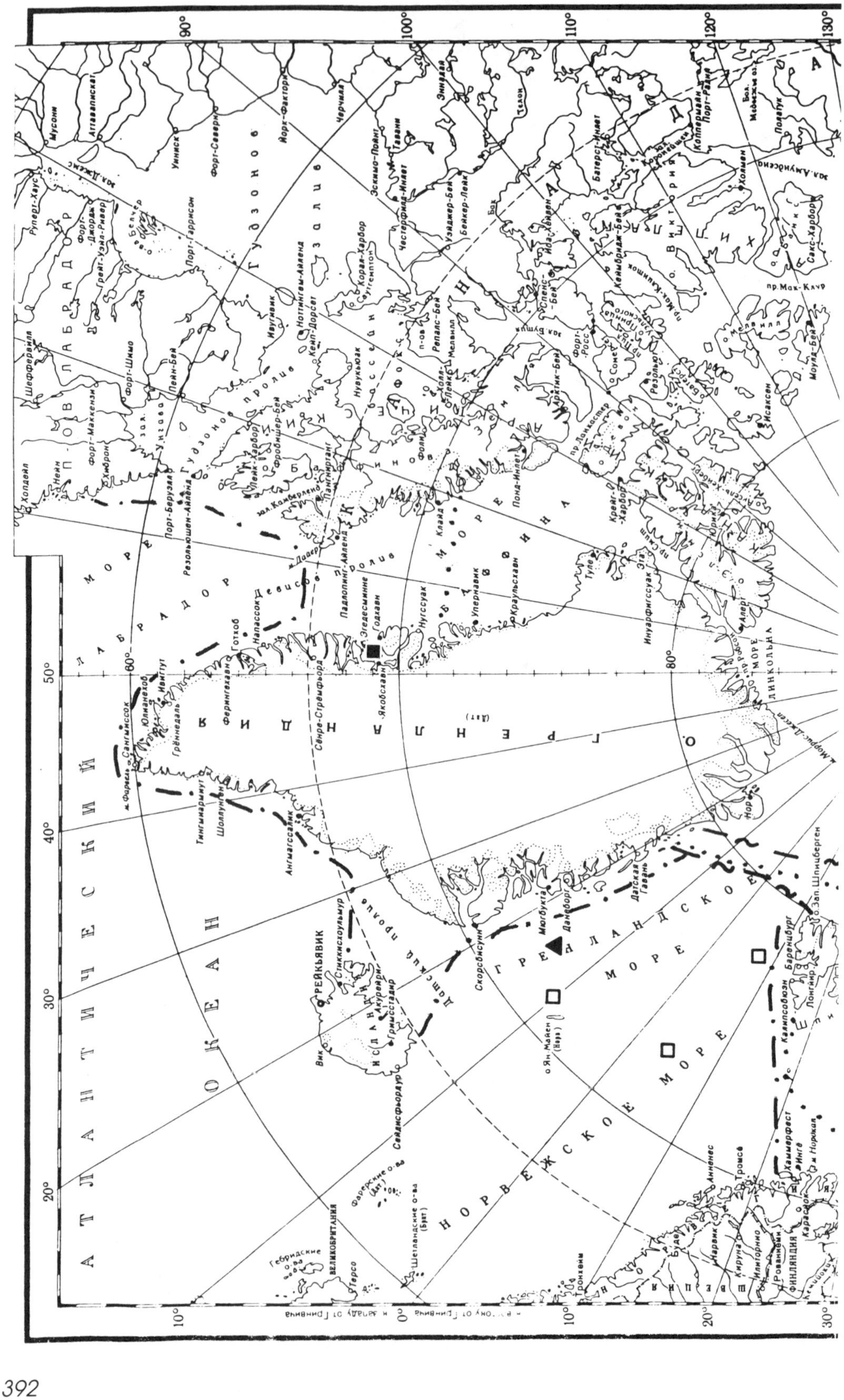
АТЛАНТИЧЕСКИЙ ОКЕАН
МОРЕ ЛАБРАДОР
п-ов ЛАБРАДОР
Гудзонов залив
Гудзонов пролив
Девисов пролив
море Баффина
ГРЕНЛАНДИЯ
ГРЕНЛАНДСКОЕ МОРЕ
НОРВЕЖСКОЕ МОРЕ
МОРЕ ЛИНКОЛЬНА
Датский пролив
ИСЛАНДИЯ
РЕЙКЬЯВИК
ВЕЛИКОБРИТАНИЯ
ФИНЛЯНДИЯ
Фарерские о-ва
Шетландские о-ва
Гебридские о-ва
о. Ян-Майен
о. Зап. Шпицберген
Годхавн
Туле
Алерт
Черчилл
Черчилл
Тромсё
Хаммерфест
Нарвик
Кируна
Тронхейм
Баренцбург
Лонгйир
Скорсбисунн
Ангмагссалик
Акурейри
Сейдисфьордур
Юлианехоб
Готхоб
Фробишер-Бей
Резольют
Эврика
90°
100°
110°
120°
130°
80°
60°
50°
40°
30°
20°
10°
0°

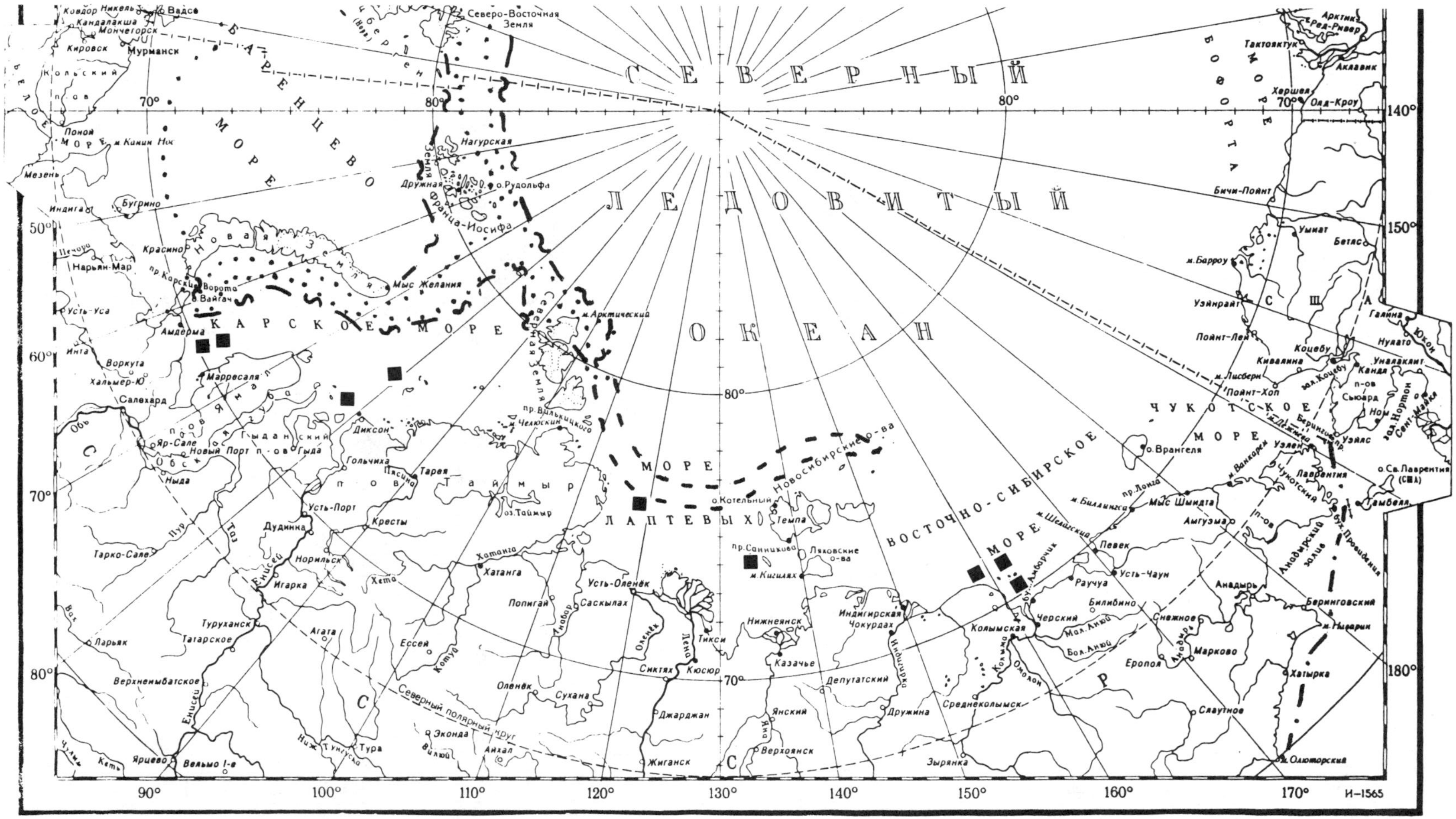

СЕВЕРНЫЙ ЛЕДОВИТЫЙ ОКЕАН
БАРЕНЦЕВО МОРЕ
КАРСКОЕ МОРЕ
МОРЕ ЛАПТЕВЫХ
ВОСТОЧНО-СИБИРСКОЕ МОРЕ
ЧУКОТСКОЕ МОРЕ
МОРЕ БОФОРТА
США
Северо-Восточная Земля
Земля Франца-Иосифа
Нагурская
о.Рудольфа
Новая Земля
Мыс Желания
Северная Земля
м.Арктический
Новосибирские о-ва
о.Котельный
Ляховские о-ва
о.Врангеля
Мурманск
Мончегорск
Кандалакша
Кировск
Кольский п-ов
Мезень
Нарьян-Мар
Воркута
Салехард
Амдерма
Марресаля
Диксон
Дудинка
Норильск
Игарка
Хатанга
оз.Таймыр
Тикси
Лена
Усть-Оленёк
Индигирка
Колымская
Черский
Билибино
Певек
Мыс Шмидта
Анадырь
Беринговский
Уэлен
Провидения
Верхоянск
Жиганск
Туруханск
Енисей
Северный полярный круг
Барроу
Ном
Коцебу
Юкон
90°
100°
110°
120°
130°
140°
150°
160°
170°
180°
И-1565


MAP 15-2

## REFERENCES


Aurivillius, C. W. S. 1896. *Das Plancton der Baffin's Bay und Davis Strait eine thiergeographische Studie.* Festschrift Lilljeborg, Uppsala, pp. 181–212, 1 plate.

Berg, L. S. 1928. O proiskhozhdenii severnykh elementov v faune Kaspiya. *Doklady AN SSSR* **14**(3):107–112. (In Russian.)

Bertelsen, E. 1937. Contributionals to the animals ecology of the fjords of Angmagssalik and Kangerdlugssuaq in east Greenland. *Medd. Grønland* **108**(3):1–58 and 2 plates.

Birula, A. 1897. Recherches sur la biologie et zoogeographie, principalement des mers Russes. III. Essai d'une faune des crustaces decapodes de la Mer Blanche et Mourmanne. *Ann. Mus. Zool. Acad. Imp. Sci., St. Petersburg* **2**:405–452 and 3 plates.

Björck, W. 1916. Die Schizopoden des Eisfjords. Zoologische Ergebnisse der Schwedischen Expedition nach Spitzbergen, 1908, Teil II. *Handl. K. Svenska Vetensk. Acad., Stockholm* **54**(6):1–10 and 1 map.

Czerniavsky, V. 1887. Monographia mysidarum imprimis imperii Rossici (marin., lacustr., et fluviatum). Fasc. III. *Trav. Soc. Imp. Nat. St. Petersburg* **18**:1–102 and 28 plates.

Dunbar, M. J. 1940. On the size distribution and breeding cycles of four marine planktonic animals from the Arctic. *J. Animal Ecol.* **9**:215–226, 5 figs.

Gerstaecker, A., and A. E. Ortmann. 1901. Malacostraca. In H. G. Bronn (ed.), *Klassen und Ordnungen des Tierreichs.* Leipzig, C. F. Winter'sche Verlagschandlung, Bd. 5, Abt. 1, Pt. 2, pp. 602–686.

Golikov, A. N. 1985*a*. Zonations and organismic assemblages: Comments on the comprehensive review by Peres (1982). *Mar. Ecol. Prog. Ser.* **23**:203–206.

Golikov, A. N. 1985*b*. Biologiya. Istoriya proishozhdeniya zhizni i biogeographicheskoe raionirovanie. *Geographiya Mirovogo Oceana-Severnyi Ledovityi i Juzhnyi Oceany, L.* pp. 114–119. (In Russian.)

Hansen, H. J. 1888. Malacostraca marine Groenlandiae occidentalis. Oversigt over det vestlige Groenlands fauna af malacostrake havkrebsdyr. *Vidensk. Meddel. Naturhist. For. Kjobenhaven* **9**(4):5–226.

Hansen, H. J. 1908. Crustacea Malacostraca. I. *Copenhagen, The Danish Ingolf Expedition* **3**(2):1–120, 4 figs., 5 plates, 1 map.

Högbom, A. G. 1917. Über die arktischen Elemente in der Aralokaspischenfauna, ein tiergeographisches Problem. *Bull. Geol. Inst. Univ. Upsala* **14**:241–260, 2 figs. and 1 plate.

Holmquist, Ch. 1949. Uber cventuelle intermediäre Formen zwischen Mysis oculata Fabr. und Mysis relicta Loven. *K. Fysiografiska Sällskapets Handlingar, N. F.* **60**(10):3–26.

Holmquist, Ch. 1958. On a new species of the genus Mysis, with some notes on *Mysis oculata (O. Fabricius). Medd. Grønland* **159**(4):3–17.

Holmquist, Ch. 1959. *Problems on Marine-Glacial Relicts, on Account of Investigations on the Genus Mysis.* Lund, 270p.

Holmquist, Ch. 1963. Some notes on Mysis relicta and its relatives in northern Alaska. *Arctic* **16**:109–128.

Holmquist, Ch. 1982. Mysidacea of the West Coast of North America. *Zoologische Jahrbücher* **109**(4):469–510.

Holt, E. W. L., and W. M. Tattersall. 1905. Schizopodous Crustacea from the north-east Atlantic slope. *Rept. Sea Inland Fisheries Ireland* pt. 2, app. 4, pp. 99–152, 11 plates.

Illig, G. 1930. Schizopoden der Deutschen Tiefsee Expedition. In Chun, *Wissenschaftliche Erbegnisse Deutsche Tiefsee—Expedition auf dem Dampfer "Valdivia" 1898–1899.* Jena, Gustav Fischer, Bd. 22, pp. 399–625.

Jaccard, P. 1901. Distribution de la flore alpine dans le Bassin de Dranses et dans quelques regions voisines. *Bull. Soc. Voudoise Sci. Natur.* **37**(140):221–272.

Jarzynsky, Th. 1870. Praemissas Catalogus crustaceorum decapodorum, inventorum, in mari albo, et in mari glacialis ad litus murmanicum auno 1869 et 1870. *Trudy Soc. Nat. Petropol.* **1**(2):317.

Jashnov, V. A. 1948. Otryad Mysidacea. *Opredelitel Fauny i Flory Severnykh Morei SSSR* pp. 224–229. (In Russian.)

Kramp, P. L. 1913. Schizopoda. In C. H. Ostenfeld (eds.), *Resume des Observations sur le Plankton des Mers Explorees par le Conseil Pendant les Annees 1902–1908.* 3 parts. A. F. Høst and Son, Copenhagen, pp. 539–556.

Kulikov, A. C. 1980. K. ecologii dvukh vidov gammarid (Amphipoda, Gammaridea) i misidy (Mysidacea) v kriopelagicheskom biotzenose Centralnogo Arcticheskogo Basseina. *Biologia Centralnogo Arcticheskogo Basseina.* pp. 111–118. (In Russian.)

Lavrova, M. A. 1960. *Chetvertichnaya Geologiya Kolskogo Poluostrova.* - M. -L., 234p. (In Russian.)

Linko, A. 1907. Schizopodes de la mer glaciale du Nord Russe. *Bull. Acad. Imp. Sci. St. Petersburg* **1**(6):491–493.

Linko, A. 1908. Schizopoda severnykh russkikh morei. *Mem. Acad. Imp. Sci. St. Petersburg* **18**(8):1–76 and 5 figs. (In Russian.)

Madsen, H. 1936. Investigations on the shore

fauna of east Greenland with a survey of the shores of other Arctic regions. *Medd. Grønland* **100**(8):1–79 and 17 figs.
Mauchline, J. 1980. The biology of mysids and euphausiids. *Adv. Marine Biol.* **18:**681p.
Mauchline, J., and M. Murano. 1977. World list of the Mysidacea, Crustacea. *J. Tokyo Univ. Fish.* **64:**39–88.
Miers, E. J. 1877. Report on the Crustacea collected by the naturalists of the Arctic expedition in 1875–1876. *Ann. Mag. Nat. Hist.* **20**(ser. 4):52–66 and 2 plates.
Monin, A. S. 1980. *Popularnaya Istoriya Zemli.* M. 224p. (In Russian.)
*Morskaya Geomorphologiya.* 1980. Terminologicheskii spravochnik. M. 279p. (In Russian.)
Murdoch, J. 1885*a*. Description of seven new species of Crustacea and one worm from Arctic Alaska. *Proc. U.S. Natl. Mus.* **7:**518–522.
Murdoch, J. 1885*b*. Marine invertebrates. *Exclusive of mollusks).* In *Report of the International Polar Expedition to Point Barrow, Alaska.* Part 4, Natural History, Washington, pp. 136–176 and 2 plates.
*Oceanographicheskaya Enciclopediya.* 1974. Pod red. Rhodes W. Fairbridge, 631p. (In Russian.) Translated from *The Encyclopedia of Oceanography,* 1966, Reinhold Publ.
Ohlin, A. 1901. Arctic Crustacea collected during the Swedish Arctic expeditions 1898, 1899 and 1900 under the direction of A. G. Nathorst and G. Kolthoff. II. Decapoda, Schizopoda. *Handl. K. Svenska Vetensk, Bihang* **27**(pt. 4):1–91.
Ortmann, A. E. 1901. Crustacea and Pycnogonida collected during the Princeton expedition to North Greenland. *Proc. Acad. Nat. Sci.* **53:**144–168.
Paulson, O. 1909. Plancton investigations in the waters round Iceland and in the North Atlantic in 1904. *Medd. Komm. Havundersøgelser, Copenhagen, ser. Plankton,* **1**(8):1–57 and 9 figs.
Ross, J. C. 1835. Marine invertebrate animals. In *Appendix to the Narrative of a Second Voyage in Search of a North-west Passage and of a Residence in the Arctic Regions during the Years 1829–1833 (Sir John Ross, captain).* A. W. Webster, London, pp. 81–150.
Rusanova, M. N. 1962. Materialy po biologii massovykh vidov misid Belogo morya. In *Gidrobiologicheskie issledovaniya. 3.* Tartu, pp. 136–140. (In Russian.)
Rusanova, M. N. 1963. Kratkie svedeniya po biologii nekotorykh massovykh vidov bespozvonochnykh raiona mysa Kartesh. *Materialy po Kompleksnomu Isucheniyu Belogo Morya* **2:**53–65. (In Russian.)
Rusanova, M. N. 1971. Genesis misid Belogo morya. In *Materialy XVI Conferencii po Isucheniyu Vnutrennikh Vodoyömov Pribaltiki.* Chapter I, Petrosavodsk, pp. 80–82. (In Russian.)
Saemundsson, B. 1937. Icelandic Malacostraca in the Museum of Reykjavik. *Reykjavik, Visendafelag Islendinga* **20:**1–32 and 1 map.
Sars, G. O. 1870. Carcinologiske Bidrag til Norges Fauna. I. Monographi over de ved Norges Kyster forekommende Mysider. Pt. I. *Christiania K. Norske Videnskabsselskab Trondhjem,* pp. 1–64 and 5 plates.
Sars, G. O. 1885. Crustacea I. In Den Norske Nordhavs Expedition, 1876–78. *Christiania* **14:**1–280.
Sars, G. O. 1886. Crustacea II. In Den Norske Nordhavs Expedition, 1876–78. *Christiania* pp. 1–96 and 1 map.
Schmitt, W. L. 1919. Schizopod crustaceans. In *Report of the Canadian Arctic Expedition 1913–18,* vol. 7. Ottawa, pp. 1–8.
Simpson, G. G. 1943. Mammals and the nature of continents. *Am. J. Sci.* **241**(1):1–31.
Sørensen, T. 1948. A new method of establishing groups of equal amplitude in plant sociology based on similarity of a species content and its application to analysis of the vegetation on Danish commons. *Kgl. Dan. Videnskab. Selskab. Biol. Skr.* **5**(4):1–34.
Stebbing, T. R. R. 1900. Arctic Crustacea: Bruce collection. *Ann. Mag. Nat. Hist.* **5**(ser. 7):1–16.
Stephensen, K. 1912. Report on the Malacostraca Pycnogonida and some Entomostraca collected by the Danmark expedition to northeast Greenland 1906–1908. *Medd. Grønland* **45:**501–630 and 5 plates.
Stephensen, K. 1913. Report on the Malacostraca collected by the "Tjalfe" expedition, under the direction of Ad. S. Jensen, especially at W. Greenland. *Vidensk. Medd. Dansk Naturhist. For. Kjøbenhavn* **64:**57–134 and 36 figs.
Stephensen, K. 1918. Grønlands Krebsdyr og Pycnogonider. *Medd. Grønland* **22:**479p.
Stephensen, K. 1933. The Godthaab expedition 1928. Schizopoda. *Medd. Grønland* **79**(9):1–20 and 6 figs.
Stephensen, K. 1938. Euphausiacea, Mysidacea, Cumacea and Nebaliacea. In A. S. Jensen et al. (eds.), *The Zoology of Iceland, Copenhagen and Reykjavik,* vol. 3, pt. 29. Levin and Munksgaard, pp. 1–24 and 1 plate.
Tattersall, W. M. 1951. A review of the Mysidacea of the United States National Museum. *Bull. U.S. Natl. Mus.* **201:**i–x, 1–292, 103 figs.
Vanhöffen, E. 1897. Die fauna und flora Grønlands. In E. von Drygalski (ed.), *Grønland Expedition der Gesellschaft für Erkunde zu Berlin, 1891–1893,* vol. 2. W. H. Kühl, Berlin, pp. 1–383, 30 figs., 8 plates, 1 map.
Walker, D. 1862. Notes on the zoology of the

last Arctic expedition under Capitain Sir F. L. M'Clintock, R. N. *J. Roy. Dublin Soc.* **3**:61–77.
Zenkevitch, L. A. 1963. *Biologiya Morei SSSR.* (In Russian.)
Zimmer, C. 1904. Die Arctischen Schizopoden. In *Römer Fritz und Fritz Schaudinn, Fauna Arctica,* vol. 3. Gustav Fischer, Jena, pp. 415–491 and 172 figs.
Zimmer, C. 1909. Die nordischen Schizopoden. In K. Brandt and C. Apstein (eds.), *Nordisches Plankton,* vol. 12. Lipsius and Tischer, Kiel and Leipzig, pp. 1–172 and 384 figs.

# 16: Hydrozoa of the Eurasian Arctic Seas

S. D. Stepanjants

## INTRODUCTION

Hydrozoans comprise one of the four classes in the phylum Cnidaria. Most members of the class have bottom sedentary (seldom pelagic) polypoid and pelagical medusoid stages of the life cycle. The polypoid generation may be an individual organism or a colony with the chitinous (rarely calcareous) skeleton. The polypoid produces the medusoid stage. The medusoid stage is not always represented by a free-swimming medusa, but rather by the medusoid, which remains attached to the colony and produces the sexual stage. Sometimes the typical medusoid or the typical polypoid stage is absent.

The majority of Hydrozoa (nearly 1,600 species) make up the subclass Hydroidea. The subclass Siphonophora comprises nearly 200 species. The siphonophores are colonial, mainly pelagic organisms, with several kinds of polypoid and medusoid zooids specialized for different functions such as feeding, movement, reproduction, and defense. The earliest Hydrozoa are believed to have evolved in the Precambrian as nonskeletal individual polyps. Skeleton-bearing fossils of Hydrozoa and jellyfish remains can be traced to the Cambrian. Recent Hydrozoa inhabit all seas and oceans, from cold polar to tropical waters. They dwell at different depths from the littoral zone to the ocean deeps (≈6.000 m). A number of Hydrozoa inhabit fresh and brackish waters.

*Note:* This chapter has been reviewed, edited, and rewritten by the Editor, Yvonne Herman.

## PREVIOUS STUDIES

The study of hydroids in the Eurasian Arctic seas commenced in 1772, with Lepechin's expedition to the White and Barents seas. In 1780, his article on several species of Sertularia was published, and *Sertularia* (= *Thuiaria*) *obsoleta* was the first Arctic species whose type specimen was discovered off Cape Svyatoi Nos (Barents Sea). The White Sea fauna was studied by Mereschkowsky (1877), Wagner (1885), Schlater (1891), Birulja (1896, 1897*a*, 1898), and Schidlowsky (1902). Also noteworthy are the following publications: M. Sars (1860,

1861, 1863), G. O. Sars (1873), Marenzeller (1878), and D'Urban (1881). They deal with the Barents Sea hydroids. Studies by Thompson (1884) describe the Barents Sea and the Kara and East Siberian seas' fauna (Thompson, 1887). Publications by Kirchenpauer (1884), Bergh (1886), Marktanner-Turneretscher (1890), Bonnevie (1898, 1899), Levinsen (1893), Hartlaub (1900), Jäderholm (1902), and Nordgaard (1904) are devoted to the hydroid fauna in several Arctic seas. We should also mention the reviews by Jäderholm (1908, 1909), the latter based on hydroids deposited in the Swedish Royal Museum.

Reviews by Broch (1910, 1916) on hydroids collected on the *R. V. Helgoland* during the Danish Expedition Ingolf represent important contributions to our knowledge of hydroids of the Arctic seas. In 1897, the study by Birulja on Hydrozoa, Polychaeta, and Crustacea collected in the Yenisei and Ob bays of the Kara Sea was published. This was followed by publications by Knipovitch (1901), Linko (1903, 1904), and Breitfus (1904) on hydroids, chiefly from the Barents Sea. A more comprehensive study of the Barents and Kara seas fauna (including hydroids) commenced in 1898 with the Murmansk Scientific Fishery Expedition (Breitfus, 1906, 1908; Derjugin, 1905, 1906, 1915). Hydroids deposited at the Zoological Museum of the Imperial Academy of Sciences in St. Petersburg collected in the above-mentioned expeditions were used for studies of various regions including the Arctic seas (Linko, 1911, 1912; Kudelin, 1914). Linko's (1912, 1913) studies dealing with the zooplankton of Western Murmansk and the Arctic Ocean mention 11 species of hydromedusae and 2 species of siphonophores. Material collected by the Russian expeditions in Spitzbergen (1899–1901), Murmansk Biological Station (1921) and Drifting Sea Research Institute (PLAVMORNIN, 1921) afforded the opportunity to increase our knowledge of the Barents Sea fauna (Rylov, 1923, 1924, 1927).

Material collected by the Murmansk Biological Station researchers on the *R. V. Alexandr Kovalevsky* (1926) and other collections of the station allowed Spassky to describe 22 new hydroid species from the Kola Bay and the southwestern part of the Barents Sea (1929), thus raising the number of species to 70 (Derjugin, 1915; Tanasijtchuk, 1927). Between 1920 and 1930, a large number of scientific expeditions were carried out in the Arctic seas. It will suffice to mention the most important ones, such as the Kara Expedition on the research vessel *Taimir* (1921), the expedition to Novaya Zemlya organized by the State Hydrological Institute (1923–1927), the expedition to Novaya Zemlya of the Institute of Study of the North on the research vessel *Elding* (1925, 1927), the Czech Expedition of the Institute on Study of the North in 1925 and 1927, the Kara Expedition of the Anglo–Russian Cooperative Union (ARCOS) on the research vessels *Malygin* (1925) and *Sedov* (1926), the expedition of the Yakut Commission of the Academy of Sciences in the Laptev Sea on the research vessel *Polarnaya Zvesda* (1927), the expedition of the Arctic Institute on the *R. V. Sedov* (1929, 1930), the expedition of the State Hydrological Institute to Wrangel Island on the *R. V. Litke* (1929), and the expedition of the Arctic Institute on the *R. V. Lomonosov* (1931). Material collected in these expeditions have added new data to the list of hydroids of the Barents and Kara seas (Vagin, 1934) and to the hydroids from the Franz Josef Archipelago as well as from the Eurasian Arctic seas (Uschakov, 1936, 1937). Uschakov (1925) made interesting observations on seasonal variations of the fauna in the littoral zone of the Kola Bay. He was the first to observe such unique phenomena as the hydroids' ability to tolerate low temperatures and even ice.

Data on the bottom population of the Novosiberian shoals (from the material collected on the *R. V. Malygin* and *Sadko,* 1935, 1937, 1938) permitted Gorbunov (1946) to present data on the vertical distributional patterns and biogeographical distributions of various groups of invertebrates, including 42 hydroids. One should mention Kramp's thorough investigations (1932, 1942, 1943, 1963) of species composition, distributional patterns, and evolution of pelagic and bottom-dwelling hydrozoans in the western Arctic region (the Greenland Coast).

The faunal composition of the Eurasian Arctic seas has been studied in the *Catalog of Fauna and Flora of the Northern Seas of the USSR* edited by Gayevskaya; the chapters on hydroids and hydromedusae were authored by Beresina (1948) and Yashnov (1948), respectively.

In the 1950s, Naumov undertook a study of the hydrozoans of the USSR waters simultaneously with investigations of the Far Eastern and Arctic hydroids and hydromedusae. He used the collections of the above-mentioned expeditions and also those collected on the *R. V. Severnyi Polus* (1946) in the East Siberian and Chukchi seas and on the *R. V. Litke* (1948, 1955), *Ob* (1956), and *Lena* (1957, 1958) in the Greenland, Barents, Kara, and Laptev seas as well as his own collections from the Barents Sea. His study resulted in a fundamental review of hydroids and hydromedusae of the USSR waters (1960) that contain 134 species from the Arctic seas, including the White Sea. In 1967, Stepanjants' review of siphonophores from USSR waters was published; 5 species were reported in the Arctic.

The following investigations discuss the distribution, growth, and feeding of zooplankton including hydromedusae and siphonophores in the Barents Sea: Zelickman (1961, 1966, 1969), Kamshilov and Zelickman (1958), Zelickman and Kamshilov (1960), Zelickman, Gelfand, and Shifrin (1969), and Zelickman and Golovkin (1972).

Collections of the Soviet Drifting Stations (SP) have added new names to the list of Hydrozoa. A new species of siphonophore belonging to the new genus *Rudjakovia plicata* (Margulis, 1982) was discovered in the central sector of the Arctic Basin.

Recently an interdisciplinary expedition undertaken by researchers of the Zoological Institute of the USSR Academy of Sciences, using scuba diving equipment, studied several shoals in the Arctic seas off the Franz Josef Archipelago (Golikov and Averintsev, 1977), off the Novosibirsk Islands (Golikov et al., 1974), and off Wrangle Island (Golikov et al., 1978).

## RESULTS

Investigations of the material collected by the Zoological Institute of the USSR Academy of Sciences, in addition to the published data, indicate that the 142 hydrozoan species inhabiting the Eurasian Arctic seas and the adjacent Central Arctic Basin belong to two subclasses: Hydroidea and Siphonophora.

All 7 orders of the class Hydrozoa are represented in this region: Athecata with 37 species, Thecaphora with 91 species, Limnomedusae with 1 species, Trachylida with 5 species, Actinulida with 1 species, Hydrida with 1 species, and Siphonantha with 6 species. A total of 27 families are represented in these regions. It is not the aim of this chapter to discuss taxonomic problems. Table 16-1 (pages 408–419) presents the list of Hydrozoa according to their systematics adapted by the author. This classification follows the opinion of most taxonomists, but includes corrections and additions to reflect data obtained in the last few decades.

All Arctic bottom-dwelling and pelagic Hydrozoa are mentioned in this chapter. Most species of hydromedusae, having a bottom-dwelling polypoid stage, follow the general distribution patterns of bottom-dwelling organisms (with pelagic larvae). Only 10 species of pelagic Hydrozoa (Trachylida and Siphonophora) do not have a bottom polypoid stage (including the trachymedusa *Ptychogastria polaris* living near the bottom). However, as will be shown, their distribution in the Arctic has the same features as all Hydrozoa discussed in this article.

As shown in Tables 16-1 and 16-2, species composition differs in the various Eurasian Arctic seas. These differences are partially accounted for by the uneven degree of knowledge of these regions. The Barents and Kara seas, which have been studied most thoroughly, have 125 and 77 species, respectively. In the other shelf seas, the results are as follows: Laptev Sea, 54 species; East Siberian Sea, 41 species; Chukchi Sea, 35 species; and the Central Arctic Basin, 12 species. The low diversity is believed to be due to the limited sampling of the deep basin. The absence of information in the basin hinders biogeographic analysis and until more information becomes available most conclusions should be considered tentative. Bearing in mind the limited data available to date we hypothesize that distributional patterns reflect adaptation to various water masses. Here we recognize 8 groups: possibly endemics, Arctic, boreal–Arctic, subtropical–Arctic (using the terminology of Golikov and Averintsev, 1977), boreal, wide-ranging warm water, bipolar, and cosmopolitan taxa.

## Possibly Endemics

Included in this group are 9 species not found outside the studied area. As will be seen all species occurred once or twice only; therefore, they are referred to as endemics by inference. Only in the Barents Sea and the adjacent area of the Central Arctic Basin did we observe *Plumularia fragilis* (depth 270 m), *Aglaophenopsis compressa* (220 m depth), and *Schizotricha polaris* (175, 196 m depth). These species inhabit temperatures above zero (1°–4°C) in the Atlantic water, and being representatives of the warm-water Plumulariidae family, they may be present in the North Atlantic also. The same is true for *Thuiaria arctica,* found off the Murmansk coast (at 86 m depth) and between the Medvezhii Island and Spitzbergen (at 38 m depth).

The new species *Rhizogeton nematophorum* (Antsulevich and Polteva, 1986) and Hydractinia (Stylactis) were found in the region of the Franz Josef Archipelago (between 3 m and 20 m water depth and temperatures of about −1°C). The latter was discovered in Kandalaksha Bay of the White Sea (at 16 m depth, in 0°C water temperature).

The Medusa *Calycopsis birulai* was found in surface plankton in all the Eurasian Arctic seas; it is tentatively regarded as an endemic form. The same applies to the recently discovered siphonophora *Rudjakovia plicata* (Margulis, 1982), which is known only from the Central Arctic Basin off the Laptev Sea, from the following depths: 950–0 m, 850–0 m, 500–0 m, 400–0 m, 350–0 m, 200–0 m, and 150–0 m, in temperatures ranging betwen −1°C and 5°C.

## Arctic Species

Thirteen species are included in this group. Their boundaries extend westward to the Canadian Arctic Archipelago and Greenland and eastward to the Arctic Alaskan Coast and Bering Sea. Of these, 10 species can be referred to as west-Arctic taxa; they extend eastward as far as the Laptev Sea (*Corymorpha glacialis, C. groenlandica, Sertularia albimaris, Marrus orthocanna*) and Kara Sea (*Acryptolaria borealis* and *Homoeonema platygonon*). Other taxa have not been found in the Arctic region east of the Barents Sea (*Hydractinia serrata, Thuiaria obsoleta, Monocoryne gigantea,* and *Rhizogeton nudum*). The remaining 3 species (*Halitholus yoldia-arctica, Ptychogastria polaris,* and *Aeginopsis laurentii*) seem to have circumpolar distribution and may occur eastward up to the Bering Sea (*Aeginopsis laurentii* found near the coast of Japan; the report by Uchida, 1928, is doubtful). No Arctic species with east-Arc-

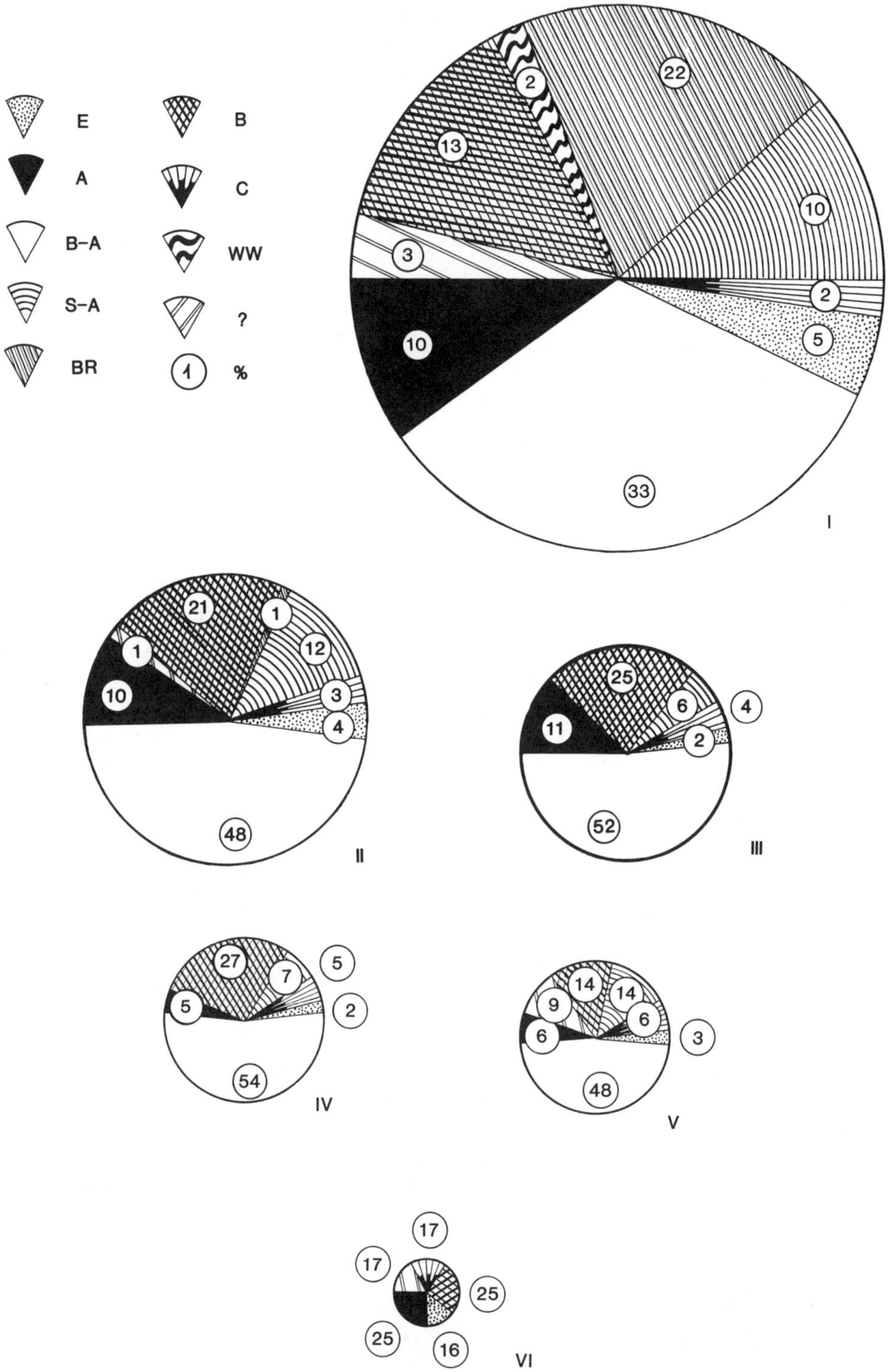


Figure 16-1. Percentage of Hydrozoa in the Eurasian Arctic seas: Barents (I), Kara (II), Laptev (III), East Siberian (IV), Chukchi (V) seas, central part of the Arctic Basin (VI). Key same as in Table 16-1.

tic distribution (Nesis, 1983*a*) are known among Hydrozoa.

A considerable part of the Arctic species were found between the upper sublittoral and the bathyal zone in a rather wide temperature range (−2°C to +4°C) (Table 16-1, Fig. 16-2). *Acryptolaria borealis* is a bathyal species inhabiting the Arctic region within temperature ranges of −1.5°C to +2.8°C. Its presence off the Japanese coasts (Hirohito, 1983) seems doubtful. *Hydractinia serrata,* known from the Greenland waters as a mainly sublittoral species (15–80 m; and temperatures of approximately 0°C), in the Barents Sea was found at a depth of 320 m (at about −1°C). The medusa *Ptychogastria polaris* inhabits the near-bottom zone, and is found in Arctic waters from Labrador up to the Novosibirsk Islands and in the Bering Sea. It lives mostly in temperatures ranging from 0°C to −1.8°C. However, according to Kramp's (1942) data, it may occur at 2.8°C also. The siphonophora *Marrus orthocanna,* found in the Barents Sea up to the central part of the Arctic Basin (the Laptev Sea sector) at depths of 2,000–300 m and 300–197 m, inhabits temperatures from 0°C to −1.8°C. According to Kramp (1963), this species was found off western Greenland in depths ranging from 1,500–1,880 m at temperatures of −0.2°C to −1.7°C. Probably, it is an indicator of the Arctic water masses.

Two Arctic species deserve special consideration: *Corymorpha groenlandica* and *Monocoryne gigantea.* They were found in the Barents Sea between 25 m and 330 m and off the Franz Josef Archipelago in water depths ranging between 25 m and 36 m, where during summer water temperatures do not rise above −0.5°C, and northeast of the Kola Peninsula at depths of 250–330 m, where the temperature is −1°C year-round (*Atlas Okeanov,* 1980). However, Spasskyi (1929) mentioned finding *C. groenlandica* in the region of 70°N and 71°N, 33°33′E at depths of 150 m and 250 m, where temperatures in August were about 3°C. In the Kara and Laptev seas, *C. groenlandica* was found once east of the Franz Josef Archipelago (542 m, and −0.5°C).

In adjacent areas of the Arctic this species has been found several times: northwest of Spitzbergen (756 m, 0.6°C) and in the Davis Strait (180 m, −1°C). During the Ingolf Expedition (Broch, 1916) it was recorded as follows: at 66°23′N, 10°26′W (1,365 m, −0.9°C); at 65°33′N, 10°28′W (895 m, 0.3°C); at 63°36′N, 7°30′W (1,260 m, −0.6°C); and 66°18′N, 25°59′W (600 m, −0.75°C); north of Iceland (795 m, 0–5°C); off Spitzbergen (45 m, approximately −1.0°C); off Liverpool Bay, and in the Beaufort Sea (0–4 m, with temperatures ranging from −1.8°C to +0.5°C) (Broch, 1916; Calder, 1972). The depth range of *Corymorpha groenlandica* extends from the sublittoral zone to the bathyal zone, mainly at negative temperatures, but apparently it tolerates summer warming up to 3°C.

*Monocoryne gigantea* was seldom encountered. The type species is from 75°12′N and 3°2′E (2,193 m, −1.57°C) (Bonnevie, 1899). In Hudson Bay it was reported at 90–100 m and about −1.5°C (Calder, 1972) and off Alaska at Stevens Pass (Fraser, 1941; Calder, 1972). According to Calder (1972), this species is known from the Barents Sea. There is reason to assume that *Monocoryne gigantea* inhabiting the shelf and abyssal depths requires subzero temperatures. Thus, almost all Arctic hydrozoans inhabit regions where temperatures range between subzero and a few degrees above zero. They can survive in the Atlantic water where the water is warm during the summer. Their distribution in the Greenland Sea, off the Canadian Arctic Archipelago and adjacent regions, suggests that the fauna has originated in the Atlantic Ocean.

It is certain that *Corymorpha groenlandica* and *Monocoryne gigantea,* and possibly *Hydractinia serrata, Ptychogastria polaris, Aeginopsis laurentii,* and *Marrus orthocanna,* inhabiting mostly temperatures below zero but having adapted to positive summer

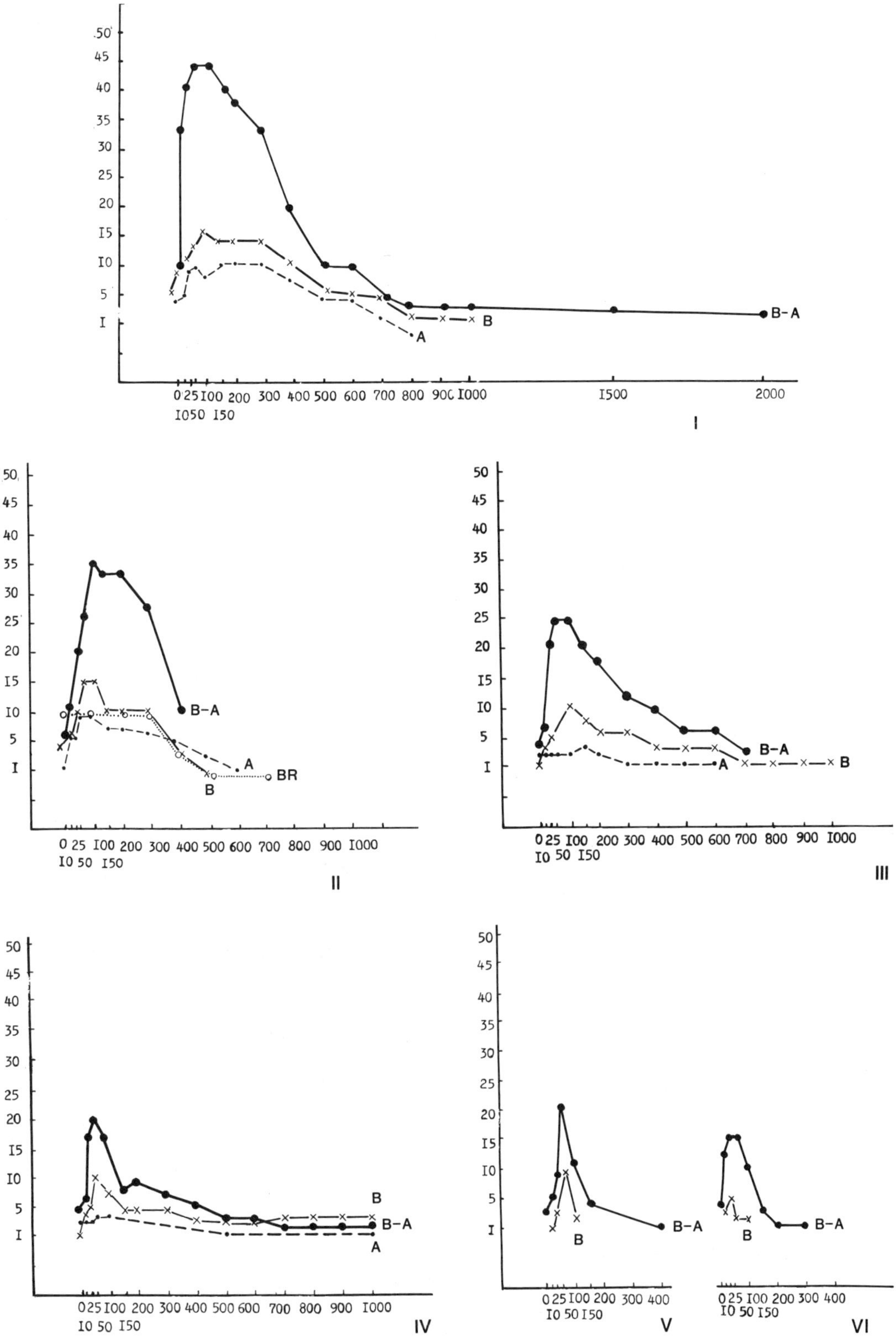


Figure 16-2. Vertical distribution of Hydrozoa (key same as in Table 16-1); all Eurasian Arctic seas (I), Barents (II), Kara (III), Laptev (IV), East Siberian (V), Chukchi (VI) seas. Abscissa: depths in m; ordinate: number of species.

temperatures, may be Arctic autochthonous taxa. Some bathyal Arctic species not found in the studied area but known from Greenland waters (Kramp, 1932) also belong to this group. These are *Hydractinia (Stylactis) arctica* (Jäderholm, 1902; 74° 41′N, 70°30′W, depth 1,200 m, −0.2°C; 72°42′N, 14°49′W, depth 2,000 m, which lives attached to the hydroid *Eudendrium planum*), *Hydractinia (Stylactis) ingolfi* (Kramp, 1932; 60°17′N, 52°05 W, depth 3,229 m, 1.4°C; 59°20′N, 40°48′W, depth 3,192 m, 1.5°C; 61°44′N, 30°29′W, depth 2,137 m, 3°C; 61°39′N, 17°10′W, depth 2,344 m, 2°C, which lives on the ophiurids *Homalophiura tesselata), Eudendrium planum* (Bonnevie, 1838; 74°4′N, 70°30′W, depth 1,200 m, −0.2°C), and *Halisiphonia arctica* (Kramp, 1932; inhabiting the same area attached to *Eudendrium planum*). It is apparent that this group of abyssal Arctic species is endemic to the Arctic and one should expect to find these species in other Arctic regions. *Hydractinia ingolfi* (Kramp, 1943) was not found in this region at shallower depths and at higher water temperatures. It is interesting that a very closely related Antarctic species, *Hydractinia vallini* (Jäderholm, 1926; depth 219–640 m), lives attached to ophiurids of the family *Ophiolepididae* (Smirnov and Stepanjants, 1980).

## Boreal–Arctic Species

The largest numbers of species (48) are the boreal–Arctic taxa. They inhabit the Arctic seas and penetrate far westward and eastward in boreal waters. Among the boreal–Arctic species the circumpolar taxa (30) are predominant, inhabiting nearly all Arctic seas (Table 16-1). They were recorded off Greenland and off the Arctic Canadian Archipelago, in the Norwegian Sea, along the North American and European coasts, in the Bering, Okhotsk, and Japan seas, along North America, and even off the Pacific Japanese coasts. In the Arctic waters these species live between the upper sublittoral zone and the bathyal zone, reaching highest abundance between 50 m and 200 m (Fig. 16-2) in water temperatures below 0°C (their temperature range is −2°C to +3°C), surviving summer warming higher than +5°C. At the boundary of their distributional range they often live in regions where annual temperatures are positive.

*Candelabrum phrygium* is known from Norwegian waters, off the Lofoten Islands, the Faeroe Islands, Iceland (at 200–1,825 m, about −1°C to +7°C); from the Greenland Sea (2,195 m, −1.57°C); from the southern sector of the North Atlantic Ocean as far as the eastern coast of Scotland; and from the rocky coast of the English Channel, near Roscoff (Korotnev, 1880), where during the summer temperatures exceed 15°C. *C. phrygium* was found along the eastern and western coasts of Greenland (10–35 m, about −1°C to +2°C), in Hudson Bay (90–100 m, about −1°C), in the Barents Sea (68–285 m, about 0°C to −1°C), and in the Kara Sea (12–15 m, about −1.5°C to +4°C). So far this species has not been found in other Arctic seas, but according to recent data (Antsulevich and Polteva, 1986) it was observed off Paramushir Island (Kurile Islands) at depths ranging between 10 m and 130 m and at temperatures from −1°C to +3°C in summer. The area of distribution of this species is boreal–Arctic. This species lives in Arctic waters at temperatures ranging between −1.5°C and +4°C; in boreal waters it may survive in high temperatures. Observations by Korotnev (1880) near Roskoff show that colonies of *C. phrygium* attained complete development in these waters only during the winter months (December–February). In late summer, their hydrorhiza that produce new polyps are found attached to rocks. This species can survive warming since it has adapted to positive temperatures and possesses a mobile larva. It is found in the coastal waters of the boreal Atlantic and Pacific oceans. According to Broch (1918), *C. phrygium* and *Corymorpha groenlandica* are Arctic relict species in the Atlantic Ocean.

*Rhizorhagium roseum, Euphysa tentaculata, Halecium corrugatum, H. marsupiale,* and *Tulpa speciosa* are not found in the Arctic east of the Kara Sea, but are known from the temperate waters of the Atlantic and Pacific oceans (*Halecium marsupiale* has not been reported from the North Atlantic). These boreal–Arctic species live in temperatures ranging from −1.5°C to +3°C. Nevertheless, their precise distributional range is difficult to determine because of the limited data. However, the medusa *Euphysa tentaculata,* being present in the Arctic region as far as the Greenland coast, may have penetrated through the Davis Strait and the Canadian Straits to the western Canadian coast (Kramp, 1942) where it has been found repeatedly (Arai and Brinckmann-Voss, 1980).

*Euphysa tentaculata* may be called a western boreal–Arctic taxon. Six other hydrozoa are western boreal-Arctic species: *Coryne lovenii, Mitrocomella polydiademata, Cuspidella grandis, Hydrallmania falcata, Cladocarpus integer,* and *Polynemertesia gracillima;* their eastward distribution in the Arctic is limited by the Laptev or Kara seas (except for the last two taxa, which are limited to east of the Barents Sea). To the west, these species are widely represented in temperate waters along the coast of North America and Europe. In the Arctic, they inhabit low temperatures above zero and can survive in temperatures as low as −1.8°C. In boreal waters they occur at temperatures higher than 6°C. Representatives of the warm-water family Plumulariidae—*Cladocarpus integer* and *Polynmertesia gracilima*—inhabit the following Arctic regions: Barents and Greenland seas; the Davis Strait; the coasts of Norway, Iceland, and Newfoundland in above zero temperatures (1° −6°C).

Seven other species are referred to as the eastern boreal-Arctic group: *Sertularia tolli, S. similis, S. cupressoides, Sertularella pellucida, Abietinaria turgida, Thuiaria cylindrica,* and *Halecium speciosum.* These species as a rule are not distributed in the Arctic west of the Chukchi and East Siberian seas, and do not occur at depths greater than 40–50 m (commonly at depths of ≈ 20m). *S. tolli* and *S. cupressoides* are found in the Kara Sea and *S. pellucida* was discovered off Spitzbergen. All these species are absent from the Atlantic Ocean but are represented in the Far East seas and in the North Pacific.

The following taxa were found in the Canadian Arctic Archipelago: *Halecium speciosum* in Hudson Bay in 20–106 m (Calder, 1970, believes that *H. ornatum* observed by Broch in 1910 off Spitzbergen is in fact *H. speciosum*); *Sertularia similis,* one of the most common species in all the straits of this archipelago (Calder, 1970); and *Abietinaria turgida,* recovered once (Calder, 1970) off Ellaf Ringnes Island (Queen Elizabeth Islands) at a depth of 46 m. Their presence in the Canadian Archipelago and absence from the Atlantic Ocean make the Canadian Arctic Archipelago the eastern boundary in their distribution.

Also found are the seven eastern boreal–Arctic species that inhabit the Arctic region chiefly in negative temperatures (from −0.5°C to −1.8°C), though they survive summer warming (e.g., *Sertularia similis* in the southern Chukchi Sea lives at 4°C). Eastern boreal–Arctic Hydrozoa discovered in the shoals of the Arctic seas at temperatures below zero cannot be regarded as indigenous to the Arctic (contrary to Nesis' opinion, 1983*b*), since many (*Sertularia similis, Sertularia tolli, Albietinaria turgida*) have been observed not only in the northern part of the Bering and Okhotsk seas but dominate in the sublittoral zone of the Kurile Islands in relatively high summer temperatures. This group is believed to be of Pacific origin (Kramp, 1963, p. 88), and these species are found off the western Greenland coast.

## Subtropical–Arctic Species

Thirteen species considered subtropical–Arctic are ecologically very similar to the previous group differing from them by their wider distributional ranges. These species occur in nearly all Arctic seas, inhabiting the upper sublittoral to upper bathyal zones. They are eurythermal, present in the Atlantic as far as Maine, New Jersey, and Florida, and off the European coast as far as Spain and the Azores. They were recovered in the Mediterranean as well. In the Pacific Ocean, representatives of subtropical–Arctic species can be found in the South Kurile Strait, off Japan, in the Yellow Sea, and off California. Within the Arctic region, species belonging to this group (e.g., *Campanulina lacerata* and *Thuiaria thuia*) were found only in the Barents Sea. The three species whose distribution in the Arctic is represented only by the medusoid stage are *Tubularia prolifer, Euphysa aurata,* and *Rathkea octopunctata.* In spite of their eurythermy, these species occur more frequently at positive temperatures than do representatives of other groups. *Lafoea grandis* occurs in the Kara Sea at depths of 200–360 m and at temperatures of 0°–5°C. Nevertheless, all species of this group live in the Laptev, East Siberian, and Chukchi seas generally at below 0°C.

## Bipolar Species

Seventeen bipolar species inhabit the Arctic Basin, the temperate waters of the Northern Hemisphere, the temperate waters of the Southern Hemisphere, the sub-Antarctic (Patagonian shelf, Crozet, Heard, Kerguelen, Southern Orkney Islands, Falkland Islands, etc.), and even the Antarctic Coast. Most (Table 16-1) inhabit all or almost all Arctic seas under conditions similar to those of pan-Arctic boreal–Arctic and subtropical–Arctic species; they are eurybiontic immigrants in the Arctic. Some forms are characteristic of the recent Arctic fauna (e.g., *Obelia longissima, Moderia plicatile, Sertularella tricuspidata,* and *Abietinaria abietina*).

In temperate waters these species inhabit positive temperatures (up to 17°C–20°C). Only a few records are known in the Southern Hemisphere. Noteworthy is *Tubularia indivisa.* It was reported in the Arctic along with *Tubularia regalis* (Jäderholm, 1909; Broch, 1910, 1916; Kramp, 1943; Calder, 1972). In the opinion of these authors, these two species can be differentiated by the presence of longitudinal ribs in the female gonophores in *T. regalis* and their absence in *T. indivisa.* The first is regarded as an Arctic taxon. It was found in the White, Barents, and Kara seas, off the eastern and western coasts of Greenland, off the Canadian Arctic Archipelago, between the Faroes Islands and Scotland, and off the Norwegian coasts. *T. regalis* was discovered in the same waters and also off the east and west coasts of North America, off South Iceland, in the Caribbean region of West Africa, in the Mediterranean Sea, in the Southern Hemisphere off the Kerguelen Islands, Crozet, and southern Georgia. Naumov (1960) lumped *T. regalis* and *T. indivisa* into one species. Until specimens with ribs on female gonophores are found in Arctic waters, it appears reasonable to adopt Naumov's point of view and regard *T. indivisa's* distributional area as bipolar.

## Boreal Species

Boreal species (28) are widely distributed in boreal waters. In the study area, they have been reported almost exclusively from the Barents Sea. Among the boreal species, 9 are amphiboreal and 19 belong to the North Atlantic group (Table 16-1). These species inhabit the Barents Sea from the littoral to the upper bathyal zone, invariably in water with positive temperatures.

## Wide-ranging Warm-water Species and Cosmopolitan Species

Three species occurring in the Barents Sea (*Obelia geniculata, Clytia gracilis,* and *Physophora hydrostatica*) can be characterized as wide-ranging warm-water taxa, since their distributional areas include, apart from boreal waters, the African, New Zealand, and Mediterranean coasts. Two species, the hydroid *Lafoea dumosa* and the siphonophora *Dimophyes arctica,* can be regarded as cosmopolitans (Stepanjants, 1975, 1980).

## DISCUSSION AND CONCLUSIONS

The distributional ranges were not established for 9 species (Table 16-1). As shown in Tables 16-1 and 16-2 and in Figure 16-1 and the maps, the majority of Arctic species consist of boreal–Arctic representatives (about 50 percent in each sea), bipolar (about 20 percent), and subtropical–Arctic (about 15 percent) forms. The bipolar term is tentative. An attempt to separate specimens of species found in the two hemispheres by traditional taxonomic methods is difficult. The percentage of boreal species is relatively high (19 percent); with the exception of one, they occur only in the Barents Sea, constituting 21 percent of the Hydrozoa Barents Sea species.

Endemic Arctic Hydrozoa, consisting of "probably" endemic forms (9 species), and 13 Arctic species contribute only 15 percent to every sea and 15 percent to the whole area. To the Arctic endemic taxa, one should probably add 4 abyssal species known so far only from the Greenland Sea Basin and the adjacent areas of the North Atlantic. Endemic genera and endemic families of Hydrozoa are unknown from the Arctic. However, there is an exception, a recently discovered genus of siphonophora *Rudjakovia* (Margulis, 1982), found so far only in the Arctic and therefore regarded possibly as an endemic form.

There are two bipolar genera, Monocoryne and Candelabrum, and 1 bipolar family, Myriothelidae, represented in the Arctic and Antarctic, as well as in the temperate waters of both hemispheres. Unlike the Arctic hydroid fauna, the Antarctic group is characterized by a high level of endemism (56 percent of the total number of hydroid species occurring in Antarctic; Stepanjants, 1979). Furthermore, 4 endemic siphonophore species are known from the Antarctic waters. Endemic genera are also known for this region: 2 Hydroidea and 3 Siphonophora. Endemic Hydroidea families are absent from both the Antarctic and the Arctic oceans. However, in the opinion of several scientists (e.g., Totton, 1965), there is one Antarctic family of siphonophore Pyrostephidae. Such low endemism of high taxonomic rank (genera and families) in the Arctic and the Antarctic is a characteristic feature of the Hydrozoa fauna. It indicates a low level of isolation of cold-water Hydrozoa and also a relatively young age of the fauna of the polar regions.

The vertical distribution of Hydrozoa is most obvious in the Barents and Kara seas, where not only shelf but also slope faunas were investigated in detail. In the eastern Arctic, mainly coastal and island shoals were studied; from deep water there are few collections: *Sadko* (1935, 1937); *Sewernyi Polus* (1940); *Litke* (1948, 1955); *Ob* (1956); *Lena* (1957, 1958); and *SP-2, SP-3,* and *SP-4* (1950–1954). However, as shown in Table 16-1 and Figure 16-2, Hydrozoa were observed from littoral to abyssal depths. As expected, species diversity is highest on the shelf between 50 m and 200 m, with maximum species diversity in the 50–100 m zone. This is seen in Figure 16-2*A* for whole area and in Figure 16-2*B,C,D* for each area. Table 16-1 shows that 60 species were found only on the shelf, 6 of them only in the upper sublittoral and 2 only in the littoral zone. Most species are boreal and 2 are endemic.

Half of the Arctic Hydrozoa species have wide depth ranges. Figure 16-2 shows that the maximum number of species occur at depths of 50–300 m; their number declines sharply with depth. This pattern is distinct in the Barents Sea. In the Kara Sea, the peak is between 50 m and

200 m. Eastward, the depth of maximum species diversity is 100 m (Laptev Sea) and 50 m (East Siberian and Chukchi seas). This pattern is in accord with Nesis' (1983*a,b*) opinion concerning the differences in the depths preference and degree of eurybathic of west-Arctic and east-Arctic species.

Thirteen Hydrozoa (bathyal) species were observed only on slopes and a few in the lower boundary of the shelf up to slope. Six of them belong to the warm-water family Plumulariidae. They live in the bathyal depth of the Barents Sea with constant low positive temperatures. The same is true for other boreal species such as *Zygophylax pinnata* and *Bonneviella grandis*. Four species were observed in bathyal depth in the studied region, though in other areas they were reported from the shelf (*Acryptolaria borealis* and *Halecium reversum*), in sublittoral zones (*Hydractinia serrata*), and near surface (*Ptychogena lactea*) at negative as well as at low positive temperatures. The typical bathyal siphonophora *Marrus orthocanna* inhabits depths with 0°C and lower temperatures. It is apparently an indicator of the Arctic waters.

True abyssal species were not observed in the studied area. However, 4 species (*Hydractinia arctica, H. ingolfi, Eudendrium planum,* and *Halisiphonia arctica*) are found only in the abyssal zone of the Greenland Sea (1,200–3,229 m). They may occur in other regions of the Arctic as well. Five species, Arctic *Corymorpha groenlandica, Monocoryne gigantea, Ptychogastria polaris, Aeginopsis laurentii* and boreal–Arctic *Candelabrum phrygium,* inhabit the Arctic from the sublittoral to the bathyal zones. Occasionally they were found at depths greater than 2,000 m. As discussed above, their depth preference in different areas is chiefly related to their preference of negative temperatures.

One hundred forty-two species of pelagic and bottom-dwelling Hydrozoa, 136 Hydroidea, and 6 Siphonophora are known from the Barents, Kara, Laptev, East Siberian, and Chukchi seas as well as from the Central Arctic Basin adjacent to these seas. Most Eurasian Arctic Hydrozoa

Table 16-1. Distribution of Hydrozoa in the Eurasian Sector of the Arctic Ocean Showing Range of Depths (m)

| Species | *Barents Sea* | | *Kara Sea* | |
|---|---|---|---|---|
| | *P* | *M* | *P* | *M* |
| | 1 | 2 | 3 | 4 |
| HYDROIDEA | | | | |
| ATHECATA | | | | |
| CLAVIDAE | | | | |
| *Clava multicornis* (Forskal, 1775) | 0–10 | — | | |
| *Rhizogeton nudum* Broch, 1909 | 35 | — | | |
| *Rhizogeton nematophorum* (Antsulevich, 1986) | | | 7–20 | — |
| BOUGAINVILLIIDAE | | | | |
| *Dicoryne conferta* (Alder, 1856) | 35–225 | — | | |
| *Bougainvillia principis* (Steenstrup, 1850) | | 250–0<br>125–65<br>10–0 | | |

P = polypoid generation; M = medusoid generation; blank spaces indicate that the species was not found in this region; dashed line (—) indicates absence of the given generation of the species; bullet (•) indicates that the given generation of the species is unknown but its presence is anticipated. In the Biogeography column A = Arctic; wA = west Arctic; B-A = boreal–Arctic; wB-

are eurybathic species, widely distributed in the Arctic and in the temperate waters of the Atlantic and Pacific oceans. They can be referred to as boreal–Arctic and bipolar taxa. In the Arctic, they live at temperatures ranging from ≈2 degrees below zero to a few degrees above zero and can survive considerable summer temperature rise and even live at constant temperatures above zero on the boundaries of their distribution ranges. There are several species with western (43) and eastern (7) distributional ranges, living in conditions similar to those of pan-Arctic species, each having well-defined bathymetric ranges. Species whose distribution to the east is limited by the Kara and Laptev seas live between the upper sublittoral and bathyal zones (occasionally deeper); however, species whose distribution is limited to the west by the above-mentioned seas do not descend below 50 m and occur mostly in the upper sublittoral zone.

In the Eurasian Arctic, 15 percent of the Hydrozoa are endemic. Most survive summer warming and can live in the Atlantic water mass at temperatures above zero. Seven species (*Rhizogeton nematophorum, Hydractinia serrata, Corymorpha groenlandica, Monocoryne gigantea, Ptychogastria polaris, Aeginopsis laurentii,* and *Marrus orthocanna*) have been found principally in waters with negative temperatures. There is only 1 possibly endemic siphonophore genus (Rudjakovia) in the Arctic and there is no endemic family in the Arctic.

The distributional ranges and ecological characteristics of Arctic Basin species suggest that almost the entire recent hydrozoan fauna consists of immigrants. As a result of their eurybiontic character these taxa adapted to the rigorous environment and became widely dispersed. Similar conclusions are drawn on the Actiniaria Arctic fauna (Grebelnyi, 1980).

The extremely low level of species endemism and lack of endemism of higher-rank taxa, as well as the low number of indigenous Arctic taxa, point to the young age of the Arctic Hydrozoa fauna. The degree and rank of endemism, as well as the composition and evolution of the arctic hydrozoan fauna, do not allow us to treat the Arctic Basin as an independent biogeographic region.

| *Laptev Sea* | | *East Siberian Sea* | | *Chukchi Sea* | | *Central Arctic Basin* | | |
|---|---|---|---|---|---|---|---|---|
| *P* | *M* | *P* | *M* | *P* | *M* | *P* | *M* | *Biogeography* |
| 5 | 6 | 7 | 8 | 9 | 10 | 11 | 12 | 13 |
| | | | | | | | | aBR |
| | | | | | | | | wA |
| | | | | | | | | E |
| | | | | | | | | aBR |
| | | | | | | | | aBR |

A = west boreal–Arctic; eB-A = east boreal–Arctic; S-A = subtropical–Arctic; BR = boreal; aBR = Atlantic boreal; B = bipolar; K = cosmopolitans; WW = widely distributed warm-water species; E = endemic taxa;? = uncertain distributional range.

Table 16-1. *(continued)*

| *Species* | *Barents Sea* P | *Barents Sea* M | *Kara Sea* P | *Kara Sea* M |
|---|---|---|---|---|
| | 1 | 2 | 3 | 4 |
| *Bougainvillia superciliaris* (L. Agassiz, 1849) | 10–20 | 10–0 | | 50–0 |
| | | 50–0 | | |
| *Rathkea octopunctata* (M. Sars, 1835) | | 175–0 | | 10–0 |
| | | 5–0 | | |
| *Rhizorhagium roseum* M. Sars, 1877 | 35–135 | — | | |
| PANDEIDAE | | | | |
| *Halitholus yoldia-arcticae* (Birula, 1897) | 12–370 | 25–0 | 6–180 | 140–0 |
| *Catablema vesicaria* (A. Agassiz, 1862) | • | 170–0 | • | 124–0 |
| | | 50–25 | | |
| *Calycopsis birulai* (Linko, 1913) | • | 25–0 | • | 18–0 |
| HYDRACTINIIDAE | | | | |
| *Hydractinia allmani* Bonnevie, 1898 | 68–320 | — | 18–628 | — |
| *Hydractinia carica* Bergh, 1887 | 4–180 | — | 34 | — |
| *Hydractinia monocarpa* Allman, 1876 | 8–60 | — | 34 | — |
| *Hydractinia echinata* (Fleming, 1828) | 15–20 | — | 34 | — |
| *Hydractinia serrata* Kramp, 1943 | 308–320 | — | | |
| *Hydractinia carnea* (M. Sars, 1846) | 7–36 | — | | |
| *Hydractinia* (Stylactis) sp. | | | 3–20 | — |
| EUDENDRIIDAE | | | | |
| *Eudendrium rameum* (Pallas, 1766) | 16–360 | — | 8–290 | — |
| *Eudendrium ramosum* (L., 1758) | 185–208 | — | 20–178 | — |
| *Eudendrium capillare* Alder, 1857 | 19–245 | — | | |
| *Eudendrium annulatum* Norman, 1864 | 42 | — | | |
| CORYMORPHIDAE | | | | |
| *Corymorpha glacialis* M. Sars, 1859 | 5–165 | — | 0.3 | — |
| *Corymorpha groenlandica* (Allman, 1876) | 25–333 | — | 542 | — |
| *Euphysa aurata* Forbes, 1848 | | 20–0 | 1–24 | |
| *Euphysa flammea* (Linko, 1904) | • | 373–0 | • | 140–0 |
| *Euphysa tentaculata* Linko, 1904 | • | 0 | | |
| TUBULARIIDAE | | | | |
| *Tubularia prolifer* (L. Agassiz, 1862) | | 100–0 | | 200–0 |
| *Tubularia simplex* Alder, 1862 | 15–23 | — | | |
| *Tubularia larynx* Ellis et Solanderf, 1786 | 35–38 | — | | |
| *Tubularia indivisa* L., 1758 | 12–288 | — | 35–360 | — |
| *Plotocnide borealis* Wagner, 1885 | • | 25–0 | • | 200–100 |
| CORYNIDAE | | | | |
| *Coryne lovenii* (M. Sars, 1846) | 45–213 | — | 6.5–17 | — |
| *Sarsia tubulosa* (M. Sars, 1835) | 58 | 29–0 | | 25–10 |
| *Sarsia princeps* (Haeckel, 1879) | | 250–0 | | 262–0 |

P = polypoid generation; M = medusoid generation; blank spaces indicate that the species was not found in this region; dashed line (—) indicates absence of the given generation of the species; bullet (•) indicates that the given generation of the species is unknown but its presence is anticipated. In the Biogeography column A = Arctic; wA = west Arctic; B-A = boreal–Arctic; wB-

| *Laptev Sea* | | *East Siberian Sea* | | *Chukchi Sea* | | *Central Arctic Basin* | | |
|---|---|---|---|---|---|---|---|---|
| *P* | *M* | *P* | *M* | *P* | *M* | *P* | *M* | *Biogeography* |
| 5 | 6 | 7 | 8 | 9 | 10 | 11 | 12 | 13 |
| | | | | | | | | S-A |
| | | | | | | | | S-A |
| | | | | 0–52 | | | | B-A |
| 51–60 | 9–30 | | 3–20 | 46 | 20 | | | A |
| • | 12–0 | • | 41–36 | | | | | B-A |
| | | | 25–0 | | | | | |
| • | ? | • | 9 | • | ? | | | E |
| 38,1542 | — | 35 | — | | | | | B-A |
| | | 35 | — | 35 | — | | | B-A |
| 19–51 | — | 35 | — | | | | | B-A |
| | | 54 | — | 50–152 | — | | | S-A |
| | | | | | | | | wA |
| | | | | | | | | aBR |
| | | | | | | | | E |
| 38 | — | 21–41 | — | | | | | B-A |
| 47–60 | — | | | | | | | B |
| | | | | 15–20 | — | | | S-A |
| | | | | 7–40 | — | | | B-A |
| 10 | — | | | | | | | wA |
| 432 | — | | | | | 756 | — | wA |
| 16 | | | | | | | | S-A |
| • | 30,51 | • | 124–0 | | | | | B-A |
| • | 0 | | | | | | | B-A |
| | | | | | | | | S-A |
| | | | | | | | | BR |
| 28–230 | — | | | | | | | B |
| 10 | — | 21 | — | 2–25 | — | | | B |
| • | 200–0 | • | 50–10 | • | 100–0 | | | B-A |
| | | | | | 30–10 | | | |
| | | | | | | | | wB-A |
| | 72–10 | | 25–10 | | | | | B |
| 186–58 | | | | | | | | B-A |

A = west boreal–Arctic; eB-A = east boreal–Arctic; S-A = subtropical–Arctic; BR = boreal; aBR = Atlantic boreal; B = bipolar; K = cosmopolitans; WW = widely distributed warm-water species; E = endemic taxa;? = uncertain distributional range.

Table 16-1. *(continued)*

| Species | Barents Sea P | Barents Sea M | Kara Sea P | Kara Sea M |
|---|---|---|---|---|
| | 1 | 2 | 3 | 4 |
| MYRIOTHELIDAE | | | | |
| *Candelabrum phrygium* (Fabricius, 1780) | 68–285 | — | 12–15 | — |
| *Monocoryne gigantea* (Bonnevie, 1898) | ? | — | | |
| THECAPHORA | | | | |
| MELICERTIDAE | | | | |
| *Melicertum octocostatum* (M. Sars, 1935) | | 40–0 | | |
| CAMPANULINIDAE | | | | |
| *Calicella syringa* (L., 1767) | 0–377 | — | 12–178 | — |
| *Staurophora mertensii* (Brandt, 1835) | 19–120 | 215–0 | 148 | |
| *Cuspidella grandis* Hincks, 1864 | 200–265 | — | 42–520 | — |
| *Mitrocomella polydiademata* (Romanes, 1876) | — | 325–0 | | |
| *Ptychogena lactea* A. Agassiz, 1865 | • | 343–74 | • | 455–520 |
| *Tiaropsis multicirrata* (M. Sars, 1835) | 10–20 | 30–0 | | |
| *Tetrapoma quadridentata* (Hincks, 1874) | 9–260 | — | 16 | — |
| *Opercularella nana* Hartlaub, 1897 | 0–66 | — | | |
| *Campanulina lacerata* (Johnston, 1847) | 0 | — | | |
| *Lafoeina maxima* (Levinsen, 1893) | 8–280 | — | 8–55 | — |
| *Lafoeina tenuis* G. O. Sars, 1874 | 33–130 | — | | |
| *Modeeria rotunda* (Quoy et Gaimard, 1827) | 50–65 | — | 102 | — |
| *Modeeria plicatile* (M. Sars, 1863) | 28–384 | — | 14–570 | — |
| *Halopsis ocellata* A. Agassiz, 1863 | • | 5–0 | | |
| CAMPANULARIIDAE | | | | |
| *Obelia flexuosa* (Hincks, 1861) | 10–0 | — | | |
| *Obelia loveni* (Allman, 1859) | 0–176 | — | | |
| *Obelia* sp. 1 aff. *flexuosa* | | | | |
| *Obelia* sp. 2 aff. *loveni* | | | | |
| *Obelia geniculata* (L., 1758) | 0–60 | 25–0 | | |
| *Obelia longissima* (Pallas, 1766) | 8–300 | 25–0 | 8–100 | |
| *Clytia gracilis* (M. Sars, 1851) | 22–30 | | | |
| *Campanularia volubilis* (L., 1758) | 5–257 | — | 16 | — |
| *Campanularia groenlandica* Levinsen, 1893 | 10–215 | — | 360 | |
| *Campanularia integra* McGillivray, 1842 | 16–308 | — | 20–120 | — |
| *Tulpa speciosa* (Clark, 1876) | 5–170 | — | 16–360 | — |
| *Rhizocaulus verticillatus* (L., 1758) | 25–343 | — | 14–185 | — |
| LAFOEIDAE | | | | |
| *Filellum serpens* (Hassal, 1848) | 15–275 | — | 34–106 | — |
| *Grammaria abietina* (M. Sars, 1851) | 10–325 | — | 20–364 | — |
| *Grammaria immersa* Nutting, 1901 | 20–271 | — | 20–57 | — |
| *Acryptolaria borealis* (Levinsen, 1893) | 192 | — | 360–570 | — |
| *Lafoea dumosa* (Flemming, 1820) | 8–477 | — | 14–570 | — |

P = polypoid generation; M = medusoid generation; blank spaces indicate that the species was not found in this region; dashed line (—) indicates absence of the given generation of the species; bullet (•) indicates that the given generation of the species is unknown but its presence is anticipated. In the Biogeography column A = Arctic; wA = west Arctic; B-A = boreal–Arctic; wB-

| *Laptev Sea* | | *East Siberian Sea* | | *Chukchi Sea* | | *Central Arctic Basin* | | |
|---|---|---|---|---|---|---|---|---|
| *P* | *M* | *P* | *M* | *P* | *M* | *P* | *M* | *Biogeography* |
| 5 | 6 | 7 | 8 | 9 | 10 | 11 | 12 | 13 |
| | | | | | | | | B-A |
| | | | | | | | | wA |
| | | | | | | | | BR |
| 38–48 | — | 38 | — | 20 | — | | | B |
| 33–48 | | 35–60 | | 76–68 | | | | B |
| 51 | — | | | | | | | wB-A |
| • | 10–0 | | | | | | | wB-A |
| • | 1455–282 | | | | | | | B-A |
| | | | | 0–9 | | | | S-A |
| 54 | — | 34 | — | | | | | B-A |
| | | | | | | | | aBR |
| | | | | 20? | | | | S-A |
| 9–220 | — | 10–50 | — | 2–68 | — | | | B-A |
| | | | | | | | | aBr |
| | | | | | | | | B |
| 38–1073 | — | 146 | — | | — | 130–280 | — | B |
| | | | | | | | | aBR |
| | | | | | | | | aBR |
| | | | | | | | | aBR |
| | | | | 5 | — | | | ? |
| | | | | 22,23 | — | | | ? |
| | | | | | | | | WW |
| 19–43 | 20–10 | 7–58 | 15 | 2–52 | | | | B |
| | | | | | | | | WW |
| 45 | — | 45 | — | | | | | S-A |
| 18.5 | — | | | | | | | B-A |
| 38–48 | — | 41–54 | — | | | | | B |
| | | | | | | | | B-A |
| 19–30 | — | 48 | — | 20 | — | | | B-A |
| 19–64 | — | 45–47 | — | | | | | B |
| 19–520 | — | 47–59 | — | | | | | B-A |
| 30–60 | — | 35 | — | | | | | B-A |
| | | | | | | | | wA |
| 19–1073 | — | 24–54 | — | 35–61 | — | 184–477 | — | K |

A = west boreal–Arctic; eB-A = east boreal–Arctic; S-A = subtropical–Arctic; BR = boreal; aBR = Atlantic boreal; B = bipolar; K = cosmopolitans; WW = widely distributed warm-water species; E = endemic taxa;? = uncertain distributional range.

Table 16-1. *(continued)*

| Species | *Barents Sea* | | *Kara Sea* | |
|---|---|---|---|---|
| | *P* | *M* | *P* | *M* |
| | 1 | 2 | 3 | 4 |
| *Lafoea grandis* Hincks, 1874 | 12–477 | — | 204–360 | — |
| *Lafoea pocillum* Hincks, 1868 | 12–59 | — | 14–19 | — |
| ZYGOPHYLAXIDAE | | | | |
| *Zygophylax pinnata* (G. O. Sars, 1874) | 130–620 | — | | |
| BONNEVIELLIDAE | | | | |
| *Bonneviella grandis* (Allman, 1876) | 161–347 | — | | |
| SERTULARIIDAE | | | | |
| *Abietinaria abietina* (L., 1758) | 2–435 | — | 12–410 | — |
| *Abietinaria filicula* (Ellis et Solander, 1786) | 9–100 | — | | |
| *Abietinaria turgida* (Clark, 1876) | | | | |
| *Abietinaria fusca* (Johnston, 1847) | 150 | — | | |
| *Abietinaria costata* (Nutting, 1901) | ? | — | | |
| *Abietinaria pulchra* (Nutting, 1904) | 9–300 | — | 12–542 | — |
| *Diphasia fallax* (Johnston, 1847) | 20–237 | — | | |
| *Diphasia rosacea* (L., 1758) | ? | — | | |
| *Dynamena pumila* (L., 1758) | 0–270 | — | | |
| *Hydrallmania falcata* (L., 1758) | 9–293 | — | 82–155 | — |
| *Tamarisca tamarisca* (L., 1758) | 60–190 | — | | |
| *Sertularella gigantea* Mereschkowsky, 1878 | 15–220 | — | 15–570 | — |
| *Sertularella rugosa* (L., 1758) | 12–58 | — | | |
| *Sertularella tenella* (Alder, 1857) | 20 | — | | |
| *Sertularella tricuspidata* (Alder, 1856) | 5–300 | — | 11–520 | — |
| *Sertularella pellucida* Jäderholm, 1908 | ? | — | | |
| *Sertularia albimaris* Mereschkowsky, 1878 | 2–605 | — | 3–6 | — |
| *Sertularia cupressina* L., 1758 | 19–120 | — | 360 | — |
| *Sertularia cupressoides* Clark, 1876 | | | 9,10 | — |
| *Sertularia mirabilis* (Verrill, 1873) | 9–187 | — | 16 | — |
| *Sertularia plumosa* (Clark, 1876) | 16–290 | — | 12–27 | — |
| *Sertularia similis* Clark, 1876 | | | | |
| *Sertularia tenera* G. O. Sars, 1874 | 6–395 | — | 17–610 | — |
| *Sertularia tolli* (Jäderholm, 1908) | | | 40 | — |
| *Thuiaria articulata* (Pallas, 1766) | 14–293 | — | 17–225 | — |
| *Thuiaria laxa* Allman, 1874 | 18–322 | — | 10–360 | — |
| *Thuiaria thuja* (L., 1758) | 12–250 | — | | |
| *Thuiaria cupressoides* (Lepechin, 1783) | 42–106 | — | | |
| *Thuiaria carica* Levinsen, 1893 | 9–256 | — | 40–360 | — |
| *Thuiaria alternitheca* Levinsen, 1893 | | | 55–64 | — |
| *Thuiaria obsoleta* (Lepechin, 1781) | 40–18 | — | | |
| *Thuiaria breitfussi* (Kudelin, 1914) | 35–62 | — | | |
| *Thuiaria arctica* (Bonnevie, 1899) | 86 | — | | |
| *Thuiaria cylindrica* Clark, 1876 | | | | |

P = polypoid generation; M = medusoid generation; blank spaces indicate that the species was not found in this region; dashed line (—) indicates absence of the given generation of the species; bullet (•) indicates that the given generation of the species is unknown but its presence is anticipated. In the Biogeography column A = Arctic; wA = west Arctic; B-A = boreal–Arctic; wB-

| Laptev Sea | | East Siberian Sea | | Chukchi Sea | | Central Arctic Basin | | |
|---|---|---|---|---|---|---|---|---|
| P | M | P | M | P | M | P | M | Biogeography |
| 5 | 6 | 7 | 8 | 9 | 10 | 11 | 12 | 13 |
| | | | | | | | | S-A |
| 64 | — | | | | | | | S-A |
| | | | | | | | | |
| | | | | | | | | |
| | | | | | | | | aBR |
| | | | | | | | | |
| | | | | | | | | |
| | | | | | | | | BR |
| | | | | | | | | |
| | | | | | | | | |
| | | 54 | — | | | | | B |
| | | | | | | | | BR |
| | | 18 | — | ? | — | | | eB-A |
| | | | | | | | | aBR |
| | | | | | | | | ? |
| 21–360 | — | 35–61 | — | 20–72 | — | | | B-A |
| | | | | | | | | aBR |
| | | | | | | | | aBR |
| | | | | | | | | aBR |
| 25 | — | | | | | | | wB-A |
| | | | | | | | | aBR |
| 19–30 | — | 54 | — | | | | | B-A |
| | | | | | | | | BR |
| | | | | | | | | BR |
| 19–45 | — | 43–45 | — | 5–53 | — | 698 | — | B |
| | | 54 | — | | | | | eB-A |
| 16–64 | — | | | | | | | wA |
| | | | | | | | | aBR |
| | | | | 6–36 | — | | | eB-A |
| 19 | — | | | | | | | B-A |
| 38–64 | — | 22–42 | — | 3–65 | | | | B-A |
| | | | | 5–47 | — | | | eB-A |
| 19–220 | — | 12–65 | — | 24–59 | — | | | B-A |
| 19 | — | | | | | | | eB-A |
| 38–64 | — | | | | | | | B |
| | | 58 | — | | | | | B-A |
| | | | | | | | | S-A |
| | | | | 36 | — | | | ? |
| 18–64 | — | 35–54 | — | | | | | B-A |
| 57 | — | 16,17 | — | | | | | B-A |
| | | | | | | | | wA |
| | | | | | | | | ? |
| | | | | | | | | E |
| | | | | 5,6 | — | | | eB-A |

A = west boreal–Arctic; eB-A = east boreal–Arctic; S-A = subtropical–Arctic; BR = boreal; aBR = Atlantic boreal; B = bipolar; K = cosmopolitans; WW = widely distributed warm-water species; E = endemic taxa;? = uncertain distributional range.

Table 16-1. *(continued)*

| Species | Barents Sea P | Barents Sea M | Kara Sea P | Kara Sea M |
|---|---|---|---|---|
| | 1 | 2 | 3 | 4 |
| HALECIIDAE | | | | |
| *Halecium beani* Johnston, 1847 | 1–297 | — | 46–207 | — |
| *Halecium curvicaule* Lorenz, 1886 | 15–192 | — | 9–360 | — |
| *Halecium speciosum* Nutting, 1901 | | | | |
| *Halecium halecinum* (L., 1758) | 55–310 | — | | |
| *Halecium corrugatum* Nutting, 1899 | 22–252 | — | 14? | — |
| *Halecium labrosum* Alder, 1859 | 18–294 | — | 16 | — |
| *Halecium tenellum* Hincks, 1861 | 30–295 | — | 72–178 | — |
| *Halecium marsupiale* Bergh, 1887 | 11–245 | — | 128 | — |
| *Halecium birulai* Spassky, 1929 | 5–160 | — | | |
| *Halecium muricatum* (Ellis et Solander, 1786) | 10–340 | — | 22–246 | — |
| *Halecium groenlandicum* Kramp, 1911 | 50–164 | — | 180 | — |
| *Halecium reversum* Nutting, 1901 | | | 125–360 | — |
| PLUMULARIIDAE | | | | |
| *Plumularia fragilis* Hamman, 1882 | 270 | — | | |
| *Schizotricha variabilis* Bonnevie, 1899 | 349 | — | | |
| *Schizotricha polaris* Naumov, 1960? | 175 | — | | |
| *Polynemertesia gracillima* (G. O. Sars, 1874) | 55–66 | — | | |
| *Nemertesia antennina* (L., 1758) | 110–220 | — | | |
| *Aglaophenopsis compressa* (Bonnevie, 1899) | 220 | — | | |
| *Cladocarpus integer* (G. O. Sars, 1874) | 149–220 | — | | |
| *Cladocarpus pourtelessii* Verrill, 1879 | 120 | — | | |
| *Cladocarpus formosus* Allman, 1874 | 270 | — | | |
| LIMNOMEDUSAE | | | | |
| OLINDIIDAE | | | | |
| *Monobrachium parasitum* Mereschkowsky, 1877 | 20–308 | — | 30 | — |
| TRACHYLIDA | | | | |
| RHOPALONEMATIDAE | | | | |
| *Aglantha digitale* (O.F. Müller, 1776) | — | 370–0<br>5–0<br>270–260 | — | 200–100 |
| *Homoeonema platygonon* Browne, 1903 | | | — | 280–200<br>50–25 |
| *Halicreas bigelowi* (Kramp, 1947) | | | | |
| PTYCHOGASTRIIDAE | | | | |
| *Ptychogastria polaris* Allman, 1878 | — | 384–25 | — | 147–120 |
| AEGINIDAE | | | | |
| *Aeginopsis laurentii* Brandt, 1835 | — | 10–0<br>325–0<br>270–360 | — | 5–0<br>260–0<br>200–100 |

P = polypoid generation; M = medusoid generation; blank spaces indicate that the species was not found in this region; dashed line (—) indicates absence of the given generation of the species; bullet (•) indicates that the given generation of the species is unknown but its presence is anticipated. In the Biogeography column A = Arctic; wA = west Arctic; B-A = boreal–Arctic; wB-

| *Laptev Sea* | | *East Siberian Sea* | | *Chukchi Sea* | | *Central Arctic Basin* | | |
|---|---|---|---|---|---|---|---|---|
| *P* | *M* | *P* | *M* | *P* | *M* | *P* | *M* | *Biogeography* |
| 5 | 6 | 7 | 8 | 9 | 10 | 11 | 12 | 13 |
| 38–1070 | — | | | | | | | B |
| | | | | | | | | B-A |
| | | | | 9–11 | — | | | eB-A |
| | | | | | | | | BR |
| | | | | | | | | B-A |
| | | 54 | — | 20 | — | | | S-A |
| | | 54 | — | | | | | B |
| | | | | | | | | B-A |
| | | | | | | | | ? |
| 20–57 | — | 35–142 | — | 5–30 | — | | | B-A |
| 180 | — | | | | | | | B-A |
| | | | | | | | | ? |
| | | | | | | | | E |
| | | | | | | | | E |
| | | | | | | 196 | — | E |
| | | | | | | | | wB-A |
| | | | | | | | | aBR |
| | | | | | | | | E |
| | | | | | | | | wB-A |
| | | | | | | | | aBR |
| | | | | | | | | BR |
| | | | | 10–72 | — | | | B-A |
| — | 1360–0 | | | — | 225–0 | | | B-A |
| | | | | | | | | wA |
| | | | | | | — | 3700–0 | ? |
| | | | | | | | 3255–990 | |
| | | | | | | | 300–0,50–0 | |
| — | 2500–35 | | | | | | | A |
| — | 0,10–0 | — | 40–25 | — | 10–0,25–0 | — | 25–0 | A |
| | 280–0 | | 25–0 | | 137–0 | | 3700–0 | |
| | | | | | | | 2000–870 | |

A = west boreal–Arctic; eB-A = east boreal–Arctic; S-A = subtropical–Arctic; BR = boreal; aBR = Atlantic boreal; B = bipolar; K = cosmopolitans; WW = widely distributed warm-water species; E = endemic taxa;? = uncertain distributional range.

Table 16-1. *(continued)*

| Species | Barents Sea P | Barents Sea M | Kara Sea P | Kara Sea M |
|---|---|---|---|---|
| | 1 | 2 | 3 | 4 |
| ACTINULIDA | | | | |
| HALAMMOHYDRIDAE | | | | |
| *Halammohydra schulzei* Remane, 1927 | — | 0 | | |
| HYDRIDA | | | | |
| HYDRIDAE | | | | |
| *Protohydra leuckarti* Greef, 1870 | 0 | — | | |
| SIPHONOPHORA | | | | |
| SIPHONANTHA | | | | |
| PHYSOPHORIDAE | | | | |
| *Physophora hydrostatica* Forskal, 1775 | 100–50 | | | |
| AGALMIDAE | | | | |
| *Marrus orthocanna* (Kramp, 1942) | ? | | | |
| *Rudjakovia plicata* Margulis, 1982 | | | | |
| PRAYIDAE | | | | |
| *Nectopyramis diomedea* Bigelow, 1911 | | | | |
| DIPHYIDAE | | | | |
| *Muggiaea bargmannae* Totton, 1954 | | | | 1360–0 |
| *Dimophyes arctica* (Chun, 1897) | 150–0 | | | 135–0 |
| | | | | 223–0 |

P = polypoid generation; M = medusoid generation; blank spaces indicate that the species was not found in this region; dashed line (—) indicates absence of the given generation of the species; bullet (•) indicates that the given generation of the species is unknown but its presence is anticipated. In the Biogeography column A = Arctic; wA = west Arctic; B-A = boreal–Arctic; wB-

Table 16-2. Biogeographic Distribution of Hydrozoa in the Eurasian Arctic Seas

| Sea | Number of Species E | wA | A | wB-A | eB-A | B-A | S-A | aBR | BR | B | K | WW | ? | Total |
|---|---|---|---|---|---|---|---|---|---|---|---|---|---|---|
| Barents | 6 | 9 | 3 | 6 | 1? | 34 | 13 | 19 | 9 | 16 | 2 | 3 | 4 | 125 |
| Kara | 3 | 5 | 3 | 3 | 2 | 32 | 9 | 1 | — | 16 | 2 | — | 1 | 77 |
| Laptev | 1 | 3 | 3 | 3 | 1 | 24 | 3 | — | — | 14 | 2 | — | — | 54 |
| East Siberian | 1 | — | 2 | — | 2 | 20 | 3 | — | — | 11 | 2 | — | — | 41 |
| Chukchi | 1 | — | 2 | — | 5 | 12 | 5 | — | — | 5 | 2 | — | 3 | 35 |
| Central Arctic Basin | 2 | 2 | 1 | — | — | — | — | — | — | 3 | 2 | — | 2 | 12 |

Key as in Table 16-1.

| *Laptev Sea* | | *East Siberian Sea* | | *Chukchi Sea* | | *Central Arctic Basin* | | |
|---|---|---|---|---|---|---|---|---|
| *P* | *M* | *P* | *M* | *P* | *M* | *P* | *M* | *Biogeography* |
| 5 | 6 | 7 | 8 | 9 | 10 | 11 | 12 | 13 |
| | | | | | | | | aBR |
| | | | | | | | | BR |
| | | | | | | | | WW |
| | | | | | | 197–300 | | wA |
| | | | | | | 2000–300 | | |
| | | | | | | 1000–0,950–0 | | E |
| | | | | | | 850–0,500–0 | | |
| | | | | | | 400–0,300–0 | | |
| | | | | | | 200–0,150–0 | | |
| | | | | | | 2000–0,800–300,260–0 | | ? |
| 993–0,360–0 | | | | | | 967–0,200–0 | | B |
| 1360–0,200–0 | | 900–200,160–100 100–0 | | 420–0,225–175 | | ? | | K |

A = west boreal–Arctic; eB-A = east boreal–Arctic; S-A = subtropical–Arctic; BR = boreal; aBR = Atlantic boreal; B = bipolar; K = cosmopolitans; WW = widely distributed warm-water species; E = endemic taxa;? = uncertain distributional range.

# APPENDIX

## Plates 16-I, 16-II, 16-III

Illustrations of representative Arctic Hydroidea and Siphonophora. (After Naumov, 1960 and Stepanjants, 1967.)

## Maps 16-1, 16-2, 16-3

Biogeographic distribution of Hydrozoa

B-A: boreal–Arctic
B: bipolar
S-A: subtropical–Arctic
BR: boreal
K: cosmopolitans
A: Arctic
E: endemic taxa
WW: widely distributed warm-water species
wB-A: west boreal–Arctic
eB-A: east boreal–Arctic

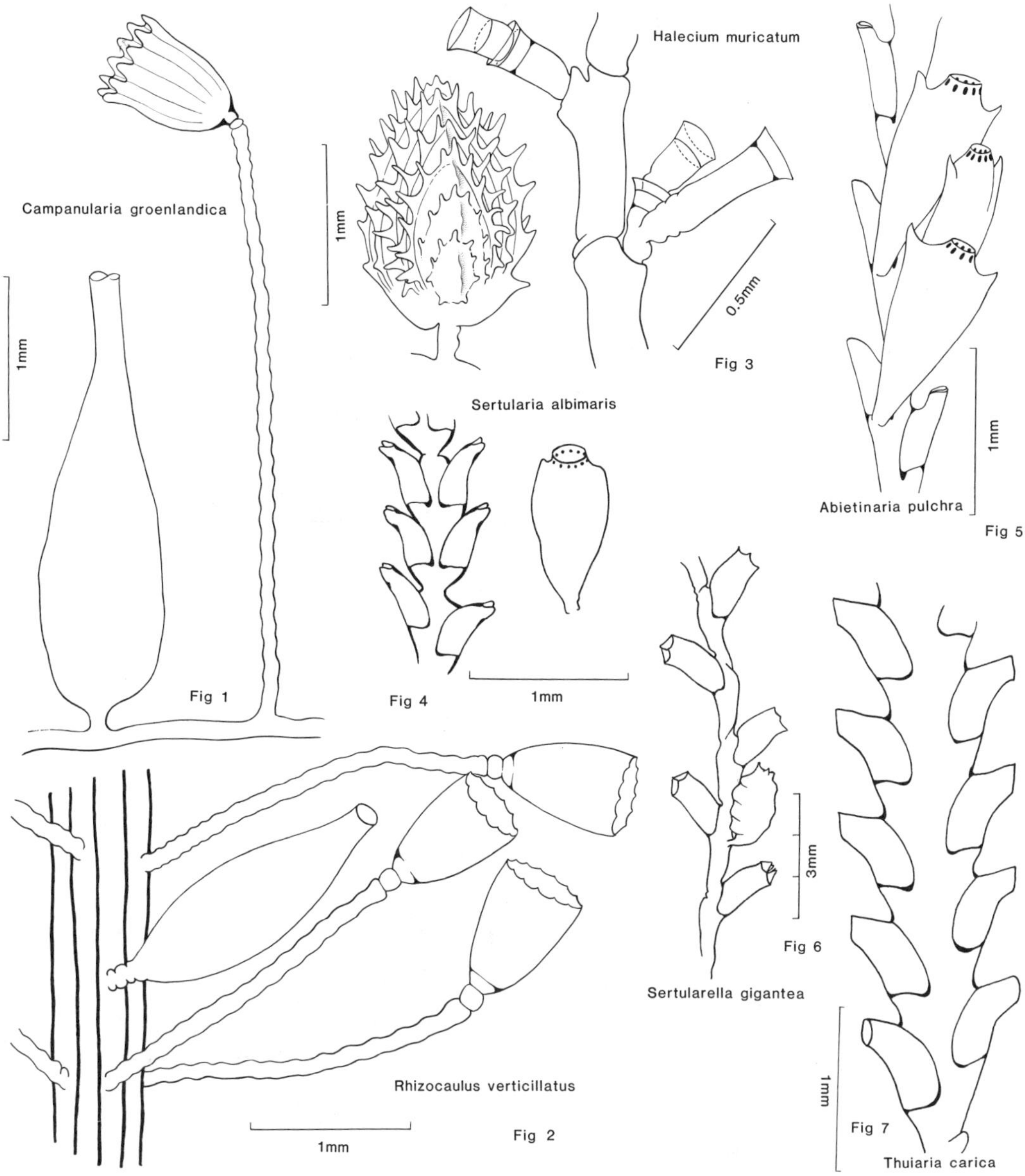
Campanularia groenlandica
1mm
Fig 1
Halecium muricatum
1mm
0.5mm
Fig 3
Abietinaria pulchra
1mm
Fig 5
Sertularia albimaris
Fig 4
1mm
Sertularella gigantea
3mm
Fig 6
Rhizocaulus verticillatus
1mm
Fig 2
Thuiaria carica
1mm
Fig 7

PLATE 16-I

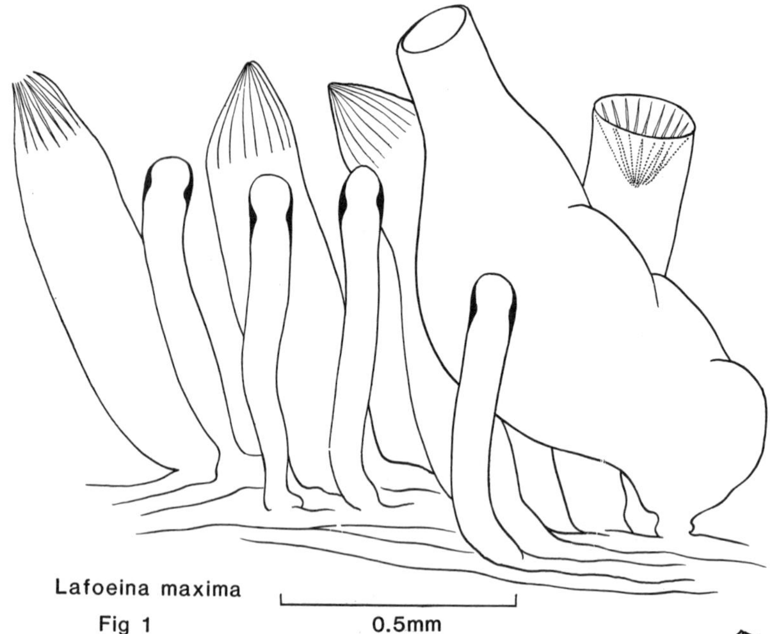


Lafoeina maxima

Fig 1

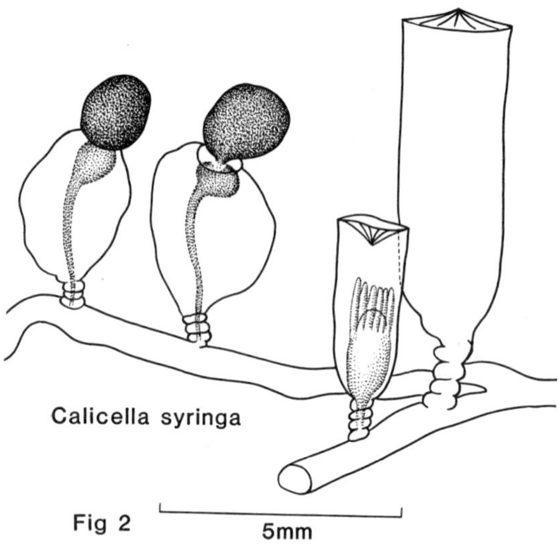


Calicella syringa

Fig 2

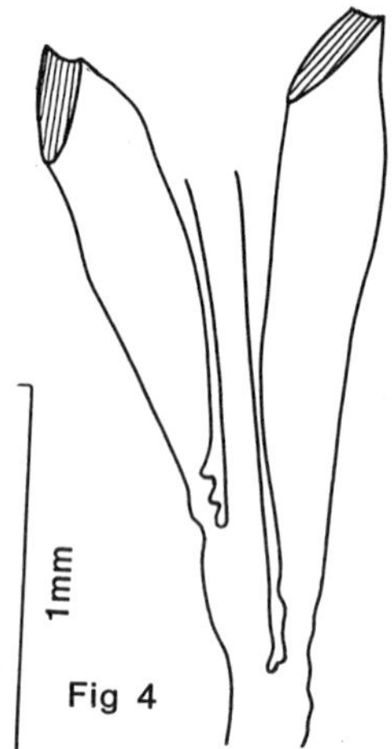


Fig 4

Modeeria plicatile

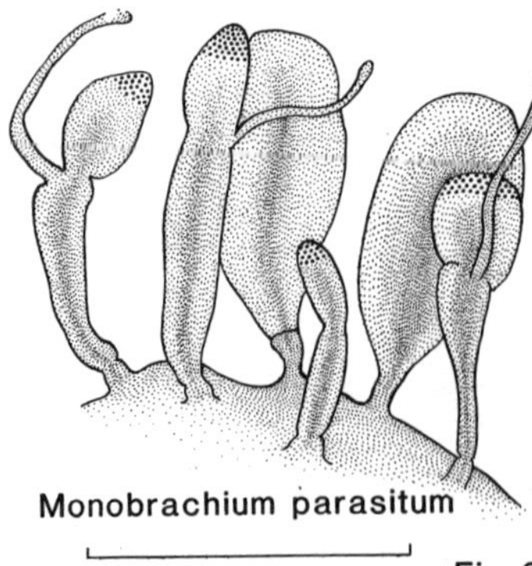


Monobrachium parasitum

Fig 3

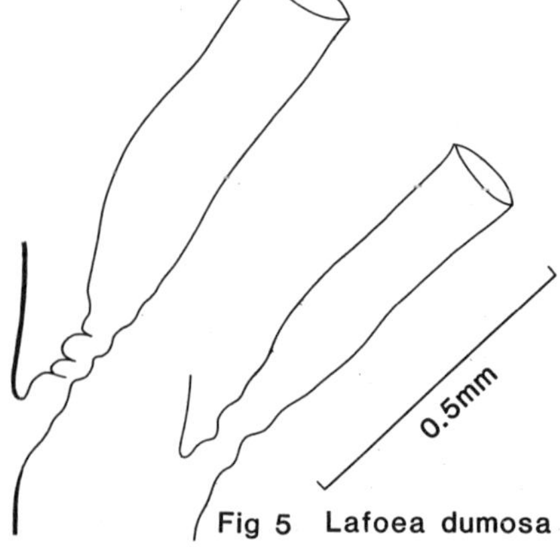


Fig 5 Lafoea dumosa

PLATE 16-II

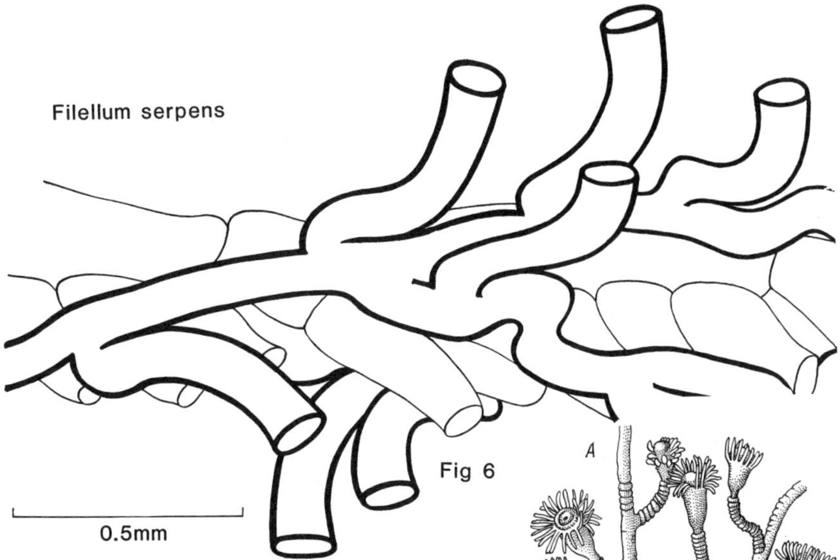


Filellum serpens

Fig 6

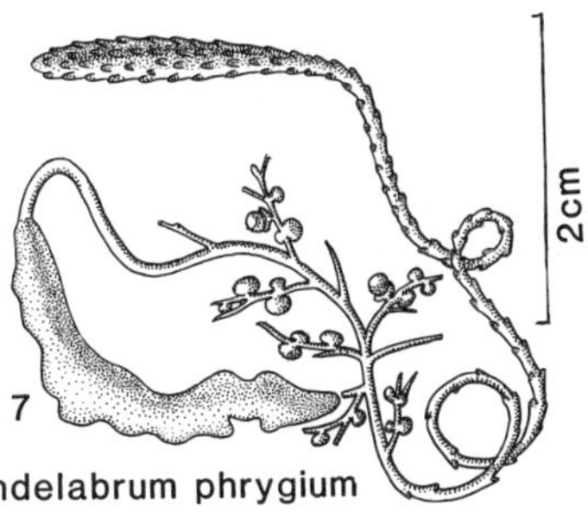


Fig 7

Candelabrum phrygium

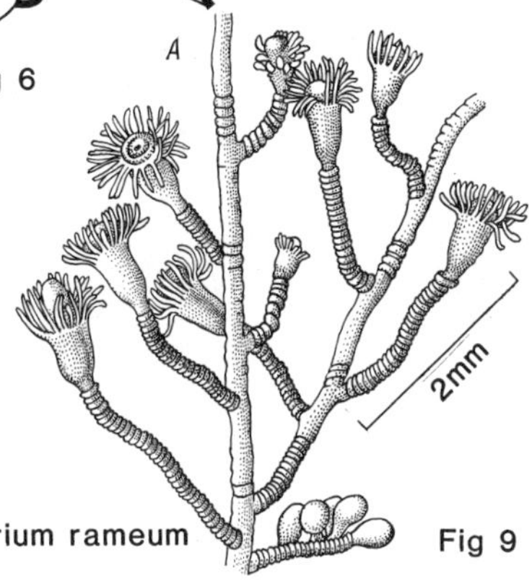


Eudendrium rameum

Fig 9

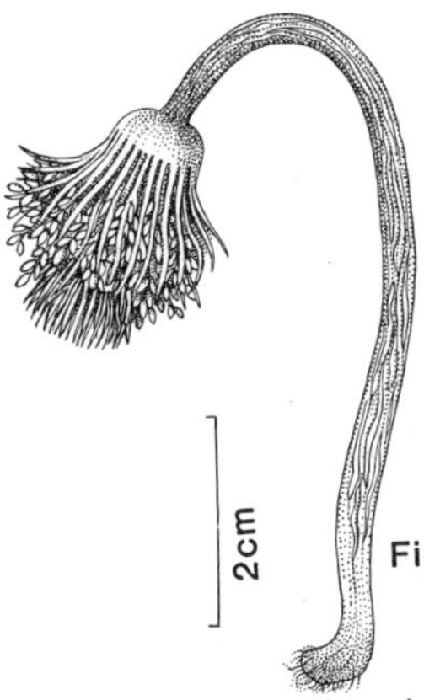


Fig 8

Corymorpha glacialis

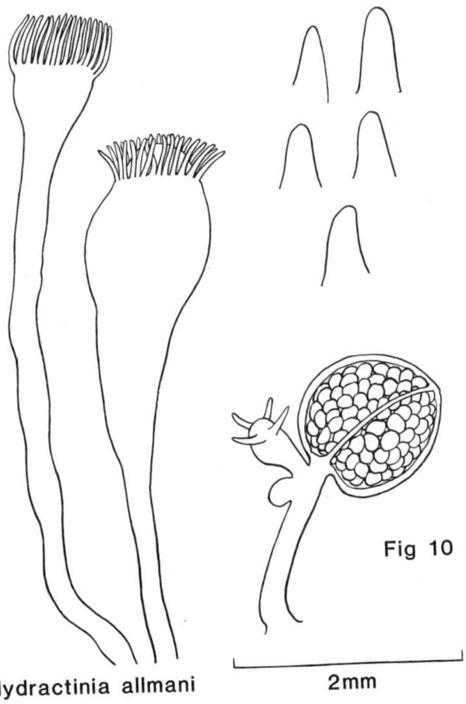


Fig 10

Hydractinia allmani

PLATE 16-II

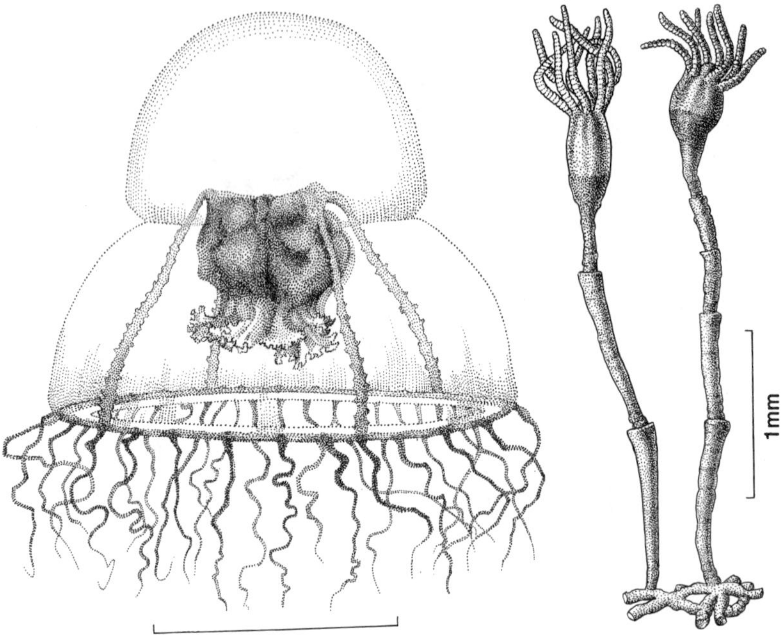


Halitholus yoldia-articae

Fig 1

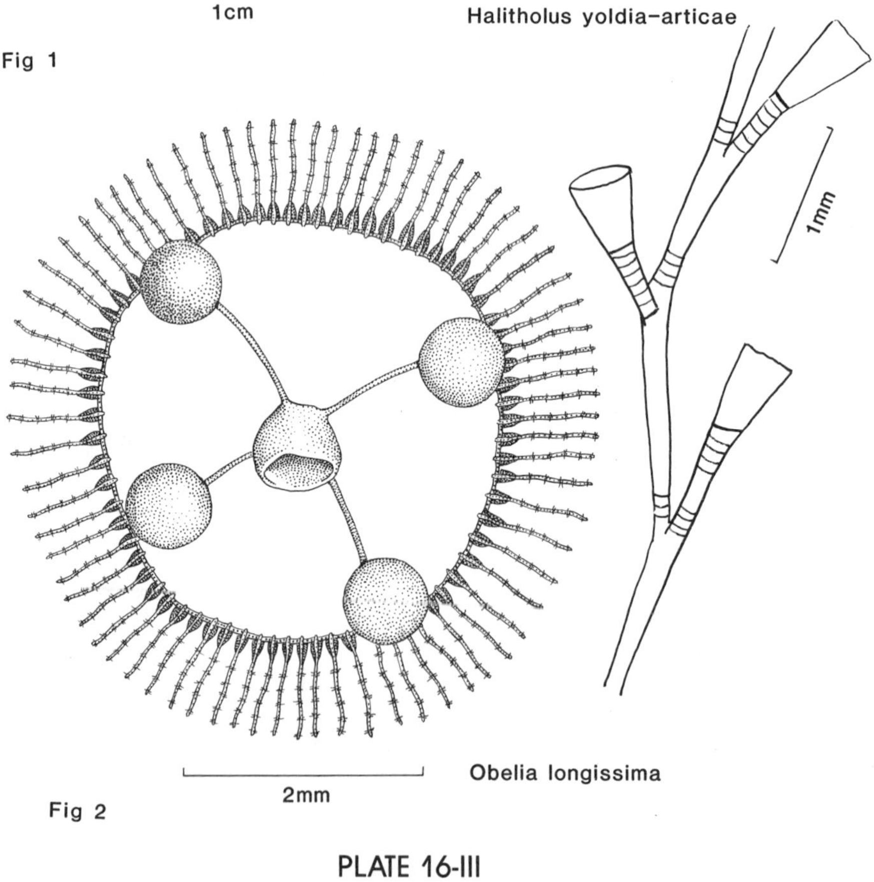


Obelia longissima

Fig 2

PLATE 16-III

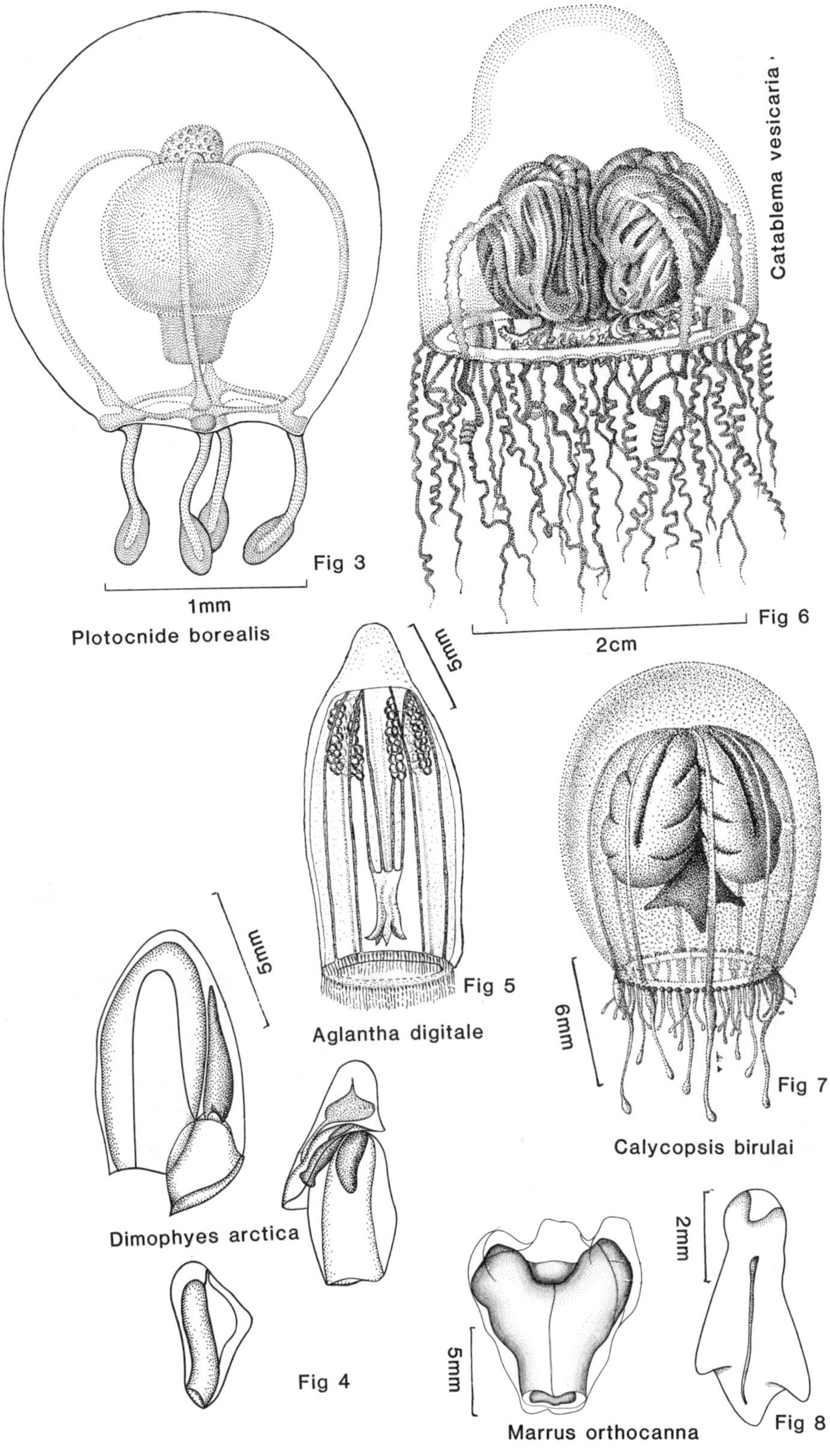
Fig 3
1mm
Plotocnide borealis
Catablema vesicaria
Fig 6
2cm
5mm
Fig 5
Aglantha digitale
5mm
6mm
Fig 7
Calycopsis birulai
Dimophyes arctica
Fig 4
5mm
2mm
Marrus orthocanna
Fig 8

PLATE 16-III

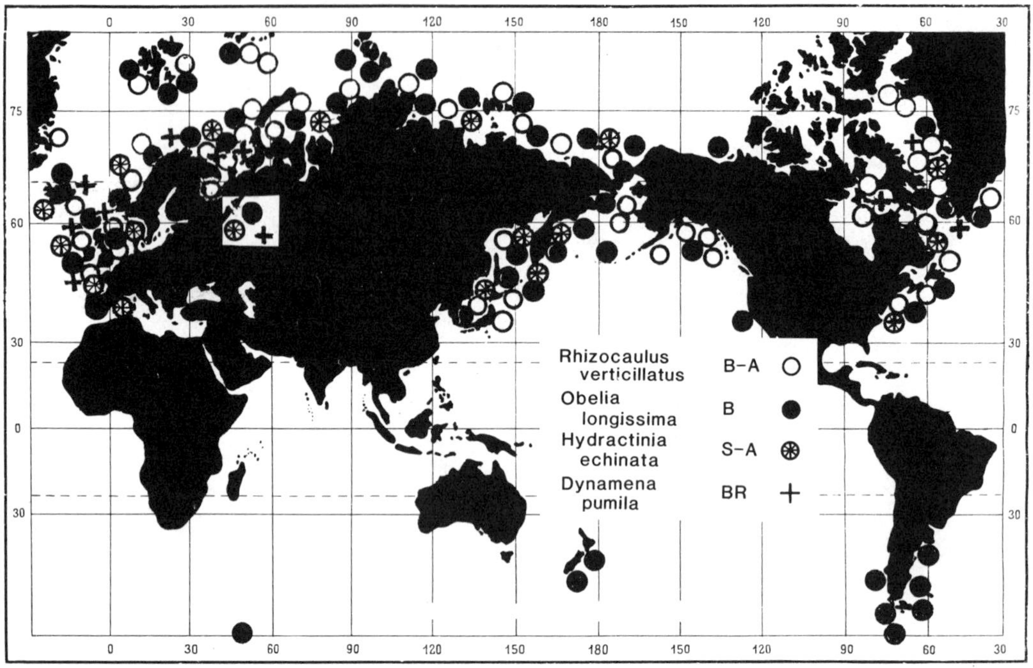


MAP 16-1

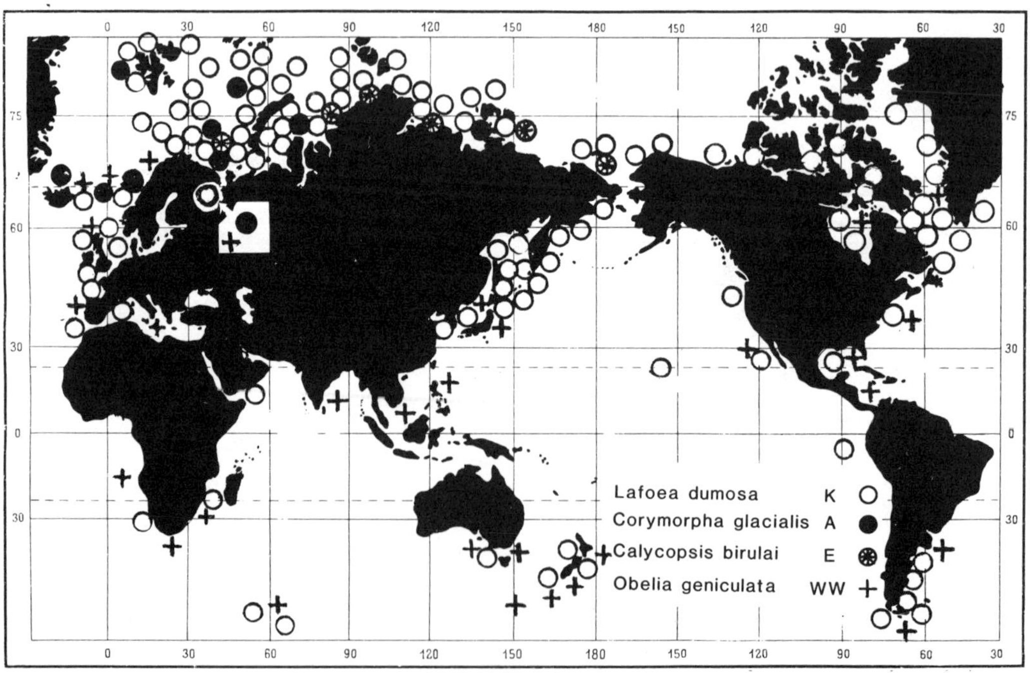


MAP 16-2

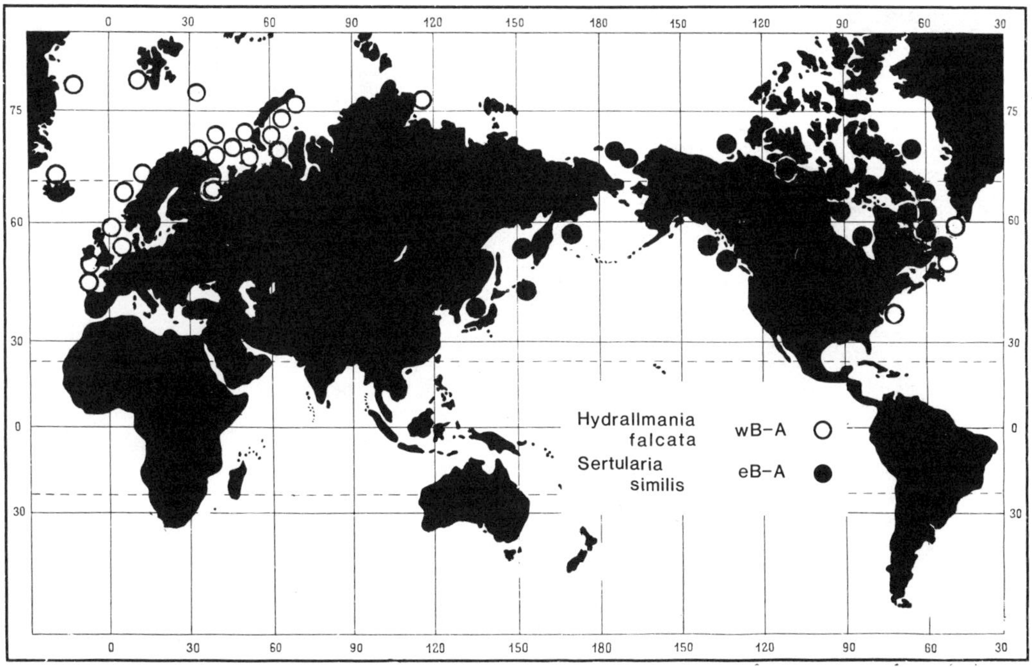


MAP 16-3

## REFERENCES


Antsulevich, A. E., and D. G. Polteva. 1986. Hydroidi roda *Rhizogeton* L. Agassiz, 1862 (*Hydrozoa, Clavidae*) v faune SSSR. *Zool. J.* **45**(7):967–972. (In Russian.)

Arai, M. N., and A. Brinckmann-Voss. 1980. Hydromedusae of British Columbia and Puget Sound. *Canad. Bull. Fish. Aquat. Sci.* **204:**192p.

*Atlas Okeanov.* 1980. Severnyi Ledovityi Ocean. Maps 118–122, pp. 133–136. (In Russian.)

Beresina, N. A. 1948. Tip *Coelenterata,* Kishechnopolostniye, Podtip *Cnidaria,* Strecajuschiye. Klass *Hydrozoa*-hydroidi. Podklass *Hydroidea*-hydroidniye. Otrjad *Leptolina.* In N. S. Gaevskaya (ed.), *Opredelit. Fauni i Vlori Sjevernich Morjei SSSR.* pp. 47–65. (In Russian.)

Bergh, R. S. 1886. Gopelepolyper (Hydroider) fra Kara-Havet. *Dijmphna-otets Zool. Botan. Udbytta Kjobenhavn.* pp. 329–337.

Birulja, A. A. 1896. K faune medus Solovetskogo zaliva. Mater. biol., zoogeogr. preimuschestv. russkich morjei, I. Yezjegodn. *Zool. Mus. Acad. Nauk,* pp. 327–354. (In Russian.)

Birulja, A. A. 1897*a*. *Cnidaria.* Otchet o sostojanii i dejatelnosti Solovetskoy stancii v 1897 g. *Tr. Imp. SPb Obstch. Estestvoisp* **28**(1):234–236. (In Russian.)

Birulja, A. A. 1897*b*. *Hydrozoa, Polychaeta* i *Crustacea,* sobrannije d-rom A. S. Botkinim v Eniseyskoy i Obskoy gubach letom 1895 g. Mater. biol., zoogeogr. preimuschestv. russkich morjei, II. Yezjegodn. *Zool. Mus. Acad. Nauk* pp. 78–115. (In Russian.)

Birulja, A. A. 1898. O savisimosti strojenija nekotorich Hydroidov poberezja Solovetskich ostrovov ot physitcheskich usloviy ich mestoobitanija. Mater. biol., zoogeogr. preimuschestv. russkich morjei, III. Yezegodn. *Zool. Mus. Acad. Nauk* pp. 203–212. (In Russian.)

Bonnevie, K. 1898. Zur Systematik der Hydroiden. *Zeitschr. Wiss. Zool.* **63:**465–495.

Bonnevie, K. 1899. Hydroida. Norske Nordhavs-Exped. 1876–1878. *Zoology* **26:**103p.

Breitfus, L. 1904. Zoologische Studien in Barents Meere auf grund der Untersuchungen der Expedition, I Liste der Fauna d. Barentsmeeres. *Izd. Komit. Pomoschi Pomoram Russkogo Severa, SPb* **47.**

Breitfus, L. 1906. Expeditcija dlja nauchno-promislovich issledovaniy u beregov Murmana. Otchet za 1903 g., SPb, 257p. (In Russian.)

Breitfus, L. 1908. Expeditcija dlja nauchno-promislovich issledovaniy u beregov Murmana. Otchet za 1904 g., SPb, 231p. (In Russian.)

Broch, H. 1910. Die Hydroiden der arktischen Meere. *Fauna Arctica* **5**(1):129–248.

Broch, H. 1916. Hydroida. Part I. *Danish Ingolf-Exped.*, 5, part 6, 66p.
Broch, H. 1918. Hydroida, Part II. *Danish Ingolf-Exped.*, 5, part 7, 200p.
Calder, D. R. 1970. Thecate Hydroids from the shelf waters of northern Canada. *J. Fish. Res. Bd. Canada* **27**:1501–1547.
Calder, D. R. 1972. Some Athecate Hydroids from the shelf waters of northern Canada. *J. Fish. Res. Bd. Canada* **29**:217–228.
Derjugin, K. M. 1905. Otchet po oborudovaniju Murmanskoy biologicheskoy stancii i letnije raboti na ney v 1904 g. *Tr. SPb Obstch. Estestvoisp* **36**(1):1–32. (In Russian.)
Derjugin, K. M. 1906. Murmanskaja biologicheskaja stancija 1899–1905 gg. *Tr. SPb Obstch. Estestvoisp.* **37**(4):176p. (In Russian.)
Derjugin, K. M. 1915. Fauna Kolskogo zaliva i uslovija jee suschestvovanija. *Zap. Acad. Nauk, Phys.-Mat. Otdelenija* **34**(ser. 8):929p. (In Russian.)
D'Urban, W. S. 1881. The zoologie of Barents Sea. *Ann. Mag. Nat. Hist. Ser.* **5**:653.
Fraser, C. M. 1941. New species of Hydroida, mostly from the Atlantic Ocean, in United States National Museum. *Proc. U.S. Nat. Mus.* **91**:77–89.
Golikov, A. N., and V. G. Averintsev. 1977. Biocenosi verchnich otdelov shelfa archipelaga Zemlja Franza-Josefa i nekotorije zakonomernosti ich raspredelenija. In A. N. Golikov (ed.), Biocenosi shelfa Zemlji Franza-Jesefa i fauna sopredelnich aquatoriy. *Issled. Fauni Morjeiy* **14**(22):5–54. (In Russian.)
Golikov, A. N., V. G. Averintsev, A. M. Scheremetevskyi, V. I. Ljuleev, and O. K. Novikov. 1974. Nekotorije zakonomernosti raspredelenija zizni na novosibirskom melkovodje i v sopredelnich aquatorijach. Otchetn. nauchn. sess. *Zool. Inst. Acad. Sci. USSR* **12**:13. (In Russian.)
Golikov, A. N., V. I. Ljuleev, O. K. Novikov, V. V. Potin, B. I. Sirenko, and A. M. Scheremetevskyi. 1978. Nekotorije zakonomernosti raspredelenija zizni na verchnich otdelach shelfa ostrova Vrangelja i m. Schmidta. In L. Y. Borkin (ed.), *Morphologija, Systematica i Evolutcija Zivotnich*, pp. 1–12. (In Russian.)
Gorbunov, G. P. 1946. Donnoe naselenie novosibirskogo melkovodja i centralnoy tchasti Severnogo Ledovitogo oceana. *Tr. Dreifujuschey Expeditsii na 1/p Sedov 1937–1940* **3**:30–138. (In Russian.)
Grebelnyi, S. D. 1980. Rasprostranenije aktiniy v. Arctice. In D. V. Naumov and S. D. Stepanjants (eds.), *Teoreticheskoje i Practicheskoje Znachenie Kishechnopolostnich*. Leningrad, pp. 20–33. (In Russian.)
Hartlaub, C. 1900. Zoologische Ergebnisse einer Untersuchungsfahrt nach Bäreninsel und Spitzbergen im 1898, I. *Einleit. Wiss. Meeresuntersuch. N. Folge, Abt. Helgoland.* Kiel, Leipzig, pp. 1–17.
Hirohito, 1983. *Hydroids from Izu Oshima and Niijima.* Biol. Lab. Imper. Household, Tokyo, 83p.
Jäderholm, E. 1902. Die Hydroiden der schwedischen zoologischen Polarexpedition 1900. Bihang till. *K. Svenska Vet. Akad. Handl.* **21**(4):1–11.
Jäderholm, E. 1908. Die Hydroiden des sibirischen Eismeeres gesammelt von der Russischen Polar-Expedition 1900–1903. *Mem. Acad. Sci. St.-Petersb.* **18**:1–26.
Jäderholm, E. 1909. IV. Hydroiden. *K. Svenska Vet. Akad. Handl.* **45**(1):124p.
Jäderholm, E. 1926. Ueber einige antarktische und subantarctische Hydroiden. *Ark. Zool.* **18**(14):1–7.
Kamshilov, M. M., and E. A. Zelickman. 1958. O vodovom sostave zooplanktona pribrezja Vostochnogo Murmana. *Tr. Murmansk. Bio. St.* **4**:41–44.
Kirchenpauer, G. H. 1884. Nordische Gattungen und Arten von Sertulariiden. *Abh. Gebiete Naturwiss., Hamburg* **8**(3):3–54.
Knipovitch, N. M. 1901. Zoologicheskije issledovanija na ledokole "Ermack" letom 1901 g. Yezjegodn. *Zool. Mus. Acad. Nauk* **6**:1–20. (In Russian.)
Korotnev, A. A. 1880. Opit sravnitelnogo isuchenija Coelenterata, II, III. *Izv. Imp. Obstch. Ljubit. Estestvozn., Antropolog. i Etnogr.* **37**(2):1–63. (In Russian.)
Kramp, P. L. 1932. Hydroids. The Gothaab Expedition 1928. *Medd. Grønland Udg. Kommis. Vidensk. Unders. Grønland* **79**(1):1–86.
Kramp, P. L. 1942. Medusae. The Gothaab Expedition 1928. *Medd. Grønland Udg. Kommis. Vidensk. Unders. Grønland* **81**(1):1–168.
Kramp, P. L. 1943. Hydroida. The zoology of east Greenland. *Medd. Grønland Udg. Kommis. Vidensk. Unders. Grønland* **121**(11):1–52.
Kramp, P. L. 1963. Summary of the zoological results of the Gothaab Expedition 1928. *Medd. Grønland Udg. Kommis. Vidensk. Unders. Grønland* **81**(7):1–115.
Kudelin, N. V. 1914. Hydroidi (Hydroidea). t. 11, 2. Fauna Rossii i sopredelnich stran. *Zool. Mus. Acad. Nauk SPb*, pp. 139–526. (In Russian.)
Lepechin, J. J. 1780. Sertulariae species duae determinatae. *Acta Acad. Sci. Petropol* **4**(1):223–225.
Levinsen, G. M. R. 1893. Meduser, Ctenopho-

rer og Hydroider fra Grønlands Vestkyst. *Vid. Medd. Nat. Foren. København for 1892* **5**:143–220.

Linko, A. K. 1903. Otchet o sostojanii i dejatelnosti biologitcheskoy stanzii Imp. SPb Obschestva Estesvoispiteteley v Ekaterininskom porte na Murmane za 1899 g. *Tr. Imp. SPb. Obstch. Estesvoisp* **1**(1):1–10. (In Russian.)

Linko, A. K. 1904. Zoologische Studien im Barents-Meere Hydromedusen. *Zool. Anz.* **28**(16):210–220.

Linko, A. K. 1911. Hydroidi (Hydroidea). T. I., Fauna Rossii i sopredelnich stran. *Zool. Mus. Acad. Nauk SPb* **1–18**:1–250. (In Russian.)

Linko, A. K. 1912*a*. Hydroidi (Hydroidea). T. II, I, Fauna Rossii i sopredelnich stran. *Zool. Mus. Acad. Nauk SPb,* pp. 1–138. (In Russian.)

Linko, A. K. 1912*b*. Plankton blis beregov Zapadnogo Murmana v svyazi s sostejaniem promislov v techenie 1903–1905 gg. *Tr. Murmansk. Nauchno-Promislovoy Exp. 1905 g.*, pp. 27–70. (In Russian.)

Linko, A. K. 1913. Zooplankton Sibirskogo Ledovitogo Oceana po sboram Russkoy Polarnoy Expeditsii 1900–1903 gg. Nauchn. Result. Russk. Polarn. Exp. 1900–1903 gg., t. II, 4. *Zap. Imp. Acad. Nauk* **19**(4):1–53. (In Russian.)

Marenzeller, E. 1878. Die Coelenteraten, Echinodermen und Würmer d.K.K. Osterreichisch-Ungarischen Nordpol. Exped. Denkschrift. d.K. *Akad. Wissen. Math. Naturwiss. Classe* **35**:357p.

Margulis, R. Y. 1982. Novaya siphonophora *Rudjakovia plicata* gen. n., sp. n. (*Coelenterata, Hydrozoa*) is Poljarnogo Basseyna i nekotorije zametchanija o drugich siphonophorach. *Zool. J.* **61**(3):440–444. (In Russian.)

Marktanner-Turneretscher, G. 1890. Die Hydroiden des k.k. *Nat. Hof-Museum Wien* **5**:195–284.

Mereschkowsky, C. S. 1877. On a new genus of Hydroids from the White Sea with a short description of other new Hydroids. *Ann. Mag. Nat. Hist.* **20**(ser. 4):220–228.

Naumov, D. V. 1960. Hydroidi i hydromedusi morskich, solonovatovodnich i presnovodnich basseinov SSSR. Opred. Faune SSSR, izd. *Zool. Inst. Acad. Nauk SSSR* **70**:585p. (In Russian.)

Nesis, K. N. 1983*a*. Hipotesa o prichine vosniknovenija zapadno-i vostochnoarcticheskich arealov morskich donnich zivotnich. *Biol. Morja,* **5**:3–14. (In Russian.)

Nesis, K. N. 1983*b*. Suschestvoval li v pleistocene panarctichesky lednikovyi pokrov? *Biol. Morja* **6**:11–20. (In Russian.)

Nordgaard, O. 1904. Hydrographical and biological investigations in Norvegian Fiords. *Bergens. Mus.* **4**:156–159, 240–244.

Rylov, V. M. 1923. Zoologische Ergebnisse der Russischen Expeditionen nach Spitzbergen. *Ann. Mag. Zool. Acad. Sci. Rus.* **24**:140–160.

Rylov, V. M. 1924. *Hydroidea-Athecata,* sobranniye expeditsiey Instituta v 1921 g. *Tr. Plov. Mor. Nauchn. Inst.* **9**:3–10. (In Russian.)

Rylov, V. M. 1927. *Hydroidea* i *Alcyonaria,* sobranniye v Barentsevom morje Murmanskoy Biologitcheskoy Stantsiey vo vremja reisov po kolskomu meridianu v 1921 g. Yezjegodn. *Zool. Mus. Acad. Nauk SSSR* **27**:33–246. (In Russian.)

Sars, G. O. 1873. Bidrag til Kundskaben om Norges Hydroider. *Vidensk. Selsk. Forhandl. 1872,* pp. 91–150.

Sars, M. 1860. Om ammeslaegten *Corymorpha,* dens arter, samt de af disse opammende meduser. *Forhandl. Vidensk. Selskab. Christiania 1859,* pp. 96–105.

Sars, M. 1861. On the Nurse-Genus *Corymorpha* and its species, together with the medusae produced from them. *Ann. Mag. Nat. Hist.* **8**(ser. 3):360–363.

Sars, M. 1863. Bemaerkinger over fire norske Hydroider. *Forhandl. Vidensk Selskab. Christiania 1862,* pp. 25–39.

Schidlowsky, A. V. 1902. Hydroidi Belogo morja u beregov Solovetskich ostrovov. Materiali po faune hydroidov arctitcheskich morjey. *Tr. Obsch. Ispitat. Prirodi Charkovsk. Univ.* **36**(1):3–268. (In Russian.)

Schlater, G. G. 1891. Otcherki hydroidnoy fauni i spisok medus poberezja Solovetskich ostrovov. *Vestn. Estestvosnan* **9**:335–342. (In Russian.)

Smirnov, I. S., and S. D. Stepanjants. 1980. Symbios hydroida *Hydractinia vallini* Jaederholm i ophiur semeystva *Ophiolepididae* v. Antarcticheskich vodach. In D. V. Naumov and S. D. Stepanjants (eds.), *Teorticheskoye i Practicheskoye Znatchenie Kishechnopolostnych.* Leningrad, pp. 105–108. (In Russian.)

Spasskyi, N. N. 1929. K faune hydroidov Kolskogo zaliva i jugosapadnoy tchasti Barentseva morja. *Tr. Murmansk. Biol. Stantsii* **3**:1–48. (In Russian.)

Stepanjants, S. D. 1967. Siphonophori morjey SSSR i severnoy tchasti Tichogo oceana. Opred. po faune SSSR, isd. *Zool. Inst. Acad. Nauk SSSR* **96**:216p. (In Russian.)

Stepanjants, S. D. 1975. Sostav i nekotoriye osobennosti raspredeleniya siphonophor v Karibskom morje, Meksikanskom zalive i v sopredelnich Rayonach Atlantici (po materialam 14-go reysa "Akademik Kurtchatov"). *Tr. Inst. Oceanologii* **100**:96–126. (In Russian.)

Stepanjants, S. D. 1979. Hydroidi vod Antarctiki i Subantarctiki. Res. Biol. Issled. Sov. Antarkt. Exp., 6. *Issled. Fauni Morjey* **22** (30):199p. (In Russian.)
Stepanjants, S. D. 1980. O. cosmopolitizme u hydroidov. In D. V. Naumov and S. D. Stepanjants (eds.), *Teoreticheskoye i Practicheskoye Znatchenie Kishechnopolostnych.* Leningrad, pp. 114–122. (In Russian.)
Tanasijtchuk, N. P. 1927. O novich i redkich dlja fauni Kolskogo zaliva (Murman) formach zivotnich. *Dokl. Akad. Nauk SSSR* **14:**213–218. (In Russian.)
Thompson, D'Arcy, W. 1884. The Hydroid zoophytes of the Willem Barents Expedition 1881. *Bijdr. Dierkunde, Amsterdam,* pp. 1–10.
Thompson, D'Arcy, W. 1887. The Hydroida of the Vega-Expedition, *Vega-Exped. Vetenskap. Jakttag.* **4:**385–400.
Totton, A. K. 1965. *A Synopsis of the* Siphonophora. Trust. Brit. Mus. Nat. Hist., London, 227p.
Uchida, T. 1928. Studies on Japanese Hydromedusae, II, *Trachimedusae* and *Narcomedusae. Jap. J. Zool., Tokyo* **2**(1):73–97.
Uschakov, P. V. 1925. Sesonnije ismenenija na litorali Kolskogo zaliva. *Tr. Leningr. Obschestva Estesvoisp.* **65**(1):47–71. (In Russian.)
Uschakov, P. V. 1936. K faune hydroidov archipelaga Frantza-Iozepha. *Tr. Arct. Inst.* **2:**141–151. (In Russian.)
Uschakov, P. V. 1937. Materiali po hydroidam arctitcheskich morjey. *Tr. Arct. Inst.* **50:**5–34. (In Russian.)
Vagin, V. L. 1934. *Hydroidea,* sobrannije expeditsiey na sudnije "Lomonosov" ljetom 1931 g., v Barentsevom i Karskom morjach. *Tr. Arct. Inst.* **9:**79–87. (In Russian.)
Wagner, N. P. 1885. Bespozvonotchnije Belogo morja. SPb., 218p.
Yashnov, V. A. 1948. Tip *Coelenterata,* Kischechnopolostnije. Podtip *Cnidaria,* Strecajuschije. Klass *Hydrozoa*-hydroidi. Podklass *Hydroidea*-hydroidnije. Hydromedusae-hydroidnije medusi. In N. S. Gaevskaja (ed.), *Opredelitel Fauni i Flori Sjevernich Morjei SSSR,* pp. 65–74. (In Russian.)
Zelickman, E. A. 1961. K Plankticheskoy charactristike jugovostochnogo sektora Barentseva morja (po materialam avgusta 1958 g.). In *Hydrologitcheskije i biologitcheskije Osobennosti Pribrjeznich vod Murmana. Murmansk,* pp. 39–58. (In Russian.)
Zelickman, E. A. 1966. Zametki o sostave i raspredelenii zooplanktona v jugo-vostochnoy tchasti Barentseva morja v avguste-oktjabre 1959 g. Sostav i raspredelenije planktona i bentosa v juznoy tchasti Barentseva morja. *Tr. Kolsk. Filiala Murmansk. Morsk. Biol. Inst.* **11**(15):34–49. (In Russian.)
Zelickman, E. A. 1969. Structurnije osobennosti massovich skopleniy medus. *Okeanologija* **9**(4):686–693. (In Russian.)
Zelickman, E. A., V. I. Gelfand, and M. A. Shifrin. 1969. Growth, reproduction and nutrition of some Barents Sea hydromedusae in natural aggregations. *Marine Biol.* **4**(3):167–173.
Zelickman, E. A., and A. N. Golovkin. 1972. Composition, structure and productivity of neritic plankton communities near the bird colonies of the northern shores of Novaya Zemlya. *Marine Biol.* **17**(N3):265–274.
Zelickman, E. A., and M. M. Kamshilov. 1960. Dinamika mnogoljetnego sostojanija biomassi planktona v juznoy tchasti Barentseva morja i faktori, opredeljajuschije jee. *Tr. Murmansk. Biol. Inst.* **2:**68–114. (In Russian.)

# 17: Arctic Ocean Cumacea

S. V. Vassilenko

## INTRODUCTION

The order Cumacea belongs to the superorder Peracarida, subclass Malacostraca, class Crustacea, according to the classification of McLaughlin, 1980. The cumaceans are primarily marine bottom-dwelling burrowing crustaceans. They live on argillaceous sands, feed on detritus, and graze on sand grains or are filter feeders. The Cumacea, like all Peracarida, have an epimorphic larval development; eggs are laid and develop in the brood pouch. This stage is followed by the manca stage and a postlarval stage (McLaughlin, 1980).

The order Cumacea comprises about 800 species widely distributed in the World Ocean, from the Arctic to the Antarctic. They occur at a wide range of depths, from 0 m to 5,000 m. Forms closely related to the lower mysids (suborder Lophogastrida, order Mysidacea) should be regarded as phylogenetic ancestors of the Cumacea (Lomakina, 1958, 1968). The fossil record of the group goes back to the Upper Permian in Germany and to the Middle Jurassic in France (Hessler, 1969).

The Cumaceans are common components of bottom communities; they are important indices of water regime and therefore a study of their distributional patterns is valuable.

*Note:* This chapter has been reviewed, edited, and rewritten by the Editor, Yvonne Herman.

## PREVIOUS STUDIES

There is a vast literature on Cumacean fauna of different regions of the North Atlantic and the Atlantic part of the Arctic Ocean (e.g., Sars, 1864, 1871, 1882, 1900; Calman, 1912, 1920; Stebbing, 1913; Hansen, 1887, 1920; Stephensen, 1936, 1937, 1938, 1943; Zimmer, 1900, 1926, 1930, 1933, 1934, 1941, 1946, 1980; Yashnov, 1948; Just, 1970). The review by Lomakina (1958) contains original data on the fauna of the Arctic seas from the Barents Sea to the Chukchi Sea. Studies by Hart (1939), Wacasey (1974, 1975), Wacasey et al. (1979), and Corey (1981) concern the faunas of the shelf of the Canadian Archipelago. In spite of the large number of publications, our knowledge of the Cumacea is incomplete. The shelf fauna has been studied to a larger extent than the continental slope and pseudoabyssal basin fauna.* The present chapter summarizes the literature data and presents the results of studies based on the vast Cumacea collections of the Zoological Institute of Leningrad and includes material from the northern seas collected since the nineteenth century.

## FAUNAL DISTRIBUTION

The cumacean fauna of the Arctic Ocean, including the Barents Sea, is represented by 56 species belonging to 14 genera and 5 families. Endemic families and genera are lacking in the Arctic. The percentage of endemic species is rather high: 34 percent of the entire Cumacea fauna (19 species). The results presented in this section are based on the author's data as well as on the vast literature listed in the References section.

Of the families inhabiting the Arctic Ocean, the Diastylidae contains the largest number of species and individuals and is the most widespread group. It has cosmopolitan distribution, but is predominant in the temperate and cold waters of the Northern and Southern hemispheres. Twenty-six species of the family Diastylidae comprise 46 percent of the entire Cumacea fauna of the Arctic Ocean and 13 of them are endemic Arctic species (62 percent of all endemic Cumacea in the Arctic). The family is regarded as being rather primitive based on a number of features (presence of 3–4 pairs of well-developed hepatic caecum, free telson not fused to the sixth abdominal somite, fusion of cephalothorax with only 3 thoracic somites, normal structure of the mouthparts) (Lomakina, 1958, 1968). The family Leuconidae is the second in species abundance in the Arctic Ocean; it has 3 genera and 16 species (28 percent of the entire Cumacea fauna of the Arctic Ocean) are known. Six of them are endemic Arctic species. The family Leuconidae belongs to the more specialized and phylogenetically younger taxonomic cumacean group (Lomakina, 1968).

The Arctic Ocean fauna contains much fewer representatives of the temperate cold-water family Lampropidae: 3 genera and 7 species (12 percent of all Arctic Cumacea). This family contains no endemic Arctic species. The predominantly warm-water Nannastacidae are represented on the Arctic Ocean shelf by 2 genera and 5 species (9 percent of the entire cumacean fauna of the Arctic Ocean); two of them are endemic to the Arctic. Only 2 genera and 2 species (3 percent of the total number of species) of the Pseudocumidae, distributed in the Mediterranean, Black, Azov, and Caspian seas, have been found in the Arctic Ocean.

The number of species in the Arctic Ocean constitute about 7% of the world cumacean fauna. In order to study the trends of changes in species composition, the Arctic Ocean was divided into 11 regions and their fauna was analyzed by estimating the degree of similarity following Jaccard (1901, 1912) and by the inclusion method (Simpson, 1943):

$$j = \frac{C \cdot 100}{D_1 + D_2 - C}$$

$$i = \frac{C \cdot 100}{D_{min}}$$

where j is the degree of similarity of a rich sample and a relatively poor or equal sample; i is the degree measure of inclusion of a smaller sample ($D_{min}$) in a larger or equal sample; C is the number of species in common; $D_1$ and $D_2$ are the numbers of species in the compared samples. The results of

*Pseudoabyssal basins have great depths (2,500–5,000 m). They are separated, however, from the ocean floor (from the true abyssal regions) by sills about 200–600 m deep. Basins of the Arctic Ocean are a typical example of these pseudoabyssal basins (Andriashev, 1977).

Table 17-1. Similarity (below the diagonal ↓) and Degree of Inclusion (above the diagonal ↑) (%) of Cumacean Fauna

| | D1 | D2 | D3 | D4 | D5 | D6 | D7 | D8 | D9 | D10 | D11 |
|---|---|---|---|---|---|---|---|---|---|---|---|
| D1 | 51 | 77 | 90 | 78 | 69 | 67 | 58 | 38 | 65 | 61 | 61 |
| D2 | 30 | 22 | 68 | 73 | 69 | 73 | 68 | 44 | 55 | 61 | 54 |
| D3 | 52 | 39 | 31 | 91 | 75 | 63 | 67 | 44 | 80 | 67 | 47 |
| D4 | 32 | 55 | 61 | 23 | 87 | 86 | 69 | 55 | 75 | 72 | 52 |
| D5 | 20 | 41 | 32 | 56 | 16 | 94 | 69 | 44 | 69 | 62 | 69 |
| D6 | 30 | 48 | 41 | 67 | 58 | 27 | 75 | 61 | 75 | 72 | 74 |
| D7 | 23 | 48 | 41 | 52 | 38 | 54 | 24 | 55 | 70 | 67 | 61 |
| D8 | 11 | 25 | 19 | 32 | 26 | 32 | 31 | 18 | 50 | 44 | 33 |
| D9 | 22 | 35 | 46 | 54 | 44 | 47 | 47 | 31 | 20 | 72 | 65 |
| D10 | 19 | 38 | 32 | 46 | 42 | 41 | 40 | 28 | 52 | 18 | 72 |
| D11 | 23 | 36 | 25 | 35 | 39 | 51 | 42 | 17 | 43 | 44 | 23 |

Norwegian Sea (D1), Spitsbergen (D2), western Barents Sea (D3), eastern Barents Sea (D4), Franz Josef Land (D5), Kara Sea (D6), New Siberian Islands (D7), Chukchi Sea (D8), Canadian Arctic (D9), Baffin Bay and northwestern Greenland (D10), northeastern Greenland (D11).

these calculations are presented in Table 17-1. As shown in this table, the number of species declines markedly in higher latitudes (Spitsbergen, Franz Josef Land). An abrupt decline in species number is observed in the Barents Sea as compared to the Norwegian Sea. Further eastward ($D_3$–$D_7$) up to the new Siberian Shoals the species number declines slightly (except the Kara Sea) and only in the Chukchi Sea and off the coasts of northwestern Greenland a considerable impoverishment of the fauna is also observed. An increase in species number is observed in the Kara Sea.

The Norwegian Sea fauna ($D_1$) is very different from fauna of other Arctic regions, having a high degree of similarity only with the fauna of the western Barents Sea ($D_3$).* The faunas of the western ($D_3$) and eastern ($D_4$) parts of the Barents Sea are very similar. The Kara Sea fauna ($D_6$) has a high degree of similarity with the fauna of the eastern Barents Sea ($D_4$) and Franz Josef Land ($D_5$), as well as with the New Siberian Shoals fauna ($D_7$), but is very different from the Chukchi Sea fauna. The New Siberian Shoals fauna ($D_7$) has closer affinities with the fauna of Spitsbergen ($D_2$), eastern Barents Sea ($D_4$), and Kara Sea ($D_6$) than with the Chukchi Sea fauna ($D_8$) and the fauna of northwestern ($D_{10}$) and northeastern Greenland ($D_{11}$). The Chukchi Sea fauna supports a number of Pacific immigrants (22 percent), thus differing from the Cumacea of other Arctic regions.

*Longitude 33°E is accepted in this paper as the boundary between the western and eastern parts of the Barents Sea. This boundary coincides partly with the boundary between the boreal and Arctic benthos (Filatova, 1938). It is variable, fluctuating between 32–36°E, depending on the changes of the North Atlantic Drift.

The fauna may be divided into 6 biogeographic groups:

**I.** Atlantic subtropical-boreal species

1. Wide-ranging, off the European and American coasts, they include the bathyal-abyssal *Platysympus typicus*, distributed east of the Brazilian coast (08°S, 34°W), off the Bermuda Islands, in the Bay of Biscay, in the Mediterranean Sea, in the Skagerrak Strait, off the coasts of Ireland, and the west coast of Norway. The northernmost locality that this taxon has been reported from

is southern Spitsbergen. This species inhabits depths of 226–4,749 m.

2. East Atlantic, primarily bathyal and sublittoral species: *Platysympus tricarinatus, Leptostylis macrura, Diastyloides serrata,* and *Pseudocuma longicornis. Platysympus tricarinatus* occurs off of the southern part of the Canary Islands; *Leptostylis macrura* inhabits the Mediterranean Sea. These species are also distributed south of Iceland, west of Ireland, off the Norwegian coast, and penetrate the southwestern part of the Barents Sea in the region of Vadso, inhabiting depths of 200–1,780 m. *Leptostylis macrura* has been found also at 73 m depth. *Diastyloides serrata* and *Pseudocuma longicornis* occur in the Mediterranean Sea, off the European coasts from Portugal up to the straits of the Baltic Sea; they are common off the English and Norwegian coasts and extend northward up to western Finmarken. *Diastyloides serrata* is a eurybathic species, inhabiting depths of 7–2,924 m. *Pseudocuma longicornis* occurs at 0–130 m depth.

**II.** Atlantic boreal species

1. Widely distributed off the European and American coasts, sublittoral-bathyal *Diastylis lucifera* is found off the American coasts from the Gulf of Maine up to Newfoundland and Davis Strait, off the European coasts north of the English Channel, in the Irish and Northern Seas, in the straits of the Baltic Sea, between the Shetland and the Faeroe Islands, off the western and northern coasts of Norway (up to Vadso), inhabiting depths of 15–791 m.
2. Near-European sublittoral and sublittoral-bathyal species: *Hemilamprops assimilis, H. rosea,* and *Lamprops fasciata* are distributed in the Irish Sea, off the coasts of England, in the Northern Sea in the straits of the Baltic Sea, along the entire western Norwegian coast, northward up to Vadso. *H. assimilis* and *H. rosea* inhabit depths of 5–970 m and *L. fasciata* are found at depths between 0 m and 71 m.

**III.** Pacific boreal, mostly sublittoral species, penetrating into the upper bathyal zone: *Diastylis alaskensis, D. aspera, D. bidentata,* and *D. dalli* are widely distributed in the Sea of Japan and the Okhotsk and Bering seas; in the Arctic Ocean they penetrate into the Chukchi Sea, where they occur in its southern part and off Wrangel Island at shallow depths. In the Arctic these species mainly inhabit the upper sublittoral zone, from 25–40 m; in the boreal waters they live between 20 m and 200 m; some species can descend to a depth of 1,000 m.

**IV.** Amphiboreal, primarily sublittoral species: *Eudorellopsis deformis* is ubiquitous in the boreal waters of the North Atlantic and Pacific; in the Arctic this species occurs off the coasts of western Greenland and northern Iceland, inhabiting depths of 5–900 m.

**V.** Boreal–Arctic species including:

1. Widely distributed circumpolar sublittoral and sublittoral bathyal species: *Lamprops fuscata, Diastylis glabra, D. rathkei, Brachydiastylis resima, Eudorella emarginata, Leucon acutirostris, L. fulvus, L. nasica,* and *L. nasicoides.* These taxa are common in boreal Atlantic water (in the western part of the Atlantic Ocean north of Cape Cod; in the eastern part of the Atlantic north of the straits of the Baltic Sea and the coast of Great Britain), in the Pacific Ocean (in the Sea of Japan and Okhotsk and Bering seas), and throughout the Arctic Ocean. Many of these species generally inhabit depths of 4–200 m, though some descend to 700 m. *Eudorella emarginata, Leucon acutirostris,* and *L. fulvus* are sublittoral-bathyal taxa inhabiting

depths of 9–2,200 m. The deepest water in which these species are found is the Davis Strait.

2. Widely distributed in boreal waters, with a hiatus in the Arctic Ocean, sublittoral and upper bathyal: *Leptostylis villosa, Petalosarsia declivis, Campylaspis costata,* and *C. rubicunda* are wide-ranging in the Atlantic, north of Cape Cod, and in the straits of the Baltic Sea; in the Pacific Ocean north of Great Peter Bay and Vancouver Island. *Leptostylis villosa, Campylaspis costata,* and *C. rubicunda* penetrate in the Arctic only up to the Kara Sea and are occasionally found in the Canadian Arctic. *Petalosarsia declivis* penetrates the Arctic up to Wilkitzkij Strait, but has not been observed north of the Davis Strait. These species live in water depths of 9–887 m.
3. Pacific, west Atlantic, primarily sublittoral species: *Eudorellopsis integra* and *Cumella carinata. Eudorellopsis integra* is widely distributed in the boreal waters of the Pacific Ocean; it is also found on the New Siberian Shoals and occurs from northwestern Greenland and Baffin Bay southward up to 40°16′N, 67°26′W. *Cumella carinata* is common in the Bering Sea and off the Canadian Arctic, western and southeastern Greenland, the Labrador Peninsula, and on the Grand Banks of Newfoundland. These species usually inhabit depths of 4–200 m, sometimes descending to 1,507 m.
4. Atlantic east-American Eurasian sublittoral-bathyal and bathyal species: *Leptostylis longimana, Eudorella hispida, Leucon pallidus, L. profundus, Leptostylis ampullacea, Hemilamprops uniplicata, Diastylis echinata,* and *D. stygia* are found in the North Atlantic east of Cape Hatteras, northward up to the Davis Strait, off Iceland, Faroes Islands, in the straits of the Baltic Sea, in Norwegian fjords, and in the Barents Sea, where they inhabit depths of 100–2,600 m (*D. stygia's* depth range is 1,134–3,300 m). In the Arctic region the first four species occur off northern Iceland, in the Kara Sea, and on the New Siberian Shoals in the sublittoral zone from 9–150 m depth. *Leptostylis ampullacea* and *Leucon pallidus* have been found off the Canadian Arctic at 72–83 m depth. *Hemilamprops uniplicata, Diastylis echinata,* and *D. stygia* have not been recorded in the sublittoral zone of the Arctic Ocean. In the Norwegian and Greenland seas they live between 325 m and 2,000 m depth. *Hemilamprops uniplicata* and *Diastylis echinata* have been recorded in the Svyataya Anna Trench and the Voronin Trench as well as in the northern part of the Kara Sea.

**VI.** Arctic species

1. Circumpolar, primarily sublittoral species: *Diastylis edwardsi, D. goodsiri, D. lepechini, D. nucella, D. oxyrhyncha, D. scorpioides, D. spinulosa, Brachydiastylis nimia,* and *Leucon nathorsti* inhabit depths of 2–200 m. *Diastylis edwardsi, D. goodsiri, D. oxyrhyncha, and Leucon nathorsti* penetrate the North Atlantic in the regions of George's Bank, Nova Scotia, Newfoundland, Davis Strait, to the Norwegian coast, descending to depths of 200–1,200 m.
2. Eurasian, primarily sublittoral species: *Diastylis sulcata, Leptostylis gorbunovi, Eudorella spitzbergensis,* and *Campylaspis stephenseni* are distributed from northeastern Greenland up to the New Siberian Shoals. *Leptostylis gorbunovi* occurs also on the outer edge of the Chukchi Sea shelf at a depth of 440 m.
3. Amerasian sublittoral species: *Eudorella arctica* and *E. groenlandica* are distributed from the New Siberian Shoals eastward up to northwestern and northeastern Greenland, inhabiting depths of 17–216 m.
4. East-American Eurasian, bathyal and bathyal-pseudoabyssal species: *Diastylis polaris, Eudorella gracilis, Leucon*

*spinulosus,* and *Campylaspis intermedia* inhabit the deep waters off the Labrador Peninsula, Davis Strait, Greenland and Norwegian seas, and penetrate into the northern part of the Kara Sea through the Svyataya Anna and Voronin trenches. *Diastylis polaris* has also been found north of Franz Josef Land and in the central part of the Arctic Ocean on the northwestern slope of the Lomonosov Ridge. The bathyal *Eurorella gracilis, Leucon spinulosus,* and *Campylaspis intermedia* inhabit depths of 200–2,700 m, and the bathyal-pseudoabyssal *Diastylis polaris* inhabits depths of 1,000–4,700 m.

The biogeographic composition of the Arctic Ocean fauna shows that it comprises primarily Arctic (34 percent) and boreal–Arctic species (41 percent). These species include three distinct groups: (1) for the most part sublittoral taxa, distributed on the shelf of the Arctic Ocean, (2) bathyal and bathyal-pseudoabyssal forms characteristic only of the bathyal and pseudoabyssal depths of the Arctic Ocean, and (3) eurybathic taxa. It should be emphasized that the Arctic shelf fauna is represented by 48 species comprising primarily of wide-ranging boreal–Arctic, mostly eurybathic and eurythermic species (21 percent of the cumacean fauna on the Arctic shelf and 56 percent of all boreal–Arctic sublittoral species of the Arctic).

Much less numerous in this region are the Atlantic boreal–Arctic species (12 percent of the cumacean fauna of the Arctic shelf) and the Pacific boreal species (4 percent of the fauna of the Arctic shelf). The impoverished fauna of the bathyal and pseudoabyssal zones of the Arctic Ocean (at depths from 200–4,700 m) is represented by 19 species; bathyal and bathyal-pseudoabyssal species of Atlantic origin constitute 42 percent of the cumaceans in the bathyal and pseudoabyssal zones of the Arctic. Comparing the biogeographic composition of the fauna of the different regions of the Arctic Ocean, certain distributional patterns emerge (Fig. 17-1). The percentage of Atlantic subtropical-boreal and boreal taxa are predominant only in the Norwegian Sea (Fig. 17-1, I). In other seas the fauna is composed primarily of boreal–Arctic and Arctic species. In the Barents Sea the relative number of Arctic species increases eastward and from the

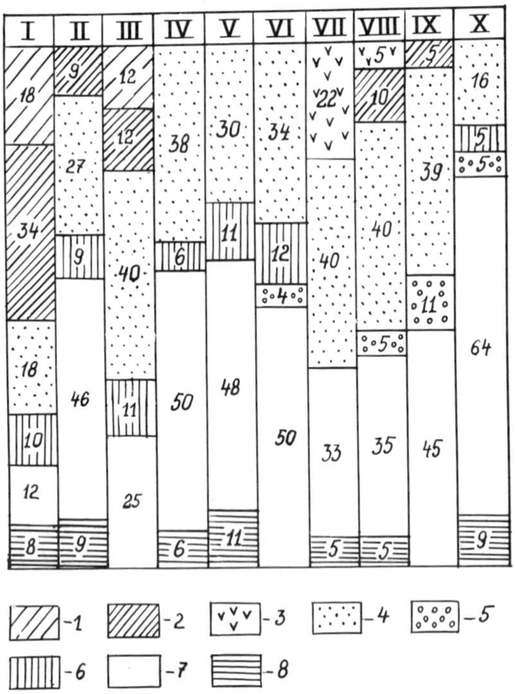


Figure 17-1. Biogeographic composition of the cumacean fauna (in %) in 10 regions of the Arctic Ocean. Regions: I, Norwegian Sea; II, Spitsbergen; III, Barents Sea; IV, Franz Josef Land; V, Kara Sea; VI, New Siberian Shoals; VII, Chukchi Sea; VIII, Canadian Arctic; IX, Baffin Sea and northwestern Greenland; X, northeastern Greenland. Biogeographic groups: 1, Atlantic subtropical-boreal species (I group in the text); 2, Atlantic boreal species and amphiboreal species (II, IV); 3, Pacific boreal species (III); 4, boreal–Arctic wide-ranging species (V, 1, 2,); 5, boreal–Arctic Pacific west Atlantic species (V, 3); 6, boreal–Arctic Atlantic east-American Eurasian species (V, 4); 7, Arctic primarily sublittoral species (VI, 1, 2, 3); 8, Arctic east-American Eurasian bathyal and bathyal-abyssal species (VI, 4).

upper to the lower part of the shelf (see Fig. 17-3, I, II). Therefore, the lower part of shelf in the western Barents Sea and on the eastern Barents Sea shelf (east of 33°E) should be regarded as a transitional zone of mixing of Arctic and boreal faunas between the Atlantic boreal and the Arctic regions.

As expected, in the seas of the Arctic Ocean the percentage of Arctic species increases northward (Fig. 17-1, II, IV) and from the Barents Sea to the New Siberian Shoals and also off the coast of northeastern Greenland (Fig. 17-1, III, V, VI, X). The New Siberian Shoals form the eastern boundary in the distribution of several Eurasian Arctic and Atlantic boreal–Arctic species (e.g., *Eudorella spitzbergensis* and *Leptostylis longimana*) and the western boundary for Amerasian Arctic and boreal–Arctic Pacific species (*Eudorella arctica, E. groenlandica,* and *Eudorellopsis integra*).

The fauna of the southern part of the Chukchi Sea includes 22 percent Pacific widely distributed boreal species, which allows for regarding this region as the zone of mixing of Arctic and boreal faunas of the Arctic and Pacific boreal regions.

## VERTICAL DISTRIBUTION OF SPECIES AND THE WATER MASSES OF THE ARCTIC OCEAN

In the Arctic Ocean the maximum number of species occur on the shelf; in the bathyal and pseudoabyssal zones the fauna is greatly impoverished and more homogenous. The maximum number of species occur at different depths in the various seas (Fig. 17-2). These depth ranges narrow from the Barents Sea to the Chukchi Sea. In the Western Barents Sea (west of 33°E) species diversity is highest between 25 m and 200 m depth, in the Eastern Barents Sea (east of 33°E) this range is more restricted, that is, from 50–150 m, in the Kara Sea from 45–100 m, on the New Siberian Shoals from 25–70 m, and on the Chukchi Sea from 2–25 m. Below 200 m, the species number falls abruptly. The second peak of increase in species number is observed in

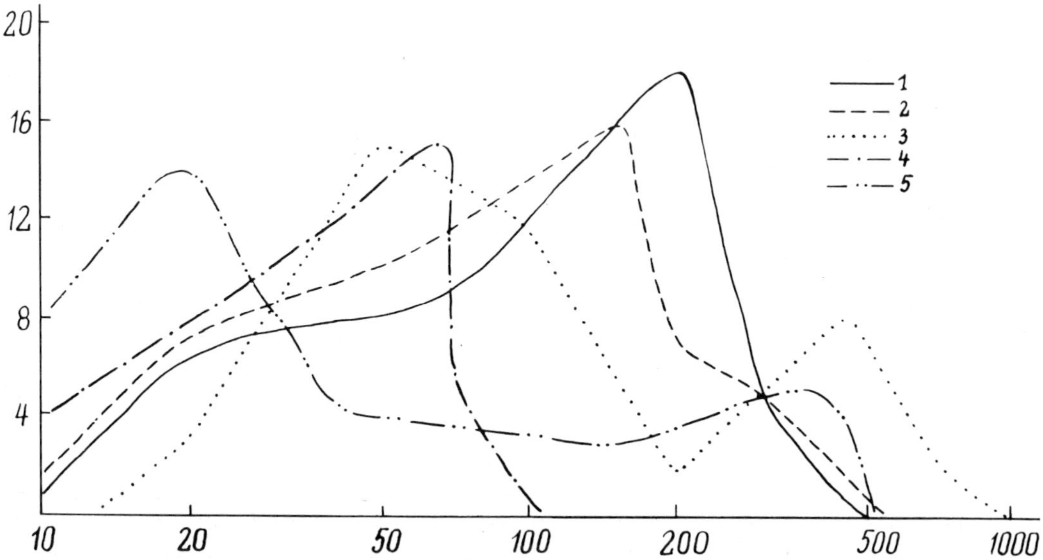


Figure 17-2. Vertical distribution of Cumacea in 5 regions of the Arctic Ocean. Ordinate: number of species; abscissa: depth on logarithmic scale. Regions: 1, Barents Sea (western part); 2, Barents Sea (eastern part); 3, Kara Sea; 4, New Siberian Shoals; 5, Chukchi Sea.

the Kara Sea at depths of 300–700 m (this peak is found in the Svyataya Anna and Voronin trenches).* A slight increase in species number is observed in the Chukchi Sea as well as on the outer edge of the shelf (200–7,400 m depth).

Four main water masses can be distinguished in the Arctic Ocean. The vertical distribution of species belonging to different biogeographic groups is related to the distribution of these water masses (Fig. 17-3).

**1.** The surface Arctic Water Mass is characterized by low salinities (30–32‰) and temperatures close to the freezing point (Coachman and Aagaard, 1974). It occupies nearly the entire shelf of the Arctic Ocean (except the southwestern part of the Barents Sea and the south part of the Chukchi Sea). Consequently, all depths of the Arctic seas' shelf are dominated by Arctic, primarily sublittoral species (group VI, 1, 2, 3) and by most eurybathic and eurythermal widespread boreal–Arctic species (group V, 1, 2,). Most of these species are so eurybiontic that the New Siberian Shoals do not impede their circumpolar dispersal.

**2.** The Atlantic Arctic Intermediate Water Mass is characterized by high salinities (about 34.9‰) and temperatures above zero (≈2.5°–3°C off Spitzbergen and 0.5°–0.6°C off the Canadian Arctic). Owing to its higher density, this water sinks below the surface Arctic water. At first this water moves along the Eurasian continental slope and later along the North American continental slope. The upper boundary of this water mass in the Atlantic part of the Arctic Ocean is at depths of 50–200 m; in the Pacific part below 250–300 m, the lower boundary is at depths of 800–900 m. The Atlantic Arctic Intermediate Water Mass penetrates into the shelf seas of the Arctic Ocean from the north through deep trenches (Kuksa, 1983). The continental slope (at depths of 200–900 m) in the Atlantic sector of the Arctic Ocean and Sv. Anna and Voronin trenches, within the Atlantic Arctic intermediate water mass, is inhabited by Arctic species, primarily by sublittoral *Diastylis goodsiri, D. oxyrhincha, D. spinulosa, D. nathorsti, Brachydiastylis nimia,* and *Leptostylis gorbunovi* (part of group VI, 1, 2); bathyal-pseudoabyssal *Leucon spinulosus* and *Diastylis polaris* (part of group VI, 4); boreal-Arctic widespread *Diastylis rathkei, Leucon nasica,* and *Petalosarsia declivis* (a small part of group V, 1, 2); and Atlantic boreal–Arctic, primarily bathyal *Hemilamprops uniplicata, Diastylis echinata,* and *Leucon pallidus* (part of group V, 4). The latter are stenothermal species inhabiting low temperatures, ranging from +4°C to −1.7°C.

**3.** The deep-water mass is characterized by high salinities and temperatures below zero. It occupies the deep basins of the Norwegian and Greenland seas and those of the Atlantic sector of the Arctic Ocean (below 1,000 m). The Atlantic boreal–Arctic, primarily bathyal *Diastylis stygia* and *Leucon pallidus* (part of group V, 40), are found here and are accompanied by Arctic bathyal-pseudoabyssal *Diastylis polarisas* and *Campylaspis intermedia* (part of group VI, 4).

**4.** The estuary–Arctic water mass with low salinities (up to 16‰) and relatively high summer temperatures (up to 4°–6°C) occupies the coastal shallow water regions of the Arctic Ocean, in the vicinity of large rivers' mouths. Here the cumaceans are represented by the estuary–Arctic *Diastylis sulcata morpha stuxbergi* and also by the widespread circumpolar boreal–Arctic *Lamprops fuscata, Diastylis glabra,* and *D. rathkei* (part of group V, 1), capable of surviving considerable freshening.

Because branches of the North Atlantic Drift penetrate into the Barents Sea, in the

*This increase in species number can be accounted for by penetration of Atlantic boreal–Arctic bathyal species with the Atlantic intermediate water mass into the Svyataya Anna and Voronin trenches. Apparently such increase in species number occurs along the whole Eurasian continental slope, but we do not possess data on other regions.

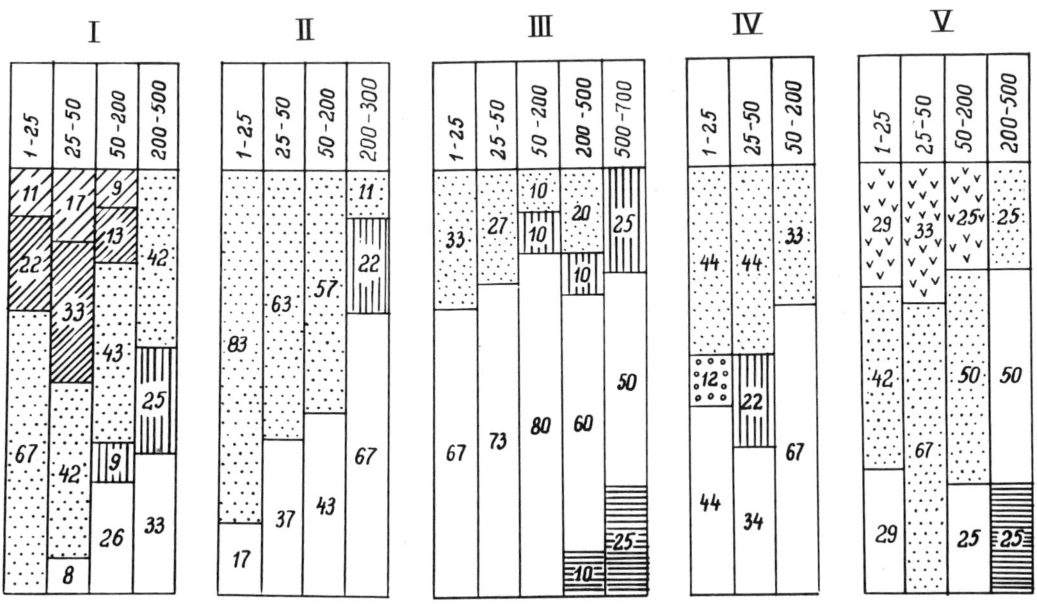


Figure 17-3. Biogeographic composition of Cumacea faunas (in %). Five regions of the Arctic seas at various depths. Regions: I, Barents Sea (western define); II, Barents Sea (eastern define); III, Kara Sea; IV, New Siberian Shoals; V, Chukchi Sea. Biogeographic key as in Figure 17-1.

western sector of the sea (west of 33°E) at depths of 25–50 m, Atlantic boreal species (group II) and Atlantic subtropical-boreal species (part of group I) comprise half of the fauna (Fig. 17-3, I). Below 50 m, the proportion of these species declines abruptly, and below 200 m, they are absent. In the Eastern Barents Sea (east of 33°E) boreal species are lacking.

Pacific water penetrates into the southern part of the Chukchi Sea, the southeastern part of the East Siberian Sea, and the western part of the Beaufort Sea. Our data indicate that a great number of Pacific species (group III) are concentrated at depths of 1–50 m in the southern Chukchi Sea and off the Wrangel Island (Fig. 17-3, V).

## ORIGIN OF ARCTIC OCEAN CUMACEA

The composition and distribution of Arctic Ocean Cumacea is related not only to the present-day structure of water masses but also to the geological history of the North Polar Ocean. The present-day structure of Arctic water masses is greatly affected by Atlantic waters that carry Atlantic taxa into the Arctic.

The origin of Arctic Cumacea is related to the geological history of the Arctic during the Cenozoic Era and to the connection between the Arctic Ocean and the Pacific and Atlantic oceans. There are two sources for the Arctic cumacean fauna: (1) Pacific fauna and (2) Atlantic fauna. The Pacific fauna penetrated the Arctic Ocean during the Pliocene and possibly later, across the Bering Strait. Judging from their morphological character, descendants of this fauna occur among endemic Arctic species (*Diastylis edwardsi, D. lepechini, D. scorpioides*), resemble the Pacific *D. bidentata* and *D. dalli*, and Arctic *D. spinulosa* has affinities with Pacific *D. paraspinulosa*. In the opinion of this author, the wide-ranging circumpolar boreal–Arctic species *Brachydiastylis resima, Eudorella emarginata, Leucon acutirostris, L. nasica*, and *L. nasicoides*, the Pacific

West Atlantic *Eudorellopsis integra* and *Cumella carinata,* and the amphiboreal *Eudorellopsis deformis* are of Pacific origin. These species are distributed most extensively in the North Pacific Ocean. Djakonov (1945) also considers these biogeographic groups of echinoderms to be of Pacific origin.

The Atlantic fauna is represented in the Arctic by the boreal–Arctic, predominantly bathyal *Leptostylis longimana, Eudorella hispida, Leucon pallidus, L. profundus, Leptostylis ampullacea, Hemilamprops uniplicata, Diastylis echinata,* and *D. stygia.* The wide-ranging circumpolar boreal–Arctic sublittoral *Diastylis rathkei* and *D. sulcata* may possibly be of Atlantic origin. The proportion of Pacific elements is particularly large in the shelf fauna, whereas the influence of Atlantic fauna is more pronounced in the bathyal and pseudoabyssal zone (below 200–500 m) of the adjacent Atlantic sector of the Arctic Ocean.

## APPENDIX

### Plate 17-I

Common Arctic Cumacea.

1. *Lamprops fuscata* Sars: female, side view. (From Sars, 1900.)
2. *Diastylis goodsiri* (Bell): female, side view. (From Sars, 1900.)
3. *Diastylis polaris* Sars: female, side view. (From Sars, 1873.)
4. *Eudorella emarginata* (Kröyer): female, side view. (From Sars, 1900.)

### Map 17-1

Distribution of Cumacea in the Arctic.

1. *Diastylis goodsiri* (Bell), Arctic circumpolar area.
2. *Eudorellopsis integra* (Smith), boreal–Arctic, Pacific west Atlantic area.
3. *Eudorella arctica* Hansen, Arctic, Amerasian area.
4. *Eudorella spitzbergensis* Zimmer, Eurasian area.
5. *Diastylis bidentata* Calman, boreal Pacific area.

### Map 17-2

Distribution of boreal–Arctic circumpolar species *Eudorella emarginata* (Kröyer).

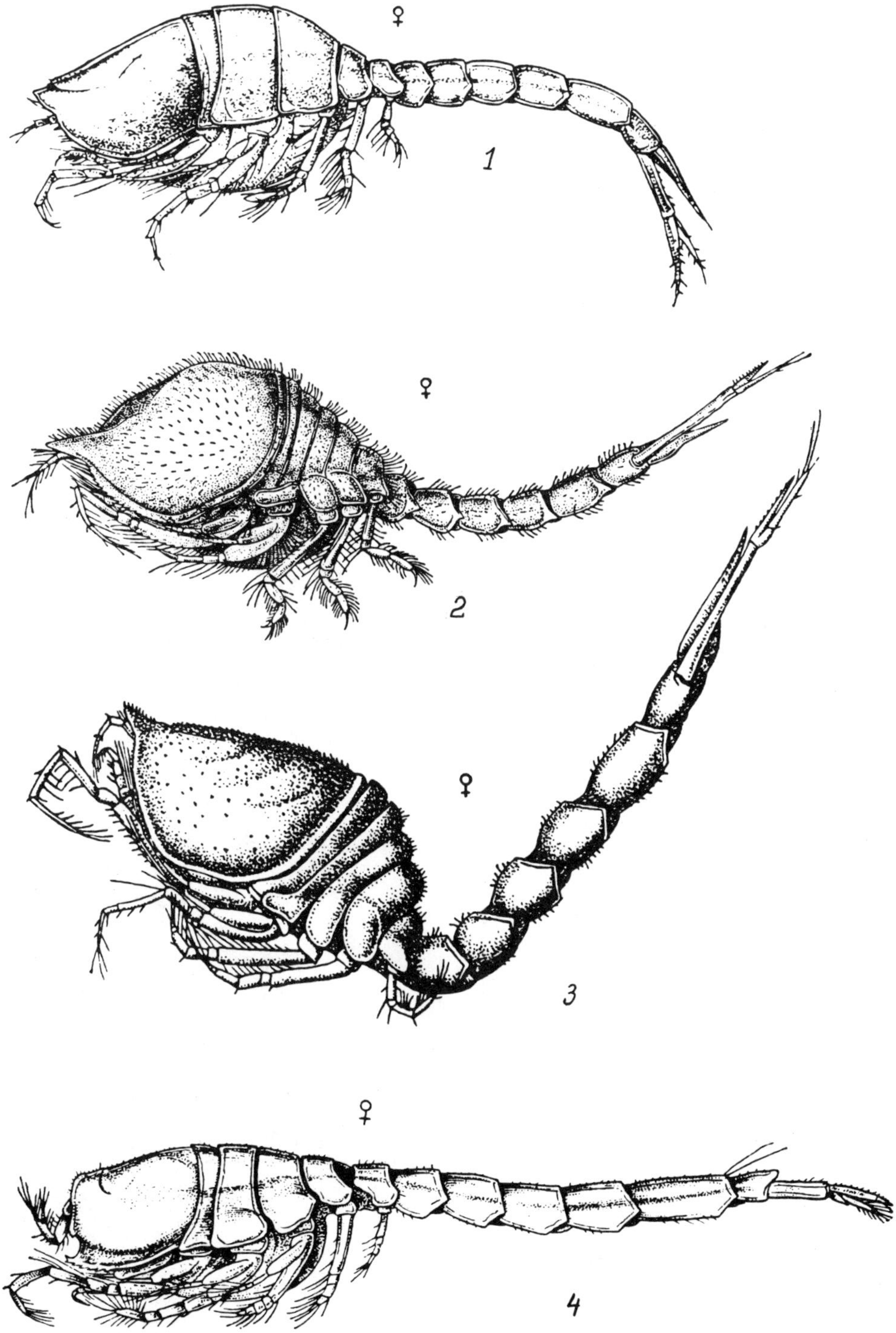


PLATE 17-I

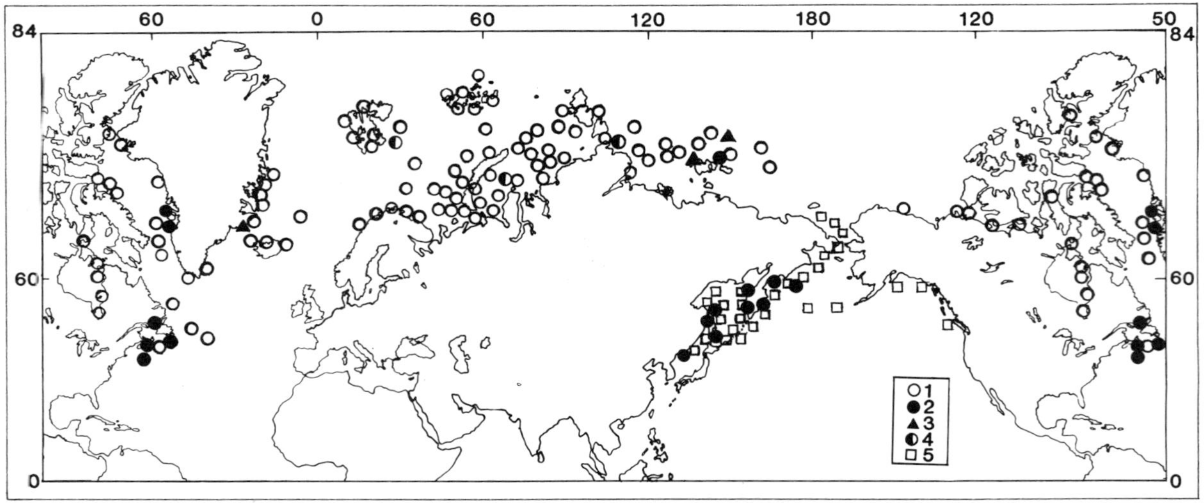


MAP 17-1

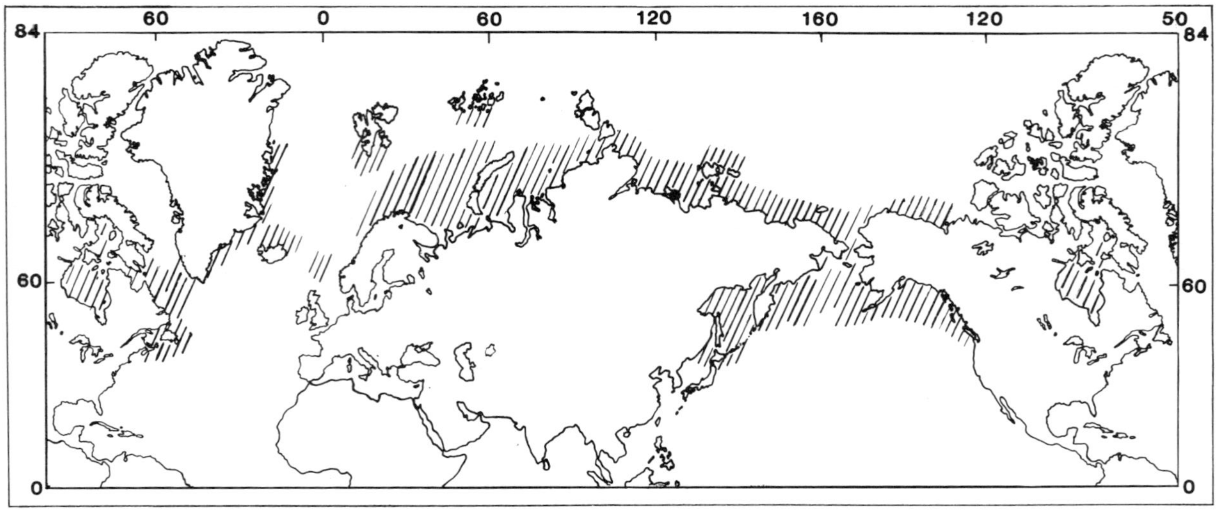


MAP 17-2

## REFERENCES


Andriashev, A. P. 1977. Some additions to schemes of the vertical zonation of marine bottom fauna. In *Adaptations within Antarctic ecosystems.* Proc. 3d SCAR Symposium on Antarctic Biology, Washington, DC, pp. 351–360.

Calman, W. T. 1912. The Crustacea of order Cumacea in the collection of the United States National Museum. *Proc. Un. St. Nat. Mus.* **41**(N1876):603–676.

Calman, W. T. 1920. Cumacea of the Canadian Arctic expedition 1913–1918. *Canadian Arctic Exp. 1913–1918,* VII, pt. C. (Cumacea), pp. 3–4.

Coachman, L. K., and K. Aagaard. 1974. Physical oceanography of Arctic and sub-Arctic seas. In Yvonne Herman (ed.), *Marine Geology and Oceanography of the Arctic Seas.* Springer, pp. 1–72.

Corey, S. 1981. Distribution of certain Arctic and sub-Arctic Cumacea in Canadian waters. *Canadian J. Zool.* **59**(9):1,726–1,733.

Djakonov, A. M. 1945. Vzaimootnoshenia arkticheskoy i tichookeanskoy morskich faun. *J. Obshei Biologii,* VI, **2**:125–155. (In Russian.)

Filatova, Z. A. 1938. Kolichestvenny uchot donnoy fauny yugozapadnoi chasti Barentseva morya. *Trudy PINRO* **23**:3–58. (In Russian.)

Hansen, H. I. 1887. Malacostraca marine Groenlandie occidentalis, VI, Cumacea. *Videnskab. Maddel.,* pp. 198–209.

Hansen, H. I. 1920. Crustacea Malacostraca, IV. *The Danish Ingolf Exped. Copenhagen, III,* **6**:1–86.

Hart, J. F. L. 1939. Cumacea and Decapoda on the western Canadian Arctic region, 1936–1937. *Can. J. Res.* **17**(sec. D):62–67.

Hessler, R. R. 1969. Peracarida. In *Treatise on Invertebrate Paleontology,* part R, Arthropoda, vol. 4. Geological Society of America and University of Kansas Press, pp. 360–393.

Jaccard, P. 1901. Distribution de la flore alpine dans le bassin de Drunses et dans quelques regions voisines. *Bull. Soc. Vandoise, Sc. Natur.* **37**:241–272.

Jaccard, P. 1912. The distribution of the flora in the alpine zone. *New Phytol.* **11**:37–50.

Just, J. 1970. Cumacea from Jørgen Brønlund. North Greenland. *Medd. Grønland* **184**:1–23.

Kuksa, V. I. 1983. *Promezhutochnye Vody Mirovogo Okeana.* Gidrometizdat, Leningrad, 290p. (In Russian.)

Lomakina, N. B. 1958. Kumovye raki (Cumacea) morei SSSR. Opredeliteli po faune SSSR izd. *Zool. Inst. AN SSSR* **66**:301. (In Russian.)

Lomakina, N. B. 1968. Stroenye pechenochnykh divertikul u kumovykh rakov (Cumacea) i ego filogeneticheskoye znachenye. *Zoologichesk. Zhurnal,* XLVII, **1**:60–72. (In Russian.)

McLaughlin, P. A. 1980. *Comparative Morphology of Recent Crustacea.* San Francisco, 177p.

Sars, G. O. 1864. Om den aberrante Krebsdyrgruppe Cumacea og dens nordiske Arter. *Forh. Vid. Selsk. Christiania,* pp. 128–208.

Sars, G. O. 1871. Cumaceer fra de store Dybder Vetensk. *Akad. Handl.* **6**:197–802.

Sars, G. O. 1882. Oversigt of Norges Crustaceer. *Forh. Vid. Selsk. Christiania* **18**:1–124.

Sars, G. O. 1900. Crustacea Norway. Cumacea, III. Bergen: 1–115.

Simpson, G. G. 1943. Mammals and the nature of continents. *Am. J. Sci.* **241**(1):1–31.

Stebbing, T. R. 1913. *Cumacea.* Das Tierreich, Berlin, 39:1–210.

Stephensen, K. 1936. Crustacea varia. The Godthaab Exp. 1928. *Medd. Grønland* **80**(2):1–138.

Stephensen, K. 1937. Marine Crustacea Mysidacea, Cumacea and Nebaliacea. *Zoology of the Faroes, Copenhagen* **2**(1):1–10.

Stephensen, K. 1938. Euphausiacea, Mysidacea, Cumacea and Nebaliacea. *Zoology of Iceland, Copenhagen* **3**(29):1–24.

Stephensen, K. 1943. Leptostraca, Mysidacea, Cumacea, Tanaidacea, Isopoda and Euphausiacea. The Zoology of east Greenland. *Medd. Grønland* **121**(10):1–82.

Wacasey, J. W. 1974. Biological oceanographic observation in the Eskimo Lakes, Arctic Canada. I. Zoobenthos data 1971–1973. *Can. Fish. Mar. Serv. Tech. Rep. 475.*

Wacasey, J. W. 1975. Biological productivity of the southern Beaufort Sea Tech. Rep. 12b, pp. 1–39.

Wacasey, J. W., E. G. Atkinson, and L. Glasspoole. 1979. Zoobenthos data from upper Frobisher Bay, 1967–1973. *Can. Data Rep. Fish. Aquat. Sci. 164.*

Yashnov, V. A. 1948. *Otryad Cumacea—Kumovye v kn.: Opredelitel Fauny, Flory Severnykh Morei SSSR, Izd.* Sov. Nauka, Moscow, pp. 223–237. (In Russian.)

Zimmer, C. 1900. Die arktischen Cumaceen. Fauna Arctica, Jena, I, 3:409–444.

Zimmer, C. 1926. Northern and Arctic invertebrates in the collection of the Swedish State Museum, X. Cumaceen. *Kunglica Svenska Vetensk. Acad. Handl.* **3**(2):1–88.

Zimmer, C. 1930. Untersuchungen an Diastyliden (Ordnung Cumacea). *Mitt. Zool. Mus. Berlin* **16**:583–658.

Zimmer, C. 1933. *Cumacea Tierwelt Nord. u. Ostse* **23**:70–120.

Zimmer, C. 1934. Die Cumaceen der norwegischen Expedition nach Ost-Grønland 1929–1932. *Zool. Results Norw. Sci. Exp. to East Greenland* **3**(61):34–40.

Zimmer, C. 1941. Cumacea. *Bronn's Klass. u. Ordn. D. Tierreich* **5**(1):1–222.

Zimmer, C. 1946. Cumacea iz Severnogo Ledovitogo okeana. *Tr. Dreif. Eksp. Glavsevmorputi na Ledok. Parokhode Sedov, 1937–1940,* pp. 264–267. (In Russian.)

Zimmer, C. (edited by T. E. Bowman and L. Walting). 1980. Cumaceans of the American Atlantic boreal coast region (Crustacea: Peracarida). *Smithsonian Contrib. Zool.* **302:**1–29.

# 18: Quaternary Calcareous Nannofossil Biostratigraphy: The Eastern Arctic Ocean Record

Gunilla Gard

## INTRODUCTION

This investigation was undertaken to study the calcareous nannofossils in sediment cores from the European Arctic Ocean, north of 80° latitude, in order to establish the biochronology and to correlate sediments over wide areas. Another objective was to decipher changes of paleoclimatic and paleoceanographic conditions in the study area.

Calcareous nannofossils constitute a heterogenous group of calcareous forms in the size range of 1–25 $\mu$m. A major part of the group, the coccoliths are calcareous plates secreted by coccolithophorid algae. Other nannofossils are discoasters, nannoconids, and various calcareous forms that sometimes are of uncertain biological affinity. This work will only briefly mention the biology of coccolithophorids and their usefulness in paleoclimatology and biostratigraphy. Comprehensive discussions of these fossils have in recent years been published by Haq (1978) and by Tappan (1980).

The first true coccoliths have been recognized from sediments of Late Triassic age (Jafar, 1983). Diverse and rich assemblages have evolved since that time, and as coccoliths and other calcareous nannofossils occur in large numbers in oceanic sediments, they are excellent as stratigraphic indicators. Biostratigraphic studies have resulted in worldwide detailed zonation schemes (Martini, 1971; Gartner, 1977; Bukry, 1973, 1975; Okada and Bukry, 1980). A comprehensive revision of Mesozoic and Cenozoic nannofossil biostratigraphy was made by Perch-Nielsen (1985*a,b*).

## BIOLOGY AND ECOLOGY

Coccolithophorids are unicellular golden brown algae with the ability to secrete shells made up of calcareous plates. They are photoautotrophic and are important primary producers in the marine food chain; however, some species are also believed to be heterotrophic. Like several other protists, the coccolithophorids have a complex life cycle that is not completely understood. Some coccolithophorids have an alternate motile phase during which they possess flagellae and a nonmotile phase. The motile phase is either naked or produces coccoliths of another type than those of the nonmotile phase (Klaveness and Paasche, 1979).

Coccolithophorids are principally marine organisms and over 150 living species have been recorded in the oceans. The highest concentrations are observed at a few meters to about 50 m below the water surface. The largest numbers are found in areas of high organic productivity, such as upwelling zones. The growth rates of coccolithophorids are relatively high; some species are able to multiply twice in less than 24 hours. Most species have relatively narrow temperature ranges but often have a wide tolerance for salinities.

Studies of coccolithophorid distribution have revealed that surface water temperature is the most important factor delineating the biogeography of various species. Generally, an increasing species diversity is observed from high latitude areas toward low latitudes (McIntyre and Bé, 1967).

## PALEOCLIMATIC APPLICATIONS

Different living climate-related assemblages of coccolithophorids have been defined for the Atlantic and Pacific oceans (McIntyre and Bé, 1967; Okada and Honjo, 1973). From seasonal sampling of living assemblages and from sediment surface assemblages, the approximate temperature ranges of selected species have been determined (McIntyre, Bé, and Roche, 1970; Okada and McIntyre, 1979). Thus fluctuations of Quaternary water masses and climates can be recognized by the different coccolithophorid assemblages preserved in the sedimentary sequences. By combining coccolithophorid and planktonic foraminiferal assemblages, McIntyre, Ruddiman, and Jantzen (1972) were able to recognize and delineate the extent of Polar Front migrations in the North Atlantic during the Pleistocene.

## TAXONOMY

Due to the diverse nature of the coccolithophorids and the unknown affinity of some nannofossils, the classification is continuously under debate. The calcareous nannofossils are today usually classified under the plant kingdom in the division Chrysophyta.

Taxa observed in this study are listed below. Taxonomy follows Perch-Nielsen (1985*b*).

Family *Coccolithaceae* Poche, 1913

Genus *Calcidiscus* Kamptner, 1950

*Calcidiscus leptoporus* (Murray and Blackman, 1898) Loeblich and Tappan, 1978; Pl. 18-I, fig. 10

1898 *Coccosphaera leptopora* Murray and Blackman, p. 439, pl. 15, figs. 1–7

1902 *Coccolithophora leptopora* (Murray and Blackman) Lohmann, p. 138

1930 *Coccolithus leptoporus* (Murray and Blackman) Schiller, p. 245, figs. 121, 122

1954 *Cyclococcolithus leptoporus* (Murray and Blackman) Kamptner, p. 23, fig. 20

1970 *Cyclococcolithina leptopora* (Murray and Blackman) Wilcoxon, p. 82

1978 *Calcidiscus leptoporus* (Murray and Blackman) Loeblich and Tappan, p. 1,391

Genus *Coccolithus* Schwarz, 1894

*Coccolithus pelagicus* (Wallich, 1877) Schiller, 1930; Pl. 18-I, figs. 4, 7

1877 *Coccosphaera pelagica* Wallich, p. 348, pl. 17, figs. 1, 2, 5, 8, 9D–11D

1930 *Coccolithus pelagicus* (Wallich) Schiller, p. 246, figs. 123–124.

Family *Helicosphaeraceae* Black, 1971

Genus *Helicosphaera* Kamptner, 1954

*Helicosphaera carteri* (Wallich, 1877) Kamptner, 1954; Pl. 18-I, fig. 11

1877 *Coccosphaera carteri* Wallich, p. 348, pl. 17.

1899 *Coccosphaera pelagica* var. *carteri* (Wallich) Ostenfeld, p. 436.
1954 *Helicosphaera carteri* (Wallich) Kamptner, pp. 21, 73, figs. 17–19.
1927 *Coccolithophora pelagica* var. *carteri* (Wallich) Kamptner, p. 174.

Family *Prinsiaceae* Hay and Mohler, 1967

Genus *Emiliania* Hay and Mohler, 1967
*Emiliania huxleyi* (Lohmann, 1902) Hay and Mohler, 1967; Pl. 18-I, figs. 1–3

1902 *Pontosphaera huxleyi* Lohmann, pp. 129–130, pl. 4, figs. 1–9, pl. 6, fig. 69
1930 *Hymenomonas huxleyi* (Lohmann) Kamptner, pp. 155–156, 159–160
1943 *Coccolithus huxleyi* (Lohmann) Kamptner, p. 44
1967 *Emiliania huxleyi* (Lohmann) Hay and Mohler in Hay et al., p. 447

Genus *Gephyrocapsa* Kamptner, 1943; Pl. 18-I, figs. 5, 6, 8, 9

Family *Syracosphaeraceae* Lemmermann, 1908

Genus *Syracosphaera* Lohmann, 1902
*Syracosphaera pulchra* Lohmann, 1902; Pl. 18-I, fig. 12

1902 *Syracosphaera pulchra* Lohmann, p. 134, pl. 4, figs. 33, 36–37

## PREVIOUS STUDIES

During the Late Pleistocene, the calcareous nannofossils have been shown to provide three biostratigraphical events that are correlated to the oxygen isotope stratigraphy: (1) The extinction of *Pseudoemiliania lacunosa* (Kamptner) Gartner in oxygen isotope stage 12, 458 kya (Thierstein et al., 1977); (2) the first appearance of *Emiliania huxleyi* in oxygen isotope stage 8 calculated to be 264 kya (Thierstein et al., 1977; Gard, 1986); (3) the increase in abundance of *E. huxleyi*. This event can also be defined as a shift in dominance between *Gephyrocapsa* spp. and *E. huxleyi*. During the earliest part of its range, *E. huxleyi* showed low relative abundances, but subsequently evolved to one of the most common species in the World Ocean. This change in abundance of *E. huxleyi* is time transgressive and correlates with oxygen isotope stages 5a and 5b in low-latitude areas, approximately 85 kya. In mid- and high-latitude areas this event occurs in oxygen isotope stage 4, approximately 61 kya (Thierstein et al., 1977; Gard, 1986).

Few studies have been devoted so far to calcareous nannofossils in the Arctic Ocean. Worsley and Herman (1980) observed rare occurrences of nannofossils in sediment cores from the Amerasian parts of the Arctic Ocean. Results on the calcareous nannofossil biostratigraphy in cores from the Fram Strait and the northern continental slope of Svalbard have been reported by Gard (1986).

## MATERIALS AND METHODS

The Swedish Arctic expedition Ymer-80 raised numerous short sediment cores from the European Arctic Ocean (Boström and Thiede, 1984). The present study includes some of the Ymer-80 cores collected from the Fram Strait between Greenland and Svalbard, the Yermak Plateau, and the continental slope and deep-sea basins north of Svalbard (Fig. 18-1, Table 18-1).

A reference core from the Norwegian Sea, V28-56, and one from the Denmark Strait, V28-14, were chosen to compare with the Ymer-80 cores (Fig. 18-1, Table 18-1). The reference cores have previously been analyzed for oxygen isotopes as well as for foraminifers and calcium carbonate content (see Kellogg, Duplessy, and Shackleton, 1978) and the stratigraphy and chronology in these cores are well established. Core V28-14 is shown to have a high sediment accumulation rate (about 4 cm per thousand years) over the past 150 thousand years, whereas V28-56 has a lower sediment accumulation rate but penetrates older oxygen isotope stages. Core

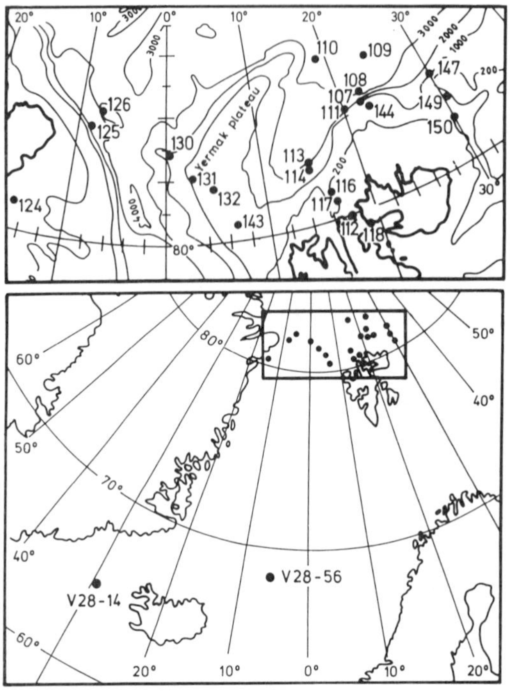


Figure 18-1. Location maps of Ymer-80 coring stations: core V28-14 in the Denmark Strait and core V28-56 in the Norwegian Sea.

V28-56 is shown to have a hiatus somewhere in stage 3 and probably also in stages 5a through 5d (Kellogg, Duplessy, and Shackleton, 1978).

The cores have been sampled at 10 cm intervals or less; smear slides were prepared for studies with the light microscope. The quantitative content of different taxa were determined as the number of specimens per unit area ($mm^2$) on the smear slides. This technique is described by Backman and Shackleton (1983) and provides a good representation of the true abundances of nannofossils in the sediment. In this study, a magnification of $\times 800$ has been used and the specimens in 25 fields of view were counted in each sample.

Using the light microscope, the observed nannofossils were grouped into different taxonomic categories: Placoliths $< 5\mu m$ (small placoliths), *Coccolithus pelagicus, Calcidiscus leptoporus, Helicosphaera carteri, Syracosphaera pulchra,* reworked Cretaceous and reworked Tertiary forms. The last category comprises rare specimens that are poorly preserved and could not be identified as belonging to any other category.

Selected levels were also studied with the scanning electron microscope (SEM) in order to specifically identify the small placoliths. The SEM studies revealed that this category was almost exclusively composed of *Emiliania huxleyi* and *Gephyrocapsa* spp. The abundance of *E. huxleyi* was determined in the SEM as the percent of this species in relation to *Gephyrocapsa* spp. For these determinations, the aim was to observe at least 200 specimens at each level. However, in many cases the abundance of nannofossils in the sediment was so low that fewer individuals were counted.

For the SEM studies, two different sample preparation techniques were used: (1) A small sediment sample was suspended in distilled water and a drop of the suspension was transferred with a capillary tube to the SEM stub and then dried; (2) a thin smear slide was prepared on a round, cover-glass, dried, and attached to the SEM stub.

## REFERENCE CORES

The results of nannofossil analyses in the reference cores from the Denmark Strait and the Norwegian Sea are shown in Figures 18-2 and 18-3 and in Tables 18-2 and 18-3. The ages of the oxygen isotope stage boundaries from the SPECMAP time scale (Imbrie et al., 1984) are shown in Table 18-4.

The nannofossil abundances vary markedly between different oxygen isotope stages. These abundance variations show consistent patterns when comparing the cores and they may be used for correlations. They are also consistent with the abundance patterns observed from oxygen

**Table 18-1. Location, Water Depth, and Thickness of Sediment in the Studied Cores**

| *Core* | *Latitude °N* | *Longitude* | *Water Depth (m)* | *Sediment Thickness (cm)* |
|---|---|---|---|---|
| V28-14 | 64°47′ | 29°03′W | 1855 | 601 |
| V28-56 | 68°02′ | 06°07′W | 2941 | 819 |
| TC 107 | 81°32′ | 22°28′E | 990 | 27 |
| PC 107 | 81°32′ | 22°28′E | 990 | 558 |
| PC 108 | 81°44′ | 22°45′E | 3155 | 103 |
| PC 109 | 82°20′ | 24°10′E | 3890 | 98 |
| PC 110 | 82°30′ | 18°41′E | 3250 | 363 |
| PC 111 | 81°37′ | 20°33′E | 2206 | 166 |
| SGC 112 | 80°00′ | 17°34′E | 395 | 379 |
| PC 113 | 81°02′ | 14°48′E | 1780 | 312 |
| PC 114 | 80°53′ | 14°31′E | 1256 | 465 |
| SGC 116 | 80°24′ | 16°04′E | 368 | 386 |
| SGC 117 | 80°14′ | 16°47′E | 270 | 164 |
| SGC 118 | 79°58′ | 17°35′E | 350 | 386 |
| SGC 124 | 80°29′ | 14°20′W | 321 | 231 |
| SGC 125 | 81°43′ | 08°52′W | 2370 | 241 |
| SGC 126 | 82°00′ | 07°05′W | 3362 | 215 |
| SGC 130 | 81°24′ | 00°52′E | 1590 | 413 |
| SGC 131 | 81°07′ | 03°12′E | 820 | 349 |
| SGC 132 | 80°47′ | 05°07′E | 680 | 470 |
| SGC 143 | 80°16′ | 07°07′E | 560 | 345 |
| SGC 144 | 81°27′ | 23°24′E | 500 | 423 |
| SGC 147 | 81°34′ | 29°30′E | 1512 | 366 |
| SGC 149 | 81°01′ | 30°00′E | 127 | 37 |
| SGC 150 | 80°44′ | 29°31′E | 473 | 404 |

TC = trigger corer; PC = piston corer; SGC = super gravity corer.

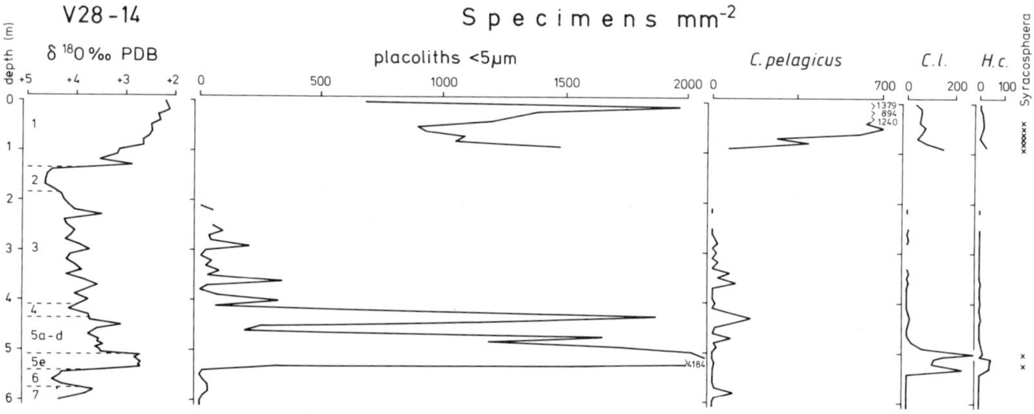


**Figure 18-2.** Plots of nannofossil abundances in V28-14. Oxygen isotope curve is from Kellogg, Duplessy, and Shackleton (1978).

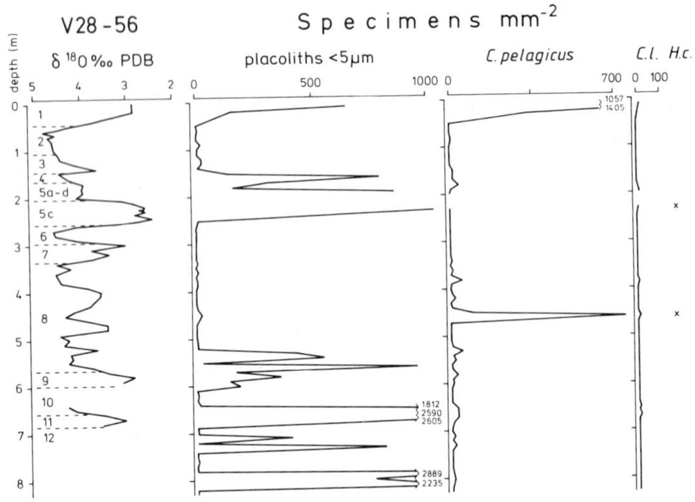


Figure 18-3. Plots of nannofossil abundances in V28-56. Oxygen isotope curve is from Kellogg, Duplessy, and Shackleton (1978).

Table 18-2. Proportional Relationship (%) Between *E. huxleyi (Eh)* and *Gephyrocapsa* spp. *(Ge)* in V28-14

| *Depth (cm)* | *Counted Specimens* | *Eh/Ge (%)* | *δ¹⁸O Stage* |
|---|---|---|---|
| 40 | 206 | 99/1 | 1 |
| 360 | 125 | 80/20 | 3 |
| 400 | 149 | 88/12 | 3 |
| 431 | 212 | 34/66 | 4 |
| 450 | 147 | 1/99 | 5a–d |
| 518 | 300 | 22/78 | 5e |

Table 18-3. Proportional Relationship (%) Between *E. huxleyi (Eh)* and *Gephyrocapsa* spp. *(Ge)* in V28-56

| *Depth (cm)* | *Counted Specimens* | *Eh/Ge (%)* | *$\delta^{18}$O Stage* |
|---|---|---|---|
| 10 | 205 | 98/2 | 1 |
| 160 | 165 | 18/82 | 4 |
| 190 | 217 | 1/99 | 5a–d |
| 420 | 124 | 22/78 | 8 |
| 540 | 100 | 0/100 | 8 |
| 560 | 157 | 0/100 | 8 |
| 650 | 125 | 0/100 | 10 |
| 790 | 200 | 0/100 | 12 |
| 809 | 51 | 0/100 | 12 |

Table 18-4. Ages of Oxygen Isotope State Boundaries 1 to 13 from the SPECMAP Time Scale

| *Boundary* | *Age (ky)* |
|---|---|
| 1–2 | 12 |
| 2–3 | 24 |
| 3–4 | 59 |
| 4–5 | 71 |
| 5–6 | 128 |
| 6–7 | 186 |
| 7–8 | 245 |
| 8–9 | 303 |
| 9–10 | 339 |
| 10–11 | 362 |
| 11–12 | 423 |
| 12–13 | 478 |

*Source:* Imbrie et al., 1984.

isotope stages 1–6 in core K-11 from the Norwegian Sea (Gard, 1986). By combining the results from V28-14, V28-56, and K-11, it has been possible to map the distribution and relative abundance pattern of calcareous nannofossils in different oxygen isotope stages in the Norwegian-Greenland seas.

*Oxygen isotope stage 1.* Stage 1 is rich in calcareous nannofossils. The assemblage is characterized by abundant *C. pelagicus* and *E. huxleyi.* Minor amounts of *Gephyrocapsa* spp., *C. leptoporus, H. carteri,* and *S. pulchra* are also present. However, a decrease in the abundance of nannofossils occurs in the lower part of the stage, which is almost barren. This decrease is particularly sharp and characteristic for *C. pelagicus.* Core V28-14 has a long postglacial section. The 105 cm level is dated by an ash layer to be 9,300 years old and the 45 cm level is dated by $^{14}C$ to be about 4,018 years old (Kellogg, Duplessy, and Shackleton, 1978). The decrease in abundance of *C. pelagicus* occurs between 80 cm and 90 cm in this core. By using the midpoint between these samples and assuming a constant sediment accumulation rate between the two dated intervals, the decrease of *C. pelagicus* is calculated to have occurred at 7,540 years B.P. Thus, common *C. pelagicus* are restricted to the Holocene (10 kya).

*Oxygen isotope stage 2.* This stage is essentially barren of calcareous nannofossils.

*Oxygen isotope stage 3.* Oxygen isotope stage 3 is characterized by barren sediments intercalated with thin horizons of small placoliths. *Emiliania huxleyi* dominates over *Gephyrocapsa* spp.

*Oxygen isotope stage 4.* The younger parts of stage 4 are barren but an abundance peak of small placoliths is present in the older part. This peak seems to have been deposited before the increase of *E. huxleyi* (61 kya), which occurs in the younger part of stage 4.

*Oxygen isotope stage 5.* High abundances of small placoliths characterize oxygen isotope stage 5. SEM studies have shown that *Gephyrocapsa* spp. is the dominant taxon.

*Oxygen isotope stage 6.* This stage is barren of calcareous nannofossils.

Core V28-14 terminates in oxygen isotope stage 7 and core K-11 was not analyzed for oxygen isotopes deeper than stage 6 (Kellogg, Duplessy, and Shackleton, 1978). Thus V28-56 alone has been used for observation of the calcareous nannofossil abundance patterns in the older oxygen isotope stages.

*Oxygen isotope stage 7.* Stage 7 is also barren of calcareous nannofossils.

*Oxygen isotope stage 8.* Stage 8 is fairly long in V28-56 and the abundances of calcareous nannofossils vary dramatically. The abundance peak of *C. pelagicus* at 460 cm depth is of interest as it is the only observed horizon below the Holocene where *C. pelagicus* is abundant. The small placoliths are abundant at 420 cm depth. Studies in SEM revealed that *Gephyrocapsa* spp. is the dominant taxon but that *E. huxleyi* is also present. In the abundance peaks of the oldest parts of oxygen isotope stage 8, no specimens of *E. huxleyi* were observed. This confirms the interpretation of the oxygen isotope curve as *E. huxleyi* has its first occurrence in mid-stage 8 at approximately 264 kya (Thierstein et al., 1977; Gard, 1986).

*Oxygen isotope stages 9–12.* This interval of V28-56 has an incomplete oxygen isotope record. Stages 9 and 10 have fairly low abundances of calcareous nannofossils whereas the small placoliths are very abundant in some layers deposited during stages 11 and 12. *Gephyrocapsa* spp. is by far the most common taxon. *Pseudoemiliania lacunosa,* whose extinction is in mid-stage 12, (Thierstein et al., 1977) was not observed.

## BIOSTRATIGRAPHY OF THE YMER-80 CORES

The calcareous nannofossil assemblages in the Ymer-80 cores are almost exclusively composed of *C. pelagicus, E. huxleyi,* and *Gephyrocapsa* spp. A few specimens of *C. leptoporus* are also present as well as occasional reworked Cretaceous and Tertiary specimens (Gard, 1986). The preservation varies from very poor to good; generally specimens in the upper core portions are better preserved than specimens in the lower parts. The content of calcareous nannofossils shows dramatic abundance variations in a pattern similar to the reference cores in the Norwegian–Greenland seas. Thus it has been possible to establish a tentative biostratigraphy and chronology based on calcareous nannofossils in the majority of the Ymer-80 cores. The abundance patterns in cores with significant amounts of nannofossils are shown in Figures 18-4 and 18-5.

## SGC 124, SGC 125, SGC 126

SGC 124, collected from the eastern shelf of Greenland, is essentially barren of calcareous nannofossils; however, minor amounts of *C. pelagicus* are present at 150–200 cm depth. This barrenness may be a result of high input of terrigenous material and dilution of nannofossils. The presence of *C. pelagicus* may indicate that the entire core represents Holocene sediments.

SGC 125 and SGC 126 were raised off the northeast margin of the Greenland continental slope. Only low frequencies of nannofossils were observed in SGC 125 whereas relatively high abundances of small placoliths were observed in SGC 126 (Fig. 18-4). Scanning electron microscope studies of samples from the 145 cm level revealed that the samples are mainly *E. huxleyi* (of 30 specimens observed, 27 were *E. huxleyi* and 3 were *Gephyrocapsa* spp.). The increase in abundance of *E. huxleyi* occurs in the younger parts of oxygen isotope stage 4 in the northern North Atlantic but this interval is barren in the Norwe-

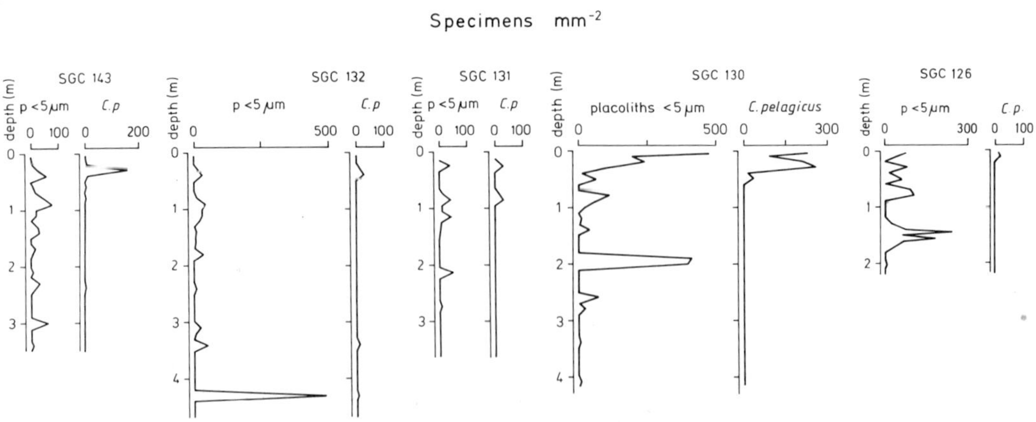


**Figure 18-4.** Plots of nannofossil abundances in SGC 143, SGC 132, SGC 131, SGC 130, and SGC 126.

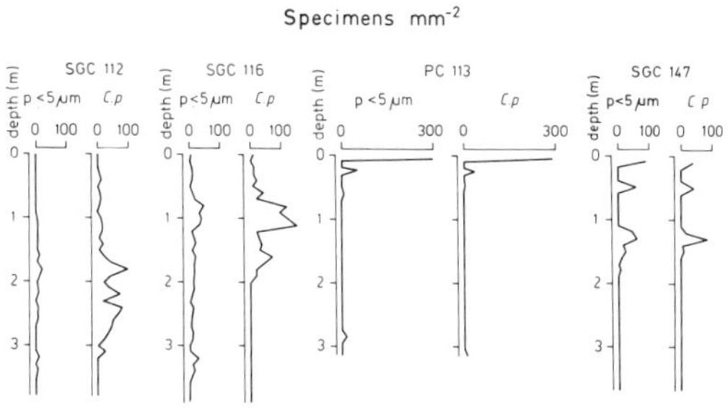


Figure 18-5. Plots of nannofossil abundances in SGC 112, SGC 116, PC 113, and SGC 147.

gian–Greenland seas. Consequently, the upper 170 cm in SGC 126 is interpreted to be younger than oxygen isotope stage 4.

## SGC 130, SGC 131, SGC 132, SGC 143

These cores were collected from various water depths on the Yermak Plateau (Fig. 18-4). SGC 130, from the northwestern slope, has intervals with relatively high abundances of nannofossils in a pattern similar to the reference cores. The Holocene extends to 55 cm depth where the abundance of *C. pelagicus* decreases sharply. The peak of small placoliths at 190–200 cm probably represents oxygen isotope stage 5 as all of the 17 specimens observed in SEM were *Gephyrocapsa* spp. Oxygen isotope stages 2–4 extend between approximately 55–185 cm depth. The interval below 200 cm is almost barren of calcareous nannofossils.

The presence of *C. pelagicus,* although in low frequencies, suggests that Holocene sediments are present at 90 cm in SGC 131, at 45 cm in SGC 132, and in at least the upper 35 cm in SGC 143. In SGC 132, 25 small placoliths from the 430 cm level were observed in SEM and all of them were *Gephyrocapsa* spp. Consequently this layer probably was deposited during oxygen isotope stage 5. Oxygen isotope stages 2–4 are at approximately 45–425 cm depth in this core. SGC 131 and SGC 143 probably represent sediments younger than oxygen isotope stage 5.

## PC 113, TC 113, PC 114, SGC 116, SGC 117, SGC 118, SGC 112

These cores were raised along a transect over the northern continental slope of Svalbard north of the Hinlopen Strait. Only low frequencies of calcareous nannofossils were observed in PC 114, SGC 117, and SGC 118. The Holocene, as indicated by abundant *C. pelagicus,* is represented in the uppermost 7 cm in PC 113, but constitutes approximately 195 cm in SGC 116 and the main part (315 cm) of SGC 112 (Fig. 18-5). Thus, sediment accumulation rates seem to have been higher in the shallower areas.

## PC 110, PC 109, PC 108, PC 111, PC 107, TC 107, SGC 144

Only very low frequencies of calcareous nannofossils were observed in these cores located in various water depths north of Svalbard. It was not possible to establish a calcareous nannofossil biostratigraphy in these cores.

## SGC 147, SGC 149, SGC 150

These cores were collected at approximately 30°E longitude along the northern continental slope of Svalbard. The Holocene comprises at least 135 cm in SGC 147 (Fig. 18-5) but only very low frequencies of calcareous nannofossils are observed in the lower parts of the core. SGC 149 and SGC 150 are essentially barren.

## Sediment Accumulation Rates

Three distinct biostratigraphic units can be distinguished in the Ymer-80 cores by the calcareous nannofossil content: (1) the topmost interval with abundant *C. pelagicus* deposited during the Holocene; (2) the essentially barren interval approximating oxygen isotope stages 2–4; (3) the interval with abundant *Gephyrocapsa* spp. deposited during oxygen isotope stage 5. The boundaries between these intervals have been estimated as the midpoint between a sample containing the key taxon and the closest sample without it. No climatically independent datings are available for the Ymer-80 cores. Therefore, it is impossible to determine whether the nannofossil events in the different parts of the study area are time-transgressive.

The length of the interval with abundant *C. pelagicus* varies from 7 cm (PC 113) to 315 cm (SGC 112). In general, cores from the shallower areas have thicker intervals with abundant *C. pelagicus* than do the cores collected from the deep-sea basins. The interval corresponding roughly to oxygen isotope stages 2–4 varies in length between 130 cm (SGC 130) and 380 cm (SGC 132). Only two cores are shown to penetrate oxygen isotope stage 5: SGC 130 and SGC 132 from the Yermak Plateau. Only 10–20 cm of sediment were recovered from this stage.

The calculated linear sediment accumulation rates for the different cores and biostratigraphical intervals are shown in Table 18-5. The interval with abundant *C. pelagicus* is estimated to correspond roughly to the Holocene. The sediment accumulation rates calculated in Table 18-5 are based on an estimated extent of maximum 10 ky for this interval. Consequently, the calculations for the barren intervals approximating stages 2–4 (12–71 ky) are based on an estimated extent of 10–71 ky.

Although the average sediment accumulation rate varies from core to core, the rates are all in the order of several cm/ky for the Holocene and oxygen isotope stages 2–4. A general trend is that sediment accumulation rates decrease in older biostratigraphical intervals and the rates calculated for oxygen isotope stage 5 are only 0.2 cm/ky (SGC 132) and 0.4 cm/ky (SGC 130).

## DISCUSSION

When correlating the calcareous nannofossil data to the oxygen isotope record, the present study shows that the presence of abundant calcareous nannofossils in the sediment reflect relatively warmer periods and that sediments poor in nannofossils

Table 18-5. Calculated Linear Sedimentation Rates (cm/ky) of Biostratigraphical Intervals in the Ymer-80 Cores

| | *Accumulation Rate (cm/ky)* | | |
|---|---|---|---|
| *Core* | *0–10 ky* B.P. | *10–71 ky* B.P. | *71–128 ky* B.P. |
| SGC 112 | 31.5 | >1.0 | — |
| PC 113 | 0.7 | >5.0 | — |
| SGC 116 | 19.5 | >3.1 | — |
| SGC 118 | 7.5 | >5.1 | — |
| SGC 126 | — | >3.5 | — |
| SGC 130 | 5.5 | 2.1 | 0.4 |
| SGC 131 | 8.5 | >4.3 | — |
| SGC 132 | 4.5 | 6.2 | 0.2 |
| SGC 143 | 3.5 | >5.1 | — |
| SGC 147 | 13.5 | >3.8 | — |

0–10 ky correspond to the interval with abundant *C. Pelagicus* approximating the Holocene. 10–71 ky correspond to an essentially nannofossil-barren interval approximating oxygen isotopes stages 2–4. 71–128 ky correspond to the interval with abundant *Gephyrocapsa* spp. aproximating oxygen isotope stage 5.

represent colder periods. These conclusions are the same as those obtained in a previous study from this area (Gard, 1986). The relative abundance of nannofossils also depends on the dilution by input of other sediment particles and/or degree of preservation of nannofossils deposited. Therefore, direct correlations between nannofossil abundances and climatic variability have to be made with caution.

The highest abundances of calcareous nannofossils are found in the Fram Strait and they decrease northward. No difference in the species composition within the investigated area has been observed. *Calcidiscus leptoporus,* which is reported to have a temperature tolerance of 5°C (Okada and McIntyre, 1979), is present in some of the most northerly located cores despite several degrees lower present-day surface water temperatures. Assuming that *C. leptoporus* has not changed its temperature tolerance range, the rare occurrences of the species may reflect incursions of warmer waters during unusually warm summers.

The intervals barren of nannofossils are believed to represent either periods of nondeposition of nannofossils or their dissolution. However, as the barren intervals in the reference cores coincide with the colder oxygen isotope stages, it seems logical to conclude that the coccolithophorids are absent due to harsh climates. Moreover, as coccolithophorids are photosynthetic algae, production and deposition of calcareous nannofossils may indicate seasonal open-water conditions. However, it was suggested that intervals barren of algae may be the result of suppressed productivity during deglaciations (Ruddiman et al., 1980; Ruddiman and McIntyre, 1981). The influx of meltwater inhibits warm-season productivity by maintaining a low-salinity surface layer with a strong basal halocline that will prevent upwelling (see Chapter 23).

Cores from the deep basins north of Svalbard (PC 108, PC 109, PC 110, PC 111, PC 113, PC 114) are almost totally barren of nannofossils, particularly below the topmost few cm. As relatively high abundances are present in cores from shallower waters from the northern continental slope of Svalbard (Gard, 1986), calcium carbonate dissolution may be the cause for their absence in the deep basins of this area.

It is not known whether coccolithophorids actually live in the Fram Strait and northward today. A possibility is that coccolithophorids do not survive the winter in this area but that living specimens are introduced by warmer water each summer or during particularly warm years.

Coccoliths are deposited mainly as fecal pellets that settle rapidly to the ocean bottom (Honjo, 1976). Therefore, it is unlikely that abundant coccoliths in the Ymer-80 cores result from transport from the Norwegian Sea. However, the composition of the assemblages in the Ymer-80 cores and in the Norwegian Sea reference cores are essentially identical, but *H. carteri* and *S. pulchra* are scarcer in the Arctic area.

The distribution of calcareous nannofossils through time and space covered by the present study is not completely understood. Further investigations are needed in conjunction with studies of the biology of living coccolithophorids and of their present-day distribution and sedimentation, before major paleoclimatic and paleoceanographic interpretations can be made from fossil assemblages of calcareous nannofossils.

## CONCLUSIONS

This study has demonstrated that calcareous nannofossils are useful for biostratigraphical studies of late Quaternary sediments in the investigated area. Most of the studied cores terminate in sediments deposited after oxygen isotope stage 5. Two cores (SGC 130, SGC 132) penetrate older strata of unknown ages.

The calcareous nannofossil biostratigraphic events in the study area are approximately correlative to oxygen isotope boundaries. Incursions of nannofossils reflect climatically favorable conditions.

The appearance of *C. pelagicus* in the study area corresponds roughly to the Holocene. The Holocene assemblages are dominated by *C. pelagicus* and *E. huxleyi*. Oxygen isotope stages 2–4 are characterized by barren intervals or layers with minor amounts of *E. huxleyi* and *Gephyrocapsa* spp.

Sediments deposited during oxygen isotope stage 5 are recognized by abundant nannofossils of which *Gephyrocapsa* spp. is the most common taxon. Oxygen isotope stages 6 and 7 are barren of calcareous nannofossils. Sediment accumulation rates during oxygen isotope stages 1–5 are highly variable in different cores and levels but are consistently in the order of several cm/ky deposited.

## ACKNOWLEDGMENTS

Jan Backman suggested and supervised this project. Valuable comments on the manuscript were made by Jason A. Crux and Yvonne Herman. University of Stockholm and Lamont–Doherty Geological Observatory provided core material. Financial support was given by the University of Stockholm, the Ymer-80 foundation, and the Swedish Natural Science Research Council.

# APPENDIX

## Plate 18-I

SEM illustrations. Length of bar equals 1 μm.

1. *Emiliania huxleyi*, coccosphere; × 9,000. V28-14, 40 cm.
2. *Emiliania huxleyi*, distal view; × 13,500. Ymer-80 SGC 126, 145 cm.
3. *Emiliania huxleyi*, distal view, broken specimen; × 13,500. V28-14, 15 cm.
4. *Coccolithus pelagicus*, distal view; × 9,000. V28-56, 10 cm.
5. *Gephyrocapsa* spp., distal view; × 18,000. V28-14, 518 cm.
6. *Gephyrocapsa* spp., distal view; × 18,000. Ymer-80 SGC 126, 145 cm.
7. *Coccolithus pelagicus*, proximal view; × 9,900. V28-56, 10 cm.
8. *Gephyrocapsa* spp., distal view, dissoluted specimen; × 18,000. Ymer-80 SGC 132, 430 cm.
9. *Gephyrocapsa* spp., distal view, overgrown specimen; × 18,000. Ymer-80 SGC 132, 430 cm.
10. *Calcidiscus leptoporus*, distal view; × 10,800. V28-14, 518 cm.
11. *Helicosphaera carteri*, distal view; × 9,000. V28-14, 40 cm.
12. *Syracosphaera pulchra*, proximal view; × 13,500. V28-14, 40 cm.

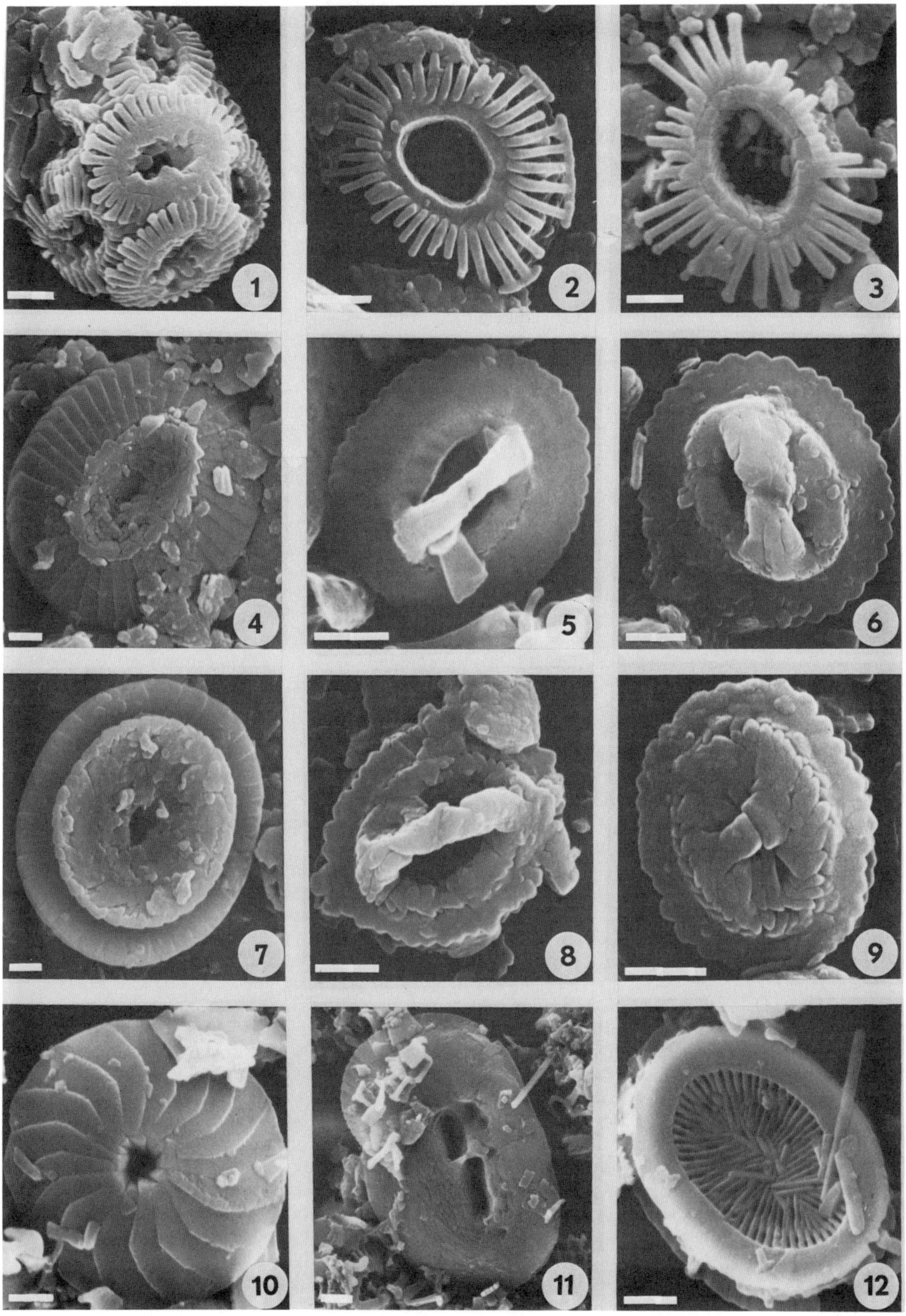


PLATE 18-I

## REFERENCES


Backman, J., and N. J. Shackleton. 1983. Quantitative biochronology of Pliocene and Early Pleistocene calcareous nannofossils from the Atlantic, Indian and Pacific oceans. *Marine Micropaleontol.* **8:**141–170.

Black, M. 1971. The systematics of coccoliths in relation to the palaeontological record. In B. M. Funnel and W. R. Riedel (eds.), *The Micropalaeontology of Oceans.* Cambridge University Press, pp. 611–624.

Boström, K., and J. Thiede. 1983. Ymer-80 Swedish Arctic expedition. Cruise report for marine geology and geophysics. Sediment core descriptions. *Medd. Stockholm Univ. Geol. Inst.* **260:**123p.

Bukry, D. 1973. Low-latitude coccolith biostratigraphic zonation. *Init. Repts. DSDP.* **15:**685–703.

Bukry, D. 1975. Coccolith and silicoflagellate stratigraphy. Northwestern Pacific Ocean. *Init. Repts. DSDP* **32:**677–701.

Gard, G. 1986. Calcareous nannofossil biostratigraphy of late Quaternary Arctic sediments. *Boreas* **15:**217–229.

Gartner, S. 1977. Nannofossils and biostratigraphy: An overview. *Earth-Sci. Rev.* **13:**227–250.

Hay, W. W., H. P. Mohler, P. H. Roth, R. R. Schmidt, and J. E. Boudreaux. 1967. Calcareous nannoplankton zonation of the Cenozoic of the Gulf Coast and Caribbean-Antillean area and transoceanic correlation. *Trans. Gulf Coast Assoc. Geol. Soc.* **17:**428–480.

Haq, B. U. 1978. Calcareous nannoplankton. In B. U. Haq and A. Boersma (eds.), *Introduction to Marine Micropaleontology.* Elsevier, New York, pp. 79–107.

Honjo, S. 1976. Coccoliths: Production, transportation and sedimentation. *Marine Micropaleontol.* **1:**65–79.

Imbrie, J., J. D. Hays, D. G. Martinson, A. McIntyre, A. C. Mix, J. J. Morley, N. G. Pisias, W. L. Prell, and N. J. Shackleton. 1984. The orbital theory of Pleistocene climate: Support from a revised chronology of the marine $\delta^{18}$O record. In A. L. Berger, J. Imbrie, J. D. Hays, G. Kukla,.and B. Saltzman (eds.), *Milankovitch and Climate, 1.* Reidel, pp. 269–305.

Jafar, S. A. 1983. Significance of Late Triassic calcareous nannoplankton from Austria and Southern Germany. *N. Jb. Geol. Palaont. Abh.* **166:**218–259.

Kamptner, E. 1927. Beitrage zur kenntnis Adriatiescher coccolithophoriden. *Arch. Protistenk.* **58:**173–184.

Kamptner, E. 1930. Die Kalkflagellaten des Süswassers and Ihre Beziehung zu jenen des Brackwassers und der Meeres. *Int. Rev. Hydrobiol. u. Hydrogr.* **24:**147–163.

Kamptner, E. 1943. Zur revision der Coccolithineen-Spezies *Pontosphaera huxleyi* Lohmann. *Anz. Akad. Wiss. Wien, Math. Naturw. Kl.* **80:**43–49.

Kamptner, E. 1950. Uber den submikroskopischen Aufbau der Coccolithen. *Anz. Østerr. Akad. Wiss., Math. Naturw. Kl.* **87:**152–158.

Kamptner, E. 1954. Untersuchungen uber den Feinbau der Coccolithen. *Arch. Protistenk.* **100:**1–90.

Kellogg, T. B., J. C. Duplessy, and N. J. Shackleton. 1978. Planktonic foraminiferal and oxygen isotopic stratigraphy and paleoclimatology of Norwegian deep-sea cores. *Boreas* **7:**61–73.

Klaveness, D., and E. Paasche. 1979. Physiology of Coccolithophorids. In *Biochemistry and Physiology of Protozoa,* vol. 1. Academic Press, New York, pp. 191–213.

Lemmermann, E. 1908. Flagellatae, Chlorophyceae, Coccosphaerales und Silicoflagellatae. In K. Brandt and C. Apstein (eds.), *Nordisches Plankton.* Lipsius and Tischer, Kiel and Leipzig, pp. 1–40.

Loeblich, A. R., and H. Tappan. 1978. The coccolithophorid genus *Calcidiscus* Kamptner and its synonyms. *J. Paleontol.* **52:**1,390–1,392.

Lohmann, H. 1902. Die coccolithophoridae, eine Monographie der Coccolithen bildenden Flagellaten, zugleich ein Beitrag zur Kenntnis des Mittelmeerauftriebs. *Arch. Protistenk.* **1:**89–165.

McIntyre, A., and A. W. H. Bé. 1967. Modern coccolithophoridae of the Atlantic Ocean I. Placoliths and cyrtoliths. *Deep-Sea Res.* **14:**561–597.

McIntyre, A., A. W. H. Bé, and M. B. Roche. 1970. Modern Pacific coccolithophorida: A paleontological thermometer. *Trans. N.Y. Acad. Sci.* **32**(ser. 2):720–730.

McIntyre, A., W. F. Ruddiman, and R. Jantzen. 1972. Southward penetrations of the North Atlantic Polar Front: Faunal and floral evidence of large-scale surface water mass movements after the last 225,000 years. *Deep-Sea Res.* **19:**61–77.

Martini, E. 1971. Standard Tertiary and Quaternary calcareous nannoplankton zonation. In A. Farinacci (ed.), *Proceedings II Planktonic Conference, Roma, 1970* **2:**739–785.

Murray, G., and V. H. Blackman. 1898. On the nature of the coccospheres and rhabdospheres. *Philos. Trans. R. Soc. London* **190B:**427–441.

Okada, H., and D. Bukry. 1980. Supplementary modification and introduction of code numbers to the low-latitude coccolith biostratigraphic zonation. *Marine Micropaleontol.* **5**:321–325.

Okada, H., and S. Honjo. 1973. The distribution of oceanic coccolithophorids in the Pacific. *Deep-Sea Res.* **20**:355–374.

Okada, H., and A. McIntyre. 1979. Seasonal distribution of modern coccolithophores in the western North Atlantic Ocean. *Marine Biol.* **54**:319–328.

Ostenfeld, C. H. 1899. Uber *Coccosphaera* und einige neue Tintinniden im Plankton des nordlichen Atlantischen Oceans. *Zool. Anz.* **22**:433–439.

Perch-Nielsen, K. 1985*a*. Mesozoic calcareous nannofossils. In H. M. Bolli, J. B. Saunders, and K. Perch-Nielsen (eds.), *Plankton Stratigraphy.* Cambridge University Press, pp. 329–426.

Perch-Nielsen, K. 1985*b*. Cenozoic calcareous nannofossils. In H. M. Bolli, J. B. Saunders, and K. Perch-Nielsen (eds.), *Plankton Stratigraphy.* Cambridge University Press, pp. 427–554.

Poche, F. 1913. Das system der Protozoa. *Arch. Protistenk.* **30**:125–321.

Ruddiman, W. F., and A. McIntyre. 1981. The North Atlantic Ocean during the last deglaciation. *Palaeogeog., Palaeoclimatol., Palaeoecol.* **35**:145–214.

Ruddiman, W. F., B. Molfino, A. Esmay, and E. Pokras. 1980. Evidence bearing on the mechanism of rapid deglaciation. *Climatic Change* **3**:65–87.

Schiller, J. 1930. Coccolithineae. In *Dr. L. Rabenhorst's Kryptogamen-Flora aus Deutschland, Østerreich und der Schweiz. Abt. Leipzig, Akad, Verlagsgesellschaft* **10**:89–267.

Schwarz, E. H. L. 1894. Coccoliths. *Ann. Mag. Nat. Hist.* **14**(ser. 6):341–346.

Tappan, H. 1980. *The Paleobiology of Plant Protists.* Freeman and Company, San Francisco, 1,028p.

Thierstein, H. R., K. R. Geitzenauer, B. Molfino, and N. J. Shackleton. 1977. Global synchroneity of Late Quaternary coccolith datum levels: Validation by oxygen isotopes. *Geology* **5**:400–404.

Wallich, G. C. 1877. Observations on the coccosphere. *Ann. Mag. Nat. Hist.* **16**(ser. 4):322–339.

Wilcoxon, J. A. 1970. *Cyclococcolithina* Wilcoxon nom. nov. (nom. subst. pro *Cyclococcolithus* Kamptner, 1954). *Tulane Stud. Geol. Paleontol.* **8**:82–83.

Worsley, T. R., and Y. Herman. 1980. Episodic ice-free Arctic Ocean in Pliocene and Pleistocene time: Calcareous nannofossil evidence. *Science* **210**:323–326.

# 19: Arctic Ocean Radiolarians

S. B. Kruglikova

## INTRODUCTION

Radiolaria are unicellular planktonic animals inhabiting all oceans and seas with salinities from 32–38‰. There is no record of their occurrence in seas with lower salinity (the Black, Baltic, and Beloye seas); in near-shore regions the radiolarian abundance and diversity drop sharply. The vertical distribution of Radiolaria ranges from surface to abyssal depths with maximum number of species between 0 m and 200–300 m.

Until recently Radiolaria included all marine protozoans possessing the characteristic intracellular structure the *central capsule*. The radiolarian group includes the Acantharia, the Spumellaria, the Nassellaria, the Phaeodaria, and the Sticholonchea. The acantharian skeleton is composed of strontium sulphate and the Phaeodaria possess siliceous shells with a large admixture of organic material. The skeletons of Acantharia and the majority of Phaeodaria dissolve immediately after the cell dies and do not reach the seafloor. Bottom sediments yield rare phaeodarian species belonging to the genera Cadium, Challengeria, Euphysetta, and Protocystis. Generally sediments contain Spumellaria and Nassellaria almost exclusively (Polycystina); they possess siliceous (opaline) skeletons. Spumellaria is represented by single and colonial organisms (Collosphaerida). Consequently, in the description of bottom sediments the term *Radiolaria* refers only to Polycystina; in the analysis of radiolarian data obtained from plankton samples our attention shall be centered on radiolarians. Due to the morphological features and chemical composition, Acantharia are at present separated from the typical Radiolaria as a kindred but separate group. The classification of Stycholonchea as Radiolaria is now dubious (Petrushevskaya, 1981; Anderson, 1983).

*Note:* This chapter has been reviewed, edited, and rewritten by the Editor, Yvonne Herman.

## PREVIOUS STUDIES

Numerous studies deal with Radiolaria; among them the pioneering works of Ehrenberg and Haeckel are considered fundamental (Ehrenberg, 1854, 1872, 1875; Haeckel, 1862, 1887). These authors described thousands of species, mostly from the tropical regions. In the early 1900s, Radiolaria were studied by Jörgensen, Cleve, and Popofsky (Jörgensen, 1900, 1905; Cleve, 1899, 1900; Popofsky, 1912, 1913). More recently important contributions were made to the systematics, morphology, and stratigraphic distribution of Radiolaria by Petrushevskaya, Riedel, Sanfilippo, Foreman, and Nazarov (Petrushevskaya, 1967, 1971, 1981, 1984; Riedel and Sanfilippo, 1970, 1971, 1977; Foreman, 1973, 1975; Nazarov, 1975), and to radiolarian cytology by Hollande, J. Cachon, M. Enjumet-Cachon, and Anderson (Hollande and Enjumet, 1960; Anderson, 1983).

Owing to the minute size of most species (a few microns) and the low biomass, Radiolaria were omitted from standard planktonic studies and there is very limited knowledge of their geographical distribution and ecology. Radiolarian distribution is primarily controlled by temperature. The majority of Radiolaria belong to the tropical fauna and their abundance and diversity is maximal in the equatorial regions of the World Ocean (Kruglikova, 1969, 1981; Petrushevskaya, 1967, 1971; Nigrini, 1967). The radiolarian data from high latitudes are of particular interest to paleoceanographic reconstructions because the skeletons of calcareous shell-bearing planktonic organisms are either absent at high latitudes or not always preserved in seafloor sediments. And yet our knowledge of Radiolaria from the Arctic Ocean and the surrounding Arctic seas is scanty.

## ARCTIC OCEAN RADIOLARIA

Brady (1878) was the first to describe ten Radiolaria in the bottom sediments of the Baffin Bay (83° 19′N). Bailey (1856) described several radiolarian species recovered from the bottom sediments of the Bering Sea near Kamchatka; he believed that they were brought there by cold currents from the north. Bailey noted the sharp difference between this complex and the associations observed off the coasts of Japan and California.

In several publications on phytoplankton and zooplankton in subpolar and polar regions, the participants of various expeditions at the beginning of this century described the discovery of radiolarians in the waters of the Northern Atlantic, Greenland, Barents, and Kara seas (Jörgensen, Cleve, Popofsky, Borgert, Schröder).

## DISTRIBUTION AND ECOLOGY OF POLYCYSTINA FROM PLANKTON SAMPLES*

The first radiolarian collections were made by Bernstein (1931, 1932, 1934) in the northern Barents Sea and in the northwestern Kara Sea (between 76° and 80°N) during the *Sedov* and *Lomonosov* expeditions. Bernstein found five species of Nassellaria, one species of Spumellaria, and three species of Acantharia in the plankton samples analyzed; she also indicated their habitat (depth, temperature, and salinity). She distinguished three groups of radiolarian species in the Barents and Kara seas: (1) the local species *Amphimelissa setosa* (Cleve) and *Acantharia ligurina* Haeckel; they compose the major part of the radiolarian plankton reproducing in these two seas; (2) species supplied from the Barents Sea to the Kara Sea: *Plectacantha oikiskos* Jörgensen and *Dictyophimus gracilipes* Bailey; and (3) newcomers from the Polar Basin: *Dictyophimus multispinus* Bernstein, *Phormacantha hystrix* Jörgensen, and *Echinomma leptodermum* Jörgensen. The species

*Figures 19-1 and 19-2 show the direction of major currents, zoogeographical zonation, and the regions of the present study area.

belonging to this group are always present in small quantities and Bernstein believed that they do not live in the Barents Sea. The "newcomers" have two origins: (1) strictly Arctic species and (2) species supplied to the Polar Basin by the North Atlantic Current flowing between Greenland and Spitzbergen.

During the *Sedov* drifting expedition in 1937–1938 (86°38′N–84°47′N, 73°02′E–28°54′E) (Fig. 19-2), Bogorov (1946) found in the zooplankton samples 59 species of animals including 4 species of Nassellaria, 4 species of Spumellaria, 1 species of Phaeodaria, and 1 species of Acantharia. The most numerous were *Haliomma beroes* Ehrenberg, and the less frequent were *Echinomma leptodermum* and *Amphimelissa setosa*.

Bogorov has established several peculiar features of the Arctic zooplankton, which apparently also apply to Radiolaria: (1) Most species are not restricted to the Arctic Ocean; they are also common planktonic taxa in the Northern Atlantic and the low-Arctic seas, particularly in the Greenland and Barents seas. (2) A large proportion of the species are rare; they are represented by a small number of specimens, which is not unusual for the Arctic Ocean. (3) The deep layers (200–750 m) are the richest in number of individuals. (4) Many zooplanktonic specimens of the Arctic Ocean are recorded even in the tropical regions at very great depths (more than 1,000 m).

Hülsemann (1963) has collected the largest "year-round" collection of Arctic Radiolaria from plankton samples of the ice-floe station T-3 (1952–55), which drifted from the vicinity of the pole to the Ellesmere Island (86°53′N–82°40′N) (Figs. 19-1 and 19-2). Hülsemann reported 32 radiolarian species, most of which were identi-

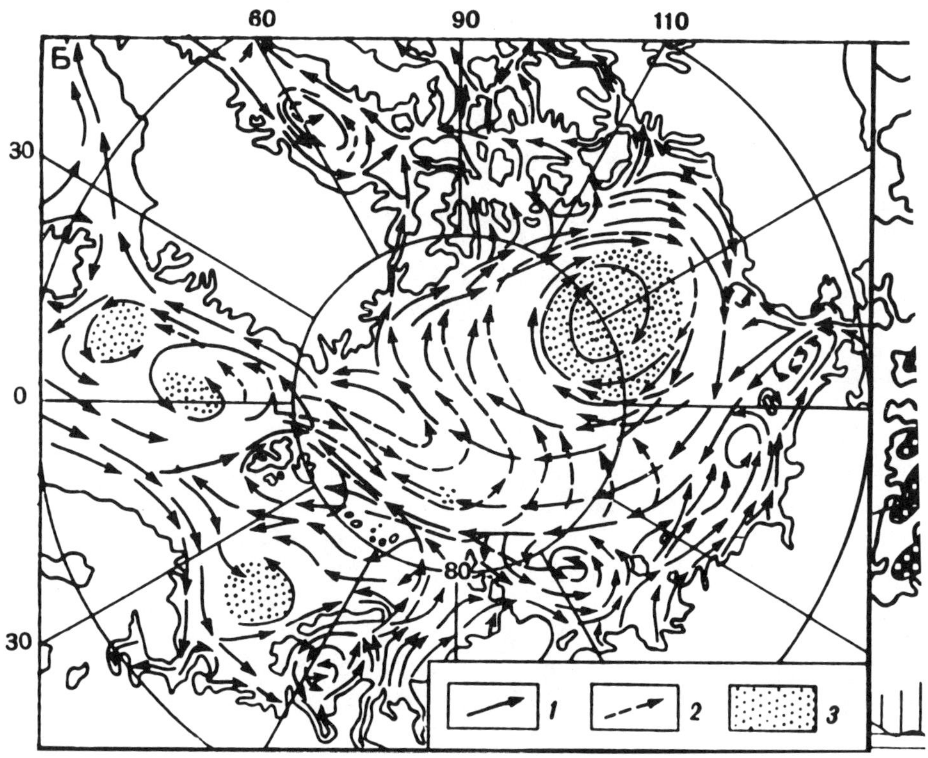


Figure 19-1. Currents in the Arctic: (1) surface, (2) subsurface, (3) variable direction. (From Lisitsyn, 1978.)

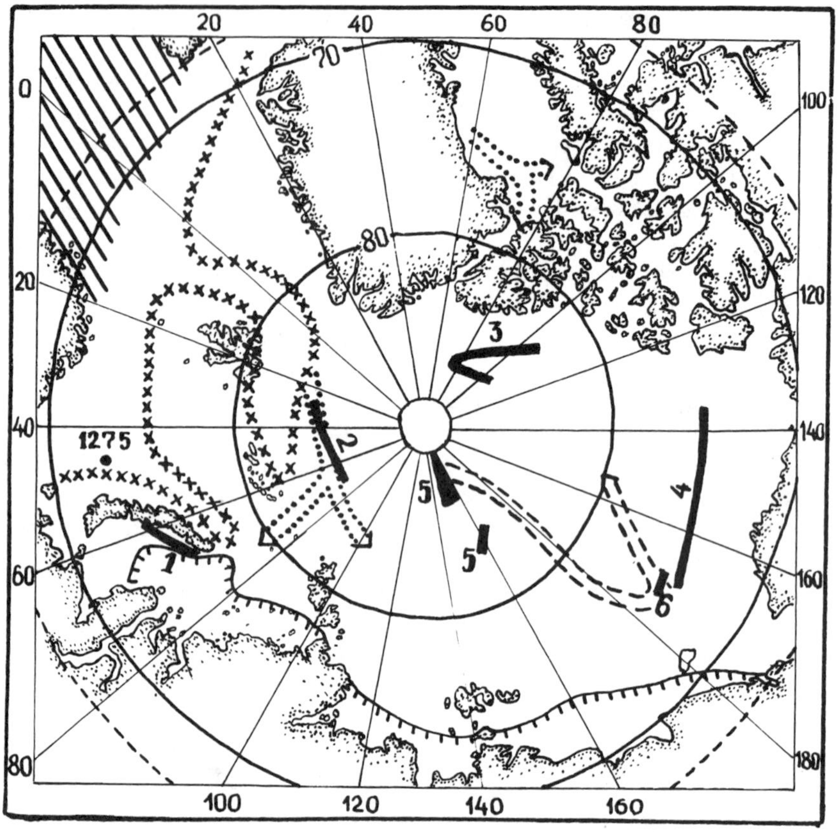

Figure 19-2. Zooplankton zonation. Distributional patterns of species: (1) glacial-neritic (*Calanus glacialis, Parathemisto libelula*); (2) marine water (*Limnocalanus grimaldii; Drepanopus bungii*); (3) zooplankton species (*Eucalanus bungii*) in the Pacific water; (4) zooplankton species in the Atlantic water (*Calanus finmarhicus*); (5) the boreal–Arctic region. (6) Transitional zone between the boreal–Arctic and tropical regions. (7) Regions of the Arctic Basin where Radiolaria were studied: (1) Bernstein, 1931, 1934; (2) Bogorov, 1946; (3) Hülsemann, 1963; (4) Tibbs, 1967; (5–6) our data (g-SP-15; 6-SP-22).

fied in the Arctic for the first time. Among them the most abundant are Phaeodaria (7 species), inhabiting depths of 1,400–1,600 m. The most frequently recorded of the 22 species of Polycystina is *Spongotrochus glacialis* Popofsky; there are about 2,000 specimens of this species in the collection.

According to Hülsemann, many radiolarian species are Arctic and boreal–Arctic. The cosmopolitan species are somewhat less numerous; 4 species are probably exclusively adapted to Arctic and Antarctic waters. Hülsemann has drawn important conclusions for the comprehensive understanding of radiolarian ecology and biogeography. (1) The size of the eurythermal species is smaller in the Arctic Ocean than in warm waters (Popofsky, 1908, mentioned this peculiarity of some of the eurythermal species in the Antarctic plankton). (2) Radiolarian distributions have well-defined depth zonation: In the upper layer (0–200 m), *Cannospaera antarctica* Haeckel (Phaeodaria) and *Spongotrochus glacialis* Popofsky (Spumellaria) are dominant; both species are found in the 200–400 m layer in considerably lesser quantities. Most Phaeodaria were collected from the deep-sea zone. The other species are too rare to be systematized. (3) Following Riedel (1958), Hülsemann suggests that two groups of species live in the polar regions; both taxa inhabit the Arctic cold upper layers and one of them has a more extensive habitat, living in the deep-water layers of warmer oceans as well. This finding confirms the regularities of the Arctic plankton established earlier by Bogorov (1946). (4) A certain proportion of radiolarian plankton is supplied to the central part of the Arctic Basin from the Atlantic (as suggested by Bernstein, 1932, 1934) and Pacific oceans. The confinement of Radiolaria to well-defined water depths allows us to identify water masses otherwise difficult to trace because of the long distances that radiolarians have to travel. The radiolarians brought alive by these water masses mix with the local fauna.

During the drift of the ice-floe station Arlis-3 in 1960–1961 (74°–75°N, 140°–170°W), (Fig. 19-2), Tibbs (1967) recorded 17 radiolarian species out of 26 protozoan species: Nassellaria, 7; Spumellaria, 3; Phaeodaria, 7, among them Polycystina, Echinomma, and Leptodermum Jörgensen, *Cromyechinus borealis* (Cleve), *Spongotrochus glacialis* Popofsky, *Plectacantha oikiskos* Jörgensen, *Phormacantha hystrix* (Jörgensen), *Dictyophimus gracilines* (Bailey), *Tetraphormis rotula* Haeckel, *T. triloba* Haeckel, *Conarachnium pronendum* Ehr., and *Botryopyle setosa* Cleve. The most abundant are *Spongotrochus glacialis* and *Botryopyle setosa* (= *Amphimelissa setosa*).

The plankton of the eastern part of the transect studied by Tibbs is richer in protozoan composition than that of the western part. The radiolarian assemblage is similar to the one established by Hülsemann, though less diversified.

Tibbs distinguishes two depth zones in the protozoan habitat where their abundance is high (70–100 m and 400m ± 70 m), and ascribes the existence of these zones to the fact that they are the zones of maximum temperature. Protozoans are able to migrate from one zone to the other. In the uppermost water layer (40 m), particularly immediately underneath the ice, their number is very low.

The above-mentioned authors have recorded in the Arctic waters a total of 41 radiolarian species: Nassellaria, 19; Spumellaria, 7; Phaeodaria, 9 (10?); Acantharia, 5; and Sticholochea, 1. A considerable number of specimens in the samples that were collected by Hülsemann and Tibbs are Phaeodaria (7 species) whereas Bogorov's records show only one Protocystis species, and Bernstein completely omits Phaeodaria from his species lists. Acantharia, however, (3 species), comprise one-third of the total number of species in the collections of Bernstein and Bogorov and considerably less (one-tenth) in Hülsemann's collection; Tibbs does not mention this group at all.

The data of the above-cited authors imply that there are a number of common species in the plankton of different (almost circumpolar) Arctic regions. Among Polycystina, on which our interest is centered, such species are *Echinomma leptodermum* Cleve, *Amphimelissa setosa* (Cleve), *Spongotrochus glacialis* Popofsky (= *Trochodiscus echiniscus* Jorgensen, Bogorov, 1946) *Plectacantha oikiscos* (Jörgensen), *Phormacantha hystrix* (Jorgensen), and *Dictyophimus gracialipes* Bailey (= *Pseudodictyophimus gracilipes*, Petrushevskaya, 1971). Among these species the most abundant and widely distributed in plankton are *Amphimelissa setosa* and *Spongotrochus glacialis.* The radiolarian plankton of the Amerasian part of the Arctic is richer in species than the Eurasian plankton. Besides the extensively distributed species of Polycystina just discussed, the species recorded both in Tibbs and Hülsemann's collections are *Cromyechinus borealis* (Cleve), *Tetraphormis rotula* (Haeckel), *T. triloba* Haeckel, and *Conarach-*

*nium profundum* (Ehrenberg). Hülsemann's collections, covering all Arctic seasons, have several other species of Polycystina.

## POLYCYSTINA FROM ARCTIC BOTTOM SEDIMENTS

Generally Acantharia and the larger part of Phaeodaria are not preserved in bottom sediments, and there are as yet no detailed data in the literature on Polycystina distribution in Arctic deposits except for the publications of the author (Kruglikova, 1982, 1984, 1985).

### Material and Methods

The author was the first to study Polycystina recovered from the surface layers of Arctic bottom sediments (Fig. 19-2). The samples were collected by "bottom-scrapers" during the drift of the Soviet ice-floe stations SP-15 from the Lomonosov Ridge and Submariners Basin in the central Arctic and the SP-22 from the outer sector of the Chukchi Sea shelf. A total of 46 samples were collected. Ten samples were studied from a 135-cm long sedimentary core recovered at station 1275 by the R/V *Shtokman* in the Central Basin of the Barents Sea (73°42'N, 46°35'W, depth 329 m). In addition, the author examined the collection of Radiolaria selected from 14 samples of cores 111, 146, 147, and 149 raised on board the R/V *Biolog* in the same basin; these specimens were kindly submitted to us by I. Sakharova.

The "bottom-scraper" samples (0–3 cm to 0–5 cm) are clays or aleurite clayey oozes, sometimes of liquid consistency. We used well-established methods for processing the sediments: The dry sediment (5–7 g) was boiled 0.5–1.5 hours in 100–150 ml of natrium tripoliphosphate, with addition of 3–5 ml of hydrogen peroxide, until the sediment was completely dispersed. The sediment was then rinsed through a 0.04 mm metal sieve. The coarse fraction was spread on a slide, dried, mounted in Canada Balsam, and covered with a cover glass. When studying Radiolaria from the tropical regions, the weight of the sediment is not more than 50–200 mg, because the number of Radiolaria in such a sample is tens or hundreds of thousands times greater than in Arctic Ocean samples.

Radiolaria were studied by light microscope at magnification × 120–300. The illustrations are drawings and photomicrograms, which were obtained by scanning electron or light microscopes. Sometimes special techniques are used to obtain photomicrograms by light microscope. Many species have volumetric skeletons with characteristic inner structures; the object is therefore photographed 2–3 times by subsequent focusing on various parts of the skeleton (edges, the most prominent middle part, etc.). Since Radiolaria are very rare in Arctic sediments, they were studied only under a light microscope.

### Results

Not all the examined samples from the central Arctic contain Radiolaria; when present their number does not exceed five individuals per gram of dry sediment. In the Chukchi Sea, the number of Radiolaria is at least 1–2 orders of magnitude larger, which means that in radiolarian abundance the Arctic is comparable with the poorest deposits of coastal regions and the halistatic zones of the World Ocean. The skeletons of some specimens in the Chukchi Sea and central Arctic sediments exhibit evidence of dissolution; preservation was excellent in only one of the samples studied, recovered from the surface sedimentary layer of the Barents Sea (St. 1275, 0–3 cm).

The samples from the central Arctic yielded only 4 Polycystina species: *Echinomma leptodermum* Cleve, *Spongotrochus glacialis* Pop., *Pseudodictyophimus gracilipes* (Bailey), and Lophophaena? sp. The most frequent was the first species; we found 12 specimens in 25 samples. The second species was found in only 4 samples. The Polycystina fauna of the Barents Sea is much more diversified, which is exemplified in 1 sample. More than 15 species were found; among them are the following:

| | |
|---|---|
| *Amphimelissa setosa* (Cleve) | 425 |
| *Botryopera setosa* (Jörg.) | 29 |
| *Plectacantha oikiskos* (Jörg.) | 29 |
| *Botryopera boreale* (Ehrenberg) | 23 |
| *Phormacantha hystrix* (Jörg.) | 16 |
| *Echinomma leptodermum* (Jörg.) | 3 |
| *Lithelius spiralis* Haeckel | 3 |
| Spongotrochus sp. | 3 |
| *Lithelius nautiloides* Pep. | ?3 |
| *Lophophaena nadegdae* Petrush. | 1 |
| *Cryomyechinus borealis* Cleve | 1 |
| *Phorticium clevei* Jörg. | ?1 |
| Euscenium sp. | 1 |
| *Hexacontium setosum* Haeckel etc. | 1 |

In the Chukchi Sea, Radiolaria were found in all 20 samples studied and the number of species in each sample did not exceed 10. A total of about 30 Polycystina species were identified:

| | |
|---|---|
| *Amphimelissa setosa* Cleve | 425 |
| *Echinomma leptodermum* (Jörg.) | 61 |
| *Plectacantha oikiskos* (Jörg.) | 54 |
| *Phormacantha hystrix* (Jörg.) | 32 |
| *Poridium longispinum* Jörg. | 17 |
| *Haloimma beroes?* Ehr. | 15 |
| *Pseudodictyophimus gracilipes* (Bailey) | 11 |
| *Spongotrochus glacialis* Pop. | 9 |
| *Pseudodictyophimus platycephalus* | 7 |
| *Mitrocalpis assansefera* Popofsky | 6 |
| *Ceratocyrtis histricosa* (Jörg.) | 3 |
| Cladoscenium sp. | 3 |
| *Sethoconus tabulatus* Cleve | 33 |
| *Artostrobus jorgenseni* Petrushevskaya | 2 |
| Spongodiscus sp. | 2 |
| *Artostrobus annulatus* (Bailey) | 1 |
| *Diplocyclas davisiana* (Ehrenberg) | 1 |
| *Lithamphora platycephala* (Ehrenberg) | 1 |
| *Arthostrobium botryocyrtum* (Haeckel) | 1 |
| *Botryopera setosa* (Jörg.) | 1 |
| *Botryopera boreale* (Ehrenberg) | 1 |
| *Artostrobium auritum?* (Ehrenberg) | 1 |
| *Arachnacorys penthacentha* Popofsky | 1 |
| *Cromyechinus borealis* (Cleve) etc. | 1 |

More than 30 Polycystina species are found in Arctic sediments, that is, more than were previously identified (Kruglikova, 1984) and many more than were found in plankton hauls. This is an indication that the Arctic zooplankton has been insufficiently studied to date. The data on Polycystina from Arctic plankton samples and from bottom sediments has demonstrated that in the Polar Ocean the most extensively distributed and numerous taxa are *Amphimelissa setosa, Echinomma leptodermum,* and *Spongotrochus glacialis* (Table 19-1, Plates 19-I, 19-III, 19-IV*), the first species being by far the most abundant (this species was not found in the sediment samples of the central Arctic). Moreover, the abundance of *S. glacialis* is considerably lower in sediments than in plankton samples. The highest diversity of radiolarian fauna was recorded in the sediments of the low-Arctic seas, whereas on the Lomonosov Ridge the samples yielded only single species.

The composition of preserved Polycystina on the seafloor, apart from its actual low diversity in this rigorous environment, is probably due to dissolution on the bottom caused by slow sedimentation rates at great depths in the central Arctic and also possibly by aggressive cold bottom water.

In agreement with Bernstein and Hülsemann, all Arctic Polycystina can be classified as boreal–Arctic, bipolar, and cosmopolitan species. There are apparently no endemic species in the present-day Arctic. The cosmopolitan species

*Plates and explanations appear as an appendix to the chapter.

Table 19-1. Distribution of Radiolaria in the Surface Layer of Bottom Sediments and in Plankton in Different Zones

| Representative Species | Arctic Basin | Pacific Boreal | Sea of Okhotsk | Bering Sea | Pacific: Northern Subtropical | Pacific: Northern Tropical | Pacific: Antarctic |
|---|---|---|---|---|---|---|---|
| 1 | 2 | 3 | 4 | 5 | 6 | 7 | 8 |
| *Lithocampe plalicephala* Ehr. | | | | | | | |
| *Artostrobus annulatus* (Bailey) | | | | | | | |
| *Diplocyclas davisiana* (Ehr.) | | | | | | | |
| *Pseudodictyophimis clevei* Jorg. | | | | | | | |
| *Pseudodictyophimis gracilipes* Bailey | | | | | | | |
| *Lithamphora furcaspiculata* Pop. | | | | | | | |
| *Lithomitra arachnea* Ehr. | | | | | | | |
| *Diplocyclas* sp. aff. *C. davisiana* Ehr. | | | | ? | | | |
| *Stylodictya validispina* Jorg. | | | | | ? | | |
| *S. stellata* Bailey | | | | | ? | | |
| *Amphimelissa setosa* (Cleve) | | | | | | | |
| *Spongotrochus glacialis* Pop. | | | | | | | |
| *Sethoconus tabulatus* (Ehr.) | | | | ? | ? | | |
| *Enneaphormis rotula* Hck. | | | | | | | |
| *Sphaeropile langii* Dryer | | | | | | | |
| *Chromyechinus borealis* Dreyer | | | | | | | |
| *Acanthodesmia micropora* (Pop.) | | | | | | | |
| *Spongurus pylimaticus borealis* Kruglikova | | | | | | | |
| *Rhizoplegma boreale* (Cleve) | | | | | | | |
| *Polysolenia arctios* Nigrini | | | | | | | |
| *Stilochlamidium venustum* (Bailey) | | | | | | | ? |
| *Saccospyris robustus* Kruglikova | | | | | | | |
| *Tholospyris spinosus* Kruglikova | | | | | | | |
| *T. borealis* (Bailey) | | | | | | | |
| *Botryocampe inflata* (Bailey) | | | | | | | |
| *Lithocampe aquilonaris* (Bailey) | | | | | ? | | |
| *Pterocorys korothevi* (Dogiel) | | | | | | | |
| *P. hirundo* Hck. | | | | | | | |
| *Diplocyclas cornuta* (Bailey) | | | | | | | |
| *Caratocytis historicosa* Jorg. | | | | | | | |
| *Spongodiscus osculosus* (Dreyer) | | | | | | ? | |
| *Echinomma delicatulum* (Dogiel) | | | | | | | |
| *Artobortys borealis* (Cleve) | | | | | | | |
| *Tholospira* sp. | | | | | | | |
| *Styptosphaera spumacea* Hck. | | | | | | | |
| *Theocorytum trachelium dianae* Hck. | | | | | | | |
| *Spongaster tetras irregularis* Nigrini | | | | | | | |
| *Actinomma medianum* Nigrini | | | | | | | |
| *Heterocantha dentata* (Hck.) | | | | | | | |
| *Lithocampe* sp. Nigrini | | | | | | | |
| *Lithopera bacca* Ehr. | | | | | | | |
| *Pterocorys junon's* (Hck.) | | | | | | | |

Table 19-1.
(continued)

| Representative Species | Arctic Basin | Pacific Boreal | Sea of Okhotsk | Bering Sea | Pacific: Northern Subtropical | Pacific: Northern Tropical | Pacific: Antarctic |
|---|---|---|---|---|---|---|---|
| 1 | 2 | 3 | 4 | 5 | 6 | 7 | 8 |
| *Heliodiscus asteriscus* Hck. | | | | | ━ | ━ | |
| *Pterocorys cranoides* Hck. | | | | | ━ | ━ | |
| *Panartus tetrathalamus* Hck. | | | | | — | ━ | |
| *Euchitonia elegans* (Ehr.) | | | | | – – | ━ | |
| *Spongaster tetras tetras* (Ehr.) | | | | | – – | ━ | |
| *Theocorythium trachelium trachelium* (Ehr.) | | | | | – – | ━ | |
| *Euchitonia furcata* Hck. | | | | | ━ | ━ | |
| *Peromelissa phalacra* (Hck.) | | | | | – – | ━ | |
| *Botryocyrtis scutum* (Harting) | | | | | – – | ━ | |
| *Pterocorys herwigii* Hck. | | | | | – – | — | |
| *Anthocyrtidium cineraria* Ehr. | | | | | | ━ | |
| *Pterocanium praetextum praetextum* Ehr. | | | | | – – | ━ | |
| *Ampirrhopalum ypsilon* Hck. | | | | | ━ | ━ | |
| *Spongocore puella* Hck. | | | | | — | ━ | |
| *Lamprocyclas maritalis* Haecker *ventricosa* Nigrini | | | | | | — | |
| *Tribonosphaera centripitalis* Hck. | | | | | | — | |
| *Centrobotrys termophila* Petruschevsk. | | | | | | — | |
| *Lophophaena hispida* (Ehr.) | | | | | – – | — | |
| *Thicolocampe cylindrica* Hck. | | | | | ━ | ━ | |
| *Lithocjrcus reticulata* Ehr. | | | | | – – | — | |
| *Nerhorospyris renilla* (Hck.) | | | | | | — | |
| *Lithelius (?) nautiloides* Pop. | | | | | | | ━ |
| *Spongurus pylomaticus pylomaticus* Riedel | | | | | | | ━ |
| *Triceraspyris antarctica* (Haecker) | | | | | | | ━ |
| *P. antarctissa* Petruschevskaya | | | | | | | ━ |
| *Sethoconus reschetnjakae* Petrusch. | | | | | | | — |
| *Saccospyris antarctica* Haecker. | | | | | | | ━ |
| *S. conitorax* Petrusch. | | | | | | | — |

*Source:* From Kruglikova, 1981.

(*Pseudodictyophimus gracilipes,* etc.) (Plate 19-II) were recorded only in the Amerasian sector (material of Hülsemann, Tibbs, and our samples from SP-22); they were not found in the Eurasian sector of the Arctic (collections of Bernstein, Bogorov, and our samples at St. 1275). Table 19-1 presents data on the distribution of the most typical Polycystina species in the sediments of the World Ocean and on a number of species found in the Arctic (Table 19-1 also includes data reported by Bernstein, Tibbs, and Hülsemann).

The Polycystina associations of the Arctic proper resemble in composition the associations found in the surrounding seas owing to the pronounced dominance of the three above-mentioned widely distributed species. These faunal compositions could possibly be ascribed to the peculiar hydrological features of the Arctic Ocean, whose circulation shows several gyres corresponding conspicuously to the circulation of Mediterranean seas (Stepanov, 1983). The Arctic Polycystina are closer to the truly oceanic associations by their high taxonomic rank composition. In the Arctic seas, and particularly in the shelf regions, the spherical and disk-shaped Spumellaria dominate; the Arctic yields not more than 1–2 Discoidea species and 2–3 Sphaeroidea species; Nassellaria are the most abundant. Nevertheless, in the seas belonging to the boreal zone Cannobotryidae are numerous.

The Polycystina identified in the plankton samples and bottom sediments belong to the families Actinommidae, Pyloniidae, Litheliidae, Spongodiscidae, Plagoniidae, Clathromitridae, Sethophormididae, Lampromitridae, Artostrobiidae, and Botryocyrtidae. There are no colonial Radiolaria in the Arctic (family Collosphaeroidae) and several other families were not recorded in the boreal–Arctic region either.

Representatives of the suborder Spyridina and of the superfamily Connobotryoidae are diverse and widely distributed in the seafloor deposits within the tropical zone, particularly in Neogene sediments. These groups appeared in the Pacific boreal zone only in the Pliocene (Kruglikova, 1979); there are not more than 1–2 Spyridina species and 2–3 Cannobotryoidae species that belong to the genus Saccospyris; both groups of species are endemic to the boreal zone. Neither our samples nor the planktonic hauls of Bernstein, Bogorov, or Tibbs have recorded Spyridina in the Arctic; Hülsemann described the new species *Tholospyris gephyristes* and Aegospyris sp. based only on a few specimens (3 and 2, respectively). The presence of this group needs to be confirmed by additional data.

*A. setosa* is by far the most widely distributed and abundant species; it belongs to the Botryocyrtidae family, a recent newcomer to the northern regions. According to our data, this species existed in the central regions of the boreal Pacific at least during the better part of the glacial Pleistocene (Kruglikova, 1976, 1977) and disappeared suddenly before the onset of the Last Glaciation, about 80 kya. This species was not recorded in the Pacific Ocean surface sediments; however, it is found occasionally in Pleistocene deposits of the Sea of Okhotsk. This taxon is now abundant in the northern sector of the Atlantic Ocean; in the Norwegian Sea, 27–30% of the radiolarian plankton is often composed of its shells (Petrushevskaya, 1971); in fjörds its relative abundance out of the radiolarian fauna sometimes reaches 80% (according to the paper by Björklund and Svanberg presented at the IV International Symposium Eurorad, 1984). In the bottom sediments of the Norwegian and North Seas this species is recorded only since the Holocene (Björklund, Thiede, and Holtedahl, 1978). The abundant standing stock of this taxon in the Arctic plankton probably means that *A. setosa* entered the boreal–Arctic Atlantic from the Pacific through the Arctic seas during a later period, perhaps during the transition from the Pleistocene to Holocene. This evidence indicates that the Arctic Ocean was at some time, during the late Pleistocene or Holocene, at least seasonally ice-free.

Another abundant species in the Arctic, *Spongotrochus glacialis,* is also of Pacific origin, but it penetrated the Arctic waters from the North Atlantic; the taxon emigrated from the Pacific Ocean to the Atlantic as early as the Miocene (Petrushevskaya, 1979).

## RADIOLARIA FROM THE BARENTS SEA CORES

In the sedimentary core recovered from St. 1275, the abundant present-day fauna is recorded only in the upper 0–3 cm; 5 cm below the top, the number of Radiolaria drop dramatically. The remainder part of the core, the 10–135 cm section, contains only a few recrystallized radiolarian skeletons, pre-Holocene species (Plate 19-IV, figs. 7–13). The rare, poorly preserved specimens ambiguously imply that the Radiolaria found in the upper part of the core are probably of Miocene age; below 55 cm in core 1275 and in cores St. 111, 146, 147, and 149 the radiolarian age determined by Vishnevskaya is possibly very early Paleocene. The following species are present: Archaespongoprunum? sp., *Stichomitra torguota?, Stichomitra? manifesta, Stichomitra livek,* and Pseudoaulophacidae gen. et ap. indet. The absence of present-day Radiolaria below the surface sediment layer can be reasonably ascribed to a sudden change in environment, which occurred not long ago, possibly in Holocene time.

## CONCLUSIONS

Several conclusions can be drawn, although we should emphasize that we consider them to be preliminary.

**1.** The composition and distribution of radiolarian fauna, the absence of species endemic to the high Arctic, we believe are evidence that the Radiolaria now living in the Arctic have only recently immigrated to this sea. This statement is in agreement with Jouse, Petrushevskaya, and others who believe that the biota of the Norwegian–Greenland seas and that of the Arctic have evolved recently.

**2.** The low faunal diversity in the central Arctic and the absence of *A. setosa*, the most abundant and widely distributed species, also imply the young age of the radiolarian population of the Arctic and their dispersal over the equator through the low-Arctic seas.

**3.** The composition of radiolarian associations is largely influenced by the influx of the Atlantic waters into the Barents Sea and of the Pacific waters via the Chukchi Sea.

**4.** The presence of cosmopolitans, the most diverse radiolarian associations, in the Amerasian sector of the ocean is an indication that probably they penetrated the Arctic mainly from the Pacific Ocean.

**5.** The true number of radiolarian species inhabiting the Arctic is apparently different from the number known at present. This is due to the limited data available on the radiolarian fauna and to the ambiguity in systematics, particularly at levels higher than the species rank. Some of the closely related species are obviously united into groups; it seems clear that other species shall be reduced to synonyms, first of all, the genera Plectacantha, Phormacantha, Arachnocorys, and Peridium.

## TAXONOMY

A review of Polycystina most frequently observed by us in bottom sediments and of their records in plankton samples reported in the literature is presented. Radiolarian systematics is rather ambiguous; there is as yet no radiolarian systematics accepted by all experts working with this group. Here we used the nassellarian system of Petrushevskaya (1981) and the spumellarian system of Riedel and Sanfilippo (1977).

Sarcodina
Actinopoda Calkins, 1909
Euradiolaria Lameere, 1931
Polycystina Ehrenberg, 1838
Spumellaria Ehrenberg, 1875
Actinommidae Haeckel, 1882 sensu Riedel, 1967

*Echinomma leptodermum* Jörgensen

*Echinomma leptodermum,* Jörgensen, 1900: 57; Jörgensen, 1905: 116, pl. 8, fig. 33 a–e; Schröder, 1909; Bernstein, 1934; Bogorov, 1946; Hülsemann, 1963; Tibbs, 1967; Björklund, 1976

This species was recorded in the northern temperate and cold waters south of the Faeroes and Shetland Islands, in the Greenland, northern Kara, and Chukchi seas. It comprises a considerable part of the radiolarian plankton of the Norwegian Sea and inhabits the deep water, 200–750 m. According to Bernstein, it does not live in the Barents Sea. Although it is the most frequently observed species in the bottom sediments of the central Arctic and was recorded in most samples from the Chukchi Sea, only single individuals of this species were found in the Barents Sea. This taxon is observed in the deep waters north and west of Norway, south and west of Spitzbergen, in the Chukchi Sea, in the high Arctic at about 84°N. In bottom sediments the species is found in the Northern Atlantic, Northern Pacific, Norwegian, Barents, Chukchi, and Bering seas and in the Sea of Okhotsk. In some of the regions of the Sea of Japan it is the dominant species.

Spongodiscidae Haeckel 1862 sensu Kozlova in Petrushevskaya and Kozlova, 1972

*Spongotrochus glacialis* Popofsky (Plate IV, photographs 1–2)
*Trochodiscus echiniscus:* Haeckel, 1887; 418; Cleve, 1899: 34, 4, fig. 4; Schröder, 1909; Bogorov, 1946
*Spongotrochus arachnius* (Haeckel); Popofsky, 1908: 227–228, pl. 26, fig. 5–7;
*Spongotrochus glacialis* Popofsky, 1908; 228–229, pl. 28, fig. 2; Riedel, 1958; Kruglikova, 1969; Petrushevskaya and Björklund, 1974

This species inhabits the Antarctica Ocean, the Northern and Southern Atlantic, and the Northern Pacific oceans. In the Arctic and Antarctic waters, the optimum depth for this taxon is less than 400 m; below this level only single individuals were found. In Arctic waters it is one of the abundant species; in bottom sediments its relative abundance is much less; in the Chukchi Sea most samples yielded single individuals of this taxon.

Order Nassellaria Ehrenberg, 1875
Fam. Plagiacanthidae Hertwig, 1879

*Plectacantha oikiscos* Jörgensen (Plate II, Photograph 1)
*Plectacantha oikiscos* Jörgensen, 1905; pl. 12, fig. 50–57; Schröder, 1929; Bernstein, 1932, 1934; Hülsemann, 1963; Tibbs, 1967; 1971; *Periplecta intricata* (Cleve) Jörgensen, 1899; *Peridium intricatum* Cleve, 1898

This species is found in the northern part of the Barents Sea and in the Kara Sea, west of Norway, in the Greenland Sea and in the high Arctic. According to Hülsemann, it is abundant only in the northern cold and temperate waters; it probably inhabits relatively deep water. This taxon is abundant in the bottom sediments of the Chukchi and Barents seas.

*Phormacantha hystrix* (Jörgensen) (Plate II, photographs 2–3)
*Peridium hystrix* Jörgensen, 1899: 76
*Phormacantha hystrix* Jörgensen, 1905: 132, pl. 14, fig. 59–63; Schröder, 1929; Bernstein, 1934; Hülsemann, 1963; Tibbs, 1967; Petrushevskaya, 1971

This species is found in the Norwegian, Greenland, northern Kara, and Chukchi seas, as well as in the Arctic Basin, and it lives at depths of 100–200 m. Atypical forms of this species were observed in the tropics at depths of 100–500 m (Petrushevskaya, personal communication). It was recovered from the bottom sediments of the Chukchi and Barents seas as well as from the Antarctic Ocean.

Fam. Lampromitridae Haeckel, 1881

*Botryopera boreale* (Ehrenberg) (Plate II, photographs 15–16)
*Lithobotrys borealis* Ehrenberg, 1872: 28: 1872, taf. 2, Fig. 3; Haeckel, 1887; Petrushevskaya, 1967; 1971; 1983; Kruglikova, 1969

This wide-ranging species was recorded from the Northern Atlantic off the coasts of Greenland, the Norwegian Sea, the Indian and the North Pacific, as well as the Antarctic oceans. It was observed in the sediments of the Chukchi and Barents seas. It is one of the dominant species in the thanatocenosis of the Barents Sea.

*Botryopera setosa* (Jörgensen) (Plate II, photographs 17–23)
*Botryopyle setosa* Jörgensen, 1900: 81; Schröder, 1929; Petrushevskaya, 1967; 1971; 1983

It is one of the frequently recorded but not abundant species living in the waters of the Norwegian Sea, North Atlantic and Antarctic oceans. It is generally observed in deep water and less frequently in surface water layers. This species was recovered from the sediments of the Barents and Chukchi seas, as well as from the Antarctic Ocean. It is one of the dominant taxa in the thanatocenosis of the Barents Sea.

Botryocyrtidae Petrushevskaya, 1983

*Amphimelissa setosa (Cleve)* (Plate III, photographs 1–25)
*Botryopyle setosa* Cleve, 1899: 27, pl. 1, fig. 10; Jörgensen, 1905; Bernstein, 1931, 1934; Bogorov, 1946; Hülsemann, 1963; Tibbs, 1967; Petrushevskaya, 1967; Kruglikova, 1969; Björklund, 1976

This species is distributed in the Norwegian, Greenland, Barents, and Chukchi seas as well as in the Arctic Basin. It is the dominant taxon in the sediments of the Barents and Chukchi seas; individual specimens are found in the sediments of the Sea of Okhotsk and in the Bering Sea.

## ACKNOWLEDGMENTS

The author wishes to express her gratitude to N. A. Belov, L. I. Moskalev, Yu. A. Pavlidis, and I. Sakharova for access to the relevant material, to V. S. Vishnevskaya for helping to identify ancient Radiolaria, to K. N. Nesis and M. S. Barash for discussions.

# APPENDIX

## Plate 19-I

1-9: *Echinomma leptodermum* Jörg.; 10–12: *Cromyechinus borealis* Jorgensen; 13–14: *Hexacontium setosa* Haeckel. (1, 2, 6, 9–10: SP-15; 3-5, 7, 8, II, 13–14: SP-22; 12: Pacific boreal zone, St. 4110, 55-56 cm, s/v *Vitjaz*). All figures × 503 (scale ''a'' 0, 05 mm); fig. 1, Plate IV × 107 (scale ''b2'' 0, 1 mm).

## Plate 19-II

1, *Plectantha oikiscos Jörgensen;* 2, 3, *Phormacantha hystrix* (Jörgensen); 4, Clathromitra? sp.; 5, Clathromitra sp.; 6, Cladoscenium sp.; 7, *Mitrocalpis aranaefera* Popofsky; 8, Lophophaena sp.; 9, *P. platycephalus* Haeckel; 11, 12, 13, *Sethoconus tabulatus* Cleve; 14, *Diplocyclas davisiana* (Ehr.); 15, 16, *Botryopera boreale* (Ehr.); 17, 18, 19, 20, 21, 22, 23, *B. setosa* (Jörgensen) (I-2. 4-6, 9, 14, SP-22; 10, SP 15; II-23, St. 1275, 0–3 cm; 7, Pacific boreal zone, St. 4110, 55-56 cm, s/v *Vitjaz*).

## Plate 19-III

1-25, *Amphimelissa setosa* (Cleve), 1-10, 12-13, St. 1275, 0-3 cm; II, 14-20, SP-22; 21-25, Pacific boreal zone, St. 4104, 96-98 cm; St. 4110, 55-56 cm, s/v *Vitjaz.*

## Plate 19-IV

1-2, *Spongotrochus glacialis* Popofsky (1, Pacific boreal zone); 3, *Streblacantha circumtexta* Jörgensen; 5-6, *Lithelius nautiloides?* Popofsky (? *Spirema haliomma* [Ehr.]); 4, 8–10, Spongodiscus spp.; 7, Euchitoniidae gen. sp.; 11, Lithelius sp.; 12, *Larnacantha polyacantha* Camp. et Clark; 13, Eucyrtidium sp. (I, St. 4110, 0-3 cm, Pacific boreal zone; 2, SP-22; S-6, St. 1275, 0-3 cm; 7-13, St. 1275, 30 cm).

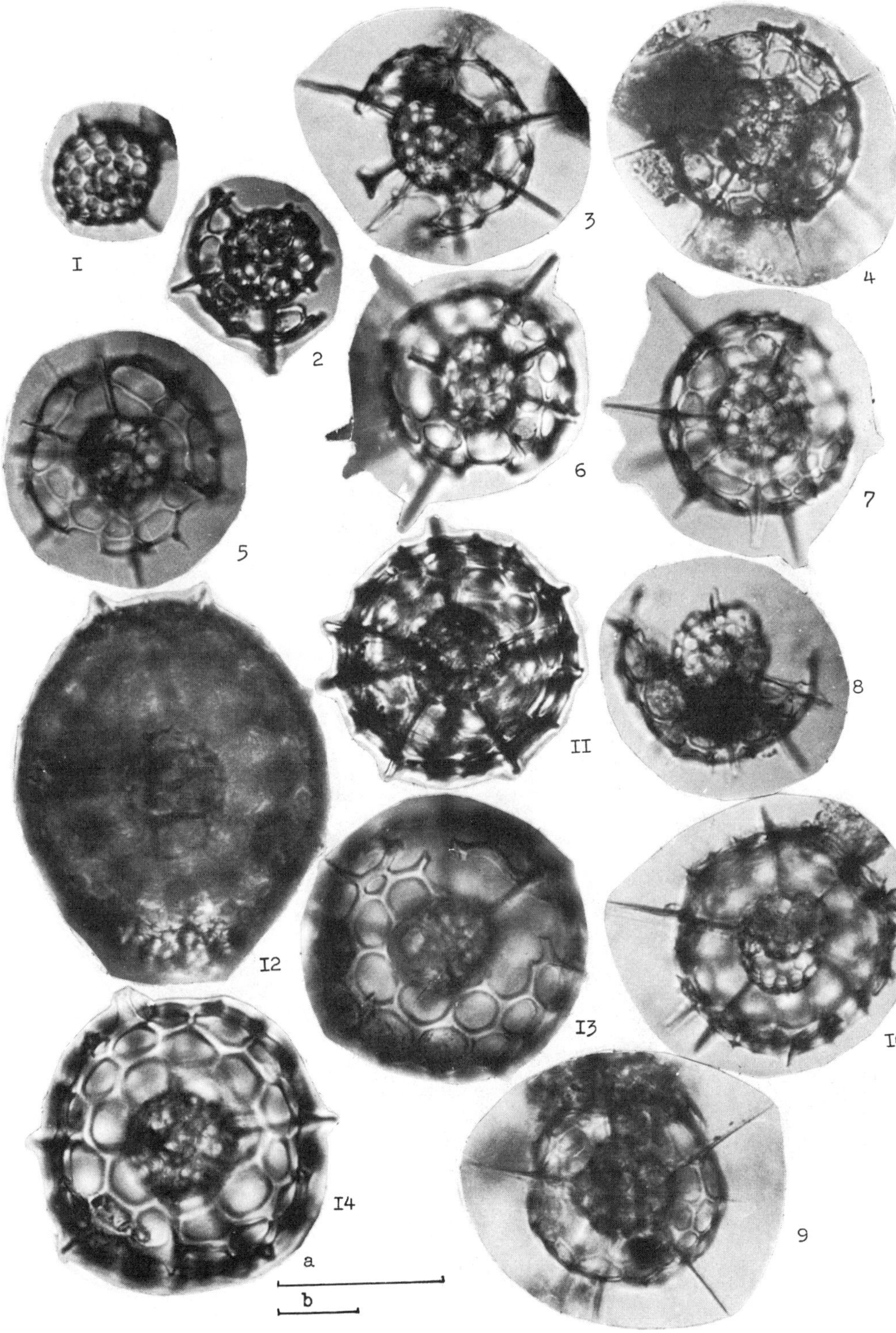


PLATE 19-1

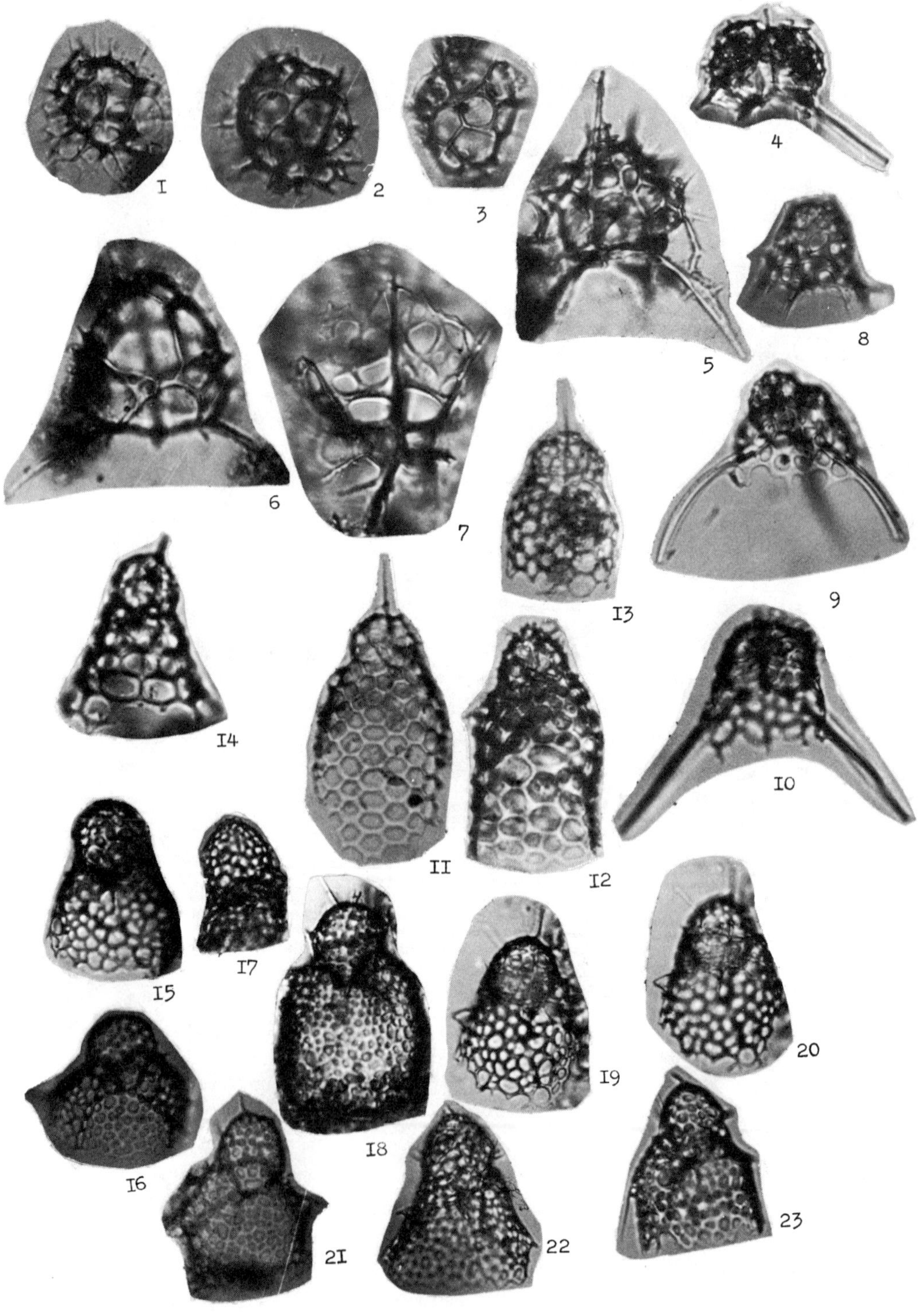


PLATE 19-II

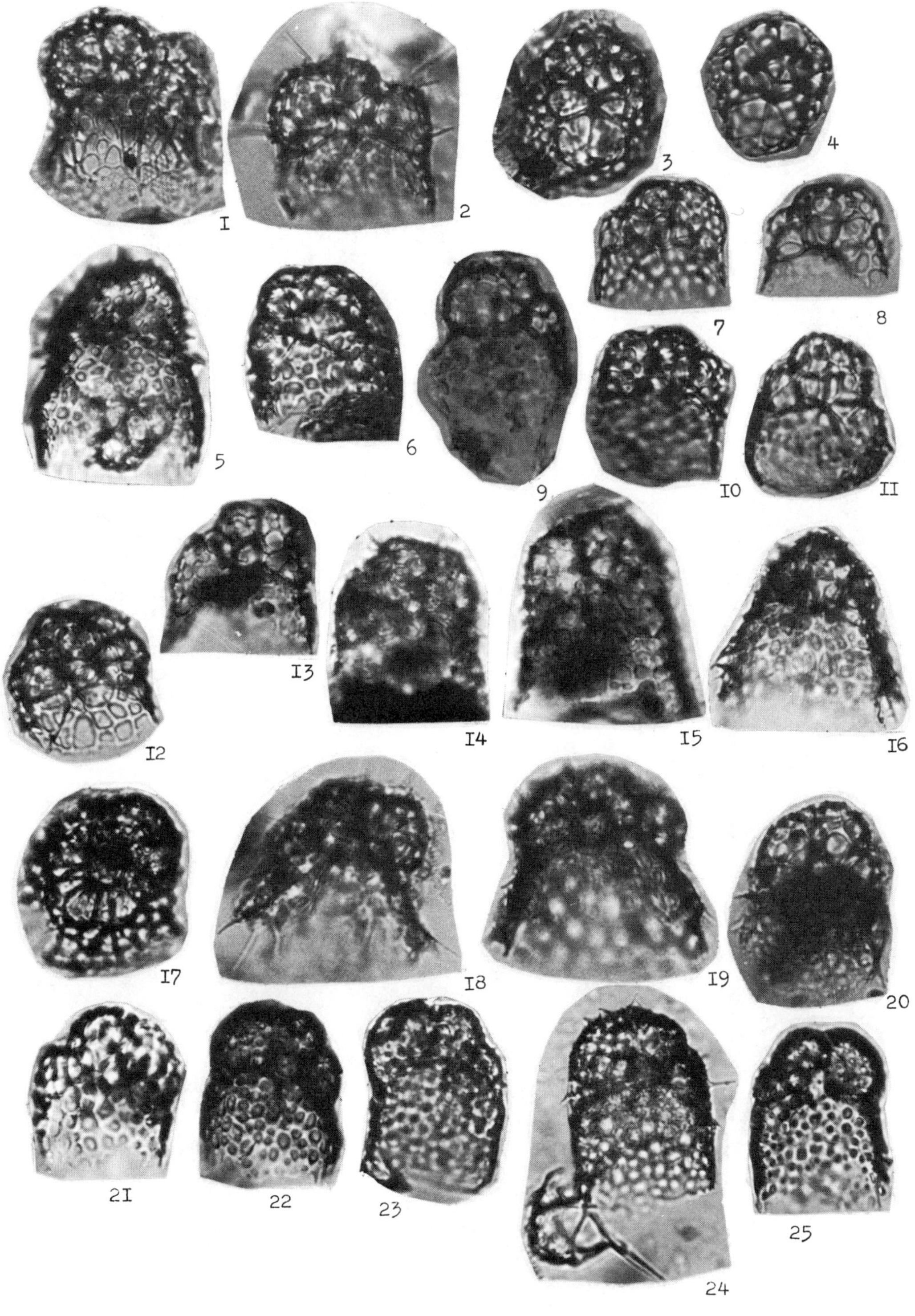


PLATE 19-III

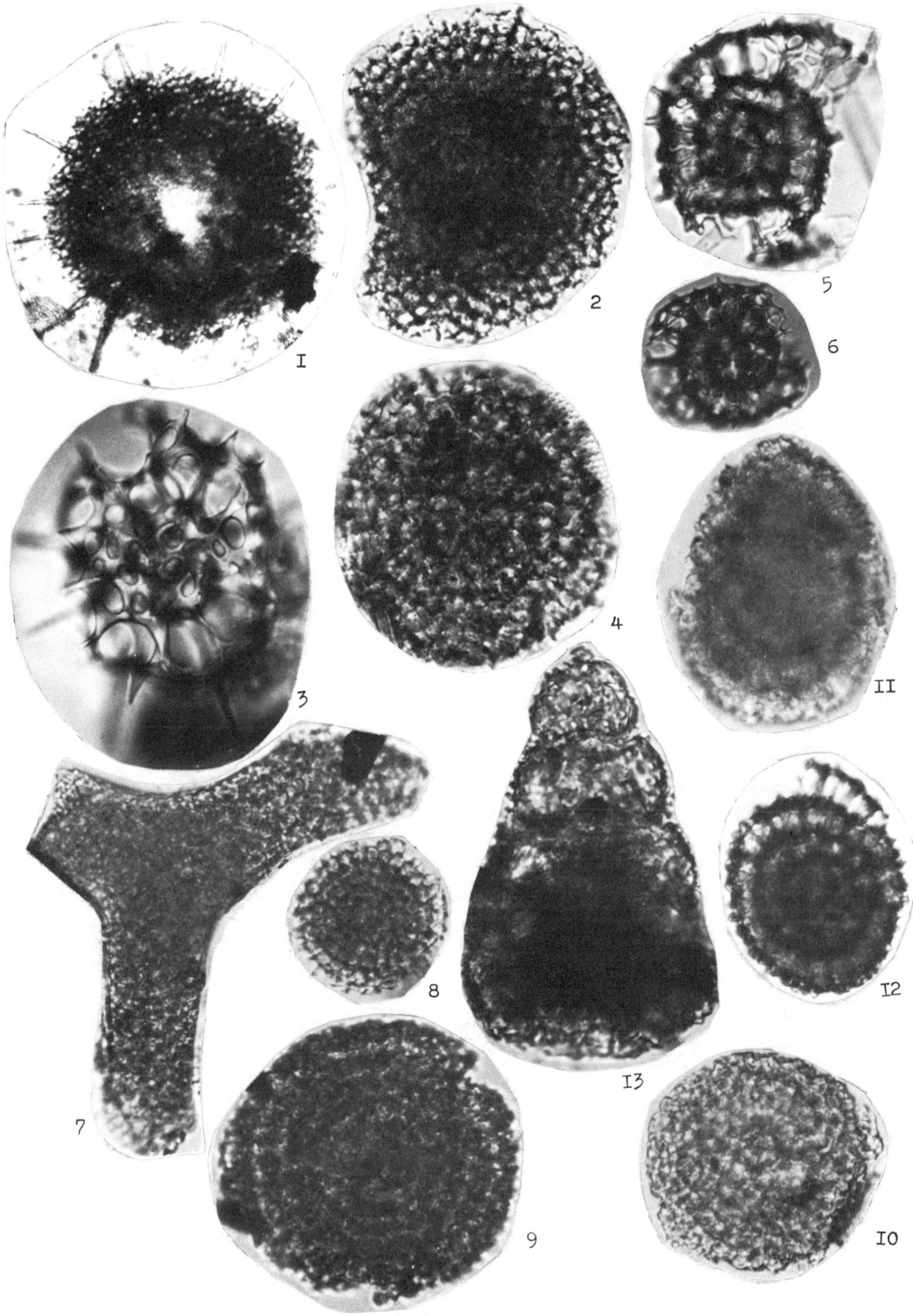


PLATE 19-IV

## REFERENCES


Anderson, R. 1983. *Radiolaria.* Springer-Verlag, Berlin, New York, Heidelberg, 355p.

*Atlas Okeanov.* 1980. Severnyi Ledovityi, GUNIO MO SSSR.

Bailey, J. W. 1856. Notice on microscopic forms found in the soundings of the Sea of Kamchatka, *Am. J. Sci. Arts* **22**(ser. 2):1–6.

Bernstein, T. N. 1931. Plankticheskie prosteishie severo-zapadnoi chasti Karskogo morya. *Tr. Arkt. Inst.* **3**(1):1–24. (In Russian.)

Bernstein, T. N. 1932. Ueber einige arktische Radiolarien. *Arch. f. Protistenk* **76:**217–227.

Bernstein, T. N. 1934. Zooplankton Karskogo morya po materialam ekspeditsii Arkticheskogo instituta na *Sedove* 1930 goda i *Lomonsove* 1931 goda. *Tr. Arkt. Inst.* **9:**3–58. (In Russian.)

Björklund, K. R. 1976. Radiolaria from the Norwegian Sea, Leg 38 of Deep Sea Drilling Project. *Init. Rpts. DSDP, Wash.* **38:**1,001–1,068.

Björklund, K. R., J. Thiede, and H. Holtedahl. 1979. Deglaciation of the southeastern Norwegian Sea towards the end of the last glacial age. *Boreas* **8:**133–135.

Bogorov, V. G. 1946. Zooplankton po sboram ekspeditsii na l/p *Sedov* 1937–1939 gg., Tr. Dreifuiuschei eksped. Glavsevmorputi na ledokolnom parokhode *G. Sedov* 1937–1940 gg. M. *Izd. Gravsevmorputi* **3:**336–370. (In Russian.)

Borgert, A. 1901. Die nordischen Tripyleen-Arten. *Nordisches Plankton* pp. 1–52.

Brady, H. B. 1878. On the Reticularien and Radiolarien (Foraminifera and Polycystina) of the Northern-Polar Expedition of 1875–76. *Ann. Nat. Hist* **1**(ser. 5):425–440.

Campbell, A., and B. Clark. 1944. Miocene radiolaria faunas from southern California. *Geol. Soc. Am. Spec. Paper 51,* pp. 1–76.

Cleve, P. T. 1899. Plankton collected by Swedish expedition to Spitsbergen in 1893. *Handl. Kgl. Svenska Vetensk Akad.* **32**(3):1–51.

Cleve, P. T. 1900. Notes on some Atlantic plankton-organisms. *Handl. Kgl. Svenska Vetensk. Akad.* **34**(1):1–22.

Ehrenberg, C. G. 1838. Ueber die Bildung der Kreidefelsen und des Kreidemergels durch unsichtbare Organismen. *Abh. Kgl. Preuss. Akad. Wiss. Berlin* pp. 1–154.

Ehrenberg, C. G. 1854. *Mikrogeologie.* Leipzig, 374p.

Ehrenberg, C. G. 1872. Mikrogeologische Studien über das kleinsteleben der Meeres-Tiefgründe aller Zonen und dessen geologischen Einfluss. *Abh. Kgl. Preuss. Akad. Wiss. Berlin* pp. 132–392.

Ehrenberg, C. G. 1875. Fortsetzung der mikrogeologischen studien als gesamtubersicht der mikroskopischen paleontologie gleichartig analysirter gebrigsarten der erde mit specieller rücksicht auf den polycystinen mergel von Barbados. *Abh. Kgl. Preuss. Akad. Wiss. Berlin,* pp. 1–223.

Foreman, H. P. 1973. Radiolaria of Leg. 10. *Init. Rpts. DSDP. Wash.* **10:**407–474.

Foreman, H. 1975. Radiolaria from the North Pacific, Deep Sea Drilling Project, Leg 32. *Init. Rpts. DSDP Wash.* **32:**579–676.

Haeckel, E. 1862. *Die Radiolarien* (Rhizopoda radiata). Berlin, 572p.

Haeckel, E. 1887. Report on the Radiolaria collected by the H.M.S. *Challenger* during the years 1873–1876. *Zoology* **18:**1,803p.

Holland, A. M., and Enjumet. 1960. Cytologie, evolution et systematique des Sphaeroides (Radiolaires). *Arch. Mus. Natl. Hist. Nat.* **1960**(7):1–134.

Hülsemann, K. 1963. Radiolaria in plankton from the Arctic Drifting Station T-3, including the description of three new species, *Arct. Inst. N. Am. Tech. Pap. Arct.* **13:**1–52.

Jörgensen, E. 1900. Protophiten und protozoen in plankton aus der norwegischen West Kuste. *Bergens Mus. Aarbog, for 1899* **6:**1–112.

Jörgensen, E. 1905. The Protist plankton and diatoms in bottom samples. *Bergens Mus. Skr.,* pp. 49–151, 195–225.

Kruglikova, S. B. 1969. Radioljarii v kolonke stantsii 4066 (severnaia chast Tikhogo okeana). In *Osnovnye problemy mikfropaleontologii i organogennogo osadkonakopleniia v okeanakh i moriakh.* Nauka, Moscow, pp. 116–126. (In Russian.)

Kruglikova, S. B. 1976. Radiolarii v verkhnepleistozenovjkh osadkakh borealnoi i severnoi subtropicheskoi zon Tikhogo okeana. *Okeanologiia* **16**(1):113–117. (In Russian.)

Kruglikova, S. B. 1977. Osobennosti raspredeleniia radioliarii v otlozheniiah borealnoi i subtropicheskoi zon Tikhogo okeana v pleistozene. *Okeanologiia* **17**(6):1,055–1,063. (In Russian.)

Kruglikova, S. B. 1979. Harakternye cherty raspredeleniia radioliarii (vysokikh taksonov) v otlozheniiakh Tikhogo okeana i okrainnyh morei. In *Iskopaemye i sovremennye radioljarii.* pp. 42–52. (In Russian.)

Kruglikova, S. B. 1981. Nekotorye cherty ekologii i rasprostraneniia sovremennykh i kainozoiskikh radioljarii. In *Sistematika, evolutsiia i stratigraficheskoe znachenie radioljarii.* Nauka, Moscow, pp. 118–139. (In Russian.)

Kruglikova, S. B. 1982. Radioljarii iz sovremennykh donnykh otlozhenii Arkticheskogo basseina po materialam SP-15 i SP-22. *Tez. dokl.*

*na 5-oi Vsesoyuznoi shkole po morsk. geol., Gelendzhik* 48. (In Russian.)

Kruglikova, S. B. 1984. Paleoekologicheskie rekonstrukzii na osnovanii izucheniia radioljarii. In *Morfologiia, Ekologiia i Evolutsiia Radioljarii.* Nauka, Leningrad, pp. 41–53. (In Russian.)

Kruglikova, S. B. 1985. Rasprostranenie radioljarii v pleistozenovikh otlozhenijakh Arktiki. Tes. dokl. na Vsesojuznom sovetschanii "Problemi chetvertichnoi paleoecologii i paleogeographii Barentzeva i Belogo morei." *Akad Nauk SSSR, Murmansk,* pp. 72–74. (In Russian.)

Lisitsyn, A. P. 1978. *Protsessy Okeanskoi Sedimentatsii.* Nauka, Moscow, 266p. (In Russian.)

Nazarov, B. B. 1975. Radioljarii nizhnego-srednego paleozoia Kazakhstana. *Tr. GIN AN SSSR* **275**:202p. (In Russian.)

Nigrini, C. A. 1967. Radiolaria in pelagic sediments from the Indian and Atlantic oceans. *Bull. Scripps Inst. Oceanogr.* **2**:1–125.

Petrushevskaya, M. G. 1967. Radioljarii otrjadov Spumellaria i Nassellaria Antarkticheskoi oblasti (po materialam Sovetskoi Antarkticheskoi ekspeditsii). In *Issled. Fauny Morei.* **4**(12). (*Rezultaty Biol. Issled. Sov. Antarkt.* Ekspeditsii (1955–1958 **3**:5–186.) (In Russian.)

Petrushevskaya, M. G. 1969. Raspredelenie skeletov radioljarii v osadkakh severnoi Atlantiki. In *Iskopaemye i sovremennye radioljarii.* Lvov, p. 127. (In Russian.)

Petrushevskaya, M. G. 1971. Radioljarii Nassellaria v planktone Mirovogo okeana. *Issled. Fauny Morei* **9**(17):3–294. (In Russian.)

Petrushevskaya, M. G. 1979. Paleobiogeografiia Norvezhsko-Grenlandskogo basseina. In *Istoriia Mikroplanktona Norvezhskogo Moria po Materialam Glubokovodnogo Bureniia.* Nauka, Leningrad. (*Issled. Fauny Morei* **23**(31).) (In Russian.)

Petrushevskaya, M. G. 1981. *Radiolarii Otr. Nassellaria Mirovogo Okeana.* Leningrad, 405p. (In Russian.)

Petrushevskaya, M. G. 1984. O klassifikatsii radioljarii Polycystina. In *Morfologiia, Ekologiia i Evolutsiia Radioljarii.* Materialy IV Simpoziuma Evropeiskikh radioljristov Eurorad. Nauka, Leningrad, pp. 124–148. (In Russian.)

Petrushevskaya, M. G., and K. R. Björklund. 1974. Radiolarinsa in holocene sediments of the Norwegian-Greenland seas. *Sarsia* **57**:33–46.

Petrushevskaya, M. G., and G. E. Kozlova. 1972. Radiolaria from Leg 14. *Init. Rpts. DSDP Wash.* **14**:495–648.

Popofsky, A. 1908. Die Radiolarian der Antarktic. *Dtsch. Südpolar-Exped.* **10**(3):183–305.

Popofsky, A. 1912. Die Sphaerellarien des Warmwassergebietes. *Dtsch. Südpolar-Exped. 1901–1903* **13**(2):73–159.

Popofsky, A. 1913. Die Nassellarien des Warmwassergebietes. *Dtsch. Südpolar-Exped. 1901–1903* **14**:217–416.

Riedel, W. R. 1958. Radiolaria in Antarctic sediments. *BAWZARE Rpts.* **6**(ser. B):217–255.

Riedel, W. R. 1967. Subclass Radiolaria. In W. B. Harland et al. (eds.), *Fossil Record.* Geological Society of London, London, pp. 291–298.

Riedel, W. R., and A. Sanfilippo. 1970. Radiolaria, Leg. 4, Deep Sea Drilling Project. In *Init. Rpts. DSDP Wash.* **6**:503–575.

Riedel, W. R., and A. Sanfilippo. 1971. Cenozoic Radiolaria from the western tropical Pacific, Leg. 7. In B. L. Winterer et al. (eds.), *Init. Rpts. DSDP Wash.* **7**:1,529–1,672.

Riedel, W. R., and A. Sanfilippo. 1977. Cainozoic Radiolaria. In A. T. S. Ramsay (ed.), *Oceanic Micropaleontology,* vol. 2. Academic Press, London, pp. 847–912.

Schröder, O. 1909. Die Nordischen Spumellarien, II: Unterlegion Sphaerellarien. In *Nordisches Plankton,* vol. 7. Kiel-Leipzig, pp. 1–63.

Schröder, O. 1911. Die Nordischen Nassellarien. In *Nordisches Plankton,* vol. 7. Kiel-Leipzig, pp. 67–140.

Schröder, O. 1929. Die Nordischen Spumellarien, I: Unterlegion Sphaerecellida. In *Nordisches Plankton,* vol. 7. Kiel-Leipzig, pp. 91–190.

Stepanov, V. N. 1983. *Okeanosfera.* Mysl, Moscow, 270p. (In Russian.)

Tibbs, J. F. 1967. On some planktonic protozoa taken from the track of Drift Station ARLIS I, 1960–1961. *Arct. Inst. N. Am.* **20**(4):247–254.

# 20: Diatoms in Arctic Shallow Seas Sediments

E. I. Polyakova

## INTRODUCTION

The research on Arctic Ocean diatoms has more than a century-long history and yet the body of information on this group is rather scarce. The data pertain mostly to species composition and seasonal development of phytoplankton (Cleve, 1873; Gran, 1904; Usachev, 1935, 1946*a,b;* Kisselev, 1935, 1937; Bursa, 1963; Belyaeva, 1980; Hsiao, 1983), or sea ice flora of diatoms (Shirshov, 1936, 1937; Usachov, 1938; Meguro, Ito, and Fukushima, 1966, 1967; Horner and Alexander, 1972; Horner, 1976). Information on diatoms in Arctic Sea sediments is incomplete and generally inventorial, as only the species composition of the benthic flora is described without reference to their habitat. No data are available so far on thanatocenoses of diatoms in the sediments of the central abyssal parts of the Arctic Ocean, or in those of some shelf seas. In fact, it is only recently that species composition and the distributional patterns of their thanatocenoses in the Arctic have commenced. The data obtained indicate that the employment of this class of algae is very promising for solving questions of Cenozoic stratigraphy as well as the paleogeographic evolution of the Arctic region.

*Note:* This chapter has been reviewed, edited, and rewritten by the Editor, Yvonne Herman.

## METHODS

Surface samples were collected with 0.25 $m^2$ ground buckets and cores were obtained with a uniflow tube. The diatoms were studied in all cores at 5–10 cm intervals and the species composition was determined in all samples. Counts were made and abundances were estimated per sample (abundance of valves per 1 gram of dry residue) as follows: (1) 1–10 valves; (2) 11–100 valves; (3) 101–500 valves; (4) 501–1,000 valves. The diatoms were concentrated by treatment with HCl and $H_2O_2$ followed by heavy liquid (KdJ + KJ) separation. Subsequently they were placed into a liquid with a high index of refraction (1,68). The valves were examined under a light microscope at × 1,000. The distributional patterns of diatoms in cores, their abundance, ecology, and biogeographic distribution are shown in Figures 20-1, 20-2, and 20-3 (see pages 484–489).

## BARENTS SEA

The benthic diatoms of the Barents Sea were first reported by Cleve (1873, 1883) and Grunow (Cleve and Grunow, 1880). They described nearly 450 species but, unfortunately, without references to their habits. Korotkevich (1960) discovered 476 species and varieties of diatoms in littoral and upper sublittoral sediments. Half are marine and brackish-water, and the remainder are indifferent* freshwater diatoms and halophiles.† The most common marine diatoms of the littoral zone are the planktonic and benthic species: *Melosira moniliformis* (O. Mull.) Ag., *Odontella aurita* (Lyngb.) Ag., *Rhabdonema arcuatum* (Lyngb.) Kutz., *R. minutum* Kutz. *Licmophora paradoxa* (Lyngb.) Ag., *Synedra investiens* W. Sm., *Navicula cancellata* Donk, *N. directa* W. Sm., *N. forcipata* var. *densestriata* A. S., *Trachyneis aspera* (Ehr.) Cl., and *Pleurosigma elongatum* W. Sm. The littoral sediments examined indicate that certain associations characterize well-defined environmental conditions and certain substrate, among which epiphytic and benthic forms are predominant. As compared to the species lists of early investigators, 786 species of benthic diatoms are known to live in the Barents Sea; among them 481 species are euhalabous and mesohalabous (Proshkina and Lavrenko, 1974).

## WHITE SEA

The benthic diatoms of the White Sea have been studied thoroughly. They were collected mostly in the wide littoral zone of the sea. The floral lists include their classification and sample location; however, no data on their habitat or on the hydrological conditions of the sampled area were presented. As the data were not published, the lists should be regarded as purely inventorial. The diatoms of the White Sea zone were first briefly reported by Merezhkovsky (1878). Later, Reinhard (1882) supplemented the list with 25 additional species discovered in Merezhkovsky's material and thus his list contained 63 species including marine, brackish-water, and freshwater taxa. Among the marine species there are Arctic and wide-ranging forms. The most complete data were presented by Meyer (1938) and his list includes 209 species. Petrov (1967) found 93 species in the upper, middle, and lower horizons of the littoral zone and indicated the substrate from which the species were collected. This list is supplementary to that of Meyer (1938). Zabelina (1939) studied diatoms in the surface layer of sediment in Kandalaksha Bay. She discovered 290 species, among which 119 are euhalabous, 38 are mesohalabous, the remainder are freshwater taxa. They included both planktonic and benthic species belonging to boreal–Arctic and boreal groups. The dominant species are *Paralia sulcata* (Ehr.) Cl., *Hyalodiscus scoticus* (Ktz.) Grun., *Rhabdonema arcuatum, R. minutum, Synedra tabulata* (Ag.) Ktz., *S. kamtschatica* Grun., *Cocconeis scutellum* Ehr., and *Trachyneis aspera.* This list did not include data on their habitat.

Studies of thanatocenoses of diatoms in the littoral zone of Onega Bay by this author established the existence of a peculiar and ecologically inhomogenous flora of diatoms including freshwater, brackish-water, brackish-marine, and truly marine taxa. The thanatocenoses are mainly composed of freshwater halophilous species of the genera Achnanthes Bory and Fragilaria Lyngb. There are several solitary marine neritic species characteristic of the White Sea plankton as well as some sublittoral

*Can tolerate low salinities but their growth is stunted.

†Halophiles are freshwater taxa that thrive when salinities increase slightly.

epiphytic and benthic forms: *Thalassiosira angustelineata* (A. Schmidt) Fryx. and Hasle, *T. eccentrica* (Ehr.) Cl., *T. lacustris* var. *septentrionalis* (Grun.) Makar., *Paralia sulcata, Cocconeis scutellum, Rhabdonema arcuatum,* and *Amphora terroris* Ehr. The content of freshwater lacustrine–alluvial diatoms in the thanatocenoses was found to decrease as the distance from the river mouths increased.

Dzhinoridze (1968, 1971), Nevesskiy, Medvedev, and Kalinenko (1977), and Vishnevskaya, Dzhinoridze, and Kuptsova (1974) studied in detail the thanatocenoses of diatoms in subfossil and Holocene sediments of the White Sea. They established the main characteristics of thanatocenoses in the sublittoral zone and reconstructed the evolution of paleogeographic conditions of the White Sea during the Holocene by using the associations of diatoms. They identified 322 species and varieties of diatoms in the sublittoral surface sediments of the White Sea. High proportions are typical for the sublittoral taxa *Paralia sulcata, Thalassiosira lacustris* var. *septentrionalis, Navicula distans* W. Sm., *Grammatophora angulosa* Ehr., *G. arctica* Cl., and neritic and panthalassic species *Melosira arctica* (Ehr.) Dickie, *Thalassiosira hyalina* (Grun.) Gran, *T. anguste-lineata, Thalassionema nitzschioides* Grun., *Coscinodiscus radiatus* Ehr., and *C. asteromphalus* Ehr. In the bay sediments the predominant species are *Thalassionema nitzschioides, Paralia sulcata, Cocconeis scutellum,* and *Rhabdonema arcuatum.* In the samples studied the content of diatoms in a gram of dry residue from the surface of the sublittoral bed ranges from 0.1 to 26 million valves. Sediments barren of diatoms occur in the sands from the vicinity of the Neck where the currents are known to have high velocities. The concentrations of diatoms, never exceeding 1 million to 5 million valves per gram, were found in sandy sediments of the shallow zones of Onega and Dvina bays. The maximum content of diatoms, 20–26 million valves, was found in the silty deposits of Kandalaksha Bay, which contain up to 6.2 percent amorphous $SiO_2$.

Dzhinoridze (1971) examined 26 cores up to 5 m long raised in different sectors of the White Sea. One 4.5-m-long core from the northern part of the White Sea was studied by Zabelina (1969). Her results are comparable with those obtained by Dzhinoridze, who distinguished several stages in the evolution of Late Pleistocene–Holocene diatom flora of the White Sea. The oldest, late glacial assemblage was found at the bottom of cores taken from Onega Bay. The assemblage is glacial lacustrine, composed of solitary freshwater cryophilic species, such as *Pinnularia lata* (Breb.) W. Sm., *P. borealis* Ehr., *Aulacosira islandica* (O. Mull.) Simons., *A. distans* (Ehr.) Simons., *Eunotia praerupta* Ehr., *Stephanodiscus dubius* (Ericke) Hust., and *Cyclotella ocellata* Pant. The time when seawater penetrated the White Sea region, in early Dryas (10,000–12,800 B.P.), is marked by the presence of marine taxa associated with both freshwater and brackish-water diatoms such as *Aulacosira islandica, Opephora martyi* Herib., *Thalassiosira lacustris* var. *septentrionalis, T. baltica* (Grun.) Ostf., and *Chaetoceros holsaticus* Schutt.

The mixed assemblages of boreal–Arctic and northern boreal diatoms of Preboreal time is characterized by the predominance of *Hyalodiscus scoticus, Rhabdonema arcuatum, Thalassiosira lacustris* var. *septentrionalis,* and *T. gravida* Cl. Boreal diatom assemblages are characterized by the abundance and widespread occurrence of the boreal–Arctic neritic species *Chaetoceros mitra* (Bail.). During Atlantic time (~5–18.5 kya) we find in White Sea sediments a large number of relatively warm-water planktonic, neritic, and panthalassic species, such as *Rhizosolenia styliformis* Bright., *Coscinodiscus asteromphalus, C. radiatus, C. perforatus* Ehr., and *Thalassiosira eccentrica;* in coastal areas there are sublittoral species such as *Grammatophora oceanica* (Ehr.) Grun., *Actinoptychus undulatus* (Bail.) Ralfs, and *Rhaphoneis surirella* (Ehr.) Grun.

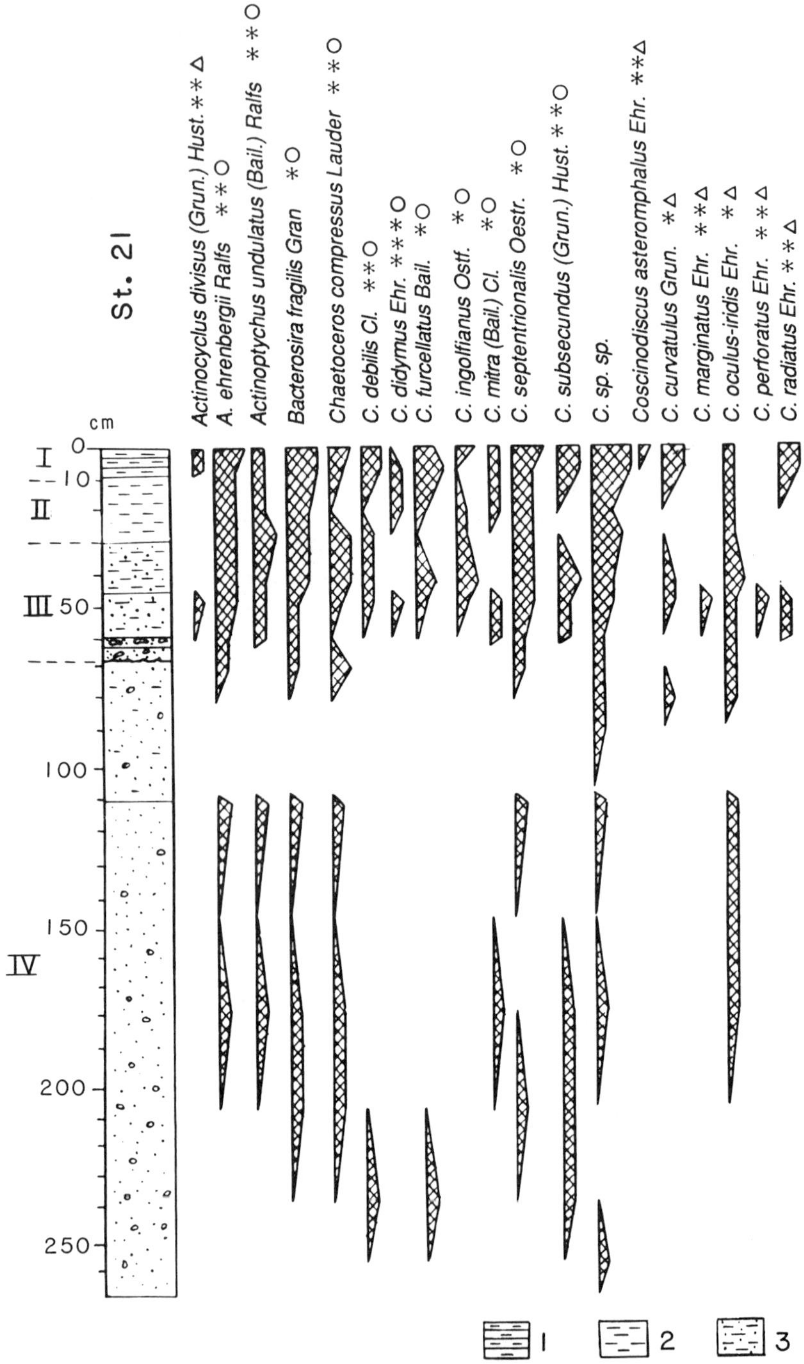


**Figure 20-1.** Diatoms in Late Pleistocene–Holocene sediments of the Chukchi Sea, Core 21. (1) silty clay; (2) silt; (3) sandy silt; (4) silty sand; (5) sand with gravel and pebbles; (6) marine diatoms; (7) freshwater diatoms; (8) silicoflagellates; (9) abundance of diatoms per sample: (a) 1–10 valves per sample, (b) 11–100, (c) 101–500, (d) 501–1,000, (e) more than 1,000; (10) stratigraphic horizons.

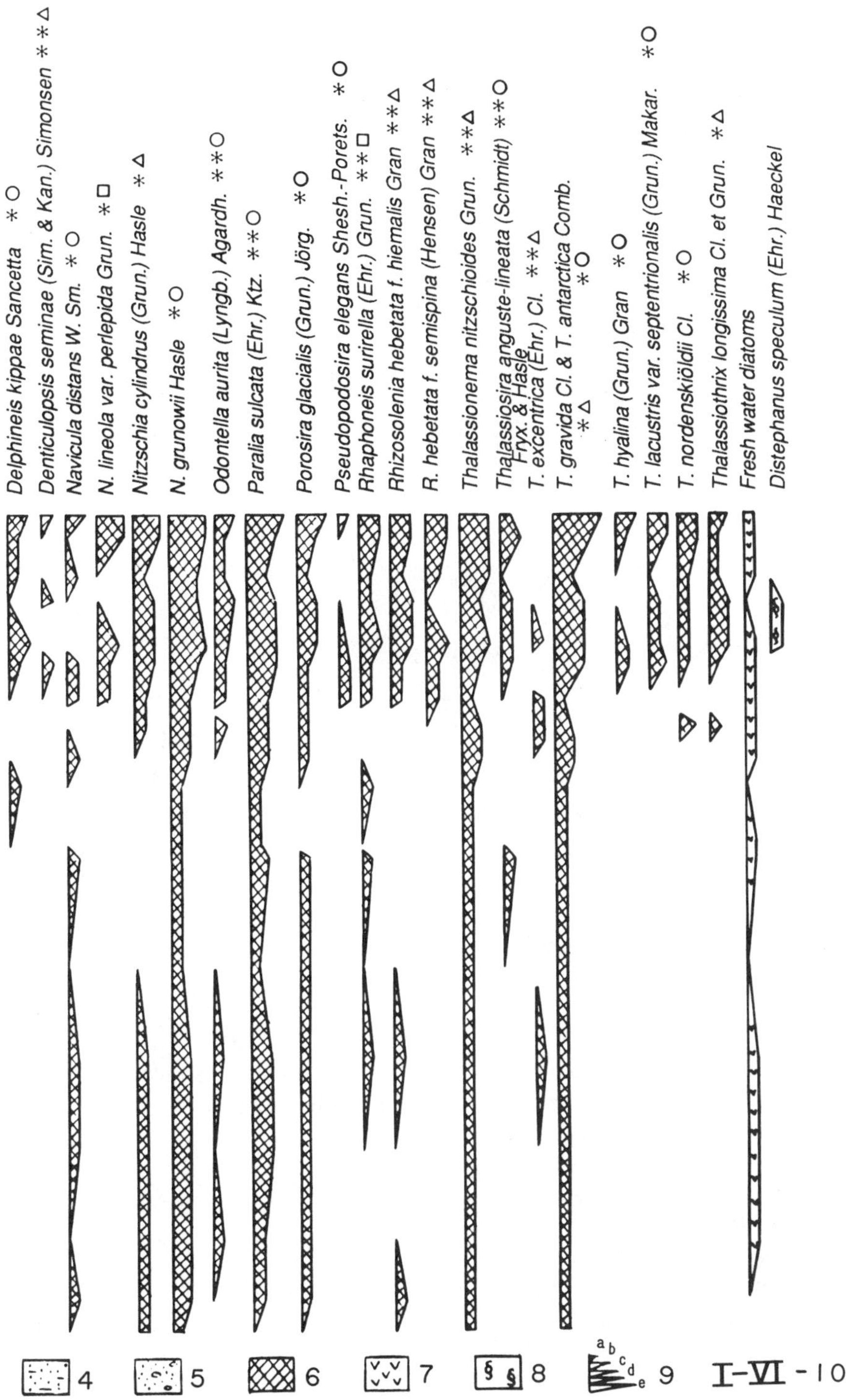


* boreal–Arctic and bipolar species
** boreal–Arctic-tropical species (or widespread species) and cosmopolitan species

□ sublittoral species
○ neritic species
△ panthalassic species

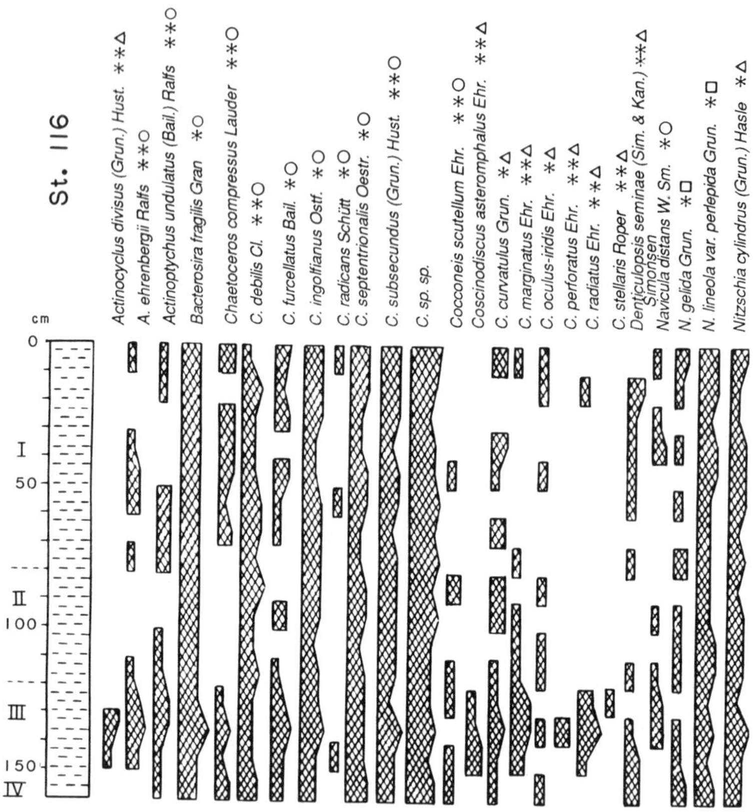


**Figure 20-2.** Diatoms in Holocene sediments of the Chukchi Sea, Core 116.

* boreal–Arctic and bipolar species
** boreal–Arctic-tropical species (or widespread species) and cosmopolitan species

Subboreal (2,700–5,000 B.P.) assemblages include boreal–Arctic and boreal species with the latter taxa dominating. In coastal regions the assemblages are chiefly composed of sublittoral species such as *Paralia sulcata, Rhabdonema arcuatum, Navicula distans, Isthmia nervosa* Kutz., and *Rhaphoneis surirella.* In the open part of the White Sea the assemblages include neritic and panthalassic species *Thalassionema nitzschioides, Thalassiosira eccentrica, T. gravida, T. hyalina* (Grun.) Gran, *T. anguste-lineata,* and *Melosira arctica,* and sometimes the relatively warm-water species *Coscinodiscus asteromphalus* and *C. radiatus.* The sub-Atlantic (~1,750–~2,700 B.P.) assemblage is represented by the recent diatom flora of the White Sea (see above).

## KARA SEA

The benthic diatoms of the Kara Sea are known from the data of Cleve and Grunow (1880) who mentioned 46 marine and 65 freshwater species on the sea coast, 58 species in the Yamal Peninsula, and 112 species near the Enisei River mouth. Later, Cleve (1883) presented an additional list of 150 marine species. All these lists should be considered as preliminary and inventorial in character. No information is avail-

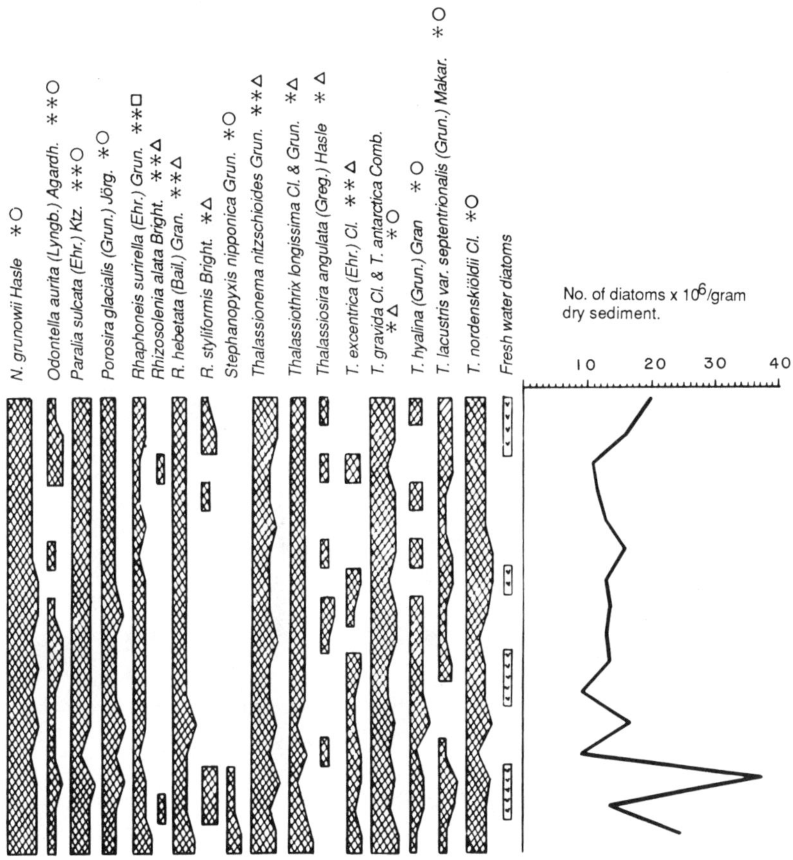


□ sublittoral species ○ neritic species △ panthalassic species

able on Laptev Sea benthic diatoms, and only its phytoplankton is listed. (Kisselyev, 1932; Usachev, 1946*b*; Shirshov, 1937).

## EAST SIBERIAN SEA

The information on East Siberian Sea diatoms is rather scarce (Polyakova, 1984). The planktonic diatoms have not been studied to date. The author's studies (Polyakova, 1984*b*) of a small number of samples from surface sediments in water depths of 14 m to 50–60 m have permitted identification of over 100 species, varieties, and forms of ecologically diverse diatoms: marine, brackish marine, freshwater, and fresh- and brackish-water taxa. The latter two groups are nonuniformly distributed in sediments, and their presence in these deposits is the result of river influx and dispersal by ice. The largest effect is attributed to the Kolyma River influx, which is detectable almost 200 km away from the river mouth.

Among freshwater diatoms carried by the Kolyma River, benthic species are predominant. They are carried from the river bed by currents. Species of the genus Thalassiosira Cl. have been found to predominate in thanatocenoses in surface sediments taken from depths of 30–60 m in the

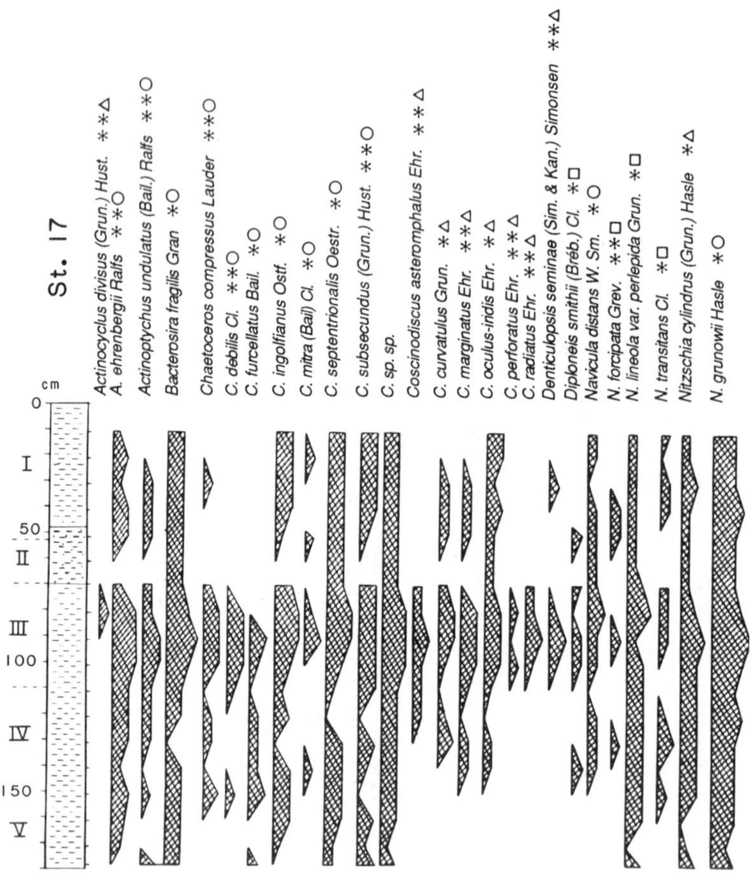


Figure 20-3. Diatoms in Holocene sediments of the Chukchi Sea, Core 17.

* boreal–Arctic and bipolar species
** boreal–Arctic-tropical species (or widespread species) and cosmopolitan species

western and central parts of the East Siberian Sea. The most abundant were the spores of the cold-water bipolar *Thalassiosira gravida* Cl. and *T. antarctica* Comb., which are indistinguishable under the light microscope (Hasle and Heimdal, 1968; Semina and Makarova, 1983). Along with these, other dominant species are *Thalassiosira lacustris* (Grun.) Hasle and var. *septentrionalis*, as well as the neritic boreal–Arctic *Thalassiosira nordenskioeldii* Cl.

Among the secondary forms the predominant species are the planktonic cold-water *Nitzschia grunowii* Hasle, *N. cylindrus* (Grun.) Hasle, *Porosira glacialis* (Grun.) Jorg, and *Bacterosira fragilis* Gran. Benthic diatoms in bottom sediments studied are not numerous and their species composition is variable. In the eastern part of the East Siberian Sea, diatoms were studied by the author in the vicinity of Chaun Bay (Polyakova, 1982). In this study, two profiles northwest of Cape Shelagski were made: 0.8–10.1 km and 0.9–10.8 km offshore, respectively. Sampling depth ranged from 14–33 m. The thanatocenoses examined included 94 species and varieties of marine, brackish-marine, freshwater, and fresh- and brackish-water diatoms.

The dominant group consists of marine

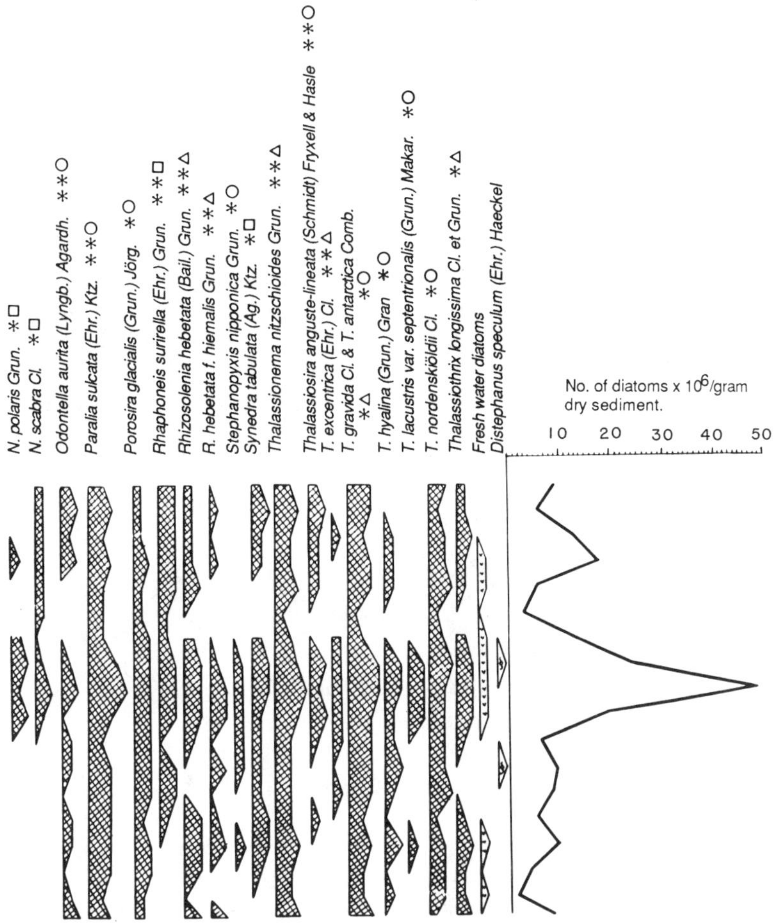


□ sublittoral species ○ neritic species △ panthalassic species

and brackish-marine taxa (65 species). Freshwater and fresh- and brackish-water diatoms were represented by 28 species and their percentage was less than one. Marine planktonic diatoms in sediments of this part of the East Siberian Sea are numerous. Numerically dominant (27–96 percent) are the spores of the genus Chaetoceros Ehr. Since most were lacking typical structural features, their identification was impossible. The spores of the following species were identified with confidence: *C. mitra* (Bail.) Cl., *C. subsecundus* (Grun.) Hust., *C. debilis* Cl., *C. furcellatus* Bail, and *C. septentrionalis* Oestr. All species except the last taxon were rather scarce. The content of Chaetoceros spores in the thanatocenoses of Chaun Bay was unusually high for the East Siberian Sea. The abundance of Chaetoceros spores in the sediments generally indicates high phytoplankton productivity (Koblentz-Mishke, Volkovinsky, and Kobanova, 1970).

Sancetta (1981, 1982) relates the high concentration of Chaetoceros spores in sediments to their high productivity in zones of active vertical displacement of water, or in zones where nutrients are supplied by river runoff. Favorable conditions for sedimentation of light and small spores of this genus are also a positive factor in their concentration. In the thanatocenoses of the East Siberian Sea, in addition to Chaetoceros spores, the dominant species are *Thalassiosira gravida* and *T. antarctica* (especially spores), *T. nordenskioeldii, Nitzschia grunowii,* and *N. cylindrus.* The geographic distribution of these species is confined to areas of cold polar and boreal–Arctic water masses (Zhuze, 1962; Semina and Makarova, 1983; Semina, 1974; Hasle and Heimdal, 1968).

The two species of the genus Nitzschia Hass. are among the basic components of the diatom flora in sea ice, forming accumulations on the lower surface of pack ice (Usachev, 1949; Gran, 1904; Horner and Alexander, 1972; Horner, 1976; Meguro, Ito, and Fukushima, 1966, 1967). *Thalassiosira gravida* spores are also present in the Arctic sea ice (Horner, 1976). The cold-water, early spring species *Thalassiosira nordenskioeldii* was found to accumulate in large masses at the edge of the ice when the latter breaks up (Schandelmeier and Alexander, 1981).

The group of secondary species in the thanatocenoses studied is composed of the neritic, bipolar, early spring taxon *Porosira glacialis,* the neritic boreal–Arctic species *Bacterosira fragilis,* the panthalassic boreal–Arctic species *Thalassiothrix longissima* Cl. et Grun., and the sea-ice, neritic (according to Makarova and Semina, 1982), boreal–Arctic species *Thalassiosira hyalina* (Grun) Gran. There are also great numbers of the sublittoral planktonic species *Paralia sulcata,* which is extremely abundant in the bottom water layers in shallow areas (Zhuze, 1962).

Marine sublittoral benthic diatoms vary in their species composition (20 species); the following taxa represent one-third of the marine diatom flora of this region: *Cocconeis scutellum* Ehr., *Navicula transitans* Cl., *N. cancellata* var. *gregorii* Ralfs, *N. forcipata* Grev., *N. lineola* var. *perlepida* Grun., and so forth. They are not numerous, probably as a result of low productivity of the benthic flora under the very severe ice conditions in the East Siberian Sea. Among the benthic diatoms, there are some species characteristic of sea ice, which continue to grow in the sublittoral zone after the ice melts (Horner, 1976). Thus, it can be stated that the thanatocenoses of the East Siberian Sea are composed of marine neritic and panthalassic planktonic species, cryophilic taxa of sea-ice diatoms, benthic diatoms living in the sublittoral zone of the sea, and freshwater species entering the latter carried by rivers. This investigation has shown that the dominant components of the thanatocenoses in the surface sediments of the East Siberian Sea and in its eastern part, in particular, are the cryophilic sea-ice diatoms.

## CHUKCHI SEA

The diatoms of the Chukchi Sea were first reported by Cleve (1883), but his results are limited chiefly to the glacial flora collected near Cape Vankarem. Mann (1907) presented a list of 62 species of diatoms from one plankton and two bottom samples taken along the American coast of the Chukchi Sea. Later papers described the planktonic flora of diatoms (Kisselyev, 1937; Shirshov, 1937; Lichti-Fedorovich, 1983) and the sea-ice flora (Horner, 1976; Meguro, Ito, and Fukushima, 1966, 1967). Not all the samples available to date to this author have been processed. The littoral (Polyakova, 1982) and sublittoral thanatocenoses of diatoms in core sequences show diatom assemblages from these two zones replacing one another in Holocene sediments (Polyakova and Vosovik, 1984).

In the western part of the Chukchi Sea, the samples studies were taken in water 1.5–2.0 m deep off the coast. The sediments are composed of gravel, pebbles, and sand. The samples contain 101 species and varieties of diatoms. The composition of thanatocenoses is diverse, as there are marine (26), brackish-marine (9), freshwater, and fresh- and brackish-water (66) species (Figs. 20-1 to 20-4). In most coastal zone samples the predominant diatoms are marine species (46–93 percent), and in only one sample this group was as low as 25 percent. The marine diatoms are mainly represented by sublittoral species (67–93 percent), namely *Diploneis subcincta, Navicula cancellata* var. *gregori*, and *N. forcipata*, as well as by a small number of plankton neritic species: *Thalassiosira nordenskioeldii, Porosira glacialis*, several species of the genus Chaetoceros Ehr., and so forth. About 70 percent of the species of marine diatoms are cold-water boreal–Arctic and bipolar,

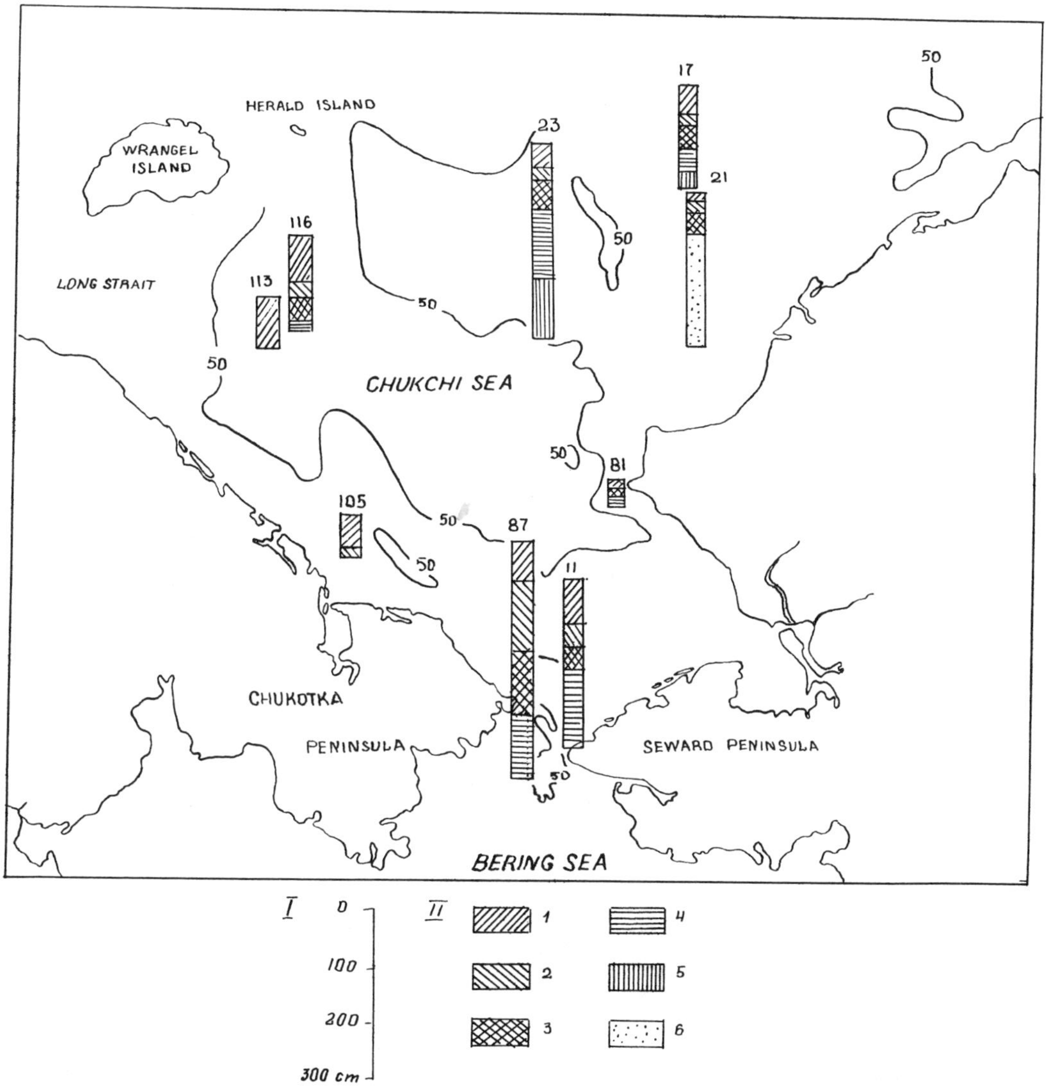


**Figure 20-4.** 1–5 refer to the five Holocene assemblages and 6 to the Late Pleistocene assemblage discussed in the text. *Notes:* Sublittoral species are planktonic and benthic species living in the sublittoral zone. Neritic species are planktonic species living in the shelf zone. Panthalassic species are marine organisms living in both neritic and oceanic waters.

while the remainder are mainly wide-ranging taxa. The percentage of brackish-marine diatoms is only 3–12.

In coastal marine thanatocenoses an important role is played by freshwater and fresh- and brackish-water species. Their percentages in some samples is as high as 68, indicating large amounts of freshwater supply to the coastal zone by river inflow.

In the sublittoral sediments of the Chukchi Sea, planktonic neritic and panthalassic species predominate (mainly cold-water forms as well as wide-ranging taxa). Among these the dominant species are *Thalassiosira nordenskioeldii, Nitzschia grunowii, N. cylindrus, Thalassiosira gravida, T. antarctica, Chaetoceros septentrionalis,* and *Bacterosira fragilis.*

The Holocene assemblages have been studied in 10 cores raised from the Chukchi Sea. The cores recovered are 30–420 cm long. Changes in species composition and the number of diatoms through the section suggest that the sediments recovered are mainly of Holocene age. Five assemblages have been distinguished, reflecting changes in paleogeography in the Chukchi Sea during the Holocene. These assemblages are not present in all cores. The shortest cores have yielded only the two upper assemblages. In some cores there is no second assemblage, or the third Holocene assemblage is missing, possibly due to erosion (Figs. 20-3 and 20-4).

In Core 21, taken in the northwestern part of the Chukchi Sea (Fig. 20-1), sandy sediments have been found underlying the deposits with the Holocene diatom assemblages. The sands contain very rare diatoms of low diversity, which are almost exclusively sea-ice neritic species. The composition of these diatom assemblages suggests that the deposits are Late Pleistocene in age. In our studies of Chukchi Sea Holocene sediments, over 100 species and varieties have been identified. In all cores neritic species (over 50 taxa) predominate; they occur mostly in the shelf zone, while panthalassic species (over 20 taxa) occur both in the shelf zone and in the open ocean. Sublittoral, generally benthic, diatoms are diverse in their species composition (about 30 taxa), but are not very abundant.

The most common assemblage is composed of cold-water boreal–Arctic planktonic species: *Nitzschia grunowii, N. cylindrius, Thalassiosira gravida, T. antarctica* (spores), *T. nordenskioeldii,* and *Bacterosira fragilis* (up to 10 million valves per gram of dry residue), and widely distributed taxa (about 6 million valves): *Thalassionema nitzschioides* and *Paralia sulcata.* The group of secondary taxa includes various neritic species, mainly of the genus Chaetoceros Ehr.: *C. septentrionalis, C. subsecundus, C. ingolfianus* Ostf., *C. debilis,* and *Porosira glacialis* and *Thalassiosira hyalina.* Their number does not exceed 1.5 million valves (per gram of dry residue). Panthalassic species *Rhizosolenia hebetata* (Bail.) Gran and *Thalassiothrix longissima* are also among the subordinate taxa. These species, much like the dominant forms, are essentially cold-water, wide-ranging, as well as cosmopolitan forms. In addition to the marine diatoms, we find solitary freshwater stenohaline species. Their presence in Chukchi Sea shelf sediments suggests transport by ice. The total number of diatoms is about 60 million valves per gram of dry residue.

The diatom flora is composed of cold-water forms; the studies of Chukchi Sea sediments have allowed us to recognize changes in their taxonomic composition and in the number of diatoms throughout the core sections and to distinguish five assemblages (Fig. 20-2 and 20-3).

In the uppermost assemblage of diatoms, the number of valves vary from 12.8 to 27 million per gram of dry residue. In addition to the dominant and subordinate, chiefly cold-water species that are mentioned above, this assemblage includes wide-ranging and cosmopolitan panthalassic forms represented by the abundant *Thalassiosira eccentrica* (Ehr.) Cl., *Coscinodiscus marginatus* Ehr., *Rhizosolenia styliformis*

Bright., *Denticulopsis seminae* (Sim. et Kan.) Simonsen, and *Thalassiosira angulata* (Greg.) Hasle. Boreal neritic species are represented by *Actinocyclus ehrenbergii* Ralfs, *Chaetoceros compressus* Lauder, and *Actinoptychus undulatus* (Bail.) Ralfs. Along with these, there are some cold-water species, not very numerous, such as *Coscinodiscus oculus-iridis* Ehr., *C. curvatulus* Grun., *Chaetoceros radicans* Schutt, and others. Almost all these species have been found in the Chukchi Sea plankton (Kisselyev, 1937), and a small number are present in the surface layer of sediment in the study area.

The second assemblage found in the underlying sediments is not as rich in species as the first assemblage. Though the species composition and the number of dominant diatoms, as well as the total number of diatoms (14.3–26.4 million), is similar to the first assemblage, it differs from the latter as warm-water species, characteristic of the first assemblage such as *Rhizosolenia styliformis, R. alata* Bright., *Actinocyclus ehrenbergii, Chaetoceros compressus,* and *Denticulopsis seminae,* have not been found in the second assemblage. In addition to wide-ranging and cosmopolitan species, this assemblage contains such solitary species as *Thalassiosira eccentrica, T. angulata, Coscinodiscus marginatus,* and *Actinoptychus undulatus.*

The third assemblage is characterized by the largest number of diatoms (up to 69 million valves per gram of dry residue) and by the highest species diversity. This increase in most subordinate taxa occurs in both wide-ranging and cold-water forms, such as *Bacterosira fragilis* (up to 3 million), *Nitzschia grunowii* (up to 9.8 million), *Thalassionema nitzschioides* (about 6 million), *Nitzschia cylindrus* (nearly 2.6 million), *Porosira glacialis* (about 1 million), *Chaetoceros subsecundus* (about 2.3 million), *C. ingolfianus* (about 3 million), *C. septentrionalis* (about 1.9 million), and *Paralia sulcata* (about 2.6 million). Among the dominant taxa, there are practically no changes in the number of species (e.g., *Thalassiosira nordenskioeldii, T. gravida,* and *T. antarctica*). Compared to the first and second assemblages, the third assemblage is characterized by a decrease by a factor of two in such panthalassic dominant forms as *Thalassiothrix longissima* and *Rhizosolenia hebetata.* At the same time, some cosmopolitan and wide-ranging species (e.g., *Coscinodiscus marginatus* and *Thalassiosira eccentrica*) increase in number and the group is more diversified. Relatively warm-water species that did not occur in the first two assemblages do occur in assemblage three; they include *Coscinodiscus radiatus* (about 375,000), and *Actinocyclus divisus* (about 75,000), as well as *Coscinodiscus perforatus, C. asteromphalus,* and *Stephanopyxis nipponica.* Most of these species have not been found in the present-day flora of the Chukchi Sea, and their presence was mentioned by Kisselev (1937) in the Bering Strait phytoplankton.

The fourth assemblage is present in the compacted sediments of the deeper sections of the cores. In its ecological and taxonomic composition this assemblage resembles the first association; however, the total number of diatom valves is higher (up to 45 million). Along with the dominant and subordinate species, we find the following species at low frequencies: *Thalassiosira eccentrica, Coscinodiscus curvatulus, C. oculus-iridis, Actinoptychus undulatus,* as well as such species as *Denticulopsis seminae* and *Stephanopyxis nipponica,* which are rare in the present-day Chukchi Sea flora.

The fifth assemblage was found in the compacted sediments of the bottom section of some cores (Fig. 20-3). The total content of diatoms is the lowest (about 1 million) and it is not so rich in species diversity. The panthalassic and neritic diatoms are represented by a small number of *Thalassiosira gravida, T. antarctica, T. hyalina, Thalassionema nitzschioides, Nitzschia grunowii, N. cylindrus,* and *Rhizosolenia hebetata,* representing the dominant and subordinate taxa.

The composition of assemblages in Chukchi Sea sediments allow the following conclusions to be drawn: (1) The cores studied represent Holocene sediments (Figs. 20-3 and 20-4). (2) The five diatom assemblages recognized suggest changes in paleoenvironments. (3) The composition of the first assemblage in core "tops" reflects present-day, Arctic conditions in the Chukchi Sea, and the influence of the relatively milder Bering Sea water. Conditions similar to those of the Recent used to prevail in the period when the fourth assemblage was laid down. (4) The second assemblage is representative of colder water than the first association. It was, most probably, deposited in a period when the connection between the Chukchi and Bering seas was more restricted than today. (5) The warmest assemblages and the highest number of diatoms was found in the third association. It suggests that water exchange between the Chukchi and Bering seas was greater than today. This assemblage is comparable to Zhuze's (1961) Complex B in the Holocene sediments of the Bering Sea and the Sea of Okhotsk, as well as in the proximal Kamchatka region of the Pacific, and is related to the time known as the Pacific thermal optimum. At that time, sea level may have been several meters higher than at present. (6) The fifth assemblage contains the coldest floral elements.

## CANADIAN ARCTIC

In the Beaufort Sea, and in the vicinity of the coast of the Canadian sector of the Arctic Ocean, the phytoplankton species composition has been studied by various authors; many articles have been published on the sea-ice diatom flora and on their seasonal development (see the References section of this chapter and Chapter 5).

Arctic Ocean diatoms have been very inadequately studied. There is no information on diatoms in the sediments of some of the shelf seas or in the abyssal regions of the Ocean. The data available to date indicate that the recent and fossil thanatocenoses could be important tools for paleogeographic reconstructions.

## REFERENCES


Belyaeva, T. V. 1980. Phytoplankton in the drift area of the station North Pole-22. In *Biology of Central Arctic Basin*. Nauka, Moscow, pp. 133–143. (In Russian.)

Bursa, A. 1963. Phytoplankton in coastal waters of the Arctic Ocean at Point Barrow, Alaska. *Arctic* **16**(4):239–262.

Cleve, P. T. 1873. On diatoms from the Arctic Sea. *Bih. Kgl. Svensk Vetensk. Akad. Handl.* **1**(13):1–28.

Cleve, P. T. 1883. Diatoms collected during the expedition of the *Vega*. In *Nordenskiold's Vega Exp.*, vol. 3. Vetensk Jaktt., Stockholm, pp. 455–517.

Cleve, P. T., and A. Grunow. 1880. Beiträge zur Kenntniss der arktischen Diatomeen. *Kgl. Svensk Vetensk. Akad. Handl.* **17**(2):1–121.

Dunbar, M. J., and J. C. Acreman. 1980. Standing crops and species composition of diatoms in sea ice from Robeson Channel to the Gulf of St. Lawrence. *Ophelia* **19:**61–72.

Dzhinoridze, R. N. 1968. Materials to study diatoms in the sediments of the White Sea. In *Materials of the Symposium "Cenozoic History of the Polar Basin."* Leningrad, pp. 50–51. (In Russian.)

Dzhinoridze, R. N. 1971. Diatoms from the bottom sediments of the White Sea as related to its Holocene history. Author's abstract of Can. Thesis, Leningrad, 23p. (In Russian.)

Foy, M. G., and S. I. C. Hsiao. 1976. Phytoplankton data from the Beaufort Sea 1973 to 1975. *Environment Canada, Fish Marine Serv. Tech. Rep. 617*, 44p.

Fukushima, H. 1965. Preliminary report on diatoms from coloured sea ice in Point Barrow, Alaska. *Antarctic Record* **24:**31–35.

Gran, H. H. 1904. Diatomacea from the ice flora and plankton of the Arctic Ocean. Norwegian North Polar Exp. 1893–1896. *Sci. Res.* **4**(11):1–74.

Hasle, G. R., and B. R. Heimdal. 1968. Morphology and distribution of the marine centric diatom *Thalassiosira antarctica* Comber. *J. Royal Micros. Soc.* **88**(3):357–369.

Horner, R. A. 1973. Biology and ecology of phytoplankton and ice organisms in coastal waters near Barrow, Alaska. *National Technical*

*Information Service (Report) AD 770 155.* Institute of Marine Science, University of Alaska, 75p.

Horner, R. A. 1976. Sea ice organisms. *Oceanogr. Marine Biol. Ann. Rev.* **14:**167–182.

Horner, R. A., and V. Alexander. 1972. Algal populations in Arctic sea ice. An investigation of heterotrophy. *Limnol. Oceanogr.* **17:**454–458.

Hsiao, S. I. C. 1976. Biological productivity of the southern Beaufort Sea: Phytoplankton and seaweed studies. *Environment Canada, Beaufort Sea Project, Tech. Rep. 12c,* 99p.

Hsiao, S. I. C. 1980. Quantitative composition, distribution, community structure and standing stock of sea ice microalgae in the Canadian Arctic. *Arctic* **33**(4):768–793.

Hsiao, S. I. C. 1983. A checklist of marine phytoplankton and sea ice microalgae recorded from Arctic Canada. *Nova Hedwigia* **37:**225–232.

Hsiao, S. I. C., M. G. Foy, and D. W. Kittle. 1977. Standing stock, community structure, species composition, distribution and primary production of natural populations of phytoplankton in the southern Beaufort Sea. *Can. J. Bot.* **55:**685–694.

Kisselev, I. A. 1932. Materials on the microflora of the southeast part of the Laptev Sea. *Research into the USSR Seas* **15:**67–94. (In Russian.)

Kisselev, I. A. 1935. Some data on the phytoplankton of the northeastern Kara Sea. *Proc. Taimyt. Hydrograph. Exp.* **2:**191–202. (In Russian.)

Kisselev, I. A. 1937. Composition and distribution of phytoplankton in the northern Bering Sea and southern Chukchi Sea. *Research into the USSR Seas* **25:**217–245. (In Russian.)

Koblentz-Mishke, O. J., V. V. Volkovinsky, and H. G. Kobanova. 1970. Plankton primary production of the world ocean. In *Scientific Exploration of the South Pacific,* W. Wooster (ed.). National Academy of Science, Washington, D. C., pp. 183–193.

Kolbe, R. W. 1927. Zur Ökologie Morphologie und Systematik der Bra Brackwasser-Diatomeen (Die Kieselalgen des Sperenberger Salzgebietes). *Pflanzenforschung* **7.**

Korotkevich, O. S. 1960. Diatom flora of the Barents Sea littoral. *Proc. Murmansk Biol. Inst.* **5**(1): 68–338. (In Russian.)

Lichti-Fedorovich, S. 1982. Contribution to the diatom flora of Arctic Canada: Report 3. On the occurrence of *Chaetoceros* spp. fo. *clavipes* in the southern Beaufort Sea. In *Current Research, Part B, Geol. Surv. Can., 82-1B,* pp. 169–171.

Lichti-Fedorovich, S. 1983. Diatoms from the southern Beaufort Sea. In *Current Research, Part B, Geol. Surv. Can, 83-1B,* pp. 267–271.

Makarova, I. V., and G. I. Semina. 1982. Some data on the morphology and distribution of *Thalassiosira hyalina* (Bacillariophyta). *Bot. J.* **67**(1):109–111. (In Russian.)

Mann, A. 1907. Report on the diatoms of the albatross Voyage in the Pacific Ocean 1888–1904. *Contrib. U.S. Nat. Herb.* **10**(5):221–419.

Mann, A. 1925. *The Marine Diatoms of the Canadian Arctic Expedition 1913–1918,* vol 4, Botany. F. A. Acland, Ottawa, pp. 1–33.

Medvedev, V. S., E. N. Nevesskiy, L. I. Govberg, E. S. Malyasova, R. N. Kirienko, and R. N. Dzhinoridze. 1970. On the structure and stratigraphic division of the bottom deposits of the White Sea. In *The Arctic Ocean and Its Coast in the Cenozoic.* pp. 253–267. (In Russian.)

Meguro, H., K. Ito, and H. Fukushima. 1966. Diatoms and the ecological conditions of their growth in sea ice in the Arctic Ocean. *Science* **152:**1,087–1,090.

Meguro, H., K. Ito, and H. Fukushima. 1967. Ice flora (bottom type), a mechanism of primary production in polar seas and the growth of diatoms in sea ice. *Arctic* **20:**114–133.

Merezhkovsky, K. S. 1878. Diatomaceae of the White Sea. *Proc. St. Petersburg Soc. of Naturalists* **9:**425–466. (In Russian.)

Meyer, K. I. 1938. Materials on the algal flora of the White Sea. *Proc. All-Union Inst. Marine Fish. and Oceanogr.* **7:**5–28. (In Russian.)

Nevesskiy, E. N., V. S. Medvedev, and V. V. Kalinenko. 1977. *The White Sea. Sedimentogenes and Development in the Holocene.* 237p. (In Russian.)

Petrov, Yu. E. 1967. Blue-green algae and diatoms of the littoral of Bolshoi Solovetskii Island (the White Sea). In *News in the Systematics of Lower Plants.* pp. 15–20. (In Russian.)

Polyakova, E. I. 1982. Diatoms of the sublittoral and lagoons of the Chukchi and East Siberian seas. *Okeanologiya* **22**(5):809–812. (In Russian.)

Polyakova, E. I. 1984. Biogeographical composition and ecology of diatoms at the recent sediments of the Chukchi Sea and the East Siberian Sea. In *Stratigraphy and Sedimentology of the Mezozoic and the Cenozoic Sediment Cover of the World Ocean.* Abstracts of reports, part 1, Stratigraphy, pp. 77–78. (In Russian.)

Polyakova, E. I., and Yu. I. Vozovik. 1984. Some paleoceanographic features of the Chukchi Sea in the Holocene (based on diatom complexes). *Okeanologiya* **24**(5):789–793. (In Russian.)

Proshkina-Lavrenko, A. I. 1974. Diatoms of the

recent seas and lakes. In *Diatoms of the USSR*. pp. 274–351. (In Russian.)

Reinhard, L. W. 1882. Zur Kenntniss der Bacillariaceen des Weissen Mures. *Bull. Soc. Nat Moscow* **1**:297–304.

Sancetta, C. 1981. Oceanographic and ecologic significance of diatoms in surface sediments of the Bering and Okhotsk seas. *Deep Sea Res.* **28A**(8):789–817.

Sancetta, C. 1982. Distributions of diatoms species in surface sediments of the Bering and Okhotsk seas. *Micropaleontology* **28**(3):221–257, plates 1–6.

Schandelmeier, L., and V. Alexander. 1981. An analysis of the influence of ice on spring phytoplankton population structure in the southeast Bering Sea. *Limnol. Oceanogr.* **26**(5):935–943.

Semina, G. I. 1974. *Phytoplankton of the Pacific Ocean*. 239p. (In Russian.)

Semina, G. A., and I. V. Makarova. 1983. On the morphology and distribution of two bipolar species of the genus *Thalassiosira* (Bacillariophyta). *Bot. J.* **5:** 611–618. (In Russian.)

Shirshov, P. P. 1936. Plankton as an indicator of the ice regime. In *Scientific results of the* Krasin *Expedition in 1935.* pp. 59–73. (In Russian.)

Shirshov, P. P. 1937. The phytoplankton life of the polar seas seasonal phenomenon as related to the sea ice regime. *Proc. Arctic and Antarctic Sci. Inst.* **82:**47–111. (In Russian.)

Usachev, P. I. 1935. Composition and distribution of phytoplankton in the Barents Sea in the summer of 1931. *Proc. Arctic and Antarctic Sci. Inst.* pp. 5–94. (In Russian.)

Usachev, P. I. 1938. Biological analysis of ice. *Trans. USSR Acad. Sci.* **19**(8):643–646. (In Russian.)

Usachev, P. I. 1946*a*. Phytoplankton collected by the *G. Sedov* expedition in 1937–1938. *Proc. Glavsevmorput Drift Exped. on the* G. Sedov *Ice Breaker in 1937–1939,* vol. 3. pp. 371–379. (In Russian.)

Usachev, P. I. 1946*b*. Biological significance of the origin of the ice at the Kara Sea, the Laptev Sea and the Straits of the Archipelago of the Franz Josef Land. *Proc. Inst. Oceanol. Acad. Sci. USSR* **1**:113–150. (In Russian.)

Usachev, P. I. 1949. Microflora of polar ice. *Proc. Inst. Oceanol. Acad. Sci. USSR* **3:**216–259. (In Russian.)

Vishnevskaya, E. M., R. N. Dzhinoridze, I. A. Kuptsova. 1974. Diatoms from Late and Post-Glacial deposits of the White Sea. In *Diatoms of the USSR*. pp. 256–259. (In Russian.)

Zabelina, E. K. 1969. Diatoms in the sediments of the northern White Sea. In *Basic Problems of Micropaleontology and Organogenic Sedimentation in Oceans and Seas.* pp. 85–95. (In Russian.)

Zabelina, M. M. 1939. Diatoms in the sediments of the White Sea in the region of Malaya Porya Bay. *Proc. St. Hydrol. Inst.* **8**(17):183–200. (In Russian.)

Zhuze, A. P. 1961. Stratigraphy of bottom deposits in the northwestern Pacific Ocean. *Proc. Inst. Geol. Estonian Acad. Sci.* **8:**183–195. (In Russian.)

Zhuze, A. P. 1962. *Stratigraphic and Paleogeographic Studies in the Northwestern Pacific Ocean.* 258p. (In Russian.)

# 21: Ecology of Recent Foraminifera on the Canadian Continental Shelf of the Arctic Ocean

G. Vilks

## INTRODUCTION

The continental margin of the Canada Basin can be divided into two major environments with respect to seafloor physiography, sedimentary processes and oceanography. To the west there is the shallow Beaufort shelf, receiving relatively warm freshwater runoff carrying fine sediments from several major rivers, and to the east there is a much deeper shelf off the Queen Elizabeth Islands. Here the shelf and channels between the islands receive very little runoff and sediments, but act as conduits for the southeasterly flow of offshore Arctic water toward Baffin Bay. Generally, the Beaufort shelf has open water or intermittent polar pack ice during the summer. Most of the channels between the Queen Elizabeth Islands are covered with multiyear pack ice throughout the melting season where the extent of open water can be limited to shore leads or occasional polynya. Notwithstanding differences in sedimentation rates and ice cover, the bottom waters in both areas are within negative centigrade ranges.

Foraminifera from these waters provide a unique opportunity to investigate the effect of low temperatures and a long winter season on very low primary production. Assemblages that are found in these extreme conditions are of particular interest to paleoceanographers attempting to reconstruct the extent and chronology of lower latitude Quaternary glaciations on the continental shelves. Elements of these present-day Arctic faunas are still found in Labrador Current waters as far south as Gulf of St. Lawrence and may be considered as remnants from Wisconsinian glaciations.

## PREVIOUS WORK

Cushman (1948) described species collected and summarized the results of early Arctic expeditions from the Bering Sea to Greenland, including Hudson Bay. Altogether 175 species are mentioned as occurring in Arctic waters.

The first detailed work was carried out by Phleger (1952) who discussed the distribution of foraminifera collected from the northern Baffin Bay and the Parry Channel system. The northern Baffin Bay samples are dominated by agglutinated species, such as *Trochammina nana* in waters deeper than 600 m and *Adecotryma glomerata* and *Spiroplectammina biformis* in waters between 180 m and 600 m. At Thule, in 30 m of water, *Eggerella advena* and *Textularia torquata* are the most abundant. In Lancaster Sound, samples collected in waters between 275 m and 660 m were also dominated by *A. glomerata* and *T. nana*, but in 30–40 m off Dundas Harbour, *Elphidium incertum* and variants (*E. excavatum*) and *Islandiella islandica* var. *minuta* (*Cassidulina reniforme*) are the dominant species. In Wellington Channel, *Cibicides lobatulus* and *Islandiella norcrossi* are prevailing between 125 m and 155 m of water, but in Melville Sound at the depth of 124 m the agglutinated species are major and dominated by *A. glomerata* and *T. nana*. Altogether 44 benthic species were identified. Thirteen species were collected living; *Saccammina atlantica* was the most common living species. Water depth was the only recorded ecological factor and seems to be important in determining species distribution.

Carsola (1952) discussed benthic foraminifera collected from 61 sediment samples of the Beaufort Shelf and Chukchi Sea between the water depth of 18 m and 1,994 m. Except for the few offshore stations, the sediments are relatively barren of foraminifera; more than 50 percent of all samples contained less then 5 specimens per gram of dry sediment. Nevertheless, 84 species were identified and the distribution of 53 species was reported. The most common species are *Cassidulina islandica* var. *minuta* (*C. reniforme*), *Islandiella norcrossi*, *Elphidium incertum* (*E. excavatum*), *Eponides frigidus*, *Spiroplectammina biformis*, *Textularia torquata*, *Trochammina nana*, and *Virgulina* cf. *V. complanata* (*Stainforthia concava*). The author reported a faunal division across longitude 155°W. The percentage of species change at this point, for example, *Nonionellina labradorica* is rare east of Coldville River. The more prominent presence of *N. labradorica* in Chukchi Sea is related to the warmer water hugging the Point Barrow area. Vilks, Wagner, and Pelletier (1979) found rare occurrences of *N. labradorica* on the Canadian Beaufort shelf and only a trace of this species in the deepest water of Hecla and Gripper Bay (Vilks, 1969).

Loeblich and Tappan (1953) studied the taxonomy of Arctic and sub-Arctic species and included a few remarks on ecology at Point Barrow. The species distribution off Point Barrow is most distinctly influenced by the type of substratum; the mixed sediments of gravel and mud contain the richest faunas. The water depth range is 3–233 m, which is not sufficient to show species depth zonation in an area where sediment redistribution is common. Thus, many of the species that occur at 3 m depths are also present at 144 m. The authors conclude that the distribution of Arctic foraminifera is basically circumpolar and the distribution of only a few species is governed by water depth.

Loeblich and Tappan's (1953) careful attention to identification has helped considerably in understanding the literature and it is still an important source of reference. Altogether 110 species were described and illustrated, including 6 new genera and 21 new species.

Anderson (1963) discussed foraminiferal ecology from the western Bering Sea, recognizing foraminiferal biotopes between 12 m and 250 m. The two dominating species in each biotope are as follows:

| | |
|---|---|
| Deltaic biotope | *Elphidium subarctium*, *E. clavatum*, (*E. excavatum*) |
| Transitional biotope | *Buccella frigida*, *Buliminella elegantissima* |
| Inner shelf biotope | *Eggerella advena*, *Reophax arctica* |
| Central shelf biotope | *Reophax arctica*, *Eggerella advena* |
| Outer shelf biotope | *Eggerella advena*, *Epistominella exigua* |

| | |
|---|---|
| Upper bathyal biotope | *Epistominella exigua, Angulogerina angulosa, (Trifarina angulosa)* |

Water mass characteristics are of primary importance for the biotope determinations.

Marlowe and Vilks (1963), and Vilks (1964, 1969, 1976) described the distribution of foraminifera in surface sediments collected from the interisland seas of the Queen Elizabeth Islands. The depth range of sample stations is 0–450 m and is entirely within the area of perennial sea ice. Foraminifera species are characteristically agglutinated in waters shallower than 200 m, except in East Bay, MacKenzie King Island. Here the agglutinated species extend to the maximum station depth of 284 m. These deep stations are located behind a sill that impedes free entrance of the deeper waters from the Arctic Ocean, which are more saline and warmer and are of Atlantic origin. It is evident that the break in faunal characteristics at 200 m is due to the differences in water mass properties rather than bathymetry.

Iqbal (1973) studied benthic foraminifera from 40 grab samples collected in M'Clure Strait. Seventy-four species were identified, described, and illustrated. Station water depths range from 230–500 m; thus distinct depth zonations could not be identified. The proximity of the Arctic Ocean water seems to influence the general characteristics of foraminiferal assemblages. Foraminifera occur in approximately ten times higher numbers on the continental shelf than in M'Clure Strait proper. The sandier and slightly shallower sediments of the Arctic Ocean continental shelf are dominated by calcareous faunas. The two most prominent arenaceous species are *Trochammina nana* and *Saccammnina sphaerica* and the dominant calcareous species are *Islandiella teretis* (*Cassidulina laevigata*), *I. norcrossi,* and *Cibicides lobatulus.* Living specimens of nine species were found based on Rose Bengal staining. *Trochammina nana* was the most abundant living species.

Smith (1974) investigated foraminifera in Resolute Bay, Northwest Territories, from short core samples collected during scuba diving operations in 10–13 m of water. The surface muds contain approximately 4 percent organic carbon and 60 percent terrigenous carbonates, which may explain the relatively rich benthic foraminiferal fauna. Fifty species were identified. The assemblage is dominated by *Elphidium* spp., *Buccella frigida,* and *Eggerella advena.* The relative abundance of *E. advena* increased in subsurface sediments between 6 cm and 18 cm, which was the maximum length of the cores. The unusually high dominance of calcareous species in the shallow Resolute Bay waters is explained by the calcareous nature of the sediments.

Vilks, Wagner, and Pelletier (1979) discussed Holocene foraminifera from 49 piston cores of the Canadian Beaufort shelf facing the MacKenzie River delta. Despite an extensive redistribution of sediments due to ice scouring of seafloor, two prominent foraminiferal zones were distinguished. The inner shelf zone is characterized by the dominance of *Elphidium excavatum* f. *clavata* and delineates the area on the shelf that is most commonly covered by the sediment and freshwater plume of the MacKenzie River runoff. The outer shelf zone is dominated by a group of forms identified as *Islandiella teretis* (Vilks, Wagner, and Pelletier, 1979) and consists of *Islandiella helenae* and *Cassidulina laevigata.* The occurrence of the planktonic foraminifer *Neogloboquadrina pachyderma* is reduced by several orders of magnitude in plankton tows taken through the ice in winter and is absent from the waters of the inner shelf year-round.

The previous reports are based on either widely spaced samples or detailed local studies. The following chapter will summarize the distribution of foraminifera on the Canadian Arctic continental shelf from samples of Beaufort shelf (Figure 21-1) (Vilks, Wagner, and Pelletier, 1979), Northwest Passage (Vilks and Joyce, 1976), M'Clure Strait (Vilks, 1970; Iqbal, 1973), Hecla and Gripper Bay and Hazen Strait (Vilks, 1969), Satellite Bay (unpublished), Byam Channel, (Vilks, 1976), East Bay (Vilks, 1963), Prince Gustaf Adolf Sea (unpublished), and Jones Sound (Hardy, 1984). The distribution of planktonic foraminifera in sediments and water column is summarized from Vilks (1973).

## QUATERNARY GLACIAL HISTORY

According to Prest (1984), the early Pleistocene glaciations may have overridden the islands and the Beaufort shelf and may have stopped somewhere on the continental shelf. On land, except for Banks Island, there is little evidence of the early glaciations, but the seafloor of the Canadian Arctic Archipelago shows geologic and physiographic features that suggest a tertiary fluvial system (Fortier and Morely, 1956). The various Quaternary ice sheets have apparently modified the Tertiary drainage pattern as summarized by Pelletier (1966).

The pre-Wisconsinan glacial history on Banks Island has been summarized by Vincent (1984). He recognized three glaciations, the oldest dating back to the Mathyama epoch. The three pre-Wisconsinan Interglacials are recorded on the basis of lake or marine deposits containing organic matter or shells.

The Late Wisconsinan glaciation is better established. According to Prest (1984), the Beaufort shelf was not glaciated by the

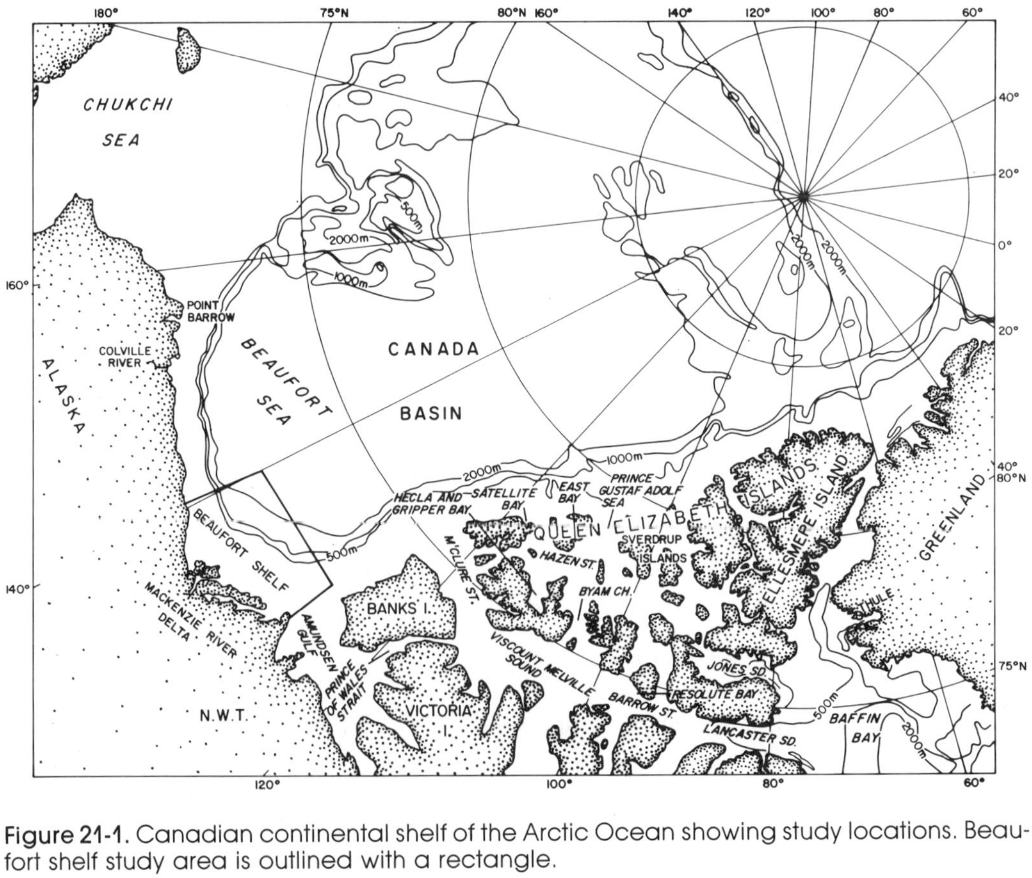


**Figure 21-1.** Canadian continental shelf of the Arctic Ocean showing study locations. Beaufort shelf study area is outlined with a rectangle.

Late Wisconsinan ice except possibly for a small lobe extending over the MacKenzie Trough. There was an ice shelf in Amundsen Gulf and possibly in M'Clure Strait (Hodgson and Vincent, 1984). Glacial ice from mainland crossed the Prince of Wales Strait and stopped along the southeastern coast of Banks Island. Viscount Melville Sound was covered with a large lobe of continental ice that emanated from the Laurentide ice sheet. The lobe reached the south coast of Melville Island, but did not reach Bathurst and Cornwallis islands. The Queen Elizabeth Islands apparently were covered with local ice sheets or ice caps.

The offshore Wisconsinan glacial history is poorly understood and the descriptions below are highly speculative. The conventional shallow seismic surveys useful in the recognition of glacial deposits have not been carried out on a regional basis in the ice-covered or ice-infested areas. Piston cores on Beaufort shelf and Viscount Melville Sound, taken during Hudson 70, and gravity cores and bottom grabs in Hazen Strait (Vilks, 1969) have not shown good evidence of glacial deposits. However, gravity cores recovered by Marlowe (1968) from Prince Gustaf Adolf Sea show distinct changes at approximately 10 cm subsurface, from mottled, yellowish oxidized sediments to stratified and dark sediments below, which are also unfossiliferous. The author suggested that these darker sediments could be representing the time when the channels were covered with a solid sheet of ice or ice shelf during the Wisconsinan maximum.

## Seafloor Morphology

The Beaufort shelf is relatively shallow and slopes gently toward the 100 m isobath where a sharp slope break characterizes the shelf edge (Fig. 21-2). The Canadian Beaufort shelf is about 150 km wide and the Alaskan shelf is only about 75 km wide. A major physiographic feature is the MacKenzie Canyon in which the 100 m isobath occurs less than 50 km from the shore.

The seafloor morphology of the Canadian Arctic Archipelago (Fig. 21-3) is characterized by straight and steep shore faces, isolated depressions, and U-shaped cross sections of the channels (Pelletier, 1966; Horn, 1967; Marlowe, 1968). The depressions may reach to depths of 500 m, which is 50 m greater than the average depth of the continental shelf. A major sill in the Parry Channel system crosses Barrow Strait with minimum depths of about 150 m. This sill is an important feature in determining the water mass characteristics entering Baffin Bay from the Arctic Ocean, because it obstructs the passage of deeper Arctic water to the southeast.

## Sediments

Sediment distribution and dynamics of Beaufort shelf are better known than the Canadian Arctic Archipelago. On the Beaufort shelf, major rivers deliver silt and sand, which is redistributed by longshore currents (Harper and Penland, 1982). On the Canadian Beaufort shelf, silt and clay are the major sediment types, with very little gravel and sand closer to the shore (Vilks, Wagner, and Pelletier, 1979). Without interference by winds, the MacKenzie River sediment plume normally migrates along the coast toward the east. Although current measurements to date do not indicate net sediment transport toward the north and west, the thickest Holocene sediments occur in the MacKenzie Trough and to the north of the Mackenzie River delta (Harper and Penland, 1982).

In the Canadian Arctic Archipelago very fine surface sediments are found in the Channels between the Sverdrup Islands (Horn, 1967) in Prince Gustaf Adolf Sea (Marlowe, 1968), in Hecla and Gripper Bay (Vilks, 1969), and in McClure Strait (Iqbal, 1973). In these areas slightly coarser sediments are present closer to the shore and on midchannel topographic highs, suggesting a weak current winnowing. Sedi-

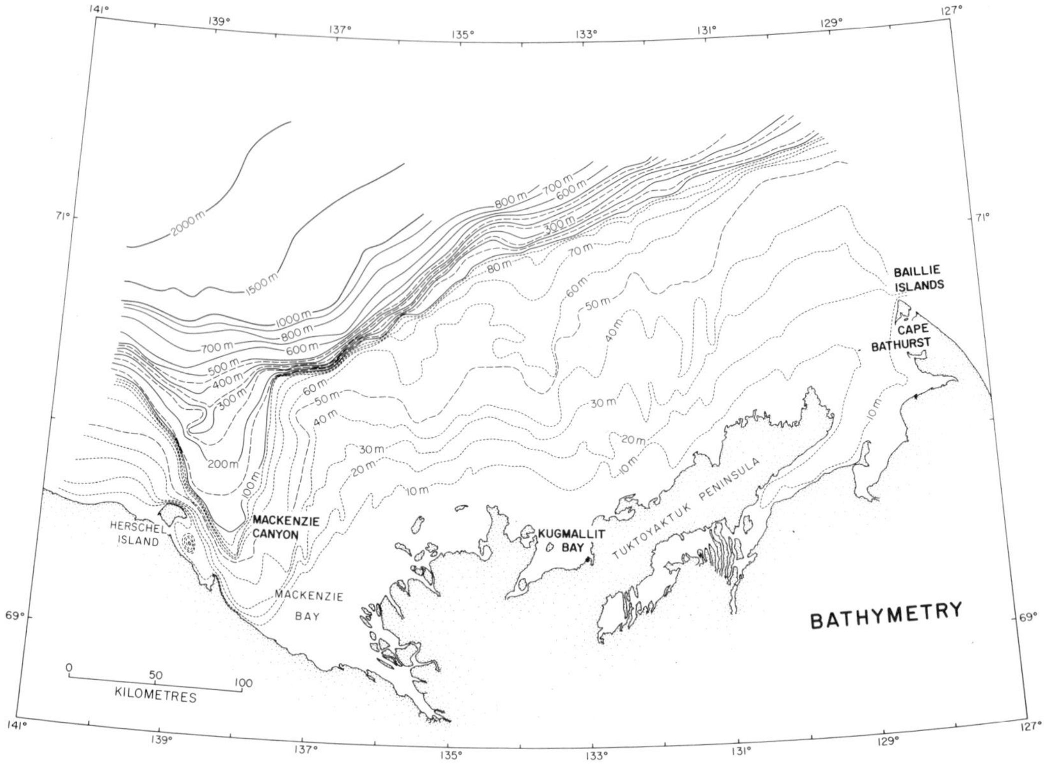


Figure 21-2. Bathymetry of the Canadian Beaufort shelf.

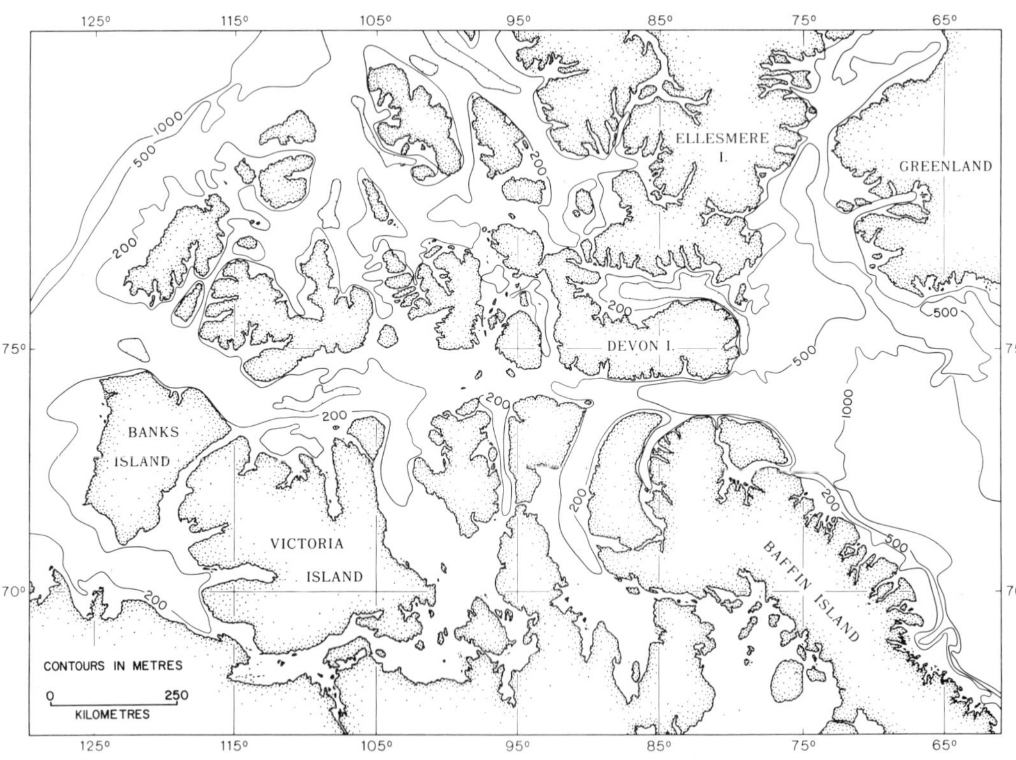


Figure 21-3. Bathymetry of the Canadian Arctic Archipelago.

ments are delivered by wind, rivers, and sea-ice rafting. The rate of sedimentation is suspected to be low, for example, 4.9 cm per 1,000 years (Vilks, 1969) due to a short runoff season, but data are very scarce.

## Water Characteristics

The waters of the Beaufort shelf and the Canadian Arctic Archipelago consist of locally modified Arctic Ocean water, which consists of three layers (Worthington, 1953; Coachman and Barnes, 1961), (Fig. 21-4):

1. The cold and relatively dilute Arctic surface layer approximately 200 m thick.
2. The warmer and more saline water mostly of Atlantic origin between 200 m and 400 m.
3. The Arctic bottom water.

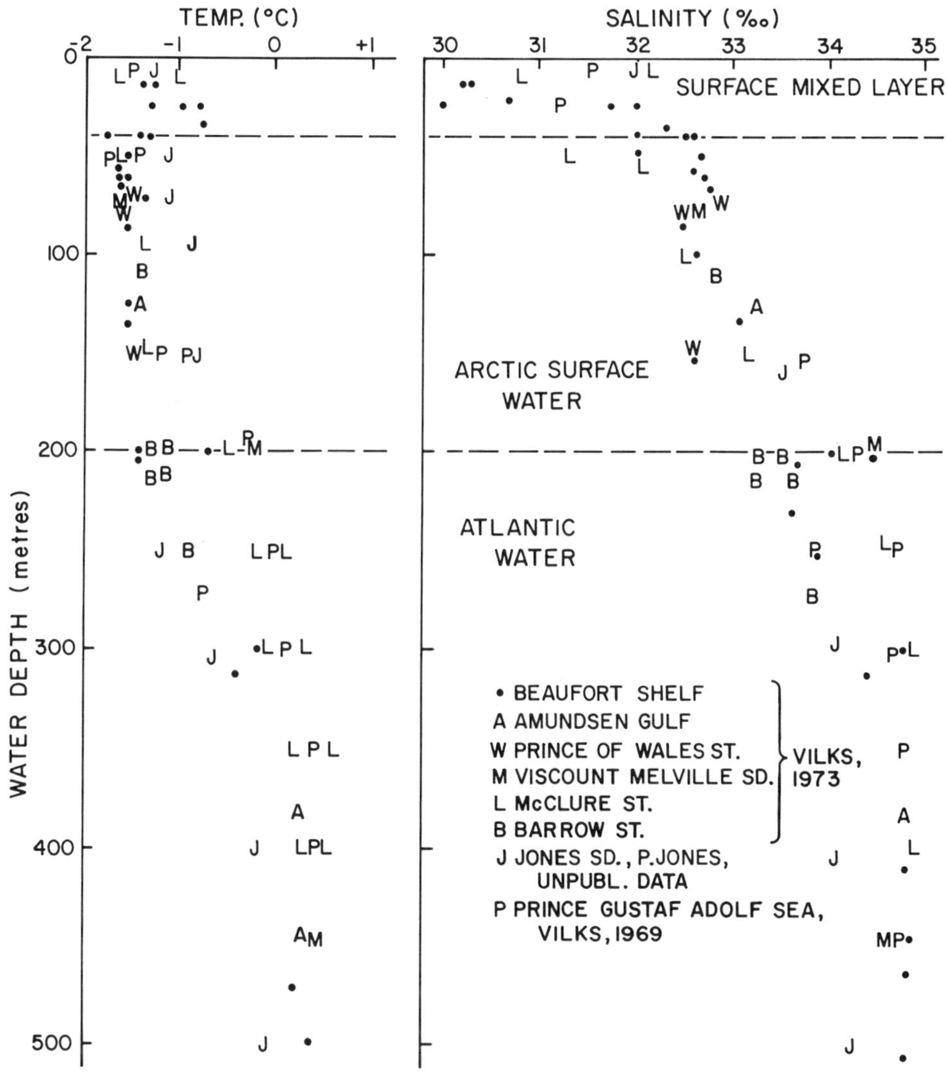


**Figure 21-4.** Temperature and salinity ranges in the water column of the Canadian Arctic Archipelago and Beaufort shelf.

A mixed seasonal surface layer to about 10–50 m is also present in the Canada Basin adjacent to the Canadian Archipelago (Melling et al., 1984). In the Beaufort Sea the Atlantic layer is slightly cooler and the Arctic surface water contains a layer of warm water entering from the Bering Strait in addition to the Colville River and MacKenzie River effluents. A typical summer profile of salinity and temperature versus depth on the southeastern Beaufort shelf shows a relatively warm and diluted surface layer down to about 20 m (Fig. 21-4). During winter the water within the same depth interval is slightly more saline and practically isothermal at −1.3°C (Vilks, 1973).

The channels of the Canadian Arctic Archipelago act as waterways for the southeastward moving water from the Arctic Ocean to Baffin Bay (Collin and Dunbar, 1964; Melling et al., 1984). The relatively deep channel entrance toward the Arctic continental shelf permits the migration of the upper two layers of the Arctic water into the channels. The southeastward movement of the deeper water is obstructed by the sills in Barrow Strait and Wellington Channel. Thus the water below 200 m in Jones Sound is colder and less saline than the water to the west of Barrow Strait (Fig. 21-4).

In summary, the water mass properties on the Beaufort shelf are influenced by the Bering Sea water and river runoff. Within the Canadian Arctic Archipelago the water mass characteristics are similar to the Arctic Ocean to the west and north of the shallow sills in the Barrow Strait area. East of the sills the water is a mixture of Baffin Bay and Arctic surface water.

## Sea Ice

Winter sea ice along the continental margin of the Beaufort Sea consists of (1) the offshore polar pack ice, (2) shore fast ice, and (3) the year-old ice between the two zones. The multiyear polar pack ice is continuous and can be 3–4 m thick; the year-old ice is 1–2 m thick and may cover about 87.5 percent of the sea surface. In the Canadian Arctic Archipelago the Parry Channel system is covered with multiyear ice cemented together by young ice. The remainder of the channels to the north are covered with a continuous multiyear pack ice.

The extent of summer ice in Beaufort Sea varies from season to season. Normally in April a polynya of open water is formed at the entrance of Amundsen Gulf and it expands toward the west. The breakup along the coast begins in mid-July. During August the open water in a normal year may extend 180–300 km off the Canadian coast. The average time of freeze-ups in the Beaufort Sea is the second week of October.

In the Parry Channel system the ice breaks up from east to west. In spring the north water polynya of the northern Baffin Bay expands and reaches Lancaster Sound in June (Dunbar and Dunbar, 1972). In Barrow Strait the ice begins to break up in July, but seldom leaves the area entirely. Normally there is a zone of open water along the north shore of Viscount Melville Sound while the remainder of the sound is covered by a fractured and mobile canopy of ice. Solid ice returns to the area in October. To the north of the Parry Channel system an almost solid cover of ice remains throughout the summer, with intermittent open leads across the channels or along the shores.

## Organic Production under Ice and Snow

Foraminifera feed on algae or organic debris (Boltovskoy and Wright, 1976) and therefore are influenced by the primary production rates. The annual production in the ice-covered Arctic waters is very low relative to other areas, because of the low average radiation from the sun and sky, and because of snow and ice cover obstructing radiation to the euphotic zone (English, 1961). The maximum daily insolation around latitude 70°N takes place during late June, but is negligible between November and February (Holmes, 1957). Thus the insolation maximum is about a month earlier than the maximum open water. About 80 percent of sunlight is reflected by snow and ice and about 40 percent by meltwater pools (Hanson, 1961). At Point Barrow, Alaska, about 5 percent of incident light reached 100 cm through snow and ice (Little, Allen, and Wright, 1972).

The phytoplankton production undergoes extensive seasonal variations in response to the fluctuations of the available energy. In the Arctic waters the algal growth season is about 120 days, usually between April and October (Subba Rao and Platt, 1984). In the Arctic Ocean chlorophyll-*a* concentrations integrated between 4 m and 64 m of water during the nonproductive winter months and the productive months of July and August are 64 times higher (English, 1961).

There are also regional differences in carbon production in the Arctic waters. For the Arctic shelf waters the primary production rate is about 17 g C $y^{-1}$ $m^{-2}$; for the slope it is 9 g C $y^{-1}$ $m^{-2}$. In these polar waters the low temperature limits the rate of photosynthesis and the regional differences may relate to differing physical conditions, such as the deepening of the euphotic zone due to the earlier removal of sea ice along the shore leads. The melting of ice also releases algae that initiate spring blooms of phytoplankton (Subba Rao and Platt, 1984).

Planktonic foraminifera also undergo drastic seasonal changes in the standing crop under Arctic conditions. On the Beaufort shelf the average numbers of planktonic foraminifera during the summer of 1970 was 7,283 tests per 1,000 $m^3$ of water (Vilks, 1973). The summer-to-winter reduction of planktonic tests was 221 to 1; a corresponding reduction in total biomass collected in the plankton tows was 34 to 1.

## Field Methods for Collecting Foraminifera

Field work on the Canadian Arctic continental shelf is carried out from vessels capable of maneuvering between ice floes or from ice as a sampling platform. Transports to sample stations are by sled or aircraft; because of load limits and because of difficulties in getting through the ice only small gravity corers and surface grabs have been used to collect samples in areas of perennial ice. The seafloor of the ice-covered areas of the Canadian Arctic Archipelago also lack the high-resolution shallow seismic surveys that are presently an integral part of sediment coring from ships in open water. Lack of long sediment cores and shallow seismic records are the main reasons for a rather poor understanding of the marine Quaternary geology of the channels.

Because of the ice constraint during most of the year, sediment sampling from the Beaufort shelf and the Parry Channel system has taken place from survey vessels on an *ad hoc* basis during a two-month open season. The first systematic sampling was carried out from *CSS Hudson* during September 1970 when close to 100 surface grabs and piston cores were taken on Beaufort shelf, and at 23 localities in the Parry Channel system and Northern Baffin Bay. In most of these localities planktonic foraminifera were also collected from the water column using a 200 mm mesh net that was towed vertically through the upper 200 m of water.

Winter plankton sampling was carried out on the Beaufort shelf during March 1972. Holes in the ice were drilled with a 12 in. auger. A plankton net with a collapsible frame was passed through the hole. The net was spread out under the ice to collect plankton. In M'Clure Strait plankton sampling during July 1968 was carried out from ice but leads and seal holes enlarged by meltwater were used to penetrate the ice with a conventional plankton net.

Field methods in the high Canadian Arctic Archipelago depended on the extent of the survey area. The Prince Gustaf Adolf Sea was surveyed with an S-55 Sikorsky helicopter. An ice auger was used to gain access holes through the ice and a light coring rig was used to take sediment samples (Marlowe and Vilks, 1963). Along the coastline of Ellef Ringnes Island and MacKenzie King Island, the field personnel and equipment were transported between camp sites and sampling stations by toboggans drawn by motor toboggans (Vilks, 1964). Sediments were taken through the 5 in. holes with a mud snapper with a 300 cc capacity. Thus, in the areas of perennial ice the samples are small and the cores rarely exceed 1 m.

Navigation in the Arctic Archipelago during the 1960s depended on a Decca Chain maintained by the Polar Continental Shelf Project. This agency of the Canadian federal government also provided logistics for the various scientific field parties operating on the islands and in the channels and seas. Modern navigation is less of a problem due to the advent of satellite navigation systems that are sufficiently versatile and light for use in medium-size helicopters.

## DISTRIBUTION OF PLANKTONIC FORAMINIFERA IN THE WATER COLUMN

*Neogloboquadrina pachyderma* has been collected with plankton tows on the Beaufort shelf and the Parry Channel (Vilks, 1973) and M'Clure Strait (Vilks, 1970). Along the continental slope of the Beaufort shelf and in the MacKenzie Canyon there are on the average 30 tests $m^3$ of water filtered, but the continental shelf water that is under the influence of the MacKenzie runoff plume is barren of planktonic tests. During winter only 0.4 percent of the summer population are present. In the Parry Channel, excluding eastern Lancaster Sound, an average of 6 tests $m^3$ were collected. At the Lancaster Sound station a high number of 175 tests $m^3$ were present in a completely mixed water column. The M'Clure Strait average summer standing crop was 0.3 tests $m^3$ of water, which is approximately four times larger than in the Arctic Ocean (Bé, 1960).

On the Beaufort shelf and in the waters of the Canadian Arctic Archipelago, the standing crop of planktonic foraminifera is adversely affected by river runoff and extensive cover of ice during the dark season. In Parry Channel, the standing crop decreases with the distance from the Arctic Ocean and Baffin Bay. The standing crop is greatest in areas of upwelling along the continental slope of the Beaufort Sea and in areas of mixing in Lancaster Sound.

## DISTRIBUTION OF FORAMINIFERA IN SEDIMENTS

The distribution of planktonic foraminifera in the sediment reflects the local rate of sedimentation and carbonate preservation rather than zones of higher plankton production. The highest number of tests found in sediments of the Beaufort continental slope is offshore the zone of the highest production. This may be due to lower allochthonous sedimentation rates to the offshore.

Sediments of the Parry Channel contain only a trace of planktonic tests, including Lancaster Sound station, where exceptionally high planktonic production rates were found. The highest planktonic test numbers are present at the western entrances to the channels; in M'Clure Strait and Prince Gustaf Adolf Sea there are large areas with more than 1,000 tests per gram of sediment. The temperature and salinity profiles do not differ significantly between the entrance to Prince Gustaf Adolf Sea, Hazen Strait, M'Clure Strait, and Viscount Melville Sound. The data suggest that the physical characteristics of the water do not change significantly before the shallow Barrow Strait. Therefore, the change in the abundance of planktonic tests on the seafloor from M'Clure Strait to Lancaster Sound may be due to different preservation rates.

## VARIATIONS IN MORPHOLOGY OF *N. PACHYDERMA*

The *Hudson-70* expedition provided a set of samples allowing a comparison of major morphological differences of *Neogloboquadrina pachyderma* between the living population in the water column and the preserved tests in the sediment. Population comparisons can also be made between the Beaufort Sea and the Northern Baffin Bay waters and Viscount Melville Sound.

In the water column the maximum diameters of planktonic tests of the Beaufort Sea are significantly larger than in Lancaster Sound (Fig. 21-5; Vilks, 1975), although the mean number of chambers per test is not significantly different between the two areas. The equal number of chambers would suggest that the two populations are equally mature; the smaller size of tests in Lancaster Sound can be explained on the basis of shorter generations due to more favorable conditions in the area of mixing waters. Also significant is the correlation between the numbers of chambers per test and its maximum diameter. This suggests that for most tests the rate of growth has not slowed down and that each new chamber added has increased in size at a constant rate.

The test size and the number of chambers are similar in the Beaufort Sea and the Parry Channel sediments. The average maximum diameter of tests in the sediment is 238 microns, which is significantly larger than the 197 microns average for

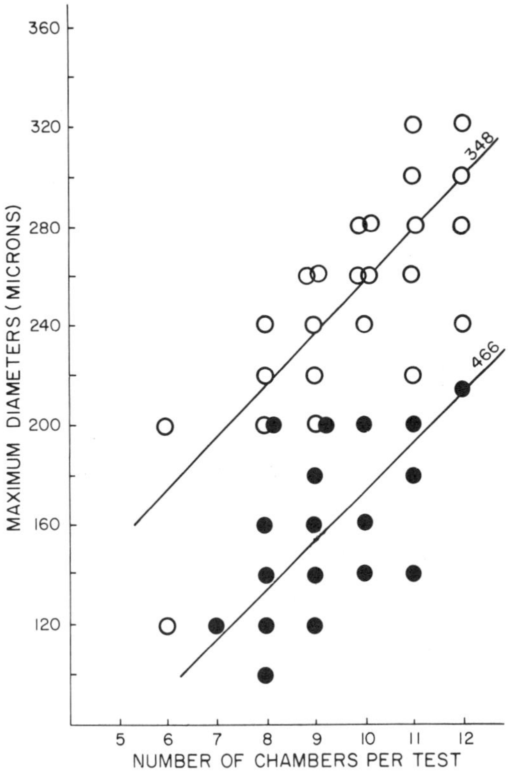


Figure 21-5. Relationship of maximum diameters to the number of chambers per test of *N. pachyderma*. Dots represent Beaufort Sea station 348, Lat. 70°6.5'N, Long. 130°31.0'W. Circles: Lancaster Sound station 466, Lat. 74°01'N, Long. 70°56'W.

tests collected from the water column. The main chamber count of the larger preserved tests is 11 in comparison to only 9 chambers in the plankton tow tests.

The morphologic difference between the tests in the water column and sediments is most likely due to preferential dissolution of the smaller tests on the seafloor. It is also possible that in the waters of the relatively shallow continental shelf and channels planktonic foraminifera live throughout the water column and continue to grow on the seafloor. If this were true in the Canadian Arctic, then areas of higher-standing crops in the water column would coincide with greater numbers of tests in the sediment. The inverse relationship exists in the Canadian Arctic Archipelago. In the very productive waters of Lancaster Sound with respect to planktonic foraminifera, the seafloor is practically barren of planktonic tests. In M'Clure Strait, where the number of tests in the water column was the smallest, the number of preserved tests in the sediment was the highest.

## DISTRIBUTION OF BENTHIC FORAMINIFERA

Foraminifera on the Arctic shelf as a rule occur in larger numbers in the finer sediments, where in addition to factors such as availability of food, preservation may be enhanced. Except for the Parry Channel, the sediments of the archipelago and the Canadian Beaufort shelf consist mainly of fine silts and clays at all water depths sampled. The distribution of species found in these fine sediments varies with water depths, most likely reflecting differences in oceanographic setting within the water column. Regional differences of species distribution could be related to ice cover, major water mass characteristics, or the presence of carbonates in the sediment.

In each sample, few species normally occur in large frequencies and a larger number of species occur in low frequencies. The low-frequency species may contribute very little toward the characterization of the environment and their presence or absence may be susceptible to chance distribution and various methodological errors. To minimize these effects, only major species are considered in this chapter. Major species are those that rank up to 90 percent of all species present in a sample or occur at least at the 2 percent level. Samples with low foraminiferal numbers are not statistically valid; therefore, only those samples that contained more than 100 tests per 35 ml of sediment, or 1.2 tests per gram of dry sediment, were considered.

The mean depth of species occurrence at a study site was calculated according to the equation (Vilks, Frances, and Pelletier, 1979)

$$\ln D = \sum_{i=1}^{N} f_{Ai} \ln D_i; \qquad D_A = e^{\ln} D_A$$

where

$D_A$ = specific depths of species $A$ in a study site
$N$ = number of stations at which the species is present
$f_A$ = frequency of species $A$
$D$ = water depth at each station

Hence the mean specific depths are weighed by frequencies at which species occur at each station.

The mean frequency of species at each sample site with its specific depth is shown in Table 21-1 (pages 510–511). For each species, the mean of the foraminiferal means from each location is shown in Table 21-2 (page 512), including the mean of the specific mean depths. In Table 21-2, the species are ranked according to their frequencies and the rank numbers are used as species codes in the dominant species distribution plots of each sample site (Figs. 21-6 through 21-14).

## BEAUFORT SHELF

The distribution of foraminiferal tests on the Beaufort shelf is patchy with large areas of depleted faunas that cannot be explained in terms of the environment. It is possible that postmortem destruction of tests takes place. From 45 core top samples, 31 contained a sufficient number of specimens (see Fig. 21-6). Twenty-five benthic species are major and dominated by calcareous assemblages; cumulative arenaceous percentage is only 6 (Table 21-1).

The distribution of dominant species in

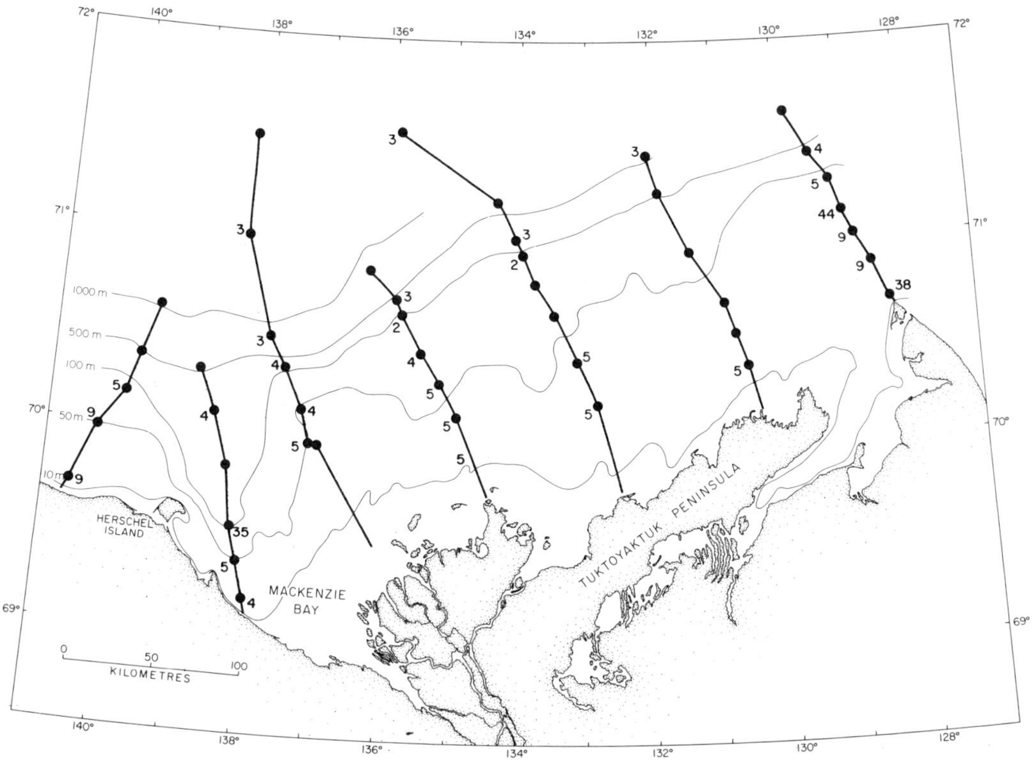


Figure 21-6. Distribution of the dominant species on Beaufort shelf. See Table 21-2 for species codes, for example, 3 is *Cassidulina laevigata* and it dominates the offshore stations.

each sample is shown in Fig. 21-6. *Elphidium excavatum* f. *clavata* (code number 5) is the most commonly dominant species on the inner shelf facing the delta. Associated with *E. excavatum* is *Cassidulina reniforme* (4) along the outer margin of the *E. excavatum* zone. *Cassidulina laevigata* (3) occurs as the dominant species along the continental slope and in the Canada Basin, presumably outside the influence of the MacKenzie River runoff plume. *Islandiella helenae* (9) occurs on the inner shelf but along the outer margins of the delta area.

The high-ranking species *Spiroplectammina biformis* (2) only dominates at two stations at the shelf break. The specific depth of *S. biformis* on Beaufort shelf is 51 m (Table 21-1) but it is basically a shallow-water species elsewhere in the Arctic; therefore, its mean depth in Table 21-1 is only 16 m. Three low-ranking Arctic species dominate one assemblage each on the Beaufort shelf: *Fursenkoina fusiformis* (35) in the fine sediments of the Mackenzie Canyon, *Haynesina orbiculare* (38) at a nearshore station, and *Haynesina nana* (44) associated with the high-diversity faunas toward the Amundsen Gulf.

The range of water depth in the Beaufort shelf study area includes Arctic bottom water at the two Canada Basin sites, Atlantic water along the continental slope, and Arctic surface water on the shelf. The range of temperatures and salinities of the upper 500 m is shown in Figure 21-4.

The distinct faunal zonation in the Beaufort shelf study area reflects the local oceanography. The Arctic bottom water and the Atlantic water depths are dominated by *C. laevigata;* on the shallow shelf *E. excavatum* dominates within the area of the runoff plume, and the other species

Table 21-1. Mean Species Frequencies at Study Areas

| | Beaufort Shelf | | N.W. Passage | | M'Clure Strait | | East Bay | | P.G.A. Sea | | H. and G. Bay | | Byam Channel | | Satellite Bay | | Jones Sound | |
|---|---|---|---|---|---|---|---|---|---|---|---|---|---|---|---|---|---|---|
| | X | D | X | D | X | D | X | D | X | D | X | D | X | D | X | D | X | D |
| *E. excavatum* | 15 | 46 | 20 | 124 | | | | | | | X | 203 | 2 | 13 | | | 2 | 349 |
| *C. reniforme* | 12 | 67 | 33 | 198 | 5 | 413 | | | 6 | 362 | 3 | 314 | 1 | 13 | | | 4 | 349 |
| *C. laevigata* | 9 | 992 | 17 | 347 | 10 | 405 | | | 63 | 386 | 11 | 335 | | | | | | |
| *I. helenae* | 9 | 53 | 2 | 84 | | | | | | | | | 1 | 13 | | | 7 | 290 |
| *H. orbiculare* | 4 | 46 | 3 | 90 | | | X | 60 | X | 382 | | | 1 | 13 | | | | |
| *S. biformis* | 4 | 51 | | | | | 53 | 16 | | | 3 | 43 | 58 | 13 | 1 | 32 | 3 | 441 |
| *B. frigida* | 2 | 49 | 1 | 93 | 1 | 443 | | | 1 | 414 | X | 146 | 1 | 13 | | | 3 | 280 |
| *E. bartletti* | 1 | 40 | | | X | 487 | | | | | X | 309 | | | | | | |
| *E. subarcticum* | 1 | 61 | 1 | 95 | 1 | 412 | | | | | | | 1 | 13 | | | | |
| *E. pulchella* | 1 | 51 | 1 | 84 | | | X | 9 | | | | | 1 | 13 | | | X | 316 |
| *H. nana* | 1 | 39 | | | | | | | | | | | | | | | | |
| *C. lobatulus* | 1 | 175 | | 84 | 5 | 436 | | | 5 | 376 | 1 | 236 | X | 13 | | | 6 | 242 |
| *F. fusiformis* | 1 | 158 | 2 | | | | | | 1 | 199 | | | | | | | X | 435 |
| *T. nana* | 1 | 199 | | 84 | 17 | 359 | 18 | 49 | | | 28 | 123 | 7 | 13 | 46 | 92 | 11 | 511 |
| *O. umbonatus* | 1 | 1634 | 7 | | 5 | 425 | | | 3 | 424 | 1 | 345 | | | | | | |
| *S. horvathi* | 1 | 1995 | | | | | | | 2 | 371 | 1 | 370 | | | | | | |
| *T. torquata* | X | 65 | | | X | 243 | X | 154 | | | 3 | 48 | 18 | 13 | | | 2 | 483 |
| *P. novangliae* | X | 32 | | | | | | | | | | | | | | | | |
| *I. norcrossi* | X | 60 | | 196 | 4 | 405 | | | 1 | 380 | X | 43 | | | | | | |
| *S. concava* | X | 125 | 5 | 118 | 1 | 383 | | | 1 | 363 | X | 422 | | | | | 1 | 314 |
| *R. turbinatus* | X | 194 | 1 | | X | 243 | 1 | 199 | | | 4 | 201 | | | 2 | 170 | | |
| *M. zaandami* | X | 417 | | | 1 | 473 | | | X | 382 | | | | | | | 2 | 334 |
| *E. advena* | X | 68 | | | | | | | | | | | 2 | 13 | | | | |
| *T. quadriloba* | X | 84 | | | 3 | 334 | X | 148 | | | 1 | 141 | | | 1 | 68 | | |
| *C. carssimargo* | X | 70 | | | 1 | 239 | 2 | 85 | | | 8 | 105 | 1 | 13 | 1 | 37 | 5 | 423 |
| *A. gallowayi* | | | | | X | 379 | | | | | | | | | | | 1 | 314 |
| *R. contortus* | | | | | 15 | 430 | | | | | 3 | 337 | | | | | | |
| *S. sphaerica* | | | | | 7 | 385 | 1 | 103 | X | 382 | 3 | 227 | | | | | 1 | 570 |
| *R. fusiformis* | | | | | 3 | 419 | X | 131 | | | 1 | 244 | 1 | 13 | | | 6 | 585 |
| *Hyperammina. sp.* | | | | | 2 | 414 | 2 | 41 | X | 382 | 1 | 178 | | | | | 1 | 424 |
| *R. guttifer* | | | | | 2 | 367 | X | 284 | | | 4 | 287 | | | | | | |
| *N. auricula* | | | | | 2 | 381 | | | | | | | | | | | | |
| *R. nodulosus* | | | | | 2 | 431 | 3 | 76 | | | 2 | 233 | | | | | | |

| | | | | | | | | | | | | | | | | | | |
|---|---|---|---|---|---|---|---|---|---|---|---|---|---|---|---|---|---|---|
| *T. trihedra* | | | | | 1 | 372 | | | | | | | | | | | | |
| *N. labradorica* | | | | | 1 | 383 | | | | | | | | | | | X | 316 |
| *S. ramosa* | | | | | 1 | 368 | | | | | | | | | | | | |
| *C. jeffreysii* | | | | | X | 240 | X | 79 | | | 2 | 95 | | | 11 | 148 | X | 641 |
| *S. tegminis* | | | | | X | 398 | | | | | | | | | | | | |
| *T. bartletti* | | | | | X | 417 | | | | | | | | | | | | |
| *Q. seminulum* | | | | | X | 329 | | | X | 394 | X | 381 | | | | | | |
| *G. arctica* | | | | | X | 381 | | | | | | | | | | | | |
| *Q. arctica* | | | | | X | 398 | | | X | 463 | | | | | | | | |
| *R. charlottensis* | | | | | X | 398 | | | | | | | | | | | | |
| *T. earlandi* | | | | | | | 6 | 70 | | | X | 70 | | | 1 | 32 | 2 | 505 |
| *S. atlantica* | | | | | | | 1 | 19 | | | 2 | 161 | 3 | 13 | 1 | 32 | 5 | 451 |
| *T. bullata* | | | | | | | 1 | 103 | | | 4 | 210 | | | 2 | 93 | | |
| *T. atlantica* | | | | | | | X | 131 | | | 4 | 114 | | | 3 | 81 | | |
| *P. variabilis* | | | | | | | X | 4 | | | | | | | | | | |
| *A. arenaria* | | | | | | | X | 226 | | | | | | | | | | |
| *R. gracilis* | | | | | | | X | 47 | | | | | | | | | | |
| *S. groenlandica* | | | | | | | X | 86 | | | | | | | | | 1 | 581 |
| *B. filiformis* | | | | | | | X | 234 | | | | | | | | | | |
| *T. inflata* | | | | | 2 | 301 | X | 201 | | | | | | | 13 | 112 | | |
| *V. arctica* | | | | | | | | | 4 | 321 | 1 | 294 | | | | | | |
| *A. glomerata* | | | | | | | | | X | 382 | | | | | | | 22 | 458 |
| *R. pilulifer* | | | | | | | | | | | X | 349 | | | | | | |
| *A. cassis* | | | | | | | | | | | | | 1 | 13 | | | | |
| *R. arctica* | | | | | | | | | | | | | 1 | 13 | | | 1 | 467 |
| *Q. agglutinata* | | | | | | | | | | | | | X | 13 | | | | |
| *R. laevigatum* | | | | | | | | | | | | | X | 13 | 11 | 117 | | |
| *T. globigeriniformis* | | | | | | | | | | | | | | | 1 | 32 | | |
| *H. subnodosa* | | | | | | | | | | | | | | | | | 3 | 364 |
| *E. vitrea* | | | | | | | | | | | | | | | | | 1 | 359 |
| *T. fluens* | | | | | | | | | | | | | | | | | X | 218 |
| *T. ochracea* | | | | | | | | | | | | | | | | | X | 531 |
| Percent arenaceous species | 28 | | 8 | | 44 | | 92 | | 18 | | 60 | | 58 | | 100 | | 54 | |
| Cumulative percent | 6 | | 8 | | 56 | | 91 | | 1 | | 74 | | 92 | | 94 | | 64 | |

X = mean frequency; percentages are less than one.
D = mean water depth of occurrence.
Northwest Passage includes Prince of Wales Strait, Viscount Melville Sound, Barrow Strait, and Lancaster Sound. H. and G. Bay is Hecla and Gripper Bay and Hazen Strait study area.

Table 21-2. Major Species of the Canadian Arctic Continental Shelf

| Species | Mean % | Mean Depth (m) | Species | Mean % | Mean Depth (m) |
|---|---|---|---|---|---|
| 1. *Trochammina nana* | 14.6 | 111 | 34. *Melonis zaandami* | 0.4 | 379 |
| 2. *Spiroplectammina biformis* | 13.6 | 16 | 35. *Fursenkoina fusiformis* | 0.3 | 199 |
| 3. *Cassidulina laevigata* | 12.2 | 382 | 36. *Eggerella advena* | 0.3 | 16 |
| 4. *Cassidulina reniforme* | 7.1 | 190 | 37. *Reophax subnodosum* | 0.3 | 364 |
| 5. *Elphidium excavatum f. clavata* | 4.3 | 80 | 38. *Haynesina orbiculare* | 0.2 | 55 |
| 6. *Textularia torquata* | 2.7 | 23 | 39. *Elphidium bartletti* | 0.2 | 94 |
| 7. *Adercotryma glomerata* | 2.5 | 457 | 40. *Astrononion gallowayi* | 0.2 | 258 |
| 8. *Cibicides lobatulus* | 2.3 | 254 | 41. *Nonionella auricula* | 0.2 | 381 |
| 9. *Islandiella helenae* | 2.1 | 97 | 42. *Reophax arctica* | 0.2 | 78 |
| 10. *Cribrostomoides crassimargo* | 2.0 | 131 | 43. *Haynesina nana* | 0.1 | 39 |
| 11. *Cribrostomoides subglobosus* | 2.0 | 413 | 44. *Triloculina trihedra* | 0.1 | 372 |
| 12. *Trochammina inflata* | 1.5 | 113 | 45. *Nonionellina labradorica* | 0.1 | 366 |
| 13. *Cribrostomoides jeffreysii* | 1.5 | 143 | 46. *Saccorhiza ramosa* | 0.1 | 368 |
| 14. *Saccammina sphaerica* | 1.4 | 313 | 47. *Quinqueloculina seminulum* | 0.1 | 368 |
| 15. *Reophax fusiformis* | 1.3 | 357 | 48. *Quinqueloculina arctica* | 0.1 | 430 |
| 16. *Saccammina atlantica* | 1.3 | 98 | 49. *Silicosigmoilina groenlandica* | 0.1 | 374 |
| 17. *Islandiella norcrossi* | 1.2 | 254 | 50. *Ammotium cassis* | 0.1 | 13 |
| 18. *Oridorsalis umbonatus* | 1.1 | 475 | 51. *Trochammina globigeriniformis* | 0.1 | 32 |
| 19. *Buccella frigida* | 1.0 | 137 | 52. *Epistominella vitrea* | 0.1 | 359 |
| 20. *Textularia earlandi* | 1.0 | 98 | 53. *Scutuloris tegminis* | 0.03 | 398 |
| 21. *Recurvoides laevigatum* | 1.0 | 110 | 54. *Trichohyalus bartletti* | 0.03 | 417 |
| 22. *Recurvoides turbinatus* | 0.8 | 193 | 55. *Gordiospira arctica* | 0.03 | 381 |
| 23. *Reophax nodulosa* | 0.8 | 172 | 56. *Robertinoides charlottensis* | 0.03 | 398 |
| 24. *Tritaxis bullata* | 0.8 | 150 | 57. *Pelosina variabilis* | 0.03 | 4 |
| 25. *Tritaxis atlantica* | 0.8 | 100 | 58. *Artrorhiza arenaria* | 0.03 | 226 |
| 26. *Hyperammina sp.* | 0.7 | 174 | 59. *Reophax gracilis* | 0.03 | 47 |
| 27. *Reophax guttifer* | 0.7 | 310 | 60. *Pseudopolymorphina novangliae* | 0.03 | 32 |
| 28. *Valvulineria arctica* | 0.6 | 319 | 61. *Reophax pilulifer* | 0.03 | 349 |
| 29. *Trochammina quadriloba* | 0.6 | 192 | 62. *Quinqueloculina agglutinata* | 0.03 | 13 |
| 30. *Stainforthia concava* | 0.5 | 263 | 63. *Bathysiphon filiformis* | 0.03 | 234 |
| 31. *Eoponidella pulchella* | 0.4 | 40 | 64. *Trifarina fluens* | 0.03 | 218 |
| 32. *Stetsonia horvathi* | 0.4 | 564 | 65. *Trochammina ochracea* | 0.03 | 531 |
| 33. *Elphidium subarcticum* | 0.4 | 75 | | | |

are prominent under various degrees of lesser runoff influence.

## THE NORTHWEST PASSAGE

Destruction of foraminifera in the Northwest Passage sediments appears to be extensive; only 4 core top samples of 11 collected contain a statistically significant number of tests (Fig. 21-7, Table 21-3). The preserved fauna is basically calcareous, containing only 8 percent of arenaceous foraminifera. The low numbers of foraminifera in Parry Channel sediment are difficult to explain, considering the expected slow sedimentation rates and the large range of water depths sampled. *Cassidulina laevigata* is the dominant species in these cores and it is possible that the species is more resistant to dissolution in comparison to the other species.

The Amundsen Gulf sediments contain a low-diversity calcareous fauna dominated by *C. laevigata* (3) (Fig. 21-7). Here the bottom water is within the relatively warm and saline Atlantic layer (Fig. 21-4). The more diverse fauna of the shallow Prince of Wales Strait and Viscount Melville Sound cores are dominated by *Cassidulina reniforme* (4) and *Elphidium excavatum* f. *clavata* (5). The water depth in these localities is within the colder and less saline Arctic surface water.

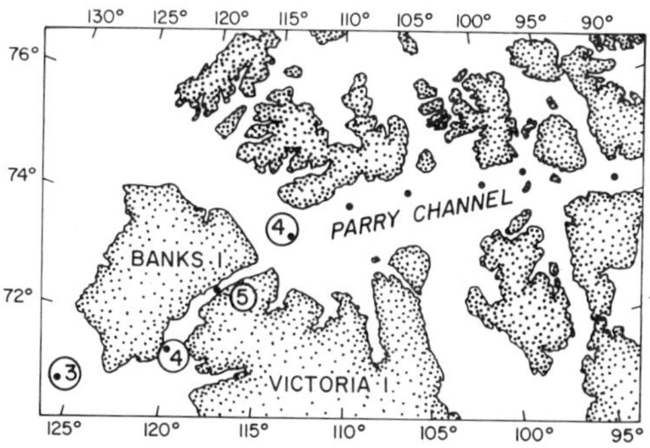


Figure 21-7. The distribution of the dominant species in Northwest Passage. Dots designate station locations and unnumbered dots indicate stations barren of foraminifera.

## M'CLURE STRAIT

Iqbal (1973) analyzed foraminifera in 40 grab samples taken from M'Clure Strait and the continental shelf at depths be-

Table 21-3. Relative Percent of Major Benthic Foraminifera from the Northwest Passage

| *Station* | *445* | *448* | *449* | *451* |
|---|---|---|---|---|
| *Water Depth (m)* | *409* | *118* | *84* | *84* |
| *Cassidulina reniforme* | 34 | 40 | 24 | 35 |
| *Cassidulina laevigata* | 58 | 10 | | |
| *Elphidium excavatum* | | 30 | 50 | |
| *Stainforthia concava* | | 2 | | |
| *Buccella frigida* | | 2 | 2 | |
| *Islandiella norcrossi* | | 2 | | 18 |
| *Elphidium subarcticum* | | 2 | | |
| *Cibicides lobatulus* | | | 8 | |
| *Eoeponidella pulchella* | | | 2 | |
| *Trochammina nana* | | | 7 | 21 |
| *Islandiella helenae* | | | | 6 |
| *Haynesina orbiculare* | | | | 11 |
| | Amundsen Gulf | Prince of Wales Strait | | Viscount Melville Sound |

Seven other core top samples taken between 475 m and 84 m of water were almost barren of foraminifera.

tween 230 m and 500 m (Fig. 21-8). The area averages 13 species per sample and 100 tests per gram of dry sediment, which is considerably higher than the Beaufort shelf and Parry Channel sediments. From a total of 74 species, 33 are major (Table 21-1). The mean cumulative percent of the arenaceous species is 56, although only 23 percent of all species present are arenaceous, indicating that most of the calcareous species occur in small numbers. Distinct species depth zonations could not be identified, but the proximity of the Arctic Ocean and distance from land influence the distribution of species. The sandier sediments from slightly shallower localities on the continental shelf are dominated by calcareous species (Fig. 21-8), and the deeper basin sediments behind the sill are dominated by arenaceous species. Oceanographic factors could not account for the major differences in faunal distribution because the bottom water of M'Clure Strait is well within the Atlantic layer of the Arctic Ocean water mass at all depths sampled. The M'Clure Strait bottom temperatures and salinities are among the highest within the envelopes that are shown in Figure 21-4.

Eight species dominate the 40 samples (Fig. 21-8). *Trochammina nana* (1), *Cribrostommoides crassimargo* (10), *Recurvoides contortus* (11), and *Saccammina sphaerica* (14) dominate the inner M'Clure Strait, the two channels, and the coastal zone. In this locality, *Trochammina nana* is dominant both at the shallowest and deepest stations. The calcareous species *Cassidulina laevigata* (3) is by far the most abundant on the Arctic shelf in association with *Oridorsalis umbonatus* (18). Both species also dominate the Arctic Ocean bottom faunas (e.g., Lagoe, 1976). The dominance of *Cassidulina reniforme* (4) at one locality at water depths close to 500 m does not agree with the general ecological concept of this species, suggesting the possibility of sediment reworking.

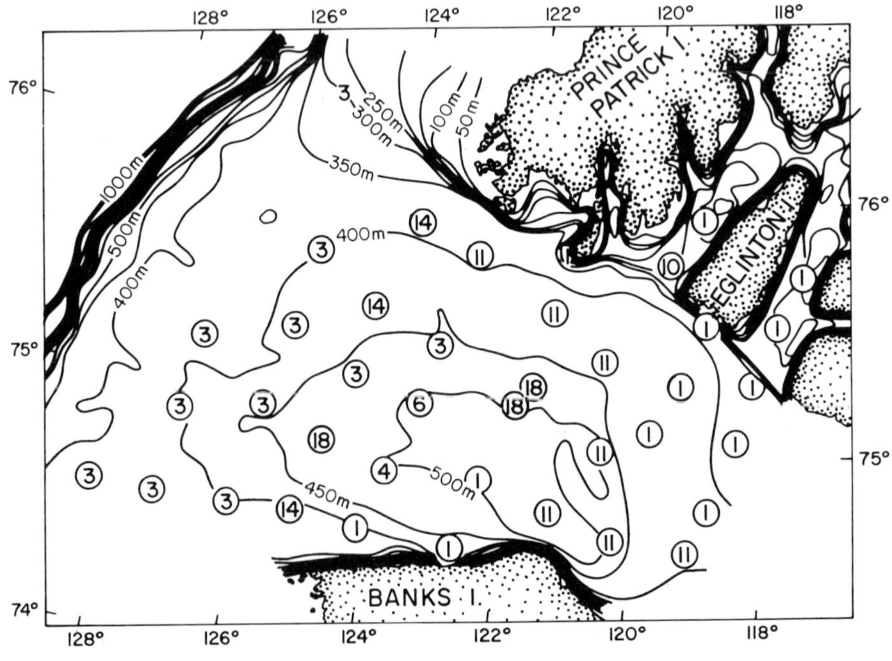


Figure 21-8. The distribution of the dominant species in M'Clure Strait.

## PRINCE GUSTAF ADOLF SEA

Marlowe (1967) collected sediment cores in Prince Gustaf Adolf Sea (Fig. 21-9) where 11 core top samples contained sufficient numbers of benthic foraminiferal tests for analysis (Table 21-1). The cores were collected from the central part of the sea at water depths between 199 m and 488 m. In comparison to M'Clure Strait, the benthic foraminiferal assemblages are less diverse and less abundant. The abundant Milliolid fragments suggest dissolution of carbonates at the seafloor. *Cassidulina laevigata* (3) is by far the most abundant species in all samples, except for the southernmost station off Bathurst Island (Fig. 21-9) where *C. reniforme* (4) is dominant.

All bottom samples came from water depths within the Atlantic layer, which in the northeastern Prince Gustaf Adolf Sea has basically similar characteristics to the M'Clure Strait water (Fig. 21-4). The water temperatures are higher than 0°C and salinities are close to 35 ‰. The polar pack enters the Prince Gustaf Adolf Sea freely and normally stays as a solid perennial cover. It disintegrates at the southern exits where annual open water may occur. Marlowe (1962 field notes) recognized the rem-

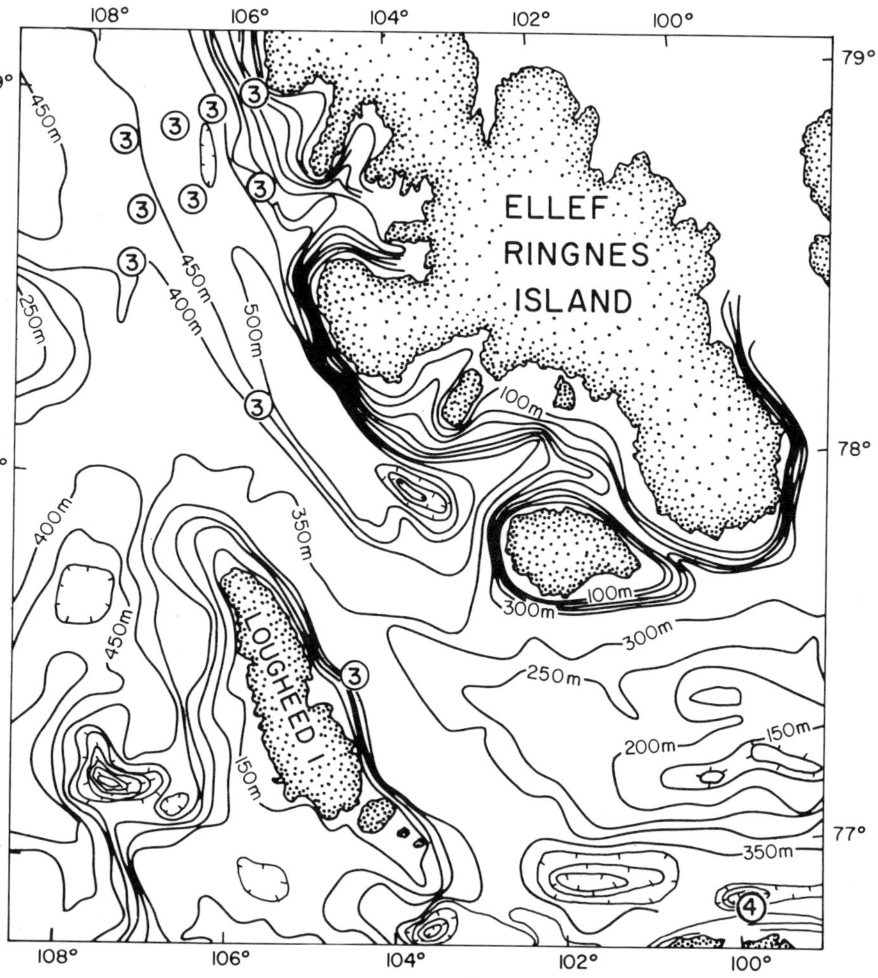


Figure 21-9. The distribution of the dominant species in Prince Gustaf Adolf Sea.

nants of an ice island between Lougheed and King Kristian islands.

The relatively solid ice cover may contribute to the low diversities of benthic foraminifera and low species abundance. Prince Gustaf Adolf Sea and Viscount Melville Sound are the two major conduits of southeasterly flow of Arctic Ocean water through the archipelago; both contain poorly preserved faunas, suggesting the efficiency of current erosion of biogenic carbonate in arctic conditions.

## HECLA AND GRIPPER BAY AND HAZEN STRAIT

Vilks (1969) studied foraminifera from 48 grab samples and 2 small cores taken at water depths between 17 m and 458 m (Fig. 21-10). The foraminiferal populations living in these fine sediments are both diverse and abundant. The average abundances are 58 tests per gram of sediment, 74 percent of which are arenaceous. From a total of 78 species, 30 are major (Table 21-1). Eleven species show a distinct preference to water depth; arenaceous species, such as *Spiroplectammina biformis, Cribrostomoides crassimargo, C. jeffreysi, Tritaxis atlantica,* and *Trochammina nana* have their specific water depths shallower than 200 m. The deep-water calcareous species are *Cassidulina laevigata, C. reniforme, Oridorsalis umbonatus,* and *Stetsonia horvathi;* the deep-water arenaceous species are *Recurvoides contortus* and *Reophax guttifer.*

Ten species dominate the 50 samples

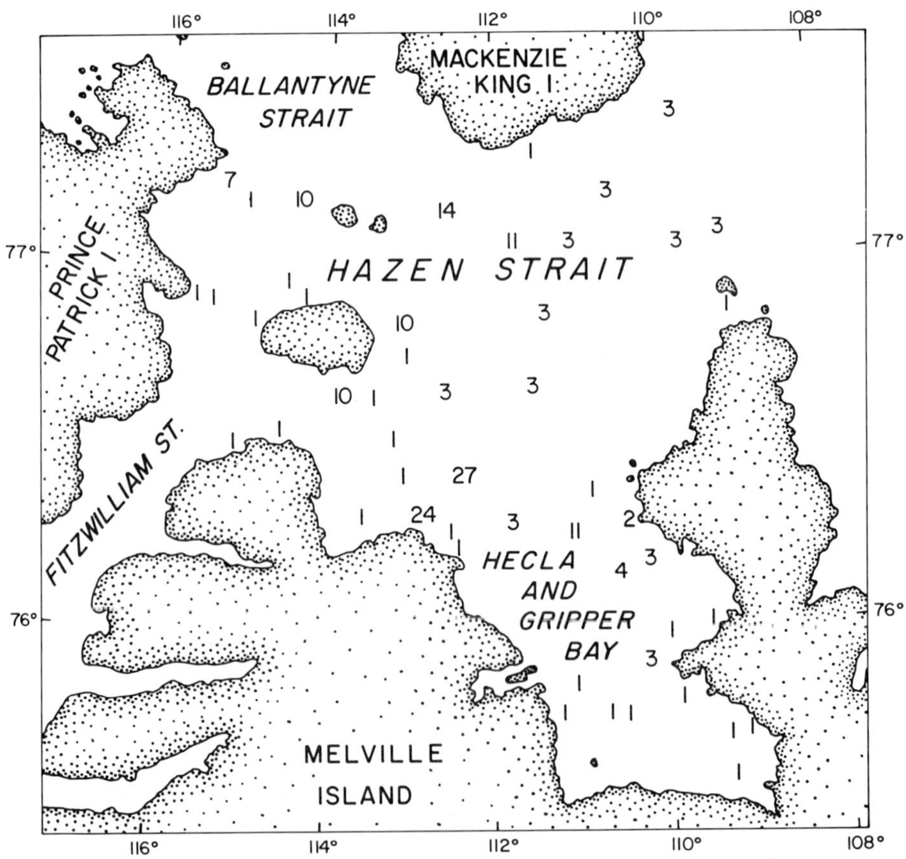


Figure 21-10. The distribution of the dominant species in Hecla and Gripper Bay and Hazen Strait.

(Fig. 21-10) with *T. nana* (1) dominant in 28 locations of the shallower waters. *C. laevigata* (3) is dominant on the deep and flat bottom of Hazen Strait in association with *R. contortus* (11). Two low-ranking species appear as dominant: *Tritaxis bullata* (24) on the shallow shoreface, and *Reophax guttifer* (27) in the deep water of Hecla and Gripper Bay.

The water mass characteristics of the area are similar to those of Prince Gustaf Adolf Sea with positive centigrade temperatures and relatively high salinities below 300 m in the Atlantic water (Fig. 21-4). The temperature is below 0°C in water shallower than approximately 250 m in the Arctic surface water and is less than −1°C above 150 m. Salinity decreases in the upper 200 m from about 34.5‰ to almost 31.5‰. The Atlantic water is characterized by *Cassidulina laevigata* and the Arctic surface water is characterized by the arenaceous Trochamminids and Textulariids.

## JONES SOUND

Jones Sound (Fig. 21-1) is a semirestricted channel to the east of the sills separating the Arctic Ocean and Baffin Bay deep waters. The maximum depths of the basin are around 800 m, with a 150 m sill in Fram Sound and a minimum depth of 450 m at the eastern entrance to the sound (Fig. 21-11). The deep basin sediments are mainly silts and clays but sand and gravel is also found along the margin of the basins and on the top of a shoreward shelf surrounding the basin (MacLean, Woodside, and Girouard, 1984). Salinity and nutrient profiles would suggest that the Arctic Ocean water entering the Jones Sound through Fram Sound overlies the Baffin Bay water as far east as Cape Hardy (Jones and Coote, 1980). Slightly higher silicate concentrations than in Baffin Bay at 600 m reflects the restricted exchange of water between Baffin Bay and Jones Sound below the sill depth. Bottom oxygen values range around 5 ml/l, which are close to bottom oxygen values of some localities in the deep water of Baffin Bay (P. Jones, unpublished data, Bedford Institute of Oceanography).

Foraminifera from surface sediments were identified by Hardy (1984). The assemblages from 11 stations between the water depth of 228 m and 640 m are more diverse than the assemblages in the western basins. The average total species per 35 mg of sample is 33, and the average number of tests per sample is 322. From the total of 28 major species, 15 are arenaceous, dominating the deep basin sediments (Table 21-1, Fig. 21-11). The cumulative mean arenaceous percent is 64.

Five species dominate the Jones Sound sediments (Fig. 21-11). *Adercotryma glomerata* (7) is the most common dominant species associated with *Trochammina nana* (1) in the deep water. The specific depth of *Islandiella helenae* (9) in Jones Sound is 290 m (Table 21-1) but it is also one of the dominant species at a locality in 540 m of water. The presence of coarse sediment in Jones Sound is reflected by the dominance of *Cibicides lobatulus* (6) at two locations.

In comparison to the western basins, the less saline and colder bottom water of Jones Sound is reflected by the dominance of *A. glomerata* and the absence of *Cassidulina laevigata.* The higher diversities of faunas suggest improved preservation of tests combined with higher productivity in an area of mixing Baffin Bay and Arctic Ocean waters.

## EAST BAY

East Bay on Mackenzie King Island (Fig. 21-12) has a maximum recorded water depth of 284 m behind a 200 m sill. The deep basin contains fine sediments along the northern margin and angular pebbles in relatively coarse and clean sand along the southern margin, suggesting current scouring of the seafloor. Close to the

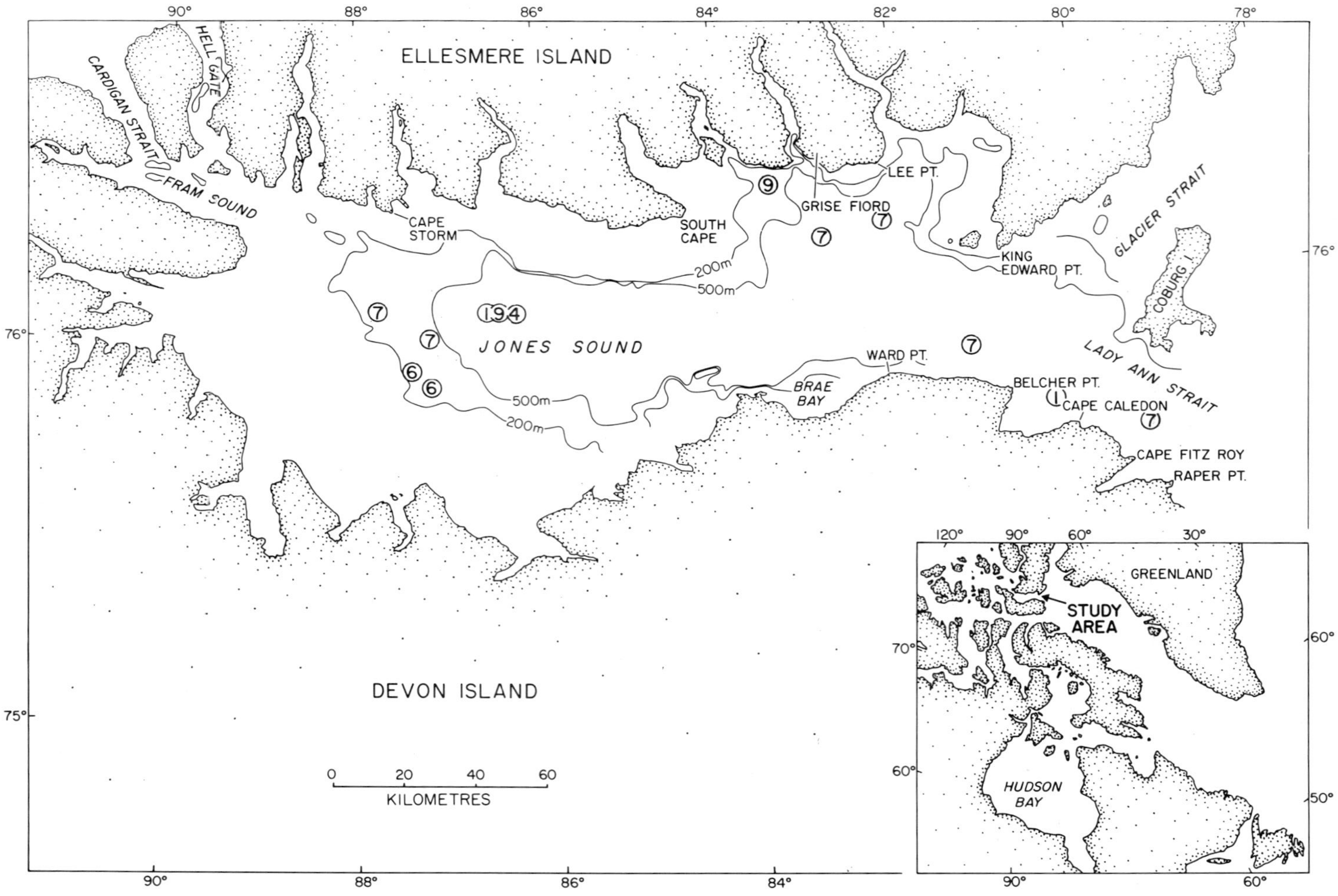


Figure 21-11. The distribution of the dominant species in Jones Sound.

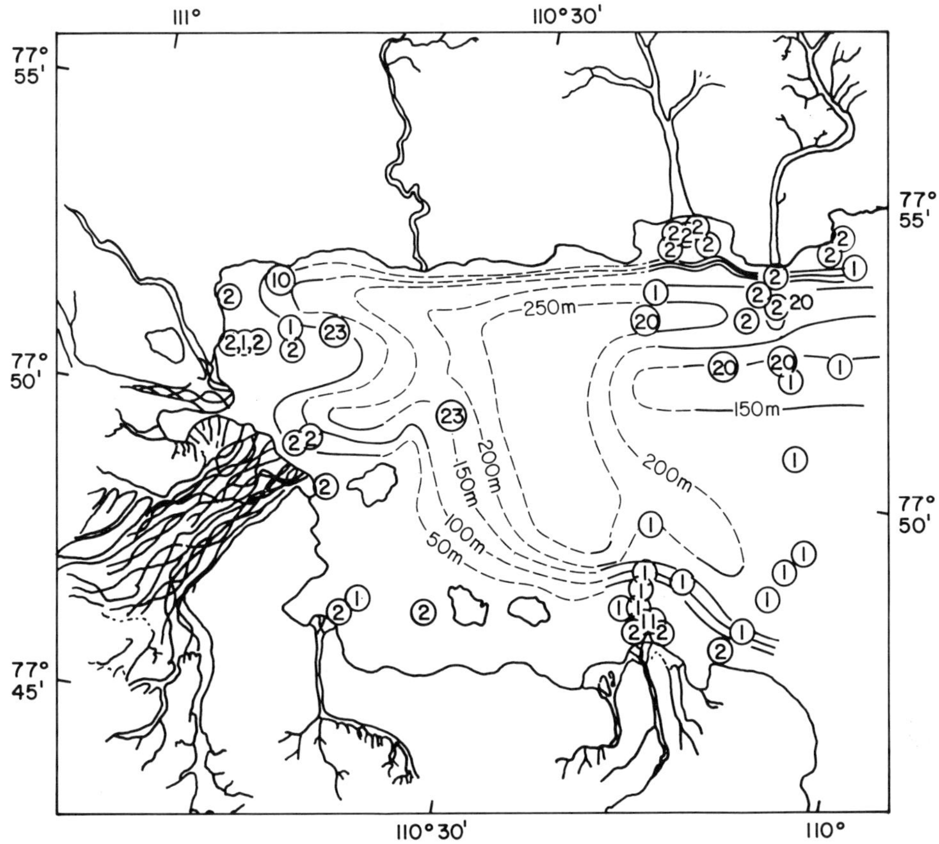


Figure 21-12. The distribution of the dominant species in East Bay, MacKenzie King Island.

shore, sediments are sandy silt and are rich in plant fragments, such as lichen, moss, lignite, and coal.

Foraminifera were studied in 78 surface grab samples with emphasis on the shallow nearshore waters (Vilks, 1963). Forty-eight species were identified, 24 of which are major (Table 21-1). The average abundance of foraminifera is 27 tests per gram of sample, with 46 tests per gram in 0–30 m nearshore zone and decreasing to 11 tests per gram in the 240–300 m zone. The average diversity is 8 species per sample in the shallow zone, increasing to 18 in the deep zone. It is evident that the shallow inshore bays that are open only as shore leads during the summer also contain large numbers of benthic foraminifera.

The foraminiferal assemblages consist almost entirely of agglutinated species, except for *Haynesina orbiculare* and *Eoeponidella pulchella,* which average less than 1 percent. There are 5 dominant species (Fig. 21-12): *Spiroplectammina biformis* (2) dominates the shallow nearshore zone, and *Trochammina nana* (1) dominates the deeper waters associated with *Textularia earlandi* (20) and *Reophax nodulosus* (23); *Cribrostomoides crassimargo* (1) also dominates one nearshore sample site.

The water depths of the sediment samples taken in East Bay are entirely within the Arctic surface water and presumably with characteristics similar to Prince Gustaf Adolf Sea (Fig. 21-4). Polar pack and heavy rafted ice were observed in the central part of the bay during the sampling (May 1963). A 3-km-wide zone of smooth ice along the shore may correspond to the extent of the shore lead. During the summer the mixing of fresh water may be sufficiently delayed in the shore leads to provide a nearshore environment having relatively low summer salinities. The low-diversity *Spiroplectammina biformis* faunas may characterize the extreme conditions in Arctic shore leads.

## BYAM CHANNEL

Foraminifera were studied in sediment cores collected by scuba divers along the shores of Byam Channel at water depths between 3 m and 26 m (Vilks, 1976) (Fig. 21-13). Short cores were taken from furrows recently excavated by sea ice and from the seafloor that did not show evidence of ice scouring. Foraminifera were investigated in 19 cores at intervals of 0–2 cm, 3–5 cm, and 6–10 cm. Table 21-1 summarizes the occurrence of the major species by combining the three layers as a single core top sample taken between 0 cm and 10 cm. This makes the comparison more meaningful of Byam Channel faunas

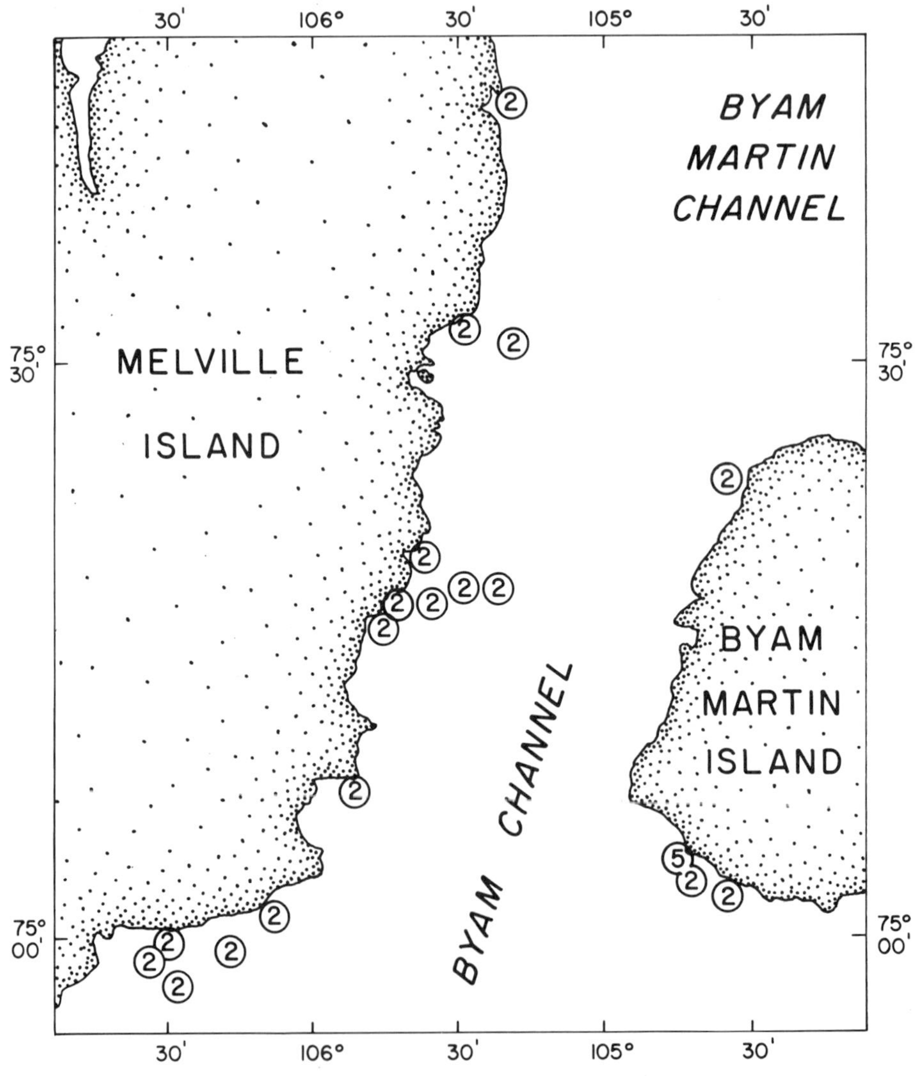


Figure 21-13. The distribution of the dominant species in Byam Channel.

with the other areas where samples were taken under less controlled conditions.

According to the divers, the nearshore zone of Byam Channel consists of bedrock and sand flats to approximately 6–8 m, and poorly sorted sand and silt with pebbles below these depths. The channel is intermittently free of sea ice during the summer.

The 57 subsamples yielded 51 species, 18 of which are listed in Table 21-1 as major species. The fauna consists mainly of arenaceous assemblages and is dominated by *Spiroplectamminia biformis.* Cluster analysis of species assemblages made a weak separation of stations facing Viscount Melville Sound from Byam Channel stations to the north. The Viscount Melville Sound cluster contained almost all the calcareous species that were found in the area. On the average, the surface 0–2 cm layer contained only 11 species, 2 of which were calcareous, and the 5–10 cm level contained 15 species with 5 calcareous species. It seems that the fauna is less diverse at the surface.

Vilks (1976) explained the higher numbers of calcareous species in the subsurface sediments in terms of more extensive sea ice cover in Byam Channel during recent times. In view of increasing evidence that calcareous species migrate in excess of 10 cm down the sediment under adverse conditions, such as deep sea sediments close to carbonate compensation depth (Corliss, 1985), the higher numbers of calcareous species in the 5–10 cm in Byam Martin Channel may be due to habitat preference rather than indicative of paleoceanographic change.

The Byam Channel samples were taken entirely within the seasonal surface layer of the Arctic water, where in addition to the meltwater dilution during summer and temperatures close to freezing point during winter, the seafloor is also disturbed by ice scouring. As in East Bay, here *Spiroplectammina biformis* (2) (Fig. 21-13) is also by far the most dominant species with one addition of *Elphidium excavatum* f. *clavata* (5) off Byam Martin Island.

## SATELLITE BAY

Satellite Bay on the northern shore of Prince Patrick Island (Fig. 21-1) faces the Arctic Ocean at a maximum water depth of 252 m. The bottom topography is irregular with two steep-sided troughs open to the ocean on both sides of the drowned headland continuing from the central peninsula (Fig. 21-14). In the western head region of the bay there is a marked depression with a local topographic difference of about 50 m.

The sediment consists chiefly of light brown mud. The black mud of the nearshore samples of the eastern head region of the bay is characterized by an $H_2S$ odor. Layers of fine black mud were also found on the platform along the east shore at depths less than 10 m.

Six sample stations, primarily from the eastern part of the bay, contain a total of 24 species, 13 of which are listed in Tables 21-1 and 21-4 as major species. All the major species are arenaceous and occur on the average of 138 individuals per gram of dry sediment. Both the numbers and diversity are highest at station 16 facing the ocean. The fauna are dominated by *Trochammina nana* with a strong presence of *Trochammina inflata,* but not by *Spiroplectammina biformis,* which is dominant in the shallow waters of the ice-covered channels of the archipelago. *Recurvoides laevigatum* is also prominent in Satellite Bay.

Satellite Bay samples are within the depth range of the Arctic Ocean surface water. Small local differences may exist, but oceanographic measurements are lacking, including summer ice observations. In May 1964, the surface of the ice in the bay was smooth, which is typical of shorefast ice around the Arctic Ocean. The field observations of the sediment suggest low ox-

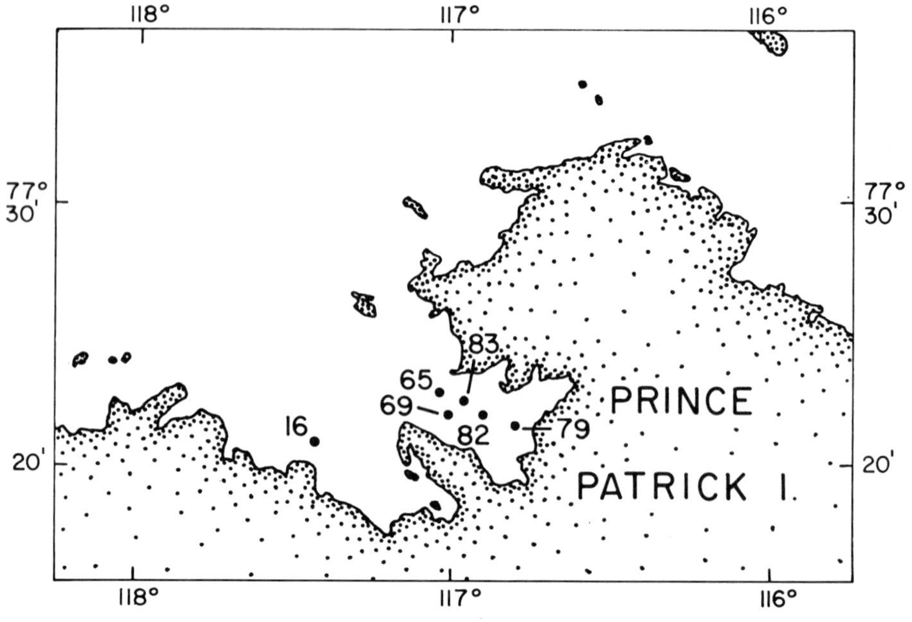


Figure 21-14. Station locations in Satellite Bay.

Table 21-4. Relative Percent of Major Species in Satellite Bay, Prince Patrick Island

| *Station* | *16* | *79* | *82* | *69* | *83* | *65* |
|---|---|---|---|---|---|---|
| *Depth (m)* | *32* | *37* | *142* | *180* | *186* | *186* |
| *Trochammina nana* | 49 | 57 | 47 | 41 | 47 | 33 |
| *Spiroplectammina biformis* | 3 | | | | | |
| *Recurvoides laevigatum* | 5 | 11 | 12 | 17 | 6 | 15 |
| *Saccammina atlantica* | 3 | | | | | |
| *Textularia earlandi* | 3 | | | | | |
| *Tritaxis atlantica* | 8 | | | | 9 | |
| *Trochammina inflata* | 6 | 15 | 21 | 8 | 16 | 14 |
| *Cribrostomoides jeffreysi* | 7 | | 5 | 18 | 6 | 27 |
| *Trochammina globigeriniformis* | 6 | | | | | |
| *Trochammina quadriloba* | 4 | | | | 3 | |
| *Tritaxis bullata* | | 5 | 3 | 5 | | |
| *Cribrostomoides crossimargo* | | 5 | | | | |
| *Recurvoides turbinatus* | | | 4 | | 4 | 4 |
| Diversity (Shannon-Wiener function) | 2.01 | 1.42 | 1.79 | 1.76 | 1.83 | 1.98 |
| Tests per gram dry sediment | 226 | 75 | 207 | 100 | 74 | 146 |

ygen levels at the sediment–water interface in semienclosed areas.

The fauna is unique with the strong presence of *T. inflata* and *R. laevigatum,* both of which occur in much lower numbers elsewhere in the Canadian Arctic. This may indicate slight differences in the environment.

## MAJOR PATTERNS OF SPECIES DISTRIBUTION

Water mass characteristics influence the distribution patterns of the major species. Locally, water mass differences in the stratified Arctic water column reflect species depth zonations. Regionally the differences can be identified with the different sources of bottom water, for example, the Atlantic water of the Arctic Ocean to the west of the major sills in the archipelago and Baffin Bay bottom water to the east of the sills. Species composition is also reflected by the physiography of the basins; semienclosed basins contain richer faunas than conduits of waters. The proximity of the Arctic Ocean also has influence. It is important to identify these major species–environment relationships in the Canadian Arctic Archipelago for the interpretation of faunal changes in sediment cores.

The Arctic deep water and the Atlantic water, regardless of locality, are characterized by calcareous faunas, and the sediments are dominated by *Cassidulina laevigata.* The Arctic surface water is characterized by calcareous faunas only in Beaufort shelf and Prince of Wales Strait sediments at the western entrance of the Northwest Passage. The dominant species are *Elphidium excavatum* f. *clavata* and *Cassidulina reniforme.* The Arctic surface water in the remainder of the western Arctic Archipelago is characterized by arenaceous species and is dominated by *Trochammina nana.* The Arctic seasonal layer of waters less than 20 m deep is dominated by *Spiroplectammina biformis.* Because of the questionable preservation of arenaceous species in sediment cores, it is useful to know which calcareous species could reflect the shallow Arctic Archipelago waters. These could be *Eoeponidella pulchella, Islandiella helenae,* and possibly *Islandiella norcrossi.*

The bottom of Jones Sound is under the influence of the Baffin Bay bottom water and is the only major basin in the archipelago to the east of the sills where the preservation of faunas is favorable. Here the deep water is characterized by arenaceous species and is dominated by *Adecotryma glomerata. A. glomerata* is not very common in the western Arctic Archipelago basins. The shallow water in Jones Sound is characterized by calcareous species and is dominated by *I. helenae. Cassidulina laevigata* was not found in surface sediments, most likely due to the colder and less saline bottom waters.

The less saline deep water in Jones Sound could also explain the much greater specific depths of species in comparison to the west of the sills. For example, the specific depth of *I. helenae* on Beaufort shelf is deeper than its specific depth in Jones Sound, but may relate to insufficient sampling at shallow depths and slumping of sediments and foraminiferal tests down the steep submarine slopes. However, deeper occurrences of shallow species are also observed on the Labrador shelf (Vilks et al., 1982) in the relatively diluted and cold inner shelf Labrador current waters.

## ACKNOWLEDGMENTS

Ms. Ann A. L. Miller, Centre for Marine Geology, Dalhousie University; Dr. B. R. Pelletier, Geological Survey of Canada, Ottawa; and Jim Syvitski, Atlantic Geoscience Centre, B.I.O., Dartmouth, reviewed and edited the manuscript. Their comments and critical suggestions are greatly appreciated.

# APPENDIX

## Taxonomy

The species (Table 21-5) are classified according to the Treatise of Loeblich and Tappan (1964), except for *Haynesina* (Banner and Culver, 1978) and *Neogloboquadrina* (Bandy, Frerichs, and Vincent, 1967). The SEM stubs containing the specimens of illustrated species (Plates 21-I–21-IX) are deposited in the SEM Specimen Repository at the Bedford Institute of Oceanography. The six-digit stub number for each specimen is included with the plate explanations. (GSC = Geological Survey of Canada).

Suborder TEXTULARIINA
Superfamily AMMODISCACEA
Family ASTRORHIZIDAE
Subfamily ASTRORHIZINAE Brady 1881
Genus ASTRORHIZA Sandahl 1858

*Astrorhiza arenaria* Carpenter: Plate 21-I, Figure 1

*Astrorhiza arenaria* Carpenter. *In:* Norman, 1877, p. 213. CUSHMAN, 1948, p. 5, pl. 1, fig. 1.

Test flat, subcircular mass of coarsely agglutinated material from which radiate short extensions, apertures at the end of extensions. Poorly preserved.

| *Locality* | *Average %* | *Average Depth (m)* |
|---|---|---|
| East Bay | X (less than 0.5%) | 226 |

Subfamily RHIZAMININAE Rhumbler 1895
Genus BATHYSIPHON M. Sars 1872

*Bathysiphon filiformis* M. Sars: Plate 21-I, Figure 2

*Bathysiphon filiformis* M. Sars, 1872, p. 251, fig. 4. COLE, 1981, p. 7, pl. 1, fig. 3.

Test walls elastic and fragile; commonly slightly flat and incomplete.

| *Locality* | *Average %* | *Average Depth (m)* |
|---|---|---|
| East Bay | X | 234 |

Superfamily HIPPOCREPININAE Rhumbler 1895
Genus HYPERAMMINA Brady 1878

*Hyperammina* sp.: Plate 21-I, Figure 3

*Hyperammina* sp. Brady, 1878, p. 433, pl. 20, fig. 2a–b.

Test large, cylindrical, consisting of bulbous proloculus and a long cylindrical second chamber; wall arenaceous. Commonly found as incomplete specimens, therefore difficult to establish species.

Table 21-5. Foraminifera

| *Suborder* | *Superfamily* | *Family* | *Subfamily* | *Genus* | *Species* |
|---|---|---|---|---|---|
| Textulariina | Ammodiscacea | Astrorhizidae | Astrorhizinae | *Astrorhiza* | *arenaria* |
| | | | Rhizammininae | *Bathysiphon* | *filiformis* |
| | | | Hippocrepininae | *Hyperammina* | sp. |
| | | | | *Hyperammina* | *subnodosa* |
| | | | | *Saccorhiza* | *ramosa* |
| | | Saccamminidae | Saccammininae | *Saccammina* | *atlantica* |
| | | | | *Saccammina* | *sphaerica* |
| | | | | *Pelosina* | *variabilis* |
| | Lituolacea | Hormosinidae | Hormosininae | *Reophax* | *arctica* |
| | | | | *Reophax* | *fusiformis* |
| | | | | *Reophax* | *gracilis* |
| | | | | *Reophax* | *guttifer* |
| | | | | *Reophax* | *nodulosus* |
| | | | | *Reophax* | *pilulifer* |
| | | Rzehakinidae | | *Silicosigmoilina* | *groenlandica* |
| | | Lituolidae | Haplophragmoidinae | *Adercotryma* | *glomerata* |
| | | | | *Cribrostomoides* | *crassimargo* |
| | | | | *Cribrostomoides* | *jeffreysii* |
| | | | | *Recurvoides* | *contortus* |
| | | | | *Recurvoides* | *laevigatum* |
| | | | | *Recurvoides* | *turbinatus* |
| | | | Lituolinae | *Ammotium* | *cassis* |
| | | Textulariidae | Spiroplectammininae | *Spiroplectammina* | *biformis* |
| | | | | *Textularia* | *earlandi* |
| | | | | *Textularia* | *torquata* |
| | | Trochamminidae | Trochammininae | *Tritaxis* | *atlantica* |
| | | | | *Tritaxis* | *bullata* |
| | | | | *Trochammina* | *globigeriniformis* |
| | | | | *Trochammina* | *inflata* |
| | | | | *Trochammina* | *nana* |
| | | | | *Trochammina* | *quadriloba* |
| | | Ataxophragmiidae | Globotextulariinae | *Eggerella* | *advena* |
| Miliolina | Miliolacea | Fisherinidae | Cyclogyrinae | *Gordiospira* | *arctica* |

Table 21-5. Foraminifera

| *Suborder* | *Superfamily* | *Family* | *Subfamily* | *Genus* | *Species* |
|---|---|---|---|---|---|
| | | Miliolidae | Quinqueloculininae | *Quinqueloculina* | *agglutinata* |
| | | | | *Quinqueloculina* | *arctica* |
| | | | | *Quinqueloculina* | *seminulum* |
| | | | | *Triloculina* | *trihedra* |
| | | | Miliolinellinae | *Scutuloris* | *tegminis* |
| Rotaliina | Nodosariacea | Polymorphinidae | Polymorphininae | *Pseudopolymorphina* | *novangliae* |
| | Buliminacea | Islandiellidae | | *Islandiella* | *helenae* |
| | | | | *Islandiella* | *norcrossi* |
| | | Buliminidae | Bulimininae | *Stainforthia* | *concava* |
| | | Uvigerinidae | | *Trifarina* | *fluens* |
| | Discorbacea | Discorbidae | Discorbinae | *Buccella* | *frigida* |
| | | | | *Eoeponidella* | *pulchella* |
| | | | | *Epistominella* | *vitrea* |
| | | | | *Stetsonia* | *horvathi* |
| | | | Bagginінae | *Valvulineria* | *arctica* |
| | Rotaliacea | Elphidiidae | Elphidiinae | *Elphidium* | *bartletti* |
| | | | | *Elphidium* | *excavatum* |
| | | | | *Elphidium* | *subarcticum* |
| | | | | *Haynesina* | *nana* |
| | | | | *Haynesina* | *orbiculare* |
| | Globigerinacea | Globigerinidae | Globigerininae | *Neogloboquadrina* | *pachyderma* |
| | Orbitoodacea | Cibicididae | Cibicidinae | *Cibicides* | *lobatulus* |
| | Cassidulinacea | Caucasinidae | Fursenkoininae | *Fursenkoina* | *fusiformis* |
| | | Cassidulinidae | | *Cassidulina* | *laevigata* |
| | | | | *Cassidulina* | *reniforme* |
| | | Nonionidae | Nonioninae | *Astrononion* | *gallowayi* |
| | | | | *Nonionella* | *auricula* |
| | | | | *Nonionellina* | *labradorica* |
| | | Alabaminidae | | *Oridorsalis* | *umbonatus* |
| | | | | *Trichohyalus* | *bartletti* |
| | | Anomalinidae | Anomalininae | *Melonis* | *zaandami* |
| | Robertinacea | Robertinidae | | *Robertinoides* | *charlottensis* |

| *Locality* | *Average %* | *Average Depth (m)* |
|---|---|---|
| M'Clure Strait | 2 | 414 |
| East Bay | 2 | 41 |
| P.G.A. Sea | X | 382 |
| H and G Bay | 1 | 178 |
| Jones Sound | 1 | 424 |

*Hyperammina subnodosa* Brady: Plate 21-I, Figure 4

*Hyperammina subnodosa* Brady, 1884, p. 259, pl. 23, figs. 11–14. COLE, 1981, p. 10, pl. 2, fig. 1.

Specimens large, may be over 1 cm in length and many appear complete. The species is characterized by relatively irregular constrictions in the tubular part of the test.

| *Locality* | *Average %* | *Average Depth (m)* |
|---|---|---|
| Jones Sound | 3 | 364 |

Genus SACCORHIZA Eilmer and Fichert 1899

*Saccorhiza ramosa* (Brady): Plate 21-I, Figure 5

*Hyperammina ramosa* Brady, 1879, p. 33, pl. 3, figs. 14–15.
*Saccorhiza ramosa* (Brady). BARKER, 1960, pl. 23, figs. 15–19.

Test large with a subglobular proloculus followed by a long dichotomously branching second chamber; wall thick and agglutinated with sand grains and sponge spicules; relatively poorly preserved.

| *Locality* | *Average %* | *Average Depth (m)* |
|---|---|---|
| M'Clure Strait | 1 | 368 |

Family SACCAMMINIDAE
Subfamily SACCAMMININAE Brady 1884
Genus SACCAMMINA M. Sars 1869

*Saccammina atlantica* (Cushman): Plate 21-I, Figure 6

*Proteonina atlantica* Cushman, 1944, p. 5, pl. 1, fig. 4. VILKS, 1969, p. 43, pl. 1, fig. 13.

Test medium in size, unilocular, flask-shaped, wall coarsely arenaceous, normally well preserved.

| *Locality* | *Average %* | *Average Depth (m)* |
|---|---|---|
| East Bay | 1 | 19 |
| H and G Bay | 2 | 161 |
| Byam Channel | 3 | 13 |
| Satellite Bay | 1 | 32 |
| Jones Sound | 5 | 451 |

*Saccammina sphaerica* Brady: Plate 21-I, Figure 7

*Saccammina sphaerica* Brady, 1871, p. 183. VILKS, 1969, p. 43, pl. 1, fig. 6.

Tests unilocular, globular with relatively smooth agglutinated surface. The aperture of well-preserved specimens consists of a neck built of fine agglutinated fragments. Occasionally several tests may be fused together.

| *Locality* | *Average %* | *Average Depth (m)* |
|---|---|---|
| M'Clure Strait | 7 | 385 |
| East Bay | 1 | 103 |
| P.G.A. Sea | X | 382 |
| H and G Bay | 3 | 227 |
| Jones Sound | 1 | 570 |

Genus PELOSINA Brady 1879

*Pelosina variabilis* Brady

*Pelosina variabilis* Brady, 1879, p. 30, pl. 3, figs. 1–3. LOEBLICH and TAPPAN, 1964, p. C200, fig. 112:7.

Tests poorly preserved and commonly found as fragments of irregular tests, consisting of characteristically light colored, finely agglutinated material.

| *Locality* | *Average %* | *Average Depth (m)* |
|---|---|---|
| East Bay | X | 4 |

Superfamily LITUOLACEA
Family HORMOSINIDAE
Subfamily HORMOSININAE Haeckel 1894
Genus REOPHAX de Montfort 1808

*Reophax arctica* Brady: Plate 21-I, Figure 8

*Reophax arctica* Brady, 1881*b*, p. 405, pl. 21, fig. 2a–b. LOEBLICH and TAPPAN, 1953, p. 21, pl.1, figs. 19–20.

The uniserial and small tests are mostly compressed with obscure suture lines in early portions. Moderately preserved.

| *Locality* | *Average %* | *Average Depth (m)* |
|---|---|---|
| Byam Channel | 1 | 13 |
| Jones Sound | 1 | 467 |

*Reophax fusiformis* (Williamson): Plate 21-I, Figure 9

*Proteonina fusiformis* Williamson, 1858, p. 1, pl. 1, fig. 1.
*Reophax curtus* Cushman, 1920, p. 8, pl. 2, figs. 2–3; LOEBLICH and TAPPAN, 1953, p. 22, pl. 2, figs. 1–4.
*Reophax fusiformis* (Williamson). VILKS, 1969, p. 44, pl. 1, fig. 8a–b.

Tests uniserial, irregular and agglutinated with grains of many sizes. Because of the large variability in test shapes, *R. curtus* and *R. fusiformis* are considered to be synonyms.

| Locality | Average % | Average Depth (m) |
|---|---|---|
| M'Clure Strait | 3 | 419 |
| East Bay | X | 131 |
| H and G Bay | 1 | 244 |
| Byam Channel | 1 | 13 |
| Jones Sound | 6 | 585 |

*Reophax gracilis* (Kiaer): Plate 21-I, Figure 10

*Nodulina gracilis* Kiaer, 1900, p. 24, text figs. 1–2.
*Reophax gracilis* (Kiaer). PARKER, 1952*a*, p. 397, pl. 2, fig. 1. VILKS, 1969, p. 44, pl. 1, fig. 9.

Test small, uniserial, agglutinated of very fine grains and slightly elastic. Poorly preserved, complete specimen very rare.

| Locality | Average % | Average Depth (m) |
|---|---|---|
| East Bay | X | 47 |

*Reophax guttifer* (Brady): Plate 21-I, Figure 11

*Lituola (Reophax) guttifera* Brady, 1881*a*, p. 49.
*Reophax guttifera* (Brady). BRADY, 1884, p. 295, pl. 31, figs. 10–15. VILKS, 1969, p. 44, pl. 1, fig. 10.

Tests small, uniserial, normally found incomplete due to fragile necks holding the chambers together.

| Locality | Average % | Average Depth (m) |
|---|---|---|
| M'Clure Strait | 2 | 367 |
| East Bay | X | 284 |
| H and G Bay | 4 | 287 |

*Reophax nodulosus* Brady: Plate 21-I, Figure 12

*Reophax nodulosa* Brady, 1879, p. 52, pl. 4, figs. 7–8. VILKS, 1969, p. 44, pl. 1, fig. 11.

Test unilocular, large, agglutinated surface normally smooth. Moderately well preserved; as a rule, tests break along the constricted necks between chambers.

| Locality | Average % | Average Depth (m) |
|---|---|---|
| M'Clure Strait | 2 | 431 |
| East Bay | 3 | 76 |
| H and G Bay | 2 | 233 |

*Reophax pilulifer* Brady: Plate 21-I, Figure 13

*Reophax pilulifera* Brady, 1884, p. 292, pl. 30, figs. 18–20. VILKS, 1969, p. 44, pl. 1, fig. 12.

Test unilocular, large, with rough agglutinated surface due to various sizes of grains. Symmetry of chambers and tests separates the species from *R. fusiformis.*

| *Locality* | *Average %* | *Average Depth (m)* |
|---|---|---|
| H and G Bay | X | 349 |

Family RZEHAKINIDAE
Genus SILICOSIGMOILINA Cushman and Church, 1929

*Silicosigmoilina groenlandica* (Cushman). Plate 21-I, Figures 14 and 15

*Quinqueloculina fusca* Brady var. *groenlandica* Cushman, 1933, p. 2, pl. 2, fig. 4.
*Silicosigmoilina groenlandica* (Cushman). LOEBLICH and TAPPAN, 1953, p. 38, pl. 4, figs. 7–9. VILKS, 1969, p. 44, pl. 1, fig. 14a–b.

Test small, elongate, chambers added at slightly larger than 180° angle, giving the test a flat appearance, surface smooth, agglutinated with very fine material.

| *Locality* | *Average %* | *Average Depth (m)* |
|---|---|---|
| East Bay | X | 86 |
| Jones Sound | 1 | 581 |

Family LITUOLIDAE
Subfamily HAPLOPHRAGMOIDINAE Maync 1952
Genus ADERCOTRYMA Loeblich and Tappan, 1952

*Adercotryma glomerata* (Brady): Plate 21-I, Figures 16 and 17

*Lituola glomerata* Brady, 1878, p. 433, pl. 20, fig. 1a–c.
*Adercotryma glomeratum* (Brady). LOEBLICH and TAPPAN, 1953, p. 26, pl. 8, figs. 1–4.
*Adercotryma glomerata* (Brady). VILKS, 1969, p. 44, pl. 1, fig. 15.

Test small, planispiral, asymmetrical, agglutinated surface rough, sutures not readily visible. Tests moderately preserved and readily break when samples are handled.

| *Locality* | *Average %* | *Average Depth (m)* |
|---|---|---|
| PGA Sea | X | 382 |
| Jones Sound | 22 | 458 |

Genus *CRIBROSTOMOIDES* Cushman 1910

*Cribrostomoides crassimargo* (Norman). Plate 21-I, Figures 18 and 19

*Haplophragmium crassimargo* Norman, 1892, p. 17.
*Alveolophragmium crassimargo* (Norman). LOEBLICH and TAPPAN, 1953, p. 31, pl. 3, figs. 4–7.
*Cribrostomoides crassimargo* (Norman). VILKS, 1969, p. 44, pl. 1, fig. 16a–b.

Test large, planispiral, surface rough, sutures moderately distinct. As a rule, tests well preserved with a readily visible areal aperture in the face of the last chamber.

| *Locality* | *Average %* | *Average Depth (m)* |
|---|---|---|
| Beaufort Sea | X | 70 |
| M'Clure Strait | 1 | 239 |
| East Bay | 2 | 85 |
| H and G Bay | 8 | 105 |
| Byam Channel | 1 | 13 |
| Satellite Bay | 1 | 37 |
| Jones Sound | 5 | 423 |

*Cribrostomoides jeffreysii* (Williamson): Plate 21-I, Figures 20 and 21

*Nonionina jeffreysii* Williamson, 1858, p. 34, pl. 3, figs. 72–73.
*Alveolophragmium jeffreysii* (Williamson). LOEBLICH and TAPPAN, 1953, p. 31, pl. 3, figs. 4–7.
*Cribrostomoides jeffreysii* (Williamson). VILKS, 1969, p. 45, pl. 1, fig. 17a–b.

Test planispiral, occasionally slightly twisted, small to moderate in size, surface smooth, tests fragile and usually poorly preserved.

| *Locality* | *Average %* | *Average Depth (m)* |
|---|---|---|
| M'Clure Strait | X | 240 |
| East Bay | X | 79 |
| H and G Bay | 2 | 95 |
| Satellite Bay | 11 | 148 |
| Jones Sound | X | 641 |

Genus RECURVOIDES Earland 1934

*Recurvoides contortus* Earland: Plate 21-II, Figure 1

*Recurvoides contortus* Earland, 1934, p. 91, pl. 10, figs. 7–9.
*Cribrostomoides subglobosus* (Cushman). VILKS, 1969, p. 45, pl. 1, fig. 18a–b.

Test large, robust, streptospiral with the coiling plane of each whorl about 90° apart. Surface moderately rough, preservation of tests moderate to poor.

| *Locality* | *Average %* | *Average Depth (m)* |
|---|---|---|
| M'Clure Strait | 15 | 430 |
| H and G Bay | 3 | 337 |

*Recurvoides laevigatum* Höglund: Plate 21-II, Figure 2

*Recurvoides laevigatum* Höglund, 1947, p. 150, pl. 11, fig. 6a–c.

Test small, streptospiral, wall finely arenaceous, smooth, aperture interio-areal above the base of last chamber with well-developed lips. Poorly preserved.

| *Locality* | *Average %* | *Average Depth (m)* |
|---|---|---|
| Byam Channel | X | 13 |
| Satellite Bay | 11 | 117 |

*Recurvoides turbinatus* (Brady): Plate 21-II, Figure 3

*Haplophragmium turbinatum* Brady, 1884, p. 313, pl. 35, fig. 9a–c.
*Recurvoides turbinatus* (Brady). PARKER, 1952*a*, p. 402, pl. 2, figs. 23–24.

Test small, streptospiral, with the last whorl coiled in a different plane. Wall finely arenaceous and smooth. Aperture interio-areal elongated slit and not readily visible. Moderate to poor preservation.

| *Locality* | *Average %* | *Average Depth (m)* |
|---|---|---|
| Beaufort Sea | X | 194 |
| M'Clure Strait | X | 243 |
| East Bay | 1 | 199 |
| H and G Bay | 4 | 201 |
| Satellite Bay | 2 | 170 |

Subfamily LITUOLINAE de Blainville 1825
Genus AMMOTIUM Loeblich and Tappan 1953

*Ammotium cassis* (Parker): Plate 21-II, Figure 4

*Lituola cassis* Parker, 1870. In DAWSON, 1870, pp. 177, 180, fig. 3.
*Ammotium cassis* (Parker). LOEBLICH and TAPPAN, 1953, p. 33, pl. 2, figs. 12–18.

Test large, flat, early portions planispiral, later uncoiling. Wall moderately to coarsely arenaceous, tests moderately to poorly preserved.

| *Locality* | *Average %* | *Average Depth (m)* |
|---|---|---|
| Byam Channel | 1 | 13 |

Family TEXTULARIIDAE
Subfamily SPIROPLECTAMMININAE Cushman 1927
Genus SPIROPLECTAMMINA Cushman 1927

*Spiroplectammina biformis* (Parker and Jones) Plate 21-II, Figures 5 and 6

*Textularia agglutinans* d'Orbigny var. *biformis* Parker and Jones, 1865, p. 370, pl. 15, figs. 23–24.
*Spiroplectammina biformis* (Parker and Jones). LOEBLICH and TAPPAN, 1953, p. 34, pl. 4, figs. 1–6.

Test small, flat, early chambers planispiral, later biserial. Wall finely arenaceous, aperture low arch at the inner margin of the last chamber. Well preserved in the channels of the Queen Elizabeth Islands, poorly preserved on Beaufort shelf and Jones Sound.

| *Locality* | *Average %* | *Average Depth (m)* |
|---|---|---|
| Beaufort shelf | 4 | 51 |
| East Bay | 53 | 16 |
| H and G Bay | 3 | 43 |
| Byam Channel | 58 | 13 |
| Satellite Bay | 1 | 32 |
| Jones Sound | 3 | 114 |

Subfamily TEXTULARIINAE Ehrenberg 1838
Genus TEXTULARIA Defrance 1824

*Textularia earlandi* Parker: Plate 21-II, Figure 7

*Textularia tenuissima* Earland, 1933, p. 95, pl. 3, figs. 21–30. HÖGLUND, 1947, p. 176, pl. 13, fig. 1a–b, text figs. 154–155.
*Textularia earlandi* PARKER, 1952*b*, p. 458b (footnote), pl. 2, fig. 4.

Test small, flat, biserial, sutures distinct, wall finely arenaceous. Moderately well preserved.

| *Locality* | *Average %* | *Average Depth (m)* |
|---|---|---|
| East Bay | 6 | 70 |
| H and G Bay | X | 70 |
| Satellite Bay | 1 | 32 |
| Jones Sound | 2 | 505 |

*Textularia torquata* Parker: Plate 21-II, Figures 8 and 9

*Textularia torquata* Parker, 1952*a*, p. 403, pl. 3, figs. 9–11.

Test small, flat, and slightly twisted, chambers biserial and increasing rapidly in size, sutures oblique and obscure. Wall finely arenaceous fragile, yet test moderately to well preserved.

| *Locality* | *Average %* | *Average Depth (m)* |
|---|---|---|
| Beaufort shelf | X | 65 |
| M'Clure Strait | X | 243 |
| East Bay | X | 154 |
| H and G Bay | 3 | 48 |
| Byam Channel | 18 | 13 |
| Jones Sound | 2 | 483 |

Family TROCHAMMINIDAE
Subfamily TROCHAMMININAE Schwager 1877
Genus TRITAXIS Schubert 1921

*Tritaxis atlantica* (Parker): Plate 21-II, Figures 10 and 11

*Trochamminella atlantica* Parker, 1952*a*, p. 409, pl. 4, fig. 17–19. VILKS, 1969, p. 45, pl. 2, fig. 1a–b.

Test small, trochospiral, chambers slightly inflated, wall finely arenaceous, aperture interio-areal, ovate with distinct lips. Moderately well preserved. *Trochamminella* Cushman 1943 is a junior synonym to *Tritaxis*.

| *Locality* | *Average %* | *Average Depth (m)* |
|---|---|---|
| East Bay | X | 131 |
| H and G Bay | 4 | 113 |
| Satellite Bay | 3 | 81 |

*Tritaxis bullata* (Höglund): Plate 21-II, Figures 12 and 13

*Trochamminella bullata* Höglund, 1947, p. 213, pl. 17, fig. 5a–c. VILKS, 1969, p. 46, pl. 2, fig. 2a–b.

Test small, trochospiral, chambers inflated forming high spire, wall finely arenaceous, aperture interio-areal, ovate slit with distinct lips. Moderately well preserved.

| *Locality* | *Average %* | *Average Depth (m)* |
|---|---|---|
| East Bay | 1 | 103 |
| H and G Bay | 4 | 210 |
| Satellite Bay | 2 | 93 |

Genus TROCHAMMINA Parker and Jones 1859

*Trochammina globigeriniformis* (Parker and Jones): Plate 21-II, Figures 14 and 15

*Lituola nautiloidea* var. *globigeriniformis* Parker and Jones, 1865, p. 407, pl. 15, figs. 46–47.
*Trochammina globigeriniformis* (Parker and Jones). CUSHMAN, 1910, p. 125, text fig. 195

Test small, trochospiral, chambers inflated, globular, wall finely arenaceous.

| *Locality* | *Average %* | *Average Depth (m)* |
|---|---|---|
| Satellite Bay | 1 | 32 |

*Trochammina inflata* (Montagu): Plate 21-II, Figures 16 and 17

*Nautilus inflatus* Montagu, 1808, p. 81, pl. 18, fig. 3.
*Trochammina inflata* (Montagu). CUSHMAN, 1910, p. 121, text fig. 188. PHLEGER, 1952, p. 86, pl. 13, figs. 27–28.

Test small, trochospiral, low-spired, sutures distinct at right angles to the test margin, wall finely arenaceous, very thin and may be flexible; thus many tests are collapsed. Preservation poor.

| *Locality* | *Average %* | *Average Depth (m)* |
|---|---|---|
| M'Clure Strait | 2 | 301 |
| East Bay | X | 201 |
| Satellite Bay | 13 | 112 |

*Trochammina nana* (Brady): Plate 21-III, Figures 1–3

*Haplophragmium nanum* Brady, 1881, p. 50. BRADY, 1884, p. 311, pl. 35, figs. 6–8.
*Trochammina nana* (Brady). LOEBLICH and TAPPAN, 1953, p. 50, pl. 8, fig. 5.

Test small, trochospiral and low-spired. Chambers inflated in the last whorl, the last chamber partly flapping over the umbilicus on ventral side. Aperture a slit on the base of the last chamber on the ventral side. Wall arenaceous of fine- to medium-size grains. Tests well preserved in the archipelago, less well preserved on Beaufort shelf and Jones Sound. The most abundant species in the Canadian Arctic.

| Locality | Average % | Average Depth (m) |
|---|---|---|
| Beaufort shelf | 1 | 199 |
| N.W. Passage | 7 | 84 |
| M'Clure Strait | 17 | 359 |
| East Bay | 18 | 49 |
| H and G Bay | 8 | 123 |
| Byam Channel | 7 | 13 |
| Satellite Bay | 46 | 92 |
| Jones Sound | 11 | 511 |

*Trochammina ochracea* (Williamson): Plate 21-III, Figures 4, 5, and 6

*Rotalia ochracea* Williamson, 1858, p. 55, pl. 4, fig. 112; pl. 5, fig. 113.
*Trochammina ochracea* (Williamson). HÖGLUND, 1947, p. 211, pl. 16, fig. 2; text fig. 190.

Test small, trochospiral, flat. Sutures distinct, forming a spiral pattern from a distinct central umbilical area on the ventral side. Ventral side concave. Preservation of tests is poor.

| Locality | Average % | Average Depth (m) |
|---|---|---|
| Jones Sound | X | 531 |

*Trochammina quadriloba* Höglund: Plate 21-III, Figures 7 and 8

*Trochammina pusilla* Höglund, 1947, p. 201, p1. 17, fig. 4a–c.
*Trochammina quadriloba* HÖGLUND, 1948, p. 46.

Test small, trochospiral, high-spired. Wall rough on dorsal side, agglutinated with medium-size grains, fine grains on ventral side. Aperture an interiomarginal arched slit at the base of the last chamber.

| Locality | Average % | Average Depth (m) |
|---|---|---|
| Beaufort Sea | X | 84 |
| M'Clure Strait | 3 | 334 |
| East Bay | X | 148 |
| H and G Bay | 1 | 141 |
| Satellite Bay | 1 | 68 |

Family ATAXOPHRAGMIIDAE
Subfamily GLOBOTEXTULARIINAE Cushman 1927
Genus EGGERELLA Cushman 1933

*Eggerella advena* (Cushman): Plate 21-III, Figures 9 and 10

*Verneuilina advena* Cushman, 1922, p. 141.
*Eggerella advena* (Cushman). CUSHMAN, 1937, p. 51, pl. 5, figs. 12–15. VILKS, 1969, p. 46, pl. 2, fig. 3.

Test small, triserial, except for the early chambers. Wall finely arenaceous, aperture a low arch at the base of the final chamber. Tests usually well preserved.

| *Locality* | *Average %* | *Average Depth (m)* |
|---|---|---|
| Beaufort Sea | X | 68 |
| Byam Channel | 2 | 13 |

| | |
|---|---|
| Suborder | MILIOLINA |
| Superfamily | MILIOLACEA |
| Family | FISHERINIDAE |
| Subfamily | CYCLOGYRINAE Loeblich and Tappan 1961 |
| Genus | GORDIOSPIRA Heron-Allen and Earland 1932 |

*Gordiospira arctica* Cushman

*Gordiospira arctica* Cushman, 1933, p. 3, pl. 1, figs. 5–7 LOEBLICH and TAPPAN, 1953, p. 49, pl. 7, figs. 1–3.

| *Locality* | *Average %* | *Average Depth (m)* |
|---|---|---|
| M'Clure Strait | X | 381 |

| | |
|---|---|
| Family | MILIOLIDAE |
| Subfamily | QUINQUELOCULININAE Cushman 1917 |
| Genus | QUINQUELOCULINA d'Orbigny 1826 |

*Quinqueloculina agglutinata* Cushman: Plate 21-III, Figure 11

*Quinqueloculina agglutinata* Cushman, 1917, p. 43, pl. 9, fig. 2. LOEBLICH and TAPPAN, 1953, p. 39, pl. 5, figs. 1–4.

Test medium to large, five chambers visible in the last whorl, wall arenaceous, tests normally well preserved.

| *Locality* | *Average %* | *Average Depth (m)* |
|---|---|---|
| Byam Channel | X | 13 |

*Quinqueloculina arctica* Cushman: Plate 21-III, Figures 12 and 13

*Quinqueloculina arctica* Cushman, 1933, p. 2, pl. 1, fig. 3a–c. VILKS, 1969, p. 46, pl. 2, fig. 8a–b.

Test medium to large in size, robust, chambers quinqueloculine with several sharp angles.

| *Locality* | *Average %* | *Average Depth (m)* |
|---|---|---|
| M'Clure Strait | X | 398 |
| P.G.A. Sea | X | 463 |

*Quinqueloculina seminulum* (Linné): Plate 21-III, Figure 14

*Serpula seminulum* Linné, 1758, p. 786.
*Quinqueloculina seminula* (Linné). PARKER, 1952*b*, p. 456, pl. 2, fig. 7a–b.
*Quinqueloculina seminulum* (Linné). VILKS, 1969, p. 47, pl. 2, fig. 10a–b.

Test medium to small, chambers quinqueloculine and rounded.

| Locality | Average % | Average Depth (m) |
|---|---|---|
| M'Clure Strait | X | 329 |
| P.G.A. Sea | X | 394 |
| H and G Bay | X | 381 |

Genus TRILOCULINA d'Orbigny 1826

*Triloculina trihedra* Loeblich and Tappan: Plate 21-III, Figures 15 and 16.

*Triloculina trihedra* Loeblich and Tappan, 1953, p. 45, pl. 4, fig. 10. VILKS, 1969, p. 47, pl. 2, fig. 11.

Test small, triangular in section and nearly flat to slightly convex sides.

| Locality | Average % | Average Depth (m) |
|---|---|---|
| M'Clure Strait | 1 | 372 |

Subfamily MILIOLINELLINAE Vella 1957
Genus SCUTULORIS Loeblich and Tappan 1953

*Scutuloris tegminis* Loeblich and Tappan

*Scutuloris tegminis* Loeblich and Tappan, 1953, p. 41, pl. 5, fig. 10.

| Locality | Average % | Average Depth (m) |
|---|---|---|
| M'Clure Strait | X | 398 |

Suborder ROTALIINA
Superfamily NODOSARIACEA
Family POLYMORPHINIDAE
Subfamily POLYMORPHININAE d'Orbigny 1839
Genus PSEUDOPOLYMORPHINA Cushman and Ozawa 1928

*Pseudopolymorphina novangliae* (Cushman): Plate 21-IV, Figures 1 and 2

*Polymorphina lactea* (Walker and Jacob) var. *novangliae* Cushman, 1923, p. 146, pl. 39, figs. 6–8.
*Pseudopolymorphina novangliae* (Cushman). CUSHMAN and OZAWA, 1930, p. 90, pl. 23, figs. 1–2. PARKER, 1952*a*, p. 410, pl. 5, fig. 1.

Test medium to large, early chambers in quinqueloculine arrangement, later biserial, aperture radiate.

| Locality | Average % | Average Depth (m) |
|---|---|---|
| Beaufort Sea | X | 32 |

Superfamily BULIMINACEA
Family ISLANDIELLIDAE
Genus ISLANDIELLA Nørvang 1958

*Islandiella helenae* Feyling-Hanssen and Buzas: Plate 21-IV, Figure 3 and 4

*Cassidulina laevigata* d'Orbigny. PARKER and JONES, 1865, p. 377, pl. 15, figs. 2–4.
*Cassidulina teretis* Tappan. LOEBLICH and TAPPAN, 1953, p. 121, pl. 24, figs. 3–4.
*Islandiella teretis* (Tappan). FEYLING-HANSSEN, 1964, p. 326, pl. 16, fig. 17. VILKS, 1969, p. 49, pl. 3, fig. 5.
*Islandiella helenae* Feyling-Hanssen and Buzas, 1976, p. 155, text figs. 1–4.

Test medium to large in size, chambers biserially arranged and coiled, in well-preserved specimens previous whorls can be seen through shell material in the umbilical region, 4 to 5 pairs of chambers in each whorl. Wall calcareous and distinctly radial when viewed in polarized light. Aperture a broad and short slit parallel to the periphery and contains a free apertural tongue.

| *Locality* | *Average %* | *Average Depth (m)* |
|---|---|---|
| Beaufort shelf | 9 | 53 |
| N.W. Passage | 2 | 84 |
| Byam Channel | 1 | 13 |
| Jones Sound | 7 | 290 |

*Islandiella norcrossi* (Cushman): Plate 21-IV, Figures 5 and 6

*Cassidulina norcrossi* Cushman, 1933, p. 7, pl. 2, fig. 7.
LOEBLICH and TAPPAN, 1953, p. 120, pl. 24, fig. 2.
*Islandiella norcrossi* (Cushman). VILKS, 1969, p. 49, pl. 3, fig. 4a–b.

Test medium to small in size, chambers biserially arranged almost symmetrically on both sides of a central plane. Always four pairs of chambers per whorl. Aperture a narrow slit symmetrical on the edge of the central plane and contains a fragile plate as lip often not preserved. Wall calcareous and radial when viewed in polarized light.

| *Locality* | *Average %* | *Average Depth (m)* |
|---|---|---|
| Beaufort shelf | X | 60 |
| N.W. Passage | 5 | 196 |
| M'Clure Strait | 4 | 405 |
| P.G.A. Sea | 1 | 380 |
| H and G Bay | X | 43 |

Family BULIMINIDAE
Subfamily BULIMININAE Jones 1875
Genus STAINFORTHIA Hofker 1956

*Stainforthia concava* (Höglund): Plate 21-IV, Figures 7 and 8

*Virgulina concava* Höglund, 1947, p. 259, pl. 23, fig. 3–4, text figs. 273–274.
*Virgulina loeblichi* Feyling-Hansen, 1964, p. 308, pl. 14, figs. 12–14.
*Virgulina complanata* Egger. PARKER, 1952*a*, p. 417, pl. 6, fig. 2.
*Cassidella complanata* (Egger). VILKS, 1969, p. 51, pl. 3, fig. 18a–b.
*Stainforthia concava* (Höglund) LOEBLICH and TAPPAN, 1964, fig. 442; 11, 12.

Test elongate, early stage triserial, later twisted biserial, chambers inflated, may have one or two apical spines. Aperture loop shaped in face, with a narrow in-curved lip at one side and folded tooth plate from opposite side partially closing the opening. The Arctic species is very similar to Höglund's 1947 figures, which are reproduced by Loeblich and Tappan, 1964.

| *Locality* | *Average %* | *Average Depth (m)* |
|---|---|---|
| Beaufort shelf | X | 125 |
| N.W. Passage | 1 | 118 |
| M'Clure Strait | 1 | 383 |
| P.G.A. Sea | 1 | 363 |
| H and G Bay | X | 422 |
| Jones Sound | 1 | 314 |

Family UVIGERINIDAE
Genus TRIFARINA Cushman 1923

*Trifarina fluens* (Todd): Plate 21-IV, Figure 9

*Angulogerina fluens* Todd. *In:* Cushman and Todd, 1947, p. 67, pl. 16, figs. 6–7.

Test fusiform, chambers triserially arranged, wall calcareous and ornamented with coastae the height of which vary with environment and preservation. Aperture on a short neck with a thickened rim. *Angulogerina* Cushman 1927 is a junior synonym to *Trifarina* Cushman 1923.

| *Locality* | *Average %* | *Average Depth (m)* |
|---|---|---|
| Jones Sound | X | 531 |

Superfamily DISCORBACEA
Family DISCORBIDAE
Subfamily DISCORBINAE Ehrenberg 1838
Genus BUCCELLA Andersen 1952

*Bucella frigida* (Cushman): Plate 21-IV, Figures 10, 11, 12

*Pulvinulina frigida* CUSHMAN, 1922, p. 144.
*Buccella frigida* (Cushman). ANDERSEN, 1952, p. 144, text figs. 4–6.
VILKS, 1969, p. 49, pl. 3, fig. 7a–b.

Test small, trochospiral, biconvex on the dorsal side, sutures distinct and curved back at the periphery, on the ventral side sutures radial and filled with granular material that also fills the umbilical region. Very common Arctic species.

| *Locality* | *Average %* | *Average Depth (m)* |
|---|---|---|
| Beaufort shelf | 2 | 49 |
| N.W. Passage | 1 | 93 |
| M'Clure Strait | 1 | 443 |
| P.G.A. Sea | 1 | 414 |
| H and G Bay | X | 146 |
| Byam Channel | 1 | 13 |
| Jones Sound | 3 | 280 |

Genus EOEPONIDELLA Wickenden 1949

*Eoeponidella pulchella* (Parker): Plate 21-IV, Figures 13–15

*Pninaella pulchella* PARKER, 1952*a*, p. 420, pl. 6, figs. 18–20.
*Asterellina pulchella* (Parker). ANDERSON, 1963, p. 314, pl. 1, figs. 5–7. VILKS, 1969, p. 50, pl. 3, figs. 10a–b.
*Eoeponidella pulchella* (Parker). HAMAN, 1973, p. 102, text fig. 1.

Test small, trochospiral, with a distinct stellate area on the ventral side of the test formed by secondary chambers. Aperture broad interiomarginal umbilical arch on the lower part of the last chamber on the ventral side with five secondary openings.

| *Locality* | *Average %* | *Average Depth (m)* |
|---|---|---|
| Beaufort shelf | 1 | 51 |
| N.W. Passage | 1 | 84 |
| East Bay | X | 9 |
| Byam Channel | 1 | 13 |
| Jones Sound | X | 316 |

Genus EPISTOMINELLA Husefima and Maruhasi 1944

*Epistominella vitrea* Parker: Plate 21-IV, Figure 16, Plate 21-V, Figure 1

*Epistominella vitrea* Parker, Phleger, and Peirson, 1953, p. 9, pl. 4, figs. 40–41.

Test small, trochospiral, sutures oblique on dorsal side, almost radial on ventral side, wall translucent. Aperture long and narrow slit parallel to the periphery and slightly curved.

| *Locality* | *Average %* | *Average Depth (m)* |
|---|---|---|
| Jones Sound | 1 | 359 |

Genus STETSONIA Parker 1954

*Stetsonia horvati* Green: Plate 21-V, Figure 2 and 3

*Stetsonia horvati* Green, 1960, p. 72, pl. 1, fig. 6a–b. VILKS, 1969, p. 49, pl. 3, fig. 8a–b.

Test small, slightly trochospiral, involute on both sides, sutures radial, four to five chambers in the last whorl. Aperture a high-arched slit at the base of the last chamber. Well-preserved specimens are transparent.

| *Locality* | *Average %* | *Average Depth (m)* |
|---|---|---|
| Beaufort shelf | 1 | 1995 |
| P.G.A. Sea | 2 | 371 |
| H and G Bay | 1 | 370 |

Subfamily BAGGININAE Cushman 1927
Genus VALVULINERIA Cushman 1926

*Valvulineria arctica* Green: Plate 21-V, Figures 4 and 5

*Valvulineria arctica* Green, 1960, p. 71, pl. 1, fig. 3a–c. VILKS, 1969, p. 49, pl. 3, fig. 9a–c.

Test small, rotoloid, last three to four chambers inflated, open umbilicus on the ventral side, aperture on the ventral side beneath a thin and transparent valvular flap, formed by the ultimate chamber.

| *Locality* | *Average %* | *Average Depth (m)* |
|---|---|---|
| P.G.A. Sea | 4 | 321 |
| H and G Bay | 1 | 294 |

Superfamily ROTALIACEA
Family ELPHIDIIDAE
Subfamily ELPHIDIINAE Galloway 1933
Genus ELPHIDIUM de Montfort 1808

*Elphidium bartletti* Cushman: Plate 21-V, Figures 6 and 7

*Elphidium bartletti* Cushman, 1933, p. 4, pl. 1, fig. 9. LOEBLICH and TAPPAN, 1953, p. 96, pl. 18, figs. 10–14. VILKS, 1969, p. 50, pl. 3, fig. 12.

Test medium in size, planispiral, involute, sides almost flat with slightly depressed umbilical regions. Sutures gently curved with 8–12 pores visible on each side, sutural bridges short and broad, umbilical area filled with granular material, aperture a row of small pores at the base of apertural face of the first chamber.

| *Locality* | *Average %* | *Average Depth (m)* |
|---|---|---|
| Beaufort shelf | 1 | 40 |
| M'Clure Strait | X | 487 |
| H and G Bay | X | 309 |

*Elphidium excavatum* forma *clavata* Cushman: Plate 21-V, Figures 8 and 9

*Elphidium incertum* (Williamson) var. *clavatum* Cushman, 1930, p. 20, pl. 7, fig. 10.
*Elphidium clavatum* Cushman. LOEBLICH and TAPPAN, 1953, p. 98, pl. 19, figs. 8–10.
*Elphidium incertum* (Williamson). VILKS, 1969, p. 50, pl. 3, fig. 13.
*Elphidium excavatum* forma *clavata* Cushman. FEYLING-HANSSEN, 1972, p. 339, pl. 1, figs. 1–9; pl. 2, figs. 1–9. MILLER et al., 1982, p. 124, pl. 1, figs. 9–12; pl. 2, figs. 3–8; pl. 4, figs. 1–7.

Test medium in size, planispiral, involute, sides slightly convex, sutures gently curved with a single row of sutural pores and short, distinct sutural bridges, wall thin and translucent, umbilical region with a single or several elevated bosses. Aperture a row of pores at the base of the apertural face of the last chamber.

| Locality | Average % | Average Depth (m) |
|---|---|---|
| Beaufort shelf | 15 | 46 |
| N.W. Passage | 20 | 124 |
| H and G Bay | X | 203 |
| Byam Channel | 2 | 13 |
| Jones Sound | 2 | 349 |

*Elphidium subarcticum* Cushman: Plate 21-V, Figure 10

*Elphidium subarcticum* Cushman, 1944, p. 27, pl. 3, figs. 34–35. LOEBLICH and TAPPAN, 1953, p. 105, pl. 19, figs. 5–7.

Test small to medium in size, planispiral, involute, sides flat, last two to three chambers moderately inflated, sutures slightly depressed, containing a row of sutural pores and a wide opaque band. Aperture a row of pores at the base of apertural face of the last chamber.

| Locality | Average % | Average Depth (m) |
|---|---|---|
| Beaufort Sea | 1 | 61 |
| N.W. Passage | 1 | 95 |
| M'Clure Strait | 1 | 412 |
| Byam Channel | 1 | 13 |

Genus HAYNESINA Banner and Culver 1978

*Haynesina nana* (Vilks, Wagner and Pelletier): Plate 21-V, Figure 11

*Protelphidium nanum* Vilks, Wagner, and Pelletier, 1979, p. 35, pl. 1, figs. 1–4, text fig. 1.

Test small, planispiral, sides gently concave, sutures gently curved, slightly depressed, and near the umbilical region granular; aperture a row of pores at the base of the final chamber.

| Locality | Average % | Average Depth (m) |
|---|---|---|
| Beaufort shelf | 1 | 39 |

*Haynesina orbiculare* (Brady): Plate 21-V, Figures 12 and 13

*Nonionina orbilculare* Brady, 1881*b*, p. 415, pl. 21, fig. 5.
*Protelphidium orbiculare* (Brady). TODD and LOW, 1961, p. 20, pl. 2, fig. 11. VILKS, 1969, p. 50, pl. 3, fig. 14.

Test medium in size, planispiral, sides gently convex, 9–11 chambers in the last whorl, sutures gently curved, slightly granular in umbilical region, aperture a series of pores at the base of the aperture face of the last chamber.

| Locality | Average % | Average Depth (m) |
|---|---|---|
| Beaufort shelf | 4 | 46 |
| N.W. Passage | 3 | 90 |
| East Bay | X | 60 |
| P.G.A. Sea | X | 382 |
| Byam Channel | 1 | 13 |

Superfamily GLOBIGERINACEA
Family GLOBIGERNIDAE
Subfamily GLOBIGERININAE Carpenter, Parker & Jones 1862
Genus NEOGLOBOQUADRINA Bandy, Frerichs & Vincent 1967

*Neogloboquadrina pachyderma* (Ehrenberg): Plates 21-VII–21-IX

*Aristospira pachyderma* Ehrenberg, 1861, pp. 276, 277, 303.
*Globigerina bulloides* d'Orbigny var. *borealis*. Brady, 1882*b*, p. 412.
*Globigerina pachyderma* (Ehrenberg) BE, 1960, p. 66, text fig. 1.
*Globorotalia pachyderma* (Ehrenberg) VILKS, 1973, pl. 1–4.
*Neogloboquadrina pachyderma pachyderma* (Ehrenberg). ROGL and BOLLI, 1973, p. 571, pl. 11, figs. 2–6, pl. 16, fig. 12.

Test trochospiral, spire flat to high, chambers inflated, aperture extra-umbilical to almost umbilical. The test wall possesses no spines, but test surface may be covered with secondary deposits in varying thicknesses from punctae on the smooth bilamellar surface to ridges between the pores. The extent of the secondary encrustations may obscure the trochospiral symmetry of the test. The number of chambers per last whorl vary from four to five, with occasional reduction in the growth of the last chamber, forming only a flap over the aperture in extreme cases.

*N. pachyderma* is very common in the surface sediments of the Arctic Ocean, offshore Beaufort shelf, outer M'Clure Strait, Hazen Strait, and northern Prince Gustaf Adolf Sea. Living specimens occur in largest numbers within the upper 200 m of water.

Superfamily ORBITOIDACEA
Family CIBICIDIDAE
Subfamily CIBICIDINAE Cushman 1927
Genus CIBICIDES de Montfort 1808

*Cibicides lobatulus* (Walker and Jacob): Plate 21-V, Figures 14 and 15

*Nautilus lobatulus* Walker and Jacob. *In:* Kanmacher, 1798, p. 642, pl. 14, fig. 36.
*Cibicides lobatulus* (Walker and Jacob). CUSHMAN, 1931, p. 118, pl. 21, fig. 3. VILKS, 1969, p. 50, pl. 3, fig. 17a–b.

Test medium to large, usually attached, plano-convex, trochospiral with the spiral side flat to excavated. Wall coarsely perforates on spiral side; aperture a low interiomarginal opening with narrow lips along the apertural face of the last chamber and up to four last chambers on the spiral side.

| *Locality* | *Average %* | *Average Depth (m)* |
|---|---|---|
| Beaufort shelf | 1 | 175 |
| N.W. Passage | 2 | 84 |
| M'Clure Strait | 5 | 436 |
| P.G.A. Sea | 5 | 376 |
| H and G Bay | 1 | 236 |
| Byam Channel | X | 13 |
| Jones Sound | 6 | 242 |

Superfamily CASSIDULINACEA
Family CAUCASINIDAE
Subfamily FURSENKOININAE Loeblich and Tappan 1961
Genus FURSENKOINA Loeblich and Tappan 1961

*Fursenkoina fusiformis* (Williamson): Plate 21-V, Figures 16 and 17

*Bulimina pupoides* var. *fusiformis* Williamson, 1858, p. 63, pl. 5, figs. 129–130.
*Virgulina fusiformis* (Williamson). PARKER, 1952*a*, pl. 6, figs. 3–6. FEYLING-HANSSEN, 1964, p. 307, pl. 14, figs. 15–18.

Test small, biserial, somewhat irregular and twisted, last chamber smaller in adult specimen. Wall very thin, aperture in adolescent tests is interiomarginal as a small semicircular opening at the inner margin of the apertural face; in adults the aperture is terminal with a collar on one side. Loeblich and Tappan, 1961, replace *Virgulina* with *Fursenkoina*.

| *Locality* | *Average %* | *Average Depth (m)* |
|---|---|---|
| Beaufort Sea | 1 | 158 |
| P.G.A. Sea | 1 | 199 |
| Jones Sound | X | 435 |

Family CASSIDULINIDAE
Genus CASSIDULINA d'Orbigny 1826

*Cassidulina laevigata* d'Orbigny: Plate 21-VI, Figures 1 and 2

*Cassidulina laevigata* d'Orbigny. BRADY, 1884, p. 428, pl. 54, fig. 1 (not d'Orbigny, 1826). SEJRUP et al., 1981, p. 290, pl. 1, fig. 5.
*Cassidulina teretis* Tappan. RODRIGUES et al., 1980, p. 59, pl. 2, figs. 1, 3, 5, pl. 5, figs. 1, 4, 7; pl. 6, figs. 7, 10.
*Cassidulina cf. teretis* Tappan. FEYLING-HANSSEN, FUNDER, and PETERSEN, 1982, pp. 90–92, pl. 1, figs. 11–13.
*Islandiella teretis* (Tappan). VILKS, 1969, p. 49, pl. 3, fig. 5.

Test medium in size, chambers biserially arranged in a coil and alternating on each side of the periphery, each reaching a central boss on one side and only extending part of the way to the boss on the other side, wall calcareous and granular in structure. Aperture an elongate slit, extending from the base of the final chamber and parallel to the periphery, with a narrow bordering lip along the lower margin, but lacking internal tooth.

| *Locality* | *Average %* | *Average Depth (m)* |
|---|---|---|
| Beaufort Sea | 9 | 992 |
| N.W. Passage | 17 | 347 |
| M'Clure Strait | 10 | 405 |
| P.G.A. Sea | 63 | 386 |
| H and G Bay | 11 | 335 |

*Cassidulina reniforme* Nørvang: Plate 21-VI, Figures 3 and 4

*Cassidulina crassa* d'Orbigny var. *reniforme* Nørvang, 1945, p. 41, text fig. 6e–h.
*Islandiella islandica* (Nørvang). VILKS, 1969, p. 49, pl. 3, fig. 3.
*Cassidulina reniforme* Nørvang. SEJRUP and GUILBAULT, 1980, p. 79, figs. 1d–e, 2f–k.

Test medium to small, chambers biserially arranged in a coil, inflated. Aperture a broad slit parallel to the periphery. A well-developed flap attached to the penultimate chamber hides part of the aperture, wall optically granular. The second most abundant species in the Arctic.

| Locality | Average % | Average Depth (m) |
|---|---|---|
| Beaufort shelf | 12 | 67 |
| N.W. Passage | 33 | 198 |
| M'Clure Strait | 5 | 413 |
| P.G.A. Sea | 6 | 362 |
| H and G Bay | 3 | 314 |
| Byam Channel | 1 | 13 |
| Jones Sound | 4 | 349 |

Family NONIONIDAE
Subfamily NONIONINAE Schultze 1854
Genus ASTRONONION Cushman and Edwards 1937

*Astrononion gallowayi* Loeblich and Tappan: Plate 21-VI, Figures 5 and 6

*Astrononion gallowayi* Loeblich and Tappan, 1953, p. 90, pl. 17, figs. 4–7. VILKS, 1969, p. 51, pl. 3, fig. 19.

Test medium in size, planispiral, umbilical region concave, sutures distinct and curved, secondary chambers surround the umbilicus, tapering toward the sutures. Wall coarsely perforate, less so of secondary chambers. Aperture a low arch at the base of the final chamber.

| Locality | Average % | Average Depth (m) |
|---|---|---|
| M'Clure Strait | X | 379 |
| Jones Sound | 1 | 314 |

Genus NONIONELLA Cushman 1926

*Nonionella auricula* Heron-Allen and Earland: Plate 21-VI, Figures 7 and 8

*Nonionella auricula* Heron-Allen and Earland, 1930, p. 192, pl. 5, figs. 68–70. PARKER, 1952*a*, p. 413, pl. 5, figs. 13a–14b.

Test medium in size, slightly trochospiral, final chamber inflated, extending further over the ventral side, sutures gently curved. Aperture at the base of the apertural face extending slightly over to the ventral side.

| Locality | Average % | Average Depth (m) |
|---|---|---|
| M'Clure Strait | 2 | 381 |

Genus NONIONELLINA Voloshinova 1958

*Nonionellina labradorica* (Dawson): Plate 21-VI, Figures 9 and 10

*Nonionina labradorica* Dawson, 1860, p. 191, fig. 4.
*Nonionellina labradorica* (Dawson). VOLOSHINOVA, 1958, p. 142. VILKS, 1969, p. 51, pl. 3, fig. 20a–b.

Test medium to large, planispiral, involute, early whorls slightly trochoid. Chambers rapidly increasing in size, ultimate chamber twice as large as penultimate; aperture a low, arched slit at the base of apertural face.

| *Locality* | *Average %* | *Average Depth (m)* |
|---|---|---|
| M'Clure Strait | 1 | 383 |
| Jones Sound | X | 316 |

Family ALABAMINIDAE
Genus ORIDORSALIS Andersen 1961

*Oridorsalis umbonatus* (Reuss): Plate 21-VI, Figures 11, 12, and 13

*Rotalina umbonata* Reuss, 1851, p. 75, pl. 5, fig. 35.
*Truncatulina tenera* Brady, 1884, p. 665, pl. 95, fig. 11.
*Eponides tener* (Brady). VILKS, 1969, p. 50, pl. 3, fig. 16a–b.
*Oridorsalis tener* (Brady). CORLISS, 1979, p. 9, pl. 4, figs. 10–15.
*Oridorsalis umbonatus* (Reuss). TODD, 1965, p. 23, pl. 6, fig. 2.

Test medium in size, trochospiral, spiral side almost flat to convex, sutures radial. Primary aperture interiomarginal, extending on umbilical side from periphery nearly to the umbilicus; secondary sutural openings on the spiral side.

| *Locality* | *Average %* | *Average Depth (m)* |
|---|---|---|
| Beaufort shelf | 1 | 1634 |
| M'Clure Strait | 5 | 425 |
| P.G.A. Sea | 3 | 424 |
| H and G Bay | 1 | 345 |

Genus TRICHOHYALUS Loeblich and Tappan, 1953

*Trichohyalus bartletti* (Cushman)

*Discorbis bartletti* Cushman, 1933, p. 6, pl. 2, figs. 3–6.
*Trichohyalus bartletti* (Cushman). LOEBLICH and TAPPAN, 1953, p. 117, pl. 23, figs. 1–7.

| *Locality* | *Average %* | *Average Depth (m)* |
|---|---|---|
| M'Clure Strait | X | 417 |

Family ANOMALINIDAE
Subfamily ANOMALININAE Cushman, 1927
Genus MELONIS de Montfort, 1808

*Melonis zaandami* (Van Voorthuysen): Plate 21-VI, Figures 14 and 15

*Nonion pompilioides* (Fichtel and Moll). CUSHMAN, 1930, p. 4, pl. 2, figs. 1–2 (not pl. 1, figs. 7–11).
*Anomalinoides barleeanum* (Williamson) var. *zaandamae*. van Voorthuysen, 1952, p. 681.
*Nonion zaandamae* (van Voorthuysen). LOEBLICH and TAPPAN, 1953, p. 87, pl. 16, figs. 11–12.
*Melonis zaandami* (van Voorthuysen). LOEBLICH and TAPPAN, 1964, p. C763, text fig. 627; 2–3.

Test medium in size, planispiral, sides slightly concave, with a distinct umbilical depression on each side; wall coarsely perforate, aperture a curved slit at the base of the apertural face of the final chamber with a slight upper lip.

| *Locality* | *Average %* | *Average Depth (m)* |
|---|---|---|
| Beaufort shelf | X | 417 |
| M'Clure Strait | 1 | 473 |
| P.G.A. Sea | X | 382 |
| Jones Sound | 2 | 334 |

Family ROBERTINIDAE
Subfamily ROBERTINACEA
Genus ROBERTINOIDES Höglund, 1947

*Robertinoides charlottensis* (Cushman): Plate 21-VI, Figure 16

*Cassidulina charlottensis* Cushman, 1925, p. 41, pl. 6, figs. 6–7.
*Robertinoides (?) charlottensis* (Cushman). LOEBLICH and TAPPAN, 1953, p. 108, pl. 20, figs. 6–7.
*Robertinoides charlottensis* (Cushman). VILKS, 1969, p. 51, pl. 3, fig. 22.

Test small to medium in size, spiral, with several chambers to a whorl, each chamber subdivided by a vertical infolding of the wall. Aperture double, consisting of a vertical slit into the face of the final chamber from its base and a second slit almost at right angles to the first.

| *Locality* | *Average %* | *Average Depth (m)* |
|---|---|---|
| M'Clure Strait | X | 398 |

## Plate Explanations

### *Plate 21-I*

I-1. *Astrorhiza arenaria*
Station 68-24 VC
Lat. 74°30′N; Long. 120°10′W
Depth, 488 m; 1164-02 (GSC 23412)

I-2. *Bathysiphon filiformis*
Station 68-28 VC
Lat. 75°00′N; Long. 118°30′W
Depth, 387 m; 1168-01 (GSC 23413)

I-3. *Hyperammina* sp.
Station 68-24 VC
Lat. 74°30′N; Long. 120°10′W
Depth, 488 m; 1162-20 (GSC 23414)

I-4. *Hyperammina subnodosa*
Station 83008-21
Lat. 75°53.0′N; Long. 83°30.0′W
Depth, 570 m; 1167-06 (GSC 23415)

I-5. *Saccorhiza ramosa*
Station 68-24 VC
Lat. 74°30′N; Long. 120°10′W
Depth, 488 m; 1162-23 (GSC 23416)

I-6. *Saccammina atlantica*
Station DH9-S4
Lat. 74°47′N; Long. 106°30′W
Depth, 20 m; 1160-17 (GSC 23417)

I-7. *Saccammina sphaerica*
Station 62-16 VC
Lat. 78°47.4′N; Long. 104°57.2′W
Depth, 57 m; 1167-05 (GSC 23418)

I-8. *Reophax arctica*
Station DH2-S3
Lat. 74°58.2′N; Long. 105°19.5′W
Depth, 13 m; 1160-33 (GSC 23419)

I-9. *Reophax fusiformis*
Station 68-24 VC
Lat. 74°30′N; Long. 120°10′W
Depth, 488 m; 1162-21 (GSC 23420)

I-10. *Reophax gracilis*
Station 62-34 VC
Lat. 78°58.9′N; Long. 104°10.2′W
Depth, 45 m; 1168-02 (GSC 23421)

I-11. *Reophax guttifer*
Station 68-24 VC
Lat. 74°30′N; Long. 120°10′W
Depth, 488 m; 1162-19 (GSC 23422)

I-12. *Reophax nodulosus*
Station 68-24 VC
Lat. 74°30′N; Long. 120°10′W
Depth, 488 m; 1162-22 (GSC 23423)

I-13. *Reophax pilulifer*
Station 68-28 VC
Lat. 74°00′N; Long. 118°30′W
Depth, 387 m; 1164-13 (GSC 23424)

I-14, 15. *Silicosigmoilina groenlandica*
Station 83008-24
Lat. 76°21.5′N; Long. 83°31.0′W
Depth, 228 m
Fig. 14: 1167-27 (GSC 23425)
Fig. 15: 1168-28 (GSC 23426)

I-16, 17. *Adercotryma glomerata*
Station DH6-S7
Lat. 75°16.4′N; Long. 105°48′W
Depth, 24 m
Fig. 16: 1160-13 (GSC 23427)
Fig. 17, apertural view: 1160-15 (GSC 23428)

I-18, 19. *Cribrostomoides crassimargo*
Station 62-46VC
Lat. 78°58.7′N; Long. 104°8.0′W
Depth, 63 m
Fig. 18, apertural view: 1168-26 (GSC 23429)
Fig. 19, side view: 1168-25 (GSC 23430)

I-20. *Cribrostomoides jeffreysii* (apertural view)
Station 63-17VC
Lat. 77°52.6′N; Long. 110° 07.5′W
Depth, 200 m; 1160-20 (GSC 23431)

I-21. *Cribrostomoides jeffreysii* (side view)
Station DH6-S1
Lat. 75°16.4′N; Long. 105°48′W
Depth, 24 m; 1160-19 (GSC 23432)

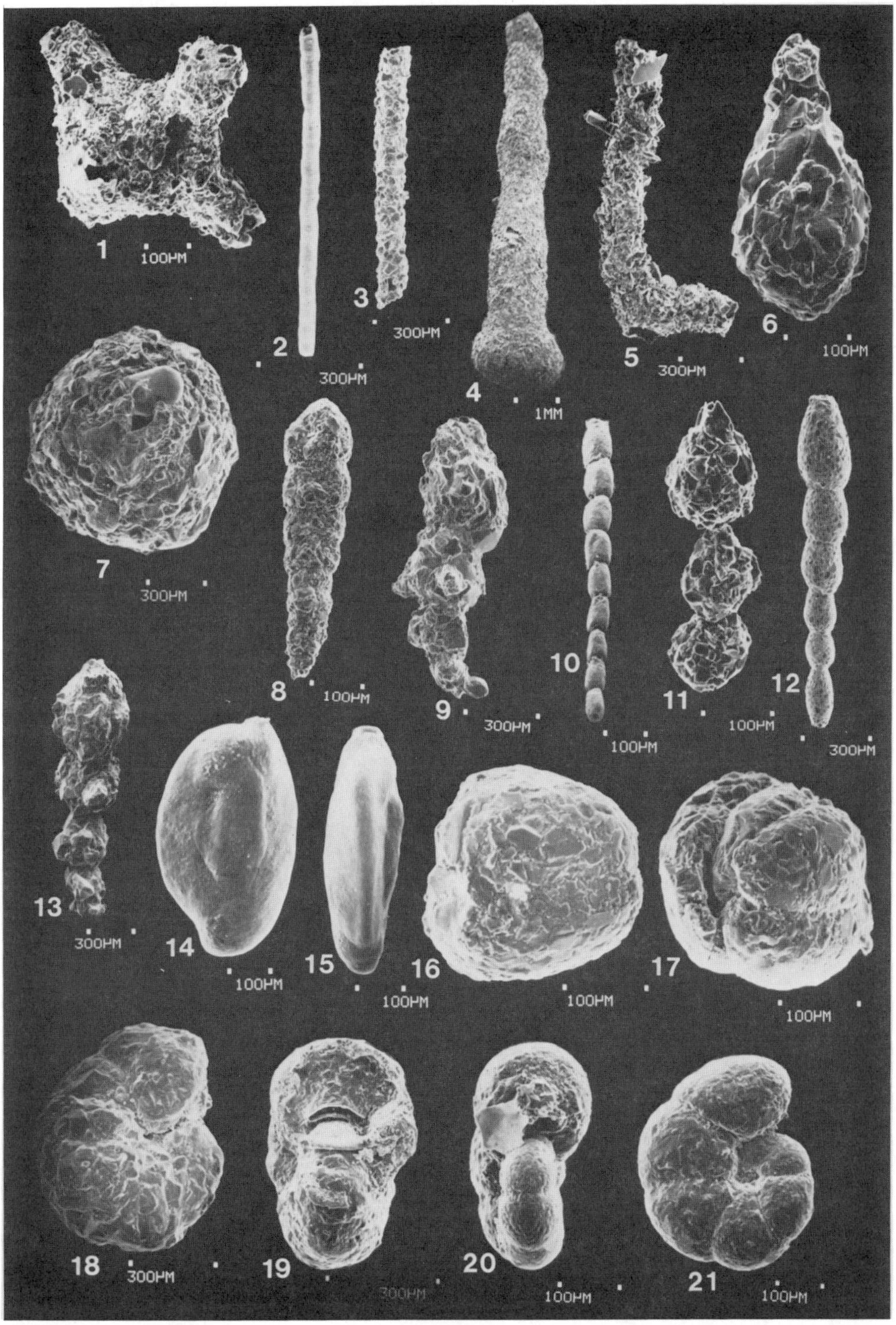
1
100HM
2
300HM
3
300HM
4
1MM
5
300HM
6
100HM
7
300HM
8
100HM
9
300HM
10
100HM
11
100HM
12
300HM
13
300HM
14
100HM
15
100HM
16
100HM
17
100HM
18
300HM
19
300HM
20
100HM
21
100HM

PLATE 21-I

## Plate 21-II

II-1. *Recurvoides contortus*
Station 66-10 VC
Lat. 76°25′N; Long. 112°24′W
Depth, 349 m; 1160-04 (GSC 23433)

II-2. *Recurvoides laevigatum*
Station 63-126 VC
Lat. 77°50.7′N; Long. 110°34.0′W
Depth, 102 m; 1160-08 (GSC 23434)

II-3. *Recurvoides turbinatus*
Station 83008-27
Lat. 75°53.5′N; Long. 87°15.5′W
Depth, 218 m; 1160-34 (GSC 23435)

II-4. *Ammotium cassis*
Station 83008-L102
Lat. 75°46.2′N; Long. 83°15.4′W
Depth, 27 m; 1168-56 (GSC 23436)

II-5. *Spiroplectammina biformis* (apertural view)
Station 66-52 VC
Lat. 77°12′N; Long. 115°02′W
Depth, 33 m; 1168-50 (GSC 23437)

II-6. *Spiroplectammina biformis* (side view)
Station 83008-2
Lat. 75°38.3′N; Long. 79°46.8′W
Depth, 532 m; 1168-45 (GSC 23438)

II-7. *Texturlaria earlandi*
Station 63-17VC
Lat. 77°52.6′N; Long. 110°07.5′W
Depth, 200 m; 1160-22 (GSC 23439)

II-8, 9. *Textularia torquata*
Station 62-7MP5
Lat. 77°29′N; Long. 104°30′W
Depth, 323 m
Fig. 8, side view: 1158-04 (GSC 23440)
Fig. 9, apertural view: 1158-02 (GSC 23441)

II-10, 11. *Tritaxis atlantica*
Station 63-17VC
Lat. 77°52.6′N; Long. 110°07.5′W
Depth, 200 m
Fig. 10, spiral view: 1160-31 (GSC 23442)
Fig. 11, apertural view: 1160-28 (GSC 23443)

II-12, 13. *Tritaxis bullata*
Station 68-28VC
Lat. 75°00′N; Long. 118°30′W
Depth, 387 m
Fig. 12, spiral view: 1165-02 (GSC 23444)
Fig. 13, apertural view: 1165-03 (GSC 23445)

II-14. *Trochammina globigeriniformis* (spiral view)
Station 68-19VC
Lat. 75°31′N; Long. 124°25′W
Depth, 417 m; 1168-05 (GSC 23446)

II-15. *Trochammina globigeriniformis* (ventral view)
Station 62-38VC
Lat. 78°58.2′N; Long. 104°10.2′W
Depth, 90 m; 1168-04 (GSC 23447)

II-16, 17. *Trochammina inflata*
Station 68-28VC
Lat. 75°00′N; Long. 118°30′W
Depth, 387 m
Fig. 16, side view: 1164-14 (GSC 23448)
Fig. 17, apertural view: 1164-16 (GSC 23449)

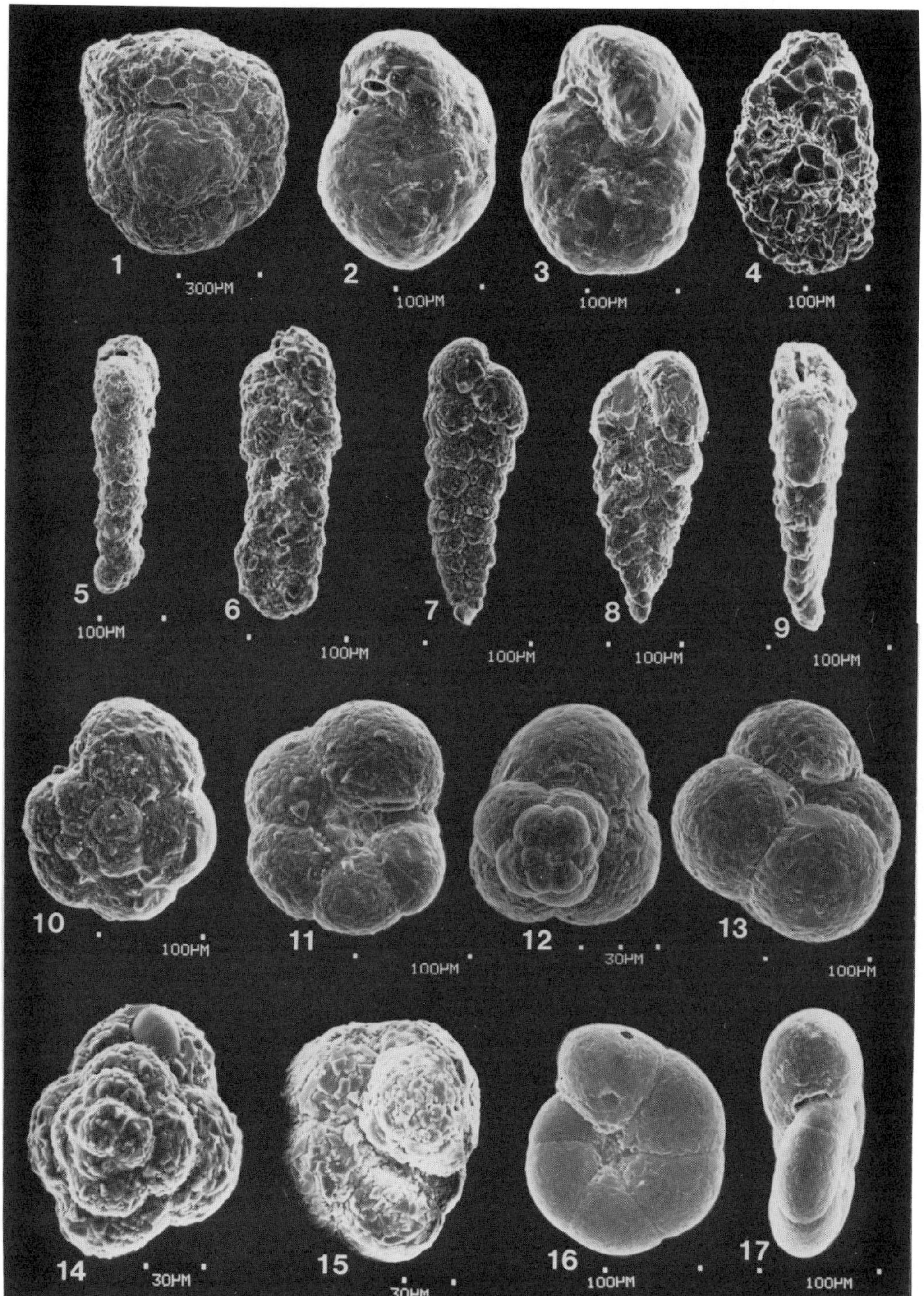
1
300μm
2
100μm
3
100μm
4
100μm
5
100μm
6
100μm
7
100μm
8
100μm
9
100μm
10
100μm
11
100μm
12
30μm
13
100μm
14
30μm
15
30μm
16
100μm
17
100μm

PLATE 21-II

## Plate 21-III

III-1–3. *Trochammina nana*
Station DH6-S1
Lat. 75°16°4′N; Long. 105°48′W
Depth, 24 m
Fig. 1, ventral view: 1159-06 (GSC 23450)
Fig. 2, spiral view: 1159-07 (GSC 23451)
Fig. 3, apertural view: 1159-09 (GSC 23452)

III-4. *Trochammina ochracea*
(spiral view)
Station 83008-L104
Lat. 75°46.7′N; Long. 83°15.3′W
Depth, 5.5 m; 1168-53 (GSC 23453)

III-5, 6. *Trochammina ochracea*
(apertural view)
Station 83008-L106
Lat. 75°48.8′N; Long. 83°15.0′W
Depth, 17 m; 1168-54 (GSC 23454)

III-7. *Trochammina quadriloba*
(spiral view)
Station 66-38VC
Lat. 76°01′N; Long. 109°40′W
Depth, 48 m; 1167-04 (GSC 23455)

III-8. *Trochammina quadriloba*
(apertural view)
Station 66-52VC
Lat. 77°12′N; Long. 115°02′W
Depth, 33 m; 1167-02 (GSC 23456)

III-9, 10. *Eggerella advena*
Station DH3-S3
Lat. 74°58.2′N; Long. 105°19.5′W
Depth, 13 m
Fig. 9, side view: 1160-03 (GSC 23457)
Fig. 10, apertural view: 1160-01 (GSC 23458)

III-11. *Quinqueloculina agglutinata*
Station 69050-849
Lat. 72°25.2′N; Long. 129°27.3′W
Depth, 69 m; 1168-10 (GSC 23459)

III-12. *Quinqueloculina arctica*
(side view)
Station 68-42VC
Lat. 75°25′N; Long. 127°00′W
Depth, 381 m; 1164-10 (GSC 23460)

III-13. *Quinqueloculina arctica*
(side view)
Station 68-45VC
Lat. 74°30′N; Long. 128°30′W
Depth, 370 m; 1160-38 (GSC 23461)

III-14. *Quinqueloculina seminulum*
Station 68-33VC
Lat. 74°50′N; Long. 126°00′W
Depth, 421 m; 1168-59 (GSC 23461)

III-15. *Triloculina trihedra*
(side view)
Station 68-33VC
Lat. 74°50′N; Long. 126°00′W
Depth, 421 m; 1164-11 (GSC 23462)

III-16. *Triloculina trihedra*
(side view showing aperture)
Station 83008-011
Lat. 75°50.5′N; Long. 80°07.0′W
Depth, 560 m; 1160-31 (GSC 23463)

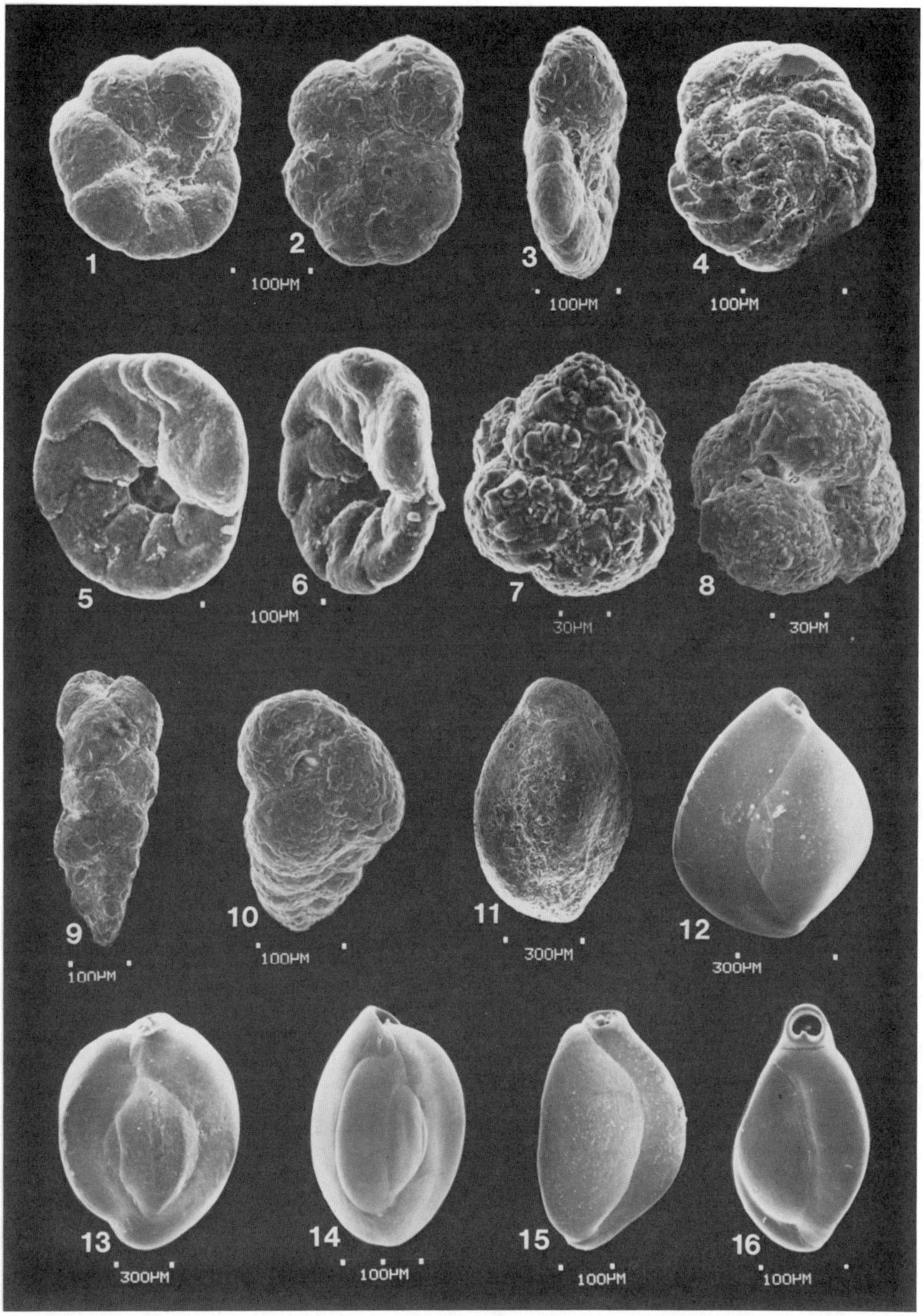
1
2
100HM
3
100HM
4
100HM
5
6
100HM
7
30HM
8
30HM
9
100HM
10
100HM
11
300HM
12
300HM
13
300HM
14
100HM
15
100HM
16
100HM

PLATE 21-III

## Plate 21-IV

IV-1. *Pseudopolymorphina novangliae* (side view)
Station 69-50-822
Lat. 69°58′N; Long. 137°00′W
Depth, 38 m; 1161-20 (GSC 23464)

IV-2. *Pseudopolymorphina novangliae* (apertural view)
Station 69-50-849
Lat. 72°25.2′N; Long. 129°27.3′W
Depth, 69 m; 1168-22 (GSC 2345)

IV-3, 4. *Islandiella helenae*
Station 69-50-849
Lat. 72°25.2′N; Long. 129°27.3′W
Depth, 69 m
Fig. 3, side view: 1168-60 (GSC 2346)
Fig. 4, edge view: 1168-58 (GSC 2347)

IV-5, 6. *Islandiella norcrossi*
Station 83008-17
Lat. 76°14.5′N; Long. 82°46.6′W
Depth, 706 m
Fig. 5, side view: 1168-33 (GSC 2348)
Fig. 6, edge view: 1168-32 (GSC 2349)

IV-7, 8. *Stainforthia concava*
Station 62-7 MP-5
Lat. 77°29′N; Long. 104°30′W
Depth, 323 m
Fig. 7, 1157-37 (GSC 2370)
Fig. 8, 1157-39 (GSC 2371)

IV-9. *Trifarina fluens*
Station 83008-82
Lat. 75°55.5′N; Long. 87°25.5′W
Depth, 247 m; 1167-03 (GSC 2372)

IV-10, 11, 12. *Buccella frigida*
Station 62-5MP-9
Lat. 77°09′N; Long. 100°56′W
Depth, 162 m
Fig. 10, spiral view: 1157-27 (GSC 2373)
Fig. 11, ventral view: 1157-28 (GSC 2374)
Fig. 12, edge view: 1157-32 (GSC 2375)

IV-13, 14, 15. *Eoeponidella pulchella*
Station 69-050-818
Lat. 69°33.0′N; Long. 138°11.8′W
Depth, 109 m
Fig. 13, spiral view: 1169-20 (GSC 2376)
Fig. 14, ventral view: 1169-17 (GSC 2377)
Fig. 15, edge view: 1169-22 (GSC 2378)

IV-16. *Epistominella vitrea*
Station 83008-27
Lat. 75°53.5′N; Long. 87°15.5′W
Depth, 218 m; 1168-30 (GSC 2379)

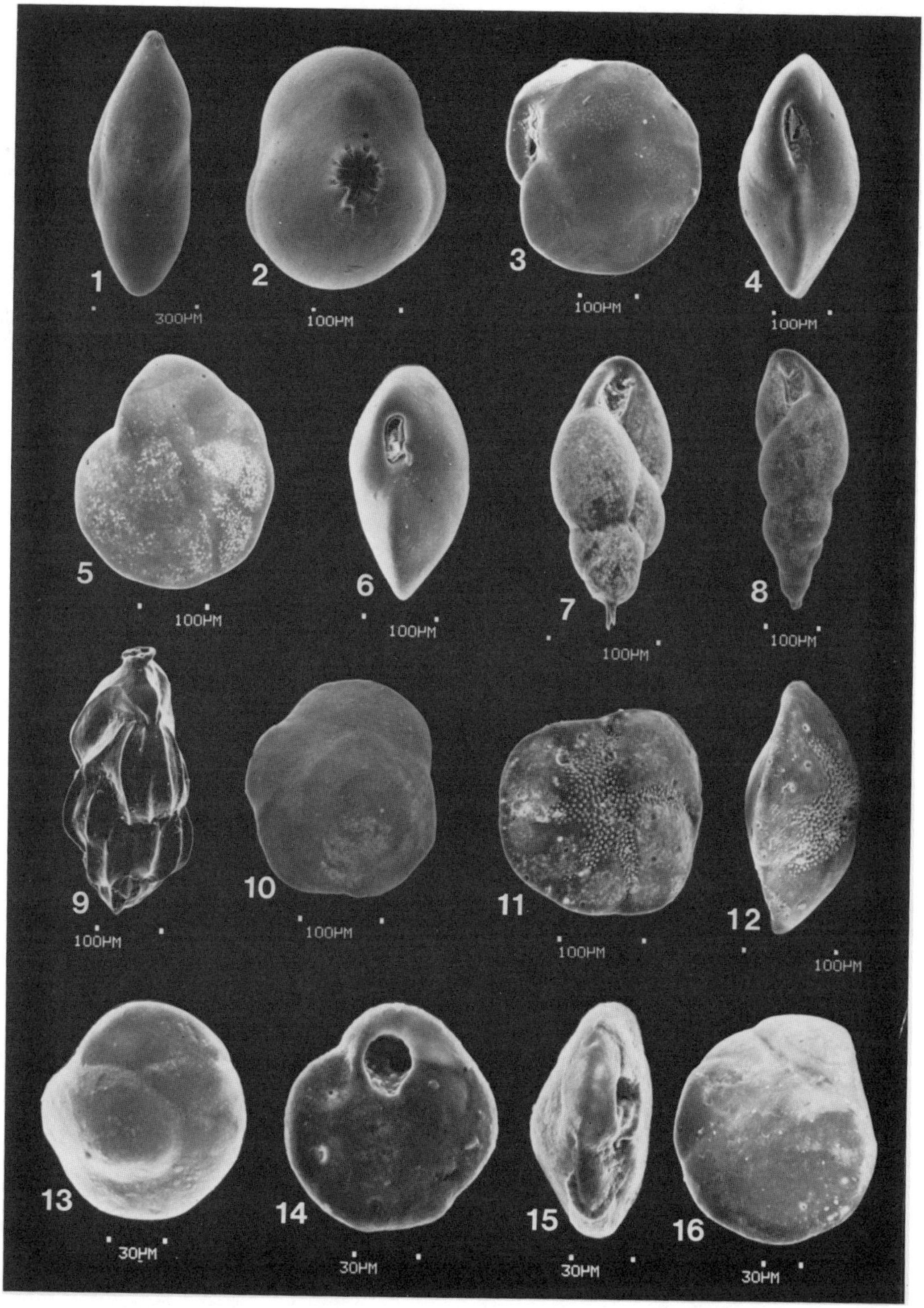
1
300µM
2
100µM
3
100µM
4
100µM
5
100µM
6
100µM
7
100µM
8
100µM
9
100µM
10
100µM
11
100µM
12
100µM
13
30µM
14
30µM
15
30µM
16
30µM

PLATE 21-IV

## Plate 21-V

V-1. *Epistominella vitrea*
Station 83008-27
Lat. 75°53.5′N; Long. 87°15.5′W
Depth, 218 m
Fig. 1, ventral view: 1168-29 (GSC 23480)

V-2. *Stetsonia horvathi*
Station 83008-34
Lat. 76°04.4′N; Long. 86°29.9′W
Depth, 543 m; 1168-18 (GSC 23481)

V-3. *Stetsonia horvathi*
(edge view)
Station 68-19VC
Lat. 75°31′N; Long. 124°25′W
Depth, 417 m; 1169-29 (GSC 23482)

V-4. *Valvulineria arctica*
(side view)
Station 62-IMP-3
Lat. 78°55′N; Long. 106°53′W
Depth, 463 m, 1157-35 (GSC 23483)

V-5. *Valvulineria arctica*
(edge view)
Station 62-7MP-5
Lat. 77°29′N; Long. 104°30′W
Depth, 323 m; 1160-35 (GSC 23484)

V-6, 7. *Elphidium bartletti*
Station H81-7
Lat. 70°25′N; Long. 135°25′W
Depth, 65 m
Fig. 6, side view: 1159-12 (GSC 23485)
Fig. 7, apertural view: 1159-11 (GSC 23486)

V-8, 9. *Elphidium excavatum* f. *clavata*
Station 69050-849
Lat. 72°25.2′N; Long. 129°27.3′W
Depth, 69 m
Fig. 8, side view: 1169-10 (GSC 23487)
Fig. 9, apertural view: 1169-12 (GSC 23488)

V-10. *Elphidium subarcticum*
Station 68-17 VC
Lat. 74°57′N; Long. 124°59′W
Depth, 473 m; 1160-39 (GSC 23489)

V-11. *Haynesina nana*
Station 69050-832
Lat. 70°08.5′N; Long. 132°47.9′W
Depth, 25 m; 1162-01 (GSC 23490)

V-12. *Haynesina orbiculare*
(side view)
Station 69050-849
Lat. 72°25.2′N; Long. 129°27.3′W
Depth, 69 m; 1168-13 (GSC 23491)

V-13. *Haynesina orbiculare*
(apertural view)
Station 69050-868
Lat. 74°53.0′N; Long. 98°3.0′W
Depth, 91 m; 1168-14 (GSC 23492)

V-14, 15. *Cibicides lobatulus*
Station 62-4.5 MP
Lat. 78°12.6′N; Long. 105°35′W
Depth, 440 m
Fig. 14, side view: 1157-45 (GSC 23493)
Fig. 15, edge view, showing aperture: 1157-44 (GSC 23494)

V-16, 17. *Fursenkoina fusiformis*
Station 69050-868
Lat. 74°53.0′N; Long. 98°3.0′W
Depth, 91 m
Fig. 16, apertural view: 1168-40 (GSC 23495)
Fig. 17, side view: 1168-42 (GSC 23496)

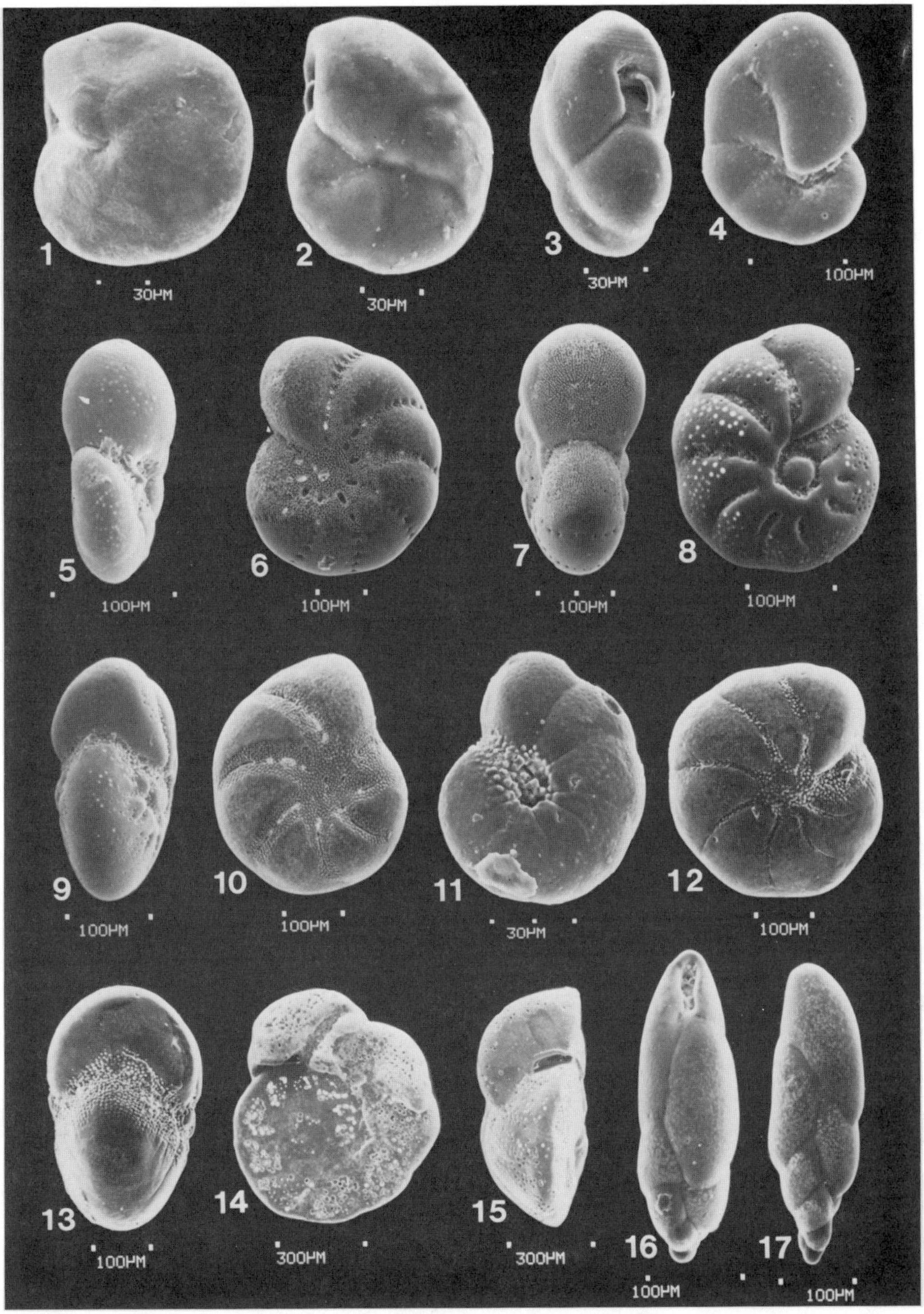
1
30µM
2
30µM
3
30µM
4
100µM
5
100µM
6
100µM
7
100µM
8
100µM
9
100µM
10
100µM
11
30µM
12
100µM
13
100µM
14
300µM
15
300µM
16
100µM
17
100µM

PLATE 21-V

## Plate 21-VI

VI-1. *Cassidulina laevigata*
(side view)
Station 62-2MP-3
Lat. 78°43′N; Long. 106°33′W
Depth, 488 m; 1157-09 (GSC 23497)

VI-2. *Cassidulina laevigata*
(apertural view)
Station 62-1MP-3
Lat. 78°55′N; Long. 106°53′W
Depth, 463 m; 1157-14 (GSC 23498)

VI-3. *Cassidulina reniforme*
(side view)
Station 69050-822
Lat. 69°58.0′N;Long. 137°00.0′W
Depth, 38 m; 1161-11 (GSC 23499)

VI-4. *Cassidulina reniforme*
(edge view)
Station 62-SMP-9
Lat. 77°09′N; Long. 100°56′W
Depth, 162 m; 1157-07 (GSC 23500)

VI-5. *Astrononion gallowayi*
(side view)
Station 69050-817
Lat. 69°22.1′N; Long. 138°4.8′W
Depth, 47 m; 1161-07 (GSC 23501)

VI-6. *Astrononion gallowayi*
(apertural view)
Station 69050-820
Lat. 70°21.5′N; Long. 137°33′W
Depth, 352 m; 1161-08 (GSC 23502)

VI-7. *Nonionella auricula*
(side view)
Station 68-19VC
Lat. 75°31′N; Long. 124°25′W
Depth, 417 m; 1168-16 (GSC 23503)

VI-8. *Nonionella auricula*
(edge view)
Station 83008-34
Lat. 76°04.4′N; Long. 86°29.9′W
Depth, 543 m; 1168-15 (GSC 23504)

VI-9, 10. *Nonionellina labradorica*
Station 69050-820
Lat. 70°21.5′N; Long. 137°33′W
Depth, 352 m
Fig. 9, side view: 1161-27 (GSC 23505)
Fig. 10, edge view: 1161-26 (GSC 23506)

VI-11–13. *Oridorsalis umbonatus*
Station 69050-810
Lat. 70°59.8′N; Long. 138°14.7′W
Depth, 1,657 m
Fig. 11, dorsal view: 1169-56 (GSC 23507)
Fig. 12, ventral view: 1169-54 (GSC 23508)
Fig. 13, edge view: 1169-57 (GSC 23509)

VI-14, 15. *Melonis zaandami*
Station 62-4.5 MP
Lat. 78°12.6′N; Long. 105°35′W
Depth, 440 m
Fig. 14, side view: 1157-46 (GSC 23510)
Fig. 15, edge view: 1157-49 (GSC 23511)

VI-16. *Robertinoides charlottensis*
Station 68-43 VC
Lat. 74°55′N; Long. 127°10′W
Depth, 398 m; 1162-04 (GSC 23512)

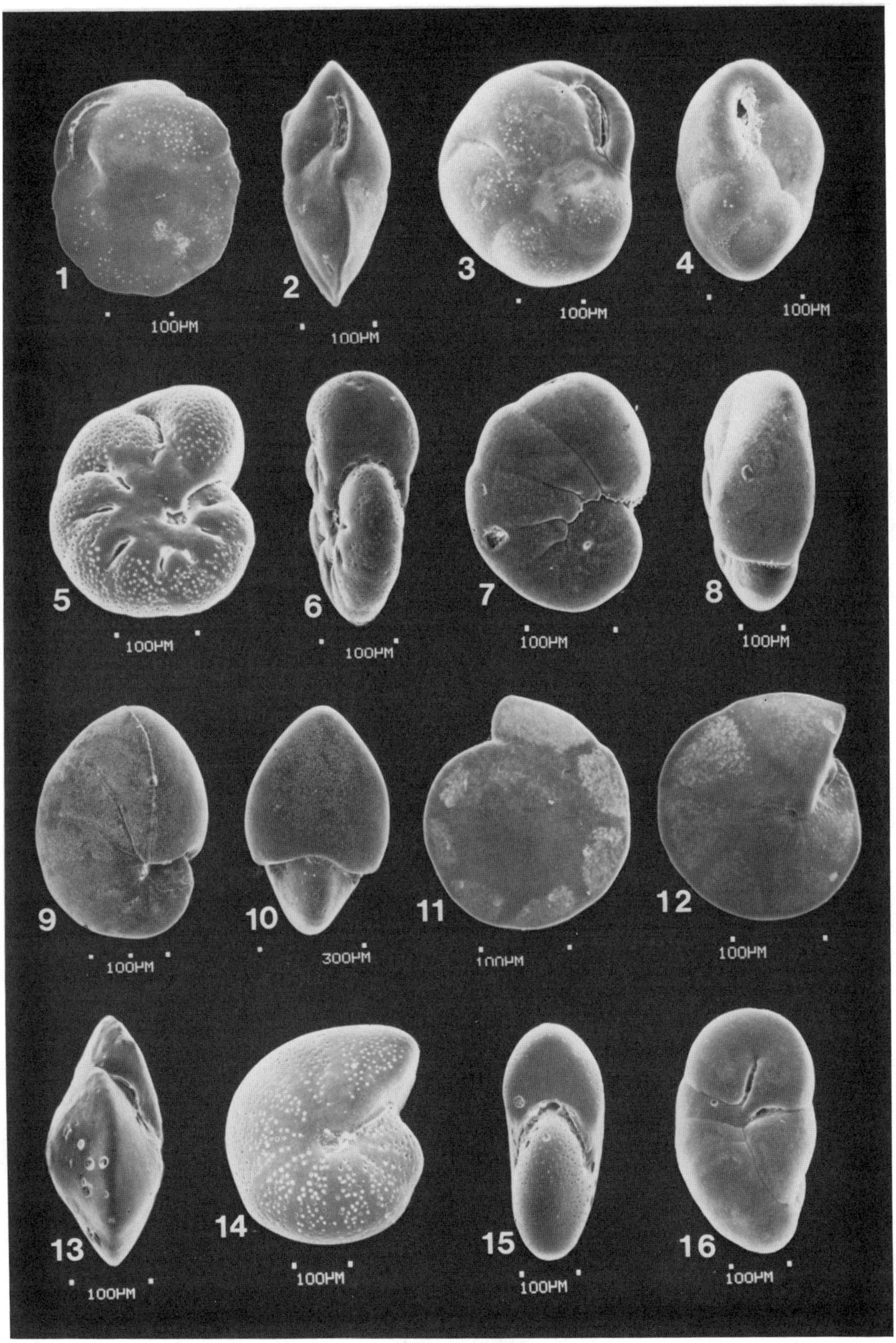
1
100PM
2
100PM
3
100PM
4
100PM
5
100PM
6
100PM
7
100PM
8
100PM
9
100PM
10
300PM
11
100PM
12
100PM
13
100PM
14
100PM
15
100PM
16
100PM

PLATE 21-VI

## Plate 21-VII

VII-1*a,b*. *Neogloboquadrina pachyderma:* Station 346, Lat. 70°19′N, Long. 138°48′W, Beaufort Sea. Depth range 0–200 m, collected Aug. 27, 1970. Magnification: Fig. 1*a* × 200; Fig. 1*b* × 3,000. (GSC 23513)

Figure 1*a*. Ventral view, sinistral, five chambers per last whorl, normalform, with an irregular and ornamented lip over the aperture. Longest diameter 400 microns, shortest diameter 325 microns.

Figure 1*b*. Surface of ultimate chamber. Secondary deposits forming irregular ridges, which begin to join around the pores. Some of the original bilamellar test surface still exposed. Average maximum diameter of pores 1.57 microns.

VII-2*a,b*. *Neogloboquadrina pachyderma:* Station 346. Magnification: Fig. 2*a* × 240; Fig. 2*b* × 3,000. (GSC 23514)

Figure 2*a*. Ventral view, sinistral, four chambers per last whorl kummerform. Longest diameter 325 microns, shortest diameter 242 microns.

Figure 2*b*. Surface of ultimate chamber. Secondary deposits in the form of semi-rounded punctae without central protuberances. Average maximum pore diameter 0.79 microns.

VII-3*a,b*. *Neogloboquadrina pachyderma:* Station 345, Lat. 70°28′N, Long. 138°57′W, Beaufort Sea. Depth range 0–200 m, collected Aug. 27, 1970. Magnification: Fig. 3*a* × 280; Fig. 3*b* × 3,000. (GSC 23515)

Figure 3*a*. Ventral view, sinistral, five chambers per last whorl, normalform, small, ornamented apertural lip. Longest diameter 271 microns, shortest diameter 214 microns.

Figure 3*b*. Surface of ultimate chamber. Secondary deposits in the form of semi-rounded punctae that may be joined by terraced ridges. A number of punctae contain central protuberances.

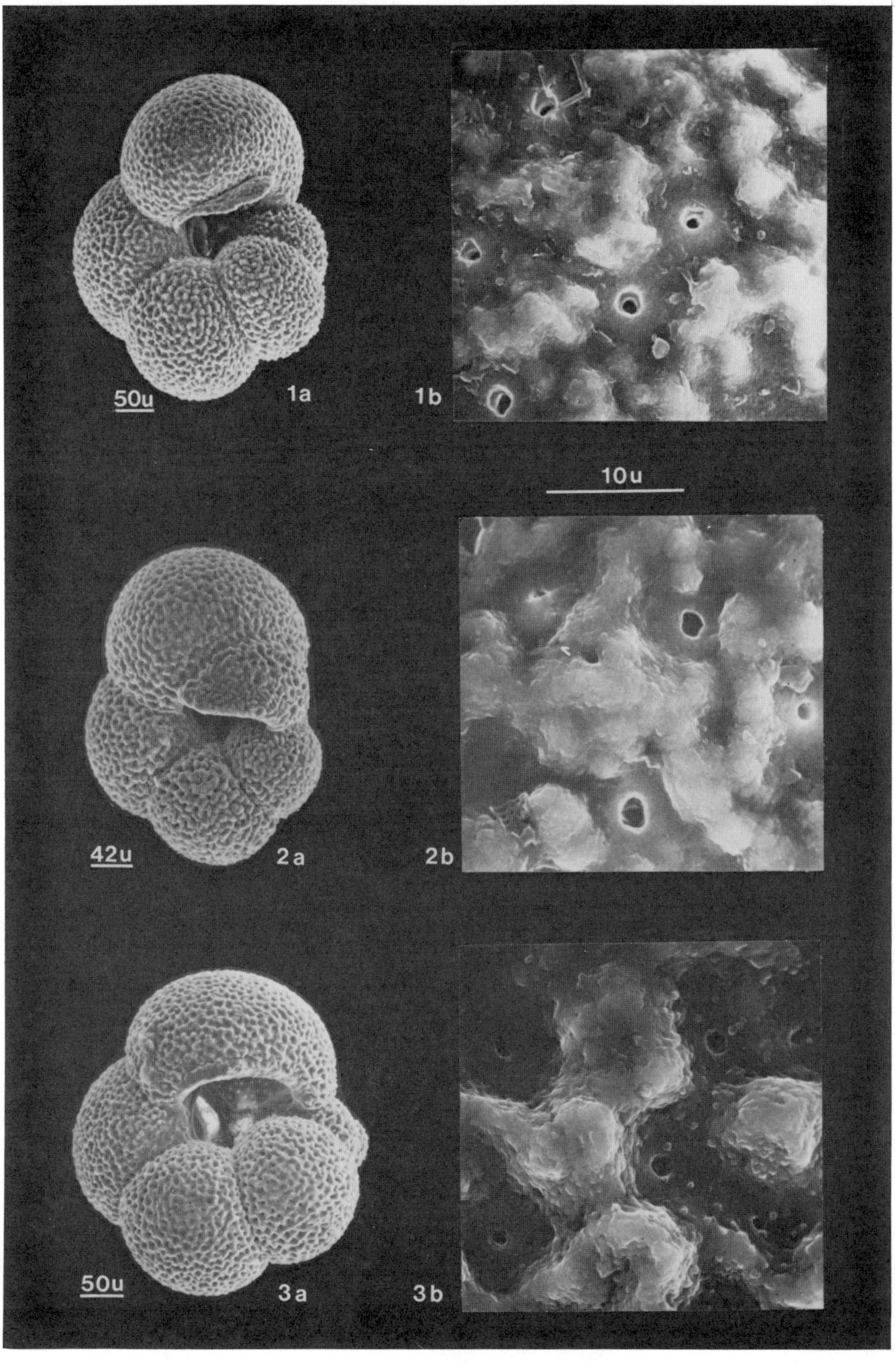


PLATE 21-VII

## Plate 21-VIII

VIII-1*a,b.* *Neogloboquadrina pachyderma:* Station 361, Lat. 70°37′N, Long. 139°29′W, Beaufort Sea. Depth range 0–200 m, collected Aug. 29, 1970. Magnification: Fig. 4*a* × 280; Fig. 4b × 3,000. (GSC 23516)

Figure 1*a*. Dorsal view, dextral, four chambers per last whorl, ten chambers total, longest diameter 292 microns, shortest diameter 278 microns.

Figure 1*b*. Surface of ultimate chamber.

VIII-2*a,b.* *Neogloboquadrina pachyderma:* Station 453, Lat. 74°39′N, Long. 108°11′W, Viscount Melville Sound. Depth range 0–200 m, collected Sept. 26, 1970. Magnification: Fig. 2*a* × 370; Fig. 2*b* × 3,000. (GSC 23517)

Figure 2*a*. Ventral view, sinistral, four chambers per last whorl, quadrate. Longest diameter 221 microns, shortest diameter 205 microns. Aperture relatively narrow with a smooth rim.

Figure 2*b*. Surface of ultimate chamber. Granular secondary deposits beginning to form irregular ridges.

VIII-3*a,b* *Neogloboquadrina pachyderma:* Station 346. Magnification: Fig. 3*a* × 300; Fig. 3*b* × 3,000. (GSC 23518)

Figure 3*a*. Ventral view, sinistral, five chambers per last whorl, normalform. Longest diameter 270 microns, shortest diameter 210 microns. Narrow and relatively smooth apertural rim.

Figure 3*b*. Surface of ultimate chamber. Secondary deposits in a form of semi-rounded protuberances or punctae. Most of the chamber consists of the original bilammellar surface. Average maximum pore diameter 1.79 microns.

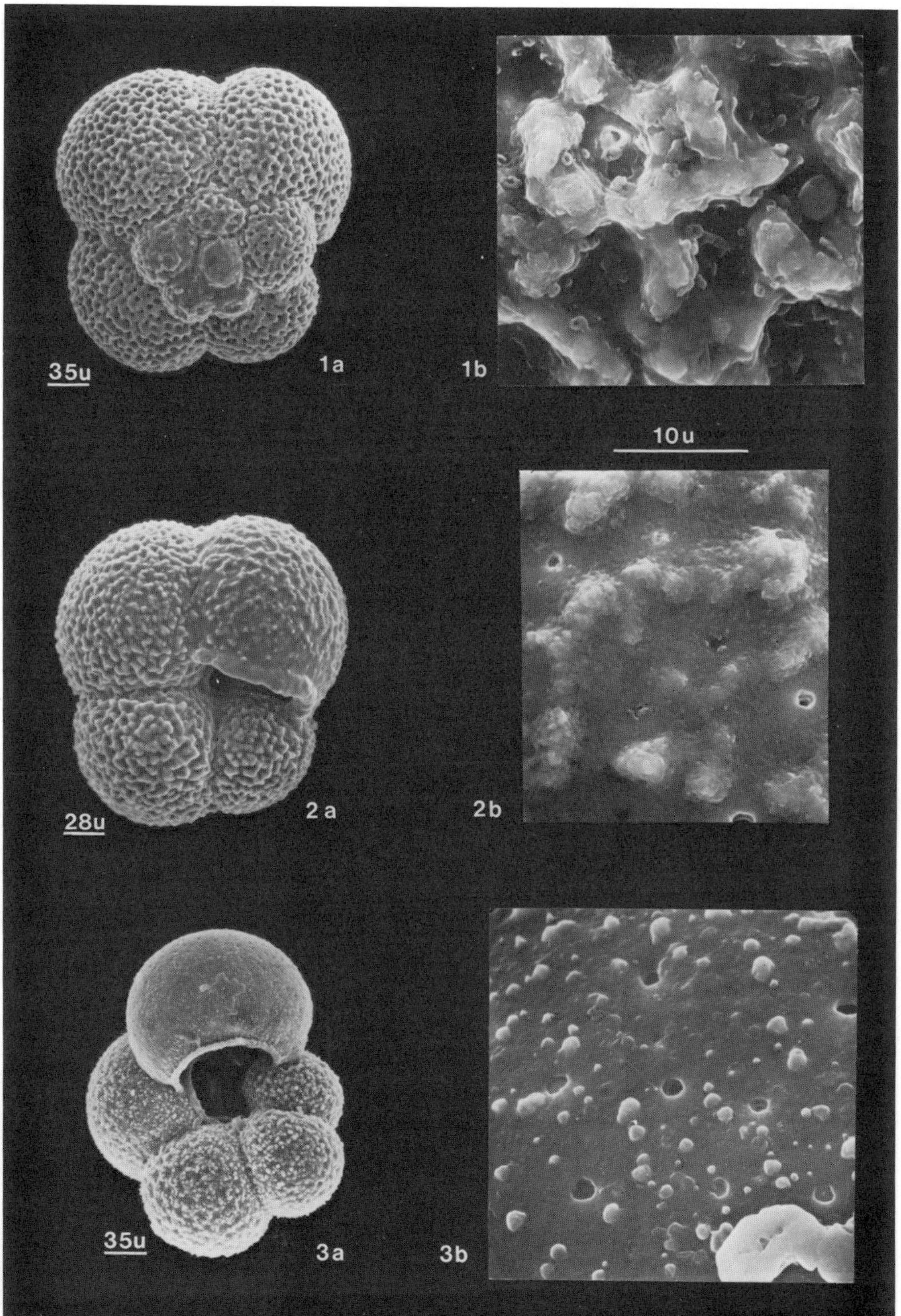


PLATE 21-VIII

## Plate 21-IX

IX-1*a,b*. *Neogloboquadrina pachyderma:* Station 469, Lat. 72°18′N, Long. 66°02′W, Baffin Bay. Depth range 0–200 m, collected Oct. 6, 1970. Magnification: Fig. 1*a* × 500; Fig. 1*b* × 3,000. (GSC 23519)

Figure 1*a*. Ventral view, sinistral, five chambers per last whorl, normalform, small nonornamental apertural lip. Longest diameter 160 microns, shortest diameter 130 microns.

Figure 1*b*. Surface of ultimate chamber. Secondary deposits in a form of punctae containing a few central protuberances that may be bases of vestigial spines. Limited ridge forming has taken place between the punctae. Average maximum pore diameter 1.21 microns.

IX-2*a,b*. *Neogloboquadrina pachyderma:* Station 466, Lat. 74°01′N, Long. 79°56′W, Lancaster Sound. Depth range 0–200 m, collected Oct. 1, 1970. Magnification: Fig. 2*a* × 600; Fig. 2*b* × 3,000 (GSC 23520)

Figure 2*a*. Ventral view, dextral, four and one-half chambers per last whorl, normalform, small and poorly ornamented apertural lip. Longest diameter 138 microns, shortest diameter 110 microns.

Figure 2*b*. Surface of penultimate chamber. The formation of secondary deposits similar to the previous example. Average maximum diameter of pores 1.57 microns.

IX-3*a,b*. *Neogloboquadrina pachyderma:* Station 346. Magnification: Fig. 3*a* × 300; Fig. 3*b* × 3,000. (GSC 23521)

Figure 3*a*. Ventral view, sinistral, four chambers per last whorl, normalform. Longest diameter 400 microns, shortest diameter 365 microns.

Figure 3*b*. Surface of ultimate chamber. Secondary deposits forming terraced ridges, which begin to join around the pores.

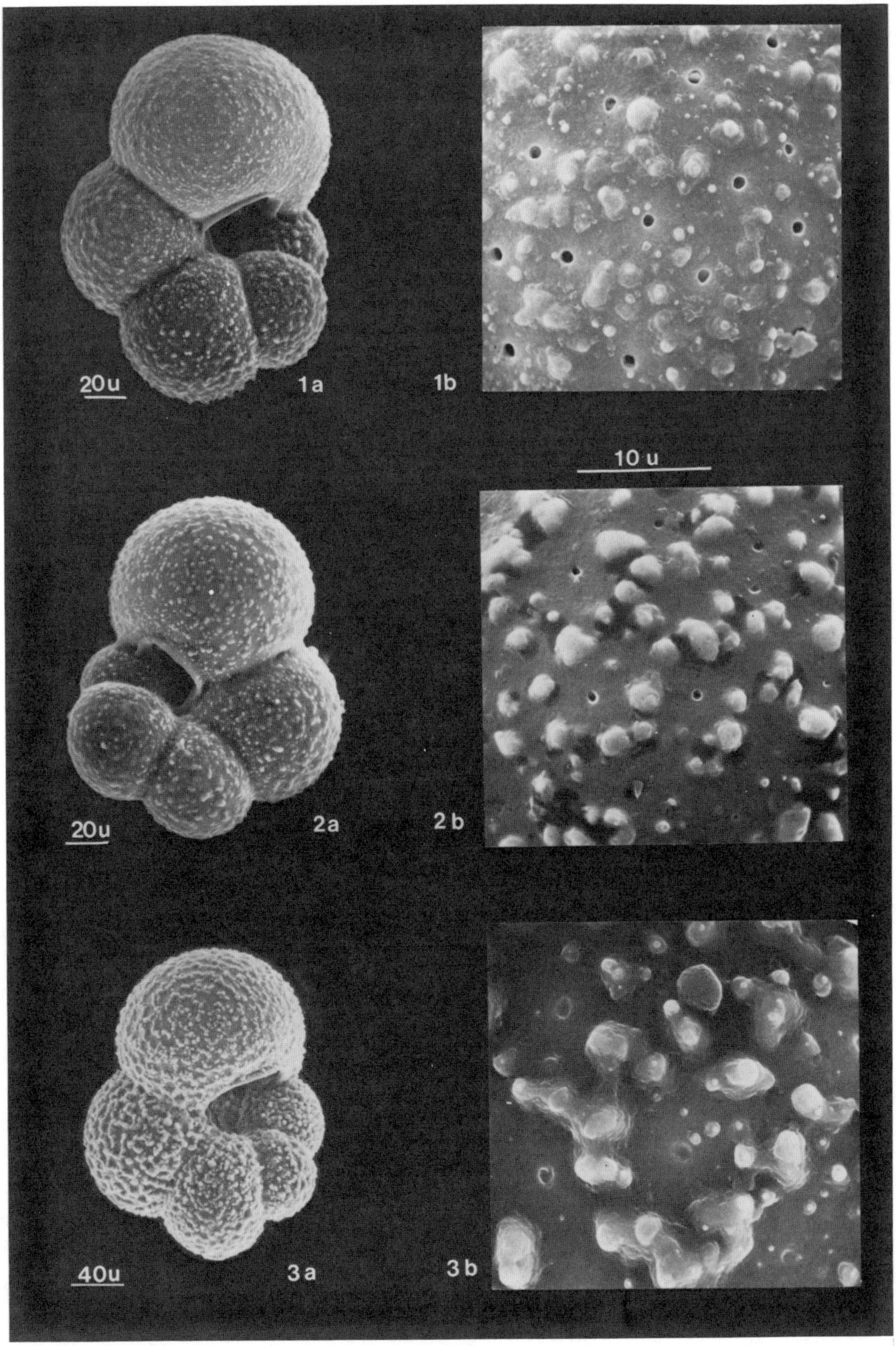


PLATE 21-IX

## REFERENCES


Andersen, H. V. 1952. *Buccella,* a new genus of the rotalid foraminifera. *J. Washington Acad. Sci.* **42**(5):143–151, figs. 1–13.

Anderson, G. J. 1963. Distribution patterns of Recent foraminifera of the Bering Sea. *Micropaleontology* **9**(3):305–317, figs. 1–12, plate 1.

Bandy, O. L., W. E. Frerichs, and E. Vincent. 1967. Origin, development and geologic significance of *Neogloboquadrina* Bandy, Frerichs and Vincent. *Cushman Found. Foram. Res., Contr.* **18**(4):152–153.

Barker, R. W. 1960. Taxonomic notes on the species figured by H. B. Brady in his report on the Foraminifera dredged by H. M. S. Challenger during the years 1873–1876. *Soc. Econ. Paleontol. Min., Spec. Publ. No. 9,* pp. 1–238, plates 1–115.

Bé, A. W. H. 1960. Some observations on the Arctic planktonic foraminifera. *Cushman Found. Foram. Res. Contr.,* **11**(part 2):64–68.

Boltovskoy, E. and R. Wright. 1976. *Recent Foraminifera.* Dr. W. Junk, b.v., The Hague, 515p.

Brady, H. B. 1871. On *Saccammina carteri,* a new foraminifera from the Carboniferous Limestone of Northumberland. *Ann. Mag. Nat. Hist.* **7**(ser. 4, no. 39):177–184, plate 12.

Brady, H. B. 1878. On the reticularian and radiolarian Rhizopoda (Foraminifera and Polycystina) of the North Polar Expedition of 1875–76. *Ann. Mag. Nat. Hist.* **1**(ser. 5): 425–440, plates 20–21.

Brady, H. B. 1879. Notes on some of the reticularian Rhizopoda of the Challenger Expedition. *Quart. J. Micro. Sci.* **19**(part 1):20–63, plates 3–5.

Brady, H. B. 1881*a*. Notes on some of the reticularian Rhizopoda of the Challenger Expedition. *Quart. J. Micro. Sci.* **31**(part 3):21–71.

Brady, H. B., 1881*b*. On some Arctic foraminifera from soundings obtained on the Austro-Hungarian North Polar Expedition of 1872–1874. *Ann. Mag. Nat. Hist.* **8**(ser. 5): 393–418, plate 21.

Brady, H. B. 1884. Report on the foraminifera dredged by H. M. S. Challenger, during the years 1873–1876. *Rept. Voy. Challenger, Zool.* **9:**1–814, plates 1–115.

Carpenter, W. B., W. K. Parker, and T. R. Jones. 1862. Introduction to the study of the foraminifera. *Ray. Soc. Publs.,* pp. 1–319, plates 1–22.

Carsola, A. J. 1952. Marine Geology of the Arctic Ocean and Adjacent Seas off Alaska and Northwestern Canada. University of California, unpublished doctoral thesis, 221p.

Coachman, L. K., and C. A. Barnes. 1961. The contribution of Bering Sea water to the Arctic Ocean. *Arctic* **14**(3):147–161.

Cole, F. E. 1981. Taxonomic notes on the bathyal zone benthonic foraminifera species off northeast Newfoundland. *Bedford Institute of Oceanography, Report Series B1-R-81-7,* 121p and 20 plates.

Collin, A. E., and M. J. Dunbar. 1964. Physical oceanography in Arctic Canada. *Oceanogr. Mar. Biol. Ann. Rev.* **2:**45–75.

Corliss, B. H. 1979. Taxonomy of Recent deep-sea benthonic foraminifera from the southeast Indian Ocean. *Micropaleontology* **25**(1):1–19, plates 1–5.

Corliss, B. H. 1985. Microhabitats of benthonic foraminifera within deep-sea sediments. *Nature* **314:**435–438.

Cushman, J. A. 1910. A monograph of the foraminifera of the North Pacific Ocean. Pt. 1, Astrorhizidae and Lituolidae. *U.S. Nat. Mus. Bull. 71,* pt. I, pp. 1–134, figs. 1–203.

Cushman, J. A. 1917. A monograph of the foraminifera of the North Pacific Ocean. Pt. 6, Miliolidae. *U.S. Nat. Mus. Bull. 71,* pp. 1–107, plates 1–38.

Cushman, J. A. 1920. The foraminifera of the Atlantic Ocean. Pt. 2, Lituolidae. *U.S. Nat. Mus. Bull. 104,* pp. 1–111, plates 1–18.

Cushman, J. A. 1922. Results of the Hudson Bay Expedition, 1920. I. The Foraminifera. *Contr. Canadian Biol. No. 9,* pp. 135–147.

Cushman, J. A. 1923. Foraminifera of the Atlantic Ocean. Pt. 4, Lagenidae. *U.S. Nat. Mus. Bull. 104,* pt. 4, pp. 1–228, plates 1–42.

Cushman, J. A. 1925. Recent foraminifera from British Columbia. *Contr. Cushman Lab. Foram. Res.* **1**(pt. 2):38–47, plates 6–7.

Cushman, J. A. 1930. The foraminifera of the Atlantic Ocean. Pt. 7, Nonionidae, Camerinidae, Peneroplidae and Alveolinellidae. *U.S. Nat. Mus. Bull. 104,* pt. 7, pp. 1–79, plates 1–18.

Cushman, J. A. 1931. The foraminifera of the Atlantic Ocean, Pt. 8. Rotaliidae, etc. *U.S. Nat. Mus. Bull. 104,* pt. 8, pp. 1–179, plates 1–26.

Cushman, J. A. 1933. New Arctic foraminifera collected by Capt. R. A. Bartlett from Fox Basin and off the northeast coast of Greenland. *Smithsonian Misc. Coll.* **89**(9):1–8, plates 1–2.

Cushman, J. A. 1937. A monograph of the foraminiferal family Valvulinidae. *Cushman Lab. Foram. Res. Spec. Publ. No. 8,* pp. 1–210, plates 1–24.

Cushman, J. A. 1944. Foraminifera from the shallow water of the New England coast. *Cushman Lab. Foram. Res. Spec. Publ. No. 12,* pp. 1–37, plates 1–4.

Cushman, J. A. 1948. Arctic foraminifera. *Cush-*

*man Lab. Foram. Res. Spec. Publ. No. 23*, pp. 1–79, plates 1–8.

Cushman, J. A., and Y. Ozawa. 1930. A monograph of the foraminifera family Polymorphinidae recent and fossil. *U.S. Nat. Mus. Proc.* **77**(6):1–195, plates 1–40.

Cushman, J. A., and R. Todd. 1947. A foraminiferal fauna from Amchitka Island, Alaska. *Cushman Lab. Foram. Res. Contr.* **23:**60–73, plates 15–16.

Dawson, G. M. 1870. On foraminifera from the Gulf and River St. Lawrence, *Canadian Nat.* **5:**172–179, 1 plate.

Dawson, J. W. 1860. Notice of Tertiary fossils from Labrador, Maine, etc. and remarks on the climate of Canada in the newer Pliocene or Pleistocene period. *Canadian Nat. Geol., Montreal, Canada* **5:**188–200, text figs. 1–5.

Dunbar, M., and M. J. Dunbar. 1972. The history of the North Water. *Proc. Roy. Soc. Edinburg* **72B:**232–241.

Earland, A. 1933. Foraminifera, Pt. 2, South Georgia. *Discovery Rpts.* **7:**27–138, plates 1–7.

Earland, A. 1934. Foraminifera, Pt. 3. The Falklands sector of the Antarctic (excluding South Georgia). *Discovery Rpts.* **10:**1–208, plates 1–10.

Ehrenberg, C. G. 1861. Elemente des tiefen Meeresgrundes im Mexikanishen Golfstrome bei Florida; uber die Tiefgrund-Verhaltnisse des Oceans am Eingange der Davistrasse und bei Island. *K. Preuss. Akad. Wiss. Berlin, Monatsber, Jahr 1861 (1862),* pp. 275–315.

English, T. S. 1961. Some biological oceanographic observations in the Central North Polar Sea; Drift Station Alpha, 1957–1958. *Arctic Inst. N. Am. Res. Paper 13,* 80p.

Feyling-Hanssen, R. W. 1964. Foraminifera in Late Quaternary deposits from Oslofjord area. *Norges Geol. Unders.* **225:**1–383.

Feyling-Hanssen, R. W. 1972. The foraminifer *Elphidium excavatum* (Terquem) and its variant forms. *Micropaleontology* **18**(3):337–354.

Feyling-Hanssen, R. W., and M. A. Buzas. 1976. Emendation of *Cassidulina* and *Islandiella helenae* new species. *J. Foram. Res.* **6**(2):154–158.

Feyling-Hanssen, R. W., S. Funder, and K. S. Petersen. 1982. The Lodin Elv Formation; a Plio-Pleistocene occurrence in Greenland. *Bull. Geol. Soc. Denmark* **31:**81–106.

Fortier, Y. O., and L. W. Morley. 1956. Geological unity of the Arctic Islands. *Roy. Soc. Can. Trans.* **1**(ser. 3):3–12.

Green, K. E. 1960. Ecology on some Arctic foraminifera. *Micropaleontology* **6**(1):57–78, plate 1.

Haman, D. 1973. Status of the foraminiferal genera *Asterellina* and *Eoeponidella. Micropaleontology* **19**(1):101–103.

Hanson, J. K. 1961. The albedo of sea-ice and ice islands in the Arctic Ocean Basin. *Arctic* **14**(3):188–196.

Hardy, I. A. 1985. Foraminifera from surface sediments of Jones Sound. Geological Survey of Canada. Open file.

Harper, J. R., and S. Penland. 1982. Beaufort Sea Sediment Dynamics. Unpublished report for the Geological Survey of Canada. Woodward-Clyde Consultants, Victoria, British Columbia, 125p.

Hodgson, D. A., and J.-S. Vincent. 1984. A 10,000 yr. B.P. extensive ice shelf over Viscount Melville Sound, Arctic Canada. *Quaternary Res.* **22**(1):18–30.

Höglund, H. 1947. Foraminifera in the Gullmar Fjord and the Skagerak. *Zool. Bidrag Uppsala* **26:**1–328, plates 1–32.

Höglund, H. 1948. New names for four homonym species described in "Foraminifera in the Gullmar Fjord and the Skagerak." *Cushman Lab. Foram. Res. Contr.* **24**(2):45–46.

Horn, D. R. 1967. Recent Marine Sediments and Submarine Topography, Sverdrup Islands, Canadian Arctic Archipelago. Ph.D. thesis, University of Texas, 362p. (unpublished).

Iqbal, J. 1973. Sedimentology and Benthonic Foraminifera in M'Clure Strait (Canadian Arctic Archipelago). M.Sc. thesis, Dalhousie University, Halifax, Nova Scotia, 265p. (unpublished).

Kanmacher, F. 1798. *Adams' Essays on the Microscope: The Second Edition, with Considerable Additions and Improvements.* Dillon and Keating, London, 724 p.

Little, E. M., M. B. Allen, and F. F. Wright. 1972. Field measurement of light penetration through sea ice. *Arctic* **25**(1):28–33.

Loeblich, A. R., Jr., and H. Tappan. 1953. Studies of Arctic foraminifera. *Smithsonian Misc. Colln.* **121**(7):1–150, plates 1–24.

Loeblich, A. R., Jr., and H. Tappan. 1961. Suprageneric classification of the Rhizopoda. *J. Paleontology* **35:**245–330.

Loeblich, A. R., Jr., and H. Tappan. 1964. Sarcodina, chiefly "Thecamoebians" and Foraminiferida. In R. C. Moore (ed.), *Treatise on Invertebrate Paleontology.* Geological Society of America, pt. c, Protista 2, vols. 1–2, pp. C1–C931, text figs. 1–653.

MacLean, B., J. M. Woodside, and P. Girouard. 1984. Geological and geophysical investigations in Jones Sound, District of Franklin. *Current Research, Part A. Geological Survey of Canada Paper 84-1A,* pp. 359–365.

Marlowe, J. I. 1968. Sedimentology of the

Prince Gustaf Adolph Sea area, District of Franklin, Polar Continental Shelf Project. *Geological Survey of Canada Paper 66-29,* pp. 1-83.

Marlowe, J. I., and G. Vilks. 1963. Marine geology, eastern part of the Prince Gustaf Adolf Sea, District of Franklin: Polar Continental Shelf Project. *Geological Survey of Canada Paper 63-22,* pp. 1-23.

Melling, H., R. A. Lake, D. R. Topham, and D. B. Fissel. 1984. Oceanic thermal structure in the western Canadian Arctic. *Continental Shelf Res.* **3**:233-258.

Miller, A. A. L., D. B. Scott, and F. S. Medioli. 1982. *Elphidium excavatum* (Terquem): Ecophenotypic versus subspecific variation. *J. Foram. Res.* **12**(2):116-144, plates 1-6.

Montagu, G. 1808. *Testacea Britannica.* Supplement. S. Woolmer. Exeter, England, 183p., plate 30.

Norman, A. M. 1877. On the Crustacea, Tunicata, Polyzoa, Echinodermata, Actinozoa, Foraminifera, Polycystina, and Spongida. In J. G. Jeffreys (ed.), Preliminary reports of the biological results of a cruise in H.M.S. Valorous to Davis Straits in 1875. *Roy. Soc. London Proc.* **25**:212-215, plates 2-3.

Norman, A. M. 1892. *Rhizopoda.* Museum Normanianum, pt. 7-8, pp. 14-21. The author (Durham).

Nørvang, A. 1945. Foraminifera. *Zoology of Iceland,* vol. 2, pt. 2. Ejnar Munksgaard, Copenhagen and Reykjavik, pp. 1-79, text figs. 1-14.

Parker, F. L. 1952*a*. Foraminifera species off Portsmouth, New Hampshire. *Bull. Mus. Comp. Zool.* **106**(9):391-423, plates 1-6.

Parker, F. L. 1952*b*. Foraminiferal distribution in the Long Island Sound—Buzzards Bay Area. *Bull. Mus. Comp. Zool.* **106**(10):425-473, plates 1-5.

Parker, W. K., and T. R. Jones. 1865. On some foraminifera from the North Atlantic and Arctic oceans including Davis Straits and Baffin Bay. *Roy. Soc. London Philos. Trans.* **155**:325-441, plates 12-19.

Parker, F. L., F. B. Phleger, and J. F. Pierson. 1953. Ecology of foraminifera from San Antonio Bay and environs, southwest Texas. *Cushman Found. Foram. Res. Spec. Publ. No. 2,* pp. 1-75, plates 1-4.

Pelletier, B. R. 1966. Development of submarine physiography in the Canadian Arctic and its relation to crustal movements. In G. D. Garland (ed.), Continental Drift. *Roy. Soc. Canada Spec. Pub. No. 9,* pp. 77-101.

Phleger, F. B. 1952. Foraminifera distribution in some sediment samples from the Canadian and Greenland Arctic. *Cushman Found. Foram Res. Contr.* **3**(pt. 2):80-89, plates 13-14.

Prest, V. K. 1984. The late Wisconsinan glacier complex. In R. J. Fulton (ed.), Quaternary Stratigraphy of Canada—A Canadian Contribution to IGCP Project 24. *Geological Survey of Canada Paper 84-10,* pp. 21-36.

Reuss, A. E. 1851. Über die fossilen Foraminiferen und Entomostraceen der Septarienthone der Umgegend von Berlin. *Deutsch. Geol. Gesell., Zeitschr.* **3**:49-91, plates 3-7.

Rodrigues, C. G., K. Hooper, and P. C. Jones. 1980. The apertural structures of *Islandiella* and *Cassidulina. J. Foram. Res.* **10**(1):48-60.

Rögl, F., and H. M. Bolli. 1973. Holocene to Pleistocene planktonic foraminifera of DSDP Leg 15, Site 147 (Cariaco Basin [French], Caribbean Sea) and their climatic interpretation. In N. T. Edgar, J. B. Saunders, et al. (eds.), *Initial Reports of the Deep Sea Drilling Project,* vol. 15. Washington, D.C. U.S. Government Printing Office, pp. 553-615.

Sars, G. 1872. Undersogelser over Hardangerfjordens Fauna. *Vidensk. Selsk. Christiania, Forhandl.* **1871**:246-255.

Sejrup, H.-P., and J.-P. Guilbault. 1980. *Cassidulina reniforme* and *C. obtusa* (Foraminifera): taxonomy, distribution and ecology. *Sarsia* **65**:79-85.

Sejrup, H.-P., T. Fjaeran, M. Hald, L. Beck, H. Jens, I. Miljeteig, I. Morvik, and O. Morvik. 1981. Benthonic foraminifera in surface samples from the Norwegian continental margin between 62°N and 65°N. *J. Foram. Res.* **11**(4):277-295, plates 1 and 2.

Smith, R. W. 1973. Sedimentological-Micropaleontological Analysis of Bottom Sediments from Resolute Bay. MacInnis Foundation, Underwater Arctic Expedition, 139p. (Unpublished manuscript.)

Subba Rao, D. V., and T. Platt. 1984. Primary production of Arctic waters. *Polar Biol.* **3**:191-201.

Todd, R. 1965. The foraminifera of the tropical Pacific collections of the "Albatross" 1899-1900. *U.S. Nat. Mus. Bull. 161,* pp. 1-139, plates 1-27.

Todd, R., and D. Low. 1961. Nearshore foraminifera of Martha's Vineyard Island, Mass. *Cushman Found. Foram. Res., Contr.* **12**(pt. 1):1-33, plates 1 and 2.

Vilks, G. 1964. Foraminiferal study of East Bay, MacKenzie King Island, District of Franklin: Polar Continental Shelf Project. *Geol. Surv. Canada Paper 64-53,* pp. 1-26.

Vilks, G. 1969. Recent foraminifera in the Canadian Arctic. *Micropaleontology* **15**(1):35-60, plates 1-3.

Vilks, G. 1970. Circulation of surface waters in parts of the Canadian Arctic Archipelago based on foraminiferal evidence. *Arctic* **23**(2):100–111.

Vilks, G. 1973. A Study of *Globorotalia pachyderma* (Ehrenberg) = *Globigerina pachyderma* (Ehrenberg) in the Canadian Arctic. Ph.D. thesis, Dalhousie University, 256p. (Unpublished.)

Vilks, G. 1975. Comparison of *Globorotalia pachyderma* (Ehrenberg) in the water column and sediments of the Canadian Arctic. *J. Foram. Res* **5**(4):313–325.

Vilks, G. 1976. Foraminifera of an ice-scoured nearshore zone in the Canadian Arctic. In C. T. Schafer and B. R. Pelletier (eds.), First International Symposium of Continental Margin Benthonic Foraminifera, Part A. Ecology and Biology, *Maritime Sediments Spec. Pub. No. 1,* pp. 267–277.

Vilks, G., and G. Joyce. 1976. Foraminifera in sediment cores from Beaufort Sea and Parry Channel, Canadian Arctic. *Bedford Institute of Oceanography Technical Report B1-D-76-1* (Microfiche).

Vilks, G., F. J. E. Wagner, and B. R. Pelletier. 1979. The Holocene marine environment of the Beaufort shelf. *Geol. Surv. Canada Bull.* **303:**43p.

Vilks, G., B. Deonarine, F. J. Wagner, and G. V. Winters. 1982. Foraminifera and Mollusca in surface sediments of southeastern Labrador Shelf. *Geol. Soc. America Bull.* **93:**225–238.

Vincent, J.-S. 1984. Quaternary stratigraphy of the western Canadian Arctic Archipelago. In R. J. Fulton (ed.), Quaternary Stratigraphy of Canada—A Canadian Contribution of IGCP Project 24. *Geol. Survey Canada Paper 84-10,* pp. 87–100.

Voloshinova, N. A. 1958. On a new systematics of the Nonionidae. *Mikrofauna SSSR.* Sbornik 9, VNIGRI, Trudy, no. 115, pp. 117–191, plates 1–16.

Voorthuysen, J. H., van. 1952. A new name for a Pleistocene foraminifer from Netherlands. *J. Paleontol.* **26**(4):680–681.

Williamson, W. C. 1858. On the Recent Foraminifera of Great Britain. *Roy. Soc.,* pp. 1–107, plates 1–7.

Worthington, L. V. 1953. Oceanographic results of project Skijump I and Skijump II in the Polar Sea, 1951–1952. *Am. Geophys. Union Trans.* **34**(4):543–551.

# 22: Thorium and Uranium Isotopes in Arctic Sediments

B. L. K. Somayajulu,
P. Sharma, and
Yvonne Herman

## INTRODUCTION

Knowledge of accumulation rates of Arctic sediments is essential to the understanding of ice dynamics and sedimentation (Herman, 1974; Herman and Osmond, 1984; Markussen, Zahn, and Thiede, 1985) as well as to the paleoclimates and paleoceanography of the region (Herman, 1974). Earlier geochronological studies involved the application of cosmogenic $^{14}C$ and $^{230}Th$, the daughter product of primordial $^{238}U$ (Ku and Broecker, 1967, and references therein). Sedimentation rates in the range of mm/$10^3$ yr were obtained. These rates are consistent with paleomagnetic data (Linkova, 1965; Hunkins et al., 1971; Clark* et al., 1980; Aksu, 1985; Chapter 24 of this volume). Since most Arctic sediments contain small to negligible amounts of foraminiferal $CaCO_3$ and other calcareous biogenic material (Herman, 1969, 1974), the most appropriate dating technique is $^{230}Th$ (Goldberg and Koide, 1962). So far only a few sediment cores were dated using this isotope whose concentrations in sediments are found to be lower than expected, based on the $^{238}U$ contents of the overlying water and the known reactivity of Th isotopes in seawater (Kaufman, 1969; Moore, 1981; see Chapter 23 of this volume for a detailed discussion concerning the Th isotope dating and reasons for the observed depletion of $^{230}Th$ in Arctic sediments.)

In order to study Arctic sedimentation rates and to understand the U-Th systematics, samples from nine cores were analyzed for Th and U isotopes. In addition to dating, Th isotopes may be used to determine the origin of certain sedimentary components such as barite (Goldberg et al., 1969). Several barite samples from one Arctic core have also been studied. Results of these measurements are presented and discussed.

*Clark's results are extremely inconsistent and therefore his data should be viewed with caution.

## MATERIAL AND METHODS

The locations, water depths, and core lengths of the cores analyzed are given in Table 22-1 and Figure 22-1. Approximately

Table 22-1. Locations, Water Depth, and Core Length

| Core | Latitude | Longitude | Water Depth (m) | Core Length (cm) |
|---|---|---|---|---|
| 19-8 | 87°15.4′N | 48°10.52′W | 2370 | 35 |
| 19–9 | 87°14.9′N | 58°17.55′W | 1192 | 23 |
| 19-10 | 87°06.6′N | 61°51.54′W | 1200 | 32 |
| 19-11 | 86°52.5′N | 62°01.51′W | 1176 | 10 |
| 21-11 | 84°40.8′N | 145°02.6′E | 840 | 65 |
| 21-13 | 84°51.0′N | 145°40.4′E | 2848 | 83 |
| T3-66-S-7 | 75°44.3′N | 157°38′W | 834 | 15 |
| T3-66-S-8 | 75°41.4′N | 158°00′W | 864 | 108 |
| T3-69-4 | 84°59.0′N | 124°52′W | 1566 | 149 |
| 3 | 77°42.81′N | 174°52.68′W | 1939 | 188 |
| 5 | 75°20.38′N | 142°55.53′W | 2105 | 241 |
| 6 | 85°25.35′N | 145°38.04′W | 2278 | 295 |

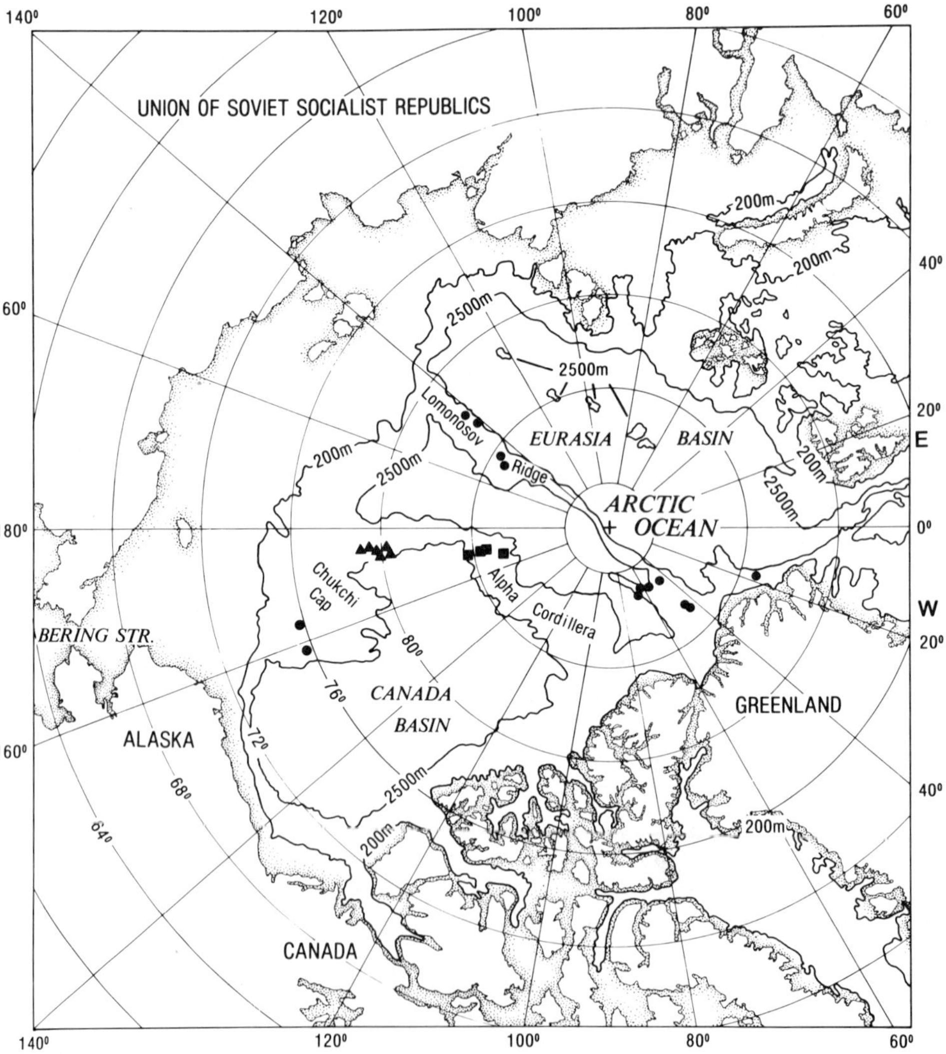


Figure 22-1. Arctic Ocean and surrounding lands, showing location of cores mentioned in the text.

2–5g of dried (at 110°C overnight) and powdered sediment was used for U-Th analysis. After a $^{232}U$–$^{228}Th$ equilibrated spike was added, samples were digested in 6M HCl until $CO_2$ effervescence ceased; they were then centrifuged. The residue was transferred into a teflon dish and brought into solution in 6M HCl after HF, $HClO_4$, and $HNO_3$ treatments. This HCl solution was combined with the 6M HCl leach and dehydrated. The residue was taken in 9M HCl, Fe was removed by di isopropyl extraction, and U and Th were separated and radiochemically purified by ion-exchange methods (Krishnaswami and Sarin, 1976). The Th and U isotopic concentrations were determined by alpha spectroscopy and the isotopic concentrations as well as activity ratios were obtained after applying standard corrections (Koide and Goldberg, 1965, Borole, 1980). The same procedure was followed for the barite samples once they were brought into solution by $Na_2CO_3$ fusion (Somayajulu and Goldberg, 1966; Borole and Somayajulu, 1977).

## RESULTS AND DISCUSSION

The $^{232}Th$, $^{238}U$, $^{230}Th_{excess}$ concentrations and the relevant isotopic ratios are given in Table 22-2. In addition to the data from the nine cores analyzed, Rosholt's (unpublished results) on two cores are also included in Table 22-2.

The $^{232}Th$ and $^{238}U$ concentrations in Arctic sediments are in the range of 3.64–29.5 ppm and 1.05–2.6 ppm, respectively. Most of the earlier data on U and Th isotopes from Arctic sediments (Ku and Broecker, 1967; Finkel, Krishnaswami, and Clark, 1977; Herman and Osmond, 1984) fall in these ranges which agree well with values found for world ocean sediments (Ku, Broecker, and Opdyke, 1968). The $^{234}U/^{238}U$ activity ratio varies from 0.83 ± 0.02 to 1.01 ± 0.03 (Table 22-2) with a mean value of less than 1.0, indicating that the U in these sediments is detrital.

## Sediment Dating Using $^{230}Th$

For details of dating using $^{230}Th$, reference is made to Goldberg and Bruland (1974), Ku (1976), and Chapter 23 of this volume. In principle, one can utilize $^{230}Th_{excess}$ activity as well as $^{230}Th_{excess}/^{232}Th$ activity ratio plots to obtain the accumulation rates that are found to be mutually consistent (Goldberg and Bruland, 1974). In the Arctic sediments there is a very large variation, by almost an order of magnitude in the $^{230}Th_{excess}$ down core and no accumulation rates can be deduced (Fig. 22-2). Only the $^{230}Th_{excess}/^{232}Th$ activity ratio shows decay with depth, from which the accumulation rates are derived (Fig. 22-3). The rates obtained in six cores are 0.8 (range 0.4–1.8), 1.3, 0.9, 3, 0.5 and 5 mm/$10^3$ yr, respectively, in cores C 19-8, C19-9, C19-10, C21-11, T3-66-S7 and T3-66-S8. In the other five cores no sedimentation rates could be deduced either due to single sample analysis per core or the absence of decrease in the $^{230}Th_{excess}/^{232}Th$ activity ratios in the deeper sample (with respect to the surface sample; Table 22-2).

It should be mentioned that the sedimentation rates obtained in the present investigation, using Th isotopes, agree with rates obtained using the magnetic reversal method (Linkova, 1965; Hunkins et al., 1971; Clark et al., 1980; Chapter 24 of this volume) and with rates deduced using $^{14}C$ (Ku and Broecker, 1967). Recently rapid sedimentation rates in the range of cm/$10^3$ yr were reported for the Arctic; they are based on epimerization of amino acids (Sejrup et al., 1984). However, these rates have been contested by Aksu and Mudie, (1985) and Darby et al. (Chapter 24 of this volume). The recently published $^{14}C$ ages using AMS on an Arctic core also yielded an accumulation rate of 2–3 mm/$10^3$ yr (Clark et al., 1986). It thus appears that

## Table 22-2. U, Th Isotopes in Artic Sediments

| Core # | Sample Depth (cm) | $^{232}Th$ (ppm) | $^{238}U$ (ppm) | $^{230}Th_{excess}$ (dpm/g) | $^{230}Th_{excess}/^{232}Th$ (activity ratio) | $^{234}U/^{238}U$ |
|---|---|---|---|---|---|---|
| 1 | 2 | 3 | 4 | 5 | 6 | 7 |
| 19-8 | 3.0-3.5 | 8.16±0.32 | 1.77±0.07 | 7.04±0.64 | 3.45±0.34 | 1.01±0.03 |
| | 3.5-5.5 | 5.72±0.24 | 2.60±0.13 | 3.80±0.21 | 2.66±0.18 | 0.94±0.04 |
| | 7.5-9.0 | 6.20±0.24 | NM | 6.65±0.22 | 4.29±0.22 | NM |
| | 9.5-10.0 | 13.16±0.96 | NM | 15.38±1.48 | 4.67±0.55 | NM |
| | 15.0-16.2 | 5.52±0.28 | NM | 2.22±0.17 | 1.61±0.15 | NM |
| | 18.5-19.0 | 10.52±0.76 | 2.20±0.09 | 3.64±0.38 | 1.34±0.14 | 0.97±0.03 |
| | 22-23 | 4.36±0.16 | 1.68±0.08 | 0.22±0.08 | 0.20±0.08 | 0.89±0.04 |
| 19-9 | 0.5-1.0 | 19.6±1.4 | 1.85±0.07 | 14.04±1.04 | 2.86±0.30 | 0.95±0.03 |
| | 5.5-6.5 | 3.92±0.17 | 1.56±0.07 | 1.92±0.13 | 1.96±0.16 | 0.95±0.04 |
| | 9.5-10.5 | 4.60±0.20 | NM | −0.20±0.06 | — | NM |
| | 12.5-13.0 | 8.88±0.64 | 1.79±0.12 | 0.43±0.16 | 0.19±0.07 | 0.86±0.06 |
| 19-10 | 3.5-4.5 | 4.76±0.17 | 1.41±0.04 | 3.12±0.12 | 2.62±0.13 | 0.91±0.02 |
| | 4.5-6.5 | 3.64±0.16 | NM | 1.80±0.13 | 1.98±0.17 | NM |
| | 9.0-10.5 | 8.60±1.16 | 2.07±0.07 | 2.36±0.50 | 1.52±0.33 | 0.88±0.03 |
| | 12.0-12.5 | 9.20±0.64 | 1.76±0.09 | 0.71±0.16 | 0.31±0.07 | 0.89±0.04 |
| 19-11 | 2-4 | 5.96±0.28 | 1.48±0.07 | 3.54±0.18 | 2.38±0.16 | 0.94±0.05 |
| 21-11 | 2-3 | 12.04±0.84 | NM | 5.32±0.49 | 1.77±0.20 | NM |
| | 6.5-8.0 | 7.01±0.24 | 2.13±0.13 | 0.78±0.13 | 0.45±0.08 | 0.97±0.06 |
| | 15.5-16.0 | 29.5±2.12 | 2.36±0.11 | 8.97±0.61 | 0.98±0.11 | 0.93±0.04 |
| | 20.5-21.5 | 5.64±0.22 | NM | 0.26±0.11 | 8.18±0.08 | NM |
| | 24.2-25.2 | 8.28±0.36 | 2.35±0.12 | 2.06±0.17 | 1.00±0.09 | 0.99±0.06 |
| 21-13 | 0.5-2.0 | 6.4±0.4 | 2.53±0.13 | 0.50±0.03 | 0.31±0.03 | 0.90±0.03 |
| 3 | 1-2 | 12.40±0.40 | 2.40±0.13 | 1.70±0.10 | 0.55±0.04 | — |
| | 7-10 | 16.40±0.80 | 2.27±0.13 | 2.8 ±0.2 | 0.68±0.02 | — |
| 5 | 1.0-2.5 | 9.2±0.13 | 1.60±0.13 | 0.4 ±0.2 | 0.17±0.09 | — |
| 6 | 1.0-2.0 | 7.28±0.20 | 1.05±0.05 | 0.67±0.06 | 0.37±0.04 | — |
| | 12-14 | 10.12±0.32 | 1.52±0.07 | 1.39±0.11 | 0.55±0.05 | — |
| T3-66-S7* | 2 | 9.39±0.23 | 2.10±0.05 | 0.73±0.08 | 0.32±0.04 | 0.90±0.02 |
| | 14 | 10.82±0.27 | 2.24±0.06 | 0.10±0.05 | 0.04±0.02 | 0.83±0.02 |
| T3-66-S8* | 2 | 8.02±0.20 | 1.91±0.05 | 0.32±0.06 | 0.16±0.03 | 0.92±0.02 |
| | 15 | 11.66±0.29 | 2.52±0.06 | 0.36±0.07 | 0.13±0.03 | 0.96±0.02 |

*Source:* Data from Rosholt (unpublished results).
*Errors quoted are due to one sigma counting statistics only.
$^{230}Th_{excess} = ^{230}TH_{total} - ^{238}U$ (all in units of dpm/g).

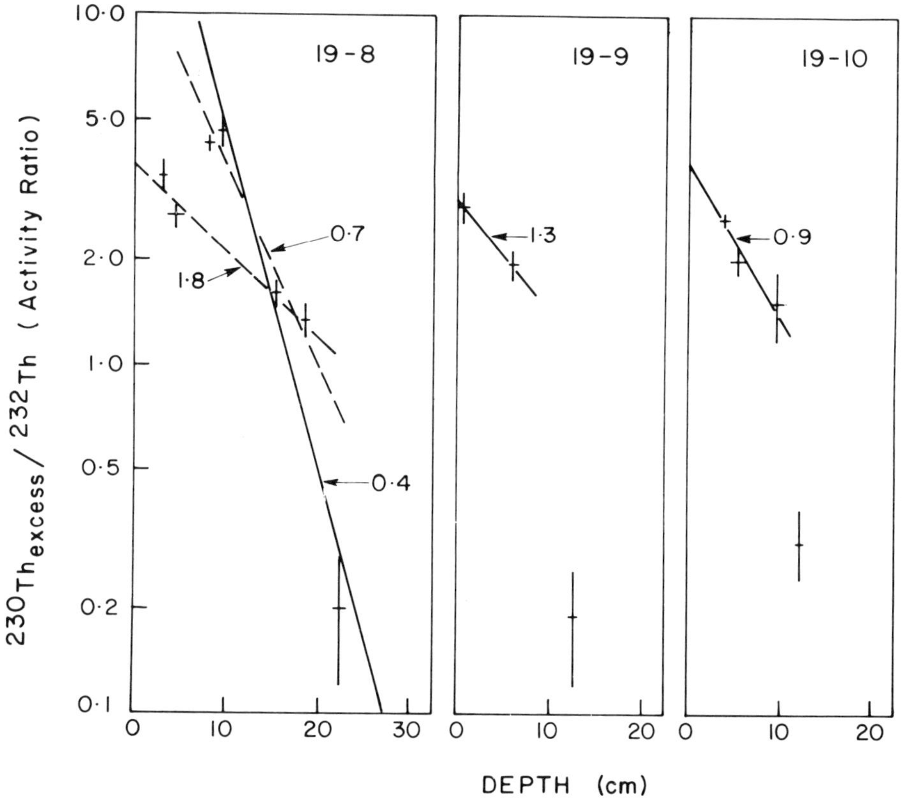


Figure 22-2. $^{230}Th_{excess}$ activity and $^{232}Th$ concentration as a function of depth in four Arctic cores (see Table 22-2 for details).

$^{230}Th$-based rates for the Arctic in our study area agree well with those measured by other nonchemical methods.

## Problems With $^{230}Th_{excess}$ Dating

As mentioned above, $^{230}Th_{excess}$ does not show a regular down core decrease in the cores analyzed (Table 22-2, Figure 22-2). Also, most of the Th and U isotope studies could be performed in samples from the uppermost 25 cm only, as there is hardly any $^{230}Th_{excess}$ in the deeper core sections. This is borne out by the fact that the accumulation rates are slow. Another reason is the fact that the $^{230}Th_{excess}$ inventories in the Arctic sediments are in general 2–20 percent of the expected inventories based on U contents of the overlying water column. This aspect has been discussed in detail by Herman and co-workers (Chapter 23). Here we wish to mention the variation in lithology of the cores, which is responsible for the observed zigzag $^{230}Th_{excess}$ profiles (Figure 22-2). Along with $^{230}Th_{excess}$, the $^{232}Th$ contents of the samples are also plotted as a function of depth for four of the cores studies (Figure 22-2). It is clear that in most cases $^{232}Th$ and $^{230}Th_{excess}$ behave in the same manner, both showing variations up to an order of magnitude. Variations are controlled by lithology and should show positive correlations with either the clay and/or Al contents. The higher the clay, the more the $^{232}Th$ should be, but to have an increase in the $^{230}Th_{excess}$, the clay has to be transported greater distances in the water column either vertically or horizontally. Being small particles with large surface area and adsorptive properties,

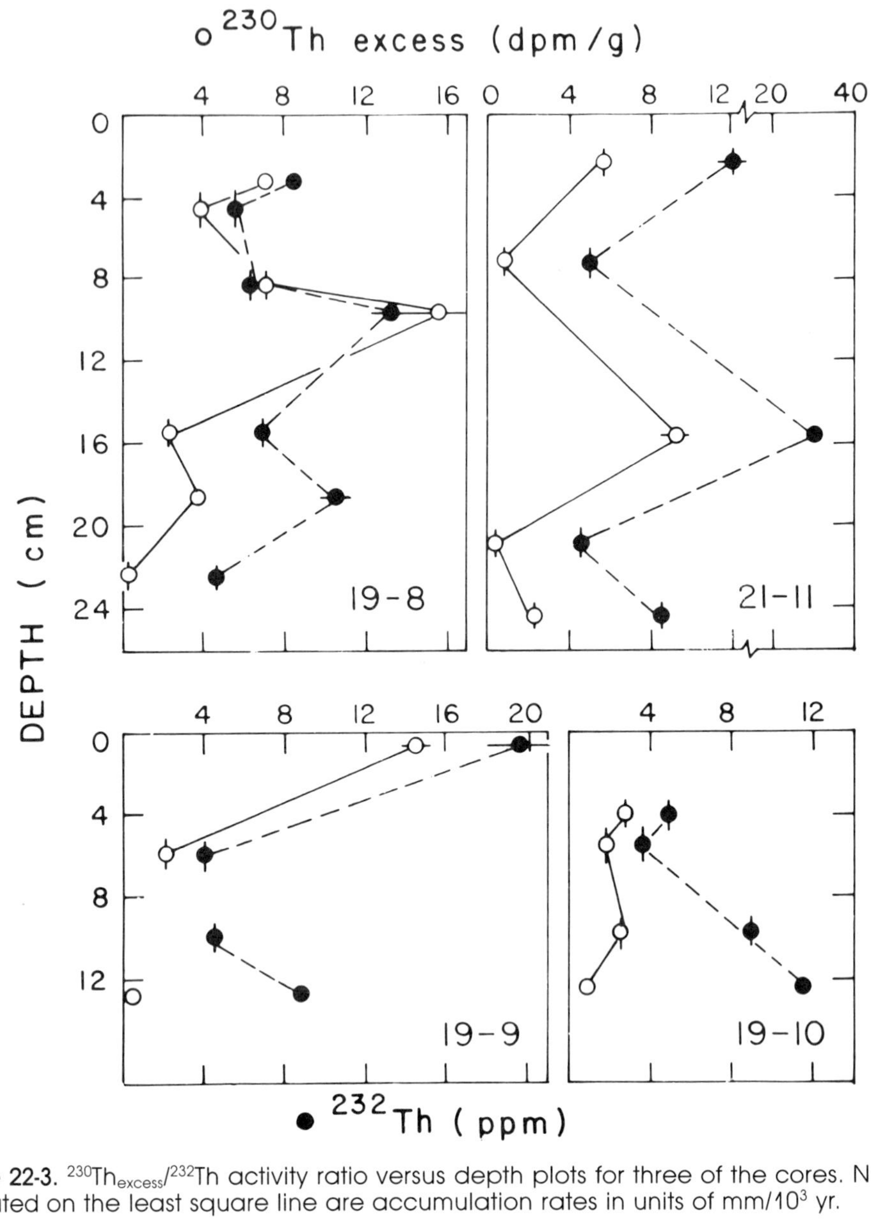


Figure 22-3. $^{230}Th_{excess}/^{232}Th$ activity ratio versus depth plots for three of the cores. Numbers indicated on the least square line are accumulation rates in units of mm/$10^3$ yr.

Table 22-3. U, Th Isotopes in Barite from Core T3-69-4

| *Sample Depth (cm)* | *$^{238}U$ (ppm)* | *$^{232}Th$ (ppm)* | *$^{230}Th$excess (dpm/g)* | *$^{230}Th$excess/$^{232}Th$ (activity ratio)* |
|---|---|---|---|---|
| 145 | — | ND | $4.1 \pm 1.2$* | $5.1 \pm 2.4$ |
| 149 | $2.1 \pm 0.4$ ($1.6 \pm 0.3$) | $3.2 \pm 0.8$ ($0.8 \pm 0.2$) | $5.5 \pm 0.6$ | $6.9 \pm 1.9$ |

*The $^{238}U$ and $^{232}Th$ concentrations of 149 cm sample are used to calculate these values. Numbers given in parentheses are activities given in units of dpm/g.
Errors quoted are due to one sigma counting statistics only.
$^{230}Th$excess $= {}^{230}Th - {}^{238}U$ (all in units of dpm/g).
ND = not detectable.

clays can absorb $^{230}Th$ efficiently from the water column during transit. We can tentatively conclude that higher clay contents indicate excess clay deposition over normal rate and the excess could have reached the site of deposition by a circuitous path. The lithology of the cores showing Th isotope variations (Figure 22-2) is given in Table 22-4.

For future $^{230}Th$ dating, it would be ideal to select cores with high and uniform clay contents as well as uniform lithology. Hopefully these would show down core decreases in both $^{230}Th_{excess}$ and $^{230}Th_{excess}/^{232}Th$.

## Th Isotopes in Barite

Deep-sea core T3-69-4, containing barite and capped by Pleistocene ice-rafted sediments above 120 cm core depth, was raised from the Alpha-Mendeleev Ridge (Fig. 22-1, Table 22-1). This core has modest amounts of barite at 145 cm and 149 cm depth. With the aim of ascertaining the origin of this barite, Th and U isotope measurements were carried out (Table 22-3).

It was shown by Somayajulu and Goldberg (1966) and Church and Bernat (1972) that authigenic marine barites contain tens of ppm Th and several ppm U. Using Sr/Ba ratio and Th, U, Sr, and S isotopic compositional measurements on barites of marine and nonmarine regions, Goldberg et al. (1969) were able to ascertain their origin. One of the diagnostic elements in this study proved to be Th.

As indicated in Table 22-3, the Th and U contents of the barites are very small (3.2 and 2.1 ppm respectively), suggesting that this mineral has not formed in situ from seawater, but rather it is probably of hydrothermal origin. The presence of $^{230}Th_{excess}$ in the two samples (4.1 and 5.5 dpm/g barite in the 145 cm and 149 cm samples respectively, Table 22-2) is surprising. Such an excess can be attributed to adsorption of $^{230}Th$ generated in the pore waters by the decay of its parent $^{234}U$ whose concentrations are in the range of 1.7–2.5 dpm/1 (Somayajulu and Church, 1973; Cochran and Krishnaswami, 1980). Borole and Somayajulu (1977) reported enrichment of $^{226}Ra$ (300–600 dpm/g) in barites separated from the deep sections of an 8-m-long piston core from the central Eastern Pacific and they explained it as due to adsorption of Ra on barite surfaces. The second possibility, that the barite could grow in the sedimentary column, was ruled out as the available data do not show increasing size of barite crystals with depth. The same explanation holds for $^{230}Th$. There is sufficient $^{230}Th$ in the pore waters (up to 0.5 dpm/kg) with $^{230}Th_{excess}/^{232}Th$ activity ratio of up to 25 (Somayajulu and Church, 1973; Cochran and Krishnaswami, 1980) to account for the observed $^{230}Th_{excess}/^{232}Th$ in the barites (Table 22-2). It is therefore suggested that the barite in core T3-69-4 is of hydrothermal origin and the excess $^{230}Th$ is due to postdepositional adsorption from the surrounding pore waters.

In conclusion, it appears that clay-rich cores collected from the deep areas of the Arctic Ocean provide the much wanted chronology by $^{230}Th$-based methods. The

**Table 22-4. Megascopic Description of Split Cores**

| | |
|---|---|
| *Core 19-8* | |
| 0–16.2 cm: | Light brown silty clay |
| 16.2–19 cm: | Light yellowish silty clay |
| 19–22 cm: | Light yellowish sandy clay |
| 22–26.5 cm: | Light brownish sandy clay |
| *Core 19-9* | |
| 0–5 cm: | Dull yellow-brown sandy clay |
| 5–16.5 cm: | Dull yellow-brown sandy clay |
| *Core 19-10* | |
| 0–27 cm: | Light yellowish sandy clay |
| *Core 21-11* | |
| 0–53 cm: | Light tan silty clay |

*Source:* Courtesy of VNII Okeanologia, Leningrad, USSR.

origin of minerals such as barite, present in Arctic sediments, can also be understood by Th isotope studies.

## ACKNOWLEDGMENTS

We thank John N. Rosholt for the T3-66-S7 and T3-66-S8 data, VNII Okeanologia, Leningrad, USSR and Lamont-Doherty Geological Observatory (LDGO) for making the cores available. Support for the curating facilities of the LDGO Deep-Sea Sample Repository was provided by the National Science Foundation through Grant OCE 85-00232 and the Office of Naval Research through Grant N00014-84-C-0132. This research was partially supported by NAS grants and Smithsonian Foreign Currency Grant 33X 0102-A20 to Yvonne Herman.

## REFERENCES


Aksu, A. E. 1985. Paleomagnetic stratigraphy of CESAR cores. In H. R. Jackson, P. J. Mudie, and S. M. Blasco (eds.), *Initial Geological Report on CESAR—The Canadian Expedition to Study the Alpha Ridge, Arctic Ocean.* Geological Survey of Canada Paper 84-22, pp. 101–114.

Aksu, A. E., and P. J. Mudie. 1985. Magnetic stratigraphy and palynology demonstrate at least 4 million years of Arctic Ocean sedimentation. *Nature* **318:**280–283.

Borole, D. V. 1980. Radiometric and Trace Elemental Investigations on Indian Estuaries and Adjacent Seas. Ph.D. thesis, Gujarat University, Ahmedabad, India, 155p.

Borole, D. V. and B. L. K. Somayajulu. 1977. Radium and lead 210 in marine barite. *Marine Chem.* **5:**291–296.

Church, T. M., and M. Bernat. 1972. Thorium and uranium in marine barite. *Earth Planet. Sci. Lett.* **14:**139–144.

Clark, D. L., R. R. Whitman, K. A. Morgan, and S. D. Mackey. 1980. Stratigraphy and glacial marine sediments of the Amerasian Basin, Central Arctic Ocean. *Geol. Soc. Am. Spec. Paper 181,* 57p.

Clark, D. L., M. Andree, W. S. Broecker, A. C. Mix, G. Bonani, H. J. Hofman, E. Morenzoni, M. Nessi, M. Suter, and W. Wolfli. 1986. Arctic Ocean chronology confirmed by accelerator $^{14}C$ dating. *Geophys. Res. Lett.* **13:**319–321.

Cochran, J. K., and S. Krishnaswami. 1980. Radium, thorium, uranium and $^{210}Pb$ in deep sea sediments and sediment pore waters from the north equatorial Pacific. *Am. J. Sci.* **280:**849–889.

Finkel, R. C., S. Krishnaswami, and D. L. Clark. 1977. $^{10}Be$ in Arctic sediments. *Earth Planet. Sci. Lett.* **35:**199–204.

Goldberg, E. D., and K. Bruland. 1974. Radioactive geochronologies. In E. D. Goldberg (ed.), *The Sea,* vol. 5. Wiley Interscience, New York pp. 451–489.

Goldberg, E. D., and M. Koide. 1962. Geochronological studies of deep sea sediments by the ionium/thorium method. *Geochim. Cosmochim. Acta* **26:**417–450.

Goldberg, E. D., B. L. K. Somayajulu, J. N. Galloway, I. R. Kaplan, and G. Faure. 1969. Differences between barites of marine and continental origins. *Geochim. Cosmochim. Acta* **39:**287–289.

Herman, Y. 1969. Arctic Ocean Quaternary microfauna and its relation to paleoclimatology. *Palaeogeogr., Palaeoclimatol. Palaeoecol.* **6:**251–276.

Herman, Y. 1970. Arctic paleoceanography in late Cenozoic time. *Science* **169:**474–477.

Herman, Y. 1974. Arctic Ocean sediments, microfauna and the climatic record in late Cenozoic time. in Y. Herman (ed.), *Marine Geology and Oceanography of the Arctic Seas.* Springer Verlag, New York, pp. 283–348.

Herman. Y., and J. K. Osmond. 1984. Late Neogene Arctic paleoceanography: Micropaleontology and chronology. In A. Berger, J. Imbrie, J. Hays, G. Kukla, and B. Saltzman (eds.), *Milankovitch and Climate.* Reidel, Dordrecht, pp. 241–250.

Hunkins, K., A. W. H. Bé, N. D. Opdyke, and G. Mathieu, 1971. The late Cenozoic history of the Arctic Ocean. In K. K. Turekiah (ed.), *The Late Cenozoic Glacial A Ages.* Yale University Press, New Haven, pp. 215–237.

Kaufman, A. 1969. The $^{232}Th$ concentration of surface ocean water. *Geochim Cosmochim. Acta* **33:**717–723.

Koide, M., and E. D. Goldberg. 1965. $^{234}U/^{238}U$ activity ratios in sea water. In M. Sears (ed.), *Progress in Oceanography,* vol. 3. Pergamon Press, London, pp. 173–178.

Krishnaswami, S., and M. M. Sarin. 1976. The simultaneous determination of Th, Pu, Ra isotopes $^{210}Pb$, $^{55}Fe$, $^{32}Si$ and $^{14}C$ in marine suspended phases. *Anal. Chem. Acta* **83:**143–156.

Ku, T. L. 1976. The uranium series methods of

age determination. *Ann. Rev. Earth Planet. Sci.* **4**:347–379.

Ku, T. L. and W. S. Broecker. 1967. Rates of sedimentation in the Arctic Ocean. In M. Sears (ed.), *Progress in Oceanography,* vol. 4. Pergamon Press, London, pp. 95–103.

Ku, T. L., W. S. Broecker, and N. D. Opdyke. 1968. Comparison of sedimentation rates measured by paleomagnetic and ionium methods of age determination. *Earth Planet. Sci. Lett.* **4**:1–16.

Linkova, T. I. 1965. Some results of paleomagnetic study of Arctic Ocean floor sediments. In *The Present and Past of the Geomagnetic Field* (trans. by E. R. Hope). Nauka, Moscow, pp. 279–291.

Markussen, B., R. Zahn, and J. Thiede. 1985. Late Quaternary sedimentation in the eastern Arctic Basin: Stratigraphy and depositional environment. *Palaeogeogr. Palaeoclimatol. Palaeoecol.* **50**:271–284.

Moore, W. S. 1981. The thorium isotope content of ocean water. *Earth Planet. Sci. Lett.* **53**:419–426.

Sejrup, H. P., G. H. Miller, J. Brigham-Grette, R. Lovile, and D. R. Hopkins. 1984. Amino acid epimerization implies rapid sedimentation rates in Arctic Ocean cores. *Nature* **310**:772–775.

Somayajulu, B. L. K., and T. M. Church. 1973. Ra, Th and U isotopes in the interstitial water from the Pacific Ocean sediment. *J. Geophys. Res.* **78**:4,529–4,531.

Somayajulu, B. L. K., and E. D. Goldberg. 1966. Th and U isotopes in sea water and sediments. *Earth Planet. Sci. Lett.* **1**:102–106.

# 23: Late Neogene Arctic Paleoceanography: Micropaleontology, Stable Isotopes, and Chronology

Yvonne Herman, J. K. Osmond, and B. L. K. Somayajulu

## INTRODUCTION

The climatic history of the Arctic has been a matter of debate ever since the systematic sampling and study of seafloor sediments commenced several decades ago. The early Soviet investigators (Sacks, Belov, and Lapina, 1955), using radium distribution in sedimentary cores, estimated that rates of sediment accumulation in the entire basin were 1.2–2 cm/$10^3$ yr. These values are an order of magnitude higher than rates based on uranium series isotope dates (Ku and Broecker, 1967; Herman and Osmond, 1984; Chapter 22 of this volume). Linkova (1965) was the first to determine the magnetic polarity of Arctic basin sedimentary cores. Her studies demonstrated conclusively that sediment accumulation rates on topographic highs, such as the Lomonosov Ridge, are extremely slow, ~1–3 mm/$10^3$ yr. Similar results were obtained for the Alpha-Mendeleev Rise (Hunkins et al., 1971; Aksu, 1985*a*; Aksu and Mudie, 1985).

Clark published several papers on the magnetic stratigraphy of the USGS cores; however, since there are numerous inconsistencies in his interpretations (Clark, 1969, 1970, 1971), which are discussed by Aksu (1985*a*), his results should be viewed with caution. The Russian geologists interpreted the uppermost 10–15-cm-thick brown foraminiferal sandy clay layer that covers vast areas of the seafloor to represent Postglacial deposits, while the underlying olive green and olive gray foraminifera-poor glacial marine sediments were thought to have accumulated during colder, glacial epochs (Sacks, Belov, and Lapina, 1955). Subsequently, Clark (1977*a*) adapted the Russian interpretation and surmised that the Arctic has been covered continuously with perennial ice since middle Cenozoic time (~35 mya) to the present. According to Clark (1977*a*, 1982*a*), the ice cover was much thicker during the deposition of the glacial marine, foraminifera-poor beds than today, when it attains thicknesses of 3–4 m at the end of winter, decreasing to about 2–3 m in summer (Maykut and Untersteiner, 1971).

A similar interpretation of a massive ice sheet filling the Arctic Ocean was given by Mercer (1970), Hughes, Denton, and Grosswald (1977), and Williams, Moore, and Fillon (1981). Hunkins et al. (1971) interpreted the foraminifera-poor beds to indicate either a more rigid ice cover than today's or dissolution of the foraminifera. Herman (1969, 1970, 1974, 1984*a,b*, 1985) suggested that during the last 0.9 my, foraminifera-poor zones were deposited during higher temperatures than those of today. These were deglacial acme periods when meltwater from surrounding continents and from ice shelves, as well as icebergs, flooded the Arctic, forming a low-density surface layer. The Matuyama foraminifera-poor beds were also deposited during periods with climates milder than those prevailing today. The scarcity of planktonic foraminifera was attributed to reduced plankton production due to low surface-water salinities and decreased nutrient levels (Herman, 1969, 1970, 1974, 1984*a*, 1984*b*, 1985).

## HOLOCENE AND MODERN DEPOSITIONAL ENVIRONMENTS

The world's largest land-locked ocean, the Arctic, occupies a broad elliptical basin, covering an area of about 12 million $km^2$ and containing 13 million $km^3$ of water. Although it constitutes over 3 percent of the world's ocean area, the Arctic holds only 1 percent of its volume; this is accounted for by the broad continental shelves that underlie almost 70 percent of its area (Beal, 1969).

The polar ice cap is the most distinctive feature of the Arctic Ocean, determining biological processes, hydrologic conditions, and sediment distribution. The maximum ice cover, which is estimated to be ~15.106 $km^2$, can be reduced by as much as 50 percent in the summer; however, the extent varies substantially from year to year (Chap. 2 of this volume; Hibler, 1980, and references therein). A detailed discussion of water masses, their sources, circulation patterns, and sea ice, is presented by Anderson and Dyrssen (Chap. 3), Hibler (Chap. 2), and Barry (Chap. 1), as well as by Nansen (1906) and Swift, Takahashi, and Livingston (1983).

### Sediment Source, Transport, and Composition

Sedimentation processes in the Arctic differ from those in other marine regions because transport of terrigenous and shallow-water sediments by surface currents and winds is inhibited during the greater part of the year by the polar ice cap. The main sedimentary components deposited today are of lithogenous, biogenous, and hydrogenous (formed in the seawater) origin. Supply rates of sediments, topography, as well as postdepositional transport and solution, control the composition of seafloor deposits.

*Ice transport:* Detrital sediments are incorporated into icebergs and sea ice; icebergs are formed from land and shelf ice. Although the Greenland ice sheet is the largest producer of icebergs, the majority of Arctic Ocean bergs originate from the Ellesmere Island ice shelves. Generally, they remain in the Beaufort Sea Gyre (Weeks and Campbell, 1973). These bergs drift westward along the Canadian Archipelago, then turn toward the pole, north of Bering Strait, and after ~10 yr, return to their point of origin. Carried by winds and currents and also under the influence of the Coriolis force, the icebergs gradually decay, releasing incorporated debris onto the seafloor (Robe, 1980). Some of the icebergs caught in the East Greenland Current (EGC) are carried south. The EGC also carries vast amounts of Arctic pack ice as far south as the southern tip of Greenland.

In shallow water, sea ice incorporates significant amounts of lithogenous and biogenous debris, mainly of silt and clay size, although sand and gravel size detritus is occasionally encountered (e.g., Carsola, 1952); sand-size particles consist predominantly of angular to subangular quartz (Herman, 1969, 1974). These poorly sorted rock and mineral fragments are of variable lithologies including metamorphic and igneous rocks as well as detrital carbonates (Schwarzacher and Hunkins, 1961). This sediment-laden ice is carried various distances, gradually melting and dropping the entrapped debris to the seafloor. The general pattern of ice-cover circulation consists of a gyre in the Beaufort Sea and a transpolar drift stream; the latter carries ice from the Bering Strait and the East Siberian Sea, crosses the North Pole, and drifts south along the east coast of Greenland (Hibler, 1980). The Beaufort Sea ice may survive in the gyre many years. Drift rates are variable, averaging 6 km $day^{-1}$.

*Currents:* With the exception of the MacKenzie and Yukon, all major rivers flow into the shallow and wide marginal seas (Barents, Kara, Laptev, East Siberian, and Chukchi), which are incised by numerous canyons (Andrew and Kravitz, 1974; Naugler, Silverberg, and Creager, 1974). Lithogenous sediments are carried by rivers and discharged onto these shelves whence they are transported further by currents via canyons to the basin floor. Shallow-water biogenous elements are transported along with the detrital sediments to the deep seafloor. Clay-size particles have a long residence time in water and are carried in suspension over great distances by currents before settling to the seafloor (Griffin, Windom, and Goldberg, 1968). They constitute important but variable percentages of the Arctic floor detrital minerals (Carsola, 1952; Berry and Johns, 1966; Carroll, 1970; Darby, 1971; Naidu, 1974; Chapter 24 of this volume). In the Canada Abyssal Plain bottom currents range from $<1$ cm/s to 2.6 cm/s (Galt, 1967; Hunkins, Thorndike, and Mathieu, 1969), while on the Alpha-Mendeleev Rise their range is $\sim 4$ cm/s to 6 cm/s. These differences obviously account for the distribution of sediments. In the vicinity of the pole, the current flows in a pulsating manner with peak speeds exceeding 12 cm/s, diagonally up the Eurasian side of the ridge over the Lomonosov Ridge, and then flows down the Canadian flank of the ridge (Aagaard, 1981).

*Eolian dust:* From a limited number of samples examined it was estimated that airborne dust contributes between 1 percent and 10 percent to the total Arctic Basin sediments (Darby, 1971).

*Lithogenous components:* Land-derived clasts transported by ice, winds, and surface and deep curents range in size from gravel to clay and are of variable lithologies as mentioned in previous paragraphs.

*Hydrogenous (authigenic) material:* Ferromanganese films cover rock and mineral fragments (Schwarzacher and Hunkins, 1961) and also fill and/or cover the empty tests of foraminifera (Herman, 1974).

*Biogenous components:* Although radiolarians and diatoms are common in the overlying water masses (Zenkevitch, 1963; Herman, unpublished data), because sea water is undersaturated with respect to silica, the skeletons of most siliceous organisms dissolve as soon as they die, and thus are rarely preserved in sediments (Herman, 1974; Chapters 19 and 20 of this volume). The calcareous tests of planktonic foraminifera constitute the bulk of faunal remains settling from the water column (Herman, 1969; Herman, 1974). Aragonitic pteropod shells belonging to one species (*Limacina helicina*) are preserved in cores shallower than 2,500 m (Herman, 1974, although occasionally they are encountered at much greater depths, where they are probably deposited during periods of rapid sediment accumulation (Herman, unpublished). Benthonic calcareous foraminifera dominate the benthonic assemblages in most Amerasian Basin sediments (Green, 1960; Fetter, 1973; Herman, 1974; Lagoe, 1977) while arenaceous tests prevail on shelves, slopes, and continental rise sediments (Cushman, 1948; Loeblich and Tappan,

1953; Vilks, 1969). Gorbunov (1946) and Shedrina (in Somov, 1955) report abundant and varied arenaceous foraminifera, comprising 41 percent of the benthonic foraminiferal fauna recovered from water depths ranging from 512–3,800 m from the Russian side of the Arctic Basin. Ostracods, pelecypods, sponge spicules, and calcareous holothurian sclerites are minor components (Fetter, 1973; Herman, 1974). Species composition and the ratio of planktonic to benthonic species are discussed in another section of the present chapter as well as in previous publications (Herman, 1969, 1974, 1984*a*, 1985).

Ice-rafted Carboniferous and Permian fossils have been dredged up from the Alpha Rise (Schwarzacher and Hunkins, 1961). Ice rafting also accounts for the scattered occurrence of shallow-water foraminifera and other skeletal remains of shelf organisms on the Alpha-Mendeleev Rise (Herman, 1970), and of rare warm-water foraminifera (Herman, 1969, 1970, 1974). Numerous other organisms are reported in several chapters of this volume. Over 2,500 seafloor photographs and many samples collected with biological trawls and cores, mainly from the Amerasian Basin, Alpha-Mendeleev Rise, and Chukchi Cap, indicate high percentages of gravel and sands on the rise, but much less on the adjacent plains (Schwarzacher and Hunkins, 1961; Carsola, 1952; Cromie, 1961; Paul and Menzies, 1973; Kitchell et al., 1978). Bottom currents, indicated by ripple marks or bending of stalked animals, can be seen on the rise. The impoverished condition of large epifaunal benthos has been discussed by Paul and Menzies (1973) and Kitchell et al. (1978). The average density of organisms is 0.02 organism/m$^2$, the lowest in the oceanic record reported from photographic surveys to date. Animal tracks and burrowers were also observed in photographs. The contrast in sediment types between the rise and adjacent plain has been attributed to the different depths and bottom current velocities in the two provinces as well as the effect of turbidity currents.

## MATERIAL AND METHODS

Piston, gravity, and multiple-barrel cores raised from the Alpha Rise, Chukchi Plateau–Northwind Escarpment, Canada Basin, Wrangel Plain, Lomonosov Ridge, Norwegian–Greenland Sea and North Pacific Ocean were examined megascopically for details of texture and structure; visual inspection of cores is of particular value for detecting anomalous layers such as slump, turbidite, and ice-rafted deposits as well as "flow-in" (including mid-core "flow in") sections (Plate 23-I).* In nearly half of the studied cores hiatuses as well as extraneous sediments have been recognized using microfaunal data and sedimentary structures. In contrast with our results, Clark claims that "only a few of the several hundred (Arctic) cores that have been studied have unconformities" (Clark et al., 1980). Obviously, the method employed by Clark is not reliable for detecting missing or extraneous sediments.

The locations, water depths, and core lengths are given in Table 23-1 and Figure 23-1. For microfaunal studies the sampling interval was determined by sedimentation rates. Initially standard-size Lamont-Doherty Geological Observatory (LDGO) equal-volume samples weighing 8–14 g were taken at 5–50 cm intervals or where lithologic changes were observed. The same procedure was followed for cores made available by other laboratories. Continuous sequences of samples were taken in the D.St.A. cores and in T3-67-3. Generally the dried and weighed sediment was washed through a 250 mesh sieve that retains particles >63 $\mu$m. Several D.St.A. cores, originally wet sieved through a 74 $\mu$m sieve, were rewashed through 38 $\mu$m

*Plates and explanations appear as an appendix to the chapter.

Table 23-1. Locations, Depths, and Lengths of Cores

| *Core* | *Latitude* | *Longitude* | *Depth (m)* | *Length (cm)* |
|---|---|---|---|---|
| D.St.A.2 | 83°52′N | 168°12′W | 1,521 | 206 |
| D.St.A.3 | 84°12′N | 168°33′W | 2,409 | 95 |
| D.St.A.4 | 84°21′N | 168°49′W | 2,041 | 116 |
| D.St.A.5 | 84°28′N | 169°04′W | 1,934 | 125 |
| D.St.A.6 | 85°15′N | 167°54′W | 1,842 | 88 |
| D.St.A.12 | 83°38′N | 161°46′W | 2,510 | 175 |
| Arlis II,1 | 81°57′N | 168°07′E | 2,816 | 44 |
| Arlis II,2 | 82°00′N | 168°04′E | 2,864 | 80 |
| T3-63-1/S | 82°56.2′N | 155°54′W | 3,548 | 18 |
| T3-63-3/SA | 82°57.8′N | 162°46′W | 3,400 | 24.5 |
| T3-63-4/SA | 82°20.7′N | 161°49′W | 3,795 | 11 |
| T3-63-4/SB | 82°20.7′N | 161°49′W | 3,795 | 13 |
| T3-63-4/SC | 82°20.7′N | 161°49′W | 3,795 | 13.5 |
| T3-63-4/SD | 82°20.7′N | 161°49′W | 3,795 | 14.5 |
| T3-66-S-W4 | 75°41′N | 157°29′W | 1,177 | 121 |
| T3-66-S-5 | 75°45.4′N | 156°47′W | 915 | 31 |
| T3-66-S-6 | 75°45.7′N | 156°50′W | 1,167 | 119 |
| T3-66-S-7 | 75°44.3′N | 157°38′W | 834 | 15 |
| T3-66-S-8 | 75°41′N | 158°00′W | 864 | 108 |
| T3-66-S-10 | 75°41.4′N | 157°53′W | 880 | 53 |
| T3-66-S-11 | 75°41.4′N | 157°50′W | 884 | 17 |
| T3-66-S-12 | 75°40.3′N | 157°36′W | 943 | 22 |
| T3-66-S-15 | 75°45.8′N | 158°22′W | 617 | 18 |
| T3-66-S-19 | 75°47.7′N | 158°25′W | 615 | 52 |
| T3-66-S-20 | 75°46.7′N | 158°28′W | 615–20 | 96 |
| T3-66-S-21 | 75°46.1′N | 158°30′W | 629 | 90 |
| T3-66-S-25 | 75°43.8′N | 160°16′W | 1,600 | 137 |
| T3-66-S-26 | 75°43.6′N | 160°18′W | 1,992 | 100 |
| T3-66-S-28 | 75°42.5′N | 161°01′W | 2,097 | 243 |
| T3-66-S-30 | 75°46.3′N | 162°02′W | 2,097 | 287 |
| T3-66-S-31 | 75°47.4′N | 162.20′W | 1,741 | 279 |
| T3-66-S-32 | 75°47′N | 162°21′W | 2,059 | 288 |
| T3-66-S-33 | 75°50.4′N | 162°59′W | 2,047 | 225 |
| T3-67-2 | 79°06.3′N | 175°34′W | 1,982 | 312 |
| T3-67-3 | 79°11′N | 175°09′W | 2,285 | 380 |
| T3-67-4 | 79°22.7′N | 174°46′W | 1,760 | 272 |
| T3-67-5 | 79°31.5′N | 174°10′W | 2,057 | 414 |
| T3-67-6 | 79°44.2′N | 173°04′W | 2,815 | 432 |
| T3-67-9 | 79°37.9′N | 172°07′W | 2,237 | 356 |
| T3-67-11 | 79°34.9′N | 172°30′W | 2,810 | 250 |
| T3-67-12 | 80°21.9′N | 173°33′W | 2,867 | 374 |
| T3-67-13 | 80°31′N | 172°42′W | 3,288 | 275 |
| T3-67-14 | 80°31′N | 172°20′W | 3,308 | 119 |
| T3-67-16 | 80°03.0′N | 170°57′W | 3,303 | 219 |
| T3-68-4 | 82°37.9′N | 158°00′W | 3,800 | 522 |
| T3-68-5 | 82°33.9′N | 157°41′W | 3,762 | 541 |
| 19-8 | 87°15.4′N | 48°10.52′W | 2,370 | 35 |
| 19-9 | 87°14.9′N | 58°17.55′W | 1,192 | 23 |
| 19-10 | 87°06.6′N | 61°51.54′W | 1,200 | 32 |
| 19-11 | 86°52.5′N | 62°01.51′W | 1,176 | 10 |
| 19-14 | 85°35.4′N | 48°51.56′W | 2,700 | 20 |
| 19-15 | 85°21.9′N | 42°02.58′W | 2,900 | 30 |
| 19-19 | 83°08.7′N | 16°17.53′W | 590 | 10 |
| 21-5 | 82°18.4′N | 142°23.3′E | 1,578 | 14 |

Table 23-1. (continued)

| Core | Latitude | Longitude | Water Depth (m) | Core Length (cm) |
|---|---|---|---|---|
| 21-6A | 82°58.8′N | 140°34.7′E | 2,085 | 36 |
| 21-6B | 84°11.3′N | 24°05′W | 590 | 6 |
| 21-11 | 84°40.8′N | 145°02.6′E | 840 | 65 |
| 21-13 | 84°51.0′N | 145°40.4′E | 2,848 | 83 |
| V-17-196TW* | 60°44′N | 57°50′W | 2,818 | 25 |
| V-27-24TW | 55°58.5′N | 56°03.5′W | 2,628 | 41 |
| V-27-25TW | 56°43.8′N | 54°41.8′W | 3,146 | 65 |
| V-27-61 | 71°19.2′N | 12°04.1′E | 2,299 | 710 |
| V-27-66 | 71°59.8′N | 30°24.5′E | 336 | 620 |
| V-27-85 | 67°21.7′N | 04°01.6′E | 1,227 | 798 |
| V-28-28TW | 73°29′N | 00°50′W | 2,288 | 38 |
| V-28-37TW | 72°04′N | 09°04′W | 2,395 | 43 |
| V-29-214TW | 72°58′W | 06°59′W | 2,608 | 42 |

*TW = trigger weight core.

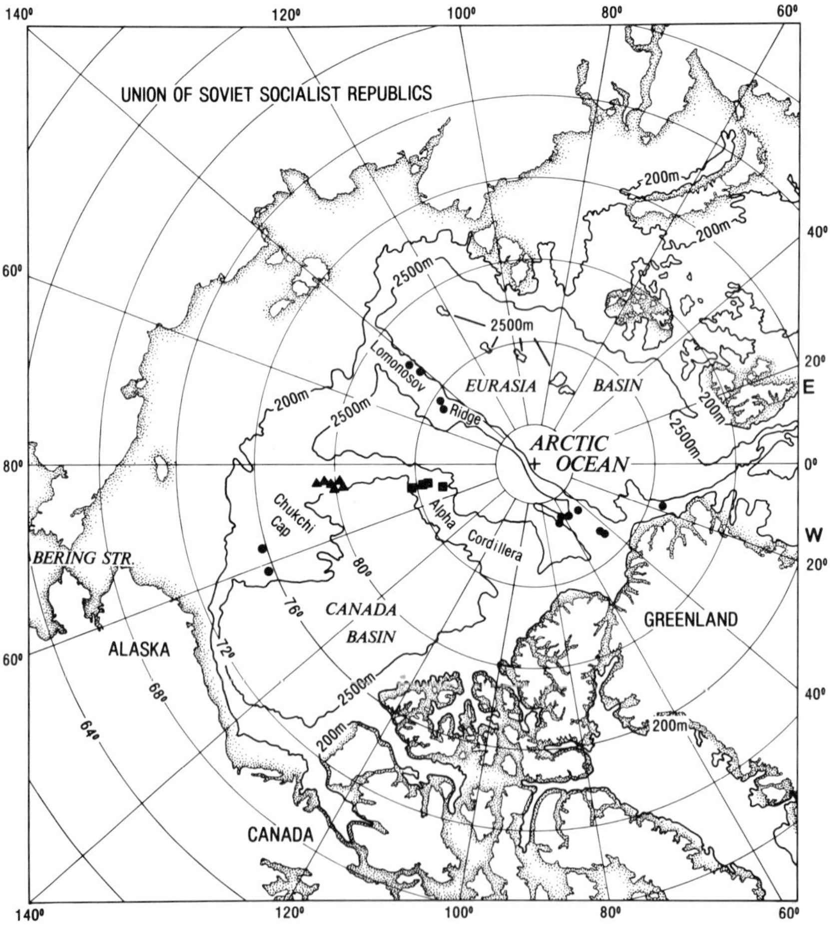


Figure 23-1. Map of the Arctic Ocean based on the geological map of the Arctic (1960), showing location of selected cores; for complete list of core locations see Table 23-1.

Table 23-2. Percentages of Fauna and Clastics in Various Size Fractions in Sample D.St.A.2, 0–0.5 cm

| 1 | 2 | 3 | 4 | 5 | 6 | 7 | 8 |
|---|---|---|---|---|---|---|---|
| 68 | 32 | 83.5 | 16.5 | 15 | 1 | 1.1 | >250 |
| 17 | 83 | 98.9 | 1.1 | 1 | — | 0.5 | >177 |
| 30 | 70 | 97.8 | 2.2 | 3 | — | 1.6 | >100 |
| 78 | 22 | 34.4 | 65.6 | 12 | 2 | 3.0 | > 63 |
| 93 | 7 | 8.3 | 91.7 | 38 | — | 4.7 | > 44 |

1. Percentage of clastics in the coarse fraction; fraction size listed in column 8.
2. Percentage of fauna.
3. Percentage of adult planktonic foraminifera of the total planktonic foraminiferal population.
4. Percentage of juvenile planktonic foraminifera of the total planktonic foraminiferal population.
5. Percentage of benthic foraminifera of the total foraminiferal population.
6. Percentage of pteropods of the total planktonic fauna.
7. Fragments of planktonic foraminifera of the total planktonic foraminiferal population.
8. Size fraction, μm.

and 53 μm screens (Table 23-2; Herman, 1974). After drying at 100°C, the coarse fraction was weighed for a second time and recorded as a percentage of the total. Curves constructed from these percentages provide valuable evidence as to processes of sedimentation and yield means for cross correlation between cores (Fig. 23-2). For faunal analyses the coarse fraction was split into subsamples and counts totaling 500–800 specimens were made; the entire population within a sample was counted when total specimens numbered less than 500. The coiling direction of planktonic and benthic foraminifera was recorded. Tests were separated from clastics in samples having low concentration of shells by flotation on carbon tetrachloride and stained using the method described by Herman and Metz (1972). In areas with low sedimentation (1–3 mm/1,000 yr), bioturbation and bottom current activity mix the sediments greatly, defocusing the stratigraphic record to a resolution of perhaps 20,000–50,000 yr. A new statistical approach of "unmixing" the sediments is being tested and will be used to obtain higher resolution of events (Herman et al., in preparation).

## LATE CENOZOIC SEDIMENTS

The pre-Pleistocene geologic evolution of the Arctic remains elusive and fraught with uncertainties as Bukry refutes all of

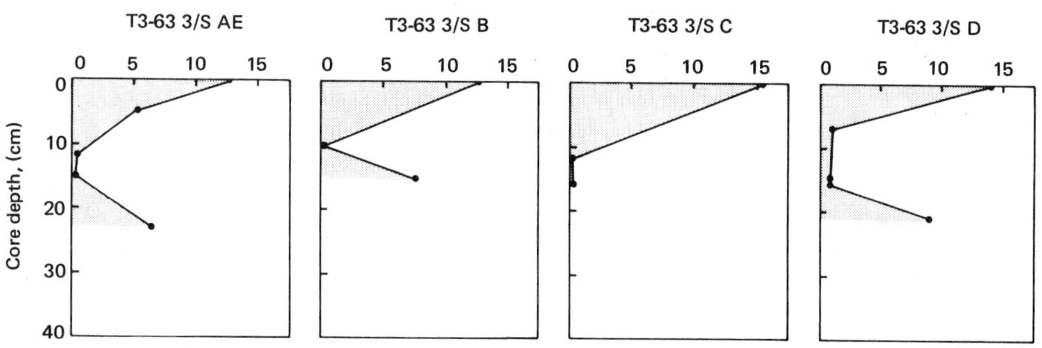


Figure 23-2. Core T3-63-3-S. Correlation between four cores based on the percentage of coarse fraction (> 63 μm) in a multiple-barrel core.

Clark's papers (e.g. 1974, 1977*b*, 1981, 1982*b*, and others), writing that Clark's "paleoceanographic inferences concerning the Cretaceous–Tertiary boundary and the timing of silica deposition in the Arctic have been erroneous" because in fact USGS core FL 422 is not Cretaceous, Paleocene, or early Eocene as claimed by Clark in the above-cited publications but rather middle–late Eocene (Bukry, 1984; see also Bukry, 1981, 1985).

Present-day sediment source, transport, and composition is discussed in the Holocene and Modern Depositional Environments section of this chapter.

## Lithostratigraphy

Lithologic criteria used for determining stratification were sediment structure, texture, color, and ferromanganese aggregate content. Using both lithologic and faunal characteristics, three major climatic units were distinguished. Climatic units are used here informally to denote important environmentally controlled depositional sequences.

In ascending order, the oldest unit (Unit III) comprising sediments deposited between approximately 2.5 mya and 4.5–5 mya consists of fairly well-sorted, oxidized, dark brown (10 YR 3/3–4/3) burrow mottled silty clays up to 140-cm-thick containing iron-manganese-rich concretions (Figs. 23-3, 23-4, 23-5). These aggregates constitute up to 70 percent of the coarse fraction (>63 $\mu$m). Small but significant amounts of ice-rafted sand-size quartz grains have also been recognized throughout Unit III. The grains exhibit surface features consistent with a cycle of glacial abrasion prior to their transport to this region from the surrounding continents and/or shelves; they also show superimposed features of subaqueous and eolian abrasion (Margolis and Herman, 1980; Plate 23-II). The cores terminate in Unit III sediments; therefore, the total thickness of this unit is unknown.

The boundary between Units II and III is defined by simultaneous lithologic and faunal changes. The sediments of Unit II consist of tan-green (5 YR 5/2) sandy clays with much fewer manganiferous micronodules and abundant coarse ice-rafted debris exhibiting a mixture of eolian and subaqueously abraded features (Figs. 23-3, 23-6 (page 592), 23-7 (page 593), 23-8 (page 594), Plate 23-III). Two brown (5 YR 3/4), mottled, silty clay layers were observed in this unit. The top of Unit II lies near the top of the Matuyama reversed epoch and is approximately 0.9 my old (Herman, 1970, 1974; Figs. 23-4, 23-5).

The boundary between Units I and II is defined by changes in the foraminiferal fauna, lithology, and appearance of sharp and frequent fluctuations in the abundance of foraminiferal tests (Herman, 1970, 1974, 1985; Figs. 23-4, 23-5). Ice-rafted coarse rock fragments are scattered throughout this unit (Figs. 23-7, 23-8, Plate 23-IV). The sediments consist of alternating brown (10 YR 5/3–3/3) mottled foraminifera-rich sandy clays and olive gray or greenish (5 YR 5/3) foraminifera-poor sandy clay layers; the thicknesses of these banded or "zebra" beds is variable (Figs. 23-4 to 23-9, pages 590–595) (Herman, 1974). Occasional pinkish (2.5 YR 6/4) and yellow-brown layers occur in this unit. The composition of clay minerals is discussed in Chapter 24.

## Ferromanganese-Rich Concretions

The mode of formation of deep-sea manganese nodules has been a topic of considerable interest since their discovery during the voyage of the *HMS Challenger* (1872–1876). They are thought to be derived from continents (Renard, in Murray and Renard, 1891), submarine volcanic activity (Murray, in Murray and Renard, 1891) and hydrothermal activity (Zelenov, 1964). Because levels of concentration in seawater are extremely low, certain mechanisms for concentrating manganese nodules on the seafloor are required. In the absence of submarine volcanic and hydrothermal ac-

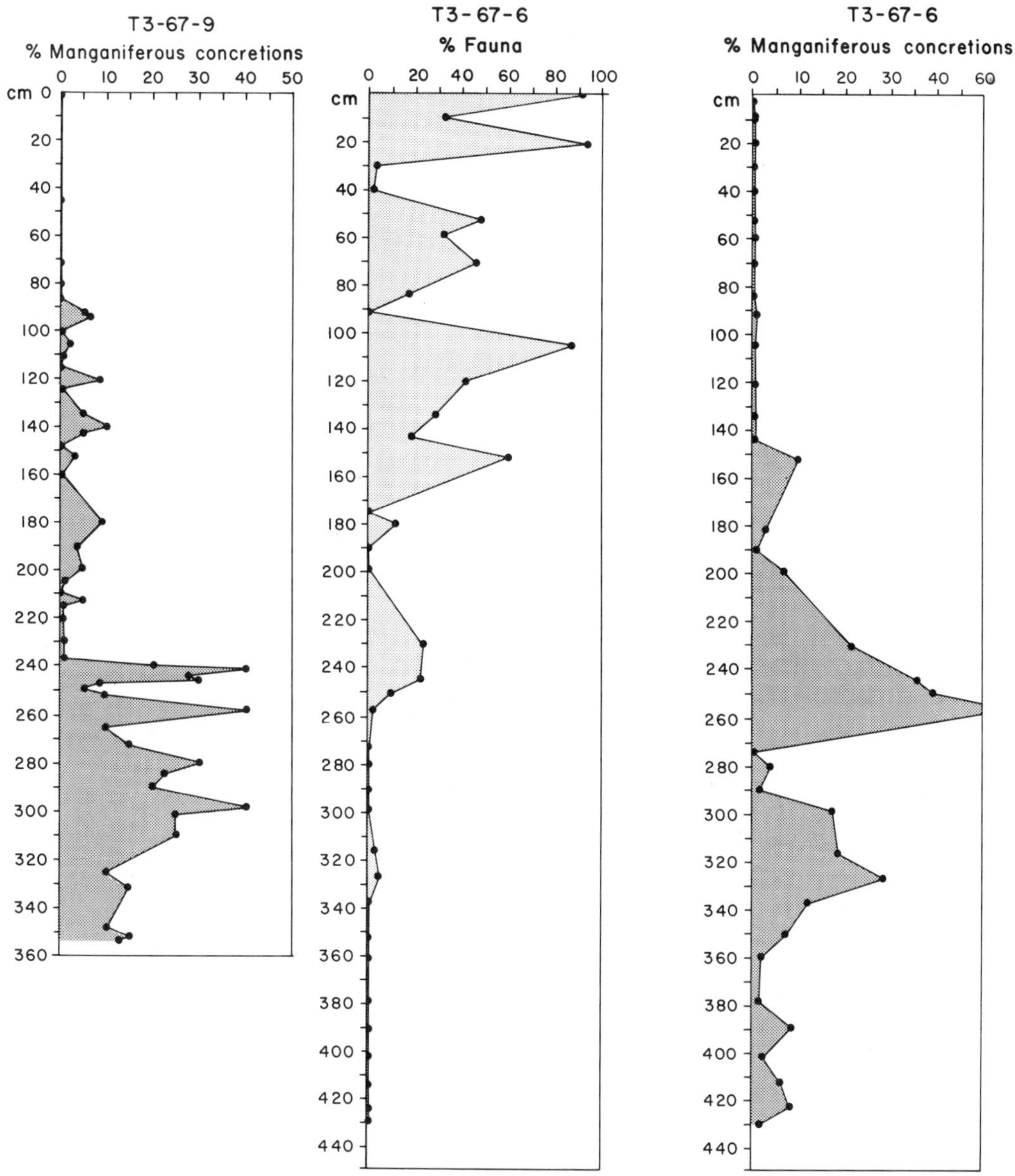


**Figure 23-3.** Ferromanganiferous concretions, percentage, in cores T3-67-9 and T3-67-6, in the coarse fraction (> 63 $\mu$m); shaded areas represent the concretions. Microfauna in the coarse fraction (> 63 $\mu$m) in core T3-67-6. Shaded areas represent microfauna; white areas represent the clastics.

tivity, biological extraction, transport, and concentration on the seafloor have been proposed by many investigators (in Fortin and Meyland, 1977).

In the Arctic sediments studied, such concretions are most abundant in Unit III sediments (Herman, 1970, 1974, 1984*a,b*; Figs. 23-3, 23-4, 23-5), constituting up to 70 percent of the fraction greater than 63 $\mu$m. They are subspherical, botryoidal, as well as irregularly shaped structures, generally 0.10–2 mm across (Plate 23-V). In reflected

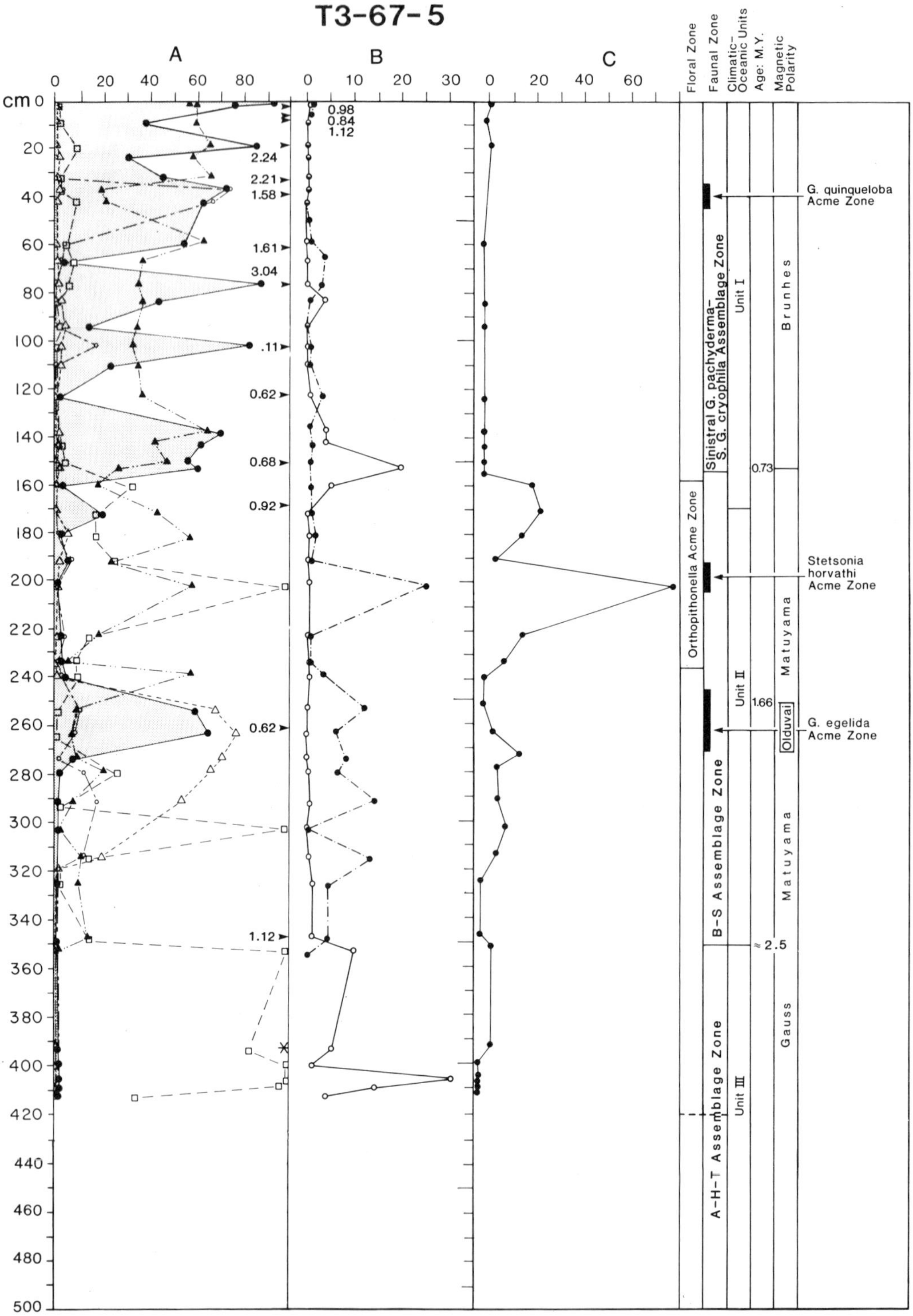


**Figure 23-4.** Core T3-67-5.

*A*

● 1. Microfauna in the coarse fraction (> 63 $\mu$m); percentage.

○ 2. *Globigerina quinqueloba* of the total planktonic foraminiferal fauna.

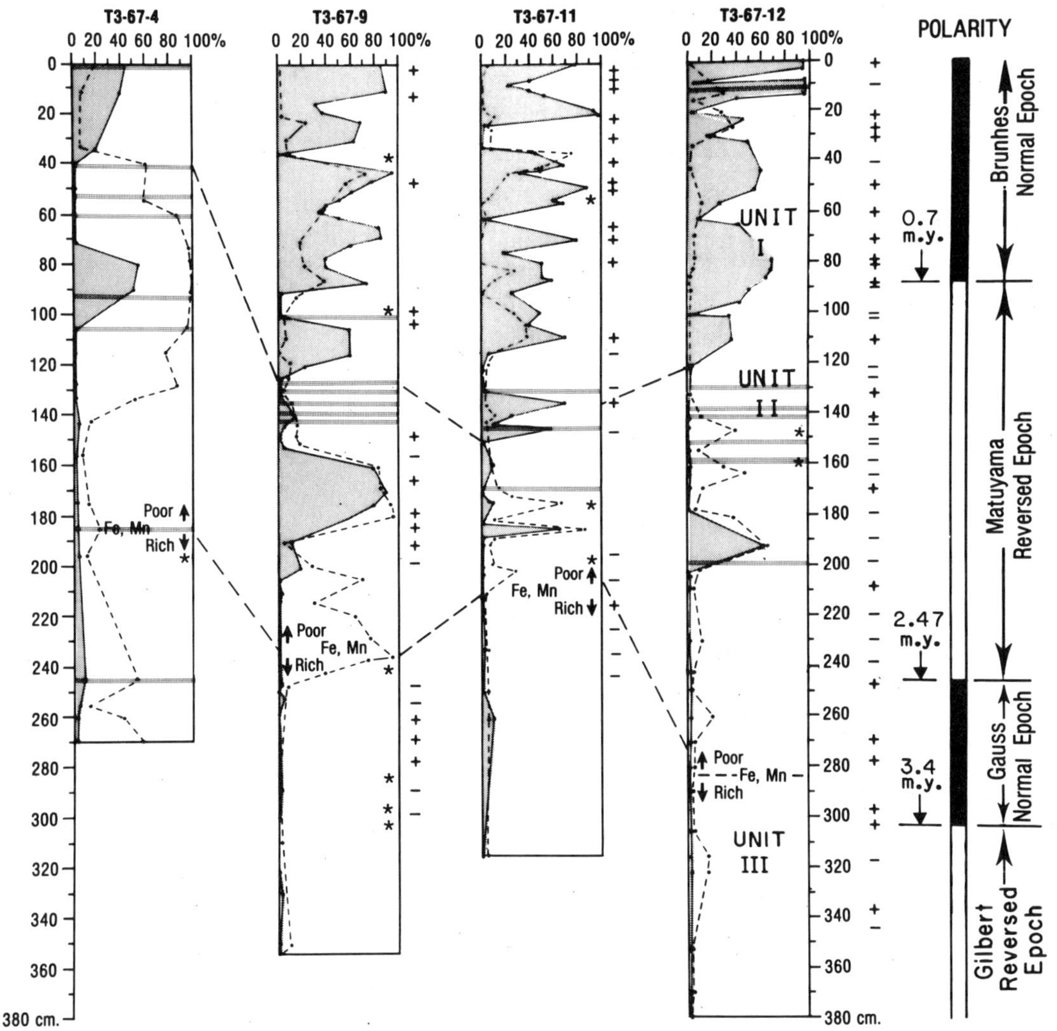


Figure 23-5. Paleomagnetic time scale (right) and percentage of microfauna in coarse fraction (> 63 $\mu$m) in cores T3-67-4, T3-67-9, T3-67-11, and T3-67-12. In core diagrams, shaded areas represent the microfauna; white areas represent the clastic particles. Dashed line represents the percentage of the total planktonic foraminiferal population composed of *G. quinqueloba* and *G. egelida*. Asterisks (*) are low-latitude planktonic foraminifera; pluses and minuses (+, −) are magnetic polarity determinations. (After Herman, 1984*a*. Magnetic stratigraphy after Hunkins et al., 1971.)

---

△ 3. *Globigerina egelida* of the total planktonic foraminiferal fauna.
▲ 4. Juvenile planktonic foraminifera of the total planktonic foraminiferal fauna.
□ 5. Benthonic foraminifera, of the total foraminiferal fauna.
* 6. Low-latitude planktonic foraminifera.

*B*

○ 1. Mn-Fe-rich aggregates.
● 2. Planktonic foraminiferal test fragments; percentage.

*C*

1. Orthopithonella abundance in the coarse fraction (> 63 $\mu$m). Numerals along columns *A* and *B* represent $\delta^{8}O$ (‰) PDB values of sinistral *G. pachyderma* shells. (From Duplessy and Herman, 1986*b*.)

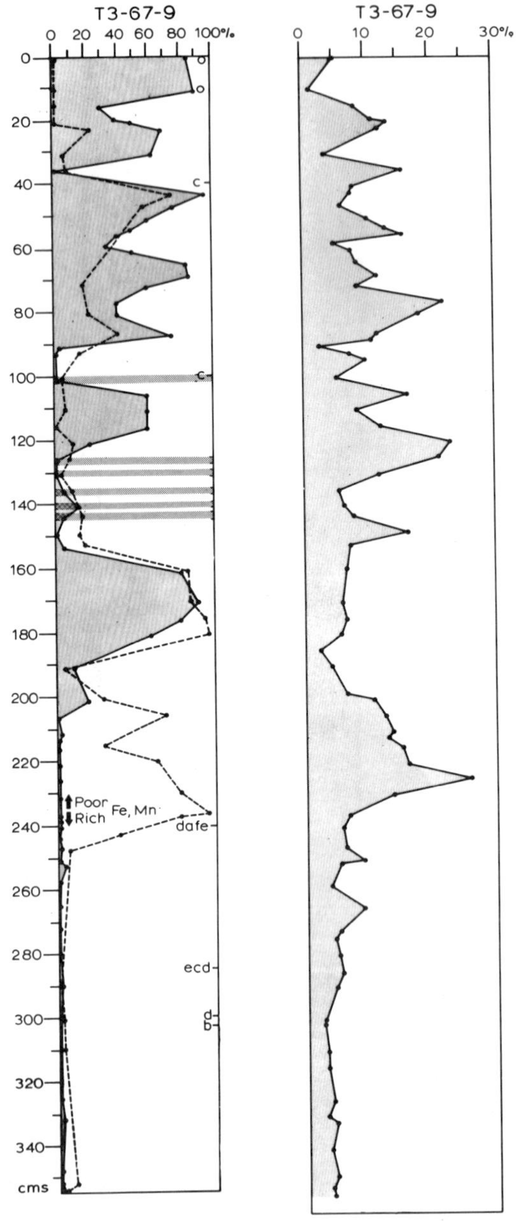


**Figure 23-6.** Core T3-67-9. Shaded areas represent the microfauna; white areas represent the clastics. Occurrence of low-latitude foraminifera: a = *Globorotalia menardii–tumida;* b = *Globorotalia scitula;* c = *Globorotalia* sp; d = *Globigerinoides* sp.; e = *Globigerinoides ruber;* f = *Sphaeroidinella dehiscens.* Circles: Pteropods. Dark horizontal lines are zones in which benthonic foraminifera are in excess of 10 percent. Dashed line represents the percentage of *G. quinqueloba* and *G. egelida* out of the total planktonic foraminiferal population. Right diagram: Shaded area represents percentage of coarse fraction (> 63 $\mu$m). (After Herman, 1984*a*.)

light they are dark red-brown. Broken tubular structures are fragments of agglutinated benthonic foraminiferal shells lined with a thin reddish film and encrusted with spherical and subspherical manganese concretions; others may be fecal pellets (Jehanno and Herman, in preparation) (Tables 23-3 to 23-6).

### Mode of Formation

On today's deep ocean floor, in areas removed from submarine hydrothermal or volcanic activity, manganese-rich micronodules occur where biological activity is high, such as the equatorial Pacific (Revelle et al., 1955), and rates of sedimentation are

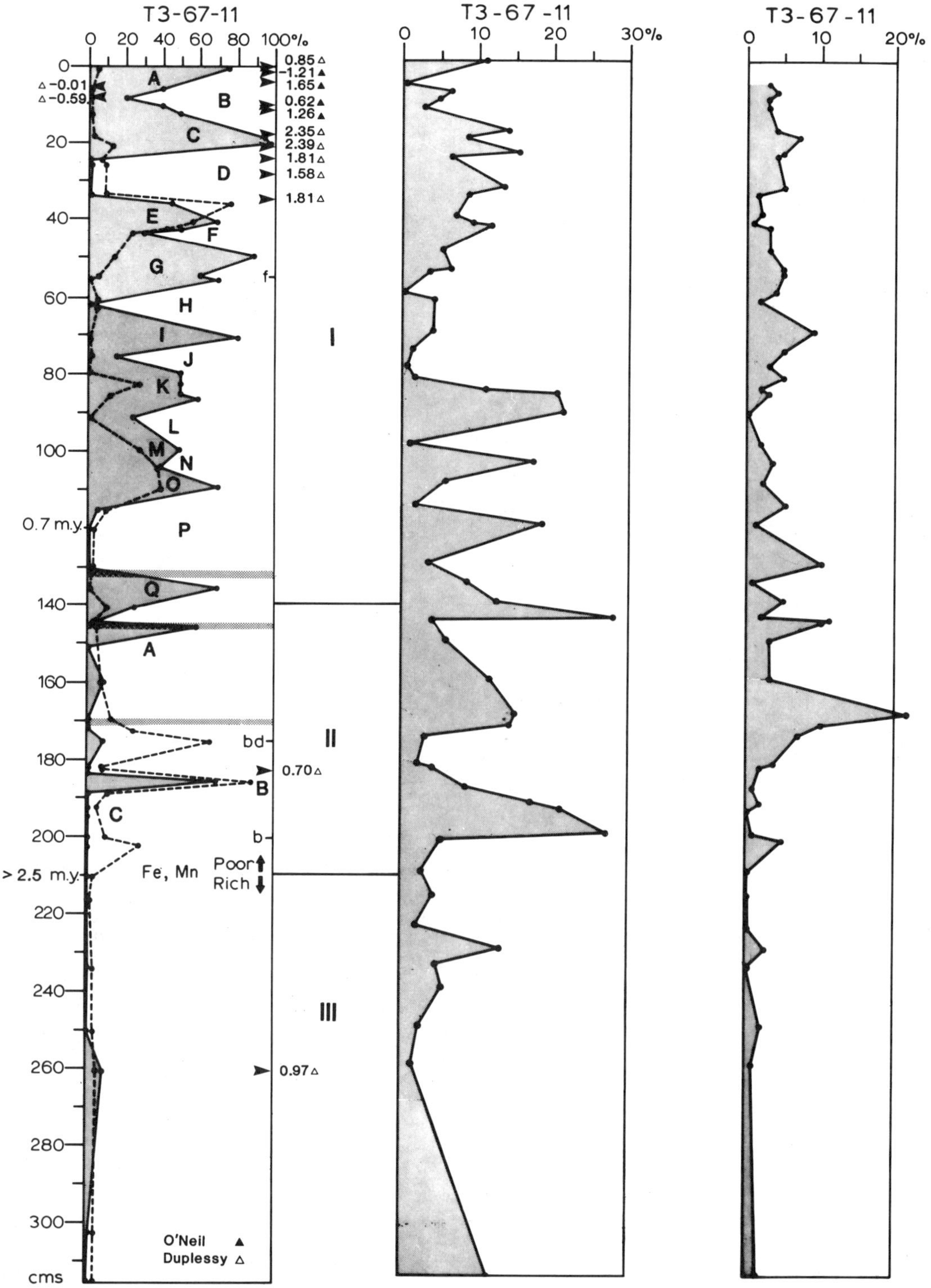


**Figure 23-7.** Core T3-67-11. Left diagram: Shaded areas represent the microfauna; white areas represent the clastics. Same legend as for Figure 23-6. Unit I: A–Q subunits; Unit II: A–C subunits. Middle diagram: percentage of coarse fraction (> 63 μm). Right diagram: percentage of benthonic foraminifera of the total foraminiferal fauna. (Modified after Herman, 1984*a*.) Numerals along column A represent $\delta^{18}O$ values of *G. pachyderma* shells; for more details see Tables 23-12, 23-13.

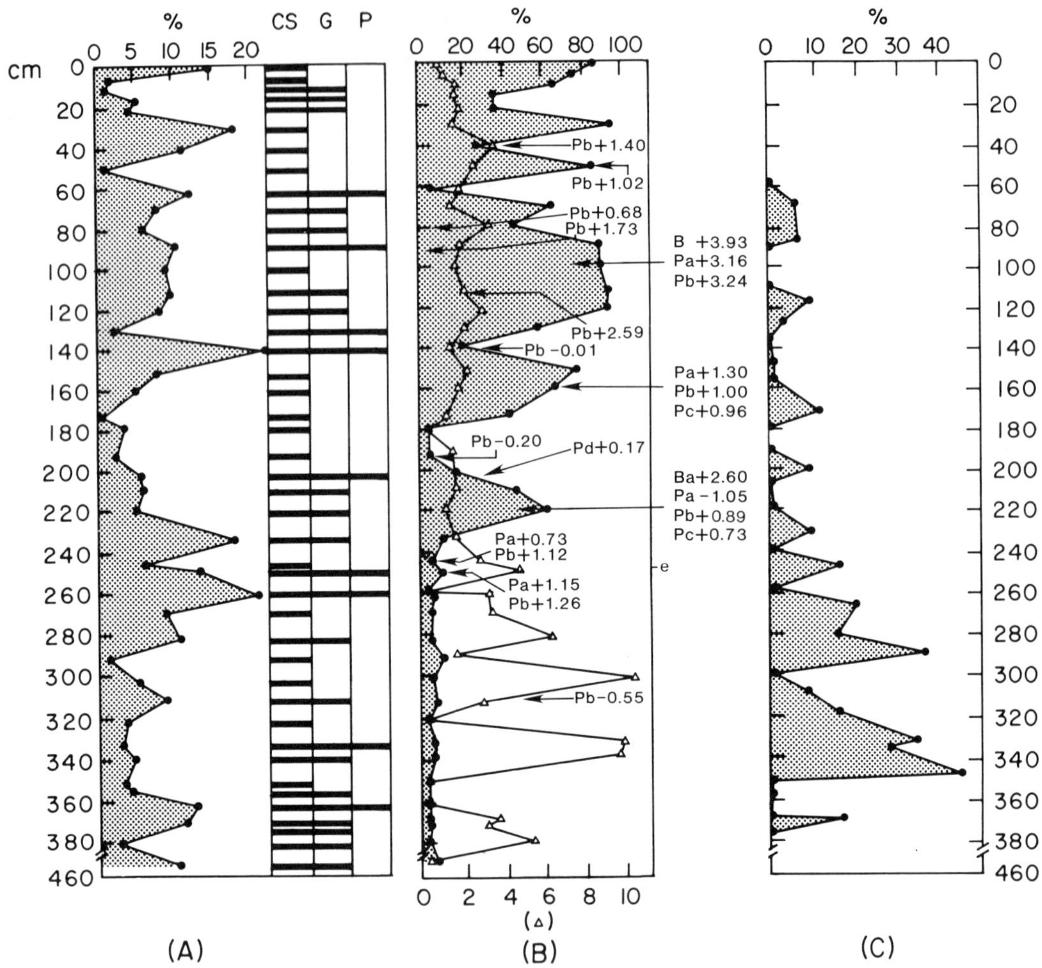


**Figure 23-8.** Core T3-67-3. *A:* Percentage of coarse fraction (> 63 μm). CS = coarse sand; G = granules; P = pebbles. *B:* Percentage of microfauna in the coarse fraction (> 63 μm). Shaded areas represent the microfauna; white areas represent the clastics; values are indicated at the top of the diagram. Triangles = percentage of dextral *G. pachyderma* out of the total *G. pachyderma* population; values are indicated on the abcissa at the bottom of the diagram; e = low-latitude planktonic foraminifera. *C:* Percentage of shallow-water Elphidium spp. out of the total benthic foraminiferal fauna. Core depth in cm. (Modified after Herman, 1984*a*.) Numerals along column *B* represent $\delta^{18}O$ values: B, benthic foraminifera; Ba, *C. wüellerstorfi;* Pa, *G. cryophila;* Pb *G. pachyderma;* Pc, *G. hermanae;* Pd, *G. quinqueloba.* (Modified after Herman and O'Neil, 1975.)

**Table 23-3. Chemical Compositions of the Outer Surface of "Micronodules"**

| | *Core T3-66-S-28: 80 cm* | | *Core T3-67-12:400 cm* | |
|---|---|---|---|---|
| MgO | 1.9 | 1.7 | 1.2 | 1.2 |
| $Al_2O_3$ | 16.6 | 16.3 | 18.2 | 19.7 |
| $SiO_2$ | 54.5 | 52.1 | 61.7 | 57.2 |
| $K_2O$ | 3.0 | 3.0 | 3.0 | 3.4 |
| CaO | 1.0 | 1.0 | 0.2 | 0.3 |
| $TiO_2$ | 0.2 | 0.5 | 0.6 | 0.4 |
| MnOOH | 14.2 | 16.6 | 6.0 | 7.3 |
| FeOOH | 8.9 | 8.8 | 9.1 | 10.5 |

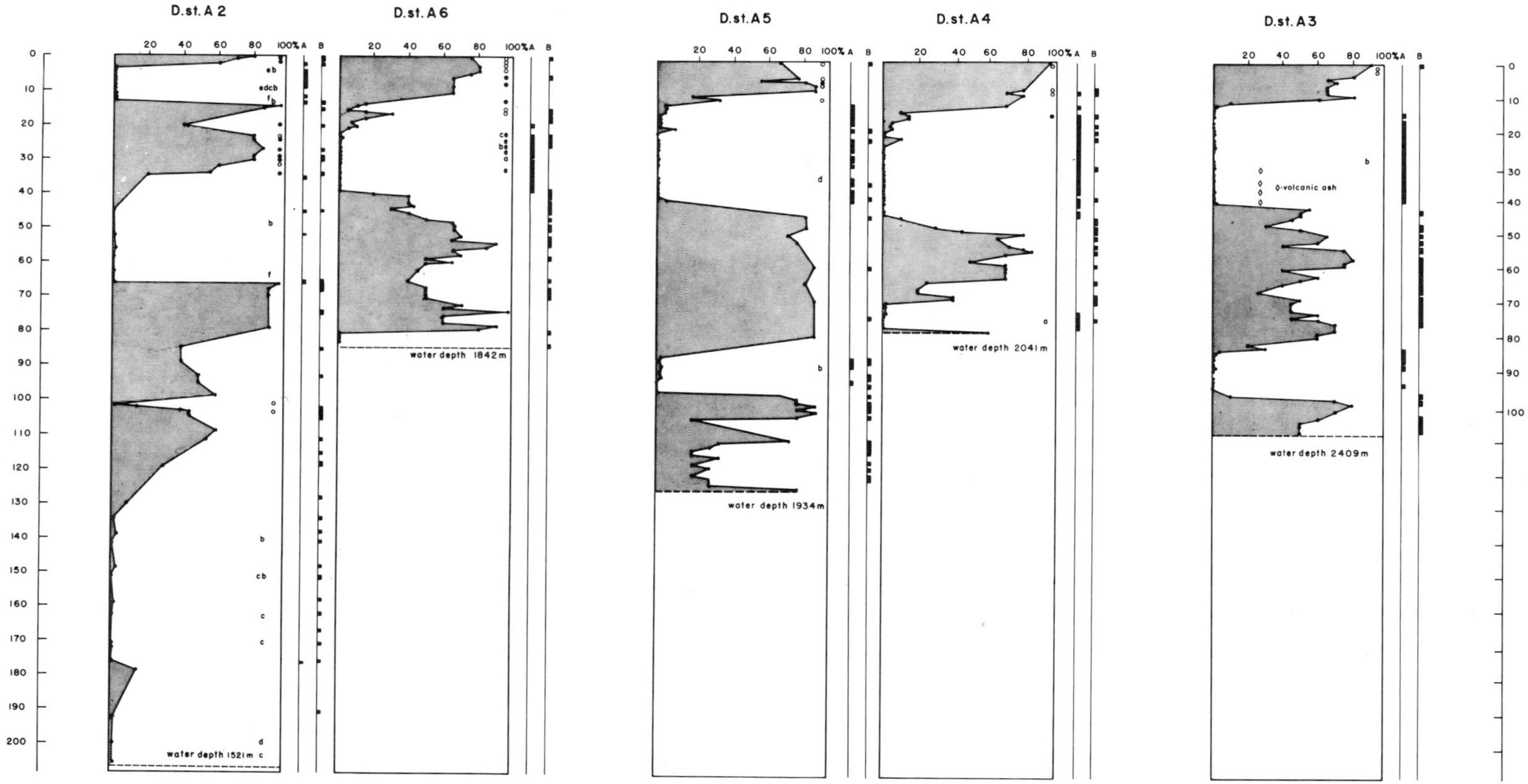


**Figure 23-9.** Cores D.St.A.2, 3, 4, 5, 6. Percentage of microfauna in the coarse fraction (> 63 μm). Occurrences of low-latitude foraminifera: a = *Globigerinoides* cf. *ruber;* b = *Globigerinoides* sp; c = *Globorotalia crassaformis;* d = *Globorotalia inflata;* e = Globorotalia sp.; f = *Globigerinoides* cf. *sacculifera.* Open circles: *Limacina helicina* = 1 percent of the total fauna. Dark circles: *Limacina helicina* = 1–5%. *A:* First column to the right of each core indicates occurrence of plant debris. *B:* Second column to the right of each core indicates occurrence of granules and pebbles. Core depth in cm. (Modified after Herman, 1974.)

**Table 23-4. Chemical Composition of the Inner Part of "Micronodules"**

| | *T3-66 S-28: 80 cm* | *T3-67-12: 400 cm* | *T3-66S-28: 80 cm* |
|---|---|---|---|
| MgO | 1.2 | N.D. | 3.2 |
| $Al_2O_3$ | 16.9 | 16.1 | 6.0 |
| $SiO_2$ | 40.0 | 45.1 | 16.8 |
| $K_2O$ | 2.9 | 4 | 1.1 |
| CaO | 0.6 | 0.2 | 1.65 |
| $TiO_2$ | 0.3 | 0.4 | — |
| MnOOH | 26.3 | 23.7 | 61.8 |
| FeOOH | 11.8 | 10.6 | 9.2 |

N.D. = not determined.

**Table 23-5. Chemical Composition of the Clay Fraction of "Micronodules" except for the MnOOH**

| | *T2-66-S-28: 80 cm* | *T3-67-12: 400 cm* |
|---|---|---|
| $Na_2O$ | N.D. | N.D. |
| MgO | 1.7 | N.D. |
| $Al_2O_3$ | 23.6 | 22.0 |
| $SiO_2$ | 55.9 | 61.8 |
| $K_2O$ | 4.0 | 3.6 |
| CaO | 0.8 | 0.3 |
| $TiO_2$ | 0.4 | 0.5 |
| FeO | 13.5 | 11.8 |
| $MnO_2$ | | |

N.D. = not determined.

**Table 23-6. Chemical Composition of the Fine Fraction (<63 μm) of Sediments Containing "Micronodules"**

| | *T3-67-13: 275 cm* | *T3-66-S-30: 60 cm* | *T3-66-S-28: 80 cm* |
|---|---|---|---|
| $Na_2O$ | 2.8 | 5.4 | 3.0 |
| MgO | 2.6 | 2.52 | 2.6 |
| $Al_2O_3$ | 21.4 | 19.4 | 21.0 |
| $SiO_2$ | 58.1 | 59.9 | 59.1 |
| $K_2O$ | 4.2 | 3.8 | 4.2 |
| CaO | 0.3 | 0.3 | 0.3 |
| $TiO_2$ | 0.8 | 0.8 | 0.8 |
| FeO | 9.5 | 7.6 | 8.8 |
| MnO | 0.25 | 0.3 | 0.3 |

low (Greenslate, Frazer, and Arrhenius, 1973). Phytoplankton and zooplankton play a major role in marine biogeochemical cycles because they have the ability to extract various trace elements such as manganese present in seawater and concentrate them in their skeletal structures, tissues, and fecal pellets. Correns (1941) has suggested the extraction of manganese from seawater by planktonic foraminifera and its incorporation into their shells. This mechanism has been confirmed by the composition of foraminiferal tests caught in plankton tows (Goldberg and Arrhenius, 1958).

Wangersky (1963), Bignot and Dangeard (1976), Ericson and Wollin (1973), Nesteroff and Heezen (1977), and Pautot and Melguen (1979) have provided evidence for the contribution of planktonic calcareous skeletal elements in the formation of micronodules in deep oceanic regions. In such areas, as a result of calcium carbonate dissolution, certain elements including manganese become concentrated at the water–sediment interface. Diatoms and radiolarians are also known to have the ability to concentrate trace metals (Boström, Joensuu, and Brohn, 1974) during their life cycle and also by organically complexing cations during decomposition in their postmortem descent to the seafloor (Martin and Knauer, 1973; Burns and Burns, 1978). Higher organisms are also able to transport elements to the seafloor by vertical migration, via fecal pellets, organic debris, and sinking of skeletal structures (Arrhenius, 1963; Osterberg, Carey, and Curl, 1963; Sorokin, 1972; Bender, Klinkhammer, and Spencer, 1977). Moulting of crustacean exoskeletons (Fowler and Small, 1967; Martin, 1970) and passage of elements to higher trophic levels (Chipman, Rice, and Price, 1958) have been suggested as a means of transporting elements to the seafloor.

On the seafloor, benthonic organisms, particularly agglutinated benthonic foraminifera, contribute actively in the growth of manganese micronodules (Graham, 1959) by collecting the micronodules with the aid of their pseudopodia and cementing these structures to their shells (Thiel, 1978) or by serving as nuclei (Dugolinsky, Margolis, and Dudley, 1977). By analogy to present-day environments conducive to the formation of manganese-rich micronodules, the Arctic Ocean Basin concretions found mainly in Pliocene deposits are also attributed to biological processes, in the absence of local submarine volcanic or hydrothermal activity. The important implications of organisms' participation in manganese micronodule formation in the Arctic Ocean during Pliocene time is that surface-water biological production must have been higher than today. This precludes the existence of a perennial ice cover. Extremely low sedimentation rates below the CCD enhanced the concentration of micronodules. The predominance of cold-water sinistral *Globigerina pachyderma* complex implies surface water nearly as cold as that of the present time. However, the presence of subpolar *Globigerina egelida* in shallow cores (Fig. 23-5; Herman, 1974) strongly suggests water temperatures at least seasonally above the freezing point.

## Fossil Record

Unaltered or virtually unaltered skeletal remains of planktonic and benthonic foraminifera constitute the bulk of microfossil remains in late Cenozoic deposits in the study area (Fig. 23-1; Table 23-1). The distribution of planktonic foraminifera in the Arctic Ocean is presented by Herman and Anderson (Chapter 8).

### *Planktonic Foraminifera*

Planktonic foraminifera are unicellular marine protozoans possessing calcitic tests generally 30–500 $\mu$m in diameter. Their protoplasmatic projections, the pseudopodia, serve for locomotion, food gathering, excretion, shell construction, respiration, and as a sensory device. The sexual phase of their life cycle has been documented (Bé and Anderson, 1976) and an asexual phase is inferred to exist by analogy with benthonic foraminifera (Loeblich and Tappan, 1964). They are omnivorous organisms feeding on bacteria, algae, other sarcodins, mollusks, and copepods. Symbiotic zooxanthellae present in their protoplasm also contribute to foraminiferal nutrition. Present-day distributional patterns show that most species are extremely sensitive to water-mass properties such as temperature, salinity, and oxygen content; hence their usefulness as oceanographic and paleoceanographic indicators (e.g., Boltovskoy, 1959; Bradshaw, 1959; Bé and Tolderlund, 1971; Parker, 1971; Cifelli and Smith, 1974). Their abundance in marine sediments in depths from 200 m to ~4,500 m, and the rapid evolution of some species, make planktonic foraminifera important in biostratigraphic zonation.

*Geologic Distribution.* Rare occurrences have been noted since Middle Jurassic when small globular forms, *Globuligerina,* were first recorded in marine sediments (Caron and Homewood, 1983). Evolving gradually, planktonic foraminifera reached their peak in the Late Cretaceous, whence several low- and high-diversity cycles are recorded; today they are represented by about 50 species (e.g., Herman, 1979). Following the terminal Cretaceous mass extinction, the morphologically simple globigerines survived, giving rise to turborotaliids and globorotaliids (Cifelli, 1969), which diversified through the Paleocene, reaching a peak by Late Paleocene (Table 23-7), approximately 57.5 mya. The 57–62 mya period was a time of fluctuating and warm-water temperatures (Table 23-7).* A

*Deep- and bottom-water paleotemperatures based on the oxygen isotopic composition of benthonic foraminifera should be viewed with caution. See section on Stable Isotopes in this chapter.

Table 23-7. Cenozoic Global Events

| *Time in my* | *Tectonic Events* | *Climatic Events* | *Oceanic Events* | *Planktonic Foraminiferal Events* | *General* |
|---|---|---|---|---|---|
| 0 | 0.9–0 Himalayan, Alpine, etc. orogenies underway (see text). | | | | |
| 0.9 | Orogeny peak. | 0.9–0 Major glacial-interglacial cycles; onset of large-amplitude climatic fluctuations. Cenozoic Threshold IV. | Artic becomes perenially ice covered; Unit I, 0.9–0. In the Arctic ice-covered/ice-free conditions alternate. | In the Arctic, initial appearance of present-day planktonic assemblages. | Onset of large-amplitude climatic fluctuations; eustatic variations. |
| 1.6 | 2–1 Increased Tibetan, Himalayan, and Sierra Nevadan orogenies. | | Northward shifts of Antarctic and sub-Antarctic water masses; Arctic CCD depressed. | PF diversity maximum; Arctic PF has no analog in present oceans: *G. egelida* zone. | |
| 2.5 | | | Arctic flooded with melt-water; significant increase in ice rafting; development of low salinity SW; climatic Units II/III boundary. | 2.5–0.9 Arctic dominated by sub-polar taxa. | |
| 2.5–2.6 | | Fluctuating temperature. | | 2.5–0.9 Arctic Ocean mostly oligotrophic. | |
| 3.5 | Uplift of Panama Isthmus; opening of Bering Strait. | Gradual temperature decline. | | | Limited NH glaciation. |

| | | | | | |
|---|---|---|---|---|---|
| 5 | | | 5–2.5 Arctic DOW oxygenated, CCD elevated. | PF diversity increases. | |
| 5.5 | Isolation of Mediterranean Sea. | | Regression. | Temporal reduction in PF diversity. | |
| 6 | | Strong global cooling. 6–4 Expansion of Antarctic ice cap. | 6.2 Cooling of the SOW: increased circulation of the BOW and fertility; $\delta^{13}C$ global negative shift; lowering of SL. | | Major expansion of Antarctic ice; limited NH glaciation. |
| 11 | | | | 11–1 Gradual increase in diversity of PF. | |
| 14 | | 14–12 Major Antarctic ice buildup. | 14–11 Dissolution climax. | | |
| 17.5 | | | 17.5–14 III $\delta^{13}C$ peak. | | |
| 18 | 18–16 Subsidence of Iceland–Faeroes Ridge. 18–14 Renewed Himalayan Tibetan orogeny. | 18–0 Numerous rapid, large-amplitude fluctuations of SOW; warming trend of SOW. | 18–14 PF diversity fluctuates. | 18–14 Short-period wide-amplitude temperature oscillations of SOW; ~38 cold–warm fluctuations in 4 my. | |
| 22 | | | Development of Antarctic convergence. | | |

Ages of events are approximate.
SOW = surface ocean water; SH = Southern Hemisphere; CCD = carbonate compensation depth; DOW = deep ocean water; BOW = bottom ocean water; SL = sea level; NH = Northern Hemisphere; PF = planktonic foraminifera.

Table 23-7. *(continued)*

| *Time in my* | *Tectonic Events* | *Climatic Events* | *Oceanic Events* | *Planktonic Foraminiferal Events* | *General* |
|---|---|---|---|---|---|
| 24 | Drake Passage opens. | 24–14 Increased glaciation of Antarctica; intensification of global climatic gradients. | 24 Regression climax followed by transgressions 23.6 to −5. 24–18 Changes in SOW temperatures. | 24–14 Gradual increase in PF diversity. Major change in global planktonic biogeography to form latitudinal belts of assemblages from tropics to poles. | |
| 25 | | | 25–22 Development of unrestricted circum-Antarctic current creating thermal isolation of Antarctica; expansion of sea ice; increased vertical and horizontal oceanic circulation. | | |
| 31 | | | | 31–29 Major extinctions followed by evolutionary appearances. | |
| 35 | | | 35–18 SOW stable regime. | | |
| 36.6 | Greenland separates from Europe; Tasman Seaway opens. | Antarctic glaciation; slow cooling; Cenozoic Threshold II. | 36–24 Slow cooling. | 36.6–24 Reduction in PF diversity; diversity minimum. Distinctive Antarctic faunal | 36.6–23.7 Lowest oceanic faunal and floral diversity for entire Cenozoic: low- |

| | | | | | |
|---|---|---|---|---|---|
| | | | | provinciality develops; Globigerines dominate. | diversity gradients between low and high latitudes. Regression, end of warm polar climate; ice growth on Antarctica. |
| 37 | 37–35 Major Himalayan and Alpine orogenies. | Cooling at high and low latitudes; mountain glaciation in Antarctica. | 37–35 II $\delta^{13}C$ peak. Short, conspicuous temp. fluctuations; DOW and SOW temp. decrease; initiation of thermohaline circulation; rapid drop of CCD: SH-sea ice. | Diversity decreases; significant changes in high-latitude species. | |
| 37.5 | | | Abrupt SOW warming. | | |
| 40 | | Gradual cooling; winter snow cover in Antarctica. | Very slow cooling. | 40–37 Diversity high, several major extinctions followed by evolutionary appearances. | |
| 44 | | | 44–40 Gradual warming of SOW. | 44–40 Diversity maxima, few major extinctions and appearances. | |
| 52 | | | | Diversity maxima. | |

Ages of events are approximate.
SOW = surface ocean water; SH = Southern Hemisphere; CCD = carbonate compensation depth; DOW = deep ocean water; BOW = bottom ocean water; SL = sea level; NH = Northern Hemisphere; PF = planktonic foraminifera.

Table 23-7. *(continued)*

| *Time in my* | *Tectonic Events* | *Climatic Events* | *Oceanic Events* | *Planktonic Foraminiferal Events* | *General* |
|---|---|---|---|---|---|
| 53 | 53–36 Labrador Sea and Baffin open. | | Flooding of Atlantic Ocean with Arctic water. | Rapid decrease. | |
| 55 | 55–38 Australia moving northward. | | 55–37 SOW gradual cooling underway. | | |
| 57 | | | | Diversity maxima. 57–53 Decline in species diversity. | |
| 60 | | | 60–59 I $\delta^{13}C$ peak; biological production maximum. | | |
| 62 | | | 62–57 Fluctuating warm temperatures. | | |
| 65 | Tasman Sea opens; Central Tethys narrows; SL drops. | 65–62 Gradual warming. | 65–62 Gradual warming. | Gradual evolutionary diversification of PF commences. | |
| 67 | Continents reach polar position; initial Himalayan orogeny, Central Tethys narrows. | Cenozoic Threshold I; abrupt, large-amplitude cooling. | Presumed change in water chemistry; regression; abrupt large-amplitude cooling. | K/P boundary event; mass extinction of taxa at the peak of their evolutionary development. | Selective mass extinctions; 140–3.5 Atlantic-Pacific communication via Panama. |
| 70 | | | Transgression. | | |

Ages of events are approximate.
SOW = surface ocean water; SH = Southern Hemisphere; CCD = carbonate compensation depth; DOW = deep ocean water; BOW = bottom ocean water; SL = sea level; NH = Northern Hemisphere; PF = planktonic foraminifera.

decline in species diversity is recorded in Early–Middle Eocene (Cifelli, 1969; Berggren, 1971), centering around 50–58 mya and following a $\delta^{13}C$ maximum, which presumably indicates high surface water biological production. Another radiation is recorded between ~37 mya and ~44 mya, interrupted by extinctions 37.3 mya, 38.2 mya, and 40 mya (Keller, 1983).

Recent foraminiferal studies indicate that with the exception of the Eocene/Oligocene (E/O) boundary, these major faunal extinctions are abrupt events being followed by first appearances. This time interval is characterized by a gradual cooling of surface ocean water (Table 23-7); the trend is interrupted by a warm pulse centering around 38 mya. The gradual E/O extinctions are much less severe than the terminal Cretaceous mass decimation of the calcareous microplankton. As in the Paleocene, morphologically simple globigerinids survive and diversify. The next major radiation commenced in Miocene, culminating in Middle Miocene ~15 mya. However, a number of fluctuations in foraminiferal diversity were recorded between 12 mya and ~17 mya. By the end of Early Miocene, fully keeled globorotaliids were diverse; orbulines and hastigerines had also evolved; the pulleniatines appear in Late Miocene. Beginning in Middle Miocene, the planktonic foraminiferal faunas are essentially modern. During the last 1.7 my numerous temperature fluctuations are recorded (Keller, 1978; Herman, 1979, and references therein).

*Modes and Tempos of Evolution.* Several modes of speciation have been suggested for oceanic microplankton. They include (1) phyletic gradualism, whereby new species arise through gradual and continuous phyletic transformation within the population of an entire species. This process is believed to proceed at a slow and constant rate; (2) punctuated equilibria, whereby evolution is concentrated in very rapid events, considered instantaneous in terms of geologic time. This second model is believed to occur in genetically isolated or semi-isolated populations at the outer fringes of the geographic range of ancestral species, followed by migration to other areas (allopatric speciation of Mayr, 1963, 1970; see also Gould and Eldredge, 1977); and (3) punctuated gradualism, described by Malmgren, Berggren, and Lohmann (1983). According to these authors, evolutionary steps could take place within a bioseries in one region, without migration of newly evolved species (Malmgren and Kennett, 1981). Furthermore, new species evolve not through lineage branching as required by punctuated equilibria through periods of rapid, but not instantaneous, phyletic transformation of entire populations (Malmgren et al., 1983).

In an interesting study Malmgren and his collaborators have analyzed the lineage of warm-water *Globorotalia tumida* through the last 10 million years of its evolutionary history in a southern Indian Ocean core. This lineage is believed to be predominantly of Indo-Pacific origin. The evolutionary sequence leads from *Globorotalia merotumida* → *Globorotalia plesiotumida* → *Globorotalia tumida*, with no evidence of divergence in this lineage. The material studied from DSDP Site 214 in water depth of 1,665 m was selected because of the shallow water, well above the lysocline; consequently, these dissolution-resistant species were very well preserved. Sampling interval was $5 \times 10^3$ yr and $15 \times 10^3$ yr across the Miocene/Pliocene boundary (5.3 my) and $2 \times 10^5$ yr in the remainder of the section. The results indicate that during the Late Miocene the *G. plesiotumida* populations were in stasis for ~5 my. About 5.6 mya, the test shape began to change, gradually completing the transformation to *G. tumida* in about 0.4 my, remaining essentially unchanged since that time. Malmgren and Kennett (1981) and Malmgren, Berggren, and Lohmann (1983) have demonstrated that at least in the *G.*

*merotumida–G. tumida* lineage punctuated gradualism is the speciation mode. In another investigation, Malmgren and Kennett (1981) recognized a case of phyletic gradualism, without branching in the evolutionary lineage of *Globorotalia conoidea* → *Globorotalia conomiozea–Globorotalia puncticulata* → *Globorotalia inflata.* Scott (1983) disputes this lineage and suggested that *G. conoidea* split and one branch gave rise to *G. inflata* in a relatively short time, thus supporting the punctuational model of speciation. Still a somewhat different ancestral-descendant relationship in these and related taxa is suggested by Berggren (1977, and references therein).

To further complicate matters, Arnold (1983) used the evolution of *Globorotalia crassaformis* from its ancestor *Globorotalia cibaoensis* via *G. puncticulata* as an example of phyletic gradualism. Different opinions concerning ancestor-descendant relationships exist in many lineages; however, lack of space precludes a detailed discussion of this problem. The state of taxonomic chaos is exemplified by *Globigerina pachyderma* referred to by various other generic names including *Aristerospira, Neogloboquadrina, Globorotalia, Turborotalia,* and *Globoquadrina.* Seven pachyderma experts believe that this species evolved either from the *Globorotalia opima-mayeri* group, or from *Globigerina angustiumbilicata,* or *Globigerina continuosa,* or a *turborotalid* ancestor (in Kennett and Srinivasan, 1980, and references therein).

Before the causes, modes, and rates of speciation can be addressed, we need to sort out taxonomical, phylogenetic, and paleoceanographic problems.

***Description of Species.*** The generic designation of species is conservative, following traditional usage (Cifelli, 1982). Thus the species *pachyderma, cryophila, hermanae, paraobesa,* and so on are retained in *Globigerina.*

*Globigerina pachyderma, Globigerina cryophila, Globigerina egelida, Globigerina quinqueloba, Globigerina hermanae,* and *Globigerina paraobesa* are the dominant taxa, generally constituting 98 percent of the planktonic foraminiferal population. Their coiling direction is given in Figures 23-10 and 23-11 and Tables 23-8 to 23-11. Low-latitude species occur sporadically and at negligible frequencies throughout the cores; however, they are more frequent in the foraminifera-poor zones (e.g., Fig. 23-12; Herman, 1963, 1969, 1974).

*Globigerina pachyderma* (Ehrenberg)
Plates 23-VI, 23-VII, Figures 1, 2, 4–12
*Aristerospira pachyderma* Ehrenberg, 1861, K. Preuss, Akad. Wiss. Berlin, Monatsber, pp. 276–277, 303.

This species is an important, often dominant element of the Arctic Basin in the Late Pleistocene (last ~0.9 my) planktonic foraminiferal population. Our counts included only typical forms characterized by their low trochospiral test with four to four and one-half chambers in the last whorl, and with a narrow low aperture, usually possessing a lip. The wall texture of this species has been discussed by several authors (e.g., Kennett and Srinivasan, 1980, and references therein). In Arctic sediments the variability in wall texture is due to the calcification stage and/or to the degree of preservation (postdepositional dissolution). Here we distinguish two stages of calcification: (1) the primary or initial calcification, which involves the formation of chambers and the multilayered wall thickening over previous chambers; during this stage sharp-edged ridges, forming polygonal (honeycombed patterns) enclosures around the pores, occur (Plate 23-VI, figs. 4, 5, 6; Plate 23-VII, fig. 7); (2) the secondary calcification stage which is characterized by broad and rounded interpore ridge formation (Plate 23-VII, fig. 10) and by euhedral crystals or irregular lumpy calcareous deposits

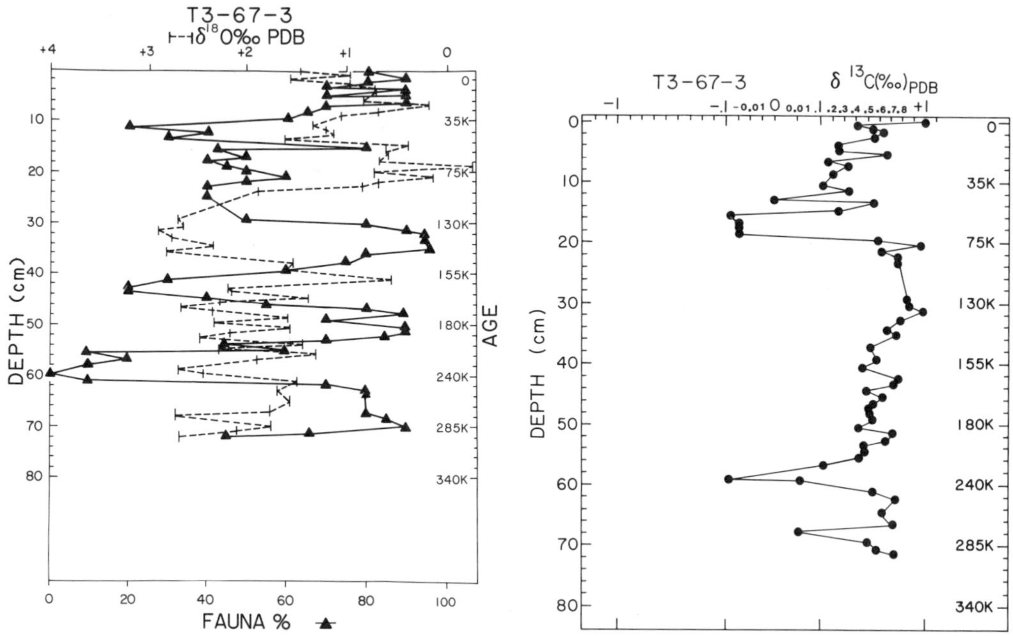


**Figure 23-10.** Core T3-67-3. Left diagram: solid line = percentage of microfauna in the coarse fraction (> 63 μm); broken line = $\delta^{18}O$ of sinistral *G. pachyderma* shells; ages based on uranium series isotopes. (From Duplessy, Herman, and Osmond, 1984.) Right diagram: $\delta^{13}C$ of sinistral *G. pachyderma*. (From Duplessy and Herman, 1986*a*.)

**Table 23-8.** Coiling Direction of *Globigerina Pachyderma*, Sinistral

| *Core* | *Foraminifera-Poor Zones—Units I, II* | | | | *Foraminifera-Rich Zones—Unit I* | | | |
|---|---|---|---|---|---|---|---|---|
| | 1* | 2* | 3* | 4* | 1* | 2* | 3* | 4* |
| D.St.A.2 | 14 | 84.7 | 60 | 100 | 16 | 99 | 96 | 100 |
| D.St.A.3 | 5 | 88 | 78 | 97 | | 98 | | |
| D.St.A.4 | 9 | 90 | 84 | 97 | 1 | 98 | | |
| D.St.A.5 | 8 | 94 | 85 | 100 | 8 | 94 | 85 | 99 |
| D.St.A.6 | | | | | | 96 | 94 | 98 |
| T3-67-9 | 10 | 95.8 | 93 | 100 | 8 | 98 | 94 | 100 |
| T3-67-12 | 13 | 93 | 82 | 100 | 11 | 96 | 91 | 98 |
| Total samples counted | 59 | 91 | | | 44 | 97 | | |

| *Core* | *Foraminifera Poor Zone—Unit III* | |
|---|---|---|
| | 1* | 2* |
| T3-67-4 | 12 | 90 |
| T3-67-9 | 7 | 95 |
| T3-67-11 | 6 | 95 |
| T3-67-12 | 8 | 96 |
| Total samples counted | 34 | 94 |

Range in Unit III samples: 82–100. Counts include levels with 2–5 specimens; thus, the range as well as the average should not be used in paleoceanographic interpretation.
1* = Samples counted; 2* = Average; 3* = Low; 4* = High.

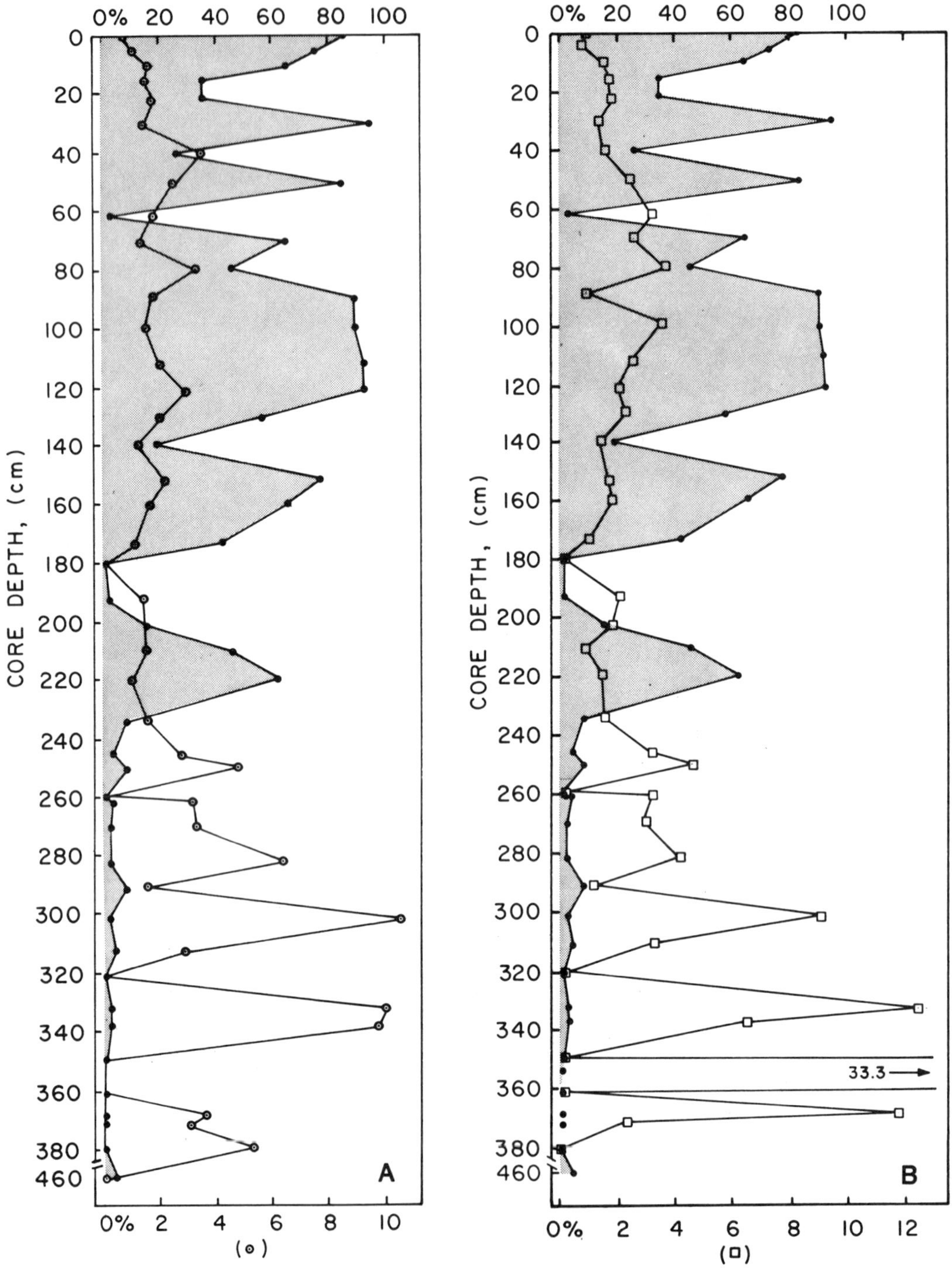


**Figure 23-11.** Core T3-67-3. *A:* Percentage of microfauna in the coarse fraction (> 63 μm). Shaded areas represent the microfauna; white areas represent the clastics; values are indicated on the abcissa at the top of the diagram. Circles represent the percentage of dextral *G. pachyderma;* values are indicated on the abcissa at the bottom of the diagram. *B:* Same legend as for diagram *A*. Squares represent the percentage of dextral *G. cryophila*. (Modified from Herman, 1984*a*.)

**Table 23-9. Coiling Direction of *Globigerina Pachyderma*, Sinistral Core T3-67-5**

| *Core Depth (cm)* | *Adults* | *Juveniles* |
|---|---|---|
| Top A | 99 (342) | 98 (605) |
| Top B | 98 (260) | 97 (588) |
| 36–37 | 95 (19) | 96 (94) |
| 42 | 96 (78) | 96 (330) |
| 50 | 95 (212) | 97 (331) |
| 66–67 | 97 (157) | 98 (158) |
| 75–76 | 96 (216) | 97 (166) |
| 100–103 | 94 (173) | 94 (144) |
| 110 | 97 (200) | 99 (162) |
| 123 | 96 (100) | 100 (61) |
| 143 | 98 (214) | 97 (218) |
| 152–155 | 97 (259) | 93 (130) |
| 160 | 93 (246) | 93 (61) |
| 181 | 92 (161) | 93 (338) |
| 233 | 95 (91) | 97 (36) |
| 239 | 94 (173) | 94 (371) |
| 273 | 90 (29) | 100 (18) |
| Averages: | 89.8 | 96 |

Numerals in parentheses indicate number of specimens counted.
Juvenile counts may include few *G. egelida* and *G. quinqueloba*, which contain high percentages of dextral specimens; thus, our juvenile counts represent minimum sinistral values.

greatly constructing or almost totally obliterating the pores (Plate 23-VII, figs. 8, 9, 12). The pores have cylindrical cross section. In fossil specimens the boundary between the primary and secondary calcification stages is accentuated because of dissolution (Plate 23-VII, fig. 11). The coiling direction of this taxon is predominantly sinistral (Figs. 23-10, 23-11; Tables 23-8, 23-9).

The highest frequencies of dextral specimens occur in the foraminifera-poor sediments of Unit II (Figs. 23-10, 23-11; Tables 23-8, 23-9; Herman, 1969, 1974). It appears that differential solution changes the coiling ratios of *G. pachyderma,* increasing the percentage of sinistral forms and consequently altering the paleotemperature estimates; the direction of bias is toward lower temperature. Clark has published data from the same USGS cores in various papers (1969, 1970, 1971); however, his results differ considerably from paper to paper and, therefore, they cannot be compared to the data and interpretation of the present study. The coiling direction in

**Table 23-10. Coiling Direction of *Globigerina Cryophila*, Sinistral**

| *Core* | *Foraminifera-Poor Zones—Units I, II* | | | | *Foraminifera-Rich Zones—Unit I* | | | |
|---|---|---|---|---|---|---|---|---|
| | 1* | 2* | 3* | 4* | 1* | 2* | 3* | 4* |
| T3-67-9 | 12 | 92 | 75 | 100 | 8 | 94 | 86 | 100 |
| T3-67-12 | 21 | 91 | 75 | 100 | 19 | 94 | 80 | 100 |
| Total samples counted | 33 | 91.5 | | | 27 | 94 | | |

1* = Samples counted; 2* = Average; 3* = Low; 4* = High.

**Table 23-11. Coiling Direction of *Globigerina Hermanae*, Sinistral**

| *Core* | *Foraminifera-Poor Zones—Units I, II* | | | | *Foraminifera-Rich Zones—Unit I* | | | |
|---|---|---|---|---|---|---|---|---|
| | 1* | 2* | 3* | 4* | 1* | 2* | 3* | 4* |
| T3-67-9 | 5 | 97 | 95 | 100 | 3 | 92 | 85 | 100 |
| T3-67-12 | 7 | 95 | 83 | 100 | 12 | 94 | 80 | 100 |
| Total samples counted | 12 | 96 | | | 15 | 93 | | |

1* = Samples counted; 2* = Average; 3* = Low; 4* = High.

adults and juveniles in Arctic Ocean cores has been determined. The percentage of dextrally coiled specimens is somewhat *higher* in adults than in juvenile forms (Table 23-9). Different results were obtained by Clark (1984; and in Steuerwald and Clark, 1972). We believe that Clark's results are based on misidentifications: Dextral *Globigerina quinqueloba* and *Globigerina egelida* were apparently included in the *G. pachyderma* counts.

In a recent study, Malmgren (1983) has obtained the following ranking of dissolution susceptibility for cold-water taxa (the species were ranked from the most resistant to the least resistant): sinistral (s) *G. pachyderma, G. inflata, Globorotalia truncatulinoides, Globigerina bulloides,* and *Globigerinita glutinata* (a tie); dextral (d) *G. pachyderma, Globigerinita uvula,* and *G. quinqueloba.*

*G. pachyderma* seems to tolerate wider ranges of temperature and salinities than *G. cryophila* (see discussion under *G. cryophila*).

The oxygen and carbon isotope values for this species are given in Figs. 23-4, 23-7, 23-8, 23-10 and Tables 23-12, 23-13, 23-14.

*Globigerina cryophila* (Herman)
Plate 23-VI, fig. 3
*Globigerina occlusa* Herman, 1974, Arctic Ocean sediments, microfauna, and the climatic record in late Cenozoic time. In Y. Herman (ed.), *Marine Geology and Oceanography of the Arctic Seas.*

This taxon ranks second in abundance after *G. pachyderma* in late Pleistocene sediments (last ~0.9 my) in the study area (Herman, 1974). It differs from *G. pachyderma* by the elongate last chamber apertural position that is umbilical–extraumbilical and the wider apertural lip. The wall texture of *G. cryophila* is similar to that of *G. pachyderma* (Plate 23-VI, fig. 3). The coiling direction of this taxon is predominantly sinistral; highest frequencies of dextral specimens occur in the foraminifera-poor sediments of Unit II (Herman, 1974; Fig. 23-11; Table 23-10). The dissolution ranking of sinistral and dextral specimens is probably similar to that of *G. pachyderma.*

S. *G. cryophila* is significantly less abundant in the relatively warm Norwegian Sea core tops V27-61 and V27-85 than in the Greenland Sea cores and Labrador Sea cores, where winter sea ice is sometimes present (V28-28, V28-37, V-17-196, V27-24, and V27-25; Table 23-15). The Labrador Sea core tops specimens are smaller than their Arctic Ocean counterparts and they are represented almost exclusively by sinistrally coiled forms.

In Arctic Ocean plankton samples the distribution of this species differs to some extent from that of *G. pachyderma* (Chapter 8). Reynolds and Thunell (1986) also noted that in the Northeast Pacific *G. cryophila* (their *N. pachyderma* Group A) is more abundant during the winter months than *G. pachyderma* (their *N. pachyderma* Group B), whereas *G. pachyderma* predominates during summers.

The oxygen and carbon isotopic analyses of the two species gave slightly different values (Table 23-14). Based on these limited data, it appears that s. *G. cryophila* is more stenothermal and stenohaline than s. *G. pachyderma.*

*Globigerina hermanae* Golbert
Plate 23-VIII, figs. 1, 2, 3, 4
*Globigerina hermanae Golbert*, 1985, Academia Nauk USSR, Sibirskoe Otdelenie, Truda Instituta Geologii I Geophyziki, 615, *Globigerina* sp., cf. *G. pachyderma* A in Herman, 1974, Arctic Ocean sediments, microfauna and climatic record in late Cenozoic time. In Y. Herman (ed.), *Marine Geology and Oceanography in the Arctic Seas.*

Table 23-12. Selected Stable Isotope Values and Inferred Paleotemperatures and Paleosalinities

| Core | Depth (cm) | Species Analyzed | $\delta^{18}O$ | $\delta^{13}C$ | Temperature (°C) | Salinity ‰ | $\delta^{18}O$ water | Meltwater (%) ($\delta^{18}O = -30$) |
|---|---|---|---|---|---|---|---|---|
| T3-67-3 | 98–101 | *Cibicides wüellerstorfi* *Eponides tener umbonatus* | +3.93 | +0.89 | −0.5 | 33.8 | −0.8 | 2.5 |
| T3-67-3 | 98–101 | *G. cryophila* | +3.16 | +0.64 | −1.7 | 32.5 | −1.9 | 6.3 |
| T3-67-3 | 98–101 | *G. pachyderma* | +3.24 | +0.78 | −1.7 | 32.6 | −1.8 | 6 |
| T3-67-3 | 160 | *G. pachyderma* | +1.00 | −2.92 | −1.0 | 30.4 | −3.7 | 12.3 |
| T3-67-3 | 160 | *G. cryophila* | +1.30 | +0.29 | −1.0 | 30.8 | −3.4 | 11.3 |
| T3-67-3 | 160 | *G. hermanae* | +0.96 | +0.30 | −1.5 | 30.0 | −4.0 | 13.4 |
| T3-67-3 | 203–205 | *G. quinqueloba* | +0.17 | −0.86 | −1.5 | 29.1 | −4.82 | 16.1 |
| T3-67-3 | 219–222 | *G. pachyderma* | +0.89 | +0.30 | −1.5 | 29.9 | −4.1 | 13.7 |
| T3-67-3 | 219–222 | *G. cryophila* | −1.05 | 1.46 | −1.0 | 26.2 | −7.3 | 24.4 |
| T3-67-3 | 219–222 | *G. hermanae* | +0.73 | +0.31 | −1.5 | 29.8 | −4.2 | 14.2 |
| T3-67-3 | 219–222 | *C. wüellerstorfi* | +2.60 | −0.04 | −0.8 | 32.2 | −2.2 | 7.3 |
| T3-67-3 | 245–248 | *G. cryophila* | +0.73 | +0.16 | +1.5 | 30.9 | −3.3 | 11.0 |
| T3-67-3 | 245–248 | *G. pachyderma* | +1.12 | +0.15 | +1.5 | 31.5 | −2.8 | 9.3 |
| T3-67-3 | 250 | *G. cryophila* | +1.15 | +0.01 | +1.5 | 31.5 | −2.8 | 9.3 |
| T3-67-3 | 250 | *G. pachyderma* | +1.26 | +1.5 | +1.5 | 31.6 | −2.7 | 9.0 |
| T3-67-4 | 80 | *G. egelida* | +1.42 | −0.42 | −1.0 | 30.6 | −3.5 | 11.9 |
| T3-67-11 | Top | *G. pachyderma* | −1.21 | −0.19 | −1.5 | 27.5 | −6.2 | 20.7 |
| T3-67-11 | 6 | *G. pachyderma* | +1.65 | +0.70 | −1.8 | 30.6 | −3.5 | 11.7 |
| T3-67-11 | 10 | *G. pachyderma* | +0.62 | +0.45 | −1.0 | 30.2 | −3.9 | 13.0 |
| T3-67-11 | 10 | *G. pachyderma* | +1.26 | +0.01 | −1.7 | 30.3 | −3.8 | 12.7 |
| T3-67-12 | 0-2 | *G. pachyderma* | +0.16 | −0.04 | −1.0 | 29.4 | −4.6 | 15.6 |
| T3-67-12 | 192 | *G. egelida* | −0.03 | −0.95 | −1.0 | 29.0 | −4.8 | 16.3 |
| D.St.A.2 | 0 | *G. pachyderma* | +1.42 | +0.56 | −1.0 | 30.6 | −3.5 | 11.9 |
| D.St.A.2 | 7 | *G. pachyderma* | +2.29 | +0.67 | −1.7 | 31.5 | −2.7 | 9.3 |
| D.St.A.2 | 39-40 | *G. pachyderma* | +1.95 | +0.82 | −1.7 | 31.0 | −3.1 | 10.5 |
| D.St.A.2 | 75 | *G. pachyderma* | +2.21 | +1.10 | −1.7 | 31.6 | −2.8 | 9.0 |
| D.St.A.2 | 180 | *G. egelida* | +1.15 | −1.26 | +1.5 | 31.4 | −2.8 | 9.5 |

*Source:* Modified after Herman and O'Neil, 1975.

Table 23-13. Stable Isotopic Composition of Sinistral *Globigerina Pachyderma*

| *Core Number* | *Depth in Core (cm)* | $\delta^{18}O$ | $\delta^{13}C$ | *Remarks* |
|---|---|---|---|---|
| T3-67-3 | 112 | +2.59 | +0.98 | Foram-rich |
| | 140 | −0.01 | −0.23 | Foram-poor |
| | 160 | +0.77 | +0.19 | Transition |
| | 193 | −0.20 | −0.26 | Foram-poor |
| | 311 | −0.55 | −0.33 | Foram-poor |
| T3-67-5 | top A | +0.98 | +0.57 | Foram-rich |
| | top C | +1.12 | +0.74 | Foram-rich |
| | 7–10 | +0.84 | −0.04 | Transition |
| | 17.5–19.5 | +2.24 | +0.86 | Foram-rich |
| | 31 | +2.22 | +0.65 | Foram-rich |
| | 36–37 | +1.58 | +0.64 | Foram-rich |
| | 59–60 | +1.61 | +0.65 | Foram-rich |
| | 75–76 | +3.04 | +1.36 | Foram-rich |
| | 100–103 | +1.11 | +0.13 | Foram-rich |
| | 123 | +0.62 | +0.34 | Foram-poor small sample |
| | 150 | +0.68 | +0.19 | Transition small sample |
| | 171 | +0.92 | +0.50 | Foram-rich |
| | 263 | +0.62 | +0.48 | Foram-rich |
| | 347 | +1.12 | +0.58 | Foram-poor red clays |
| T3-67-11 | top | +0.85 | +0.62 | Foram-rich |
| | 18 | +2.35 | +0.81 | Foram-rich |
| | 20 | +2.39 | +0.73 | Foram-rich |
| | 24 | +1.81 | +0.79 | Top, Foram-poor layer |
| | 25 | +1.58 | +0.50 | Foram-poor |
| | 35 | +1.81 | +0.75 | Foram-rich |
| | 183 | +0.70 | +0.56 | Foram-poor |
| | 260 | +0.97 | +0.37 | Foram-poor red clays |

*Source:* From Duplessy and Herman, 1986*a*.

Table 23-14. Stable Isotopic Composition of Planktonic Foraminifera

| *Species* | *Core Number* | *Core Depth (cm)* | $\delta^{18}O/PDB$ (‰) | $\delta^{13}C/PDB$ (‰) |
|---|---|---|---|---|
| | T3-67-11-TW | | | |
| *G. cryophila* (dextral) | | 0–3 | +0.03 | +0.60 |
| *G. cryophila* (dextral) | | 3–5 | −0.45 | +0.27 |
| *G. cryophila* (sinistral) | | 0–3 | +0.55 | +0.56 |
| *G. cryophila* (sinistral) | | 3–5 | +0.30 | +0.21 |
| *G. cryophila* (sinistral) | | 5–7 | −0.17 | −0.06 |
| *G. pachyderma* (dextral) | | 0–3 | +0.14 | +0.59 |
| *G. pachyderma* (dextral) | | 3–5 | −0.04 | +0.16 |
| *G. pachyderma* (dextral) | | 5–7 | −0.50 | +0.07 |
| *G. pachyderma* (sinistral) | | 0–3 | +0.27 | +0.71 |
| *G. pachyderma* (sinistral) | | 3–5 | −0.01 | +0.21 |
| *G. pachyderma* (sinistral) | | 5–7 | −0.59 | +0.14 |
| | T3-74-1-TW | | | |
| *G. cryophila* (sinistral) | | top | +3.42 | +0.73 |
| | T3-74-3-TW | | | |
| *G. pachyderma* (sinistral) | | top | +1.30 | +0.38 |

*Source:* From Duplessy and Herman, 1986*b*.

Table 23-15. Sea Surface Temperature and Salinities[1]

| | *SSWT*[a] | *SSWS*[b] | *WSWT*[c] | *WSWS*[d] |
|---|---|---|---|---|
| Norwegian Sea | 7° to 11°C | 34 to 35‰ | −1° to 5°C | 34.5 to 35‰ |
| Greenland Sea | 5°C | 34.6‰ | −1.4°to −1.7°C | 34.7‰ |
| Labrador Sea | 5° to 6°C | 33 to 34‰ | 0.5° to −1.7°C | 34 to 34.2‰ |

[a]Average summer surface water temperatures in the region where the cores were raised.
[b]Average summer surface water salinities.
[c]Average winter surface water temperatures.
[d]Average winter surface water salinities.

The test is trochospiral with subrounded periphery; it has five chambers per whorl, and five to six chambers in the last whorl visible ventrally. The shell texture is similar to that of *G. pachyderma* (Plate 23-VIII, figs. 3, 4). The coiling direction of this taxon is predominantly sinistral (Table 23-16). *G. hermanae* is present at low frequencies in the studied sediments (Herman, 1974; Table 23-16).

*Globigerina paraobesa* Herman
*Globigerina paraobesa* Herman, 1974, Arctic Ocean sediments, microfauna and the climatic record in late Cenozoic time. In Y. Herman (ed.), *Marine Geology and Oceanography of the Arctic Seas.*

The fairly large and robust test (0.40–0.50 mm average diameter) consists of five chambers per whorl with five to six chambers in the last whorl visible ventrally, arranged in two and one-half to 3 whorls, in a high, loosely coiled trochospire. The chambers are globular; a modified, larger, grotesque, last chamber is generally present. The aperture possesses a well-developed lip extending from umbilicus to periphery. The wall is thick, perforate; its surface texture is very similar to that of *G. pachyderma* and the sutures are radial and narrow.

This species resembles *G. pachyderma* but can be distinguished from the latter by its larger size, number of chambers, outline of test, which is circular in *G. paraobesa* and subquadrate in *G. pachyderma,* and its modified, larger, last chamber. The coiling direction of this taxon is predominantly sinistral (average: 98 percent sinistral in core T-67-3, based on 19 sample counts and 99.4 percent sinistral in core T3-67-5 based on 9 sample counts; Table 23-17;

Table 23-16. Percentage of *Globigerina Hermanae* of the Total Planktonic Foraminifera Population in Core T3-67-5

| *Foraminifera-Rich* | | *Foraminifera-Poor* | |
|---|---|---|---|
| *Unit I* | | | |
| 1.5 | (94) | 2.2 | (80) |
| 0.7 | (98) | 1.4 | (100) |
| 0.6 | (100) | 1.8 | (100) |
| 1.2 | (91) | 1.8 | (83) |
| 1.9 | (100) | | |
| 2 | (100) | | |
| 1.4 | (91) | | |
| 0.4 | (100) | | |
| 1.1 | (100) | | |
| 1.8 | (100) | | |
| 0.6 | (100) | | |
| 2.1 | (90) | | |
| 1.1 | (100) | | |
| 2.5 | (100) | | |
| 2.5 | (77) | | |
| 0.4 | (100) | | |
| 0.9 | (100) | | |
| 1.6 | (100) | | |
| 1.1 | (100) | | |
| 1.0 | (83) | | |
| *Unit II* | | | |
| | | 1.7 | (83) |
| | | 0.8 | (100) |
| | | 3.8 | (100) |
| | | 4.00 | (93) |

Numbers in parentheses indicate percentage sinistral.

Table 23-17. Percentage of *Globigerina Paraobesa* of the Total Planktonic Foraminiferal Population in Core T3-67-5

| *Foraminifera-Rich* | | *Foraminifera-Poor* | |
|---|---|---|---|
| *Unit I* | | | |
| 0.3 | (100) | 0.2 | (100) |
| 0.06 | (100) | 0.2 | (100) |
| 0.1 | (100) | | |
| 0.4 | (100) | | |
| *Unit II* | | | |
| | | 0.3 | (100) |
| | | 0.3 | (100) |
| | | 4.3 | (95) |

Numbers in parentheses indicate percentage sinistral.

Herman, 1974). *G. paraobesa* is present throughout the studied Arctic cores in Units I, II, and III sediments, constituting 0.06–4.3 percent of the planktonic foraminiferal population (Herman, 1974; Table 23-17).

*Globigerina quinqueloba* (Natland)
Plate 23-IX, figs. 1, 2
*Globigerina quinqueloba* Natland, 1938, University of California, Scripps Inst. Oceanography, Bull., Tech. Ser., vol. 4, no. 5.

This spinose species has an elongate last chamber that usually extends across the umbilical area. There are four and one-half to five chambers in the last whorl. The coiling direction of *G. quinqueloba* is random (Herman, 1969, 1974).

*G. quinqueloba* is considered an opportunistic taxon, able to occupy environments unfavorable to other species (Colbourn et al., 1980); it is abundant in subpolar oceans. In the North Pacific, *G. quinqueloba* is one of the dominant taxa (Bradshaw, 1959; Reynolds and Thunell, 1985). The latter authors found it in automated time series sediment traps at highest frequencies in April when temperatures in the upper water column are coldest for the year. *G. quinqueloba* appears to prefer the period of high phytoplanktonic productivity that characterizes the winter water mass. Its Arctic Ocean distribution pattern is discussed by Herman and Andersen in Chapter 8.

In Arctic sediments this solution-susceptible species is present in all studied cores, attaining highest frequencies (up to 90 percent of the planktonic foraminiferal population) in relatively shallow-water cores (< ~2,000 m) such as D. St. A.-2 (Fig. 23-12; Herman, 1974). Several *G. quinqueloba* acme zones were observed in the Alpha Rise cores (Table 23-18). These intermittent abundance maxima may represent periods of high sea level, when water exchange with the Pacific was unrestricted and this North Pacific species was carried into the Arctic. Its periodic absence from sediments may be due to interruption in water exchange between the two oceans

Table 23-18. *Globigerina Quinqueloba* Acme Zones in Selected Cores

| *Core* | *1* | *2* | *3* | *4* | *5* | *6* |
|---|---|---|---|---|---|---|
| T3-67-2 (1,982) | top missing | | | | 30 | 70 |
| T3-67-3 (2,285) | | 60–80 | | | | 150–160 |
| T3-67-5 (2,057) | | 40–48 | | | 100 | |
| T3-67-6 (2,815) | 10 | 58–60 | | | | |
| T3-67-9 (2,237) | 25 | 40 | | | | 88 |
| T3-67-11 (2,810) | 20 | 35–53 | | | 82–86 | 105–110 |
| T3-67-12 (2,867) | 12 | 28 | | | | |
| D.St.A.2 (1,521) | | 14–18 | 45 | 60–70 | | 125–130 |

Numbers in parentheses indicate water depth in meters.

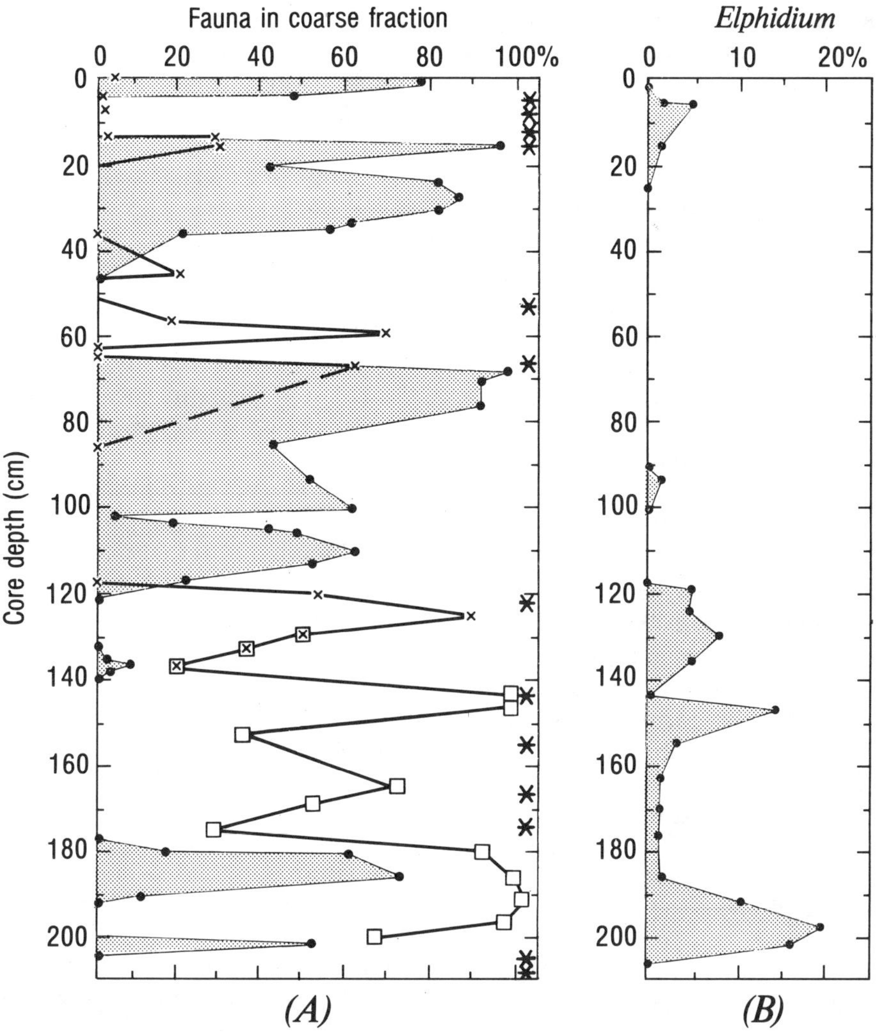


**Figure 23-12.** Core D.St.A.2. *A:* Percentage of microfauna in the coarse fraction (> 63 μm) Shaded areas represent the microfauna; white areas represent the clastics. Crosses (X) represent percentages of *G. quinqueloba* out of the total planktonic foraminiferal fauna. Open squares (□) represent percentages of *G. egelida* of the total planktonic foraminiferal fauna. Asterisks (*) represent low-latitude planktonic foraminifera. *B:* Percentages of Elphidium spp., a shallow-water genus of the total benthonic foraminiferal fauna. Core depth in cm. (Modified after Herman, 1984*a*.)

during low sea levels, or to temperatures below their tolerance range and/or to their dissolution.

*Globigerina egelida* (Cifelli and Smith)
Plate 23-IX, figs. 3, 4
*Globigerina quinqueloba egelida* Cifelli and Smith, 1970, Distribution of planktonic foraminifera in the vicinity of the North Atlantic Current, *Smithsonian Contributions to Paleobiology No. 4.*

The small trochospiral test has a rounded, lobate, periphery; chambers are spherical, enlarging gradually, arranged in three to three and one-half whorls. Generally there are four and one-half to five chambers in the last whorl; the final chamber is spherical and possesses a thin but distinct lip. The umbilicus is deep; it is surrounded by umbilical teeth. The aperture of the last chamber is a low arch, umbilical–extraumbilical. The sutures are distinct, narrow, and depressed. Surface texture: There are no interpore ridges (Plate 23-IX, fig. 4) and the spine bases are rounded (Plate 23-IX, fig. 4).

Cifelli and Smith (1974) and Cifelli (1974) described the distribution of *G. egelida* from Atlantic Ocean and Mediterranean Sea plankton tows. Other authors included this species with *G. quinqueloba;* therefore, its distributional patterns mentioned in the present chapter are incomplete. In the North Atlantic it attains peak abundances in the cold western slope waters, between 36° and 45°N lat., during the winter when temperatures may drop below 2°C in the most northerly parts. In the summer, temperatures may exceed 24°C and *G. egelida* exhibits a marked minimum. Along the west coast of Africa, near Cape Verde, where upwelling occurs it constitutes 6 percent of the planktonic foraminifera during April. In the sub-Arctic region, *G. egelida* maxima were recorded during summer, when surface temperatures range from 12°–15°C; it is also common in the Transitional Zone (Bé and Hamlin, 1967). It was found in Mediterranean plankton tows collected during the summer. In the Eastern Mediterranean it constitutes up to 17 percent of the planktonic foraminiferal population in a region where surface temperatures were 19°C and salinities were 39‰. In Arctic plankton tows where surface temperatures were −1.7°C and salinities were 31.9‰ it is rare (Chapter 8).

In Arctic sediments eurythermal–euryhaline, solution-susceptible *G. egelida* is abundant in the shallow ($< \sim 2000$ m) Northwind Escarpment and in Alpha Rise sediments deposited during the Matuyama epoch. The *G. egelida* acme zone, a unique event during the early Matuyama epoch (Figs. 23-4, 23-5, 23-12), is interpreted to represent an environment unfavorable to other species. This species is scarce in Brunhes and Gauss epoch sediments. In earlier publications (Herman, 1970, 1971) this species was included with the counts of *G. quinqueloba;* it was also referred to *Globigerina* sp., cf. *G. quinqueloba* and *G. exumbilicata* (Herman, 1974).

*Dissolution Effects.* Differential solution of calcareous foraminiferal tests alters the ratio of various species enriching the seafloor assemblages in the resistant forms (Berger, 1976; Malmgren, 1983, 1987, and references therein). In our sediments the bias introduced by dissolution is toward colder temperatures; species tolerant of lowest temperatures (s. *G. pachyderma* and s. *G. cryophila*) are also the most resistant to dissolution.

Berger (1968, 1970, 1973, 1976, 1982, and references therein) has considered various factors that influence the dissolution of planktonic foraminiferal tests including (1) the degree of undersaturation of water with respect to $CaCO_3$; (2) the rate of supply of calcareous versus noncalcareous particles to the seafloor; (3) the length of time tests are exposed to the cold corrosive bottom water; (4) the rates of sedimentation, and (5) the shell wall thickness, porosity, and microstructure. In our study the following three criteria were used to establish the degree of dissolution: (1) percentage of planktonic foraminiferal test fragmentation; (2) percentage of benthic foraminifera, which are generally more resistant to dissolution than planktonic species (varying with the degree of fragmentation); and (3) wall textures. Typically test dissolution results in chalky texture, thinning, increase in pore size, and surface etching. As shown in Figure 23-4, dissolution effects are marked in Units II and III sediments at 160–220 cm and 290–410 cm core depths.

### Pteropoda

Pteropods are marine gastropods adapted to pelagic life. Some species possess delicate external aragonitic shells while others are devoid of it. Pteropods are ubiquitous in the world oceans. After the animals' death their empty shells settle to the seafloor and, being more susceptible to solution than the calcitic skeletons of coccolithophores and foraminifera, they are generally preserved between about 200 m and 3,000 m. Preservation depths differ in various basins and change with time, depending in part on productivity, predation, temperature, and rates of sedimentation. As with other calcareous shells, dissolution increases with decreasing temperature and increasing depth. While important water mass indicators in present oceans and useful paleoceanographic tools for the Pleistocene, their fragile tests are rarely preserved in older sediments (Herman, 1971, 1978, and bibliography therein). The geologic range of pteropods is Cretaceous–Recent, the earliest records being from the Tethyan Province (Herman, 1978, and bibliography therein).

In the Arctic, pteropod shells are generally preserved in the uppermost 30–40 cm in cores raised from depths shallower than 2,500 m (Herman, 1970, 1974). Their distribution in plankton samples is given in Chapter 8.

### *Coccolithophores*

Coccoliths, which are the skeletal remains of planktonic unicellular photosynthetic golden brown algae, occur very sparsely (~1 per $10^4$ silt-size grains) in central Arctic Basin deposits (Worsley and Herman, 1980) in cores D.St.A.-2, T3-67-2 and T3-67-3 (Fig. 23-13; Herman, in preparation). Their distribution in eastern Arctic sediments is discussed by Gard (Chapter 18) and in the Arctic Ocean by Heimdal (Chapter 7).

### *Radiolaria and Diatoms*

The siliceous skeletons of radiolaria and diatoms are rarely preserved in sediments although they abound in the Arctic Ocean (Herman, unpublished). They were observed in cores D.St.A.2, D.St.A.3, and D.St.A.6, associated with colorless volcanic glass shards in the uppermost foraminifera-poor layer (Herman, 1974). The dis-

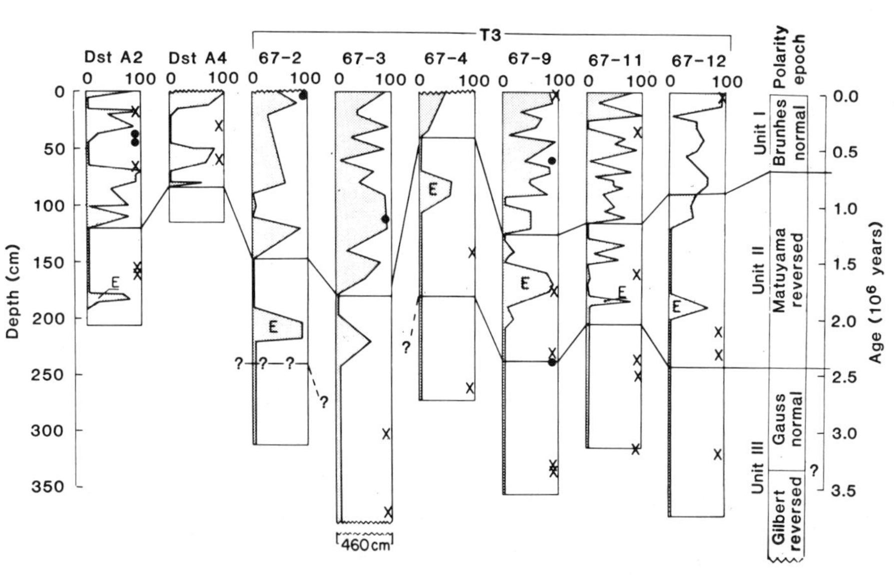


**Figure 23-13.** Shaded areas represent percentages of the microfauna; white areas represent the clastics in the coarse fraction (> 63 $\mu$m). Crosses = coccoliths looked for and not found. Dark circles = coccoliths found. E = *G. egelida* zone. Core depth in cm on left margin. (Modified after Worsley and Herman, 1980.)

tributional patterns of both groups in Arctic sediments is presented by Kruglikova (Chapter 19) and Polyakova (Chapter 20) and in the Arctic Ocean by Horner (Chapter 5) and Heimdal (Chapter 7).

### Dinoflagellates

Dinoflagellates are minute unicellular algae 2–5 μm in size, predominantly planktonic; most taxa are autotrophic (capable of photosynthesis); others are heterotrophic or saprophytic. Today they are the second most abundant phytoplankton group in the world oceans, following the diatoms; the apparent restriction of certain assemblages to specific water masses make dinoflagellates important tools in paleoenvironmental reconstructions (Williams, 1971*a,b*). They are found in all environments from fresh water to open ocean. The known stratigraphic range of dinoflagellates is Silurian to Recent.

The Arctic Ocean sediments' calcareous sphaeroids are cysts of peridinian dinoflagellates belonging to the genus Orthopithonella (family Calciodinellaceae; Herman and Keupp, 1987; Plate 23-X, figs. 1, 2, 3). They were mistakenly classified as *Thoracosphaera arctica* by Clark (1984) and Gilbert and Clark (1983); therefore, Clark's interpretation based on this taxon should be viewed with caution. Thoracosphaera is a name occupied for the vegetative calcareous stage only (Tangen et al., 1982). Their stratigraphic range within the studied cores is late Pliocene to Holocene (Unit III–Unit I, Fig. 23-4).

### Tasmanitids

Tasmanitids are unicellular green algae varying in size from less than 100 μm to more than 600 μm, believed to represent the cysts of marine planktonic organisms. Their yellow or red-brown sphaeroidal organic walls (more rarely ovoidal, reniform, biconvex, or discoidal) are characterized by the presence of numerous fine radial canals (Plate 23-X; Figs. 23-4, 23-5; Williams, 1978). The geologic record of tasmanitids suggests that they were deposited in littoral environments, where freshwater runoff from rivers and *melting glaciers* had lowered the ocean water salinity. The known stratigraphic range of Tasmanaceae is Precambrian to Holocene (Tappan, 1980). In Arctic Ocean sediments their occurrence is as follows: in Northwind Escarpment cores (T3-66-S-7, 15, 19, 20, and 21) they are scarce in the foraminifera-rich layers and abundant in the foraminifera-poor beds. In Alpha Rise cores (e.g., T3-67-5) they are rare throughout the entire core length.

### Benthonic Skeletal Elements

Calcareous foraminifera generally dominate the benthonic assemblages; however, arenaceous tests occur sporadically in the study area (Table 23-19; Herman, 1974; Chap. 21). Ostracods, pelecypods, gastropods, tubeworms, and echinoderm fragments are minor components; sponge spicules are generally sparse; however, they occur at high frequencies in the foraminifera-poor zones of the Northwind Escarpment cores. The distribution, ecology, paleoecology, geological range, and modes of evolution of numerous taxa are discussed by several authors in this volume.

### Biostratigraphy

Our informal biostratigraphic zonation was based on the presence or dominance of certain taxa. The boundaries of these zones are not defined by first or last appearances of any of their constituents. Three major zones, coinciding with lithostratigraphic and climatic units, are recognized.

In ascending order, the oldest, corresponding to the climatic Unit III is the *Alveolophragmium–Haplophragmoides–Trochammina* assemblage zone. It is characterized by the dominance of several arenaceous benthonic foraminifera including *Alveolophragmium subglobosum,* species of

Table 23-19. Distribution of Benthonic Foraminifera in Arctic Ocean Sediments

| | |
|---|---|
| *Alveolophragmium subglobosum:* Unit III | *Fissurina longispina:* Units I, II |
| *Bolivina arctica:* Units I, II | *Glomospira gordialis:* Units I, II, III |
| *Bolivina* cf. *B. pseudopunctata:* Units I, II | Haplophragmoides sp.: Units II, III |
| Bolivina sp.: Units I, II | Hyperammina spp.: Units I, II, III |
| *Buccella frigida,* Units I, II | Lagena spp.: Units I, II |
| *Bulimina aculeata:* Units I, II | Miliolids: Unit II |
| *Cassidulina islandica:* Units I, II | Nodosaria sp.: Units I, II |
| *Cassidulina laevigata:* Units I, II | Nonion sp.: Unit II |
| *Cassidulina norcrossi:* Units I, II | *Oolina melo:* Units I, II |
| *Cassidulina teretis:* Units I, II | Oolina sp.: Units I, II |
| Chilostomella sp.: Units I, II | *Patellina corrugata:* Unit I |
| *Cibicides lobatulus:* Units I, II | Pyrgo spp.: Units I, II |
| *Cibicides wüellerstorfi:* Units I, II | *Quinqueloculina akneriana:* Units I, II |
| Cibicides sp.: Units I, II, III | Quinqueloculina sp.: Units I, II |
| *Cyclammina pusilla:* Unit III | Reophax spp.: Units II, III |
| *Dentalina frosbisherensis:* Units I, II | *Sphaeroidina bulloides:* Units I, II |
| *Dentalina pauperata:* Units II, III | *Stetsonia horvathi:* Units I, II |
| Elphidium spp.: Units I, II | Triloculina sp.: Units I, II |
| *Epistominella arctica:* Units I, II | *Trochammina inflata:* Unit III |
| *Eponides tener umbonatus:* Units I, II, III | *Valvulineria arctica:* Units I, II |
| *Eponides tumidulus horvathi:* Units I, II | *Virgulina davisi:* Units I, II |
| Enthoselenia sp.: Units I, II | *Virgulina loeblichi:* Units I, II |
| *Fissurina cuculatta:* Units I, II | Virgulina sp.: Units I, II, III |
| *Fissurina kerguelensis:* Units I, II | |

Haplophragmoides, and *Trochammina inflata* (Table 23-19). The top of this assemblage zone coinciding with the sharp lithologic contact between Units II and III is placed at 185 cm in core T3-67-4, at 350 cm in core T3-67-5, at 235 cm in core T3-67-9, at 210 cm in core T3-67-11, and at 285 cm in core T3-67-12 (Figs. 23-4, 23-5). Since the cores terminate in this zone, the zone's total thickness is unknown. Calcareous benthonic and planktonic species are rare; the former are listed in Table 23-19 and are shown in Figure 23-4. Robust, solution-resistant, mostly corroded s. *G. pachyderma* and s. *G. cryophila* dominate the planktonic assemblages. Solution-susceptible taxa are rare or absent and the benthonic/planktonic foraminiferal ratio is very high (Fig. 23-4).

The *Bolivina–Stetsonia* assemblage zone, corresponding to climatic Unit II, is characterized by the dominance of several calcareous benthonic species including *Bolivina arctica* and *Stetsonia horvathi* (Table 23-19). In cores shallower than 2,000 m (e.g. D.St.A.2. T-3-67-4), solution-susceptible subpolar *G. egelida* and *G. quinqueloba* dominate the planktonic assemblages. Dextral *G. pachyderma* and d. *G. cryophila* attain highest frequencies in this unit (Tables 23-8, 23-10; Figs. 23-10, 23-11). The *G. egelida* and the Orthopithonella acme zones provide important datum planes in this assemblage zone (Fig. 23-4), which is divided in subzones (Fig. 23-7). The double peak of *G. egelida* in core T3-67-11 is due to mid-core "flow in" (Fig. 23-5).

The s. *Globigerina pachyderma*–s. *G. cryophila* assemblage zone, corresponding to climatic Unit I, is characterized by the dominance of these two polar species. Several *G. quinqueloba* acme zones (Table 23-18) and six to ten alternating foraminifera-rich and foraminifera-poor beds characterize this assemblage zone, which is divided into several subzones (A–Q, Fig. 23-7). The benthonic foraminifera are listed in Table 23-19. As discussed in the Material and Methods section, in nearly half of our cores hiatuses as well as extraneous sediments have been recognized using microfaunal data.

For example, missing sediments are indicated by the distribution of *G. quinqueloba*. In the continuously studied shallow D.St.A.2 core there are four *G. quinqueloba* acme zones (Table 23-18; Fig. 23-12). Two *G. quinqueloba* zones are absent from this and other cores.

*Sediment Mixing.* Bottom currents and biological activity mix the surface layers of marine sediments (Arrhenius, 1963; Bramlette and Bradley, 1941; Berger and Heath, 1968), redistributing their constituents and defocusing the stratigraphic record. Sediment mixing depends primarily on two parameters: depositional rates and the rate of volume mixing below the seafloor. Benthonic faunal density, being extremely low in the Arctic Basin (Paul and Menzies, 1973), offsets the effect of slow sedimentation on topographic highs (< 1mm–~3mm/1,000 yr). Consequently, in spite of the low sedimentation rates, the abrupt climatic changes are well recorded. In cores for which the continuous sedimentary record is available (D.St.A. and T3-67-3), extremely sharp changes in lithology, fauna, and stable isotopes occur within <0.5 cm intervals (Figs. 23-9, 23-10).

## STABLE ISOTOPES AS PALEOCEANOGRAPHIC INDICATORS

Emiliani (1955, 1964, 1966*a*, 1966*b*, 1967, 1971) was the first to analyze the oxygen isotopic composition of foraminiferal tests in deep-sea cores, establishing that the isotopic variations reflected Pleistocene climatic changes. Subsequent work in several laboratories on deep-sea cores from major oceans and adjacent seas has shown that the $^{18}O/^{16}O$ ratio in the carbonate tests of planktonic and benthonic foraminifera provide the combined record of past glacial ice volumes and ocean water temperatures (e.g., Duplessy, 1978; Berger, 1981, and references therein). The $\delta^{18}O$ values of shells depend on both temperature and the isotopic composition of the ocean (Williams and Healy-Williams, 1980, and references therein). Studies of planktonic foraminifera collected with plankton tows, in sediment traps and core tops, as well as of benthonic foraminifera, indicate that some species do not deposit their shells in isotopic equilibrium with the seawater in which they grow. Different explanations have been suggested to account for these isotopic disequilibria (e.g., Duplessy, Lalou, and Vinot, 1970; Duplessy, 1978; Deuser, 1979; Erez, Honjo, and Shackleton, 1979; Kahn, 1979; Kahn and Williams, 1981; Belanger et al., 1981; Graham et al., 1981; Berger, 1981, and references therein; Durazzi, 1981; Curry and Matthews, 1979; Grossman, 1987).

In a recent study, Keigwin (1987) measured the $\delta^{18}O$ separately in benthonic foraminifera *Planulina wüellerstorfi* and *Globocassidulina subglobosa* from North Atlantic cores at DSDP Site 606. At numerous levels at which $\delta^{18}O$ values were measured, when one species registered enrichment the other showed depletion in $^{18}O$. Furthermore, the trend in these two taxa differs, frequently changing in opposite directions: When *P. wüellerstorfi* indicates warming, *G. subglobosa* registers cooling and vice versa. These two species are widely used in stable isotope studies (see references in Keigwin, 1987) and are often combined for measurements. Therefore, the average value apparently does not reflect temperature trends, ice volume, or the isotopic composition of ocean water. Another problem is that within the same core these species occur at different levels and paleotemperature curves are plotted on one diagram using the values derived from monospecific taxa.

The carbon isotopic variations in planktonic and benthonic foraminifera reflect the combined effects of ''(1) global changes in the rates of exchange of the ocean's carbon reservoir with biosphere, soil, and sediments; (2) global and regional changes in surface water productivity; (3) internal shifts in water-mass structure and circulation (basin–basin fractionation, oxygen minimum development); and (4) organism-specific fractionation effects that are due to changes in microhabitat and/or ontogenic fractionation (vital effects). Additional complications arise from differential preservation. It is impossible to entirely isolate these various factors'' (Berger and Vincent, 1986; see also Williams, Sommer, and Bender, 1977; Kroopnick, Margolis, and Wong, 1977). $\delta^{13}C$ for both groups generally show species-dependent nonequilibrium fractionation, although some exceptions were recently reported (Grossman, 1984).

## The Arctic Ocean Record

Among the factors conditioning and modifying surface-water salinities in the Arctic, precipitation, river runoff, glacial meltwater, evaporation, inflow and subsequent mixing of Atlantic and Pacific water, convective mixing, and the presence of drift ice are considered important. The oxygen isotopic composition of Arctic surface water is determined by the same factors; evaporation results in an increase of $^{18}O$ whereas precipitation and river inflow causes its decrease (Craig and Gordon, 1965). At the present time the ranges of temperatures and salinities of Arctic surface water are $-1.6°$–$0°C$ and $<29$–$\sim 34‰$, respectively (Coachman and Aagaard, 1974). As a result of slow vertical mixing, present-day surface salinities are about 4.8‰ lower than those of the bottom water. Oxygen isotopic analysis of Arctic water indicates that the $^{18}O/^{16}O$ ratio closely follows the salinity profile with $\delta^{18}O$ decreasing by 0.8‰ with a 1‰ decrease in salinity in the upper 350 m (Van Donk and Mathieu, 1969). The $\delta^{18}O$ value of this surface seawater is about 4‰ lower than at greater depth, where the $\delta^{18}O$ approaches values close to 0.

Paleotemperatures and paleosalinities in representative cores were estimated in an earlier study (Herman and O'Neil, 1975); up to 5,000 specimens of *G. egelida* and *G. quinqueloba* were hand-picked for each analysis. For paleotemperatures we used the known correspondence between the coiling direction of *G. pachyderma* and present-day temperatures in the world ocean (Ericson, 1959; Bandy, 1959, 1960). With these temperature estimates, and assuming a $\delta$ value of $-30‰$ for glacial meltwater (Herman and O'Neil, 1975), we calculated the paleosalinities using the oxygen isotope paleotemperature expression of Epstein et al. (1953; Table 23-12).

In a more recent study the $^{18}O/^{16}O$ ratios of s. *G. pachyderma* were obtained from a continuously sampled core (T3-67-3) spanning the last 0.3 my (Duplessy, Herman, and Osmond, 1984; Fig. 23-10). Variations in $\delta^{18}O$ range from $-0.01$ to 2.90 $\delta^{18}O‰$ PDB (Figs. 23-4, 23-7, 23-8, 23-10). Twenty-five to fifty specimens were analyzed at each level (for laboratory procedures, see Herman and O'Neil, 1975 and Duplessy, 1978). In the continuously sampled core there are five major foraminifera-rich and four foraminifera-poor zones; most zones can be further subdivided into subzones. For example, between 24 and 14 cm there are three peaks and two valleys (Fig. 23-10). There are more variations in $^{18}O$ than in faunal abundance because the isotopic composition of tests is a more sensitive oceanographic indicator than the ratio of foraminifera to clastics. $^{18}O$ changes are not consistently in phase with the faunal abundance variations (Fig. 23-10).

Before the oxygen isotope data became available, the main differentiation of Unit I sediments (the last 0.9 my) was based on the alternation of foraminifera-rich and foraminifera-poor zones (Figs. 23-4 to 23-9), which are accompanied by lithological changes (Figs. 23-6 to 23-9). When comparing the oxygen isotope composition and the faunal content in the same sample, we noticed that in some foraminifera-rich layers the $\delta^{18}O$ of s. *G. pachyderma* is near 1–

1.5‰ and that of benthonic foraminifera is 2.60–3.93‰ (Fig. 23-8, Table 23-12), while in other foraminifera-rich zones planktonic and benthonic $\delta^{18}O$ values are very close to each other (Fig. 23-8, Table 23-12). Generally the most negative $\delta^{18}O$ values were measured in the foraminifera-poor zones (Figs. 23-4, 23-7, 23-8, 23-10).

In the Arctic the $\delta^{13}C$ in s. *G. pachyderma* is variable (Fig. 23-10, Tables 23-12, 23-13, 23-14). The lowest values were recorded in a foraminifera-poor zone (Fig. 23-10) and in an $^{18}O$ depleted layer (Fig. 23-10) whereas carbon enrichments of more than 1‰ were measured in foraminifera-rich zones (Fig. 23-10). This relationship is not consistent throughout the studied cores (Fig. 23-10, Tables 23-12, 23-13). When comparing the three measured parameters—foraminiferal abundance, $\delta^{13}C$, and $\delta^{18}O$—a complex pattern emerges (Fig. 23-10, Tables 23-13, 23-14). The sedimentary and stable isotope record of the Eastern Arctic differs from that of our study area (Markussen, Zahn, and Thiede, 1985; Zahn, Markussen, and Thiede, 1985).

## DISCUSSION

Despite the very low sedimentation rates (1.5–4.5 mm/1.000 yr), remarkably sharp oxygen isotope shifts reflecting important oceanographic changes were observed within a few mm in the continuously sampled core T3-67-3 (Fig. 23-10). This is partially the result of the low rate of sediment mixing in a nutrient-poor region with an extremely low density of burrowing organisms (Paul and Menzies, 1973).

Based on the oxygen isotopic composition of foraminifera, three major regimes are distinguished: (1), represented by foraminifera-rich zones (abundance > 40 percent) and planktonic foraminiferal $\delta^{18}O$ values higher than 2.5‰ (e.g., Fig. 23-10: T3-67-3, 29–36 cm; Fig. 23-8: T3-67-3, 98–101 cm; Fig. 23-4: T3-67-5, 75–76 cm, etc.). The foraminifera are believed to have been deposited in an ice-free nearly isopycnal ocean; (2) foraminifera-poor zones (< 40 percent with planktonic $\delta^{18}O$ values lower than 1‰ (e.g., Fig. 23-10: T3-67-3, 40 cm, 15–22 cm; Fig. 23-8: T3-67-3, 140 cm, 193 cm, 311 cm), interpreted to represent periods of marked density stratification and at least a seasonally ice-free ocean. These are deglacial or interglacial intervals. The $\delta^{18}O$ decreases until values reach a negative plateau (e.g., Fig. 23-10: T3-67-3, 18–19 cm) whence they rebound. The $^{18}O$ data support Herman's hypotheses (Herman, 1984*a,b*) contending that foraminifera-poor zones represent deglacial acme events during the deposition of Unit I sediments, not the presence of a thick, rigid ice cover (Clark, 1982*a* and his earlier papers); (3) foraminifera-rich zones (> 40 percent with planktonic foraminiferal $\delta^{18}O$ values ranging between ~0.6‰ and <2‰ (e.g., Fig. 23-8: T3-67-3, 160 cm; Fig. 23-10: T3-67-3, 12 cm, 50 cm, 64–67 cm), representing oceanographic conditions intermediate between regimes 1 and 2. These periods characterized by a low-salinity surface mixed layer and perennial ice cover represent conditions similar to those prevailing today in the Arctic Ocean.

## RADIOMETRIC CHRONOLOGIES

The actual dating, by radiometric methods, of Arctic Ocean sediments is not commonplace. Only a limited number of cores have been studied, with an applicable time span, thus far, of only a few hundred thousand years. However, these studies have been particularly important in defining sedimentation rates and in indicating the contemporary conditions of sediment accumulation. Furthermore, the applicability of paleomagnetic, stable isotopic, and faunal time scales depend in part on confirmation by radiometric dating.

Deep-sea sediments accumulate at rates averaging 1–100 m per million years, and so radionuclides with half-lives in the range of 10,000 to 10 million years are needed to measure sediment accumulation rates. The two principal modes of origin for suitable radionuclides are (a) intermediate daughters generated by decay of uranium dissolved in seawater, and (b) spallation products formed by cosmic ray bombardment of nitrogen, oxygen, and argon in the atmosphere. Among the former, $^{234}U$ has an ideal half-life of 250,000 years, but seldom occurs in a state sufficiently out of equilibrium to be useful; $^{210}Pb$ (T1/2 = 222 years) and $^{226}Ra$ (T1/2 = 1,622 years) are too short-lived to define accumulation rates in slowly depositing deep-sea sediments, whereas $^{230}Th$ (T1/2 = 75,200 years and $^{231}Pa$ (T1/2 = 34,300 years) both occur in significant excess in deep-sea sediments and have appropriate half-lives. Among the latter cosmogenic nuclides, $^{10}Be$ has an ideal half-life (1.5 my) but until recently has been very difficult to measure at the low levels of its occurrence in sediments, while $^{32}S^{i}$ (T1/2 = 108 years) is too short-lived.

Carbon-14, the most commonly studied cosmogenic nuclide worldwide, also has a half-life too short (5,730 years) to be used for dating deep-sea sediments generally. It is true that several $^{14}C$ analyses by Ku and Broecker (1967) were among the first indications of the exceptionally slow sediment accumulation rates in the Arctic Ocean. For these reasons the topic of Arctic sediment chronologies tends to focus on just two radionuclides: $^{230}Th$ (and its sister nuclide $^{231}Pa$), which has a relatively uniform supply and inert geochemical behavior but has a useful time range limited to about 300 Ka, and $^{10}Be$, which has a promising half-life except there are questions regarding supply rate and, until recently, measurement. This synthesis, therefore, is concerned with these two nuclides primarily: $^{230}Th$, for which useful Arctic data are available (see previous sections), and $^{10}Be$, for which considerable potential application is foreseen.

Whether generated in the atmosphere or in the water column, the successful application of radiometric dating depends on some consistent model of incorporation of radioelement into the sediment, so that some form of the dating equation can be applied:

$$\frac{A}{A_o} = \exp(-\lambda t) \qquad (23\text{-}1)$$

where $A$ and $A_o$ are the observed and assumed original radioactivity respectively, $\lambda$ is the decay constant, and $t$ is the age of the sediment horizon.

Both $^{230}Th$ and $^{10}Be$ are generally insoluble and have short residence times in seawater, being scavenged by particulate matter sinking to the seafloor. Both nuclides are thought to have fairly uniform rates of production, $^{230}Th$ in the water column and $^{10}Be$ in the atmosphere above. The most logical model of accumulation holds that the flux of radionuclide onto the seafloor is constant, and that the concentration in the sediment varies with the rate of diluting particulate accumulation. However, because it is the particulates that are thought to scavenge both nuclides, the more commonly applied model holds that the sediment radioactivity (i.e., the ratio of nuclide to particulates) remains constant (and equation 23-1 can be applied directly).

In practice, neither model, nor proposed hybrid models (Osmond, 1979), work consistently well enough to permit the assignment of accurate ages to individual sediment horizons. Rather, the overall pattern of decay in the core usually permits a determination of a reliable average sediment accumulation rate.

There have been suggestions that biological detritus and fecal matter are the primary means of transport of radionuclides to the seafloor (Spencer et al., 1978). In this respect, the rate of scavenging in Arctic waters may be less than elsewhere; thus, the possible effects of an enhanced residence time on Arctic sediment dating is also a focus of this chapter.

## Thorium-230 Studies

### Theory

The basic model of $^{230}Th$ dating of deep-sea sediments depends on (a) the soluble, conservative, uniform distribution of uranium in seawater, and (b) the insolubility of its daughter thorium. As a consequence, $^{230}Th$ generated by $^{234}U$ is quickly scavenged by particulate matter and settles to the seafloor, so that an unsupported excess of $^{230}Th$ is found in deep-sea sediments. This excess is evident to varying degrees, especially in clays and microfossil oozes, provided that the influx of detrital sediment is not so great as to dilute the effect. In such detritus, the equilibrium condition approximately pertains, $A(^{230}Th) = A(^{234}U)$, so that if accumulation rates exceed several cm per Ka the determination of sedimentation rates by disequilibrium is usually not possible.

For the oceans as a whole, and for many individual basins studied, an approximate balance between uranium dissolved in the ocean and daughter thorium deposited in sediments has been demonstrated (Ku, 1966). Such is not the case in the Arctic Ocean. All Arctic cores studied have considerably less $^{230}Th$ than expected, based on the depth of water column above, and relative to similar cores elsewhere (Ku and Broecker, 1967; Finkel, Krishnaswami, and Clark, 1977; Herman and Osmond, 1984; Chapter 22 of this volume; Rosholt, personal communication).

The equation that describes the $^{230}Th$ balance between seawater and sediment is

$$P_{230} = D + E + F$$

where $P$ is the production rate of $^{230}Th$ in dpm/cm$^2$/Ka in a 1 cm$^2$ column of seawater, as a result of decay of dissolved $^{234}U$ ($P_{230} = Q_{234}\lambda_{234}$); $D$ is the fraction of $^{230}Th$ that decays before reaching the seafloor, in dpm/cm$^2$/Ka, which is a function of the residence time ($T_{230}$) of the $^{230}Th$ in solution or suspension relative to the half-life of $^{230}Th$; $E$ is the fraction of $^{230}Th$, in dpm/cm$^2$/Ka, which escapes from the immediate water column system and ends up elsewhere in unsampled sediments or sinks; and $F$ is the flux of $^{230}Th$ into the sediment at the base of the water column, in dpm/cm$^2$/Ka.

The normal situation is that the residence time of $^{230}Th$ is very short, much less than 1 Ka, so that $D$ is insignificant; $E$ is also not appreciable, on average, although considerable variation from site to site in a given basin is commonly observed. The usual relationship, therefore, is that $F = Q_{234}\lambda_{234}$.

There are only a few possible explanations for the observation that in the Arctic sediment water system $F < Q_{234}\lambda_{234}$:

1. The uranium concentration in Arctic waters is unusually low, and therefore $P_{230}$ is unusually low.
2. The value for $T_{230}$ is unusually high in Arctic waters, either because of lack of scavengers as well as adsorption sites on suspended particulates or because the sediment particles are so small that settling rates are extremely slow.

   By our definition, a long $T_{230}$ can result from either true solution or semipermanent suspension, and would result in a decrease in the rate of accumulation of $^{230}Th$ onto the seafloor. (Although $^{230}Th$ is not in excess in the water column, all $^{230}Th$ atoms decay, and the $D$ term refers to that fraction that decays before sedimentation.)
3. There are significant $^{230}Th$ sinks in Arctic Ocean sediments that have yet to be sampled.
4. There is a systematic loss of sediment particles from the tops of cores raised from the Arctic. By this we mean either natural winnowing by bottom currents or losses resulting from the impact of the coring device.

Of these explanations, there is no evidence for item 1 or item 3.

There are data reported by Bacon and Huh (1984) that the residence time for $^{230}Th$ may be greater in Arctic waters due to low scavenging rates. The same idea was proposed earlier by Ku and Broecker (1967). However, in order to explain the shortage of $^{230}Th$ on the seafloor, a residence time of $\approx 10^5$ yr would be required. This seems unlikely, given the usual estimate in normal ocean water of $10–10^2$ yr (Bacon and Huh, 1984).

We are left with item 4, that winnowing, either by bottom currents or loss resulting from impact of the coring device (McCoy, 1980, 1985) during coring operation, is the main reason for the deficiency of $^{230}Th$ excess in Arctic sediments. In effect, this is an explanation similar to item 3, that there are $^{230}Th$ sinks—in this case, very fine particulate matter—that have not been sampled, due to winnowing processes.

### *Methods*

Although the application of uranium series disequilibrium methods to the study of deep-sea sediments dates from early in this century, the most systematic treatment of the subject was that of Ku (1966), and most investigators since have followed the general methods he developed. The procedure employed for cores 19-8, 19-9, and 19-20 (Table 23-20, Fig. 23-14) is described by Somayajulu, Sharma, and Herman (Chapter 22).

An alternative procedure, described in

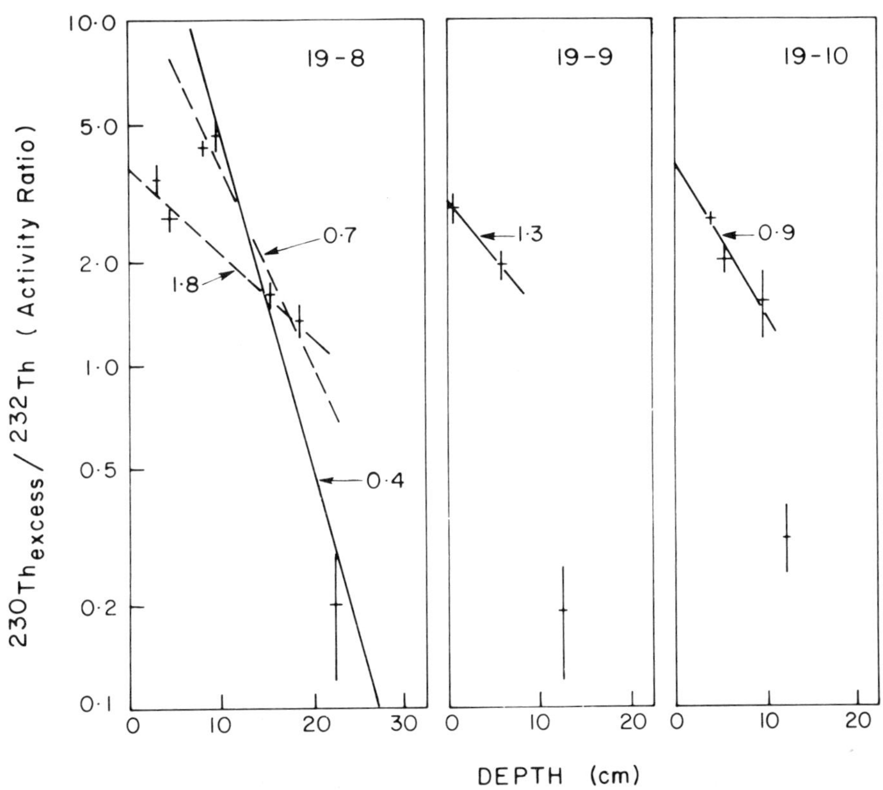


**Figure 23-14.** Examples of plots of log $^{230}Th/^{232}Th$ activity ratio vs. core horizon, showing that excess $^{230}Th$ decays with depth, but not necessarily along a strictly logarithmic line. The best-fit line slope equals the sediment accumulation rate in cm/Ka. Three possible fits are shown for 19-8, though the median rate is most likely, and is shown on Figure 23-15.

Table 23-20. Thorium Isotope Data from Arctic Ocean Sediments

| 1. *Core Number* | 2. *Depth (m)* | 3. *Sampled Horizons* | 4. $^{232}Th$ *dpm/g (Ave.)* | 5. $\frac{^{230}Th}{^{232}Th}$ *(Initial)* | 6. *Sed. Rate cm/Ka (Ave.)* | 7. $^{230}Th_{excess}$ dpm/gm³ (Initial) | 8. $^{230}$Th Acc. Rate dpm/cm²/Ka(Ave.) | 9. $^{230}$Th Prod. dpm/cm²/Ka | 10. $^{230}$Th Accum. % | 11. Ref. |
|---|---|---|---|---|---|---|---|---|---|---|
| T3-63-1 | 3,437 | 11 | 2.4 | 8.8 | .21 | 10.6 | 2.22 | 7.9 | 23 | a |
| T3-67-3 | 2,285 | 9 | 1.9 | 3.0 | .27 | 2.9 | 0.77 | 5.3 | 15 | b |
| T3-67-12 | 2,867 | 9 | 2.4 | 2.8 | .16 | 3.4 | 0.54 | 6.6 | 8 | b |
| 19-8 | 2,370 | 7 | 1.4 | 6.2 | .08 | 4.4 | 0.35 | 5.5 | 6 | c |
| 19-9 | 1,192 | 4 | 1.7 | 3 | .13 | 2.5 | 0.33 | 2.7 | 12 | c |
| 19-10 | 1,200 | 4 | 1.6 | 4 | .09 | 3.2 | 0.29 | 2.8 | 10 | c |
| 21-11 | 840 | 5 | 3.3 | 2 | <0.3 | < 3.3 | <1.0 | 1.9 | <53 | c |
| FL-277 | 2,871 | 4 | 2.6 | 9 | <0.3 | <11.7 | <.35 | 6.6 | < 5 | d |
| FL-106 | 1,856 | 3 | 0.7 | 2 | .23 | 0.7 | .16 | 4.3 | 4 | d |
| T3-66-S7 | 834 | 2 | 2.4 | 0.44 | .05 | 0.67 | .03 | 1.9 | 2 | e |
| T3-66-S8 | 864 | 2 | 2.4 | 0.17 | .50 | 0.2 | .10 | 2.0 | 5 | e |

*Notes:* Cols. 1, 2, 3, 4: Original authors' data. Col 5: From Figs. 23-14 and 23-15. Col. 6: From Figs. 23-14, 23-15, and 23-16. Col. 7: Col. 4 × Col. 5 divided by density (0.5g/cm³). Col. 8: Col. 6 × Col. 7. Col. 9: Assumes 3.1 ppb U in seawater. Col. 10: Col. 8/Col. 9 × 100. Col. 11 (a) Ku, 1966; (b) Herman and Osmond, 1984; (c) Chapter 22 of this volume; (d) Finkel and Krishnaswami, 1977; (e) Rosholt, unpublished.

detail by Cochran and Osmond (1974), uses gamma-ray spectrometry rather than alpha-ray spectrometry. Cores T3-67-3 and T3-67-12 (Table 23-20) were studied by this method (Herman and Osmond, 1984). Although individual data points are probably less reliable (short-lived daughter isotopes are measured, rather than $^{230}$Th and $^{232}$Th themselves, and the supported fraction of $^{230}$Th must be inferred indirectly), data can be collected more easily, and the results of the two methods are usually comparable (Cochran and Osmond, 1974).

The use of $^{231}$Pa, daughter of $^{235}$U, with a half-life of 35 Ka, sometimes has been used to confirm $^{230}$Th-derived sedimentation rates (Ku, 1966). This is the case with core T3-67-1 (Ku and Broecker, 1967; Table 20); in this core $^{230}$Th and $^{231}$Pa decay plots yield similar sediment accumulation rates.

There are two approaches to the determination of $^{230}$Th flux in sediments: one based primarily on the summation of $^{230}$Th excess in the core, and the other based primarily on the sediment accumulation rate as deduced from a best-fit decay curve. Both approaches assume steady-state conditions.

In the first approach, the rate of deposition of $^{230}$Th, in activity units, is assumed to be equal to the rate of decay of the total excess $^{230}$Th in the sediment column ($F = Q_{230}\lambda_{230}$). To make this summation, the $^{234}$U in the sediment and the sediment density must be known. In some cases, sampling is continuous in the core; in others, estimates must be made for unsampled intervals. If the $^{230}$Th excess data do not form a smooth curve, such estimates may be difficult to make. For this reason, the summation approach is not used here as a means of calculating *F* values.

The second approach, illustrated in Table 23-20, is based on the relationship

$$F(\text{dpm/cm}^2/\text{Ka}) = A_o(\text{dpm/cm}^3) \times S(\text{cm/Ka}) \qquad (23\text{-}2)$$

This states that the average rate of deposition of unsupported $^{230}$Th excess (flux) is equal to the $^{230}$Th excess in contemporary sediment times the average rate of sediment accumulation. The advantage of this approach is that most $^{230}$Th studies are aimed at determining the average sediment accumulation rate, and this is usually the most reliable output of such studies. Often, the value of *S* is determined by plotting $^{230}$Th(xs)/$^{232}$Th ratio or $^{230}$Th(xs)/$^{231}$Pa(xs) ratio versus depth in the core (Fig. 23-15). Such ratio data tend to produce smoother decay curves than do the plots of $^{230}$Th excess (Fig. 23-16) itself (Osmond, 1979; Chapter 22). All of the data points are used to deduce both the average sedimentation rate and the zero-age value of the plotted parameter. If, as in most of the cases cited in Table 23-20, the ratio $^{230}$Th/$^{232}$Th is used, a representative $^{232}$Th dpm/g value is necessary to convert to the desired $^{230}$Th excess value. As in approach 1 above, a value for average sediment density is also required (to convert dpm/g to dpm/cm$^3$).

### *Results*

Eleven Arctic cores have been analyzed in five studies in sufficient detail to permit useful estimates of sediment accumulation rates and $^{230}$Th fluxes (Tables 23-1, 23-20). For these cores, representative values for *F* and *S* can be calculated, by method 2 above (Table 23-20 and Figure 23-17).

Reduced to these basic parameters, there seem to be some systematic variations in *S* and *F* in Arctic Ocean cores. In general, higher $A_o$ values are associated with higher sedimentation rates (the data for two inconsistent cores, Fl-106 and T3-66-S8, are based on only three and two sample points, respectively). This means that high flux rates of $^{230}$Th result from both increased sediment accumulation rates and higher $^{230}$Th activity in the sediments; conversely, slowly accumulating sediments seem to have lower concentrations of $^{230}$Th.

The pattern noted above is consistent with the winnowing scenario in which the

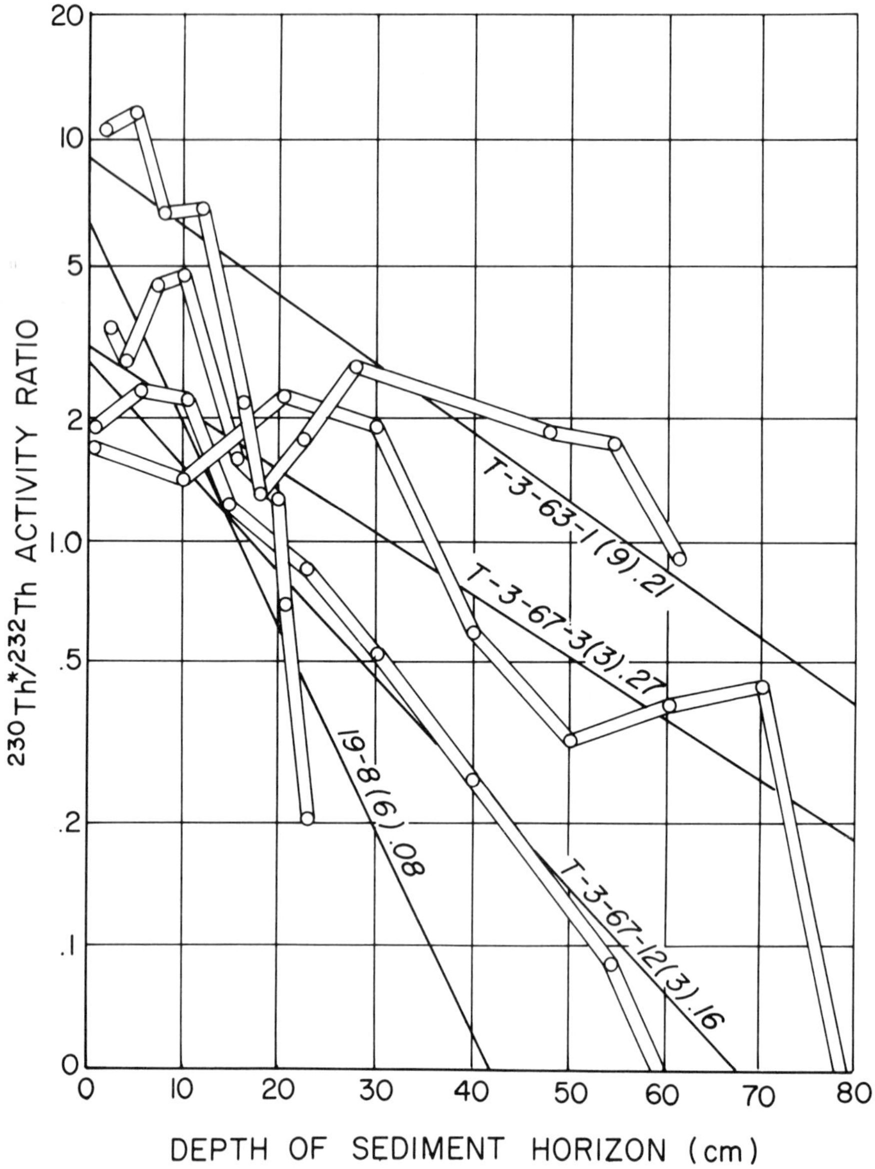


**Figure 23-15.** $^{230}Th/^{232}Th$ activity ratio vs. depth in four well-studied cores. The asterisk (*) denotes excess $^{230}Th$, unsupported by $^{234}U$ in the sediment. For each core, segmented stick graphs show the actual pattern of decay, and the associated straight line is the best-fit logarithmic decay function. The latter is used to determine the contemporary activity ratio (in parentheses) and the average sediment accumulation rate (last figure, in cm/Ka).

$^{230}Th$ excess, associated with fine particulate matter, may be reworked from place to place by currents. Sites susceptible to winnowing will exhibit both low *S* and *A* values. If such is the case, other sites downcurrent and/or at greater depth should exhibit higher values of *F*, *A*, and *S*. In fact, the deepest core studied, T3-63-1, has the highest of these values, although its flux of 2.2 dpm/cm²/Ka is still only 28 percent of the expected 7.98 dpm/cm²/Ka, based on water depth. A plot of all cores studied

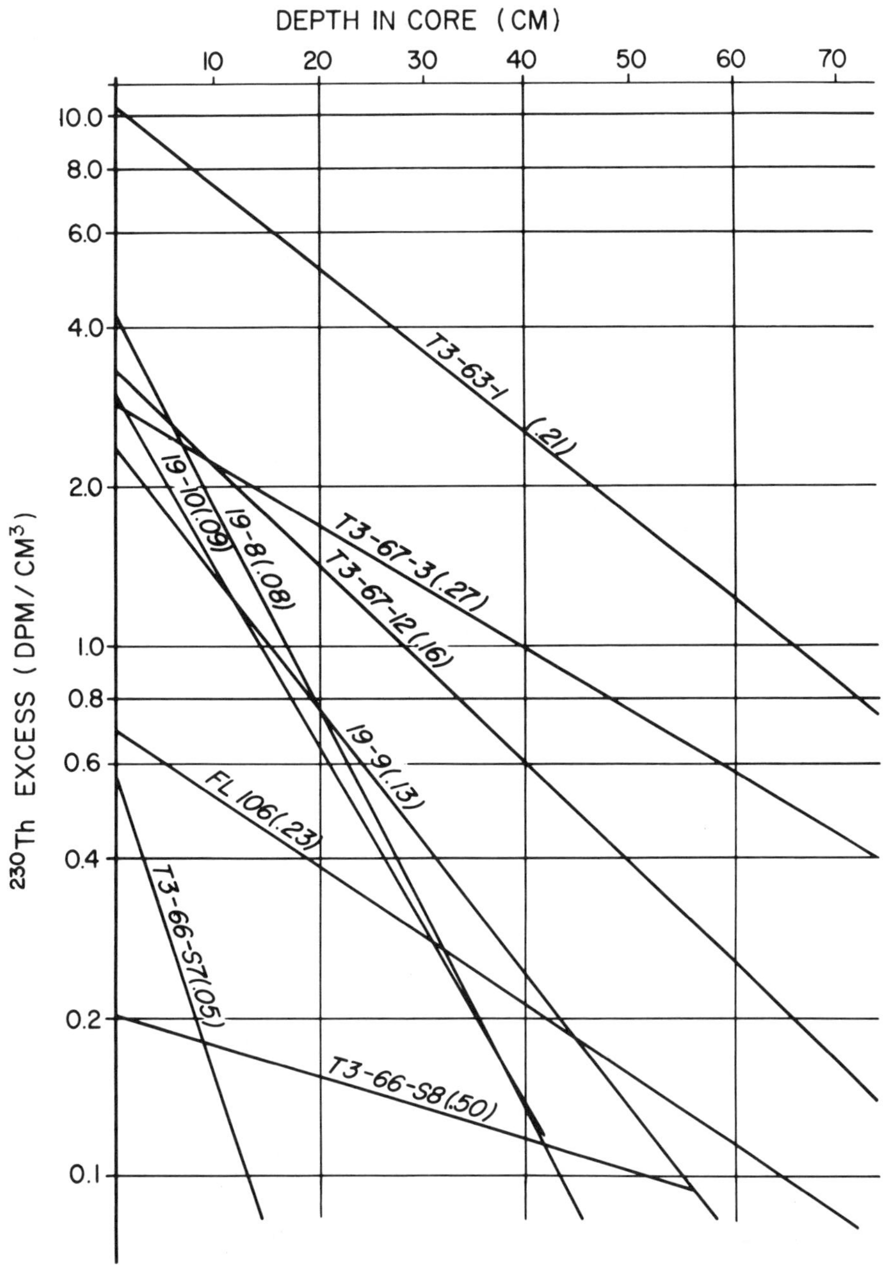


Figure 23-16. Best-fit logarithmic plots for nine cores with more or less regular decay patterns. The zero intercept for each is the $^{230}Th$ excess activity in contemporary sediment, and the slope is the average sediment accumulation rates in cm/Ka (in parentheses).

does show a pattern of increasing $F$ with depth (Fig. 23-18). The plotted line, $F = KD^3$, is not a best-fit line, but merely suggests that a sort of exponential relationship might pertain.

Four of the Arctic cores studied have been analyzed in sufficient detail (7–11 data points) so that the regularity of sediment accumulation and $^{230}Th$ flux at a given site can be investigated. In Figure 23-

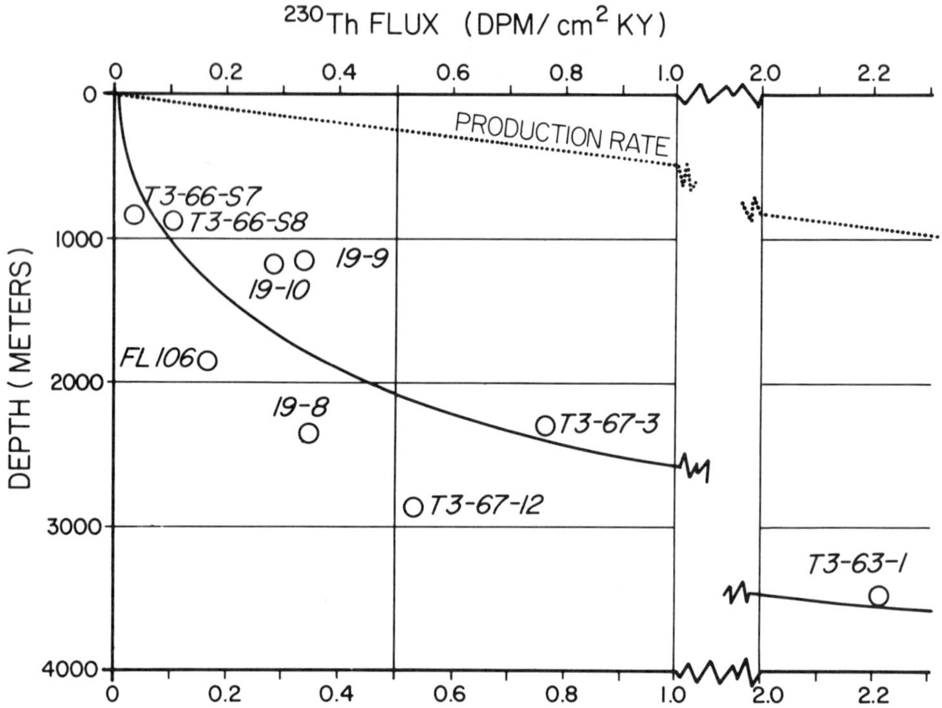


**Figure 23-17.** Computed $^{230}Th$ flux in nine cores relative to ocean depth (from Table 23-20 and Fig. 23-16). The dotted line shows the expected flux based on the $^{230}Th$ production rate by decay of $^{234}U$ dissolved in the water column. The curved line has the equation $F = KD^3$.

16 the authors' $^{230}Th/^{232}Th$ data versus core horizon are plotted. One core, T3-67-12, displays a fairly smooth logarithmic decay curve, but the other three exhibit a more irregular pattern. In general, the observed variations exceed analytical uncertainty.

Equation 23-2 can be written in the form $S = F/A$, so that it can be used to deduce the sediment accumulation rate at any given core horizon (Osmond, 1979). To do this, we convert the ratio data to $^{230}Th_{excess}$, and noting that $F = Q_{230}\lambda_{230}$, integrate the total amount of $^{230}Th$ below a given core horizon. Inasmuch as both $Q^{230}$ and $A^{230}$ decay at the same rate, the ratio of $F_i$ to $A_i$ at any horizon yields a value for $S_i$, the sedimentation rate at a given horizon at the time of deposition. The assumption is that $F$ is more or less constant, and that the changing $A$ values reflect changes in $S$ rate. Low values of $A.R$ (Fig. 23-18) are produced by dilution of $^{230}Th$ in times of fast sedimentation. The age equation of the horizon, by this model, is

$$Q_{230i}/Q_{230t} = e^{-\lambda T} \qquad (23\text{-}3)$$

Figure 23-18 shows the results of computation of $S$ versu $T$ for the four most studied cores. There appears to be no systematic basinwide correlation of sedimentation rates; for example, the high $S$ excursions of T3-63-1 at 100 Ka and those of T3-67-3 at 200 Ka are not duplicated by the others.

On the other hand, it may be significant that in all four cores, the apparent $S$ rate at 0–50 Ka is greater than at 250–350 Ka. Such a constant flux interpretation may not seem convincing when the flux observed in the sediment is only a small fraction of the $^{230}Th$ production rate in ocean water above. Nevertheless, the curves in Figure 23-18 represent, at least, the relative accu-

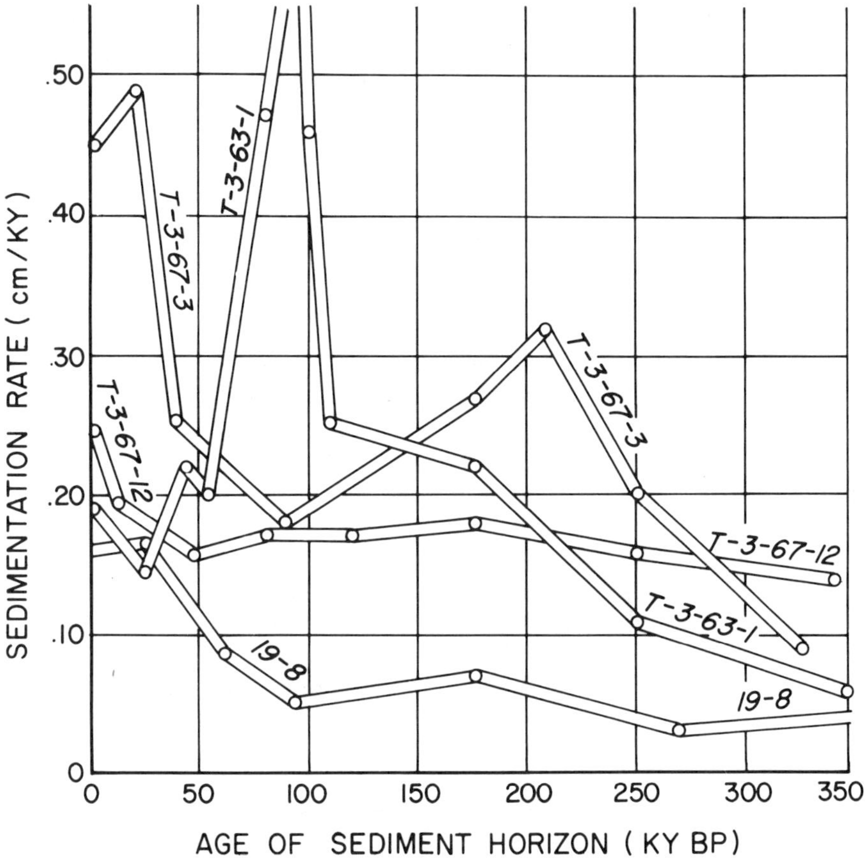


**Figure 23-18.** Graph of inferred sedimentation rate vs. age of sediment horizon for four well-studied cores. *S* (sediment accumulation rate) and *T* (age of deposition) are calculated on the basis of a constant flux model.

mulation rates of $^{230}Th$ (or $^{230}Th$-bearing fine sediment) compared to diluting inert sediments. In the period before 250 Ka, relative to the last 50 Ka, the accumulation rate of diluting sediment was less relative to $^{230}Th$, or the amount of winnowing of the radioactive fines was diminished. Conversely, there has been a greater dilution by inert sediments and/or enhanced winnowing of $^{230}Th$-bearing fine sediment.

## Beryllium-10 Studies

### *Theory*

In the past few years, sufficient data on $^{10}Be$ has accumulated to provide a fairly clear picture of the atmospheric production and the geochemistry of this important nuclide with its significant 1.5 million year half-life. The production rate of $^{10}Be$ was originally calculated by Lal and Peters (1967) to be $4.5 \times 10^{-2}$ atoms $cm^{-2}s^{-1}$. It was subsequently revised to $1.8 \times 10^{-2}$ atoms $cm^{-2}s^{-1}$ by Amin, Lal, and Somayajulu (1975), with the availability of new cross-section data (Amin et al., 1972; Raisbeek and Yiou, 1974).

Cosmic-ray-produced $^{10}Be$ has a residence time of only a few years in the atmosphere; it is brought down primarily by rain and snow. Its residence time in the ocean is also very short (<1,000 years) compared to its half-life. Thus, most of the $^{10}Be$ inventory on earth is in deep-sea sediments. Settling particles, both mineral and biogenic in origin, are the most likely scavengers of $^{10}Be$ from the seawater column.

On the assumption that the $^{10}Be$ production rate is constant, which in turn implies that the cosmic ray and earth's magnetic field intensities also remain constant over the dating interval, and that there is no postdepositional migration of $^{10}Be$ in the sediment column, it should in principle be possible to date ocean sediments. In fact, many of the sediment cores used for $^{10}Be$ measurements, though limited in number, do show a decrease of $^{10}Be$ with depth, yielding accumulation rates, despite *a* ± 30 percent variation (Amin, Lal, and Somayajulu, 1975; Tanaka and Inoue, 1979; Tanaka, Inoue, and Huang, 1982). Sharma, Bhattacharya, and Somayajulu (1983) show that the observed variation/scatter is attributable to cosmic ray and geomagnetic field intensity variations. Somayajulu (1977) suggests that it is also possible, in the Arctic or the Antarctic, where intense ice melting took place in the past, that a noticeable $^{10}Be$ "melting" spike could be present in the sediment deposited in those areas.

The principle of dating marine sediments using $^{10}Be$ is the same as that using $^{230}Th$. The logarithm of $^{10}Be$ activity ($\alpha$) versus sample depth (Z) is plotted and the slope of the best-fit through the data points is

$$\frac{d\ln\alpha}{dZ} = \frac{-\lambda}{S} \qquad (23\text{-}4)$$

where $\lambda$ is the decay constant of $^{10}Be$ ($4.62 \times 10^{-7}$ $yr^{-1}$), and $S$ is the sedimentation rate.

## Methods

Rigorous radiochemistry followed by low background beta counting were employed to measure the beta radiation emitted by $^{10}Be$ ($E_{max}$ $\beta^-$ = 0.555 MeV) by early workers (Arnold, 1956, and Goel et al., 1957). Large samples, of the order of 100 g (dry weight), are required for this method in order to be able to detect one decaying atom every minute. Consequently, this method found very limited application in dating marine sediments. Even a refined beta counting system (Fujita et al., 1975) could not decrease the sample size to less than a few tens of grams. To date, only about a dozen cores have been analyzed for $^{10}Be$. With the advent of accelerator mass spectrometry (Raisbeck et al., 1978) the sensitivity of detection of cosmogenic $^{10}Be$ has increased by as much as a factor of 1 million, and it is now being used for dating long piston cores and deep sea drilling cores (Ku et al., 1984; Mangini et al., 1984).

The accelerator mass spectrometer consists of an ion source that generates $BeO^-$ ions by sputtering the BeO target (~ 0.5 mg BeO recovered from a few grams of sediment) with a $Cs^+$ beam. The $^9BeO^-$ or $^{10}BeO^-$ beam is selected and injected into an accelerator (linear accelerator or cyclotron can be used). The beam is accelerated to the high-potential terminal where it is stripped to $Be^{2+}$, $Be^{3+}$, or $Be^{4+}$ state, depending on the terminal voltage of the accelerator. Boron, which is found to interfere with the $Be^{2+}$ and $Be^{3+}$ states, can be removed chemically as well as by Al or Ni absorber foils (Nelson et al., 1984). The Be ions are then accelerated and sent to the detection system, which consists of a Faraday cup for $^9Be$ detection and $\Delta$E-E detector for $^{10}Be$ (Turekian et al., 1979). $^{10}Be$ measurements can now be made routinely on a few grains of marine sediment to a precision of ≤ 10 percent (Mangini et al., 1984). Reference is made to Brown (1984) for an up-to-date review on the applications of $^{10}Be$ to earth sciences.

## Results

Finkel, Krishnaswami, and Clark (1977) attempted to measure $^{10}Be$ in two Arctic sediment cores using the decay counting method. In the 0–30 cm section of Fl-277 (88°35′N; 149°27′W; water depth 2,871 m) they found measurable $^{10}Be$, $2 \times 10^9$

atoms/g. In three sections of the second core, Fl-106 (77°38′N; 173°43′W; water depth 1,856 m) only upper limits $< 10^9$ atoms/g could be set, inasmuch as the $^{10}Be$ activities were not statistically significant above their detector backgrounds. However, the deposition rate of $^{10}Be$ calculated from their data agrees very well with that deduced by McCorkell, Fireman, and Langway (1967) based on Greenland ice measurement, namely $0.6 \times 10^{-2}$ atoms $cm^{-2}\,s^{-1}$. Finkel, Krishnaswami, and Clark (1977) also found that in general the $^{10}Be$ fallout pattern resembles that of $^{90}Sr$. The near-polar $^{10}Be$ deposition rate is much lower than the global average fallout rate of $5 \times 10^{-2}$ atoms $cm^{-2}\,s^{-2}$ obtained recently by Somayajulu et al. (1984) and the assessed global mean production rate of $(3.8 \pm 0.8) \times 10^{-2}$ atoms $cm^{-2}\,s^{-1}$ (Monaghan, Krishnaswami, and Turekian, 1986).

More recently, Somayajulu et al. (1985) measured six levels of an argillaceous core located north of Svalbard (80°35′N; 10°04′E; water depth 916 m). The $^{10}Be$ concentration varied from $2.46 \times 10^8$ to $7.34 \times 10^8$ atoms/g. The two extreme variations are believed to have been caused by the presence and absence (melting) of ice, respectively.

## DISCUSSION

Our synthesis of new and previously reported $^{230}Th$ data confirms the radiometric observations of Ku and Broecker (1967), Finkel, Krishnaswami, and Clark (1977), and Herman and Osmond (1984), also supported by magnetic stratigraphy and palynology (Linkova, 1965; Hunkins et al., 1971; Aksu, 1985*a*; Aksu and Mudie, 1985, and references therein; Herman and Kusumgar, in preparation), as well as by amino acid epimerization (Macko and Aksu, 1986, and references therein), that sediment accumulation rates in our study area of the Arctic Ocean are lower than the norm for other oceans. We also conclude that the flux of $^{230}Th$ onto the Arctic seafloor is about 1/8, by weighted average, of the production rate $^{230}Th$ in seawater. Review of the few available beryllium data suggests that Arctic sediments are similarly deficient in $^{10}Be$ relative to calculated fallout rates.

Coarse particulates, perhaps fecal matter, are the normal scavengers of $^{230}Th$ and $^{10}Be$ (Spencer et al., 1978). Lacking such scavengers in Arctic waters, the radionuclides probably remain in solution longer and attach to finer suspended particulates. The result is that they would have longer effective residence times in the Arctic Ocean, so that lateral transport becomes a more important factor. Once deposited, these fine particulates would be vulnerable to winnowing by bottom currents, or by the impact of a coring tool. There is an indication (Figure 23-18) that both sediment and radionuclide accumulation rates are greater in deeper water. It seems likely that further study of the deepest parts of the Arctic basins might tend to restore some of the ''lost'' $^{230}Th$ to the regional equilibrium balance.

Higher than average $^{10}Be$ concentrations in certain sediment horizons may be due to episodic melting of the Greenland ice cap and other circum-Arctic ice. Higher than average $^{230}Th$ concentrations in younger parts of the cores may be due to increased sediment accumulation rates during the last 50,000 years.

## SUMMARY AND CONCLUSIONS

The evolution of Arctic oceanic climatic regimes during the last 4.5–5 my appears to have been linked to major global climatic, oceanographic, and tectonic events. The terminal Miocene temperature decline culminating in the marked expansion of the Antarctic ice sheets and coeval sea-level lowering (Hayes and Frakes, 1975), as well as in the isolation of the Mediterranean

Sea, approximately 6.2 million years ago, may have been synchronous with the onset of small-scale high-latitude northern hemispheric glaciation. This event was followed by a warm interval. North Pacific microfaunal data suggest 6°C higher surface water temperatures between ≈4 mya and 5 mya than today preceded by an extended cold period (Ingle, 1973). North Atlantic paleotemperatures follow a similar trend; the 5.5–6.5-my cold interval is replaced by ≈ 0.6 my of fluctuating temperatures. Approximately 4.9 mya a warm period punctuated by short colder intervals ensued; it persisted for nearly 1 my (Hooper and Weaver, 1987; Keigwin, Aubry, and Kent, 1987). The oxygen isotope data for the 2.5–4-my interval indicate numerous cold–warm cycles with a $\delta^{18}O$ amplitude of ≈ 1‰ (0–1‰) recorded by planktonic foraminifer *G. bulloides* (Keigwin, 1987).

The Arctic sedimentary record commences ≈4.5–5 mya, if we accept the "long" chronology based on our uranium series isotopes, Macko and Aksu's amino acid data (1986), $^{14}C$, and the magnetic stratigraphy (Linkova, 1965; Hunkins et al., 1971; Aksu, 1985*a*; Herman and Kusumgar, in preparation). We did not use Clark's magnetic stratigraphy because of the many inconsistencies in his interpretation. For example, core 224, with one of the best polarity records according to Clark, was interpreted to represent more than 5.5 my (1969, in Steuerwald, Clark, and Andrew, 1968). However, in his 1970 and 1971 papers as well as in later publications the same core ends in Gauss, thus representing only 3 my. Still a different time frame is given for core 224 by Darby et al. (Chapter 24; see also Aksu, 1985*a*). This is a time of global warming (Ingle, 1973; Hooper and Weaver, 1987). Even if the Arctic was covered with sea ice prior to the deposition of Unit III's red clays, a warming of 6°C of the North Pacific and Atlantic Surface Water must have required much higher atmospheric temperatures. According to Budyko (1977), thermal conditions in the Arctic are closely related to those in upper water layers at high latitudes in other oceans and polar sea ice will melt in several years when atmospheric temperatures increase by 4°C during the warm season. The Arctic sedimentary record for the last ≈2.5–5 my is scanty (Unit III); it is contained in the bioturbated, FeMn concretions bearing rich red clays.

On today's deep ocean floor, in areas removed from submarine hydrothermal or volcanic activity, manganese-rich micronodules occur where biological activity is high, such as the equatorial Pacific (Revelle et al., 1955), and where rates of sedimentation are low (Greenslate, Frazer, and Arrhenius, 1973), in regions below the CCD. In these areas dissolution of calcareous and siliceous oozes results in the formation of red clays rich in iron and manganese. By analogy with present-day environments conducive to the formation of manganese-rich micronodules, the Arctic Ocean Basin concretions are also attributed primarily to biological processes in the absence of local submarine volcanic or hydrothermal activity. The important implications of organism's participation in manganese micronodule formation in the Arctic is that surface-water biological production must have been higher than today. This precludes the existence of a perennial ice cover.

The predominance of s. *G. pachyderma* complex implies cold surface-water temperatures. However, it should be noted that among the high-latitude species this taxon is the most resistant to dissolution (Malmgren, 1983) and, therefore, our paleotemperature estimates are biased toward lower temperatures. Furthermore, the presence of subpolar *G. egelida* in a shallow core (T3-67-4, 1,760 m) strongly suggests water temperatures at least seasonally above freezing point. This regime continued through the time of the opening of the Bering Strait, the emergence of the Panamanian Isthmus, and a reorganization of oceanic circulation to a more vigorous south–north current system 3.2–3.5 mya (Woodring, 1966; Crowell and Frakes, 1970; Keigwin, 1978). The Gulf Stream as we know it today is speculated to have evolved following the elevation of the Isthmus of Panama. This change in oceanic circulation may have resulted in the penetration of the warm Norway Current deep into the Arctic via the Barents Shelf. Increased transport of moist air over the North Atlantic to adjoining subpolar and polar highlands enhanced precipitation, possibly triggering the development of local ice sheets, winter coastal ice, and ice shelves over the Arctic continental shelves (Crowell and Frakes, 1970). The Barents ice sheet is also believed to have formed about 3.5 mya (Kvasov and Blazhchishin, 1978). The lower-latitude temperature fluctuations that characterize the late Miocene–middle Pliocene are not detectable in the Arctic red clays. Quartz grains in Unit III sediments exhibit surface features consistent with a cycle of glacial abrasion (Margolis and Herman, 1980). Occurrence of ice-rafted quartz sand grains, a minor but persistent component, as well as shallow-water benthonic foraminifera endemic to continental shelves, such as Elphidium (Figs. 8, 12) on the crest and upper flank provinces of the Alpha Rise, firmly establish that ice rafting was underway in the Arctic Ocean at least as early as 4.5–5 mya (Herman, 1970; Margolis and Herman, 1980).

A major oceanographic and climatic threshold was crossed about 2.5 mya at the Units II/III boundary when the residual, oxidized red silty clays, deposited below the CCD, were replaced by tan-green sandy clays with abundant coarse ice-rafted debris (Figs. 23-6, 23-7; Herman, 1970, 1974). The much greater percentages of coarse ice-rafted debris in Unit II and the sporadic high abundances of shallow-water benthic Elphidium in cores raised from isolated bathymetric highs (Figs. 23-8, 23-12) indicate that glaciers gauged material from the continental shelves and dropped it far from their source more frequently than during the earlier climatic regime (Herman, 1974, 1984*b*; Herman and Hopkins, 1980). Unit II records the inception of sharp density–salinity stratification in the Arctic, a consequence of the sudden dilution of surface water by the influx of large quantities of thawing icebergs and ice shelves coupled with intensified river runoff during an early deglaciation at high northern latitudes (Herman, 1970, 1974, 1984*a,b*). The lack of manganese micronodules and the tan-green sediment color indicate poorer oxygenation of the bottom water and decreased vertical mixing of water masses. The scarcity of planktonic microfauna during the deposition of most of Unit II's sediments (subunits IIa and IIc, Fig. 23-7) is attributed to reduced zooplankton production, due to low availability of nutrients and to decreased surface water salinities (the latter indicated by $\delta^{18}O$ depleted values) (Figs. 23-4, 23-7, 23-8) to which very few open-ocean planktonic foraminifera are adapted. Density stratification was most pronounced during the summer and early fall, when ice melting and river runoff are most intense. Nutrient recycling by upwelling must have been minimal and, consequently, plankton production was very low (Herman, 1970, 1974).

The *G. quinqueloba–G. egelida* dominance may also be related to the availability of certain nutrients. *G. quinqueloba* is considered an "opportunistic" species able to occupy environments unfavorable to other taxa (Colbourn, Parker, and Berger, 1980). The water temperatures indicated by the dominant planktonic foraminifera, subpolar *G. egelida* and *G. quinqueloba*, and peak frequencies of d. *G. pachyderma* (Figs. 23-4, 23-5, 23-8, 23-12) are incompatible with the presence of a perennial ice cover. Seasonal ice would enhance salinity stratification by reducing mixing due to wind shear and by diluting the surface low-salinity layer when the ice melted in summer (Herman and Hopkins, 1980). The influx through the Bering Strait of the low-salinity water may have contributed to the persistence of a salinity-density stratification throughout most of the time represented by Unit II.

The northern sector of the Baffin Bay was suggested as a present-day analog of the Arctic during the deposition of Unit II sediments (Herman, 1983, 1984*c*). Herman's interpretation of an ice-free Arctic during most of the time represented by Unit II is supported by the discovery of warm-water cells during the winter in the north water of the Baffin Bay (Steffen, 1986).

Interpretation of global temperatures is not unequivocal. The oxygen isotopic composition of *G. bulloides* in North Atlantic cores taken at DSDP Site 606 indicates frequent large amplitude fluctuations ranging from 0–1.6‰ commencing ≈ 2.5 mya (Keigwin, 1987). The rise in abundance of temperate–subpolar d. *G. pachyderma* between 1.6 mya and 2.35 mya in cores taken at DSDP Site 609 is remarkably similar to the late Miocene–early Pliocene warm interval recorded in the same DSDP core (Raymo, Ruddiman, and Clement, 1987; Hooper and Weaver, 1987). Plio-Pleistocene climates in eastern North Greenland were humid and much warmer than today in that region (Bennike, 1987). Shallow-water marine sediments from Northern Greenland contain abundant well-preserved invertebrate and vertebrate fauna as well as terrestrial plant remains deposited approximately 2 mya, indicating much warmer water than those prevailing today in the same region. According to Funder, Petersen, and Simonarson (1987) and Funder et al. (1985), these data preclude the existence of a perennial sea-ice cover over the Arctic. South Atlantic oxygen isotopic composition of *G. sacculifer* indicates a mild period between around 1.8 my and 2.2 my (Leonard, Williams, and Thunell, 1983). Southwest Pacific DSDP data for the 2.1–6.3 mya period suggest a cold interval between ≈ 4.3 mya and 6.3 mya followed by warm surface-water temperatures from 2.1 mya to ≈ 4.3 mya (Elmstrom and Kennett, 1985).

Judging from the oxygen isotope data and the fossil record, the Northern Hemisphere and Antarctic ice sheets appear to have reached their maximum extent and temperatures dived to lowest values about 0.9 mya (e.g., Emiliani, 1955; Selli, 1967; Herman, 1970; Bandy, Casey, and Wright, 1971; Olsson, 1971; Repenning, 1984; Repenning and Brouwers, 1987). Oceanic surface-water temperatures dropped approximately 12°C in the North Pacific (Ingle, 1973). Linked with the Pleistocene global refrigeration trend—the increased Tibetan, Himalayan, and Sierra Nevadan orogenies that reached a peak about 1 mya (Charlesworth, 1957; Valdiya, 1984)—another climatic threshold was crossed in the Arctic with the inception of a perennial sea-ice cover approximately 0.9 mya. As discussed previously, the Arctic Ocean density stratification was established ≈ 2.5 mya and a decrease in temperature of the magnitude mentioned above was more than sufficient to cover the ocean with a 2–4-m-thick ice lid (Budyko, 1977).

The development of perennial sea-ice cover, expressed by the earliest record of sediments similar to those being deposited today, with the fauna dominated by s. *G. pachyderma*, s. *G. cryophila*, and planktonic foraminifera $\delta^{18}O$ values of about 1‰, was dated about 0.9 my. This event marks the period of deposition of sediments representing the third and most recent climatic-oceanic regime, Unit I. The nearly complete disappearance of *G. egelida* and lower percentages of d. *G. pachyderma* (Herman, 1970, 1974, 1985) suggest that water temperatures dropped to present-day levels. Between ≈ 0.9 mya and 0 Ma ago foraminifera-rich and foraminifera-poor zones alternate. Based on oxygen isotope analyses three regimes are distinguished within this unit: (1) Foraminifera-rich zones (>40%), with planktonic foraminifera $\delta^{18}O$ higher than 2‰, interpreted to have been deposited in an ice-free nearly isopycnal ocean, (2) foraminifera-poor zones, with planktonic foraminiferal $\delta^{18}O$ lower than 1‰, representing periods of marked density stratification and (at least) a seasonally ice-free ocean; these are deglacial or interglacial intervals, and (3) foraminifera-rich zones with planktonic foraminiferal $\delta^{18}O$ values that range between $\approx 0.6‰$ and <2‰, representing oceanographic conditions intermediate between regimes 1 and 2 characterized by a low salinity surface layer and perennial ice cover, similar to conditions prevailing today in the Arctic Ocean. The chronology and stable isotope data of core T3-67-3 suggest that the Arctic was ice free and the pycnocline was greatly reduced during periods apparently corresponding to isotope/faunal stages 12, 8 and 6 (Fig. 23-10). As discussed in the Radiometric Chronologies section, many Arctic core "tops" are missing, including perhaps the uppermost part of T3-67-3. Therefore, the ages determined for this core should be considered tentative and correlations with paleoceanographic events at lower latitudes must await confirmation by additional absolute age determinations.

## ACKNOWLEDGMENTS

The first author is indebted to many colleagues who have been willing to share their knowledge and have helped her to formulate the ideas put forward in the present chapter as well as in previous publications listed in the References section. Philip E. Rosenberg read the manuscript and offered valuable suggestions. The cores were raised by the LDGO and Sevmorgeologia, Leningrad, USSR. I thank R. Lotti for making the Lamont-Doherty Geological Observatory (LDGO) samples available and I. S. Gramberg for the Sevmorgeologia cores. Particularly valuable were discussions with P. E. Rosenberg, G. L. Johnson, I. S. Gramberg, V. S. Zarchidze, L. V. Poliack, V. D. Dibner, M. S. Barash, P. A. Lisitzin, A. Juzé, V. Gudina, S. Kruglikova, L. P. Zonenshain, A. N. Golikov, V. T. Frolov, J. C. Duplessy, S. Krishnaswami, C. Jehanno, C. A. Repenning, and E. Brouwers. We also thank P. Sharma and J. Rosholt for radiometric data for cores T3-66-S7 and T3-66-S8, and the Sevmorgeologia cores, and R. E. Garrison for making the scanning electron microscope available for some of the foraminifera SEM, W. A. Berggren for several foraminifera SEM, S. V. Margolis for the SEM of quartz sand grains, C. Jehanno for the SEM of the ferromanganese concretions and for the chemical analyses of the concretions presented in Tables 23-3, 23-4, 23-5, and 23-6, H. Keupp for the SEM of dinoflagellates and tasmanitids and for identifying these taxa. Support for the curating facilities of the LDGO Deep-Sea Sample Repository was provided by the NSF through Grant OCE85-00232 and the ONR through Grant N00014-84-C-0132. Radiometric analyses at the Physical Research Laboratory (PRL) were supported by the Department of Space, Government of India.

## APPENDIX

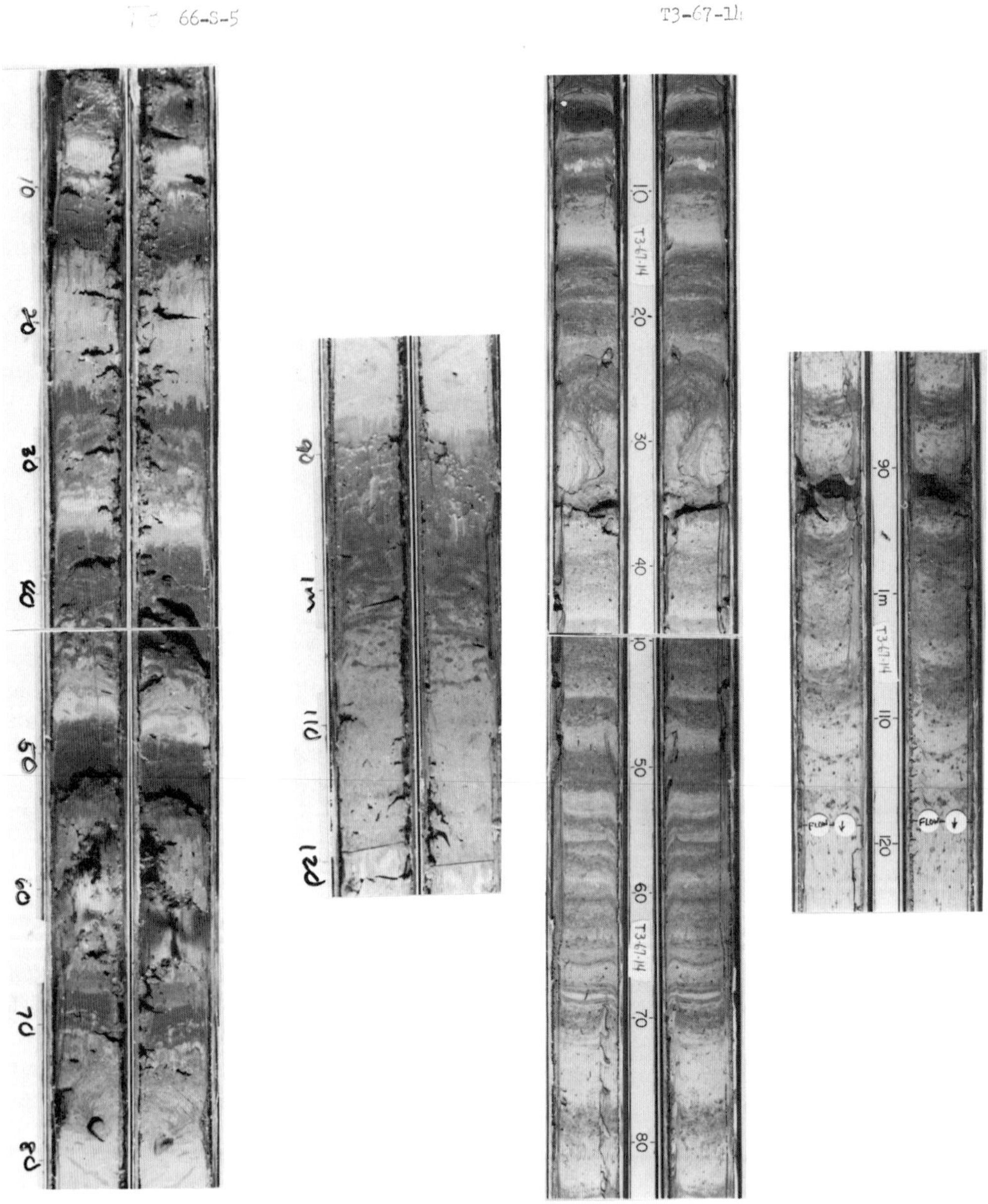


**Plate 23-I**

A. Photograph of core T3-66-S5 showing diagnostic features such as layering and burrow mottling.
B. Photograph of core T3-67-14. (Courtesy of LDGO Core Laboratory.)

## Plate 23-II

Scanning electron micrographs of quartz sand grains from Unit III.

A. Sand grain from sample 1 showing sharp angular outline, step-shaped fractures, and conchoidal fractures, typical of grains subject to glacial action; from core T3-67-9, 330 cm.
B. Edge of quartz grain from sample 2 showing conchoidal fractures, semiparallel steps, striations, and other features shown in *A*; from core T3-67-12, 375 cm.
C. Surface of grain from sample 3 exhibiting features of glacial origin, with some rounding of edges; from core T3-67-11, 250 cm.
D. Grain showing glacial features modified by eolian features along right-hand margin, and on center ridge (sample 1).
E,F. Rounded grains with subdued features of glacial origin (striations, breakage blocks, fractures). Rounding is believed to be caused by a cycle of eolian and/or subaqueous abrasion.

## Plate 23-III

Scanning electron micrographs of quartz sand grains from Unit II.

A. Grain from sample 4 showing angular outline, sharp edges, striations, steplike fractures, all of glacial origin; from core T3-67-11, 200 cm.
B. Rounded grain from sample 5; note randomly oriented ''V''-shaped pits of subaqueous origin; from core sample 7; from core T3-67-11, 188 cm.
C. Striations and steplike fractures on grain from sample 7; from core T3-67-11, 144 cm.
D, E, F. Grains showing features typical of glacial action from samples 13 (*D*) from core D.St.A.4 27 cm, 14 (*E*), from core D.St.A.4 29 cm and 5 (*F*).

## Plate 23-IV

Scanning electron micrographs of quartz sand grains from Unit I.

A, B. Grain from sample 8 showing high relief, large breakage blocks, and conchoidal fractures. *B* is higher magnification photo of upper right-hand portion of *A*; from core T3-67-11, 85 cm.
C, D, E. Grains from sample 9 showing rounded outline and smooth overgrowths similar to grains of eolian origin. These are from a foraminifera-rich zone; from core T3-67-12, 42 cm.
F. Grain from sample 12 exhibiting typical glacial surface features; from core T3-67-11, top. (Courtesy of S. V. Margolis.)

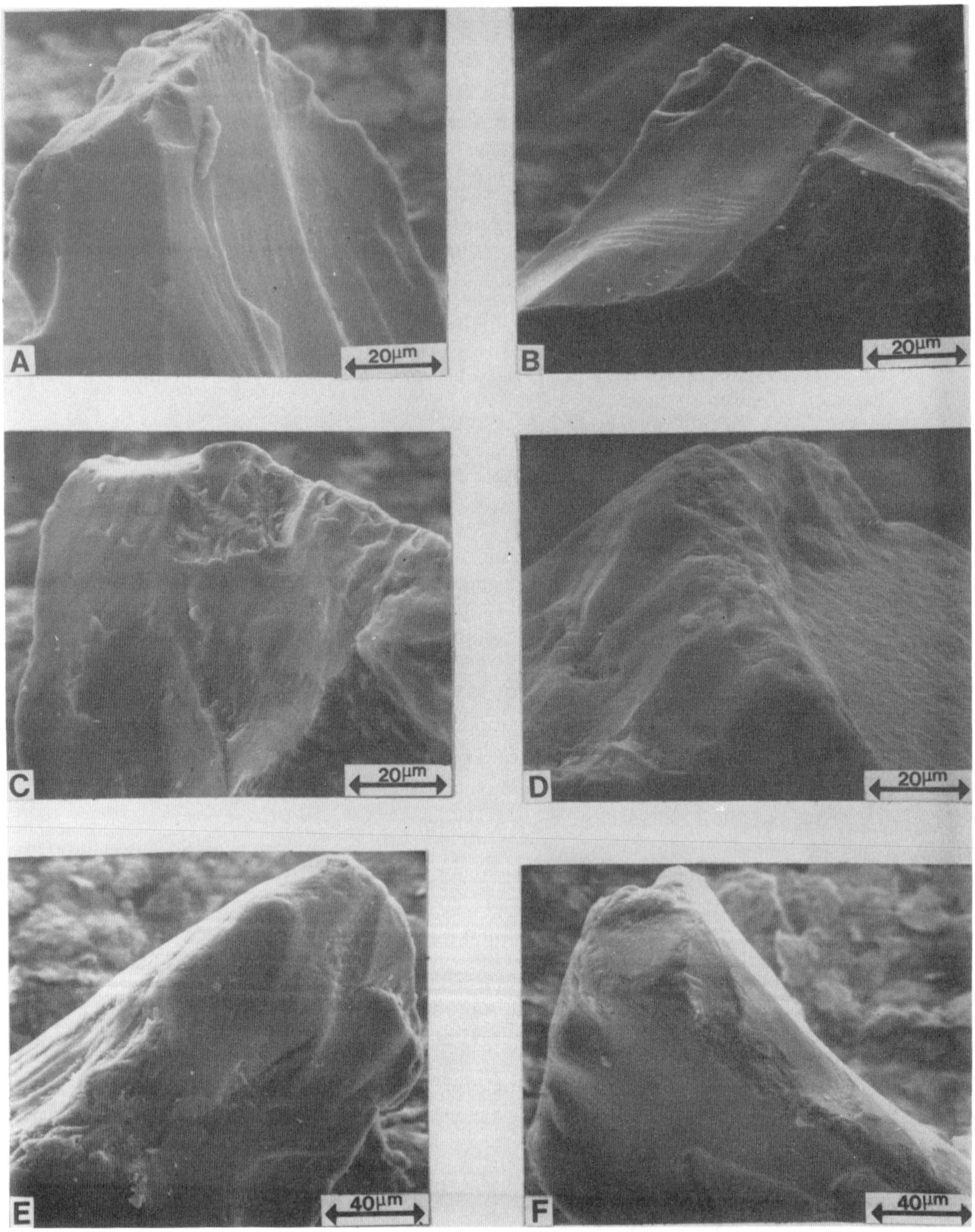


PLATE 23-II

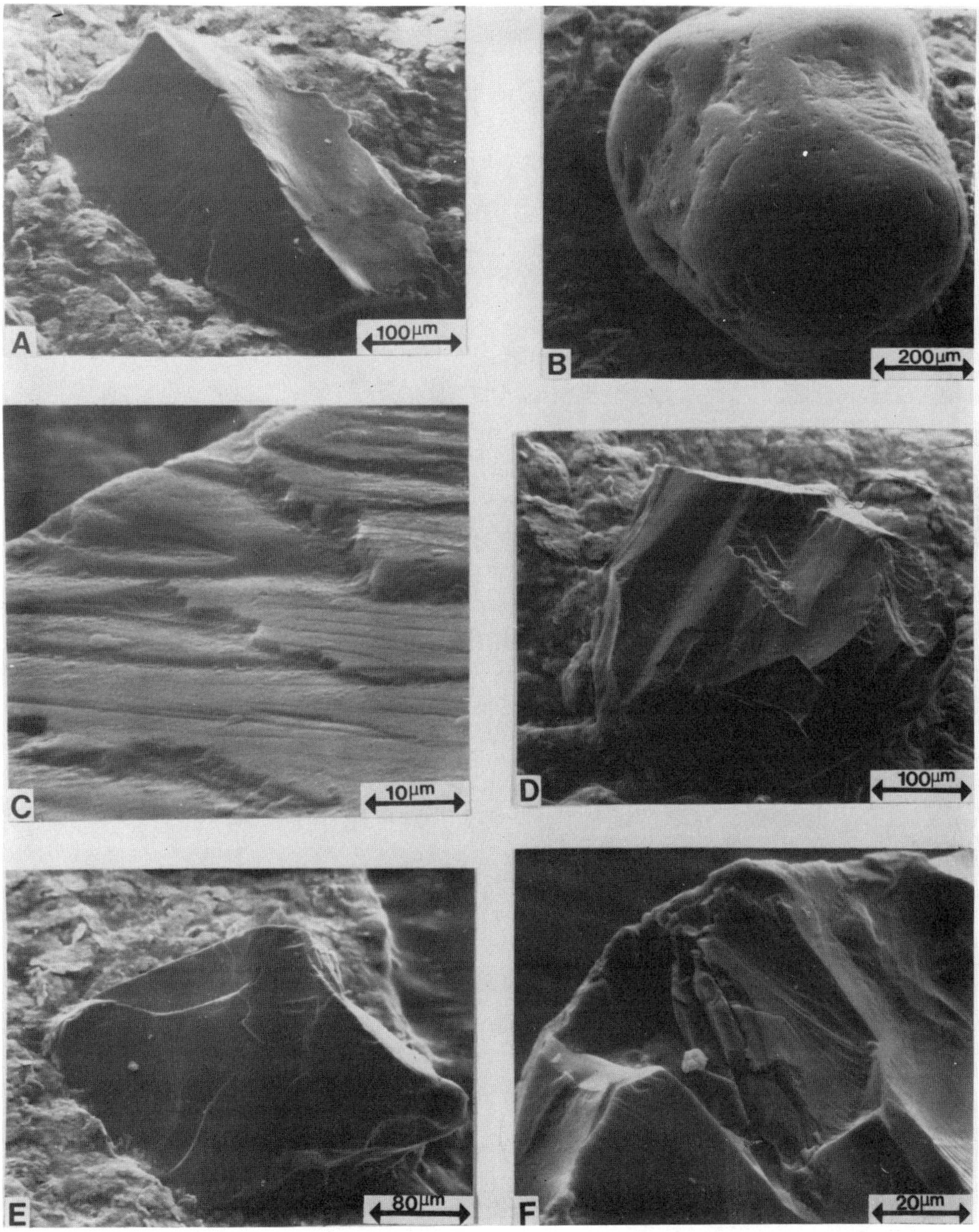


PLATE 23-III

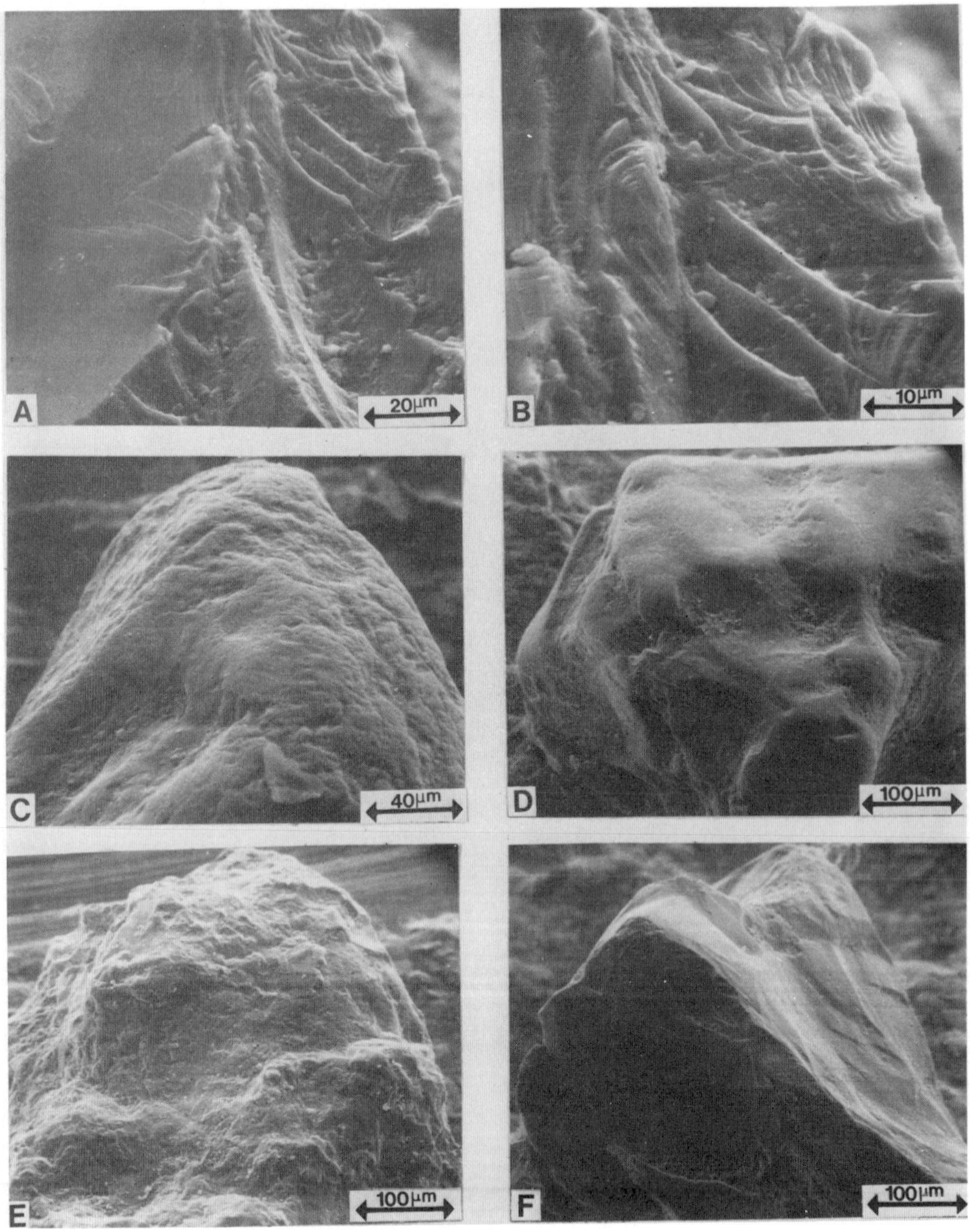


PLATE 23-IV

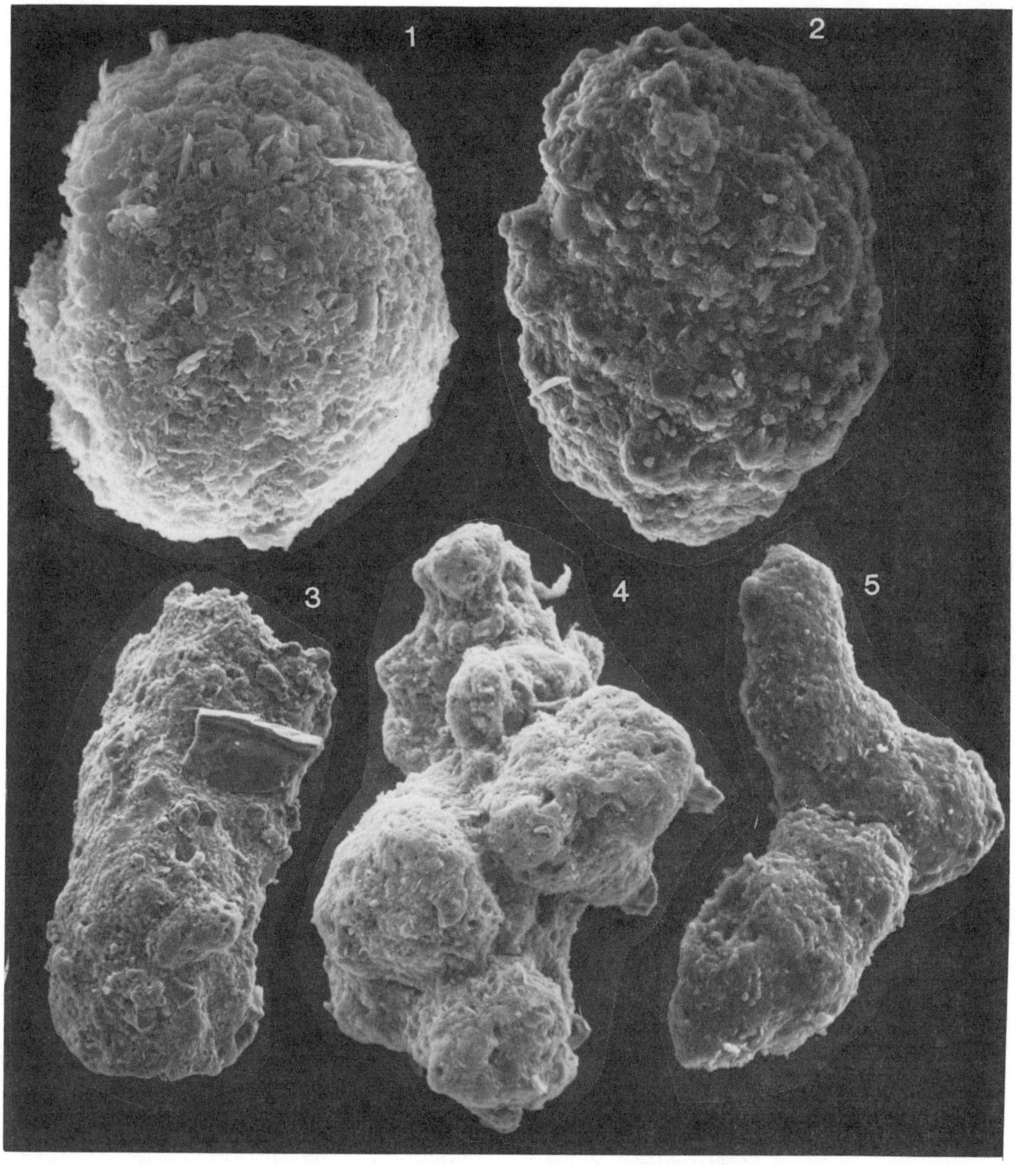


## Plate 23-V

1. Photomicrograph of ferromanganese micronodule from core T3-67-16, 99 cm; SEM × 675.
2. Photomicrograph of ferromanganese micronodule from core T3-66-S-28, 80 cm; SEM × 225.
3. Photomicrograph of ferromanganese micronodule from core T3-67-13, 252 cm; SEM × 75.
4. Photomicrograph of ferromanganese micronodule from core T3-67-16, 70 cm; SEM × 150.
5. Photomicrograph of ferromanganese micronodule from core T3-66-S-28, 80 cm; SEM × 150. (Courtesy of C. Jehanno.)

## Plate 23-VI

1. G. *pachyderma,* dextral; umbilical view, × ≈ 272, SEM from core T3-67-9, 220 cm.
2. G. *pachyderma,* sinistral; umbilical view, × ≈ 408, SEM from core T3-67-11, top.
3. G. *cryophila,* sinistral; umbilical view, × ≈ 245, SEM from core T3-66-S-8, 5 cm.
4. G. *pachyderma,* side view, × ≈ 190, SEM from core T3-66-S-8, 5 cm; sharp-edged ridges forming polygonal enclosures around the pores.
5. G. *pachyderma,* × ≈ 272, SEM from core T3-66-S-8, 5 cm, showing sharp-edged ridges.
6. G. *pachyderma,* × ≈ 374, SEM from core T3-66-S-8, 5 cm, showing sharp-edged ridges.

## Plate 23-VII

7. G. *pachyderma,* last chamber, ≈ 1,360, SEM from core T3-66-S-8, 5 cm, showing sharp-edged ridges.
8. G. *pachyderma,* dextral; dorsal view, × ≈ 408, SEM from core T3-67-11, 175 cm, showing euhedral crystals.
9. G. *pachyderma,* dextral, × ≈ 408, SEM from core T3-67-11, 20 cm, showing euhedral crystals.
10. G. *pachyderma,* sinistral, × ≈ 408, SEM from core T3-67-11, 109 cm, showing broad and rounded interpore ridges.
11. G. *pachyderma,* sinistral, × ≈ 408, SEM from core T3-67-11, 160 cm, showing partially dissolved interpore ridges.
12. G. *pachyderma,* sinistral, × ≈ 408, SEM from core T3-67-11, 160 cm, showing irregular calcareous deposits that constrict and/or obliterate the pores.

## Plate 23-VIII

1. G. *hermanae,* dextral; dorsal view, × ≈ 216, SEM from core T3-66-S-8, 5 cm.
2. G. *hermanae,* sinistral; umbilical view, × ≈ 130, SEM from core T3-66-S-8, 5 cm.
3. G. *hermanae* (same as fig. 2), × ≈ 351, SEM, showing sharp-edged ridges forming polygonal enclosures around the pores.
4. G. *hermanae* (same as fig. 2), × ≈ 540, SEM, showing same features as fig. 3.

## Plate 23-IX

1. G. *quinqueloba,* dextral; umbilical view, × ≈ 410, SEM from core T3-67-S-7, top.
2. G. *quinqueloba,* last whorl penultimate chamber, × ≈ 1,080, SEM from core T3-67-12, 192 cm, showing rounded spine bases.
3. G. *egelida,* sinistral; umbilical view, × ≈ 243, SEM from core T3-67-12, 192 cm.
4. G. *egelida,* × ≈ 2,700, SEM from core D.St.A.2 showing broken, rounded spine bases.

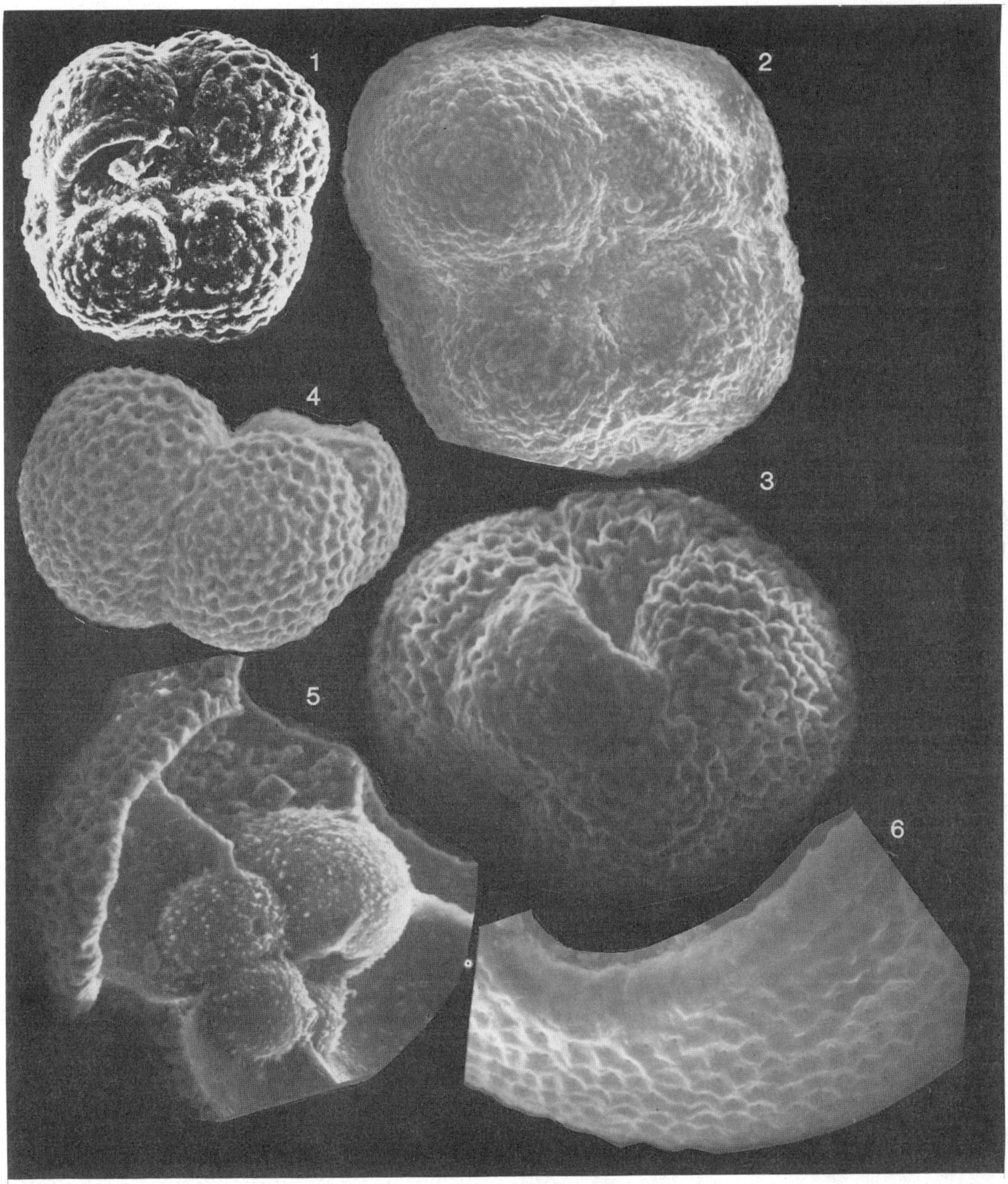


PLATE 23-VI

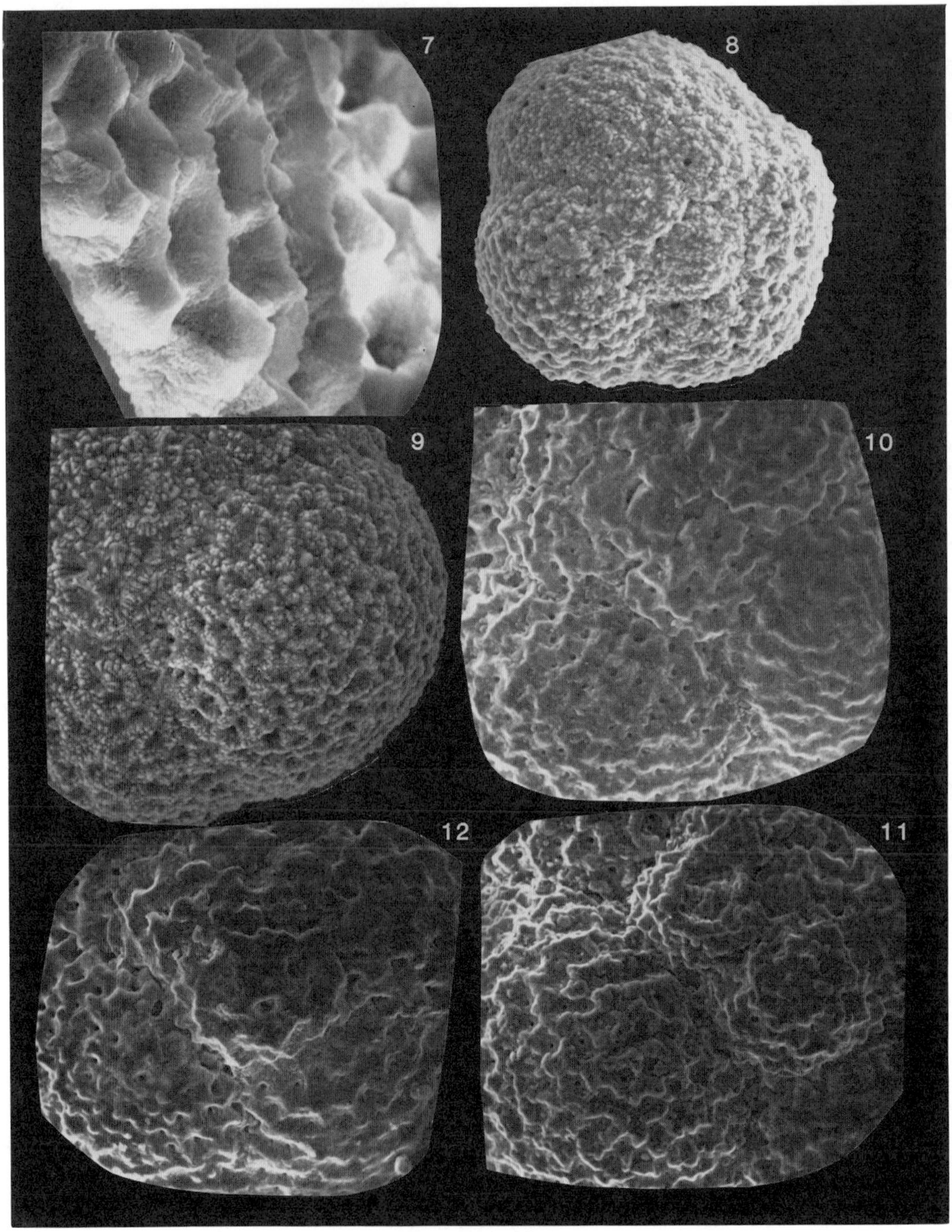


PLATE 23-VII

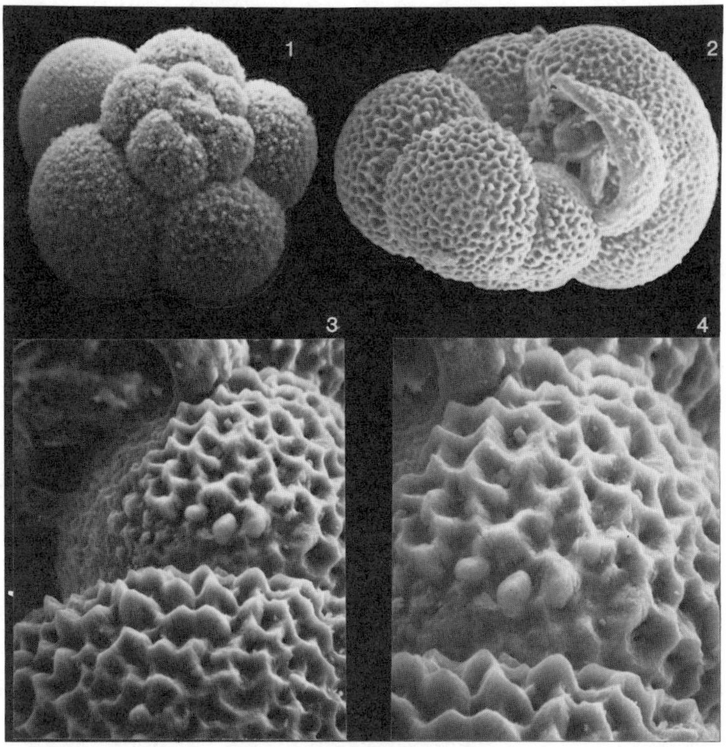


PLATE 23-VIII

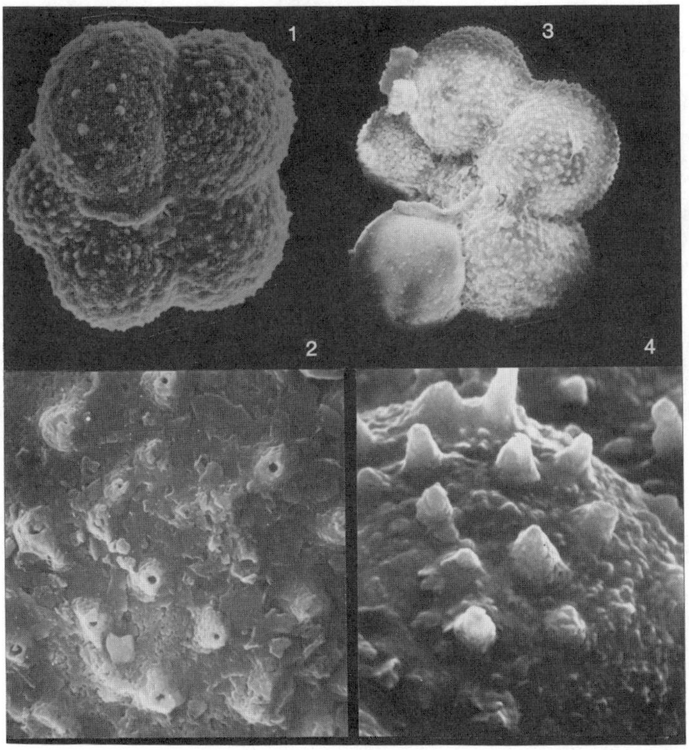


PLATE 23-IX

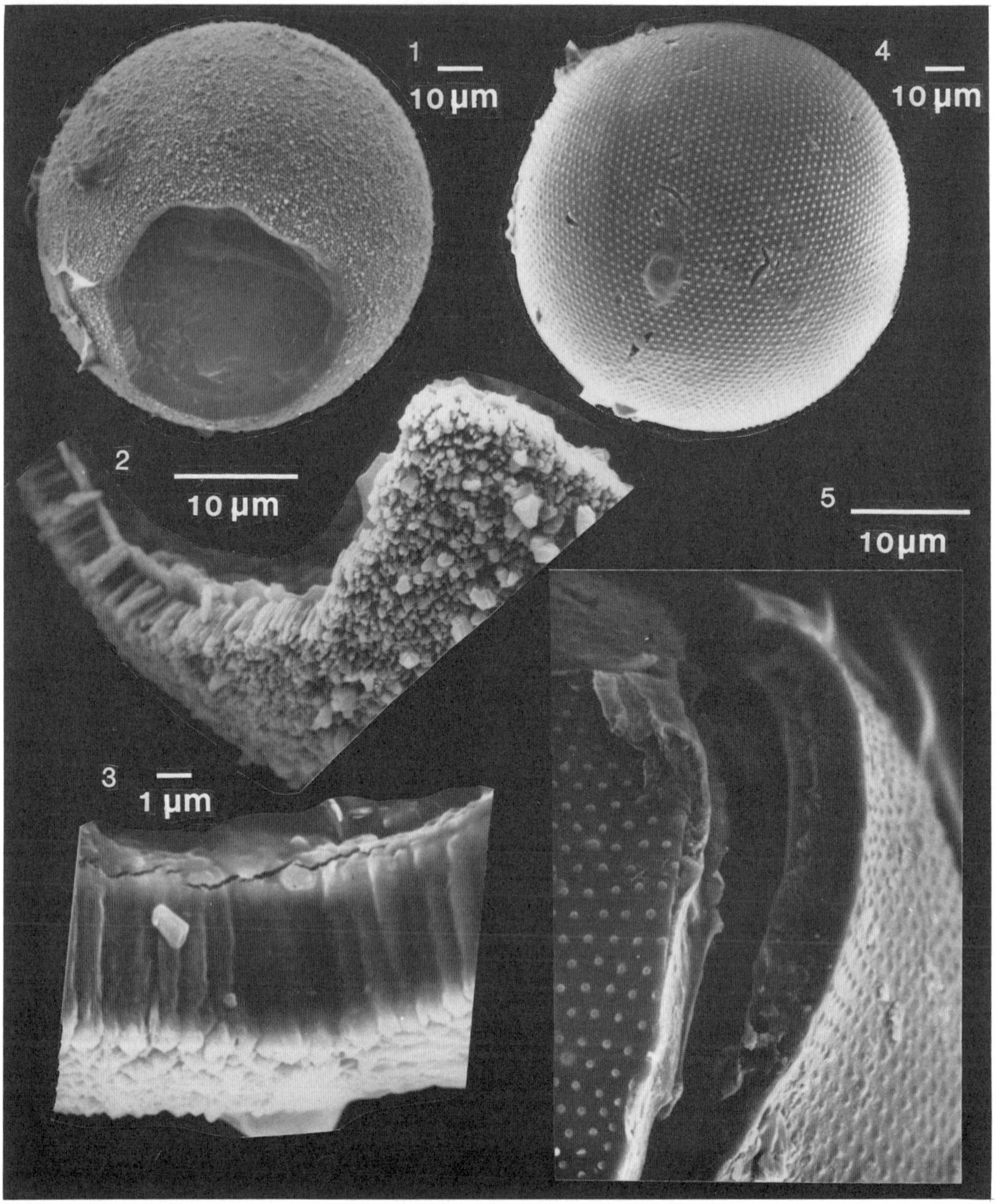


## Plate 23-X

1. Orthopithonella Keupp 1984, with cryptotabulation, from core T3-67-5, 200 cm × 600, SEM.
2. Same specimen as above, showing organic periphragm/endophragm inside; × 1,725, SEM.
3. Same specimen as above: single-layered wall with radially arranged calcite crystallites, × 4,500, SEM.
4. Tasmanites sp. (yellowish spheroid), × 525, SEM, thick-walled organic cyst from core T3-66-S-21, 70 cm.
5. Same specimen as above, showing thickness of organic wall, × 1,650, SEM. (Courtesy of H. Keupp.)

## REFERENCES


Aagaard, K. 1981. On the deep circulation in the Arctic Ocean. *Deep-Sea Res.* **28**(3A):251–268.

Aksu, A. E. 1985*a*. Paleomagnetic stratigraphy of the CESAR cores. In H. R. Jackson, P. J. Mudie, and S. M. Blasco (eds.), *Initial Geological Report on CESAR, The Canadian Expedition to Study the Alpha Ridge, Arctic Ocean.* Geol. Survey Canada Paper 84-22, pp. 101–114.

Aksu, A. E. 1985*b*. Planktonic foraminiferal and oxygen isotopic stratigraphy of CESAR cores 102 and 103: Preliminary results. In H. R. Jackson, P. J. Mudie, and S. M. Blasco (ed.), *Initial Geological Report on CESAR, The Canadian Expedition to Study the Alpha Ridge, Arctic Ocean.* Geol. Survey Canada Paper 84-22, pp. 115–124.

Aksu, A. E., and P. J. Mudie. 1985. Magnetostratigraphy and palynology demonstrate at least 4 million years of Arctic Ocean sedimentation. *Nature* **318**:280–283.

Amin, B. S., S. Biswas, D. Lal, and B. L. K. Somayajulu. 1972. Radiochemical measurements of $^{10}$Be and $^{7}$Be formation cross sections in oxygen by 135 and 550 MeV protons. *Nucl. Phys.* **195A**:20.

Amin, B. S., D. Lal, and B. K. L. Somayajulu. 1975. Chronology of marine sediments using $^{10}$Be method: Intercomparison with other methods. *Geochim. Cosmochim. Acta* **39**:1,187–1,192.

Andrew, J. A., and J. H. Kravitz. 1974. Sediment distribution in deep areas of the Northern Kara Sea. In Y. Herman (ed.), *Marine Geology and Oceanography of the Arctic Seas.* Springer-Verlag, New York, pp. 231–256.

Arnold, A. J. 1983. Phyletic evolution in the *Globorotalia crassaformis* (Galloway and Wissler) lineage: A preliminary report. *Paleobiology* **9**:390–397.

Arnold, J. R. 1956. Beryllium-10 produced by cosmic rays. *Science* **124**:584–585.

Arrhenius, G. 1963. Pelagic sediments. In M. N. Hill (ed.), *The Sea: Ideas and Observations on Progress in the Study of the Seas,* vol. 3. Interscience, New York, pp. 655–722.

Bacon, M. P., and C-A. Huh. 1984. Water column distributions of $^{230}$Th and $^{231}$Pa at three oceanographic sites (abstract). *EOS, Trans. AGU* **65**(16):227.

Bandy, O. L. 1959. Geologic significance of coiling ratios in the foraminifer *Globigerina pachyderma* (Ehrenberg). *Geol. Soc. Am. Bull.* **70**:1,708.

Bandy, O. L. 1960. The geologic significance of coiling ratios in the foraminifer *Globigerina pachyderma* (Ehrenberg). *J. Paleontol.* **34**:671–681.

Bandy, O. L., R. E. Casey, and R. C. Wright. 1971. Late Neogene planktonic zonation magnetic reversals, and radiometric dates, Antarctic to the tropics. *Biol. Antarctic Seas* (4, Antarctic Res. Ser.) **15**:1–26.

Bé, A. W. H. 1960. Some observations on Arctic planktonic foraminifera. *Contrib. Cushman Found. Foram. Res.* **11**(2):64–68.

Bé, A. W. H., and O. R. Anderson. 1976. Gametogenesis in planktonic foraminifera. *Science* **192**(4,242):890–892.

Bé, A. W. H., and W. H. Hamlin. 1967. Ecology of recent planktonic foraminifera, III: Distribution in the North Atlantic during the summer of 1962. *Micropaleontology* **13**:87–106.

Bé, A. W. H., and D. S. Tolderlund. 1971. Distribution and ecology of living planktonic foraminifera in surface waters of the Atlantic and Indian oceans. In B. M. Funnel and W. R. Riedel (eds.), *Micropalaeontology of Oceans.* Cambridge University Press, Cambridge, pp. 105–149.

Beal, M. A. 1969. Bathymetry and Structure of the Arctic Ocean. Ph.D. thesis, Oregon State University, Corvallis, 204p.

Belanger, P. E., W. B. Curry, and R. K. Matthews. 1981. Core-top evaluation of benthic foraminiferal isotopic ratios for paleo-oceanographic interpretations. *Palaeogeog., Palaeoclimatol., Palaeoecol.* **33**:205–220.

Bender, M. L., G. P. Klinkhammer, and D. W. Spencer. 1977. Manganese in sea-water and marine manganese balance. *Deep-Sea Res.* **24**:799–812.

Bennike, O. 1987. Plio-Pleistocene environment at Kap-København, Eastern North Greenland. *XII INQUA Congress, Ottawa, Abs. Vol.,* p. 127.

Berger, W. H. 1968. Planktonic foraminifera: Selective solution and paleoclimatic interpretation. *Deep-Sea Res.* **15**:31–43.

Berger, W. H. 1970. Planktonic foraminifera: Selective solution and the lysocline. *Marine Geol.* **8**:111–138.

Berger, W. H. 1973. Deep-sea carbonates: Pleistocene dissolution cycles. *J. Foram. Res.* **3**:187–195.

Berger, W. H. 1976. Biogenous deep-sea sediments: Production, preservation and interpretation. In J. P. Riley and R. Chester (eds.), *Treatise on Chemical Oceanography,* vol. 5. Academic Press, London, pp. 265–387.

Berger, W. H. 1981. Paleoceanography: The deep-sea record. In C. Emiliani (ed.), *The Oceanic Lithosphere: The Sea,* vol. 7. Wiley & Sons, New York, pp. 1,437–1,520.

Berger, W. H. 1982. Deep-sea stratigraphy:

Cenozoic climatic steps and the search for chemo-climatic feedback. In G. Einsele and A. Seilacher (eds.), *Cyclic and Event Stratification.* Springer-Verlag, Berlin, pp. 121–157.
Berger, W. H ., and G. R. Heath. 1968. Vertical mixing in pelagic sediments. *J. Marine Res.* **26**(2):134–143.
Berger, W. H., and E. Vincent. 1986. Deep-sea carbonates: Reading the carbon-isotope signal. *Geol. Rundsch.* **75**(1):249–269.
Berggren, W. A. 1971. Multiple phylogenetic zonation of the Cenozoic based on planktonic foraminifera. In A. Farinacci (ed.), *Proceedings of the II Planktonic Conference, Roma 1970.* Edizioni Tecnoscienza, Rome, pp. 41–56.
Berggren, W. A. 1977. Late Neogene planktonic foraminiferal biostratigraphy of the Rio Grande Rise (South Atlantic). *Marine Micropaleontol.* **2**:265–313.
Berry, R. W., and W. D. Johns. 1966. Mineralogy of the clay-sized fractions of some North Atlantic-Arctic Ocean bottom sediments. *Geol. Soc. Am. Bull.* **77**(2):183–195.
Bignot, G., and L. Dangeard. 1976. Contribution à l'étude de la fraction biogene des nodules polymétalliques des fonds océaniques actuels. *C.R. Somm. Soc. Geol. Fr.* **18**(3):96–99.
Boltovskoy, E. 1959. Foraminifera as biological indicators in the study of ocean currents. *Micropaleontology* **5**(4):473–481.
Boltovskoy, E., and R. Wright. 1976. *Recent Foraminifera.* W. Junk, The Hague, 515p.
Boström, K., O. Joensuu, and I. Brohm. 1974. Plankton: Its chemical composition and its significance as a source of pelagic sediments. *Chem. Geol.* **14**:255–271.
Bradshaw, J. S. 1959. Ecology of living planktonic foraminifera of the north and equatorial Pacific Ocean. *Cushman Found. Foram. Res. Contr.* **10**(2):25–64.
Bramlette, M. N., and W. H. Bradley. 1941. Geology and biology of North Atlantic deep sea cores between Newfoundland and Ireland, I: Lithology and geologic interpretations, *U.S. Geol. Surv. Prof. Paper 196A,* pp. 1–34.
Brown, L. 1984. Applications of accelerator mass spectrometry. *Ann. Rev. Earth Planet. Sci. Lett.* **12**:39–59.
Budyko, M. I. 1977. *Climatic Changes.* Waverly Press, Inc., Baltimore, 261p.
Bukry, D. 1981. Cretaceous Arctic silicoflagellates. *Geo-Marine Lett.* **1**(1):57–63.
Bukry, D. 1984. Paleogene paleoceanography of the Arctic Ocean is constrained by the Middle or Late Eocene age of USGS Core Fl-422: Evidence from silicoflagellates. *Geology* **12**(4): 199–201.
Bukry, D. 1985. Correlation of Late Cretaceous Arctic silicoflagellates from Alpha Ridge. In H. R. Jackson, P. J. Mudie, and S. M. Blasco (eds.), *Initial Geological Report on CESAR, The Canadian Expedition to Study the Alpha Ridge, Arctic Ocean.* Geol. Survey Canada Paper 84-22, pp. 125–135.
Burns, V. M., and R. G. Burns. 1978. Diagenetic features observed inside deep-sea manganese nodules from the north equatorial Pacific. *Scanning Electron Microscopy* **1**:245–252.
Caron, M., and P. Homewood. 1983. Evolution of early planktonic foraminifers. *Marine Micropaleontol.* **7**:453–462.
Carroll, D. 1970. Clay minerals in Arctic Ocean sea floor sediments. *J. Sed. Petrol.* **40**(3):814–821.
Carsola, A. J. 1952. Marine Geology of the Arctic Ocean and Adjacent Seas off Alaska and Northwestern Canada. Ph.D. thesis, University of California, Los Angeles, 226p.
Charlesworth, J. K. 1957. *The Quaternary Era.* E. Arnold, Ltd., London, 2 vols., 1,700p.
Chipman, W. A., T. R. Rice, and T. J. Price. 1958. Uptake and accumulation of radioactive zinc by marine plankton, fish, and shellfish. *U.S. Fish Wild. Serv. Bull.* **58**:279–292.
Cifelli, R. 1969. Radiation of Cenozoic planktonic foraminifera. *Syst. Zool.* **18**:154–168.
Cifelli, R. 1974. Planktonic foraminifera from the Mediterranean and adjacent Atlantic waters (cruise 49 of the *Atlantis II,* 1969). *J. Foram. Res.* **4**(4):171–183.
Cifelli, R. 1982. Textural observations on some living species of planktonic foraminifera. *Smithsonian Contr. Paleobiol.* **45**:1–45.
Cifelli, R., and R. K. Smith. 1970. Distribution of planktonic foraminifera in the vicinity of the North Atlantic Current. *Smithsonian Contr. Paleobiol.* **4**:1–52.
Cifelli, R., and R. K. Smith. 1974. Distribution patterns of planktonic foraminifera in the western North Atlantic. *J. Foram. Res.* **4**(3):112–125.
Clark, D. L. 1969. Paleoecology and sedimentation in part of the Arctic Basin. *Arctic* **22**(3):233–245.
Clark, D. L. 1970. Magnetic reversals and sedimentation rates in the Arctic Ocean. *Geol. Soc. Am. Bull.* **81**(10):3,129–3,134.
Clark, D. L. 1971. Arctic Ocean ice cover and its Late Cenozoic history. *Geol. Soc. Am. Bull.* **82**(12):3,313–3,323.
Clark, D. L. 1974. Late Mesozoic and early Cenozoic sediment cores from the Arctic Ocean. *Geology* **2**(1):41–44.
Clark, D. L. 1975. Geological history of the Arctic Ocean Basin. In C. J. Yorath, E. R. Parker, and D. J. Glass (eds.), *Canada's Continental*

*Margins and Offshore Petroleum Exploration.* Can. Soc. Petrol. Geol. Mem. 4, pp. 501–524.

Clark, D. L. 1977*a*. Climatic factors of the late Mesozoic and Cenozoic Arctic Ocean. In M. Dunbar (ed.), *Polar Oceans.* Arctic Institute of North America, Calgary, Alberta, pp. 603–615.

Clark, D. L. 1977*b*. Paleontologic response to post-Jurassic crustal plate movements in the Arctic Ocean. In R. M. West (ed.), *Paleontology and Plate Tectonics.* Milwaukee Pub. Museum, Spec. Publ. Biol. Geol. 2, pp. 55–76.

Clark, D. L. 1981. Geology and geophysics of the Amerasia Basin. In A. E. M. Nairn, M. Churkin, Jr., and F. G. Stehli (eds.), *The Ocean Basins and Margins, 5, The Arctic Ocean.* Plenum, New York, pp. 599–634.

Clark, D. L. 1982*a*. The Arctic Ocean and post-Jurassic paleoclimatology. In W. Berger and J. Crowell (eds.), *Climate in Earth History: Studies in Geophysics.* National Academy Press, Washington, D.C., pp. 133–138.

Clark, D. L. 1982*b*. Origin, nature and world climate effect of Arctic Ocean ice cover. *Nature* **300**(5,890):321–325.

Clark, D. L. 1984. Determining the age of Arctic Ocean sediment cores. *13th Ann. Arctic Workshop.* University of Colorado, Boulder, pp. 46–47.

Clark, D. L., R. R. Whitman, K. A. Morgan, and S. D. Mackay. 1980. Stratigraphy and glacial-marine sediments of the Amerasia Basin, central Arctic Ocean. *Geol. Soc. Am. Spec. Paper 181,* 57p.

Coachman, L. K., and K. Aagaard. 1974. Physical oceanography of the Arctic and Subarctic seas. In Y. Herman (ed.), *Marine Geology and Oceanography of the Arctic Seas.* Springer-Verlag, New York, pp. 1–72.

Cochran, J. K., and J. K. Osmond. 1974. Gamma spectrometry of deep sea cores and sediment accumulation rates. *Deep-Sea Res.* **21**:721–737.

Correns, C. W. 1941. *Nachr. Ges. Wiss. Göttingen Math-Phys. Kl.* **5**:219p.

Colbourn, W. T., F. L. Parker, and W. H. Berger. 1980. Faunal and solution patterns of planktonic foraminifera in surface sediments of the North Pacific. *Marine Micropaleontol.* **5**:329–399.

Craig, H., and L. I. Gordon. 1965. Isotopic oceanography. In *Symposium in Marine Geochemistry, 3.* Narragansett Marine Laboratory, University of Rhode Island, pp. 277–374.

Cromie, W. J. 1961. Preliminary results of investigations on Arctic drift station "Charlie." In G. O. Raasch (ed.), *Geology of the Arctic,* vol. 1. University of Toronto Press, Toronto, pp. 690–708.

Crowell, J. C., and L. A. Frakes. 1970. Phanerozoic glaciation and the causes of ice ages. *Am. J. Sci.* **268**:193–224.

Curry, W. B., and R. K. Matthews. 1979. Isotopic fractionation in recent fossil planktonic foraminifera from the Indian Ocean: Analysis of equilibrium and disequilibrium patterns. *Geol. Soc. Am. Abstr.* **11**:408.

Cushman, J. A. 1948. Arctic foraminifera. *Cushman Lab. Foram. Res. Spec. Pub. 23,* 79p.

Darby, D. A. 1971. Carbonate Cycles and Clay Mineralogy of Arctic Ocean Sediment Cores. Ph.D. thesis, University of Wisconsin, Madison, 117p.

Deuser, W. G. 1979. Seasonal changes in isotopic and species composition of foraminifera collected in a deep-water sediment trap. *Geol. Soc. Am. Abstr.* **11**:412.

Dugolinsky, B. K., S. V. Margolis, and W. C. Dudley. 1977. Biogenic influence on growth of manganese nodules. *J. Sediment. Petrol.* **47**:428–445.

Duplessy, J. C. 1978. Isotopic studies. In J. Gribbin (ed.), *Climatic Change.* Cambridge University Press, Cambridge, pp. 46–67.

Duplessy, J. C., and Y. Herman. 1986*a*. Stable isotopic and faunal variations in planktonic foraminifera from the Arctic Ocean. In *Fifteenth Annual Arctic Workshop and Symposium.* Univ. Colorado, Boulder, Proc. Vol. 10–12.

Duplessy, J. C., and Y. Herman. 1986*b*. Foraminifera and pteropoda beneath the Arctic sea-ice: New distributions and paleoceanographic implications. In *Fifteenth Annual Arctic Workshop and Symposium.* Univ. Colorado, Boulder, Proc. vol. 13–15.

Duplessy, J. C., Y. Herman, and J. K. Osmond. 1984. Late Quaternary Arctic paleoceanography: Stable isotopes, micropaleontology and chronology (abstract). *Geol. Soc. Am., Abstracts with Programs 497.*

Duplessy, J. C., C. Lalou, and A. C. Vinot. 1970. Differential isotopic fractionations in benthic foraminifera and paleotemperatures reassessed. *Science* **168**:250–251.

Durazzi, J. T. 1981. Stable isotope studies of planktonic foraminifera in North Atlantic core tops. *Palaeogeog., Palaeoclimatol. Palaeoecol.* **33**:157–172.

Elmstrom, K. M., and J. P. Kennett. 1985. Late Neogene paleoceanographic evolution of site 590: Southwest Pacific. In J. P. Kennett, C. C. von der Borch, et al. (eds.), *Initial Rpts. DSDP 90.* U.S. Government Printing Office, Washington, D.C., pp. 1,361–1,381.

Emiliani, C. 1955. Pleistocene temperatures. *J. Geol.* **63**(6):538–573.

Emiliani, C. 1964. Paleotemperature analysis of the Caribbean cores A254-BR-C and CP-28. *Geol. Soc. Am. Bull.* **75:**129–143.

Emiliani, C. 1966*a*. Isotopic paleotemperatures. *Science* **154:**851–857.

Emiliani, C. 1966*b*. Paleotemperature analysis of Caribbean cores P6304-8 and P6304-9 and a generalized temperature curve of the past 425,000 years. *J. Geol.* **74**(2):109–124.

Emiliani, C. 1967. The Pleistocene record of the Atlantic and Pacific oceanic sediments; correlations with the Alaskan stages by absolute dating; and the age of the last reversal of the geomagnetic field. *Progr. Oceanogr.* **4:**219–224.

Emiliani, C. 1971. The amplitude of Pleistocene climatic cycles at low latitudes and the isotopic composition of glacial ice. In Karl K. Turekian (ed.), *The Late Cenozoic Glacial Ages.* Yale University Press, New Haven, Conn., pp. 183–197.

Epstein, S., R. Buchsbaum, H. Lowenstam, and H. C. Urey. 1953. Revised carbonate-water isotopic temperature scale. *Geol. Soc. Am. Bull.* **64:**1,315–1,325.

Erez, J., S. Honjo, and N. J. Shackleton. 1979. Isotopic composition of planktonic foraminifera in plankton tows, sediment traps and sediments. *Geol. Soc. Am. Abstr.* **11:**421.

Ericson, D. B. 1959. Coiling direction of *Globigerina pachyderma* as a climatic index. *Science* **130:**219–220.

Ericson, D. B., and G. Wollin. 1973. Precipitation of manganese oxide in deep-sea sediments. *Inter-University Program of Research on Ferromanganese Deposits of the Ocean Floor Phase I Report.* National Science Foundation, Washington, D.C., pp. 99–103.

Ericson, D. B., W. M. Ewing, and G. Wollin. 1964. Sediment cores from the Arctic and sub-Arctic seas. *Science* **144**(3,623):1,183–1,192.

Fetter, F. C. 1973. Recent Deep-Sea Benthic Foraminifera from the Alpha Ridge Province of the Arctic Ocean. M.S. thesis, Florida State University, Tallahassee, 121p.

Finkel, R., S. Krishnaswami, and D. L. Clark. 1977. $^{10}$Be in Arctic Ocean sediments. *Earth Planet. Sci. Lett.* **35:**109–204.

Fortin, L., and M. A. Meyland. 1977. *Bibliography and Index to Literature on Manganese Nodules.* Hawaii Institute of Geophysics Publ., Hawaii, 180p.

Fowler, S. W., and L. F. Small. 1967. Moulting as a possible mechanism for vertical transport of zinc-65 in the sea. *J. Oceanol. Limnol* **1:**237–245.

Fujita, Y., Y. Taguchi, M. Imamura, T. Inoue, and S. Tanaka. 1975. A low-level needle counter. *Nucl. Instrum. Meth.* **128:**523–524.

Funder, S., K. S. Petersen, and L. A. Simonarson. 1987. The early Pleistocene Arctic Ocean: View from the beach. *XII INQUA Cong., Ottawa, Abs.* **170.**

Funder, S., N. Abrahamsen, P. Bennike, and R. W. Feyling-Hanssen. 1985. Forested Arctic: Evidence from North Greenland. *Geology* **13:**542–546.

Galt, J. A. 1967. *Current Measurements in the Canada Basin of the Arctic Ocean, Summer 1965.* University of Washington, Department of Oceanography, Technical Report 184, 17p.

Gilbert, M. W., and D. L. Clark. 1983. Central Arctic Ocean paleoceanographic interpretations based on Late Cenozoic calcareous dinoflagellates. *Marine Micropaleontol.* **7**(5):385–401.

Goel, P. S., D. P. Kharkar, D. Lal, N. Narasappaya, B. Peters, and V. Yatirajam. 1957. Beryllium-10 concentration in deep sea sediments. *Deep-Sea Res.* **4:**202–210.

Golbert, T. A. 1985. *Globigerina hermanae* Golbert. In V. I. Gudina and A. V. Kanegin (eds.), *Microfauna Phanerozoi Sibiri Sibirskoe Otdelenie.* Truda Instituta Geologii i Geophyziki, 615, pp. 68–70, 123–125.

Goldberg, E. D., and G. Arrhenius. 1958. Chemistry of Pacific pelagic sediments. *Geochim. Cosmochim. Acta* **13:**153–212.

Gorbunov, G. 1946. Bottom life of the Novosiberian shoal-waters and the central part of the Arctic Ocean, Vol. 3, *Compendium of Results of Drifting Expedition of Ice-Breaker "Sedov" 1937–1940.* Chief Office of North Road, Moscow, 138p.

Gould, S. J., and N. Eldredge. 1977. Punctuated equilibria: The tempo and mode of evolution reconsidered. *Paleobiology* **3:**115–151.

Graham, D. W., B. H. Corliss, M. L. Bender, and L. D. Keigwin, Jr. 1981. Carbon and oxygen isotopic disequilibria of recent deep-sea benthic foraminifera. *Marine Micropaleontol.* **6:**483–497.

Graham, J. W. 1959. Metabolically induced precipitation of trace elements from seawater. *Science* **129:**1428–1429.

Green, E. 1960. Ecology of some Arctic foraminifera. *Micropaleontology* **6**(1):57–78.

Greenslate, J. L., J. Z. Frazer, and G. Arrhenius. 1973. Origin and deposition of selected transition elements in the seabed. In M. Morgenstein (ed.), *The Origin and Distribution of*

*Manganese Nodules in the Pacific and Prospects for Exploration*. Honolulu, pp. 45–69.

Griffin, J. J., H. Windom, and E. D. Goldberg. 1968. The distribution of clay minerals in the World Ocean. *Deep-Sea Res.* **15**:433–459.

Grossman, E. L. 1984. Carbon isotopic fractionation in live benthic foraminifera—comparison with inorganic precipitate studies. *Geochim. Cosmochim. Acta* **48**:1,505–1,512.

Grossman, E. L. 1987. Stable isotopes in modern benthic foraminifera: A study of vital effect. *J. Foram. Res.* **17**(1):48–61.

Hayes, D. E., and L. A. Frakes. 1975. General synthesis, Deep Sea Drilling Project, Leg 28. In D. E. Hayes et al. (eds.), *Initial Repts. DSDP 28*. U.S. Government Printing Office. Washington, D.C., pp. 914–942.

Herman, Y. 1963. Temperate water planktonic foraminifera in Quaternary sediments of the Arctic Ocean. *Nature* **20-1**(4,917):386–387.

Herman, Y. 1969. Arctic Ocean Quaternary microfauna and its relation to paleoclimatology. *Palaeogeogr., Palaeoclimatol. Palaeoecol.* **6**:251–276.

Herman, Y. 1970. Arctic paleoceanography in late Cenozoic time. *Science* **169**:474–477.

Herman, Y. 1971. Vertical and horizontal distribution of pteropods in Quaternary sequences. In B. M. Funnel and W. R. Riedel (eds.), *The Micropalaeontology of Oceans*. Cambridge University Press, Cambridge, pp. 463–486.

Herman, Y. 1974. Arctic Ocean sediments, microfauna and the climatic record in late Cenozoic time. In Y. Herman (ed.), *Marine Geology and Oceanography of the Arctic Seas*. Springer-Verlag, New York, pp. 283–348.

Herman, Y. 1978. Pteropods. In B. U. Haq and A. Boersma (eds.), *Introduction to Marine Micropaleontology*. Elsevier, New York and Oxford, pp. 151–160.

Herman, Y. 1979. Plankton distribution in the past. In S. van der Spoel and A. C. Pierrot-Bults (eds.), *Zoogeography and Diversity of Plankton*. Bunge Scientific Publisher, Utrecht, pp. 29–49.

Herman, Y. 1983. Baffin Bay: Present day analog of the Central Arctic during late Pliocene to mid-Pleistocene time. *Geology* **11**:356–359.

Herman, Y. 1984*a*. Arctic paleoceanography in late Neogene time and its relationship to the evolution of Northern Hemisphere glaciations. *Proc. XXVIIth Int'l. Geol. Cong., Moscow, Colloquium O* **4**:162–173.

Herman, Y. 1984*b*. Arctic ice cover history and chronology in late Neogene time. *Proc. XXVIIth Int'l. Geol. Cong., Moscow, Colloquium O* **3**:156–162.

Herman, Y. 1984*c*. Comment and reply on "Baffin Bay: Present day analog to the Central Arctic during late Pliocene to mid-Pleistocene time." *Geology* **12**:379–380.

Herman, Y. 1985. Arctic paleoceanography in late Neogene time and its relationship to global climates. *Inter-Nord* **17**:9–14.

Herman, Y., and D. M. Hopkins. 1980. Arctic oceanic climate in Late Cenozoic time. *Science* **209**(4,456):557–562.

Herman, Y., and H. Keupp, 1987. Biochronology and geochronology of late Neogene Central Arctic deep-sea cores. *Proc. Arctic Symposium and Workshop, Polar Research*.

Herman, Y., and M. C. Metz. 1972. Staining technique for recent and fossil calcareous invertebrates. *J. Paleontol.* **46**:152.

Herman, Y., and J. R. O'Neil. 1975. Arctic paleosalinities during Late Cenozoic time. *Nature* **258**(5,536):591–595.

Herman, Y., and J. K. Osmond. 1984. Late Neogene Arctic paleoceanography: Micropaleontology and chronology. In A. Berger, J. Imbrie, J. Hays, G. Kukla, and B. Saltzmann (eds.), *Milankovitch and Climate*. Reidel, Dordrecht, pp. 241–250.

Hibler, W. D., III. 1980. Sea ice growth, drift and decay. In Samuel C. Colbeck (ed.), *Dynamics of Snow and Ice Masses*. Academic Press, New York, pp. 141–209.

Hooper, P. W. P., and P. P. E. Weaver. 1987. Paleoceanographic significance of late Miocene to early Pliocene planktonic foraminifers at Deep Sea Drilling Project Site 609. In W. F. Ruddiman, R. B. Kidd, E. Thomas, et al. (eds.), *Initial Rpts. DSDP 94*. U.S. Government Printing Office, Washington, D.C., pp. 925–934.

Hughes, T., G. Denton, and M. Grosswald. 1977. Was there a late-Würm Arctic ice sheet? *Nature* **266**:596–602.

Hunkins, K. L., E. M. Thorndike, and G. Mathieu. 1969. Nepheloid layers and bottom currents in the Arctic Ocean. *J. Geophys. Res.* **74**:6,995–7,008.

Hunkins, K., A. W. H. Bé, N. D. Opdyke, and G. Mathieu. 1971. The late Cenozoic history of the Arctic Ocean. In K. K. Turekian (ed.), *The Late Cenozoic Glacial Ages*. Yale University Press, New Haven, Conn., pp. 215–237.

Ingle, J. C., Jr. 1973. Summary comments on Neogene biostratigraphy, physical stratigraphy, and paleoceanography in the marginal north-eastern Pacific Ocean. In L. D. Kulm and R. von Huene (eds.), *Initial Rpts. DSDP*

*18*. U.S. Government Printing Office, Washington, D.C., pp. 949–960.

Kahn, M. I. 1979. Non-equilibrium oxygen and carbon isotopic fractionation in tests of living planktonic foraminifera. *Oceanol. Acta* **2:**195–208.

Kahn, M. I., and D. F. Williams. 1981. Oxygen and carbon isotopic composition of living planktonic foraminifera from the northeast Pacific Ocean. *Palaeogeogr., Palaeoclimatol., Palaeoecol.* **33:**47–69.

Keigwin, L. D., Jr. 1978. Pliocene closing of the Isthmus of Panama, based on biostratigraphic evidence from nearby Pacific Ocean and Caribbean Sea cores. *Geology* **6:**630–634.

Keigwin, L. D. 1987. Pliocene stable isotope record of Deep Sea Drilling Project Site 606: Sequential events of $^{18}O$ enrichment beginning at 3.1 Ma. In W. F. Ruddiman, R. B. Kidd, E. Thomas, et al. (eds.), *Initial Rpts. DSDP 94.* U.S. Government Printing Office, Washington, D.C., pp. 911–920.

Keigwin, L. D., M.-P. Aubry, and D. V. Kent. 1987. North Atlantic late Miocene stable-isotope stratigraphy, biostratigraphy, and magnetostratigraphy. In W. F. Ruddiman, R. B. Kidd, and E. Thomas et al. (eds.), *Initial Rpts. DSDP 94.* U.S. Government Printing Office, Washington, D.C., pp. 935–964.

Keller, G. 1978. Late Neogene biostratigraphy and paleoceanography of DSDP Site 310 central North Pacific and correlation with the southwest Pacific. *Marine Micropaleontol.* **3:**97–119.

Keller, G. 1983. Biochronology and paleoclimatic implications of Middle Eocene to Oligocene planktonic foraminiferal faunas. *Marine Micropaleontol.* **7:**463–486.

Kennett, J. P. 1978. The development of planktonic biogeography in the Southern Ocean during the Cenozoic. *Marine Micropaleontol.* **3:**301–345.

Kennett, J. P., and M. S. Srinivasan. 1980. Surface ultrastructural variation in *Neogloboquadrina pachyderma* (Ehrenberg): Phenotypic variation and phylogeny in the Late Cenozoic. *Cushman Found. Spec. Pub.* **19:**134–162.

Kitchell, J. A., J. F. Kitchell, G. L. Johnson, and K. L. Hunkins. 1978. Abyssal traces and megafauna: Comparison of productivity, diversity and density in the Arctic and Antarctic. *Paleobiology* **4:**171–180.

Krishnaswami, S. A., and M. M. Sarin. 1976. The simultaneous determination of Th, Pu, Ra isotopes $^{210}Pb$, $^{55}Fe$, $^{32}Si$, and $^{14}C$ in marine suspended phases. *Anal. Chem. Acta* **83:**143–156.

Kroopnick, P., S. V. Margolis, and C. S. Wong. 1977. $\delta^{13}C$ variations in marine carbonate sediments as indicators of the $CO_2$ balance between atmosphere and oceans. In N. R. Anderson and A. Malahoff (eds.), *The Fate of Fossil Fuel $CO_2$ in the Oceans.* Plenum, New York, pp. 295–321.

Ku, T. L., and W. S. Broecker. 1967. Rates of sedimentation in the Arctic Ocean. In M. Sears (ed.), *Progress in Oceanography,* vol. 4. Pergamon Press, New York, pp. 95–104.

Ku, T. L., M. Kusakabe, J. R. Southon, J. S. Vogel, and D. E. Nelson. 1984. $^{10}Be$ distributions in Deep Sea Drilling project sediments studied by accelerator mass spectrometry (abstract). *EOS, Trans. AGU* **65:**949.

Kvasov, D. D., and A. I. Blazhchishin. 1978. The key to sources of the Pliocene and Pleistocene glaciation is at the bottom of the Barents Sea. *Nature* **273:**138–140.

Lagoe, M. B. 1977. Recent benthic foraminifera from the central Arctic Ocean. *J. Foram Res.* **7**(2):106–129.

Lal, D., and B. Peters. 1967. Cosmic ray produced radioactivity on the earth. *Handbuch der Physik* **46:**551–612.

Leonard, K. A., D. F. Williams, and R. C. Thunell. 1983. Pliocene paleoclimatic and paleoceanographic history of the South Atlantic Ocean: Stable isotopic records from leg 72 Deep Sea Drilling Project holes 516A and 517. In P. F. Barker, R. L. Carlson, and D. A. Johnson (eds.), *Initial Rpts. DSDP 72.* U.S. Government Printing Office, Washington, D.C., pp. 895–906.

Linkova, T. I. 1965. Some results of paleomagnetic study of Arctic Ocean floor sediments. In *The Present and Past of the Geomagnetic Field.* Nauka, Moscow, Directorate of Scientific Info. Serv. CRB Canada, pp. 279–291.

Loeblich, A. R., Jr., and H. Tappan. 1953. *Studies of Arctic Foraminifera.* Smithsonian Miscellaneous Collections, vol. 121, 150p.

Loeblich, A. R., Jr., and H. Tappan. 1964. Foraminiferida. In R. C. Moore (ed.), *Treatise on Invertebrate Paleontology, Part C, Protista,* 2(2 vols.). University of Kansas Press, Lawrence, 900p.

McCorkell, R. H., E. L. Fireman, and C. C. Langway, Jr. 1967. Aluminum-26 and beryllium-10 in Greenland ice. *Science* **150:**1,690–1,692.

McCoy, F. W. 1980. Photographic analysis of coring. *Marine Geol* **38:**263–282.

McCoy, F. W. 1985. Mid-core flow-in: Implications for stretched stratigraphic sections in piston cores. *J. Sed. Petrol.* **55**(4):608–610.

Macko, S. A., and A. E. Aksu. 1986. Amino acid epimerization in planktonic foraminifera suggests slow sedimentation rates for Alpha Ridge, Arctic Ocean. *Nature* **322**:730–732.

Malmgren, B. A. 1983. Ranking of dissolution susceptibility of planktonic foraminifera at high latitudes of the South Atlantic Ocean. *Marine Micropaleontol.* **8**:183–191.

Malmgren, B. A. 1987. Differential dissolution of Upper Cretaceous planktonic foraminifera from a temperate region of the South Atlantic Ocean. *Marine Micropaleontol.* **11**:251–271.

Malmgren, B. A., and J. P. Kennett. 1981. Phyletic gradualism in a Late Cenozoic planktonic foraminiferal lineage: DSDP Site 284, southwest Pacific. *Paleobiology* **7**:230–240.

Malmgren, B. A., W. A. Berggren, and G. P. Lohmann. 1983. Evidence for punctuated gradualism in the Late Neogene *Globorotalia tumida* lineage of planktonic foraminifera. *Paleobiology* **9**:377–389.

Mangini, A., M. Segl, G. Bonani, H. J. Jofman, E. Morenzoni, M. Nessi, M. Suter, W. Wolfli, and K. K. Turekian. 1984. Mass spectrometric $^{10}$Be dating of deep sea sediments applying the Zürich tandem accelerator. *Nucl. Instrum. Meth. Phys. Res.* **B5**:353–358.

Margolis, S. V., and Y. Herman. 1980. Northern Hemisphere sea-ice and glacial development in the Late Cenozoic. *Nature* **286**(5,769):145–149.

Markussen, B., R. Zahn, and J. Thiede. 1985. Late Quaternary sedimentation in the Eastern Arctic basin. Stratigraphy and depositional environment. *Palaeogeog., Palaeoclimatol. Palaeoecol.* **50**:271–284.

Martin, J. H. 1970. The possible transport of trace metals via moulted copepod exoskeletons. *Limnol. Oceanogr.* **15**:756–761.

Martin, J. H., and G. A. Knauer. 1973. The elemental composition of plankton. *Geochim. Cosmochim. Acta* **37**:1,639–1,653.

Maykut, G. A., and N. Untersteiner. 1971. Some results from a time-dependent thermodynamic model of sea ice. *J. Geophys. Res.* **76**:1,550–1,575.

Mayr, E. 1963. *Animal Species and Evolution.* Harvard University Press, Cambridge, Mass., 797p.

Mayr, E. 1970. *Population, Species, and Evolution.* Harvard University Press, Cambridge, Mass., 453p.

Mercer, J. H. 1970. A former ice sheet in the Arctic Ocean? *Palaeogeog., Palaeoclimatol., Palaeoecol.* **8**:19–27.

Monaghan, M. C., S. Krishnaswami, and K. K. Turekian. 1986. The global average production rate of $^{10}$Be. *Earth Planet. Sci. Lett.* **76**:279–287.

Mudie, P. J. 1985. Palynology of the CESAR cores, Alpha Ridge. In H. R. Jackson, P. J. Mudie, and S. M. Blasco (eds.), *Initial Geological Report on CESAR, The Canadian Expedition to Study the Alpha Ridge, Arctic Ocean.* Geol. Survey Canada Paper 84-22, pp. 149–174.

Murray, J., and A. Renard. 1891. Manganese nodules. In C. W. Thomson (ed.), *Report of the Scientific Results of the Voyage of the H.M.S. Challenger, 5, Deep Sea Deposits.* Eyra and Spottiswoode, London, pp. 341–378.

Naidu, A. S. 1974. Sedimentation in the Beaufort Sea: A synthesis. In Y. Herman (ed.), *Marine Geology and Oceanography of the Arctic Seas.* Springer-Verlag, New York, pp. 173–190.

Nansen, F. 1906. *Northern Waters: Captain Roald Amundsen's Oceanographic Observations in the Arctic Seas in 1901.* Vitensk.-Selsk. Skr., I, Math.-Naturv. Kl., 3, 145p.

Naugler, P. F., N. Silverberg, and J. S. Creager. 1974. Recent sediments of the East Siberian Sea. In Y. Herman (ed.), *Marine Geology and Oceanography of the Arctic Seas.* Springer-Verlag, New York, pp. 191–210.

Nelson, D. E., J. R. Southon, J. S. Vogel, and R.G. Korteling. 1984. Progress in $^{14}$C and $^{10}$Be dating at SFU Nucl. Instrum. Meth. *Phys. Res.* **B5**:139–143.

Nesteroff, W. D., and B. C. Heezen. 1977. Du mode de formation de quelques nodules et encroûtements de ferromanganèse des grands fonds océaniques. *C.R. Acad. Sci. Paris* **284**(D):799–801.

Olsson, R. K . 1971. Pliocene-Pleistocene planktonic foraminiferal biostratigraphy of the Northeastern Pacific. In A. Farinacci (ed.), *Proceedings of the Second International Planktonic Conference,* vol. 2. Edizioni Tecnoscienza, pp. 921–928.

Osmond, J. K. 1979. Accumulation models of $^{230}$Th and $^{231}$Pa in deep sea sediments. *Earth Sci. Rev.* **15**:95–150.

Osmond, J. K. 1981. Quaternary deep sea sediments: Accumulation rates and geochronology. In C. Emiliani (ed.), *The Sea, vol. 7, The Oceanographic Lithosphere.* Wiley, New York, pp. 1,329–1,371.

Osterberg, C. L., A. G. Carey, Jr., and H. Curl, Jr. 1963. Acceleration of sinking rates of radionuclides in the ocean. *Nature* **200**:1,276–1,277.

Parker, F. L. 1971. Distribution of planktonic foraminifera in Recent deep-sea sediments. In B. M. Funnel and W. R. Riedel (eds.), *The Micropalaeontology of Oceans.* Cambridge University Press, Cambridge, pp. 289–308.

Paul, A. Z., and R. J. Menzies. 1973. *Benthic*

*Ecology of the High Arctic Deep-sea.* Technical Rept., Florida State University, Department of Oceanography, Tallahassee, 337p.

Pautot, G., and M. Melguen. 1979. Influence of deep water circulation and sea floor morphology on the abundance and grade of central South Pacific manganese nodules. In J. L. Bischoff and D. Z. Piper (eds.), *Marine Geology and Oceanography of the Pacific Manganese Nodule Province.* Plenum Press, New York, pp. 621–645.

Raisbeck, G. M., and F. Yiou. 1974. Cross sections for the spallation production of $^{10}$Be in targets of N, Mg, and Si and their astrophysical applications. *Phys. Rev. C.* **9**(4):1,383–1,395.

Raisbeck, G. M., F. Yiou, M. Lieuvin, and J. M. Loiseaux. 1978. Measurement of $^{10}$Be in 1000 and 5000 year old Antarctic ice. *Nature* **275:**731–733.

Raymo, M. E., W. F. Ruddiman, and B. M. Clement. 1987. Pliocene-Pleistocene paleoceanography of the North Atlantic at Deep Sea Drilling Project Site 609. In W. F. Ruddiman, R. B. Kidd, E. Thomas, et al., *Initial Rpts. DSDP 94.* U.S. Government Printing Office, Washington, D.C., pp. 895–901.

Repenning, C. A. 1984. Quaternary rodent biochronology and its correlation with climatic and magnetic stratigraphies. In W. E. Mahaney (ed.), *Quaternary Chronologies.* GeoBooks, London, pp. 105–118.

Repenning, C. A., and E. M. Brouwers. 1987. Mid-Pliocene to late Pleistocene changes in the Arctic Ocean borderland ecosystem. *XII INQUA Cong., Ottawa, Abs.* **251.**

Revelle, R., M. Bramlette, G. Arrhenius, and E. D. Goldberg. 1955. Pelagic sediments of the Pacific. *Geol. Soc. Am. Spec. Paper 62,* pp. 221–236.

Reynolds, L., and R. C. Thunell. 1985. Seasonal succession of planktonic foraminifera in the subpolar North Pacific. *J. Foram. Res.* **15**(4):282–301.

Reynolds, L. A., and R. C. Thunell. 1986. Seasonal production and morphologic variation of *Neogloboquadrina pachyderma* (Ehrenberg) in the Northeast Pacific. *Micropaleontology* **32**(1):1–18.

Robe, R. Q. 1980. Iceberg drift and deterioration. In S. G. Colbeck (ed.), *Dynamics of Snow and Ice Masses.* Academic Press, New York, pp. 211–259.

Sacks, V. N., N. A. Belov, and N. N. Lapina. 1955. Our present concept of the geology of the central Arctic. *Priroda* (trans.) *Defense Research Board (DRB), Canadian Transl. No. T 196 R, 7, 13, 1–14.*

Schwarzacher, W., and K. L. Hunkins. 1961. Dredged gravels from the Central Arctic Ocean. In G. O. Raasch (ed.), *Geology of the Arctic, Proceedings of the International Symposium on Arctic Geology, 1960.* University of Toronto Press, Toronto, vol. 1, pp. 666–677.

Scott, G. H. 1983. Divergences and phyletic transformations in the history of the *Globorotalia inflata* lineage. *Paleobiology* **9:**422–426.

Selli, R. 1967. The Pliocene-Pleistocene boundary in Italian marine sections and its relationship to continental stratigraphies. In M. Sears (ed.), *Progress in Oceanography.* Pergamon Press, London, vol. 4, pp. 67–86.

Sharma, P., S. K. Bhattacharya, and B. L. K. Somayajulu. 1983. Beryllium-10 in deep sea sediments and cosmic ray intensity variations. *Proc. 18th ICRC, Bangalore, India* **2:**337–340.

Shedrina, Z. G. 1956. Results of the study of foraminifera in the Soviet seas. *Voprosy Mikropaleontologii* **1:**23–36. (In Russian.)

Somayajulu, B. L. K. 1977. Analysis of causes of $^{10}$Be variations in deep sea sediments. *Geochim. Cosmochim. Acta* **41:**909–913.

Somayajulu, B. L. K., P. Sharma, J. Beer, G. Bonani, H. J. Hofman, E. Morenzoni, M. Nessi, M. Suter, and W. Wolfi. 1984. $^{10}$Be annual fallout in rains in India. *Nucl. Instrum. Meth. in Phys. Res.* **B5:**398–403.

Somayajulu, B. L. K., P. Sharma, A. S. Naidu, B. Markussen, J. Thiede, G. Bonani, M. Nessi, M. Suter, and W. Wolfli. 1985. $^{10}$Be variation in an Arctic sediment core: Implication to climate. *International Symposium on Concepts and Techniques of Applied Climatology.* Andhara University, Waltair, India, March 18–21.

Somov, M. M. 1955. *Observational Data of Scientific Research, Drifting Station of 1590–1951.* Vol. 1–3, Vol. 1, Sec. 4 (translated by the American Meteorological Society), p. 9.

Sorokin, Y. I. 1972. Role of biological factors in the sedimentation of iron, manganese, and cobalt and in the formation of nodules. *Oceanology* **12:**1–11.

Spencer, D. W., P. G. Brewer, A. Fleer, S. Honjo, S. A. Krishnaswami, and Y. Nozaki. 1978. Chemical fluxes from a sediment trap in the deep Sargasso Sea. *J. Marine Res.* **36:**493–523.

Steffen, K. 1985. Warm water cells in the North Water, Northern Baffin Bay during winter. *J. Geophys. Res.* **90:**9,129–9,136.

Steuerwald, B. A., and D. L. Clark. 1972. *Globigerina pachyderma* in Pleistocene and Recent Arctic Ocean sediment. *J. Paleontol* **46**(4):573–580.

Steuerwald, B. A., D. L. Clark, and J. A. Andrew. 1968. Magnetic stratigraphy and faunal patterns in Arctic Ocean sediments. *Earth Planet. Sci. Lett.* **5**(2):79–85.

Swift, J. H., T. Takahashi, and H. D. Livingston. 1983. Contribution of the Greenland and Barents seas to the deep water of the Arctic Ocean. *J. Geophys. Res.* **88**(C10):5,981–5,986.

Tanaka, S., and T. Inoue. 1979. Beryllium-10 dating: $^{10}$Be concentration and $^{10}$Be/$^{9}$Be ratios in north Pacific sediment cores up to 2.5 million years B.P. *Earth Planet. Sci. Lett.* **45:**181–187.

Tanaka, S., T. Inoue, and Z. Y. Huang. 1982. $^{10}$Be and $^{10}$Be/$^{9}$Be in near Antarctic sediment cores. *Geochem. J.* **16:**321–325.

Tangen, K., L. E. Brand, P. L. Blackwelder, and R. R. Guillard. 1982. *Thorascosphaera heimii* (Lohmann) Kamptner is a dinophyte: Observations on its morphology and life cycle. *Marine Micropaleontol.* **7:**193–212.

Tappan, H. 1980. *The Paleobiology of Plant Protists.* Freeman & Co., San Francisco, 1,028p.

Thiel, H. 1978. The faunal environment of manganese nodules and aspects of deep sea time scales. In W. E. Krumbein (ed.), *Environmental biogeochemistry and Geomicrobiology,* vol. 3. Ann Arbor Science Publ., Inc., Ann Arbor, Mich., pp. 887–896.

Turekian, K. K., J. K. Cochran, S. Krishnaswami, W. A. Lanford, P. D. Parker, and K. Bauer. 1979. The measurement of $^{10}$Be in manganese nodules using a tandem Van de Graaf accelerator. *Geophys. Res. Lett.* **6:**417–420.

Valdiya, K. S. 1984. Evolution of the Himalaya. *Tectonophysics* **105:**229–248.

van Donk, J., and G. Mathieu. 1969. Oxygen isotope compositions of foraminifera and water samples from the Arctic Ocean. *J. Geophys. Res.* **74**(13):3,396–3,407.

Vilks, G. 1969. Recent foraminifera in the Canadian Arctic. *Micropaleontology* **15**(1):35–60.

Wangersky, P. J. 1963. Manganese in ecology. In V. Schultz and A. W. Klement (eds.), *Radioecology.* pp. 499–508.

Weeks, W. F., and W. J. Campbell. 1973. *CRREL Research Report 200.* U.S. Army Cold Regions Research and Engineering Laboratory, Hanover, N. H.

Weyl, P. K. 1968. The role of the oceans in climatic change: A theory of the Ice Ages. *Meteorological Monographs* **8**(30):37–62.

Williams, D. B. 1971*a*. The distribution of marine dinoflagellates in relation to physical and chemical conditions. In B. M. Funnel and W. R. Riedel (eds.), *The Micropalaeontology of Oceans.* Cambridge University Press, Cambridge, Mass., pp. 91–96.

Williams, D. B. 1971*b*. The occurrence of dinoflagellates in marine sediments. In B. M. Funnel and W. R. Riedel (eds.), *The Micropalaeontology of Oceans.* Cambridge University Press, Cambridge, pp. 231–244.

Williams, D. F. and N. Healy-Williams. 1980. Oxygen isotopic-hydrographic relationships among recent planktonic foraminifera from the Indian Ocean. *Nature* **283**(5,750):848–852.

Williams, D. F., W. S. Moore, and R. H. Fillon. 1981. Role of glacial Arctic Ocean ice sheets in Pleistocene oxygen isotope and sea level records. *Earth Planet. Sci. Lett.* **56:**157–166.

Williams, D. F., M. A. Sommer, and M. L. Bender. 1977. Carbon isotopic compositions of Recent planktonic foraminifera of the Indian Ocean. *Earth Planet. Sci. Lett.* **36:**391–403.

Williams, G. L. 1978. Dinoflagellates, acritarchs and tasmanitids. In B. U. Haq and A. Boersma (eds.), *Introduction to Marine Micropaleontology.* Elsevier, New York and Oxford, pp. 293–326.

Woodring, W. P. 1966. The Panama land bridge as a sea barrier. *Am. Philos. Soc. Proc.* **110:**425–433.

Worsley, T. R., and Y. Herman. 1980. Episodic ice-free Arctic Ocean in Pliocene and Pleistocene time: Calcareous nannofossil evidence. *Science* **210**(4,467):323–325.

Yiou, F., and G.M. Raisbeck. 1972. Half-life of $^{10}$Be. *Phys. Rev. Lett.* **29:**372–375.

Zahn, R., B. Markussen, and J. Thiede. 1985. Stable isotope data and depositional environments in the late Quarternary Arctic Ocean. *Nature* **314**(6,010):433–435.

Zelenov, K. K. 1964. Iron and manganese in exhalations from the submarine volcano Bano Vuku (Indonesia). *Dokl. Akad. Nauk SSSR, Earth Science Sections* **155:**91–94.

Zenkevitch, L. A. 1963. *Biology of the Seas of the U.S.S.R.* (translated by S. Botcharskaya). Interscience Publishers, New York, 955p.

# 24: Sediment Composition and Sedimentary Processes in the Arctic Ocean

D. A. Darby, A. S. Naidu,
T. C. Mowatt,
and G. Jones

## INTRODUCTION

Sedimentologic investigations in the central Arctic Ocean have been limited primarily to the western half, that is, west of the Lomonosov Ridge and Fletcher Abyssal Plain (Fig. 24-1). The most detailed sampling was obtained from Fletcher's Ice Island T-3 between 1963 and 1974. This sampling involves the areas within the Canada Basin, defined herein as the Canada Abyssal Plain and surrounding ridge and continental slopes. More recently, temporary ice stations have sampled bottom sediments east of the Alpha Ridge to the Lomonosov Ridge (for example, the Canadian CESAR and LOREX Projects, Fig. 24-2). The sediments of the eastern half of the Arctic Ocean remain an enigma with little or no sediment cores or grab samples having been studied to date. The probing in the western Arctic is not very uniform and areas such as the Wrangel Abyssal Plain have yet to be sampled. Even large parts of the Canada Abyssal Plain have not been studied.

With such a paucity of available sediments due partially to logistic problems inherent with deep-sea sampling in the Arctic, an understandably incomplete picture of Arctic sedimentary processes has emerged. While the general nature of the central Arctic sediments has been reported along with a remarkably consistent lithostratigraphy throughout the western Arctic (e.g., Clark et al., 1980), several problems remain. Foremost among these is the lack of an equally detailed chronostratigraphy for these widely correlated Plio-Pleistocene lithostratigraphic units. Such a sediment chronology is essential for answering many questions concerning Arctic depositional history and linking these events to global paleoclimatic events. Although this problem remains beyond our grasp, this chapter will reevaluate the existing sediment data, and incorporate new results. For example, most Arctic scientists agree that ice rafting is important in this ocean but the source(s) of this sediment and the relative quantitative importance of this process have yet to be adequately de-

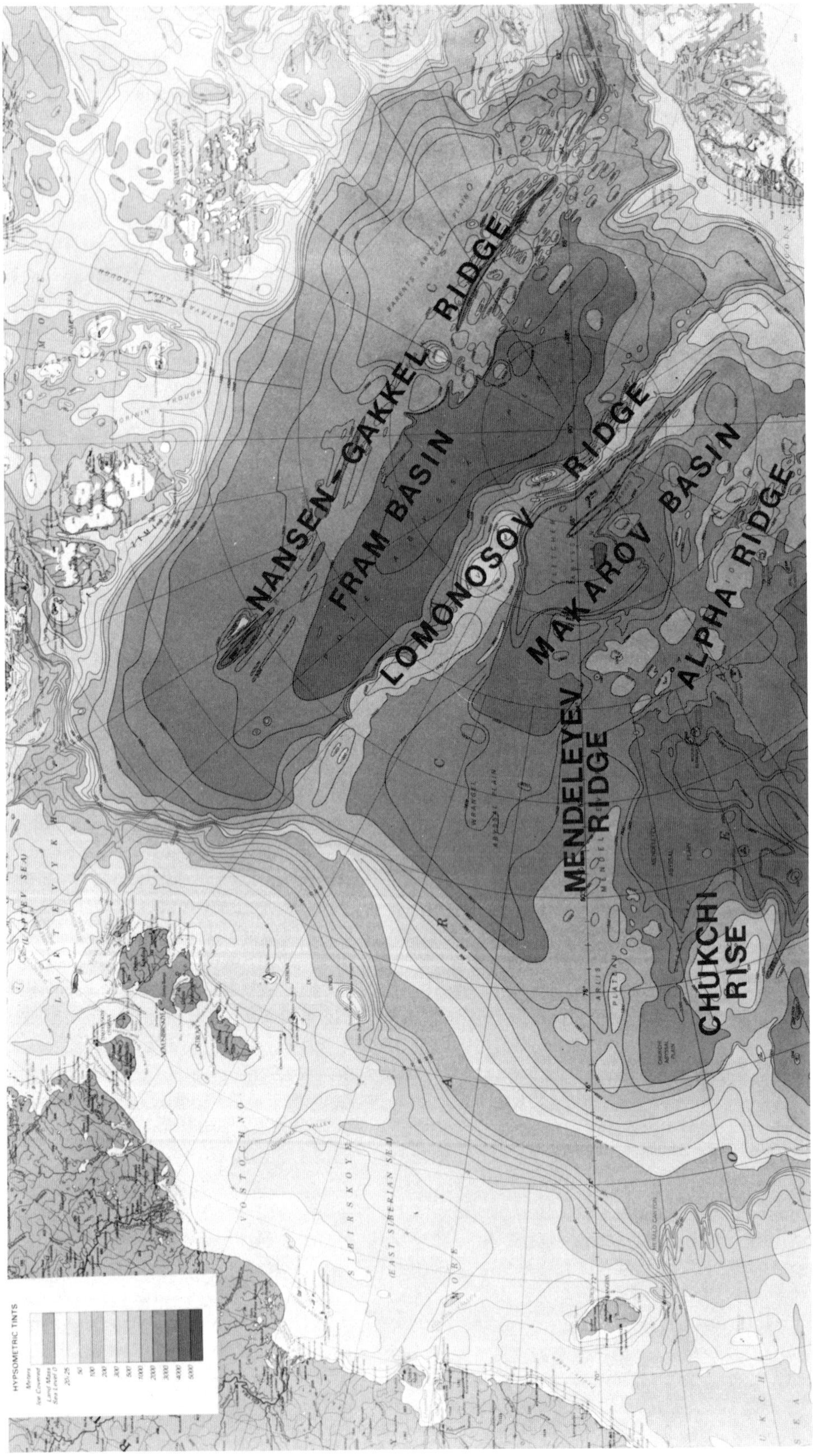
NANSEN–GAKKEL RIDGE
FRAM BASIN
LOMONOSOV RIDGE
MAKAROV BASIN
ALPHA RIDGE
MENDELEYEV RIDGE
CHUKCHI RISE
HYPSOMETRIC TINTS
LAPTEV SEA
EAST SIBERIAN SEA

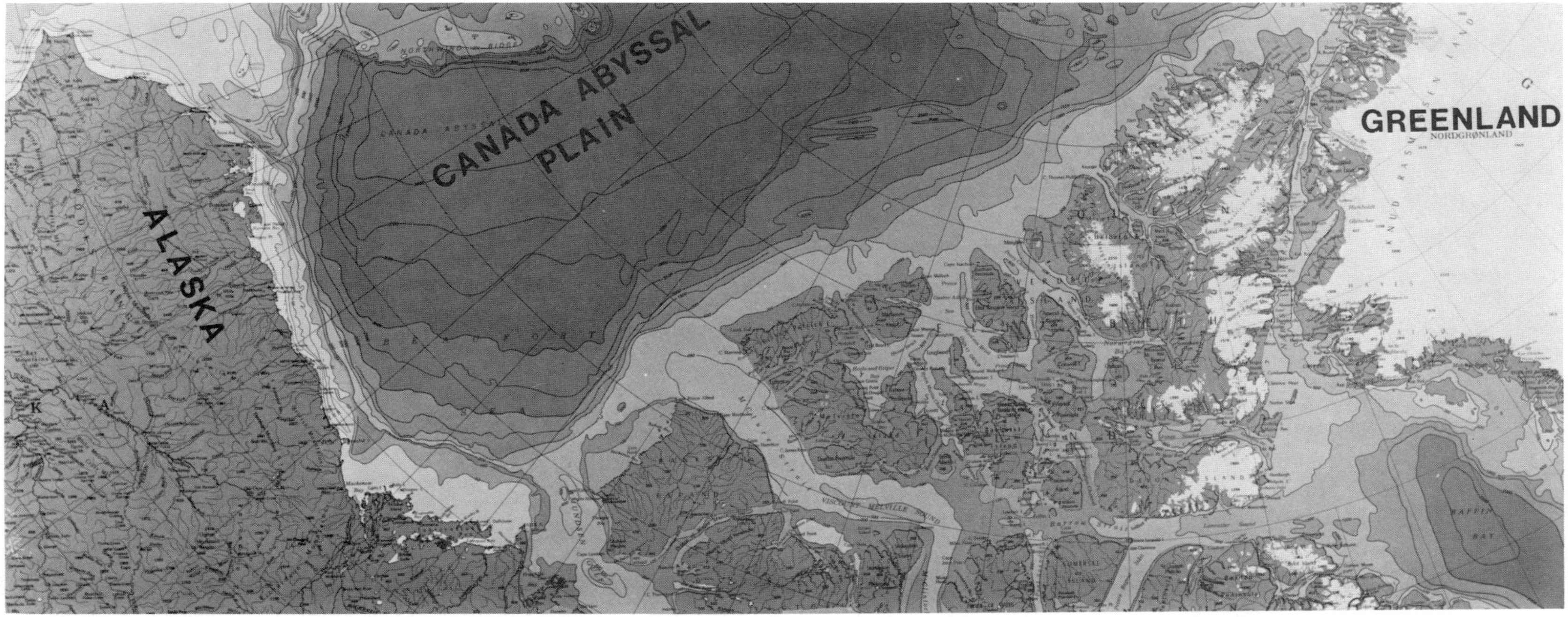


Figure 24-1. Bathymetric map of the Arctic Ocean showing the major physiographic features. (From Perry et al., 1985.)

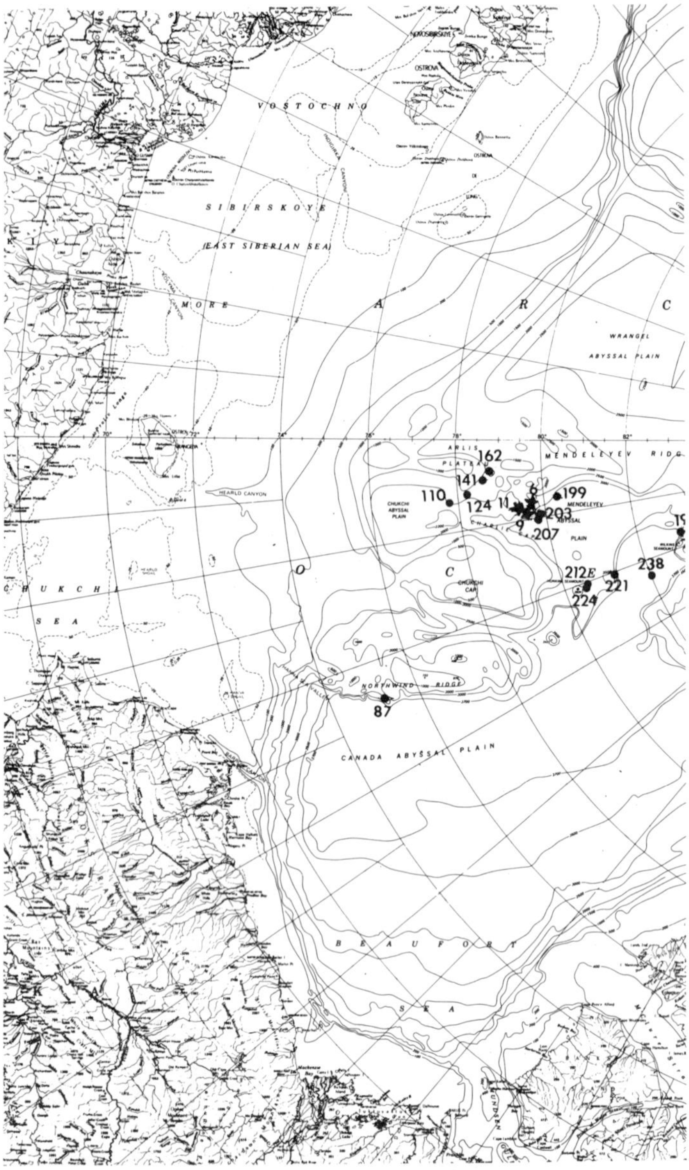


Figure 24-2. Map of the Arctic Ocean showing important core and sampling locations discussed in the text. Numbered cores are those from the University of Wisconsin collection (FL-), except those numbered 6, 9, and 11, which are

termined. The source(s) of turbidite sediments of the Canada Abyssal Plain also need to be more precisely defined. Furthermore, the geochemistry of Arctic sediment is largely unknown.

## MATERIALS AND METHODS

Samples for mineralogic and chemical analyses were taken either from grab samples or from the top 0–5 cm of gravity cores

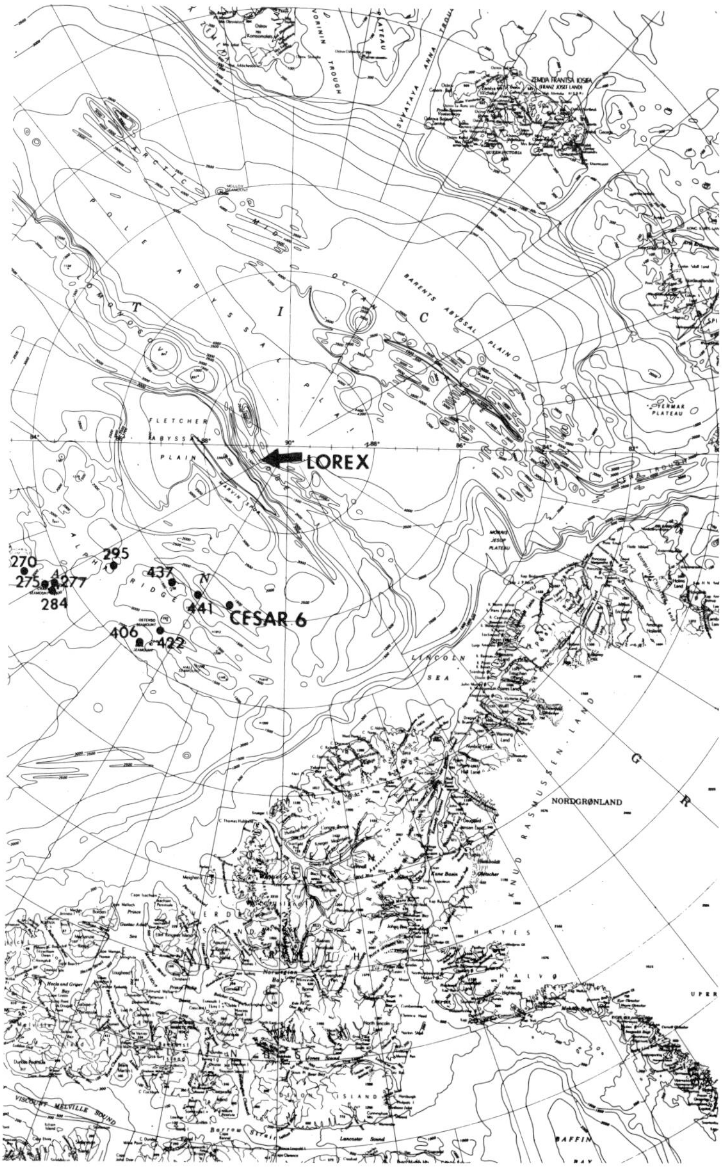


Lamont-Doherty Geological Observatory (T-3) cores. Also shown are CESAR core 6 and the general location of the LOREX cores taken across the Lomonosov Ridge. (After Perry and Fleming, 1983.)

collected from the Canada Basin and the Beaufort Sea. These argillaceous sediment samples were collected between 1959 and 1977 either from U.S. Coast Guard icebreakers or from Fletcher's Ice Island T-3.

The procedures for the clay mineral and chemical analyses are discussed elsewhere (Naidu, Creager, and Mowatt, 1982; Naidu and Hood, 1972; Naidu and Mowatt, 1974, 1983); therefore, only a brief description follows. Clay mineral assemblages in the $<2\mu m$ fraction of sediments were analyzed by X-ray diffraction. Each sample was treated initially with $H_2O_2$ to remove organic matter and the $<2\mu m$ clay fraction was separated by suspension settling. The clay was collected following centrifugation and smeared onto glass slides (Gibbs, 1965). Clay mineral abundances were estimated following Biscaye's (1965) method.

For chemical analyses, an initial 10–20 g sample split was oven dried and then lightly disaggregated. A portion of this sample was pulverized into a fine powder. Analyses for carbonate, organic carbon, and the concentrations of a suite of elements (Si, Al, P, Fe, Mn, N, Zn, V, Cu, Cr, Co, and Ni) were carried out on these powders. Organic carbon (OC) in the sediments was calculated from the difference between total carbon and carbonate carbon. Total carbon was analyzed by a LECO, TC-12 automatic carbon analyzer, whereas the carbonate contents ($CO_3$) were measured gasometrically (Hulsemann, 1966). Precisions for the total carbon and carbonate analyses were each better than 10 percent. Total nitrogen (N) was analyzed using a Coleman nitrogen analyzer. Acetanilide was used as a standard and the precision of the nitrogen analysis was better than 8 percent. The OC/N ratios were calculated on a weight basis. The weight percent of carbonate minerals was determined from the gasometric analysis for carbonate ($CO_3$) using the X-ray diffraction mineral percents (Griffin, 1971) to convert $CO_3$ weight percent to calcite ($CaCO_3$) or dolomite ($CaMg(CO_3)_2$).

The other elements, except P, were analyzed by atomic absorption (AA) spectrophotometry. Aluminum and Si were analyzed by AA from $LiBO_2$ fusions of ground sediment (Medlin, Suhr, and Bodkin, 1969). The total concentrations of each of the other elements were analyzed from $HF\text{-}HNO_3$ digestions of the pulverized sediment (Rader and Grimaldi, 1961). Total P from these solutions was analyzed calorimetrically (Murphy and Riley, 1962). To acquire information on the partitioning patterns of eight of the elements (Fe, Mn, Zn, V, Cr, Co, Cu, and Ni), portions of a selected number of lightly disaggregated sediment samples were leached with 1*M* hydroxylamine hydrochloride and 25 percent acetic acid solutions (HHAA), following the method described by Chester and Hughes (1967). Vanadium in this leachate was analyzed by neutron activation, following the method described by Weiss et al. (1977). The other elements were analyzed by AA using a graphite furnace.

This treatment presumably mobilizes into solution all portions of the metals contained in a sediment, except those that are held tightly in the crystal lattice. Therefore, the metals in the leachates are presumed to represent metals essentially bound with sediment fractions such as ferromanganese oxides/hydroxides, carbonates, or variously adsorbed constituents. Based on an operational definition, these leachate metal fractions are commonly termed nonlithogenous (NL) fractions of sediments as opposed to the lithogenous fractions (lattice-held or detrital) (Chester and Hughes, 1967).

Analytical precision for the gross elemental analysis was better than 12 percent and the accuracy was checked with reference to USGS standard rock powders G-2, AGV-1, and BCR-1. Precision for the analysis of the metals in the leachates varied between 7 percent and 17 percent.

Heavy mineral ($>2.89$ specific gravity) counts were made on grain mounts using either the fleet method (central Arctic samples) or the line method (Galehouse, 1971). No more than 100 opaque grains were counted because of their abundance. No attempt was made to identify the various opaque minerals. A minimum of 300 nonopaque grains were counted provided that this number existed on each slide.

## GENERAL NATURE AND DISTRIBUTION OF CENTRAL ARCTIC SEDIMENTS

The thickness of unconsolidated Arctic Ocean sediments increases dramatically from ridges to basins (Demenitskaya et al., 1968). The greatest accumulation occurs beneath the abyssal plains of the oldest abyssal plain in the Arctic, the Canada Basin, where up to 2–3 km has been reported (Hunkins and Kutschale, 1967). In the Fletcher and Pole abyssal plains (Markov and Fram basins), on either side of the Lomonosov Ridge, the thickness of sediments is about 1 km (Blasco, Bornhold, and Lewis, 1979; Sweeney, Weber, and Blasco, 1982). In contrast, the sediment on the ridges is less than 0.1 to 0.5 km. In general, the thinnest accumulation, that of 0–40 m, was measured on the Lomonosov Ridge (Blasco, Bornhold, and Lewis, 1979).

The Alpha Ridge is a series of troughs and peaks with depths between 1,000 m and 2,500 m (Fig. 24-1; Weber and Jackson, 1985; Hall, 1979). This bottom topography has been attributed to the presence of a series of grabens (Hall, 1979). The sediment thickness in the troughs is 300–1,200 m while thinner deposits occur on topographic highs and in some places no unconsolidated sediment occurs. The oldest-known sediment in the Arctic Ocean (Cretaceous age) has been recovered from these areas of thin deposits (Ling, McPherson, and Clark, 1973; Clark, 1974). These sediments consist of siliceous oozes quite different in appearance and composition from Quaternary Arctic sediments (Mudie and Blasco, 1985; Clark, Byers, and Pratt, 1986).

Photographs of the Arctic seafloor reveal pebble- to cobble-size clasts, usually sediment draped, scattered among occasional benthic organisms and animal tracks (Schwarzacher and Hunkins, 1961; Hunkins, Thorndike, and Mathieu, 1969; Hunkins et al., 1970; Clark and Hanson, 1983; Amos, 1985). The abundance of these clasts decreases from ridges to abyssal plains. Due to their size and scattered occurrence, these coarse clasts are largely excluded from sediment cores and thus from textural descriptions of Arctic sediments. All Arctic surface sediments, as well as Plio-Pleistocene sediments, are sandy or silty lutites with variable proportions of gravel-size clasts.

Clark et al. (1980) described four textural types based on Coulter counter analysis of the $<62\mu m$ fraction. Type I is an unsorted clay and silt with a generally low but variable sand component (Fig. 24-3). Type II is a bimodal lutite with modes at 1.26–2.52 $\mu m$ and 6.35–63 $\mu m$ with moderately high sand fractions averaging 35.6 percent. Type III is a lutite with low sand percentages and type IV is a coarse silty to sandy lutite with high sand percentages. Gradations exist from type I to type IV and they have been attributed to ice rafting, by sea ice, glacial ice, or transport by turbidite currents (Clark et al., 1980; Clark and Hanson, 1983; Goldstein, 1983). Although types I and II are most commonly restricted to the ridges, types III and IV are common to both ridges and abyssal plains (Clark et al., 1980, p. 40). In the abyssal plains, type III contains less silt and more clay with very little, if any, sand. Likewise, type IV contains finer sand and generally lower sand percentages than similar textural types on the ridges (Clark et al., 1980). These textural differences for types III and IV between ridges and abyssal plains are primarily due to the abundance of fine-grained turbidite deposits on the abyssal plains (Campbell and Clark, 1977; Clark et al., 1980).

The central Arctic sediments differ from other oceans in that the former contain nearly equal percentages, 10–30 percent, of both kaolinite and chlorite in the clay fraction ($< 2$ $\mu m$) and abundant detrital dolomite and calcite in the silt-size and coarser fractions. The clay fraction is domi-

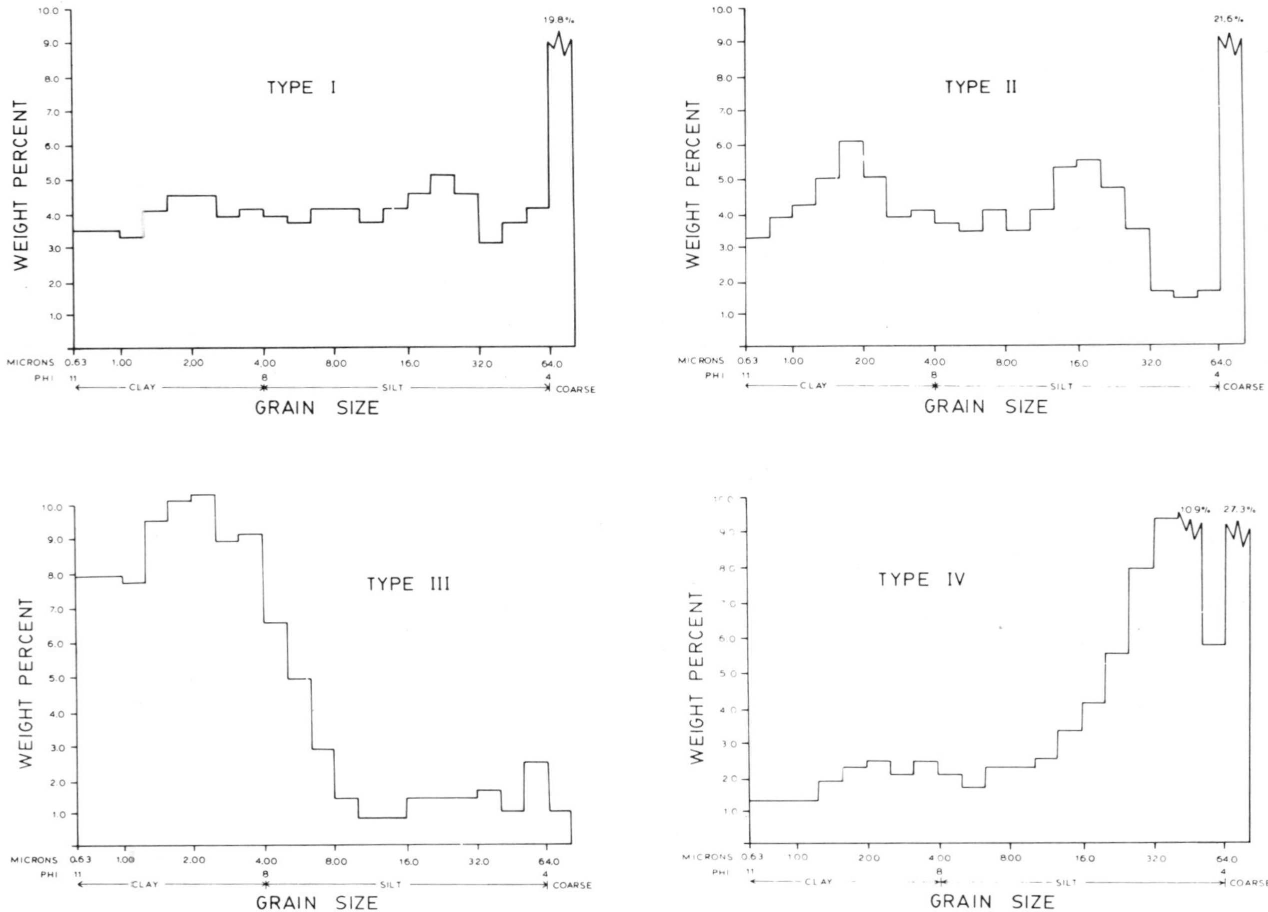


**Figure 24-3.** Four sediment-size types in central Arctic Ocean sediment based on Coulter counter analysis of the silt and clay fractions. (From Clark and Hanson, 1983.)

nated by illite (30–60 percent) with minor but variable amounts of smectite and mixed-layered illite and smectite besides the kaolinite and chlorite (Darby, 1975*a*; Clark et al., 1980). The abyssal plain sediments contain more illite (up to 85 percent) relative to other clay minerals (Campbell and Clark, 1977). In spite of the variation expected from slightly different methodologies, these differences appear significant.

The silt and sand mineralogy is primarily quartz and feldspar with variable amounts of carbonate, micas, and heavy minerals (Darby, 1971; Campbell and Clark, 1977). Modal analyses of sand grains from abyssal turbidites indicated that monocrystalline quartz dominates (47–63 percent) this size fraction with 6–21 percent chert; 8–15 percent detrital carbonate grains (75–95 percent dolomite); 1–9 percent K-feldspar; < 1 percent plagioclase; 3–10 percent limonite; and less than 1 percent of amphibole, pyroxene, and various rock fragments including sandstone, dolostone, limestone, shale, and igneous and metamorphic rock fragments (Darby, 1971; Campbell and Clark, 1977).

Dredged gravels (>2 cm) and erratics (>8 mm) from sediment cores have been described from the Alpha Ridge (Schwarzacher and Hunkins, 1961; Mullen, Darby, and Clark, 1972; Amos, 1985). Carbonate rocks and sandstones were usually the most abundant, accounting for more than half of the erratics. More than half the carbonates were dolostone. Metamorphic and igneous (granitic) rock fragments are about 15 percent of the erractics. Chert comprises about 15 percent also. Thus the dominant provenance is a sedimentary terrane, primarily carbonate (dolostone). Based on the rounded nature of most of the carbonate clasts recovered from the seafloor at CESAR sites, source area was interpreted to be a beach or deltaic environment on the adjacent Canadian landmass (Amos, 1985).

The biogenic component of central Arctic sediment is less than about 6 percent by weight of the total sediment. In most cases it is less than 1 or 2 percent. In spite of this, several fluctuations in foraminifera abundance accounting for as much as 90 percent of the coarse fraction (>62 $\mu$m) have been reported for Plio-Pleistocene Arctic sediment and interpreted as significant paleoclimatic events (Clark, 1969, 1971, 1977; Herman, 1969, 1974; Herman and O'Neil, 1975; Worsley and Herman, 1980; Herman and Hopkins, 1980). The most prolific biogenic contributor to central Arctic sediments is the planktonic foraminifera; however, their genera are limited to *Globigerina* and *Globorotalia* (Herman, 1969, 1970; Clark, 1977). There are several less important benthic foraminifera genera (Herman, 1974; Lagoe, 1976, 1977). Of minor importance are ostracods (Joy and Clark, 1977), microscopic mollusks, echinoderms, pteropods, holothurians, and sponges (Gamber and Clark, 1978; Paul and Menzies, 1973), as well as calcareous dinoflagellates (Gilbert and Clark, 1983; see Chapter 23 of this volume concerning the taxonomy of this group), crinoids, gastropods, radiolaria, tintinnids, and silicoflagellates (Aksu, 1985; Clark, 1977). Several trace fossil types (tracks, trails, and burrows) have been described from bottom photographs and Arctic cores (Kitchell et al., 1978; Clark et al., 1980).

The weight percent of organic carbon in the central Arctic Ocean is generally less than 2 percent with occasional values up to 7 percent (Darby, 1971; Campbell and Clark, 1977). Like other oceans, the organic carbon appears more abundant in finer-grained sediments (Lisitzin, 1972). Campbell and Clark (1977) found an average of 1 percent (standard deviation 1.5%) organic carbon in silty turbidite layers and 1.2 percent (standard deviation 0.5 percent) in clayey nonturbidite layers in the same cores from the Canada Basin. While these means are not significantly different,

85 percent of the clay-rich layers in these cores had higher organic carbon than silty layers.

## DEPOSITIONAL PROCESSES IN PLIO-PLEISTOCENE CENTRAL ARCTIC SEDIMENTS

### Ice Rafting

The Plio-Pleistocene olive brown to gray lutites in the central Arctic Ocean have long been thought to have an important ice-rafted component due to the presence of occasional erratics(>8 mm) and several sandy lutite layers in the upper 3–4 m of cores (Schwarzacher and Hunkins, 1961; Clark, 1969). Ice rafting is more apparent on the ridges where the seafloor is strewn with pebble- to cobble-size clasts and where sandy lutite (especially textural types I and IV, Fig. 24-3) is most abundant. Small, pelletlike, irregularly shaped mud clasts have been used as further evidence of an ice-rafted glacial origin for sandy lutite (Clark et al., 1980). These clasts are similar to those described by Ovenshine (1970) as typically glacial in origin and are thought to have formed between ice crystals in foliation bands in glacier ice. In addition, the microsurficial features of quartz grains are typically glacial in origin for many grains (Clark et al., 1980; Margolis and Herman, 1980).

More recently, even much of the finer silty lutite from the ridges and abyssal plains, where turbidite bedding is absent, has been attributed to ice rafting (Clark and Hanson, 1983). Because of the absence of appreciable sand, especially coarse sand, in these silty lutites (primarily textural type III, Fig. 24-3), their origin has been attributed, in part, to rafting by sea ice instead of glacial ice. Apparently significant quantities of silt and clay can be stirred up by waves and currents on the shallow shelves and incorporated into sea ice during the fall freeze (Clark and Hanson, 1983). Coarser sediment, which is incorporated into grounded sea ice, apparently is not often transported into the central Arctic Ocean because this ice is not as mobile and is melted before significant transport occurs.

One might argue that sea-ice rafting is too limited in extent or occurrence to be volumetrically important. Yet, the perennial sea-ice cover over most of the Arctic Ocean coupled with the small volume of glacial ice in this ocean today would favor low biogenic productivity, low glacial input, and thus low sedimentation rates. The weak ocean currents (Galt, 1967; Hunkins, Thorndike, and Mathieu, 1969) and lack of turbidites except in the abyssal plains (Campbell and Clark, 1977) further supports low sedimentation rates today and makes sea-ice rafting relatively more significant.

Volumetrically, ice-rafted sediment (primarily glacial) might constitute more than 70 percent or 80 percent of the sediment on the ridges and shallower depths (<3,800 m) in the central Arctic Ocean (Mullen, Darby, and Clark, 1972; Clark et al., 1980; Clark and Hanson, 1983). Even the abyssal plains and continental slopes, where turbidity deposits have been recognized as volumetrically important, may contain up to 60–70 percent of ice-rafted sediment (Campbell and Clark, 1977; Goldstein, 1983).

Sources of glacial ice are rather limited today. Some of the larger blocks of glacial ice have been attributed to Ellesmere Island and northern Greenland (Stoiber et al., 1960). Other potential glacial sources today are Spitsbergen, Franz Joseph Land, and Severnaya Zemlya. Most, if not all, of the ice from these islands along the eastern margin of the central Arctic Ocean would be swept south into the North Atlantic by surface currents and thus contribute little sediment to the central Arctic Ocean.

During periods of glacial maxima throughout the Plio-Pleistocene, the occurrence of glacial ice peripheral to the Arctic

Ocean should increase. Based on the distribution of preserved glacial features in Canada, a large lobe of the Laurentide ice moved northwestward and most likely into the Arctic Ocean near the MacKenzie River–Amundsen Gulf area (Prest, Grant, and Rampton, 1968; Dyke and Prest, 1987). In addition, there most likely would have been increased glaciation in the Queen Elizabeth Islands and northern Greenland at these times. The narrow shelves adjacent to these potential glacial sources contrast markedly with the East Siberian–Laptev shelf, which is among the widest of the world's oceans. With lowered sea level accompanying glacial maxima, these wide shelves would impose a formidable obstacle to significant calving into the central Arctic Ocean. Furthermore, the Eastern Siberia and the north slope of Alaska were not glaciated for the most part (Flint, 1971). Thus, northern Canada and Greenland might also have supplied most of the glacial ice-rafted sediment to the central Arctic Ocean during glacial as well as interglacial periods.

## Turbidity Flows

Turbidite deposits have been recognized only in the Canada abyssal plain on the basis of Bouma sequences (Ericson, Ewing, and Wollin, 1964; Hunkins and Kutschale, 1967; Campbell and Clark, 1977). Several other Arctic workers have concluded that important volumes of sediment deposited in the Canada Basin and probably other abyssal plains in the Arctic Ocean were deposited from turbidity currents (Clark, 1969; Hunkins, Thorndike, and Mathieu, 1969; Hunkins et al., 1970, 1971; Demenitskaya and Hunkins, 1970; Vilks, Hall, and Piper, 1977; Goldstein, 1983). Campbell and Clark (1977) concluded that up to 68 percent (avg. 27 percent) of the cored sediments in the southeastern portion of the Canada Abyssal Plain adjacent to the narrow Canadian shelf and the Mackenzie cone were deposited by turbidity currents. The abundance of turbidites decreases markedly northward away from the continent with a maximum of 6.5 percent turbidite deposits of the cored sediments in the northwestern Canada Abyssal Plain, adjacent to the Mendeleyev-Alpha Ridge. This suggests that the major source of turbidity deposits in this basin is the western Canadian shelf, in particular the Mackenzie River Delta. The Northwind Ridge, Chukchi Cap, and Mendeleyev-Alpha Ridge system (Fig. 24-1) contribute little, if any, to turbidity flows in the Canada Abyssal Plain (Hunkins and Kutschale, 1967).

Turbidity currents are thought to be channeled down continental slopes by shallow ($<5$ m deep) but wide ($>5$ km) deep-sea channels with poorly developed levees (Hunkins, 1968; Goldstein, 1983). Submarine canyons such as the Barrow Canyon and the Herald Canyon–Charlie Gap might also play a role in funneling turbidity currents into the Canada Abyssal Plain and Mendeleyev Abyssal Plain, respectively (Fig. 24-1).

The presence of chert, detrital carbonate, and metamorphic quartz led Campbell and Clark (1977) to conclude that the lower Paleozoic metamorphic and carbonate sequence of the Canadian Archipelago (Queen Elizabeth Islands) were the most probable source for the Canada Basin turbidites. This agrees with the general decrease in the abundance of turbidite deposits northward in this basin away from Canada.

The texture of these turbidite deposits is primarily that of silty or sandy lutites where the coarsest fraction is usually coarse silt to medium sand with the very fine sand fraction (3-$\phi$) dominating (Campbell and Clark, 1977; Goldstein, 1983). Although all five Bouma units are found, unit A (massive graded) is relatively uncommon. Thus most turbidite sediments consist of finely parallel-laminated silty sandy lutite (Bouma unit B), finely cross-laminated silty sandy lutite (unit C), and parallel-laminated silty lutite (unit D) or massive silty lutite (olive gray lutite of Goldstein, but most likely Bouma unit E). Each unit varies from 1 mm to more than 15 cm in thickness. The general absence of the graded Bouma A units might be due to the location of the cored sediments examined thus far. Most of these are located near the toe of the continental rise and at some distance from likely source areas for turbidity flows on the edge of the continental shelf. Because the coarser sediment associated with the graded unit is most likely deposited closer to the source, few of these units have been recognized in the cored sediments (Campbell and Clark, 1977; Goldstein, 1983).

The olive gray lutites (Bouma unit E) in the Canada Abyssal Plain lack erratics or coarse sand, are gradational with Bouma D units below, are low in pelagic foraminifera, have a fine-tailed size distribution indicative of turbidites (Bouma and Hollister, 1973), and have a mud-rich texture with moderately poor sorting similar to mud turbidites (Rupke, 1975; Stow and Shanugan, 1980). These observations led Goldstein to conclude that these units were deposited by turbidity current instead of normal pelagic deposition.

In the Canada Abyssal Plain, turbidite deposits are interbedded with ice-rafted sediment (Goldstein, 1983). This sediment is either a sandy lutite similar to that found on the ridges with an abundance of coarse sand and pebbles (>20 percent coarser than 62$\mu$m) or a silty lutite that is similar to the turbidite sediment of the Bouma unit E (olive gray) layers. The ice-rafted silty lutite has a darker brown or pinkish color and is more bioturbated than the turbidite sediment of similar texture. These ice-rafted lutites contain small amounts of coarse sand and pelletoids that are of a different color than the silty lutite matrix. Many pelletoids are irregularly shaped, and sometimes highly angular. Their till-like texture in the sandy lutite sediment led Goldstein (1983) to conclude that they were ice rafted, similar to glacial pelletoids found by Ovenshine (1970), rather than fecal pellets. Reworking by currents or winnowing was ruled out as an origin for these pelletoids due to their unconsolidated nature and the lack of grading or tractive current structures associated with them.

Similar silty lutites from the ridges, but lacking pelletoids, were interpreted as ice-rafted sediment from sea ice during times of decreased glacial ice rafting (Clark, 1981*a*). The detailed analysis of quartz grain-surface texture and other sedimentologic features described above by Campbell and Clark (1977) and Goldstein (1983) indicates that much of the nonturbidite sediment in the Canada Abyssal Plain is also ice rafted, either by glacial ice (sandy lutite) or sea ice (silty lutite).

The average sedimentation rate for the Canada Abyssal Plain depends on the frequency of turbidites as well as the depositional rates of ice-rafted and "normal" pelagic sediments. The upper 6 cm sediments from the northern Canada Abyssal Plain average sedimentation rates were 83 mm/1,000 yr with a range of 4–462 mm/$10^3$ yr, based on a $^{14}C$ age date of 700 ± 100 years B.P. (Hunkins and Kutschale, 1967) for a unique foraminifera-rich turbidite, correlated between nine cores in this area (Campbell and Clark, 1977). These rates are more than an order of magnitude higher than the 0.5–5.0 mm/$10^3$ yr, calculated for the adjacent ridge sediments (Clark et al., 1980, and references therein). Assuming that the entire 2.5–3 km sediment thickness in the Canada Abyssal Plain was deposited in 100 my (Clark, 1981*b*), the average sedimentation rate would have been 25–30 cm/$10^3$ yr. This is 3.3–3.6 times the average rate calculated for surface sediments in the northern Canada Basin. This data (300 vs. 83 mm/$10^3$ yr) suggest that earlier Quaternary sedimentation rates might have been higher than late Pleistocene/Holocene rates.

## Current Transported Sediment

As of yet, there is no direct evidence for nonturbidite, current-transported sediment in the modern or Plio-Pleistocene sediments of the central Arctic Ocean. In fact, bottom photographs show well-preserved trace fossils at all depths (Kitchell et al., 1978). Clark et al. (1980) have examined hundreds of sediment cores and concluded that none of the typical sediment characteristics of fluidized sediment flows, grain flows, debris flows, contourites, and density current deposits occur in the central Arctic Ocean. Furthermore, the occurrence of similar sandy lutite units in depths of more than 3,500 m at the toe of the Alpha Ridge and at the ridge crest (<2,000 m) argues against a turbidite origin for these sandy lutites in this area (Clark et al., 1980).

The sparse current velocity data for the Canada Basin and adjacent ridges seem to support these conclusions. For example, Hunkins, Thorndike, and Mathieu (1969) measured bottom currents on the Mendeleyev Ridge of 4–6 cm/s and Galt (1967) found even weaker bottom currents in the Canada Abyssal Plain (1.5–2.6 cm/s). The abundant clay-size fraction (<2 μm) in central Arctic sediments rules out bottom winnowing even of noncohesive sediments such as discussed by Domack (1984) for glaciomarine sediments. Instead, Clark et al. (1980) explained the bimodal silt texture of type II sediments (Fig. 24-3) by portions of the fine-to-medium silt-size particles being separated from the coarser ice-rafted sediments while they settle through mid-ocean depths (200–900 m) and are sorted by highly variable currents that might reach velocities of 51 cm/s (Galt, 1967). This explanation fails to account for the finer sediment that is associated with type II sediment. Apparently the clay-size sediment accumulates when currents are very weak, a problematic coincidence. The explanation also does not consider the variable floc size of fine sediment that might be rafted into the central Arctic Ocean.

If ice rafting is more or less random throughout the central Arctic as Clark et al. (1980) and other workers have concluded, then these mid-ocean currents would have to be much more consistent and widespread than suspected in order to remove some of the fine silt and prevent its deposition over the vast areas in which type II sediment has been found (Clark et al., 1980, p. 40). Instead, the bimodal nature of some of the central Arctic sediment (namely, textural type II), which is not significantly different from the ice-rafted type I sediment, might be better attributed to the original size distribution of the ice-rafted sediment. Glaciers are known to erode weathered as well as fresh rock, grinding the latter into rock flour that might have a coarse silt-size mode (Dreimanis and Vagners, 1971; Vincent, 1983). Weathered rocks, carbonates, and shales would produce a clay-size mode (Flint, 1947). The combination of these two sources might produce the bimodal type II sediment observed by Clark et al. (1980).

Seismic reflection data on the crest of the Alpha Ridge indicate that structures interpreted to represent paleosediment waves occur over an estimated area of 30,000 km$^2$ buried beneath a few meters of Plio-Pleistocene lutite (Hall, 1979). These waves, presumably of sand, have amplitudes of up to 55 m and wavelengths of about 1 km and were formed at a time when there was an assumed net sediment transport from north to south across the Alpha Ridge. While the age and origin of these waves is uncertain, Hall speculated that they resulted prior to the development of the present ice cover, possibly in the mid-Pliocene (3.5–4 my B.P.) when stronger circulation might have existed due to the freshening winds accompanying the onset of glacial conditions.

Indirect evidence for currents sufficiently strong to prevent sediment accumulation or to cause possible erosion has been cited by Hall for parts of the Mendeleyev Fracture Zone. Currents funneled from north to south through the Cooperation Gap near the junction of the Mendeleyev and Alpha ridges were thought to be responsible for this erosion. Bottom nepheloid layers over the ridges and rises surrounding the Canada Basin, but absent on the abyssal plain, attest to the swifter tractive currents (4–6 cm/s vs. 1 cm/s) over the shallower seafloor (Hunkins, Thorndike, and Mathieu, 1969). While these currents are too weak to erode the bottom, they are capable of maintaining fine particles in suspension (Heezen and Hollister, 1964). Hunkins, Thorndike, and Mathieu concluded that these bottom currents were due to deep water flowing into the Canada Basin over a sill in the Lomonosov Ridge and ultimately came from the Norwegian Sea.

Similar evidence exists for current transport of sediment across the Lomonosov Ridge and into the Fletcher Abyssal Plain (Blasco, Bornhold, and Lewis, 1979; Clark and Morris, 1984). Bottom photographs suggest active current erosion on this ridge (Blasco, Lewis, and Bornhold, 1980). Tractive current velocities up to 12 cm/s were measured over the Lomonosov Ridge as a series of pulses directed toward the Canada Basin during the LOREX 79 project (Aagaard, 1980). While these velocities are probably sufficient to erode noncohesive silt and to maintain fine silt-size particles in suspension, cohesive lutite would not be eroded unless the tractive velocities were more than twice those measured (Sundborg, 1956; Mignoit, 1968).

### Aeolian Sediment Transport

Based on 32 snow samples from the Arctic pack ice at three locations, an estimated fallout rate of 3.3–14.0 $\mu g/cm^2$ yr was calculated for the southern Canada Basin (Mullen, Darby, and Clark, 1972; Darby, Burckle, and Clark, 1974). Assuming that all of this dust reaches the seafloor by summer melting and runoff into fissures in the pack ice, only 0.02–0.09 mm/$10^3$ yr would be deposited. Depending on the sedimentation rate from other processes (generally 1–2 mm/$10^3$ yr on the ridges), airborne dust could account for 1–10 percent of the Arctic sedimentation in these areas. Obviously airborne dust would be relatively negligible in the abyssal plains where depositional rates are much higher due to turbidity flows.

## MINERALOGY OF THE CONTINENTAL MARGIN AND BASIN SEDIMENTS

### Clay Mineralogy

Despite the rather poor sample coverage for the Arctic Ocean as a whole, the clay mineral types reflect the source mineralogies of adjacent landmasses. Typically, illite dominates with subequal amounts of chlorite and kaolinite (Darby, 1971, 1975*a*; Naidu, Burrell, and Hood, 1971; Silverberg, 1972; Mowatt, Naidu, and Veach, 1974; Campbell and Clark, 1977; Clark et al., 1980; Naidu, Creager, and Mowatt, 1982; Naidu et al., 1982; Naidu and Mowatt, 1983). Expandable clay minerals and mixed-layered clays are variable but generally occur in minor amounts.

#### *North Bering, Chukchi, and Siberian Seas*

The nature of the clay mineral suites from the larger potential source rivers for the western Arctic seas is fairly well known. Thus dispersal patterns and possible authigenic or diagenetic influences can be interpreted based on the areal distribution of clay minerals in these seas. The average compositions of the clay mineral suites characteristic of each of the major rivers of the west Arctic region indicate subtle but important differences among potential source rivers (Table 24-1). The Colville and Yukon rivers contain expandable clay mineral in excess of 20 percent, while the Kuparuk and Sagavanirktok rivers, adjacent to and east of the Colville River in northern Alaska, contain lesser but appreciable amounts (up to 16 percent and 10 percent, respectively). The clay mineral suites from the other rivers studied are characterized by low contents of expandable clay minerals (generally six percent or less).

The occurrence of appreciable amounts of kaolinite is surprising in these high-latitude rivers, based on previously held broad generalizations regarding the global distribution of clay minerals as related to latitude and resulting weathering regimes (Biscaye, 1965; Griffin, Windom, and Goldberg, 1968). In northern Alaska, the presence of kaolinite-bearing rocks in nearly all of the drainage basins readily explains the observed kaolinite contents of these fluvial sediments (Table 24-1). Similarly, the contents of the other clay minerals (illite, chlorite, expandable minerals) recognized in the fluvial sediments analyzed can be readily reconciled with the known geologic materials presently being eroded within the respective drainage basins (Mowatt, Naidu, and Veach, 1974; Naidu and Mowatt, 1983).

The greater relative abundance of expandable clay minerals in two of the larger Alaskan rivers, the Colville and the Yukon, suggests that their assemblages ought to be useful in delineating fine-grained sediment dispersal patterns. The Yukon River is the predominant source of terrigenous clay minerals for the northern portion of the Bering Sea, and more northerly transport through the Bering Strait results in significant contributions to the Chukchi Sea as well (Naidu et al., 1982; Moser and Hein, 1984). Apparently, the central Chukchi Sea is a major depositional repository for clay minerals derived from the Yukon River (Fig. 24-4). Expandable clay mineral abundances in the Chukchi Sea generally decrease toward the adjacent coasts of Siberia and Alaska. In general, kaolinite and chlorite are present only in relatively minor proportions in sediments from the Chukchi Sea. Kaolinite/chlorite ratios are relatively higher adjacent to the Alaskan coast, occurring areally as tonguelike seaward projections. These relatively high

Table 24-1. Clay Mineralogy of Some Major Arctic Rivers

| *Fluvial or Deltaic Sediment* | *No. of Samples Analyzed* | *Expandable Minerals* | *Illite* | *Chlorite* | *Kaolinite* | *Kaolinite/ Chlorite Ratios* |
|---|---|---|---|---|---|---|
| Kobuk River | 2 | 4 | 66 | 27 | 3 | 0.1 |
| Noatak River | 2 | trace | 66 | 34 | 0 | <0.1 |
| Yukon River | 6 | 21 | 41 | 26 | 12 | 0.5 |
| Kolyma River | 1 | 5 | 59 | 27 | 9 | 0.3 |
| Kolyma-Indigirka Delta | 5 | 3 | 71 | 21 | 5 | 0.2 |
| Colville River | 20 | 26 | 46 | 19 | 9 | 0.5 |
| Kuparuk River | 10 | 16 | 47 | 22 | 15 | 0.7 |
| Sagavanirktok River | 10 | 10 | 58 | 20 | 12 | 0.6 |
| Canning River | 6 | 0 | 82 | 15 | 3 | 0.2 |
| Mackenzie Delta | 33 | 5 | 64 | 19 | 12 | 0.6 |
| Mackenzie River | 7 | 6 | 66 | 15 | 13 | 0.9 |

*Note:* The deltaic sediments are associated with the marine facies of river deltas and were collected off the river mouths. The fluvial samples were from the estuarine-channel bed load (Naidu and Mowatt, 1974, 1983; Naidu, Creager, and Mowatt, 1982; Naidu et al., 1975, 1982; Silverberg, 1972).

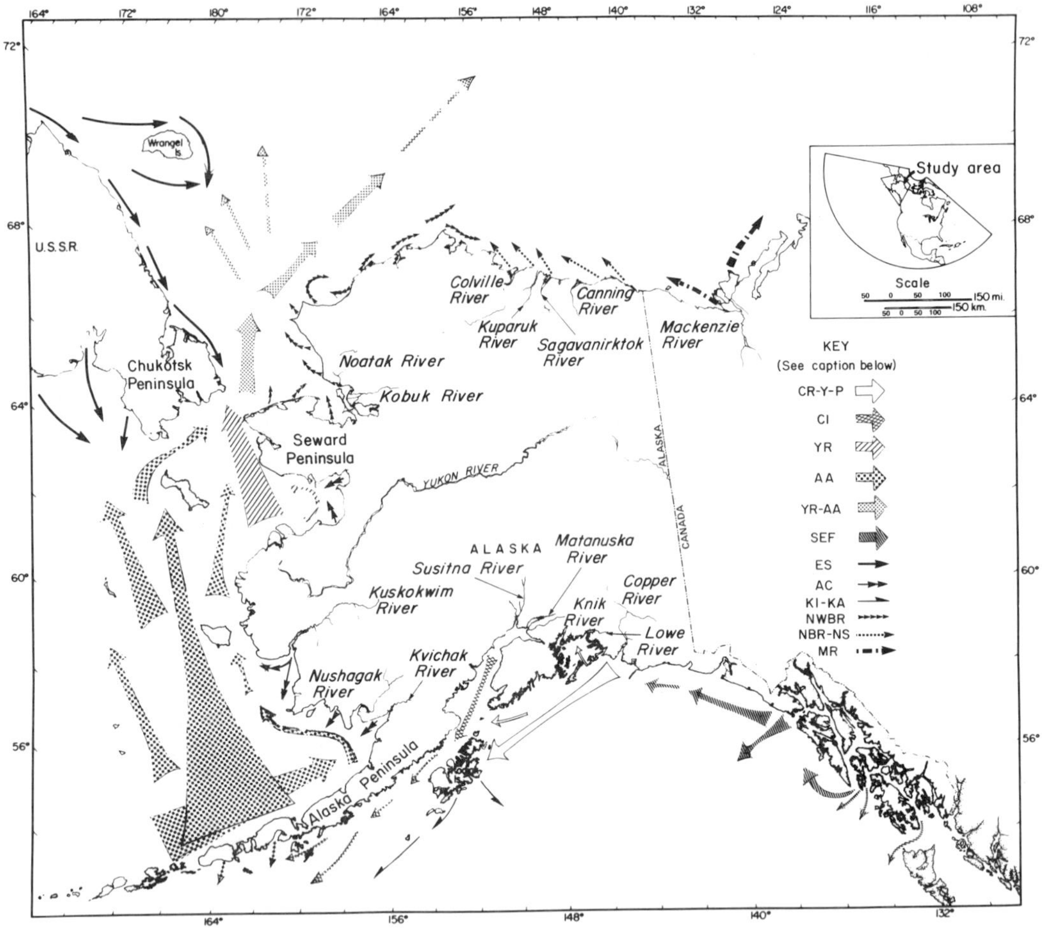


Figure 24-4. Inferred major transport trajectories and depositional sites of clay-size particles in the marginal seas of Alaska (From Naidu and Mowatt, 1983). The clay sources are CR-Y-P, Copper River–Yakataga–Poul Creek Formations; CI, Cook Inlet; YR, Yukon River; AA, Alaska Peninsula; SEF, southeast fjords; ES, east Siberia; AC, southeast coastal Alaska; KI-KA, Kodiak Island-Katmal Ash of Kodiak Shelf; NWBR, West Brooks Range; NBR-NS, North Brooks Range–North Slope; MR, Mackenzie River.

kaolinite/chlorite ratios are readily attributable to the nature of the sedimentary rocks exposed on adjacent land areas, some of which contain significant proportions of kaolinite.

The expandable clay mineral distributions show a concentration in the central portion of the Chukchi Sea, with a generalized bifurcation into two trends (Naidu and Mowatt, 1983), one extending generally westward from Point Hope toward the USSR and the other extending generally northeastward beyond Point Barrow toward the Canada Basin (Fig. 24-4). This latter trend is of particular interest in that it suggests a possible source for the minor expandable clay mineral assemblages of the Canada Abyssal Plain as reported by Campbell and Clark (1977).

Illitic sediments with relatively low proportions of expandable clay minerals are moving into the East Siberian Sea and apparently the adjacent northwestern portion of the Chukchi Sea from the Kolyma and Indigirka rivers of northern Siberia (cf. Table 24-1; Naugler, 1967; Silverberg,

1972). Coastal currents also transport these illitic clays to the Wrangel Island area and the western Chukchi Sea to the southeast.

Naidu, Creager, and Mowatt (1982) presented remote-sensing imagery to define the trajectory of the Alaskan Coast–Bering Strait current component in the Chukchi Sea (Fig. 24-4). This influx is composed chiefly of the warmer Yukon River discharge (Coachman and Aagaard, 1974). Thus, net northward transport of Yukon River clays through the Bering Strait and significant deposition in the central portions of the Chukchi Sea is well established. The presence of turbid bottom waters in the central Chukchi Sea is consistent with this finding (McManus and Smyth, 1970), as well as additionally suggestive that some as yet undetermined proportion of these Yukon River-derived clays may be transported further, possibly to a final site of deposition in the Canada Basin. Thus, transport via Barrow Canyon into the adjacent Canada Basin is a distinct possibility.

### *Beaufort Sea*

In the nearshore portion of the Beaufort Sea, the influx and dispersal of clays from fluvial sources, largely by water currents, are clearly delineated. On the outer continental shelf, clear patterns of clay mineral distributions have not been discerned. This absence of well-defined patterns might be due to sediment transport by ice rafting and currents, as well as by redistribution resulting from ice gouging of bottom sediments and eccentric drainage of fluvial materials onto and through sea ice (Barnes, Schell, and Reimnitz, 1984).

Appreciable regional variations in the abundance of expandable clay minerals have been documented for the bottom sediments across the Beaufort Sea. The marine facies of the Colville River Delta contains the greatest concentrations (about 30 percent), while there is a net seaward decrease to less than 10 percent away from the delta. The adjacent nearshore regions north of the Kuparuk and Sagavanirktok river deltas, as well as the outer continental shelf of the Alaskan portion of the Beaufort Sea and the entire shelf off the Mackenzie River Delta, are characterized by rather low proportions of expandable clay minerals. A narrow isolated area extending east–west in the central portion of the continental shelf north of the Canning River contains appreciable amounts (20–30 percent) of expandable clay minerals, with somewhat less illite than the rest of the adjacent shelf.

The clay mineral assemblages typical of the eastern Chukchi Sea, containing high concentrations of illite, persist eastward of Point Barrow into the Beaufort Sea, but are confined generally to the nearshore, as far east as the western margin of the Colville River Delta. Farther east, the illite variations close to shore can be related to the major river systems.

As in the Chukchi Sea, concentrations of kaolinite of up to 10–15 percent are common in the Beaufort Sea. Chlorite content across the Beaufort Sea generally ranges from 10–20 percent, although a few samples contain up to 37 percent chlorite. The kaolinite/chlorite ratios in individual samples are generally high ($>0.5$) throughout the Beaufort Sea, although appreciable lateral variations are observed. For example, this ratio decreases seaward of the Colville River Delta and the highest average ratios ($>0.7$) extend westward from the Kuparuk River mouth.

The presence of relatively low illite contents, accompanied by relatively high proportions of expandable clay minerals and high kaolinite/chlorite ratios off the Canning River, is somewhat enigmatic. Because of their isolated occurrence relative to the Colville River, these sediments have been considered as relict (Naidu and Mowatt, 1983). A possible source of these materials from the east, in particular from the Mackenzie River, is considered unlikely. The sediments of the western Beaufort

Sea, in general, do not appear to be significantly influenced by contributions from the Mackenzie River. This is evidenced by the presence in this river of relatively higher illite contents (averaging 60–70 percent) and illite/expandable clay mineral ratios compared to western Beaufort Sea sediments.

There is some question as to the fate of at least some of the expandable clay minerals from Alaska subsequent to their introduction into the marine environment, especially those derived from the Yukon and Colville rivers (Fig. 24-4). Apparently much of the expandable clay minerals characteristic of the Yukon River remain essentially unchanged, and can be traced into the Bering and Chukchi seas. But further distribution of these clay minerals into the central Arctic Ocean is somewhat more problematic.

### Central Arctic Ocean

Analyses of clay mineral assemblages in the central Arctic Ocean indicate a relatively homogeneous distribution with most of the sample variance due to expandable clay percents (Table 24-2; Naidu et al., 1975; Darby, 1975*a*; Clark et al., 1980). Perusal of these clay mineral results indicates little in the way of characteristics

Table 24-2. Clay-Mineral Abundances in the <2 μm Fraction of the Central Arctic Deep-Sea Sediments

| *Sample Number* | *Latitude North* | *Longitude West* | *Water Depth (m)* | *Expandable Minerals* | *Illite* | *Kaolinite* | *Chlorite* |
|---|---|---|---|---|---|---|---|
| T3-1 | 74°19′ | 141°51′ | 3,637 | 11 | 62 | 12 | 15 |
| T3-2 | 74°35′ | 142°35′ | 3,650 | 7 | 57 | 13 | 23 |
| T3-3 | 75°18′ | 146°57′ | 3,795 | 14 | 57 | 11 | 18 |
| T3-4 | 75°24′ | 146°57′ | 3,792 | 13 | 56 | 13 | 18 |
| T3-5 | 75°18′ | 146°37′ | 3,761 | 14 | 57 | 13 | 16 |
| T3-6 | 75°60′ | 151°20′ | 3,827 | 8 | 62 | 12 | 18 |
| T3-7 | 75°45′ | 151°14′ | 3,830 | 15 | 55 | 13 | 17 |
| T3-8 | 75°49′ | 153°36′ | 3,833 | 12 | 52 | 20 | 16 |
| T3-9 | 75°25′ | 156°4′ | 2,860 | 6 | 58 | 14 | 22 |
| T3-10 | 75°12′ | 156°35′ | 1,705 | 3 | 65 | 16 | 16 |
| T3-11 | 75°12′ | 156°52′ | 1,160 | 6 | 64 | 11 | 19 |
| T3-12 | 75°45′ | 156°42′ | 3,835 | 5 | 62 | 16 | 17 |
| BSS-51 | 72°51′ | 156°11′ | 2,477 | 7 | 64 | 11 | 18 |
| BSS-54 | 72°1′ | 145°4′ | 3,458 | 4 | 59 | 16 | 21 |
| BSS-97 | 71°52′ | 152°41′ | 1,500 | 11 | 55 | 11 | 23 |
| BSS-92 | 71°20′ | 150°3′ | 1,600 | 5 | 60 | 10 | 25 |
| BSS-91 | 71°20′ | 150°0′ | 1,800 | 5 | 60 | 10 | 25 |
| GLA7122 | 71°9′ | 148°0′ | 1,006 | 20 | 53 | 9 | 18 |
| GLA7158 | 71°15′ | 149°3′ | 1,011 | 18 | 50 | 11 | 21 |
| GLA7185 | 71°23′ | 150°34′ | 1,053 | 25 | 49 | 10 | 16 |
| GLA7157 | 71°20′ | 149°7′ | 1,829 | 12 | 55 | 11 | 22 |
| GLA7120 | 71°14′ | 147°23′ | 2,000 | 10 | 64 | 6 | 20 |
| GLA7186 | 71°30′ | 150°31′ | 2,150 | 19 | 48 | 10 | 23 |
| FL441 | 85°03′ | 121°13′ | 2,431 | 1 | 51 | 24 | 24 |
| FL264 | 82°34′ | 158°1′ | 3,820 | 6 | 58 | 14 | 22 |
| GLA775 | 72°56′ | 146°36′ | 3,593 | 9 | 56 | 17 | 18 |
| GLA776 | 72°46′ | 146°22′ | 3,566 | 10 | 56 | 21 | 23 |
| GLA777 | 72°54′ | 142°5′ | 3,566 | 9 | 56 | 16 | 19 |
| GLA7184 | 71°17′ | 150°24′ | 1,000 | 12 | 52 | 14 | 22 |

*Note:* Abundances are expressed as weighted peak area percents.

definitive of any particular source or mode of origin for the clay minerals in the Canada Basin. The wide ranges of kaolinite (0–20 percent), chlorite (0–23 percent) and illite (20–70 percent) encompass all possible terrigeneous source areas, while the observed range for expandable clay minerals or smectite (0–15 percent) suggests that those sediments with the higher proportions of these clays must be derived from a fluvial source(s), such as the clay mineral suites of the Yukon and Colville rivers where relatively high expandable clay mineral abundances occur.

Carroll (1970) indicated some diagenetic modification of the clay assemblage with depth in 3-m-long cores from the Alpha Ridge. In particular, there appeared to be recognizable increases in the mixed-layered character of the micas, with some development of chlorite and vermiculite. Darby (1975*a*) found no such trend in sediment cores from the same areas. Detailed clay mineral analysis from core FL-441 seems to confirm this lack of change in clay mineral types or abundance with depth, except for the expandable mineral component, which varies considerably (Table 24-3). Lithostratigraphic unit F contains the highest percent of expandable clay, and Darby (1975*a*) found a similar increase at this interval in two other widely spaced cores from the Alpha Ridge. Because expandable clays, such as smectite, commonly are derived from altered volcanic ash, fluctuations in this clay type might reflect times of greater ash input into this part of the Arctic Ocean. A reworked fluvial or glacial source is less probable because of the limited stratigraphic occurrence of these elevated smectite values, but cannot be ruled out.

Few analyses have been reported on the mineral nature of suspended particulate materials from the water column within the central Arctic Ocean. Preliminary results from suspended particle samples off of Ice Island T-3 at 84°30′N latitude and 84°30′W longitude show predominantly a zeolitic component—as yet to be defined more rigorously—with relatively minor proportions of expandable clay and a possible trace of illite in both the surface and 500 m deep water. The suspended materials from 1,500 m depth were similar to shallower samples but exhibited relatively poor crystallinity.

The apparent presence of zeolitic material is noteworthy. No terrigenous source can be identified. Possibly this zeolite could be an alteration product from volcanic ash or siliceous biogenic material within the water column. Zeolitic components have not been reported from Arctic bottom sediments, though such materials are not uncommon in bottom sediments from marine environments elsewhere, particularly in regions of appreciable volcanic activity (Boles, 1981). If the zeolites of the Arctic-suspended sediments are volcanogenic, then the question remains as to the source of the parent volcanic detritus.

## Heavy Minerals and Sand Provenance

Lisitzin (1972) summarized the Russian heavy-mineral studies of surficial sediments, showing a more or less random pattern of heavy mineral provinces in the Arctic Ocean. The dominant provinces were epidote–amphibole and ore (opaque minerals)–epidote–pyroxene. Analyses of heavy minerals from three widely spaced sediment cores along the Alpha Ridge indicate that amphibole dominates with lesser but significant number frequencies of pyroxene, opaques, and garnet (Table 24-4, pages 678–679). The percent of heavy minerals in the sand faction of these three cores averaged 1.45 ± 1.10 percent) (Zauderer, 1982).

The majority of clinopyroxene grains were augite although other varieties were identified. Most of these pyroxenes displayed some degree of etching and some contained delicate coxcomb features that could not have survived tractive transport.

Table 24-3. Stratigraphic Variations of the Percentages of Clay Minerals in the <2 μm Fraction of Core FL-441, Collected from the Alpha Ridge, Arctic Ocean

| *Lithostratigraphic Units*[a] | *Core Section* | *Expandable Minerals*[b] | *Illite* | *Kaolinite* | *Chlorite* | *Kaolinite/ Chlorite Ratio* | *Illite/ Exp.Min. Ratio* |
|---|---|---|---|---|---|---|---|
| | 1 | 1 | 51 | 24 | 24 | 1.0 | 51 |
| | 2 | 12 | 40 | 27 | 21 | 1.3 | 3 |
| | 3 | 10 | 37 | 33 | 20 | 1.7 | 4 |
| M | 4 | 4 | 60 | 17 | 19 | 0.9 | 15 |
| | 5 | 8 | 55 | 18 | 19 | 1.0 | 9 |
| | 6 | 13 | 54 | 17 | 17 | 1.0 | 4 |
| | 7 | 7 | 44 | 24 | 25 | 1.0 | 6 |
| | 8 | 9 | 52 | 21 | 18 | 1.2 | 6 |
| L | 9 | 8 | 46 | 27 | 19 | 1.4 | 6 |
| $K_1$ | 10 | 6 | 56 | 17 | 21 | 0.8 | 9 |
| K | 11 | 5 | 53 | 22 | 20 | 1.1 | 11 |
| $K_2$ | 12 | 7 | 50 | 22 | 23 | 1.0 | 7 |
| | 13 | 3 | 56 | 23 | 18 | 1.3 | 19 |
| J | 14 | 3 | 74 | 11 | 12 | 0.9 | 11 |
| | 15 | 7 | 58 | 16 | 19 | 0.8 | 8 |
| $I_1$ | 16 | 4 | 61 | 15 | 20 | 0.8 | 15 |
| $I_2$ | 17 | 4 | 59 | 17 | 20 | 0.9 | 15 |
| | 18 | 11 | 54 | 19 | 16 | 1.2 | 5 |
| | 19 | 7 | 53 | 19 | 21 | 0.9 | 8 |
| H | 20 | 7 | 49 | 23 | 21 | 1.1 | 7 |
| | 21 | 5 | 57 | 18 | 20 | 0.9 | 11 |
| | 22 | 5 | 61 | 13 | 21 | 0.6 | 12 |
| G | 23 | 5 | 58 | 17 | 20 | 0.9 | 12 |
| | 24 | 5 | 57 | 18 | 20 | 0.9 | 11 |
| | 25 | 12 | 52 | 13 | 21 | 0.7 | 5 |
| | 26 | 12 | 54 | 17 | 19 | 0.9 | 4 |
| F | 27 | 27 | 40 | 19 | 15 | 1.3 | 2 |
| | 28 | 5 | 60 | 15 | 20 | 0.8 | 12 |
| E | 29 | 4 | 62 | 13 | 21 | 0.6 | 16 |
| | 30 | 6 | 62 | 13 | 19 | 0.7 | 10 |
| D | 31 | 10 | 58 | 12 | 20 | 0.6 | 6 |
| | 32 | 8 | 55 | 15 | 22 | 0.7 | 7 |
| C | 33 | 10 | 51 | 18 | 21 | 0.9 | 5 |
| | 34 | 9 | 59 | 11 | 21 | 0.5 | 7 |
| B | 35 | 7 | 62 | 12 | 18 | 0.7 | 9 |
| A | 36 | 7 | 64 | 13 | 16 | 0.8 | 9 |

[a]See Figure 24-5 for lithostratigraphic units.
[b]Includes all minerals that expand upon glycolation and display a peak in the vicinity of 17Å.

Similar etching was described for hornblende grains in Arctic soils of Baffin Island (Locke, 1979). Yet the hornblende and other amphiboles identified in the central Arctic Ocean cores did not generally display etching features.

A diverse assemblage of heavy minerals exists in the Beaufort Sea (Dygas, Tucker, and Burrell, 1972; Naidu and Sharma, 1972; Burrell, Dygas, and Tucker, 1974; Naidu and Mowatt, 1974; Luepke, 1975). This assemblage is dominated by clinopyroxene (Table 24-4). Luepke concluded that the source of these heavy minerals was the Alaskan North Slope and that only garnet, which increased westward, and iron-stained aggregates appeared to have source-related distribution patterns. The percentages of each heavy mineral in the major North Slope rivers are variable but generally compatible with those of the Beaufort Sea.

The heavy mineral suite of the East Siberian Sea is dominated by amphibole (44 ± 8 percent), epidote (19 ± 5 percent) and clinopyroxene (18 ± 8 percent) with lesser amounts of apatite, sphene, orthopyroxene, zircon, tourmaline, and garnet (Naugler, Silverberg, and Creager, 1974). Three heavy-mineral provinces were identified for the East Siberian and Laptev seas: clinopyroxene–garnet (west), amphibole–clinopyroxene–epidote (central), and clinopyroxene–epidote–amphibole–opaque minerals (east) (Silverberg, 1972). The different provinces were attributed to the influences of the major rivers emptying into these seas (the Lena in the west and the Kolyma in the east) as well as reworking by strong currents in the shallower areas such as the New Siberian Shoal region.

Apparently the Kara Sea is dominated by an opaque minerals–epidote–pyroxene province and an epidote–amphibole province (Lisitzin, 1972). The western margin of the Barents Sea contains a pyroxene–amphibole or pyroxene subprovince.

Because of the high frequency of amphibole (primarily hornblende) relative to pyroxene in the central Arctic Ocean sediments, several of these surrounding seas can be eliminated as probable direct sources of sand-size sediment; namely, the Beaufort Sea with its very high pyroxene frequency and low amphibole can be eliminated as a source. Also, the Barents Sea and probably the Kara Sea are unlikely sources for the heavy minerals in the central Arctic due to the higher pyroxene to amphibole ratios and high number frequency of opaque minerals in these areas. The Laptev and East Siberian seas have more compatible heavy-mineral abundances with the central Arctic Ocean; however, sluggish currents over much of the Laptev Sea as well as the small amount of ice rafting of sand-size sediment in this sea (Holmes and Creager, 1974) make it an unlikely sediment source for the central Arctic, at least during modern high sea-level conditions.

The abundance of well-preserved, delicate etching features on the pyroxene grains in the central Arctic indicates a glacial source and transport by ice rafting. Since much of east Siberia and the Alaskan North Slope were unglaciated during the Pleistocene (Flint, 1971), they can be discounted as likely sources for these etched grains and probably the entire sand fraction in the sandy lutites of the central Arctic Ocean. This leaves northern Canada and Greenland as the most probable source areas. The diverse geology of metamorphic, volcanic, and plutonic rocks outcropping along the Arctic coast of Ellesmere and the other Queen Elizabeth Islands as well as northern Greenland could account for the amphibole–pyroxene–garnet–epidote province found in the three Alpha Ridge sediment cores (cf. Douglas and MacLean, 1963; Trettin, 1972). Associated with these igneous and metamorphic rocks are abundant sedimentary rocks, especially early Paleozoic dolo-

**Table 24-4. Number Frequencies of Heavy Minerals (2-4$\phi$) from Lithostratigraphic Unit J of Central Arctic Ocean Sediment Cores (Zauderer, 1982), Beaufort Sea Sediments (Luepke, 1975), and North Alaskan Rivers**

| *Sample* | *Opaque* | *Clino-pyroxene* | *Ortho-pyroxene* | *Altered Pyroxene* | *Amphibole* | *Epidote Group* |
|---|---|---|---|---|---|---|
| Core FL-212 (112–114 cm) | 11.7 | 30.1 | .07 | NC | 45.7 | 9.3 |
| Core FL-284 (67–68 cm) | 19.8 | 18.3 | — | NC | 44.3 | 13.7 |
| Core FL-406 | 2.0 | 21.1 | — | NC | 61.8 | 2.2 |
| Central Arctic | NC | 17.3 | 0.7 | NC | 27.5 | 13.5 |
| Sediments[b] $n=43$ | | ±9.7 | ±1.1 | | ±11.9 | ±10.0 |
| Beaufort Sea $n=42$ | 17.0 | 28.5 | 2.8 | NC | 10.0 | 6.6 |
| | ±6.5 | ±16.1 | ±3.2 | | ±9.5 | ±5.7 |
| Colville River $n=3$ | 50.3 | 44.3 | 6.8 | 17.0 | 4.2 | 1.8 |
| | ±21.5 | ±5.4 | ±4.7 | ±11.6 | ±1.0 | ±2.7 |
| Canning River $n=2$ | 56.0 | 3.9 | 0.7 | 26.7 | 0.5 | 2.2 |
| | ±5.6 | ±0.4 | ±0.9 | ±23.0 | ±0.4 | ±1.8 |
| Kuparuk River $n=2$ | 67.4 | 18.5 | 4.6 | 32.7 | 3.6 | 7.8 |
| | ±11.6 | ±0.7 | ±2.2 | ±12.6 | ±1.0 | ±3.0 |
| Sagavanirktok River $n=1$ | 69.8 | 4.3 | 2.4 | 9.1 | 1.4 | 3.3 |

NC = not counted
[a]Includes apatite, spinel, staurolite,kyanite, andalusite, monazite, and chloritoid.
[b]Average (±1$\sigma$) of all sandy lutite in all three of the above cores.
[c]Includes iron-stained aggregates and altered opaque minerals.

stones and limestones of the northwest Canadian Shield (Hoffman, 1974). This large carbonate source area, which was frequently glaciated, explains the abundant detrital dolomite and calcite in central Arctic sediments discussed earlier. Thus both heavy-mineral and light-mineral data point to northern Canada as a dominant source area for the sands of the central Arctic Ocean.

The heavy minerals derived from northern Canada and probably northern Greenland appear to have changed little during the past 2.5 my based on stratigraphic studies of their distributions in three sediment cores from the Alpha Ridge (Table 24-5). These minerals and probably the entire sand fraction were supplied from the same sources throughout this interval. While the dominant minerals remain essentially constant within the limits of the sample variance, tourmaline increases in lithological units H and I. Likewise monazite, zircon, kyanite, and rutile are slightly more abundant in the older sandy lutite units (C through I). Most of these minerals, which are more abundant in older horizons, are relatively more stable than the other heavy minerals in the central Arctic sediments. This slight increase in the more stable minerals might reflect more prolonged or intense weathering prior to some of the earlier pulses of continental glaciation in northern Canada.

## CENTRAL ARCTIC OCEAN LITHOSTRATIGRAPHY

### Early Developments in Lithostratigraphy

Much of the geological materials recovered by the early expeditions in the Arctic Ocean prior to the mid-1950s were either grab samples or short (less than 15 cm) gravity cores. As a result, not until the late 1950s were sufficiently large numbers of long (greater than 1 m) piston and gravity cores available to even attempt the devel-

| *Tour-maline* | *Zircon* | *Garnet* | *Chlorite Group* | *Sphene* | *Rutile* | *Other*[a] | *Unknown* | *Total Count Nonopaque* |
|---|---|---|---|---|---|---|---|---|
| 0.7 | 0.7 | 12.5 | — | — | — | — | — | 130 |
| 1.5 | 1.5 | 16.8 | 3.8 | — | — | — | — | 131 |
| 0.3 | — | 12.7 | 1.9 | — | — | — | — | 361 |
| 7.5 | 5.6 | 16.6 | 8.7 | Tr | — | 1.0 | — | 36– |
| ±6.4 | ±5.5 | ±8.5 | 4.7 | | | ±1.8 | | 383 |
| 1.0 | 1.0 | 6.2 | 3.3 | 0.4 | 0.1 | 11.1 | 32.1[c] | 300– |
| ±1.0 | ±1.5 | ±10.6 | ±0.4 | ±1.1 | ±0.2 | ±15.3 | ±12.3 | 600 |
| 1.0 | 0.6 | 8.2 | 3.3 | 0.3 | 0.2 | 8.0 | 6.5 | 180– |
| ±0.1 | ±0.6 | ±9.7 | ±2.9 | ±0.3 | ±0.3 | ±4.8 | ±0.6 | 318 |
| 1.4 | 0.1 | 0.1 | 36.1 | 0.1 | — | 18.1 | 2.0 | 131– |
| ±0.4 | ±0.1 | ±0.1 | ±42.7 | ±0.1 | | ±4.1 | ±2.0 | 531 |
| 18.0 | 2.0 | 3.7 | 8.3 | 0.6 | 1.0 | 4.8 | 2.5 | 233– |
| ±2.3 | ±2.2 | ±3.9 | ±0.1 | ±0.4 | ±1.0 | ±3.1 | ±0.1 | 374 |
| 7.7 | 19.7 | 12.0 | 25.5 | — | 3.4 | 5.8 | 4.8 | 208 |

opment of the first Arctic lithostratigraphy.

Belov and Lapina (1962) and Lapina and Belov (1960) studied more than 500 cores from the central Arctic Ocean. These cores averaged 1.0–1.5 m in length, with the longest core approaching 4 m. They recognized that central Arctic Ocean sediments were characterized by dark brown, foraminifera-rich intervals that alternated with light brown, foraminifera-poor intervals. Unfortunately, characteristic sedimentologic features that could distinguish one dark (or light) layer from another were not identified. Thus, it was impossible to make firm correlations between cores in the absence of any unique lithostratigraphic markers or chronostratigraphic datums. Lin'kova (1965) reported the first paleomagnetic stratigraphy from Lomonosov Ridge cores. Unfortunately, no attempt was made to place the Lapina and Belov lithologic units within this paleomagnetic framework.

American scientists occupied several Arctic drift stations between 1952 and 1963, and a number of short cores were collected. Ericson and Wollin (1959) described seven free-fall gravity cores collected from the Ice Island T-3 between 1952 and 1955. These cores ranged from 9 cm to 81 cm in length. Due to the limited number of cores and their short lengths, no attempt was made to correlate these records. Although the data did not exist at that time, Ericson and Wollin pointed out that the way to obtain a usable Arctic lithostratigraphy would be to construct a typical section based on analyses of a large number of cores from widely spaced regions throughout the Arctic.

By the late 1960s, the United States had obtained cores from various regions within the Arctic Basin. It was becoming clear that despite the widespread occurrence of the alternating light and dark units, there were regions that did not follow this pattern. Hunkins and Kutschale (1967) indicated that cores collected from the rises and some marginal areas of the abyssal

Table 24-5. Heavy Mineral Frequencies from Correlated Sandy Lutite Horizons in Three Alpha Ridge Sediment Cores (Zauderer, 1982)

| *Core* | *Litho-stratigraphic Unit* | *Amphibole* | *Clino-pyroxene* | *Garnet* | *Epidote Group* | *Chlorite Group* | *Tour-maline* |
|---|---|---|---|---|---|---|---|
| FL-212 | M | 36 | 22 | 16 | 10 | 3 | 4 |
| FL-284 | M | 41 | 9 | 13 | 15 | 17 | 4 |
| FL-406 | M | 8 | 13 | 38 | 20 | 5 | 1 |
| FL-212 | L | 39 | 33 | 5 | 12 | 4 | trace |
| FL-284 | L | 30 | 26 | 15 | 10 | 7 | 1 |
| FL-406 | L | 18 | 22 | 18 | 28 | 8 | 0 |
| FL-212 | J[a] | 26 | 30 | 17 | 7 | 11 | 4 |
| FL-284 | J[a] | 28 | 13 | 16 | 13 | 19 | 9 |
| FL-406 | J[a] | 18 | 15 | 25 | 22 | 7 | 1 |
| FL-212 | J[b] | 44 | 30 | 12 | 9 | 0 | 1 |
| FL-284 | J[b] | 44 | 15 | 18 | 13 | 8 | 2 |
| FL-496 | J[b] | 50 | 23 | 13 | 2 | 11 | 2 |
| FL-212 | I | 18 | 41 | 7 | 7 | 7 | 11 |
| FL-284 | I | 11 | 3 | 26 | 5 | 5 | 26 |
| FL-406 | I | 12 | 17 | 30 | 2 | 12 | 14 |
| FL-212 | H | 24 | 28 | 11 | 13 | 7 | 8 |
| FL-284 | H | 25 | 17 | 12 | 12 | 9 | 13 |
| FL-406 | H | 18 | 17 | 23 | 11 | 6 | 11 |
| FL-212 | F | 25 | 5 | 20 | 15 | 12 | 14 |
| FL-284 | F | 24 | 13 | 22 | 15 | 12 | 8 |
| FL-406 | F | 23 | 14 | 21 | 12 | 6 | 7 |
| FL-212 | C | 22 | 7 | 22 | 5 | 10 | 5 |
| FL-284 | C | 39 | 11 | 8 | 13 | 17 | 1 |
| FL-406 | C | 39 | 8 | 11 | 13 | 12 | 4 |

[a]White layer in lithostratigraphic unit J.
[b]Pink layer in lithostratigraphic unit J.

plains did follow this pattern; however, some of the abyssal plain sediments contained completely different sediment textures. For example, the Canada Abyssal Plain sediments were composed primarily of olive-gray lutites.

In 1963, the United States began a decade-long occupation of the Ice Island T-3. A regular coring program was undertaken, and by 1973, the USGS recovered over 600 cores for heat-flow studies. Most cores are stored at the University of Wisconsin. Increased experience and improvements in coring technique resulted in the routine recovery of 2.5–3.5-m cores, with some cores nearly 6 m long. Early paleomagnetic studies of these cores (Steuerwald, Clark, and Andrew, 1968; Hunkins et al., 1971) indicated that sediments as old as 3–5 my had been recovered. Sedimentological studies of these cores showed that these longer records were not entirely composed of alternating light and dark brown lutites.

Herman (1970) classified the Arctic sediments into three climatic units. The most recent unit (Unit I) was laid down during the last 0.9 my. This unit consists primarily of the alternating light and dark layers previously described (Herman, 1969). Unit II consists of sediments that are poor in Fe and Mn oxides as well as foraminifera deposited between $\approx$0.9 mya and $\approx$2.5 mya. Some of the coarsest sediments occur within this unit. The oldest unit, Unit III, deposited between $\approx$2.5 mya and 4.5

| *Kyanite* | *Monazite* | *Zircon* | *Rutile* | *Sphene* | *Xeno- time* | *Ortho- amphibole* | *Ortho- pyroxene* | *Other* |
|---|---|---|---|---|---|---|---|---|
| 0 | 3 | 0 | 0 | 0 | 0 | 0 | 1 | 5 |
| 0 | 0 | 0 | 0 | 0 | 0 | 0 | 0 | 0 |
| 0 | 10 | 2 | 1 | trace | 3 | 0 | 0 | 0 |
| 0 | 4 | trace | trace | trace | trace | 1 | 1 | trace |
| 0 | 4 | 1 | 0 | 2 | 1 | 1 | 1 | 0 |
| 0 | 1 | 0 | 0 | 0 | 1 | 1 | 1 | 1 |
| 0 | 2 | 2 | 0 | 0 | 0 | 0 | 0 | 0 |
| 0 | 3 | 0 | 0 | 0 | 0 | 0 | 0 | 0 |
| 0 | 8 | 1 | 1 | trace | 0 | 1 | 0 | 0 |
| 0 | 1 | 0 | 0 | 0 | 0 | 0 | 0 | 0 |
| 0 | 0 | 2 | 0 | 0 | 0 | 0 | 0 | 0 |
| 0 | 0 | 0 | 0 | 0 | 0 | 0 | 0 | 0 |
| 0 | 3 | 2 | 1 | 0 | 1 | 0 | 2 | 0 |
| 0 | 5 | 16 | 3 | 0 | 0 | 0 | 0 | 0 |
| 2 | 5 | 6 | 6 | 0 | 0 | 0 | 0 | 0 |
| 1 | 3 | 2 | trace | 1 | 0 | 0 | trace | 1 |
| 2 | 1 | 4 | trace | 2 | trace | 1 | trace | trace |
| 3 | 2 | 1 | 1 | 1 | 0 | 2 | 2 | trace |
| trace | 1 | 3 | 0 | 0 | 0 | 1 | 1 | 2 |
| 0 | 7 | 0 | 1 | 0 | 1 | 0 | 1 | 0 |
| 1 | 2 | 9 | 9 | 0 | 0 | 1 | 0 | 0 |
| 4 | 7 | 8 | 1 | 1 | 0 | 0 | 1 | 5 |
| trace | 7 | 1 | 1 | trace | trace | 0 | 0 | 0 |
| 1 | 7 | 2 | 2 | 1 | 1 | 0 | 0 | 0 |

mya, consists of sediments that are rich in Fe and Mn oxides and poor in foraminifera (Chap. 23).

This study was based on material recovered from the Ice Island T-3 between 1963 and 1967 by the Lamont-Doherty Geological Observatory. The collection consists of 74 piston cores, 54 trigger weight cores, 1 Phleger core, and 30 stations where small camera tripod or Gatlin gravity cores were taken as well as several sediment dredges. The data used to define these climatic units are from the longest cores (Herman, 1974; Chap. 23 of this volume).

The vast majority of the T-3 cores were collected by the USGS for heat-flow studies of the Arctic Basin (Lachenbruch and Marshall, 1966). The sedimentology of these cores was studied by Clark and associates. Darby (1971, 1975*b*) correlated 7 detrital carbonate maxima among 11 cores with a wide geographic distribution in the central Arctic Ocean. Thirteen lithostratigraphic units (Clark et al., 1980) are correlatable throughout most of the Amerasian Basin west of the Lomonosov Ridge. The lithology of all the LDGO cores is routinely described by the Core Laboratory's technical staff. Lithostratigraphy has proven to be the quickest method of correlation.

## Central Arctic Ocean Ridge and Ridge Flank Lithostratigraphic Units

Nearly all of the longest sediment cores (3–5 m in length) from the ridges and ridge flanks of the central Arctic Ocean reveal a more or less complete sequence of 13 lithostratigraphic units, initially described by Clark et al. (1980), refined by Minicucci and Clark (1983), and extended to older units by Mudie and Blasco (1985). The key to this lithostratigraphic sequence is the weight percent of sand-size material; however, the occurrence of several obvious sandy lutites interbedded with silty lutites and the presence of two pink and white (Munsell color 10YR5/3 and 10YR8/2, respectively), coarse, sandy lutite layers (0.5–2-cm-thick) are important to the recognition of the lithostratigraphic units (Fig. 24-5).

Other sediment characteristics useful for distinguishing each lithostratigraphic unit are (1) amount of bioturbation, (2) nature of unit contacts, that is, whether contacts are distorted, burrowed, gradational, or sharp, (3) abundance of Mn micronodules (microaggregates), (4) abundance of foraminifers, (5) thickness and nature of laminae, (6) paleomagnetic reversals, (7) abundance of erratics, and (8) color. The key lithostratigraphic units are the sandy lutites labeled C, F, H, J, L, and parts of M in Fig. 24-5. For example, unit C is the only sandy lutite with a median silty lutite and with light brown to light olive-brown silty lutite above and below (units B and D). Unit H can be subdivided into as many as five subunits (Fig. 24-5). Besides the key sandy units, each lithostratigraphic unit contains features that aid in recognition and correlation (cf. Clark et al., 1980; Mudie and Blasco, 1985).

The lithostratigraphy described by Clark et al. (1980) was based on sediment cores from the western part of the Alpha Ridge, the Mendeleyev Ridge, and the Chukchi Cap. Minicucci and Clark (1983) extended the correlations to cores from the eastern Alpha Ridge with minor modifications. In general, there is a gradual increase in the frequency of additional sandy lutite layers within units A, D, G, I, and K from the western and central Alpha Ridge area to the eastern Alpha Ridge nearest to Greenland (Minicucci and Clark, 1983, p. 346) (Fig. 24-5). Thus the lithologic sequence becomes coarser as one moves closer to land.

A similar lithologic sequence exists in LOREX cores (Fig. 24-2), differing only in detail from the Alpha Ridge sequence (Morris, Clark, and Blasco, 1985). The lithostratigraphic units recognized in cores from the Lomonosov Ridge represent the same major sedimentary events as those of stratigraphic units K, L, and M of the Canada Basin (Fig. 24-5; Chap. 23). Thus rafting of glacial bergs covered more than 500,000 km$^2$ of the western Arctic Ocean with correlatable thin deposits.

### *Cyclicity*

Units A through F represent two cycles of burrowed silty lutite that change upward into nonburrowed silty lutite and then to sandy lutite (units A to C and D to F). Units G and H might also represent a third cycle of similar sedimentation. Each cycle begins with an interval of slow deposition (bioturbated units A and D), grading upward into a more rapidly deposited silty lutite (units B and E) characterized by much less burrowing and fewer ferromanganese particles. Each cyle is climaxed by rapidly (?) deposited coarse sediment (units C and F) that might represent deposition of glacial ice-rafted detritus during the deglacial/interglacial part of the climatic curve of Clark and Morris (1985).

**Figure 24-5.** Lithostratigraphic units $A_3$ to M in ridge and ridge flank sediments from the central Arctic Ocean. Thickness and other sediment parameters are averaged for 67 Arctic cores studied in detail (Clark et al., 1980) except for lithostratigraphic units $A_1$, $A_2$, and $A_3$, which are based on CESAR core 14 (Mudie and Blasco, 1985). The paleomagnetic interpretation is based on Aksu (1985), Clark (1970), Clark et al. (1980), and Jones (1987), using the paleomagnetic time scale of Mankinen and Dalrymple (1979).

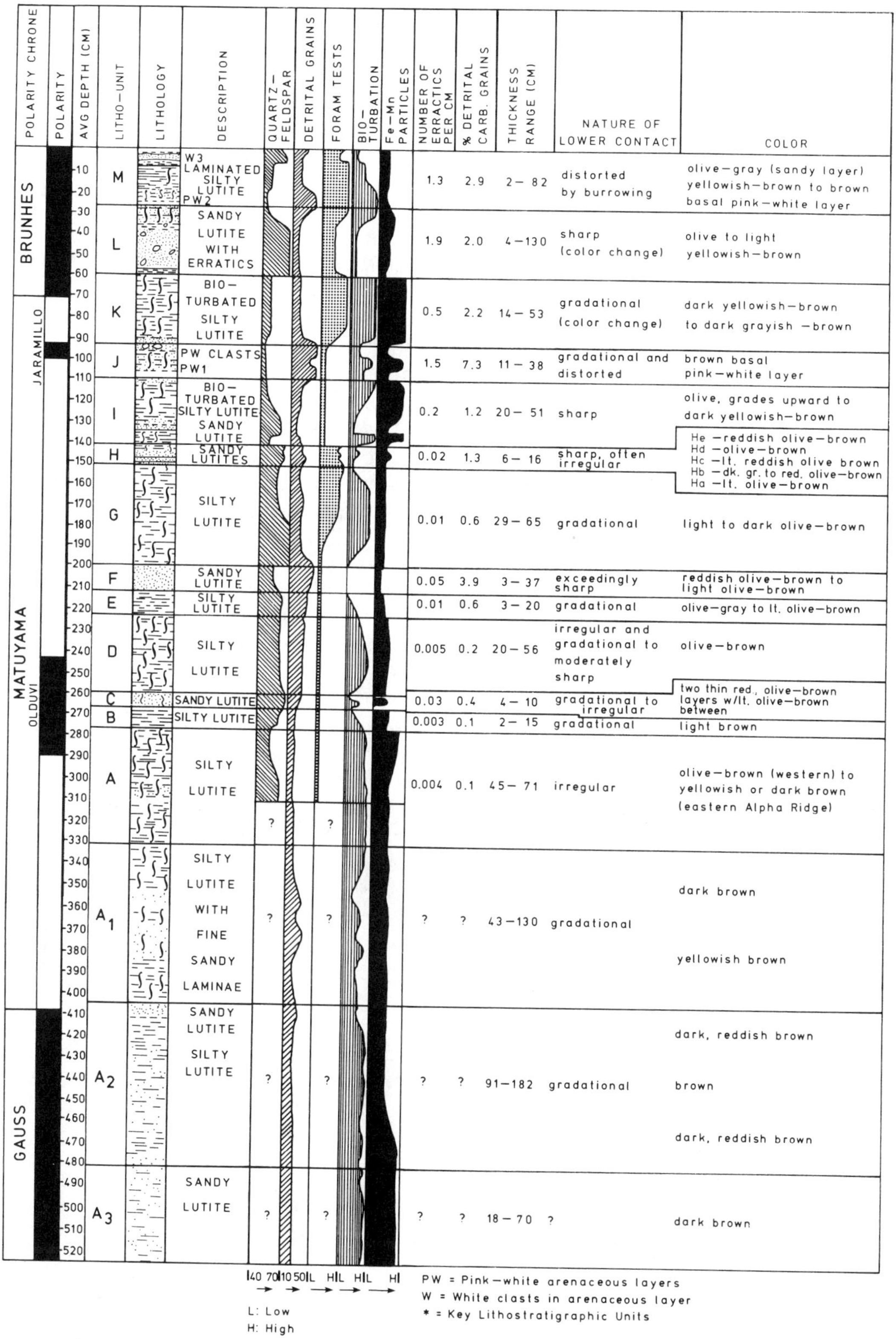
POLARITY CHRONE
POLARITY
AVG DEPTH (CM)
LITHO-UNIT
LITHOLOGY
DESCRIPTION
QUARTZ-FELDSPAR
DETRITAL GRAINS
FORAM TESTS
BIO-TURBATION
Fe-Mn PARTICLES
NUMBER OF ERRACTICS PER CM
% DETRITAL CARB. GRAINS
THICKNESS RANGE (CM)
NATURE OF LOWER CONTACT
COLOR
BRUNHES
JARAMILLO
MATUYAMA
OLDUVI
GAUSS
M W3 LAMINATED SILTY LUTITE PW2 1.3 2.9 2- 82 distorted by burrowing olive-gray (sandy layer) yellowish-brown to brown basal pink-white layer
L SANDY LUTITE WITH ERRATICS 1.9 2.0 4-130 sharp (color change) olive to light yellowish-brown
K BIO-TURBATED SILTY LUTITE 0.5 2.2 14- 53 gradational (color change) dark yellowish-brown to dark grayish -brown
J PW CLASTS PW1 1.5 7.3 11- 38 gradational and distorted brown basal pink-white layer
I BIO-TURBATED SILTY LUTITE SANDY LUTITE 0.2 1.2 20- 51 sharp olive, grades upward to dark yellowish-brown
H SANDY LUTITES 0.02 1.3 6- 16 sharp, often irregular
He -reddish olive-brown
Hd -olive-brown
Hc -lt. reddish olive brown
Hb -dk. gr. to red. olive-brown
Ha -lt. olive-brown
G SILTY LUTITE 0.01 0.6 29- 65 gradational light to dark olive-brown
F SANDY LUTITE 0.05 3.9 3- 37 exceedingly sharp reddish olive-brown to light olive-brown
E SILTY LUTITE 0.01 0.6 3- 20 gradational olive-gray to lt. olive-brown
D SILTY LUTITE 0.005 0.2 20- 56 irregular and gradational to moderately sharp olive-brown
C SANDY LUTITE 0.03 0.4 4- 10 gradational to irregular two thin red. olive-brown layers w/lt. olive-brown between
B SILTY LUTITE 0.003 0.1 2- 15 gradational light brown
A SILTY LUTITE 0.004 0.1 45- 71 irregular olive-brown (western) to yellowish or dark brown (eastern Alpha Ridge)
A1 SILTY LUTITE WITH FINE SANDY LAMINAE ? ? 43-130 gradational dark brown yellowish brown
A2 SANDY LUTITE SILTY LUTITE ? ? 91-182 gradational dark, reddish brown brown dark, reddish brown
A3 SANDY LUTITE ? ? 18-70 ? dark brown
-10 -20 -30 -40 -50 -60 -70 -80 -90 -100 -110 -120 -130 -140 -150 -160 -170 -180 -190 -200 -210 -220 -230 -240 -250 -260 -270 -280 -290 -300 -310 -320 -330 -340 -350 -360 -370 -380 -390 -400 -410 -420 -430 -440 -450 -460 -470 -480 -490 -500 -510 -520
40 70 10 50 L H L H L H
L: Low
H: High
PW = Pink-white arenaceous layers
W = White clasts in arenaceous layer
* = Key Lithostratigraphic Units

The upper units (I through M) might represent another type of cyclic deposition similar to that represented by units A through H. Each cycle begins with a silty lutite such as unit I and grades up to a sandy lutite such as unit J. Three such cycles can be identified (units I to J, K to L, and middle M to upper M [W3]). In all, at least six cycles are represented by the lithostratigraphic units A through M.

Each of the older cycles (lithostratigraphic units A through H) includes an average sediment thickness of about 60 cm. Using an average sedimentation rate of 2 mm/$10^3$ yr, each cycle is approximately 0.3 my or about three times the 0.1-my frequency found in many $\delta^{18}O$ records (e.g., Hays, Imbrie, and Shackleton, 1976; Ruddiman and McIntyre, 1981). With each younger cycle, the average sediment thickness decreases (45 cm in I through J, 32 cm in K through L, and 22 cm in lower M to upper M). Using a slower deposition rate of 1 mm/$10^3$ yr for these intervals, as dictated by the paleomagnetic interpretation of Jones, Clark, and King (1987), these younger cycles range from 0.22 my to 0.45 my. Examination of the lithology in Figure 24-5 or in individual cores, such as done by Boyd et al. (1984), shows more frequent sandy lutite layers than the more general pattern of cycles described here. Boyd and co-workers were able to resolve a 0.1-my frequency in Arctic deep-sea cores. Yet the crude cyclicity of silty lutite to sandy lutite is discernable and might suggest a lower-frequency climatic driving force (0.2–0.4 my) for Arctic Ocean sedimentation with higher-frequency (0.02, 0.04, or 0.1 my) cycles superimposed.

### Lithologic Units and Textural Types: Implications for Depositional Processes

As discussed earlier, if the silty lutites were largely rafted by pack ice and the sandy lutites by icebergs, then the sediment textural types that should dominate the sandy lutite layers are those interpreted as representing glacial ice sedimentation such as types I and IV as well as type II. Clark et al. (1980) interpreted type II sediment as current winnowed but it is also probably glacial in origin. The textural type III representing reduced glacial input but active sea-ice sedimentation should dominate the silty lutites. This is especially evident in the sandy lutite units where type III comprises only 0–25 percent of the sediment. The silty lutite units contain a higher proportion of type III sediment as expected (21.5 percent ± 12.6 percent but are not dominated by this textural type). The same holds for the eastern Alpha Ridge area (Minicucci and Clark, 1983). Obviously either the interpretation of sedimentary processes based on the clay and silt-size distribution for one or more of these textural types is in error or sea-ice transport of sediment accounts for only about one-fifth of the silty lutite deposits and a maximum of 42.8 percent in unit K (cf. Clark et al., 1980).

Because the sand-size component and occurrence of erratics and mud clasts support a glacial iceberg origin for textural types I, II, and IV, the importance of sea-ice transport of shelf sediment into the central Arctic is probably reflected by the small proportion of type III sediment in the silty lutite units (20 percent). The majority of the silty lutite sediment is probably derived from glacial iceberg rafting, supplemented by other marine processes such as biogenic accumulation, aeolian input, and current-wafted fines. Of course, a small (<2 percent) portion of the silty lutite sediment is made up of authigenic ferromanganese particles that typically form in slowly accumulating sediment such as the silty lutite.

### Unconformities and Pre-Glacial Arctic Sediments

Only four cores from the western Alpha Ridge and Chukchi Cap areas were found to contain unconformities in the lithostratigraphic sequence (Clark et al., 1980). The unconformity in one core (FL-278) involves the three youngest units that might be missing due to the coring process. Unit L is missing along with parts of M or I, J, and K in the other three cores (FL-218,FL-268 and FL-283). All of these cores are located less than about 30 km to the east of seamounts in 2,600–3,600 m water depths. While nearby cores lack unconformities, the role that local topography might have played in slumping or concentrating tractive currents, which might explain the missing units, is uncertain.

Unconformities are more common in cores from the eastern Alpha Ridge (Minicucci and Clark, 1983). Of the 48 cores studied from this area, 29 contained unconformities, most involving the units above H. Minicucci and Clark (1983) suspected that stronger bottom currents in this area were responsible for the increase in missing units relative to more western ridge areas.

Four cores thus far revealed older sediment. Core FL-437 contained a Late Cretaceous (Campanian–Maestrichtian?) "slumped" silicoflagellate ooze, overlain by 46 cm of Pleistocene gray-brown lutite (Ling, McPherson, and Clark, 1973; Bukry, 1981). Core FL-422 contained similar slumped sediment of Paleogene age overlain by 12 cm of Pleistocene lutite (Clark, 1974; Kitchell and Clark, 1982). Bukry (1984) discussed this core and Kitchell and Clark's paleoceanographic interpretations. CESAR Core 6 was taken from the northern flank of the Alpha Ridge and also contained siliceous ooze of Late Cretaceous age (Jackson and Mudie, 1984; Mudie and Blasco, 1985; Bukry, 1985). The oldest sediment recovered thus far is 67 cm of black, organic-rich mud of late Campanian or Maestrichtian age from core FL-533, raised from the eastern Alpha Ridge (Clark, Byers, and Pratt, 1986). Overlying this black mud with an average organic content of 15 percent is 20 cm of yellowish banded, siliceous ooze that is highly distorted. Unconformably overlying this sediment is 261 cm of Pleistocene sediment, representing units F to M.

At least two of the FL cores (422 and 437) were taken from steep slopes leading into what Hall (1979) interpreted as graben features near the crest of the Alpha Ridge. These steep slopes coupled with possible earthquakes might have played a significant role in the above-described sediment slumping. Another interpretation is that the unconformities resulted from erosion during intervals of stronger tractive currents than exist today in these areas. "Normal" Arctic deposition blanketed this older sediment subsequent to the slumping or erosional events that apparently occurred during the Pliocene or Pleistocene.

## Central Arctic Ocean Abyssal Plain Lithostratigraphy

The lithostratigraphic units defined for the central Arctic ridge areas cannot be extended into the deep abyssal plain areas. Although Goldstein (1983) recognized ice-rafted deposits in the Canada Abyssal Plain, no correlation between the ice-rafted silty lutite in this basin and the Alpha Ridge was possible. He did note that in both areas, the uppermost ice-rafted sediment was a silty lutite underlain by a sandy lutite. Thus the possibility of such a correlation between the ridge and abyssal areas is open for future work.

A sequence of 20 lithostratigraphic units consisting of turbidites and ice-rafted layers was correlated among many cores in the northern part of the Canada Abyssal Plain and a sequence of 16 units was correlated within the southern part. Correlation between these two areas, separated by more than 300 km, was very tenuous. Apparently, unique marker beds, such as the foraminifer-rich turbidite unit correlated among several cores in the northern Canada Abyssal Plain by Hunkins and Kutschale (1967), are rare and limited in geographic extent.

## PALEOMAGNETIC GEOCHRONOLOGY

While the Clark et al. (1980) Arctic lithostratigraphy is useful in correlating sediments throughout a wide area of the Arctic Ocean, the more difficult task of establishing accurate ages for all of these lithostratigraphic units still remains. Early attempts to establish such a chronology have focused on paleomagnetics (Lin'kova, 1965; Hunkins and Kutschale, 1967; Steuerwald, Clark, and Andrew, 1968; Clark, 1970; Hunkins et al., 1971). More recently, researchers have actively used a wide range of new methods and approaches in order to obtain a more detailed and accurate chronology than was previously possible. A few examples of these new approaches include cryogenic magnetometer paleomagnetic analyses, radiocarbon dating by accelerator mass spectrometer, cosmogenic $^{10}$Be, chemically fingerprinting and correlating volcanic ash found in the Arctic sediments with volcanic eruptions of known age, and planktonic and benthic foraminiferal stable isotope stratigraphy (Chap. 23). The results of these efforts are still being evaluated.

### Paleomagnetic Record

Recently the most fundamental data base of Arctic chronology has been questioned. For nearly 20 years it has been assumed that Arctic ridge and ridge flank sediments have accumulated at rates that are 10–20 times slower than those found in similar areas of other ocean basins. The Arctic rates have been assumed to be 0.1–0.3 cm/1,000 yr based on the published polarity interpretations of paleomagnetic analysis of Arctic sediments.

Sejrup et al. (1984) recently questioned these slow sedimentation rate assumptions and suggested rates 10-50 times higher based on amino acid racemization rates of planktonic foraminifers. The workers reconcile these differences by pointing out that actual paleomagnetic data have been published from only one core, and it is possible that the published paleomagnetic polarity interpretations could be in error. The lack of published paleomagnetic data prevents these polarity interpretations from being unambiguously supported or refuted. In light of this recent controversy some of the unpublished paleomagnetic data that have been used previously to infer low rates of sedimentation will be evaluated.

The first Arctic paleomagnetic data were obtained by Lin'kova (1965). The only data presented were the polarity of each sample, and none of the inclination, declination, or intensity measurements were given. A transition from normal to reversed polarity was shown to occur at an approximate depth of 110 cm in a number of central Arctic cores. If this transition marks the Brunhes/Matuyama boundary (730,000 B.P.), it suggests an average Brunhes sedimentation rate of only 0.1–0.2 cm/1,000 yr. Subsequent studies by Hunkins and Kutschale (1967) and Steuerwald, Clark, and Andrew (1968) followed Lin'kova's interpretation.

Despite these interpretations, the data have been published from only two of the more than 180 Arctic Ocean cores that have been analyzed for paleomagnetics (Hunkins et al., 1971; Witte and Kent, 1988). Interpretations of the polarity records of only 32 other cores have been published. This very limited published data set is the major source of evidence used to support the assumption of very low rates of sediment accumulation in the central Arctic Ocean.

The largest existing paleomagnetic data base is from the University of Wisconsin.

Between March 1968 and July 1976, Clark and his colleagues analyzed nearly 7,500 paleomagnetic samples from 146 cores. All data were obtained with a Princeton SM2 spinner magnetometer. None of the paleomagnetic data have been published and only 20 interpretive polarity records, out of a possible 146, have been published.

The second largest Arctic Ocean paleomagnetic data set is from the Lamont–Doherty Geological Observatory (LDGO) collection. All of the samples from some 30 cores were demagnetized in a 50 Oe field, and inclinations, declinations, and intensities were calculated. The data from these cores are comparable to the University of Wisconsin data, and since only one LDGO core (T-3-67-9) has been incorporated into the currently accepted lithostratigraphic framework of Clark et al. (1980) (see Fig. 24-12), we will not discuss the data at this time.

In order to evaluate the contention of Sejrup et al. (1984) that the published paleomagnetic polarity interpretations might be wrong, the unpublished University of Wisconsin paleomagnetic data that were used for polarity interpretations were examined. None of the inclination, declination, or intensity values had been calculated for any of these nearly 7,500 samples. Only 8 of these 146 cores had been analyzed with six- or three-spin measurements, and only 57 of them had been demagnetized.

In order to obtain as much sedimentologic and paleomagnetic data as possible from many cores shortcuts were made. It was reasoned that since these cores were collected from such high latitudes, the inclination vectors would be nearly straight up or down. Therefore, just the sign of the in-phase value from a one-spin measurement would be sufficient to determine if a sample was of normal or reversed polarity. Additional time was saved by not having to manually calculate the paleomagnetic parameters from the data obtained with the nonautomated spinner magnetometer. The recent claims of Sejrup et al. (1984) required that this raw data be reduced and evaluated.

## Cautions with the Arctic Paleomagnetic Data

A number of caveats must be raised when studying the University of Wisconsin's paleomagnetic data set, and to some extent, the data set of the LDGO. First, true inclinations cannot be calculated from the Wisconsin data set since only a one-spin measurement was made on each sample. However, these data are sufficient to calculate a "pseudo" inclination. The high latitudes from which these cores were recovered produce calculated one-spin "pseudo" inclinations that are within $\pm 5°$ of the true inclinations calculated from three- or six-spin measurements.

Second, 60 percent of these 146 cores were never demagnetized, and another 30 percent were only demagnetized at 50 Oe. Jones (1987) has demonstrated that for most Arctic Ocean samples a demagnetization of at least 150 Oe and preferably 250 Oe is needed to remove the viscous remanent magnetization (VRM) overprint before the true stable remanent magnetization can be obtained (Fig. 24-6). Without demagnetization there are numerous normal polarity values. When demagnetized at 150 Oe, the VRM overprint is removed and spurious reversals are significantly eliminated. This is a common characteristic of Arctic sediments as well as for most all paleomagnetic data of marine and lake sediments (Opdyke, 1972). Because of insufficient demagnetization intensities, only 10 percent of the cores previously analyzed for paleomagnetics yield useful results.

Third, these cores are composed largely of glaciomarine sediments and one would expect the Arctic paleomagnetic records to be noisier than records from regions with-

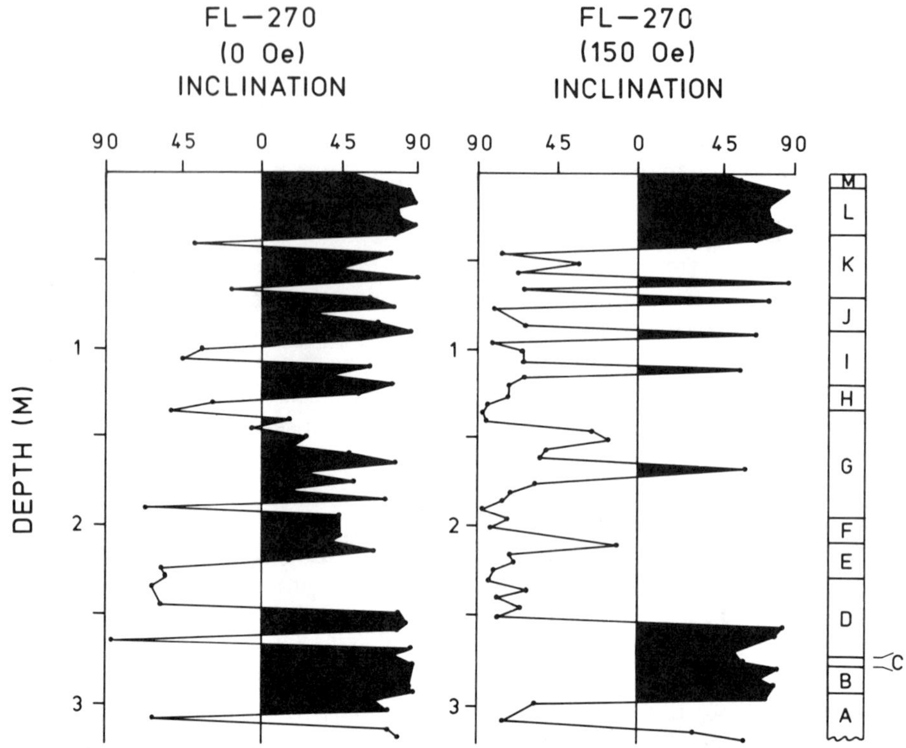


Figure 24-6. Effect of demagnetization at 150 Oe on the paleomagnetic inclinations from core FL-270. Note the removal of many spurious reversals.

out glaciomarine sediments. The main reason is that ice-transported pebbles settle too rapidly to the seafloor to align with the Earth's magnetic field. Any pebble containing a sufficient amount of magnetic material to quickly align with the Earth's field would result in erroneous paleomagnetic measurements due to subtle deviations from this magnetic alignment upon impact on the seafloor. Previous Arctic paleomagnetic studies never quantitatively addressed this problem, so the occasional spurious data due to coarse magnetic grains are difficult to recognize. Many of the anomalous paleomagnetic samples contain a disproportionately large amount of the coarse-grained, pebble-size sediment (Jones, 1987). Considerable improvement in the overall quality of a paleomagnetic record results when any sample containing gravel is rejected from the data set. On the average, this criteria rejects 10 percent of the samples from any one core.

Fourth, Jones (1987) has found that the same demagnetization treatment for all Arctic sediment core samples is inappropriate. The lithostratigraphic units of Clark et al. (1980) are primarily defined by sediment texture. Each of the 13 units are characterized by different grain sizes, compositions, and so forth. Some lithostratigraphic units are optimally demagnetized at 150 Oe, whereas others require demagnetization intensities of nearly 400 Oe.

Because of these four caveats, we can conclude, in the most general sense, that Sejrup et al. (1984) are correct in stating that there really is not a large body of existing good-quality paleomagnetic data from the central Arctic Ocean. However, with the established lithostratigraphy in the western Arctic Ocean, only a very few good-quality records are needed to evaluate the paleomagnetics from this region. As Sejrup et al. contend, if Arctic sedimentation rates are much higher than previously thought, then all of the sediments recovered from this basin, even from the longest core, would have to be of Brunhes age (i.e., all sediments would have to be of normal polarity). Any reversed intervals would have to be interpreted to be one of the smaller magnetic field excursions such as Biawa, Laschamp, or Emperor and would be of very short duration relative to the normal intervals.

## Paleomagnetic Records and Lithostratigraphy

The lithostratigraphic units M and L are characterized by steep (>70°) normal polarity inclination (Figs. 24-7 and 24-8). Within unit K the inclinations change from normal to reverse polarity. The magnetic reversal is not always a clear-cut change from normal to reversed inclination values. Clark et al. (1980) have described the polarity transitions as being gradual. Sejrup et al. (1984) interpreted this statement as meaning the sedimentation rates had to have been very high in order to preserve a gradational polarity transition. However, Clark et al. (1980) had only looked at whether the inphase value was either positive or negative and they did not calculate the inclination vector. Thus their data could not determine if the transition was gradational or not. More commonly, the transition is preserved as an interval of alternating normal and reversed polarity values. This type of noisy behav-

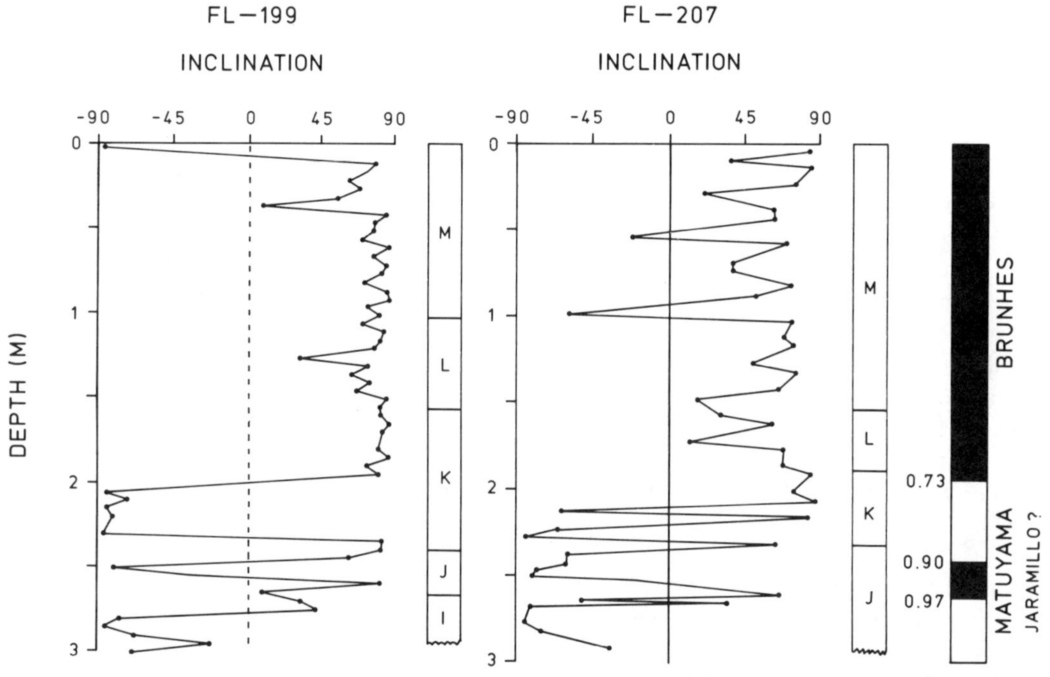


Figure 24-7. The relationship between paleomagnetic inclination values and the lithostratigraphy of two central Arctic Ocean cores (see Figure 24-2 for core locations). The Jaramillo event ((0.90–0.97 mya) is uncertain in these cores. Note the different thickness of the lithostratigraphic units, presumably representing different deposition rates. Yet the Brunhes-Matuyama reversal occurs at about the same depth, representing the same average deposition rate during the past 0.73 my.

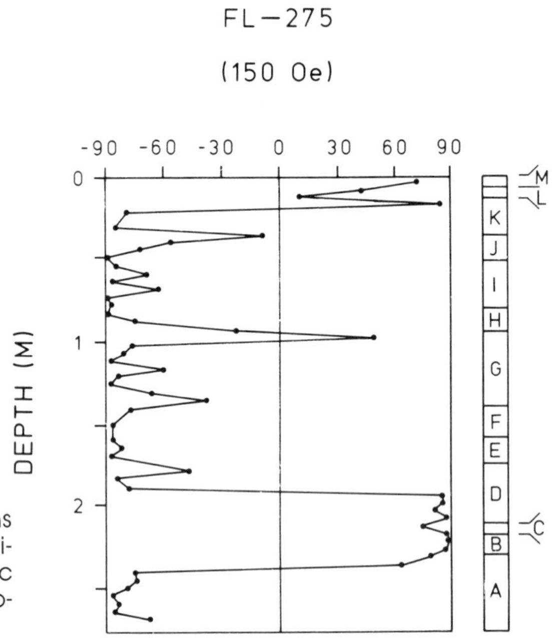


Figure 24-8. The paleomagnetic inclinations in three cores with nearly complete stratigraphic records back to lithostratigraphic unit C or A. Major reversals occur in lithostratigraphic units K, D, and AB.

ior is commonly observed in regions characterized by low sedimentation rates (J. King, personal communication, 1985), and is more the result of sedimentological processes rather than paleomagnetic processes associated with a field reversal.

One possible approach in more precisely establishing the location of a reversal is assuming nonhomogeneous mixing of these sediments by benthic organisms. A comparable example is commonly seen in foraminiferal stable isotope records. The noisiest isotope data occur at the climatic transition, and not within those sediments above or below the transition (Boyle, 1984). Although it is difficult to identify the exact level of the magnetic reversal in many of these cores, it can be confidently stated that the first magnetic reversal occurs within unit K.

The lithostratigraphic units that are characterized by reversed polarity are noisier than the normal polarity units. The most reasonable explanation for this is the incomplete removal of the VRM. As discussed earlier, the demagnetization intensity needed to remove the VRM was rarely obtained for these cores. During an interval of normal polarity an incomplete removal of the normal polarity VRM overprint will not dramatically change the measured inclination values. However, an incomplete removal of the normal polarity VRM overprint during an interval of reversed polarity can dramatically affect the measured inclination values (Fig. 24-6).

The University of Wisconsin, Lamont-Doherty, Russian (Lin'kova, 1965), and CESAR (Aksu and Mudie, 1985) paleomagnetic data, as they have been interpreted, indicate that a magnetic reversal occurs near a depth of 100 cm in many Arctic cores. This reversal consistently occurs *within* lithostratigraphic unit K and is not found at a lithostratigraphic boundary. This is important, because if found at a boundary, the reversal could be interpreted as a major hiatus, and regional rates of sediment accumulation could be higher than they appear. The fact that this reversal occurs within a lithostratigraphic unit supports the hypothesis that it mea-

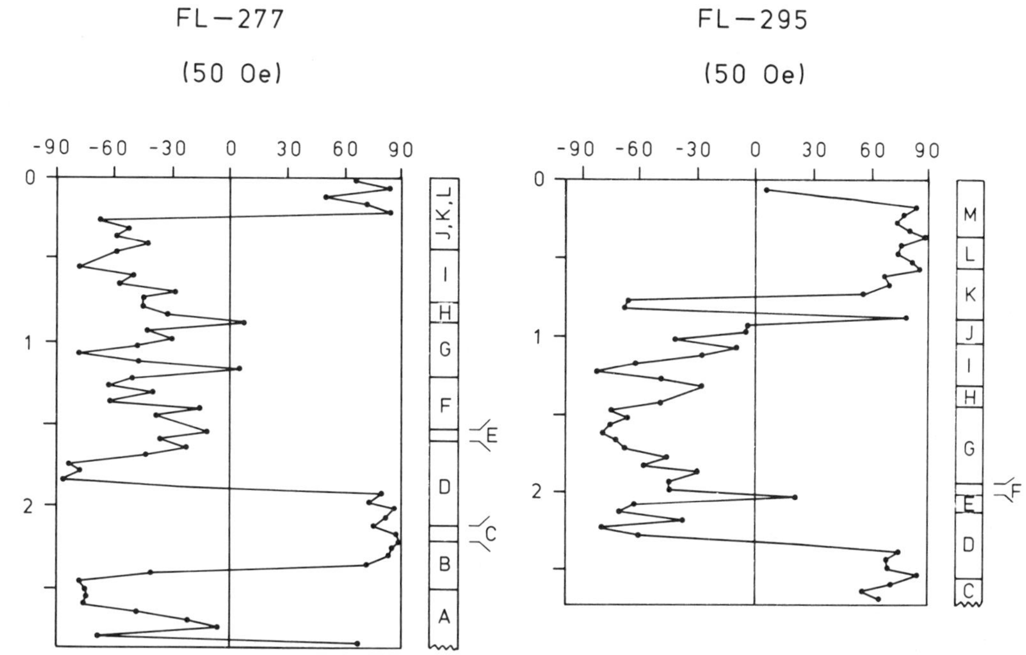


sures a true reversal of the Earth's field and is not the result of sediment processes (Figs. 24-7 and 24-8).

Given the thickness of the sediments below this polarity change that are characterized by reversed polarity, the data would have to be forced into a high-sedimentation rate scheme where this reversed interval is interpreted as one of the short-duration magnetic excursions that have been found within the Brunhes. The only reasonable conclusion is that this polarity transition is indeed the Brunhes/Matuyama boundary, and that Brunhes sedimentation rates are less than 0.2 cm/1,000 yr. This slow rate has also been confirmed by accelerator $^{14}C$ dating of the upper 5 cm of core FL-124 (Clark et al., 1986) and new amino acid data (see Chaps. 22 and 23 of this volume).

Below this major reversal event, there is a hint of a short interval of normal polarity occurring within unit J. The evidence is not overwhelming and is primarily based on a statistical averaging of all the available data from all cores demagnetized at 50 Oe or greater. If the Brunhes sedimentation rates are approximately 0.2 cm/1,000 yr, and we extrapolate these rates downcore, the age of this interval of normal polarity closely corresponds to the age of the Jaramillo (0.9 - 0.97my).

A number of factors contribute to this interval being poorly defined paleomagnetically. A major problem is that if this normal interval is the Jaramillo, and Arctic sedimentation rates are approximately 0.2 cm/1,000 yr, then the average thickness of the sediments recording this event would be 10–15 cm. Most of the cores used for paleomagnetics were sampled at a 5 cm interval. Under optimal conditions there would be no more than two or three samples of normal polarity. Combining this with the problems of less than sufficient demagnetization intensities to remove all VRM effects, and the fact that these are glaciomarine sediments, it is easy to see that this short interval could very easily become absorbed within the noise of the available data. We conclude that where sedimentation rates are of the order of 0.2 cm/1,000 yr, it is impossible to confidently identify the Jaramillo with the existing data, and it will be difficult to identify this event with new data as well.

The next distinctive event below the Jaramillo is an interval of normal polarity encompassing lithostratigraphic units D, C, and B. This data is of very good quality, with inclination values uniformly exceeding 70°, and the event occurs in all cores containing these lithostratigraphic units. The event previously has been interpreted to be the Gauss (Clark et al., 1980; Aksu and Mudie, 1985). However, after studying the existing unpublished paleomagnetic data, as well as recently analyzed cores, the event has been reinterpreted to be the Olduvai (Jones, 1987). Assuming that the overall sedimentation rates for the Arctic Basin have remained relatively constant, as averaged over long periods of time, and extrapolating the Brunhes sedimentation rates, the age of this interval of normal polarity is much more comparable to that of the Olduvai (1.67–1.87 my) than to the Gauss (2.48–3.40 my). In addition, the thickness of this interval is more appropriate for the Olduvai than for the Gauss. Assuming a sedimentation rate of 0.2 cm/1,000 yr, one would expect this interval to be approximately 40 cm long, and based on the 5 cm sampling interval that was used, there should be at least eight samples recording this event. Typically, the zone of normal polarity within units D, C, and B is 30 to 40 cm long, and there are five to ten consecutive samples of normal polarity (Fig. 24-8).

The final polarity change that is observed in these cores is a transition from reversed to normal polarity. This transition occurs approximately 130 cm below the B/A boundary. Jones (1987) has interpreted this polarity transition as the Matuyama/Gauss boundary, dated at approximately 2.48 my (Mankinen and Dalrymple, 1979). Thus, the oldest conformable sediments that have been recovered from Arctic Basin cores, except possibly for CESAR core 14, are no older than 3 my, and are not of Miocene age as has been interpreted previously (Clark et al., 1980). The paleomagnetic and palynological data from CESAR 14 suggest at least a late Pliocene age for lithostratigraphic unit A2 (Aksu and Mudie, 1985).

## CARBONATES IN ARCTIC SEDIMENTS

### Nature of Carbonate Materials

Unlike the carbonate sediment fraction of every other ocean, the dominant carbonate mineral in the central Arctic Ocean is dolomite (Darby, 1971, 1975*b*; Fig. 24-9). Generally calcite constitutes the major portion of deep marine carbonates, reflecting their biogenic nature (Blatt, Middleton, and Murray, 1980). In the Arctic sediment, low Mg calcite occurs with the more abundant dolomite. In some sediment samples the calcite equals or exceeds the dolomite (Tables 24-6 and 24-7). The average dolomite to calcite ratio for nearly 50 carbonate mineral analyses (gasometric combined with X-ray diffraction) from core-top samples where both minerals are present is 3.2 (Table 24-6). As expected, the variance about the mean is high ($\sigma = 4.5$) and 29 percent of the 89 samples in Tables 24-6 and 24-7 contain only one carbonate mineral, most commonly dolomite (80 percent of the cases).

The samples in Tables 24-6 and 24-7 are from both abyssal plain and ridge locations, representing both turbidite and ice-rafted sediment. Thus carbonate is widespread in the Arctic Ocean. In addition, the samples represent both modern (core tops) and Pleistocene sediment. Thus carbonate is common throughout this period, averaging $14 \pm 8$ percent (weight percent calcite + dolomite).

High Mg-calcite is rare but occasionally it is more common than low Mg-calcite (Table 24-7). Aragonite is virtually nonexistent except for an occasional pteropod

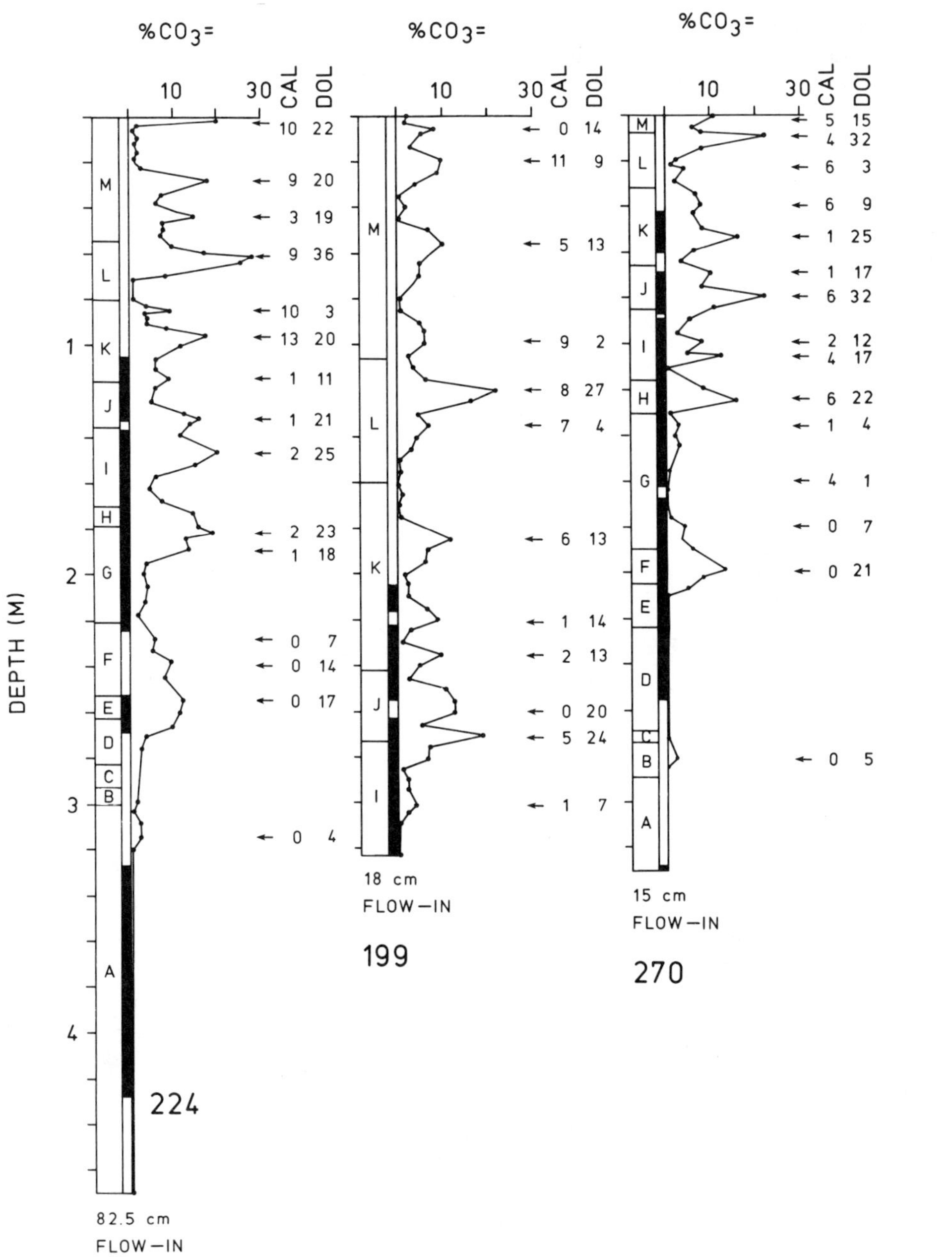


Figure 24-9. Weight percent calcite and dolomite compared to the total weight percent carbonate ($CO_3$) for three Mendeleyev-Alpha Ridge cores. Lithologic units of Clark et al. (1980) and paleomagnetic signature are shown (dark core intervals are magnetically reversed).

shell (Herman, 1970, 1974). It was never detected on the XRD analysis.

Microscopic analysis of the sand-size fraction ($>62$ $\mu$m) indicates that most of the carbonate occurs as microcrystalline (micrite) grains and less commonly as individual crystals of dolomite (rhombs) and calcite. Carbonate lithofragments are occa-

Table 24-6. Ratio of Dolomite to Calcite in Arctic Sediment Core-Top Samples Based on X-ray Diffraction with Available Gasometric Analysis of Carbonate Reported as Weight Percent $CaCO_3$ Plus $CaMg(CO_3)_2$

| Core Number | Latitude | Longitude | Water Depth (m) | Dolomite/ Calcite | % Carbonate |
|---|---|---|---|---|---|
| FL-1 | 82°31′ | 156°29′ | 3,747 | 1.3 | |
| FL-4 | 82°39′ | 157°15′ | 3,738 | ∞ | |
| FL-16 | 83°01′ | 158°55′ | 2,247 | 0.6 | 14 |
| FL-17 | 82°59′ | 159°07′ | 2,215 | 1.0 | 26 |
| FL-19 | 83°03′ | 162°49′ | 3,417 | 0.6 | 29 |
| FL-21 | 83°03′ | 163°31′ | 3,494 | 2.3 | 25 |
| FL-23 | 83°02′ | 163°00′ | 3,748 | 1.0 | 11 |
| FL-24 | 82°23′ | 162°07′ | 3,743 | 1.0 | 11 |
| FL-25 | 82°27′ | 160°37′ | 3,750 | 2.0 | 12 |
| FL-32 | 80°58′ | 136°13′ | 3,698 | 1.0 | 11 |
| FL-33 | 80°55′ | 136°48′ | 3,679 | ∞ | |
| FL-34 | 80°48′ | 136°55′ | 3,675 | 9.0 | 14 |
| FL-35 | 80°46′ | 136°49′ | 3,683 | ∞ | |
| FL-36 | 80°38′ | 137°24′ | 3,683 | ∞ | |
| FL-37 | 80°41′ | 137°14′ | 3,680 | 0 | 11 |
| FL-39 | 80°22′ | 136°47′ | 3,680 | 9.0 | 12 |
| FL-40 | 80°26′ | 136°22′ | 3,668 | 6.1 | 13 |
| FL-57 | 76°38′ | 138°19′ | 3,616 | ∞ | 17 |
| FL-60 | 70°09′ | 141°47′ | 3,702 | ∞ | |
| FL-61 | 76°08′ | 142°09′ | 3,707 | ∞ | |
| FL-68 | 75°29′ | 140°24′ | 3,684 | 13.3 | |
| FL-69 | 75°35′ | 140°30′ | 3,690 | ∞ | |
| FL-92 | 77°38′ | 174°44′ | 2,078 | 2.3 | |
| FL-93 | 77°48′ | 175°25′ | 1,567 | ∞ | 14 |
| FL-94 | 77°53′ | 175°52′ | 1,611 | 0.7 | |
| FL-99 | 77°43′ | 174°51′ | 1,931 | 4.6 | |
| FL-104 | 77°37′ | 173°46′ | 1,848 | 3.0 | 9 |
| FL-106 | 77°38′ | 173°44′ | 1,856 | ∞ | |
| FL-107 | 77°30′ | 173°43′ | 1,873 | 0 | 8 |
| FL-116 | 78°05′ | 174°49′ | 1,570 | 4.0 | |
| FL-122 | 78°08′ | 174°23′ | 1,837 | 1.4 | |
| FL-125 | 78°15′ | 175°23′ | 1,288 | 1.5 | |
| FL-126 | 78°22′ | 175°53′ | 1,181 | 1.9 | |
| FL-129 | 78°34′ | 175°56′ | 1,453 | 1.0 | |
| FL-138 | 78°37′ | 176°24′ | 1,360 | 1.0 | |
| FL-143 | 78°34′ | 175°44′ | 1,519 | 1.0 | |
| FL-149 | 78°34′ | 176°08′ | 1,422 | 3.8 | |
| FL-165 | 78°48′ | 176°08′ | 1,490 | 1.6 | |
| FL-167 | 78°40′ | 175°58′ | 1,672 | ∞ | |
| FL-171 | 78°41′ | 175°54′ | 1,678 | ∞ | |
| FL-179 | 78°52′ | 176°41′ | 1,645 | 4.0 | |
| FL-199 | 80°12′ | 172°48′ | 2,988 | 4.0 | |
| FL-204 | 79°44′ | 170°57′ | 3,157 | 0.8 | |
| FL-207 | 79°42′ | 170°10′ | 3,262 | 1.0 | |
| FL-217 | 80°36′ | 157°59′ | 3,713 | 2.3 | |
| FL-218 | 80°41′ | 158°22′ | 3,654 | 2.1 | |
| FL-221 | 80°33′ | 159°39′ | 3,638 | 1.6 | |
| FL-224 | 80°28′ | 158°49′ | 3,467 | 0.5 | |
| Average[a] | | | | 3.2 | |
| Standard deviation | | | | 4.5 | |

[a]Excluding ∞ values.

Table 24-7. Weight Percent of Carbonate Minerals Based on Gasometric and XRD Analysis of the Sediment in Two Cores from the Alpha Ridge

| *Core Number* | *Depth (cm)* | *% Dolomite (2.88 Å)* | *% Low Mg-Calcite (3.03 Å)* | *% High Mg-Calcite (2.98 Å)* |
|---|---|---|---|---|
| T3 67-9 | 0 | 5.0 | 4.4 | 0 |
| | 10 | 18.0 | 15.3 | 0 |
| | 15 | 5.8 | 4.9 | 0 |
| | 20 | 14.7 | 15.9 | 0 |
| | 35 | 1.6 | 1.7 | 0 |
| | 40 | 7.0 | 10.5 | 0 |
| | 42 | 6.3 | 9.8 | 0 |
| | 55 | 0 | 8.2 | 0 |
| | 75 | 12.6 | 0 | 6.5 |
| | 90 | 5.2 | 5.7 | 0 |
| | 110 | 8.2 | 4.4 | 0 |
| | 115 | 2.7 | 5.7 | 0 |
| | 120 | 10.8 | 4.3 | 0 |
| | 125 | 16.1 | 0 | 0 |
| | 185 | 1.2 | 0.5 | 0.8 |
| | 205 | 23.1 | 0 | 0 |
| | 225 | 18.1 | 0 | 0 |
| | 230 | 11.4 | 0 | 0 |
| | 295 | 0.8 | 0.6 | 0.9 |
| T3 67-11 | 15 | 7.5 | 15.1 | 0 |
| | 17 | 5.6 | 15.6 | 0 |
| | 20 | 0.4 | 8.4 | 0 |
| | 25 | 0 | 0.7 | 0 |
| | 35 | 0 | 0.3 | 0 |
| | 42 | 18.4 | 0 | 0 |
| | 62 | 0.8 | 0 | 0 |
| | 67 | 0.4 | 9.6 | 0 |
| | 90 | 46.2 | 1.6 | 0 |
| | 135 | 9.2 | 8.4 | 0 |
| | 137 | 1.5 | 15.0 | 0 |
| | 140 | 0.4 | 3.9 | 0 |
| | 141 | 2.0 | 2.2 | 0 |
| | 148 | 15.8 | 4.0 | 0 |
| | 150 | 19.3 | 0.1 | 0.1 |
| | 167 | 25.8 | 0.6 | 0 |
| | 170 | 6.2 | 0.1 | 0 |
| | 173 | 3.1 | 0 | 0.5 |
| | 195 | 16.6 | 0 | 0 |
| | 197 | 10.7 | 0.1 | 0 |
| | 200 | 10.8 | 0 | 0 |
| | 202 | 2.3 | 0 | 0 |

*Note:* See Figure 24-2 for core locations.

sionally encountered with fossils (Paleozoic) such as those found in core FL-221 (Fig. 24-10; cf. Mullen, Darby, and Clark, 1972, p. 209).

Most carbonate clasts are angular. This fact, combined with their largest size (boulders; Aksu, 1985), suggests an ice-rafted transport mode into the central Arctic. If most of the carbonate in the Arctic sediments is deposited from glacial ice-rafted bergs, then there should be a good correlation between carbonate and sand content because of the demonstrated glacial origin of the detrital sand in these sedi-

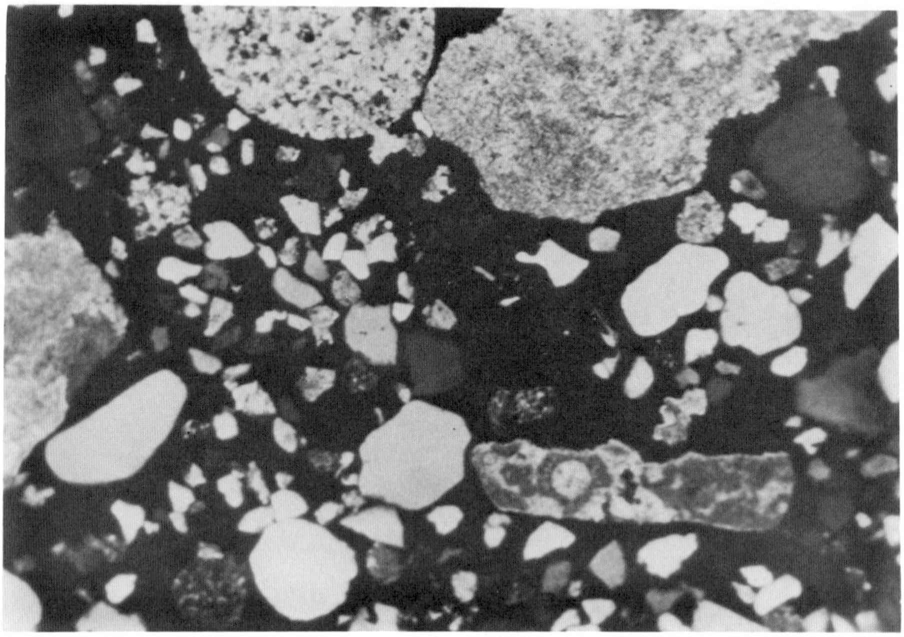

Figure 24-10. Microphotograph of sectioned grain mounts (>62 μm). The four largest grains are carbonate lithofragments showing a microcrystalline to sparry calcite matrix. The elongated grain with the coral is 1.8 mm long. Sample is from 62–66 cm depth in core FL-221.

ments. It turns out that this is the case and the high carbonate abundance corresponds to sandy lithologic units (M, L, J, H, F, and C) (Fig. 24-11). In some cases carbonate comprises the bulk of the sand fraction. For example, at a depth of 45 cm in core FL-275 (lithostratigraphic unit J) carbonate accounts for about 70 percent of the sand fraction (Darby, 1971). Mullen, Darby, and Clark (1972) reported that 30 percent of the glacial erratics in 44 Arctic cores were carbonate lithofragments. The occurrence of coarse detritus and carbonate percentages is not perfect (Figs. 24-11 and 24-12). For example, lithostratigraphic unit M often has a sandy interval with low carbonate (cf. cores 110 and 19). However, except for very few occasions, high carbonate contents are associated with elevated percentages of coarse detritus.

An examination of the carbonate content in the sand and mud fractions separately shows that the bulk of the carbonate is mud-size (Fig. 24-12). This is easily explained by the softness of carbonate rocks and the abundance of rock flour generated by glaciers.

Evidence for partial dissolution of certain calcitic foraminifera tests in the Arctic (Herman, 1969; Hunkins et al., 1971; Herman and Hopkins, 1980; Chapter 23 of this volume) might also affect the nondolomitic portion of the sediment carbonate. The evidence for dissolution is largely circumstantial, based primarily on the decrease in foraminifera with water depth in some cases. The probable origin of Arctic bottom water in the North Atlantic and the 4,500–5,000 m carbonate compensation depth (Li Takahashi, and Broecker, 1969) suggests that dissolution should be minor in the Arctic Ocean, which is no deeper than 5,000 m. Care must be taken not to include turbidite sediments when relating foraminifera abundance to water depth. Some of the sediments deeper than about 3,500 m in the abyssal plains are turbidite deposits in which the biogenic pelagic content is

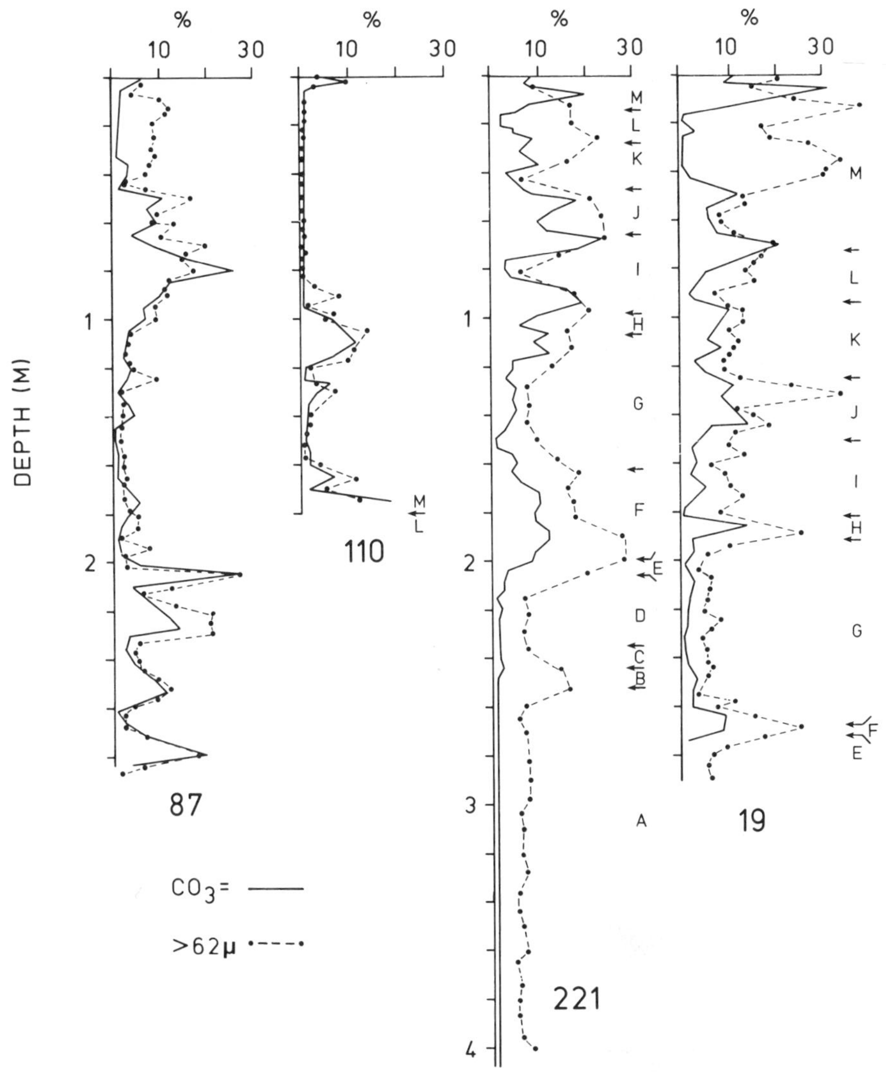


Figure 24-11. Relationship between carbonate and coarse fraction in central Arctic cores. The lithostratigraphic unit boundaries are shown where known.

greatly diluted relative to the ridge and ridge-flank sediment. Morris and Clark (1986) reported the disappearance of foraminiferal-rich zones in some nonturbidite lithostratigraphic units in deeper ridge-flank cores, suggesting a fluctuating lysocline throughout the Pleistocene. Despite this finding, the abundance of dolomite in the detrital fraction of central Arctic sediment still argues against significant dissolution of carbonates.

## Biogenic Carbonate

Planktonic foraminifera, primarily *Globigerina* and *Neogloboquadrina* sp., dominate the calcareous fauna and the biogenic component of Arctic sediments (Clark, 1969, 1977; Herman, 1969, 1974). Herman (1969, 1974) estimated that faunal maxima accounted for more than 90 percent by volume of the greater than 62 $\mu$m fraction. Clark (1971) found that foraminifera com-

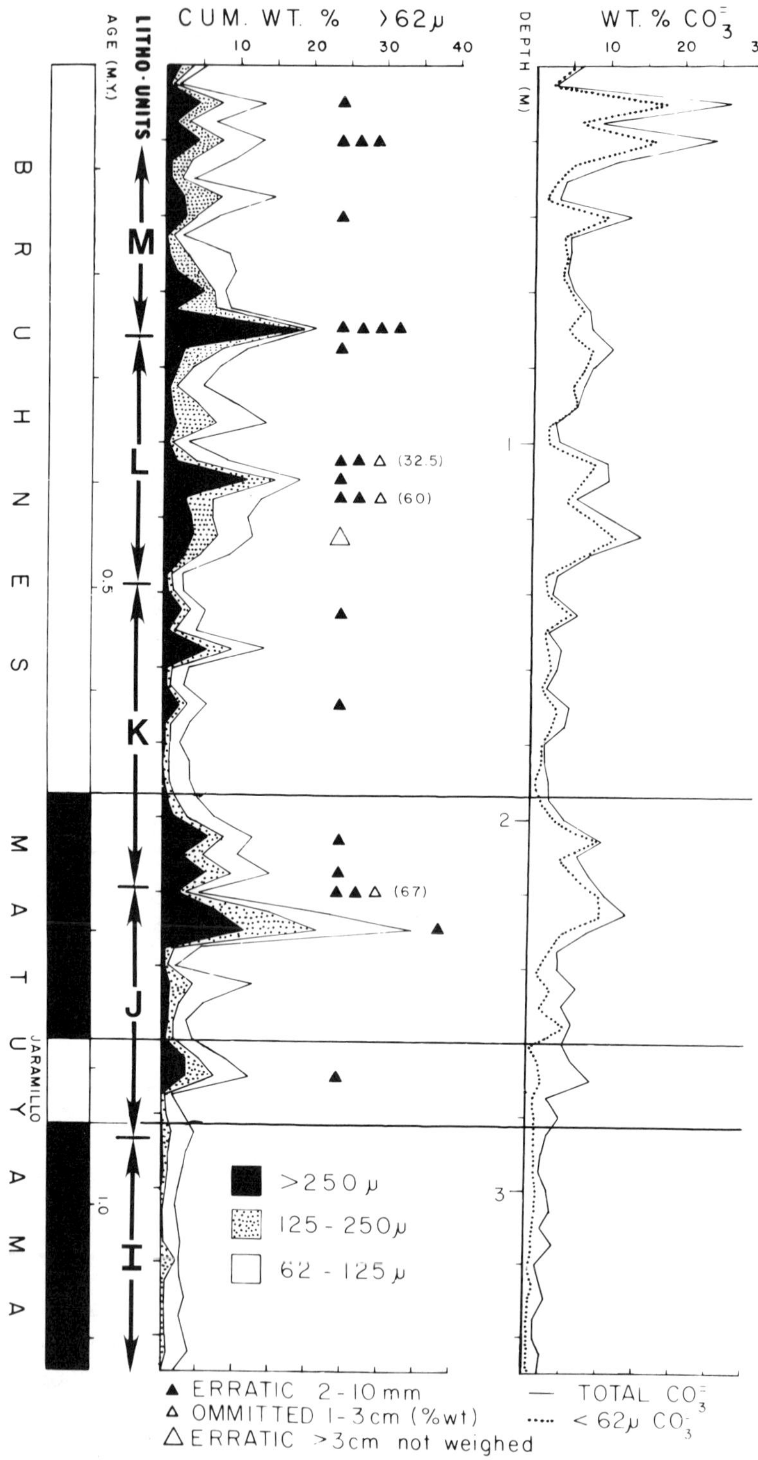


**Figure 24-12.** Cumulative weight percent of sand and gravel compared to cumulative weight percent carbonate ($CO_3$) in the less than 62 μm and greater than 62 μm fractions for core T3-67-9. Note that most of the carbonate is finer than 62 μm.

prised up to fifteen percent by weight of this coarse fraction in Alpha Ridge cores. Core intervals between faunal maxima contained less than one percent (by weight) foraminifera. Actual counts on foraminifera in the > 62 μm fraction show about $10^4$ to $3 \times 10^4$ tests per gram of bulk sediment (all sizes less than 2mm) in faunal maxima (Boyd et al., 1984; Clark, personal communication, 1985; Chapter 23). These foraminiferal counts are similar to those of North Atlantic marine sediments above the carbonate compensation depth (Kellogg, 1977).

Regardless of the apparent abundance of foraminifera in certain layers of central Arctic sediment, their weight percent contribution to the total sediment is relatively minor. For example, the maximum of 15 percent by weight of foraminifera in the > 62 μm fraction amounts to less than 8 percent of the total sediment in the same cores and usually less than 3 percent based on the abundance of coarse sediment as reported by Clark et al. (1980).

Likewise, the contribution of foraminifera to the total carbonate is relatively minor, with a few exceptions. Nearly all of the foraminifera are larger than 62 μm, yet the bulk of the carbonate is less than this size (Fig. 24-12). Also, the foraminiferal abundance maxima generally do not correspond to carbonate maxima (Fig. 24-13). In core T3-67-11, no foraminifera maxima occur with less than about 4 percent carbonate; however, where high foraminifera and low carbonate content coincide, the foraminifera might account for more than half of the carbonate. For example, at 67 cm in this core, such a coincidence occurs and the XRD mineral data (Table 24-7) shows that this horizon is primarily calcite instead of dolomite. Other sampled horizons where calcite dominates in this core also correspond to foraminiferal abundance maxima (17 cm and 137 cm). While the foraminiferal tests are calcitic, detrital calcite and dolomite also contribute to the carbonate in these horizons.

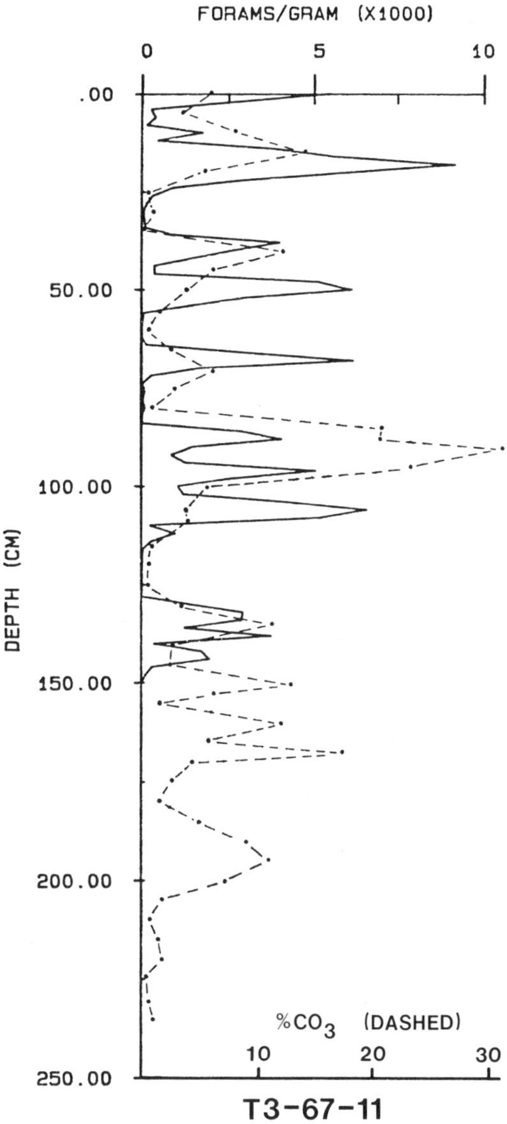


Figure 24-13. Number of foraminifera (> 150 μm) per gram of bulk sediment compared to weight percent $CO_3$. Foraminifera samples are at a 2 cm interval and those for $CO_3$ are at a 5 cm interval except for a few closer-spaced samples.

## Carbonate Cycles

Darby (1971, 1975*b*) demonstrated that seven or eight carbonate maxima occurred in Plio-Pleistocene sediment from the Arctic Ocean and that they were correlatable over hundreds of kilometers except between ridge or ridge flank and abyssal plain areas. More detail is evident when samples are taken at 1 cm instead of 5 cm intervals, but the general patterns of carbonate fluctuations remain the same (Fig. 24-14).

Because these carbonate fluctuations are correlatable and because several occur in the same lithostratigraphic units, they are potentially useful in establishing a more detailed stratigraphy than the lithostratigraphy alone. For example, there are four carbonate maxima below the upper 10 cm in lithostratigraphic unit M (Fig. 24-15). The oldest maxima in these cores occurs at the boundary between units M and L. While both the range in carbonate percent and the shape of each carbonate maxima are similar within localized areas, such as the Chukchi Rise (Fig. 24-15), only the sequence of peaks can be correlated over larger distances. For example, cores FL-224 and FL-270 in Fig. 24-9 are approximately 300 km apart, and despite the poor paleomagnetic signature for lithostratigraphic unit F, the two cores can be correlated by the sequence of carbonate maxima and lithostratigraphic units with greater detail than by lithostratigraphic criteria alone.

The primary mode of origin for most of the Arctic carbonate is glacial ice rafting. While minor outcrops of limestone and dolostone occur on Ellesmere Island and northern Greenland, the only adjacent areas with active glaciers today, the probable primary sources for the carbonate in the central Arctic basin are the widespread outcrops of early Paleozoic carbonates in the Northwest Territories of Canada. This statement is supported by maps of Canada

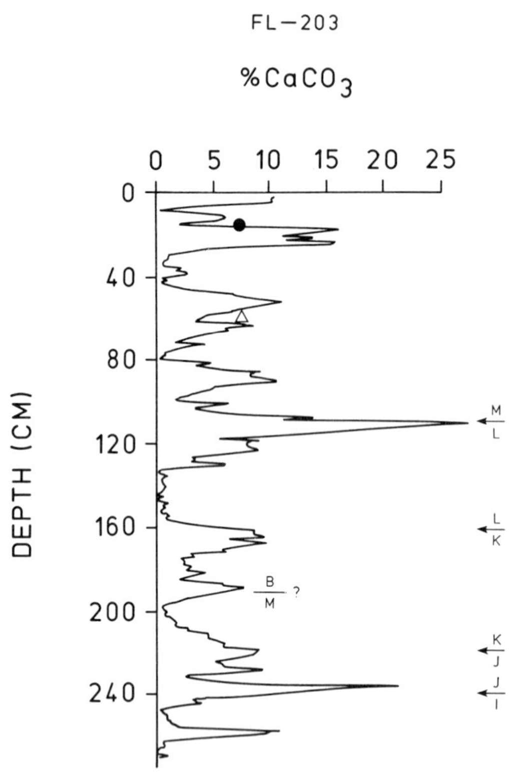


Figure 24-14. $CaCO_3$ records from Chukchi Rise core FL-203. Interval shown represents lithostratigraphic units I to M. Lithostratigraphic unit boundaries are shown at right. Note that $CaCO_3$ record offers possibility of increased resolution in core-to-core correlations over the currently used lithostratigraphic correlations alone. Also shown are predicted occurrences of Old Crow Tephra based on paleomagnetics (∘) and amino acid (△) age models. For convenience, all $CO_3$ is assumed to be $CaCO_3$.

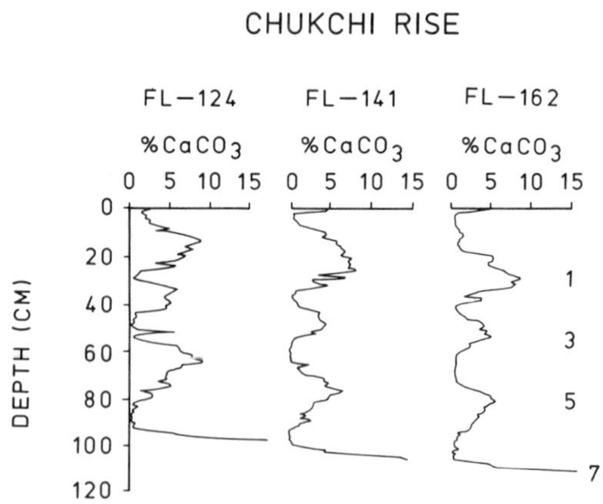


Figure 24-15. Core-to-core correlations within Unit M (0–390,000 yr B. P.) using detailed (1 cm) and precise (±0.25 percent) $CaCO_3$ analyses. All three records are from the Chukchi Rise. For convenience, all $CO_3$ is assumed to be $CaCO_3$.

showing the extent and paths of Laurentide ice movements northward toward the Amundsen Gulf–Mackenzie Delta area and the Arctic Ocean (Prest, Grant, and Rampton, 1968; Dyke and Prest, 1987).

Based on spectral analysis of carbonate cycles, coarse detritus percents, and foraminiferal abundance, Boyd et al. (1984) found a tendency for variance to be concentrated in three cores at powers near 100,000 yr using linear paleomagnetic age constraints. This is similar to a 100,000-yr power found in many $\delta O^{18}$ records (e.g., Hays, Imbrie, and Shackleton, 1976; Ruddiman and McIntyre, 1981). In two cores, Boyd et al. found an additional periodicity near 41,000 yr and 29,000 yr. These observations, which need further testing when a better chronology is established for Arctic cores, add to the evidence for an orbital climatic response in the carbonate, coarse fraction, and foraminiferal abundance signatures.

## SEDIMENT GEOCHEMISTRY

### Organic Carbon and Nitrogen

The average organic carbon (OC) and the average total nitrogen (N) in surface sediment from the southern Canada Abyssal Plain are significantly higher than that reported for sediments having a similar composition (e.g., argillaceous and 1–10 percent $CaCO_3$) in the Atlantic or Pacific oceans (Table 24-8; cf. Muller, 1977; Lew, 1981*a*,*b*). Yet the average OC/N ratio in this Arctic sediment is notably lower than the ratios for the Atlantic Ocean and Pacific Ocean sediments. The higher OC and N contents in the Arctic Ocean sediments are apparently unrelated to the overall biological production and primary productivity rates.

Numerous studies have pointed out that the net yearly production in the Arctic Basin, as defined by the abundance of the faunal and floral standing crops in the water columns and in the benthic substrate as well as the productivity rates, is generally among the lowest found in the world's oceans (English, 1961; Tibbs, 1967; Mohr and Geiger, 1968; Dunbar, 1968). Presumably as a result of this low productivity, there is very low (2–13 $\mu$g/cc) particulate organic carbon (POC) in the Canada Basin waters, as compared to the Pacific and Atlantic deep waters (Kinney, Loder, and Groves, 1971).

In contrast, the average content of dissolved organic carbon (DOC) in the Arctic waters is unexpectedly high (1.03 $\mu$g/l). This is attributed to the relatively slow decomposition rate of organic matter compared to the turnover rate in the Arctic Ocean (Kinney, Loder, and Groves, 1971, p. 135). We presently have no ready explanation for the relatively higher contents of OC and N in the Canada Basin sediment. Obviously the physicochemical depositional conditions (Trask, 1939) prevalent in the Arctic Ocean are such that they are more conducive to the preservation, and hence, relative enhancement of organic matter in the sediments.

A number of possible factors acting independently or together determine the OC/N ratios in marine sediments (e.g, Lew, 1981*b*). This ratio decreased from 14 to 5 in the nearshore Beaufort Sea to the contiguous Canada Basin sediments possibly because of dilution of terrigenous organic detritus by marine biogenous particulates (Naidu, 1985). This contention is supported by the seaward increase in $\delta^{13}C$ values in Beaufort Sea sediments (Gearing, Plucker, and Parker, 1977). Furthermore, the presence of relatively high (0.19–0.55) particulate organic carbon/particulate matter (POC/PM) ratios of central Arctic seawater samples presumably results from the relatively small input of terrigenous detritus into this basin by way of eolian, water current, or ice transport today (Kinney, Loder, and Groves, 1971).

An alternative explanation for the observed seaward decrease in the OC/N ra-

Table 24-8. Location and Partial Chemical Analysis of 28 Surface Sediments of the Canada Basin

| *Sample Number* | *Latitude North* | *Longitude West* | *Water Depth (m)* | *Si* | *Al* | $CO_3$ | *OC* | *N* |
|---|---|---|---|---|---|---|---|---|
| T3-1 | 74°19′ | 141°51′ | 3,637 | 25.0 | 7.5 | 4.2 | 0.74 | 0.13 |
| T3-2 | 74°35′ | 142°35′ | 3,650 | 24.0 | 7.4 | 3.1 | 0.71 | 0.11 |
| T3-3 | 75°18′ | 146°57′ | 3,795 | 23.0 | 8.4 | 0.6 | 0.73 | 0.22 |
| T3-4 | 75°24′ | 146°57′ | 3,792 | 21.0 | 8.1 | 2.9 | 0.74 | 0.24 |
| T3-5 | 75°18′ | 146°37′ | 3,761 | 24.5 | 8.2 | 1.2 | 0.75 | 0.20 |
| T3-6 | 75°60′ | 151°20′ | 3,827 | 21.5 | 8.2 | 1.1 | 0.79 | 0.35 |
| T3-7 | 75°45′ | 151°14′ | 3,830 | 23.0 | 8.4 | 0.9 | 0.77 | 0.35 |
| T3-8 | 75°49′ | 153°36′ | 3,833 | 23.0 | 8.6 | 1.2 | 1.16 | 0.35 |
| T3-9 | 75°25′ | 156°4′ | 2,860 | 23.5 | 6.9 | 5.3 | 0.39 | 0.15 |
| T3-10 | 75°12′ | 156°35′ | 1,705 | 24.0 | 6.3 | 8.0 | 0.39 | 0.11 |
| T3-11 | 75°12′ | 156°52′ | 1,160 | 23.5 | 6.7 | 4.1 | 0.47 | 0.10 |
| T3-12 | 75°45′ | 151°42′ | 3,835 | 24.0 | 8.4 | 0.8 | 0.79 | 0.09 |
| BSS-51 | 72°51′ | 156°11′ | 2,477 | 28.0 | 7.6 | 1.25 | 0.91 | 0.13 |
| BSS-54 | 71°1′ | 145°4′ | 3,458 | 28.0 | 7.5 | 2.65 | 0.88 | NA |
| BSS-97 | 71°52′ | 152°41′ | 1,500 | 31.0 | 6.1 | 3.15 | 1.34 | 0.17 |
| BSS-92 | 71°20′ | 150°3′ | 1,600 | 35.5 | 6.9 | 1.75 | 1.11 | NA |
| BSS-91 | 71°20′ | 150°0′ | 1,800 | 29.0 | 6.8 | 0.15 | 1.41 | NA |
| GLA7122 | 71°9′ | 148°0′ | 1,006 | 28.0 | 6.6 | 2.61 | 0.66 | 0.15 |
| GLA7158 | 71°15′ | 149°3′ | 1,011 | 28.0 | 6.4 | 1.87 | 1.00 | 0.15 |
| GLA7185 | 71°23′ | 150°34′ | 1,053 | 26.0 | 6.2 | 1.98 | 1.33 | 0.16 |
| GLA7157 | 71°20′ | 149°7′ | 1,829 | 25.5 | 6.8 | 1.41 | 1.04 | 0.18 |
| GLA7120 | 71°14′ | 147°23′ | 2,000 | 23.5 | 6.2 | 2.36 | 1.05 | 0.16 |
| GLA7186 | 71°30′ | 150°30′ | 2,150 | 24.5 | 6.3 | 3.61 | 1.53 | 0.23 |
| GLA7121 | 71°23′ | 147°55′ | 2,200 | 25.5 | 6.8 | 1.12 | 1.35 | 0.17 |
| GLA7184 | 71°17′ | 150°24′ | 1,000 | NA | NA | NA | NA | NA |
| GLA775 | 72°56′ | 146°36′ | 3,593 | NA | NA | 1.78 | 0.78 | 0.10 |
| GLA776 | 72°46′ | 146°22′ | 3,566 | NA | NA | 2.48 | 0.56 | 0.11 |
| GLA777 | 72°54′ | 142°5′ | 3,566 | NA | NA | 2.97 | 0.58 | 0.09 |
| Average | | | | 25.64 | 7.22 | 2.39 | 0.87 | 0.18 |

*Note:* The concentrations of Si, Al, organic carbon (OC), total N (N), carbonate ($CO_3$), Fe, and Mn are expressed on a dry weight % basis, whereas those of other elements are in μg/g.

tios in the Arctic sediments is the possible enrichment basinward of inorganic N (fixed + exchangeable) in the sediments with organic carbon values remaining unchanged. However, at present we have no regional estimate of the inorganic N contribution to the total N in Arctic Ocean sediment. Generally, the inorganic N (expressed as ammonium) in marine sediments is quantitatively unimportant as it constitutes less than 5 percent of the total N (Muller, 1977), but occasionally it can be as high as 95 percent (Stevenson and Cheng, 1972; Muller, 1975; Lew 1981*b*).

## Elemental Composition of Sediments

The elemental composition of southern Canada Abyssal Plain sediment samples is generally uniform throughout the area (Table 24-8). As expected, most (average 90 percent) of the Mn is associated with the easily reducible sediment phase represented by the hydroxylamine hydrochloride–acetic acid extract (Table 24-9). Approximately 15–40 percent of the other metals (Fe, Cu, Ni, Zn, Co, Cr, and V) are released from the sediment in this extract.

| *OC/N* | *P* | *Fe* | *Mn* | *Cu* | *Ni* | *Zn* | *Co* | *Cr* | *V* |
|---|---|---|---|---|---|---|---|---|---|
| 6 | 650 | 5.15 | 0.48 | 60 | 90 | 120 | 55 | 130 | 360 |
| 7 | 588 | 4.90 | 0.40 | 52 | 90 | 93 | 53 | 110 | 305 |
| 3 | 800 | 5.45 | 0.32 | 125 | 80 | 94 | 41 | 130 | 305 |
| 3 | 850 | 5.25 | 0.48 | 80 | 100 | 110 | 55 | 145 | 274 |
| 4 | 713 | 5.20 | 0.36 | 52 | 60 | 104 | 44 | 140 | 243 |
| 2 | 469 | 5.40 | 0.33 | 55 | 60 | 95 | 41 | 145 | 260 |
| 2 | 463 | 5.25 | 0.38 | 50 | 70 | 84 | 52 | 145 | 260 |
| 3 | 875 | 5.10 | 0.49 | 49 | 90 | 94 | 74 | 145 | 305 |
| 3 | 713 | 6.05 | 0.58 | 85 | 110 | 130 | 121 | 160 | 285 |
| 4 | 625 | 5.10 | 0.37 | 48 | 70 | 102 | 74 | 100 | 185 |
| 5 | 525 | 4.60 | 0.50 | 46 | 80 | 104 | 87 | 90 | 210 |
| 9 | 950 | 5.10 | 0.50 | 52 | 90 | 88 | 78 | 120 | NA |
| 7 | 625 | 4.55 | 0.86 | 80 | 68 | 83 | 34 | 80 | 245 |
| NA | 1,000 | 4.88 | 0.33 | 61 | 58 | 103 | 62 | 135 | 320 |
| 8 | 219 | 4.06 | 0.03 | 60 | 41 | 83 | 35 | 135 | NA |
| NA | 800 | 3.65 | 0.03 | 92 | 65 | 122 | 19 | 90 | 215 |
| NA | 1,075 | 4.06 | 0.08 | 53 | 50 | 65 | 30 | 100 | 230 |
| 4 | 1,875 | 5.10 | 0.26 | 32 | 52 | 129 | 46 | 90 | 208 |
| 7 | 1,500 | 5.00 | 0.09 | 30 | 50 | 90 | 44 | 80 | 195 |
| 8 | 1,750 | 5.80 | 0.09 | 37 | 55 | 111 | 42 | 90 | 208 |
| 6 | 1,406 | 5.20 | 0.53 | 36 | 54 | 111 | 46 | 85 | 220 |
| 7 | NA | 4.80 | 2.58 | 70 | 50 | 103 | 47 | 80 | 215 |
| 7 | 1,750 | 5.20 | 1.86 | 40 | 52 | 106 | 44 | 90 | 195 |
| 6 | 875 | 5.20 | 1.20 | 30 | 54 | 156 | 44 | 85 | 233 |
| NA | 1,538 | 3.68 | 0.04 | 35 | 60 | 88 | 15 | 113 | 210 |
| 8 | NA | 5.40 | 0.47 | 50 | 55 | 121 | 28 | 103 | 220 |
| 5 | NA | 6.05 | 0.40 | 57 | 63 | 144 | 28 | 117 | 252 |
| 6 | NA | 5.39 | 0.31 | 53 | 58 | 116 | 26 | 110 | 234 |
| 5 | 943 | 5.02 | 0.51 | 56 | 67 | 105 | 49 | 112 | 246 |

Thus the bulk of these metals is presumably associated with crystalline phases other than Fe or Mn oxides presumably extracted with the hydroxylamine wash (cf. Chester and Hughes, 1967).

Comparing the elemental composition of the Arctic pelagic sediment with that from tropical to temperate oceans, the Arctic Ocean contains significantly lower Al, Fe, Mn, Cu, Ni, Zn, and Co (Table 24-10). Only Si, Cr, and V were not significantly different in Arctic and low-latitude ocean sediments. The sediments used in these comparisons were lithologically similar to those of the Arctic. Belov and Lapina (1970) and Belov et al. (1970) also observed a similar paucity of Fe and Mn in surficial Arctic deep-sea clays.

There are two sets of average abundance values listed for most elements in the tropical to temperate ocean sediments in Table 24-10. One set of values is computed from analysis of the total sediment, including the contents of micro Fe-Mn nodules and concretions (Wedepohl, 1956; Goldberg and Arrhenius, 1958; El Wakeel and Riley, 1961; Krishnaswami, 1976). The second set of values is from Pacific pelagic

Table 24-9. Concentrations of Heavy Metals in the Total Sediment (T) and in the Hydroxylamine Hydrochloride-Acetic Acid Extract of Sediment (E)

| *Sample Number* | *Water Depth (m)* | *Fe* | | | *Mn* | | | *Cu* | | |
|---|---|---|---|---|---|---|---|---|---|---|
| | | *T* | *E* | *% E* | *T* | *E* | *% E* | *T* | *E* | *% E* |
| GLA71-84 | 1,000 | 3.68 | 0.84 | 23 | 0.04 | 0.03 | 75 | 35 | 2 | 6 |
| GLA77-7 | 3,566 | 5.39 | 0.89 | 17 | 0.31 | 0.28 | 90 | 53 | 20 | 38 |
| GLA77-5 | 3,593 | 5.40 | 1.09 | 20 | 0.47 | 0.45 | 96 | 50 | 24 | 48 |
| T3-1 | 3,637 | 5.15 | 1.00 | 19 | 0.48 | 0.42 | 88 | 60 | 26 | 43 |
| T3-2 | 3,650 | 4.90 | 0.98 | 20 | 0.40 | 0.27 | 68 | 52 | 23 | 44 |
| T3-3 | 3,795 | 5.45 | 1.08 | 20 | 0.32 | 0.29 | 91 | 125 | 8 | 6 |
| T3-4 | 3,792 | 5.25 | 1.13 | 22 | 0.48 | 0.47 | 98 | 82 | 10 | 12 |
| T3-5 | 3,761 | 5.20 | 1.10 | 21 | 0.36 | 0.36 | 100 | 52 | 13 | 25 |
| T3-6 | 3,827 | 5.40 | 1.13 | 21 | 0.33 | 0.29 | 88 | 55 | 10 | 18 |
| T3-7 | 3,830 | 5.24 | 1.25 | 24 | 0.38 | 0.37 | 97 | 50 | 17 | 34 |
| T3-8 | 3,833 | 5.10 | 1.18 | 23 | 0.49 | 0.48 | 98 | 49 | 14 | 29 |
| T3-9 | 2,860 | 6.05 | 0.70 | 12 | 0.58 | 0.40 | 69 | 85 | 23 | 27 |
| T3-10 | 1,705 | 5.10 | 0.70 | 14 | 0.37 | 0.30 | 81 | 48 | 16 | 33 |
| T3-11 | 1,160 | 4.60 | 0.90 | 20 | 0.50 | 0.50 | 100 | 46 | 15 | 33 |
| T3-12 | 3,835 | 5.10 | 1.25 | 25 | 0.50 | 0.49 | 100 | 52 | 14 | 27 |
| Average | | 5.13 | 1.01 | 20 | 0.40 | 0.36 | 90 | 60 | 16 | 27 |

*Note:* The concentrations of Fe and Mn are expressed in wt. % and those of the other metals are in μ g/g. The proportion of extractable metals from the total sediment is expressed as % E. See Table 24-8 for sample locations.

clays (Cronan, 1969) and these exclude the proportion of elements that were contributed by Fe-Mn nodules associated with the clays. As expected, the values for these Pacific clays, excluding micro Fe-Mn nodules, are, in general, significantly lower than the other values from low-latitude deep-sea sediments. Because the sampled Arctic sediments were inclusive of any Fe-Mn nodules, only the results from the low-latitude sediments inclusive of nodules were statistically compared to the Arctic sediments in Table 24-10.

## Patterns of Metals in the Arctic Ocean Sediments

Typically deep-sea, tropical to temperate, marine sediments contain higher concentrations of several heavy metals compared with continental margin sediments (Wedepohl, 1956; Chester, 1971). A comparison of the chemical data for the Arctic deep-sea sediments (Table 24-8) with published data on Arctic continental margin sediments (Naidu and Hood, 1972; Naidu et al., 1975, 1982; Naidu, 1979; Sweeney, 1984) shows that the overall regional trend of total metal partitioning in the Arctic Ocean is generally in agreement with that displayed in lower-latitude oceans (Table 24-11). Chester (1971) reviewed the various factors that might lead to this observed concentration of heavy metals in deep-sea sediments relative to adjacent shelf sediment. In the Arctic, the primary factor is thought to be the concentration of silt and clay-size particles in the Canada Basin sediments (Naidu and Hood, 1972; Naidu et al., 1975; Naidu, 1979). These fine-grained sediments generally have larger adsorptive/exchange capacities than coarser shelf sediment. Presumably such concentrations resulted from selective transport of fine particles from the shelves to the adjacent abyssal plain, either by winnowing action of currents or more likely by sea-ice transport.

| *Ni* | | | *Zn* | | | *Co* | | | *Cr* | | | *V* | | |
|---|---|---|---|---|---|---|---|---|---|---|---|---|---|---|
| *T* | *E* | *% E* | *T* | *E* | *% E* | *T* | *E* | *% E* | *T* | *E* | *% E* | *T* | *E* | *% E* |
| 60 | 3 | 5 | 88 | 33 | 38 | | | | 113 | 2.7 | 2 | | | |
| 58 | 8 | 14 | 116 | 31 | 27 | 26 | 9 | 35 | 110 | 2.1 | 2 | 234 | 36 | 15 |
| 55 | 8 | 15 | 121 | 34 | 28 | 28 | 10 | 36 | 103 | 1.4 | 1 | 220 | 56 | 26 |
| 90 | 18 | 20 | 120 | 36 | 30 | | | | | | | 360 | 68 | 19 |
| 90 | 13 | 14 | 93 | 34 | 37 | | | | | | | 305 | 53 | 17 |
| 80 | 9 | 11 | 94 | 42 | 45 | | | | | | | 305 | 47 | 15 |
| 100 | 17 | 17 | 110 | 40 | 36 | | | | | | | 274 | 64 | 23 |
| 60 | 13 | 22 | 104 | 56 | 54 | | | | | | | 243 | 51 | 21 |
| 60 | 12 | 20 | 95 | 45 | 47 | | | | | | | 260 | 4 | 16 |
| 70 | 18 | 26 | 84 | 46 | 55 | | | | | | | 260 | 71 | 27 |
| 90 | 19 | 21 | 94 | 38 | 40 | | | | | | | 305 | 69 | 23 |
| 110 | 16 | 15 | 130 | 24 | 19 | | | | | | | 285 | 39 | 14 |
| 70 | 12 | 17 | 102 | 24 | 24 | | | | | | | 185 | 39 | 21 |
| 80 | 21 | 26 | 104 | 29 | 28 | | | | | | | 210 | 47 | 22 |
| 90 | 22 | 24 | 88 | 42 | 48 | | | | | | | | | |
| 78 | 14 | 18 | 103 | 37 | 36 | 27 | 10 | 37 | 109 | 2 | 2 | 265 | 52 | 20 |

Table 24-10. Differences in the Average Concentrations of Some Elements in Deep-Sea Sediments of the Arctic Ocean (Canada Basin) and the Tropical to Temperate Locales in the Atlantic and Pacific Oceans

| | *Arctic Ocean*[a] | *Tropical-Temperate Oceans* | *Student T Test Results (95% Confidence)* |
|---|---|---|---|
| Si% | 25.64 | 24.22[c,d] | insignificant |
| Al% | 7.22 | 9.55[c,d] | significant |
| Fe% | 5.02 | 5.57[b,c,d,e]; 5.07[f] | significant |
| Mn% | 0.40 | 0.69[b,c,d,e]; 0.48[f] | significant |
| P | 943 | 634[d]; 1,052[f] | significant |
| Cu | 60 | 391[b,c,d,e]; 323[f] | significant |
| Ni | 78 | 222[b,c,d,e]; 211[f] | significant |
| Zn | 103 | 125[b,c,d] | significant |
| Co | 27 | 100[b,c,d,e]; 101[f] | significant |
| Cr | 109 | 99[b,c,d]; 102[f] | insignificant |
| V | 265 | 274[b,c,d]; 215[f] | insignificant |

*Note:* The concentrations are in μg/g dry weight, unless stated otherwise.
[a]This study (Table 24.8)
[b]Wedepohl (1956).
[c]Goldberg and Arrhenius (1958).
[d]El Wakeel and Riley (1961).
[e]Krishnaswami (1976).
[f]Cronan (1969); samples of Pacific Ocean clays without Fe-Mn nodules.

Table 24-11. Comparison of the Averages of Total and Nonlithogenous (NL) Metal Contents, and the Contribution of the Nonlithogenous Fraction (%NL) to the Total Measured Metal Content in Argillaceous Sediments of the Alaskan Shelf (Beaufort Sea) and Basins of the Arctic, Pacific, and Atlantic Oceans

| | *Arctic Ocean*[a] | | | *Beaufort Sea Clays*[b] | | | *Pacific Ocean*[c] | | | *Atlantic Ocean*[d] | | |
|---|---|---|---|---|---|---|---|---|---|---|---|---|
| *Element* | *Total* | *Nonlithogenous (NL)* | *% NL* | *Total* | *Nonlithogenous (NL)* | *% NL* | *Total* | *Nonlithogenous (NL)* | *% NL* | *Total* | *Nonlithogenous (NL)* | *% NL* |
| Fe% | 5.13 | 1.01 | 20 | 2.85 | 0.34 | 12 | 5.44 | 1.00 | 18 | 4.83 | 0.26 | 5 |
| Mn% | 0.40 | 0.36 | 90 | 0.034 | 0.017 | 51 | 0.74 | 0.58 | 78 | 0.21 | 0.12 | 57 |
| Cu | 60 | 16 | 27 | 19 | 4 | 21 | 338 | 145 | 43 | 49 | 18 | 37 |
| Ni | 78 | 14 | 18 | 33 | 4 | 12 | 244 | 146 | 65 | 82 | 33 | 40 |
| Co | 27 | 10 | 37 | 9 | 3 | 33 | 87 | 64 | 73 | 31 | 21 | 68 |
| Cr | 109 | 2 | 2 | 61 | 2 | 3 | | | | 98 | 25 | 25 |
| Zn | 103 | 37 | 36 | 94 | 22 | 23 | | | | | | |
| V | 265 | 52 | 20 | 111 | 8 | 7 | | | | 143 | 11 | 8 |

*Note:* The element concentrations are in $\mu g/g$ dry weight, unless indicated otherwise.
[a] This study; based on the analysis of 15 samples (Table 24.9).
[b] Naidu and Hood (1972), Naidu and Mowatt (1975), and Sweeney (1984).
[c] Krishnaswami (1976).
[d] Chester and Messiha-Hanna (1970); based on data from 9 argillaceous samples.

Chester and Messiha-Hanna (1970) demonstrated that the metal content in the lithogenous (lattice-held, detrital) fraction were similar in the near-shore and deep-sea sediments of the Atlantic Ocean. On this basis they concluded that this detrital component in the Atlantic deep-sea sediment was essentially derived from adjacent near-shore sediments. By implication they also concluded that the content of a metal in the detrital fraction of deep-sea sediments constituted the "background" (via terrigenous input) to the deep ocean. Any contents of metals in excess of background values in deep-sea sediments must be authigenic in origin and acquired from within the pelagic system. As for the Arctic sediments analyzed, all the heavy metals except Mn in the lithogenous fractions (Total minus NL in Table 24-11) are clearly different between Beaufort Sea clays and the adjacent Canada Basin. Following Chester and Messiha-Hanna's reasoning, this would imply that the Alaskan Beaufort Sea is evidently not a major source of the fine-grained sediments in the Canada Basin. This concurs with our earlier discussion on the fact that the Beaufort Sea contributes little, if any, to the Canada Basin clays based on clay-mineral and heavy mineral patterns in these sediments.

## Possible Implications of Low Heavy-Metal Values in Canada Basin Sediments

Naidu et al. (1975) considered the various possible factors, including analytical techniques, sedimentation rates, input of submarine volcanogenic components, and source-terrane lithology, that might possibly explain the relatively low heavy-metal values in the southern Canada Basin (Table 24-10). They hypothesized that because of a general lack of contemporary submarine volcanic exhalations in the west Arctic, the sediments there are deprived of a larger proportion of Fe and Mn as well as other metals that tend to coprecipitate with them. Regardless of its origin, that is, rifting with volcanism or structural due perhaps to subduction (Vogt and Avery, 1974; Herron, Dewey, and Pitman, 1974; Weber and Sweeney 1985), the Alpha Ridge system that dominates the west Arctic Ocean has not experienced volcanism during the entire Plio-Pleistocene to the present.

The close association of deep-sea metalliferous sediments locally with some form of submarine volcanism has been reported by several investigators (Dymond and Veeh, 1975; Bischoff and Rosenbauer, 1977; Edmond et al., 1981; Rona et al., 1983; and Aplin and Cronan, 1985). Whether volcanogenic elements are dispersed laterally (in solution and/or adsorbed to various suspended materials) over great distances by currents and/or by advective-diffusive processes is speculative (Chester, 1971). In support of this theory is the presence of Fe- and Mn-rich sediments in the Bauer Basin, which are thought to have originated on the East-Pacific Rise (EPR) about 1,000 km away. In addition, hydrothermal input of Mn, Fe, and several other metals along the EPR from 2°N to 42°S has been noted and clearly shown to progressively decrease from the volcanic source (Walter and Stoffers, 1985). Because of the distances involved in this example, currents rather than advective-diffusive processes are responsible (Kennett, 1982; Walter and Stoffers, 1985). Assuming that this type of metal transport is possible in the Arctic Ocean, either the distant sources such as the Arctic Mid-Ocean Ridge are relatively inactive or the currents are inadequate.

Another possible explanation for the relatively low concentrations of several metals in the southern Canada Basin sediment is the somewhat unusual character of the sediments. Unlike the argillaceous sediment of the temperate oceans, the sediments of the Canada Basin are predominantly glaciomarine. They are derived principally from regoliths of glacial flour, the result of essentially mechanical rather than chemical weathering. Consequently, such deposits tend to have chemical and mineralogical characteristics that are generally quite similar to their source rocks (Naidu and Mowatt, 1983).

The landmass surrounding the Canada Basin (especially Greenland and northern Canada) has extensive outcrops of igneous, high-grade metamorphic, and carbonate rocks. Presumably, the clay- and silt-size minerals in the glacial flour derived from such rocks have very low exchange and adsorptive capacities for Fe. This was clearly demonstrated for glaciomarine and glacial deposits of southeast Alaska; the cation-exchange capacities of these deposits range from less than 1 me/100g to 14 me/100 g (Kunze, Knowles, and Kitano, 1968; Slatt, 1970). In this context it is important to note that compared to deep-sea sediment in lower latitudes the glaciomarine sediment of the Arctic Ocean contains significantly lower concentrations of a number of metals (Angino, 1966). The fact that most of the metals extracted from Arctic sediments are associated with crystalline phases rather than adsorbed or Fe-Mn oxide phases (Tables 24-9 and 24-11) would seem to support the contention that glaciomarine rock flour has a low capacity to scavenge metals.

In attempting to account for the differences observed in the concentrations of metals between the surficial deep-sea deposits of the Canada Basin and tropical to temperate oceans, previous studies have not considered the possibility of fundamental differences in the postdepositional fluxes of metals from subsurface to surficial sediments. During diagenesis a number of metals (especially Fe and Mn) can be mobilized upward in solution. Their subsequent reprecipitation at the sediment–water interface can lead to anomalously high concentrations of metals in some deep-sea deposits. Such metal enrichments primarily occur as disseminated, encrusted, or nodular oxides and hydroxides of Fe and Mn, with several other metals coprecipitated (scavenged) with them. This diagenetic process appears to be a major contributing factor to the regional generation of metalliferous deep-sea sediments in extensive parts of the tropical to temperate oceans.

In the Arctic Basin diagenetic recycling of Mn to surface sediments has been well documented by Li, Bischoff, and Mathieu (1969) and some additional supportive data on Fe and Mn have been provided by Naidu (1974). However, the quantitative importance of this process is difficult to assess (Chester, 1971), especially where there is a distinct paucity of geochemical data such as in the Arctic. Although diagenetic recycling occurs in the Arctic surface sediments, large-scale deposits of Fe-Mn nodules have yet to be found there. Thus there is probably a low diagenetic flux of Fe and Mn to the surface sediments in the Canada Basin, presumably resulting from immediate subsurface sediments that in turn contain relatively low concentrations of Fe and Mn. This low diagenetic flux, coupled to the glaciomarine nature of most of the Plio-Pleistocene sediments and the lack of volcanogenic components, results in the general paucity of Fe, Mn, and several other metals in the deep-sea deposits of the southern Canada Basin.

## SUMMARY AND CONCLUSIONS

Significant differences exist between the ridges and abyssal plain sediments of the central Arctic Ocean. There is essentially no change in the sediments from the ridge crests to the toe of the ridges except perhaps for some carbonate dissolution in the deeper areas (Morris and Clark, 1986). Either glacial- or sea-ice rafted sediment dominate the ridge areas. The abyssal plains also contain appreciable ice-rafted sediment but turbidity flow deposits become more significant and result in much higher rates of deposition in these deeper regions (>3,800 m). Sediment thickness is generally less than 300 m on the ridges and ridge flanks and up to 2–3 km on the abyssal plains. Presumably the oldest basin in the central Arctic Ocean, the Canada Basin contains the thickest unconsolidated sediment accumulation.

Current deposited sediment (other than turbidity flows) is of minor importance today although there is some evidence to suggest that older sand waves, now buried, on the ridges were deposited by currents. Presumably this occurred before the perennial ice cover was established because tractive currents in the central Arctic today are inadequate to transport sand.

Aeolian input into the modern Arctic Ocean is also of minor importance, probably accounting for less than 1 percent of the total sediment, even in areas of slow accumulation from other processes.

Pelagic sedimentation, primarily biogenic, is less than 6 percent (dry weight) of the total sediment even in areas of slow deposition rates. Most of these pelagic sediments are composed of the shells of planktonic foraminifera but nearly all other biogenic components typically found in deep-sea sediments occur in the Arctic.

The abundance and distribution of turbidite deposits and their composition suggests that the major source of both ice-rafted and turbidite sediment in the Amerasian Basin (west of the Lomonosov Ridge) is northern Canada and Greenland. Some sediment might move into the central Arctic Ocean from the Chukchi Sea but the most important source of the thicker basin deposits is from the Mackenzie Delta region of northern Canada. Little, if any, sediment is presently moving from the ridges to the abyssal plains.

Clay mineral assemblages of contemporary ocean-bottom sediments in the study area are dominated by terrigenous components. These materials appear to have been essentially unchanged, in terms of their fundamental mineralogic characteristics, attendant to their sojourns in the marine environment. In fact, the Yukon River clay-mineral suite can be traced into the Chukchi Sea, where it contributes a major portion of the clay fraction. Likewise, these Yukon River clays might even reach the Canada Abyssal Plain via the Barrow Canyon. There is some evidence of the possible authigenic or diagenetic development of zeolitic as well as expandable mineral phases in a few samples of suspended materials acquired in the Canada Basin. The overall impression however is one of clay-mineral suites dominated by terrigenous detrital components, randomly distributed in contemporary bottom sediments across the region primarily as the result of ice rafting.

The dominant heavy minerals in the Arctic seas and central Arctic Ocean are amphibole, clinopyroxene, opaque minerals, garnet, and epidote. The heavy minerals and probably the entire sand-size fraction on the Arctic continental shelves or seas are derived from nearby rivers and locally important glaciers. The heavy-mineral suite from the central Arctic Ocean and, in particular, the Alpha Ridge is most likely derived from northern Canada, the Queen Elizabeth Islands, and northern Greenland via glacial ice rafting. Small parts of this vast source area are presently glaciated but most of the sampled sandy lutites from the Alpha Ridge were deposited probably from glacial bergs calved from much more extensive continental glaciers. Despite the slight increase in more stable heavy minerals below lithostratigraphic unit J, there is no evidence to indicate that the sources of heavy minerals in the central Arctic have changed over the past 3.5 my; however, weathering and soil development may have been more pronounced prior to about 1 mya (below lithostratigraphic unit J) and the increase in frequency of glacial pulses.

Thirteen lithostratigraphic units can be correlated over 100,000 sq. km covering the central Arctic Ocean ridges and ridge flanks. This stratigraphic framework reveals that ice-rafted sediment has been deposited throughout the central Arctic at least since the late Pliocene. The nature of this sedimentation has been cyclic. Although glacial ice-rafted sediment comprises the bulk of the Arctic deposits, intervals of relatively reduced glacial activity and increased sea-ice rafting of fine-grained sediment have occurred at least six times throughout the Plio-Pleistocene interval represented by the longest sediment cores in the central Arctic.

This sea-ice-rafted sediment is incorporated into the pack ice during the fall freeze along the margins of the permanent pack ice where storm waves stir up the fine-grained shelf sediment just before the ice forms. The overall input rate of sediment into the central Arctic Ocean during these times, however, is quite low. This input is supported by the intense burrowing and greater abundance of ferromanganese particles in the silty lutite deposited at these times. The slow deposition allows for even small amounts of sea-ice-rafted sediment to account for at least 20 percent of the sediment while infrequent bergs from local glaciers, such as those existing today along the coast of Ellesmere Island and northern Greenland, might contribute up to 70–80 percent of the remaining accumulation. Thus silty lutites, which are typified by this slow deposition, are being deposited today; however, the amount of sediment that has accumulated since the end of the last glacial maxima is so small (probably less than 3–5 cm) that often this thin surface blanket of fine sediment is overlooked or missing due to the coring process. In the abyssal plains where sedimentation rates are higher and even the weak currents suspected of maintaining nepheloid layers are absent (Hunkins, Thorndike, and Mathieu, 1969), this modern silty lutite is more evident (Goldstein, 1983).

The influx of glacial icebergs from local glaciers like those of Ellesmere Island into the central Arctic during intervals of silty lutite deposition is further supported by the increase in frequency of sandy layers within the silty lutites deposited on the eastern Alpha Ridge closest to Ellesmere or Greenland. These sandy layers probably represent glacial detritus from local valley glaciers during their more active phases or surges.

The fact that these local glaciers appear to supply the bulk of the sediment for the silty lutites, which are the thickest lithostratigraphic units in the central Arctic, is significant. This finding suggests that a large portion of the Plio-Pleistocene Arctic Ocean sediment, perhaps including much of the tens of meters thickness deposited in the deep basins, came from a relatively small geographic area. Less than 30 such valley glaciers exist today on the Arctic coasts of the Queen Elizabeth Islands and of northern Greenland, where icebergs would drift into the central Arctic. Of course, the number and extent of potential glacial sources might increase dramatically with even a slight global cooling, culminating in Laurentide-type glaciation and deposition of a sandy lutite unit such as units C, F, H, J, L, or parts of M.

The correlation of such thin lutite units over a vast area such as the central Arctic Ocean is unusual for deep marine sediments. Several quantitative methods have been used to correlate discrete horizons in deep-sea cores (e.g., paleomagnetics, oxygen isotopes, fauna, and ash). The successful lithostratigraphic framework for the Arctic Ocean suggests that other glaciomarine deposits could possibly contain a widely correlatable lithostratigraphy. Pleistocene ice-rafted sediments from the North Pacific have been correlated over more than 1,300 km using paleomagnetic reversals and faunal extinctions (Kent, Opdyke, and Ewing, 1971). Kellogg, Duplessy, and Shackleton (1978) correlated cores over a wide area of the Norwegian Sea that contain Pleistocene ice-rafted coarse layers using oxygen isotopes, fauna, and carbonate curves. In these sediments and other glaciomarine sediments of the world's oceans, no lithostratigraphy has yet been developed except for the Arctic Ocean.

The sediment geochemistry of the west Arctic Ocean is unique in several respects. Compared to similar low-latitude deep-sea sediments, the Arctic is high in organic carbon and nitrogen and low in the ratio of organic carbon to nitrogen as well as suspended particulate organic carbon (POC). POC is low probably because of the low biological productivity in the Arctic. Dissolved organic carbon is unexpectedly high in the Arctic relative to other oceans due probably to the slow decomposition rate compared to the turnover rate in the Arctic. At present there is no good explanation for the relatively high sediment organic carbon and nitrogen content.

The Arctic Ocean is also somewhat unique in that its deep-sea argillaceous sediments are generally low in Al, Fe, Mn, Cu, Ni, Zn, and Co relative to similar sediments of low-latitude oceans. Only Si, Cr, and V concentrations were similar to those of other oceans. The lower metal concentrations in the Arctic are probably due to a combination of factors, the most significant of which is the general paucity of submarine volcanic exhalations in the west Arctic today. This finding is a reflection of the cessation of most tectonic seafloor activity in the area since approximately the Cretaceous. Also a factor is the glaciomarine origin of most of the Arctic deep-sea sediment. Because of its origin, primarily via mechanical weathering, the terrigenous particulate matter has lower exchange capacity than particulates derived from more intense chemical weathering conditions. Furthermore, there is probably a low net diagenetic flux of metals from the sediment subsurface in the Arctic that results in a low production of Fe-Mn nodules and low values of other metals commonly associated with these nodules.

The heavy-metal partitioning patterns in sediments of the Canada Basin and the Alaskan shelf support the conclusions based on heavy-mineral analysis that the Beaufort Sea off Alaska is not an important source of sediment (sand or mud) for this basin.

The carbonate component in central Arctic Ocean sediments is mostly ice-rafted detrital grains from Laurentide-type glacial events with minor amounts of biogenic carbonate (primarily foraminifera). On the average, more than 85 percent of the detrital carbonate is dolomite; however, low Mg-calcite dominates in some horizons, especially in the upper meter of Arctic sediment. High Mg-calcite occurs rarely in Arctic cores and its precise source is unknown.

Carbonate cycles also have the potential of contributing to a refinement of the established Arctic lithostratigraphy. Several carbonate maxima occur in some lithostratigraphic units such as unit M and these can be correlated over the same expanses as the lithostratigraphic units themselves. Thus, in some cases, an eightfold increase in lithostratigraphic detail can be realized by correlating the carbonate peaks and troughs from core to core.

The paleomagnetic record in central Arctic Ocean sediment cores remains the best geochronologic data available despite the problems and limitations of these data (see Chap. 23). While some amino acid racemization analyses of Arctic planktonic foraminifera suggest much younger dates and higher rates of sedimentation, more recent amino acid, accelerator $^{14}C$, and paleomagnetic data, as well as a thorough reexamination of previous paleomagnetic data, contradict the younger dates of Sejrup et al. (1984).

We conclude that the early interpretations of low sedimentation rates were essentially correct, although the older polarity events may have been misidentified. Palynological data such as provided by Aksu and Mudie (1985) and Mudie (1985) are helpful in dating some of the lithologic units.

The central Arctic sedimentation regime

was dominated throughout the Plio-Pleistocene by glacial ice rafting. Because of this glacial influence, sedimentation rates fluctuate between times of iceberg melt when large amounts of both coarse and fine sediment were introduced and times of sea-ice rafting such as today when sedimentation rates are probably reduced substantially and only a few glacial bergs from local sources are rafted into the central Arctic. Even turbidity-flow frequency and coincident sedimentation rates in the abyssal plains could be increased during intervals of widespread and rapid glacial melting. These higher rates would be due to enhanced runoff from the proximal glaciated landmass in northern Canada and the accompanying high rates of deposition on the narrow Canadian continental shelf in the Arctic from whence sediment might be remobilized into the central Arctic Ocean.

Confirmation of this model of fluctuating Arctic sedimentation rates awaits the development of a more detailed geochronology for the Plio-Pleistocene lithostratigraphic units. The widely correlated sandy lithostratigraphic units (units C, F, H, J, L, and parts of M) are key marker laminae in the lithostratigraphy but their age and duration can only be interpolated from paleomagnetic records. However, these records do not provide the necessary detail in order to confirm whether sandy lutites represent much faster rates of sedimentation than silty lutites. Indirect evidence such as lower bioturbation in the sandy units supports this paradigm but does not confirm their more rapid deposition nor does it indicate how fast the sandy units were deposited.

Equally significant is the complicated relationship between deposition of coarse ice-rafted detritus and the fluctuating biomass represented primarily by planktonic foraminifera. Although correlations between the distributions of the two are poor (cf. Herman, 1984), spectral analysis of their variance suggests a similar periodicity, perhaps related to global climatic events (Boyd et al., 1984). Often the foraminifera maxima occur on the shoulder of coarse detritus maxima but there is no consistency in this relationship. Obviously, several factors control the abundance of the planktonic and even the benthic species in the Arctic, and these include fluctuating detrital sedimentation rates, pack-ice thickness that might control their food source, rapid changes in surface salinity due to freeze/thaw processes, and possible changes in circulation that could affect nutrient supplies. The relative importance of these processes is unknown.

The existence of open Arctic conditions free of seasonal pack ice during part of the Plio-Pleistocene (prior to 0.9 my) as proposed by Herman and Hopkins (1980) has been questioned by Clark (1977) who favors a continuous history of sea-ice cover throughout the last 3.2 my. Again, resolution of this controversy might hinge in part on a more detailed chronostratigraphy as well as a close sampling interval (1 cm), which is warranted by the generally slow sedimentation rates (1–3 mm/$10^3$ yr). The chronostratigraphy would enable better correlation between time-series changes in Arctic sediments and global climatic changes and the close sampling would better resolve relationships among sediment parameters such as biomass and ice-rafted detritus. Apparently, some of the differences between the paleoceanographic interpretations of Herman and Clark (see, for example, Herman, 1974; Clark and Hanson, 1983; Chap. 23) are due to differences in the sediment cores analyzed by each group and the data available for their interpretations.

More certain is the source of much of the central Arctic sediment. Heavy-mineral and clay-mineral abundances throughout the Plio-Pleistocene sediment indicate a uniform, unchanging source, most likely northern Canada and northern Greenland. The abundant dolomite in the silt- to gravel-size fractions points to a long history of periodic glacial ice movement from the above source areas into the Arctic Ocean. More locally derived glacial detritus from small-valley glaciers such as those present today on Ellesmere Island need to be contrasted from detritus derived from regional glacial masses in order to resolve the generally complex relationship between coarse detritus and other significant paleoceanographic parameters such as foraminiferal abundances.

In conclusion, major advances have been made in understanding Arctic sedimentary processes; however, each new data set has raised a number of questions. A paucity of sediment samples, limited core penetrations, and restricted distribution of samples have hindered any progress. The Arctic Ocean holds potentially significant direct links between a remarkably detailed lithostratigraphy and global climatic events. These links await the necessary chronostratigraphy to be fully exploited.

### ACKNOWLEDGMENTS

We gratefully acknowledge the support of the Department of Geological Sciences of Old Dominion University in preparing the figures for this chapter. Drafting was done by Susan Hoebeke at the Center for Instructional Development at ODU and Jeanne Ashmore typed the manuscript.

## REFERENCES

Aagaard, K. 1980. Moored current measurements over the Lomonosov Ridge. *EOS* **61**:279.

Aksu, A. E. 1985. Paleomagnetic stratigraphy of the CESAR cores. In H. R. Jackson, P. U. Mudie, and S. M. Blasco (eds.), *Initial Geological Report on CESAR— The Canadian Expedition to Study the Alpha Ridge, Arctic Ocean.* Geological Survey of Canada Paper 84-22, pp. 101–114.

Aksu, A. E., and P. J. Mudie. 1985. Magnetostratigraphy and palynlogy demonstrate at least 4 million years of Arctic Ocean sedimentation. *Nature* **381**:280–283.

Amos, C. L. 1985. Bottom photography and sediment analyses on CESAR. In H. R. Jackson, P. J. Mudie, and S. M. Blasco (eds.), *Initial Geological Report on CESAR—The Canadian Expedition to Study the Alpha Ridge, Arctic Ocean.* Geological Survey of Canada Paper 84-22, pp. 25–45.

Angino, E. E. 1966. Geochemistry of Antarctic pelagic sediments. *Geochim. Cosmochim. Acta* **30**:939–961.

Aplin, A. C., and D. S. Cronan. 1985. Ferromanganese oxide deposits from the Central Pacific Ocean, I. Encrustations from the Line Islands Archipelago. *Geochim. Cosmochim. Acta* **49**:427–436.

Barnes, P. W., D. M. Schell, and E. Reimnitz (eds.). 1984. *The Alaskan Beaufort Sea: Ecosystems and Environments.* Academic Press, New York, 466 p.

Belov, N. A., and N. N. Lapina. 1962. Geologic investigations of the Arctic Ocean bottom during 25 years. *Problemy i Antarktiki Sbornik Statei* **11**:97–104. (In Russian.)

Belov, N. A., and N. N. Lapina. 1970. Climatic variations of the Arctic in the light of analysis of bottom depositions. In *The Arctic Ocean and Its Shores During the Cenozoic.* Hydrometeorological Publishers, Leningrad, pp. 27–33. (English translation by L. A. Hutchinson, UCSD, La Jolla, Calif.)

Belov, N. A., N. N. Kulikov, N. N. Lapina, and Y. P. Semenov. 1970. Distribution of iron, manganese, and carbonates in Arctic Ocean sediments. In I. M. Belov (ed.), *Problems of Polar Geography.* National Science Foundation, Washington, D. C., pp. 71–77. (Translated from Russian.)

Biscaye, P. E. 1965. Mineralogy and sedimentation of Recent deep-sea clay in the Atlantic Ocean and adjacent seas and oceans. *Geol. Soc. Am. Bull.* **76**:803–832.

Bischoff, J. L., and R. J. Rosenbauer. 1977. Recent metalliferous sediment in the North Pacific manganese nodule area. *Earth Planet. Sci. Lett.* **33**:379–388.

Blasco, S. M., B. D. Bornhold, and C. F. M. Lewis. 1979. *Preliminary Results of Surficial Geology and Geomorphology Studies of the Lomonosov Ridge, Central Arctic Basin.* Current Research Paper 79-1C, Geological Survey of Canada, pp. 73–83.

Blasco, S. M., C. F. M. Lewis, and B. E. Bornhold. 1980. Surficial geology and geomorphology of the Lomonosov Ridge. *EOS* **61**:176.

Blatt, H., G. Middleton, and R. Murray. 1980. *Origin of Sedimentary Rocks,* 2nd ed. Prentice-Hall, Englewood Cliffs, N.J., 782 p.

Boles, J. R. 1981. Zeolites in deep-sea sediments. In F. A. Mumpton (ed.), *Mineralogy and Geology of Natural Zeolites.* Reviews in Mineralogy, vol. 4, Mineralogical Society of America, Washington, D.C., pp. 137–163.

Bouma, A. H., and D. C. Hollister. 1973. Deep ocean basin sedimentation. In H. A. Bouma and G. V. C. Middleton (eds.), *Turbidites and Deep Water Sedimentation.* Short Course Notes, Society of Economic Paleontologists and Mineralogists, Tulsa, Okla., pp. 79–118.

Boyd, R. F., D. L. Clark, G. Jones, W. F. Ruddiman, A. McIntyre, and N. F. Pisias. 1984. Central Arctic Ocean response to Pleistocene earth-orbital variations. *Quat. Res.* **22:**121–128.

Boyle, E. A. 1984. Sampling statistic limitations on benthic foraminifera chemical and isotopic data. *Marine Geol.* **58:**213–244.

Bukry, D. 1981. Cretaceous Arctic silicoflagellates. *Geo-Marine Lett.* **1:**57–63.

Bukry, D. 1984. Paleogene paleoceanography of the Arctic Ocean is constrained by the Middle or Late Eocene age of USGS Core Fl-422: Evidence from silicoflagellates. *Geology* **12**(4): 199–201.

Bukry, D. 1985. Correlation of Late Cretaceous Arctic silicoflagellates from Alpha Ridge. In H. R. Jackson, P. J. Mudie, and S. M. Blasco (eds.), *Initial Geological Report on CESAR—The Canadian Expedition to Study the Alpha Ridge Arctic Ocean.* Geological Survey of Canada Paper 84-22, pp. 125–135.

Burrell, D. C., J. A. Dygas, and R. W. Tucker. 1974. Beach morphology and sedimentology of Simpson Lagoon. In V. Alexander, et al. (eds.), *Environmental Studies of an Arctic Estuarine System.* Technical Report 74-1 (Sea Grant Report 73-116), University of Alaska Institute of Marine Science, pp. 45–143.

Campbell, J. S., and K. D. Clark. 1977. Pleistocene turbidites of the Canada Abyssal Plain of the Arctic Ocean. *J. Sed. Petrol.* **47:**657–670.

Carroll, D. 1970. Clay minerals in Arctic Ocean sea-floor sediments. *J. Sed. Petrol.* **40:**814–821.

Chester, R. 1971. The geochemistry of deep-sea sediments. In J. P. Riley and R. Chester (eds.), *Introduction to Marine Chemistry.* Academic Press, London, pp. 388–421.

Chester, R., and M. J. Hughes. 1967. A chemical technique for the separation of ferro-manganese minerals, carbonate minerals and adsorbed trace elements from pelagic sediments. *Chem. Geol.* **2:**249–262.

Chester, R., and R. G. Messiha-Hanna. 1970. Trace element partitioning patterns in North Atlantic deep-sea sediments. *Geochim. Cosmochim. Acta* **34:**1,121–1,128.

Clark, D. L. 1969. Paleoecology and sedimentation in part of the Arctic Basin. *Arctic* **22:**233–245.

Clark, D. L. 1970. Magnetic reversals and sedimentation rates in the Arctic Ocean. *Geol. Soc. Am. Bull.* **81:**3,129–3,134.

Clark, D. L. 1971. Arctic Ocean ice cover and its late Cenozoic history. *Geol. Soc. Am. Bull.* **82:**3,313–3,324.

Clark, D. L. 1974. Late Mesozoic and Early Cenozoic sediment cores from the Arctic Ocean. *Geology* **2**(1):41–44.

Clark, D. L. 1977. Climatic factors of the late Mesozoic and Cenozoic Arctic Ocean. In M. H. Dunbar (ed.), *Polar Oceans.* Arctic Institute of North America, Calgary, Canada, pp. 603–613.

Clark, D. L. 1981*a*. Iceberg and glacial-marine sediment of central Arctic Ocean. *Am. Assoc. Petroleum Geologists Bull. (Abstract)* **65:**911.

Clark, D. L. 1981*b*. Geology and geophysics of the Amerasian Basin. In A. E. M. Nairn, M. Churkin, Jr., and F. G. Stehli (eds.), *The Ocean Basins and Margins,* vol. 5. Plenum Press, New York, pp. 599–634.

Clark, D. L., and A. Hanson. 1983. Central Arctic Ocean sediment texture: A key to ice transport mechanisms. In B. F. Molnia, (ed.), *Glacial-Marine Sedimentation.* Plenum Publishing Corporation, New York, pp. 301–330.

Clark, D. L., and T. H. Morris. 1984. Cenozoic sedimentation patterns in the evolving Arctic Ocean: Canada Basin and Alpha-Chukchi region to Lomonosov Ridge. In I. S. Gramberg (ed.), *Arctic Geology,* vol. 4. 27th International Geological Congress, Moscow, pp. 173–178.

Clark, D. L., and T. H. Morris. 1985. Arctic Ocean sediment texture and the Pleistocene climate cycles. *Geol. Soc. Am. Abstracts Programs* **17**(7):547.

Clark, D. L., C. W. Byers, and L. M. Pratt. 1986. Cretaceous black mud from the Central Arctic Ocean. *Paleoceanography* **1**(3):265–271.

Clark, D. L., R. R. Whitman, K. A. Morgan, and S. D. Mackey. 1980. Stratigraphy and glacial-marine sediments of the Amerasian Basin, central Arctic Ocean. *Geol. Soc. Am. Spec. Paper 181,* 57p.

Clark, D. L., M. Andree, W. S. Broecker, and A. C. Mix. 1986. Arctic Ocean chronology confirmed by accelerator $^{14}C$ dating. *Geophys. Res. Lett.* **13**(4): 319–321.

Coachman, L. K., and K. Aagaard. 1974. Physical oceanography of Arctic and sub-Arctic seas. In Y. Herman (ed.), *Marine Geology and Oceanography of the Arctic Seas.* Springer-Verlag, New York, pp. 1–72.

Cronan, D. S. 1969. Average abundance of Mn, Fe, Ni, Co, Pb, Mo, V, Cr, Ti and P in Pacific pelagic clays. *Geochim. Cosmochim. Acta* **33:**1,562–1,565.

Darby, D. A. 1971. Carbonate cycles and clay mineralogy of Arctic Ocean sediment cores. *Technical Report 8, Arctic Ocean Sediments.* University of Wisconsin, Madison, 117p.

Darby, D. A. 1975*a*. Kaolinite and other clay minerals in Arctic Ocean sediments. *J. Sed. Petrol.***45**(1):272–279.

Darby, D. A. 1975*b*. An evaluation of Arctic Ocean paleomagnetic stratigraphy. In R. M. Fisher, M. Fuller, V. A. Schmidt, and P. J. Wasilewski (eds.), *Proceedings of the Takesi Nagata Conference.* University of Pittsburgh, pp. 230–244.

Darby, D. A., L. H. Burckle, and D. L. Clark. 1974. Airborne dust on the Arctic pack ice, its composition and fallout rate. *Earth Planet. Sci. Lett.* **24:**166–172.

Demenitskaya, R. M., and K. Hunkins. 1970. Shape and structure of the Arctic Ocean. In A. E. Maxwell (ed.), *The Sea, New Concepts of Seafloor Evolution,* vol. 4. Wiley-Interscience, New York, pp. 223–249.

Demenitskaya, R. M., A. M. Karasik, G. Kiselev Yu, I. V. Litvinenko, and S. A. Ushakov. 1968. The transition zone between the Eurasian continent and the Arctic Ocean. *Canadian J. Earth Sci.* **5:**1,125–1,129.

Domack, E. W. 1984. Rhythmically bedded glaciomarine sediments on Whidbey Island, Washington. *J. Sed. Petrol.* **54**(2):589–602.

Douglas, R. J. W., and B. MacLean. 1963. Geology of Yukon Territory and Northwest Territories. *Geol. Survey Canada Map 30-1963.* Scale 1:3,000,000.

Dreimanis, A., and U. J. Vagners. 1971. Bimodal distribution of rock and mineral fragments in basal tills. In R. P. Goldthwait (ed.), *Till: A Symposium.* Ohio State University Press, Columbus, pp. 237–250.

Dunbar, M. J. 1968. *Ecological Development in Polar Regions: A Study in Evolution.* Prentice-Hall, New Jersey, 119 p.

Dygas, J. A., R. Tucker, and D. A. Burrell. 1972. Geologic report of the heavy minerals, sediment transport and shoreline changes of the barrier islands and coast between Oliktok Point and Beechey Point. In P. J. Kinney, et al. (eds.), *Baseline Data Study of the Alaskan Arctic Aquatic Environment.* Report R72-3, University of Alaska Institute of Marine Science, Fairbanks, pp. 61–121.

Dyke, A. S., and V. K. Prest. 1987. Paleogeography of northern North America, 18,000–5,000 years ago. *Geol. Survey Canada Map 1703A.* Scale 1:12,500,000.

Dymond, J., and H. H. Veeh. 1975. Metal accumulation rates in the southeast Pacific and the origin of metalliferous sediments. *Earth Planet. Sci. Lett.* **28:**13–22.

Edmond, J. M., C. Measures, B. Mangum, B. Grant, F. R. Sclater, R. Collier, A. Hudson, L. I. Gordon, and J. B. Corliss. 1981. On the formation of metal-rich deposits at ridge crests. *Earth Planet. Sci. Lett.* **46:**19–30.

El Wakeel, S. K., and J. P. Riley. 1961. Chemical and mineralogical studies of deep-sea sediments. *Geochim. Cosmochim. Acta* **25:**110–146.

English, T. S. 1961. Some biological oceanographic observations in the central North Polar Sea. Drift Station Alpha, 1957–1958. *Arctic Institute of North America Scientific Report 15,* 80p.

Ericson, D. B., M. Ewing, and G. Wollin. 1964. Sediment cores from the Arctic and sub-Arctic seas. *Science* **144:**1,183–1,192.

Ericson, D. B., and G. Wollin. 1959. Micropaleontology and lithology of Arctic sediment cores. *Geophysical Research Papers 63, Scientific Studies at Fletcher's Ice Island, T-3, 1952-5511,* vol. 1, pp. 50–58.

Flint, R. F. 1947. *Glacial Geology and the Pleistocene Epoch.* Wiley, New York, 589p.

Flint, R. F. 1971. *Glacial and Quaternary Geology.* Wiley, New York, 892p.

Galehouse, J. S. 1971. Point counting. In R. E. Carver (ed.), *Procedures in Sedimentary Petrology.* Wiley-Interscience, New Jersey, pp. 385–408.

Galt, J. A. 1967. *Current Measurements in the Canada Basin of the Arctic Ocean, Summer 1965.* Technical Report 184, University of Washington, 17p.

Gamber, J. H., and D. L. Clark. 1978. Distribution of microscopic molluscs, echinoderms and sponges in the central Arctic Ocean. *Micropaleontology* **29:**422–431.

Gearing, P., F. E. Plucker, and P. S. Parker. 1977. Organic carbon stable isotope ratios of continental margin sediments. *Marine Chem.* **5:**251–266.

Gibbs, R. 1965. Error due to segregation in quantitative clay mineral X-ray diffraction mounting techniques, *Am. Mineralogist* **50:**741–751.

Gilbert, M. W., and D. L. Clark. 1983. Central Arctic Ocean paleoceanographic interpretations based on late Cenozoic calcareous dinoflagellates. *Marine Micropaleontol.* **7:**385–401.

Goldberg, E. D., and G. O. S. Arrhenius. 1958. Chemistry of Pacific pelagic sediments. *Geochim. Cosmochim. Acta* **13:**153–212.

Goldstein, R. H. 1983. Stratigraphy and sedimentology of ice-rafted and turbidite sediment, Canada Basin, Arctic Ocean. In B. F. Molnia (ed.), *Glacial-Marine Sedimentation.* Plenum Publishing Corp., New York, pp. 367–400.

Griffin, G. M. 1971. Interpretation of X-ray diffraction data. In R. E. Carver (ed.), *Procedures in Sedimentary Petrology.* Wiley-Interscience, New York, pp. 541–572.

Griffin, J. J., H. Windom, and E. D. Goldberg. 1968. The distribution of clay minerals in the world ocean. *Deep-Sea Res.* **15:**433–459.

Hall, J. K. 1979. Sediment waves and other evidence of paleo-bottom currents at two locations in the deep Arctic Ocean. *Sed. Geol.* **23:**269–299.

Hays, J. D., J. Imbrie, and N. J. Shackleton. 1976. Variations in the Earth's orbit: Pacemaker of the ice ages. *Science* **194:**1,121–1,132.

Heezen, B. C., and C. D. Hollister. 1964. Deep-sea current evidence from abyssal sediments. *Marine Geol.***1:**141–174.

Herman, Y. 1969. Arctic Ocean Quaternary microfauna and its relation to paleoclimatology. *Palaeogeog., Palaeoclimatol., Palaeoecol.* **6:**251–276.

Herman, Y. 1970. Arctic paleo-oceanography in late Cenozoic time. *Science* **169:**474–477.

Herman, Y. 1974. Arctic Ocean sediments, microfauna, and the climatic record in late Cenozoic time. In Y. Herman (ed.), *Marine Geology and Oceanography of the Arctic Seas.* Springer-Verlag, New York, pp. 283–348.

Herman, Y. 1984. Arctic paleoceanography in Late Cenozoic time and its relationship to global climates. In *Arctic Geology.* Colloquium 4, 27th International Geological Congress, Moscow, pp. 161–173.

Herman, Y., and D. M. Hopkins. 1980. Arctic Oceanic climate in Late Cenozoic time. *Science* **209:**557–562.

Herman, Y., and J. R. O'Neil. 1975. Arctic paleosalinities during late Cenozoic time. *Nature* **258:**591–595.

Herron, E. M., J. F. Dewey, and W. C. Pitman. 1974. Plate tectonics model for the evolution of the Arctic. *Geology* **2:**377–380.

Hoffman, P. 1974. Shallow and deepwater stromatolites in Lower Proterozoic platform—basin facies change, Great Slave Lake, Canada. *Am. Assoc. Petroleum Geol. Bull.* **58:**856–867.

Holmes, M. L., and J. S. Creager. 1974. Holocene history of the Laptev Sea continental shelf. In Y. Herman (ed.), *Marine Geology and Oceanography of the Arctic Seas.* Springer-Verlag, New York, pp. 211–230.

Hulsemann, J. 1966. On the routine analysis of carbonates in unconsolidated sediments. *J. Sed. Petrol.* **36:**622–625.

Hunkins, K. 1968. Geomorphic provinces of the Arctic Ocean. In *Arctic Drifting Stations.* Arctic Institute of North America, Washington, pp. 365–376.

Hunkins, K., and H. Kutschale. 1967. Quaternary sedimentation in the Arctic Ocean. In M. Sears (ed.), *Progress in Oceanography,* vol. 4. Pergamon Press, New York, pp. 89–94.

Hunkins, K., E. M. Thorndike, and G. Mathieu. 1969. Nepheloid layers and bottom currents in the Arctic Ocean. *J. Geophys. Res.* **74**(28):6,995–7,008.

Hunkins, K., G. Mathieu, S. R. Teeter, and A. Gill. 1970. The floor of the Arctic Ocean in photographs. *Arctic* **23**(3):175–189.

Hunkins, K., A. W. H. Be, N. D. Opdyke, and G. Mathieu. 1971. The late Cenozoic history of the Arctic Ocean. In K. K. Turekian (ed.), *The Late Cenozoic Glacial Ages.* Yale University Press, New Haven, Conn., pp. 215–237.

Jackson, R., and P. Mudie. 1984. CESAR cores: Geological time capsules. *Geoscience* **13**(2):15–18.

Jones, G. A. 1987. The Central Arctic Ocean sediment record: Current progress in moving from a litho- to chronostratigraphy. *Polar Res.* **5:**309–311.

Joy, J. A., and D. L. Clark. 1977. The distribution, ecology and systematics of the benthic Ostracoda of the central Arctic Ocean. *Micropaleontology* **23:**129–154.

Kellogg, T. B. 1977. Paleoclimatology and paleo-oceanography of the Norwegian and Greenland seas: The last 450,000 years. *Marine Micropaleontol.* **2:**235–249.

Kellogg, T. B., J. C. Duplessy, and M. H. Shackleton. 1978. Planktonic foraminiferal and oxygen isotopic stratigraphy and paleoclimatology of Norwegian Sea deep-sea cores. *Boreas* **7:**64–73.

Kennett, J. 1982. *Marine Geology.* Prentice-Hall, New Jersey, 813p.

Kent, D., N. D. Opdyke, and M. Ewing. 1971. Climate change in the North Pacific using ice-rafted detritus as a climatic indicator. *Geol. Soc. America Bull.* **82:**2,741–2,754.

Kinney, P. J., T. C. Loder, and J. Groves. 1971. Particulate and dissolved organic matter in the Amerasian Basin of the Arctic Ocean. *Limnol. Oceanogr.* **16:**132–137.

Kitchell, J. A., and D. L. Clark. 1982. Late Cretaceous–Paleogene paleogeography and paleocirculation: Evidence of North Polar upwelling. *Paleogeog. Palaeoclimatol. Palaeoecol.* **40:**135–165.

Kitchell, J. A., J. F. Kitchell, D. L. Clark, and L. Dangeard. 1978. Deep-sea foraging behavior: Its bathymetric potential in the fossil record. *Science* **200:**1,289–1,291.

Krishnaswami, S. 1976. Anthigenic transition

elements in pelagic clays. *Geochim. Cosmochim. Acta* **40:**425–434.

Kunze, G. W., L. E. Knowles, and Y. Kitano. 1968. The distribution and mineralogy of clay minerals in the Taku Estuary of southeastern Alaska. *Marine Geol.* **6:**439–448.

Lachenbruch, A. H., and B. V. Marshall. 1966. Heat flow through the Arctic Ocean floor: The Canada Basin–Alpha Ridge boundary. *J. Geophys. Res.* **71:**1,223–1,248.

Lagoe, M. B. 1976. Species diversity of deep-sea benthic foraminifera from the central Arctic Ocean. *Geol. Soc. America Bull.* **87:**1,678–1,683.

Lagoe, M. B. 1977. Recent benthic foraminifera from the Central Arctic Ocean. *J. Foraminiferal Res.*7(2):106–129.

Lapina, N. N., and N. A. Belov. 1960. Peculiarities of sediment formation in the Arctic Ocean. In N. M. Strahov, P. L. Bezrukov, and V. S. Yablokov (eds.), *Recent Sediments of the Seas and Oceans.* Reports of the 21st International Geological Congress, 24 to 27 May 1960, pp. 88–93. (English translation by the U.S. Navy Electronics Lab., 1 August 1965.)

Lew, M. 1981*a*. The distribution of some major and trace elements in sediments of the Atlantic Ocean (DSDP Samples). 1. The distribution of sulphur isotopes, and Mn, Fe, Zn and Cu. *Chem. Geol.* **33:**205–224.

Lew, M. 1981*b*. The distribution of some major and trace elements in sediments of the Atlantic Ocean (DSDP Samples). 2. The distribution of total, fixed and organic nitrogen. *Chem. Geol.* **33:**225–235.

Li, Y. H., J. Bischoff, and G. M. Mathieu. 1969. The migration of manganese in the Arctic Basin sediment. *Earth Planet. Sci. Lett.* **7:**265–270.

Li, Y. H., T. Takahashi, and W. S. Broecker. 1969. Degree of saturation of calcium carbonate in the oceans. *J. Geophys. Res.* **74:**5,507–5,526.

Ling, H. Y., L. M. McPherson, and D. L. Clark. 1973. Late Cretaceous (Maestrichtian?) silicoflagellates from the Alpha Cordillera of the Arctic Ocean. *Science* **180:**1,360–1,361.

Linkova, T. I. 1965. Some results of paleomagnetic study of Arctic Ocean floor sediments. In E. R. Hope (trans.), *The Present and Past of the Geomagnetic Field.* Nauka, Moscow, pp. 279–291.

Lisitzin, A. P. 1972. *Sedimentation in the World Ocean.* Special Publication 17. Society of Economic Paleontologists and Mineralogists, Tulsa, Okla. 218p.

Locke, W. W., III. 1979. Etching of hornblende grains in Arctic soils: An indicator of relative age and paleoclimate. *Quaternary Res.* **11:**197–212.

Luepke, G. 1975. *Heavy Mineral Trends in the Beaufort Sea.* Open File Report 75-667. U.S. Geological Survey, 16p.

McManus, D. A., and C. S. Smyth. 1970. Turbid bottom water on the continental shelf of northern Bering Sea. *J. Sed. Petrol.* **40:**869–873.

Mankinen, E. A., and G. B. Dalrymple. 1979. Revised geomagnetic polarity time scale for the interval 0–5 mybp. *J. Geophys. Res.* **84**(B2):615–626.

Margolis, S. V., and Y. Herman. 1980. Northern hemisphere sea-ice and glacial development in the late Cenozoic. *Nature* **286**(5,769):145–149.

Medlin, J. H., N. H. Suhr, and J. B. Bodkin. 1969. Atomic absorption analysis of silicates employing $LiBO_2$ fusion. *Atomic Absorption Newsletter* **8:**25–29.

Mignoit, C. 1968. Etude des proprietes physiques de differents sediments tres fin et leur comportement sous des actions hydrodynamiques. *La Houille Blanche* **7:**59–62.

Minicucci, D. A., and D. L. Clark. 1983. A late Cenozoic stratigraphy for glacial-marine sediments of the eastern Alpha Cordillera, central Arctic Ocean. In B. F. Molnia (ed.), *Glacial-Marine Sedimentation.* Plenum Publishing Corp., New York, pp. 331–365.

Mohr, J. L., and S. R. Geiger. 1968. Arctic Basin faunal precis. Animals taken mainly from Arctic drifting stations and their significance for biogeography and water-mass recognition. In J. E. Sater (coordinator), *Arctic Drifting Stations.* Arctic Institute of North America, pp. 297–313.

Morris, T. H., and D. L. Clark. 1986. Pleistocene calcite lysocline and paleocurrents of the central Arctic Ocean and their paleoclimatic significance. *Paleoceanography* **1:**181–195.

Morris, T. H., D. L. Clark, and S. M. Blasco. 1985. Sediments of the Lomonosov Ridge and Makarov Basin: A Pleistocene stratigraphy for the North Pole. *Geol. Soc. America Bull.* **96:**901–910.

Moser, F. C., and J. R. Hein. 1984. Distribution of clay minerals in the suspended and bottom sediments from the Northern Bering Sea shelf area, Alaska. *U.S. Geol. Survey Bull.* 1,624, 19p.

Mowatt, T. C., A. S. Naidu, and N. Veach. 1974. Detailed Clay Mineralogy of the Colville River and Delta, Northern Arctic Alaska. Open File Report No. 45, Alaska Division of Geology and Geophysical Survey, 36p.

Mudie, P. J. 1985. Palynology of the CESAR cores, Alpha Ridge. In H. R. Jackson, P. J.

Mudie, and S. M. Blasco (eds.), *Initial Geologic Report on CESAR—The Canadian Expedition to Study the Alpha Ridge, Arctic Ocean.* Geological Survey of Canada Paper 84-22, pp. 149–174.

Mudie, P. J., and S. M. Blasco. 1985. Lithostratigraphy of the CESAR cores. In H. R. Jackson, P. J. Mudie, and S. M. Blasco (eds.), *Initial Geologic Report on CESAR—The Canadian Expedition to Study the Alpha Ridge, Arctic Ocean.* Geological Survey of Canada Paper 84-22, pp. 59–100.

Mullen, R. E., D. A. Darby, and D. L. Clark. 1972. Significance of atmospheric dust and ice rafting for Arctic Ocean sediment. *Geol. Soc. America Bull.* **83**:205–212.

Muller, P. J. 1975. Diageneses stickstoffhaltiger organischer Substanzen in oxischen and anoxischen marinen Sedimenten. "Meteor" Forsch.-Ergebn. C, No. 22, Berlin, pp. 1–60.

Muller, P. J., 1977. C/N ratios in Pacific deep-sea sediments: Effects of inorganic ammonium and organic nitrogen compounds sorbed by clays. *Geochim. Cosmochim. Acta* **41**:765–776.

Murphy, J., and J. P. Riley. 1962. A modified single solution method for determination of phosphate in natural waters. *Analytica Chim. Acta* **27**:31–36.

Naidu, A. S. 1974. Sedimentation in the Beaufort Sea: A synthesis. In Y. Herman (ed.), *Marine Geology and Oceanography of the Arctic Seas.* Springer-Verlag, New York, pp. 173–190.

Naidu, A. S. 1979. *Sources, Transport Pathways, Depositional Sites and Dynamics of Sediments in the Lagoon and Shallow Marine Region, Northern Arctic Alaska.* Annual Report to BLM-NOAA-OSCEAP Office, Boulder, Colo. 81p.

Naidu, A. S. 1985. Organic carbon, nitrogen, and C/N ratios of deltaic sediments, north Arctic Alaska. In *SCOPE/UNEP Sonderband, Hefd 58: Mitt. Geol.-Paleont. Inst.* University of Hamburg, pp. 311–321.

Naidu, A. S., and D. W. Hood. 1972. Chemical composition of bottom sediments of the Beaufort Sea, Arctic Ocean. In *Proceedings of the 24th International Geological Congress, Montreal.* Section 10, 307–317.

Naidu, A. S., and G. D. Sharma. 1972. *Texture, Mineralogy and Chemistry of Arctic Ocean Sediments.* Technical Report R72-12, University of Alaska Institute of Marine Science, 31p.

Naidu, A. S., and T. C. Mowatt. 1974. Aspects of size distributions, mineralogy, and geochemistry of deltaic and adjacent shallow marine sediments, north Arctic Alaska. In G. L. Hufford et al. (eds.), *An Ecological Survey in the Beaufort Sea.* U.S. Coast Guard Oceanographic Unit, Oceanographic Report CG373-064, Washington, D. C., pp. 238–262.

Naidu, A. S., and T. C. Mowatt. 1975. Clay mineralogy and geochemistry of continental shelf sediments of the Beaufort Sea. In J. C. Reed and J. W. Sater (eds.), *The Coast and Shelf of the Beaufort Sea.* Arctic Institute of North America, Arlington, Va., pp. 493–510.

Naidu, A. S., and T. C. Mowatt. 1983. Sources and dispersal patterns of clay minerals in surface sediments from the continental-shelf areas off Alaska. *Geol. Soc. America Bull.* **94**:841–854.

Naidu, A. S., D. C. Burrell, and D. W. Hood. 1971. Clay mineral composition and geological significance of some Beaufort Sea sediments. *J. Sed. Petrol.* **41**:691–694.

Naidu, A. S., J. S. Creager, and T. C. Mowatt. 1982. Clay mineral dispersal patterns in the north Bering and Chukchi seas. *Marine Geol.* **47**:1–15.

Naidu, A. S., T. C. Mowatt, D. B. Hawkins, and D. W. Hood. 1975. Clay minerals and geochemistry of some Arctic Ocean sediments: Significance of paleoclimate interpretation. In G. Weller and S. A. Bowling (eds.), *Climate of the Arctic.* Geophysical Institute, University of Alaska, Fairbanks, pp. 59–67.

Naidu, A. S., L. H. Larsen, T. C. Mowatt, M. D. Sweeney, and H. V. Weiss. 1982. *Aspects of Size Distributions, Clay Mineralogy and Geochemistry of Sediments of the Beaufort Sea and Adjacent Deltas, North Arctic Alaska.* Final Report to NOAA-OCSEAP Office, Boulder, Colo. 114p.

Naugler, F. P. 1967. Recent Sediments of the East Siberian Sea. Unpublished masters thesis, University of Washington, Seattle, 71p.

Naugler, F. P., N. Silverberg, and J. S. Creager. 1974. Recent sediments of the East Siberian Sea. In Y. Herman (ed.), *Marine Geology and Oceanography of the Arctic Seas.* Springer-Verlag, New York, pp. 191–210.

Opdyke, N. D. 1972. Paleomagnetism of deep-sea cores. *Rev. Geophys. Space Phys.* **10**:213–249.

Ovenshine, A. T. 1970. Observations of iceberg rafting in Glacier Bay Alaska, and the identification of ancient ice-rafted deposits. *Geol. Soc. America Bull.* **81**:891–894.

Paul, A. Z., and R. J. Menzies. 1973. *Benthic Ecology of the High Arctic Deep-Sea.* Technical Report, Department of Oceanography, Florida State University, 337p.

Perry, R. K., and H. S. Fleming. 1983. *Bathymetry of the Arctic Ocean.* Naval Research Laboratory, Washington, D.C. Scale 1:4, 704, 075.

Perry, R. K., H. S. Fleming, J. R. Weber, Y.

Kristoffersen, J. K. Hall, A. Grantz, and G. L. Johnson. 1985. *Bathymetry of the Arctic Ocean.* Geological Society of America, Boulder, Colo. Scale 1:4, 704, 675.
Prest, V. K., D. R. Grant, and V. N. Rampton. 1968. Glacial Map of Canada. *Geol. Survey Canada Map 1253A.* Scale 1:5,000,000.
Rader, L. D., and R. S. Grimaldi. 1961. Chemical analysis for selected minor elements in Pierre Shale. *U.S. Geol. Survey Prof. Paper 391-A,* pp. A1–A45.
Rona, P. A., K. Bostrom, L. Laubier, and K. L. Smith, Jr. (eds.). 1983. *Hydrothermal Processes at Seafloor Spreading Centers.* Plenum Press, New York, 796p.
Ruddiman, W. F., and A. McIntyre. 1981. Oceanic mechanisms for amplification of the 23,000-year ice volume cycle. *Science* **212:**617–627.
Rupke, M. A. 1975. Deposition of fine-grained sediments in the abyssal environment of the Algero-Balearic Basin, Western Mediterranean Sea. *Sedimentology* **22:**95–109.
Schwarzacher, B. A., and K. Hunkins. 1961. Dredged gravels from the central Arctic Ocean. In G. O. Raasch (ed.), *Geology of the Arctic,* vol. 1. University of Toronto Press, pp. 666–677.
Sejrup, H. P., G. H. Miller, J. Brigham-Grette, R. Lovlie, and D. Hopkins. 1984. Amino-acid epimerization implies rapid sedimentation rates in Arctic Ocean cores. *Nature* **310:**772–775.
Silverberg, N. 1972. Sedimentology of the Surface Sediment of the East Siberian and Laptev Seas. Unpublished Ph.D. Dissertation, University of Washington, 184p.
Slatt, R. M. 1970. Sedimentological and Geochemical Aspects of Sediments and Water from Ten Alaskan Valley Glaciers. M. S. thesis, University of Alaska, Fairbanks, 125p.
Steuerwald, B. A., D. L. Clark, and J. A. Andrew. 1968. Magnetic stratigraphy and faunal patterns in Arctic Ocean sediments. *Earth Planet. Sci. Lett.* **5:**79–85.
Stevenson, F. J., and C. N. Cheng. 1972. Organic geochemistry of the Argentine Basin sediments: C-N relationships and Quaternary correlations. *Geochim. Cosmochim. Acta.* **36:**653–671.
Stoiber, R. E., J. B. Lyons, W. T. Elbertz, and R. H. McCrehon. 1960. Petrographic evidence on the source area and age of T-3. In V. Bushnell (ed.), *Scientific Studies at Fletcher's Ice Island, T-3 (1952–1955),* vol. 3. Air Force Cambridge Research Center, Cambridge, Mass., p. 114.
Stow, D. A. V., and G. Shanugan. 1980. Sequence of structures in fine-grained turbidites: Comparison of recent deep-sea flysch sediments. *Sed. Geol.* **25:**23–42.
Sundborg, A. 1956. The River Klaralven: A study of fluvial processes. *Geogr. Annalr.* **38:**127–316.
Sweeney, J. F., J. R. Weber, and S. M. Blasco. 1982. Continental ridges in the Arctic Ocean: LOREX contraints. *Tectonophysics* **89:**217–237.
Sweeney, M. D. 1984. Heavy Metals in the Sediments of an Arctic Lagoon, Northern Alaska. M.S. thesis, University of Alaska, Fairbanks, 256p.
Tibbs, J. F. 1967. On some planktonic protozoa taken from the tract of drift station ARLIS I, 1960–61. *Arctic* **20:**247–254.
Trask, P. D. 1939. Organic content in recent marine sediments. In P. D. Trask (ed.), *Recent Marine Sediments.* Thomas Murby and Co., London, pp. 428–453.
Trettin, H. P. 1972. The Innuitian Province. In R. A. Price and R. J. W. Douglas (eds.), *Variations in Tectonic Styles in Canada.* Special Paper 11. The Geological Association of Canada, Toronto, pp. 83–179.
Vilks, G., J. M. Hall, and D. J. W. Piper. 1977. The natural remanent magnetization of sediment cores from the Beaufort Sea. *Canadian J. Earth Sci.* **14:**2,007–2,012.
Vincent, J.-S. 1983. La geologie du Quaternaire et al geomorphologie de L'ile Banks. *Arctique Canadien Mem. 405.* Geological Survey of Canada, Ottawa, 118p.
Vogt, P. R., and O. D. Avery. 1974. Tectonic history of the Arctic Basins: Partial solutions and unsolved mysteries. In Y. Herman (ed.), *Marine Geology and Oceanography of the Arctic Seas.* Springer-Verlag, New York, pp. 83–117.
Walter, P., and P. Stoffers. 1985. Chemical characteristics of metalliferous sediments from eight areas on the Galapagos Rift and East Pacific Rise between 2°N and 42°S. *Marine Geol.* **65:**271–287.
Weber, J. R., and H. R. Jackson. 1985. CESAR bathymetry. In H. R. Jackson, P. J. Mudie, and S. M. Blasco (eds.), *Initial Geological Report on CESAR—the Canadian Expedition to Study the Alpha Ridge, Arctic Ocean.* Geological Survey of Canada Paper 84-22, pp. 15–18.
Weber, J. R., and J. F. Sweeney. 1985. Reinterpretation of morphology and crustal structure in the central Arctic Ocean Basin. *J. Geophys. Res.* **90:**663–677.
Wedepohl, K. H. 1956. Spurenanalytische Unterschuchungen an Tiefseetonen aus dem Atlantic. *Geochim. Cosmochim. Acta* **18:**200–231.
Weiss, H. V., M. A. Guttman, J. Korkisch, and I. Steffan. 1977. Comparison of methods for the determination of vanadium in sea-water. *Talanta* **24:**509–511.

Witte, W. K., and D. V. Kent. 1988. Revised magnetostratigraphies confirm low sedimentation rates in Arctic Ocean cores. *Quaternary Res.* **29**:43–53.

Worsley, T. R., and Y. Herman. 1980. Episodic ice-free Arctic Ocean in Pliocene and Pleistocene time: Calcareous nannofossil evidence. *Science* **210**:323–325.

Zauderer, K. 1982. Analysis of Heavy Minerals in Arenaceous Lutites from the Northern Canada Basin, Arctic Ocean. Unpublished masters thesis, Old Dominion University, Norfolk, 83p.

# 25: Late Cenozoic Stratigraphy and Paleoceanology of the Barents Sea

V. S. Zarkhidze
and Yu. G. Samoilovich

## INTRODUCTION

Until recently stratigraphical studies of the Arctic seas were based on surface seafloor sediments collected with gravity cores. We now have data from sparker profiles and drill cores of the entire unconsolidated sediment sequence above the upper regional unconformity (average 25–30 m).

Barents Sea Neogene–Quaternary sediment thicknesses were determined by seismoacoustic (sparker) methods. Maximum thickness occurs at the edge of the Spitsbergen shelf (up to 500 m) in the proximity of the continental slope in the western Bjørnoyrenna Trough and Storfjordrenna (Solheim and Kristoffersen, 1984). Generally, Late Cenozoic sediment thicknesses greater than 50 m are rare on the Barents and Norwegian shelves. On the Kola shelf in areas with a maximum thickness of 50–250 m, the shelf topography is very rugged and buried paleoriver valleys can be traced over long distances; they develop into branched paleodelta channels. The sediments that fill and overlie these landforms are a rhythmical alternation of continental and marine lilthofacies.

Within the circumpolar Arctic Basin and surrounding continents, there are Pliocene–Pleistocene glacial marine diamictons yielding abundant fossils, mainly mollusks and foraminifers. Pollen and diatoms, however, do not lend themselves very well for correlations, particularly in coastal sections.

We have made the ''rhythmostratigraphic'' correlations and have integrated biostratigraphic data from the Northern Norwegian and Barents seas with the coastal sections of the Kanin Peninsula as well as with regions located to the east of this peninsula. Such distant areas as the North Sea can be very easily correlated to our sections (Feyling-Hanssen, 1980). Some of our foraminiferal biozones can be considered as marker horizons on the Barents Sea shelf. Most important is the ''boreal transgression maximum'' horizon. It represents the Eemian, Miculina, and Kazantsevo horizons of western Europe, eastern Europe, and northern Siberia, re-

*Note:* This chapter was reviewed, edited, and rewritten by the Editor, Yvonne Herman.

spectively. The age of the horizon is 50–100,000 yr. On the Kola-Kanin shelf there are faunal assemblages typical of the middle Pleistocene Rogovskaya Formation (125,000–400,000 yr ago) of the Pechora Lowland. Strictly speaking, by the "epoch of boreal transgressions" we are referring to the Middle–Late Pleistocene. Figure 25-1 shows the main relationships between Plio-Pleistocene seismofacies and their inferred age and genesis (Kola shelf provides a typical example of the southern Barents Sea).

In our discussion of paleogeography and paleoceanology we will not be able to confine ourselves to the Barents Sea and adjacent areas because during the Cenozoic, mainly during the Plio-Pleistocene, the Barents Sea shelf was often (but not always) connected with the Norwegian–Greenland Basin as well as with the Eurasian subbasin of the Arctic Ocean. Furthermore, the Barents Sea together with the Kara Sea form a single morphostructural subdivision named the Barents–Kara sector of the Arctic Ocean (Dibner, 1978).

During the beginning of the Late Cenozoic, tectonic activity increased in the southern margins of the Barents–Kara region. The Polar Urals and Novaya Zemlya were the scenes of mountain-building activity, while the oceanic abyssal regions became deeper, causing the incisions of the Late Miocene drainage systems. These incised areas are shown in Figure 25-1.

## EARLY–MIDDLE PLIOCENE (2.4–5.2 mya)

The marine transgression in the Arctic Ocean began during the Early–Middle Pliocene, at a time when the territory of northeastern Europe and western Siberia were covered by a shallow sea (50–100 m). The thickness of marine Pliocene sediments in boreholes reached 100–500 m in the northern Timan-Ural region. In the northeastern part of the Western Siberian Lowland it is somewhat thinner. A peculiar feature of the tectonic environment of this transgression was in the isolation of marine epicontinental basins from the North Atlantic (Gramberg and Pogrebitsky, 1984). A vast landmass seems to have occupied most of the present Barents Sea shelf. This assumption is based on the following data:

1. Marine facies of the Prosundui and Kolva formations in the Timan-Ural region and those of the Ust' Soleninskaya Formation in Western Siberia do not contain boreal North Atlantic faunal as-

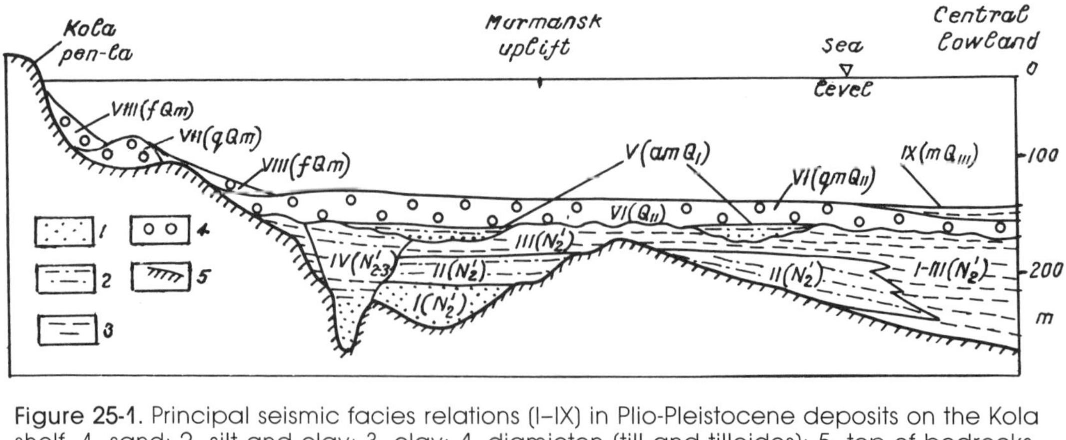


Figure 25-1. Principal seismic facies relations (I–IX) in Plio-Pleistocene deposits on the Kola shelf. 1, sand; 2, silt and clay; 3, clay; 4, diamicton (till and tilloides); 5, top of bedrocks. $N_2$, Upper Neogene, Pliocene; Q, Quaternary. g, glacial, f, fluvioglacial; a, alluvial; m, marine; gm, glaciomarine.

semblages. (They appear in the Arctic in younger deposits.)

2. Along with the relatively monotonous lithology of deep-sea deposits with molluskan faunas, which do not change even in the vicinity of the Urals, shallow, near-shore facies were found on the shores of the Pechora Sea (borehole 100 Tobseda Village). Here a 70 m sequence of shallow-water deposits with *Macoma baltica,* interlayered with relatively deep-water sediments containing the diagnostic mollusks Propeamussium and Nucula, were observed (Gramberg and Kulakov, 1983). However, no Pliocene deposits of that period have been found farther north on Kolguev Island.
3. Spore, pollen, and foraminiferal assemblages in the marine Pliocene deposits of Novaya Zemlya also suggest paleoenvironments quite different from the present. Representative foraminiferal assemblages do not contain species typical of this island shelf. Baranovskaya (personal communication, 1981) suggested that they represent a continental shelf biozone. The pollen spectra are dominated by boreal flora with the Asteraceae, Caryophylaecae, and representatives of the Artemisia families. Malyasova (personal communication, 1984) believes that the palinoflora has no equivalents in the Arctic. At the same time marine Pleistocene deposits of the southern areas contain pollen spectra dominated by fir and pine trees.
4. The Pliocene paleogeographic environment when a land occupied the site of the Western Barents Sea was discussed earlier (Fig. 25-1). Thus, the assumption concerning the existence of a hypothetical land separating the Kolva Pliocene marine basin from the Atlantic Ocean seems very likely.

The marine deposits and the monotonous fauna in the areas that were subjected to Pliocene marine transgression suggest a transgression on a peneplained surface on the background of which even the Urals were small uplifts (300–500 m) separated by wide transverse depressions. Deposits of the Prosundui Formation in the northern Timan-Ural region that filled the depressions were eroded mainly from Mesozoic rocks. The Kolva Formation was more widespread: The transgression spread over the Malozemelskaya and the Bolshezemelskaya tundras (Pechora Lowland). Thus, at the end of the first major Pliocene cycle, within most western sectors of the Soviet Arctic, the development of an accumulative plain took place over an extensive zone that spread over the southern margin of the Barents Sea and most Kara Sea shelves.

## LATE PLIOCENE (1.8–2.4 mya)

This stage began with a new orogenic cycle including the uplift of the Polar Urals, Novaya Zemlya, Taimyr, and Severnaya Zemlya. The uplift extended over regions adjacent to the shelf. The marine Pliocene deposits were partially eroded even in the previously submerged areas (e.g., the Khoreiver Trough with the Pechora Lowland). The preexisting shape of shelf basins changed. The land areas in the western Barents Sea shelf, which prevented the water exchange between the Arctic and the Atlantic oceans, were subsiding simultaneously with the northward shift of the coastal line (Zarkhidze, unpublished). This resulted in the deposition of shallow-water sediments represented by sands and silts with calcareous concretions containing *Cyrtodaria angusta* (Gramberg and Pogrebitsky, 1984). The deposits are rich in warm-water mollusks and foraminifera and are predominated by Atlantic species. This period was characterized (judging from the pollen spectra) by a considerable climatic warming. Spiridonova notes the predominance of arborescent pollen grains (fir, pine, alder, and birch) and the presence of pollen grains of broad-leaved trees (elm, oak, norhnbeam, and hazelnut) (Yakhimovich, 1981). The penetration of the warm Atlantic current into the central Arctic Ocean qualitatively changed the faunal composition of the North Polar Sea. In this period (2.4 my B.P.) the Pacific Ocean mollusks migrated into the North Atlantic via the Arctic Ocean shelf.

## EARLY-MIDDLE PLEISTOCENE (0.32–1.8 mya)

This period is characterized by a unique paleoceanographic environment in the Arctic with the accumulation of thick (up to 200–250 m), complex, sedimentary sequences; Sacks (1953) called it the ''age of maximum glaciation,'' but it is also known as the ''period of maximum transgression'' (Gramberg and Pogrebitsky, 1984). Sedimentation occurred mainly in northeastern Europe and the Western Siberian Lowlands, where a number of large sedimentary basins, such as the Timan-Ural and Western Siberian, formed during the Early(?)–Middle Pleistocene. These deposits are composed mainly of pebbly mudstones (known as tilloides, or diamicton). Successions such as the Rogovskaya Formation in the Pechora Lowland and the Samarov Horizon in western Siberia are examples of such types of deposits.

In the Kola Peninsula, most investigators mention the lack of the Middle Pleistocene sediments. Marine Late Pleistocene facies overlay directly crystalline basement or a thin sedimentary cover of continental origin (alluvial, lacustrine-alluvial, and other). Middle Pleistocene deposits have not been found on the Kola shelf either. In the central Barents region, as on the Kola shelf, there are no Middle Pleistocene sediments. Gravity cores have not recovered sediments older than Late Pleistocene Polyak, 1982).

On the Northern Kolguev Island the early Middle Pleistocene deposits crop out in typical glacial-marine facies (tilloides). Their thickness does not exceed 20–30 m. Thus, the central Barents region could not have been an area of high sedimentation. The drastic decrease in sediment thickness during Middle Pleistocene suggests that the region either had an intermittent depositional regime or that it is a denudation area. The Novaya Zemlya elevations (400–500 m) suggest the presence of a denudation area. The southern and central parts of the Kara shelf are also believed to have been denudation areas.

Pleistocene deposits are absent on the Arctic Institute Islands and the Sverdrup Island (Samoilovich et al., 1981). We believe that this absence is evidence for active erosion, primarily from northern areas of denudation on the West Arctic shelf, the main regions of Cretaceous and Cenozoic deposits. The presence of a northern band of denudation accounts for the nature of the climatic and paleohydrological environment with shelf sedimentary basins, which despite their large size had limited connections with oceanic basins to the west through the Baltic Sea (Afanasiev, 1979), and to the east along the North Siberian Lowland. Reconstructions of Middle Pleistocene paleogeography show dissected paleobasin shorelines, presence of islands and archipelagos, relatively low water temperature, and seasonal ice fields (Zarkhidze, unpublished).

## MIDDLE-LATE PLEISTOCENE (0.05–0.32 mya)

At the end of Middle Pleistocene and beginning of Late Pleistocene, a major tectonic reorganization took place within the Barents-Kara region. This resulted in a change in the areas of denudation and accumulation. The sedimentary basins of northeastern Europe and western Siberia were uplifted. The Pre-Kazantsevo incision, with a typical example being the Yenisei paleovalley, provides a comparison of the position of coastlines during the late stages of Middle Pleistocene paleobasins (Rogovskaya Formation and others) and the early stages of Late Pleistocene basins. Table 25-1 shows that both the Kola Peninsula and the Southern Novaya Zemlya Island were denudation areas during Middle Pleistocene time. However, at the beginning of Late Pleistocene, they underwent a significant submergence and the sea ingressed. A sharp change in the hydrobiological environment took place on the Barents-Kara shelf: With increased temperatures warm Atlantic fauna penetrated this region. The Kazantzevo deposits form a terrace 140 m above sea level (in northeastern Europe) and 170–180 m (in northwestern Siberia). The post-Kazantsevo regression (the Zyryan) was a cold time interval.

During the Kazantsevo–Zyryan subphase the Barents–Kara region underwent significant changes that led to the accumulation of marine sediments over the extensive western Barents and central Barents areas, most of the Kara Sea shelf, much of the Kola Peninsula, and Novaya Zemlya, as well as along the northern coasts of Europe and northeastern and northern Western Siberia. These are mainly near-shore marine facies (Kazantsevo horizon) and glaciomarine facies (Zyryan horizon). The Kazantsevo horizon contains the largest and most diversified Late Cenozoic fauna (molluska, foraminifera, and ostracoda).

Table 25-1. Comparison of Terrace Levels for the Middle and Upper Pleistocene Transgressions in Northern Eurasia

| | *Maximum Transgression Level (m)* | |
|---|---|---|
| *Location* | $Q_{II}$ | $Q^1_{III}$ |
| Kola Pen-la | 20–30 | 160 |
| Novaya Zemlya | 60 | 200–220 |
| Northeastern Europe | 200 | 60 |
| Western Urals | 230–250 | 60 |
| Northwestern Siberia | 230–250 | 60 |

The relatively deep-water facies (mainly clays) of the Kazantsevo deposits occur on the Kola Peninsula, in Novaya Zemlya, in boreholes on the coast of the Pechora Sea, as well as on Spitsbergen (Semevsky, 1967).

## LATE PLEISTOCENE (KARGA) (0.02–0.05 mya)

Tectonic movements continued within the Barents–Kara region during the Karga period. Spitsbergen, Franz Josef Land, and Severnaya Zemlya became the arena of marine terrace formation (20 to 100–200 m). At this time the above-mentioned archipelagos acquired their present-day topography. Karga sediments on this perioceanic shelf margin consist of sands and silts. During the Karga period tectonic activity was very intense on the Kola Peninsula and Novaya Zemlya. According to Gudina and Evzerov (1973), the coastline of the Karga transgression on the Kola Peninsula was 140 m high. An abrasion bench of the Karga transgression on Novaya Zemlya occurs at the same level of 140 m. The Karga deposits formed a terrace at 40 m in the maritime lowland of Eurasia (Zarkhidze, unpublished).

## LATE PLEISTOCENE (SARTANIAN) (0.01–0.02 mya)

A distinctive environment is related to global sea-level drop by 100–120 m and continued orogeny, mainly in the perioceanic areas of the Barents–Kara shelf. Geological data suggest active uplift and drainage of large areas within the Barents and Kara seas. Fragments of young subsided drainage systems are identifiable in the topography of the Kara seafloor. The Sartanian substage witnessed the initiation of depression of future trenches. Simultaneously, in northern Eurasia, Sartanian incisions of drainage networks are observed. There are radiocarbon dates of peat giving an age of 15 kya B.P. from the coastal sequences of Novaya Zemlya, Yamal Peninsula, and Severnaya Zemlya (Zarkhidze and others, unpublished).

## HOLOCENE (0.001 mya)

Tectonic activity during the Holocene was as prevalent as during the Late Pleistocene. There is evidence of rejuvenation of morphostructures, such as the Franz Victoria, St. Anna, and Voronin Novaya Zemlya oceanic troughs, and associated uplifts within the Barents-Kara shelf.

Traces of the Holocene marine transgression are observed at 20–60 m elevation on the Kola Peninsula and Novaya Zemlya. The Holocene sea-level changes were accompanied by the formation of coastal terrace accumulative units on Novaya Zemlya. However, there are no synchronous monolevel terraces. Marine sediments with ice-rafted material were deposited during the Holocene on the Barents-Kara shelf. Our climatic-stratigraphic scale based on pollen data (Blytt-Sernander scale) is very detailed and our pollen, foraminiferal, and diatom biozones for the Barents shelf are comparable to this scale (Zarchidze and others, unpublished).

Holocene sandy and argillaceous deposits occur ubiquitously on the Barents Sea floor, except for neotectonically active sites of recent uplift where we find submarine erosion of the upper layers of sediments.

## CONCLUSIONS

Our paleoceanographic reconstructions are based on tectono-eustatic models for the Pliocene-Pleistocene-Holocene periods. The rhythmic global eustatic sea-level fluctuations superimposed upon regional tectonic movements can be recognized on islands in the Arctic seas as abrasion terraces in bedrocks. However, on the continent, on low coastal plains such as the eastern European and western Siberian plains, the sea ingression resulted in submergence and intense transport of unconsolidated deposits (alluvial, marine). Many geologists consider these reworked sediments moraines; however, we do not share this viewpoint. Terrigenous ice-rafted material was deposited everywhere on the vast offshore of the Arctic seas during the Pleistocene glacial periods. The deposits of the so-called ''boreal transgression'' consist mainly of clays, silts, and pebbles and yield marine well-preserved fossils.

## REFERENCES


Afanasiev, B. L. 1979. *Istoria Geologicheskogo Razvitia Pribaltiki v Pliotsen-Chetvertichnoe Vremya.* Riga, 70p. (In Russian.)

Dibner, V. D. 1978. *Morfostruktura Shel'fa Barentseva Morya.* Nedra, Leningrad, 211p. (In Russian.)

Feyling-Hanssen, R. W. 1980. Foraminiferal indication of Eemian interglacial in northern North Sea. *Bull. Geol. Soc. Denmark* **29:**175–189.

Gramberg, I. S., and Yu. N. Kulakov (eds.). 1983. *Osnovnye Problemy Paleogeografii Pozdnego Kainozoya Arktiki.* Nedra, Leningrad, 263p. (In Russian.)

Gramberg, I. S., and Yu. E. Pogrebitsky (eds.). 1984. Geologischeskoe stroenie SSSR i zakonomernosti razmeshchenia poleznykh iskopaemykh. T. 9 *Morya Sovetskoi Arktiki.* Nedra, Leningrad, 280p. (In Russian.)

Gudina, V. I., and V. Yu. Evzerov. 1973. *Stratigrafia i foraminifery Pozdnego Pleistosena Kol'skogo Poluostrova.* Nauka, Novosibirsk, 145p. (In Russian.)

Polyak, L. V. 1982. Vertikal'noe rasprostranenie foraminifer v otlozheniayh kraevykh zhelobov Barentsevo-karskogo shel'fa. In V. S. Zarchidze (ed.), *Stratigrafia i Paleogeografia Pozdnego Kainozoya Arktiki.* Sevmorgeologia, Leningrad, pp. 19–27. (In Russian.)

Sacks, V. N. 1953. *Chetvertichnyi period v Sovetskoi Arktike.* Nedra, Moscow, 627p. (In Russian.)

Samoilovich, Yu. G. 1982. O pogrebennoi rechnoi sisteme na akvatorii Kol'skogo shel'fa. In V. S. Zarkhidze (ed.), *Stratigrafia i Paleogeo-*

*grafia Pozdnego Kainozoya Arktiki.* Sevmorgeologia, Leningrad, pp. 5–9. (In Russian.)

Semevsky, D. V. 1967. Osnovyne stadii razvitia arkhipelaga Spitsbergen v pliotsen-chetvertichnoe vremya. In V. N. Sokolov (ed.), *Materialy po Stratigrafii Spitsbergena.* NIIGA, Leningrad, pp. 84–98. (In Russian.)

Solheim, A., and Y. Kristoffersen. 1984. *The Physical Environment of Western Barents Sea 1:1,500,000.* Norsk Polarinstitutt Skrifter, Norway, 179B, 25p.

Yakhimovich, V. L. (ed.). 1981. *Pliotsen i Pleistotsen Volgo-Ural'skoi Oblasti.* Nauka, Moscow, 162p. (In Russian.)

Zarkhidze, V. S. 1970. Istoria razvitia fauny morskiky molluskov priatlanticheskogo sektora Arktiki v pozdnem kainozoe. In B. A. Yurtsev (ed.), *Severnyi Ledovityi Okean i Ego Poberezhie v Kainozoe.* Leningrad, pp. 186–194. (In Russian.)

# 26: The Last Glaciation of Eurasia

A. A. Velitchko,
L. L. Isayeva, D. B. Oreshkin,
and M. A. Faustova

## SCANDINAVIAN ICE SHEET

The southwestern margin of the Scandinavian Ice Sheet was on the Jutland Peninsula where the preglacial Jutland ridge appears to have been an obstacle to ice movement. The marginal zone crosses the peninsula in a submeridional direction and is incomplete. It consists of individual terminal moraines and associated outwash plain and glacial tunnel valleys. Most researchers refer to it as the maximum stage of the Last Glaciation. The marginal zone formed approximately 16 kya. This date was derived from varve analysis (Hansen, 1965). The glacier is believed to have had the maximum extent during deglacial times.

Eastward, on the northern European lowland, marginal ice forms have the shape of a ridge oriented southward between the Elba and the Oder rivers. They are expressed by small terminal moraines and glacial complexes of combined type occasionally including push moraines. To the east of the Oder River the limit of the maximum extension of glaciation is deferred by a straighter line and is defined weakly morphologically. The age of maximum ice advance was dated 20 ky by radiocarbon dating the Lausitz marginal valley sediments (Cepek, 1972). Sometimes marginal deposits of a younger stage of glaciation approach quite closely the limit of glacier maximum development and overlap it. On the Russian Plain marginal formations associated with maximum advance of ice occur along the margins of the Grodno Highlands whence they extend northeastward. They are represented by terminal moraines on the northern slopes of the Grodno and Lida Highlands. The age of ice advance maximum was dated here to be 18–25 ky by radiocarbon dating of plant debris of siltstones that are overlain by till in the Niemen River valley near the village of Gozha.

Eastward, near the slopes of Oshmyany and Minsk highlands on the right bank of the Viliya River, the boundary of maximum ice advance is marked by individual terminal moraines and by well-developed angular interlobate hummocks. On the

*Note:* This chapter was reviewed, edited, and rewritten by the Editor, Yvonne Herman.

northern slopes of the Smolensk highland ice stagnation landforms are sporadically developed. The age of the maximum ice boundary is 18–24 ky, bracketed by radiocarbon dates of lacustrine–alluvial deposits that underly the moraine in the Zapadnaya Dvina River Basin known as the *usvyatcha suite* (Arslanov et al., 1971). Further east an active glacial tongue existed in the upper reaches of the Volga River. It left well-defined ice-marginal formations in the Mologo–Sheksna and Kubenskoye lowlands.

In the Vaga River Basin marginal landforms are of a more intermittent nature. The age of maximum ice advance was dated here on submoraine lacustrine sediments with peat interbeds in the depression of Lake Kubenskoye (Putchka River) as ≈21 ky while to the east, in the Vaga River Valley, it is ≈24 ky. Ice-marginal landforms are generally less pronounced eastward. The northern boundary of the Scandinavian Ice Sheet is not as well defined. Possible marginal glacial forms have been observed on the continental shelf. Diamictons have been reported from these localities. These are considered to be in the zone of ancient glaciation by most researchers. However, the chronology of marginal zones and moraines remains problematic; therefore, the ages of marginal moraines on the seafloor are estimated according to morphostratigraphy and correlation with dated marginal zones on the coast.

The Norwegian shelf contains a system of ridges that are most clearly defined along the northwestern shores of Scandinavia. Based on analysis of glacial features and a series of radiocarbon dates on marine and lacustrine sediments on islands and along the coast, Andersen and other researchers associated this system with the maximum advance of ice cover (Andersen, 1981). A series of ridges indicating ice extent on the shelf are traced along the northern and northeastern coast of the Kola Peninsula on the shelf of the Barents Sea. The aforementioned marginal forms are believed to belong to the former Scandinavian Ice Sheet combined with the local Lofoten and Ponoj ice domes.

## NOVAYA ZEMLYA ICE SHEET

A system of marginal forms corresponding to the Novaya Zemlya Ice Sheet joins that of the Scandinavian Ice Sheet to the east. On land these marginal forms are distinctly different in morphology from those of the Scandinavian Ice Sheet. The marginal ridges of the ice maximum occur in the basins of the Mezen, Kuloj, Pechora, and Kolva rivers and in the Varsh depression, stretching from the upper reaches of the Mezen in the Timan area to the basin of the upper Pechora and Usa rivers and the foothills of the Polar Urals. These ridges are push moraines and glacial thrust masses. Drumlin-like hills and fluted moraines occur southward of the outer marginal formations though they were not observed in Europe at this time. They were deposited by active outlet glacier tongues penetrating further than the main line of the ice front. According to existing radiocarbon dates (Lavrov and Arslanov, 1977; Arslanov, Lavrov, and Nikiforova, 1981) outer marginal landforms and the associated tills are underlain by deposits of older age than those of the Scandinavian ice cover.

Evidence of glacial activity such as moraines and erratics present on the Jamal Peninsula are present only in its southwestern part. These moraines can be compared with those of the Polar–Ural Ice Sheet. Eastward in the lower reaches of the Ob River a moraine ridge of the Salechard Urals defines probably the southern limit of the Polar-Ural ice cover that converged with the Novaya Zemlya ice cover. Radiocarbon dates obtained from submoraine deposits are contradictory. According to initial dates, the ages of submoraine alluvial–lacustrine deposits range from 25–

28 kya (two dates) to 47 kya (Zubakov and Levkovskaya, 1969). New dates on samples collected by Makeev provide infinite age for the sediments.* For this reason we favor the concept of an older glaciation of this region.

Ice-marginal landforms on the Kanin Peninsula and or the Polar Urals also appear to have been deposited by ice masses whose centers were both on the Polar Urals and on Novaya Zemlya. This is confirmed not only by the orientation of terminal moraines but also by the petrographic composition of the moraines' debris, which was most likely transported from the Polar Urals and Novaya Zemlya (Velitchko, 1979). Furthermore, according to Lavrov's data (1977), the areas west of the Pechora lowland and Timan are characterized by an abundance of boulders that originated from the Scandinavian center and that belong to the youngest of the two Valdai moraines distinguished in this region.

The distribution of glacial features on the Kara shelf is limited in the north by the St. Anna trench slope and the eastern margin of the Jamal-Gydan bank (Dibner, 1970). West of Novaya Zemlya, marginal glacial formations occur at a distance of 150–180 miles from the island according to Matishov (1980, 1984). East of the Novaya Zemlya trench on the Kara seafloor narrow deep valleys are observed; their structure differs from the relief of glacial troughs.

## NORTHERN SIBERIAN ICE SHEET

In interpreting the data on the distribution of former glacial covers of the more eastern regions of western and central Siberia one is confronted with major difficulties. For a long time the viewpoint prevailed that the last ice cover centered on the Jamal and Gydan peninsulas and affected the northern part of western Siberia (Astakhov, 1978, 1980; Arslanov, Lavrov, and Potapenko, 1983). This concept rested on the hypothesis that glaciers moved from the Kara center toward the mainland. The edge of this ice cover followed the Haduttey ridges crossing the Tazovksy Peninsula in a sublatitudinal direction (Maps 26-1 and 26-2).* In fact, these ridges are composed of thrusted Paleogene clays and do not contain glacial deposits per se. Thus the age of the Haduttey range seems to be unresolved.

A Late Valdai (Sartan) age for a residual frozen moraine layer on the Yamal Peninsula is rejected (Kaplyanskaya and Tarnogradsky, paper presented at the meeting Climate and Glaciations in the Anthropogene, 1983). The entire series of dates within the range of 26–40 ky B.P. is known from peat deposits and mammoth remains that are not overlapped by moraines on the Tazov and Gydan peninsulas. Dates of 15,830 B.P. and 15,360 B.P. were obtained from peat in the terrace alluvium on the Gydan Peninsula. These dates indicate ice-free conditions (Avdalovich and Bidzhiev, 1984). Avdalovich and Bidzhiev concluded that the northern territory of western Siberia was a periglacial area—a polar desert with firn fields and snow beds and different forms of eolian and frost relief with a poorly developed hydrological network.

Data on the distribution of the last ice cover in Central Siberia is presently undergoing thorough review.

Until recently, former glacial complexes that reflected a rather wide distribution of ice extent were associated with the Last (Sartan) Glaciation (Table 26-1). Thus from the Haduttey ridges the boundary of the Last Glaciation passed the Yenisey valley where on the right bank (in the Turukhan River Basin) pronounced glacial deposits are observed. The latter were considered

*Paper presented at the All-Union meeting Climates and Glaciations in the Anthropogene, Moscow, 1983.

*Maps and explanations appear as an appendix to the chapter.

Table 26-1. Late Pleistocene Geologic Climatic Units (USSR, West and Central Europe, Alaska)

| | *West and Central Europe* | EPOCH | *East Europe (USSR)* | *ky B.P.* | | *Siberia ky B.P.* | *North-East USSR* | *Alaska* |
|---|---|---|---|---|---|---|---|---|
| Late Würm (Late Vistulian) Glaciation | Junger Dryas (Salpausselka) | Late Valday Glaciation | Salpausselka | 10.8–9.8 | Sartan Glaciation | Norilsk stage 11.4–10.3 | Iskaten', Khaymikino Glaciation (several stage) | Naptown Glaciation |
| | Alleröd | | Alleröd | 11.9–10.8 | | | | |
| | Older Dryas | | Neva Stage | 12.0 | | | | |
| | Bölling | | Bölling | 12.7–12.2 | | | | |
| | Oldest Dryas | | Luga stage | 13.5–13.0 | | Warming 11.8–11.8 | | |
| | not expressed | | Raunis warming | 14.0–13.5 | | Cooling | | |
| | Pomeranian Stage | | Vepsovo stage | 16.0–14.0 | | Warming 13.0–12.0 | | |
| | not expressed | | Nonglacial Episode (dry, cool climate) | 17.0–16.0 | | Nyapan stage 15.0–13.0 | | |
| | Frankfurt, Poznan stage | | Edrovo regressive phase | 18.0–17.0 | | Warming 16.0–15.0 | | |
| | Brandenburg, Leszno (maximum) stage | | Bologoye (maximum) stage | 24.0–18.0 | | Gydan, Oneka stage 22.0–16.0 | | |

| Epoch | Stage | Substage |
| --- | --- | --- |
| WÜRM (VESTULIAN) GLACIAL EPOCH | Middle Würm Megainterstadial | Denekamp |
| | Early Würm Glaciation | Hengelo |
| | | Brörup |
| | | Amersfort |
| | | Early Würm (Early Vistulian) Glaciation |
| Eemian Interglacial | | |

| Epoch | Stage | Substage | Age |
| --- | --- | --- | --- |
| VALDAY GLACIAL | Middle Valday Megainterstadial | Dunaevo (Bryansk) warm interval | 32.0–25.0 |
| | | Cooling<br>Warming<br>Cooling<br>Leningrad Warming | 40.0–32.0 |
| | | Cooling<br>Warming<br>Cooling<br>Warming | 54.0–40.0 |
| | Early Valday Glaciation | Cooling | |
| | | Upper Volga warming | |
| | | Cooler–warmer climate fluctuations | |
| | | Early Valday Glaciation | |
| Mikulino Interglacial | | | |

| Epoch | Stage | Substage |
| --- | --- | --- |
| ZYRYANKA GLACIAL EPOCH | Karginsky Interstadial | Warming 30.0–22.0 |
| | | Cooling (Zsigansk Glaciation in Verkhoyansk Ridge) |
| | | Warming 43.0–33.0 |
| | | Cooling ~45 |
| | | Warming 55.0–45.0 |
| | Early Zyryanka (Ermakovo, Muructa) Glaciation | Two stages of glaciation in Central Siberia |
| Kazatsevo Interglacial | | |

| Stage |
| --- |
| Karginsky Interstadial |
| Vankarem, Tylkhoy Glaciation |
| Interglacial |

| Epoch | Stage |
| --- | --- |
| WISCONSIN GLACIATION | Interstadial (Cape Barrow type section) |
| | Knik Glaciation |
| Sangamon Interglacial | |

to be of the same age as the Haduttey on the Tazovsky Peninsula. However, the age of lacustrine sediments associated with marginal glacial landforms in the Turukhan River Basin are dated 34–41 ky (Kind, 1974), which does not allow to include these marginal formations to the Sartan Glaciation. In earlier studies attention was paid to data that do not support the viewpoint advocating wide distribution of ice cover in Central Siberia. In particular, within the boundary marked by the abovementioned ridges, there are regions where glacial deposits are absent and sediments 30–40 ky old occur at the very surface. Furthermore, in the central part of Taimyr, in a section on Cape Sabler, from a series of sediments where till is absent radiocarbon dates were obtained for the interval of 2–34 ky (Kind and Leonov, 1982; Map 26-2).

Further information on the maximum extent of glaciation at the end of the Pleistocene in Siberia was provided by the discovery of buried glacier ice that was considered to be the remains of this glacial advance. The state of preservation of this ice suggests a young age for the glaciation responsible for its deposition. Detailed study of such an exposure, located in the low reaches of the Yenisey River downward of the Kureika River mouth known as Ledyanaya Gora, showed that this interpretation should be accepted with caution and even critically.

In sediments formed during ice thawing and considered by some researchers to be flow till there are peat inclusions, wood debris, and other heterogeneous materials. Radiocarbon dates of organic inclusions suggest that these deposits are more than 40 ky old. Their accumulation continued up until the beginning of the Holocene. Thus even supporters of widespread glacier distribution in Late Pleistocene (e.g., Astakhov, 1978, 1980; Isayeva, 1972; Isayeva and Kraush, 1979; Isayeva et al., 1976, 1981) have to reconsider their idea that the above-mentioned marginal moraine complexes were deposited during the Sartan Glaciation. It seems more likely that they are Early Valdai deposits. The authors mentioned above believe that traces of Sartan glaciation in the low reaches of the Yenisey River exist northward of Dudinka (Map 26-3).

Nevertheless, the new data suggest a young age for the Sartan deposits of Central Siberia. This minimum age agrees with the ideas of Saks (1953). According to Saks, in the Central Siberian plateau in the valley of the Putorana plateau, young marginal moraines delimit the boundaries of a small ice cover from which glaciers flowed westward, southward, and partially eastward. On the Anabar plateau there are also fresh marginal moraines in the valleys. Based on their distribution we think that mountain–valley glaciation developed only in the uppermost part of the plateau (with elevations of 800–900 m). The existence of a large ice sheet with the center on the Kara shelf during the Last Glaciation has not been substantiated. This hypothesis is in conflict with the results of the investigations mentioned above concerning Arctic islands and archipelagos.

## THE LAST GLACIATION IN THE ARCTIC

Recent work carried out in the area of the Arctic shelves and its archipelagos indicate limited shelf glaciation. The Severnaya Zemlya Archipelago has been thoroughly studied during the 1980s. A series of dates ranging between 19 ky and 21 ky has been obtained here on mammoth bones that lie directly on contemporary surface or from a till older than the Late Valdai till (Makeev, Arslanov, and Garrut, 1979). According to Makeev and co-workers, during Sartan time the glaciers were even less extensive than they are today.

For Novaya Zemlya Island, even more limited dimensions of the Sartan Glaciation are envisaged than those previously discussed. On the basis of glaciological and geological–geomorphological studies,

Tchizhov (Tchizhov et al., 1968; Krasnozhen et al., 1982). Baranovskaya et al. (1982) suggest only mountain-valley glaciation on the island. It is quite possible that the two variants do not contradict each other since wider distribution of the Novaya Zemlya ice cover is dated to have occurred somewhat earlier than the time of maximum Late Pleistocene cooling.

For the Franz-Josef Land Archipelago there are no direct data on the dimensions of the Sartan ice dome. Geomorphological observations that include the tracing of a continuous system of terraces with elevations ranging from 10–160 m lead us to conclude that the dimensions of the ice dome were small, even if the possibility of advances of ice onto the shelf is accepted (Danilov, 1979; Velitchko and Faustova, 1982; Makkaveev et al., 1980). On the other hand, Grosvald's (1983*a*) reconstructions suggest a large ice cover in this region.

It is on the glaciation of the Spitsbergen Archipelago that the largest amount of data have been obtained since investigations in the 1970s. During the Late Pleistocene a multiple expansion of glaciers took place on Spitsbergen. The traces of maximum development of the Late Valdai Glaciation are observed at only a few sites on the archipelago (Boulton, 1979; Boulton et al., 1982; Salvigsen, 1979).

In the northwestern sector of the island ice marginal landforms suggest that glaciers advanced down fjords no further than 20 km from the mainland. Dates obtained on the moraine ridges are 26 ky B.P. (thermoluminescens) and 28.5 ky B.P. ($^{14}C$; Salvigsen, 1979).

Sections in the western and southern margins of the island contain sediments deposited during an interstadial 40–50 kya. There are no overlying glacial sediments. Shorelines at the mouth of Isfjorden were dated at about 40 ky B.P. by $^{14}C$. Sediments in the inner part of the fjord not overlain by till and dated 30–50 ky B.P. (Mangerud et al., 1983) may be from the same interval. In the southern coast of Hornsund, 25–30 kya a sea terrace 80–100 m high was being formed (Piotr and Leszek, 1981). These new data agree well with earlier results on the structure of sea terraces dated in detail by radiocarbon method (Salvigsen and Osterholm, 1982; Troitsky et al., 1979). From these data it follows that interstadial conditions characterized Spitsbergen 25–45 kya and consequently ice could not spread far from the centers of outflow.

On the islands of the archipelago evidence of the Late Pleistocene glacial event is not clear. Twenty-three radiocarbon dates obtained for Kong Karls Land (Salvigsen, 1981) reflect the Holocene history of the island and the interstadial period of 45–50 ky B.P. Deposits of the latter are occasionally observed under sediments whose origin is only assumed. Only on one of the Kong Karls Land islands were there traces of glacial striae on boulders. On the island of Prins Karl Forland where Boulton (1979) thought a wider expansion of ice during the period of 12–32 kya possible, Salvigsen (1979) discovered merely frost-weathered material.

These data indicate a restricted style for Late Pleistocene glaciation in the Spitsbergen Archipelago, as assumed by most researchers. The ice dispersed from two or three local ice domes in the northwestern (between Smerenburgfjorden and Lifdefjorden), eastern, and possibly southern margins of the archipelago, leaving nunataks and leaving parts of the coast ice-free. Thus the largest western fjords of the island (Isfjorden and Mijenfjorden) were covered with ice in their eastern sections only, according to most researchers. The largest ice mass evidently was the eastern dome, which probably occupied some part of the shelf.

The late Valdai ice sheet of the archipelago was thin, and it disintegrated before the beginning of the Holocene Epoch. Judging by the fact that the sea terrace deposits are 20–25 ky old, the expansion of the Late Valdai ice was probably short-lived and occurred independently of the Scandinavian Ice Sheet. Out-of-phase glaciations are not uncommon: During the younger Dryas when rejuvenation of glaciers (Salpausselka pulsations) was observed on the continent, most active ice disintegration characterized Spitsbergen (Salvigsen and Osterholm, 1982). In contrast to continental Europe, maximum development of the Late Pleistocene glaciation on the archipelago took place prior to 45–50 kya.

## RECONSTRUCTION OF THE LATE PLEISTOCENE GLACIATION OF THE ARCTIC SHELF AND ADJACENT LAND

According to the latest reconstructions of shorelines and former ice directions, the center of the Scandinavian Ice Sheet is believed to have been positioned not above the Gulf of Bothnia as was thought before but 100–150 km west of the Gulf, on land within Sweden. There were probably at least two coalescent ice caps. The main features of the ice cover have been described in a number of articles (Aseev, 1974, 1977; Gerasimov and Velitchko, 1982*a*): They indicate the asymmetry of slopes and radial structures caused by the existence of a series of ice flows. The southwestern, southern, and southeastern slopes of the ice cover that were on land were characterized by long extension especially in the southern regions. On these slopes ice streams were confined by valleys. They were nourished by the Swedish–Finnish–Karelian ice caps. The main sources of boulders were the eastern slopes of the Scandinavian mountains, the basin of the Baltic Sea, and the highlands of Finland and the Kola Peninsula. The northwestern and western slopes of the ice cover were shorter and steeper. Westward, the Scandinavian ice did not spread beyond the limits of the shelf zone where further advance was restricted by the Egga I ranges at the end of the Lofoten shelf.

Calculations of the amount of ice in the North Sea by Mangerud (1981) confirm that the ice did not reach far onto the Norwegian shelf. The limits of ice extension in the northern sector of the Barents Sea shelf have not been estimated.

Similarities between glacial morphostructures and marginal ice deposits along the coast of northern Norway and on the Barents Sea shelf suggest that the maximum extent of the Scandinavian ice cover was from the ridges of the terminal moraine Egga I at the end of the Lofoten shelf northward through ridges with depths of 300–400 m in the Bjornoya lowland and submarine ridges near the western coast of Spitsbergen (Matishov, 1984), joining the Scandinavian and Spitsbergen ice covers. However, another reconstruction is possible, with the ice sheet limited in the north by the coastal regions of the Kola Peninsula following the configuration of marginal formations near the northern coast of Norway and the USSR. In favor of such a reconstruction are data from the island and the raised parts of the shelf between Spitsbergen and the Scandinavian Peninsula.

On Björnöya Island located south of Spitsbergen only local glaciation is evident. The same is true for the archipelago of Hopen Islands (Hoppe, 1981) and for the Björnöya-Hopen submarine elevation. Peculiarities of its relief make us doubt the Valdai age of this glaciation, though other researchers think that an independent ice cover may have existed here (Makkaveev et al., 1980). The ice cover on the Spitsbergen Archipelago may have been composed of many small domes. The same is true for glacial covers of the archipelagos of Franz Josef Land and Severnaya Zemlya.

The Novaya Zemlya ice cover had an extended domelike shape roughly following the configuration of the island. Its slopes were asymmetrical. The eastern slope was most probably steep and short. Here the ice extent was limited by the Novaya Zemlya trench and the margin was parallel to the island. The northern slope was also steep and was limited by the St. Anna trench. The western and southern slopes were more extensive and gently sloping. The ice margin extended out to the adjacent parts of the shelf. Lobes of the southern slope flowed to the northeast of the European sector of the USSR (Malaya Zemlya and Bolshaya Zemlya tundras). This territory was a zone of highly active ice streams.

The Barents Sea–Pechora and Novaya Zemlya–Kolvinsky flows originated from the Novaya Zemlya center. To the east they interacted with ice flows of the Ural outflow. To the west of this area, in the zone of the Kola–Mezen ice flows, there is evidence of ice penetration from Scandinavia (Lavrov, 1977). This ice penetration suggests the existence of a convergence zone here, between the Scandinavian and the Novaya Zemlya ice covers. However, this reconstruction should be considered preliminary. The Novaya Zemlya ice cover was not a precise chronological equivalent of the Scandinavian Ice Sheet (Velitchko and Faustova, 1982). Radiocarbon dates indicate that the Novaya Zemlya Ice Sheet reached its maximum extent about 33 kya. We assume that in the northeastern part of the Russian plain these ice covers expanded at different times—first the Novaya Zemlya ice, later the Scandinavian ice. Accordingly these ice caps represent heterochronous ice advances.

The question as to whether the ice flows from Novaya Zemlya reached the Yamal Peninsula remains unanswered. It is more likely that the region south of Yamal was nourished by glaciers that originated in the lower reaches of the Ob River. Glaciers migrated down from the Ural's eastern slope. Eastward, in the Arctic sector on the Taimyr Peninsula and in Central Siberia, the areal distribution of glaciers was of limited dimensions. The size of the glacial ice caps on Severnaya Zemlya Island was smaller than those of the present day. In Taimyr only in the highest part of the Byrranga Mountains reticulate glaciation developed, filling the valleys with ice up to 500–700 m thick. The glaciation on the Anabar highland in Central Siberia was of similar nature. On the Putorana Plateau, in its highest part (1,500–1,700 m), in the dividing zone of Yenisey, Kheta, Kotui, and Lower Tunguska tributaries, an ice dome more than 100 km in diameter and 700 m thick existed. Valley glaciers 50–80 km long flowed from the dome at heights of 600 m.

There were several types of mountain glaciation in the northeastern mountains of the USSR. The widest scale glacial phenomena could be observed within the Verkhoyansk ridge. Valley glaciers formed in the highest part of this mountain system and at times they coalesced. Firn fields were the main area accumulating solid precipitation. From them glaciers flowed down the valleys, filling their upper reaches. Reticulated glaciation could have been formed in restricted areas. Different types of glaciations also characterized the Suntar–Khayata, Chersky, and several ridges of the Koryak upland where cirque and mountain–valley glaciation prevailed, while reticulated glaciation was developed to a lesser extent. Radiocarbon dates indicate that glacial pulsations occurred four times during the period between 11 kya and 33 kya.

Glaciation was extensive on the Kamchatka Peninsula where mountain–valley and reticulated glaciation occupied significant parts of the Median and Eastern ridges. In the European and Asian Arctic, several independent ice sheets existed during the maximum of the Last Glaciation. Their characteristic structure and dynamic regime were related to regional climatic conditions (Map 26-4). In spite of the fact that all types of ice cover existed during the Last Glaciation their development was not simultaneous.

In the European sector of the Arctic, independent ice sheets existed in Scandinavia, Iceland, Faeroes, Britain, Franz Josef Land, and Spitsbergen. From the Asian Arctic the Polar-Ural, Novaya Zemlya, and Severnaya Zemlya ice sheets are known, and the Putorana and Byrranga ice caps extended on the continent. Based on recent data the reconstruction of ice covers of the Last Pleistocene Glaciation in Eurasia, including the Arctic, suggests that this glaciation was characterized by its limited size with outflow centers on high elevations of land and islands. Such reconstruction differs radically from earlier reconstructions (Grosvald, 1983*a,b*). One must conclude that the existence of major ice sheets with centers of outflow on the bottom of the Barents and Kara seas during the Last Pleistocene Glaciation is not supported by the geological record. A glacial system consisting of many, often isolated, ice domes existed. This confirms the assumption of the small size of glaciation (Saks, 1958; Velitchko, 1979; Velitchko and Khodakov, 1981; Velitchko and Faustova, 1982).

Our reconstruction of ice covers supports the concept of glaciation centers tied to high elevations as centers of solid precipitation. This concept proposed by Gerasimov and Markov (1939), Saks (1958), and Strelkov (1968) was developed by Velitchko (1979), and confirmed by calculations based on the development of the Scandinavian Ice Sheet (Khodakov, 1979). These calculations showed that there was no sufficient time to form a single pan-Arctic ice cover in the central sector of the shelf, which would have covered the islands' high elevations and would have flowed out through trenches and depressions. Furthermore, data on glacioeustatic land rise in the Arctic sector suggest rates of 1–2 m per hundred years; such data do not support the hypothesis of a single ice cover over the Barents Sea shelf. The data on crustal deformation is from the periphery of this hypothetic ice sheet and could not have been caused by a single ice sheet. One of its centers, for instance, was believed to have been more than 2,000 m thick over the Novaya Zemlya (Grosvald, 1983*b*, fig. 13). However, in the data mentioned in the same work, Postglacial terraces have an elevation of 4–8 m above sea level, while the 50–75 m terraces on Novaya Zemlya are more than 30 ky old.

The suggestion of a major ice cap on the shelf contradicts not only the above-stated data but the record from other high-latitude regions of the Northern Hemisphere. Thus, of great significance for the solution to the problem of the Scandinavian Ice Sheet extent onto the shelf to the southwest of the Scandinavian Peninsula and the dimensions of the ice domes of the British Islands, were studies off the coastal regions of Norway within the Norwegian shelf, on the Shetland Islands, Orkney-Shetland Channel, and the North Sea. Supporters of a ''large'' British Ice Sheet included in their reconstruction the British, Shetland, Hebrides, and other islands, the ice coalescing with the Scandinavian Ice Sheet, which in turn extended partially on the Norwegian shelf during the maximum stage of glaciation. They assumed that the ice moving from Scandinavia crossed the Norwegian Sea Basin and filled the depression of the North Sea, flowing mainly westward and northwestward across Cape Caithness, Shetland, and Orkney Islands toward the continent.

However, calculations of the Vistulian ice volume in the North Sea (Mangerud, 1981), as well as reconstructions of the temperature regime of seawater (Sejrup et al., 1984; Dupplessy et al., 1980), material of glacial-marine origin on the Norwegian shelf, and the presence of nunataks near shore suggest that the extent of Scandinavian ice westward on the Norwegian Sea shelf was not large and was limited by the coastal regions along Norway. Detailed studies by Flinn (1978) and Price (1975) also showed that the ice cover on the Shetland highlands had its own center and was much smaller than had been previously estimated. Eastward of the Moray Firth Bay, the sea was ice-free. Glacial striae from Caithness Bay ice were stated to move toward the bay and not from the sea. The outer Hebrides had an independent ice cap. Consequently, the ice did not cross the Norwegian Basin and the confluence zone was very narrow as shown by the Boulton model (1979). The data obtained support Reinhard's opinion (1974) concerning an ice-free shelf strip stretching from the Shetland Islands southeastward.

In the Canadian Arctic studies by Loriol and Gray (1982), Scott (1976), Miller (1980), and Andrews, Shilts, and Miller (1983) question the assumption of a single stable Laurentide Ice Sheet with the center of outflow in Hudson Bay. Flint's concept (1943) of a major ice sheet over Hudson Bay was used by Hughes, Denton, and Grosvald (1977) and Walcott (1975) in their reconstructions. The conclusions of these authors was not confirmed by the Geological Survey of Canada studies. First of all, it appears that the maximum thicknesses and the main ice mass were concentrated east of Hudson Bay, in New Quebec where the main ice divide occurs. Ice advanced from northeast toward James and Hudson bays. This is clearly indicated by data on rock debris transport and by the direction of ice advance (Shilts, 1980). Amino acid dates also showed that during the glaciation Hudson Bay was filled with seawater two or three times and therefore must have been ice-free.

Many researchers (e.g., Hillaire, 1981; Dyke, Dredge, and Vincent, 1982) developed the concept of a multidome Laurentide Ice Sheet; this concept has been accepted by most glaciologists. The largest domes in northeastern Canada were in New Quebec, Ellesmere, and Baffin Land. The Ellesmere dome was not confluent with the Greenland Ice Sheet and was separated from Baffin Land according to England and Bradley (1978). It is remarkable that ice domes formed under arid Arctic conditions when some islands remained largely ice-free. These active domes were partially situated in Kevateen and near the boundary with the Cordilleras. Their dynamics, growth, and development were related to the subglacial relief and local climates. The zone of ice divides and the contact between domes were changing continuously (Shilts, 1980). Possibly some domes were isolated from the Laurentide Ice Sheet.

The Greenland ice cover was probably stable: Its size during the Last Glacial maximum was only slightly larger than today (Paterson, 1977; Hjort, 1973) and decreased gradually during the glacial epoch. Large nunataks characterize the maximum stage of Late Pleistocene glaciation and according to Hakansson's data (1982) some areas were ice-free. According to the majority of Canadian researchers, a recent glacial event, 18–20 kya, was of limited dimensions on the Baffin Land, Canadian Archipelago, and northeastern part of the continent. This is indicated by geological–geomorphological data and by radiocarbon dates (England, 1976; Ives, 1978; Vilks and Mudie, 1978) showing that the Late Pleistocene glaciers did not leave the shorelines limits.

The widespread extent of tundra during the last 21 ky in the coastal zone of the Labrador Peninsula (Vilks and Mudie, 1978) and the results of palynologic and faunal studies of sediments from boreholes in the southern part of the Labrador shelf indicate the absence of glaciers.* The extreme northwest of the North American continent was ice-free. The Yukon was glaciated only in restricted regions and most areas were occupied mainly by forest tundra and

*Evidence of deep weathering and widespread development of block fields (Ives, 1978) indicate weak glaciation; these traces were observed on the high plateau of the Canadian Arctic. However, these features can be interpreted differently (Sugden, 1977).

occasionally by boreal forests. The large glacial complexes of mountain and ice sheets adjacent to the Laurentide Ice Sheet are referred to as the Appalachians and Cordilleras. The latter did not adhere completely to the Laurentide ice cover. Both complexes consisted of independent ice sheets. The maximum development of the Cordilleras and Laurentide ice were not simultaneous (Rutter, 1981; Stalker and Mac, 1980); in the Cordilleras the development began first. Therefore, the "corridor" between the above-mentioned ice covers remained free of ice for a significant time. Only during a short time did the Cordilleras valley glaciers advance to their central part. The northern and southern part of the corridor was ice-free.

In the central and southern parts of western Alberta the width of the ice-free zone 22 kya is estimated to have been 20–40 km, while 17 kya it was 80–100 km. The presence of refugia on the islands off British Columbia and nunataks in the zone of the coastal ridge should be mentioned.

New data were used to model "limited" glaciations (Boulton, 1979). The hypothesis of synchronous glaciations of the Northern Hemisphere in Late Pleistocene are supported by Blake (1970), Hughes, Denton, and Grosvald (1977), and Grosvald (1983*a,b*), while the asynchronous concept of Scandinavian and Laurentide ice covers are supported by Andrews (1982).

During the Last Glaciation (Valdai, Sartan, Wisconsin) there were outflow centers on continents and islands that advanced toward the shelf edge 18–20 kya. This dome glacial system characterized even the largest North American and Scandinavian ice covers.

## EXTENT OF THE LAST GLACIATION

The geochronology of the entire Last Glacial (Valdai, Zyrianka and Sartan, Vistula, Wisconsin Ice Age) has been well documented in many publications, and most scholars recognize the Early and Late Valdai. Differences in the concepts concern interpretation of climatic conditions of the interval between these two epochs, embracing the period between 25 kya and 50 kya. Detailed studies of many sections in the European USSR have been used for paleogeographical reconstructions of this interval. Except for Raukas (1976) and Serebryannyi (1978), there is unanimity in defining it as a large interstadial (or as an ice-free period in the Russian Plain) characterized by complex temperature fluctuations. Persistence of periglacial flora during the whole interval (even in times of rising temperature) indicates that the whole period was much cooler than the interglacials.

In contrast to the Late Valdai, the Early Valdai Glaciation is not as well known. Several reconstructions have been made of the Early Valdai Glaciation. They are based on the following data: In the European USSR, on the Kola Peninsula, in a number of sections (for instance, in the Kovdor Basin) above the Middle Valdai deposits, there are two moraines, the lower one considered to be of Early Valdai age (Yevzerov and Koshechkin, 1982). However, the Middle Valdai age and the interstadial rank of the intermoraine deposits have a till that has been the subject of discussions. In the same region a washout surface has been discovered, which divides two marine strata: Ponoian (boreal) layers, corresponding to the Eemian of Europe, and the younger, Strelnin layers, which have formed under conditions cooler than today's. At this level, gravel, pebbles, and moraines were present (Apukhtin et al., 1977).

According to Ekman et al. (1977), glacial deposits of Early Valdai age are found on the slopes of the Baltic sheet, in Karelia. In the northern and northwestern parts of the Russian plain reliable features of the Early Valdai glacial deposits have not been detected. Most researchers believe that there is no conclusive evidence of sections where Mikulino deposits are overlain by Early Valdai moraines, which in turn would be covered by Middle and Late Valdai deposits and characterized paleobotanically. There is also a different interpretation on the limits of the Early Valdai ice within the Russian plain. According to a group of specialists from Leningrad (Krasnov, Zarrina, and others, unpublished), the moraine of this glaciation is present in a number of sections, including the one in Kileshino in the Upper Volga Basin (Malakhovsky and Spiridonova, 1981). But unfortunately not one of the above-cited sections represents indisputably the Early Valdai. The concept of a small ice sheet in the Early Valdai period supports data obtained in the periglacial zone of the Russian plain (Velitchko, 1975).

In contrast, in western Siberia and on the Central Siberian tableland, reconstructions indicate a large spatial development of the first Late Pleistocene Zyrianka (Murutkin) Ice Age dated at >60 ky B.P. Its extent is deduced from the location of marginal glacial forms, which are unequivocally identified from aerial and Sputnik photographs and by moraines on deposits of the Kazantsevo (Eem) Interglacial age. This moraine contains deposits with spore and pollen spectra and malacofauna different from the Kazantsevo ones dated 30 ky and 45 ky by $^{14}C$ (Bardeeva and Isayeva, 1980).

The maximum ice extent in western Siberia was along the northern slope of the Sibirskie Uvaly, protruding significantly southward along the Ob and Yenisei valleys (Fig. 26-2) (Isayeva et al., 1981). The province reconstructed within these limits that has formerly been covered by ice completely embraces the north of the western Siberian plain, the Taimyr Peninsula, and the northern Siberian lowland, as well as the northwestern sector of the Central Siberian tableland. The centers of glaciation have probably been north of Taimyr or on the Kara shelf, on Novaya Zemlya, in the north and Polar Urals, on the Putorana plateau, and on the Anabar upland. During the glacial maximum, ice merged from all these centers, forming a single cover, which occupied a considerable part of the Arctic shelf except for the dry land. The larger size of this glaciation, as compared with the Late Zyrianka Sartan Glaciation, are well documented by data from Severnaya Zemlya Island (Arslanov et al., 1980; Makeev, Arslanov, and Garrut, 1979).

The more modest dimensions of the Early Zyrianka Glaciation in western Siberia (the second alternative) may be reconstructed from the marginal formations between the Ob and the Yenisei rivers approximately along the Polar Circle. However, one may assume that large ice lobes moved along the river valleys of the Ob, Nadym, Pur and Taz, and Yenisei. In the mountainous regions of the northeastern USSR, the Early Zyrianka Glaciation is also believed to have been more extensive than the Late Zyrianka Glaciation.

According to most researchers in western Eurasia, the Early Valdai Ice Sheet was much more restricted than the Late Valdai Ice Sheet. Its southern edge did not extend outside the limits of the Baltic Sea, leaving even the northern regions of the Russian plain ice-free. At that time glaciation of the Siberian sector of Eurasia was much more pronounced. Most probably its size was much greater than the size of younger glaciers. It covered the north of the western Siberian lowland and the greater part of the Taimyr and the Central Siberian tableland.

It is believed that in northeastern Eurasia the Early Valdai Glaciation was also larger than the Late Valdai Glaciation. In the early phases of the Valdai, the most favorable conditions for ice sheet growth were probably in the central and in the eastern parts of Eurasia. The climate was sufficiently cold and quite humid. Westward in the vicinity of the Atlantic Ocean the humidity was even greater; however, the warming effect of the ocean and Gulf

Stream retarded the growth of glaciers. With progressive climatic cooling and drying, the western Siberian ice covers became deprived of moisture, as the influence of anticyclonic air masses became more pronounced. The continental Siberia glaciers began to retreat and the centers of glacier growth gradually migrated westward. As cooling proceeded, the Atlantic Ocean became the principal moisture source (Table 26-2). During the Late Pleistocene glacial maximum of Europe (18–20

**Table 26-2. Radiocarbon Dates of Submoraine Deposits Indicating the Age of the Maximum Extent of the Last Glaciation in the Northeast and Central Regions of the Russian Plain**

| *Section* | *Age (ky)* | *Sample Number* |
|---|---|---|
| Soz'va I, Soz'va River, right tributary of Pechora | 33,520 ± 470 | LU-513 A |
| | 35,540 ± 1,570 | LU-513 B |
| Tyrybei I, basin of Soima River, Pechora's left tributary | 38,670 ± 870 | LU-512 B |
| | 39,840 ± 570 | LU-512 A |
| Shapkina II, right tributary | 40,650 ± 790 | LU-550 |
| | 40,860 ± 1,260 | LU-517 B |
| | 38,400 ± 640 | LU-517 A |
| Shapkina I | 42,660 ± 970 | LU-519 |
| | 43,240 ± 1,090 | LU-394 C |
| | 45,280 ± 1,200 | LU-515 A |
| Urdyuga I, basin of Soima Rivers, left tributary of Pechora | 42,810 ± 1,200 | LU-533 |
| Velt I, Timan shore, Malaya Zemlya tundra | 43,250 ± 1,110 | LU-677 |
| Chyornaya I, Bolshaya Zemlya tundra | 47,210 ± 1,270 | LU-674 |
| Soz'va II, right tributary of Pechora | over 51,200 | LU-514 B |
| Drechaluki, vicinity of the town of Surazh, Byelorussia | 17,770 ± 170 | LU-95 A |
| | 17,900 ± 160 | LU-95 B |
| | 18,370 ± 180 | LU-96 A |
| | 23,630 ± 370 | LU-97 A |
| Gozha River, vicinity of Grodno, Byelorussia | 18,730 ± 1,230 | LU-76 A |
| | 22,740 ± 1,870 | LU-76 B |
| | 23,200 ± 520 | LU-76 C |
| | 22,950 ± 440 | LU-89 |
| | 25,100 ± 240 | LU-90 A |
| | 24,860 ± 230 | LU-90 B |
| Pokrovskoye, Puchka River, basin of Kubenskoye Lake, Vologda oblast | 21,410 ± 150 | LU-18 B |
| | 21,880 ± 110 | LU-18 A |
| Shapurovo, vicinity of the town of Surazh, Byelorussia | 22,500 ± 210 | LU-91 |
| | 29,150 ± 850 | LU-78 A |
| Town of Shenkursk, valley of Vaga River | 24,900 ± 470 | U-40 |
| Dunaevo, Lovat' River, Novgorod oblast | 27,500 ± 1,500 | LU-28 A |
| | 25,600 ± 360 | LU-28 C |
| | 25,440 ± 270 | LU-28 B |
| Borisova Gora, vicinity of the town of Surazh, Byelorussia | 28,170 ± 750 | LU-105 |

kya) the ice of the easternmost sectors of the continent was decaying due to decreased moisture.

## PALEOCLIMATES DURING THE LAST GLACIATION MAXIMUM

For the paleometeorological reconstructions of the 18–20-ky period we used fossil plants (Grichuk, Gurtovaya, Zelikson, and Borisova, unpublished), mammals (Markova, unpublished), paleoentomological methods (Nazarov, unpublished), and paleocryogenic methods (Velitchko, 1983, 1985). These results have shown that the greatest deviations from modern values occurred in winter. The maximum temperature decrease was observed in the western (the margin of the ice sheet) as well as the southwestern USSR and in the far eastern sectors of Eurasia (Map 26-5). The temperature decrease as compared to present values was 9°C, 15°C, and 10°C respectively.

In the central parts of the continent the temperature decrease did not exceed 5°C or 6°C. Reconstructions for glacial summers are quite different. In the west, July temperatures were 5°–7°C lower than at present. In the central region and in the northeast there were practically no temperature changes. An insignificant fall of temperature (−2°C) is recorded for the south of the Far East.

The frost-free period was two or three months shorter in the European USSR, about a month shorter in central Eurasia, and about 1.5–2 months shorter in its eastern periphery. The annual amplitude of temperature changes increased in Europe and in the Far East by 7°–10°C and in western Siberia by 5°C. Precipitation decreased everywhere, except in the mountainous regions of the country's southeast. Estimates of winter temperatures using several methods (e.g., small mammals) gave values close to −30°C for the territory of Bryansk Oblast and the Ukraine, suggesting winter temperatures were ≈20°C lower than today.

Buried pseudomorphs in 3- or 4-m-deep fossil ice, discovered in the western Ukraine, suggest that during the Last Glaciation maximum mean annual temperatures were not lower than −7° or −8°C, while the mean January temperature was probably −30°C. The reliability of this estimate (−26°C to −30°C) is also confirmed by paleoentomological studies (Nazarov, 1984) carried out in the neighboring regions of Byelorussia. There are also extensive data that indicate considerable (although not as sharp) fall of winter temperatures in Siberia. Increase in ice-sheet thickness brought about a strong cooling of the overlying air masses. It is believed that the mean annual temperatures over large ice sheets fell to −30°C to −32°C (July values were −15°C to −16°C and January temperatures were −40°C to −50°C, Maps 26-5 and 26-6). The sharpest climatic changes took place in the vicinity of the ocean.

Recent data on moisture content of the atmosphere over the North Atlantic were collected on board Soviet research vessels (Nikonov, 1984). The winter moisture content of the air above the sea on the average is 2 or 2.5 times lower than in summer, and the minimal values (on the average from 6.84–8.0 mm) were recorded in the highest latitudes.

With the growth of sea ice and decreased frost-free periods (Velitchko, 1973), the air masses' regime was similar to present winter conditions during the greater part of the year. The cold months were characterized by high stability of cold and extremely dry air masses over a considerable part of the extratropical belt. This situation was enhanced by the southward deflection of the Gulf Stream.

Reconstructions by Gates (1976) suggest that the atmospheric pressure over the center of the Scandinavian glacier could reach 1,045 mb even in summer, and in winter it increased by another 35 mb or 40 mb. Similar figures characterize the Siberian anticyclone. In winter, in the center of the anticyclonic systems, the atmospheric pressure may have reached values that have no present-day analogue.

Air masses from Siberia and central Asia may have moved far to the west, joining the masses of Arctic air that were cooled over the surface of the Scandinavian glaciers and brought about extremely cold and dry European winters.

A more complicated atmospheric situation evolved in the short warm period of the year. Meteorological observations show that the minimal amplitude of the annual moisture content (up to 11 mm or 12 mm) is found in the zone of sea and continental air interaction. There is no direct relationship with latitude or with circulation systems. This intermediate zone existed during the maximum glaciation as well. Of significance is the fan-shaped spreading of summer temperature anomalies from east to west (Fig. 26-1).

It is doubtful that humid air masses coming from the ocean could have moved far into the middle of the continent, the thick Scandinavian Ice Sheet (3,000 m) and the area of high pressure over the ocean being the major barriers. Estimated summer temperatures were −16°C over glaciers and 11°–14°C in the extraglacial area. Increased circulation at the contact between these two masses led to the development of a system of air-mass movement, which could be called *summer glacial monsoon* (Fig. 26-2 and Map 26-7).

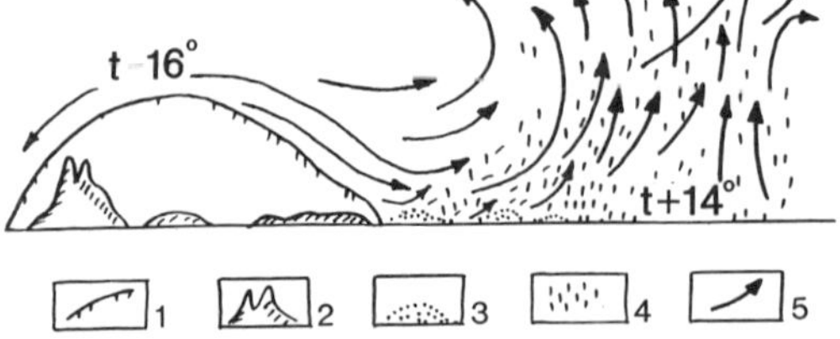


Figure 26-1. Diagram of summer glacial monsoon and its impact on the lithosedimentational processes in the periglacial zone. 1. Ice sheet. 2. Under-ice relief. 3. Sand accumulation. 4. Loess accumulation. 5. Direction of air mass.

The wedge of a dense air flowing down from the glacier squeezed out the warm air masses from the south. Being warmed adiabatically on the way from the glacier and further over the periglacial plain, the glacial air masses moved away from their center. In the loess areas of the Russian plain, the amount of precipitation decreased by 40–60 percent, resulting in not more than 300 mm/yr (according to floral and lithological data).

## DECAY OF THE LAST GLACIAL ICE

Three major stages of deglaciation can be distinguished in Europe and in North America. The first occurred between 18–20 kya. The mean rate of glacial retreat was between 20 m/yr and 30 m/yr, while the amplitude of individual glacier surges did not surpass a few tens of kilometers. In Europe, marginal ice of the maximum (Brandenburg–Leshno–Bologovo) glaciation was in the form of a chain of arcs extending from Shverin through Frankfurt to Poznan, and in the Russian plain it can be traced in the direction of Snyardva and the Svir lakes across the upper reaches of Berezina and Zapadnaya Dvina. Further to the east, the marginal belt loses its continuity.

The second stage of deglaciation (12 kya to 15–16 kya) is characterized by the high mobility of glacial margins and by a combination of glacier retreat accompanied by surges. In Europe this stage began with the large Vepsovo (Pomorsky) advance, which is traced by a belt of massive marginal forms with a great number of "glaciodislocations" and other features indicative of glacial stress.

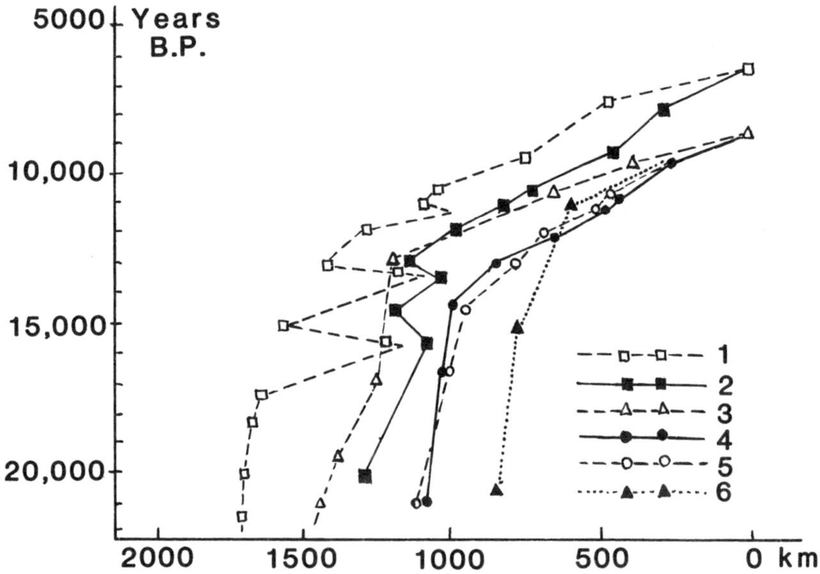


Figure 26-2. Rates of retreat of the Laurentide and Scandinavian ice sheets during the Last Glaciation. 1. The sector of Great Lakes of the Laurentide Ice Sheet. 2. The Atlantic sector of the Laurentide Ice Sheet. 3. The western European sector of the Scandinavian Ice Sheet. 4. The central sector of the Scandinavian Ice Sheet, the Onega–Karelia Stream. 5. The central sector of the Scandinavian Ice Sheet, the Ladoga Stream. 6. The Arctic sector of the Scandinavian Ice Sheet.

The glacial relief of this stage is characterized by complicated structures. It was formed by numerous active tongues and microtongues of ice. This stage includes the well-known ranges to the south of Neubrandenburg in the upper reaches of the Ikker River, in the Middle Oder, as well as the Baltic range and the glacial complexes of Velikie–Luki–Toropets, Valdai, Vepsovo, and other uplands. Within North America the surges of Port-Guron and Kary probably correspond to this advance. The characteristic feature of surges of this kind is probably the proximity of glaciers to the moisture sources (the oceans). In Europe the amplitude of Vepsovo surges diminishes considerably from west to east.

The end of the second deglaciation stage is comparable to the Old and Middle Dryas and to the corresponding surges of the late Glacial, the Luga and Neva phases. Marginal formations of this age are situated close to the Baltic Sea, within the depressions of the Onega and Ladoga lakes.

The third stage of deglaciation, which began after 12 kya, is characterized by the irreversible breakdown of ice sheets. In Young Dryas, between 10 kya and 11 kya, glacial movement was limited by Fennoscandia. The continuation of the Salpausselka ranges was discovered in northern Karelia and in the Kola Peninsula (Ilyin, Lukashev, and Ekman, 1978). The Scandinavian Ice Sheet ceased to exist about 9.4–9.5 kya. The decay of the North American ice went on for another 3 ky.

In summary, the first stage of deglaciation of Europe was similar to that of North America: the ice retreating slowly and undergoing small-amplitude surges. The mobility of the glacier's margin was hindered by very low temperatures. Beginning 15 kya or 16 kya, a rise in temperature caused a qualitative change in the heat and moisture balance. In the zone of oceanic climate the growth of glaciers was mainly limited by temperature, while in areas with continental climate a decisive role was played by humidity.

In the third stage of deglaciation the leading role was again played by temperature. Because of the global rise in temperature, glaciers disintegrated rapidly both in areas with humid climate and in those with dry climate.

It is probably not by chance that in Young Dryas glaciers advanced not in the areas with minimum temperature (continental Eurasia) but in the zone influenced by humid air masses. It is possible that the growth of ice sheets in coastal zones was not caused by the cooling of the sea but by its capacity to provide moisture for the growth of glaciers.

## APPENDIX

### Map 26-1

The Late Pleistocene glacial morphosculpture of the north of Siberia.

*Elements of the Morphosculpture*

1. Narrow ridges of push and fill terminal moraines. 2. Massive ridges of push and fill moraines with the core of dislocated preglacial deposits. 3. Massive interlobate and intertongue uplands in the northern Siberian lowland and distinct from western Siberia. 4. A contrasting relief of hummock in the zone of melting of buried (glacier?) ice. 5. Water divides ridges and uplands, built by lacustrine, fluvioglacial, and marine (?) deposits, which accumulated between the blocks of dead ice (fissure-kame massives with a moraine core). 6. Hummock relief of an ablation moraine. 7. Glacial thrust masses: (a) parallel ridges built by dislocated Meso-Cenozoic rocks and systems of linear fissures; (b) swell-like uplands built by Meso-Cenozoic rocks—''glacioanticlinals.'' 8. The hilly kame relief of western Siberia. 9. (a) Troughs cutting the Byrranga Ridge; (b) drumlinoids on valley bottoms between the basins of the Nizhnyaya Tunguska and Bakhta rivers; (c) fields of drumlinoids. 10. Assumed boundary glacial formation in northern Taimyr Peninsula. 11. Regions of accumulation of thick glacial deposits. 12. Areas of development of mountain-valley and reticulated glaciation in the Sartan Ice Age.

*Boundaries*

13. The assumed boundaries of the advance of the first Late Pleistocene glaciation (Early Zyrianka Ice Age, the Murkutin Ice Age): (a) assumed maximum; (b) other alternatives. 14. Boundaries of the second Late Pleistocene glaciation (Sartan Ice Age): (a) assumed maximum; (b) other alternatives for western Siberia. 15. Limits of local glaciation of an ice cap. 16. Comparison of terminal moraines on the Putoran plateau. 17. Assumed minimum extent of the Sartan Ice Age. 18. Glacial landforms: (1) Belogorskaya Upland, (2) Sibirskie Uvaly, (3) Kellog–Yeloguy Ridge, (4) Bakhta Ridge, (5) Nadym Marginal Zone, (6) Upper Tax Ridges, (7) Tchernyi Ostrov Ridge, (8) Salekhardski Uvaly, (9) Hadutte Ridges, (10) Turukhan Amphitheatre, (11) Igarka Amphitheatre, (12) Bolshaya Kheta Ridge, (13) Potapovo Amphitheatre, (14) Malyshevskaya (Zyrianka) Ridge, (15) Karaul Ridges, (16) Urdakhskaya Ridge, (17) Sampesa Ridge, (18) Djangodo-Syntabul Ridge–Severokokorskaya Ridge, (19) Gydan Ridge, (20) Mokoritto Ridges, (21) Upper Taimyra Ridge, (22) North Taimyra Ridge, (23) Onyoka Marginal Belt, (24) Lower Tunguska Marginal Belt. 19. (a) Cape Sabler section, (b) Ice Hill section.

Assumed centers of glaciations: I, Novaya Zemlya; II, Kara Sea–North Taimyr; III, Polar Urals; IV, Putoran; V, Anabar.

The data have been compiled by L. L. Isayeva from the following sources: Kind and Leonov, 1982; Astakhov, 1978, 1980; Isayeva et al., 1976; Isayeva and Kraush, 1979; Makkaveev et al., 1980.

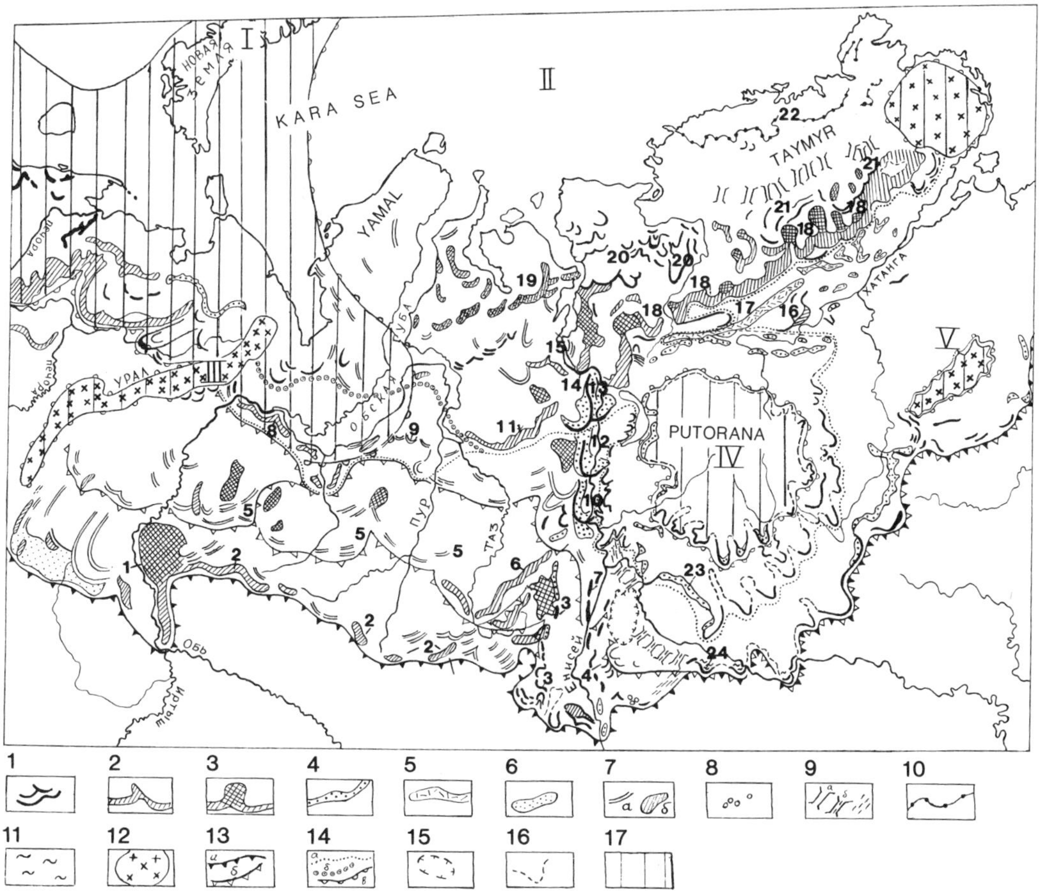
KARA SEA
YAMAL
TAYMYR
PUTORANA
I
II
III
IV
V
1
2
3
4
5
6
7
8
9
10
11
12
13
14
15
16
17

MAP 26-1

## Map 26-2

Cross sections of Severokokorskaya (North Kokora) and Verkhnetaimyrskaya (Upper Taimyra) marginal ridges. 1, Push moraine; 2, sand, gravel, boulders; 3, sand; 4, find sand; 5, clay and silt; 6, varved clay and silt; 7, rock debris; 8, till; 9, peat; 10, bedrock surface stated by geophysics; 11, marine mollusk shells; 12, plant remains: wood fragments, tree trunks, and organic detritus; 13 (a) $^{14}C$ dated plant remains and (b) mollusk shells; 14, concave cross bedding; 15, inclined bedding; 16, disturbed stratification; 17, ice wedges.

Stratigraphic level: IV, Holocene; $III_{sr}$, Sartan Glaciation (Late Würm, Late Valda); $III_{mr}$, Murukta Glaciation (Early Würm, Early); $III_{kz}$, Kazantsevo Valday (Interglacial, Mikulino, Eem).

## Map 26-3

Boundaries of the Late Pleistocene glaciations of Siberia: possible alternatives. For symbols see Map 26-2.

## Map 26-4

Reconstruction of the glacial systems of the Northern Hemisphere during the maximum of the Last Glaciation (18–20 kya). Compiled by M. A. Faustova. 1. Ice sheets. 2. Mountain ice sheets. 3. The main ice divides. 4. Areas of ice sheet convergence. 5. Areas of heterochronous advances of ice sheets. 6. Ice-flow lines. 7. The principal ice flows and lobes. 8. Limits of ice sheets: (a) documented; (b) assumed. 9. Maximum limit of ice sheets (assumed) 10. Shore line. 11. Boundaries of sea ice (marine pack ice and seasonal marine ice). 12. Key sections ($^{14}C$ dated).

## Map 26-5

The Late Valdai maximum temperature fall. Deviations of January temperatures from present-day values, based on floral data. 1. Isoanomalies. 2. Areas with radiocarbon datings. Numerical symbols: in the numerator—reconstructed temperature; in the denominator—present-day temperature; in brackets—deviation. 3. The limit of glaciation: (a) "maximum" alternatives; (b) "minimum" alternatives. 4. Mountain–valley glaciation. 5. Permafrost boundary. 6. Shorelines.

## Map 26-6

Late Valdai maximum temperature fall. Deviations of mean July temperature from present-day values. For symbols see Map 26-5.

## Map 26-7

Cross-section of Quaternary deposits of northwest Russian plain. 1, Valdai (Late Pleistocene) till; 2, Middle Pleistocene till; 3, Early Pleistocene unspecified tills; 4, glaciofluvial gravel and sand; 5, coastal sands of ice-dammed basins; 6, sand of valley terraces; 7, glaciolacustrine clay and fine-grained sand; 8, bog-lacustrine deposits (gyttja and peat); 9, cover loams; 10, bedrocks; 11, stratigraphic boundaries; 12, $^{14}C$ dates.

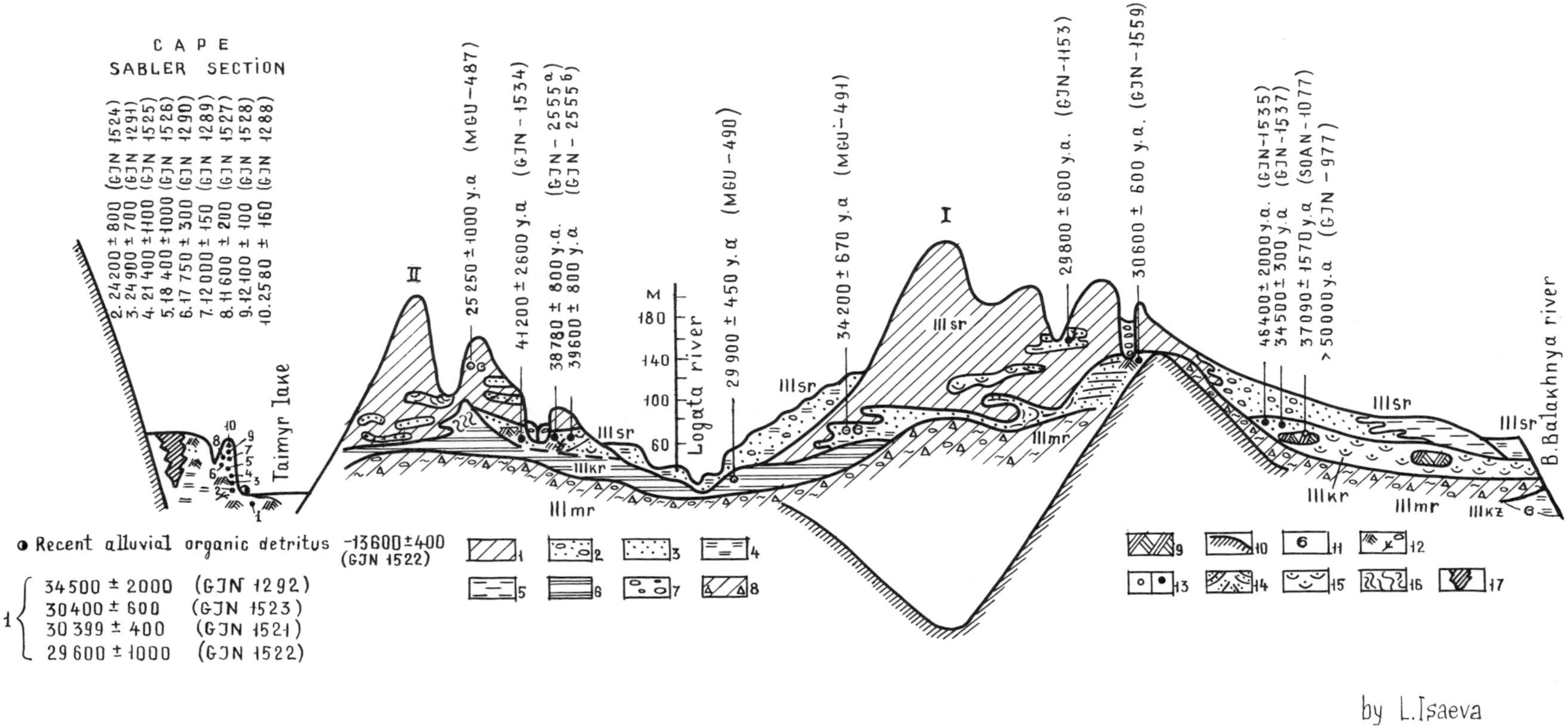

CAPE
SABLER SECTION
2. 24200 ± 800 (GJN 1524)
3. 24900 ± 700 (GJN 1291)
4. 21400 ± 1100 (GJN 1525)
5. 18400 ± 1000 (GJN 1526)
6. 17750 ± 300 (GJN 1290)
7. 12000 ± 150 (GJN 1289)
8. 11600 ± 200 (GJN 1527)
9. 12100 ± 100 (GJN 1528)
10. 2580 ± 160 (GJN 1288)
Taimyr lake
II
25 250 ± 1000 y.a (MGU-487)
41200 ± 2600 y.a (GJN-1534)
38780 ± 800 y.a. (GJN-2555a)
39600 ± 800 y.a (GJN-2555б)
IIIsr
IIIkr
IIImr
M
180
140
100
60
Logata river
29 900 ± 450 y.a (MGU-490)
34200 ± 670 y.a (MGU-491)
I
IIIsr
29800 ± 600 y.a. (GJN-1153)
30600 ± 600 y.a. (GJN-1559)
46400 ± 2000 y.a. (GJN-1535)
34500 ± 300 y.a (GJN-1537)
37090 ± 1570 y.a (SOAN-1077)
> 50000 y.a (GJN-977)
IIIsr
IIIkr
IIImr
IIIkz
B. Balakhnya river
Recent alluvial organic detritus -13600 ± 400 (GJN 1522)
1 { 34500 ± 2000 (GJN 1292)
30400 ± 600 (GJN 1523)
30 399 ± 400 (GJN 1521)
29 600 ± 1000 (GJN 1522) }
1 2 3 4 5 6 7 8 9 10 11 12 13 14 15 16 17
by L.Isaeva


MAP 26-2

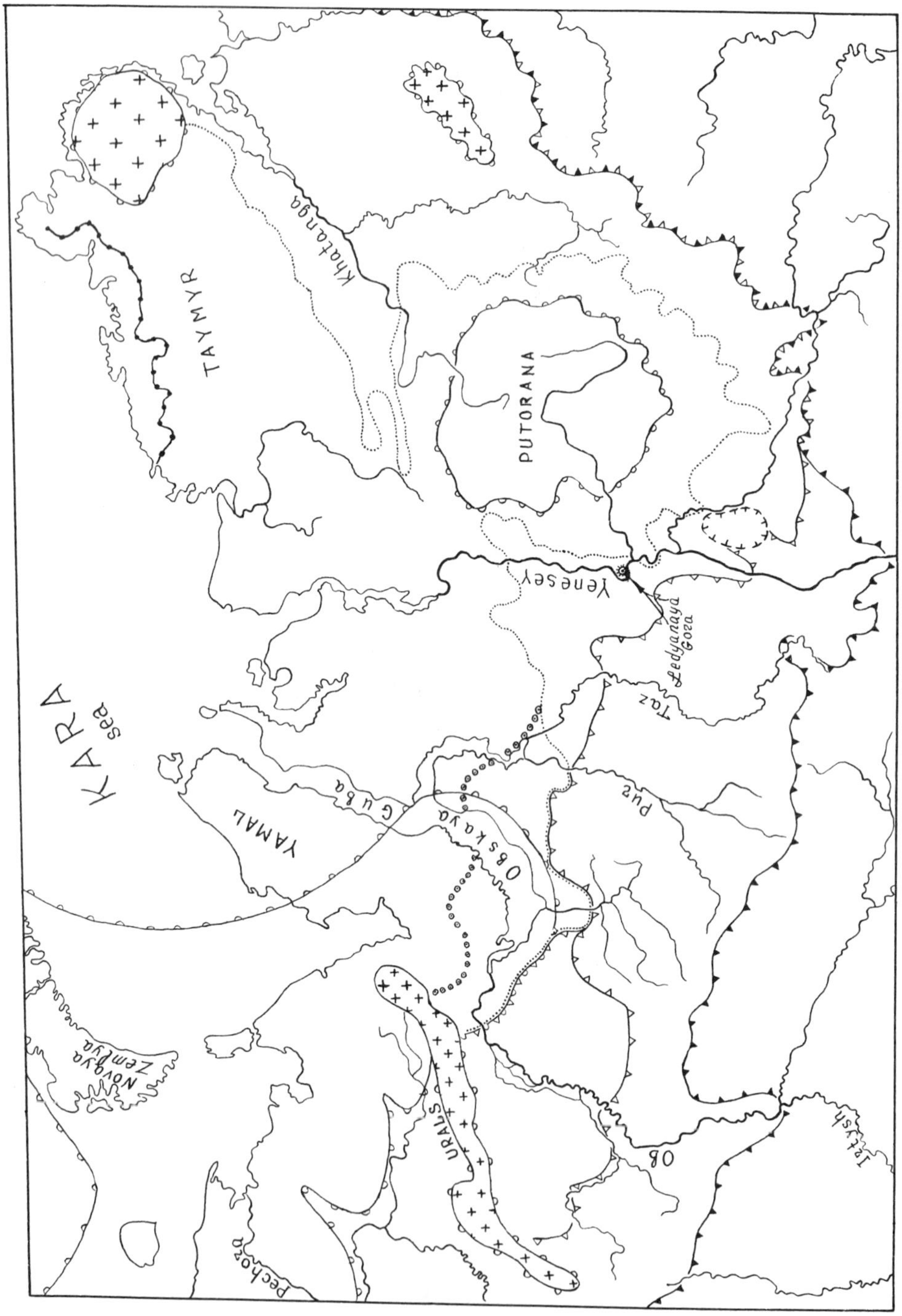
KARA sea
TAYMYR
Khatanga
PUTORANA
Yenesey
Ledyanaya Goza
Taz
Pur
YAMAL
Obskaya Guba
Novaya Zemlya
URALS
Ob
Irtysh
Pechora

MAP 26-3

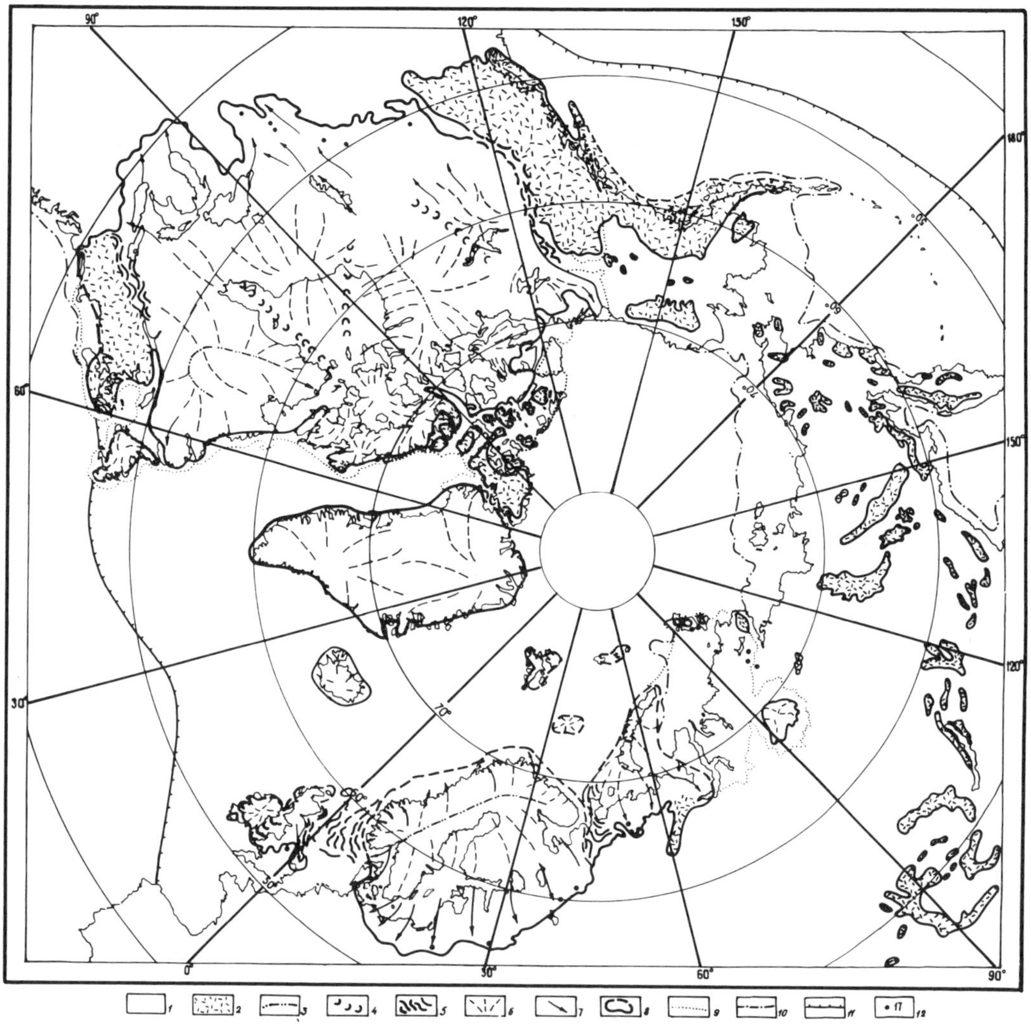


MAP 26-4

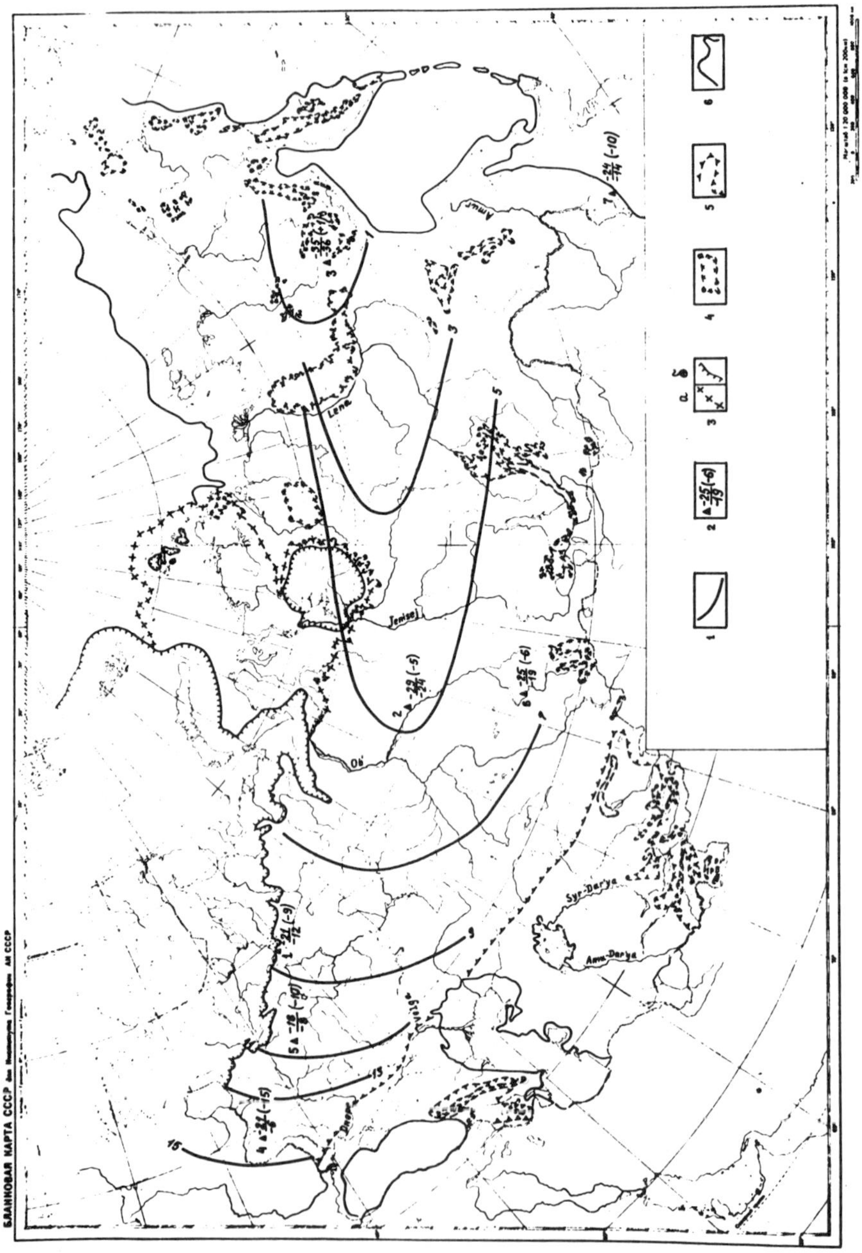


MAP 26-5

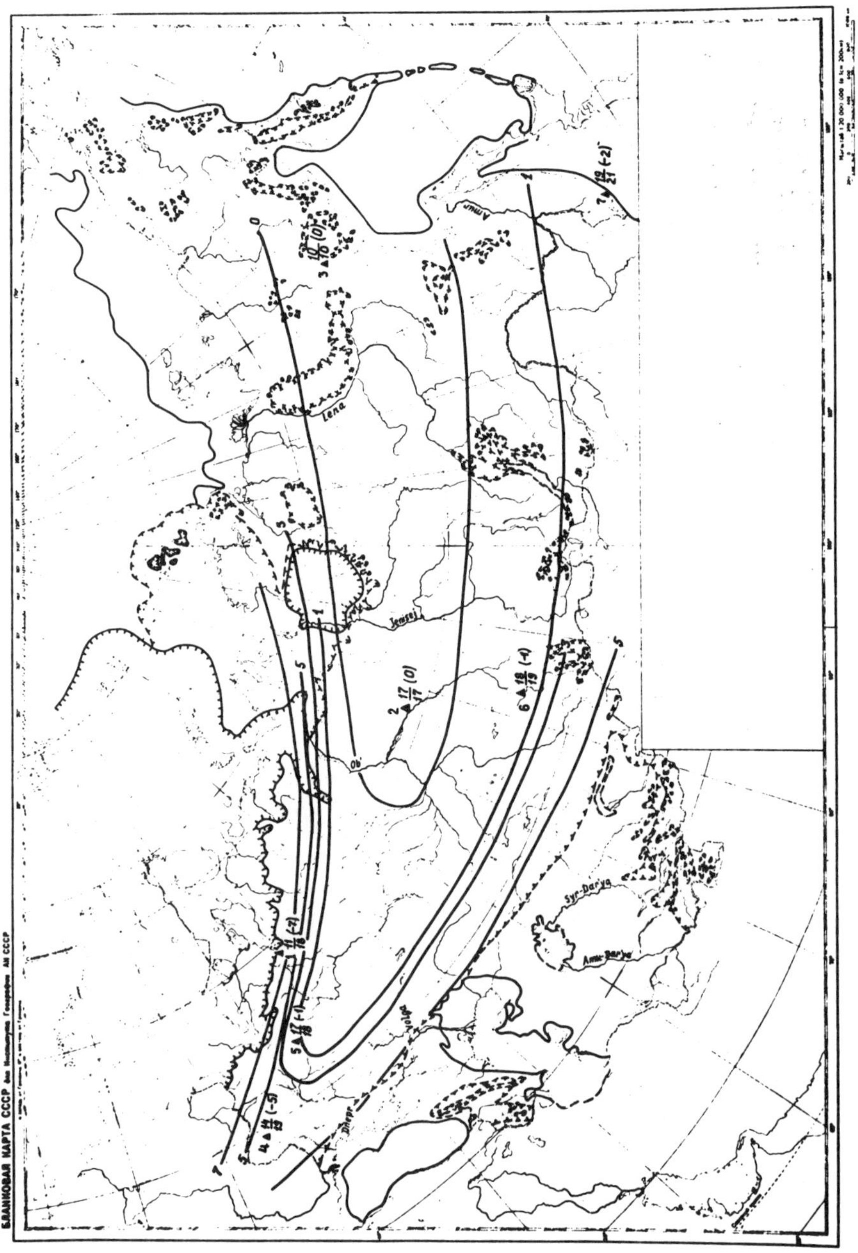


MAP 26-6

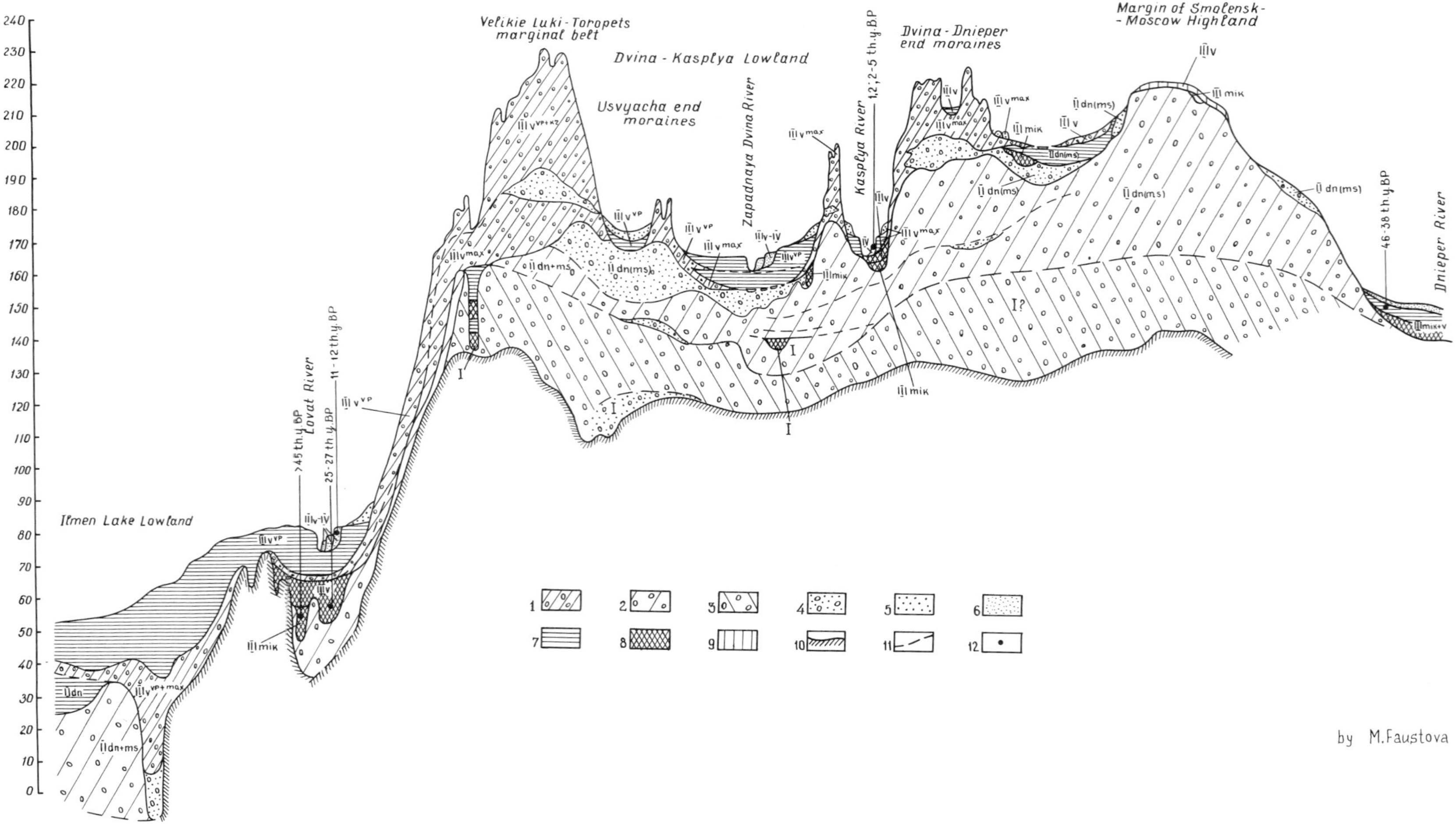
Velikie Luki-Toropets marginal belt
Dvina - Kasplya Lowland
Usvyacha end moraines
Zapadnaya Dvina River
Kasplya River
1,2; 2-5 th.y.BP
Dvina - Dnieper end moraines
Margin of Smolensk- - Moscow Highland
46-38 th.y.BP
Dnieper River
Ilmen Lake Lowland
Lovat River
>45 th.y.BP
25-27 th.y.BP
11-12 th.y.BP
240
230
220
210
200
190
180
170
160
150
140
130
120
110
100
90
80
70
60
50
40
30
20
10
0
1
2
3
4
5
6
7
8
9
10
11
12
by M.Faustova

MAP 26-7

## REFERENCES


Andersen, G. B. 1981. Late Weichselian ice sheets in Eurasia and Greenland. In G. H. Denton and T. J. Hughes (eds.), *The Last Great Ice Sheets.* Wiley, New York, pp. 1-65.

Andrews, J. T. 1982. On the reconstruction of Pleistocene ice sheets: A review. *Quat. Sci. Reviews* **1**(1):1-30.

Andrews, J. T., W. W. Shilts, and G. H. Miller. 1983. Multiple deglaciations of the Hudson Bay lowlands, Canada, since deposition of the Missinaibi (Last Interglacial) formation. *Quat. Res.* **19**(1):18-37.

Arkhipov, S. A., M. R. Voktakh, A. V. Gol'bert, et al. 1977. Poslednee oledeneniye v Nizhnem Priobiye (The Last Glaciation in the Lower Ob region). Nauka, Novosibirsk, 216 p. (In Russian.)

Arslanov, Kh. A., A. S. Lavrov, and L. M. Potapenko. 1983. Novye dannye o pozdne-pleistotsenovom oledenenii Zapadnoi Sibiri. In *Oledeneniya i Paleoclimaty Sibiri v Pleistotsene.* Nauka, Novosibirsk, pp. 27-36. (In Russian.)

Arslanov, Kh. A., L. N. Voznyachuk, F. Yu. Velichkevich, L. V. Kurierova, and G. S. Petrov. 1971. Vozrast maksimal'noi stadii poslednego oledeneniya na mezhdurechie Zapadnoi Dviny i Dnepra. *Doklady AN SSSR* **1**:901-909. (In Russian.)

Arslanov, Kh. A., V. M. Makeev, O. F. Baranovskaya, D. B. Malakhovsky, and T. V. Tertychnaya. 1980. Geokhronologiya i nekotorye voprosy paleogeografii vtoroi poloviny pozdnego pleistotsena Severnoi Zemli. In *Geochronologiya Chetvertichnogo Perioda.* Nauka, Moscow, pp. 168-176. (In Russian.)

Arslanov, Kh. A., A. S. Lavrov, and L. D. Nikiforova. 1981. O stratigrafii, geokhronologii i izmeneniyakh klimata srednego i pozdnego pleistotsena na severo-vostoke Russkoi ravniny. In *Pleistotsenovye Oledeneniya Vostochno-Yevropeiskoi Ravniny.* Nauka, Moscow, pp. 12-37. (In Russian.)

Aseev, A. A. 1974. Drevniye materikovye oledeneniya Yevropy. Nauka, Moscow, 319p. (In Russian.)

Astakhov, V. I. 1978. O Karskom tsentre pokrovnogo oledeneniya Zapadnoi Sibiri. In *Sovremennoye i Drevnee Oledeneniye Ravninnykh i Gornykh Rayonov SSSR.* Nauka, Moscow, pp. 29-45. (In Russian.)

Astakhov, V. I. 1980. Struktura severnogo pleistotsena po dannym kosmicheskikh i vysotnykh s'yomok. *Issledovaniya Zemli iz Kosmosa* **5**:22-25. (In Russian.)

Astakhov, V. I., and Yu. B. Fainer. 1979. Lednikovyi relief i osobennosti pleistotsenovogo oledeneniya na zapade Srednei Sibiri. *Geomorfologiya* **3**:45-55. (In Russian.)

Avdalovich, S. A., and R. A. Bidzhiev. 1984. Karginskiye morskie terrasy na severe Zapadnoi Sibiri i problema sartanskogo oledeneniya. *Izvestiya Akad. Nauk SSSR,* ser. geograf., **1**:70-73. (In Russian.)

Bardeeva, M. A., and L. L. Isayeva. 1980. O vydelenii Muruktinskogo gorizonta v chetvertichnykh otlozheniyakh Srednei Sibiri. *Doklady Akad. Nauk SSSR* **251**(1):169-173. (In Russian.)

Blake, W., Jr. 1970. Studies of glacial history in Arctic Canada. I. Pumice, radiocarbon dates and differential postglacial uplift in the eastern Queen Elizabeth islands. *Canadian J. Earth Sci.* **7**(2):634-664.

Boulton, G. S. 1979. Glacial history of the Spitsbergen Archipelago and the problem of a Barents shelf ice sheet. *Boreas* **8**(1):35-57.

Boulton, G. S., C. T. Baldwin, J. D. Peacock, A. M. McCabe, J. Miller, J. Jarvis, B. Horsefield, P. Worsley, N. Eyles, P. N. Chroston, T. E. Day, P. Gibbard, P. E. Hare, and V. Von Brunn. 1982. A glacioisostatic facies model and amino-acid stratigraphy for Late Quaternary events in Spitsbergen and the Arctic. *Nature* **298**(5,873): 437-441.

Cepek, A. J. 1972. Zum Stand der Stratigraphie der Weichsel-Kaltzeit in der DDR. *Wissenschaft. Zeitschr. Universitat Grcifswald. Mat.-Naturn. Reihe* **21**(1).

Chebotareva, N. S. (ed.) 1977. *Struktura i Dinamika Poslednego Yevropy.* Nauka, Moscow, 142p. (In Russian.)

Chizhov, O. P., V. S. Koryakin, N. V. Davidovich, et al. 1968. *Oledeneniye Novoi Zemli.* Nauka, Moscow, 338p. (In Russian.)

Chuklenkova, I. N. 1982. *Morfometricheskiye Issledovaniya Drevnelednikovoi Morfoskul'ptury.* Nauka, Moscow, 76p.

Danilov, I. D. 1979. O gipoteze posdnevyurmskogo arkticheskogo lednikovogo shchita. *Vestnik MGU,* ser 5, geografiya, **5**:18-26. (In Russian.)

Dibner, V. D. 1970. Geomorfologiya. In *Sovetskaya Arktika.* Nauka, Moscow, pp. 59-107. (In Russian.)

Dupplessy, J. C., J. Moyes, and C. Pujol. 1980. Deep water formation in the North Atlantic Ocean during the last Ice Age. *Nature* **286**:479-482.

Dyke, A. S., L. A. Dredge, and J. S. Vincent. 1982. Configuration and dynamics of the Laurentide Ice Sheet during the Late Wisconsin Maximum. *Georg., Phys. et Quatern.* **36**:(1-2):5-14.

Ekman, I. M., Yu. M. Levkin, and G. K. Morozov. 1979. Glyatsiotektonika pozdnego pleistotsena rayona g. Petrozavodska. In *Paleogeografiya Oblasti Skandinavskikh Materikovykh Oledeneniy.* Leningrad, pp. 11–18. (In Russian.)

England, J. 1976. Late Quaternary glaciation of the eastern Queen Elizabeth Islands, NWT, Canada: Alternative models. *Quat. Res.* **6**(2):185–202.

England, J., and R. S. Bardley. 1978. Past glacial activity in the Canadian High Arctic. *Science* **200**(44,399):265–270.

Flinn, D. 1978. The most recent glaciation of the Orkney-Shetland Channel and adjacent areas. *Scot. J. Geology* **14**(2):109–123.

Flint, R. F. 1943. Growth of North American ice sheet during the Wisconsin Age. *Bull. Geol. Soc. Am.* **54:**325–362.

Gates, W. L. 1976. Modeling the ice-age climate. *Science* **191:**1,137–1,144.

Gerasimov, I. P., and K. K. Markov. 1939. Lednikovyi period na territorii SSSR. *Trudy Instituta geografii.* vyp. 33, Izdatelstvo AN SSSR, Moscow and Leningrad, 462p. (In Russian.)

Gerasimov, I. P., and A. A. Velitchko (eds.). 1982*a*. *Paleogeografiya Yevropy za Posledniye sto Tysyach Let. Atlas Monografiya.* Nauka, Moscow, 259p. (In Russian.)

Gerasimov, I. P., and A. A. Velitchko. 1982*b*. Sovremennye vedushchiye napravleniya issledovaniy po paleogeografii antropogena v SSSR. *Izv. AN SSSR,* ser. geograf., **4:**6–17. (In Russian.)

Grant, D. R. 1977. Glacial style and ice limits, the Quaternary stratigraphic record and changes of land and ocean level in the Atlantic Provinces, Canada. *Geogr. Phys., Quat.* **31**(3–4):247–260.

Grosvald, M. G. 1983*a*. Pokrovnye ledniki kontinental'nykh shel'fov. Nauka, Moscow, 210p. (In Russian.)

Grosvald, M. G. 1983*b*. Oledeneniye kontinental'nykh shel'fov. *Itogi Nauki i Tekhniki,* ser. paleogeografiya, **1:**145p. (In Russian.)

Hakansson, S. 1982. University of Lund radiocarbon dates XV. *Radiocarbon* **24**(2):194–213.

Hansen, S. 1965. The Quaternary of Denmark. In *The Quaternary,* vol. 1, New York.

Hillaire, M.-C. 1981. Continental ice complexes: Distinction between centres of outflow and of loading of the crust. Example of the eastern Laurentide Ice-Sheet. *Zeitschr. Geomorphologie* **25**(2):225–228.

Hjort, G. 1979. Glaciation in northern East Greenland during the Late Weichselian and Early Flandrian. *Boreas* **8**(3):281–296.

Hoppe, G. 1981. Glacial traces on the island of Hopen, Svalbard: A correction. *Geogr. Ann.* **63**(1–2):67–68.

Hughes, T. J., G. H. Denton, and M. G. Grosvald. 1977. Was there a Late-Wurm Arctic ice sheet? *Nature* **266**(5,603):596–602.

Ilyin, V. A., A. D. Lukashov, and I. M. Ekman. 1978. Kraevye lednikovye obrazovaniya Zapadnoi Karelii i ikh korrelyatsiya s gryadami finskikh Sal'paussel'kya. In *Kraevye Obrazovaniya Materikovykh Oledeneniy.* Kiev, pp. 96–108. (In Russian.)

Isayeva, L. L. 1972. Kraevye lednikovye obrazovaniya Sredne-Sibirskogo ploskogoriya. In *Kraevye Obrazovaniya Materikovykh Oledeneniy.* Nauka, Moscow, pp. 205–211. (In Russian.)

Isayeva, L. L., and M. A. Kraush. 1979. Lednikovye morphoscul'ptury Severo-Sibirskoi nizmennosti. In *Regional'naya Geomorfologiya rayonov Novogo Osvoyeniya.* Izdaniye MFGO SSSR, Moscow, pp. 98–109. (In Russian.)

Isayeva, L. L., N. V. Kind, M. A. Kraush, and L. D. Sulerzhitsky. 1976. O vozraste i stroyenii kraevykh lednikovykh obrazovaniy severnogo podnozhiya plato Putorana. *Byull. Komissii po Izucheniyu Chetvertichnogo Perioda* **46:**117–123. (In Russian.)

Isayeva, L. L., N. V. Kind, S. M. Andreeva, and M. A. Bardeeva. 1981. Chetvertichnye oledeneniya severa Srednei Sibiri. In *Chetvertichnye Oledeneniya Zapadnoi Sibiri i Drugikh Oblastei Severnogo Polushariya.* Nauka, Novosibirsk, pp. 98–105. (In Russian.)

Ives, J. D. 1978. The maximum extent of the Laurentide Ice Sheet along the east coast of North America during the Last Glaciation. *Arctic* **31**(1):24–53.

Kaplina, T. N., and I. L. Kuznetsova. 1975. Geotemperaturnaya i klimaticheskaya model' epokhi nakopleniya osadkov edomnoi svity Primorskoi nizmennosti Yakutii. In *Problemy Paleogeografii Lyossovykh i Periglyatsial'nykh Oblastei.* Moscow, pp. 170–173. (In Russian.)

Keigwin, L. D. 1982. An Arctic Ocean ice sheet in the Pleistocene? *Nature* **296**(5,860):808–809.

Khodakov, V. G. 1979. Paleoglyatsiologicheskiye rekonstruktsii dlya epokhi maksimuma posdnevyurmskogo oledeneniya territorii SSSR i nekotorye paleoklimaticheskiye sledstviya. *Izv. An SSSR,* ser. geograf., **6:**27–32. (In Russian.)

Kind, N. V. 1974. *Geokhronologiya Pozdnego Antropogena po Izotopnym Dannym.* Nauka, Moscow, 255p. (In Russian.)

Kind, N. V., and B. N. Leonov (eds.). 1982. *Anthropogen Taimyra.* Nauka, Moscow, 179p. (In Russian.)

Klimanov, V. A. 1982. Paleogeograficheskiye rekonstruktsii kak zadacha raspoznavaniya obrazov. *Izv. An SSSR,* ser. geograf., **5:**98–105. (In Russian.)

Krasnozhen, A. S., O. F. Baranovskaya, V. S.

Zarkhidze, E. S. Malyasova, and O. M. Lev. 1982. Verkhnechetvertichnye otlozheniya Yuzhnogo ostrova Novoi Zyemli. In *Stratigrafiya i Paleogeografiya Pozdnego Kainozoya Arktiki.* Leningrad, pp. 108–114. (In Russian.)

Lavrov, A. S. 1977. Kol'sko-Mezenskiy, Barentsovomorsko-Pechorskiy, Novozemel'sko-Kolvinskiy lednikovye potoki. In *Struktura i Dinamika Poslednego Lednikovogo Pokrova Yevropy.* Nauka, Moscow, pp. 83–100.

Lavrov, A. S., and Kh. A. Arslanov. 1977. Vozrast i genezis terras Pechorskoi nizmennosti: Novye geokhronologicheskiye i radiouglerodnye dannye. In *Rechnye Sistemy i Melioratsiya.* Chast I, Materialy XIV plenuma geomorfologicheskoi Komissii AN SSSR. Novosibirsk, pp. 52–58. (In Russian.)

Lebuis, I., and P. P. David. 1977. La stratigraphie et les evenments du Quaternaire de la partie occidentale de la Gaspesie, Quebec. *Geogr. Phys., Quat.,* 3-e Coll. Quat. du Quebec, **31**(3–4):275–296.

Levkov, E. A. 1980. *Glyatsiotektonika.* Minsk, Nauka i teknika, 280p. (In Russian.)

Loriol, B., and J. Grey. 1982. Evolyutsiya posdneviskonsinskogo lednikovogo shchita v sectore Ungava: Dannye po podnyatiyu morskikh i ozernykh beregovykh liniy. In *XI Congress of INQUA. Abstracts of Reports, vol. 1.* Moscow, pp. 157–158.

Makeev, V. M., Kh. A. Arslanov, and V. E. Garrut. 1979. Vozrast mamontov Severnoi Zemli i nekotorye voprosy paleogeografii pozdnego pleistotsena. *Doklady AN SSSR* **245**(2):1,114–1,118. (In Russian.)

Makkeveev, A. N., A. A. Aseev, V. I. Astakhov, V. V. Bronguleev, L. L. Isayeva, A. S. Lavrov, and G. G. Matishov. 1980. Geomorfologicheskiye kriterii rekonstruktsii pozdnepleistotsenovykh lednikovykh pokrovov Severnoi Evrazii. *Geomorfologiya* **4:**13–26. (In Russian.)

Malakhovsky, D. B., and E. A. Spiridonova. 1981. O nizhnevaldaiskikh otlozheniyakh i nekotorykh voprosakh paleogeografii poslednego oledeneniya Severo-Zapada Russkoi ravniny. In *Geologiya Pleistotsena Severo-Zapada SSSR.* Apatity, pp. 83–90. (In Russian.)

Mangerud, J. 1981. The Early and Middle Weichselian in Norway: Review. *Boreas* **10**(4).

Mangerud, J., A. Elgersma, D. Helliksen, J. Landvik, and O. Salvigsen. 1983. The Late Weichselian (25–10 KA B.P.) Glacial Maximum in Isfjorden and Van Mijenfjorden, Spitsbergen, Svalbard. *Abstracts 13th Annual Arctic Workshop INSTAAR.* University of Colorado, pp 67–68.

Matishov, G. G. 1980. Geomorfologicheskiye priznaki vozdeistviya Skandinavskogo, Novozemel'skogo, i Shpitsbergenskogo lednikovykh pokrovov na poverkhnost' dna Barentseva morya. *Okeanologiya* **20**(4):669–680. (In Russian.)

Matishov, G. G. 1984. *Dno Okeana v Lednikovyi Period.* Nauka, Leningrad, 175p. (In Russian.)

Miller, G. H. 1980. Late Foxe glaciation of southern Baffin Island, NWT, Canada. *Bull. Geol. Soc. America* **91**(7):1.

Nazarov, V. N. 1984. *Rekonstruktsiya Landshaftov Byelorussii po Paleoentomologicheskim Dannym.* Nauka, Moscow, 95p. (In Russian.)

Nikonov, V. N. 1984. Srednee integral'noye vlagosoderzhaniye atmosfery severnoi chasti Atlanticheskogo okeana po dannym aerologicheskogo zondirovaniya korablei pogody s 1958 po 1970 g. *Problemy Arktiki i Antarktiki* **58:**85–90. (In Russian.)

Paterson, W. S. 1977. Extent of the Late-Wisconsin glaciation in northwest Greenland and northern Ellesmere Island: A review of the glaciological and geological evidence. *Quat. Res.* **8**(2):180–190.

Piotr, K., and L. Leszek. 1981. Development of glaciers on the southern coast of Hornsund in Spitsbergen during the Wurm (Vistulian) Glaciation. *Acta Geologica Polonica* **31**(1–2).

Price, R. J. 1975. The Glaciation of West Central Scotland—a review. *Scot. Geogr. Mag.* **91**(3):134–145.

Raukas, A. V. 1976. Korrelyatsiya verkhnepleistotsenovykh otlozhenii i krayevykh lednikovykh zon Syevernoi Ameriki i Yevropy. *Abstracts of Reports of the Vsesoyuznoe Soveshchaniye po Izucheniyu Krayevykh Obrazovaniy Materikovykh Oledeneniy.* Naukova Dumka, Kiev, pp. 141–143. (In Russian.)

Reinhard, H. 1974. *Genese des Nordseeraumes im Quartar.* Fennia, Helsinki, Bd. 129.

Rutter, N. W. 1981. Relationship between Late Pleistocene Laurentide and Cordilleran glaciations, Canada, IGCP Program 73-1-24. *Quat. Glaciations North. Hemisph. Rept. N 6, Ostrava, 1979.* Praga, pp. 205–218.

Saks, V. N. 1953. *Chetvertichnyi Period v Sovetskoi Arktike.* Vodtransizdat, Moscow and Leningrad, 177p. (In Russian.)

Saks, V. N. 1958. Nekotorye soobrazheinya o geologicheskoi istorii Arktiki. In *Problemy Severa,* vol. 1. Izdatelstovo AN SSSR, Moscow, pp. 7–31. (In Russian.)

Salvigsen, O. 1979. The last deglaciation of Svalbard. *Boreas* **8**(2):229–231.

Salvigsen, O. 1981. Radiocarbon dated raised beaches in Kong Karls Land, Svalbard, and their consequences for the glacial history of the Barents Sea area. *Geogr. Ann.* **63A**(3/4):283–291.

Salvigsen, O., and H. Osterholm. 1982. Radiocarbon dated raised beaches and glacial history of the northern coast of Spitsbergen, Svalbard. *Polar Res.* **1**:97–115.

Scott, J. S. 1976. Geology of Canadian tills. In R. F. Leggett (ed.), *Glacial Till.* Spec. Publ. No. 12. Royal Society of Canada, Ottawa, pp. 50–66.

Sejrup, H. P., E. Jansen, H. Erlenkenser, and H. Holtedahl. 1984. New faunal and isotopic evidence of the Late Weichselian–Holocene oceanographic changes in the Norwegian Sea. *Quat. Res.* **21**:74–84.

Serebryannyi, L. R. 1978. *Dinamika Pokrovnogo Oledenleniya i Glyatsioevstaziya v Pozdnechetvertichnoye Vremya.* Nauka, Moscow, 270p. (In Russian.)

Shilts, W. W. 1980. Flow patterns in the central North America ice sheet. *Nature* **286** (5,770):213–218.

Stalker, A., and S. Mac. 1980. The geology of the ice-free corridor: The southern half. *Can. J. Anthropol.* **1**(1):11–16.

Strelkov, S. A. 1968. Tsentry oledeneniya na severe Sibiri, usloviya i stadii razvitiya. In *Neogenovye i Chetvertichnye Otlozheniya Zapadnoi Sibiri.* Nauka, Moscow, pp. 31–37. (In Russian.)

Sugden, D. E. 1977. Reconstruction of the morphology, dynamics and thermal characteristics of the Laurentide Ice Sheet at its maximum. *Arctic Alpine Res.* **9**(1):21–47.

Troitsky, L., J.-M Punning, G. Hutt, and R. Rajamae. 1979. Pleistocene chronology of Spitsbergen. *Boreas* **8**(4):401–407.

Velichko, A. A. 1973. *Prirodnyi Protsess v Pleistotsene.* Nauka, Moscow, 241p. (In Russian.)

Velichko, A. A. 1975. Problemy korrelyatsii pleistotsenovykh sobytiy v lednikovoi, periglyatsial'no-lyossovoi i primorskoi oblastyakh Vostochno-Yevropeiskoi ravniny. In *Problemy Regional'noi i Obshchei Paleogeografii Lyossovykh i Periglyatsial'nykh Oblastei.* Moscow, pp. 7–26. (In Russian.)

Velichko, A. A. 1979. Problemy reconstruktsii pozdnepleistotsenovykh oledeneniy i lednikovykh pokrovov na territorii SSSR. *Izvestiya AN SSSR*, ser. geograf., **6**:21–37. (In Russian.)

Velichko, A. A. 1980. Shirotnaya asimmetriya v sostoyanii prirodnykh komponentov lednikovykh epokh v Severnom polusharii. *Izv. AN SSSR*, ser. geograf., **5**:5–14. (In Russian.)

Velichko, A. (ed.) 1983. *Late Quaternary Environment of the Soviet Union.* Minnesota, 327p.

Velichko, A. A. 1985. Problemy rekonstriktsii Klimata lednikovykh epokh. *Materialy glyatsiologicheskikh Issledovaniy* **52**:120–129. (In Russian.)

Velichko, A. A., and M. A. Faustova. 1982. Problemy obosnovaniya maksimal'noi granitsy pozdnepleistotsenovogo oledeneniya na severe Yevrazii. In *Razvitiye Prirody Territorii SSST v Pozdnem Pleistotsene i Golotsene.* Nauka, Moscow, pp. 72–84. (In Russian.)

Velichko, A. A., and V. G. Khodakov. 1981. Voprosy rekonstruktsii pozdnepleistotsenovykh lednikovykh pokrovov arktischeskogo shel'fa na osnove konkretnykh dannykh i paleoglyatsiologicheskogo modelirovaniya. In *Chetvertichnye Oledeneniya Zapadnoi Sibiri i Drugihk Oblastei Severnogo Polushariya.* Nauka, Novosibirsk, pp. 25–34. (In Russian.)

Vilks, G., and P. Mudie. 1978. Early deglaciation of the Labrador shelf. *Science* **202**(4,373):1,181–1,183.

Walcott, R. I. 1975. Characteristics of recent uplift in North America. In *Problems of Recent Crustal Movements.* Valgus, Tallin, pp. 146–155.

Yevzerov, V. Ya., and B. N. Koshechkin. 1980. Paleogeografiya pleistotsena zapadnoi chasti Kol'skogo poluostrova. Nauka, Leningrad, 170p. (In Russian.)

Zubakov, V. A., and G. M. Levkovskaya. 1969. Stratigrafiya noveishikh otlozhenii nizoviev r. Obi. In *Chetvertichnaya Geologiya i Geomorphologiya Sibiri,* chast I. Novosibirsk, 215p. (In Russian.)

# 27: Geological and Paleoclimatic Evolution of the Arctic During Late Cenozoic Time

I. D. Danilov

Three main stages are recognized in the development of the Arctic Ocean and its marginal seas: They occurred during the Miocene, Pliocene–Early Pleistocene and Late Pleistocene. The Late Cenozoic climatic deterioration was the result of increased area and elevation of continents simultaneously with the deepening, isolation, and cooling of the Arctic Ocean.

The Late Cenozoic fluctuations of climate resulted from changing paleogeographic conditions in the Northern Hemisphere that were caused by tectonically controlled transgressions and regressions of the Arctic water, which in turn determined the variations in the degree of its isolation from the World Ocean. During transgressions the climate was milder and more humid and the exchange of fauna between the Arctic and boreal regions was pronounced. During regressions the Arctic Basin became more isolated and acquired a stable ice cover that merged with continental ice sheets widely developed on the adjoining land.

*Note:* This extended abstract was reviewed, edited, and rewritten by C. A. Repenning and the Editor, Yvonne Herman.

The boundary between the Paleogene and Neogene is marked by a gradual deepening and isolation of the epicontinental seas of the Arctic. This was a consequence of uplift and Alpine folding in many areas circling the Arctic Basin. The modern basin of the deep Arctic Ocean began to develop at that time. The change was of great significance and resulted in the distinct difference between the Arctic Ocean of the Early Cenozoic and that of the Late Cenozoic.

In the latest Miocene, about 5.2–6.5 mya (Messinian), global lowering of sea level caused maximal isolation of the Arctic Ocean from the World Ocean. The Arctic shelf was completely drained and became a deeply dissected plain of erosion. This shallowing of the Arctic Ocean could have led only to greater warmth and salinity of its waters and the sea was ice-free.

In Early Pliocene, about 3.5–5 mya, and during the Akchagyl transgression, sea level rose and a major transgression took place throughout the Arctic Basin. Its waters covered vast expanses of northern Europe and western Siberia. The Mediterranean also experienced a transgressive stage following the Messinian salinity crisis. Thus expansion of the cold-water marine fauna to the south was not due to planetary cooling of the subtropical regions of Europe but to the southward transgression of the waters of the Arctic Ocean.

In Late Pliocene, about 2–3 mya, and beginning with the opening of the Bering Strait and the introduction of mollusks from the Pacific to the Arctic and Atlantic oceans (Gladenkov, 1981), the World Ocean records indicate that ice began to accumulate on the continents of the Northern Hemisphere (Shackleton and Opdyke, 1977) and by 2.4 mya glacial till was deposited as far south as central North America (Iowa; Boellstorff, 1978). At this time, about 2.4 mya, Sher and others (1979) report that forest vegetation near the coast of the Arctic Ocean had disappeared and was replaced by tundra in the Kolyma Lowlands of northeastern Siberia and the deposits show the first evidence of permafrost.

However, evidence from Alaska indicates that the Arctic Ocean was not yet frozen at that time. This evidence is from marginal marine beds of a transgressive stage of the Arctic Ocean, although contemporaneous with a major accumulation of continental ice in central North America and, thus, suggesting the subsidence of the North Slope of Alaska at that time (C. A. Repenning, personal communication, 1986). By 2 mya, the North American glaciation had ended and records from northern Greenland indicate that forests had returned to the shore of the Arctic Ocean; the ocean was ice-free (Herman, 1970; Funder et al., 1985).

The Eopleistocene (about 0.85–2 mya) glacial advances, including the Nebraskan, commenced about 850 kya simultaneously with the inception of perennial sea-ice cover in the Arctic Ocean (Unit I of Herman, 1970; Herman and Hopkins, 1980). This coincides in time with the end of the northwestern Europe Waalian warm period.

## REFERENCES


Boellstorff, J. 1978. North American Pleistocene stages reconsidered in light of probable Plio-cene–Pleistocene glaciation. *Science* **202:**305–307.

Funder, S., N. Abrahamsen, O. Bennike, and R. W. Feyling-Hanssen. 1985. Forested Arctic: Evidence from North Greenland. *Geology* **13:**542–546.

Gladenkov, Yu. B. 1981. Marine Plio-Pleistocene of Iceland and problems of its correlation. *Quat. Res.* **15:**18–23.

Herman, Y. 1970. Arctic paleoceanography in Late Cenozoic time. *Science* **169:**474–477.

Herman, Y., and D. M. Hopkins. 1980. Arctic Ocean climate in late Cenozoic time. *Science* **209:**557–562.

Shackleton, J. J., and N. D. Opdyke. 1977. Oxygen isotope and paleomagnetic evidence for an early Northern Hemisphere glaciation. *Nature* **270:**216–219.

Sher, A. V., T. N. Kaplina, R. E. Giterman, A. V. Lozhkin, A. A. Arkhangelov, S. V. Kiselyov, Yu. V. Kouznetsov, E. I. Virina, and V. S. Zazhigin. 1979. Scientific excursion on problem: Late Cenozoic of the Kilyma Lowland. *Acad. Sci. USSR, XIV Pacific Science Congress,* Moscow.

# 28: Organic Geochemistry of Barents Sea Sediments

A. N. Belyaeva,
A. I. Daniushevskaya,
and E. A. Romankevich

In the present chapter, factors that control the distribution and transformation of organic matter (OM) in the system living organic matter—particulate matter—bottom sediments during sediment genesis and diagenesis are discussed. Based on our data, it is possible to characterize the OM (seawater, particulate matter, bottom sediments) and offer a tentative model of organic matter balance in the Barents Sea.

## FACTORS CONTROLLING ORGANIC MATTER DISTRIBUTION

A characteristic feature of the Barents Sea, as compared to the other Arctic seas is the low river discharge (163 $km^3/yr$), the Pechora River being the most important freshwater contributor. Other small rivers, flowing across the Kola Peninsula, Novaya Zemlya, and other islands are fed by glaciers and freeze in winter. Therefore, their contribution to the river discharge of the Barents Sea is not significant. Particulate matter carried mainly by the Pechora River is left within the river's liman. This factor along with low rates of coastal erosion of the predominantly igneous rocks is responsible for transparency of the Barents Sea waters.

*Note:* This chapter was reviewed, edited, and rewritten in part by the Editor, Yvonne Herman.

### Hydrological and Hydrochemical Elements

The Barents Sea water structure and its hydrological regime are determined primarily by the water exchange with the adjacent seas and with the Arctic Basin, which supply about $125 \cdot 10^3$ $km^3$ of water annually to the Barents Sea; the entire initial volume of water is completely renewed nearly every 12 years (Tantsyura, 1973). Based on temperature and salinity, typical water masses, and hydrochemical parameters, distribution can be distinguished for major areas (Fig. 28-1) (Nikiforov and Shpaiker, 1980; *Biogeochimia organicheskogo veshchestva arkticheskich morei,* 1982).

The Nordcap area I is mainly influenced by Atlantic waters flowing via the Norwegian Sea to form the water mass from surface to the seafloor. The total influx of the

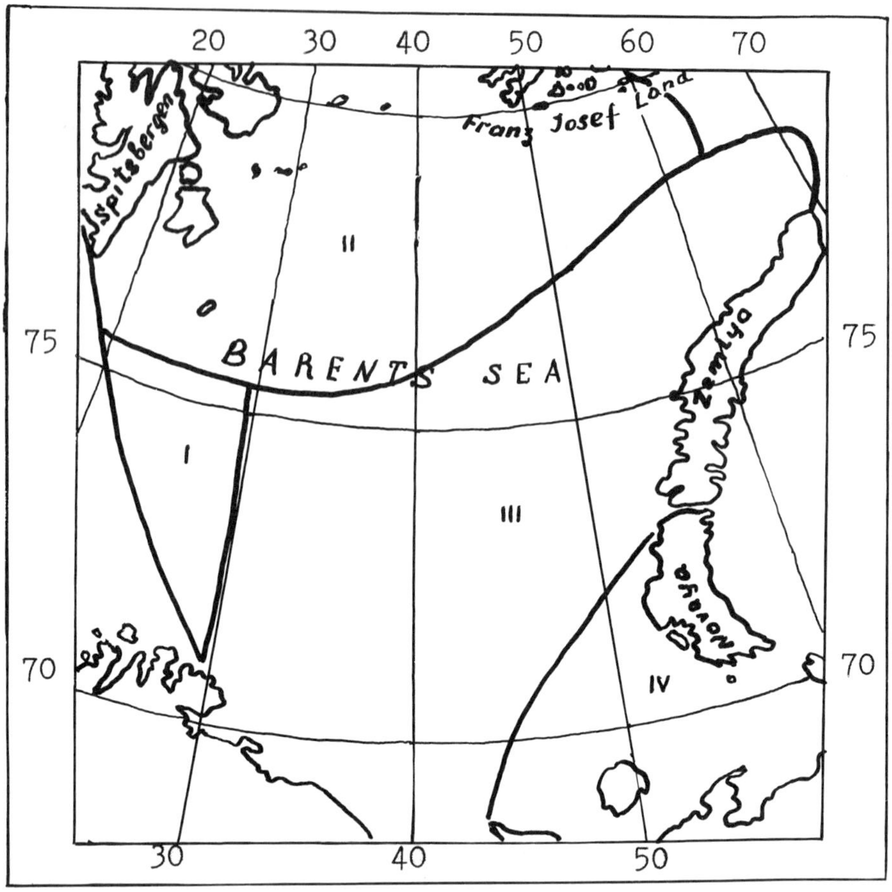


Figure 28-1. Barents Sea water masses. Regions: I, Nordcap area; II, Spitsbergen area; III, Central area, IV, Pechora area.

Atlantic water along the Nordcap–Bear Island section ranges from 392.7 $km^3$ per day to 268.6 $km^3$ per day. Seasonal and annual variations of the Atlantic water influx have a wide range and during some periods the discharge can be negative due to strong counter currents. However, according to Timofeev (1960), the volume of these waters in the Barents Sea accounts for about $60 \cdot 10^3$ $km^3$ or about 20 percent of the sea volume.

The water structure and the hydrological regime are strongly influenced by the water exchange with the Arctic basin. The influx of Arctic water in the 0–100 m layer between Victoria Island and Alexandra Land is estimated at 0.44 km $s^{-1}$. Therefore, the surface layer in the area is represented mainly by the Arctic surface water underlain by the Barents Sea water. The volume of the Barents Sea water totals $32 \cdot 10^3$ $km^3$ or 11 percent of the sea volume while the Barents Sea water accounts for about 64 percent of the sea volume.

The Arctic and Barents sea water form the water mass structure in the central area III as well. The surface layer in the area consists of the Arctic water underlain by Barents Sea water transformed by Atlantic water in winter. The southern part of the area, off the Murman coast stretching as a narrow band, is apparently affected by river discharge. The annual discharge of all Norwegian and northwestern USSR rivers totals 153 $km^3$, accounting for 1.5 percent of the water (together with the Nordcap branch of the coastal current). The effect of river runoff is most pronounced within the Pechora area IV, where a coastal water mass underlain by the Barents Sea water is formed during the spring–summer period. The depth of the coastal water sharply increases in the White Sea; however, the volume of river runoff does not exceed 5 percent (about $14 \cdot 10^3$ $km^3$).

The pattern of water mass circulation in the Barents Sea is affected by the hydrological regime, which is influenced by the water exchange with the adjacent basin, predominant winds, density, freshwater discharge, currents, and so on. In general, the sea is characterized by a stable cyclonic water movement intermittent with numerous cyclonic and anticyclonic gyrals in different areas of the sea. Redistribution of water masses and consequently hydrochemical parameters depend on year to year and monthly changes of the water circulatory system. Since the Barents Sea is located within the confines of the continental shelf with depths of 50–100 m predominating, winter convection sometimes reaches the seafloor, introducing changes in hydrochemical conditions at depth. This convection is responsible for oxygen supply to the deep and bottom water and for the enrichment of the surface water with biogenic elements. Thus, conditions become favorable for high photosynthetic activity in early spring, which in turn provides high biological productivity in summer.

The content of dissolved oxygen in the Barents Sea surface waters increases northward due to the gas solubility at low water temperature and the weaker effect of the warmer Atlantic water whose oxygen content is much lower than that of the Barents Sea or Arctic water. In the Atlantic water entering via the Norwegian Sea to area I, the oxygen content of the surface water averages 105 percent, but at a depth of 50 m it decreases to 100 percent and to 90 percent in the bottom water. The highest oxygen content is found in the layer with maximum photosynthesis; thus, in the southern part it is present within the 50 m layer; however, farther north, in the vicinity of drift ice, it is found near the surface. The most favorable conditions for unicellular algal growth and suppressed development of their consumers in low-temperature water causes oversaturation with respect to oxygen in the photosynthesis zone in the northern parts of the sea (area III) (sometimes reaching 30 percent above average). A fairly high oxygen content is caused by dynamic processes in the Barents Sea deep waters.

Seasonal changes in the content of biogenic elements throughout the water column occur. These changes are extremely pronounced in the surface water. During the spring–summer period the active photosynthesis zone shows rapid depletion in biogenic elements accompanied by higher oxygen content and higher pH. Almost simultaneously there is a gradual increase in biogenic elements with increased water depth reaching a maximum in the bottom layers in early winter. As a result, we see a decrease in oxygen content, an increase in $pCO_2$, and a decrease in pH. A clear-cut stratification of biogenic element distribution stabilized by the end of summer is absent due to surface water cooling and winter mixing in October–November. The content of biogenic elements becomes gradually evenly distributed throughout the entire water column during the winter convection. The vertical distribution of hydrochemical parameters becomes completely homogeneous in the convective layer between February and April and, as a result, biogenic elements are replenished to the surface layer by the onset of the spring photosynthesis.

## Biology

### Phytoplankton

The qualitative composition and distribution of phytoplankton are determined by physical-geographical, hydrochemical, and hydrodynamic conditions. Lack of light in winter, dynamics of biogenic elements, and low water temperatures shorten the phytoplankton vegetation period in the Barents Sea. Unlike other seas of the Arctic Ocean, in the Barents Sea the bloom period of phytoplankton, its qualitative and quantitative parameters, are subject to change due to the complex hydrodynamic and hydrochemical regime. In general, because of low mean annual insolation, low water temperatures, and increased ice cover from south to north, the phytoplankton bloom period decreases from 9 months to 4 months in the coastal waters of Murman off Spitsbergen and Franz Josef Land.

Phytoplankton bloom starts at a water temperature of about 0°C, insolation of 100–200 cal/cm$^2$ per day, when vertical stability of water masses is highest in the photic zone and is coincident with the time of maximum discharge of biogenic matter. The depth of the photosynthesis boundary changes throughout the entire phytoplankton bloom season but in general the isolume of 1 percent illumination lies at a depth of 50 m. About 0.1–2.1 percent of the total phytoplankton is found below the 50 m layer in the southern coastal regions (Roukhiyainen, 1966). In spring, the photosynthesis maximum is at the water surface, while in summer it is at 10–15 m depth, due to light reduction and in autumn it rises again to the 0–5 m layer (Solovieva, 1969; Bobrov, 1978*b*).

The major OM producers in the Barents Sea are the diatoms, constituting more than a half of the total phytoplankton species (Zenkevich, 1956) and up to 79 percent of the unicellular algal biomass (*Sovetskaya Arctica,* 1974). In spring, apart from such diatom genera as Fragilaria, Thalassiosira, and Chaetoceros, the yellow-green algae *Phaecystis pouchelii,* showing photosynthetic ability at a temperature of −0.22°C to −0.98°C can also grow (Bliznichenko et al., 1976). In summer and in autumn peridinian algae are more abundant.

Phytoplankton diversity and composition change from year to year and depend on abiotic factors, mainly temperature. A decrease in temperature leads to a decrease in total number of phytoplankton species: Diatoms and yellow-green algae dominate, while peridineans become less abundant. The total number of phytoplankton cells in a coastal area may vary greatly (Roukhiyainen, 1961*b*).

The amount of pigments capable of detecting light in the plankton algal communities of the Barents Sea is fairly constant during the phytoplankton bloom period despite variation in their number. The phytoplankton assimilation off Murman ranges between 0.1 and 5 mg C/mg chl. hr during the phytoplankton bloom period, chiefly depending on mineral nutrients (nitrogen in particular) (Vedernikov and Solovyeva, 1972; Solovieva, 1973; Solovieva, 1976).

At present, the annual primary phytoplankton production in the Barents Sea can be estimated only tentatively. To estimate the annual production we have assumed that three peaks occur in the phytoplankton production process in all the offshore regions such as the coastal areas. They are the spring, summer, and autumn maxima, the latter being the lowest. In addition, the correlation of OM synthesized during various biological seasons in the Barents Seas was assumed to be generally constant. The fact that the maximum to medium primary production ratio based on numerous data varies slightly lends support to this assumption.

According to estimates available (Table 28-1), the primary production in various parts of the Barents Sea at different times ranges from 17 to 170 g/m$^2$ per year, aver-

Table 28-1. Biomass and Primary Productivity of Phytoplankton in the Barents Sea

| Study Area (yr) | Time of Investigation (water depth, m) | Primary Biomass mg/m³ | Primary Productivity mg C/m³ | Primary Productivity mg C m² d | Primary Productivity Annual[a] g C m² yr | Reference |
|---|---|---|---|---|---|---|
| Dalnezelenetskaya Bay (1960) | April–November (0–5) | — | 4.3–218.6 | 10.8–868.0 (202.4) | 55 | Roukhiyainen (1961*a,b*) |
| Dalnezelenetskaya Bay (1967) | April–October (0–10) | 40–50 | 0.2–31.3 | 7.4–262.0 (94.2) | 25 | Sokolova and Solovieva (1971) |
| Dalnezelenetskaya Bay (1976) | Annual cycle (0–50) | | — | 1.0–2,500.0 (563.0) | 152[b] | Bobrov (1978*a,b*) |
| Dalnezelenetskaya Bay Yarnyshskaya | Vegetation season | | 0.2–36.1 | — | 28 | Solovieva (1976) |
| Murman Coast (1967, 1970) | June–August (0–50) | — | 0.8–57.8 | 44.6–593.9 | 42 | Vedernikov and Solovieva (1972) |
| Western part of the sea (Atlantic waters) | May | — | — | 280.0–320.0 | 17 | Corlett (1958) (edited after Vedernikov and Solovieva, 1972) |
| Arctic waters off Spitsbergen | May | — | — | 1,330.0 | 56 | Corlett (1958) (edited after Vedernikov and Solovieva, 1972) |
| Barents Sea waters | Spring bloom period | — | 1,350.0 | — | 40 | Fedorov and Ermachenko (1961) |
| Atlantic waters | Spring bloom period | | 1,120 | — | 29 | Fedorov and Ermachenko (1961) |
| Southwestern, Central Nordcap current branch | Spring (0–100) | 36.9 | — | — | 16 | Ryzhov and Syuzeva (1974) |
| Southeastern part | Spring (0–100) | 334.6 | — | — | 48 | Ryzhov and Syuzeva (1974) |
| Southeastern part | Summer (0–100) | 140.5–433.4 | — | — | 170 | Ryzhov and Syuzeva (1974) |
| Southwestern part | | 1.4[c] | — | — | 56 | Bliznichenko et al. (1976) |
| The sea as a whole | Phytoplankton bloom season | — | — | — | 55–180[c] | Marti (1971) |

[a]Annual primary productivity estimated on the basis of data presented in the references cited.
[b]Authors' data given in the last column.
[c]Biomass mg/l.

aging 156 mg C/m$^2$ or 77 million tons per year. These results compared to the phytoplankton production reported from other seas of the USSR (Maksimova, 1971; Sorokin, 1964; Kondratieva and Finenko, 1971) show that the primary production of the Barents Sea per 1 m$^2$ is two and a half times less than that of the Baltic Sea and two and six times less than that of the Black Sea and the Sea of Azov, respectively. However, the total amount of carbon resulting from photosynthesis in the Barents Sea is much higher than that of the southern seas because of the larger area of the Barents Sea.

### Phytobenthos

Multicellular algae form well-developed mats of differing width within the drained and the sublittoral zones of the Barents Sea. Their contribution to the coastal primary production is essential and this is a bioproduction feature of particular importance distinguishing the Barents Sea among other seas of the USSR. The floral composition of macrophytes and their production are controlled by the distribution and species composition of unicellular algae.

The general effect of abiotic and biotic environmental factors of the Barents Sea on bottom phytocenoses is responsible for the concentration of most macrophytes off Murman. The sublittoral area of the southern Barents Sea is zoned by fine green filamentous algae. Four principal species of fucus algae are common on the littoral zone. Below the fucus zone we find red algae while laminaria occur in the sublittoral zone (Kuznetsov, 1956; Blinova and Vozzhinskaya, 1971). Macrophytes occur at a depth greater than 15 m, but the width of the algal zone may reach 1.5 km.

At present, the total amount of bottom algae has been reliably estimated for most of the southern coast of the Barents Sea (Blinova and Vozzhinskaya, 1971), except for the southeast part of the sea, Novaya Zemlya, and Spitsbergen, which are dominated mainly by laminaria. Algae inhabiting the East Murman area account for 205.5 thousand tons; *Fucus vesiculosus* and *Ascophyllum nodosum* amount to 70 thousand tons and 60 thousand tons, respectively. The total annual macrophyte production can reach values as high as 215.5 thousand tons (of wet weight). Based on general algal mass (in the Barents Sea) of 600 thousand tons per year and P/B coefficient of 1.05 (Blinova, 1969), annual macrophyte production accounts for 630 thousand tons of wet weight or about 66 thousand tons of organic carbon per year. Macrophytes synthesize only a small part of OM as compared to the phytoplankton primary production; however, phytobenthos production per unit area in coastal regions (12.0–23.1 kg/m$^2$) can exceed phytoplankton production several hundred times. Therefore, in the coastal areas of the sea, macrophytobenthos is of prime importance for OM production (this must be considered in the general OM balance of the sea).

According to Kuznetsov (1956), about 35 percent of the production of coastal macrophytes is lost as a result of grazing by phytophagous invertebrates forming heterotrophic links of the food chain. About 21 percent of the algal biomass decomposes; consequently, about 13 thousand tons of organic carbon is discharged annually into the seawater and bottom sediments of the coastal zone.

### Bacterioplankton

The bacterial flora is the principal agent in biotransformation of dissolved and particulate OM in the upper layer of bottom sediments. Most microorganisms are hemotrophs, bacterial types with phototrophic exchange, forming only a small proportion of microbial population in the ocean (Kriss, 1976). Hemotrophs are much more common than photosynthesizing organisms but their distribution and production are controlled by abiotic factors and bioproduction parameters of the total ecosystem.

Bacterioplankton of the Barents Sea is relatively poor, chiefly due to the low ambient temperature. Bacteria composing the periphyton contain 64 percent coccoid rod bacteria and the remaining 36 percent are also cocco-like organisms (Rossova, 1971). A peculiar feature of the high-latitude ocean microflora is a sharp increase in the proportion of rod bacteria in the macrophyte composition (Kriss, 1976).

The number and production characteristics of bacterioplankton in the southwest Barents Sea water and adjacent bays is fairly well known (Rossova, 1977; Bliznichenko et al., 1976). There is a general trend of spatial and temporal variability in the microperiphyton quantitative parameters. The bacterial biomass increases northward. In the cold Bear Island current it is three times higher than in the warm Nordcap current (58.2 $mg/m^3$). In the southwestern part of the sea the highest values of bacterial biomass are reported from the Bear Island shallow-water areas (167.7–315.8 $mg/m^3$).

In the open sea, bacteria are concentrated mainly in the upper 0–50 m layer (averaging 112.7 $mg/m^3$). The amount decreases with depth and the 100–200 m layer contains half of the upper layer biomass. Abiotic factors such as temperature and salinity do not affect the vertical distribution of bacteria in seawater, but similarly to the lateral and seasonal distribution, it is controlled mainly by the intensity of OM production as well as by phytoplankton and zooplankton.

Bacterial production in the Barents Sea water varies between 0.3 and 11.9 $mgC/m^3$ (Rossova, 1977; Bliznichenko et al., 1976). In some of the offshore regions there is not only a seasonal variation but also a yearly variation in production.

The amount of bacteria in the bottom sediments of the Barents Sea is determined by the content of organic carbon and soil adsorption. In littoral oozes it ranges between $50.8 \cdot 10^5$ and $111.1 \cdot 10^5$ per 1 g of dry sediment (Ruban, 1961). The number of bacteria decreases sharply with depth (in cores) and depends on grain size distribution of sediments.

Bacterial production contributes only slightly to the general production system of seas and oceans as compared to the phytoplankton primary production. Hemosynthesis should be considered as a primary OM source only when allochthonous OM or volcanic gases or products of hot springs are involved. Chemosythesizing microorganisms use mainly energy of compounds formed as a result of decomposition of photosynthesized OM and, therefore, the bacterial synthesis may be considered as a secondary process. At present the allochthonous OM component does not allow us to estimate bacterial production; however, a priori one may suggest a small proportion of the bacterial production of the sea as a whole.

### *Zooplankton*

The main consumer of phytoplankton in the Barents Sea is *Calanus finmarchicus,* which accounts for 98 percent of the zooplankton biomass. Macroplankton is composed of abundant euphausides, dominated by *Thysanoessa inermis* and *T. raschii,* belonging to Arctic–boreal neritic species (Zelikman, 1964). The Atlantic–boreal euphauside *Meganyctyphanes norvegica* is found in the Barents Sea in warmer years, brought by the Gulf Stream. In general, east of the Kola meridian, autochthonous communities account for 70 percent of the planktonic population (Zelikman, 1977).

Copepods and euphausides are phytoplankton consumers and therefore their reproduction cycle is closely related to the dynamics of autotrophous OM production. These zooplankters winter at a depth of about 150 m and rise to the upper layers in the early biological spring during phytoplankton bloom. The time of calanus and euphausides rise to the surface is determined by water illumination (Zelikman, 1964; Fomin, 1978; Digby, 1961), which is a major indicator of nutrient availability at the surface.

Zooplankton production in the Barents Sea follows mainly the same trend as the primary production. Therefore, the production of zooplankton in the Arctic and coastal waters is six times greater than in the Atlantic Ocean. If we assume that the mean annual production in the Barents Sea is 45 tons per km$^2$, the synthesis of organic carbon amounts to 6.3 million tons per year. This is one order of magnitude lower than phytoplankton primary production.

### *Zoobenthos*

The bottom-dwelling fauna is the best known of the Barents Sea ecosystems (Filatova, 1938; Brotskaya and Zenkevich, 1939; Turpaeva, 1953, 1954; Kuznetsov, 1970, 1980).

The most important factors controlling the benthos trophic structure are food availability and abundance, composition of bottom sediments, and bottom water dynamics. Based on food source and mode of nutrition, bottom dwellers can be subdivided into the following feeding categories: immobile and mobile sestonophages, detritus eaters, and carnivores indiscriminately gathering and ingesting soil.

The biomass of bottom-dwelling organisms of the Barents Sea averages 78–112 g/m$^3$, which is equivalent to 110–160 million tons, wet weight, or 7.7–11.0 million tons organic carbon. The annual zoobenthos production can be estimated using a mean value of 0.4 for P/B coefficient, accepted for the World Ocean (Bogorov, 1971; Antipova, 1978). It amounts to about 3.1–4.4 million tons of organic carbon per year. This value is slightly lower compared to that of zooplankton, because bottom-dwelling organisms have a longer life span and show an inverse relation to the rate of OM turnover (Golikov, 1970).

Of interest is the production of the main links in the Barents Sea ecosystem as compared to that of the World Ocean (Table 28-2).

The values mentioned above suggest that the Barents Sea is responsible for the synthesis of 37 percent organic carbon production, created by phytoplankton and phytobenthos in the World Ocean, indicating the high productivity of the Barents Sea. The bottom-dwelling fauna of the Barents Sea accounts for the highest percentage of the total benthos production of the World Ocean. This is likely to be due to the relatively high phytoplankton production, short food chains in the water masses, low mean annual temperatures of OM decomposition, and, finally, to the long life span of benthic organisms.

## Sediments

Morphostructurally, the Barents Sea is mainly represented by shelf plateaus cov-

Table 28-2. Production Parameters of the Ecosystem in the Barents Sea and World Ocean

| *Source of Organic Carbon* | *Production $10^6$ Tons of Organic Carbon in the Barents Sea* | *Total production in the World Ocean* | *Production in the Barents Sea, % of Total Value* |
|---|---|---|---|
| Phytoplankton | ~80.0 | 30.0[a] | 0.3 |
| Phytobenthos | 0.066 | 0.1[b] | 0.07 |
| Zooplankton | 5.4 | 2.0[b] | 0.3 |
| Zoobenthos | 7.7–11.0 | 0.6[b] | 1.3–1.8 |

[a]Romankevich, 1984
[b]Koblents-Mishke and Vedernikov, 1977

ered by Cenozoic unconsolidated and poorly consolidated deposits of various thicknesses underlain by Mesozoic and Upper Paleozoic deposits. The plateaus are dissected by shelf and infra-shelf trenches. These trenches are geomorphological markers of fracture zones (Polkin, Romanovich, and Zatsepin, 1975; Dibner, 1978).

During the Holocene, as a result of low sediment supply, the Barents Sea, unlike other shelf seas of Eurasia, was characterized by fairly slow sedimentation rates (less than 30 mm per 1000 yr). The abrasion of coasts and floor, along with the silt load, have supplied most of the sediments. These sources have supplied 59.0, 7.7, and 8.6 million tons/yr, respectively (totaling 75.3 million tons/yr) (Suzdalsky, 1974). The Barents Sea waters owe their transparency and low sedimentation to a fairly small supply of POM from land.

There are few descriptions of Quaternary successions (Suzdalsky, 1976); opinions on ''ancient'' clay genesis differ. Most workers assign them to glacial formations (Klenova, 1961; Dibner, 1978). Apart from typical till, cores contain clay sediments characterized by ''mottled'' facies, suggesting a glacial-marine origin. Based on lithology, the deposits may be divided into an upper Holocene sequence, consisting mainly of poorly consolidated sediments, 0 to 1.0 m thick, underlain by lower Pliocene–Pleistocene sequences, composed chiefly of the so-called ''ancient'' clays, 2–6 km thick. The lower part of the succession is composed of gray-blue argillaceous silts. They contain a high proportion of clay fraction.

## Dissolved and Particulate Organic Matter

Dissolved OM (DOM) is represented by compounds in solution (particle size <0.001 mkm), colloids, and forming particles of 0.001–0.1 mkm as well as coarser particles, passing through filters with 0.45–1 mkm pore size.

DOM is formed by bacterial decomposition products and autholysis of plant and animal cells, their inlife products and OM supplied from land. Both the absolute mass of DOM and its residence time in ocean are much higher than those of particulate (POM) (Romankevich, 1984).

### *Carbon of Dissolved Organic Matter (C-DOM)*

The dissolved organic carbon distribution (C-DOM) in August–September 1976, along the Murmansk–Franz Josef Land section, is shown in Figure 28-2. The C-DOM content along the section ranges between 1.0 and 4.8 mgC-DOM/1 (2.2 average). The highest concentrations of the C-DOM are reported from the upper water layers in the southern part of the section. On the average, the C-DOM content decreases both with depth and toward Franz Josef Land.

The Gulf Stream water is characterized by high concentrations of C-DOM. The cold polar current waters entering the Barents Sea from the northwest, south of Franz Josef Land, form a distinct tongue of low C-DOM concentrations. The content of DOM increases markedly near the coasts (Rybachy Peninsula) and in bottom water layers. The Gulf Stream waters bring a large proportion of anthropogenic OM from the North Atlantic into the Barents Sea (Levy and Walton, 1976; Smith, 1976). In general, the C-DOM content in the southern part of the section is twice as high as weighted average of C-DOM concentration in the Atlantic; they are 1.8 and 1.6 mgC/l for 0–100 m and 100–200 m layers (Romankevich, 1984). For the Barents Sea as a whole, the average C-DOM concentration is nearly twice as high as in the 0–200 m layer of the World Ocean, mainly due to high bioproductivity.

### *Carbon of the Particulate Organic Matter (C-POM)*

Particulate OM is more labile as compared to dissolved OM distribution; it is more in-

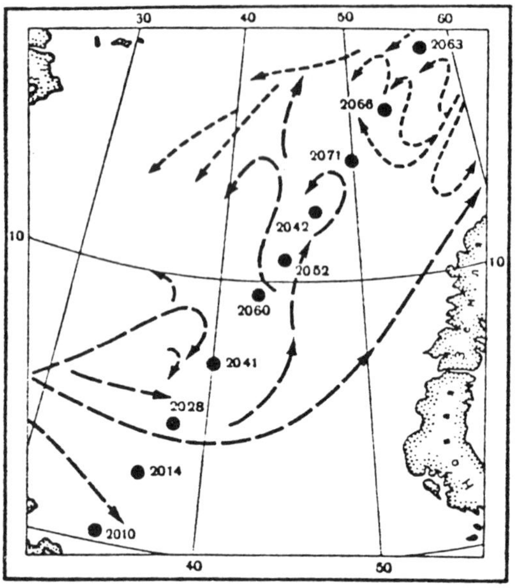


Figure 28-2. Dissolved organic carbon distribution along the Murmansk–Franz Josef Land section.

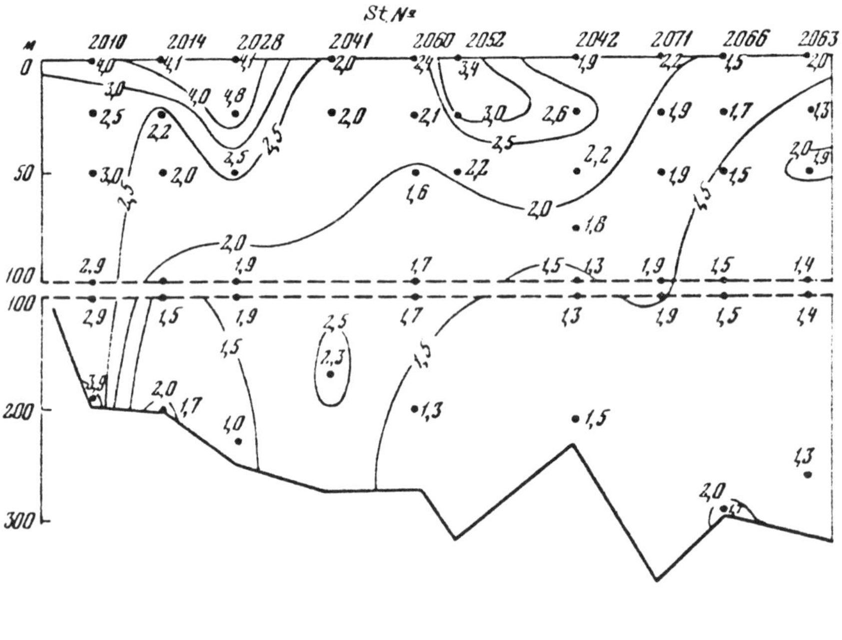


timately related to authotrophic biosynthetic processes and also to land efflux of allochthonous OM. Thus, its temporal and spatial distribution is more heterogeneous than that of C-DOM.

The Murmansk–Franz Josef Land section showed C-POM variation from 19 to 314 mkgC/l, averaging 87 mkgC/l in August–September 1976; these values considerably exceed the average of C-POM concentrations in the 0–200 m layer of the World Ocean.

In the coastal regions of the sea, mean yearly C-POM concentrations are even higher, for example, averaging 250 mkg of C-POM/l in Dalnozelenetskaya Bay (Myasnikov, 1976). Maximum C-POM content in dry sediments is positively correlatable with maxima phytoplankton abundance and river discharge.

Seasonal changes of C-POM are also a peculiar feature of open sea areas. In winter, mean C-POM concentrations are expected to be lower than those in autumn, when phytoplankton growth commonly shows a second maximum.

In the Barents Sea the distribution of C-POM and C-DOM are similar, that is, the C-POM concentrations decrease from south to north and from the surface to bottom (Fig. 28-3). However, unlike C-POM, the distribution of organic carbon is more complex.

The C-POM content amounts to 1.1–12.6 percent of the C-DOM content. The weight average C-POM content in organic carbon of the seawater (sum of C-DOM + C-POM) decreases with depth (Table 28-3), reaching a minimum value in the central part of the section considered, which is consistent with regular trends of phytoplankton distribution. The C-POM content in organic carbon of the seawater averages 3.5 percent in this section. Actually this value is consistent with the C-POM/C-DOM ratio for the Baltic Sea (about 3.8) (Ehrchardt, 1969), but higher than values of the Sea of Japan (1.0–2.4 percent) (Lyutsarev and Romankevich, 1975) and of open areas of the World Ocean.

The high mean concentrations of particulate and dissolved OM and POM/DOM ratio in the Barents Sea may be attributed to two main reasons. First, despite a short phytoplankton bloom period, a large amount of OM is synthesized in the Barents Sea. Second, the OM transformation rate here is lower than in marginal and inland low-latitude seas. In the Arctic seas the change in kinetics of OM transformation is mainly due to low water temperature, which impairs the activity of most microorganisms (Petrov, 1978) and, hence, the process of OM biodegradation.

The fact that OM degradation kinetics is mainly temperature dependent is evidenced by labile, newly synthesized organic compounds. It was shown (for example, Bikbulatov and Bikbulatov, 1979) that the rate of OM aerobic degradation is one-half of the initial rate with a 10°C temperature decrease. This rate means that the degradation rate of dead phytoplankton cells in the Arctic seas is one-half of

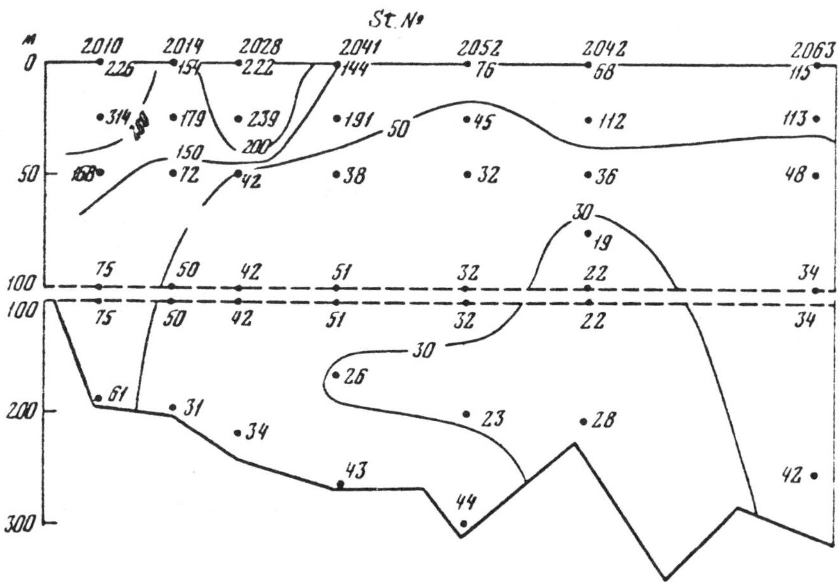


**Figure 28-3.** Particulate organic carbon distribution along the Murmansk–Franz Josef Land section.

Table 28-3. Weighted Average Concentrations of Dissolved and Particulate Organic Carbon at the Section Murmansk-Franz Josef Land

| Water Depth, m | St. 2010 | | | St. 2014 | | | St. 2028 | | |
|---|---|---|---|---|---|---|---|---|---|
| | $C_{DOM}$, mg C/l | $C_{POM}$, mkg C/l | $C_{POM}/C_{DOM}$, % | $C_{DOM}$, mg C/l | $C_{POM}$, mkg C/l | $C_{POM}/C_{DOM}$, % | $C_{DOM}$, mk C/l | $C_{POM}$, mkg C/l | $C_{POM}/C_{DOM}$, % |
| 0–50 | 3.0 | 256 | 8.5 | 2.6 | 146 | 5.6 | 4.0 | 186 | 4.7 |
| 50–100 | 3.0 | 122 | 4.1 | 1.8 | 61 | 3.4 | 2.2 | 42 | 1.9 |
| 100–200 | 3.4 | 68 | 2.0 | 1.6 | 41 | 2.6 | 1.5 | 38 | 2.5 |
| 0–200 | 3.2 | 131 | 4.1 | 1.9 | 72 | 3.8 | 2.2 | 88 | 4.0 |
| Rate of $C_{POM}$ decrease with depth in layer 0–50 m ($K_{50}\cdot 10^2$) | | 2.5 | | | 3.6 | | | 6.9 | |
| The same in layer 0–200 m ($K_{200}\cdot 10^3$) | | 9.9 | | | 10.0 | | | 10.0 | |

that in the southern seas, and affects the transformation kinetics of POM as a whole.

The variations in the POM transformation rate occur within the Barents Sea as well. The vertical decrease in C-POM concentration within the upper 300 m layer, beyond the oligotrophic regions, is shown by the equation (Nakajima, 1973)

$$C_i = C_{io} \cdot e - k\,(Z_i - Z_{io})$$

where $C_{io}$ is the highest concentration at a depth of $Z_{io}$ in the interval ($i$); $C_i$ concentration is at a depth of $Z_i$.

Using the equation one can estimate the rate of vertical decrease of C-POM concentration ($K_i$) along the Murmansk–Franz Josef Land section. The obtained $K$ values for the 50 m layer average $3.9\cdot 10^{-2}$ at the upper limit and account for $K_i(0.2–4)$, $10^{-2}$, the value reported by Nakajima (1973) for the northwestern Pacific Ocean. Below the photic layer, the rate of C-POM concentration decreases considerably. Both $K_{50}$ and $K_{200}$ have a direct relationship to the C-POM maximum concentration in the surface water layers despite a considerable scatter (Fig. 28-4). Therefore, in the surface water the maximum C-POM concentrations are formed chiefly by labile organic compounds, indicating high biodegradation rates. Station 2010, located in the coastal zone, is an exception. Here $K_{50}$ and $K_{200}$ are much lower than those expected, using maximum C-POM values in the surface water layers. Such an anomaly may be attributed to local influx of allochthonous OM, containing a small proportion of easily degradating compounds. It also may be due to degradation-resistant anthropogenic OM, supplied by the Gulf Stream.

The $K_{200}$ distribution implies a low rate of OM concentration decrease from surface to bottom in the northern part as compared to the southern sector of the section; the difference in temperature is probably the reason. However, POM formed by

Table 28-3.
Continued

| St. 2041 | | | St. 2052 | | | St. 2042 | | | St. 2063 | | |
|---|---|---|---|---|---|---|---|---|---|---|---|
| $C_{DOM}$, mg C/l | $C_{POM}$, mkg C/l | $C_{POM}/C_{DOM}$, % | $C_{DOM}$, mg C/l | $C_{POM}$, mkg C/l | $C_{POM}/C_{DOM}$, % | $C_{DOM}$, mg C/l | $C_{POM}$, mkg C/l | $C_{POM}/C_{DOM}$, % | $C_{DOM}$, mg C/l | $C_{POM}$, mkg C/l | $C_{POM}/C_{DOM}$, % |
| — | 142 | — | — | 50 | — | 2.3 | 82 | 3.6 | 1.6 | 97 | 6.1 |
| — | 44 | — | — | 32 | — | 1.8 | 24 | 1.3 | 1.6 | 41 | 2.6 |
| — | 36 | — | — | 31 | — | 1.4 | 25 | 1.8 | 1.3 | 38 | 2.9 |
| — | 57 | — | — | 34 | — | 1.7 | 38 | 2.2 | 1.4 | 50 | 3.6 |
| | 6.4 | — | — | 1.7 | | | 4.5 | | | 1.7 | |
| | 6.1 | | | 5.4 | | | 7.5 | | | 4.2 | |

melting ice is more resistant than the newly synthesized OM in this part of the sea.

A comparison of the maximum concentrations of the surface water layers, with the $K_{200}$ and organic carbon content of the upper layer of bottom sediments, shows a correlation between these parameters, namely, the lower the $K_{200}$, the higher the carbon concentration in sediments (Fig. 28-5). Based on these data, the average concentration for C-DOM and C-POM of

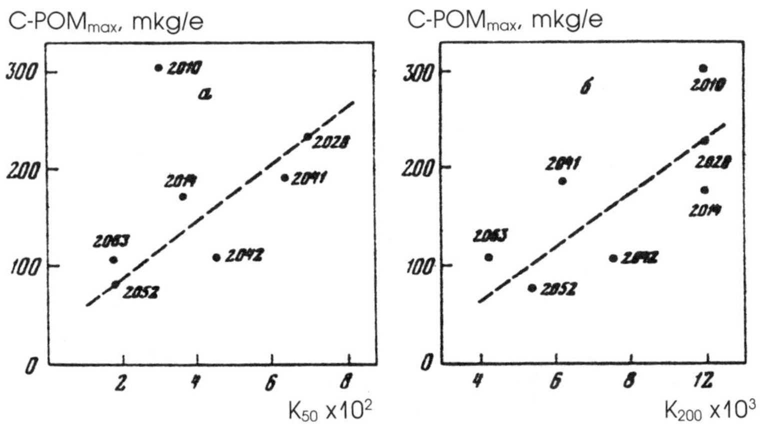


Figure 28-4. The relationship between the rate of C-POM concentration decrease with depth and its maximum concentration in the surface water layers.

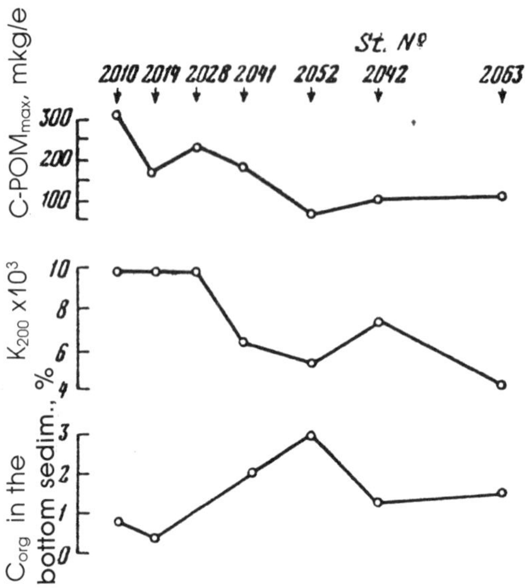


Figure 28-5. Comparison of maximum concentrations in the surface water layers with rate of C-POM concentration decreasing with depth (0–200 m) and the sediment organic carbon concentration along the Murmansk–Franz Josef Land section.

2.2 mg/l and 87 mkg/l, respectively, their amount in the Barents Sea, was estimated to be 320 million tons and 10 million tons.

### Composition of OM in the Barents Sea

Information concerning the OM composition of the Barents Sea water is limited to determinations of amino acids from the coastal area and of lipids on the Murmansk–Franz Josef Land.

Both combined and free amino acids were determined in organic compound hydrolysates that were precipitated from the filtered seawater on zirconium phosphates (Zlobin, Perlyuk, and Orlova, 1975). Water samples taken from Dalnozelenetskaya Bay in May 1972 showed amino acid concentrations ranging between 1.28 and 3.41 (2.62 average) mg/l, that is, higher than those in the water. In the coastal parts of the Barents Sea C-DOM concentration is 4 mg/l and the sum of dissolved amino acids averages about 50 percent OM (Romankevich, 1984).

The marine waters of Dalnozelenetskaya Bay are dominated by asparagine acid, serine, and glycine whose content reaches 30 percent; glutaminic acid and treonine make up 34 percent and metonine and valine constitute 20 percent of the total amino acids. Glutamine, alanine, leucine, and isoleucine were also found in the analyzed samples. Unicellular and multicellular algae provide the main source for amino acids in Dalnozelenetskaya Bay water. Among macrophytes, *Fucus inflatus* have an amino acid composition similar to that of the seawater (Zlobin, Perlyuk, and Orlova, 1975).

Dissolved and particulate lipids along the Murmansk–Franz Josef Land section were quantitatively determined using the carbon of lipids in aliquats extracted from filtered seawater (1.5–3 l, pH 2) and suspension. The suspension was collected on "Whatman" GF/C fiber-glass filters by filtration of 20–25 l of seawater. The same techniques were used for comparing the lipid distribution in different OM forms, extraction and determination of their concentration in seawater, in suspension, and in bottom sediments (Belyaeva, 1980*a,b*).

Concentrations of dissolved lipids along the Murmansk–Franz Josef Land section ranges betwen 200 and 600 mkg c/l, averaging 400 mkg c/l (Table 28-4). Unlike the decrease of dissolved carbon toward Franz Josef Land, the absolute concentration of lipids increases in this direction, which is a distinctive feature of dissolved lipid distribution. The vertical distribution of the latter in the Barents Sea water also differs from that of C-DOM. In the dissolved carbon, concentration decreases with depth; the dissolved lipid content in a few cases increases toward the seafloor surface. This fact in particular holds true for the northern part of the section, which intersects water masses of various origin. The surface water layer (about 30–50 m) here is formed by the sub-Arctic water masses, while the lower layer is composed of the transformed Atlantic water with typically

Table 28-4. Lipid Content in a Water Mass of the Barents Sea

| Station Number | Horizon m | Lipid Suspension mkg C/l | Lipid Suspension % $C_{POM}$ | Dissolved Lipids mkg C/l | Dissolved Lipids % $C_{DOM}$ |
|---|---|---|---|---|---|
| 2,010 | 0 | 25 | 11.2 | — | — |
| | 25 | 22 | 7.9 | — | — |
| | 50 | 31 | 18.2 | — | — |
| | 100 | 17 | 22.4 | — | — |
| | 190 | 18 | 30.8 | — | — |
| 2,014 | 0 | 53 | 34.2 | 200 | 4.9 |
| | 25 | 36 | 20.1 | 217 | 9.9 |
| | 50 | 31 | 43.0 | — | — |
| | 100 | 8 | 16.0 | — | — |
| | 200 | 9 | 29.0 | — | — |
| 2,028 | 0 | 25 | 11.3 | 370 | 9.0 |
| | 25 | 19 | 7.9 | 611 | 12.7 |
| | 50 | 7 | 16.5 | — | — |
| | 100 | 8 | 19.1 | 343 | 18.0 |
| | 220 | 6 | 17.5 | 354 | 35.4 |
| 2,041 | 0 | 48 | 33.4 | 387 | 19.4 |
| | 25 | 48 | 25.0 | — | — |
| | 50 | 24 | 63.2 | — | — |
| | 100 | 30 | 58.6 | 395 | — |
| | 170 | 33 | — | — | — |
| | 270 | 25 | 58.7 | 426 | — |
| 2,060 | 0 | — | — | 522 | 21.8 |
| | 100 | — | — | 532 | 31.3 |
| | 200 | — | — | 340 | 26.2 |
| 2,071 | 0 | — | — | 610 | 27.7 |
| | 25 | — | — | 455 | 23.9 |
| | 50 | — | — | 661 | 34.8 |

low C-DOM content and maximum concentration of dissolved lipids as compared to that of the surface water.

In the Barents Sea the distribution of dissolved lipids within the C-DOM composition shows a distinct pattern not reported earlier for seawater from other areas of the World Ocean, namely, inverse correlation between the C-DOM concentration in the water mass and lipid carbon content in the carbon of dissolved OM (Fig. 28-6).

Such an inverse relationship previously found in bottom sediments characterizing thoroughly transformed OM is known as the Uspensky–Vassoevich relationship. It shows an increase in lipid-bitumoid (or stable organic compounds) proportion in the OM of bottom sediments, as labile OM components are being destroyed. This relationship was first established for the Barents Sea waters and it may indicate a high degree of DOM transformation throughout the entire offshore area. Most likely it may be attributed to the influx of transformed "old" water of the Gulf Stream with compositionally conservative OM into the Barents Sea. The influence of river-discharged OM may be excluded as a major source inasmuch as the volume of river water and contribution to OM is small.

Allochthonous strongly transformed OM is partially supplied by melting sea ice

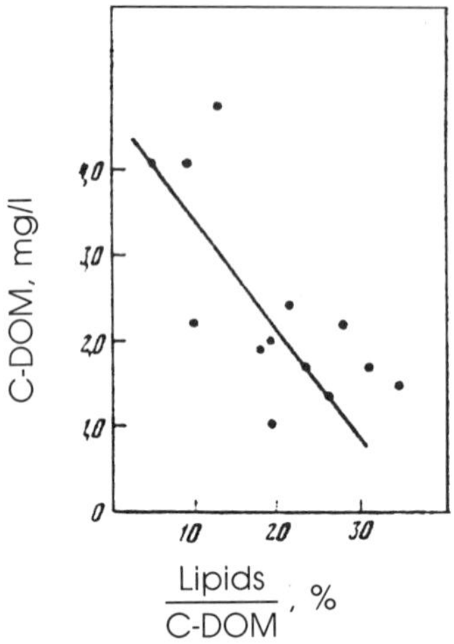


Figure 28-6. The relationship between the C-DOM concentration in the water body and lipid carbon content in the carbon of the dissolved OM.

or by eolian transport. Owing to eolian transport, OM is supplied as particulate matter and the compounds are transformed into dissolved form, only as a result of POM transformation, which in turn increases the age of seawater OM.

The content of suspended lipids along the Murmansk–Franz Josef Land section ranges from 6 to 53 mkg C/l (averages 23 mkg C/l). This amount is twice as much as the mean values for the suspension in the southeastern Pacific (Belyaeva and Romankevich, 1976). Absolute concentrations of lipids in suspension decreases with depth (Table 28-4). The higher the lipid concentration in surface water layers, the more marked the decrease, and the process is the most rapid in the upper 0–50 m. The vertical gradient of lipid concentrations decreases below the photic layer, possibly due both to a high rate of unstable POM degradation (including complex lipids of the suspension) in the surface water layers and to low concentration of living OM.

The lipid carbon of the suspension constitutes 7.9–63.2 percent of C-POM ($K_{lip}$), with increased proportion of OM of deep-water suspension, as compared to that of the surface layers. The importance of stable lipid compounds containing mainly simple lipids—hydrocarbons, fatty acids, alcohols—in POM, from a layer deeper than 50 m, is illustrated in Figure 28-7. The illustration shows different $K_{lip}$ for the same absolute concentration of particulate lipids in 0–50 m and in deeper layers, $K_{lip}$ increasing with water depth. The increase of lipid content in POM with depth reflects regular features in OM transformation in the course of sedimentogenesis. There is a decrease in the number of labile compounds and an increase in components most resistant to microbial and geochemical decomposition, such as simple lipids, which accumulated selectively in the OM.

The atomic H/C ratio, averaging 2.74 (6 samples) and 2.07 (13 samples), respectively, for the suspension and water, was determined using the elemental analysis (C, H, N Kovo-analyzer, CSSR) of lipids of seawater and the suspended matter of the Barents Sea. This is an indication of higher content of oxygen-bearing compounds in dissolved lipids as compared to lipids of the suspension and is consistent with earlier data (Belyaeva and Romankevich,

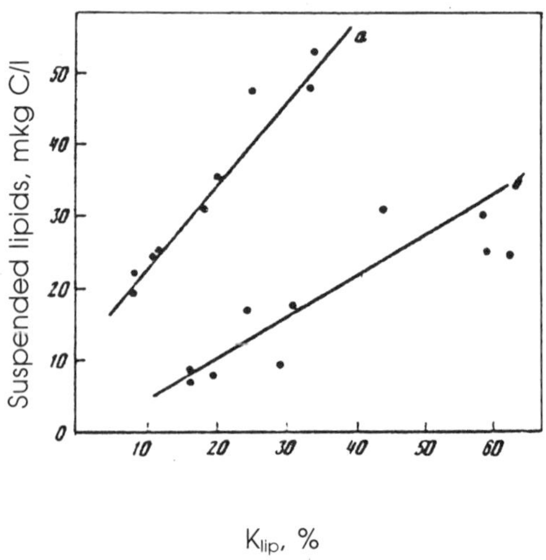


Figure 28-7. The relationship between the lipid carbon concentration in the suspension and lipid content in POM (Klip) in the layer 0–50 m (*a*) and deeper layers (*b*).

1976). Such lipid groups as compound ethers occur mainly in dissolved form.

In the upper 0–5 cm layer of bottom sediments along the Murmansk-Franz Josef Land section, the content of lipids using cold extraction with chloroform-methanol mixture in 2:1 ratio is 0.0137–0.0480 mg C/g, dry weight. The lipid content in sediments depends on weighted average concentrations in suspension in a water mass from the surface to the seafloor (Table 28-5). The absence of direct correlation between these values may be due to different degrees of OM transformation at the water–sediment interface, where the change in OM composition is as active as in the seawater photic layer.

Unlike the suspended matter, the lipid content in the OM (of carbon) of bottom sediments decreases by a factor 6–30 and is inversely proportional to the organic carbon concentration in sediments. This decrease confirms the global occurrence of the Uspensky–Vassoevich relationship. In the upper sediment layer (0–5 cm) OM diagenetic transformations are modest; this may be attributed to the fact that the relationship is underway during sedimentogenesis. In this case, the higher the rate of decrease in the OM concentration in the suspended matter, the higher the lipid content in the OM of the upper sediment layer.

In summary, the data indicate that the longer the residence time of OM in the forms studied (solution, suspension, sediments), the higher the degree of its transformation; the concentration of all OM components decreases and is the lowest for the lipid compounds, which results in their selective presence in OM.

The Uspensky–Vassoevich relationship holds true not only for bottom sediments but for seawater where the residence time of OM is 1,500–3,000 yr. POM is one of the most labile forms in which organic compounds occur in seas and oceans and the above relationship does not apply to them. The latter is a distinctive feature of stable OM geopolymers.

## Organic Matter in Bottom Sediments

Geochemical studies of organic matter in Barents Sea shelf bottom sediments commenced in 1970, and three main aspects were investigated. First, the relationship of OM content versus the morphostructural framework of the offshore area was examined; second, its relationship to the original organic material composing OM and finally its relationship to the environment and intensity of early diagenetic transformations was studied.

The average content of organic carbon

**Table 28-5. Lipid Content in Suspension in the Upper Layer of Bottom Sediments through the Murmansk–Franz Josef Land Section**

| *Station Number and Depth* | *Lipid Concentration in Suspension*[a] *mkg C/l* | *Content in a Dry Sediment*[b] *Organic Carbon, %* | *Lipids mg C/g Dry Matter* | *Organic Lipid/ Organic Carbon, %* |
|---|---|---|---|---|
| 2,010 190 m | 21 | 0.79 | 0.040 | 5.06 |
| 2,014 200 m | 18 | 0.46 | 0.014 | 3.05 |
| 2,041 270 m | 32 | 1.68 | 0.048 | 2.86 |

[a]Weighted average concentrations in a layer O-floor.
[b]Sediment layer 0–5 cm.

in the upper sediment layer of the Barents Sea is approximately 1.28 percent, which is higher than values obtained in the White (1.14 percent), Kara (0.83 percent), East Siberian (0.79 percent), and Chukchi (0.98 percent) seas (Gorshkova, 1975). Thus, owing to high primary productivity (compared to other northern seas), the Barents Sea has the highest content of OM in sediments.

Geochemical studies of bottom sediments of the Barents Sea carried out at stations and sections shown in Figure 28-8 suggest that the content of organic carbon in bottom sediments depends on their lithology and bottom topography. The organic carbon content in argillaceous sediments of shelf plateaus is generally higher than in sandy deposits (1.07 percent versus 0.42 percent), but less than in silty sediments of trenches (which average 1.35 percent). The concentration of organic carbon in argillaceous sediments in trenches ranges from 1.0 to 1.4 percent (Fig. 28-9). In clay sediments in plains, the distribution of organic carbon is bimodal and it shows two peaks of 0.6–0.8 percent and 1.2–1.4 percent, respectively. Sandy sedi-

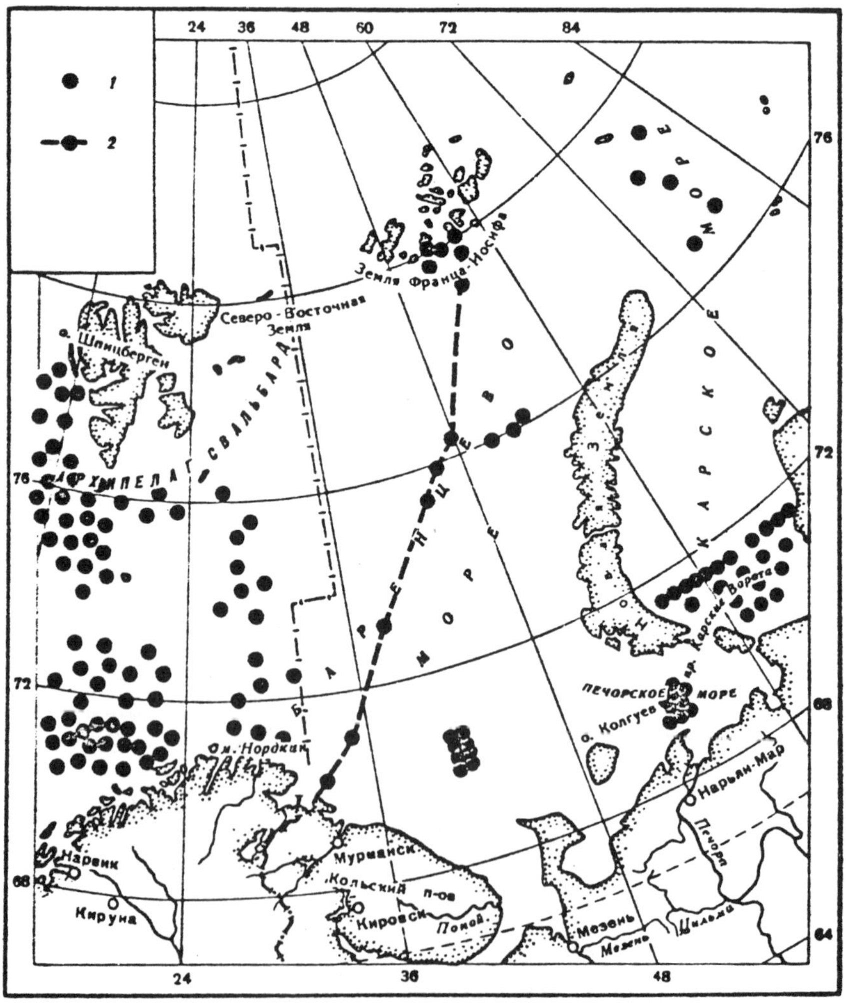


Figure 28-8. Studied regions in the Barents–Kara shelf. 1, Sample sites; 2, profile.

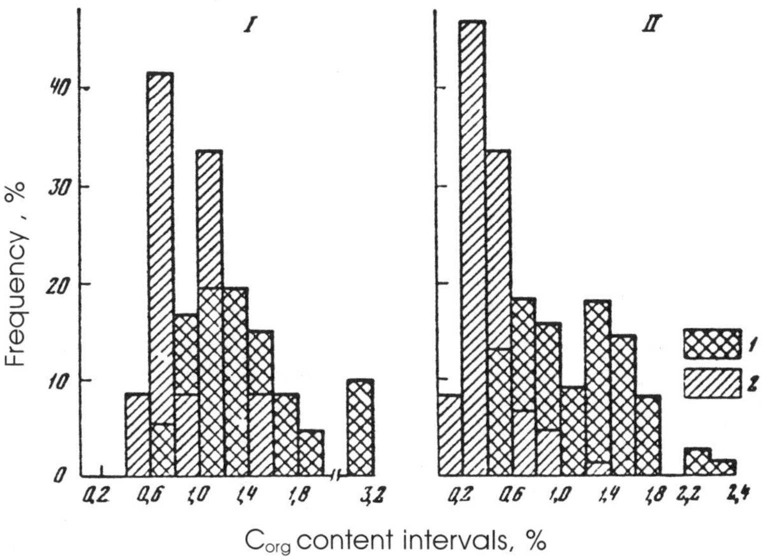


Figure 28-9. Histogram of organic carbon distribution in the sediments of shelf trenches (I) and shelf plateaus (II) of the Barents Sea. 1, Clayey sediments; 2, silty sediments.

ments in trenches contain more organic carbon than their counterparts in plains (0.6–0.8 percent as compared to 0.2–0.4 percent) (*Biogeochimia Organicheskogo Veshchestva Arkticheskich Morei,* 1982).

The distribution of organic carbon is indicative of the importance of OM of terrigenous genesis. For example, in the Kanin-Kolguev part of the offshore area where evacuated and precipitated particulate matter constitutes 25–30 percent of the runoff of the Pechora River and of the rivers of Murmansk (Klenova, 1961), the organic carbon content in bottom sediments is <0.5 percent, that is, two to three times less than that of recent sediments of the Pechora Sea and Kola shelf. Furthermore, off Spitsbergen where coal-bearing strata are very common, the average content of organic carbon is fairly high (1.88 percent) and drops to 0.92 percent toward Scandinavia where the deposits are underlain mainly by igneous rocks depleted in OM. Consequently, the amount of OM resulting from diagenesis, used for reduction of reactable forms of iron and manganese and for the formation of a gaseous phase, changes at the transition from the Holocene to Pliocene–Pleistocene sediments.

Judging from the iron forms in bottom sediment cores along the profile from Rybachii Peninsula to Franz Josef Land, the environment changes from oxidizing to strongly reducing ($Fe^2/Fe^3$, 0.5–3.0). The change in the content and composition in cores taken from different morphostructural zones of the Barents Sea and other Arctic seas suggests intense oxidizing-reducing processes. Estimates show that the discharge of organic carbon in the Barents Sea during anaerobic stages of sediment diagenesis averages 7–23 percent of organic carbon. Diagenetic losses of organic carbon at the same depth vary in cores taken from shelf plateaus as compared to those from shelf trenches. The latter exhibit very rarely oxidizing conditions; $Fe^2/Fe^3$ ratio is stable and close to 1.0, while on shelf plateaus it varies from 0.4–0.5 and 2.0–3.0 in the upper and lower horizons, respectively. Therefore, diagenetic organic carbon losses required for reduction of ferrous iron and manganese in cores taken from shelf trenches is higher than that on shelf plateaus, reaching 29–30 percent with respect to the original organic carbon. The highest present-day content of organic carbon (3.2 percent) was found in shelf trenches, and with the allowance made for the above-mentioned diagenetic organic carbon losses, it can be assumed that the original organic carbon content in sediments of these morphostructural offshore zones reached 10 percent (the scaling factor 1.82 corresponds to a level of OM transformation at silt stage; OM losses for life activity of saprodent microflora were not taken into account). Thus, the contents of buried OM in shelf trench sediments were similar to some of the Norwegian fjord sediments (Strom, 1955).

The OM genesis in sediments can be surmised from N content and the $C/N_{at}$ ratio. The latter increase proportionally to the terrigenous organic components. In general, the N content of the Barents Sea sediments varies from 0.05 to 0.2 percent. The $C/N_{at}$ ratio equals 10–40. These values suggest a mixed genesis of OM in sediments. Thus, the distribution of organic carbon in the recent sediments of the Barents Sea seems to be affected by phytoplankton, phytobenthos, and terrigenous OM. The relationship between them is governed by a peculiar polar lithogenesis; processes such as mechanical denudation and terrigenous deposition are the most important elements.

The study of OM content in the Barents Sea bottom sediments was based on the analysis of group composition (determination of bitumoids in chloroform and then in benzol and alcohol-benzol extracts, determination of humic acids) and on the analysis of molecular structure of bitumoids using IR and UV spectrometry; molecular composition of alkans was studied using gas liquid chromatography; amino acid composition was determined using quantitative paper chromatography. Free amino acids were extracted directly from sediments while combined acid from hydrolysate of humic acids formed by hydrolysis of the latter by 6N HCl solution at 100°C.

In general, the Barents Sea exhibits a direct relationship between an organic carbon content and bitumoids in sediments ($r = 0.67$; 93 samples). Bitumoids constitute 2–2.5 percent in the composition of sedimentary OM, and 0.4–0.6 percent of the chloroform bitumoid (Ach1) along the Rybachii–Franz Josef Land profile.

Elemental composition of bitumoids in bottom sediments containing less than 1 percent organic carbon is poor in carbon and rich in heteroelements. The value of $H/C_{at}$ is higher as well as the aliphatic level. In general, elemental and group compositions of bitumoids varies along the profile within wide ranges. For example, carbon content varies from 78 percent to 84 percent, $H/C_{at}$ from 1.36 to 1.83. The bitumoids of the Barents Sea differ from those of other offshore areas by their higher carbon content (Gorskaya, 1960; Bordovsky, 1974; Romankevich, 1977). The amount of oils in bitumoids of sediments varies from 17 percent to 57.5 percent. The content of hydrocarbons in sediments and in OM is unstable and changes with depth, being the highest in trench sediments. Maximum HC contents are $(2 \cdot 10^{2}) - (2 \cdot 10^{-2})$ percent and $(9 \cdot 10^{-2}) - (7 \cdot 10^{-1})$ percent in sediment and in OM, respectively. The ratio of methane–naphthene to aromatic HC varies within a wide range (0.4–2.3).

The comparison of n-alkanes distribution in bottom sediments in different parts of the Barents Sea (Figs. 28-10 and 28-11) implies a similar OM genesis. Alkanes are dominated by high-molecular C23–C33 components (accounting for 46.3–77 percent). Concentration maxima were found for odd n-alkanes $C_{23}$, $C_{25}$, $C_{27}$, $C_{29}$ typical of waxes of higher plants; they suggest the importance of terrigenous plant material within the OM of sediments. This conclusion is also confirmed by the presence of consolidated aromatic structures (naphtalene, phenanthrene, chrizen, benzopyrene, and especially, perilene) that may be

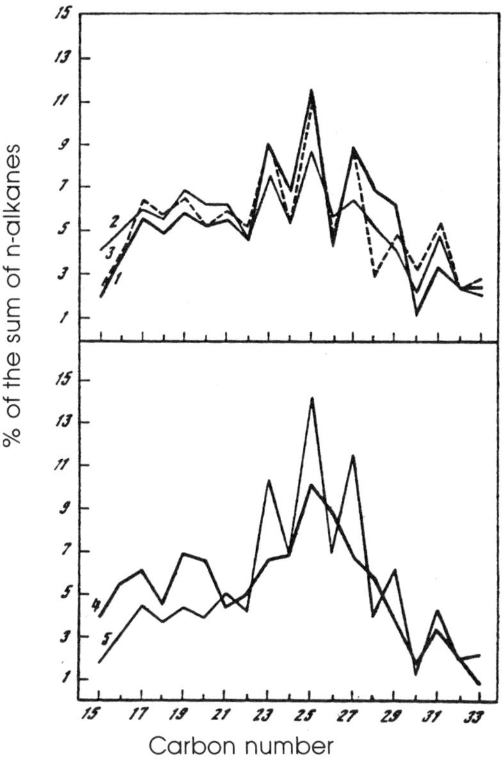


Figure 28-10. n-Alkanes distribution in the OM in the shelf trench sediments of the Barents Sea along the Rybachii Peninsula to Franz Josef Land section. Horizons from bottom surface, cm: 1, 100–135; 2, 275–300; 3, 420–450; 4, 338–362; 5, 630–670.

used as indicators of terrigenous OM remains of higher plants (Aizenschtadt, Baedecker, and Kaplan, 1973). Furthermore, the predominance of pristane over phytane suggests the occurrences of zooplankton remains in the sediments.

The odd to even alkanes ratio (CPI) in sediments through the Rybachii Peninsula–Franz Josef Land profile varies from 1.08 to 2.47 and decreases with increased degree of diagenesis. The extent of sediment humification (content of humic acid in the OM) shows a direct correlation with the organic carbon. The relationship is not obvious in trench sediments (0.38; 48 samples), but becomes more pronounced within the entire offshore area ($r$ = 0.56; 77 samples). Most humic acids in the samples on the Rybachii Peninsula–Franz Josef Land constitute about 1/3 OM. Their content varies from 12 percent to 38 percent, suggesting a general OM instability.

The analysis of nonhydrocarbon components of the OM shows that HA elemental composition changes along the profile as follows: 60–68 percent C, 5–6.6 percent H, 1.6–3.2 percent N. High C content found in the HA of bottom sediments (48–60 percent) (Romankevich, 1977) may be due to greater abundance of higher plant remains. This is consistent with variations in N content in humic acid. The amount of hydrolized material in humic acid varies from 16 percent to 36.5 percent. HA hydrolysate contains about 11–15 combined amino acids (free amino acids have not been found in sediments), dominated by neutral and acidic ones. The content of combined amino acids range from 15 to $67 \cdot 10^3$ mgk per 1 g of OM (or 100 g of sediment). The data on nonhydrocarbon OM components show changes in their composition both in the surface sediment layer in different sectors of the offshore area and vertically within the cores studied. Variations of HA content and composition suggest a mixed nature of OM formed both by terrigenous plant detritus and hydrobiont remains.

The analysis of OM composition in sediments of different morphostructural zones suggests important differences in bitumoid molecular structure, amino acid composition, and other parameters for trenches and shelf plateaus, respectively (Tables 28-6, 28-7, Fig. 28-12). For example, Ach1 bitumoids of sediments display low oxygen C-O group content (1,740, 1,720 $cm^{-1}$). The content of long-chain paraffins (720 $cm^{-1}$) is also low, which points to their presence in a single molecular structure. However, the intensity of aromatic structure absorption bands (1,605, 880, 809, 750 $cm^{-1}$) does not increase, suggesting the predominance of methane–naphthene structures (970 $cm^{-1}$) with shorter methane chains. Such bitumoids are typi-

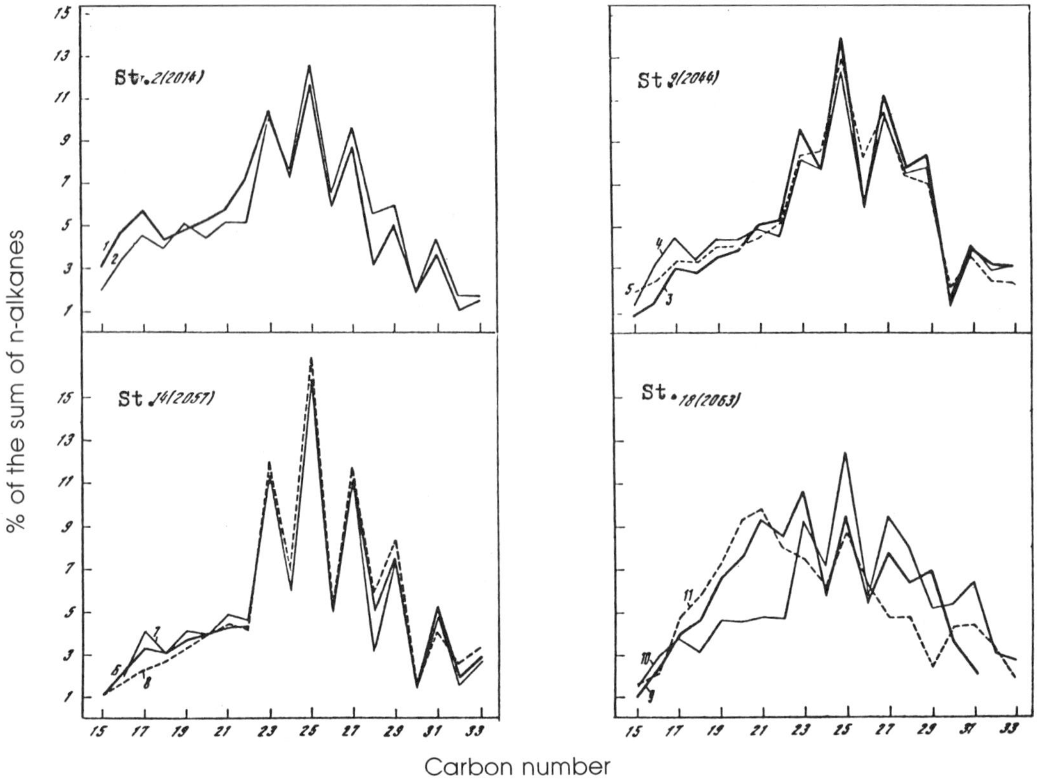


**Figure 28-11.** n-Alkanes distribution in the OM of the shelf plateaus of the Barents Sea. Horizons from bottom surface cm.: 1, 53–66; 2, 268–290; 3, 268–290; 4, 316–326; 5, 670–675; 6, 67–93; 7, 257–280; 8, 360–363; 9, 90–105; 10, 265–280; 11, 330–340.

cal of OM formed by mixed autochthonous and allochthonous organic material, indicating their epigenetic nature (Danyushevskaya et al., 1973, 1978). IR bitumoid absorption spectra of shelf plateau bottom sediments are rich aromatic structures (intense absorption bands of aromatic rings at 1,605 $cm^{-1}$ and replaced structures at 880, 809, 750 $cm^{-1}$) of ketones, acids, and ethers. Oxygen-bearing compounds contain mainly C-O groups (1,740 and 1,720 $cm^{-1}$; 1,170, 1,250–1,270 $cm^{-1)}$ and OH-acid group (2,700–2900 $cm^{-1}$ and 3,200–3,400 $cm^{-1}$) of aliphatic and aromatic structures forming part of aromatic acids, ketones, ethers, and aldehydes. All spectra contain strong 1,650–1,680 $cm^{-1}$ absorption bands belonging to compounds whose C-O groups are linked with unsaturated bonds (enol $\beta$-ketoethers, O-oxybensoates, chinoid structures). These molecular spectra are very common. They are known in bitumoids from different parts of offshore areas especially on shelf plateaus where sedimentary OM consists mainly of terrigenous detritus.

The OM of trenches and shelf plateaus sediments differ also with respect to amino acid composition. In trenches the total amino acid content is 7–8 percent higher (Fig. 28-12), owing to high concentration of arginine, proline, gistidine, lysine, glycine, leucine, and asparagine acid. However, the content of alanine, proline, serine, and phenylalanine is low in the OM of trench sediments. The content of other amino acids is almost the same in shelf plateau and trench sediments.

## Table 28-6A. Geochemical Properties of the OM in Bottom Sediments of Various Morphostructural Zones of the Barents Sea for the Rybachi–Franz Josef Land Profile

| | | | Content in a Dry Sediment (%) | | | | | | | Group Composition of OM (%) Bitumoid A | | |
|---|---|---|---|---|---|---|---|---|---|---|---|---|
| | | | | | | Bitumoid A | | | | | | |
| Morpho-type | Number of samples | Sedi-ment Type | $C_{carb}$ | Organic Carbon | OM ($C_{org}$ × 1.82) | Chlor. | Benzol | Alcohol–benzol | Humic Acid | Chlor. | Benzol | Alcohol-Benzol |
| Trench | 4 | Clay | 0.08 | 1.05 | 1.91 | $\frac{0.01}{2}$ | $\frac{0.02}{2}$ | 0.01 | $\frac{0.47}{2}$ | 0.5 | 1.0 | 0.5 |
| | 10 | Clayey silt-stone | 0.19 | 0.76 | 1.38 | $\frac{0.01}{8}$ | $\frac{0.02}{6}$ | 0.01 | $\frac{0.15}{6}$ | 0.7 | 1.4 | 0.7 |
| Shelf plateau | 30 | Clay, clayey silt-stone | 0.21 | 1.14 | 2.07 | $\frac{0.01}{23}$ | $\frac{0.03}{11}$ | 0.01 | $\frac{0.57}{11}$ | 0.5 | 1.4 | 0.5 |

## Table 28-6B. Physical-Chemical Characteristics of Bitumoids A

| | Group Composition of OM, % | | Elementary Composition, (%) | | | Group Composition, (%) | | | | | Hydrocarbons (%) | |
|---|---|---|---|---|---|---|---|---|---|---|---|---|
| Morpho-type | Humic Acid | ROM | C | H | $H/C_{at}$ | Oil | Resin | Asphaltene Asphaltogene Acids | Hydrocarbon ratio $\frac{Me + Nf}{Ar}$ | $\Sigma Ar$ in OM $10^{-4}$% | In Sedi-ment | In OM |
| Trench | 24.6 | 73.4 | $\frac{82.1}{2}$ | 10.9 | 1.59 | 38.2 | 52.8 | 6.3 | 1.20 | 22.7 | 0.004 | 0.19 |
| | 10.9 | 86.3 | $\frac{80.7}{6}$ | 11.1 | 1.65 | 32.0 | 60.6 | 6.0 | 1.11 | 13.7 | 0.003 | 0.22 |
| Shelf plateau | 27.5 | 70.1 | $\frac{81.2}{11}$ | 10.4 | 1.54 | 30.4 | 58.5 | 9.8 | 0.98 | 16.5 | 0.003 | 0.15 |

Table 28-7. Quantitative Changes of Amino/Acid Composition from Cores in Different Morphostructural Zones of the Barents Sea (Averaged Data from Rybachii Peninsula–Franz Josef Land Profile)

| *Depth of Sampling (m)* | *Amino Acid Content mkg per 100 g of Dry Sediment* | | | | | | *Loss of Amino Acid (%) Compared to Original Content* | | | | | |
|---|---|---|---|---|---|---|---|---|---|---|---|---|
| | *Basic* | *Neutral* | *Acid* | *Aromatic* | *Sulfur* | *Total* | *Basic* | *Neutral* | *Acid* | *Aromatic* | *Sulfur* | *Total* |
| Shelf Trench | | | | | | | | | | | | |
| 0.5–1.3 | 14,277 | 16,798 | 15,644 | 12,497 | — | 59,216 | — | — | — | — | — | — |
| 2.7–7.7 | 10,121 | 14,758 | 12,643 | 3,488 | — | 41,010 | 29.2 | 19.2 | 12.1 | 72.1 | — | 30.7 |
| Shelf Plateau | | | | | | | | | | | | |
| 0.6–1.1 | 9,422 | 19,941 | 15,927 | 7,671 | 1,134 | 54,095 | — | — | — | — | — | — |
| 3.7–7.6 | 11,046 | 15,366 | 12,065 | 7,656 | 302 | 46,435 | Not found | 24.2 | 22.9 | 0.2 | 73.4 | 14.2 |

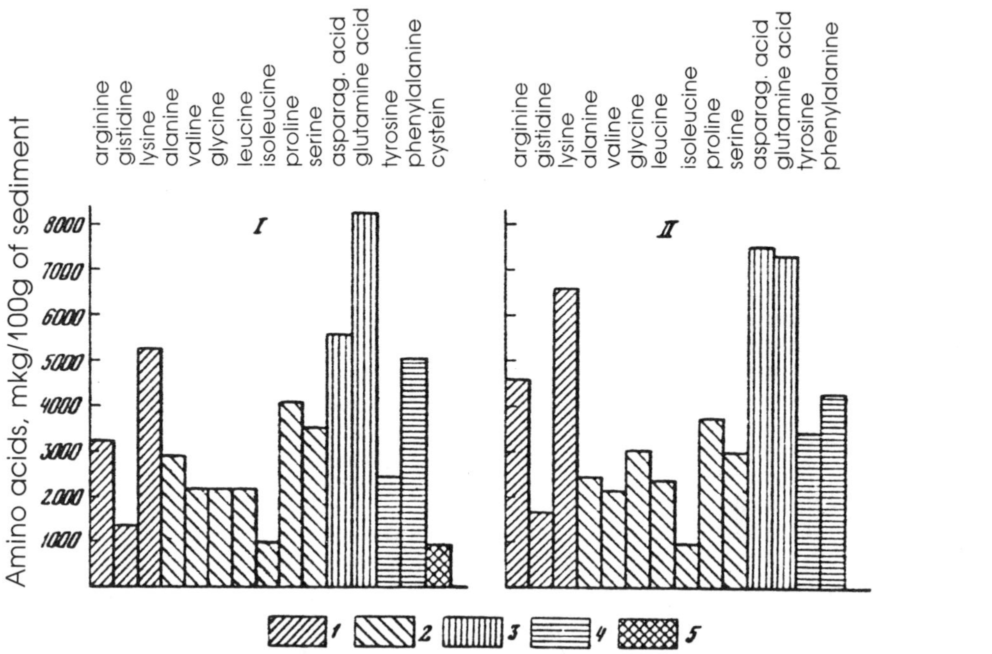


Figure 28-12. Amino acid composition of bottom sediments of different morphostructural zones of the Barents Sea. I, Shelf plateaus; II, shelf trenches; 1, basic acids; 2, neutral acids; 3, acidic acids; 4, aromatic acids; 5, sulfur-containing acids.

About 15 amino acids, totaling 50–670 mkg/g of dry sediment, have been found in HA hydrolysates. The amino acids seem to have been formed as a result of the decomposition of hydrobiont and higher plant proteins. The amount of amino acids in sediments is one to two orders of magnitude higher than that in the Mesozoic deposits of the adjacent Arctic landmass. Trace contents, if any, of free amino acids in the sediments show that OM biopolymers have suffered sedimentogenetic and diagenetic changes and were transformed into geopolymers. A relatively high content of residual OM (80–90 percent) supports the above statement.

Humification observed in bottom sediments during the earliest oxydizing stages of OM transformation at 1.0–1.5 m depth is frequently marked by higher content of combined amino acids. At this stage, the content of dissolved OM components depends on biochemical degradation of organic material affected by bacterial activity.

At great depths biochemical processes of OM polymerization become more intense as evidenced by the transformation of its constituents and lower solubility. As a result, we see a decrease in the content of humic amino acids and an increase in the concentration of residual organic matter with increasing depth in bottom sediments (2.5–7.0 m interval was studied). OM formed from hydrobionts and those rich in allochthonous components display similar trends, the former being more intense. For example, the comparison of amino acid distribution within sediments of upper and lower horizons in trenches and on shelf plateaus shows that the tendency toward the decrease of amino acids with increasing depth is independent of the morphostructural setting of sediments. However, the total loss of amino acids (as compared to the original content) is larger in trench sediments (30.7 percent); on shelf plateaus it averages 14.2 percent. A high proportion of finely dispersed clay component in trench sediments favors the OM polymerization and accelerates the transition of amino acids into an undissolved state. OM (consisting mainly of hydrobionts remains, rich in aliphatic structures) in trench sediments may be more labile.

A comparison of amino acid groups shows that the content of aromatic (by 72 percent) and basic (29 percent) amino acids drastically decreases with depth in trench sediments. Shelf plateau sediments display a different trend: the content of basic amino acids gistidine and lysine increases with depth. Furthermore, the content of sulfur amino acids, in particular, cystein, decreases by 73 percent (of importance is that cystein in surface sediments may be due to pollution) as do asparagine and glutamine acids (on the average by 24 percent).

Thus, despite a general trend toward a decrease in amino acid content with depth in sedimentary strata, the dynamics of these changes are different in bottom sediments of various morphostructural zones.

The change in OM group composition in sediment cores is mainly due to a decrease in the content of dissolved bituminous and especially humic acid components, resulting from OM disintegration and polymerization and its attendant transformation (of oxydizing and reducing stages of diagenesis). Therefore, the OM composition of the Barents Sea bottom sediments is fairly diverse, mainly due to OM genesis, oxydizing-reducing conditions of diagenesis and morphostructural setting of sediments. The OM in sediments at all stations shows a mixed allochthonous-autochthonous genesis. The allochthonous component consists of higher plant remains and marine hydrobionts. The importance of the latter increases in the trench zones stations.

The analysis of the results of OM studied from bottom sediments of the Barents Sea allowed the recognition of three major types of OM (Fig. 28-13) (*Biogeochimia Organicheskogo Veshchestva Arkticheskich Morei,* 1982). Type I OM consists mainly of decomposition products of marine organisms. Sediments containing type I OM have lower (as compared to background) organic carbon concentrations (0.5–0.8 percent in clay sediments). Bitumoids are characterized by low C content (up to 79 percent), large proportion of oxygen-bearing and aliphatic structures, and low content of aromatic compounds. These bitumoids are most typical of recent shelf plateau sediments. The oils in bitumoids varies from 10 percent to 20 percent; hydrocarbons are dominated by methane-naphthene compounds Me − Nf/Ar >1.0. Mass-spectrometer analyses indicate that the latter are dominated by naphthene fraction (76 percent represented mainly by bitricyclic naphthenes, 24 percent and 21 percent, respectively. The distribution of n-alkanes displays a $C_{15}$–$C_{17}$ maximum of hydrobiont lipids. In n-alkanes, $C_{12-22}$ fraction totals 45 percent, $C_{23-33}$ totals 55 percent; pristane and phytane with a low ratio of 1.37 have been found in type I sediments (Table 28-8).

A distinctive feature of type II OM found only in shelf trench sediments is the high content of bitumoids, oils, and methano-naphthene hydrocarbons. The HC content is ten times higher than that of the remaining OM types. Judging from IR spectra, bitumoids are more neutral in composition due to drastic relative decrease of the C-O group. The proportion of long paraffin chains decreases along with the C-O groups, suggesting that they belong to a single molecular structure. However, the relative content does not increase sufficiently, indicating the predomi-

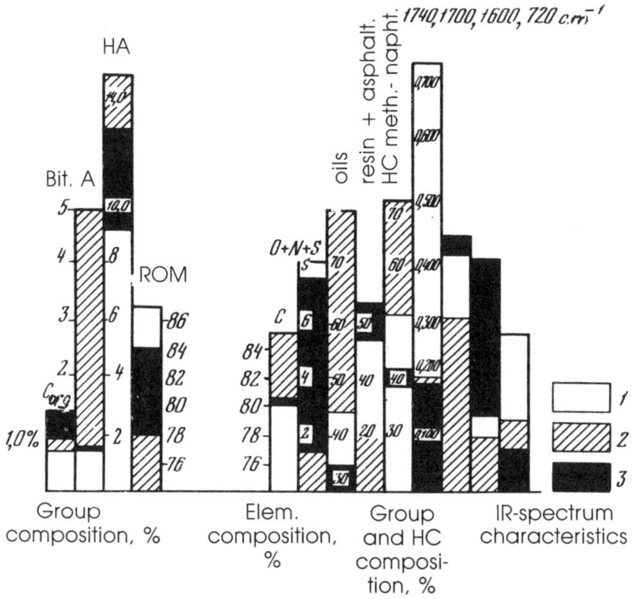


Figure 28-13. Geochemical characteristics of major OM types in the sediments of the Arctic Seas. 1, OM formed mainly by marine organisms; 2, OM of autochthonous–allochthonous origin; 3, OM of allochthonous origin.

nance of methane–naphthene structures (they seem to have shorter methane chains).

Type III OM, most common on the shelf, is rich in aromatic compounds as expressed in bitumoid composition and their spectral parameters. Bitumoids are enriched in carbon and depleted in hydrogen; the content of resinous–asphaltene components and aromatic HC increases.

Table 28-8. Typical Distribution (%) of n-Alkanes Within the OM of Barents Sea Bottom Sediments of the Sum

| | *Geochemical Types of OM* | | |
|---|---|---|---|
| *Index* | *I* | *III* | *IIIa* |
| $C_{12}$–$C_{17}$ | 12.5 | 6.3 | 6.2 |
| $C_{18}$–$C_{24}$ | 46.7 | 37.3 | 43.9 |
| $C_{25}$–$C_{28}$ | 27.6 | 34.2 | 32.3 |
| $C_{29}$–$C_{35}$ | 13.2 | 22.2 | 17.6 |
| $\Sigma C_{12}$–$C_{22}$ | 44.7 | 27.5 | 32.6 |
| $\Sigma C_{23}$–$C_{35}$ | 55.3 | 72.5 | 67.4 |
| CPI | 1.48 | 2.0 | 1.28 |
| $iC_{19}/iC_{20}$ | 1.37 | 1.29 | 1.15 |

*Note:* Average values for type I (9 samples), III (15 samples), IIIa (8 samples) are shown.

1-12-benzoperylene, perylene, and other polycyclic arens, naphtalin, phenantren, chrysen structures have been found in the aromatic fraction. Unlike other OM types, the composition of methane-naphthene HC changes and they are dominated by solid methane HC. The latter contain n-alkanes represented mainly by long-chain compounds: Lipid remains of higher terrestrial plants with $C_{25,27,29,30}$ maxima, $C_{12-22}$, and $C_{23-33}$ fractions account for 24–30 percent and 71–76 percent, respectively. $C_{15-19}$ typical of hydrobiont lipids maxima are absent in the composition of n-alkanes. The pristane to phytane ratio does not exceed 1.0 and on the average it is lower than that of type I OM.

Genetically, IR spectra are typical and suggest a considerable increase of the content of aromatic structures in bitumoids because there is a maximal value of 1,600 $cm^{-1}$ absorption bands. Thus, the geochemical data indicate high levels of terrigenous organic material in type III OM. Diagenetic vertical changes in type III OM through cores are evidenced by decrease of CPI from 2.0 to 1.28.

The distribution of OM types (Table 28-9) indicates the presence of allochthonous organic material in all the Barents Sea morphostructural zones, especially in sediments of the continental slope and rise. This peculiar feature is consistent with a distinctive polar lithogenesis determining the leading role of terrigenous sedimentation. However, a fairly high bioproductivity of the Barents Sea is reflected in OM composition. The effect of hydrobionts on the OM nature of sediments (composition and structure) is stronger here than in the Kara Sea.

## Organic Matter Balance

The absolute mass of dissolved particulate OM and OM in bottom sediments are in dynamic equilibrium with the introduced masses of autochthonous and allochthonous OM and OM lost during mineralization and transport of water masses.

The OM input incorporates also the OM production of phytoplankton and macrophytes and the allochthonous OM from different sources (river and ice discharge, coastal wave abrasion, and eolian material as well as organic pollution). As a result of phytoplankton photosynthesis in the Barents Sea, annually about 77 millions tons of organic carbon are formed. The annual macrophyte production is much lower: 66

Table 28-9. Distribution of OM Types According to Morphostructure Occurrences in Barents Sea Sediments

| *Morphotype* | *Sea Depth (m)* | *Total Number of Samples Studied* | *Distribution with Respect to OM Types, % of the Number of Types Studied* | | |
|---|---|---|---|---|---|
| | | | *I* | *II* | *III* |
| Socle plains (shallow water) | 20–70 | 13 | 30.8 | — | 69.2 |
| Shelf plateaus | 50–400 | 81 | 22.3 | 4.9 | 72.8 |
| Coastal-shelf and inner-shelf trenches | 100–475 | 68 | 35.3 | 11.8 | 52.9 |
| Continental slope zones | 200–2,400 | 65 | 12.3 | — | 87.7 |
| Deep-water zones | 2,400–3,500 | 4 | 25.0 | — | 75.0 |
| Linear uplifts of the sedimentary cover | 60–230 | 10 | — | — | 100 |

thousand tons of organic carbon. Autotropic OM forms most of OM supplied into the Barents Sea (Table 28-10).

Among allochthonous sources of OM, the most important is liquid river discharge, mainly the Pechora River: According to mean multiyear data the river carries 1.44 million tons of organic carbon into the Barents Sea annually (Maltseva, 1980).

The silt load was estimated using the suspension concentration of the Pechora at 0.016–0.038 g/l (Konovalov, Ivanova, and Kolesnikova, 1966) and organic carbon content in dry particulate matter of the northern rivers at 2 percent (Skopintsev and Krylova, 1955). The proportion of liquid and solid river discharge with respect to carbon is much higher in the former because rivers flowing into the Barents Sea contain small amounts of particulate matter (Norina and Borisova, 1965) and have a fairly high content of dissolved OM.

The total wave abrasion of the Barents Sea shores is estimated at 59 millions tons/yr (Suzdalsky, 1974). If we take the average organic carbon content in this material to be 0.4 percent (Romankevich, 1984), then annual supply of organic carbon (with wave abrasion material) totals 236 thousand tons per year. Assuming that about half of the Barents Sea discharge is derived from Spitsbergen, Franz Josef Land, and Novaya Zemlya glaciers (48.2 $km^3$/yr) (Mirovoi Vodny Balans i Vodnye Resursy Zemli, 1974), and the organic carbon content in liquid ice discharge is 0.5–1.0 mg/l (Romankevich, 1984), the annual supply of organic carbon is 20 thousand tons 0.03 percent of the total organic carbon supply. The value is higher by factor 3 than that for the Atlantic Ocean as a whole (Emelianov and Romankevich, 1979). Based on the same data, ice debris discharge accounts for 24 million tons and by organic carbon content of 0.1 percent (Romankevich, 1984) it is 24 thousand tons carbon per year.

The predominantly land to sea wind direction and processes of freezing and weathering of rocks contribute high proportions of eolian suspension into the Barents Sea. At present, we can give only an approximate value of total organic carbon supply from this source. The data available

Table 28-10. Annual Supply and Deposition of OM into the Barents Sea

| *Sources of Organic Carbon* | *Supply of Original Matter of Organic Carbon Carrier* | *Content of Organic Carbon in Original Matter* | *Supply of Organic Carbon $10^6$ tons* | *Different Sources Supplying Organic Carbon % of the Sum* |
|---|---|---|---|---|
| Phytoplankton | — | — | 77.00 | 97.03 |
| Phytobenthos | — | — | 0.066 | 0.05 |
| River runoff | 163 $km^3$/yr | 13.0 mg C/l | 2.10 | 2.64 |
| Sediment load | $3.5 \cdot 10^6$ tons | 2.0% | 0.07 | 0.09 |
| Eolian discharge | $4.2 \cdot 10^6$ tons | 2.5% | 0.11 | 0.14 |
| Material of wave abrasion | $88 \cdot 10^4$ tons | 0.4% | 0.01 | 0.01 |
| Ice discharge (without debris) | 24.1 $km^3$ | 0.5–1 mg C/l | 0.02 | 0.02 |
| Debris of ice discharge | $12 \cdot 10^6$ tons | 0.1% | 0.01 | 0.01 |
| Anthropogenic pollutants | $2 \cdot 10^3$ tons | 75% | 0.0015 | 0.002 |
| Total | | | 79.36 | 100.0 |
| autochthonous | | | 77.04 | 97.08 |
| allochthonous | | | 2.32 | 2.92 |

show that during freezing 100 tons of terrigenous eolian material, mainly of sand size with organic carbon content of 1.0–0.5 percent, precipitates in an area of 2 $km^2$ off the Barents Sea (Tarasov, 1979). With allowance made for the coastline (excluding Novaya Zemlya and other islands) from shore to sea, during freezing, $4.2 \cdot 10^5$ tons of eolian suspension or $2.1–4.2 \cdot 10^3$ tons of organic carbon are carried annually. Petrographical examination shows that the spore and pollen content in bottom sediments contains 6–15 percent OM. At an annual burial rate of 106 tons of carbon, excluding plant debris supplied by river discharge, the supply of organic carbon with eolian suspension may account for no less than 60 thousand tons annually.

The water circulation system of the World Ocean accounts for 30 percent of organic pollutants transported by the Gulf Stream into the Barents Sea. Oil is the most important pollutant. Based on Levy's data (1980), there is an influx of about 2.5 thousand liters of oil per day of which 0.6 thousand liters is transported by the Nordcap Current into the Barents Sea. Thus, the annual formation of carbon in autochthonous OM in the Barents Sea is estimated at about 77 million tons or 96.86 percent of the total carbon content supplied into the sea from different sources. The supply of allochthonous OM from different sources totals about 2.5 million tons or 3.14 percent of the total organic carbon supply into the Barents Sea. Because the Barents Sea is not a closed sea, in addition to the OM influx and synthesis within the sea, some of the OM is carried by water exchange with adjacent offshore area. Various estimates show that about 9 million tons C-DOM and C-POM is annually discharged due to different concentrations of dissolved and suspended organic carbon in various parts of the Barents Sea and in areas where water exchange takes place. Absolute masses of DOM and POM organic carbon in this sea amounts to 320 and 10 million tons, respectively. A comparison of weighted mean concentrations of organic carbon in suspension from a 0–50 m layer and bottom suspension suggests that 18–24 percent of C-POM occurring in the photic layer reaches the bottom or 2.1–2.7 million tons of organic carbon annually. Most of this OM is not buried in bottom sediments; it is dissolved and decomposed by aerobic saprophytes.

The rates of organic carbon accumulation in Quaternary deposits and the total amount of organic carbon deposited on the Barents seafloor during the last 1,000 yr may be obtained by averaging sedimentation rates of 10–20 cm per 1,000 yr and the volume weight of sediments at 1.5 $g/cm^3$. With an average moisture of sediments of 60 percent and an average organic carbon content of 0.84 percent, the rate of carbon accumulation is estimated to be $7.55 \cdot 10^{-2}$ $g/cm^3$ per 1,000 yr and the total carbon content is $10.5 \cdot 10^5$ g C yr. Hence, the fossilization coefficient of OM or a fraction of the OM buried of the entire primary production and influx from land into the Barents Sea equals 1.3 percent. This value is twice as high as the average fossilization coefficient for the World Ocean and twice that of submarine continental margins (Romankevich, 1984); this is due to low OM transformation at the sedimentogenesis stage, low water temperature, and low bacterial activity.

Comparing the amount of carbon that reaches the floor in the form of particulate matter and the amount annually buried in sediments we obtain about 1–1.5 million tons of organic carbon (or 50–57 percent of POM) that is decomposed, dissolved, or comes into the deep OM reserve at the water–sea bottom interface in the Barents Sea. The value obtained is 1.7–1.8 times less than that for the entire World Ocean (Romankevich, 1984). The low OM mineralization on the floor is due to slow biogeochemical processes of OM transformation in the Barents Sea. The annual amount of OM mineralized in the Barents Sea water was estimated at 118 million tons (65 million tons of organic carbon) or 82 percent OM discharged into the sea (Fig. 28-14). It is noteworthy that in the entire World Ocean water, up to 95 percent of the OM discharged is mineralized (Romankevich, 1977). Therefore, the general balance of OM production–decomposition system in the Barents Sea is much lower than that of the World Ocean and its submarine margins. This may be explained by the effect of biological, physical-geographical and lithogeochemical processes responsible in general for high OM influx to the seafloor.

In summary, our study shows that the hydrochemical and hydrodynamic regimes of the sea, its biological structure, and distinctive polar lithogenesis, favor

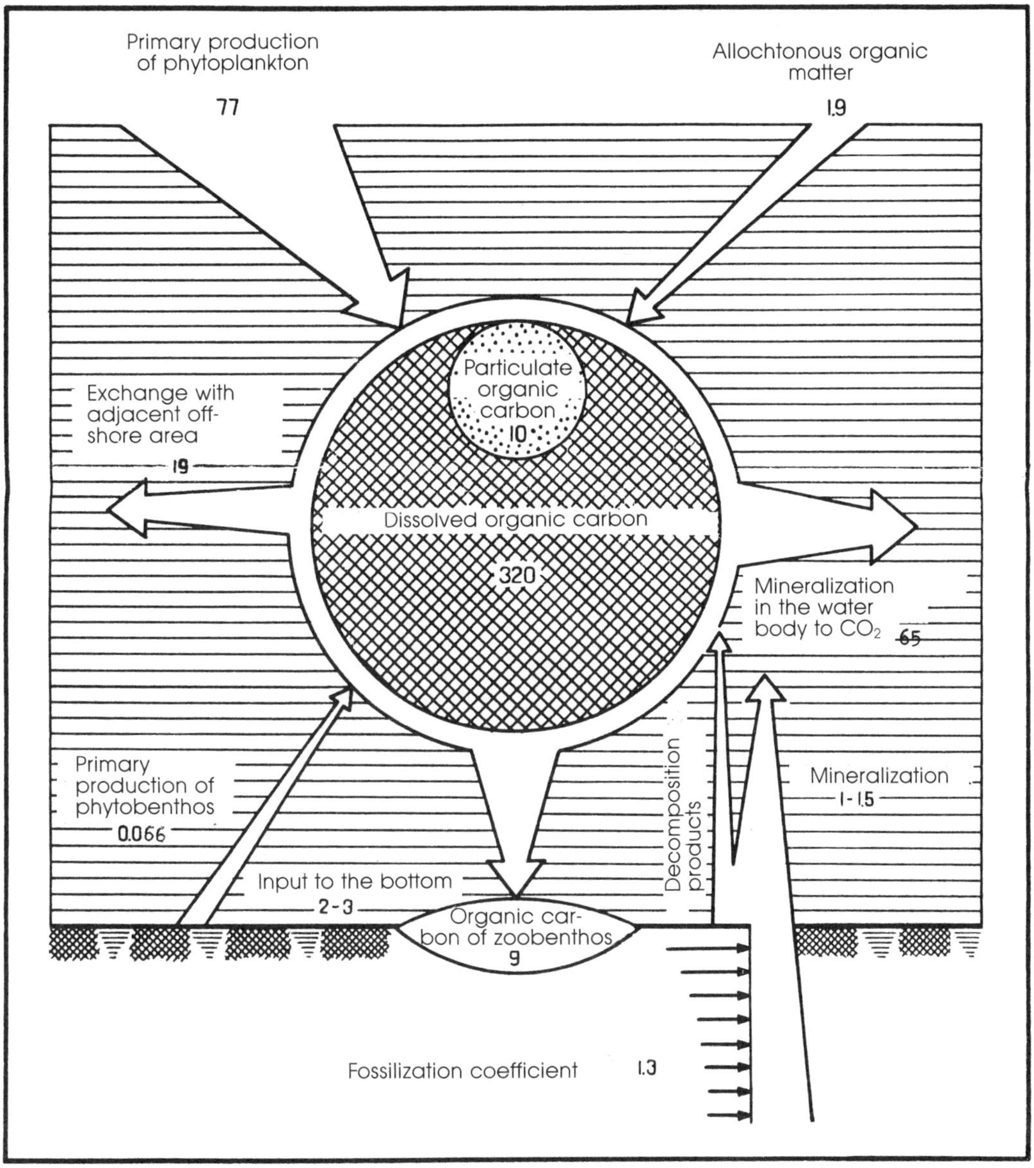


Figure 28-14. The balance of organic matter in the Barents Sea.

the formation of a peculiar OM turnover in the Barents Sea. The high productivity and relative stability of DOM and POM compounds at low temperatures (along with their influx from offshore areas and the ice rafting of less labile OM as compared to newly synthesized forms) provide conditions favorable for the formation of higher OM reserves as compared to those in the surface layer of the World Ocean.

The differences in the rates of POM transformation in water masses of various parts of the sea are mainly due to temperature gradients of the water masses. Variations in kinetic parameters of OM sedimentogenesis, the amount of OM influx to the floor and its burial, in addition to a variable genesis, determine the concentration and content of OM in bottom sediments.

Based on composition the OM in the bottom sediments may be subdivided into three major types. Type I OM (autochthonous OM) consists mainly of marine hydrobionts, and occurs in sediments of all morphostructural zones; however, this type of OM is not very common. Type II OM (mixed autochthonous-allochthonous OM) occurs in trench and shelf bottom sediments off Franz Josef Land. The most common is the type III OM dominated by higher plant remains. This OM type also dominates the shelf sediments of the study area.

In surface sediments OM of all three types is at a stage of transition from biopolymers to geopolymers. The absence of chlorophyll in most samples suggests that they are at a final stage of biopolymer transformation.

The study of diagenetic changes of OM in bottom sediments suggests that during transition from oxidizing to reducing environments in sediments, carbon consumption increases from 3–4 percent to 24 percent as compared to the inferred original values, while in humic acids and bitumoids there is an increase of carbon and condensed aromatic structures. However, the amount of hydrolyzable material and acid amino acids decreases in humic acids.

## REFERENCES


Antipova, T. V. 1957. Raspredelenie biomassy bentosa Barentseva morya. *Tr. PINRO* **35:**121–124.

Antipova, T. V. 1978. Produktsia populyatsii nektorych vidov dvustvorchatykh mollyuskov yugo-vostochnoi chasti Barentseva i yuzhnoi chasti Karskogo morei. *Okeanologia* **18**(4):737–741.

Aizenschtadt, Z., M. Baedecker, and I. Kaplan. 1973. Distribution and diagenesis of organic compounds in JOIDES sediment from Gulf of Mexico and Western Atlantic. *Geochim. Cosmochim. Acta* **37:**1881–1898.

Belyaeva, A. N. 1980*a*. Ekstraktsia lipidov iz morskoi i ilovoi vody, vzvesi i donnykh osadkov. In *Metody Issledovania Organicheskogo Veshchestva v Okeane*. Nauka, Moscow, pp. 129–139.

Belyaeva, A. N. 1980*b*. Metod opredelenia kontsentratsii lipidov v morskoi vode, vzvesi i donnykh osadkakh. In *Metody Issledovania Organicheskogo Veshchestva v Okeane*. Nauka, Moscow, pp. 139–142.

Belyaeva, A. N., and E. A. Romankevich. 1976. Chimicheskie prevrashchenia lipidov pri sedimentogeneze. In *Issledovania Organicheskogo Veshchestva Sovremennykh i Iskopaemykh Osadkov*. Nauka, Moscow, pp. 81–103.

Bikbulatov, E. S., and E. M. Bikbulatov. 1979. Skorost raspada organischeskogo veshchestva otmershego fitoplanktona. In *Mikrobiologicheskie i Khimicheskie Protsessy Destruktsii Organicheskogo Veshchestva v Vodoemakh*. Nauka, Leningrad, pp. 213–224.

*Biogeochimia Organicheskogo Veshchestva Arkticheskich Morei*. 1982. Nauka, Moscow, 239p.

Blinova, E. I. 1969. Raspredelenie, zapasy i godovaya produktsia dominiruyushchikh vidov vodorosley litorali Murmanskogo poberezhya Barentseva morya. *Rastitelnye Resursy* **2:**47–59.

Blinova, E. I., and V. B. Vozzhinskaya. 1971. Morskie makrofity i rastitelnye resursy. In *Osnovy Biologicheskoi Produktivnosti Okeana i ee Ispolzovanie*. Nauka, Moscow, pp. 137–172.

Bliznichenko, T. E., A. A. Degtereva, S. S. Drobysheva, et al. 1976. Results of plankton investigation in the Norwegian and Barents seas in 1976. *Ann. Biol. Cons. Int. Explor. Mer.* **33:**54–56.

Bobrov, Yu. A. 1978*a*. Sezonnaya dinamika khlorofilla fitoplanktona Barentseva morya. In *Zakonomernosti Bioproduktsionnykh Protsessov v Barentsevom More*. Apatity, pp. 27–36.

Bobrov, Yu. A. 1978*b*. Sezonnye izmenenia nekotoryph pokazatelei produktivnosti fitoplanktona Barentseva morya. In *Zaknomernosti Bioproduktsionnykh Protsessov v Barentsevom More*. Apatity, pp. 37–52.

Bordovsky, O. K. 1974. *Organicheskoe Veshchestvo Morskikh i Okeanskokh Osadkov v Stadii Rannego Diagenesa*. Nauka, Moscow, 91p.

Brotskaya, V. A., and L. A. Zenkevich. 1939. Kolichestvennii uchet donnoi fauny Barentseva morya. *tr. VNIRO* **4**:1–126.

Danyushevaskaya, A. I., E. A. Glebovskaya, E. A. Rogozina, and E. P. Shkatov. 1973. Geokhimicheskie issledovania organicheskogo veshchestva osadkov shelfa Barentseva morya. *Geologia Nefti i Gaza* **1**:33–40.

Danyushevskaya, A. I., V. D. Dibner, E. P. Shkatov, and E. A. Rogozina. 1978. Kompleksnye geokhimicheskie issledovania na Arkticheskom shelfe. *Okeangeologia* **1**:70–78.

Dibner, V. D. 1978. Morfostruktura shelfa Barentseva morya. Nedra. Leningrad, 197p.

Digby, P. S. B. 1961. The vertical distribution and movements of marine plankton under midnight sun conditions in Spitsbergen. *J. Animal. Ecol.* **30**(1).

Ehrchardt, A. 1969. A new method for the automatic measurement of dissolved organic carbon in sea water. *Deep Sea Res.* **16**(4):393–397.

Emelianov, E. M., and E. A. Romankevich. 1979. Geokhimica Atlanticheskogo okeana. *Organicheskoe Veschchestvo i Fosfor*. Nauka, Moscow, 219p.

Fedorov, M. V., and I. A. Ermachenko. 1961. Intensivnost obrazovania i raspada organicheskogo veshchestva v severnykh moryakh. In *Pervichnaya Produktsia Morey i Vnutrennikh Vod. Minsk: Usdatelstvo Ministerstva Vysshego i Srednego Spetsialnogo i Professionalnogo Obrazovania BSSR*, pp. 40–46.

Filatova, Z. A. 1938. Kolichestvenny uchet donnoi fauny yugozapadnoi chasti Barentseva morya. *Tr. PINRO* **2**:3–58.

Fomin, O. K. 1978. Zhiznennii tsyki i sezonnoe vertikalnoe raspredelenie vozrastnykh kategorii Kalyanus Finmarchikus v pribrezhye Vostochnogo Murmana. In *II Vsesoyuznaya Konferentsia po Biologii Shelfa, 1978*. Tex. Dokl. Kiev, pp. 115–116.

Golikov, A. N. Metod opredelenia produktsionnykh svoistv populyatsii po razmernoi strukture i chislennosti. *DAN USSR* **93**(3):730–733.

Gorshkova, T. I. 1975. Organicheskoe veshchestvo sovremennych shelfovykh osadkov severnych morey SSSR. In *Proplemy Geologii Shelfa*. Nauka, Moscow, pp. 66–72.

Gorskaya, A. I. 1960. Issledovanie organicheskogo veshchestval chetvertichnykh i sovremennykh osadkov razlichnykh fatsii. In *Bitumo-Obrazovanie v Chetvertichnykh Osadkakh i Genezis Nefti*. Gostophechizdat, Moscow.

Klenova, M. V. 1961. Sovremennoye osadkoobrazovanie v Barentsevom more. In *Sovremennye Osadki Morei i Okeanov. Izd. AN USSR*, pp. 419–436.

Kondratieva, T. M., and Z. Z. Finenko. 1971. Pervichnaya produktsia i nekotorye aspekty fiziologii planktonnykh vodorosley. In *Problemy Morskoi Biologii*. Naukova Dumka, Kiev, pp. 197–194.

Konovalov, G. S., A. A. Ivanova, and T. Kh. Kolesnikova. 1966. Redkie i rasseyannye elementy (mikroelementy) v vode i vo vzveshennykh veshchestvakh rek evropeiskoi territorii SSSR. *Gidrokhim. Mater.* **42**:94–111.

Kriss, A. E. 1976. *Mikrobiologicheskaya Okeanografia*. Nauka, Moscow, 269p.

Kudlo, B. P. 1973. Mnogoletnie izmenenia solensoty u poberzhya Murmans. *Tr. PINRO* **34**:113–122.

Kuznetsov, A. P. 1970. Zakonomernosti raspredeleniya pishchevych gruppirovok donnykh bespozvonochnykh v Barentsevom more. *Tr. IOAN SSSR* **88**:5–80.

Kuznetsov, A. P. 1980. *Ekologia Donnykh Soobshchestv Mirovogo Okeana. Troficheskaya Struktura*. Nauka, Moscow, 243p.

Kuznetsov, V. V. 1956. Rose pribreshnykh morskikh maktofitov i potreblenie imi pitatelnykh veshchestv v Barentsevom i Belom more. *Bot. zh.-l.* **41**:510–521.

Levy, E. M. 1980. Fonovye urovni i krugovorot v okruzhayushchei srede neftyanych uglevodorodov i trebovania k mnogosrednemu monitoringu. In *Kompleksny globalny Monitoring Zagryaznemia Okruzhayushchei Prirodnoi Sredy*. Tr. Mezhdunar. Simposiuma, Riga, 1978. Gidrometeoizdat, Leningrad, pp. 277–293.

Levy, E. M., and A. Walton. 1976. High seas oil pollution: Particulate petroleum residues in the North Atlantic. *J. Fish. Res. Board Canada* **33**:2,781–2,791.

Lyubarsky, A. M. 1976. O kolebaniakh ledovitosti Barentseva morya i ikh vozmoznykh fizicheskikh prichinakh. *Tr. GGO* **380**:81–88.

Lyutsarev, S. V., and E. A. Romankevich. 1975. Organicheskii uglerod, fosfor i azot v vodakh Yaponskogo morya. *Okeanologia* **15**(2):261–265.

Maksimova, M. P. 1971. Osobennosti balansa

organicheskogo veshchestva v yuzhnykh i severnykh vnutrennikh moryakh SSSR. *Tr. VNIRO* **80:**7–17.

Maltseva, A. V. 1980. Srednii mnogoletnii stok organicheskikh veshchestv s territorii SSSR i ego izmenenie vo vremeni. *Gidrokhim. Mater.* **68:**14–21.

Marti, Yu. 1971. Osobennosti formirovania biologicheskoi produktsii v severnykh moryakh i troficheskoi oblasti Mirovogo okeana. In *Materialy Rybokhozyaistvenykh Issledovanii Severnogo Basseina. Pischchepromizdat,* Moscow, pp. 155–166. (*Tr. PINRO,* vol. 17).

Myasnikov, V. L. 1976. Vzves v pribrezhnykh vodakh Vostochnogo Murmana. In *Biologia Barentseva i Belogo Morei.* Apatity, pp. 19–24.

Nakajima, K. 1973. Suspended particulate matter in the Western North Pacific Ocean. *Mem. Fac. Hokkaido Univ.* **20**(1/2):1–106.

Nikiforov, E. G., and A. O. Shpaiker. 1980. *Zakonomernosti Formirovania Krupnomasshtabnykh Kolebanii Gidrologicheskogo Rezhima Severnogo Ledovitogo Okeana,* A. F. Treshnikova (ed.). Gidrometeoisdat, Leningrad, p. 269.

Petrov, G. N. 1978. Nekotorye fizicheskie protsessy samoochishchenia ot nefti. *Gidrobiol. Zhurnal* **14**(4):52–54.

Polkin, Ya. I., B. S. Romanovich, and E. N. Zatsepin. 1975. Tektonika shelfov Barentseva i Karskogo morei. In *Problemy Geologii Shelfa.* Nauka, Moscow, pp. 73–76.

Romankevich, E. A. 1977. *Geokhimia Organicheskogo Veshchestva v Okeane.* Nauka, Moscow, 256p.

Romankevich, E. A. 1984. *Geochemistry of Organic Matter in the Ocean.* Springer-Verlag, Berlin, 334p.

Rossova, E. Ya. 1971. Kolichestvennaya kharakteristika bakterialnogo naselenia Severo-Vostochnoi Atlantiki. In *Materialy Rybokhozyaistvennykh Issledovanii Severnogo Basseina.* Pishchepromizdat. Moscow, pp. 63–67. (*Tr. PINRO,* vol. 17.)

Rossova, E. Ya. 1977. Nekotorye dannye o chislennosti i produktsii bakterioplanktona v yugo-zapadnoi chasti Barentseva morya. *Okeanologia* **17**(5):878–882.

Roukhiyainen, M. I. 1961*a*. K voprosu o pervichoi produktsii fitoplanktona gub Barentseva morya. *DAN SSSR* **141**(1):205–207.

Roukhiyainen, M. I. 1961*b*. Osobennosti vesennego razvitia fito planktona v 1955–1957. In *Gidrologicheskie i Biologicheskie Osobennosti Pribrezhnykh vod Murmana.* Murmansk, pp. 98–108.

Roukhiyainen, M. I. 1966. Vertikalnoe raspredelenie fitoplanktona v yuzhnoi chasti Barentseva morya. In *Sostav i Raspredelenia Planktona i Bentosa v Yuzknoi Chasti Barentseva Morya* (Tr. Murman. Mor. Biol. Inst.). Nauka, Moscow and Leningrad, vyp. II (15), pp. 24–33.

Ruban, E. L. 1961. Mikrobiologicheskoe razlozhenie organicheskogo veshchestva v gruntakh litorali Barentseva morya. In *Gidrologicheskie i Biologicheskie Osobennosti Pribreznykh vod Murmana.* Murmansk knizhnoye izd., pp. 229–237.

Ryzhov, V. M., and N. G. Syuzeva. 1974. Fitoplankton yugo-zapadnoi chasti Barentseva morya. In *Gidrobiologia i Biogeografia Shelfov Knolodnykh i umerennykh vod Mirovogo Okeana.* Tez. Dokl. Nauka, pp. 102–103.

Sedykh, K. A. 1960. O sezonnykh i mnogoletnikh izmeneniakh solenosty v yuzhnoi chasti Barentseva Morya. *Tr. Okeanograf. Komissii* **9**(I):90–95.

Skopintsev, B. A., and L. P. Krylova. 1955. Vynos organicheskogo veshchestva krupneishimi rekami. *DAN SSSR* **5**(4):770–773.

Smith, G. B. 1976. Pelagic tar in the Norwegian coastal current. *Mar. Pollut. Bull.* **7**(4):70–72.

Sokolova, S. A., and A. A. Solovieva. 1971. Pervichnaya produktsia na gube Dalnozelentskoi (pribrezhie Murmana) v. 1967. *Okeanologia* **3:**460–470.

Solovieva, A. A. 1973. Pervichnaya produktsia fitoplanktons zalivov Vostochnogo Murmana. *Gidrobiol. Zhurnal* **9**(4):14–19.

Solovieva, A. A. 1975. Dinamika chislennosti fitoplanktona i khlorofilla v gube Dalnozelenetskoi (Barentsevo more). *Gidrobiol. Zhurnal* **4:**26–31.

Solovieva, A. A. 1976. Pervichnaya produktsia i fitoplankton v pribrezhnykh vodakh Barentseva morya. In *Biologia Barentseva i Belogo Morey.* Apatity, pp. 25–32.

Sorokin, Yu. I. 1964. Produktsia fotosinteza v Chernom more. *Izd. AN SSSR, Ser. Biol.* **5:**749–759.

*Sovetskaya Arctica* (Morya i ostrova Severnogo Ledovitogo okeana). 1974. Nauka, Moscow, 526p.

Strom, K. M. 1955. Land-locked waters and the deposition of black muds. In *Recent Marine Sediments.* Tulsa, Okla., pp. 356–372.

Suzdalsky, O. V. 1974. Litodinamika melkovodia Belogo, Barentseva i Karskogo morei. In *Geologia Morya,* vol. 3. NIIGA, Leningrad, pp. 27–33.

Suzdalsky, O. V. 1976. *Paleogeografia Arkticheshikh Morei SSSR v Neogene i Pleistocene.* Nauka, Moscow, 112p.

Tantsyura, A. I. 1973. O sezonnykh ismeneniakh techenii Barentseva morya. *Tr. PINRO* **3:**108–112.

Tarasov, G. A. 1979. Vlianie eolovogo vynosa na osadkonakoplenia v Barentsevom more. *DAN SSSR* **244**(3):728–730.

Timofeev, V. T. 1960. *Vodnye Massy Arkticheskogo Basseina*. Gidrometeoizdat, Leningrad, 191p.

Turpaeva, E. P. 1953. Pitanie i pishchevye gruppirovki morskikh donnykh bespozvonochnykh. *Tr. IOAN SSSR* **7**:259–299.

Turpaeva, E. P. 1954. Tipy morskikh donnykh biotsenozov i zavisimost ikh raspredelenia ot abioticheskikh faktorov sredy. *Tr. IOAN SSSR* **2**:36–55.

Yashnov, V. A. 1940. *Palnkticheskaya Produktovnost Severnykh Morei SSSR*. MOIP, Moscow, 85p.

Zelikman, E. A. 1964. K ekologii razmnozheniz massovykh vidov Euphausiacea yugo-vostochnoi chasti Barentseva morya. In *Novye Issledovania Planktonia i Bentosa Barentseva Morya*. Nauka, Moscow and Leningrad, pp. 12–21.

Zelikman, E. A. 1977. Soobshchstva arkticheskoi pelagiali. In *Biologicheskaya Produktivnost Okeana*. Nauka, Moscow, pp. 43–55.

Zenkevich, L. A. 1956. Morya SSSR, Ikh Fauna i Flora. Uchpedgiz, Moscow, 422p.

Zlobin, B. S., M. F. Perlyuk, and T. A. Orlova. 1975. Aminokislotny sostav rastvorennogo organicheskogo veshchestva v atlanticheskikh vodakh. *Okeanologia* **15**(4):643–648.

# 29: Physiography and Bathymetry of the Arctic Ocean Seafloor

J. R. Weber

## PHYSIOGRAPHICAL OVERVIEW

The Arctic Ocean consists of a deep ocean basin, submarine ridges, continental shelves, and marginal plateaus (Fig. 29-1). Five major epicontinental seas—the Barents, Kara, Laptev, East Siberian, and Chukchi seas—are located on the very wide European and Siberian continental shelves. The much smaller Beaufort and Lincoln seas overlie part of the generally narrow continental shelves that fringe the North American and Greenland margins of the Arctic Ocean. The regular pattern of continental shelves is interposed by five marginal plateaus: the Morris Jesup and Yermak plateaus off the Greenland and Svalbard shelves, and the Northwind Ridge, Chukchi Cap, and Arlis Plateau off the Chukchi shelf, which together comprise the Chukchi Borderland. The Lomonosov Ridge, which crosses near the North Pole, divides the Arctic Basin into Eurasia and Amerasia basins.

### Eurasia Basin

The Eurasia Basin (Fig. 29-1) is bounded by the Lomonosov Ridge and the Laptev, Kara, Barents, and North Greenland continental shelves; it is bisected by the active Nansen–Gakkel Ridge into Nansen and Fram basins and joins the Greenland Sea. The Lomonosov Ridge is believed to be a continental sliver that broke off the Barents and Kara shelves some 60 mya and was transported to its present position by seafloor spreading centered about the Nansen–Gakkel Ridge that opened up the Eurasia Basin (Weber and Sweeney, 1985).

### Amerasia Basin

The Amerasia Basin (Fig. 29-1) is bounded by the Lomonosov Ridge, the Canadian and Alaskan continental shelves, the Chukchi Borderland, and the East Siberian shelf. The Alpha and Mendeleev ridges divide it into Canada Basin and Makarov Basin. The deepest part of the Canada Basin is the remarkably flat, 3,800 m deep Canada Plain (Fig. 29-2). A smaller, 2,900 m deep abyssal plain, the Mendeleev Plain lies between Chukchi Cap and Mendeleev Ridge.

Between the Lomonosov Ridge and the Alpha–Mendeleev Ridge lies the Makarov Basin. It is wedge-shaped, in excess of 500

Figure 29-1. Overview map of Arctic Ocean with names used in text. The Lomonosov Ridge divides the Arctic Ocean Basin into Eurasia and Amerasia basins.

km wide along the East Siberian Shelf, and narrows to a point toward Ellesmere Island. The basin contains two abyssal plains, Wrangel Plain that adjoins the East Siberian shelf, and Siberia Plain that connects via Arlis Gap with the deeper Siberia Plain.

The narrow Marvin Spur (indicated by a line of dots in Fig. 29-2) juts out some 120 km (at the 3,000 m isobath) into the Siberia Plain. It extends in the opposite direction toward Ellesmere Island for at least 400 km paralleling the 40°W/140°E meridian.

The Alpha–Mendeleev Ridge is the largest submarine mountain complex in the Arctic Ocean. In areal extent it exceeds that of the Alps. The ridge is about the same length, slightly deeper, much wider, and morphologically much more complex than the Lomonosov Ridge.

## HISTORICAL BACKGROUND

### Early Explorations

From the Middle Ages until the end of the nineteenth century geographers believed

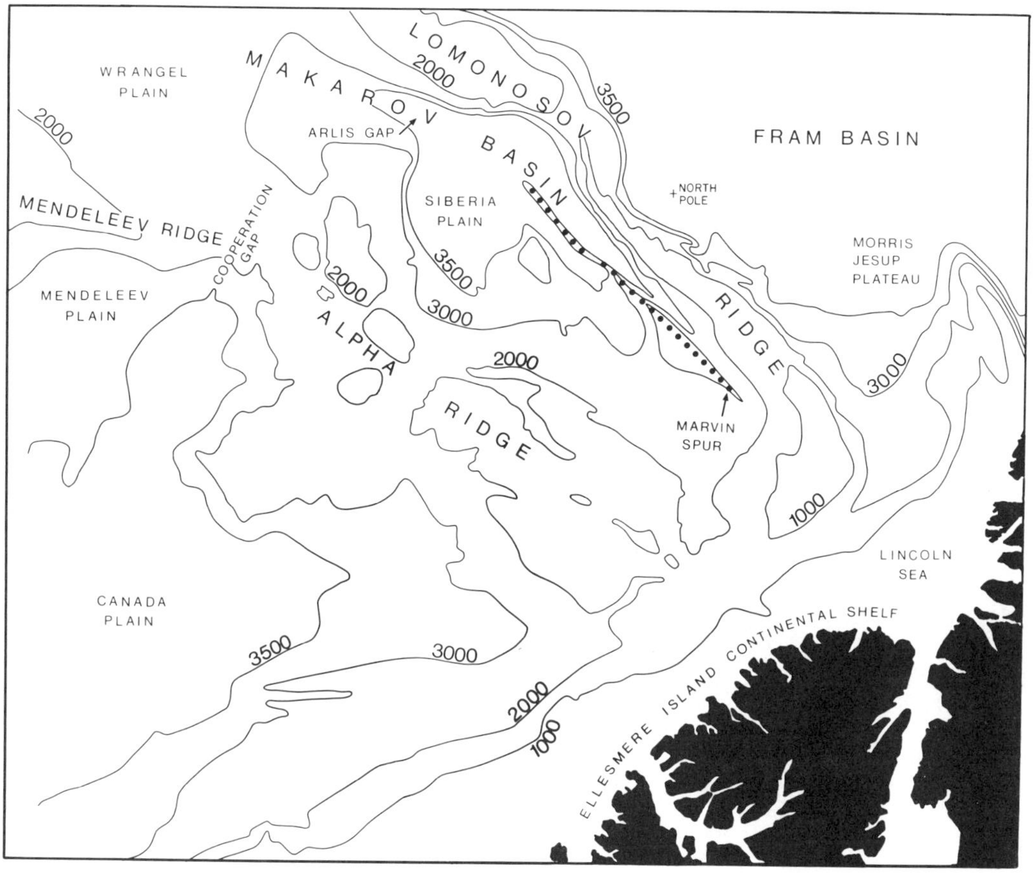


Figure 29-2. Generalized bathymetric map of the Alpha Ridge. Dotted line shows location of Marvin Spur. Broken line labeled SP indicates the location of the depth profile along the submarine tracks of the *Skate* and *Seadragon* shown in Figure 29-9. Depth contours are from *Bathymetry of the Arctic Ocean* by Perry et al. (1985).

that the North Pole was surrounded by land or by an archipelago. This concept lost credibility after Nansen's ship *Fram* drifted with the Arctic pack ice from the New Siberian Islands to Svalbard between 1893 and 1896 (Nansen, 1897). It was found that the ocean along the entire length of *Fram's* drift path was more than 3,000 m deep. This led many geographers to believe that the whole Arctic Ocean Basin was a single oceanic deep.

During the first decade of the twentieth century, Bob Bartlett, supporting Peary's attempts on the North Pole, made a number of soundings off the northern coast of Ellesmere Island that outlined the general shape of part of its continental shelf. In 1918, Storkerson (Stefansson, 1921), a member of Vilhjalmur Stefansson's Canadian Arctic Expedition 1913–1918, took numerous soundings in the Beaufort Sea.

In 1927, Sir Hubert Wilkins, flying from Alaska, landed his aircraft on the ice some 1,300 km north of Bering Strait. By setting off an explosive charge and using a stopwatch to measure the time it took for the echo to return from the ocean floor, he obtained a depth of 5,440 m (Wilkins, 1928). This sounding later turned out to be erroneous, but for the next twenty years it was much quoted in support of the opinion of a single deep Arctic Ocean Basin. The opinion was seemingly confirmed by 38 depth soundings made in 1937 between the North Pole and the East Greenland coast, along the nine-month drift path of the Soviet research station now known as North Pole 1 (Papanin, 1946). Further support came from the Soviet icebreaker *Sedov,* which made 46 soundings while drifting in the Eurasia Basin, following *Fram's* path, between 1937 and 1940 (Armstrong, 1958).

During April 1941, personnel from the Soviet All-Union Arctic Institute, in three successive flights in a ski-equipped four-engine aircraft based at Wrangel Island, established three stations in the regions of the Wrangel and Mendeleev abyssal plains between latitudes 78°N and 81°N and between longitudes 175°E and 170°W. They stayed four to five days at each station, taking soundings, measuring temperature and salinity, and sampling the water column and the seabed. Their depth measurements of 2,427 m, 1,856 m, and 3,370 m covered a large range, but neither confirmed nor disproved the depth determined by Wilkins (Treshnikov, 1966).

The Soviet airborne expeditions were highly successful. They demonstrated the advantages of using aircraft over letting a ship freeze into the ice or occupying an ice station for a long period of time. Aircraft provided choice of location, simplified the logistics, and, in some cases, decreased the cost.

## Postwar Explorations

### The Soviet Contribution

The war interrupted Arctic scientific explorations, but in 1948 the All-Union Arctic Institute started a program of systematic exploration of the whole of the Arctic Ocean, which they called High-Latitude Air Expeditions (HLAE). Several aircraft were used to land parties at a series of points on the floating ice on the Arctic Ocean during a period of approximately six weeks in spring (Armstrong, 1958).

During the 1948, 1949, and 1950 HLAE, groups of scientists were landed to carry out short-term observations, each lasting from a few hours to a few days, at 87 points distributed over the Arctic Ocean (Somov, 1955). Discoveries included the basic features of relief of the ocean floor and information on the scale and limits of the distribution of the Atlantic water layer. The surveys revealed that, far from being a flat abyssal plain, the Arctic Basin has a very complicated structure with both depressions and submarine ridges and plateaus. The most significant discovery was made on 17 April 1948, at 85°26′N, 145°E, where a relatively shallow depth of 1,290 m was recorded. Further soundings at smaller intervals, the same year and again the following year, revealed the outline of a massive submarine mountain range, rising 3,000 m above the seafloor and extending 1,800 km from Ellesmere Island to the New Siberian Islands. It was named after M. V. Lomonosov, the eighteenth-century Russian scientist, grammarian and poet. These annual springtime High-Latitude Air Expeditions, later renamed NORTH-series, took place every year from 1948 to the present.

In 1950, the Soviets initiated the NORTH POLE-series expeditions (Gordienko and Laktinov, 1960; Treshnikov, 1960). In contrast to the annual, short-period NORTH-series operations, the NORTH POLE (NP) stations remain occupied from one to several years until they either "exit" via the Greenland Current into the Atlantic or otherwise lose their usefulness.

With the exception of 285 depth soundings released as a contribution to the International Geophysical Year (Somov, 1955), Soviet bathymetric data remains classified, unfortunately, and is unavailable to the public.

### The U.S. Contribution

It was not until April 1951 that U.S. scientists joined in the exploration of the Arctic Ocean by mounting two airborne expeditions to the Beaufort Sea (Crary, Cotell, and Oliver, 1952). From oceanographic observations made at 23 stations, Worthington (1953, pp. 550–551) concluded that "there is a submarine ridge, running roughly from Ellesmere to the New Siberian Islands, which separates the deepest water of the Beaufort Sea from the remainder of the basin," and the "sill depth of the ridge should not exceed 2,300 m." Ironically, the Americans were unaware of the Soviet NORTH- and NORTH POLE-series operations and did not know that the existence of the Lomonosov Ridge had in fact been confirmed five years earlier.

The brief-landing type of airborne survey was replaced in 1952 by year-round occupation of drifting ice islands. The best known of these ice islands, Fletcher's Ice Island T-3, has been occupied at intervals from 1952 to 1974 (Hunkins, 1977). Other drifting ice islands that were occupied were Alpha (1957/1958), Charlie (1959), and Arlis II (1961–1965). T-3 and Alpha served as U.S. stations during the International Geophysical Year.

Throughout its occupation, Alpha drifted over a ridge that was parallel to the Lomonosov Ridge. Although identified as a general feature on the 1954 Soviet chart, it was mapped in comparative detail for the first time from Station Alpha, from which it derives its name. Research on these ice islands embraced a broad range of oceanographic, geophysical, atmospheric, and glaciological studies (Hunkins, 1960*a,b,c,d;* Hunkins et al., 1962).

In 1960, the Americans resumed their airborne reconnaissance surveys under the sponsorship of the Office of Naval Research (ONR) and the Geophysical and Polar Research Centre of the University of Wisconsin. Using Cessna 180 aircraft based along the north Alaskan coast and on drifting stations, they landed on the ice of the Beaufort Sea, Chukchi Shelf, and Canada Basin to take depth soundings and gravity measurements. This program continued every spring until 1969, by which time over 800 stations had been established (Wold, 1973).

Since 1957, the U.S. Navy has operated nuclear submarines in the Arctic Ocean. Echograms from the ocean floor obtained from the *USS Nautilus* (1957, 1958), *Skate* (1958, 1959, 1962), *Sargo* (1960), and *Seadragon* (1960, 1962) have been analyzed by Beal (1969). Figure 29-3 shows the cruise tracks of the four submarines.

For lack of other depth measurements the data from these echograms were used extensively in the compilation of bathymetric maps although the positioning of these early submarines was very inaccurate (Beal, 1983), resulting in displacement errors as high as 70 km (Weber, 1983). Unfortunately, since 1963 all U.S. submarine-collected bathymetry has been classified and is unavailable to cartographers.

In 1951, the icebreaker *USS Burton Island,* and in 1957, the icebreakers *USS Staten Island* and *USCGC Northwind* carried out continuous echo soundings across the continental slope into the Beaufort Sea north of Point Barrow (Fisher, Carsola, and Shumway, 1958).

During the course of conducting seismic reflection surveys across the continental shelf into the Beaufort Sea, bathymetric data were acquired by the U.S. Geological Survey between 1972 and 1978, and by the Geophysical Corporation of Alaska in 1973 (Greenberg, Hart, and Grantz, 1981). Based on all the data available, including Canadian soundings east of longitude 141°W, Greenberg, Hart, and Grantz (1981) published a bathymetric map of the continental shelf, slope, and rise of the Beaufort Sea north of Alaska at a scale of 1:5 000 000.

From 1979 to 1983, ONR carried out four springtime expeditions, which were named FRAM I to IV, into the part of the Eurasia Basin adjoining the Yermak and Morris Jesup plateaus. Studies conducted from these drifting ice stations included bathymetry (Kristoffersen, 1982*a*; Manley et al., 1982; Kristoffersen, 1982*b*).

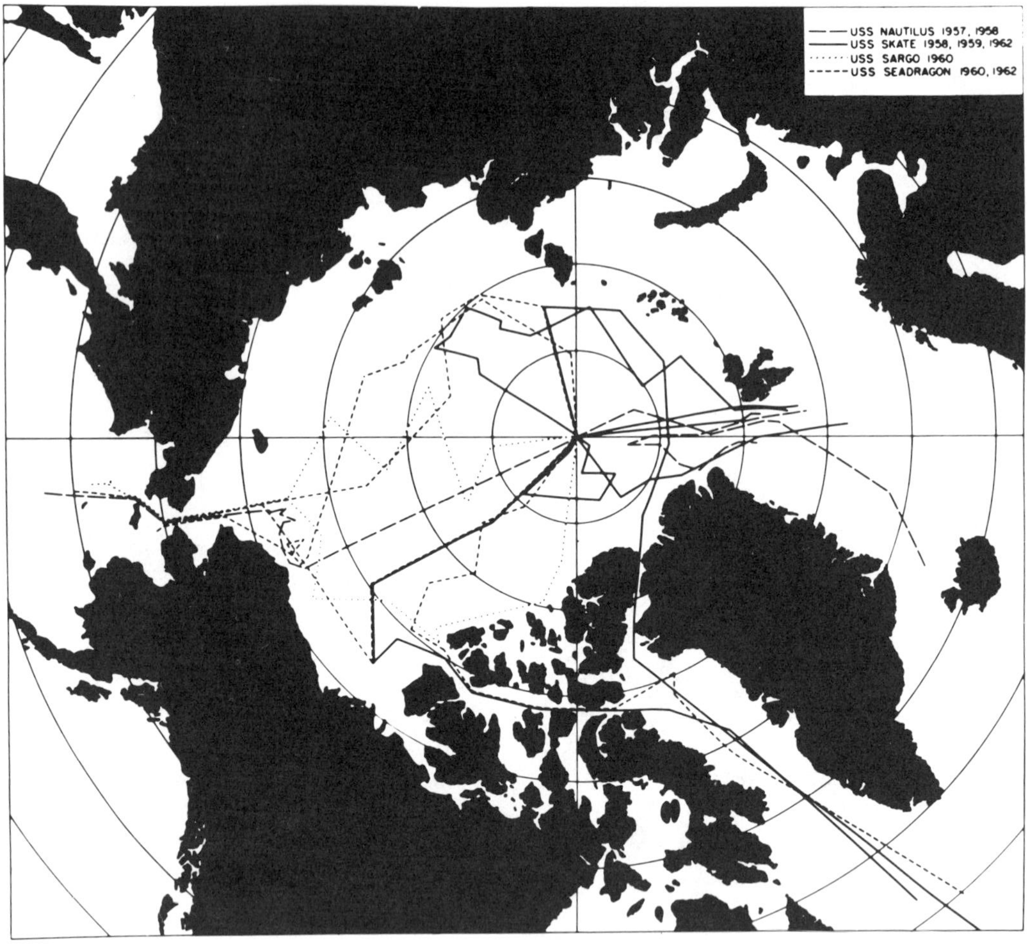


**Figure 29-3.** U.S. nuclear submarine tracks in the Arctic Ocean 1958–1962. (From Committee on Polar Research, 1970, p. 51.)

During the MIZEX (Marginal Ice Zone Experiment) 1983 and 1984 operations, sponsored by ONR, some bathymetric data were collected in the Fram Strait.

### *The Canadian Contribution*

Systematic exploration of the Arctic Ocean between the 48°W meridian (Lincoln Sea), the 141°W meridian (Yukon–Alaska border), and the north polar region began as a direct result of the creation, in 1958, of the Polar Continental Shelf Project (PCSP). The PCSP is under the jurisdiction of the Canadian Department of Energy, Mines and Resources. Its broad mandate (Weber, 1983) includes the charting of the polar continental shelf by the Canadian Hydrographic Service, a task started in 1960 and now almost complete except for a gap off Axel Heiberg Island. Figure 29-4 shows the locations of all the bathymetric and gravity stations that were established in the Arctic Ocean with the support of PCSP between 1960 and 1985. PCSP also supported four polar expeditions: the Canadian North Pole Expedition 1967 and 1969, LOREX 79, and CESAR 83. The studies conducted during the first two small-scale expeditions, the forerunners of

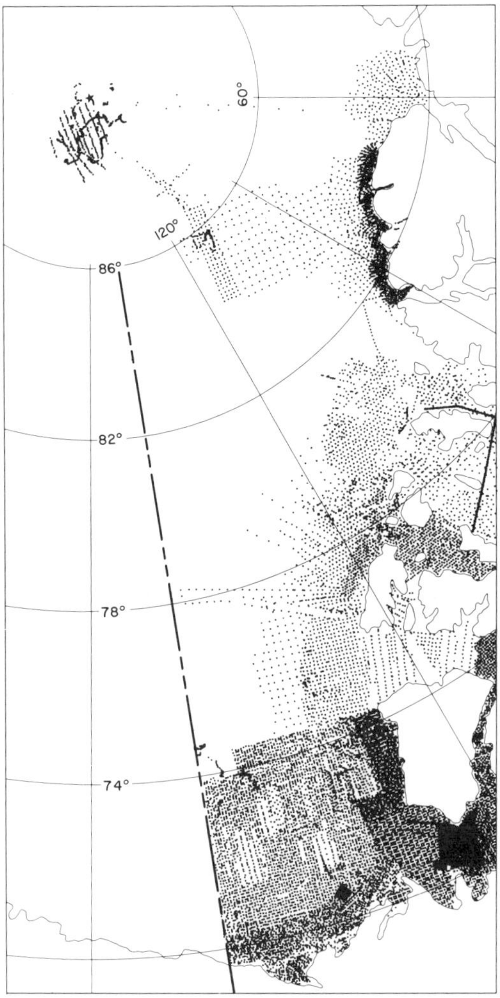


Figure 29-4. Plot of Canadian bathymetric and gravity stations established on the Arctic Ocean and adjoining interisland waters as of 1985. Stations near the North Pole were established primarily on LOREX in 1979 and those off Ellesmere Island on CESAR in 1983. Lines of stations off Axel Heiberg Island are located along seismic lines established from an ice island by EPB and Atlantic Geoscience Centre, EMR, in 1985. The ice island, used as forward base, has been occupied by PCSP since 1984.

LOREX and CESAR, included the testing of modern navigational aids (precision celestial observations using computers, Omega radio navigation, and Transit satellite navigation) and the establishment of some 70 depth soundings and gravity measurements north of 89°20′ latitude as well as a line of bathymetric and gravity stations between Lincoln Sea and the North Pole.

The Lomonosov Ridge Experiment, code named LOREX 79, was a large-scale, multidisciplinary project to study the nature and origin of the Lomonosov Ridge (Weber, 1979). Bathymetric information was obtained from echo soundings along some 800 km of drift track of the main camp and two satellite camps and at 270 helicopter landing sites.

During CESAR 83 (a code name for Canadian Expedition to Study the Alpha Ridge), an operation of similar scale and nature as LOREX (Weber, 1986), some 1,300 spot soundings were taken over the Ellesmere Island continental shelf and Alpha Ridge (Fig. 29-4).

In 1984, PCSP established a permanent camp on an ice island, measuring some 35 $km^2$ in areal extent, that had broken off the Ward Hunt Ice Shelf the previous year (Jeffries, 1985). In 1985, using the ice island as a forward base when it was located off Nansen Sound, some 77 bathymetric and gravity stations were established over the continental shelf in conjunction with a seismic refraction survey (Asudeh, Green, and Forsyth, 1985) (Fig. 29-4).

### The Norwegian, British, and Swedish Contributions

During the last two decades Norwegian fishing boats, equipped with echo sounders and radio navigational aids, have collected bathymetric data in the Barents Sea as far east as Novaya Zemlya. Kristoffersen from the Norwegian Polar Institute assembled these data and compiled an as yet unpublished bathymetric chart of the Barents Sea. This information has been incorporated in Perry and coworkers' (1985) bathymetric chart of the Arctic Ocean.

In 1975 and 1976, the British nuclear

submarine *Sovereign* made a number of cruises into the Arctic Ocean. Some of its nonclassified bathymetric data have been used in the compilation of the 1979 GEBCO map (Sobczak, pers. comm., 1980.

In 1980, in commemoration of Nordenskjold's first voyage through the Northeast Passage, the Swedish navy's icebreaker *Ymer* carried out oceanographic and geophysical studies in Fram Strait and in the Barents Sea, which included bathymetric measurements over the Yermak Plateau.

## The Maps and Charts of the Arctic Ocean

### *First Soviet Map (1954)*

By 1954, the Soviets had established over 2,000 soundings, resulting in the compilation of a map of the Arctic seafloor that revealed the existence of the Lomonosov Ridge for the first time (Burkhanov, 1956). The map caused a sensation among Arctic experts outside the Soviet Union, who until 1954 had been unaware of the scope and magnitude of Soviet exploration.

### *DRB Chart (1956)*

In 1956, the Defense Research Board of Canada (DRB) compiled a bathymetric chart of the Arctic Ocean (Fig. 29-5). It was compiled by using the Soviet map as a base and adding U.S. airborne and T-3 data, and by recontouring Somov's NP2 soundings (Greenaway, 1983). Here, for the first time, the Alpha-Mendeleev Ridge is outlined as separating the Makarov Basin from the Canada Basin. For many years the DRB map remained the best bathymetric chart of the Arctic available to the public outside the Soviet Union.

### *Second Soviet Map (1960)*

By 1960, the Soviets had established over 20,000 soundings. These data resulted in the publication of a second Soviet map (Gordienko and Laktionov, 1960; also fig. 9 in Weber, 1983). Although generalized because of its small scale and because depth soundings are highly classified in the Soviet Union, it clearly delineates the Canada Basin, the Alpha-Mendeleev Ridge, and the Makarov Basin, but the Nansen-Gakkel Ridge is not yet clearly defined. No Soviet maps have since been published that show significantly more detail.

### *Heezen and Ewing (1961)*

Based on inference from the location of earthquake epicenters and by reexamining Soviet data, Heezen and Ewing (1961) produced a revised chart of the Eurasia Basin that shows the midocean ridge extending from the Lena Trough to the Laptev shelf and bisecting the Eurasia Basin roughly halfway between the edge of the Barents continental shelf and the Lomonosov Ridge. The existence of the Nansen-Gakkel Ridge, as the Arctic Mid-Ocean Ridge was later named, was not unequivocally confirmed until 1969 with the publication by Beal of the U.S. submarine bathymetry. Virtually no new unclassified depth data from the Eurasia Basin have become available until 1979 to 1983 when detailed soundings were made near the Atlantic end of the ridge during the FRAM expeditions. This leaves most of the Nansen-Gakkel Ridge still poorly mapped.

### *Ostenso's Bathymetric Map (1961)*

In 1961, Ostenso compiled a new map of the Arctic Ocean seafloor based on all available unclassified material (Ostenso, 1962, 1963). The map is contoured at 200, (500), 1,000, 2,000, 3,000, (3,800), 4,000 and 5,000 m and as published is plotted at a scale of about 1:18 000 000.

### *Canadian Bathymetric Chart (1967)*

In 1967, the Canadian Hydrographic Service published two charts of the Arctic

Ocean north of 72° from longitude 0° to 90°W (map sheet 896) and from 90°W to 180°W (map sheet 897). Beyond the 200 m isobath the chart is contoured at 500-m intervals; the scale is 1:2 000 000. The bathymetry is color coded with six color ranges and land masses are left blank. Except for the continental shelf off the Queen Elizabeth Islands, which had been charted by PCSP, and the Chukchi Rise, Alpha Ridge, and Canada Basin, where the University of Wisconsin had been carrying out sounding operations, the chart does not differ significantly from Ostenso's.

### *GEBCO Map (1968)*

The Canadian Bathymetric Chart of 1967 served as the basis for the General Bathymetric Chart of the Oceans (GEBCO), 3rd edition, sheets CI and CII, which were published in Paris in 1968 by the Institute Géographique National at a scale of 1:3 000 000. The land topography is contoured at 1000-m intervals and color coded. Other than the scale, and the fact that the GEBCO bathymetry shows more soundings, it is identical to Canadian Charts 896 and 897.

### *The British Admiralty North Polar Chart (1969)*

In 1969, The United Kingdom Hydrographic Department published the North Polar Chart, Admiralty Chart No. 4006. It was compiled by Ritchie (1969) at an approximate scale of 1:7 500 000. Beyond the 100-fathom isobath the depth is contoured at 500-fathom intervals. The chart does not present any new information, and the contouring is essentially an interpolation, in the old depth units, of the Canadian Hydrographic Charts 896 and 897.

### *Heezen and Tharp's Map of the Arctic Region (1975)*

In 1975, the American Geographical Society published *Map of the Arctic Region,* sheet 14 of The World Series. The colored map, scale 1:5 000 000, measures approximately 120 × 150 cm; 18 color steps cover an elevation range from a depth of 8,000 m in the Aleutian Trench to a height of over 5,000 m in the St. Elias Range. The map was compiled by Bruce C. Heezen and Marie Tharp of the Lamont-Doherty Geological Observatory, based on all available unclassified soundings and on extensive use of Beal's (1969) submarine depth profiles. The ocean floor is depicted in great detail everywhere with contour intervals as small as 10 m, without regard to how much or how little data are available. Unfortunately, many of the details are largely a fiction of the imagination, and because no distinction is made between fact and speculation it is the least reliable of the modern charts.

### *Sobczak's Arctic Bathymetry (1978)*

In 1978, the Earth Physics Branch of EMR published as part of *Arctic Geophysical Review* (Sweeney, 1978) a new map of the Arctic Ocean Basin. It was compiled by Sobczak and Sweeney (1978) at a scale of 1:7 500 000 and is contoured at 500-m intervals beyond the 200-m isobath. Although nearly 250 000 digitized soundings were used, the distribution is very uneven. For example, not a single sounding is available from the Laptev shelf side of the Eurasia Basin south of 85°. The data used by Sobczak, and earlier by Heezen and Tharp, originate from the same sources, yet the interpretation is quite different. Where the contouring is necessarily speculative, the isobaths are represented by broken lines.

### *GEBCO Map (1979)*

In 1979, the Canadian Hydrographic Service published Chart No. 517 of the General Bathymetric Chart of the Oceans under the authority of the International Hydrographic Commission. It encompasses the Arctic regions north of 64° and is printed at a scale of 1:6 000 000. The topo-

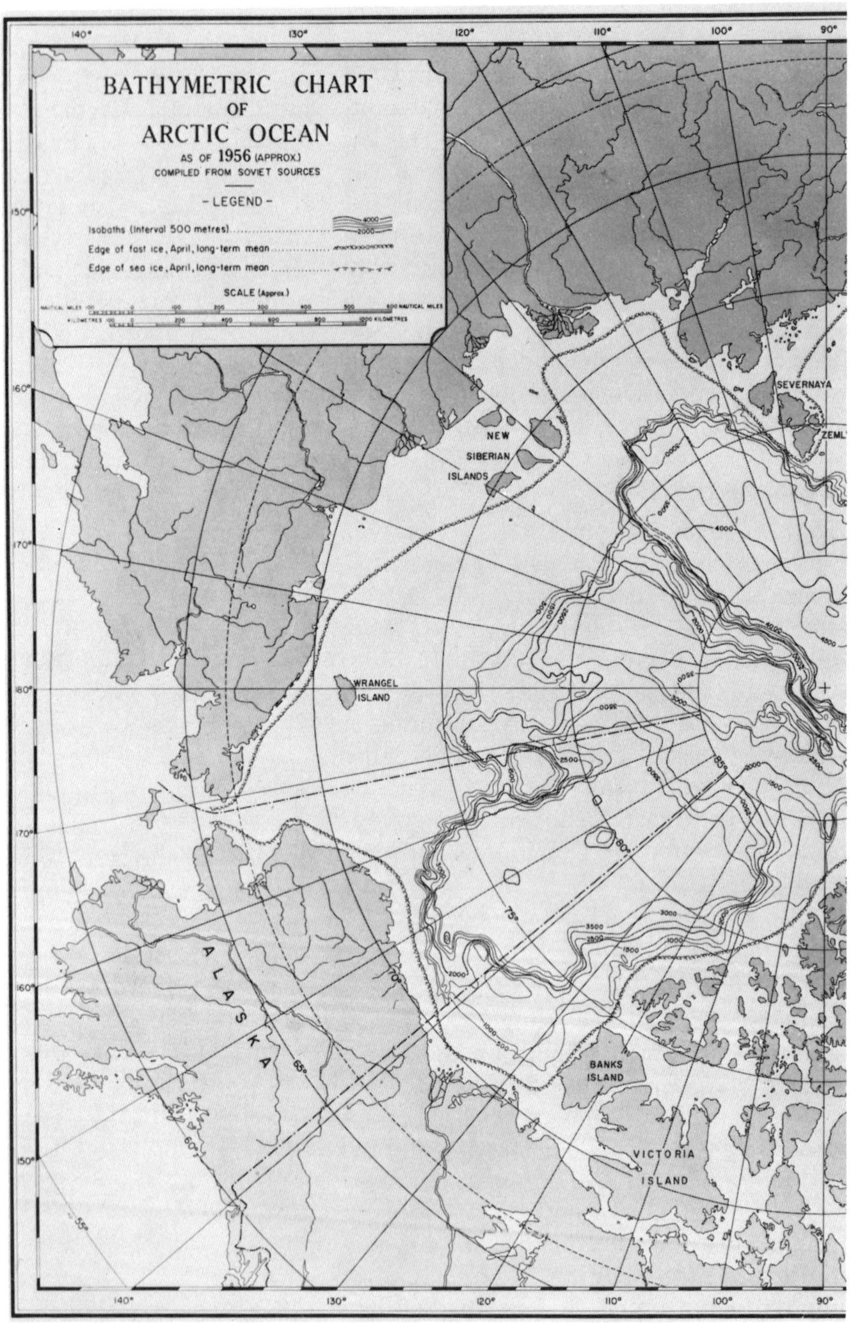


Figure 29-5. Defense Research Board of Canada bathymetric chart (1956) of the Arctic Ocean. This is the first chart that delineates the Alpha-Mendeleev Ridge complex including Marvin Spur; it was compiled by combining Soviet (inset) and T-3 data. The chart shows Nansen's Sill, which was believed to connect Greenland with Svalbard and to impede

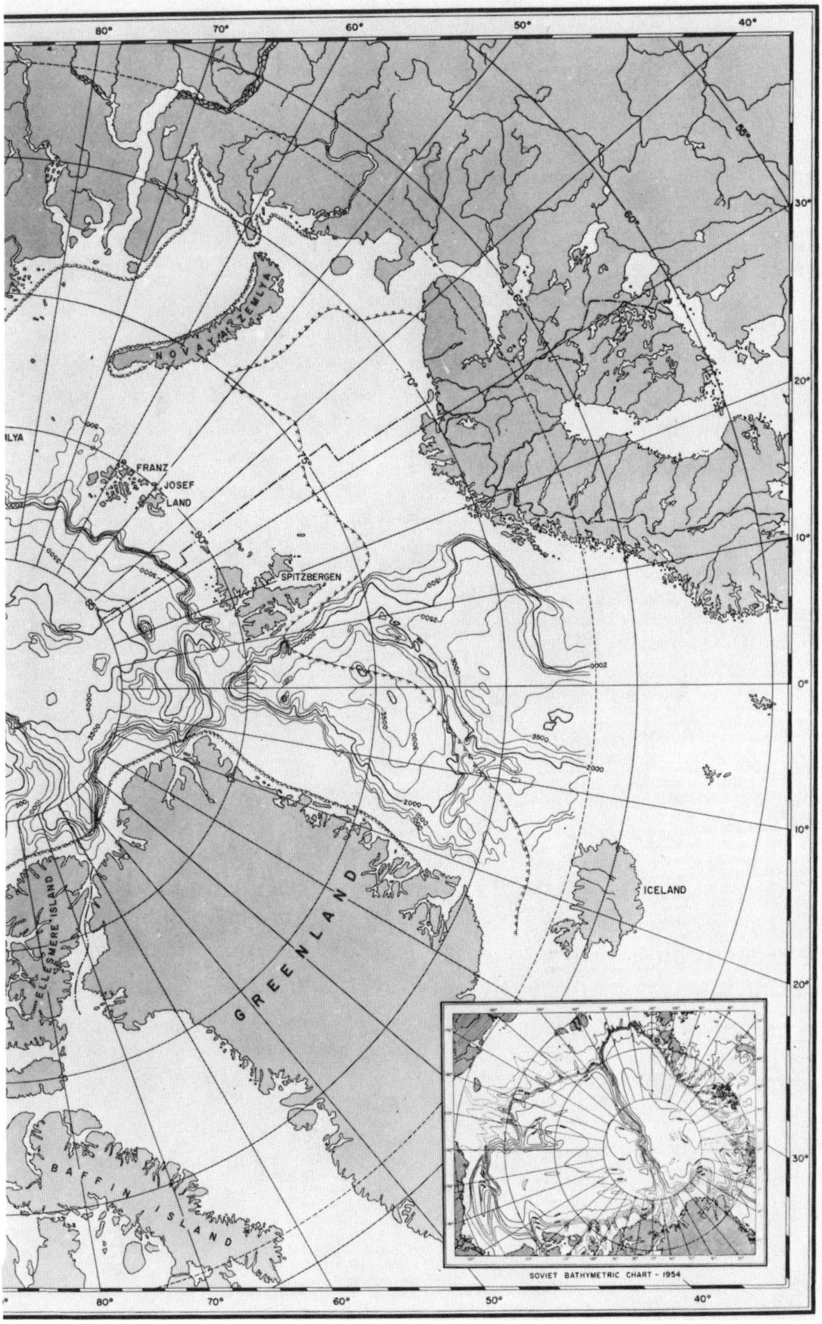


water exchange between Arctic and Atlantic oceans; its existence was disproved with the discovery in 1957 (Laktionov, 1959) of the Lena Trough (Fig. 29-7). The chart also still shows the Beaufort Sea as an epicontinental sea; it exists in name only and not as a separate physiographic feature of the Canada Basin (Fig. 29-6).

graphic and bathymetric color scheme is the same as for Heezen and Tharp's chart. The contour intervals range from 1,000 m to 100 m depending on the sounding density. Unlike Heezen and Tharp's chart, however, the location of surface ship, submarine, and drifting station tracks, as well as individual soundings, are plotted, and areas with high-density soundings, such as over the Canadian Polar Continental Shelf, are delineated. This allows the user to gauge the reliability of the chart for any particular area. Contouring is conservative, and in the Arctic Ocean is generally the same as Sobczak's 1978 chart.

### *Perry et al.'s Bathymetry of the Arctic Ocean (1985)*

In 1985, the U.S. Naval Research Laboratory published a preliminary bathymetric chart of the Arctic Ocean at a scale of 1:4 704 475 (Perry et al., 1985). The 14-color coded chart measures 107 cm by 149 cm and contour intervals range from 20 m to 1,000 m depending on the available data density. It was compiled as a base map for the Decade of North American Geology series publication and includes all the data collected until 1984. The 500 m interval contours over the uncharted areas of the Arctic Ocean have been updated on the basis of classified U.S. submarine data (G.L. Johnson, pers. comm., 1984). For the general public, of course, it is impossible to estimate the accuracy of these contours. Except for a few minor modifications the bathymetric maps of the Amerasia Basin (Fig. 29-6) and Eurasia Basin (Fig. 29-7) have been compiled from Perry et al.'s chart.

### *CESAR Bathymetric Map (1986)*

During the CESAR expedition in 1983, in cooperation with the Canadian Hydrographic Service, some 1,300 spot soundings were taken from the coast of Ellesmere Island over the continental shelf and Alpha Ridge to the 120°W meridian (Fig. 29-4). A 100-m contour map was compiled incorporating all publicly available sounding data from U.S. stations T-3, Alpha, Arlis II, U.S. airlifted stations, CHS stations, Canadian North Pole Expeditions 1967 and 1969, LOREX 79, and CESAR 83 (Weber, 1987*a*). The chart extends from the Ellesmere Island continental shelf across Alpha Ridge, Makarov Basin, Lomonosov Ridge, to the North Pole area. It is divided into three maps as illustrated in Figure 29-8*a–c* (pages 812–814). Dots indicate the sounding locations used for contouring. Blank areas, where no data are available, were filled in with 500-m contour lines (broken) taken from the latest preliminary *Bathymetric Map of the Arctic Ocean* (Perry et al., 1985).

## CONTINENTAL MARGINS

### Continental Shelves, Slopes, and Rises

Two-thirds of the Arctic Ocean is underlain by continental shelves, the largest of which are the very wide and shallow Siberian and European shelves. In contrast, the Greenland and North American continental shelves are much narrower and deeper.

### *Greenland*

Following the coast of Greenland westward its continental shelf narrows from nearly 300 km along the Greenland Sea to 20 km off the northeast coast, then widens again to some 90 km off the north coast, where it merges with the Morris Jesup Plateau (Fig. 29-7). The depth of the continental shelf break ranges between 350 m and 450 m.

The Greenland continental shelf is characteristic of shelves along glaciated coasts and includes an inner shelf—the strandflat of Nansen (1922)—which, seaward of Greenland, is generally a submerged continuation of the Paleozoic and Precambrian

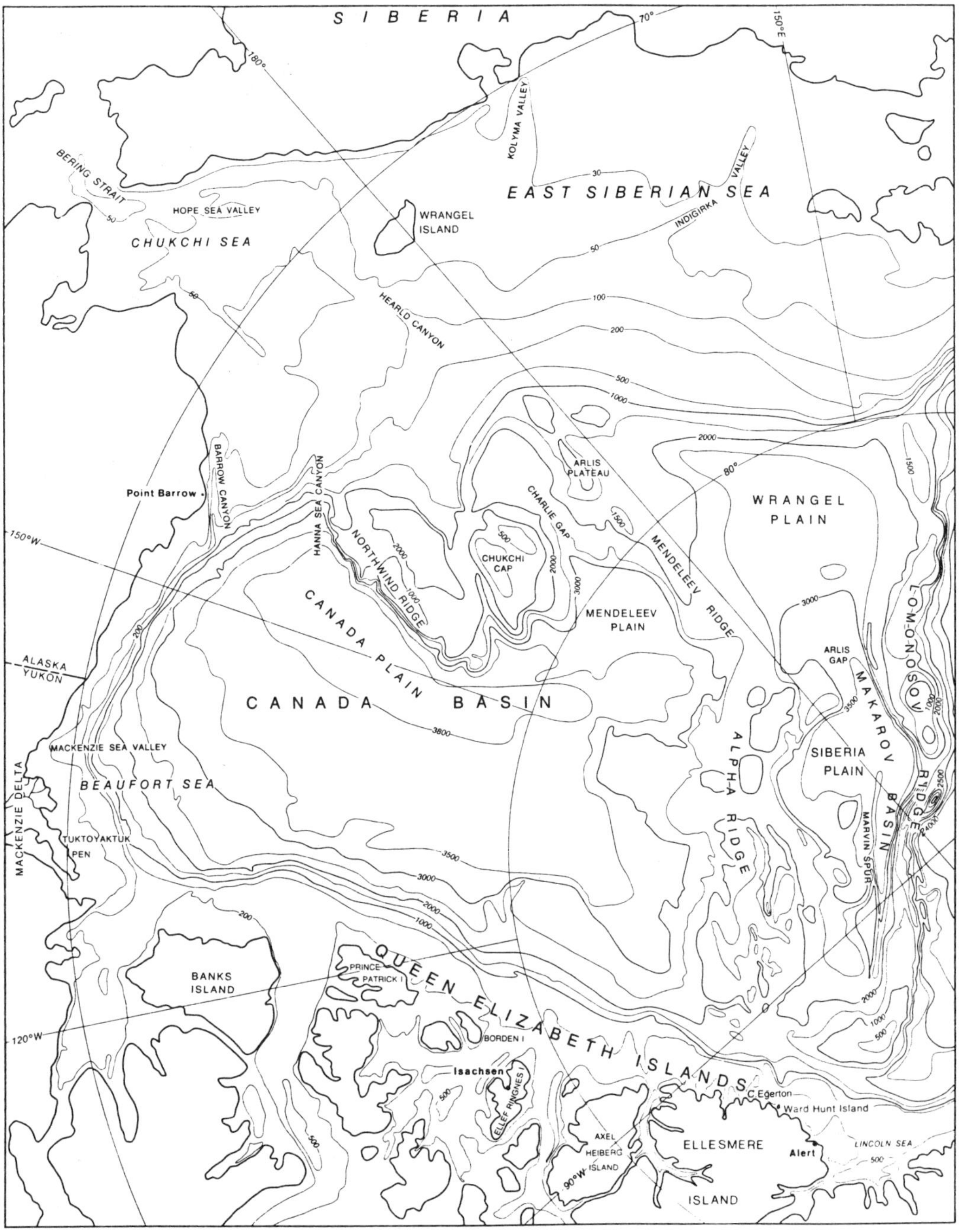


Figure 29-6. Bathymetric map of the Amerasia Basin. Except for some minor modifications of the contouring over the Alpha Ridge, based on the CESAR bathymetry (Fig. 29-8*b*), the contours agree with those on Perry et al.'s 1985 chart.

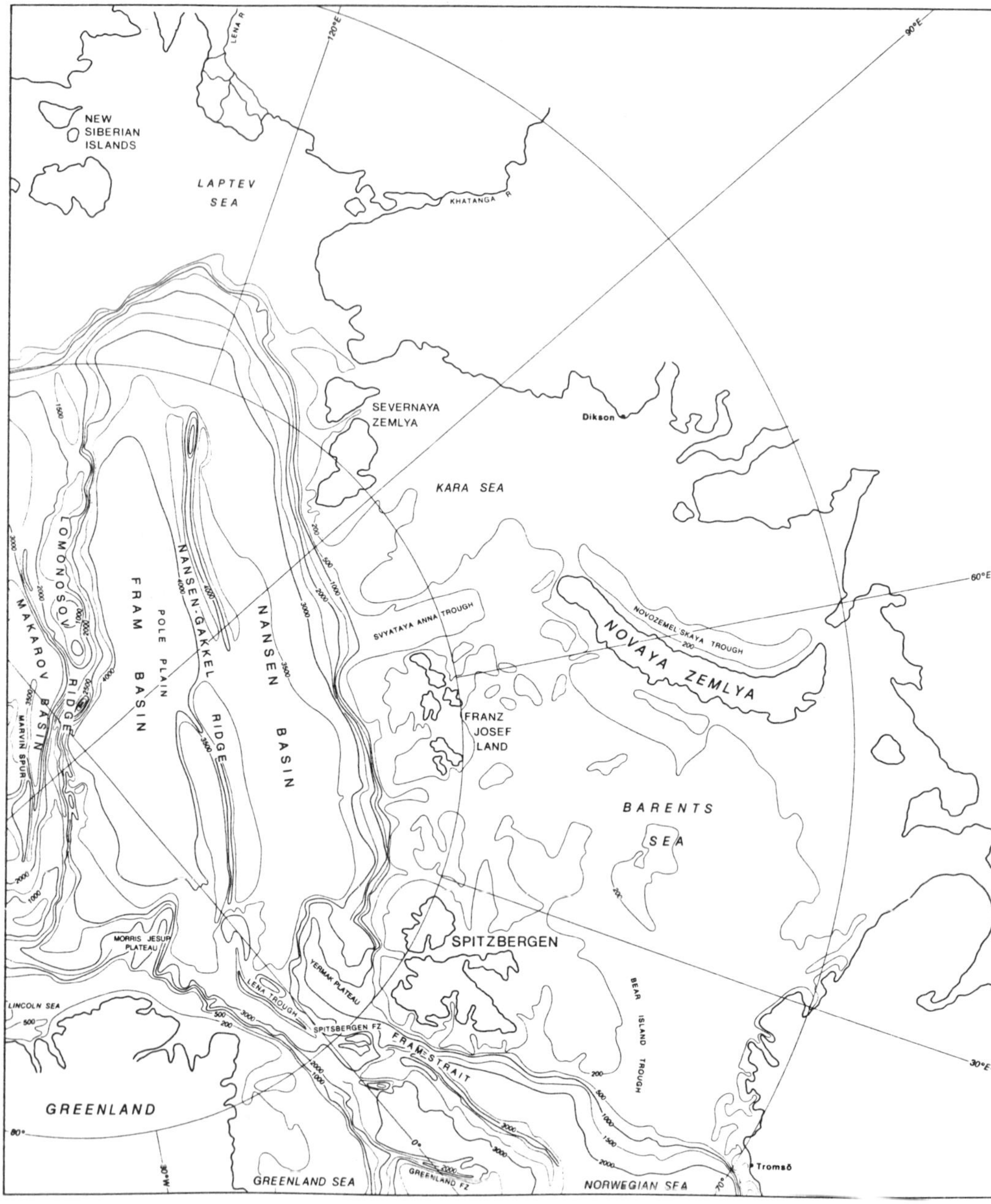


Figure 29-7. Bathymetric map of the Eurasia Basin compiled from Perry et al.'s 1985 chart. The contouring of the Nansen–Gakkel Ridge has been simplified.

continental basement. The outer banks and strandflat are usually separated by relatively long, narrow depressions several hundred meters deep. These marginal channels are generally parallel to the coastline (Holtedahl and Holtedahl, 1961). During glacial periods, glaciers tended to flow along these shallow channels, breaking seaward locally to form transverse valleys. The advancing glacial ice abraded the channels, which originally were quite shallow, as well as shallower parts of the outer shelf. In many places the outer shelf is characterized by banks that were presumably constructed by glaciers. These banks, which are thought to be terminal or lateral moraines of glaciers that grounded or terminated on the shelf, have been notched or beveled by wave action during lower sea-level stands (Johnson et al., 1975; Johnson, Vanney, and Hayes, 1982; Johnson, Grantz, and Weber, 1986).

The Greenland continental slope is well developed, with gradients ranging from 20 m/km to 65 m/km. North of the Greenland fracture zone, the base of the continental slope generally lies below 2,000 m. South of the fracture zone, the base of the slope is found at depths of 1,500–3,000 m. The continental rise is about 75 km wide on the south side of the Greenland fracture zone. It narrows abruptly at the fracture zone, and averages 55 km in width to the north of it. This narrowing is probably a reflection of increasing distance from the sediment source. The rise pinches out at about 80°30′N, where the Spitsbergen fracture zone impinges upon the Greenland continental block (Johnson, Grantz, and Weber, 1986).

### *Canada*

The Canadian continental shelf ranges in width from 65 km to 180 km (Fig. 29-6). Except for the northernmost part of Ellesmere Island, where the shelf is shallower, the depth of the continental shelf break seaward of the Queen Elizabeth Islands and Banks Island varies between 400 m and 650 m. This is much deeper than the world average of 200 m (Worzel, 1968), suggesting that the present day subsidence rate there is greater than the sedimentation rate. Perhaps glacial isostatic rebound is incomplete. At the Mackenzie Delta, where sedimentation is intense, the shelf is much shallower and the inner shelf break lies between 60 m and 100 m.

The Canada continental shelf is bisected by a number of canyons that trench toward glaciated fiords (Ellesmere and Axel Heiberg islands) or interisland channels. Generally, water depths are greatest in the fiords and interisland channels and shoal as much as several hundred meters toward the continental edge. The nonglaciated coasts of the Queen Elizabeth Islands and mainland do not have the coast-parallel marginal channels found off Greenland, but they have been subjected to scour and gouging by sea ice.

The character of the continental slope in the Lincoln Sea is unknown. To the southwest, however, the continental shelf off Ellesmere Island is flanked by a saddle as much as 1,800–2,000 m deep that separates it from the Alpha Ridge (Weber, 1987*a*). Opposite the Alpha Ridge, the depth of the continental shelf break ranges from 250 m at Cape Albert Edward to 400 m at Cape Egerton (Fig. 29-6). The slope gradient in this area ranges from 10 m/km to 20 m/km, and the slope bottoms out in a saddle at depths of 1,300–1,650 m. The continental slope seaward of Ellef Ringnes, Borden, and Prince Patrick islands consists of an inner slope with a gradient of 12–30 m/km and an outer, more gentle slope with a gradient of 5–15 m/km. Off Banks Island, the slope has a gradient that averages 20 m/km, and the base of the slope is slightly deeper than 3,000 m. Seaward of Tuktoyaktuk Peninsula the gradient is 12 m/km, and the base of the slope rises to a depth of only 1,500 m. The southward decrease in the depth to the base of the continental slope is a consequence of the abutment of the large Mackenzie Cone continental rise sedimentary prism against the continental slope. The surface of this large depositional prism commences in the Mackenzie Delta and merges with the Canada Abyssal Plain 400–450 km to the west at depths of about 3,600 m. The Mackenzie Sea Valley that cuts the continental shelf and slope is the major conduit of debris to the Canada Basin (Grantz et al., 1981). Ice

Figure 29-8. CESAR 83 and LOREX 79 bathymetric maps. Contour interval 100 m. Consecutive dots are drift tracks of ice stations; individual dots are locations of spot soundings. Uncharted areas have been filled in by 500 m dashed contour lines taken from Perry et al.'s 1985 chart.

a

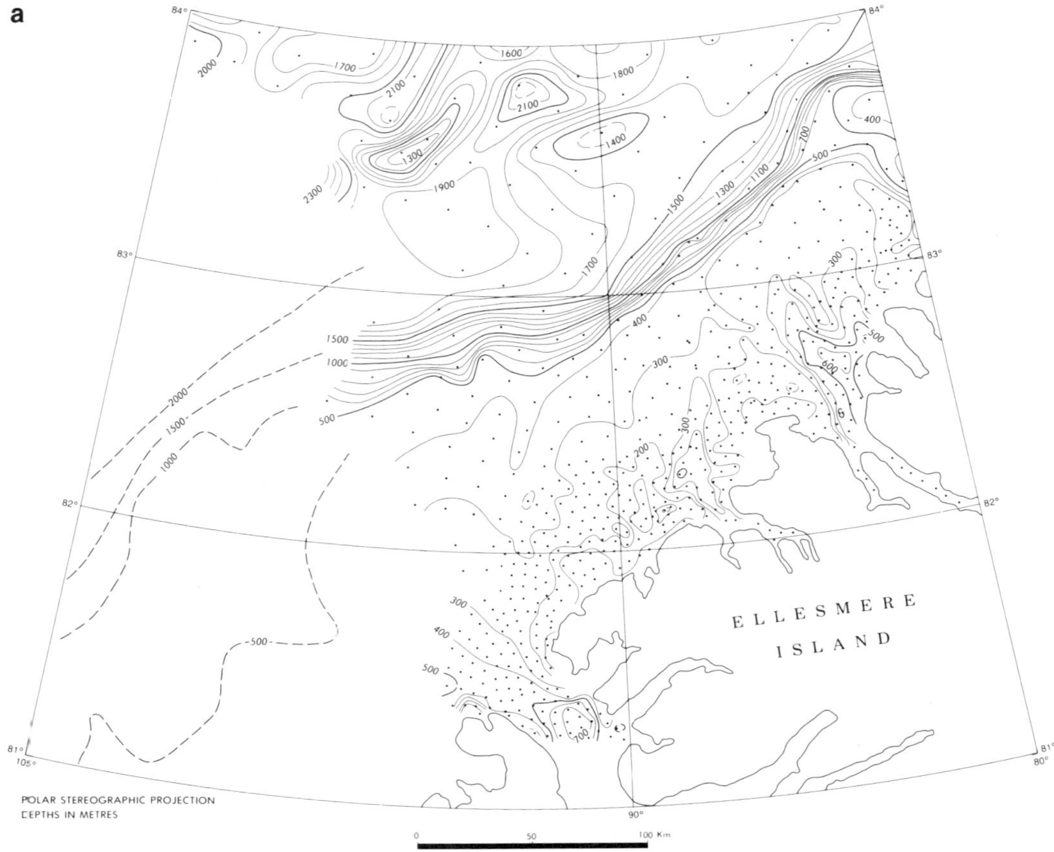

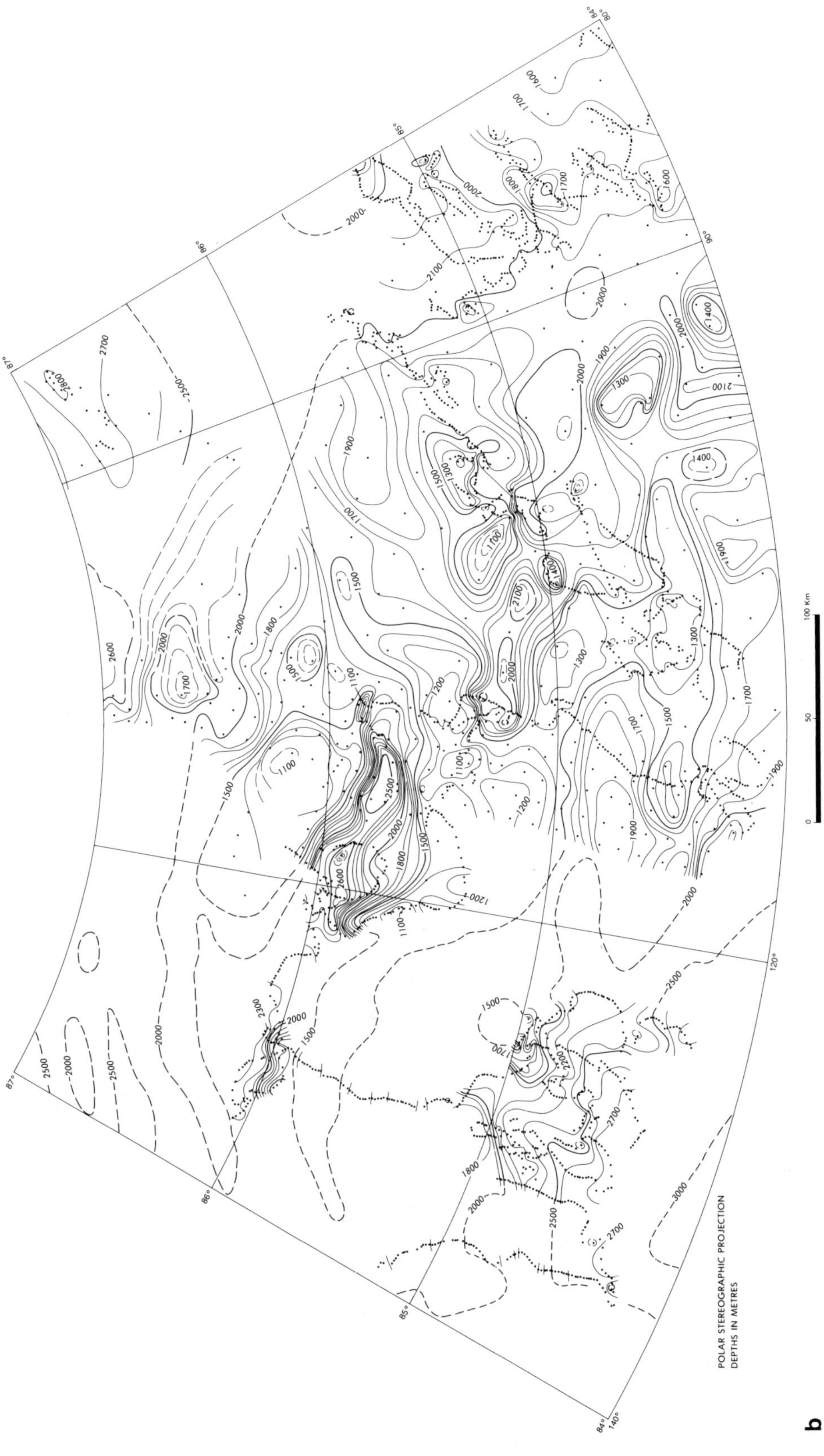
POLAR STEREOGRAPHIC PROJECTION
DEPTHS IN METRES
0
50
100 Km
80°
84°
85°
86°
87°
90°
120°
140°

b

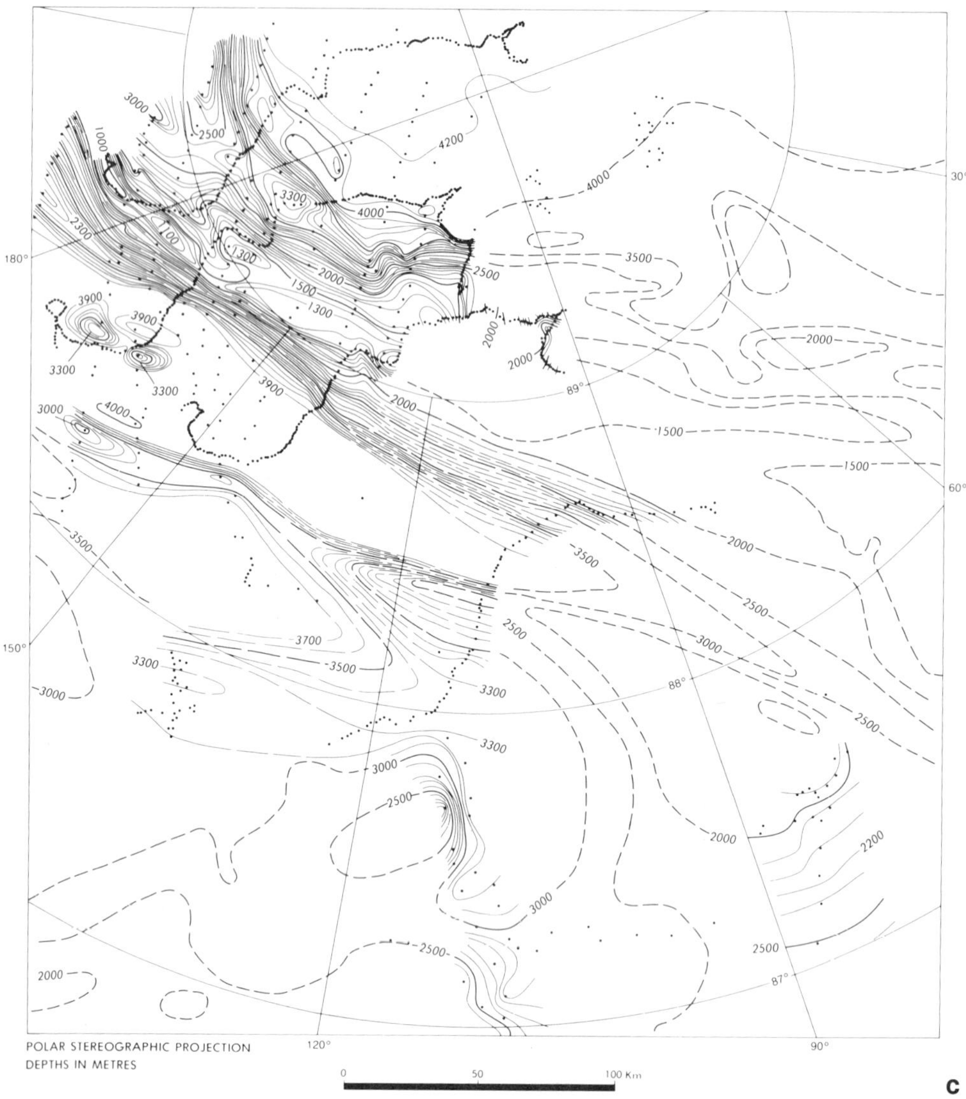


Figure 29-8. *(continued)*

lobes and meltwater of the Pleistocene Laurentide ice sheet were major factors in carving this sea valley.

### Alaska

The continental shelf north of the Alaska facing the western Beaufort Sea is narrow with complex morphology. The shelf varies in width between 70 km and 120 km. It extends seaward with a gentle gradient from the barrier islands or low bluffs at the coast to a complex outer continental shelf break at depths of 200–800 m. East of 147°W the shelf consists of an inner section with a gradient of about 1 m/km that extends from shore to a slump-controlled break in slope near the 60 m isobath, and an outer section with a gradient of about 16 m/km that extends from the 60 m isobath to the outer shelf break.

The inner shelf west of the 151°W longitude is basically a wave-cut surface, although predominantly fine-grained Quaternary clastic sediments, locally as much as 140 m thick, have been deposited upon it. East of 151°W Neogene and Quaternary clastic sediments have been deposited on the inner shelf, which has experienced very little erosion (Grantz et al., 1981). However, west of 151°W, fine-grained Quaternary clastic sediments forming pockets up to 140 m thick have been deposited on the inner shelf even though the surface has been wave-cut.

The inner shelf morphology has been shaped by glacier and sea-ice movements. Pleistocene glaciers apparently reached the shelf east of 143°W but to the west the effect of the glaciers on the inner shelf was minimal. The ice gouges, produced by grounding of deep-draft sea-ice pressure ridge and ice islands that are driven westward along the shelf by a wind-driven clockwise current, are generally parallel to the coast and extend from nearshore to at least the 75 m isobath (Reimnitz and Barnes, 1974). Those related to the present stand of sea level are abundant to water depths of 30 m. These ice gouges are usually 1–3 m deep (Grantz et al., 1981), but some as deep as 5.5 m have been observed (Reimnitz and Barnes, 1974).

The continental slope north of Alaska follows a gentle arc from the Mackenzie Bay to the Northwind Ridge of the Chukchi Borderland. Its junction with the continental rise lies at depths of 1,100–2,000 m and its slope ranges generally from 7 m/km to 22 m/km. Numerous slumps, landslides, and incised channels characterize the continental slope. The largest slide masses originate at or near the outer shelf break. Along many profiles the vertical drop at the headwall of these slides is 900–1,200 m, and the head-to-toe length of some large slides exceeds 35 km (Grantz et al., 1981).

Seismic and bathymetric profiles show that many canyons incise the Alaska continental slope. Some appear to be slump scars; others contain bedded sediments, are bordered by levees, and appear to be turbidity current channels (Johnson, Grantz, and Weber, 1986). The Barrow Sea Valley and Barrow Canyon, which cut across continental shelf and slope, delineate the western boundary of the continental margin. The axis of Barrow Sea Valley is incised into the continental shelf about 20 km off the northeast-trending coast. It is a broad, flat-bottomed channel 200 km long and 2–8 km wide with an overall axial gradient of about 1 m/km. About 20 km south of the continental shelf break, the sea valley merges into Barrow Canyon. The canyon attains its maximum relief of 900 m where it crosses the projection of the shelf break. Its average axial gradient is 30 m/km but a steep upper segment slopes at 48 m/km. Barrow Sea Valley is thought to have been incised into the shelf and slope by subaerial streams during Pleistocene regressions in sea level and by the northeast-flowing Alaskan coastal water during interglacial times. Barrow Canyon is presumed to have been cut by turbidity currents.

The continental rise north of Alaska is much narrower (about 35 km) than along the Canadian sector.

### Siberia

The continental shelf off the Siberian and Alaskan northwest coast, including the Chukchi, East Siberian, Laptev, and Kara shelves is one of the widest in the world and very shallow.

The Chukchi shelf lies between Bering Strait on the south, Barrow Sea Valley on the east, and Wrangel Island (180°) on the west. This broad, 900 km square shelf was eroded far into the continent and transects the principal mountain ranges of northern Alaska. The shelf is characterized by low bathymetric gradients (Grantz et al., 1981). Hope Sea Valley and its continuation, Herald Canyon, cut across the Chukchi shelf and slope from Bering Strait to the Chukchi Abyssal Plain between 170°W and 180°W (Grantz et al., 1981). Hanna Sea

Canyon cuts across the Chukchi slope between 165°W and 158°W and enters the Canada Abyssal Plain at the foot of the Northwind Ridge.

Nearshore sand bars, ice gouges, and large sediment dunes characterize the Chukchi shelf. Well-defined ice gouges have been found where the seabed is as deep as 60 m, but they are most abundant in depths of 25–40 m (Toimil, 1978). Most are 1–3 m deep. The deepest ice keels that have been measured in the western Arctic do not extend more than 50 m, and most do not extend more than 25 m below sea level. Thus the gouges found at depths of 60 m in the north-central Chukchi Sea were probably made prior to 3,000–4,000 years ago, when sea level was either eustatically lower than at present (Grantz et al., 1981), or the shelf was higher in elevation and has subsided since.

The bathymetry of the East Siberian, Laptev, and Kara seas is poorly known. The continental shelf of the East Siberian Sea, up to 800 km wide, is a flat, very shallow plain. Only the Indigirka and Kolyma submarine valleys, which reflect submerged river channels, provide relief. The extreme shoalness is demonstrated by the fact that 350 km from shore west of the Kolyma Valley the shelf is less than 30 m deep. West of Indigirka Valley water depths 250 km from shore range from 10 m to 15 m. The major topographic features of the shelf floor were shaped by grounded ice and subaerial erosion during glacial low sea-level stands, as well as by river sedimentary discharges and tectonic movements. The Laptev Sea is cut by numerous transverse submarine valleys of erosional and tectonic origin that can be traced into landward extensions (Holmes and Craeger, 1974). Seaward of the shallow shelf a prominent bench is shown between 100 m and 200 m (Perry et al., 1985). Its origin is uncertain, but it may be structurally controlled in a manner similar to parts of the boundary between the inner and outer shelf seaward of Alaska (Johnson, Grantz, and Weber, in press). Sediments originating mainly from the Lena and Khatanga rivers are transported into the Fram and Nansen basins by turbidity currents, thereby slowly burying the eastern end of the Nansen–Gakkel Ridge.

The East Siberian and Laptev continental slopes, although scarcely known, appear to be typical of other continental slopes as inferred from a number of U.S. submarine echograms (Johnson, Grantz, and Weber, in press). All show irregularities indicative of either canyons or slumps. The continental rise is well developed landward of the Wrangel Abyssal Plain, and the 2,500 m depth of this plain is at least partially a function of sediment fill behind a bedrock sill (Kutschale, 1966).

The floor of the Kara Sea, with an average depth of 100 m and a maximum of 640 m, is occupied by a series of broad terraces stepping down from east to north and west (Fairbridge, 1966; Andrew and Kravitz, 1974). The up to 400 m deep Novozemel'skaya Trough borders the east coast of Novaya Zemlya. The trough is separated by a saddle located at the northeast tip of the island from the wider and deeper Svyataya Anna Trough west of Franz Josef Land.

### *Europe*

In contrast to the East Siberian, Laptev, and Kara seas the bathymetry of the Barents Sea is better known. It is bounded to the east by Novaya Zemlya and Svyataya Anna Trough and to the west by Fram Strait and the Norwegian Sea. The Barents Sea includes Franz Josef and Svalbard archipelagos. The depths of the seafloor vary for the most part from 100–350 m except in Bear Island Trough where the depth reaches 500 m. Bottom relief is complex with many submarine ridges, banks, and depressions. The continental break lies between 150 m and 200 m.

The structural relief of the Barents Sea is marked by two distinct terrace levels at about 200 m and 70 m. They correspond to two stages of the Quaternary and are connected with the two glaciations most developed in Russia (Klenova, 1966). A submerged shore line at about 200 m forms a terrace escarpment. Abundant rocks, boulders, and other rock fragments of local origin are dredged here. Outcrops of the underlying older clay sediments are frequent. The rock fragments at depths of more than 200 m are very altered and covered by thick Fe-Mn oxide crusts. This crust testifies to their long exposure on the seafloor. The submarine slopes above 200 m are very dissected, possibly because of their long history of subaerial erosion. The younger level (70 m) has smoother expression in the relief, is very flat, and contains many well-rolled fresh pebbles (Klenova, 1966). The bathymetry of the continental slope and rise off the Barents Sea is largely unknown.

## Marginal Plateaus

### *Morris Jesup and Yermak Plateaus*

The Morris Jesup Plateau is an aseismic spur that juts out 240 km into the Fram Basin from the north coast of Greenland in a north-northeasterly direction as outlined by the 1,000 m isobath. It was discovered from Ice Island ARLIS II from which echo soundings, shallow seismic reflection, and gravity measurements were obtained in 1964 (Ostenso and Wold, 1977). It was also crossed by the British submarine *Sovereign*. The bathymetry of the plateau is poorly known.

The Yermak Plateau is an arcuate, aseismic plateau located seaward of Spitzbergen. In contrast to the Morris Jesup Plateau its bathymetry is better known. The plateau is divided into two zones, one lies north of latitude 82°N, where the bathymetric contours trend northeast, and the other lies to the south where they strike north-northwest. Jackson, Mudie, and Blasco (1985) inferred from magnetic and seismic refraction measurements that the northern portion is of continental origin while the southern portion is oceanic. Feden, Vogt, and Fleming (1979) suggested that the Yermak and Morris Jesup plateaus once formed a continuous Iceland-like mass before breaking apart some 35 mya.

### *Chukchi Borderland*

The Chukchi Borderland consists of Northwind Ridge, Chukchi Cap, and Arlis Plateau and encompasses an area 600 km by 600 km. The Northwind Ridge or Northwind Escarpment rises to within 600 m of sea level. To the west it is characterized by a steep continental slope and rise that abuts the Canada Basin at depths near 3,900 m. To the west it is separated from the Chukchi Cap by a deep basin and deep slope canyon. The Chukchi Cap has flat shelves at 270 m and at 400–500 m below sea level rising locally to within 246 m of sea level (Hunkins et al., 1962). The highest shelf is marked by small-scale relief of 5–30 m, which may be the result of ice gouging during a period when the sea level was lower than at present (Johnson, Grantz, and Weber, 1986). The Arlis Plateau is separated from the Chukchi Cap by the Chukchi Abyssal Plain and Charlie Gap. It is commonly considered part of the Mendeleev Ridge but appears, morphologically, to more closely resemble the Chukchi Cap and Northwind Ridge (Johnson, Grantz, and Weber, 1986). It has been suggested (Hunkins et al., 1962; Shaver and Hunkins, 1964) from bathymetric and geophysical considerations that the high-standing parts of the borderland consist of continental crust.

# BASINS AND ABYSSAL PLAINS

## Eurasia Basin

The Nansen–Gakkel Ridge divides the Eurasia Basin (Fig. 29-7) into the Fram Basin to the north and Nansen Basin to the south.

### Nansen Basin

The Nansen Basin lies subparallel to the Kara and Barents shelves. It is about 1,800 km long, 360 km wide at its widest part, and deepest (3,850 m) at the western end where the relatively small Barents Abyssal Plain is wedged between Barents shelf, Yermak Plateau, and Nansen–Gakkel Ridge. Sediments originating mainly from the Kara and Laptev shelves have filled the east end of the Nansen Basin, thereby burying the eastern end of the Nansen–Gakkel Ridge. The Lena River is the major contributor to this sedimentary fill (Johnson, 1969).

### Fram Basin

The Fram Basin is similar in size to the Nansen Basin and lies subparallel to the Lomonosov Ridge. It is floored by the Pole Abyssal Plain which is enclosed by the 4000 m isobath and is much larger and deeper than the Barents Abyssal Plain. It is deepest (4400 m) halfway between North Pole and Nansen-Gakkel Ridge.

Publicly available bathymetric data in the Eurasia Basin are very scarce except for the area of the western end near Yermak and Morris Jesup plateaus.

## Amerasia Basin

The Alpha-Mendeleev Ridge divides the Amerasia Basin (Fig. 29-6) into Makarov Basin and Canada Basin.

### Makarov Basin

The wedge-shaped Makarov Basin lies between the Lomonosov and Alpha-Mendeleev ridges. It is more than 500 km wide along the East Siberian Shelf and, at the 2,500 m isobath, narrows to a point between Lomonosov Ridge and Ellesmere Island. The Makarov Basin includes two abyssal plains.

The Wrangel Abyssal Plain, about 2,800 m deep, lies between the East Siberian shelf and Arlis Gap. Kutschale (1966) reported that a prominent basement ridge underlies Arlis Gap and acts as a dam behind which the sediments of Wrangel Abyssal Plain are ponded, and over which they spill onto the deeper (4,000 m) Siberian Abyssal Plain. The bathymetry suggests that the sediments that incompletely fill the Makarov Basin were deposited from turbidity currents that flowed from the East Siberian shelf across Wrangel Plain, through Arlis Gap, and onto Siberia Plain.

### Canada Basin

The Canada Basin is bounded by the Alpha–Mendeleev Ridge, the Canadian and Alaskan continental shelves, and the Chukchi Borderland. Five major abyssal plains lie within its boundary. The deepest area of the Canada Basin, the Canada Abyssal Plain, is remarkably flat with a uniform depth of some 3,850 m over a great part of its surface. On the east it shallows gradually and uniformly into the Mackenzie Cone and the Canadian continental rise toward the Mackenzie Delta and the Canadian Arctic Islands. In contrast, this abyssal plain abuts the Alaskan and Northwind continental slopes and rises at marked changes in slope. These features indicate that most of the sedimentary fill beneath the Canada Abyssal Plain has probably come from the area of the Mackenzie River Basin and Arctic Islands region (Hunkins, 1968; Johnson, Grantz, and Weber, in press).

The Chukchi and Mendeleev abyssal plains lie between Chukchi Cap and Mendeleev Ridge. The Chukchi Plain (2,100 m depth) is connected via Charlie Gap to the lower-lying Mendeleev Plain (2,900 m), and a 500 m drop separates the Mendeleev

Plain from the north end of the Canada Abyssal Plain, which at its center is a little more than 3,800 m deep. Herald Canyon on the western Chukchi shelf was at least a minor source of sediment to the northern Canada Abyssal Plain via the Chukchi Plain, Charlie Gap, and the Mendeleev Plain (Hunkins, 1968; Johnson, Grantz, and Weber, in press).

## SUBMARINE RIDGES

### Mid-Ocean Ridge

#### *Nansen-Gakkel Ridge*

The Nansen–Gakkel Ridge is part of the North American–Eurasian plate boundary. Based on (1) the location of the line of earthquake epicenters that bisects the Eurasia Basin and (2) on the recently discovered over 4,000-m-deep Lena Trough in Fram Strait, Heezen and Ewing (1961) speculated that the seismic belt signifies the continuation of the Mid-Atlantic Ridge, and that the Lena Trough is a rift feature of the ridge. The existence of the Nansen–Gakkel Ridge, as it was later named, was later confirmed by echograms obtained by U.S. submarines between 1958 and 1962 (Beal, 1969) and by the British submarine *Sovereign* in 1975 and 1976. Except for the western end of the ridge in the vicinity of the Greenwich meridian, which was investigated during the FRAM-series operations (Kristoffersen, 1982*a*), the bathymetry of the Nansen–Gakkel Ridge is based mainly on echograms obtained from early submarine cruises with unreliable navigation (Weber, 1983).

The Nansen–Gakkel Ridge is a 1,500 km long continuous feature. Within 10° of the Greenwich meridian, at its western end, the ridge has a width of approximately 270 km and the ridge crest rises 2 km above the medial valley floor, which lies 5 km below sea level (Johnson, Grantz, and Weber, in press). The ridge valley is well developed and appears to be between 30 km and 40 km wide at least to 90°E longitude. To the east the ridge disappears under the sediments of the Laptev continental slope.

The contouring of the ridge on the 1979 GEBCO and earlier bathymetric charts shows a complex topography suggestive of numerous transform faults and fracture zones. Sobczak (1977, 1979) pointed out that this Mid-Atlantic Ridge-type diagrammetric representation is highly speculative and not based on published bathymetric soundings. Perry et al.'s (1985) chart, although the contouring is less complex, still shows this diagrammetric representation of the Nansen–Gakkel Ridge, although, except for the western end of the ridge, no new bathymetric data has since become available. Because this detailed bathymetry is not based on publicly available, and therefore verifiable, data, the contouring of the ridge in Figure 29-7 has been simplified.

### Aseismic Ridges

#### *Lomonosov Ridge*

The Lomonosov Ridge is an aseismic submarine mountain range. It extends over a distance of 1,700 km from the Lincoln Sea continental shelf to the continental shelf off the New Siberian Islands (Fig. 29-7). The Lomonosov Ridge rises over 3 km above the adjacent Fram Basin that flanks the ridge on the Eurasia side and 2 km above the Makarov Basin on the Amerasia side. The ridge is a relatively narrow, linear, steep-sided, flat-topped feature paralleling the 43°W/137°E meridians. Above the 3,500 m isobath it is some 200 km wide over most of its length except near the North Pole where it narrows to 65 km. Toward the Lincoln Sea the ridge widens considerably and appears to veer to the southwest toward Ellesmere Island merging with the Alpha Ridge.

Before the LOREX expedition in 1979 cartographers mapping the Lomonosov Ridge had to rely on generalized contours of early small-scale Soviet maps and on information obtained from early U.S. submarines that cruised in the polar area between 1958 and 1962. The only reliable early bathymetric data available were from the U.S. drifting ice station ARLIS II that in 1964 crossed the Lomonosov Ridge obliquely from 88°30′N to 86°30′N latitude between the 100°W and 40°W meridians (Ostenso and Wold, 1977). One of the LOREX accomplishments was the compilation of a 100 m contour map of the central part of the Lomonosov Ridge (Fig. 29-8*a*). The map, and the results of seismic air-gun profiling revealed that the Lomonosov Ridge consists of several slightly tilted en echelon fault blocks with scarps facing the Makarov Basin (Blasco, Bornhold, and Lewis, 1979).

The Makarov-facing slope (up to 14°) is much steeper than the Fram-facing slope (not exceeding 6°). The shallowest depth recorded on top of the ridge was 955 m. Publicly available bathymetric data on the remainder of the Lomonosov Ridge toward Siberia and toward Lincoln Sea, including the junctions with the continental shelves, are very scarce. On contemporary maps the contours in areas with little or no sounding data are based partly on early Soviet maps (Burkhanov, 1956; Gordienko, 1960) and partly on unpublished U.S. submarine data. The reliability of the maps in these areas is therefore difficult to assess. Contemporary maps suggest that the Siberian end of the Lomonosov Ridge is an extension of the continental shelf north of the New Siberian Islands while the North American end of the Lomonosov-Alpha ridge complex appears to be separated from the continental shelf off Ellesmere Island by a trough of between 1,500 m and 2,000 m depth (Fig. 29-8*c*).

### *The Alpha-Mendeleev Ridge Complex*

Although isolated areas of the Alpha-Mendeleev ridge complex had first been discovered earlier by the Soviets (Fig. 29-5; Gordienko, 1960), the Alpha Ridge was first recognized as a distinct submarine feature by U.S. scientists aboard the drifting station ALPHA during the IGY (Hunkins, 1961). Much of the topography of the ridge complex is known from the soundings carried out from U.S. drifting ice stations, ALPHA 1957–1958 (Cabaniss, 1962), T-3 1953–1957, 1962–1963, 1966–1974 (Hunkins and Tiemann, 1977) and ARLIS II 1962–1964 (Wold, 1973), and from early submarine echograms. Hall (1970) carried out a comprehensive study of the Alpha and Mendeleev ridges based on the sounding, seismic, gravity, magnetic and coring results collected on board these drifting stations.

Both the Lomonosov Ridge and Alpha Ridge were traversed by a number of early U.S. submarines (Fig. 29-3). Echograms from two of these, from the *Skate* and *Seadragon* that sailed from the North Pole across the Lomonosov and Alpha ridges along the 135°W meridian in 1962, are reproduced from Beal (1969) in Figure 29-9, top; their cruise tracks are shown in Figure 29-2. These submarines were equipped with an early version of an inertial navigation system that was quite inaccurate, and they had to surface from time to time to obtain celestial position fixes (Beal, 1983). In these echograms echo travel time is converted to depth without correction for the variation of water velocity with depth, so that the basins appear some 100 m too shallow. The bathymetric profile at the bottom of Figure 29-9 has been obtained by adjusting the *Skate* echogram vertically and horizontally to be compatible with the newly compiled CESAR bathymetric map (Fig. 29-8*a–c*).

Examination of the bathymetric map and bathymetric profile show that the

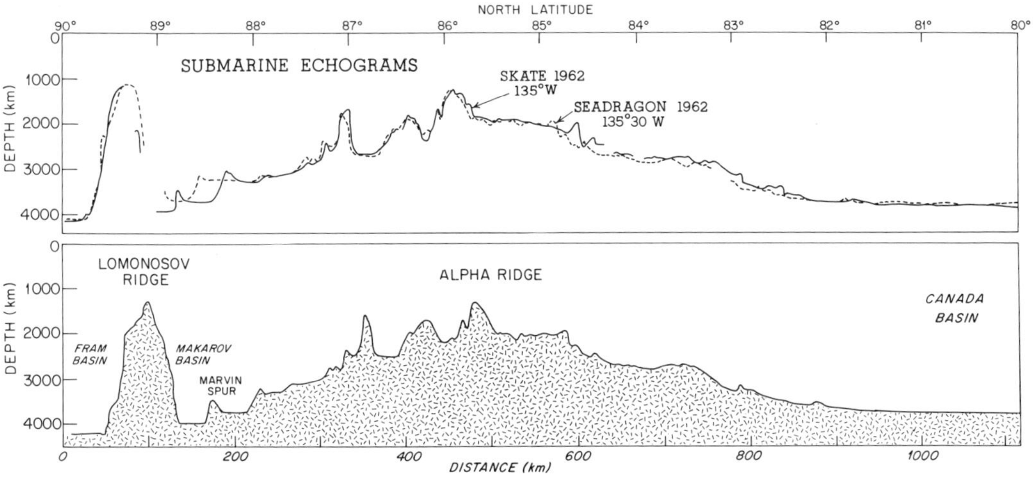


**Figure 29-9.** Top: Echograms of the U.S. nuclear submarines *Skate* and *Seadragon* that sailed from the North Pole along the 135°W meridian in 1962. Track is shown in Figure 29-2. Bottom: Depth profile compiled from the *Skate* echogram corrected vertically and horizontally.

Alpha Ridge is a very rugged, broad, fractured arch consisting of seamounts, depressions, and long ridges and valleys striking parallel to its axis. Hall (1970) reported sediment thicknesses from 100–1,200 m on the ridge. A similar range of values was obtained by the CESAR expedition, confirming that relief on the basement is rugged. Shallow seismic profiles recorded from CESAR (Jackson, Mudie, and Blasco, 1985) as well as from T-3 show many short wavelength bathymetric features that were not revealed by the CESAR regional bathymetric survey used to map the ocean floor. Numerous graben-like troughs bounded by steep faults were reported from the Alpha Ridge by Hall (1970). A similar structure, a prominent east-southeast trending valley centered along 86°N latitude between 110°W and 125°W longitude (Fig. 29-8), was traversed by ice station CESAR (Weber and Jackson, 1985). It was named Cesar Trough and the adjoining ridges were named Cesar North Ridge and Cesar South Ridge (Weber, 1987*a*). These ridges are illustrated in Figure 29-10, which shows a 100 m contour map of the CESAR Camp and surrounding area compiled from the spot soundings and main and remote camp echo soundings. The bottom topography in this area is complex, ranging in depth from 2,536 m in the Cesar Trough to less than 1,100 m on the Cesar North and South ridges. It appears roughest where the sounding density is greatest. There are many short wave-length topographic features visible on the continuous echo sounder and shallow seismic records along the drift track that remain unresolved in the 10 km grid of spot soundings, the smallest grid interval that could be surveyed during the time available (Weber and Jackson, 1985).

Along the eastern end of the Makarov Basin the Alpha Ridge is bounded by the Marvin Spur (dotted line in Figure 29-2). As the spur trends out of the Siberia Plain toward Greenland it becomes strongly asymmetric and its northern (Lomonosov Ridge-facing) flank becomes a steep scarp. Ice Station ARLIS II found a steep scarp with 1,650 m of relief at the north flank of the spur when it drifted north on the 150°W meridian in 1964 (Ostenso and

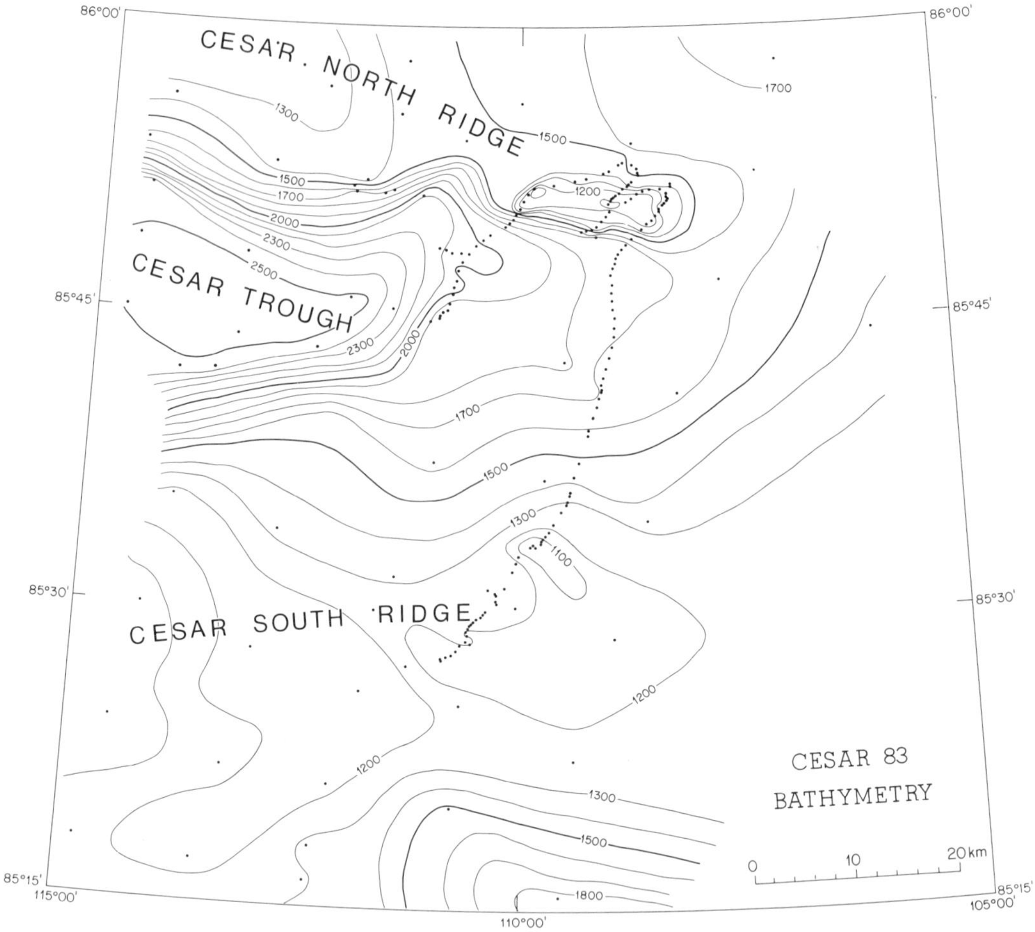


Figure 29-10. Bathymetric map of the CESAR area with 100 m contour interval. Drift track of main camp and remote camp is shown as consecutive dots at three-hour intervals. Also shown are locations of spot soundings and gravity observations. The prominent valley and the two ridges were named Cesar Trough and Cesar North and South ridges.

Wold, 1977). Spot soundings taken during the LOREX and CESAR expeditions along the 160°W and 125°W meridians showed elevation changes of 748 and 554 m, respectively, at the scarp. This long, steep, and continuous escarpment is identified on older maps (e.g., Heezen and Tharp, 1975) as the Lomonosov Ridge-facing flank of the Marvin spur, and Beal et al. (1966) proposed that it is an independent feature that bisects the Makarov Basin. Submarine echograms (fig. 5 in Weber and Sweeney, 1985), ARLIS II data as reinterpreted by Weber and Sweeney (1985), and preliminary CESAR seismic refraction results (Forsyth, Asudeh, and Green, 1985) suggest that this escarpment is probably the Lomonosov Ridge-facing flank of the Alpha Ridge. The part of the Siberia Plain that lies between the escarpment and Lomonosov Ridge, the eastern end of the Makarov Basin, consists of a wedge-shaped graben that is 70 km wide at its junction with Siberia Plain but narrows to a point where, morphologically, the Alpha and Lomonosov ridges merge near the Lincoln Sea.

A number of cones protrude through

the sediments of the eastern end of the Makarov Basin. These cones or diapirs, which form seamounts that rise as much as 700 m above the abyssal plain (Sobczak and Sweeney, 1978), were referred to as the Marvin Seamounts by Sobczak (1977). Air-gun seismic reflection profiles (Blasco, Bornhold, and Lewis 1979) show that these diapirs were formed before the basin fill was deposited, and an analysis of the gravity observations indicate that the cones consist of rocks higher in density (3.15 $Mg/m^3$) than those exposed in the escarpment (2.5 $Mg/m^3$) (Weber, 1980). These diapirs may represent mafic extrusions in the form of volcanos that originated at the time that the Makarov Basin was first formed.

## TECTONIC IMPLICATIONS AND CONSTRAINTS

### Eurasia Basin

The origin and evolution of the Eurasia Basin is quite well understood. Since Heezen and Ewing (1961) and Wilson (1963) recognized that the Eurasia Basin is the extension of the North Atlantic seafloor into the Arctic, the Lomonosov Ridge has been commonly assumed to be originally part of the outermost Barents and Kara continental shelves and that it was split off and transported to its present position in the Arctic Ocean by seafloor accretion centered along the Nansen-Gakkel Ridge. The Lomonosov Ridge, therefore, should be a sliver of continental crust.

The continental nature of the ridge is well supported by geophysical and compositional constraints. The approximate symmetry of the magnetic anomaly pattern about the Nansen-Gakkel Ridge and the present seismicity along the ridge indicate that it is an active accreting margin. There is evidence that the Fram Basin consists of a thin oceanic crust (Duckworth, Beggeroer, and Jackson, 1982). The relatively flat and smooth crest and steep flanks, the en echelon fault block nature of the Lomonosov Ridge (Blasco, Bornhold, and Lewis, 1979; Overton, 1982), the nearly 30 km deep crustal root (Mair and Forsyth, 1982; Weber, 1980), the relatively low average crustal density (Lillestrand and Weber, 1974; Sweeney, Weber, and Blasco, 1982; Weber and Sweeney, 1985), the similarity between the seismic velocity structure below the Kara Sea and that of the Lomonosov Ridge (Forsyth and Mair, 1984), the low-amplitude magnetic anomalies, except for local areas (Vogt et al., 1979; Coles, 1985), are all constraints indicative of the continental nature of the Lomonosov Ridge.

Given that the Lomonosov Ridge is a continental sliver, when did the separation occur? Anomaly 24 is the oldest magnetic reversal that can be positively identified in the Eurasia Basin. However, there is room for anomalies 25-28 on the deep seafloor beyond anomaly 24, which could have been suppressed or erased (Vogt et al., 1979). This room puts the onset of the separation of the Lomonosov Ridge from the Barents and Kara shelves within the interval between magnetic anomalies 24 and 29, or sometime between 56 mya and 65 mya.

### Amerasia Basin

Unlike the Eurasia Basin the origin and evolution of the Amerasia Basin is speculative at best, and the relationship of the basin to the surrounding land is poorly understood.

#### *Canada Basin*

Based largely on aeromagnetic evidence (Vogt et al., 1982) and seismic refraction studies (Mair and Lyons, 1981) it appears that the southern Canada Basin consists of oceanic crust. The North American polar continental shelf is probably an Atlantic-type passive margin as implied by elliptically shaped free-air gravity anomalies centered over the shelf edge (Weber, 1987*b*).

Some authors favor the hypothesis that the Canada Basin opened when a crustal block containing northern Alaska and northeastern Siberia rifted from Arctic Canada and rotated anticlockwise to its present position (Carey, 1958; Hamilton, 1970; Tailleur, 1973; Grantz, Eittreim, and Whitney, 1981; Sweeney, 1985). The authors proposed that the Canada Basin opened in Early and mid-Cretaceous time with a continental breakup phase between 113 mya and 131 mya and a seafloor formation phase between 79 mya and 118 mya (Sweeney, 1985). Another possibility is that the Alaskan–East Siberian margin was a transform (shear) margin along which a continental fragment moved away from Arctic Canada. In this scenario Vogt et al. (1982) propose, on the basis of magnetic lineations in the southern Canadian Basin, that the Canada Basin must have opened between the Late Jurassic and late Early Cretaceous (110–115 mya).

### *Alpha Ridge*

The origin and nature of the Alpha Ridge has been an enigma since its discovery nearly 40 years ago. Many hypotheses have been proposed for its formation, and they range from subsided continental crust, rafted continental fragment, extinct spreading center, extinct island arc, trace of hot spot, to aseismic ocean ridge or plateau. The present concensus among CESAR investigators is that at least the Alpha Ridge part of the Alpha-Mendeleev Ridge Complex is oceanic in character. The high-amplitude, short wavelength geomagnetic anomaly pattern (Taylor et al., 1981), the highly magnetized basement rock below sediment cover (Vogt et al., 1979), the extreme Magsat anomaly high over the Alpha Ridge (Coles, 1985), the extraordinary homogeneity of density, seismic velocity, and magnetization (Weber, 1987*b*), the similarities of the seismic velocity–depth relationship functions with those of Iceland (Forsyth et al., 1985), and the higher average crustal density than for the Lomonosov Ridge (Weber, 1987*b*) suggest that the Alpha Ridge may consist of a large accumulation of basaltic-type material. CESAR coring results (Jackson, Mudie, and Blasco, 1985) indicate that the Alpha Ridge in the Cesar Trough area has not undergone any significant deformation since late Cretaceous time. Based on the hypothesis of Alaska having rotated away from Arctic Canada, Sweeney (pers. comm., 1986) estimates that the formation of the Alpha Ridge took place between about 79 mya and 125 mya.

### *Makarov Basin*

Much of what is known about age and origin of the Makarov Basin is derived from studies of the adjacent submarine ridges. The creation of the Canada Basin and of the Alpha Ridge may be broadly contemporary and their development could be genetically linked. The Makarov Basin crust, where bounded by Alpha and Lomonosov ridges, appears to have formed by crustal stretching and thinning (Weber and Sweeney, 1985) and may have included seafloor spreading west of 160°W longitude (Taylor et al., 1981).

## SUMMARY AND CONCLUSIONS

A systematic exploration program of the Arctic Basin by the Soviets started in 1948, and by 1959 their reconnaissance mapping of the whole basin and large parts of the continental shelf areas had been completed. Exploration of the Arctic Basin by U.S. scientists started in 1951, followed in 1967 by Canadian researchers, and was mostly confined to working from drifting ice islands and ice floes, or to the systematic survey of specific areas. Mapping of the Arctic seafloor by U.S. nuclear submarines started in 1957, but since 1963 all U.S. submarine data have been classified. Mapping of the Siberian and European continental shelves along the northeast passage began early in the 1900s, while systematic mapping of the Alaskan and Canadian continental shelves did not start until the 1950s and 1960s.

The Arctic Ocean occupies a 3,800–4,300 m deep basin surrounded by continental shelves varying in depth from less than 50 m to over 600 m, and in width from less than 60 km to 900 km. If sea level were to drop 200 m, the Arctic Ocean could be classified as a Mediterranean sea. It would connect with the North Atlantic via a 300 km wide opening through Fram Strait, and with Baffin Bay via a 15 km wide access through Nares Strait. Access to the Norwegian Sea via Barents shelf would be impeded by an archipelago similar to the present-day Canadian Arctic Islands.

The Arctic Ocean Basin is divided by the aseismic Lomonosov Ridge into Eurasia and Amerasia basins, which contain the actively spreading Nansen–Gakkel Ridge and the aseismic Alpha–Mendeleev ridge complex, respectively. These various ridges and other structures create a very complex seafloor morphology.

The tectonic history of the Eurasia Basin and origin and nature of the Lomonosov Ridge, which is believed to be a sliver of continental crust originating from the Kara and Barents continental shelves, is fairly well known. The evolution of the Amerasia Basin, on the other hand, is poorly understood. Hypotheses on the opening of the Canada Basin are speculative and include counterclockwise rotation of Alaska away from the Canadian Arctic Islands. Very little is known about the Makarov Basin. Understanding of the nature of the enigmatic Alpha–Mendeleev Ridge complex is beginning to emerge. The concensus among investigators from the 1983 CESAR expedition to the Alpha Ridge is that at least the Alpha Ridge part of the complex consists of a massive accumulation of basaltic rocks. Hypotheses on its age and evolution, however, are still speculative.

## ACKNOWLEDGMENTS

Some of the material included in this chapter has been freely extracted from the chapter Bathymetry and Physiography in *The Arctic Ocean Region,* volume 5, of The Decade of American Geology series, authored by G. L. Johnson, Arthur Grantz, and myself. I wish, therefore, to acknowledge the indirect contribution of my co-authors. Many thanks also to Michael D. Thomas and Larry W. Sobczak of the Earth Physics Branch for reviewing the manuscript. This is a contribution of the Earth Physics Branch No. 1216.

## REFERENCES

Armstrong, T. 1958. *The Russians in the Arctic. Aspects of Soviet Exploration and Exploitation of the Far North, 1937–57.* Methuen, London. 182p.

Asudeh, I., A. G. Green, and D. A. Forsyth. 1985. The uniform waveforms of Arctic crustal data from CESAR 1983: An extensive study of multiple reflections and P to S converted phases. *Abs. Geophys.* **50:**1,362.

Beal, M. A. 1969. Bathymetry and structure of the Arctic Ocean. Ph.D. thesis, Oregon State University, 205p.

Beal, M. A. 1983. Letters to the editor. *Arctic* **36:**372–373.

Beal, M. A., F. Edvalson, K. Hunkins, A. Molley, and N. A. Ostenso. 1966. The floor of the Arctic Ocean. Geographic names. *Arctic* **19:**215–219.

Blasco, S. M., B. D. Bornhold, and C. F. M. Lewis. 1979. Preliminary results of surficial geology and geomorphology studies of the Lomonosov Ridge, Central Arctic Basin. Current Research. *Geol. Survey Canada Paper 79-1C,* pp. 73–83.

Burkhanov, V. 1956. *New Soviet Discoveries in the Arctic.* Foreign Languages Publishing House, Moscow, 60p.

Cabaniss, G. H. 1962. Geophysical data from U.S. Arctic Drifting Stations, 1957–60. *Air*

*Force Cambridge Res. Lab. U.S.A.F., Res. Note AFCRL-62-683,* 234p.

Carey, S. W. 1958. The orocline concept in geotectonics. *Royal Soc. Tasmania Pap. Proc.* **89:**255–288.

Coles, R. L. 1985. Magsat scalar magnetic anomalies at northern high latitudes. *J. Geophys. Res.* **90:**2,576–2,582.

Committee on Polar Research. 1970. Geology and solid-earth geophysics of the polar regions. In *Polar Research—A Survey.* Washington, D.C.: U.S. National Academy of Sciences.

Crary, A. P., R. D. Cotell, and J. Oliver. 1952. Geophysical studies in the Beaufort Sea, 1951. *Trans. Am. Geophys. Union* **33:**211–216.

Duckworth, G. L., A. B. Baggeroer, and H. R. Jackson. 1982. Crustal structure measurements near FRAM II in the Pole Abyssal Plain. *Tectonophysics* **89:**173–216.

Fairbridge, R. W. 1966. Kara Sea. In R. W. Fairbridge (ed.), *Encyclopedia of Oceanography,* vol. 1. Reinhold Publishing Corporation, New York, pp. 430–432.

Feden, R. H., P. R. Vogt, and H. S. Fleming. 1979. Magnetic and bathymetric evidence for the "Yemark Hot Spot" northwest of Svalbard in the Arctic Basin. *Earth Planet. Sci. Lett.* **44:**18–38.

Fisher, R. L., A. J. Carsola, and G. Shumway. 1958. Deep-sea bathymetry north of Point Barrow. *Deep-Sea Res.* **5:**1–6.

Forsyth, D. A., A. Asudeh, and A. Green. 1985. Crustal structure of the Alpha Cordillera. *Am. Geophys. Union Trans.* **66:**957.

Forsyth, D. A., and J. A. Mair. 1984. Crustal structure of the Lomonosov Ridge and the Fram and Markarov basins near the North Pole. *J. Geophys. Res.* **89:**473–481.

Gordienko, P. A., and A. F. Laktinov. 1960. Principal results of the latest oceanographic research in the Arctic Basin. *Akad. Nauk. SSSR., Isvestiya, Geographical Series* **5:**22–23.

Grantz, A., S. Eittreim, and D. A. Dinter. 1979. Geology and tectonic development of the continental margin north of Alaska. *Tectonophysics* **59:**263–291.

Grantz, A., S. L. Eittreim, and O. T. Whitney. 1981. Geology and physiography of the continental margin north of Alaska and implications of the origin of the Canada Basin. In A. E. M. Nairn, M. Churkin, Jr., and F. G. Stehli (eds.), *The Ocean Basins and Margins,* vol. 5, *The Arctic Ocean.* Plenum Press, New York, pp. 439–492.

Greenaway, K. R. 1983. Letters to the editor. *Arctic* **35:**372–373.

Greenberg, J., P. E. Hart, and A. Grantz. 1981. *Bathymetric map of the Continental Shelf, Slope and Rise of the Beaufort Sea North of Alaska.* Miscellaneous Investigations Series, U.S. Geological Survey. 5p. and map.

Hamilton, W. 1970. The Uralides and the motion of the Russian and Siberian platforms. *Geol. Soc. Am. Bull.* **81:**2,253–2,576.

Hall, J. K. 1970. *Arctic Ocean Geophysical Studies: The Alpha Cordillera and Mendeleev Ridge.* Office of Naval Res., Tech. Report No. CU-2-70, Washington, D.C.

Heezen, B. C., and M. Ewing. 1961. The Mid-Oceanic Ridge and its extension through the Arctic Basin. In G. O. Raasch (ed.), *Geology of the Artic.* University of Toronto Press, pp. 622–642.

Holmes, M. L., and J. S. Creager. 1974. Holocene history of the Laptev Sea continental shelf. In Yvonne Herman (ed.), *Marine Geology and Oceanography of the Arctic Seas.* Springer-Verlag, New York, pp. 211–230.

Holtdahl, O., and H. W. Holtdahl. 1961. On "marginal channels" along continental borders and the problem of their origin. *Bull. Geol. Inst. Univ. Uppsala* **40:**183–187.

Hunkins, K. 1960*a*. *Seismic Studies of the Arctic Ocean Floor.* Air Force Cambridge Research Laboratory, AF 19 (604)-2030, 15p.

Hunkins, K. 1960*b*. *Geophysical and Oceanographic Research in the Arctic Ocean.* Lamont-Doherty Geological Observatory, Columbia University. Final Report AF 19 (604)-2030, ASTIA 253807.

Hunkins, K. 1960*c*. *Results of Depth Soundings Taken from Drifting Station Charlie in the Arctic Ocean.* Lamont-Doherty Geological Observatory, Columbia University.

Hunkins, K. 1960*d*. *Results of Depth Soundings Taken at Drifting Station Bravo (T-3) in the Arctic Ocean.* Lamont-Doherty Geological Observatory, Columbia University.

Hunkins, K., and W. Tiemann. 1977. *Geophysical Data Summary for Fletcher's Ice Island (T-3) May 1962–Oct. 1974.* Office of Naval Res., Tech. Report No. CU-1-77, Washington, D.C.

Hunkins, K., T. Herron, H. Kutschale, and G. Peter. 1962. Geophysical studies of the Chukchi Cap, Arctic Ocean. *J. Geophys. Res.* **67:**235–247.

Jackson, H. R., P. J. Mudie, and S. M. Blasco. 1985. Initial geological report of CESAR: The Canadian Expedition to Study the Alpha Ridge, Arctic Ocean. *Geol. Survey Canada Paper 84-22.*

Jeffries, M. O. 1985. Ice shelf studies off Northern Ellesmere Island, Spring 1983. *Arctic* **38:**147–177.

Johnson, G. L. 1969. Morphology of the Eurasian Arctic Basin. *Polar Rec.* **14:**619–628.

Johnson, G. L., A. Grantz, and J. R. Weber (ir

press). Bathymetry and physiography. In G. L. Johnson, A. Grantz and J. F. Sweeney (eds.), *The Decade of North American Geology,* vol. L. Geological Society of America, Boulder, Colo.

Johnson, G. L., J. R. Vanney, and D. Hayes. 1982. The Antarctic Continental shelf. In C. Craddock (ed.), *Antarctic Geoscience.* University of Wisconsin Press, pp. 995–1,002.

Johnson, G. L., N. J. McMillan, M. Rasmussen, U. Campsie, and F. Dittmar. 1975. Sedimentary rocks dredged from the southwest Greenland continental margin. In C. Yorath, E. R. E. R. Parker and D. J. Glass (eds.), *Mem. Can. Soc. Petrol. Geol. 4,* pp. 391–410.

Klenova, M. V. 1966. Barents Sea and White Sea. In R. W. Fairbridge (ed.), *Encyclopedia of Oceanography,* vol. 1. Reinhold Publishing Corporation, New York, pp. 95–101.

Kristofferson, Y. 1982*a. U.S. Ice Drift Station FRAM-IV: Report on Norwegian Field Program.* Norsk Polar Institutt, Rapportserie No. 11, Oslo, 60p.

Kristofferson, Y. 1982*b.* The Nansen Ridge, Arctic Ocean. Some geophysical observations of the rift valley at slow spreading rate. *Tectonophysics* **89:**161–172.

Kutschale, H. 1966. Arctic Ocean geophysical studies: The southern half of the Siberia Basin. *Geophysics* **21:**683–709.

Laktionov, A. F. 1959. Bottom topography of the Greenland Sea in the region of Nansen's Sill. *Priroda* **10:**95–97. (Translated from the Russian by E. R. Hope. Defense Research Board translation T333R, Nov. 1959, 5 p.)

Lillestrand, R. L., and J. R. Weber. 1974. Plumbline deflection near the North Pole. *J. Geophys. Res.* **23:**3,347–3,352.

Mair, J. A., and D. A. Forsyth. 1982. Crustal structures of the Canada Basin near Alaska, the Lomonosov Ridge and adjoining basins near the North Pole. *Tectonophysics* **89:**239–254.

Mair, J. A., and J. A. Lyons. 1981. Crustal structure and velocity anisotropy beneath the Beaufort Sea. *Can. J. Earth Sci.* **18:**724–747.

Manley, T. O., L. A. Codispoti, K. Hunkins, H. R. Jackson, E. P. Jones, V. Lee, S. Moore, J. Morison, T. T. Packard, and P. Wadhams. 1982. The FRAM III Expedition. *EOS* **63:** 35.

Nansen, F. 1897. *Farthest North.* Archibald Constable & Co., London, vol. 1, 510p. and vol. 2, 671p.

Nansen, F. 1922. The strandflat and isostasy. Vidinsk. Skr., I. Math. *Nature* **2:**313p.

Ostenso, N. A. 1962. Geophysical investigations of the Arctic Ocean Basin. Res. Report Series 62-4, University of Wisconsin, Geophysical and Polar Research Center, 124p. (Unpublished manuscript.)

Ostenso, N. A. 1963. Physiography of the Arctic Ocean Basin. In *13th Alaskan Science Conference Proceedings.* Alaska Division, American Association for the Advancement of Science, pp. 92–114.

Ostenso, N. A., and R. J. Wold. 1977. A seismic and gravity profile across the Arctic Ocean Basin. *Tectonophysics* **37:**1–24.

Papanin, I. 1946. *Life on an Icefloe.* Translated from the Russian by Fanny Smitham. Hutchinson & Co., London, 240p.

Perry, R. K., H. S. Fleming, J. R. Weber, Y. Kristoffersen, J. K. Hall, A. Grantz, and G. L. Johnson. 1985. *Bathymetry of the Arctic Ocean.* Bathymetric chart, 1.49 m by 1.06 m, scale 1: 4 704 075. Naval Research Laboratory, Washington, D.C.

Reimnitz, E., and P. W. Barnes. 1974. Sea ice as a geologic agent on the Beaufort Sea shelf of Alaska. In J. C. Reed, and J. E. Sater (eds.), *The Coast and Shelf of the Beaufort Sea.* Arctic Institute of North America, Arlington, Va., pp. 301–354.

Ritchie, G. S. 1969. *North Polar Chart No. 4006.* Hydrographer of the Navy, Taunton, United Kingdom.

Shaver, R., and K. L. Hunkins. 1964. Arctic Ocean geophysical studies: Chukchi Cap and Chukchi Abyssal Plain. *Deep-Sea Res.* **11:**905–916.

Sobczak, L. W. 1977. Bathymetry of the Arctic Ocean north of 85° latitude. *Tectonophysics* **42:**T-27-33.

Sobczak, L. W. 1979. Bathymetry of the Arctic Ocean north of 85° latitude—reply. *Tectonophysics* **60:**297–302.

Sobczak, L. W., and J. F. Sweeney. 1978. Bathymetry of the Arctic Ocean. In J. F. Sweeney (ed.), *Arctic Geophys. Rev.* Publications of the Earth Physics branch **45**(4):7–14.

Somov, M. M. 1955. Observational data of the scientific research drifting station of 1950–1951. vol. I, sec. 4, Moroskoy Transport, Leningrad. (Translated by D. Kraus. American Meterological Society, Boston ASTIA Document No. AD 117135, 59p.)

Stefansson, V. 1921. *The Friendly Arctic.* New York, Macmillan, 784p.

Sweeney, J. F. (ed.) 1978. Publication of the Earth Physics branch. *Arctic Geophy. Rev.* **45:**108p.

Sweeney, J. F., J. R. Weber, and S. M. Blasco. 1982. Continental ridges in the Arctic Ocean. LOREX constraints. *Tectonophysics* **89:**217–238.

Tailleur, I. L. 1973. Probable rift origin of Canada Basin. In M. G. Pitcher (ed.), *Arctic Geology. Am. Assoc. Pet. Geol. Mem.* **19:**526–535.

Taylor, P. T., L. C. Kovacs, P. R. Vogt, and G. L. Johnson. 1981. Detailed aeromagnetic investigation of the Arctic Basin. 2. *J. Geophys. Res.* **86:**6,323–6,333.

Toimil, L. J. 1978. *Ice-Gouged Morphology on the Floor of the Eastern Chukchi Sea, Alaska: A Reconnaissance Survey.* U.S. Geol. Surv. Open-File Rep. 78-693, 113p.

Treshnikov, A. F. 1960. The Arctic discloses its secrets. *Priroda* **2:**25–32. (Translated from the Russian by E. R. Hope. Defense Research Board Translation T357 R. Aug. 1961, 12p.)

Treshnikov, A. F. 1966. *At the Poles of the Earth.* Sovetskaya Rossiya Publishing House, Moscow. (Translated from the Russian. Secretary of State Translation Bureau No. 123842, 30p.)

Vogt, P. R., P. T. Taylor, L. C. Kovacs, and G. L. Johnson. 1979. Detailed aeromagnetic investigation of the Arctic Basin. *J. Geophys. Res.* **84:**1,071–1,089.

Vogt, P. R., P. T. Taylor, L. C. Kovacs, and G. L. Johnson. 1982. The Canada Basin: Aeromagnetic constraints on structure and evolution. *Tectonophysics* **89:**295–336.

Weber, J. R. 1979. The Lomonosov Ridge Experiment. LOREX 79. *EOS* **60:**715–721.

Weber, J. R. 1980. Exploring the Arctic seafloor. *GEOS* **9**(3):2–7.

Weber, J. R. 1983. Maps of the Arctic Basin seafloor: A history of bathymetry and its interpretation. *Arctic* **36:**121–142.

Weber, J. R. 1987*a*. Maps of the Arctic Basin seafloor. Part II: Bathymetry and gravity of the Alpha Ridge. The 1983 CESAR Expedition. *Arctic* **40:**1–15.

Weber, J. R. 1987*b*. The Alpha Ridge: Gravity, seismic and magnetic evidence for a homogeneous, mafic crust. *J. Geodynamics* **6:**117–136.

Weber, J. R., and H. R. Jackson. 1985. Cesar bathymetry. In *Initial Geological Data Report of CESAR: Canadian Expedition to Study the Alpha Ridge.* Geological Survey of Canada Paper 84–22:15–17.

Weber, J. R., and J. F. Sweeney. 1985. Reinterpretation of morphology and crustal structure in the Central Arctic Ocean Basin. *J. Geophys. Res.* **90:**663–677.

Wilkins, G. H. 1928. *Flying the Arctic.* G. P. Putnam Sons, New York, 222p.

Wilson, J. T. 1963. Hypothesis of the earth's behaviour. *Nature* **198:**925–929.

Wold, R. J. 1973. *Gravity Surveys of the Arctic Ocean Basin.* Final Report. University of Wisconsin, Dept. of Geological Sciences, Milwaukee, 222p.

Worthington, L. V. 1953. Oceanographic results of Project Skijump I and Skijump II in the Polar Sea, 1951–1952. *Am. Geophys. Union Trans.* **34:**543–551.

Worzel, J. L. 1968. Advances in marine geophysical research of continental margins. *Can. J. Earth Sci.* **5:**963–983.

# 30: Tectonic History of the Arctic Region from the Ordovician Through the Cretaceous

L. P. Zonenshain
and L. M. Natapov

## INTRODUCTION

The Arctic region has been inadequately studied to date. However, many attempts have been undertaken to reconstruct the history of its development.

The existing concepts were reviewed by Eardley (1961), Churkin (1973), Pushcharovsky (1976), Churkin and Trexler (1981), as well as in a number of articles published in a volume of the *Proceedings of the 27th International Geological Congress* in Moscow (Karasik, Ustritsky, and Khramov, 1984; Burke, 1984). Two opposing concepts concerning the tectonic evolution of the Arctic were formulated over a half century ago. The discovery of Precambrian metamorphic rocks and slightly deformed Paleozoic and Mesozoic sequences in the Novosibirsk, Wrangel, and De Long Islands permitted Shatsky (1935) to distinguish the Hyperborean Platform, which had split and subsided along fault lines to form deep basins. This idea was representative of the ''fixist'' concept. The existence of the ancient massif seemed indisputable. The second or ''mobilist'' concept was originally developed by Wegener (Koppen and Wegener, 1924), who conjectured that the divergence of North America from Eurasia resulted in the genesis of the Arctic basins. This viewpoint was supported by Du Toit (1937). Carey (1955, 1958) applied his orocline concept of continental drift to the Arctic region and suggested that the Arctic basins replaced the Alaskan orocline.

The center of the Alaskan orocline was in the Mackenzie Delta, while the angle of rotation of Alaska and the adjacent Eurasian blocks were displaced with respect to North America by 28°. Although some geologists (Pushcharovsky, 1960, 1976; Udintsev, 1984; Pogrebitsky et al., 1984) still support the concept of the Arctic basins' evolution by continental crust subsidence and destruction, the plate tectonic theory has made it obvious that the Arctic oceanic basins were formed by rifting (Heezen and Ewing, 1961; Karasik, 1968; Churkin, 1973; Ostenso and Wold, 1973; Tailleur, 1973) as well as by the divergence of the North American from the Eurasian lithospheric

*Note:* This chapter was reviewed, edited, and rewritten by the Editor, Yvonne Herman.

plate. The earliest plate tectonic models were proposed by Churkin (1972) and Herron, Dewey, and Pitman (1974). Apart from the Eurasian and North American plates, these authors also considered the small Kolyma, or the Omolon plate, which collided with Eurasia and America to form the Verkhoyansk and South Anui fold belts. In the 1970s and 1980s, a considerable amount of data have been obtained on the bottom topography of the Arctic basins, their structure and magnetic lineations, as well as on Recent and Mesozoic–Cenozoic plate tectonics (Karasik, 1968, 1970; Savostin and Karasik, 1981; Karasik, Savostin, and Zonenshain, 1983; Savostin, Natapov, and Stravsky, 1984; Zonenshain et al., 1978; Vogt et al., 1979, 1982, 1984; Taylor et al., 1981; Jackson and Johnson, 1984; Jackson, Forsyth, and Johnson, 1985; Burke, 1984; Grantz and May, 1983, 1984).

The relative positions of North America/Eurasia are accepted as a basis for all plate tectonic reconstructions of the Arctic basins. In the period spanning the Devonian, to the middle of the Cretaceous, these two continents were united into Euramerica and later into Laurasia. At the end of the Mesozoic, they were broken up and moved apart to create the North Atlantic Ocean between them. The relative position of these two main continental plates may be reliably established using the North Atlantic magnetic anomalies as well as the initial fit of continents. Geoscientists agree that the Eurasian Basin began to open ≈60 mya. The detailed reconstruction of this basin has been based on magnetic lineation patterns (Karasik, Savostin, and Zonenshain, 1983; Savostin, Karasik, and Zonenshain, 1984; Srivastava, 1985).

The Lomonosov Ridge, considered a fragment of the former Eurasian continental margin that split from the continent, was rafted to its present location by seafloor spreading prior to the beginning of the Tertiary. A controversy exists concerning the time of formation of the Canadian Basin. The magnetic lineations mapped in the Canadian Basin were previously interpreted as produced by spreading and identified from anomaly M25 to anomaly M12, suggesting that the ocean floor was created in the 125–155 my time interval (Vogt et al., 1982, 1984). However, Sweeney (1985) and Jackson, Forsyth, and Johnson (1986) believe that the magnetic pattern is related to the topography and does not reflect seafloor spreading. According to Sweeney (1985), the Canadian Basin began to open in Late Jurassic–Early Cretaceous (from 113–144 mya) judging from the appearance of basalts on the passive margins and the sedimentary complexes of the Canadian Artic Archipelago. Seafloor spreading and the creation of new oceanic crust in the Canadian Basin took place only in Late Cretaceous time, from 80–118 mya. We shall discuss this problem later.

The nature of the Alpha–Mendeleev Ridge is also in dispute. Seismic refraction data obtained during the CESAR experiment in 1983 show a very thick (up to 40 km) crust under the Alpha Ridge (Jackson and Johnson, 1984; Jackson, Forsyth, and Johnson, 1986), which seems to exclude an interpretation of this feature as an extinct spreading center, as proposed earlier by Hall (1977). Jackson, Forsyth, and Johnson (1986) suggest that the Alpha Ridge, similar to the Manihiki Plateau and to the Shatsky Rise, was formed by hot spot magmatism near an extinct triple junction along a previous spreading center from which the Canadian Basin crust was created.

The reconstruction of the locations of minor continental blocks in the eastern part of the Arctic region, particularly of Alaska, Chukotka, and the Novosibirsk Islands, point out the differences in opinion. There are two viewpoints: (1) Alaska and Chukotka either occupied approximately the same position relative to North America today or drifted along the Tintina Fault from the southeast (Churkin and Trexler, 1980, 1981; Jones, 1980). In the latter case, the Canadian Basin with its Mesozoic anomalies may be considered part of the Pacific Kula plate that wedged out into Laurasia and was separated from the ocean by the Alaska–Chukotka block and the Kolymian Massif when they arrived from the south. (2) The second viewpoint, following an earlier idea, is that the Canadian Basin was opened up as a result of the rifting and divergence of Chukotka, Alaska along with the Novosibirsk Islands from the Canadian Arctic Archipelago (Heron, Dewey, and Pitman, 1974; Fujita, 1978; Freeland and Dietz, 1978; Burke, 1984; Zonenshain, 1984). Simultaneously with the opening of the Canadian Basin and shifting of the Alaskan-Chukotka block from the Canadian Arctic Archipelago the South Anui oceanic bay in the Paleo-Pacific (Panthalassa) was being closed to form the South Anui Suture. The second viewpoint is more likely to be correct; therefore, it is developed in this chapter. A series of 11 maps showing plate reconstruction of the modern Arctic region from 65–460 mya, that is, from Late Ordovician to Late Cretaceous time, are presented herein. The maps have been compiled following conventional methods of the plate tectonic analyses. These analyses include the use of kinematic, paleomagnetic, paleoclimatic, and geological data sources.

The motions of the four continents (Eastern Europe, North America, Siberia, and Arctida) are considered when studying the history of the Arctic region. Eastern Europe collided with North America about 400 mya to form Euramerica. Arctida joined Euramerica at the end of the Devonian and Siberia at the end of the Paleozoic, when Laurasia was formed as a northern part of Pangea. During the Mesozoic continental drift, Arctida was split into isolated blocks, which later joined different continents localized at the margins of the Arctic Basin.

All the reconstruction maps were compiled on an oblique Mercator projection with a pole at 0°, 180°E. Eastern Europe was taken as a fixed frame of reference.

## GEOLOGICAL DATA

The main structural units of the Arctic region are shown in Figure 30-1 within the circle limited by 60°N. latitude. Three types of structures are distinguished in the Arctic region: (1) young oceanic basins formed in Late Mesozoic and Cenozoic time; (2) large shelves and low terrains on land, with deep sedimentary basins subsiding since the Late Paleozoic; and (3) continental basement consisting of ancient shields, overlain by late Precambrian and Paleozoic platforms, as well as fold belts separating the platforms.

The oceanic crust of the Norwegian-Greenland and the Eurasian basins was formed during the last 55 my from the active spreading systems of the Mohns, Knippovich, and Gakkel ridges.

Recent geodynamic processes are the result of the interaction of the Eurasian and North American plates, separated by a divergent plate boundary. The pole of rotation is either at 61°9′N; 143°1′E (Zonenshain et al., 1978) or at 59°5′N, 141°2′E (Savostin, Natapov, and Stavsky, 1984), that is, close to the map's frame. Therefore, the Eurasian and North American plates are moving apart along the entire boundary.

The spreading midoceanic ridge extends from Iceland along the center of the Norwegian–Greenland and the Eurasian basins and further into the continent via the Laptev Sea. Here the plate boundary becomes less distinct, though it may be traced by the diffuse seismicity of the Chersky Ridge, the Moma Rift grabens, and the recent Balagan-Tas and Urga-Khaya volcanoes (Zonenshain et al., 1978).

The other oceanic basins of the Arctic region (the Labrador, Makarov, Canadian [Amerasian] Basins) were formed at an earlier time. The Labrador Basin was formed between 44 mya and 80 mya (between anomalies 19 and 34) (Kristoffersen and Talwani, 1977; Srivastava, 1978), the Makarov Basin between 55 mya and 76 mya (between anomalies 23 and 33),

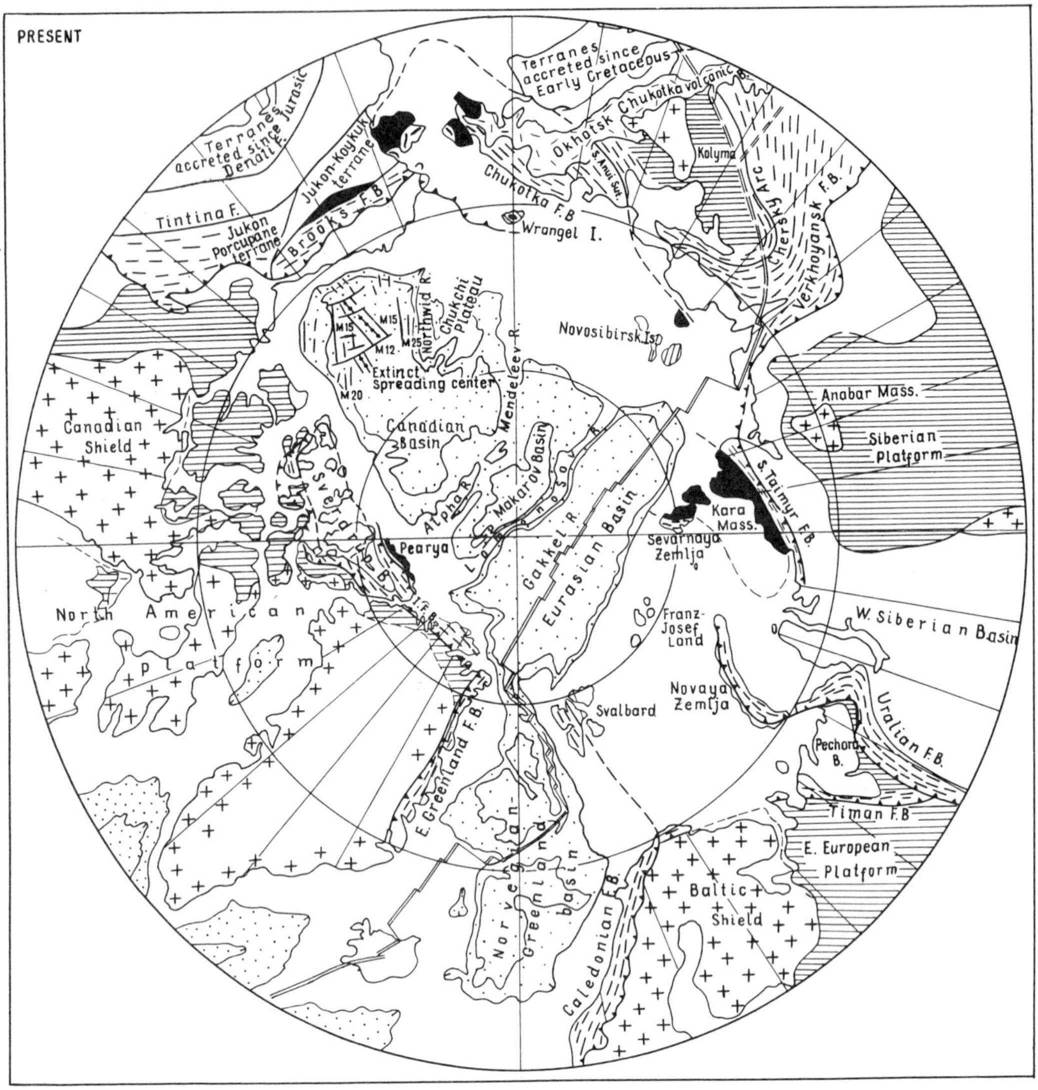


Figure 30-1. Main geological structures of the Arctic. Dotted lines are oceanic basins deeper than 2,000 m; 2,000 m and 3,000 m isobaths are shown. The active spreading center is drawn as a double line, dashed where diffused or uncertain. On the continents, shields (crosses), platform (horizontally hatched), and orogenic belts (dashed lines) are shown. Ancient massifs, remnants of the Arctida continent, are black. Mesozoic and Cenozoic sedimentary basins are blank. Toothed lines indicate fronts of thrusting. Thick lines correspond to main sutures.

and the Canadian Basin between 125 mya and 155 mya (between anomalies M12 and M25) (Vogt et al., 1984; Jackson and Johnson, 1984). The diachroneity in the evolution of these basins and their partial overlap in time suggests the existence of several spreading systems of different ages active simultaneously in the 45–60 my interval.

The Arctic oceanic basins are rimmed by extremely wide shelves (500–1,000 km). Only a minor part of these shelves was formed as a result of initial extension and thinning of the crust, caused by the formation of young oceanic basins. These shelves are narrower bands (100–200 km wide) near the margins of the Norwegian-Greenland and Eurasian basins. Major portions of the shelf depressions were formed prior to the breakup of Laurasia in the Mesozoic. The first minor extensional phase in the Devonian caused the formation of a sedimentary basin on the shelf of the Barents Sea and in grabens on Svalbard (Birkenmajer, 1981) as well as the basaltic eruptions in the Pechora Basin and on the Kola Peninsula (Schwarz, 1981; Krasilshchikov, 1977).

Major extensional phases preceded the opening of oceanic basins including: (1) the Late Carboniferous when a large sedimentary basin was formed, extending from the Urals via the Voring Plateau and the Viking Graben to the North Sea and Middle Europe; (2) the Triassic and Middle–Late Jurassic, when the sedimentary basin was considerably deepened covering nearly the entire Arctic region (Shulgina and Mikhailov, 1979; Eldholm, Myhre, and Sundvor, 1984; Gramberg et al., 1984); and (3) the beginning of the Early Cretaceous, when basaltic eruptions occurred on Svalbard (Birkenmajer, 1981) and on Franz Josef Land (Tkachenko and Egiazarov, 1970). The Kara Sea shelf sedimentary basin with its continuation into the West Siberian Basin existed at least since the Triassic. This existence is confirmed by drilling data from the Jamal Peninsula, in the northern part of the Western Siberian Basin (Nesterov et al., 1984). The sedimentary basins on the Chukchi Sea shelf and the Beaufort Sea, located along the southern edge of the Canadian oceanic basin, started to develop in Carboniferous time (Mississippian) (Grantz and May, 1983). They continue to the north of Alaska, where they were contoured by drilling (Brosque and Dutro, 1973; Detterman, 1973; Dutro, 1981; Grantz and May, 1983, 1984).

In Late Paleozoic and Early Mesozoic time, prior to the opening of the Canada Basin, the basins of the Beaufort and Chukchi seas were connected with the Sverdrup Basin in the Canadian Arctic Archipelago. The Sverdrup Basin was formed between Late Mississippian (Visean) and Late Cretaceous time (Balkwill, 1978; Kerr, 1981). During its development, there were several additional extensional phases, with basaltic eruptions during the Carboniferous, Permian, Jurassic, and Early Cretaceous (Balkwill, 1978). Remnants of ancient shelf basins are preserved in the marginal zones of Greenland. The Wandel Sea Basin off northeast Greenland is one of these remnants. This basin evolved in the Mississippian and is probably filled with a sedimentary sequence spanning the Carboniferous and Jurassic periods (Dawes and Peel, 1981). It is apparent that the shelf basins of the Arctic region started to form mainly after the Devonian and earlier orogenic movements and thus corresponded to the Caledonian and Ellesmerian orogenies. The presence of these vast shelf basins surrounding the Arctic region indicate that since the Devonian the main constituents of the Arctic region were located in close proximity, and the continental margins of this area belonged to the passive margin type and were facing a stable ocean (probably the Panthalassa).

The basement underlying the sedimentary basins both on the shelf and on land consists of ancient massifs, platforms, and shields separated by orogenic belts. While stable regions coincide with continents, orogenic belts developed in the adjacent oceanic basins that separate them and mark the continental collision zones.

The northern margins of the three large Precambrian platforms are found to be localized in the Arctic region (the Eastern European Platform with the Baltic shield, the Siberian Platform with the Anabar Massif or shield, and the North American Platform with the Canadian shield). The Eastern European and Siberian platforms are separated by the Uralian orogenic belt, the folding of which was completed at the beginning of the Triassic in the south and at the end of the Triassic in the north (Novaya Zemlya). The Uralian paleoocean existed between Eastern Europe and Siberia during the Paleozoic (Hamilton, 1970; Zonenshain et al., 1984). The continuation of the Uralian Belt is not known. According to Hamilton (1970), the Uralian Belt was connected via a transform fault with the South Taimyr fold belt separating Siberia from the Precambrian basement of the Kara Massif (North Taimyr and Severnaya Zemlya). In this reconstruction dextral strike slip with an amplitude of 450 km is assumed. The folding in the south Taimyr fold belt had been completed by the end of the Jurassic (Pogrebitsky, 1971); therefore, the strike slip motion had taken place by that time.

The boundary between the Eastern European and the North American platforms extends along the Caledonian fold belt. It is known that at present this fold belt is separated by the Norwegian-Greenland oceanic basin. Its major part with well-developed oceanic complexes occurs in Scandinavia, where the Caledonides were overthrusted on the crystalline basement of the Eastern European Platform (Strand and Kulling, 1972; Holtedahl and Wolff, 1978). In the Varanger Fjord the Caledonian belt bends northward continuing to Bear Island and to Svalbard Island (Hjelle and Ohta, 1978; Berkenmajer, 1981). The folding was completed prior to the Devonian (400–420 mya). The Caledonian movements also took place in the East Greenland Belt (the western side of the Atlantic). Here the ancient basement and the Late Precambrian–Ordovician miogeosynclinal sedimentary sequences overlying this basement were involved in the deformation (Haller, 1976; Peel, 1981). These deformations had been completed prior to the Devonian, similarly to the other side of the Atlantic. The Caledonian fold belts replaced the Paleo-Atlantic or the Iapetus Ocean. They mark the suture of the Eastern European and North American convergence, which resulted in the formation of the very large Euramerican continent.

In the northern part of the Eastern European Platform, the Timan fold belt, deformed at the end of the Precambrian (Getzen, 1975), extends from the Scandinavian Caledonides up to the Urals. It separates the Barents Sea and the Pechora Basin from the East European Platform. According to Ershov et al. (1974) a Preriphean massif is localized in the southern part of the Barents Sea. The Timan fold belt may be considered as a possible suture, marking the collision between the small Barentzia continent and Eastern Europe, which occurred at the end of the Precambrian.

The Innuitian fold belt extends along the northern margin of Greenland and the Canadian Arctic Archipelago passing further west into the Brooks Ridge. This fold belt is composed of Precambrian and Lower Paleozoic deformed rocks. The southern margin of the fold belt is formed by the Arctic Platform, which is part of the North American craton. The northern part of the basin is a clastic wedge derived from the north (Dawes and Soper, 1973; Dawes and Peel, 1981; Kerr, 1981; Dutro, 1981). The deformations united into the Ellesmerian orogeny, which occurred from the Devonian to the Early Mississippian, that is, later than in the Caledonian fold belts. At that time, the region of the fold belt was uplifted and eroded. A possible continuation of the Innuitian Belt to the east may be in the Severnaya Zemlya fold zone, where the Lower Paleozoic–Ordovician complex was deformed prior to the Silurian. At the end of the Devonian this complex underwent refolding together with the Silurian and Devonian sequences (Tkachenko and Egiazarov, 1970). Traces of possible Devonian tectonic movements are observed in the northern part of Novaya Zemlya and are expressed by variegated unsorted sandstones and by granitic intrusions (Schwartz, 1981).

The Innuitian Belt and its equivalents

on Severnaya Zemlya may be considered a collisional suture of Arctida and Euramerica. The remnants of the Arctida basement are preserved in some ancient massifs within the Arctic region (Fig. 30-1). The Kara Massif surrounding the northern part of Taimyr and the southern part of Severnaya Zemlya is the largest (Pogrebitsky, 1971; Tkachenko and Egiazarov, 1970). The massif contains crystalline complexes that are overlain by Cambrian fossiliferous beds. The radioisotopic age of the gneisses is 2,300–2,400 my, while that of the granites protruding these gneisses is 850–1,100 my (Khain, 1979).

The Precambrian metamorphic rocks of the Bolshoi Lyakhovsky Island compose another massif (the Novosibirsk Island) (Tkachenko and Egiazarov, 1970; Genin, Lipkov, and Piskarev, 1977). Precambrian rocks compose the basements of the Chukchi fold belt, the Seward peninsula, and the Wrangel Island. The Precambrian rock outcrops in the eastern part of Chukotka and on Seward Peninsula constitute a single crystalline massif. The biotite and amphibole crystalline schists, gneiss, and marble in the eastern part of Chukotka are 650–1,570 my old (radioisotopic ages; Drabkin, 1970). The Rb-Sr age on the Seward Peninsula is 650 my (Churkin and Trexler, 1981). The Precambrian rocks of the Wrangel Island, composed of crystalline schists and amphibolites of the Gromov Formation, belong to the same Chukotka–Alaska group of ancient massifs.

East of the Seward peninsula, the Precambrian crystalline rocks outcrop in the Schist Belt of the Brooks Ridge, where they are cut by granite whose radioisotopic age is 970 my (Churkin and Trexler, 1981). On Chukotka, Alaska, the Wrangel Island, and in the Schist Belt of the Brooks Ridge, Precambrian basement is overlain by sequences of predominantly carbonate rocks of Ordovician, Silurian, Devonian, and Carboniferous age. Finally, Pearya, the northern geoanticline of the Innuitian fold belt supplying detrital material to the clastic wedge in the south, also belongs to the Arctic massifs. The isotopic age of the Pearya metamorphic sequences is 800–1,000 my (Kerr, 1981).

At present, the relicts of the Arctida basement are spread along the periphery of the Arctic oceanic basins. If these basins would be closed, as shown in Fig. 30-2, the separated massifs would be united. Approximate contours of Arctida are shown in Figure 30-2. The ancient massifs constitute the core of this Early Paleozoic continent, whereas the Innuitian and Severnaya Zemlya fold belts represent zones of collision between Arctida and Euramerica. This figure also shows that Arctida was not attached to Siberia in Late Paleozoic and Early Mesozoic time near the Kara Massif (Taimyr), Chukotka, and Alaska.

The Verkhoyansk–Kolyma Belt consists of the following structural elements: (1) the Verkhoyansk fold belt, (2) the magmatic arc of the Chersky Ridge, (3) the Kolymian–Omolon Massif, and (4) the South Anui suture belt.

The Verkhoyansk Belt is composed of a thick (10–12 km) sedimentary complex formed during the Vendian (Upper Jurassic). No structural disconformities were observed in this complex. A significant environmental change took place at the end of the Early Carboniferous. Prior to this time, shallow-water carbonate sediments with occasional evaporites prevailed. In Early Carboniferous they are replaced by shallow-water clastic sediments derived from the eroded uplifted basement of the Siberian continent. The complex has the shape of a sedimentary prism of the passive margin of the Siberian continent. Approaching the inner part of the belt, the thickness of the sedimentary cover increases and marine deposits prevail. The prolonged existence of the passive margin (about 500 my) suggests that it was located in the vicinity of a large ocean. The folding in the eastern parts of the Verkhoyansk

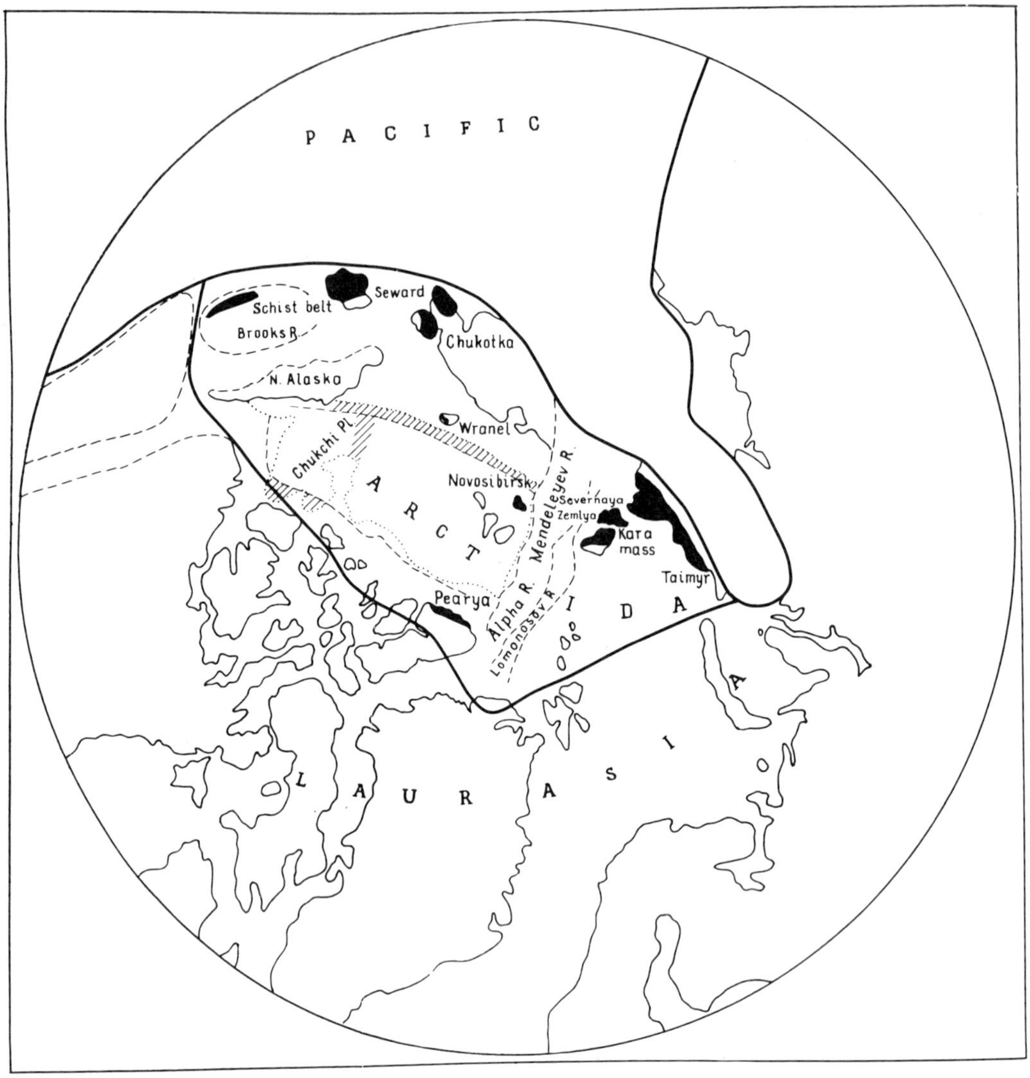


Figure 30-2. Early Jurassic reconstruction showing ancient Arctic massifs surrounding Arctida continent attached to Laurasia.

Belt commenced in Volgian time and in Early Cretaceous encompassed the major part of the belt. The linear folding in major parts of the belt was of the decollement type. Thrusts and nappes displaced in the direction of the Siberian continent arose in the frontal part of the belt. Intrusions of granites took place 90 mya, which indicates that the folding had been completed by then.

The magmatic arc of the Chersky Ridge consists of two structural elements: the Ilintass volcanic belt and the Kolymian batholitic belt. The Ilintass Belt began to form during the Callovian (Jurassic). It is composed of calcalkaline volcanics alternating with turbidites and olistostromes.

The South Anui Belt is composed of ophiolites and island arc assemblages of Upper Jurassic and Lower Cretaceous age (Seslavinsky, 1970; Natalin, 1979). It is a suture representing the collision between the Kolymian-Omolon Massif and Chukotka. During the Triassic and Jurassic, an oceanic basin replaced this belt. This basin was closed at the end of Jurassic-beginning of Cretaceous, during the subduction of the oceanic crust, mainly beneath the Kolymian-Omolon Massif. The fold belt was formed in a very short period of time, between the Neocomian and Aptian.

During the Early Cretaceous, the Verkhoyansk-Kolymian Belt evolved during a very short period. The Okhotsk-Chukotka continental volcanic belt was overlain with a sharp unconformity by a basement composed of rocks of different ages. This belt was superimposed on all the structures formed in Middle Cretaceous (commencing in the Aptian).

The southern marginal parts of the Chukotka Peninsula (Koryakia) and Alaska are occupied by tectono-stratigraphic terranes, which accreted the Eurasian and the North American margins, since Late Jurassic and Early Cretaceous times. The majority of these terranes, such as Wrangelia, Khatyrka, and Talovo-Mainsky, had moved thousands of kilometers to collide with the Laurasian continental margin (Churkin and Trexler, 1980, 1981). They were formed in the tropical zone and are exotic to the Arctic region.

The geological data support our interpretation concerning the evolution of the Arctic, discussed in the Introduction of this chapter. The following sequence in the development of the Arctic region is proposed:

1. Collision of East Europe and North America prior to the Devonian, the Caledonian orogeny, and the development of Euramerica.
2. Collision of Arctida with Euramerica, the Ellesmerian orogeny.
3. Collision of Siberia (including Kazakhstan) with Euramerica in Late Paleozoic, the Hercynian orogeny and the development of Laurasia.
4. Continuous development at the end of the Paleozoic and Early Mesozoic; Triassic rifting in the western part of Siberia, formation of deep shelf sedimentary basins.
5. Beginning of splitting and breakup of Laurasia in Late Jurassic, rifting in the North Sea and in the Norwegian-Greenland Basin, opening of the Canada Basin and the simultaneous collision of Chukotka and the Novosibirsk Islands with Siberia and the Kolymian Massif. This process was completed by the beginning of Early Cretaceous.
6. The breakup of Laurasia is still underway in Middle and Late Cretaceous with the subsequent formation of the Labrador and the North Atlantic (south of Iceland) oceanic basins.
7. Opening of the Norwegian-Greenland and Eurasian basins.

## PALEOMAGNETIC DATA

The apparent polar wandering paths were established for the three main continental blocks of the Arctic region: Eastern Europe (Khramov, 1982), North America (Irving, 1977, 1983), and Siberia (Khramov, 1982). The coordinates of paleomagnetic poles are given in Table 30-1 and the apparent polar wandering paths of Eastern Europe, North America, and Siberia are shown in Figure 30-3. Although there is an inconsistency between the Triassic and Late Paleozoic paleomagnetic poles of these three continents (Van der Voo, Peinade, and Scotese, 1984), generally the paleomagnetic data indicate the following sequence of events in the Arctic evolution of these three plates: Eastern Europe and North America were separated prior to the Devonian and were joined after the Devonian. Eastern Europe and Siberia were separated until Early Triassic time. Devonian paleomagnetic data indicate that Siberia was rotated 60° clockwise and moved toward Europe 3,000 km to join with Eastern Europe at the end of the Paleozoic.

Table 30-1. Paleomagnetic Pole Positions for the Arctic Continents

| | *North America* | | | *East Europe* | | | *Siberia* | | | *Kolyma* | | |
|---|---|---|---|---|---|---|---|---|---|---|---|---|
| | *(Ref.)* | | | *(Ref.)* | | | | *(Ref.)* | | *(Ref.)* | | |
| Late Cretaceous | 70 | 206 | (2) | 78 | 188 | (3) | the same as for East Europe | | | | | |
| Early Cretaceous | 70 | 184 | (2) | 75 | 189 | (3) | the same as for East Europe | | | | | |
| Late Jurassic | 66 | 187 | (2) | 71 | 150 | (3) | the same as for East Europe | | | 66 | 153 | (1) |
| Early Jurassic | 68 | 140 | (2) | 70 | 121 | (3) | | | | 70 | 195 | (4) |
| Triassic | 53 | 107 | (2) | 51 | 153 | (1) | 49 | 149 | (1) | 45 | 279 | (5) |
| | | | | | | | | | | 58 | 233 | (4) |
| Late Permian | 52 | 114 | (2) | 45 | 162 | (1) | | | | | | |
| Early Permian | 44 | 118 | (2) | 40 | 167 | (1) | 45 | 141 | (1) | | | |
| Late Carboniferous | 37 | 124 | (2) | 41 | 169 | (1) | | | | | | |
| Early Carboniferous | 34 | 128 | (2) | 32 | 163 | (1) | 30 | 150 | (1) | 34 | 232 | (6) |
| Late Devonian | 44 | 122 | (2) | 24 | 160 | (1) | 20 | 140 | (1) | 24 | 222 | (6) |
| Early Devonian | | | | 31 | 161 | (1) | 15 | 116 | (1) | | | |
| Silurian | 29 | 123 | (2) | 38 | 155 | (1) | −10 | 104 | (1) | | | |
| Ordovician | 33 | 138 | (2) | 30 | 143 | (1) | −21 | 131 | (1) | | | |

(1) Khramov (1982).
(2) Irving (1977).
(3) Westphal et al. (1985).
(4) Pechersky (1970).
(5) Lozhkina (1981).
(6) Kolesov (1981).

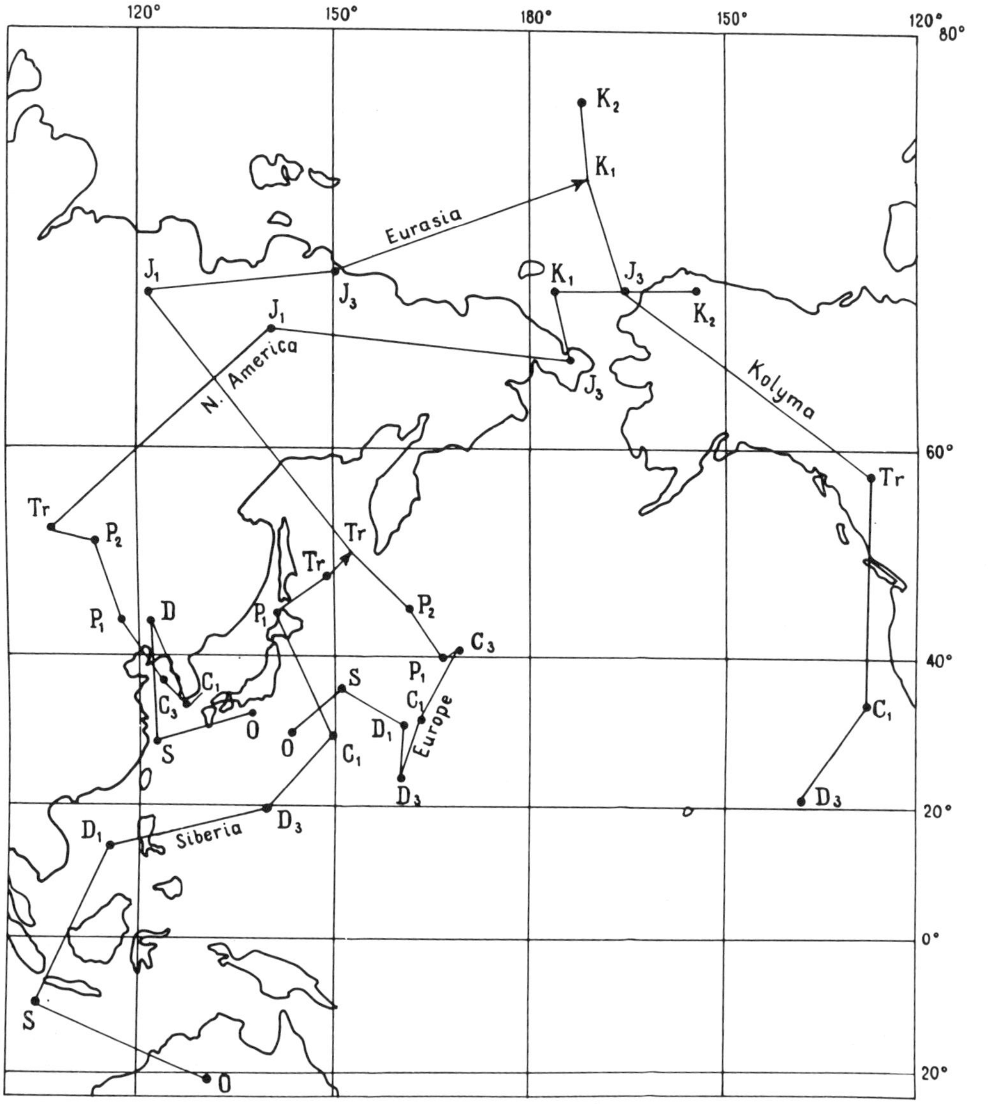


Figure 30-3. Apparent polar wandering paths for Europe, Siberia, North America, and Kolyma derived from Table 30-1.

Paleomagnetic data suggest that the three continents were in the equatorial zone in Early Paleozoic time and moved slowly northward. Siberia was the first to reach the Polar circle; its southern margin was in the vicinity of Baikal at the end of the Devonian. Other continents reached the Polar Region only at the end of the Permian and Triassic. Greenland, however, was at that time only at 30°N. latitude. At the end of the Cretaceous, these continents had reached their present-day position.

Paleomagnetic data were used in our reconstruction of paleolatitudes. Generally, the paleomagnetic data are in agreement with the paleoclimatic reconstructions (Zonenshain and Gorodnitsky, 1977). The widespread Ordovician and Silurian reef limestones outline the position of continents in the equatorial zone. The Devonian red beds extending across northwestern Europe, eastern Greenland, Spitzbergen, and Severnaya Zemlya delineate the extent of the low-latitude arid zones. Furthermore, the Permian evaporites and limestones extending over the vast areas of the Urals, Svalbard, and the Sverdrup and Zechstein basins of Europe were formed in arid zones. The northern humid zone was characterized by the deposition of clastic sequences and coals in eastern Siberia and the Verkhoyansk Belt.

Unfortunately no paleomagnetic data were obtained for the Arctida massifs. Therefore, the latitudinal position of Arctida is based tentatively on sedimentary rocks. The Ordovician and Silurian Arctida rock units are represented essentially by shelf reef limestones suggesting a low latitudinal position of Arctida. Devonian sequences consist in part of red beds and of carbonate rocks, indicating that Arctida was displaced together with Euramerica to higher latitudes. Limestones and evaporites prevail among the Carboniferous and Permian rocks, for examples, on the Novosibirsk Islands (Tkachenko and Egiazarov, 1970), Chukotka (Drabkin, 1970), and Wrangel Island (Kameneva, 1977). Hence these areas were located in the arid zone, isolated from Siberia and the Verkoyansk region. The latter two areas were situated at that time in the northern humid zone.

Paleomagnetic determinations were obtained only for the Late Paleozoic and Mesozoic sequences of the Kolymian Massif (Pechersky, 1970; Kolesov, 1981; Lozhkina, 1981) (Table 30-1). Table 30-2 shows the calculated values of paleolatitudes and declinations toward the pole for an arbitrarily chosen point at 65°N and 165°E located on the Kolymian Massif. The table shows the separation of the Kolymian Massif from Siberia in Late Paleozoic and Triassic time (latitudes 30°–38°), and if we assume a point on Siberia, it would have shifted from 43° to 74°N latitude. The difference of the paleomagnetic vectors indicate that the Kolymian Massif should have rotated no less than 100° clockwise to occupy its present position relative to Eurasia. In Jurassic there are no longer paleomagnetic discrepancies, suggesting the approach of the Kolymian Massif to Siberia.

## KINEMATIC DATA

Heron, Dewey, and Pitman (1974) have shown that the tectonic history of the Arctic region was determined first by the interaction of the Eurasian and North American plates. Based on the calculated parameters of the motion of these plates, Pitman and Talwani (1972) and Heron and co-workers (1974) came to the conclusion that the plates should have converged simultaneously with the opening of the North Atlantic in the Arctic region between Chukotka and Alaska–Canada. Based on the above, Heron, Dewey, and Pitman (1974) assumed that the Alpha Ridge was a structure arisen above the compressional boundary.

Table 30-2. Paleolatitudes (ϕ) and Paleodeclinations (λ) for the Point 65°N, 165°E Computed from the Siberia and Kolyma Poles Given in Table 30-1

| | *Late Devonian* | | *Lower Carboniferous* | | *Triassic* | | *Jurassic* | |
|---|---|---|---|---|---|---|---|---|
| | ϕ | λ | ϕ | λ | ϕ | λ | ϕ | λ |
| Siberia | 43 | 206 | 54.5 | 195 | 74 | 204 | 81 | 12 |
| Kolyma | 30 | 106 | 38 | 90 | 45 | 30(1) | 87 | 112(1) |
| | | | | | 56.5 | 67(2) | 76 | 54(2) |

(1) From the Khramov's (1980) pole.
(2) From the Pechersky's (1970) pole.

Table 30-3. Parameters of the North America/Eurasia Motion

| *my* | *Pole Position* N. Lat. | *Pole Position* E. Long. | *Angle of Rotation* | *Ref.* |
|---|---|---|---|---|
| 0–20 | 64.5 | 136.5 | 3.76 | (1) |
| 0–35 | 73 | 133 | 8.16 | (1) |
| 0–54 | 63.16 | 139.13 | 12.48 | (1) |
| 0–80 | 81.6 | 150.3 | 23.8 | (2) |
| 0–110 | 89.14 | −103.14 | 29.6 | (2) |
| 0–130 (initial fit) | 86.4 | 158.5 | 30.21 | (3) |

(1) Karasik, Savostin, and Zonenshain (1983).
(2) Olivet et al. (1985).
(3) Savostin et al. (1985).

For the reconstruction maps, the authors of the present article used the relative motion parameters of Eurasia and North America published by Karasik, Savostin, and Zonenshain (1983), Savostin et al. (1985), and Olivet et al. (1985). The data are given in Table 30-3. We assume that no movements took place between Eurasia and North America prior to Early Cretaceous time. Therefore, all the events associated with the opening of the Canada Basin in the Late Jurassic and Early Cretaceous occurred within a rigid frame formed by the single Laurasian continent. The initial breakup of Laurasia is obviously related to the Early Cretaceous, although the plate separation started only in the middle of the Cretaceous. By the end of the Cretaceous, several new oceanic basins developed: (1) Iberia and Newfoundland; (2) the Rockall Basin, and (3) the Labrador and Baffin basins between Greenland and North America.

The parameters of differential motion of the Eurasian and North American plates computed from Table 30-3 are given in Table 30-4. They explain the interaction of these plates within the Arctic region. Figure 30-4 shows the motion trajectories of selected points of the North American plate relative to Eurasia. According to the alteration of the North America/Eurasia pole of rotation, three phases of interaction of these plates may be distinguished in the Arctic region. During the first phase (the initial breakup in Early Cretaceous time) the pole of rotation was located far south. This location caused general extension in the Arctic region. The amount of extension was 200 km in the Polar area and 300 km in the Charly Gibbs Fault. The Early Cretaceous basaltic widespread vol-

Table 30-4. Relative Motion of North America and Eurasia

| *my* | *Eurasia Fixed* N. Lat. | *Eurasia Fixed* E. Long. | *North America Fixed* N. Lat. | *North America Fixed* E. Long. | *Angle of Rotation* |
|---|---|---|---|---|---|
| 0–20 | 64.5 | 136.5 | 64.5 | 136.5 | 3.11 |
| 20–35 | 79.86 | 122.88 | 80.16 | 128.54 | 4.47 |
| 35–54 | 45.77 | 146.3 | 45.33 | 140.80 | 4.66 |
| 65–80 | 79.80 | −36.85 | 78.30 | −76.30 | 12.60 |
| 80–110 | 59.35 | −21.75 | 58.95 | −51.06 | 7.03 |
| 110–130 | 15.24 | 160.64 | 15.64 | 131.00 | 2.06 |

*Note:* Computed from Table 30-3.

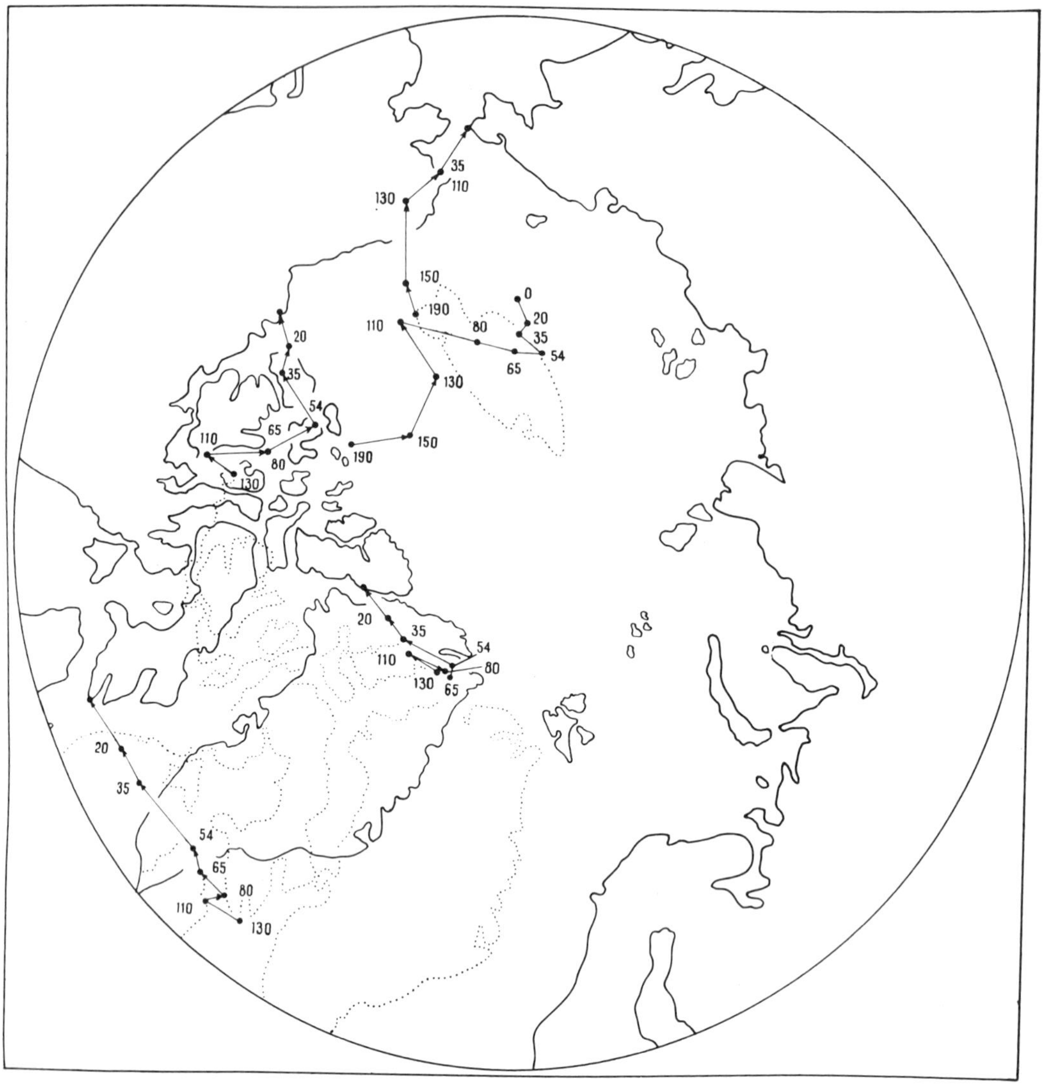

Figure 30-4. Relative motion history of North America with respect to Eurasia within the Arctic, derived from Table 30-4. Numbers indicate ages in million years. Dotted lines indicate the positions of North America, Greenland, and Chukotka 190 mya.

canism extending from Greenland to De Long Island and the formation of the North Sea rifts are most likely associated with this phase (Zonenshain, unpublished).

During the second phase (late Cretaceous to Paleocene, between 55 mya and 110 mya) the pole of relative rotation of North America and Eurasia was located between Greenland and Ellesmere Island. The plates diverged along the Atlantic, south of the pole of rotation, converging toward the Bering Strait. In the Chukchi Sea area the plate converged 900 km, and north of Ellesmere Island it converged 350 km. Based on plate kinematics, it is shown that in Late Cretaceous–Paleocene time in the Arctic there should have existed structures that resulted from the convergence of North America and Eurasia. The fold front from the Brooks Ridge via Point Hope Peninsula to Herald Island and

Wrangel Island is taken as a compressional structural belt. The continuation of this compressional belt is unclear: It should be somewhere between Wrangel Island and the point of the North American/Eurasian pole position, at 59°35′N, 21°75′E (Table 30-3). The best candidate for such a continuation is in our opinion the Mendeleev–Alpha Ridge, although this contradicts the Jackson, Forsyth, and Johnson (1986) interpretation of the Alpha Ridge.

Jackson and colleagues believed that the Alpha–Mendeleev Ridge was generated by hot spot magmatism. To obtain a 1,000 km convergence it is necessary to assume that oceanic crust consumption had taken place apart from continental shortening. Therefore, we assume that the Canada Basin, from Alaska to the Alpha–Mendeleev Ridge, was approximately 900 km larger than it is today. As almost all Arctic regions were during Late Cretaceous time under compression, induced by the North American/Eurasian convergence, it is very difficult to envision the main opening stage for the Canada Basin just at that time (Sweeney, 1985). Therefore, we prefer to follow the previously identified magnetic lineations in the Canada Basin indicating that oceanic crust accretion was completed before Middle Cretaceous time.

The third phase occurred during the Cenozoic. At that time the pole of opening of the North Atlantic and the Arctic basins was located mainly within Eastern Siberia. The convergence of North America with Eurasia was manifested as the oroclinal bending of the Brooks–Herald Islands orogen, and as the deformations diffused in the Kolymian Ridge in the southwestern USSR, which in Tertiary was uplifted to form mountain ranges.

## PLATE RECONSTRUCTIONS

Figures 30-5 through 30-15 show the reconstruction of the Arctic region from 65–460 mya. The above-mentioned geological, paleomagnetic, and kinematic data were used in these reconstructions. The relative location of the main plates (Europe, Siberia, and North America) serves as a basis for these reconstructions. In all these reconstructions Europe has been taken as a fixed reference frame.

In the 130–140 mya reconstructions North America was integrated with Europe into Euramerica. Siberia (Kazakhstan included) is shown together with Europe up to 235 mya. The relative position of Siberia and Europe given in the 280–460 my reconstructions as well as the positions of North America and Europe in the 460 my reconstruction are shown according to global reconstructions compiled by Zonenshain et al. (1984).

When reconstructing the Arctic region history the most difficult problem was to assemble the separate Arctic continents with the continents of Early Mesozoic Pangea. It has been mentioned earlier that there are at least two alternative viewpoints: either to close the Canada Basin by clockwise rotation of the Alaska–Chukotka block or to converge the Canada Basin with the Pacific Ocean moving the Alaska–Chukotka block southeastward along the Tintina Fault. The occurrence in the Canada Basin of a Mesozoic system of magnetic anomalies with a northward direction from the Mackenzie Delta supports the first viewpoint, which is adapted in the present article.

Figure 30-2 shows the assumed pattern of Arctic region continents in Early Mesozoic time. According to this reconstruction the Lomonosov Ridge is moved to the shelf edge of Franz Josef Land and Severnaya Zemlya. The Alpha–Mendeleev Ridge whose crust is most likely subcontinental is joined with the Lomonosov Ridge along the 2,000 m isobath in order to close the Makarov Basin. The northern part of the Taimyr along with Severnaya Zemlya, Franz Josef Land, the Lomonosov, and the Alpha–Mendeleev ridges or the Kara Massif are displaced some 600 km north–

northeast along the assumed transform fault connecting the northern extremity of Novaya Zemlya with the South Taimyr fold belt. Such displacement is necessary to obtain the convergence between the northern part of Taimyr and Siberia, which resulted in the deformation of the South Taimyr fold belt. The displacement complies with the matching of the fold structures of Novaya Zemlya and the southern part of Taimyr and with the closing of the oceanic gap between Greenland and Ellesmere Island, which is due to the initial fit of Greenland with Scandinavia and North America.

The Mesozoic magnetic anomalies in the Canada Basin include two systems: the old system, from M25 to M20, parallel to the Canadian continental shelf and the Northwind Ridge on one side of the Chukchi Plateau, and the younger system, from M20 to M12, fan shaped, northward of the Alaskan shelf near the Mackenzie Delta. These two series are observed in the magnetic lineation maps (Vogt et al., 1979, 1982, 1984; Karasik and Sochevanova, 1981). These series indicate that there were two stages in the opening of the Canada Basin. In order to fit the anomalies of the young fan-shaped system, the Alaska–Chukotka block has been rotated 45° clockwise around the pole near the Mackenzie Delta. Between the edge of the Chukchi Plateau (the Northwind Ridge) and the Canadian shelf there is a 350 km gap. This gap is closed if the Chukchi plateau is moved to the Canadian shelf parallel to the edge of the Alaskan shelf. The edge of the block of the Novosibirsk Islands along with the Chukchi Plateau is displaced east of Chukotka. As a result the gap between the Kara Massif and Canada is filled, although there is a 100 km overlap of the Novosibirsk block and the Chukchi Plateau and a 50 km overlap of the Novosibirsk block with the Chukchi block in the area of the Wrangel Island. The first overlap is due to the spreading and formation of a deep (>3,000 m) basin between the Chukchi Plateau and the southern part of the Mendeleyev Ridge, while the second is most likely due to diffused spreading.

This reconstruction of the Arctic region blocks in Early Mesozoic time applies to both new reconstructions, for which kinematic data are available, and for old reconstructions, for which we do not possess kinematic data. The Early Mesozoic reconstructions given in Figure 30-2 demonstrate convincingly the contours of the Arctida continent joined to Laurasia. These reconstructions show the diverging and converging plate boundaries and the transform faults. The boundaries for the post 155 my period are based on the linear magnetic anomalies and kinematic data. The distribution of key geological assemblages showing plate boundaries have been used to a lesser extent. The main criteria for identifying plate boundaries for the period older than 155 my are geological data, especially the magmatic record. These geological assemblages include the calcalkaline volcanic and plutonic series, which allow us to delineate the converging plate boundaries or the bimodal volcanic series, indicating continental rift.

## 460 my Reconstruction (Fig. 30-5)

Based mainly on paleomagnetic data, the 460 my location of the continents is given with a certain degree of confidence. At that time there were four separate continents: East Europe, Siberia, North America, and Arctida; all were located in the equatorial zone. Europe and North America were separated by the Iapetus Ocean, which had a width of not less than 2,500 km. The Uralian Ocean separated East Europe from Siberia and Kazakhstan. It began to open in the Ordovician (Zonenshain et al., 1984); however, at that time it had not reached its maximum width. Its width during the Ordovician was about 1,000 km. Arctida was an independent continent 4,000 km long and 2,000 km wide. This continent was most likely sepa-

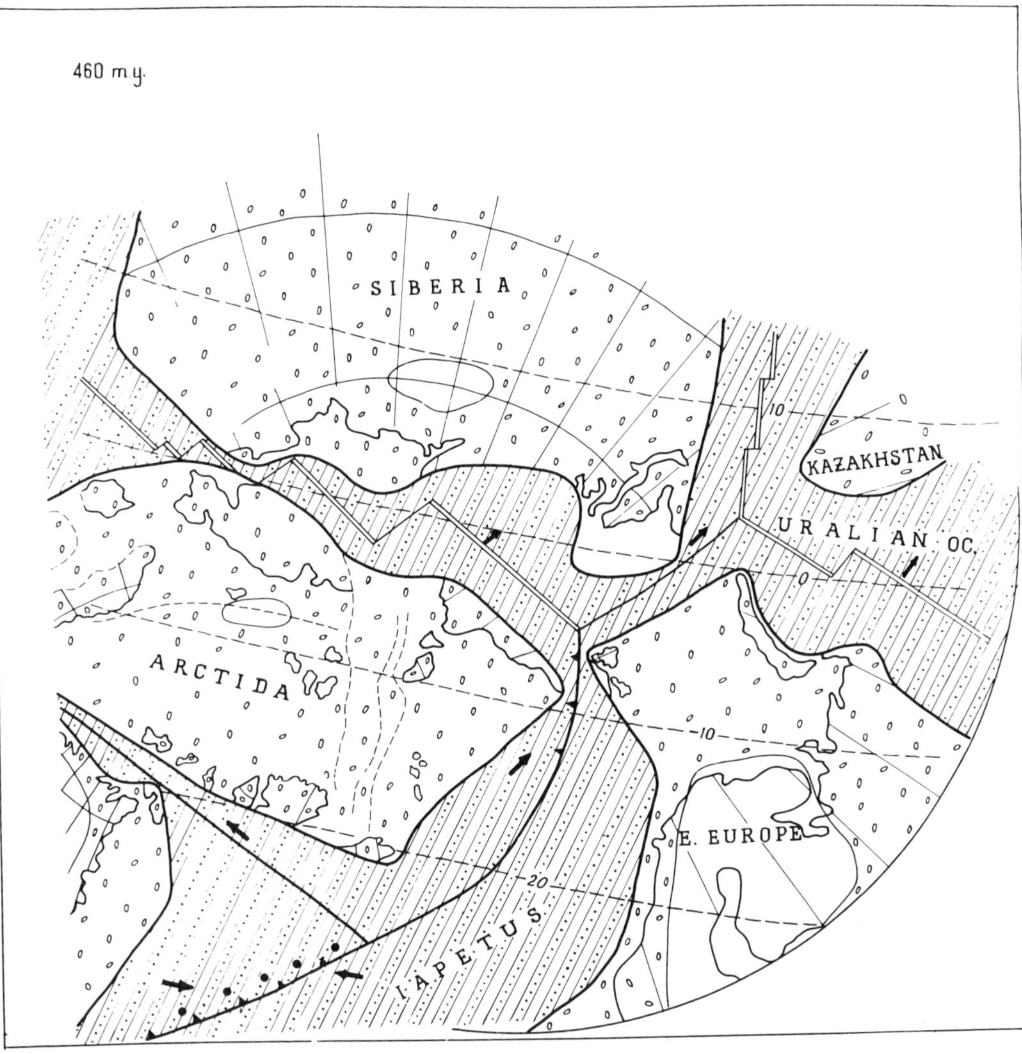


Figure 30-5. The 460 my reconstruction. Thin double line indicates diverging plate boundary, thick toothed line indicates converging plate boundary, and solid line indicates transform fault. Arrows show direction of relative plate motion. Black dots indicate calcalkaline volcanism. Toothed dashed lines show thrusting and folding. Basaltic volcanism on continents is shown as r-s. Oceanic basins are crosshatched, shallow seas are circled, and land is blank. Oblique Mercator projection with pole at $\lambda^{0°}$, 180°. Eurasia is fixed.

rated from Siberia and North America by relatively narrow oceanic straits extending along transform faults. Sweeney (1982) proposed a 2,000 km left lateral motion of Arctic–Alaska along a transform fault in the Mid-Paleozoic. The continents were almost submerged. Land existed in the central part of Siberia at the site of the Anabar Massif, in East Europe at the site of the Baltic shield, and in North America at the site of the Canadian shield. Arctida was most probably below sea level.

Widespread undisturbed carbonate sequences, occasionally with gypsum, were deposited during the Ordovician on the Arctic continents. They represent a major marine transgression and suggest that these continents were located in the low latitudes. Ordovician clastic sequences were recognized only in fold belts. On the western slope of the Urals these sequences are represented by quartzite and quartz conglomerates associated with bimodal volcanics. They are confined to the grabens that resulted from the separation of the East European margin and the formation of the Uralian Ocean. In the Innuitian Belt the clastic sequences composing the Hazen Formation are represented by graptolitic shales, often with radiolarites, passing into turbidites (Trettin, 1973). They were derived from Arctida and mark the continental rise, stretching along the Arctida slope. The Ordovician turbidites of the Scandinavian Caledonides most likely should also be attributed to continental rise deposits (Strand and Kulling, 1972). The shallow-water shelf limestones are replaced on the Novosibirsk Islands by a thick sequence of Ordovician clastics that trend in a southwest direction, toward the oceanic basin, separating Arctida from Siberia (Genin, Lipkov, and Piskarev, 1977). A possible source area is believed to have been the Chukchi Sea in the area of the Wrangel Island.

It is not known whether the pre-Ordovician Arctida was an independent continent or was split from another continent. A considerable part of the Arctida basement is composed of Late Precambrian fold complexes: on Severnaya Zemlya, the Bolshoi Lyakhovskoi, the Wrangel Islands and the Seward Peninsula, and in the Schist Belt of the Brooks Ridge (as mentioned above). It is obvious that the basement was formed during the Baikalian orogeny most likely as a result of accretional tectonics. In this respect Arctida strongly differs from the neighboring continents with their old crystalline cores (shield). Therefore, Arctida was probably an independent continental block formed at the end of the Precambrian Era. Its equivalent is Barentsia, a part of Eastern Europe joined to the Baltic shield along the Timan fold belt at the very end of the Precambrian Era.

## 400 my Reconstruction (Fig. 30-6)

Two events took place by the beginning of the Devonian: (1) North America had collided with East Europe; this collision resulted in the closure of the Iapetus Ocean; (2) the Uralian Ocean opened widely between East Europe and Siberia. The first event gave rise to an extended fold belt along the collision zone. The fold belt included the Caledonides of Scandinavia, Svalbard, and eastern Greenland. As a result, the so-called Old Red continent was formed. In the grabens and intermontane depressions nonmarine coarse clastic sediments accumulated on Svalbard (for instance, a sequence of up to 6 km thick was deposited; Birkenmajer, 1981).

The Uralian Ocean reached its maximum width of 2,500–3,000 km. It served as an oceanic strait connecting the Panthalassa Ocean with the Paleo-Tethys Ocean (Zonenshain et al., 1984). Within the Arctic region the Uralian Ocean was fringed to the south by East Europe and Arctida and to the north by Siberia. The ocean fringed by the passive continental margins was of the Atlantic type. The southern passive margin may be traced from Novaya Zemlya to South Taimyr and Chukotka. On this margin carbonate rocks were deposited in places with gypsum-bearing sequences, suggesting arid depositional environments. The bituminous facies possibly indicate stagnant environments rich in organic matter. The northern passive margin (northern part of Siberia) was the site of shallow-water carbonate deposits. In some areas, such as Novaya Zemlya, we can follow the change from shelf deposits to abyssal turbidites and radiolarites of the continental rise (Bondarev et al., 1979).

The ocean margin becomes active only near the boundaries of the territory considered within the Urals itself. The Devonian

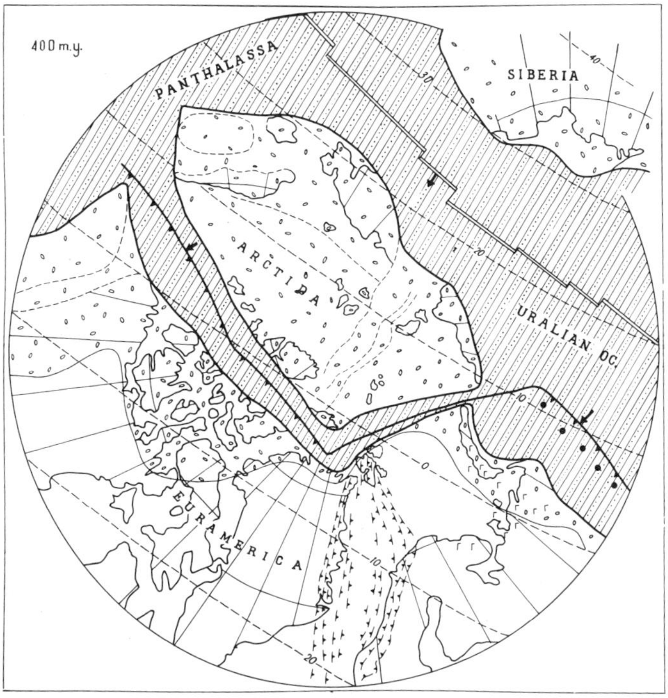


Figure 30-6. The 400 my reconstruction. For explanation see Figure 30-5.

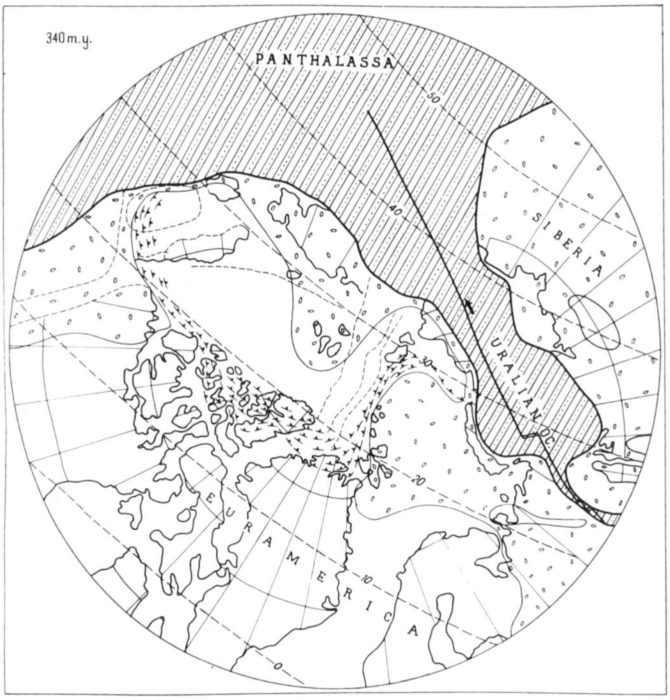


Figure 30-7. The 340 my reconstruction. For explanation see Figure 30-5.

calcalkaline volcanics are widely represented in the Uralian fold belt (Perfiliev, 1979). These rocks allow us to reconstruct the island arc beneath which the Uralian Ocean floor was consumed.

At the beginning of the Devonian, Arctida was still an independent continent separated from Euramerica by the basins of the Innuitian Belt. To the east the collision with Euramerica was underway in the middle of the Devonian. This collision is suggested by the Devonian folding on Severnaya Zemlya and in the north of Novaya Zemlya. Gradually, the collision migrated eastward, embracing the Innuitian Belt by the end of the Devonian. In the Devonian, Arctida was flooded by sea except probably insignificant land areas in the Kara and Pearya massifs, which supplied the clastic material to the adjacent basins.

The Arctic continents were located, as during the Ordovician, in the tropical zone, having drifted slightly northward (Fig. 30-6). The arid zone of the Southern Hemisphere is characterized by sediments of the Old Red continent. The equatorial zone is delineated by the organogenic limestones (including reefs) distributed along the passive margins of the Uralian Ocean.

## 340 my Reconstruction (Fig. 30-7)

By the beginning of the Carboniferous, Arctida had collided with Euramerica and had become part of it by Jurassic time. We assume that the Innuitian fold belt stretching from the Brooks Ridge to the northern part of Greenland extends further toward Caledonian structures of Severnaya Zemlya. Simultaneously, the Uralian Ocean had been considerably reduced and the contours of the Late Paleozoic Pangea were outlined.

Considerable parts of the continents located in the vicinity of the collision zones rose above sea level. A very large land mass stretched from the Baltic shield to Alaska. On continents the sea extended only along the margins in the vicinity of the oceanic basins. Carbonate sedimentation prevailed in response to the location of continents in the relatively low latitudes of the Northern Hemisphere. The vast coal accumulations are attributed to the Early Carboniferous. At this time, there were two well-developed passive continental margins along the Uralian Ocean and Panthalassa. The remnants of these margins occur in the Carboniferous rock successions of the Verkhoyansk Belt, the nearby Urals, the southern part of Taimyr, the Novosibirsk Islands, and Chukotka.

The extension and the formation of the sedimentary basins of the Barents Sea shelf are believed to have occurred during the Devonian. The Devonian basalt flows recognized on Novaya Zemlya, on the Pechora Basin (Bondarev, 1973), and on the Kanin Peninsula (Krasilshchikov, 1977), as well as the alkaline rock complexes of the Khibin Mountains, support this theory.

By the beginning of the Carboniferous, a vast depression had formed 1,500 km across, from the Wandel Sea (eastern part of Greenland) via Svalbard and Franz Josef Land to the Pechora Basin. This depression subsided and was filled with sediments until the Triassic, that is, until the continental collision occurred near Novaya Zemlya. In the Carboniferous the continents shifted to the low and moderate latitudes of the Northern Hemisphere.

## 280 my Reconstruction (Fig. 30-8)

The Pangea supercontinent was formed at the end of the Carboniferous and the beginning of the Permian. Collision zones were developed between Siberia and East Europe. The thrusting and nappe displacement for many tens and hundreds of kilometers are attributed to this period (Perfiliev, 1979). Simultaneously, in the middle of the Carboniferous, the Visean, and later extension embraced the northern part of Euramerica. The Sverdrup and Wandel

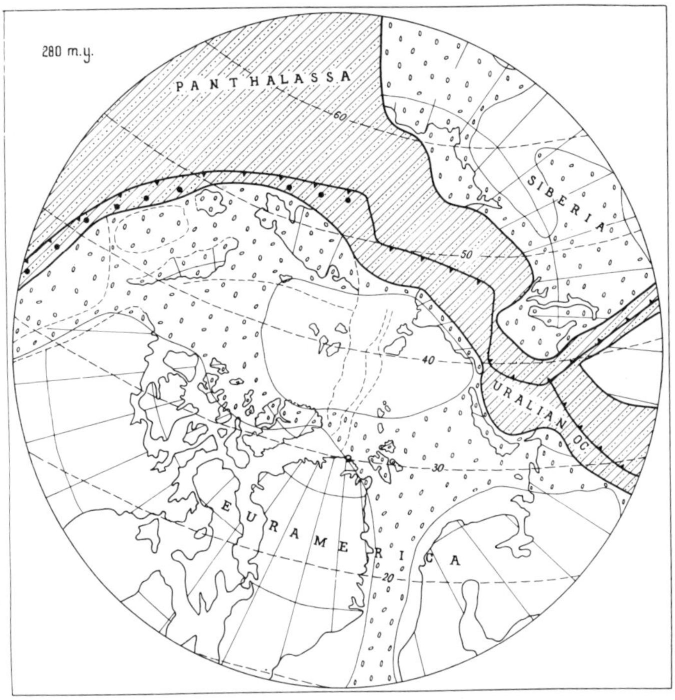


**Figure 30-8.** The 280 my reconstruction. For explanation see Figure 30-5.

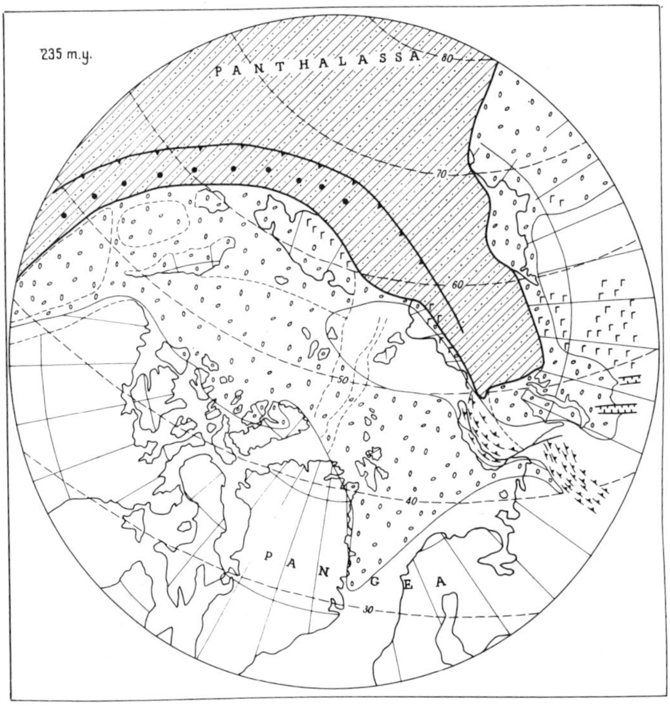


**Figure 30-9.** The 235 my reconstruction. For explanation see Figure 30-5.

Sea sedimentary basins were formed. They were connected with the Barents Sea Basin. At the base of the section of the Sverdrup Basin there are basalt flows (Balkwill, 1979), which suggest indirectly that the extension preceded the basin subsidence. It is possible that this newly formed basin extended between Greenland and Scandinavia to the North Sea and further toward western Europe.

The paleogeography of the Late Carboniferous and Early Permian included the Panthalassa Ocean, an oceanic strait located in the place of the Uralian Ocean, sea basins along the continental margins, and the Sverdrup and Barents Sea intercontinental basins. Vast land areas existed in the region occupied by the Baltic and Canadian shields. The sea withdrew from a considerable portion of Siberia for the first time, thus giving rise to extended erosion from the Anabar and Aldan shields. This area supplied the adjacent Verkhoyansk and Tunguska sedimentary basins with clastic material.

In the place of the former Arctida the Kara land was formed; this land included the northern part of Taimyr, Severnaya Zemlya, Pearya, and occasionally, the Novosibirsk Islands and Wrangel Island. The Kara land was a clastic source area for the Sverdrup, Barents Sea, and South Taimyr basins. The character of the sedimentation in the Verkhoyansk passive margin of Siberia had sharply changed by the end of the Visean: The steady carbonate accumulation that prevailed from Late Precambrian to Early Carboniferous changed to clastics of the Verkhoyansk series, which continued to deposit until the Jurassic. This change may be associated either partly with the withdrawal of the sea from the Siberian Platform and increased dissection of the Siberia relief, or with the shifting of Siberia from the arid to the humid zone and the origination of large rivers.

The northward motion of the Arctic continents continued in the Late Paleozoic (Fig. 30-8). For the first time Siberia was in the area of the Polar Circle. The Siberian fauna is attributed to the boreal province, whereas the fauna of East Europe and North America (including Chukotka) belongs to an independent province (Ustritsky and Stepanov, 1976). The data available on the distribution of the sedimentary complexes confirm the latitudinal location of the continents determined from paleomagnetic data. The rocks formed in the arid zone of the Northern Hemisphere were as follows: Late Carboniferous limestones, Early Permian reefs, red beds and evaporites of the Uralian margin of East Europe, Late Carboniferous and Permian limestones and evaporites of the Sverdrup basin, Carboniferous limestones of the Lisburn group on Alaska and their equivalent on the Wrangel Island in eastern Siberia. The coals of the Tunguska Basin mark the humid zone of the Northern Hemisphere. In the Verkhoyansk Belt, there are marine glacial deposits that confirm the location of that part of Siberia within the Polar Circle (Ustritsky, 1975).

## 235 my Reconstruction (Fig. 30-9)

All the continents at this time were integrated into Laurasia. The formation of the Uralian fold belt was completed in the Early Triassic. Its continuation through Novaya Zemlya connected the Kara Massif with Eurasia. The general configuration of the oceans and continents during the Triassic was simple. The Panthalassa (or the Paleo-Pacific) occupied a major part of the Polar region. To the south, the ocean wedged out as a narrow (700–1,200 km wide) embayment into Laurasia. It separated Chukotka and the Kara massif from Siberia. Its paleogeography was the same as during the Permian and followed the configuration of the oceanic embayment. The sea basins were located in the Laurasian margin. The Verkhoyansk and Chukotka margins were the widest. Siberia, North America, and the Kara massif supplied these margins with clastic material. Furthermore, the vast Barents Basin continued to subside.

The Permian and Triassic calcalkaline volcanic complexes of the Alazeya Plateau, in the Verkhoyansk–Kolymian region (Natapov, Zonenshain, and Shulgina, 1977), and the tectono-stratigraphic terrains of Alaska, particularly in the Yukon–Koyukuk zone (Churkin, 1981), permit the reconstruction of the island arc and subduction zone along the northern fringe of the continent. The oceanic floor of the Panthalassa was consumed in this zone. An important event of this period was the intensive trapp volcanism of Siberia. Flood basalts poured out over major parts of Siberia. Basalts flowed into the western Siberian grabens. The extension gave rise to the West Siberian Basin. Basalt occurrences stretching as a band along the Taimyr–Chukotka continental margin define the additional extension. The Triassic basalt volcanism and extension observed in the Arctic region are coeval with the Triassic extensional phase in the Central Atlantic and in the Tethys Belt.

## 190 my Reconstruction (Fig. 30-10)

During the Early Jurassic the same oceanic embayment (the South Anui oceanic basin) existed as during the Triassic and the continental margins were flooded with shallow-water seas. The oceanic embayment was also separated from the Pacific by a volcanic island arc. The remnants of this arc were preserved in the Jurassic calcalkaline volcanics of the eastern margin of Siberia. Parfenov et al. (1981) believe that this arc (they named it the Udsko–Murgal arc) stretched along the Siberian margin, from the Sikhote–Alin Ridge to Alaska. Judging by the close location of the Kolyma and the Siberian paleomagnetic poles we assume that in Early Jurassic time the Kolymian microcontinent approached the subduction zone from the Pacific accompanied by the Udsko–Murgal arc.

Another stage of the development of the Barents Sea Basin began during the Jurassic. According to seismic reflections (Golmshtok, personal communication, 1983), a sharp reflector is traced on the Barents Sea shelf along which Jurassic sequences rest unconformably on Triassic deposits. The thickness of the Jurassic–Triassic and Cenozoic deposits does not exceed 1,000 m, in contrast with the 10,000–12,000 m thickness of rocks accumulated over the period spanning the Devonian and Triassic. Therefore, we believe that the subsidence rate decreased sharply. This decrease may be due to the fact that the sedimentation compensated for the basin subsidence. A new basin started to develop between Greenland and Scandinavia. In Jurassic the West Siberian Basin subsided rapidly at the site of the Triassic grabens.

## 150 my Reconstruction (Fig. 30-11)

From mid-Jurassic on, a major change occurred in the Arctic region caused by the breakup of Pangea. The Canada Basin was opening along the near-American fringe of the former Arctida. During the initial separation, approximately 350 km of new oceanic crust were accreted. The breakup most likely included all Laurasia from the Canada Basin to the central part of the Atlantic but resulted in the separation of continents only in the Arctic region. In other regions, particularly in the Norwegian–Greenland Basin and in the North Sea, rifts and grabens developed. In the middle of the Jurassic a shallow-water seaway was formed joining the Atlantic to the Pacific via the Arctic; this shallow sea remained a permanent feature.

In the course of the opening of the Canada Basin, Arctida (or the Chukotka/Alaska block) was moving from America northward (the Jurassic coordinates) toward Siberia. As a result, the South Anui oceanic basin started to decrease in size and along the northern margin of Arc-

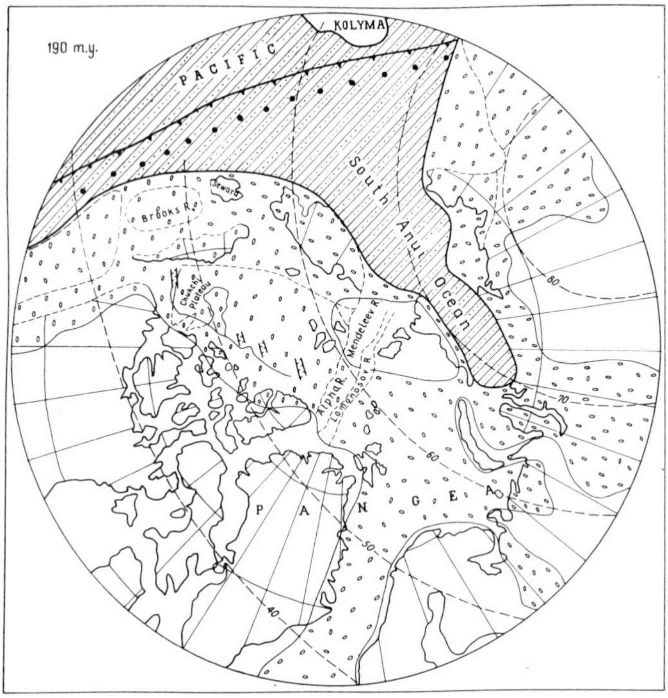


Figure 30-10. The 190 my reconstruction. For explanation see Figure 30-5.

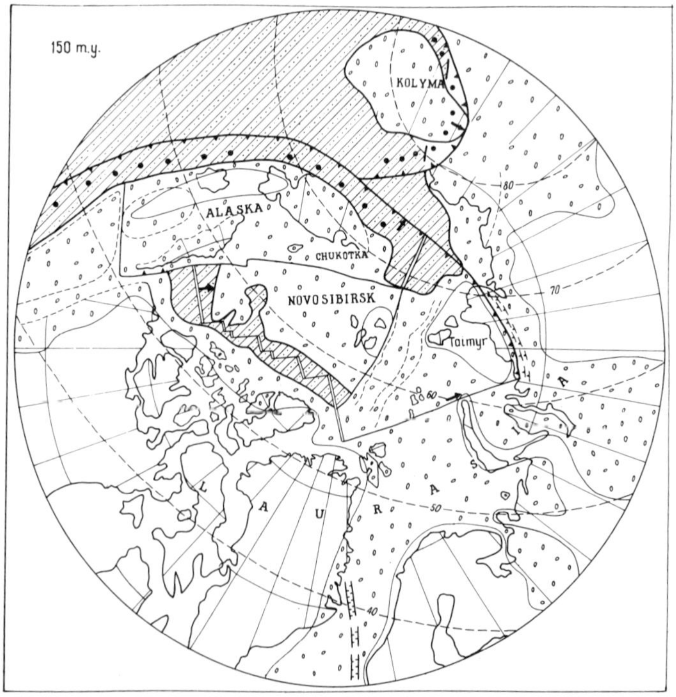


Figure 30-11. The 150 my reconstruction. For explanation see Figure 30-5.

tida (facing the Pacific) a subduction zone originated. The intensive Jurassic calcalkaline volcanism of the South Anui suture zone suggests this zone as a subduction zone.

We assume that during its separation from America, Arctida was broken up into three blocks: the Chukotka/Alaska, the Novosibirsk, and the Kara blocks. The Kara block drifted approximately 600 km relative to East Europe. It was the first to reach the Siberian margin and collide with it. By the end of the Jurassic, the folding within the South Taimyr Belt had been completed. According to accepted reconstructions the Novosibirsk block should have undergone sinistral displacement ($\approx$500 km) relative to Chukotka/Alaska.

During the Jurassic, Kolyma moved along with the Kula plate toward Siberia and in mid-Jurassic it blocked the subduction zone and collided with the Udsk–Murgal arc. As a result a new subduction zone developed along the leading edge of Kolyma and the Ilintass volcanic belt started to be formed with the consumption in the trench of the Verkhoyansk passive margin in Siberia. Therefore, Kolyma was found to be upthrusted on the thinned Siberian margin and the orogenic belt advanced toward Siberia. The collision of Kolyma with Siberia probably started in Kimmeridgian time.

## 130 my Reconstruction (Fig. 30-12)

In Early Cretaceous time, during the second phase of the opening of the Canada Basin, the Arctida blocks rotated as a whole anticlockwise relative to North America around the pole located near the Mackenzie Delta, Arctida, most likely, was broken up into two small plates: Alaska and Chukotka. It is quite possible that there was a third small plate: the Novosibirsk plate. Chukotka moved away from Alaska and in Early Cretaceous the gap between them was approximately 450 km. According to kinematic data, the separation of Alaska and Chukotka was due to the reduction in distance between them, during the Late Cretaceous and Tertiary convergence of Eurasia and North America in the Arctic region. The tensional axis between Alaska and Chukotka gave rise to the basin that exists today between the Chukchi Plateau and the Mendeleyev Ridge, as well as to the assumed basin between Alaska and Chukotka (Fig. 30-12), which was consumed later during the convergence of North America and Eurasia.

By the beginning of the Cretaceous, the South Anui oceanic basin was closed and the collision of Chukotka with Kolyma and Siberia began along the South Anui suture. Nappes were emplaced onto both sides (onto Kolyma and Chukotka), though the last of the subduction zones was inclined beneath Chukotka (southward according to Early Cretaceous coordinates).

The sinistral displacement relative to Siberia prevailed in the Novosibirsk locality of the Chukotka plates. Therefore, no considerable nappes, except for the southern portion of the Bolshoi Lyakhovsky Island, were formed in this region. The tectonic slices, ultramafics, and spilites of the southern portion of this island were described by Genin, Lipkov, and Piskarev (1977).

At that time Kolyma was connected with the Kula plate, and continued to converge with Siberia. As a result, the folding of the Verkhoyansk passive margin of Siberia started and the Kolymian granitic belt was formed above the subduction zone.

The opening of the Canadian oceanic basin caused the subsidence of the thinned continental margins and the formation of sedimentary basins. The best example of sedimentary sequences of a reformed passive margin is the Brookian complex in North Alaska (Grantz and May, 1984). Both here and along all the northern margin of Canada this complex started to accumulate in the Early Cretaceous; it is sepa-

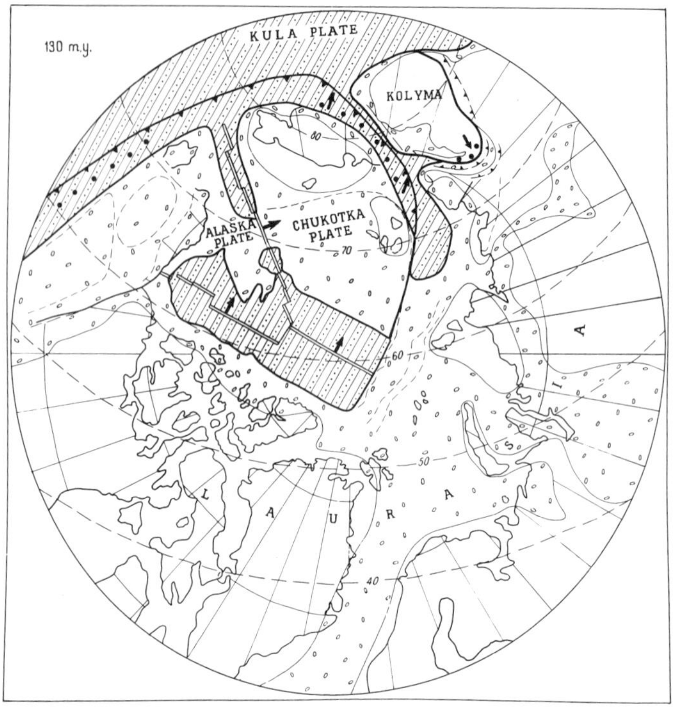


Figure 30-12. The 130 my reconstruction. For explanation see Figure 30-5.

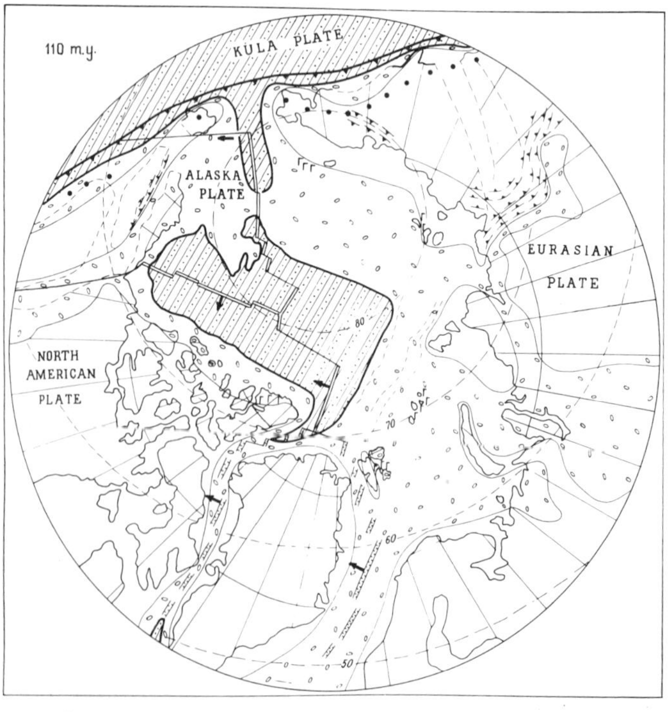


Figure 30-13. The 110 my reconstruction. For explanation see Figure 30-5.

rated by an unconformity from older rocks and is characterized by a change in the source areas of clastics. For the first time in north Alaska the source area was not within the Canadian Shield but in the adjacent Brooks Ridge. The sedimentary sequence on the islands of the eastern sector of the Soviet territory of the Arctic region also began to accumulate in the Cretaceous (Tkachenko and Egiazarov, 1970).

## 110 my Reconstruction (Fig. 30-13)

By the middle of Early Cretaceous the Canadian oceanic basin had attained its maximum size. It was 2,500 km long (from Alaska to Severnaya Zemlya) and 1,200 km wide. The Middle Cretaceous Canadian Basin was 900 km longer than the present basin. The important extension phase in the Arctic region that produced numerous basaltic eruptions is believed to have occurred during this time. The extension gave rise to the initial separation of Greenland (essentially with Eurasia) from North America. At the site of the Labrador Sea and Baffin Bay a sea strait originated that had partly oceanic and partly thinned continental crust.

By Aptian time the South Anui Basin had completely closed and the South Anui fold belt had been formed. It was the time of culmination of the Verkhoyansk fold belt formation. During that period the nappe emplacement toward Siberia occurred and the Verkhoyansk foredeep formed.

We assume that the Chukotka plate was connected to Eurasia, though the Alaska plate still existed; it rotated anticlockwise relative to North America and by the middle of the Early Cretaceous it occupied its present position. As a result of the Alaskan motion the fold front of the Brooks Ridge originated; it supplied clastics to the Colville trough toward the Canada Basin.

Since Chukotka and Kolyma had collided with Siberia, and the Kula/Eurasia convergence continued along the reformed Eurasian margin, a new subduction zone originated. This subduction zone is marked by the Okhotsk–Chukotka volcanic belt that consists of calcalkaline subaerial lavas and granitic plutons protruding them. This belt marks the formation of Andian-type active margin. The volcanic polarity has permitted the determination that the Benioff zone dipped under the continental margin at a 30° angle (Zonenshain, Kuzmin, and Moralev, 1976). The continental volcanic belt stretched eastward as far as the Seward Peninsula, while further on it was most likely replaced by a volcanic island arc, whose remnants may be recognized in the Cretaceous volcanics of the Yukon–Koyukuk zone of Alaska (Lathram, 1973; Churkin and Trexler, 1981).

## 80 my Reconstruction (Fig. 30-14)

The compressional phase recognized in the Arctic Basin is attributed to the Late Cretaceous, because the Eurasian/North American pole of rotation was located at the southern end of Greenland (see Kinematic Data section). The most probable location of the plate convergence boundary is shown in Figure 30-14. The boundary most likely extended along the fold front of the Brooks Ridge to the Herald–Wrangel fold zone and further along the shelf of the Chukchi and East Siberian seas via the Canada Basin margin along the Eurasian shelf edge, and continued southward to the Nares Strait between Greenland and Ellesmere Islands as well as between Greenland and Svalbard. The Cretaceous rock deformations in the eastern portion of the Sverdrup Basin and in northern Greenland may be attributed to plate interactions along this boundary.

In the sector of the plate boundary between Eurasia and the Canada Basin consumption of the oceanic floor occurred

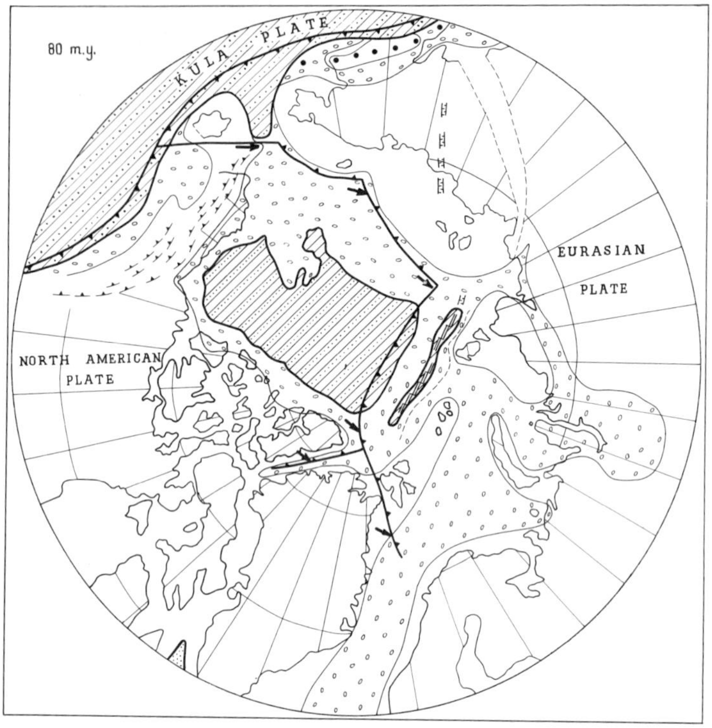


**Figure 30-14.** The 80 my reconstruction. For explanation see Figure 30-5.

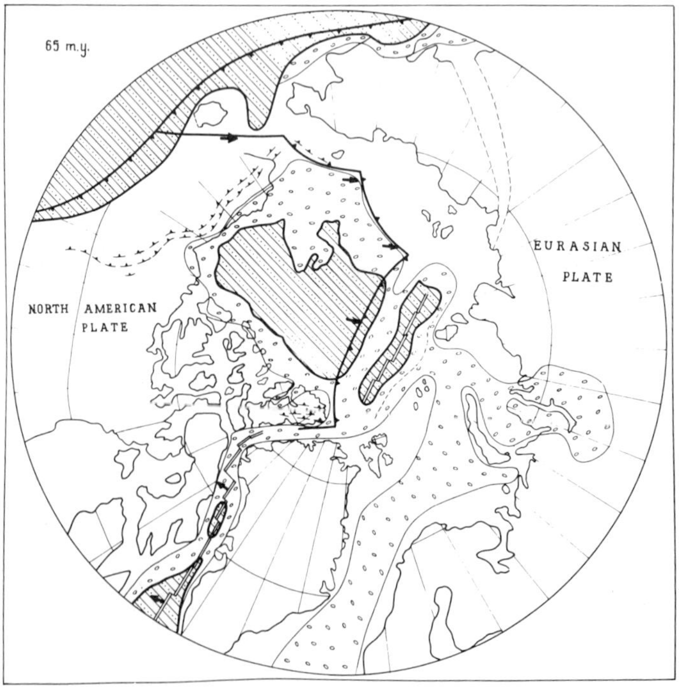


**Figure 30-15.** The 65 my reconstruction. For explanation see Figure 30-5.

and was accompanied by reduction in the basin's size by approximately 450 km. This process should have caused formation of a subduction zone and a volcanic arc. Proceeding from this event it is necessary to assume the presence of Late Cretaceous calcalkaline volcanics in the Alpha–Mendeleyev Ridge. In this interpretation the Makarov is a marginal basin, originated at the expense of the back arc spreading. By this time the basin that caused the splitting of the Alpha–Mendeleyev Ridge from Eurasia had started to open. The Lower Indigirka rift composed of bimodal volcanics, basalts, and rhyolites may be considered as a contribution of the Makarov Basin spreading zone to East Siberia (Savostin, Natapov, and Stavsky, 1984; Stavsky, 1982).

In Late Cretaceous, the oceanic strait between Chukotka and Alaska was most likely nearly closed. The Kula plate localized in the Pacific was subducted under Eurasia and North America. It was at that time that the Talovo–Mainsk, Pekulnei, and Khatyrka terrains shifted from the Pacific into the subduction zone. As a result, the margin of Eurasia had accreted and prograded southward to the Pacific.

### 65 my Reconstruction (Fig. 30-15)

In the 65–80 my interval plate interactions show close resemblance to those of the previous period. The main difference is that the pole of Eurasia/North America motion shifted northward to the Ellesmere Island. Therefore, the convergence of Eurasia and North America took place between Greenland and Alaska. During 15 my approximately 300 km of the Canada Basin oceanic crust was consumed. South of the pole, between Greenland and North America, a plate separation that gave rise to the oceanic crust of the Labrador and Baffin seas took place. By the end of the Cretaceous the width of the Labrador oceanic basin was 350 km.

## DISCUSSION AND CONCLUSIONS

The Arctic region history involves the interaction of four continents: East Europe, North America, Siberia, and Arctida. These four continents were rifted apart in Early Paleozoic and, by the end of the Paleozoic, they were integrated during successive collisions, forming Laurasia, as part of Pangea. Because the Arctic portion of Laurasia was for a long time (since the Devonian and Carboniferous) facing the Panthalassa or the Paleo-Pacific, well-developed passive continental margins were formed on which thick sedimentary wedges accumulated. This accumulation makes the Arctic region promising for oil and gas exploration. The Mesozoic history involves the separation of North America and Eurasia on some portions of their common boundary and the convergence of these continents along other portions, depending on the position of the rotation pole. The opening of the Canada Basin in Late Jurassic and in the beginning of Early Cretaceous caused the closure of the South Anui oceanic basin and the formation of the Verkhoyansk and South Anui fold belts. The opening of the North Atlantic and Labrador basins in Late Cretaceous was accompanied by partial consumption of the Canada Basin crust and the formation of fold zones in the eastern sector of the Arctic region. At that time the former Arctida broke up and its separate fragments ended up on different sides of the newly formed oceanic basins of the Arctic region.

Since the Ordovician, the Arctic continents were in comparative proximity relative to one another, from time to time either converging or diverging. In the Early Paleozoic, however, the continents were located in the equatorial zone. During the Paleozoic and Early Mesozoic they drifted gradually northward, reaching polar latitudes only in the Jurassic.

## ACKNOWLEDGMENTS

We would like to thank K. Burke, M. Churkin, and R. L. Thiessen for reviewing the manuscript in June 1985, and especially G. L. Johnson for many helpful discussions and for sending new preprints on the Arctic tectonics.

## REFERENCES

Balkwill, H. R. 1978. Evolution of Basin Arctic Canada. *Am. Assoc. Petroleum Geol. Bull.* **62**(6):1,004–1,028.

Birkenmajer, K. 1981. The geology of Svalbard, the western part of the Barents Sea and the continental margin of Scandinavia. In A. E. M. Nairn, M. Churkin, Jr., and F. Cr. Stehli (eds.), *The Ocean Basins and Margins,* vol. 5, *The Arctic Ocean.* Plenum Press, N.Y. and London, pp. 265–329.

Bondarev, V. I., S. V. Cherkesova, V. S. Enokyan, and B. S. Romavich. 1973. Geologic structure of Novaya Zemlya, Vaygach, Pay-Khoy and Northern Pechora. In M. C. Pitcher (ed.), *Arctic Geology.* American Association of Petroleum Geologists, Tulsa, Okla., pp. 301–308.

Bondarev, V. I., Ju. P. Ershov, B. S. Ipatov, and I. A. Andreeva. 1979. The history of tectonic development of North Novaya Zemlya. In V. I. Bondarev (ed.), *Geology and Stratigraphy of Novaya Zemlya.* NIIGA, Leningrad, pp. 5–17. (In Russian.)

Brosge, and J. T. Dutro. 1973. Paleozoic rocks of northern and central Alaska. In M. G. Pitcher (ed.), *Arctic Geology.* American Association of Petroleum Geologists, Tulsa, Okla., pp. 351–375.

Burke, K. 1984. Plate tectonic history of Arctic. In *Proceedings of the 27th International Geological Congress.* Coll. 04, Geology of the Arctic, Moscow, pp. 159–167.

Carey, S. W. 1955. Orocline concept in geotectonics, I. Paper A. *Proc. Roy. Soc. Tasmania* **89**:255–288.

Carey, S. W. 1958. The tectonic approach to continental drift. In *Continental Drift. A Symposium.* University of Tasmania, pp. 177–355.

Churkin, M., Jr. 1972. Western boundary of the North American continental plate in Asia. *Geol. Soc. Am. Bull.* **83**:1,027–1,036.

Churkin, M., Jr. 1973. Geologic concepts of Arctic Ocean Basin. In M. G. Pitcher (ed.), *Arctic Geology.* American Association of Petroleum Geologists, Tulsa, Okla., pp. 485–499.

Churkin, M., Jr., and J. H. Trexler, Jr. 1981. Continental plates and accreted oceanic terranes in the Arctic. In A. E. M. Nairn, M. Churkin, Jr., and F. C. Stehli (eds.), *The Ocean Basins and Margins,* vol. 5, *The Arctic Ocean.* Plenum Press, New York and London, pp. 1–20.

Dawes, P. R., and J. S. Peel. 1981. The northern margin of Greenland Sea. In A. E. M. Nairn, M. Churkin, Jr., and F. C. Stehli (eds.), *The Ocean Basins and Margins,* vol. 5, *The Arctic Ocean.* Plenum Press, New York and London, pp. 201–264.

Dawes, P. R., and N. J. Soper. 1973. Pre-Quaternary history of North Greenland. In M. G. Pitcher (ed.), *Arctic Geology.* American Association of Petroleum Geologists, Tulsa, Okla., pp. 117–134.

Deherman, R. L. 1973. Mesozoic sequence in Arctic Alaska. In M. G. Pitcher (ed.), *Arctic Geology.* American Association of Petroleum Geologists, Tulsa, Okla., pp. 376–387.

Drabkin, I. E. (ed.) 1970. *Geology of the USSR,* vol. XXX, *Northeast of the USSR.* Book I, 548p. (In Russian.)

Du Toit, A. L. 1937. *Our Wandering Continents.* Hafner Publish. Co., 366p.

Dutro, J., Jr. 1981. Geology of Alaska bordering the Arctic ocean. In A. E. M. Nairn, M. Churkin, Jr., and F. C. Stehli (eds.), *The Ocean Basins and Margins,* vol. 5, *The Arctic Ocean.* Plenum Press, New York and London, pp. 21–36.

Eardley, A. J. 1961. History of geologic thought on the origin of the Arctic Basin. In G. O. Raash (ed.), *Geology of the Arctic,* vol. 1. University of Toronto Press, pp. 607–621.

Eldholm, O., A. M. Myhre, and E. Sundvor. 1984. Structure and evolution of the margins around the Norwegian-Greenland Sea. In *Proceedings of the 27th International Geological Congress.* Coll. 04, Geology of the Arctic, Moscow, pp. 37–52.

Ershov, Ju. P., A. A. Krasilshchikov, V. E. Volk, and V. N. Shimarajev. 1974. Geotectonic description of the south part of the Barents Sea shelf. In A. A. Krasilshchikov (ed.), *Geotectonic Conditions for Mineral Deposit Exploration on the Arctic Ocean Shelf.* Nauka, Leningrad, pp. 34–50. (In Russian.)

Freeland, G. L., and R. S. Dietz. 1973. Rotation history of Alaskan tectonic blocks. *Tectonophysics* **18**:379–389.

Fujita, K. 1978. Pre-Cenozoic tectonic evolution of north-east Siberia. *J. Geology* **86**:159–172.

Genin, B. L., L. Z. Lipkov, and A. L. Piskarev. 1977. Basement structure of the East Siberian Sea exemplified by Novosibirsk Archipelago. In Yu. E. Pogrebitsy and M. K. Kos'ko (eds.), *Tectonics of the Shelf Sedimentary Basins.* NIIGA, Leningrad, pp. 86–97. (In Russian.)

Getzen, V. G. 1975. *Structure of the Basement of North Timan and Kanin Peninsula.* Nauka, Leningrad, 142p. (In Russian.)

Gramberg, I. S., V. E. Volke, V. S. Zarkhidze, Ju. N. Kulakov, I. V. Shkola, and D. S. Jashin. 1984. Geological structure of the Arctic continental margin of the USSR. *Proceedings of the 27th International Geological Conference.* Coll. 04, Geology of the Arctic, Moscow, pp. 3–11.

Grantz, A., S. Eittreim, and O. T. Whitney. 1981. Geology and physiography of the continental margin of the Canada Basin. In A. E. M. Nairn, M. Churkin, Jr., and F. C. Stehli (eds.), *The Ocean Basins and Margins,* vol. 5, *The Arctic Ocean.* Plenum Press, New York and London, pp. 439–492.

Grantz, A., and S. D. May. 1983. Rifting history and structural development of the continental margin north of Alaska. In J. S. Watkins and C. C. Drake (eds.), *Studies in Continental Margin Geology.* Am. Assoc. Petroleum Geol. Mem. 34, pp. 77–100.

Grantz, A., and S. D. May. 1984. Sedimentary basins and geologic structure of the continental margin north of Alaska. *Proceedings of the 27th International Geological Conference.* Coll. 04, Geology of the Arctic, Moscow, pp. 105–118.

Hall, J. K. 1973. Geophysical evidence for ancient sea floor spreading from Alpha Cordillera and Mendeleev Ridge. In M. G. Pitcher (ed.), *Arctic Geology.* American Association of Petroleum Geologists, Tulsa, Okla., pp. 542–561.

Haller, J. 1976. *Geology of the East Greenland Caledonides.* Interscience Publishers, London, 413p.

Hamilton, W. 1970. The Uralides and the motion of the Russian and Siberian platforms. *Geol. Soc. Am. Bull.* **81**:2,553–2,576.

Heezen, B. C., and M. Ewing. 1961. The mid-ocean ridge and its extension through the Arctic Basin. In G. O. Raash (ed.), *Geology of the Arctic.* University of Toronto Press, pp. 622–642.

Heron, E., J. E. Dewey, and W. C. Pitman. 1974. Plate tectonic model for the Arctic. *Geology* **2**(8):377–380.

Hjelle, A., and I. Ohta. 1978. Tectonics of Spitzbergen. In A. V. Peive, V. E. Khain, and M. V. Muratov (eds.), *Tectonics of Europe and Adjacent Areas. Ancient Platforms, Baikalides, Caledonides.* Nauka, Moscow, pp. 355–366. (In Russian.)

Holtedahl, O., and C. Wolff. 1978. Norwegian Caledonides. In A. V. Peive, V. E. Khain, and M. V. Muratov (eds.), *Tectonics of Europe and Adjacent Areas. Ancient Platforms, Baikalides, Caledonides.* Nauka, Moscow, pp. 355–366. (In Russian.)

Irving, E. 1977. Drift of the major continental blocks since the Devonian. *Nature* **270**:304–307.

Irving, E. 1983. Fragmentation and assembly of continents, mid-Carboniferous to present. *Geophys. Surv.* **5**:299–333.

Jackson, H. R., D. A. Forsyth, and G. L. Johnson. 1986. Oceanic affinities of the Alpha Ridge (in preparation).

Jackson, H. R., and G. L. Johnson. 1984. Structure and history of the Amerasian Basin. In *Proceedings of the 27th International Geological Congress.* Coll. 04, Geology of the Arctic, Moscow, pp. 105–118.

Jones, P. B. 1980. Evidence from Canada and Alaska on plate tectonic evolution of the Arctic Ocean Basin. *Nature* **5,762**:377–380.

Kameneva, G. I. 1977. The problem of the Wrangel Island tectonic position and its structural relation with Alaska. In Yu. E. Pogrebitsky and M. K. Kos'ko (eds.), *Tectonics of the Arctic. Fold Basement of the Shelf Sedimentary Basins.* NIIGA, Leningrad, pp. 122–131. (In Russian.)

Karasik, A. M. 1968. Magnetic lineation of the Gakkel Ridge and the origin of the Eurasian subbasin in the Arctic Ocean. *Geophysical Methods of Exploration in the Arctic,* no. 5. NIIGA, Leningrad, pp. 8–25. (In Russian.)

Karasik, A. M. 1980. Main features of the history and structure of the Arctic Basin bottom from acromagnetic data. In *Marine Geology, Sedimentology, Sedimentary Petrology and Geology of the Oceans.* Nedra, Leningrad, pp. 178–193. (In Russian.)

Karasik, A. M., L. A. Savostin, and L. P. Zonenshain. 1983. Motion parameters for lithosphere plates in Eurasian Basin of the Arctic Ocean. *Doklady Akad. Nauk USSR* **273**(5):1,191–1,196. (In Russian.)

Karasik, A. M., and N. A. Sochevanova. 1981. *World Magnetic Lineation Map.* IZMIRAN, Leningrad.

Karasik, A. M., V. I. Ustritsky, and A. N. Khramov. 1984. The history of the Arctic ocean formation. In *Proceedings of the 27th International Geological Congress.* Coll. 04, Geology of the Arctic, Moscow, pp. 151–159.

Kerr, J. W. 1981. Evolution of the Canadian Arctic Islands, a transition between the Atlantic and Arctic oceans. In A. E. M. Nairn, M. Churkin, Jr., and F. C. Stehli (eds.), *The Ocean Basins and Margins,* vol. 5, *The Arctic Ocean.* Plenum Press. New York and London, pp. 105–190.

Khain, V. E. 1979. *Regional Geotectonics. Outalpine Asia and Australia.* Nedra, Moscow, 279p. (In Russian.)

Khramov, A. N. (ed.) 1982. *Paleomagnetology.* Nedra, Leningrad, 312p.

Kolesov, E. V. 1981. Paleomagnetic characteristic of Middle Paleozoic rocks in the Omolon massif. In Yu. Ya. Vashchilov (ed.), *Rock Magnetism and Paleomagnetic Stratigraphy of Eastern and Northeastern Asia.* DVNC, Magadan, pp. 68–674. (In Russian.)

Koppen, W. P., and A. Wegener. 1924. *Die Klimate der Geologisch Vorzeit.* Gebrueder Borntraegger, 255p.

Krasilshchikov, A. A. 1974. Tectonic structure of basement of the Kanin Peninsula. In Yu. E. Pogrebitsky and M. K. Kos'ko (eds.), *Tectonics of the Arctic: the Basement of the Shelf Sedimentary Basins.* NIIGA, Leningrad, pp. 5–14. (In Russian.)

Kristoffersen, J., and M. Talwani. 1977. Extinct triple junction south of Greenland and the Tertiary motion of Greenland relative to North America. *Geol. Soc. Am. Bull.* **88:**1,037–1,049.

Lathram, E. 1973. Tectonic framework of northern and central Alaska. In M. G. Pitcher (ed.), *Arctic Geology.* American Association of Petroleum Geologists, Tulsa, Okla., pp. 351–360.

Lozhkina, N. V. 1981. Magnetic properties of the reference section of Middle-Upper Triassic in the Okhotsk region. In Yu. Ya. Vashchilov (ed.), *Rock Magnetism and Paleomagnetic Stratigraphy of Eastern and Northeastern Asia.* DVNC, Magadan, pp. 59–67.

Natal'in, B. A. 1979. Eugeosynclinal zones of Chukotka and Alaska. Abstracts of the XIY Pacific Congress. Comm. B, Section BII. *Continental Margins, Island Arcs and Oceanic Floor.* Khabarovsk, pp. 34–35.

Natapov, L. M., L. P. Zonenshain, and V. S. Shulgina. 1977. Geological evolution of the Kolyma–Indigirka region and the problem of the Kolymian massif. *Geotectonica* **N4:**18–31.

Nesterov, I. I., N. Kh. Kulakhmetov, V. S. Bochkarev, M. I. Mishunsky, and Ju. N. Kulakov. 1984. The geology and oil–gas-bearing potential of the Jamal–Gydan Syneclise (West Siberia). In *Proceedings of the 27th International Geological Congress.* Coll. 04, Arctic Geology, Moscow, pp. 80–88.

Olivet, J. Z., J. Bonnin, P. Beazart, and J. M. Auzende. 1985. Cinematique de L'Atlantique nord et central. In *Rapports Scientifiques.* CNEXO, Paris.

Ostenso, N. A., and R. J. Wold. 1973. Aeromagnetic evidence for origin of Arctic Ocean Basin. In M. G. Pitcher (ed.), *Arctic Geology.* American Association of Petroleum Geologists, Tulsa, Okla., pp. 506–516.

Parfenov, L. M., B. A. Natal'in, I. R. Voinova, and L. I. Popeko. 1981. Tectonic evolution of the active continental margins on the North-West Pacific. *Geotectonica* **1:**85–104.

Pechersky, D. M. 1970. Paleomagnetic study of Mesozoic rocks in the northeastern USSR. *Izvestiya Akad. Nauk, s. Physics of the Earth,* nb, 69–83.

Perfil'ev, A. S. 1979. *Formation of the Earth Crust of the Uralian Eugeosyncline.* Nauka, Moscow, 188p. (In Russian.)

Pitman, W. C., III, and M. Talwani. 1972. Sea floor spreading in the North Atlantic, *Geol. Soc. Am. Bull.* **83:**619–646.

Pogrebitsky, Yu. E. 1971. *Paleotectonic Analysis of the Taimyr Foldsystem.* Nedra, Leningrad, 248p. (In Russian.)

Pogrebitsky, Yu. E., A. N. Vishersky, S. A. Gubin, G. I. Kavardin, E. A. Korago, M. K. Kos'ko, M. I. Mitroshin, A. I. Trukhalev, and N. K. Shanurenko. 1984. Endogenic differentiation of matter in the geodynamic system of Arctic Ocean. In *Proceedings of the 27th International Geological Congress.* Coll. 04, Geology of the Arctic, Moscow, pp. 11–26.

Pushcharovsky, Yu. M. 1960. Some general problems of the tectonics of the Arctic. *Izvestiya Akad. Nauk, Ser. Geol.* **9:**15–28. (In Russian.)

Pushcharovsky, Yu. M. 1976. Tectonics of the Arctic Ocean. *Geotectonics* **2:**3–14. (In Russian.)

Savostin, L. A., and A. M. Karasik. 1981. Recent plate tectonics of the Arctic Basin and of north-eastern Asia. *Tectonophysics* **74:**111–145.

Savostin, L. A., A. M. Karasik, and L. P. Zonenshain. 1984. The history of opening of the Eurasian Basin in the Arctic. *Doklady Akad. Nauk USSR* **275**(5):1,156–1,161. (In Russian.)

Savostin, L. A., L. M. Natapov, and A. P. Stavsky. 1984. Mesozoic paleogeodynamics and paleogeography of the Arctic region. In *Proceedings of the 27th International Geological Conference.* Coll. 03, Paleoceanography, Moscow, pp. 172–187.

Savostin, L. A., J. C. Sibuet, L. P. Zonenshain, X. Le Pichon, and M. J. Roulet. 1985. Kinematic evolution of the Tethys belt from the Atlantic Ocean to Pamyr since Triassic. *Tectonophysics* **123:**1–36.

Schwartz, I. V. 1981. Main phases in the development of sedimentary basins of the Barents Sea during Phanerozoic. In D. S. Sorokin and J. N. Kulakov (eds.), *Lithology and Paleogeography of the Barents and Kara Seas.* NIIGA, Leningrad, pp. 5–33. (In Russian.)

Seslavinsky, K. B. 1970. Structure and development of the South Anui suture. *Geotectonics* **5:**73–85.

Shatsky, N. S. 1935. On tectonics of the Arctic. In *Proceedings of the 1st Geological Conference, Glavsevmorput,* vol. 1. Moscow-Leningrad, pp. 149–169. (In Russian.)

Shulgina, N. I., and Yu. A. Mikhailov. 1979. New data concerning stratigraphy of Mesozoic in the Franz Josef Land. In N. I. Shulgina (ed.), *Upper Paleozoic and Mesozoic of the USSR Arctic Seas and Coast.* NIIGA, Leningrad, pp. 5–10. (In Russian.)

Srivastava, S. P. 1978. Evolution of the Labrador sea and its bearing on the early evolution of the North Atlantic. *Geophys. J. Roy. Astr. Soc.* **52:**313–357.

Srivastava, S. P. 1985. Evolution of the Eurasian Basin and its implication to the motion of Greenland along Nares Strait. *Tectonophysics* **114:**29–53.

Stavsky, A. P. 1982. Nizhne-Indigirskaya rift zone—a new element in the structure of the north-east USSR. *Doklady Akad. Nauk USSR* **262:**1,443–1,446. (In Russian.)

Strand, T., and O. Kulling. 1972. *The Scandinavian Caledonides.* Interscience Publishers, London, 380p.

Sweeney, J. F. 1982. Mid-Paleozoic travers of Arctic-Alaska. *Nature* **298:**647–649.

Sweeney, J. F. 1985. Comments about the age of the Canada Basin. *Tectonophysics* **114:**1–10.

Tailleur, I. L. 1973. Probable rift origin of Canada Basin, Arctic Ocean. In M. G. Pitcher (ed.), *Arctic Geology.* American Association of Petroleum Geologists, Tulsa, Okla., pp. 526–535.

Taylor, P. T., L. C. Kovacs, P. R. Vogt, and G. L. Johnson. 1981. Detailed aeromagnetic investigation of the Arctic Basin. 2. *J. Geophys. Res.* **86:**6,223–6,333.

Tilman, S. M. 1973. *Comparative Tectonics of Mesozoides on the North Pacific Belt.* Nauka, Moscow, 326p. (In Russian.)

Tkachenko, B. V., and B. Kh. Egizarov (eds.) 1970. *Geology of the USSR,* vol. IVLMC, *Islands of the Soviet Arctic.* Nedra Moscow, 547p. (In Russian.)

Trettin, H. P. 1973. Early Paleozoic evolution of northern parts of Canadian Arctic Archipelago. In M. G. Pitcher (ed.), *Arctic Geology.* American Association of Petroleum Geologists, Tulsa, Okla., pp. 57–75.

Udintsev, G. B. 1984. Norwegian-Greenland Basin: Rifting and oceanization. In *Proceedings of the 27th International Geological Congress.* Coll. 04, Geology of the Arctic, Moscow, pp. 51–58.

Ustritsky, V. I. 1975. Development history of the north-east USSR in Upper Paleozoic. In V. I. Ustritsky (ed.), *Upper Paleozoic of the North-east USSR.* NIIGA, Leningrad, pp. 54–75. (In Russian.)

Ustritsky, V. I., and D. L. Stepanov. 1976. Paleobiography and climate of Eurasia in Permian. In *Proceedings of the 25th International Geological Congress.* Reports of Soviet Geologists. Sect. Paleontology: Marine Geology. Nauka, Moscow, pp. 103–108.

Van der Voo, R., J. Peinade, and C. R. Scotese. 1984. A paleomagnetic reevaluation of Pangea reconstructions. In R. R. Van der Voo, C. R. Scotese, and N. Bonhommet (eds.), *Plate Reconstruction from Paleozoic Paleomagnetism.* Geodyn. ser., vol. 12, American Geophysical Union, Washington, D.C., pp. 11–26.

Vogt, P. R., P. T. Taylor, J. Kovacs, and G. L. Johnson. 1979. Detailed aeromagnetic investigation of the Arctic Basin. I. *J. Geophys. Res.* **84:**1,071–1,089.

Vogt, P. R., P. T. Taylor, L. C. Kovacs, and G. L. Johnson. 1982. The Canada Basin: Aeromagnetic constraints on structure and evolution. *Tectonophysics* **89:**295–336.

Vogt, P. R., J. Kovacs, R. K. Perry, and P. T. Taylor. 1984. Amerasian Basin in the Arctic: Magnetic lineation and its identification. In *Proceedings of the 27th International Geological Conference.* Coll. 04, Geology of the Arctic, Moscow, pp. 128–135.

Westphal, M., M. C. Bazenov, J. P. Lauer, D. M. Pechersky, and J. C. Sibuet. 1985. Paleomagnetic implications on the evolution of the Tethys belt from the Atlantic Ocean to Pamir since Triassic. *Tectonophysics* **123:**37–82.

Wilson, J. T. 1966. Did the Atlantic close and then reopen. *Nature* **211:**676–681.

Zonenshain, L. P. 1984. Tectonics of innercontinental foldbelts. In *Proceedings of the 27th International Geological Conference.* Sect. 07, Tectonics, Moscow, pp. 48–59.

Zonenshain, L. P., M. I. Kuzmin, and M. V. Kononov. 1985. Absolute reconstructions of the Paleozoic oceans. *Earth Planet. Sci. Lett.* **74:**103–116.

Zonenshain, L. P., M. I. Kuzmin, and V. M.

Moralev. 1976. *Global Tectonics, Magmatism and Metallogeny*. Nedra, Moscow, 230p. (In Russian.)

Zonenshain, L. P., L. M. Natapov, L. A. Savostin, and A. P. Stavsky. 1978. Present plate tectonics of North-East Asia in relation with opening of the North Atlantic and Arctic oceans. *Oceanology* **5**:38–47.

Zonenshain, L. P., M. I. Kuzmin, M. V. Kononov, and A. M. Gorodnitsky. 1984*a*. Paleozoic oceans. An attempt of absolute reconstructions. In *Proceedings of the 27th International Geological Conference*. Coll. 03, Paleoceanography, Nauka, Moscow, pp. 35–46.

Zonenshain, L. P., V. G. Korenevski, V. G. Kazmin, D. M. Pechersky, V. V. Khain, and V. V. Matveenkov. 1984*b*. Plate tectonic model of the South Urals development. *Tectonophysics* **109**:95–135.

# Index